The Handbook of Science and Technology Studies

The Handbook of Science and Technology Studies

Fourth Edition

Edited by
Ulrike Felt
Rayvon Fouché
Clark A. Miller
Laurel Smith-Doerr

Published in cooperation with the Society for Social Studies of Science

The MIT Press
Cambridge, Massachusetts
London, England

This book was set in Stone Serif and Stone Sans by Toppan Best-set Premedia Limited. Printed and bound in the United States of America.

Library of Congress Cataloging-in-Publication Data

Names: Felt, Ulrike, editor.
Title: The handbook of science and technology studies / edited by Ulrike
    Felt, Rayvon Fouché, Clark A. Miller, and Laurel Smith-Doerr.
Description: Fourth edition. | Cambridge, MA: The MIT Press, [2017] |
    Includes bibliographical references and index.
Identifiers: LCCN 2016023009 | ISBN 9780262035682 (hardcover: alk. paper)
Subjects: LCSH: Science. | Technology.
Classification: LCC Q158.5 .H36 2017 | DDC 303.48/3—dc23 LC record available at
    https://lccn.loc.gov/2016023009

10 9 8 7 6 5 4 3 2 1

Contents

Preface

Welcome to the fourth edition of the *Handbook of Science and Technology Studies*! We live in a world in which the technosciences have become a dominant mode of knowledge-making, with wide-ranging material and conceptual consequences. If you are relatively new to science and technology studies (STS), this substantial book and the interdisciplinary field that it landscapes (an apt metaphor adopted by the editors and elaborated in their introduction) can begin to answer your questions. Are you interested in understanding how the technical dimensions of science and technology are always enmeshed with their nontechnical dimensions? Might you be someone who feels compelled to analyze sciences and technologies critically, with the goal of questioning received assumptions and exploring new conceptualizations? Do you find yourself asking what the technosciences are for, and for whom, and imagining how they might be otherwise? Then this book is for you. Alternatively, if you are already deeply engaged with STS, this extraordinary collection of thematic articulations of the field will serve as an invaluable reference and further resource for your ongoing work.

Officially, it is the Society for Social Studies of Science (4S) that offers this handbook to all readers interested in the scholarship of STS, including what STS researchers and practitioners write, make, and do and the implications of our scholarly practices. Yet as the editors' thoughtful and reflexive introduction makes clear, 4S does so in recognition of its own limitations and with the collegial support of sister societies, researchers, and practitioners beyond its boundaries and across the planet. This handbook invites readers to find inspiration and, we hope, new directions in which to take STS projects, mobilizing relevant scholarly practices and infrastructures both already made and in the making.

On behalf of the 4S membership, we ask you to join us in thanking all of those scholars whose collective work brought this remarkable volume into being. Thanks go first to Kelly Moore, who as chair of the 4S Publications Committee translated formal procedural oversight into many days of managing complexity and offering generous personal and intellectual support to the editors of this edition of the *Handbook*. Many thanks also to members of the 4S Handbook Committee, who provided invaluable and timely contributions to the editorial process: Linda Layne, Trevor Pinch, Judy Wajcman, and Stephen Zehr (chair). Hearty thanks as well to all of the authors who made the effort to share their scholarly insights and to document the projects that compel their attention, doing so in ways that might best reach a diverse readership. As the editors explain, a total of 121 scholars writing in 36 chapters demonstrate that science, technology, and society are made in the doing; that engagements in worlds of science and technology must come with critical reflection on those interventions; and that, increasingly, STS scholarship traces and follows practices of science and technology into broader societal territories.

Finally, we congratulate and thank Ulrike Felt, Rayvon Fouché, Clark A. Miller, and Laurel Smith-Doerr for their commitment to defining "newcomers to the field" as the handbook's primary audience, and for the tenacity, integrity, and years of intensive labor entailed in realizing that vision. The result of their efforts is a collection that renders the varied field of STS scholarship as an open but definable terrain on which all of our future work can build.

Gary Downey, President, Society for Social Studies of Science, 2014–2015
Lucy Suchman, President, Society for Social Studies of Science, 2016–2017

Acknowledgments

Thanks are due first and foremost to all the authors for their remarkable patience, responsiveness, generosity, and perseverance. All chapters went through several rounds of revisions and exchanges among reviewers, editors, and authors. Sometimes deadlines were tight and authors had to put up with us invading their inboxes with reminders. We're not sure what the authors would say, but from our perspective it was a very good experience to work together and to struggle over how to bring ideas to paper and participate in advancing our field. Our special thanks go to Trevor Pinch and Gary Downey, who accompanied us in the core phases of shaping this handbook when most of the key choices had to be made, and to Lucy Suchman who was with us in the final phase of the project, preparing the launch of this new handbook, forty years after the appearance of the first edition. All three in their function as 4S presidents generously lent their support to the project, encouraged us, and offered valuable feedback. In particular, we are indebted to Kelly Moore (chair of 4S Publications Committee), who was our interlocutor all along the project, housed us for an editorial meeting in Chicago, and skillfully handled the contract negotiations with MIT Press for us. The handbook committee, chaired by Steve Zehr, offered constructive critique and encouragement in particular during the earlier phases of the process.

About 120 referees generously gave us their time, shared their critical but very constructive reflections on the chapters, and pointed to missing elements. We owe them much for their contributions to making the handbook what it is. Our profuse thanks go to Victoria Neumann, a research assistant at the Department of Science and Technology Studies in Vienna, who accompanied us for nearly two years in this process, keeping track of the flows of communication, participating in some of our Skype night sessions, continuously reminding authors, helping with a first round of copyediting, and much more. It has been a delight working with her. Finally, Katie Helke from MIT Press as well as Nancy Benjamin not only were always supportive but also guided us through the production process in a very efficient and competent manner.

We shared editorial responsibility for the chapters, with at least two members of the editorial team being responsible for accompanying the writing process of each chapter. The distribution of chapters into the five sections changed several times as the process of producing the chapters advanced—in our introduction to the handbook, we compare this work of continuously reimagining the whole volume out of the chapters with landscaping. This "cross-reading" effort, we hope, has given as much coherence as possible to the handbook as a whole. In providing peer reviewers with a standard questionnaire when refereeing the chapters, and through continuous exchange with authors there were more than 5,000 e-mails exchanged, numerous phone calls, and some face-to-face meetings with authors during the entire production cycle of this volume. The editors aimed at maintaining common standards as well as a shared vision of where the handbook should go. Finally, in imagining that many of our readers were newcomers to the field or researchers who wanted to enter a new area of specialization, we reminded authors to explain the concepts they used and guide the readers through the argument, giving them a sense of familiarity with the unknown, attracting their attention to previously unexplored ideas and connections, and creating a heightened sensitivity for the many ways in which science, technology, and society are entangled.

We met face to face several times, for a few days at a time, and spent many hours in Skype conversations crossing three time zones. We thank each of our universities (Vienna, Purdue, Arizona State, University of Massachusetts) for their support in funding our travel to editorial meetings. At the end of the day, we remain highly skeptical of the claim that communication technologies will one day replace our physical meetings. So San Diego, Washington, Chicago, Buenos Aires, Denver, and finally again Chicago became our physical meeting places. Many of our editorial debates are captured in one way or another in the introductory chapter to the handbook. The landscape the handbook cultivates is surely different from the one imagined when we began; yet we are confident that the multiple displacements have actually enriched the outcome. As editors we hope that the landscape that emerged out of this collective effort is inviting, even if sometimes still a bit rough to navigate as a newcomer. Welcome to STS thinking! We look forward to an expanding community of scholars and people concerned with how we live in and engage with ever-changing technoscientific worlds.

Ulrike Felt
Rayvon Fouché
Clark A. Miller
Laurel Smith-Doerr

Introduction to the Fourth Edition of *The Handbook of Science and Technology Studies*

Ulrike Felt, Rayvon Fouché, Clark A. Miller, and Laurel Smith-Doerr

Science and technology studies—STS, for short—is an interdisciplinary field that investigates the institutions, practices, meanings, and outcomes of science and technology and their multiple entanglements with the worlds people inhabit, their lives, and their values. As a dynamic and innovative intellectual field, STS explores the transformative power of science and technology to arrange and rearrange contemporary societies. Over the past two hundred years, science and technology have evolved to be among the most significant forms of human activity and inseparable from social, political, and economic organization. They were and continue to be instruments of military power, economic innovation, democratic governance, moral judgment, political imagination, and cultural difference. Increasingly, science and technology permeate the social and material fabric of everyday life via, for example, the explanatory power of scientific models, the quantification of metrics of individual and organizational performance, and the globalization of information, communications, energy, transportation, and other technological infrastructures. Ultimately, science and technology shape how humans experience, imagine, assemble, and order the worlds they live in.

For STS, understanding science and technology means interrogating not only how science and technology shape social life and the world around us but also how the latter in turn shape developments in science and technology. Fundamentally, STS views science and technology as historical products of human labor, investments, choices, and designs. People construct and perform science and technology. At the same time, STS scholarship emphasizes that in the process of making science and technology, people also make and remake themselves, their bodies and identities, their societies, and their material surroundings. Thus, the idea that epistemic, technological, and social orders are *co-produced* has become commonplace in the field (Jasanoff 2004; Latour 1993); so, too, has reflexivity toward the *situatedness* of knowledge claims and technological developments (Haraway 1988). In this sense, all knowledge is local and reflects the specific historical moment, cultural context as well as the networks within which it is made. STS is sensitive to the moral economies that guide scientific research and techno-logical development as well as to the various sociotechnical modalities through which ways of knowing and living get arranged. Classificatory practices, boundary-drawing activities—for example, between science and nonscience or between disciplinary spe-cializations, specific entanglements of the social and the technical, and the complex processes of inclusions and exclusions created through these practices—all come in for careful attention within the work of the field (Bowker and Star 1999; Gieryn 1999).

STS aims to position science and technology alongside, intertwined with, and inte-gral to other important arenas of human activity. It explores the particularities of where, when, and how people do science and technology and put them to work in making and changing the worlds they inhabit. STS research and pedagogy seek to open up science, technology, and society to critical assessment and interrogation. STS schol-arship broadly asks, why have contemporary societies centered their imaginings of the past, present, and future on science and technology—and why do those visions differ so starkly from one another across times and cultures? Why do societies make science and technology—and, along the way, themselves—in one way as opposed to another? What and how do communities, from smaller groups to nations, choose to know, and how does that knowledge intersect with how they choose to live and govern? How do new sociotechnical arrangements come into being, get deliberated, and stabilize? How do people create diverse assemblages of cognitive, social, material, and technological realities, and what outcomes do these configurations have for how people live, work, and play? What happens when these arrangements are contested, changed, or even fall apart, potentially reconfiguring the worlds they have helped to constitute? STS has always asked *cui bono*—Who benefits from specific configurations of science and tech-nology? Increasingly, STS also asks, how can our insights be put to work in ways that improve outcomes for people and the planet?

Give Me a Handbook and I Will Raise a Field—What Handbooks Do

The title of this section alludes to an early article by Bruno Latour (1983, 141), "Give Me a Laboratory and I Will Raise the World," that reflected on what was then the STS innovation of the moment: detailed studies of the work and life of scientists in the laboratory. Latour reminded us that what scientists do and how they know is always performative, in respect to both defining their field and its relations to wider society. This particularly holds for the process of editing a handbook. From an STS perspective, a handbook—commonly understood as a reference book in a particular field—does not merely reference or reflect "a reality"; rather, in purporting to describe a field, it also helps to call it into being, to shape its relations with both neighboring disciplines and society more broadly, to make it real for its readers. In the next few paragraphs, we reflect on this idea, followed by a short review of the three previous handbook editions, exploring how they mirror important parts of the history and the concerns, engagements and contexts of the field in time and space. Finally, we end the chapter by discussing what this fourth edition of this handbook tries to achieve and which features of STS it highlights.

A handbook tries to capture a specific moment and its trajectory; it draws boundaries and makes selections. Over the past four decades, the three previous editions have done just this. They accompanied and revealed the gradual establishment of STS; they observed the field's dynamic evolution and demonstrations of its significance; they charted the continuities and shifts in its sensibilities; and they showcased the contributions made by its manifold ways of analyzing technoscientific worlds. Handbooks thus always reflect the invisible timescapes (Adam 1998) specific to a particular moment. Individual handbooks bear witness—and invite attentiveness—to the multiple, culturally rooted time-related practices: recalling and mapping past endeavors; projecting potential roads to take; showing disruptions, continuities, and accelerations in the development; capturing experiences of the pressing problems of a time; and imagining technoscientific and societal developments as well as how they connect pasts and futures. Handbooks—and their individual chapters—thus perform a specific reality of the field.

Media such as handbooks have another important role for academic fields. They are deeply entangled with the memory practices (Bowker 2005), forming a kind of repository of past futures while creating future pasts. They are arranging the past in specific ways, attributing central actor positions, identifying essential innovations, highlighting key questions to be asked, and shaping them in prospective and retrospective moves into developmental trajectories. And handbooks are directed toward the future; they point to issues to be taken up, sensitivities to be developed, and new approaches to be imagined.

Thus, making a handbook is in no way a straightforward mapping exercise. While the metaphor of a map may seem attractive and something to strive for, as it promises to put some order in place, to draw boundaries, and to indicate directions, STS has been very good at resisting clear-cut mapping enterprises so far. The field, if that term can even capture the richness and diversity of work adopting the label STS, has always been much more a complex choreography of elements and identities. Texts often "belong" to multiple categories and classifications. They claim and are claimed as STS; yet they frequently also belong to sociology, political science, anthropology, or other academic territories. STS works draw on a complex hinterland of methodological and theoretical repertoires from a broad array of social scientific and humanistic disciplines; yet it also has its own stance on method(s) (e.g., Law 2004). In particular locations and at particular times, STS seems to have a strong core and clear boundaries; at other moments, its strength lies in its capacity to be an integral part of other intellectual and social environments.

The dynamics of STS remind us that we live in a world of multiplicities, in which naming, defining, and mapping are acts of ontological politics (Mol 1999), which give or take explanatory power and authority. Each handbook, including this one, is a specific site of production and performance, with distinct practices of assemblage, where meanings are negotiated, and where a particular version of STS gets done, across the different chapters and in their conversation, whether following the same line or expressing dissenting voices, as they are written and read, individually and together.

Altogether, the contents of our handbooks aim to stand for STS as a field, with everyone fully knowing that this is an impossible task. Many chapters—and their corresponding ideas and lines drawn through the intellectual territory—inevitably remained unwritten. Choices are made about what to include in limited space. The chapters that make it into any volume participate in giving shape to a landscape populated by many important concepts, by a broad range of intellectual traditions, by diverse authors and institutions, by multiple training programs, and by imagined and real readers and seekers of expertise. It is worth reflecting, therefore, on the stories each edition of the handbook tells us.

Getting to Know STS through the Lens of Handbooks

The 1977 Handbook

Each previous edition of the STS handbook has generated a rich set of performances, both retrospective and prospective, documenting aspects of a moment in the field's historical development. For the editors of the first handbook, Ina Spiegel-Rösing and Derek de Solla Price (1977), the struggle was to conceive and bound a field that was not yet born, that existed only in the imagination of possibilities yet to come, while seeming nonetheless already to share some territory, or at least some concerns. Produced under the aegis of the International Council for Science Policy Studies, the process of envisioning the volume began in 1971, when a group of scholars identified a need "for some sort of cross-disciplinary mode of access to this entire spectrum of scholarship" (ibid., 1) addressing issues of science, technology, and society.

Spiegel-Rösing (1977, 2) described the difficulty of "select[ing] a team of authors from all the different disciplines and fields we felt had to be incorporated." One of the contributors later shared some nostalgic reflections on the excitement of the moment: a small group sat together and sketched the outlines of a field that was there and not yet there at the same time, unclear about its future or even whether it would have one. Ultimately, sixteen researchers authored individual chapters, grouped into three sections: "The Normative and Professional Contexts," "Social Studies of Science: The Disciplinary Perspectives," and "Science Policy Studies: The Policy Perspective." Classical disciplines (e.g., psychology, sociology, economics, etc.) were the ordering principle, especially in section II, something that has disappeared altogether from more recent handbooks (although traces remain). Some of the authors and their institutions were key in building programs in the following decades; others are today no longer visible actors in the landscape of STS.

The introductory chapter of the first edition of the handbook offered a depiction (or, perhaps more accurately, a prediction) of a field in the making, highlighting that "SSTS" (the acronym used in the first handbook, for the study of science, technology, and society) "exhibits all the signs of an institutional field, such as specialized research institutions and teaching programs, organization at a national and international level, and specialized media of communication" (ibid., 9). The introduction described an array of institutions that in one way or another studied science and technology from a humanities and social sciences perspective. The chapter argued that the growing study of science, technology, and society emerged out of "a changed relationship of science to power" after World War II. "A certain *laissez-faire* attitude toward the development of science, alongside government support for enormous growth in research funding and activities," had led to a "new experience in the scale of use of science and technology" (ibid., 7–8). Spiegel-Rösing thus portrayed SSTS as triggered by "war, economic expediency and political power," combined with "the misdirection of science and technology, [and] the subsequent widespread disillusionment about their unconditional usefulness and desirability" (ibid., 8). She depicted SSTS as having a clear policy orientation and an important potential future voice in processes of technoscientific development.

Central to the map of the emerging field was an "East-West" ideological, Cold War–inspired divide: the "East" being socialist and seeing planning and prediction in science and technology at its core, while "the West" was described as nonsocialist or bourgeois (Spiegel-Rösing 1977, 12). Today, seen from hindsight, we recognize the deep problems attached to such differentiations. While their nuances are beyond the scope of this introduction, the first edition serves as a crucial reminder of just how rooted the field's origins are in an environment that took such divisions as essential—indeed, that emphasized the significance of science and technology to the East-West divide. In the end, as expressed in the first edition of the handbook, what held the western version of SSTS together was a concern for how science and technology came to matter in a world full of tensions and conflicts.

We also see in the first edition the effort to sketch what could hold the field together. Spiegel-Rösing (1977, 20–25) described the core of STS as having five major "tendencies": (1) the humanistic tendency to "get back the 'actor' into the picture"; (2) the relativistic tendency focusing "on the role of the specific historical moment in which knowledge and technologies are created"; (3) the reflexive tendency to analyze "science and technology as taking place within a situation at a given time"; (4) the desimplifying tendency that fosters "movement away from black-boxism"; and (5) the normative tendency which points at "increasing readiness to take the normative aspects of science and technology into account" (Spiegel-Rösing 1977, 20–25). This last led to critical discussions about "the social responsibility of scientists" that cut across a number of chapters in the first edition of the handbook.

The 1995 Edition of the Handbook

Nearly twenty years later, the 1995 publication of the handbook illuminated a field that had developed considerably and a world that had changed markedly. Edited by Sheila Jasanoff, Gerald Markle, James Petersen, and Trevor Pinch (1995), the second edition was published in collaboration with the Society for Social Studies of Science, by then a burgeoning professional society that had grown rapidly after its 1975 founding (the establishment of the European Association for the Study of Science and Technology followed in 1981). Using the metaphor of mapmaking—precisely, "constructing a map of a half-seen world"—the editors sought to capture the struggles and difficulties encountered in the representational work that any handbook must do. They described this work as a form of "imaginative risk taking and diligent codification" (ibid., xi). What and where are divisions and boundaries? Where are the blank spots that need filling in? Where has work occurred that may not have been seen?

For the editors "the project [was conceived] as something more than the traditional, treatise-like handbook that would clinically describe the world of STS" (ibid., xi). The field had, in their view, "not yet achieved the hoary respectability that merits such dispassionate, and unimaginative, treatment" (ibid., xi). Yet institutional developments were pushing change. New doctoral programs in the field, established in the 1980s and early 1990s, created new demands for graduate training. The handbook thus "wanted to compile scholarly assessments of the literature that could be presented to neophyte graduate students as the state of the art in STS" (ibid., xi). The aim was to guide an incoming generation of young scholars, attracting them to a field that was still very much described as "in the making" but that nonetheless now aspired to present a clear intellectual vision rather than a sum of different disciplinary approaches.

The editors described the process of assembling the handbook, which in the end encompassed twenty-eight chapters. They observed the push by the field to define itself, with suggestions for contributions flowing in from many unexpected areas. As they wrote, "We decided to accept this movement towards self-definition … [and redrew] the boundaries [of the initial map] to include more of the topics that authors wished to address" (ibid., xiii). The second edition thus clearly reflected the growth and diversity of the field, even as it still showed traces of the disciplines from which scholars had come and STS had taken inspiration. Yet section titles no longer referenced traditional social science disciplines as in the first edition of the handbook. The name of the field and its acronym changed from SSTS to STS, as the editors of the second edition saw that the time had come "to adopt the newer guise of S&TS—'science and technology studies'" (ibid., xi).

Across the chapters, the second edition displayed methodological innovations in the field as being entangled with emerging theoretical understandings, including "controversy studies," "laboratory studies," and the beginnings of "actor-network theory." The situatedness of knowledge, the (social) constructivist approach to understanding science and technology in the making, an emerging attention to practices of knowledge and technology making, the realization that what we perceive as science is the outcome of complex boundary work, as well as the very idea that natural and social orders have to be seen as co-produced: all were presented as starting to form important basic understandings driving STS research. The chapters illuminated a merging of science and technology in the vision of the field—also reflected in the chapter headings—expressed through the more frequent appearance of the notion "technoscience." The chapters highlighted the shared concerns about science and technology emerging in other closely aligned new fields of interdisciplinary scholarship, such as feminist theory, gender studies, and science communication, which were absent in the first edition. Concerns about gender in the second edition focused on the exclusion of women from science and on the relationships between knowledge production and the ordering of gender dynamics in society—obvious topics in the face of the success of the women's movement in the two decades since the first edition. But also issues of feminism and technology were debated. Yet, gender issues remained mostly confined to the chapters explicitly addressing them. Postcolonial theory or reflections on race were only present at the margins, if at all. Growing attention to the rhetoric of science and to the communication practices of science, both internally and between science and society, as well as issues of public participation marked an important shift in the field toward understanding how science and society entangle. There were initial signs of what would become a very important theme in the third edition of the handbook: the interweaving of public participation, engagement, and governance.

It is also relevant to ask which questions were less clearly visible in the second edition. Indeed, while the role of values in science and technology were addressed, explicit questions of the social responsibility of scientists, which were quite prominent in the first edition, were much less clearly visible, as were rapidly growing broader concerns about ethics in scientific research. A handful of chapters addressed broad topics such as the military, environment, and biotechnology, yet there is a sense that, in a post–Cold War world, STS seemed to be freer to address the immediate societal concerns that had driven earlier investments in the field in favor of reflecting on the field's development and deepening its theoretical core. Interestingly, reflections on method were sparse.

The 2008 Edition of the Handbook

More than a decade later, in 2008, Edward Hackett, Olga Amsterdamska, Michael Lynch, and Judy Wajcman edited a third edition of the handbook. It presented STS as a broad, accomplished field "that is creating an integrative understanding of the origins, dynamics, and consequences of science and technology" (ibid., 1). The third edition was substantially larger than its predecessors, with thirty-eight chapters encompassing a significantly wider breadth of topics and authorship. By underlining in the introduction that the field should not be seen as a "narrow academic endeavor," the editors pointed to the double profile STS developed from the start and that has become an explicit part of its identity. On the one hand, the field should become a successful academic endeavor comparable to a discipline, offering a somewhat distanced (even though reflexive) analysis in order to gain intellectual authority and credibility. On the other hand, it should be engaged with a whole range of societal actors and, as the editors described, to take a clear stance on "matters of equity, policy, politics, social change, national development, and economic transformation" (ibid., 1).

First and foremost, the third edition's chapters paid growing attention to the notion of practice. Science and technology were thus investigated not as stable entities but rather through how they are done and the multiple complex and situated arrangements in which this doing happens. This shift also focused research on processes of meaning making, how they work, and the multiple outcomes they produce. "Hybrids and ambiguities, tensions and ambivalences," were thus at the center of attention for most authors. The editors stressed that "sets are fuzzy; categories are blurred; singulars become plurals (sciences, not science; publics, not public, for example)" (ibid.,

4). Attending to multiple entities and identities in the practices and performances of science and technology required different forms of analyses. Perhaps this critical complexity and the challenges it raises for STS scholars is related to the increasing collaboration and multiple authorship of chapters that began to appear in the third edition; this expanding authorship is even more apparent in this current edition's chapters, as we discuss below. The third edition also spelled out very clearly another way to conceptualize the complex multiplicity of practice, which was already visible in the second edition but only just becoming known: the co-production of epistemic, social, and technoscientific orders. Co-production, as a number of chapters in the third edition emphasized, moved STS analyses away from thinking in terms of clear-cut causalities and directionality in the relationship between knowledge and society. Rather, knowledge (how the world is understood) and norms (how people expect to live in the world) are co-produced, that is, come into being together, through dynamic interactions, and constitute one another.

The geography of concerns visibly shifted in the third edition as well: "global" perspectives, combined with a heightened attention to place and history, are palpable in many chapters, as is sensitivity to comparison. The field's greater attention to these topics and perspectives in 2008 was perhaps not surprising given the rise of globalization on the international agenda by the beginning of the twenty-first century and the growing reconceptualization of many facets of the social and natural worlds as planetary in scope (i.e., markets, networks, systems, ecologies). Yet STS has insisted on attending to the very real differences that occur as nominally universal entities or ideas move across national and cultural boundaries, thus giving rise to a productive comparative sensibility. STS as a field had also become more international—even though this was not yet significantly present in the authorship of chapters in 2008—with growing communities of scholars active in Asia, India, and Latin America from the turn of the century onwards.

Gender issues as well as postcolonial perspectives started to appear across a number of chapters in the 2008 edition. Rather than being confined to separate chapters, these perspectives brought concerns about inequalities into conversation with many other topics. This integration is in line with a heightened attention to agendas of social change at a moment in history when a series of financial crises threw global inequalities into a harsh new light and created international dissent around institutions of global knowledge and power. It is also in line with a renewed attention to questions of values and politics in knowledge-making, which were again explicitly addressed in the third edition in chapters centered on questions of ethics and responsibility in science and technology, after these themes had largely disappeared in the second edition.

Readers of the third edition encountered numerous chapters in which close analysis of fields or cases came with clearly normative conclusions, but in a much broader sense than those in the first edition. For example, the third edition included multiple chapters examining the relationships between industrial and academic science, critical reflections on the commercialization of universities, and related trends toward privatization and patenting in research that began in the 1980s and accelerated through the 1990s and 2000s. The very notion of politics also changed in the third edition, going beyond classical ideas from political science and policy toward a <u>view of power as imbued and embedded in the mundane practices and arrangements of day-to-day work</u> and life of scientists, engineers, and publics. The notion of governance made its appearance with growing concerns about future-making practices in contemporary democracies, as well as societies' engagement and participation in these processes.

A topic virtually absent from the third edition of the handbook, as from the previous ones, was explicit broader attention to research methods (as opposed to using STS methods to make theoretical points, which appeared from the second edition onwards). Editors sought to walk a fine line by acknowledging that STS "may have been 'against method'" (ibid., 5) while still trying to use their edition's focus on practices to urge readers to be attentive to methods and epistemic assumptions. As discussed below, this current edition in your hand or on your screen takes up this challenge by including a section dedicated to methods.

Handbook as Reflexive Landscaping

Bringing to life the fourth edition of *The Handbook of Science and Technology Studies*, forty years after the first STS handbook appeared (Spiegel-Rösing and de Solla Price 1977), is a fascinating and exciting endeavor. Like its predecessors, this edition is a snapshot of a particular intellectual moment. Today, the social, institutional, and epistemic landscapes of the field continue to develop in both breadth and depth. STS scholarship has proliferated in the humanities and social sciences, assisted by the creation of diverse subfields focused on science and technology in sociology, anthropology, economics, rhetoric, history, philosophy, and political science. The number of undergraduate and graduate students trained in STS continues to grow, as universities invest in hiring new STS faculty across many departments and programs. Independent undergraduate and graduate programs have developed in Europe and North America and are starting to appear in Asia and Latin America. Ph.D. programs have been established in the field, under a variety of names, even as the majority of researchers continue to learn their trade from STS faculty teaching in disciplinary graduate programs. And STS continues its long tradition of providing intellectual and institutional space for projects and collaborations that extend well beyond academia.

The historical development of STS has been accompanied by continuous reflections on the tensions between the agenda to form an autonomous field and the wish to remain entangled with traditional social science disciplines. While this tension is sometimes rightfully problematized, it might also be seen as supporting the field's diversity and creativity. The field thrives in part from a shared identity, motivated by a perceived need to promote critical analyses of science, technology, and society in research and public discourse. For many, STS is and remains a necessary antidote to the over-promises expressed on behalf of science and technology by scientists, engineers, entrepreneurs, business leaders, politicians, and even publics. The work of building a strong "brand," setting standards for rigorous scholarship as well as expanding opportunities for high-quality training in research and its applications, stem from a commitment to have a voice in the shaping of technoscientific worlds we live in. The relationships of STS to science and engineering, business, and policy institutions range from antagonistic to unnoticed to highly productive partnerships. By attending not only to the execution of good research but also to the effective engagement with multiple audiences, these interactions help the field continuously innovate beyond its borders as well as in its own work. This work has resulted in important accomplishments in recent years. STS has proven in multiple contexts the richness and intellectual strengths of its contributions to the understanding of complex technoscientific worlds, a fact that is increasingly acknowledged not only through new funding for research in the field but also through the uptake of concepts in other research fields. The field's expertise is recognized on many occasions and in many places as an important building block in policy deliberations, especially in Europe.

Yet while STS has become a significant field, at times it still feels more open to question than many others, perhaps mirroring the fluidity of the technoscientific worlds it studies. The perceived fragility also derives from the field's commitment to plurality and openness to neighboring fields as well as from the relative partiality of its institutionalization. STS, unlike more classical disciplinary approaches, is not everywhere taken for granted as part of the institutions of higher education. Nonetheless, the field thrives in the sheer variety of people who have conducted exciting research projects in the humanistic and social studies of science and technology. At many universities, even though STS may not have an institutional home, it still has numerous practitioners, perhaps even more than some traditional departments. The field inhabits and draws upon the insights of many disciplines, flourishing amid an enormously varied array of research methods and approaches, demanding its practitioners read and engage across numerous intellectual traditions. STS practitioners call home a remarkably diverse array of institutional settings, from academic departments in the humanities, social sciences, natural sciences, and engineering, to business, policy, and nongovernmental organizations. This diversity challenges the field's unity, opening it up to myriad forms of inquiry, reward systems, institutional norms, and imaginative endeavors that pull it in many directions, all the while making possible new avenues for fruitful and productive research. Yet the upshot is a highly diverse workforce with the necessary knowledge and skills to tackle an impressive range of problems and to put the resulting insights to work in solving them.

The field's diversity and dynamism created a challenge as we tried to put together a handbook matching the richness of the field. As editors, our principal goal in constructing this edition was to invite newcomers to the field. Our own trajectories into and through the field helped us reflect on what it would take to make this vision work. Unlike fields with longer traditions and more thorough institutionalization, STS has no single or preferred entry point. As is typical, each of the editors entered the field through individual pathways: one via one of the field's long-standing Ph.D. programs, one from an early science career, one after doctoral training in engineering, and one from sociology. Helpful, too, in understanding the challenge of creating a volume designed for newcomers to the field, has been the diversity of our current work: one leads an STS department within a social science faculty with designated STS master's and doctoral programs open to students across all disciplines; one leads an STS doctoral program within the context of an interdisciplinary school that bridges STS with science, engineering, and policy; one leads an interdisciplinary research institute that encompasses all of the social sciences while holding a faculty appointment within a sociology department; and one leads an interdisciplinary humanities program that sits alongside STS in the epistemic landscape of U.S. academic institutions. We hope that we have created a rich representation of the field, making visible its productive diversity. This handbook might become for newcomers a first step on a much longer journey of exploration, encounter, and investigation. We also aspire to introduce those already working in the field to their neighbors and to neighboring terrains and landscapes.

In the editorial work of crafting this edition, the metaphor of reflexive landscaping has taken on an important role. Landscapes are never stable. They emerge out of a blend of material elements and cultural arrangements, subject to individual and collective perceptions. They have a built-in temporality, growing and evolving over time as multiple elements and actors interact. Landscaping thus demands an effort to develop more or less complex choreographies, bringing together diverse elements, imagining how they might co-evolve, reflecting what gets left out, always aiming at the creation of a meaningful whole. Handbook editors and authors individually and collectively imagine "the field," reflecting where it comes from but also projecting where it ideally might go.

For us as editors, the work of assembling this handbook is akin to that of landscape architects, which is not always an easy task. Authors provided us with "seeds"—chapter proposals in answer to our call for contributions—out of which we created an imagination of what the landscape might look like, what each chapter could contribute, what was missing, and who the readers might be who would stroll through it. We debated, shifted, deliberated, and created a draft blueprint. Then, of course, authors wrote and rewrote draft chapters, contributing with their unique understandings of STS literatures, their personal and intellectual histories, and their ways of engaging technoscientific worlds. The chapters developed in new and exciting directions that often differed from early ideas. Reviewers weighed in, as did we, leaving traces via multiple feedback loops. Chapters started to interact, overlapping, complementing, and diverging. The editors struggled to decide how much landscaping to do and how much to let chapters simply grow. Some rearrangement of the overall landscape—expressed in the table of contents—became necessary as chapters took more concrete form. Finally, the text went to press, and our job was done. The landscape is now open to the readers and will surely take ever-new shapes through their experiences, interpretations, and engagements.

A Political Geography This Edition Speaks and Thinks From/With

Landscapes are also material places that matter. In STS, political geography is widely recognized as a key element in the making of knowledge, both in natural science and in our own field. Authors inevitably speak from and about particular cultural contexts, experiences, institutions, and positions. Already, in the early phase of conceptualizing this edition, the council of the Society for Social Studies of Science and others expressed the importance of ensuring the "inclusion of non-Western science and technology" and this remained a concern during the design and production process of this edition.

Being attentive to place is not embracing determinism but rather reflecting how such unequal spatial distributions might matter to what and how we (can) know. Like everyone else, chapter authors and editors circulate within narratives, discourses, paradigms, and problem framings—as well as social and material sites of work—that vary across geographies. As Gieryn (2000, 464) points out, STS does not worry "that the particularities of discrete places might compromise the generalizing and abstracting ambitions of the discipline." Quite the contrary, STS has been very much concerned with the geographic asymmetries inherent to the field that require particular care with regard to the situatedness of particular knowledge claims (Haraway 1988). When chapters describe certain places, regions or whole continents, do authors speak from that location and with voices present there, or simply about the locale? To what extent does our scholarship, more broadly speaking, tend to address the experiences of a particular place as if it were universal? And, if we integrate place into our frame of reflexivity, do we study places of alternative knowledge-making (e.g., Watson-Varran and Trunbull 1995)? Do we fully acknowledge that postcolonial sensitivities should also be present in reflecting on our own field? Or do we tacitly embrace the languages and paradigms of development and of catching up when speaking about STS?

STS has shown in its own studies of different scientific and technological fields how place matters and how knowledge is touched by local conditions. Livingstone offers a reminder that "space is far from a neutral 'container' in which social life is transacted, [...] is not [...] simply the stage on which the real action takes place. Rather, it is itself constitutive of systems of human interaction" (Livingstone 2003, 7)—and thus also of the knowledge we (can) create. He continues to underline the crucial role of the positions we speak from for what can be spoken, pointing at the intimate connection between "location and locution" (ibid., 7).

The places where knowledge is created, as well as the localities where investigations are performed and where financial support originates, matter; and so does the distribution of places where people receive their education and first socialization in STS, as well as the "maps" that locate academic centers and peripheries and that are used to guide researchers in their academic career trajectories (e.g., Felt and Stöckelová 2009). Our efforts to be attentive to different places where ideas and perspectives develop are tempered by the acknowledgment that resources and ways of seeing the world are distributed very unevenly in the field.

In a world of manifold mobilities, however, knowledge geographies are more complex than straightforward physical maps of institutions. Figure 0.1 maps a specific indicator: where each author works or worked while writing the chapter. It sheds light on the prevailing institutional geography of authors and shows a quite uneven distribution, which mirrors the specific history and politics of the institutionalization of STS.

In this fourth edition of the *Handbook*, two large clusters of author affiliations in the United States and in Europe are visible (see figure 0.1), making up approximately 90% of the number of authors. About 50 percent of the authors have U.S. affiliations. The United States and the United Kingdom together provide the institutional home for two-thirds of the authors. Including the Netherlands brings the total coverage of the three countries to roughly 75 percent of the authors' institutions. However, it is important to highlight that for the first time six authors (5 percent) have their institutional

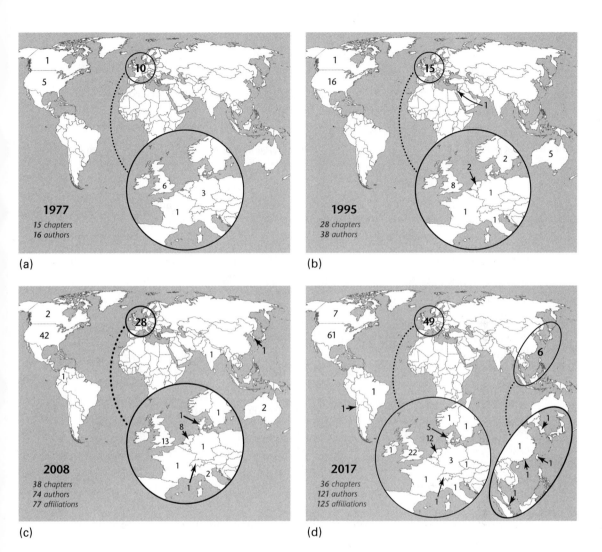

Figure 0.1
Geography of institutional affiliations of authors in the fourth (2017), third (2008), second (1995), and first (1977) editions of the *Handbook of Science and Technology Studies*. (See Color Plate 1.) Map of World with Countries (http://freevectormaps.com/world-maps/WRLD-EPS-01-0013)— Outline by FreeVectorMaps.com.

affiliations in Asia and two authors in South America. Neither Australia nor Africa is present in the institutional geography of this edition.

While this could be read as a story of a developing yet rather uneven distribution of who gets voice, it is essential to stress that each author's personal geographic history is inevitably much richer and reflects a much broader range of experiences. Each author's location may not accurately reflect cultural origins, political leanings, social perspectives, or sites of investigation. Often we live and work very far from our place of origin and development or the locations we call home. Furthermore, within national borders there often exists much regional variation in experience and inequalities. Still, the maps tell us something; these locations and the distribution of authors across them surely shape academic experiences and express specific academic traditions and valuation regimes as well as dominant research approaches. Locations also reflect distributions of power and wealth within the field: wealth to attract the best from all over the world, power to set standards and frame problems. Situatedness and reflexivity are key concepts of STS—therefore it is essential to understand this short reflection on the field's own geography as an expression of these core traditions.

No fewer than 121 authors (with 125 institutional affiliations at the time of publication) were involved in the 36 chapters of this edition. This author list is very large compared to the first three editions, which listed 16, 38, and 74 authors, respectively. The editors, early in the process of calling for submissions and negotiating the orientation of chapters, invited the building of author consortia in order to include more scholars earlier in their careers, to expand the geography of authorship, and thus to integrate a greater diversity in the situated perspectives of authors on the interactions of science, technology, and society. This effort was only partially successful. We are pleased to have a greater number of younger scholars present. The success of our pushing to increase the geographic spread of authors' institutional affiliation was less effective. Though we may not want to overgeneralize, the maps are nonetheless telling about the current configuration of the field.

When comparing the fourth edition's author map to those of the previous editions, we do observe some signs of gradual change (see figure 0.1). The first edition (1977) had only 16 authors, all from institutions in Northern America and Western Europe (United Kingdom, Germany, and France). This distribution changed slightly for the second edition. The focus on the United States and Europe remained, but the U.S. presence increased, the diversity of European countries hosting authors grew, and Australia emerged as an important site of authorship (5 authors). East Asian and Latin American authors were still completely absent in the second edition (1995). In the third edition (2008), we observe a further shift in the institutional geographic of authors. Fifty-five percent of the authors worked in U.S. institutions when writing their chapters, which is the highest value of all four editions. The United States, the United Kingdom, and the Netherlands together hosted more than 80 percent of the authors in the third edition. The number of Australian contributors had decreased compared to the second edition, and there would be no contributors from Australia in the fourth edition. However, it is important to see that Asian institutions began to appear in small numbers (2) in the third edition (2008), as well as Latin American institutions (1).

In this fourth edition, more authors work in these regions, reflecting the field's expansion on the international level. In 2001, for example, the Japanese Society for Science and Technology Studies (JSSTS) was founded, holding regular meetings and producing publications. In 2007 the first issue of the international journal *East Asian Science, Technology and Society*, sponsored by the Taiwanese Ministry of Science and Technology, was published on the topic of public participation in science and technology. STS also acquired new salience within India's social science and environmental research communities, even though not reflected in the contributions to this edition. Similarly, with the establishment of ESOCITE (Latin American Association for the Social Study of Science and Technology), STS in Latin America became more clearly visible in the field's terrain from the mid-1990s onwards. These changes are also visible in the locations of the annual conferences of the Society for Social Studies of Science (4S), which were held in Tokyo in 2010 and in Buenos Aires in 2014.

Changing Coauthorship Practices

Coauthorship practices across the editions of the STS handbooks have undergone considerable change. The growth in numbers of participating authors and in the authorship teams of individual chapters expresses not only the ways in which we perform research but also a stronger phenomenon of network building in the field. It also reflects our explicit invitation and encouragement to form collaborations to try to broaden the representation of the field. Figure 0.2 shows this growth in the number of collectively written chapters. While the first edition's chapters each had a single author, the second edition already showed a number of coauthored chapters. This pattern only changed slightly for the third edition, with a small number of chapters having three or more authors. This fourth edition of the handbook has shifted the pattern insofar as single-authored chapters have become rare—only four of thirty-six; nearly one-third of the chapters have three authors.

Examining these author partnerships, we find that geographic proximity remains important, as does inclusion in recognized academic systems and networks, to processes of group formation in the field. This helps explain our failure, described above,

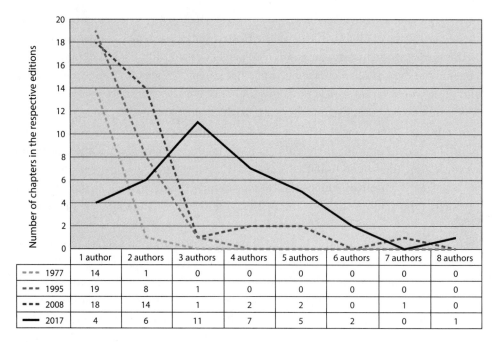

	1 author	2 authors	3 authors	4 authors	5 authors	6 authors	7 authors	8 authors
--- 1977	14	1	0	0	0	0	0	0
--- 1995	19	8	1	0	0	0	0	0
--- 2008	18	14	1	2	2	0	1	0
— 2017	4	6	11	7	5	2	0	1

Figure 0.2

Changing coauthorship patterns in the editions of *The Handbook of Science and Technology Studies*. (See Color Plate 2.)

to expand the geography of authorship more successfully, although it does not excuse it. Writing in a more collective manner also carries its drawbacks. Increases in the size of author groups may enrich the writing process by bringing in additional perspectives, but it also can make drafting a coherent essay more challenging. In an era of strict word and time limits, teams had to find time to negotiate common languages, arguments, and voices. Nonetheless, nearly 60 percent of the chapters had coauthors working at institutions in different national and regional settings, a considerable increase over the 20 percent of chapters coauthored internationally in the third edition. As the first two editions had predominantly single-authored chapters, it is not surprising that international author collaborations were rare in their cases (e.g., two such chapters in the second edition).

The Epistemic Landscape of the Fourth Edition

This edition is organized into five major sections. The first section covers methods and participatory practices in STS research. For our field, methods can never simply be a straightforward application of tools and techniques. In the field's history, the empirical observation and analysis of other fields' research methods is one of the most important topics of study. That body of work highlights that all methods are partial, are contingently and contextually applied, and yield inevitably value-laden data. Reflecting those findings back onto STS research creates a unique stance toward method in the field, the subject of John Law's opening chapter, which has the potential to become a classic as an introduction to STS as method/STS and method. An important facet, as described by Law, is that for an STS researcher, method is more than mere data collection and analysis; it is a form of ongoing reflexive practice and an exercise in systematic engagement with her interlocutors and the world she and they inhabit. The rest of the chapters in the section illustrate that reflexivity. Chapters 2 through 4 present particular examples of diverse research techniques and materials in the field, and the subsequent four chapters explore engagements of STS research with other communities of practice. This short collection of chapters should in no way be considered a comprehensive treatment of method and practice in STS, a topic that could fill an entire volume by itself. We hope, however, that it will help spur fruitful conversation on the topic.

The second section of chapters explores the field's preoccupation with the subject of knowledge. How do scientists, societal actors, and collectives like states know? How does the resulting knowledge get constructed and shaped in diverse social contexts? How does that construction process ultimately reconfigure social institutions, identities, and relationships? Each chapter examines, via different methods in diverse settings, the mechanisms through which knowledge, people, and societies are co-produced. The chapters explore the politics of making and organizing knowledge in contemporary societies, its relationship with the exercise of power and the institutions of governance, and its intersection with the dynamics of inequality and exclusion. Inspired by feminist scholarship and postcolonial studies, several of the chapters critically reflect on how social orders are constructed and performed through diverse science-based classifications such as race, gender, sexuality, and ethnicity. Many of the chapters explicitly address scientific knowledge. Other chapters examine less formalized, more diffuse forms of knowing and imagining that contribute to the self-understanding of individuals and communities, entanglements with science and technology, and the ways futures are envisioned. In all of the chapters, one senses a clear call to enrich our stories of how knowledge and worlds get made together, thus opening possibilities for new ways of accounting for—and holding to account—how technoscientific societies and the people who inhabit them come to be.

The third section shifts focus to the field's other great fixation: technology. For STS, societies make themselves not only through knowledge but also through the design, construction, and use of material devices and infrastructures. Collectively, the chapters in this section reflect anew on and update some of the field's oldest and most important questions. Who designs and uses technologies, for example, to what ends, and with what implications for their identities, values, relationships, networks, and larger societal contexts? How do social configurations and relationships shape and get shaped by technological and informational infrastructures and landscapes, whether in individual locales, like cities, or in far-flung global networks and markets? How can STS—and people more broadly—achieve a critical purchase on the interdependencies and relationships among technologies and societies that enables them to direct future developments toward greater equality and improved social, political, and economic outcomes? In answering these questions, the chapters in this section draw on highly diverse methods and intellectual traditions to examine the complexity of societies' intertwining with technology. They reveal how humanity's symbiotic relationship with material artifacts is continually remade, reconfigured, and reassembled. At the same time, the chapters also hint at the limitations of allowing the experiences of the United States and Western Europe to define and delimit the human relationship to technology in STS scholarship. They press us, when thinking, writing, and theorizing about our technological pasts, presents, and futures, not only to unravel the ways technology informs and supports diverse forms of inequality and exclusion but also to contest and destabilize how the design and operation of technologies embed and reinforce concepts like race, gender, and nation within social and cultural imagination and identity.

The fourth section examines the organization and governance of science, looking both at the structures and institutions of scientific research and the values that are inscribed in and around this significant form of human activity. Science is one of the most important institutions in contemporary societies, and as such STS has always insisted that it deserves careful attention. The chapters in this section focus on academic science as a form of work and track persistence as well as change in the organization and structures of that work. The chapters also explore the ways that the funding and institutionalization of scientific research and training creates and reinforces inequalities, for example, across race and gender, among disciplines, between elite scientists and other research workers, across different kinds of research topics, and more broadly in society. Finally, the chapters delve into the ways values are laden into knowledge production and circulation through a wide range of pathways, including the organization of research priorities; the development of legal, regulatory, and ethical requirements for scientific work and practice; the incorporation of scientific research in larger processes of technological innovation; and the development of new practices and expectations in the communication of science with broader public and policy audiences.

Last, but not least, the final section of the handbook looks at the contributions of the field to addressing some of the most important challenges confronting today's societies. From the first edition to the present one, a defining characteristic of the field has been its focus on science and technology as both causes of and imagined solutions to what today are called grand challenges. At the same time, as funding for the field has grown in recent decades, especially through the incorporation of STS into highly visible research programs such as the European Union's *Horizon 2020* funding schemes or the U.S. National Nanotechnology Initiative, many STS scholars have increasingly found it valuable to organize their work around problems defined by these programs as key areas of societal need. The final chapters of the handbook reflect on six such challenges: aging, agriculture, security, disasters, environmental justice, and climate change. In reviewing STS scholarship in these fields, the chapters highlight three critical ideas. First, design matters enormously with regard to the human outcomes that flow from the application of science and technology. Hence, critical analyses and reflections on the design of science and technology are essential if societies are to meet their challenges in ways that create real improvements rather than re-create inequalities. Second, the current understanding of innovation—generally defined as a process of technological creativity—is inadequate. New technologies take shape and are put to work only through society-wide processes of change. Third, the politics of expertise—or who is authorized to speak to and for what is known about an issue—is critical to the organization of governance arrangements that identify, frame, and solve problems both effectively and justly. Together, these chapters review a wide breadth of scholarship. Even so, they merely scratch the surface of relevant STS research, testifying to the ongoing significance of the field's work to societies' efforts to confront grand challenges.

Cross-Cutting Issues and Emergent Themes

One of the great joys of collaborative projects—and part of what makes them worthwhile, despite their complexity—is that shared themes and concerns emerge out of the process. Reflecting on the past decades and previous editions, it is clear that STS is a field in motion. Across the five sections in this edition, for example, we observe a shift in what we might call the narrative infrastructure of the stories STS tells. By narrative infrastructure, we refer to characteristic ways that the field describes both the problems it explores and the insights it develops. We see, for example, the continuation of trends first identified in the third edition toward investigating problems through the lens of practices and processes rather than fixed entities. There is a clear sense that science, technology, and society are made in the doing and that they are constantly being remade and redone, that performances are thus important, as are the multiplicities that come along with such an understanding. Likewise, STS practitioners are increasingly committed not only to engaging in the world—as they have always done—but also in reflecting on the field's capacity to develop engagements with diverse societal actors in order to make a difference in the world. Finally, there is attention not only to the formal institutions and work practices of science and engineering but also to broader imaginaries and entanglements of science, technology, and society. The chapters in this fourth edition add nuances to notions of performance, widen the scope of reflections on imaginaries, and point at intersections of different forms of practices. They thus draw out a more fluid, dynamic, and perhaps unstable picture of the relations among science, technology, and society—a picture that reveals the frequent reinscription of inequalities through science and technology but that is perhaps also more open to proactive redesign toward greater justice.

How does this move toward practices manifest in recent STS scholarship? While the importance of practices has been acknowledged since its early phases—for example, in the laboratory (Knorr Cetina 1983) and in other activities where knowledge is created and distributed or technologies are shaped—the contributions to this edition have enlarged the focus. Explicit reflections on how concrete knowledge generation happens through the micropractices of laboratory researchers are virtually absent. By contrast, an abundance of chapters review scholarship examining the practices of doing science and technology across wider terrains—for example, in democratic institutions, social movements, cities, information infrastructures, art studios, scenario exercises, public participation experiments, markets, development projects, and security agencies. Increasingly, STS finds its objects of inquiry scattered throughout contemporary societies, wherever people are engaged in creating knowledge and rearranging materiality.

In this edition there is also new attention to STS as a form of reflexive method. This attention is particularly visible in the first section but traces run throughout the volume. Reflexivity, as already mentioned, has long been a key theme in the field. Yet there is a sense in many of the chapters that reflexivity has settled in, not as an add-on but as something to be practiced daily. A number of chapters deliver short reflections on how to approach issues of concern while also pointing at the field's blind spots. Other chapters explore collective experiments, such as in the field of participation, highlighting many new configurations of science, technology, and society, critically unpacking questions of authority and expertise, and pointing out how it is often unclear who can speak in the name of science or technology.

Many of the chapters express a renewed attention to questions of responsibility and ethics as well as to valuation practices. In recent years, new ways of valuing—and evaluating—science and technology have emerged. These evaluations range from initiatives to measure research productivity and use the results to redistribute science funding to new forms of venture capital decisions for investing in technological innovation. These evaluative frameworks embed and enshrine certain kinds of values, with deep impacts on the practice and structure of research. Consideration of ethics and responsibility in science and technology have also become more programmatic and institutionalized, as seen in the rise of ethics advisory boards, review mechanisms, and funding programs for "ethical, legal and social aspects/implications" and "responsible innovation." These new governance tools often accompany research in the life sciences and other emerging fields of research (e.g., nanotechnology and geoengineering), with the goal of managing problems of moral and political disagreements or ambivalences.

Critically examining the rise of new ways of governing science and technology, while also reflecting on the role of STS in their generation and propagation, emerged as an important theme for many authors. A number of chapters highlight a pressing need to specifically examine representations and performances of the future in scientific and societal discourses. Key questions include how particular patterns of imagination, analysis, and calculation bring about specific visions of the future; how these images of the future get constructed, contested, and stabilized; how they are put into practice and help to realize specific social and technological arrangements; and how they become rooted more deeply in social thought and practice through the development and stabilization of (sociotechnical) imaginaries. These concerns have become a lively field of STS inquiry and permeate this edition.

The growing presence and integration of feminist, gender, and postcolonial studies into the body of STS knowledge is a further specificity of this edition. Not only does it include chapters explicitly devoted to these subjects, as in the second and third editions, but most other chapters also incorporate theoretical framings, sensitivities, and insights from these fields as well. Some of our authors argue there is a need for still further integration; nonetheless, the degree of penetration of key ideas is significant. For one, there is a heightened sensitivity toward how we classify the contents of the world, the ontological politics implicated in such ordering work, the epistemic and material infrastructures built to establish new social orders, and, in all this, which categories come to matter. Likewise, in many chapters we encounter the field's widespread attention to questions of inequality, exclusion, and injustice, which are deeply entangled in powerful ways with key categories of gender, race, class, ethnicity, sexuality, and coloniality, and which may be attributed to the ongoing influence of feminist, gender, and postcolonial studies. In the process of reviewing the chapters, we tried to draw authors' attention to still existing blind spots, sometimes with success, sometimes not.

Finally, while the third edition carried considerable debate around emerging techno-scientific fields like nanotechnology, this focus is much less present here. Concerns for new technologies are still important, evident in discussions of ethics and responsible innovation as well as of futures and (sociotechnical) imaginaries. What has shifted is the breadth of frame. Many of the chapters address the sense that it is not just technologies but the world as a whole that is in transition or needs to be reordered to respond to current problems. When crafting the first proposal, we explicitly reflected on the fact that, more than ever, societies are calling on STS and other academics to make an active contribution to solving societal challenges. While the field can and does critically reflect on this language, recent STS work in these areas has opened up new opportunities to engage with the framing of problems and the development of solutions, as illustrated not only by the chapters in the final section but by a number of others as well.

Indeed, the long-standing concern that the field can and should seek to make a difference in the world is present across virtually all chapters in this edition. The attention given this topic in the chapters that follow is not so much on how the field's work can be made to have lasting practical impact in areas of business, policy, or engineering; in public participation; or as a new form of expert advice. Rather, the chapters focus on how the field's diverse modes of thinking and analysis create the potential to ask critical questions about the social responsibility of science and technology as well as STS itself. Some chapters remind the reader that the birth and rise of the field in the 1970s clearly emerged out of political and ethical concerns over sociotechnical developments and accompanying practices in areas such as the environment, warfare, and gender relations. The first edition is in many ways a reflection of those concerns. And so is this one, which we find interesting.

As science and technology have evolved over the past forty years, so have their entanglements with problems of environment, security, and social inequality, and so has STS. The field's capacity has grown enormously in that time, both in the quality and depth of its ideas and in the diversity and numbers of its adherents. Yet, the problems are no less significant, raw, or immediate, even if their material and moral specificities have shifted. Global terrorism is not the same as the Cold War. Climate change differs from concerns about the net primary productivity of the biosphere. Inequalities today take different forms than forty years ago. Yet STS tells us much about how these problems remain "the same"—and why—as well as which kinds of differences are consequential for developing meaningful responses. Cutting across the volume is a renewed call to make the field matter, to better integrate past STS experiences into the development and learning of the field itself, to collaborate more closely and learn together with diverse societal actors, as well as to <u>better connect the different strands</u> <u>of thinking within the field and neighboring disciplines</u> in order to enrich our ways of capturing, describing, and intervening in the world. We invite you to explore the landscapes of STS through engaging with the ideas, observations, and analysis presented in this handbook and to find ways to bring them into conversation with the concerns that matter to you.

References

Adam, Barbara. 1998. *Timescapes of Modernity: The Environment & Invisible Hazards*. London: Routledge.

Bowker, Geoffrey C. 2005. *Memory Practices in the Sciences*. Cambridge, MA: MIT Press.

Bowker, Geoffrey, and Susan L. Star. 1999. *Sorting Things Out: Classification and Its Consequences*. Cambridge, MA: MIT Press.

Felt, Ulrike, and Tereza Stöckelová. 2009. "Modes of Ordering and Boundaries That Matter in Academic Knowledge Production." In *Knowing and Living in Academic Research: Convergence and Heterogeneity in Research Cultures in the European Context*, edited by Ulrike Felt, 41–124. Prague: Institute of Sociology of the Academy of Sciences of the Czech Republic.

Gieryn, Thomas F. 1999. *Cultural Boundaries of Science*. Chicago: University of Chicago Press.

___. 2000. "A Space for Place in Sociology." *Annual Review of Sociology* 26: 463–96.

Hackett, Edward J., Olga Amsterdamska, Michael Lynch, and Judy Wajcman, eds. 2008. *The Handbook of Science and Technology Studies*. 3rd ed. Cambridge, MA: MIT Press.

Haraway, Donna. 1988. "Situated Knowledges: The Science Question in Feminism and the Privilege of Partial Perspectives." *Feminist Studies* 14 (3): 575–99.

Jasanoff, Sheila, ed. 2004. *States of Knowledge: The Co-production of Science and Social Order*. New York: Routledge.

Jasanoff, Sheila, Gerald E. Markle, James G. Petersen, and Trevor Pinch, eds. 1995. *Handbook of Science and Technology Studies*. 2nd ed. Thousand Oaks, CA: Sage.

Knorr Cetina, Karin. 1983. "The Ethnographic Study of Scientific Work: Towards a Constructivist Sociology of Science," In *Science Observed: Perspectives on the Social Study of Science*, edited by Karin Knorr Cetina and Michael J. Mulkay, 115–40. London: Sage.

Latour, Bruno. 1983. "Give Me a Laboratory and I Will Raise the World." In *Science Observed: Perspectives on the Social Study of Science*, edited by Karin Knorr Cetina and Michael J. Mulkay, 141–70. London: Sage.

___. 1993. *We Have Never Been Modern*. Cambridge, MA: Harvard University Press.

Law, John. 2004. *After Method: Mess in Social Science Research*. London: Routledge.

Livingstone, David N. 2003. *Putting Science in Its Place: Geographies of Scientific Knowledge*. Chicago: University of Chicago Press.

• Mol, Annemarie. 1999. "Ontological Politics: A Word and Some Questions." *The Sociological Review* 47: 74–89.

Spiegel-Rösing, Ina. 1977. "The Study of Science, Technology and Society (SSTS): Recent Trends and Future Challenges." In *Science, Technology and Society: A Cross-disciplinary Perspective*, edited by Ina Spiegel-Rösing and Derek de Solla Price, 7–42. London: Sage.

Spiegel-Rösing, Ina, and Derek de Solla Price, eds. 1977. *Science, Technology and Society: A Cross-disciplinary Perspective*. London: Sage.

• Watson-Verran, Helen, and David Turnbull. 1995. "Science and Other Indigenous Knowledge Systems." In *Handbook of Science and Technology Studies*. 2nd ed., edited by Sheila Jasanoff, Gerald E. Markle, James G. Petersen, and Trevor Pinch, 115–39. Thousand Oaks, CA: Sage.

I Doing, Exploring, and Reflecting on Methods

Laurel Smith-Doerr

Reflections on the methods and design practices of science and technology studies (STS) offer a valuable point of entry and one deserving of greater attention for a community that has placed reflexivity at the forefront of discussions about scholarly practices. STS methods have not had their own dedicated section in prior editions of this handbook. We place this discussion up front to highlight the importance of how STS is done, to describing what STS is. In fact, STS as a method is one way to envision the field, through close and in-depth attention to practices is a unifying approach in much STS research.

This section of the current *Handbook* includes chapters that reflect on some of the field's sensibilities about method and practice, positioning STS in relation to broader debates about method in cognate fields and disciplines. These chapters display the variety and multiplicity of epistemologies and methods that STS scholars use through expositions of emerging arenas of research practice, such as digital, video, and art and participatory design as methods of STS. This section also discusses some increasingly common STS practices in designing our research—including engagements with public participation in science as well as collaboration with natural scientists and engineers. There are eight chapters in this section, which begins with a cohesive view of STS method in the field and works outward toward looking at the ways STS engages with making and doing science and technology.

Method

Chapter 1, by John Law, employs classic case studies to illustrate how STS examines practices and narratives in a contextualized, multivalent, and symmetrical way. Law notes that STS methods involve attending to practices and performativity, conducting case studies, critiquing epistemologies, and documenting social constructions in their multiplicities. Chapter 2, by Kalpana Shankar, David Hakken, and Carsten Østerlund, argues for a new approach to documents as a basis for STS methods, with an ethnographic sensibility toward in-depth and thick understanding of documents and their contexts. Shankar et al. make a case that thinking digitally will allow STS to improve documenting methods by attending to features such as the fixity or fluidity of documents and the practices of ownership and control of documents. Chapter 3, by Sally Wyatt, Staša Milojević, Han Woo Park, and Loet Leydesdorff, aims to demonstrate the utility of greater integration of scientometrics into STS. Scientometrics uses quantitative methods to analyze scientific literature. In this chapter, Wyatt and her coauthors present new data on co-citation of scientometric and qualitative STS publications to show there are gaps between these knowledge communities that are ripe for cross-fertilization. Chapter 4, by Philippe Sormani, Morana Alač, Alain Bovet, and Christian Greiffenhagen, describes the role of ethnomethodology, especially video analysis of interactions and conversations, in its close connection to STS methods. Ethnomethodology studies mundane practices, and video recording contributes to understanding of local accomplishments of action in everyday situations, including the mustering of scientific evidence and development of epistemology in situ.

Participation

Chapter 5, by Chris Salter, Regula Valérie Burri, and Joseph Dumit, argues that art and design are methodological as well as dissemination practices for STS and can develop more politically engaged scholarship. Salter et al. call for greater engagement with art and design in order for STS scholars to see how art and design are enrolled in the production of facts, to expand our methods through the complexity of perspectives offered by art, to explore new ways to communicate research, and to engage in political action through the intersection of science, technology, and art. Chapter 6, by Janet Vertesi, David Ribes, Laura Forlano, Yanni Loukissas, and Marisa Leavett Cohn, discusses the ways that STS designs IT as well as analyzes it. Vertesi et al. present the engagement of STS with the design of information technology in four ways: corporate employment of scholars who consult on human-computer interaction, critical analysis by scholars who analyze the values embedded in IT design, inventive collaboration by academics who focus on IT design, and inquiry-based studies of participatory design. Chapter 7, by Javier Lezaun, Noortje Marres, and Manual Tironi, describes the multivalent experiments in participation in science with goals for demonstrating and democratizing technoscience. Lezaun et al. discuss the proliferation of experiments, from reality TV to the gold standard in social science results as reported in the media. The authors ask whether experiments allow more meaningful public participation in science and technology or whether more democratic approaches are undermined by agendas that are set through experimental method and design. Chapter 8, by Gary Downey and Teun Zuiderent-Jerak, argues for putting STS ideas to work in the world. The authors reflect on how STS scholars are creating initiatives and projects for "making and doing," including public engagement products from op-eds and documentary films to pedagogical tools, supporting social movement activism with scholarship on controversies, and doing experiments in participation often in collaboration with natural scientists and engineers.

Reflection

Thus, two major ways we do STS are developed in this section of the *Handbook*: first, discussion about research methods (and research design) in the field and second, the methods for interacting with and engaging in design of science and technology with communities outside the field. To invoke boundaries implicit in these categories of "inside" and "outside" the field invites a reflexive analysis of how STS scholars—in writing of methods and design—are constructing, maintaining, and working on the territory of "our field." We feel this sort of boundary work is appropriate for a handbook and provides a kind of blueprint for those new to STS. It is of course charted at a particular time, with a particular set of authors and editors, marked up and redrawn in documenting the lay of the land.

Research methodology is a topic that has had less purchase, as yet, in STS, but we hope that this handbook, particularly the chapters in this section, will open further explicit discussion in the field about methodology. The chapters that focus on research methods each take a different approach; some are closely focused on a particular method, such as Sormani et al. on ethnomethodology and Wyatt et al. on scientometrics, while the chapter by Shankar et al. focuses on a particular object/method in interrogating documents. Other methods chapters in this section have a wider, more panoramic lens, such as Law's chapter. Taken as a whole, these four methods complement each other, and transition well into the participatory design chapters.

At this particular moment in STS, there is much attention to participation in science by various publics and the role of STS in democratizing science and technology; chapters in other sections of this handbook also incorporate these themes. The chapters in this section that discuss participatory design take two basic approaches. Some examine participatory design in particular technoscience contexts, such as Vertesi et al. on information technology and computational design and Salter et al. on art and design. Other participatory design chapters discuss particular ways that STS engages with publics, such as Lezaun et al. on experiments in participation and Downey and Zeiderent-Jerak on making and doing alongside journalists, activists, and scientists and engineers.

And while the eight chapters in this section provide very helpful introductions to the methodologies that STS scholars use to conduct research and to engage in design work with publics, it is important to note that any focus is partial and incomplete. Each of the chapters points to work yet to be done. We can also point to gaps in the section as a whole; for example, more work on thinking (and writing) about organizational and international-level methods used to conduct STS analyses would be welcome. An STS approach to methods is not about getting it right, however, but rather about thinking and reflecting. If we consider the performativity of methods, it may sensitize us to the ways that our chosen method shapes our insights and indeed works as a set of blinders.

The overall aim of this section of the handbook is to develop an argument for increasing attention to our methods in STS. If we think of *STS as method*, we may attend to the ways that an STS focus goes beyond data collection and analysis to reflexive practices and engagements with interlocutors. Or if we think of *methods of STS* in a more traditional research methods framework, perhaps we may improve our insights by attending to our methods.

Consider a reflexive interlude that combines STS methods and STS *as* method concerns: normative claims about methods like the sentence above on "improvement" may be one reason why STS methods have been avoided in previous editions of the *Handbook*. If arguing for rigor in methods is a normative argument, however, at least it does not have to be an optimized "one best way" argument; certainly problematizing "scientific" approaches to STS is to be encouraged. At the risk of making a "should do" argument by including this section in the *Handbook*, we feel that readers new to STS looking for guidance and learning about the field will be aided by clarity in presenting different methods that we use and how they are used. The absence of discussion about method looks like showing a finished product and expecting a neophyte just to figure STS out and "make it work." As STS research has long shown, iterative learning and tacit knowledge construction requires apprenticeship in a social context; exposure to a handbook will not suffice in teaching STS methods of understanding and intervening. Rather, we hope this section of text becomes a point of conversation about STS methods.

1   STS as Method

John Law

Introduction

How do science and technology shape the world? Or medicine and engineering? How does the world in turn shape them? And how, if at all, might we intervene in these processes? These are core questions for science and technology studies, and STS authors tackle them by asking how science (and technology) work in practice. This means that they operate on the assumption that "technoscience" is a set of social and material practices. Then they note that those practices work in different ways in different locations: laboratories, firms, and hospitals, and also (since STS interests are wide) for financial traders, farms, care homes, environmental movements, and indigenous ways of knowing. They look at how theories, methods, and materials are used in practice in specific social, organizational, cultural, and national contexts—and they look at the effects of those practices. So the first lesson is this: _STS attends to practices_.

How did this arise? One answer is that STS started by looking at "the scientific method." It showed that scientists don't usually follow philosophers' rules. Science is powerful, but in practice the scientific method is material and messy. (The same is true for social science method.) More than forty years have passed, but still STS looks at messy methods, scientific and otherwise, at how they get shaped, and also at what they actually _do_. It argues that technoscience practices are methods that shape and reproduce the social world. Indeed, I want to suggest, more strongly, that STS may be understood as the study of method in practice—that method, broadly conceived, lies at the heart of STS. This chapter therefore has a double focus. I both describe STS's own methods and explore the methods in the practices that it studies. Necessarily I do this from a particular "situated" point of view, that of a material-semiotics in which materials and meanings are woven together. Others in STS will understand method differently.

In one way or another STS almost always works through _case studies._ These evoke, illustrate, disrupt, instruct, and help STS to craft and recraft its theory (Heuts and Mol 2012; Yates-Doerr and Labuski 2015). Though the status of case studies is a source of controversy (Beaulieu, Scharnhorst, and Wouters 2007; Gad and Ribes 2014; Jensen 2014), nonetheless, if you want to understand STS—and STS theory—you need to read it through its cases. Such cases include how fishermen and scallops interact in practice (Callon 1986), how engineers and military chiefs create a warplane, how work in a laboratory generates new theories about physical forces, how primatology reproduces patriarchy, and how environmental scientists misunderstand aboriginal Australians. Some outside the discipline find this difficult: they think of theory as abstract. But the STS focus on practice means that theory, method, and the empirical get rolled together with social institutions (and sometimes objects). They are all part of the same weave and cannot be teased apart.

In the next section, "Shaping," I describe how early STS rejected philosophers' stories about scientific method and argue that science is shaped by social interests. In the "Structuring" section I use feminist STS to show that technoscience methods also in turn shape or format the social. The "Methods at the Center" section pushes this further by suggesting that everything, social and natural, is shaped in practices. The section on "Difference" shows how this varies in different practices and opens up the possibility of a politics of things. In "Knowing Spaces" I briefly review how methods link with subjects, objects, expressions and representations, and institutions; and the "Conclusion" reviews the argument of the chapter as a whole. Along the way boxes give a flavor of particular STS case studies.

Shaping

What's Wrong with Epistemology: How STS Started

Technoscience is shaped by society: scientific ideas or technologies reflect social interests. Many in STS say this. But where does the argument come from? I foreshadowed an answer in the introduction. Fifty years ago most of those who thought about science believed that science is special because it uses the scientific method. Philosophers debated its character, but the general consensus was that the scientific method is especially good at collecting accurate data, generating logical generalizations which explain that data, and testing those generalizations. Philosophers generally said that scientific knowledge—good, true, or accurate knowledge—grows if people adopt the scientific method (Popper 1959). In response to Nazi and Soviet political interference, they said prejudice distorts how scientists observe phenomena, erodes logical reasoning, and undermines objectivity.

The earliest sociology of science shared this view (Merton 1957), but STS came into being by reacting against it in two quite different ways. Some said that this was a nice picture in theory but that in practice scientific methods in a class or gendered society cannot escape social power (Slack 1972). This means that scientific knowledge is irredeemably ideological. Others argued that science is necessarily social. Scientists are trained to see the world in particular disciplinary (and therefore social) ways. They work with appropriate experimental arrangements and theories to identify core scientific puzzles and what will count as appropriate solutions to those puzzles. They also learn whom they can trust. This "sociology of scientific knowledge" (SSK) (Barnes 1977; Bloor 1976; Collins 1975) drew on the work of historian Thomas Kuhn (1970b). It said that science, its knowledge, its methods, and its practices are disciplinary cultures and that scientific knowledge is shaped in interaction between the world on the one hand and the culture of science, including its methods, on the other.

Note three points before moving on. First, in SSK it doesn't matter whether scientific knowledge is true or false. Since the same kinds of social processes are at work in each, we need the same methods to explain both (David Bloor [1976] called this the "principle of symmetry"). Second, scientific knowledge doesn't reflect nature. Instead it is a practical *tool* for handling and making sense of the world. In this philosophically pragmatist position either scientific tools do the job or they don't. Knowledge that works in practice is taken to be true. That which doesn't is taken to be false. And there is no other way of knowing truth. And then third, following Kuhn, SSK added that theories, methods, perceptions, practices, and institutional arrangements are all mixed together: that methods are not simply techniques but carry personal, skill-related, theoretical, and other agendas (Kuhn 1970a; Polanyi 1958; Ravetz 1973). This means that the STS concern with methods spills over into much that is not obviously methodological because methods cannot be separated from their social context. These become cultural, practical, materially based, theoretically implicated, institutionally located, and socially shaped routines or procedures, all raveled up with everything else.

Social Shaping

SSK opened up space for laboratory studies which ethnographically explored the construction of knowledge (Knorr Cetina 1981; Latour and Woolgar 1986; Lynch 1990; Traweek 1988). It also, and a little differently, asked how science, its methods, and its findings are *shaped*. SSK answered this question in two closely related ways. First it said, as I have just noted, that scientists work with cultural tools. And second, it said that

---

Box 1.1

Statistics: Case Study 1

Correlation is a way of measuring how two variables relate to one another. Here's an example. Vaccination (or not) is one (nominal) variable, and catching a disease (or not) is a second. If none of those vaccinated catch the disease and all those unvaccinated do, then the two variables are highly inversely correlated. We tend to take statistics for granted. But measures like correlation are invented. They are tools for handling data that can be quantified. And since they are invented, they may be constructed in different ways, and statisticians may get into disputes about them.

SSK writer Donald MacKenzie (1978) looked at one such dispute. In 1905 the protagonists—George Udny Yule and Karl Pearson—had invented two different ways of measuring correlation. Yule's approach was straightforward. Pearson's was more complicated: he assumed that variables reflected normal distributions. Why? MacKenzie makes two arguments about how interests shaped Pearson's approach. Pearson had previously worked on normal distribution (the "bell curve") so he found it natural to think about correlations in this way. It was in his *cognitive* interest to do so. But his approach to correlation also fitted his social agendas. A bell-curve way of thinking about correlation made it easier to think about the supposed superiority of middle-class over working-class people. The middle classes (including Pearson) were toward the top of the curve, and the working classes (in need of eugenic improvement) lay toward the bottom. MacKenzie suggests that his complex way of calculating correlation was also in Pearson's *social* interests.

---

scientific knowledge is shaped by social interests. Donald MacKenzie (1978) explored this for a controversy about statistical correlation (see box 1.1).

Statistical procedures (like other scientific theories or methods) are tools for making sense of the world. But (this is the new move) how those tools are constructed depends on the tasks that we set them. Yule's method for calculating correlation was useful in many ways, but it was never going to do the kind of work that Pearson sought to do with his tetrachoric coefficient of correlation, namely, to help to show whether some kinds of people are superior to others.

MacKenzie's case study is much more sophisticated than this brief account suggests. For instance, he argues that both professional and broader social class interests are at work. It's also important to understand both that interests may shape science in ways invisible to those involved and that the fact that interests are at work tells us nothing about the validity of the science involved. Knowledge that works, "good" knowledge, is necessarily shaped, and sometimes by social interests of which we disapprove (Barnes 1977). And finally, MacKenzie's study is just one example of SSK at work. For instance, Jonathan Harwood (1976) wrote on race and intelligence, and the same approach was developed to explore technology in the *social construction of technology* (SCOT). Why did the penny-farthing bicycle give way to the safety bicycle? Wiebe Bijker (1995) showed the penny-farthing was linked to macho forms of masculinity. Women—and many men—couldn't or weren't supposed to ride it. But since this meant that the market for bicycles was small, it was in the interests of manufacturers to create a bicycle that was safer and more modest. Here gender and commercial interests together shaped a technology. And (another example) Cynthia Cockburn (1999) argued that the technologies of the precomputer print trade expressed and reproduced both class and gender interests: the creation of heavy manual work was just one of the mechanisms working to exclude women. (For further SCOT studies, see Bijker, Hughes, and Pinch [1987], and Bijker and Law [1992].)

Objectivity, Nature, and Culture

In contemporary STS the idea that science can be separated from the social has almost disappeared. But the insight that technoscience and society are woven together also came from feminism. For instance, Donna Haraway (1988) talked of the "God trick" to describe the mistaken and self-serving claim that science speaks impartially. The idea—or the ideal—of objectivity has a long and varied history. (See Daston [1999] and the case study in box 1.2 below.) Usually this implies impartial detachment from local prejudices, blinkers, and idiosyncrasies. But Haraway argued that knowledges and methods are irredeemably *situated*. The stories they tell about the world always reflect their location and reproduce social agendas. Achieving the God-like status of being above everything is impossible, though the myth that this can be achieved retains a powerful grip.

To say this is not to object to science. We are all located. But does it also mean that everything is subjective? Haraway's response is that we can hang on to objectivity by making two methodological and political moves. First, scientists and social scientists need to acknowledge their own social location. And second, they need to treat that location, its prejudices, and its blind spots as matter of critical inquiry in their own right. For Haraway, objectivity is doubly "partial" because it knows that it is one-sided and because it also recognizes that it is incomplete. Her argument is that to achieve objectivity, scientists and social scientists need to be *accountable* for what they write rather than hiding behind the fiction that what they are reporting comes direct and unmediated from nature. Sandra Harding makes a similar argument. What she calls "strong objectivity" grows out of a self-critical examination of the social basis of knowing—a way of doing science or social science that explores the position (and questions the assumptions) of those producing knowledge (Harding 1993). The idea

---

Box 1.2

Separating Science from Society: Case Study 2

In London in the 1660s in the newly created Royal Society Robert Boyle was wrestling with the question: how can we reliably learn about nature? The answer wasn't obvious. For instance, the Bible was full of powerful stories about the creation of the world. Boyle was interested in air pressure. A devout Anglican and a royalist, he also wanted to divide facts about the world (or nature) from politics and God. He did this by making a radical proposal. We can learn about nature, he said, if we do three things. First, we need to conduct reliable experiments. We need an experimental apparatus, an air pump. It was large, complex, expensive, and difficult to run. But the very idea of an experiment was a novelty. This is the first innovation. Historians Shapin and Schaffer (1985) call this a *material technology*. But more was needed. The experiments needed witnesses, but not everyone could come to London to see those experiments for themselves. They needed to be told about them. This led to the creation of a *literary technology,* in which experimental accounts were written in a modest and matter-of-fact way, sticking to the facts, and excluding opinions and speculations. This was the second big innovation. But there was a third question: who could be trusted as a reliable witness? To answer this question, Boyle drew from the English legal system. In a court of law reliable witnesses were independent. Servants could not be trusted because they were not independent of their masters. Neither could women: they were beholden to husbands, fathers, or brothers. And this is the third innovation: the creation of what Shapin and Schaffer call a *social technology.*

This is the foundation of contemporary technoscience. Nature is separated from the social. It is imagined that facts can be described in ways that separate them from opinions and social contexts. And only disinterested specialists can decide about those facts. This is where what Haraway calls the "God trick" came from (Shapin 1984; Shapin and Schaffer 1985).

---

is that knowledge-makers are part of what they study and that their methods should reflect this.

But where did the idea of objectivity as impartiality come from? Steven Shapin and Simon Schaffer (1985) suggest that this was created in very particular social circumstances in London in the 1660s and the 1670s (see the case study in box 1.2). At this historically important moment "nature" was separated from "the social" and "the political" and this separation was successfully institutionalized. Natural science came into being in Europe—and later across the world. A passive nature that might be known and mastered was divided from people who were active—and male. At the same time, objectivity was separated from subjectivity and opinion and impartiality from partiality.

Structuring

Haraway (1997) raises questions about parts of Shapin and Schaffer's account, but most, including Haraway, accept its overall significance. This is the moment when the God trick was embedded in science, and the methods of the latter appeared to step outside the social. But the stories about Boyle, Pearson, and class, gender, and technology in the print trade hint at something more. They suggest that technoscience is not simply shaped by the social but helps in turn to shape it. Indeed many in STS argue that knowledges and methods are often shaped in ways that are gendered, racist, class-based, and/or imperialist and also that they help to reproduce such inequalities. But how?

Feminist Cultural Studies of Science

The third case study (see box 1.3) shows how one version of primatology carried and reproduced a whole range of social concerns (and horrors), including sadism, masculinist self-birthing, patriarchy, anxieties about child-rearing, and assumptions about functional nuclear families.

It was shaped by concerns that could not be separated from those of educated, middle-class, mid-twentieth-century America. But, at the same time, it helped to give shape to those concerns and reproduce them. Social concerns fed into technoscience practice, and technoscience fed these back into social agendas. These were "structuring" practices—methods—that give simultaneous form to science experiments, structures or forms of knowing and social structures. Removed from concerns about nuclear families, child-rearing, and gender roles, it is very difficult to make sense of the Wisconsin experiments at all.

But how to study that structuring? Haraway draws on feminist cultural studies. The key term here is *narrative*. Narratives are embedded in texts, materials, and methods, and in turn draw on *tropes*. Tropes are figures of speech or metaphors. Think, for instance, of phrases like "society is an organism" or the notion of "scientific discovery" and contrast these with "society is a machine," or "scientific invention." They do different kinds of work. Such tropes shape our narratives and carry clouds of connotations. This is not a complaint: tropes make up the weave of language and culture. They help to make us what we are. But they also carry political and social agendas. And this has been the insight of feminist STS: that formatting work is done in storylines and the practices in which these are embedded. So, for instance, technoscience stories may naturalize sex-gender differences. Anthropologist Emily Martin talks about metaphors of bounded bodies in pregnancy (Martin 1998) and immune system discourse (Martin

---

Box 1.3

Primatology: Case Study 3

Primatology is the study of the great apes. But how should they be studied? Sometimes—think of Jane Goodall—scientists live with their subjects. This allows primatologists to observe natural behavior. Others place their apes in laboratories, which makes it easy both to observe behavior and to control important variables. Donna Haraway (1989) tells the story in her book *Primate Visions*.

In the 1950s and the 1960s Harry Harlow's Primate Research Laboratory at the University of Wisconsin-Madison was an important center for primate research. Harlow was a master communicator who told media-savvy stories. But he was also working on topics and questions that reflected the anxieties of post–World War II America. The focus was the nuclear family. The big question was: how was this holding up in an era of stress? How was it responding to the demands of consumerism? So the focus was gender roles and child-rearing. Were children suffering, as middle-class American women came under pressure to go out to work? What was the importance of maternal love? How might children be brought up happily and healthily given such pressures?

Where better to study these questions experimentally than with those close relatives of human beings, the primates? Harry Harlow's laboratory ran experiment after experiment. These were well planned and managed—and often sadistic. One example is the surrogate mother experiment. This was designed to find out what infant primates needed to be secure. Perhaps something to hold on to of a vaguely simian shape, a surrogate mother? A wire shape with something like a face? What was the minimum needed to secure a version of maternal love? Harlow and his team reduced many young simians to psychosis (there was a freezing "ice mother") but in the end created what one might think of as the minimally functional version of the mother. This was the "cloth mother": a frame covered by a blanket with a caricature of a face and a feeding teat. Infants, it turned out, survived with the latter.

1994). Cultural analyst Jackie Stacey (1997) explores the role of the monstrous in cancer. The insight that power generates silences has also been explored in feminist writing (Ryan-Flood and Gill 2010), while Lorraine Daston (1999) describes the shifting history of objectivity. But if stories structure common sense and science alike, then how can we narrate and create better alternatives? Haraway's answer is to create alternative tropes that interfere with those that are dominant. So, for instance, she creates a feminist cyborg (Haraway 1985). No longer a Cold War–created destructive masculinist military machine-human enhancement, this is a set of partial connections that blurs boundaries, including the distinction between fact and fiction. It offers a path to emancipation in which alternative non-militarist futures might be imagined. It makes a difference, politically, theoretically, and methodologically. And so too should STS (Haraway 1997).

Performativity

So technoscience practices are shaped by but also shape the social. They help to format the world. This means that they are *performative*. "Performativity," a term from linguistic philosophy, says that words are sometimes also actions (like "I do" in a Christian wedding ceremony [Austin 1962]). We can link this to the dramaturgical idea that social life may be understood as a performance and to its corollary that performances may have real effects that order the social (Goffman 1971). This double move suggests a new focus for STS: to explore *how methods are staged*.

Think, for instance, about the performativity of social surveys. These stage and structure people as respondents (see box 1.4). In practice this means that people have telephone lines, speak the appropriate language, are willing to answer questions, are willing to be classified as men or women, understand ordinal scales, and are willing to admit that they "don't know" (which suggests that they are buying into something like a "knowledge society"). None of this is exceptionable, but neither is it given in the order of things. The survey works because people are being made to fit, even if they don't. (What of transsexual people, or those who don't work with ordinal scales?)

Unsurprisingly, it turns out that survey research is an historical social science achievement (Igo 2006; Savage 2010). It didn't exist until the twentieth century, when

---

Box 1.4

Surveys: Case Study 4

No doubt social research methods are socially shaped, but what do they *do*?

I asked this question for a Europe-wide survey, the Eurobarometer (Law 2009). In 2007 around 29,000 people were interviewed in the different EU countries. The sample was stratified by country—about 1,000 respondents were interviewed by phone in each. They were asked about their attitudes to farm animal welfare, and how (or whether) this influenced them when they bought meat products. The survey concluded that farm animal welfare was seen as important by European consumers. "Please tell me on a scale from 1 to 10 how important it is to you that the welfare of farmed animals is protected." This was one of the questions, and the mean score was 7.8. There were significant country differences. For instance Scandinavian respondents trusted the state to look after animals more than people from southern Europe. And many said that they took farm animal welfare into account when they went shopping. But what was the survey *doing*? The answer is lots of things, but here are two. It was shaping *interview subjects*. (A person is not necessarily an interviewee. You need to be formatted right.) And it shaped *collectivities* in particular and specific ways. (A collectivity is not necessarily a country, for instance, and a country is not necessarily a collection of people.)

people learned that it is acceptable for strangers to ask them questions. But if surveys perform people in their methods of data collection, they also stage them in their findings. For instance, in the Eurobarometer people are formatted as sets of attitudes that are seeking information to decide whether to buy animal-sourced products. And collectivities are being done too. These become collections of individuals: collections of isomorphic social atoms in a homogeneous conceptual and geographical space.

To say this is not necessarily to criticize. There is no God trick and all methods narrate and format the world (Waterton and Wynne 1999). This means that general complaints about the performativity of methods miss the point. Any criticism needs to focus on particular forms of performativity.

So, for instance, Eurosceptics correctly argue that the Eurobarometer stages the European Union as a collectivity, and sociologists are right to say that people don't necessarily have stable attitudes shaping how they behave (Shove 2010). Perhaps we simply need to say that the Eurobarometer is flawed. But there is a less obvious and more interesting STS argument. This ties validity to location. So in shops the survey is probably wrong: the extent to which attitudes shape what people buy is limited. But in other places the survey is (taken to be) right. Pragmatically, for instance, in the European Commission the figure of "the consumer-with-attitude" is successfully staged. It becomes real because it is epistemologically and politically performative. The conclusion? STS tells us truths are practice-embedded, but as Bruno Latour (1988) showed when he explored why Pasteur was so successful, it also tells us that truths are location-dependent. If French farms were to be "Pasteurised" they needed to be reformatted as laboratories. And if people are to be treated as attitude-carrying decision makers, then receptive administrative and political audiences similarly need to be created.

A final point. The Eurobarometer says that "Hungarians" believe this, whereas "Italians" believe that. This tells us that it is staging the nation-state as well as national citizens. But how? Note that national terms are used unproblematically. The survey makes no argument for the nation-state, but does this mean that its performative effects of nationality are weak? I want to suggest, on the contrary, that formatting is often most powerful when it is almost incidental. Nationality is being done strongly precisely because it is built unproblematically into the survey's frame, since it is simply taken for granted. My suggestion is that methods, social scientific and otherwise, powerfully enact such incidental "collateral realities" (Law 2011a) by assuming them. Surely Haraway is right. It is one of STS's tasks to scrape away the self-evident to understand and question how methods structure the world.

Methods at the Center

SSK author Harry Collins (1975) long ago showed that knowledge and methods and scientific authority may all be negotiated together. In a different idiom Thomas Hughes (1983) made a related argument about system building. Hughes argued that when Thomas Edison created the New York public electricity system he generated a heterogeneous web of social, legal, political, economic, geographical, scientific, and technical relations. Everything was raveled up together. But what is the best way of thinking about such interconnectedness? STS has tackled this question in various ways. For Hughes, system builders were specially gifted at fitting together heterogeneous components. One of the successor projects to SSK and SCOT, *co-construction* or *co-production*, explores how the social and the scientific are constructed together, for instance, in the form of regulatory frameworks (Jasanoff 2004; Shackley and Wynne 1995). As we have seen, feminist material semiotics uses narrative analysis to understand the forms taken by heterogeneous relations. Differently again, actor-network theory (ANT) has also tackled interconnectedness in ways that put methods at the center.

Actor-Network Theory

Actor-network theory is radically relational. So Michel Callon (see box 1.5) (drawing from post-structuralism [Deleuze and Guattari 1988; Serres 1974] and innovation studies [Callon 1980]) created a conceptual tool kit for talking about heterogeneous relationality, a method for mapping how every object or actor is shaped in its relations.

Box 1.5
Scallops: Case Study 5
In 1986 Michel Callon published what may be the most cited article in STS. This was on the scallops, the fishermen, and the scientists of Saint Brieuc Bay. The story is about the decline of the scallop population, the attempts by three scientists to understand that decline, and efforts to create zones protected from fishing where scallops might breed and mature. The story traces the successful attempts by the scientists to create collectors for scallop larvae. It details the negotiations between the scientists and the fishermen to create non-fishing zones, and it concludes with the dramatic moment when the agreement broke down and the fishermen scraped the protected areas clean of scallops. However, the success or the notoriety of Callon's article has little to do with the scallops themselves. Instead it arises because he treats the fishermen, the scientists, and the scallops in the same terms. *All* are actors. *All* are strategists and tacticians. *All* seek to enroll others in their schemes. At Callon's hands, there is no difference in principle between scallops, fishermen, and scientists.

Here nothing has a given form. The differences between scallops and fishermen grow in the web of relations and don't preexist those relations. So scallops and people might be different elsewhere, and it is important to explore specificities without pre-judging their form or shape (he calls this the principle of "generalized symmetry"). This is radical in explanatory terms: it represents a substantial shift from SSK. For Callon, the social doesn't shape or explain anything. Society and nature, humans and nonhumans, people and technologies—essential divisions have simply disappeared (Law and Mol 1995). So the macrosocial doesn't explain anything either—like everything else the "macro" and the "micro" are relationally generated (Callon and Latour 1981).

Ordering Methods

These conclusions are controversial. Many in STS remain attached to macro-micro dis-tinctions. But if we follow its logic, we need to study relations, networks, and webs of practice. We need to look at how webs assemble themselves to stage effects such as actors and objects, and binaries such as nature and culture, human and nonhuman, or indeed macro and micro. But this is a profound methodological shift, because with it STS moves from explanations (like social interests) which lie behind events to attend instead to methods for assembling. Whatever is going on is seen as an expression of strategies or tactics. Indeed the case studies of ANT and its related projects can be seen as a list of methods for assembling, stabilizing, or undoing realities. These methods include delegation into durable materials (Latour 1987), the creation of circulating immutable mobiles (Law 1986) or fluid and mutable objects (de Laet and Mol 2000; Yates-Doerr 2014), inscription devices (Latour 1998), and the preformatting of distant locations (Latour 1988). They also include the logic of tactics (Callon 1986) and mul-tiple "modes of ordering," which together secure temporarily robust human and non-human arrangements (Latour 2013; Law 1994, 2002; Thévenot 2001).

It is easy to see why the critics say that actor-network theory is a Machiavellian description of ruthlessly successful political tactics. Sometimes it is guilty as accused, but not, I think, always, for it is not necessarily cynical to explore how power is done. On the contrary, if we want to undo power, it may help if we understand its methods. Here the similarities between ANT, feminist material semiotics, and Michel Foucault's (1979) history of the present are instructive. Despite differences, all attend to mate-rial and linguistic heterogeneities, and how these generate effects including asymme-tries and dualisms. All insist that these are not given in the order of things (Foucault's phrase) and might be otherwise. And all argue that patterns recur: that the world isn't a different place every morning. Perhaps (early ANT excepted) they are also saying that there are sustained patterns of inequality. At any rate, they are all assuming that

a methodological microphysics of power is systematically at work that is both productive and excludes alternatives. (Think of Haraway on primates and Foucault on judicial
torture [Foucault 1979].) And crucially, none works on the assumption that strategies
are inevitably explicit or cynical. The argument, then, is that ANT is not necessarily
Machiavellian. An analysis of the methods of power and their productivity—a history
of the present—may, instead, be used to make a political difference.

Difference

So in a material semiotic way of thinking everything is radically relational. Essential differences disappear. Everything is endowed with a "variable geometry," and it becomes
crucial to explore the tactics and strategies—the methods—embedded in practices. No
assumptions are made about what will be found. But there is a knock-on effect. Since
practices may vary, so too may the entities that they are formatting. This means that
"the same" object may be one thing in one place and another somewhere else. In STS
this is called the *problem of difference.*

Multiplicity

Mol (2002) explores multiplicity for lower-limb atherosclerosis. She shows (see box 1.6)
that the practices that perform this condition are different in different places.

Then she makes the claim that I just mentioned: that the objects being enacted in
those relations are being differently shaped too. Her counterintuitive conclusion takes
us to the problem of difference. She says that in practice there isn't a single atherosclerosis; there are four. But the practices that format atherosclerosis aren't independent of
one another. This means that atherosclerosis is a complex pattern of intersections, an
object that is more than one but less than many. The different atherosclerosis may line
up, contradict, include one another, never meet up—or combine some mix of these.
Like Haraway's cyborg, atherosclerosis isn't a unity but a set of partial connections
(Haraway 1988; Strathern 1991). We live in a world of *ontological multiplicity.*

Philosophers use the term *ontology* to talk of what there is in the world, or what
reality out there is made of. Most Western philosophers assume that the stuff of reality
is constant, that we share the same reality-world, and that we disagree about reality
because we have different perspectives on it. But recent STS is pushing back against
this. In the way I have just suggested, it is saying that ontologies are relational effects
that arise in practices (Barad 2007; Law 2002) and that since practices vary, so too do
objects. This softens realities—it means that they are not given (Abrahamsson et al.
2015). It also means that we might imagine better alternative realities. A "politics of

---

Box 1.6

Disease: Case Study 6

What is atherosclerosis? Annemarie Mol explored this in an ethnographic study of lower-limb atherosclerosis in a Dutch town (Mol 2002). She visited GPs' surgeries and listened to patients worrying about leg pain when they walked. In the hospital she watched technicians taking radiographs which showed the circulatory system in the form of a tracery of curves and lines. She visited the ultrasound department and watched the specialists looking for Doppler differences reflecting changes in the speed of blood flows. And then she watched surgeons opening up blood vessels and scraping out white, puttylike, arterial plaque.

Four practices, each about lower-limb atherosclerosis, but what *is* this condition? The standard story says that long-term changes in the blood lead to the buildup of arterial plaque, which limits the blood flow which in turn starves the muscles of oxygen and causes pain. In practice Mol found that sometimes these signs and symptoms fitted together nicely, but sometimes they didn't. If this happened, then the differences were hammered out at a case conference. Mol notes that this worked because everyone assumed that there is an object out there, and the specialists had different perspectives on it. However, her own argument is quite different and very far from the common sense of this standard story. She says that *different practices enact different atheroscleroses*. These practices and their atheroscleroses relate to one another in theory but not necessarily in practice.

---

things," an ontological politics (Mol 1999) or a cosmopolitics (Stengers 2005) becomes possible because different normativities and realities are being woven together in what Mol calls "ontonorms" (Mol 2012). So a feminist cyborg may be better than one that is militarist, or the atherosclerosis of physiotherapy might have advantages over the one performed in surgery.

Two further points. First, a caution. Performing objects is tough, even in this relational world. It is difficult and costly (think of Mol's hospital departments). We can't just dream up new realities. (Latour and Woolgar 1986; Law 2011a). Second, we need to ask where we might find difference. We can debate, but the intuition that underpins Mol's intervention is that we will always discover it if we go looking for it, and that doing so is an analytical and normative choice. But this implies a methodological rider: we need to be wary of stories about consistency and coherence. Instead it might be better to cultivate a sensibility for mess (Law 2004). Though, of course, there is also an art in distinguishing between mess that is politically and methodologically important and that which is not. There are no rules here, but simply noting that the world is noncoherent is not a discovery. We also need to know what kind of a difference we are hoping to make.

Method and Difference

In this version contemporary STS asks questions that are simultaneously about realities and politics or normativities. Recognizing its own performativity, it understands that it makes a difference. But what kind of difference does it make? The answer is that it typically tries to find ways of living together well. It does this in many ways, but here are two.

In a world in crisis economically, socially, and environmentally, it is clear that we urgently need to find better ways of living together. STS tells us that technoscience in its present form is part of the problem. Separated from the political, it is destructive because it takes reality to be fixed. So how to think about this? One answer draws on democratic political theory and practice. Democracy is about living together well in a common world. Perhaps the old ways of reconciling difference democratically— parliaments and their analogues—have failed because they reproduce the nature-culture divide, fix nature and exclude it from politics. The task, then, is to invent new methods for softening realities, reworking social collectivities, and melding these productively and democratically together. Many have wrestled with this, but none more systematically than Bruno Latour. He has talked of non-modern constitutions, of parliaments of things, of matters of concern, of new forms of political ecology, of the importance of due process, and of the need for diplomacy to hold together different conditions of felicity or modes of existence (Latour 1993, 2004a, 2013). Throughout, his urgent task has been to imagine ways of generating common responses to common problems in a common world. Less ambitious but related concerns inform work on publics (Marres 2007) and the work of Michel Callon, Pierre Lascoumes, and Yannick Barthe (2009), who experiment with hybrid forums which mix experts, nonexperts, and politicians. The object is to melt the categories of nature that were previously hardened and fixed in professional silos. As a part of this they undertake experiments "in the wild" to secure collective learning and recompose a better common world. Again the interference is procedural and methodological.

How can we go on together well in difference? This question—adapted from Helen Verran (2013)—takes us to the second strategy. Though similar to the first, it is more modest because it makes no assumption about common frameworks. If democracy wants to reconcile difference overall, then the second strategy is not about democracy. Neither does it try to generalize. Rather it is about detecting and handling difference well, case by case (Law et al. 2014). So Mol (2010) argues that the atherosclerosis of

---

Box 1.7

Eutrophication: Case Study 7

In Western ways of thinking "nature" is divided from "culture." Nature is taken to have particular attributes. Science seeks to reveal these. In contrast, culture is known to be variable: different groups of people believe different things. But what to make of the STS idea that science is cultural too? Situated? Potentially revisable? How to think about intractable problems that are both natural and social?

Between 2007 and 2010 Claire Waterton and Judith Tsouvalis (2015) brought together farmers, residents, social scientists, environmental scientists, and administrators to discuss the persistence of blue-green algal bloom in Loweswater in the English Lake District. Everyone was clear that something needed to be done. The issue was what? Waterton and Tsouvalis looked for ways of opening up scientific and social uncertainties. How were the scientific findings produced? Might these be discussed and questioned? Was it possible to situate them alongside other kinds of framings, economic, social, or recreational? Could scientific findings be softened (Latour 2004b) from stabilized "matters of fact" to situated "matters of concern"? The answer, it turned out, was yes, at least within limits.

Was it possible to appreciate that there are limits to all forms of knowing, those of technoscience included? Was the group capable of working with the idea that human-nonhuman relations are complex, that there were no definitive solutions, and that humility in the face of complexity might be what was needed? Again the answers to these questions were yes. There were many frustrations too, but a more relationally fluid collectivity, one that attended to the importance of context and process, was provisionally tinkered into being.

---

physiotherapy may be better for some patients than that of surgery. Michel Callon and Vololona Rabeharisoa (2004) explore the intersection of different forms of morality and humanity and the role of silence for the case of a patient with muscular dystrophy. Waterton and Tsouvalis (2015) (see box 1.7) work locally on the environmental problem of algal bloom to soften scientific and social categories. And Ingunn Moser's (2008) work on dementia care suggests that Marta Meo care methods enact patient competences that don't fit textbook medical science. Like Mol, she is chipping away at the dominance of biomedical realities and treatment régimes by talking up processes of care (see also Pols [2006] and Singleton [2010]). Importantly, none of these authors offer general prescriptions.

Postcoloniality

Similar power-asymmetrical encounters across difference are common in North-South relations where alternative Southern realities about land, gods, animals, people, bodies, and social ordering are typically turned from realities into mistaken beliefs. So people are not visited by spirits: they are psychotic (Bonelli 2012). The land is not a living thing: it is empty (Verran 1998, 2002). The mountain is not a god: it is a mineral-rich resource (de la Cadena 2010). The forest does not depend on shamans: it is a place to mine gold (Kopenawa and Albert 2013). A food additive does not reduce children's malnutrition: this is a fabrication (Marques 2014). Bodies don't have meridians: they are neuromuscular entities (Kuriyama 1999). Glaciers don't take offense: they are ice floes (Cruickshank 2012). Chinese medicine is not experiential: it is theory-deficient (Zhan 2014). In all these encounters two realities are being staged, but one is refusing the other (Law 2015). The issue then becomes how to discover techniques for undoing this refusal and going on well together in difference. As I earlier noted, this phrase comes from Helen Verran (1998, 2013), who charts how the Australian legal system and aboriginal people have learned how to respond to one another across difference. Is land an area, or is it part of a continuing creation? The solutions are far from perfect, but Australian law has created practices which recognize ownership in both senses. Such techniques for living well with difference do not always work and they need to be crafted case by case. Perhaps a task for STS—working here with postcolonialism—is to chart differences, articulate these, and help to craft ways of going on well together in difference (Blaser 2009; Feit 2004; TallBear 2014; Turnbull 2000; Verran 2002).

Knowing Spaces

STS suggests that methods are never simply techniques. Theories, methods, the empirical, modes of writing, disciplinary structures, audiences, authorities, and realities—all are staged together. Other candidates are jostling to join this list, including organizational structures, career concerns, social, economic, technical and publishing infrastructures, and imaginaries, national and otherwise (Felt 2015; Jasanoff and Kim 2015). The argument (as in the empirical examples above) is that knowing and its methods are materially complex and performative webs of practice that imply particular arrays of subjects, objects, expressions or representations, imaginaries, metaphysical assumptions, normativities, and institutions. Perhaps we might think of these heterogeneous arrays as "knowing spaces" (Law 2011b) because they set more or less permeable boundaries to the possible and the accessible; they are defined by patterns of relations which enact those gradients of possibility and accessibility; and they intersect with and are implicated in the generation of alternative knowing spaces that cannot be included (think of Darwinism and Creationism).

Now think about the power and the obduracy of these knowing spaces. In any given location it is easier to know in some ways than in others. It may be challenging to publish in major academic journals, but at least the appropriate literary conventions, procedures, competences, topics, and theoretical frameworks are reasonably clear. Together they enact academic knowing spaces within which it is comparatively easy to operate, and they substantially define what is possible in an academic career. But if we shift beyond these conventions, knowing becomes progressively more difficult for an academic. The wrong topic? A case study that is not of interest to an international (a U.S.?) audience? The wrong language? A strange theory? Inappropriate methods? Excessive commitment to activism? The "wrong kind" of activism? Writing that doesn't look like a standard journal article? This is getting risky. So here is the question. Is it possible to imagine alternative STSs?

In practice, the answer is yes: hybrid or unconventional knowing spaces are indeed possible. Some have successfully worked through exhibitions (Latour and Weibel 2006), or by writing poetry (Cole 2002), or poetry in combination with other media (Watts, Ehn, and Suchman 2014), or semi-popular texts (Raffles 2010), or in simulations (Guggenheim, Kräftner, and Kröll 2013), reciprocal human-animal interactions (Despret 2013), activism (Haraway 2008; Wynne 1996), artworks (Jones 2011; Neuenschwander 2008), art-science intersections (Gabrys and Yusoff 2011; Kräftner et al. 2010), or in participative methods (Waterton and Wynne 1999). Others have done so in dance (Cvejic 2010; Myers 2012) or by consulting with the spiritual realm (Smith 2012)—a way of knowing important in some postcolonial contexts. Such efforts represent brave efforts to experiment with hybrid knowing spaces. But creating different knowing spaces is slow, hazardous, and often lonely and uncertain. And, to pick up a theme touched on in the previous section, the unwitting "Northern" character of STS knowing spaces sets stark limits to alternative "Southern" forms of STS. So, for instance, in a "Chinese"-inflected STS, theories and methods might look quite different (Lin and Law 2014).

Conclusion

In this chapter I have argued that methods are shaped *by* the social; that they *also* shape, stage, and structure the social; that they are performative and heterogeneously enact objects, worlds, and realities; that they are situated, productive, essentially political, and normative; and that they might be otherwise. Then I have argued that with the decline of larger explanatory schemes, STS has increasingly attended to the tactics and strategies of practice, to methods, and to how these stage the world. I have also suggested that since practices vary between locations, they generate different realities and normativities; that the relations between these are uncertain; and that much STS is currently struggling in one way or another to generate methods that recognize, properly attend to, or stage better ways of handling difference.

The story I have told has been both about the methods in the processes that we study and those that make up our own STS practices. As is obvious, the two are intertwined. What we detect in the world arises in the interference between our own practices and those of the world. And this is why this chapter should be understood as its own situated intervention. Evenhandedness is not possible, and the God trick is out. Coming from a space between actor-network theory, feminist material-semiotics, and postcolonialism, I have staged relationality, specificity, difference, binary breakdowns, and politics or normativities in ways which others might not. I have reinterpreted essential categories and realities as relational effects and searched for multiplicity rather than causal explanations. As a part of this, I have adopted an expansive or generous understanding of method and sought noncoherences as a matter of both taste and politics. My object has been to suggest, both implicitly and explicitly, that it is the urgent task of STS first to attend to difference and second to craft specific but multiple ways of going on well together in difference. There are no single solutions. What it means to go on well together in difference is necessarily contested. Though we need to remind ourselves that the world is not open and that not everything is possible, this does not mean that we cannot try, just a little, to open up and enact alternative and better possibilities. The hope is that in this way we can avoid giving comfort to a politics that denies that it is political, and resist the claim that reality is destiny. So perhaps in the end, the enemy is hubris. Things never have to be the way they are. Such is the point of this STS of method.

Acknowledgments

Thanks to Wiebe Bijker, Ivan da Costa Marques, Ulrike Felt, Bernd Kräftner, Judith Kröll, Marianne Lien, Maureen McNeil, Annemarie Mol, Wen-yuan Lin, Lucy Suchman, Heather Swanson, Claire Waterton, Vicky Singleton, and the *Handbook* referees.

References

Abrahamsson, Sebastian, Filippo Bertoni, Annemarie Mol, and Rebeca Ibañez Martin. 2015. "Living with Omega-3: New Materialism and Enduring Concerns." *Environment and Planning D: Society and Space* 33 (1): 4–19.

Austin, John L. 1962. *How to Do Things with Words*. Edited by James O. Urmston and Marina Sbisà. Oxford: Oxford University Press.

Barad, Karen. 2007. *Meeting the Universe Halfway: Quantum Physics and the Entanglement of Matter and Meaning*. Durham, NC: Duke University Press.

Barnes, Barry. 1977. *Interests and the Growth of Knowledge*. London: Routledge and Kegan Paul.

Beaulieu, Anne, Andrea Scharnhorst, and Paul Wouters. 2007. "Not Another Case Study: A Middle-Range Interrogation of Ethnographic Case Studies in the Exploration of E-science." *Science, Technology, & Human Values* 32 (6): 672–92.

Bijker, Wiebe E. 1995. *Of Bicycles, Bakelite, and Bulbs: Toward a Theory of Sociotechnical Change*. Cambridge, MA: MIT Press.

Bijker, Wiebe E., Thomas P. Hughes, and Trevor J. Pinch, eds. 1987. *The Social Construction of Technical Systems: New Directions in the Sociology and History of Technology*. Cambridge, MA: MIT Press.

Bijker, Wiebe E., and John Law. 1992. *Shaping Technology, Building Society: Studies in Sociotechnical Change*. Cambridge, MA: MIT Press.

Blaser, Mario. 2009. "The Threat of the Yrmo: The Political Ontology of a Sustainable Hunting Program." *American Anthropologist* 111 (1): 10–20.

Bloor, David. 1976. *Knowledge and Social Imagery*. London: Routledge and Kegan Paul.

Bonelli, Cristóbal. 2012. "Ontological Disorders: Nightmares, Psychotropic Drugs and Evil Spirits in Southern Chile." *Anthropological Theory* 12 (4): 407–26.

Callon, Michel. 1980. "The State and Technical Innovation: A Case Study of the Electric Vehicle in France." *Research Policy* 9: 358–76.

___. 1986. "Some Elements of a Sociology of Translation: Domestication of the Scallops and the Fishermen of Saint Brieuc Bay." In *Power, Action and Belief: A New Sociology of Knowledge?*, edited by John Law, 196–233. London: Routledge and Kegan Paul.

Callon, Michel, Pierre Lascoumes, and Yannick Barthe. 2009. *Acting in an Uncertain World: An Essay on Technical Democracy*. Cambridge, MA: MIT Press.

Callon, Michel, and Bruno Latour. 1981. "Unscrewing the Big Leviathan: How Actors Macrostructure Reality and How Sociologists Help Them to Do So." In *Advances in Social Theory and Methodology: Toward an Integration of Micro- and Macro-Sociologies*, edited by Karin Knorr Cetina and Aaron V. Cicourel, 277–303. Boston: Routledge and Kegan Paul.

Callon, Michel, and Vololona Rabeharisoa. 2004. "Gino's Lesson on Humanity." *Economy and Society* 33: 1–27.

Cockburn, Cynthia. 1999. "The Material of Male Power." In *The Social Shaping of Technology*, edited by Donald MacKenzie and Judy Wajcman, 177–98. Buckingham, PA: Open University Press.

Cole, Peter. 2002. "Aboriginalizing Methodology: Considering the Canoe." *International Journal of Qualitative Studies in Education* 15 (4): 447–59.

Collins, Harry. 1975. "The Seven Sexes: A Study in the Sociology of a Phenomenon, or the Replication of Experiments in Physics." *Sociology* 9: 205–24.

Cruickshank, Julie. 2012. "Are Glaciers 'Good to Think With'? Recognising Indigenous Environmental Knowledge." *Anthropological Forum* 22 (3): 239–50.

Cvejic, Bojana. 2010. "Xavier Le Roy: The Dissenting Choreograph of One Frenchman Less." In *Contemporary French Theatre and Performance*, edited by Clare Finburgh and Carl Lavery, 2011, 188–99. Basingstoke: Palgrave Macmillan, http://www.xavierleroy.com/page.php?id=4b530eff07 7090c4cdd558852f04f24fb0840bae&lg=en.

Daston, Lorraine. 1999. "Objectivity and the Escape from Perspective." In *The Science Studies Reader*, edited by Mario Biagioli, 110–23. New York: Routledge.

de la Cadena, Marisol. 2010. "Indigenous Cosmopolitics in the Andes: Conceptual Reflections beyond 'Politics.'" *Cultural Anthropology* 25 (2): 334–70.

de Laet, Marianne, and Annemarie Mol. 2000. "The Zimbabwe Bush Pump: Mechanics of a Fluid Technology." *Social Studies of Science* 30 (2): 225–63.

Deleuze, Gilles, and Félix Guattari. 1988. *A Thousand Plateaus: Capitalism and Schizophrenia.* London: Athlone.

Despret, Vinciane. 2013. "Responding Bodies & Partial Affinities in Human-Animal Worlds." *Theory, Culture & Society* 30 (7–8): 51–76.

Feit, Harvey A. 2004. "James Bay Crees' Life Projects and Politics: Histories of Place, Animal Partners and Enduring Relationships." In *In the Way of Development*, edited by Mario Blaser, Harvey A. Feit, and Glenn McRae, 92–110. London: Zed Books, also available at http://www.mtnforum.org/sites/default/files/publication/files/1372.pdf.

Felt, Ulrike. 2015. "Keeping Technologies Out: Sociotechnical Imaginaries and the Formation of Austria's Technopolitical Identity." In *Dreamscapes of Modernity: Sociotechnical Imaginaries and the Fabrication of Power*, edited by Sheila Jasanoff and Sang-Hyun Kim, 103–25. Chicago: University of Chicago Press.

Foucault, Michel. 1979. *Discipline and Punish: The Birth of the Prison.* Harmondsworth: Penguin.

Gabrys, Jennifer, and Kathryn Yusoff. 2011. "Arts, Sciences and Climate Change: Practices and Politics at the Threshold." *Science as Culture* 21 (1): 1–24.

Gad, Christopher, and David Ribes. 2014. "The Conceptual and the Empirical in Science and Technology Studies." *Science, Technology, & Human Values* 39 (2): 183–91.

Goffman, Erving. 1971. *The Presentation of Self in Everyday Life.* Harmondsworth: Penguin.

Guggenheim, Michael, Bernd Kräftner, and Judith Kröll. 2013. "'I Don't Know Whether I Need a Further Level of Disaster': Shifting Media of Sociology in the Sandbox." *Distinktion: Scandinavian Journal of Social Theory* 14 (3): 284–304.

Haraway, Donna J. 1985. "Manifesto for Cyborgs: Science, Technology and Socialist Feminism in the 1980s." *Socialist Review* (80): 65–108.

___. 1988. "Situated Knowledges: The Science Question in Feminism and the Privilege of Partial Perspective." *Feminist Studies* 14 (3): 575–99.

___. 1989. *Primate Visions: Gender, Race and Nature in the World of Modern Science*. London: Routledge and Chapman Hall.

___. 1997. *Modest_Witness@Second_Millenium.FemaleMan©_Meets_OncoMouse™: Feminism and Technoscience*. New York: Routledge.

___. 2008. *When Species Meet*. Minneapolis: University of Minnesota Press.

Harding, Sandra. 1993. "Rethinking Standpoint Epistemology: What Is 'Strong Objectivity'?" In *Feminist Epistemologies*, edited by Linda Alcoff and Elizabeth Potter, 352–84. New York: Routledge. Accessed at http://www.msu.edu/~pennock5/courses/484%20materials/harding-standpoint -strong-objectivity.pdf

Harwood, Jonathan. 1976. "The Race-Intelligence Controversy: A Sociological Approach I— Professional Factors." *Social Studies of Science* 6 (3–4): 369–94.

Heuts, Frank, and Annemarie Mol. 2012. "What Is a Good Tomato? A Case of Valuing in Practice." *Valuation Studies* 1 (2): 125–46.

Hughes, Thomas P. 1983. *Networks of Power: Electrification in Western Society, 1880–1930*. Baltimore: Johns Hopkins University Press.

Igo, Sarah E. 2006. "A Gold Mine and Tool for Democracy: George Gallup, Elmo Roper, and the Business of Scientific Polling." *Journal for the History of the Behavioral Sciences* 42 (2): 109–34.

Jasanoff, Sheila, ed. 2004. *States of Knowledge: The Co-production of Science and the Social Order*. London: Routledge.

Jasanoff, Sheila, and Sang-Hyun Kim, eds. 2015. *Dreamscapes of Modernity: Sociotechnical Imaginaries and the Fabrication of Power*. Chicago: University of Chicago Press.

Jensen, Casper Bruun. 2014. "Continuous Variations: The Conceptual and the Empirical in STS." *Science, Technology, & Human Values* 39 (2): 192–213.

Jones, Kristin M. 2011. "Signs of Life." *Frieze* 122, also available at http://www.frieze.com/article/ signs-life.

Knorr Cetina, Karin D. 1981. *The Manufacture of Knowledge: An Essay on the Constructivist and Contextual Nature of Science*. Oxford: Pergamon Press.

Kopenawa, Davi, and Bruce Albert. 2013. *The Falling Sky: Words of a Yanomami Shaman*. Cambridge, MA: The Belknap Press of Harvard University Press.

Kräftner, Bernd, Judith Kröll, Gerhard Ramsebner, Leo Peschta, and Isabel Warner. 2010. "A Pillow Squirrel and Its Habitat: Patients, a Syndrome, and Their Dwelling(s)." In *New Technologies and Emerging Spaces of Care*, edited by Michael Schillmeier and Miquel Domenech, 169–95. Burlington: Ashgate.

Kuhn, Thomas S. 1970a. "Reflections on My Critics." In *Criticism and the Growth of Knowledge*, edited by Imre Lakatos and Alan Musgrave, 231–78. Cambridge: Cambridge University Press.

___. 1970b. *The Structure of Scientific Revolutions*. Chicago: University of Chicago Press.

Kuriyama, Shigehisa. 1999. *The Expressiveness of the Body and the Divergence of Greek and Chinese Medicine*. New York: Zone Books.

Latour, Bruno. 1987. *Science in Action: How to Follow Scientists and Engineers through Society*. Milton Keynes: Open University Press.

___. 1988. *The Pasteurization of France*. Cambridge, MA: Harvard University Press.

___. 1993. *We Have Never Been Modern*. Brighton: Harvester Wheatsheaf.

___. 1998. "Circulating Reference: Sampling the Soil in the Amazon Forest." In *Pandora's Hope: Essays on the Reality of Science Studies*, edited by Bruno Latour, 24–79. Cambridge, MA: Harvard University Press.

___. 2004a. *Politics of Nature: How to Bring the Sciences into Democracy*. Cambridge, MA: Harvard University Press.

___. 2004b. "Why Has Critique Run Out of Steam? From Matters of Fact to Matters of Concern." *Critical Inquiry* 30: 225–48.

___. 2013. *An Inquiry into Modes of Existence: An Anthropology of the Moderns*. Cambridge, MA: Harvard University Press.

Latour, Bruno, and Peter Weibel, eds. 2006. *Making Things Public: Atmospheres of Democracy*. Karlsruhe: ZKM, Centre for Art and Media; Cambridge, MA: MIT Press.

Latour, Bruno, and Steve Woolgar. 1986. *Laboratory Life: The Construction of Scientific Facts*. 2nd ed. Princeton, NJ: Princeton University Press.

Law, John. 1986. "On the Methods of Long Distance Control: Vessels, Navigation and the Portuguese Route to India." In *Power, Action and Belief: A New Sociology of Knowledge?* Sociological Review Monograph 32, edited by John Law, 234–63. London: Routledge and Kegan Paul.

___. 1994. *Organizing Modernity*. Oxford: Blackwell.

___. 2002. *Aircraft Stories: Decentering the Object in Technoscience*. Durham, NC: Duke University Press.

___. 2004. *After Method: Mess in Social Science Research*. London: Routledge.

___. 2009. "Seeing Like a Survey." *Cultural Sociology* 3 (2): 239–56.

___. 2011a. "Collateral Realities." In *The Politics of Knowledge*, edited by Fernando Domínguez Rubio and Patrick Baert, 156–78. London: Routledge.

___. 2011b. "The Explanatory Burden: An Essay on Hugh Raffles' Insectopedia." *Cultural Anthropology* 26 (3): 485–510.

___. 2015. "What's Wrong with a One-World World." *Distinktion: Journal of Social Theory* 16 (1): 126–39.

Law, John, Geir Afdal, Kristin Asdal, Wen-yuan Lin, Ingunn Moser, and Vicky Singleton. 2014. "Modes of Syncretism: Notes on Non-coherence." *Common Knowledge* 20 (1): 172–92.

Law, John, and Annemarie Mol. 1995. "Notes on Materiality and Sociality." *The Sociological Review* 43: 274–94.

Lin, Wen-yuan, and John Law. 2014. "A Correlative STS? Lessons from a Chinese Medical Practice." *Social Studies of Science* 44 (6): 801–24.

Lynch, Michael. 1990. "The Externalized Retina: Selection and Mathematization in the Visual Documentation of Objects in the Life Sciences." In *Representation in Scientific Practice*, edited by Michael Lynch and Steve Woolgar, 153–86. Cambridge, MA: MIT Press.

MacKenzie, Donald. 1978. "Statistical Theory and Social Interests: A Case Study." *Social Studies of Science* 8 (1): 35–83.

Marques, Ivan da Costa. 2014. "Ontological Politics and Latin American Local Knowledges." In *Beyond Imported Magic: Essays on Science, Technology, and Society in Latin America*, edited by Eden Medina, Ivan da Costa Marques, and Christina Holmes, 85–107. Cambridge, MA: MIT Press.

Marres, Noortje. 2007. "The Issues Deserve More Credit: Pragmatist Contributions to the Study of Public Involvement in Controversy." *Social Studies of Science* 37 (5): 759–80.

Martin, Emily. 1994. *Flexible Bodies*. Boston: Beacon Press.

___. 1998. "The Fetus as Intruder: Mother's Bodies and Medical Metaphors." In *Cyborg Babies: From Techno-Sex to Techno-Tots*, edited by Robbie Davis-Floyd and Joseph Dumit, 125–42. New York: Routledge.

Merton, Robert K. 1957. "Science and Democratic Social Structure." In *Social Theory and Social Structure*, edited by Robert K. Merton, 550–61. New York: Free Press.

Mol, Annemarie. 1999. "Ontological Politics: A Word and Some Questions." In *Actor Network Theory and After*, edited by John Law and John Hassard, 74–89. Oxford: Blackwell.

___. 2002. *The Body Multiple: Ontology in Medical Practice*. Durham, NC: Duke University Press.

___. 2010. "Actor-Network Theory: Sensitive Terms and Enduring Tensions." *Kölner Zeitschrift für Soziologie und Sozialpsychologie* 50 (1): 253–69.

___. 2012. "Mind Your Plate! The Ontonorms of Dutch Dieting." *Social Studies of Science* 43 (3): 379–96.

Moser, Ingunn. 2008. "Making Alzheimer's Disease Matter: Enacting, Interfering and Doing Politics of Nature." *Geoforum* 39: 98–110.

Myers, Natasha. 2012. "Dance Your PhD: Embodied Animations, Body Experiments, and the Affective Entanglements of Life Science Research." *Body & Society* 18 (1): 151–89.

Neuenschwander, Rivane. 2008. *Contingent.* Accessed at http://www.youtube.com/watch?v=gurlpLOyubA.

Polanyi, Michael. 1958. *Personal Knowledge: Towards a Post-Critical Philosophy.* London: Routledge and Kegan Paul.

Pols, Jeannette. 2006. "Accounting and Washing: Good Care in Long-Term Psychiatry." *Science, Technology, & Human Values* 31 (4): 409–40.

Popper, Karl R. 1959. *The Logic of Scientific Discovery.* London: Hutchinson.

Raffles, Hugh. 2010. *Insectopedia.* New York: Pantheon Books.

Ravetz, Jerome R. 1973. *Scientific Knowledge and Its Social Problems.* Harmondsworth: Penguin.

Ryan-Flood, Róisín, and Rosalind Gill, eds. 2010. *Secrecy and Silence in the Research Process: Feminist Reflections.* Abingdon: Routledge.

Savage, Mike. 2010. *Identities and Social Change in Britain since 1940: The Politics of Method.* Oxford: Oxford University Press.

Serres, Michel. 1974. *La Traduction, Hermes III.* Paris: Les Éditions de Minuit.

Shackley, Simon, and Brian Wynne. 1995. "Global Climate Change: The Mutual Construction of an Emergent Science-Policy Domain." *Science and Public Policy* 22 (4): 218–30.

Shapin, Steven. 1984. "Pump and Circumstance: Robert Boyle's Literary Technology." *Social Studies of Science* 14: 481–520.

Shapin, Steven, and Simon Schaffer. 1985. *Leviathan and the Air Pump: Hobbes, Boyle and the Experimental Life.* Princeton, NJ: Princeton University Press.

Shove, Elizabeth. 2010. "Beyond the ABC: Climate Change Policy and Theories of Social Change." *Environment and Planning A* 42 (6): 1273–85.

Singleton, Vicky. 2010. "Good Farming: Control or Care?" In *Care in Practice: Tinkering in Clinics, Homes and Farms*, edited by Annemarie Mol, Ingunn Moser, and Jeannette Pols, 235–56. Bielefeld: Transcript.

Slack, Jonathan. 1972. "Class Struggle among the Molecules." In *Counter Course: A Handbook for Course Criticism*, edited by Trevor Pateman, 202–17. Harmondsworth: Penguin Educational Specials.

Smith, Linda Tuhiwai. 2012. *Decolonizing Methodologies: Research and Indigenous Peoples*. 2nd ed. New York: Zed Books.

Stacey, Jackie. 1997. *Teratologies: A Cultural Study of Cancer*. London: Routledge.

Stengers, Isabelle. 2005. "The Cosmopolitical Proposal." In *Making Things Public: Atmospheres of Democracy*, edited by Bruno Latour and Peter Weibel, 994–1003. Karlsruhe: ZKM, Centre for Art and Media; Cambridge, MA: MIT Press.

Strathern, Marilyn. 1991. *Partial Connections*. Savage, MD: Rowman and Littlefield.

TallBear, Kim. 2014. "Standing with and Speaking as Faith: A Feminist-Indigenous Approach to Inquiry [Research note]." *Journal of Research Practice* 10 (2): Article N7.

Thévenot, Laurent. 2001. "Which Road to Follow? The Moral Complexity of an 'Equipped' Humanity." In *Complexities: Social Studies of Knowledge Practices*, edited by John Law and Annemarie Mol, 53–87. Durham, NC: Duke University Press.

Traweek, Sharon. 1988. *Beamtimes and Lifetimes: The World of High Energy Physics*. Cambridge, MA: Harvard University Press.

Turnbull, David. 2000. *Masons, Tricksters and Cartographers: Comparative Studies in the Sociology of Scientific and Indigenous Knowledge*. Amsterdam: Harwood Academic Publishers.

Verran, Helen. 1998. "Re-Imagining Land Ownership in Australia." *Postcolonial Studies* 1 (2): 237–54.

___. 2002. "A Postcolonial Moment in Science Studies: Alternative Firing Regimes of Environmental Scientists and Aboriginal Landowners." *Social Studies of Science* 32: 729–62.

___. 2013. "Engagements between Disparate Knowledge Traditions: Toward Doing Difference Generatively and in Good Faith." In *Contested Ecologies: Dialogues in the South on Nature and Knowledge*, edited by Lesley Green, 141–61. Cape Town: HSRC Press.

Waterton, Claire, and Brian Wynne. 1999. "Can Focus Groups Access Community Views?" In *Developing Focus Group Research: Politics, Theory and Practice*, edited by Rosaline Barbour and Jenny Kitzinger, 127–43. London: Sage.

Waterton, Claire, and Judith Tsouvalis. 2015. "On the Political Nature of Cyanobacteria: Intra-active Collective Politics in Loweswater, the English Lake District." *Environment and Planning D: Society and Space* 33 (3): 477–93.

Watts, Laura, Pelle Ehn, and Lucy Suchman. 2014. Prologue. In *Making Futures: Marginal Notes on Innovation, Design, and Democracy*, edited by Pelle Ehn, Elisabet M. Nilsson, and Richard Topgaard, ix–xxxix. Cambridge, MA: MIT Press.

Wynne, Brian. 1996. "May the Sheep Safely Graze? A Reflexive View of the Expert-Lay Knowledge Divide." In *Risk, Environment and Modernity: Towards a New Ecology*, edited by Scott Lash, Bronislaw Szerszynski, and Brian Wynne, 44–83. London: Sage.

Yates-Doerr, Emily. 2014. "The World in a Box? Food Security, Edible Insects, and 'One World, One Health' Collaboration." *Social Science & Medicine* 129: 106–12.

Yates-Doerr, Emily, and Christine Labuski. 2015. "The bookCASE: Introduction." In *Somatosphere: Science, Technology and Medicine.* Accessed at http://somatosphere.net/2015/06/the-bookcase -introduction.

Zhan, Mei. 2014. "The Empirical as Conceptual: Transdisciplinary Engagements with an 'Experiential Medicine'." *Science, Technology, & Human Values* 39 (2): 236–63.

2   Rethinking Documents

Kalpana Shankar, David Hakken,* and Carsten Østerlund

Introduction: Why an STS of Documents

In this chapter we assess the place of documents in science and technology studies (STS) scholarship. As a first approximation, by *document* we mean any artifact that includes substantial references to the social processes through which it was produced and reproduced. Whether paper based or electronic, wall mounted or etched in stone, material or virtual, documents document. It is this property that makes them of great use to STS scholars. Hence, scholars should be reluctant to separate the artifact (a text) from the practices (which are documented), the noun from the verb.

Acknowledging this property of documents is the starting point for our overall thesis—that STS needs a robust, reflective discourse on documents. In outline, we make the following points: First, STS already depends strongly on interpretation of *artifacts that document* and is likely to continue to do so. Such artifacts, like an article in a scientific journal, a blueprint, or an archaeological simulation, suggest much about the techniques that were used in creating the things on which they focus. STS scholars have made substantial contributions to our understanding of documents and their roles in social settings. Second, there have recently been important changes in how documents are created and the forms they take. Many such changes are related to digital mediation, of either preexisting documents or the creation of new ones. Technoscience practices have come to include computing, as in the move from paper to electronic lab notebooks or from wetware experiments to simulations. Hence, most documents with which STSers now interact are digital. Third, these changes often make interpreting documents substantially more difficult. While such mediations do provide important additional resources for STS research, they also make interpretation more complex. Some digitizations, like making ancient documents available online,

*In memory of David Hakken.

make scholarship easier, but others, like frequent changes in the content of websites, make it harder. Fourth, there are additional justifications for STS, giving more explicit attention to the digitization of documents. They include how the massively expanded use of digital technologies also undermines existing knowledge technologies. Consider the partial displacement of peer-reviewed journals by multiple, less formal, web-based communications, such as preprints or blog posts and other alternatives to the scholarly publication process. This displacement complicates the tracing of "invisible college" disseminations of knowledge.

In sum, since so many of these changes to knowledge production are digital in nature, STSers need to learn to think digitally. To make our case, we suggest ways the STS community might improve its use of documents. One is to increase awareness of the practices and debates of other document scholars. Another is to add to the community's capacity for reflection by incorporating research techniques that increase sensitivity to the properties (especially the new ones) of documents. Increased awareness of such techniques, our primary methodological contribution, should also make the STS discourse on digitization of documents more accessible to new researchers.

The significance of our argument emerges when one considers some implicit assumptions about documents too present in STS research. Perhaps the most important of these are (1) that the meaning of documents is contained within them and (2) that this meaning is largely straightforward and self-evident—in a phrase, that a document "means what it says." For example, an STS scholar doing fieldwork is likely to bracket an informant's verbal comments as personal opinion, but she may treat a written document or a ceremonial object as a collective representation whose very existence establishes its representativeness. The mutability of digital documents amplifies the dangers to interpretation lurking in such assumptions. We need to understand how they came to acquire their form and content.

In this chapter we do not aim to present a theory of current documents and documenting, which is a long-term project and something we STS scholars need to create together. Rather, we see ourselves as presenting initial steps that could justify this broader project. We argue for a more explicit, self-conscious, and occasionally even critical approach to documents and documenting in STS. We do so by focusing heavily on recent changes and what they suggest about how STS scholars, especially new ones, should deal with documents in the future.

Defining Documents

Documents do document. However, a definition that suggested this is all they do would imply that everything that exists now can be connected to what went before, and thus everything is documenting. Perhaps not surprisingly, we do find a number of such all-encompassing definitions in the literature on documents. For instance, Martinez-Comeche (2000, 6), who describes documents as "humanity's materialized memory," argues that anything, as long as it is considered a document, can be a document. To Levy (2001, 23), documents "are, quite simply, talking things." Invoking Latour's notion of delegation, Levy continues: "They are bits of the material world—clay, stone, animal skin, plant fiber, sand—that we've imbued with the ability to speak."

In contrast, our concept of documents focuses on a narrower set of phenomena, documental properties of artifacts that overtly document and that point toward the collection of processes that resulted in the artifact's creation. Documents that document have discernable histories. They have gone through several forms that make it possible to infer things from present instantiations. They have a palimpsestic quality.

The long history of use of the notion "document" reveals important tensions that reverberate in the contemporary literature on documents. Lund (2009) provides a helpful conceptual archaeology. In the Middle Ages, *documentum* referred to a model, a teaching method, or a demonstration. An oral lecture could be a document. Only with the emergence of European state bureaucracies in the seventeenth century did "document" come to refer to something written or inscribed, whether a title deed, coin, picture, or tombstone, that furnished evidence or information about a subject (Lund 2009). Over the next centuries, scholars increasingly developed, shared, used, and stored mostly paper documents as empirical evidence of their research. Around the turn of the twentieth century, American and European scholars collaborated to create an infrastructure of international scientific associations for the many scholarly journals being founded, all to increase the sharing of documents. The proliferation of scientific materials created a need for tools to locate colleagues' work, find publications, create and share collections of data, and coordinate collaborations. A notion of "document" emerged as the organizing concept for these efforts, as did a new metascience of *documentalism*. This latter's agenda mostly deals with written documents' material manifestations, temporal and spatial production, distribution, inventory, statistical properties, preservation, and use (Lund 2009; Otlet 1903, 1907).

For those in the developing field of information science, documentalism raises a number of important questions germane to contemporary STS scholarship (also affording our definition of documents). The French documentalist Suzanne Briet (1894–1989) asked: "Is a star a document? Is an antelope on the savanna a document?" The rhetorical answers were "no," but a consensus developed that a picture of the star or a report on the antelope is a document. What about an antelope in the zoo, caged up side by side with other classes of antelopes? Briet (2006, 11) claimed that it was only when the very concrete animal is "cataloged" that it becomes a document. That is, the process of documenting is seen as a constitutive part of a document itself. She argued that a document must have been "preserved or recorded towards the ends of representing, of reconstituting, or of proving a physical or intellectual phenomenon" (Briet 2006, 10). In light of this pioneering work, it is not surprising that the contemporary literature on documents in anthropology, sociology, and STS often debate how or to what extent the power of a document is associated with its capacity to represent, stand for, or orient us toward some particular aspect of the world (Hull 2012).

More recently, document scholars have tended to move beyond representations and content as their central empirical and theoretical categories and concerns. Hull (2012) and others (Garfinkel 1967; Hetherington 2008; Reed 2006) argue that there is a tendency to supplement or even deemphasize their information, content (Elyachar 2006; Li 2009), knowledge (Strathern 2006), facts (Hull 2003), and meaning (Riles 1998). Instead, documents are analyzed as more than mere instruments of representation, denotation, reference (Hull 2012), or texts to be interpreted (Riles 1998). Deemphasizing the representational does not mean taking the content of documents lightly. It is rather an acknowledgment of the rich literature on discourse that has emerged at least since Foucault (1971).

Documents do more than represent the world; they often also refer to the practices, objects, rules, knowledge, and organizational forms that produced them. While documents may stand for something else, they are more than haulers, carrying preformed ideas or abstract semiotic constructions through space and time. Engaging with the constitutive dimensions of documents has already led STS scholars to the more self-conscious use of documents to address a number of major issues in the field.

Documents Already in STS Scholarship

In this section we address several issues of general concern to STS scholars and others that follow from the more self-conscious approach to documents that we advocate. These include the tension between fixity and fluidity in sociotechnical processes of production and circulation, materiality, coordination and control, and indexical range. For the reasons outlined initially, we give particular attention to additional difficulties introduced by digitization.

Fixity and Fluidity

One example of a perennial issue in STS is the tension between fixity and fluidity in sociotechnical processes of production and circulation. The elements of a set of documents can often be arranged from the highly permanent to the very temporary. Where a document lies on this continuum can be an important indication of its social uses. Permanence may be associated with relatively developed group structures, like organizations and governments, while ephemerality may be associated with social movements or waves of popular culture.

STS scholars can learn from the evolution of documentalists' efforts to deal with this property of documents. Early documentalist and archival studies tended to concentrate on the permanent end, highlighting document aspects that implied stable "facts" that could be fixed to index cards (Otlet 1907). These could then be sorted, combined, retrieved, and indexed into a larger network of knowledge. The presumptions of permanence and objective capture of knowledge were born of the positivism of the nineteenth century (Otlet 1903). However, subsequent documentalists, such as Suzanne Briet (2006), argued for seeing documents as parts of more fluid networks of processes (although the document itself remained relatively fixed in time and space. Levy (1994) has asked what "fixed versus fluid" means in the context of new media. What matters more, the fixity of the physical document or the stability of content from one document to another?

For archivists—those emphasizing the practical organization and maintenance of repositories—an ability to fix both the medium and the message of a particular document is the foundation of their profession. A version of this approach, one that takes the "fixed" document as the core of information studies, has been resurrected in recent years (Buckland 2007; Lund 2009). Some archival scholars argue for an *archival wall* approach: once documents are committed to an archive, their form and content should be considered fixed (Duranti 1997).

However, while documents (and archives) are artifacts, they also result from ongoing processes. As they have both fluid and fixed aspects, there is substantial benefit from approaching *document* not solely as a noun but also as a verb—*to document*. Focusing on documenting as a process leads to probing for a document's origin, the changes in it from version to version, and its provenance. Where did it come from, as in where did a bit of data or artifact first appear, but also, what is the "genealogy" of the artifact (e.g., an artwork or an idea), and who or what (person or organization) had control of it (Baudoin 2008; Jensen et al. 2010; Sweeney 2008)? This processual approach has also been adopted by scholars in information management and computer science (Blanc-Brude and Scapin 2007; Lynch 2001; Shen, Fitzhenry, and Dietterich 2009). They stress how an *original* text may be authored by one person or it may be assembled from multiple sources. Also, content often flows from application to application and document to document, constantly recycled, reworked, and repackaged.

When applied to STS, this flow means we can no longer expect to say definitively what created a document and why. We should expect instead to deal with the multiple appearances of documents. How do they get placed in particular locations for a moment in all their heterogeneity? How do we attend to the ways documents keep changing in the processes that enact or perform realities (Law 2004; Suchman 2007)? While some documents may leverage their power through an immutability even while mobile, as Latour (1990) has argued, such fixity is an accomplishment that is best understood by studying the practices and materials that have gone into stabilizing them in time and space. Current workplace studies, like those referenced above, engage each document's form as a point in ongoing work practices, so one document's multiple forms might be read as moments in ongoing conversations and processes.

Attention to fluidity/fixity dynamics greatly affords the basic STS idea that science and technology are contingent social constructions. Keeping this in mind will help STS scholars deal more effectively with documents.

Materiality

Recently, a general appreciation and rehabilitation of artifacts has developed in the social sciences and humanities. Much of this has followed from attention to their materiality. Increased understanding of documents will also follow from their being included in this project. Documents are knowledge-made material, taking material form in journal articles, patents, conference presentations, whiteboards, computer screen images, graphics, recordings, and databases of all kinds (Hull 2012).

Nonetheless, some implications drawn from the stress on materiality are contradictory. For example, material cultural studies highlight how documents' material symbols communicate through representations. Emerging in the late 1970s as a humanistic reaction to neo-functionalist theories dominating archaeology and anthropology (Hodder 1982), material culture studies explore how material objects, including documents, in practice carry meanings as vehicles of symbolic expressions. For instance, in "The Interpretation of Documents and Material Culture," Hodder (1994) distinguishes two types of material symbolism. On the one hand, we have documents that are intended to communicate through representations and language, like syntax does. On the other hand, we find material symbols that are evocative or implicative. They come to have abstract meaning through association and practice, as in the ways certain foods or sports become embedded within social conventions and thus come to have common meanings. (For example, the Magna Carta—although the charter long ago lost its practical use and significance, invoking it has certain symbolic notions of resistance to tyranny.) They become objects that provide mirrors in whose reflection contemporaries and successors can fashion themselves (Ingold 2012).

Such material culture studies, however, are at odds with many other contemporary studies of materiality. Ingold (2012) points out that material culture studies tend to prioritize finished artifacts over attending to process perspectives. In doing so, Hodder (1994) and his contemporaries assume fixed, not fluid, documents. What flows in material culture studies is not the document but its interpretation, and the material becomes a mere attribute. Instead, Ingold (2012) argues that material properties should be approached as having histories. To understand the materiality of a document is to be able to tell its history, what it does, and what happens to it when treated in particular ways.

In the broader literature on documents we find a similar shift away from just the "objectness" of documents toward the material flows and formation processes through which documents come into being. Brown and Duguid (1994) illustrate how the materiality of a simple document, a movie ticket, is carefully chosen to support particular practices of authentication. A movie ticket sold at the door needs little extra authentication and is, consequently, quite insubstantial: *Admit One* printed on a small slip of paper (Brown and Duguid 1994, 15). But when tickets are sold at a distance in time or space, their authorial force must be performed. Many different forms can come into play in redeeming such demands on documents, but the solutions are often manifest in the material of their composition. Solutions include things like watermarks, letterheads, engravings, and embossings. Even if the ticket is electronic, one needs a way of corroborating the document's history, tracing it back through any earlier transformations. These transformations might involve encrypted messages sent to the eager moviegoer, including unique ticket numbers issued by the ticket agency. Arriving at a theater, the unique identifier number is electronically corroborated with a ledger hosted on the ticket agency's server. Emerging payment systems, like Bitcoin, raise the required complexity demanded by hosting transaction traces distributed widely across a public network rather than at some centralized and private organization.

In the work on documents' materiality, a vocabulary is emerging that is sufficient to the entwined nature of sociomateriality. Influential terms, among others, include *bricolage* (Fujimura 1992), *actor networks* (Callon 1986), *assemblages* (Law 2004), *entanglements* (Barad 2003), *practice-order bundles* (Schatzki 2002), *mangles of practice* (Pickering 1995), and even the *double mangling of the human and material* (Jones 1998). The image of something knotty, evoked in both common empirical observations and in analytically developed relational ontologies, unites all these terms.

Important laboratory and workplace studies show us how to pay keen attention to the documenting aspects of the material "mangle" with which organizational members surround themselves (Pickering 1995). For instance, in their classic study of an airport operations room, Suchman (1997) and her colleagues at Xerox Palo Alto Research Center described the work setting as resembling archaeological strata. Here, the layerings were of documents and communication technologies in various material instantiations: paper, whiteboards, networked workstations, video monitors, and clocks, all acquired in bits and pieces over time. The operations staff constantly reconfigured this heterogeneous collection to make available the remote locations across the airport that they are managing. To be made useful, these documents' particular material instantiations need to be read in relation to one another and to the unfolding work at hand. The documents involved are constituted materially through and inseparably from the specifically situated practices of their use (Blomberg, Suchman, and Trigg 1996).

Subsequent studies have used this material perspective to explain why attempts to introduce new systems often fail (e.g., Berg 1999). Iconic examples include the consequences of replacing medical charts with electronic medical record systems and the implementation of large enterprise systems (e.g., Bjørn and Østerlund 2014).

Drawing on Haraway (1987), we can describe documents as sociomaterial realities that resemble messy bundles of yarn. It takes careful effort to trace the string and thus give the mess some organization. To get to a comfortable order, like that of a knitted sweater, STS scholars have to attend to the practices of shaping and reshaping the bundle as the yarn is used in cat's cradle–like games. Creators continuously let the document "string" slide from hand to hand, in the process creating new patterns. Similarly, Latour and Woolgar (1986) show how laboratory practices could be explained by looking at the transformation of rats and chemicals into papers. Instruments produce small documents containing various signs and inscriptions that can later be integrated into databases, combined or superimposed as figures and charts into conference presentations, patents, slides, and published articles. Documents' specific material instantiations make it possible for scientists and others to accumulate, compare, combine, contrast, manipulate, and evaluate work (Latour 1990).

Classic actor-network theory likewise places great importance on documents and their materiality in the accumulation of work and the enrollment of allies (Callon 1986). In the mustering of power by increasing the effective alignment or promoting the fidelity of new allies, documents help. Latour (1990) highlights the advantages of paperwork, including how inscriptions are mobile and yet immutable when they move; are made flat and the scale of the inscriptions may be modified at will; can at little cost be reproduced, spread, reshuffled, and recombined; and how it is possible to superimpose several images of totally different origins and scale and be made part of the same written text.

In these analyses, Latour (1990) assumes the use of print on paper. He does not explore the challenges and possibilities that come with other material instantiations (e.g., electronic, as in the movie tickets discussed above). Moving beyond paperwork, we find a lively debate over how much of an ontological role materiality plays in documenting work. At the "highly ontological" end of a spectrum, we find positions that assume artifacts are associated with relatively stable and pre-given material orders. For instance, Leonardi (2011) introduces the notion of imbrication to describe human and material agencies arranged as distinct elements in overlapping patterns so that they function interdependently. Here, particular materials are associated with specific affordances that allow them to be imbricated into particular organizational routines. The relative rigid ordering associated with the verb *imbricate* derives from the Greek and Roman name for roof tiles neatly interlocked into a visible pattern.

At the "low ontological" end of the spectrum, one finds a position often associated with Barad's work (2007) and promoted in studies by Scott and Orlikowski (2012). Four basic assumptions characterize their perspective. First, the world of documents is already sliced and diced by practitioners into galleries of doings and discourses. In order to perform, practitioners bind the world into actionable bundles. Second, we should not regard these bundles as pre-given or timeless. Their boundaries are not fixed (Barad 2003; Knorr Cetina 1997; Suchman 2007). Third, documents are always also enacted as parts of smaller and larger entanglements (Barad 2003). For instance, we can choose to approach a medical record as a single entity. Simultaneously, by focusing solely on the sociomaterial practice associated with a single sheet of the record, we could break the medical record apart. We could also choose to see the record as part of a larger health information system. To do so depends in part on how the practitioners have sliced and diced their world. Fourth and finally, in the absence of predetermined boundaries of what constitutes the document and its materiality, much analytical work has to go into distinguishing the relations relevant for the particular situation at hand.

The preceding sections illustrate how documentary practices are often at the center of debates over how materiality is to be made manifest in STS analyses. The openness and uncertainty associated with documents' materiality, as well as their extensive use, interpretation, and impacts, make them very relevant to post-essentialist debate in STS (Law 2004; Woolgar and Lezaun 2013; Woolgar et al. 2008). Focusing in even more self-conscious, deeper, and more systematic ways on documents will generally benefit the materiality project (Robichaud and Cooren 2013).

Coordination and Control: Accounts of and Accounts For

Studies of bureaucracy often treat documents as the central image of formal organizational practices, or even as the central technology for coordination and control (Harper 1998; Hull 2012). Yet precisely how and how much documents contribute to organizational order varies a great deal. Weber (1978) saw files and written documents as a central source of stability and the expression of norms that allowed bureaucratic institutions to remain stable despite regime changes. While avoiding Weber's focus on the normative, Goody (1977) and later Yates (1993) offer wonderful descriptions of how new technologies and document genres differ in the role they play in new organizational forms. They do this through how they store and transmit information, becoming instruments of control (an issue taken up by Hull 2012) or political technologies that can introduce disciplinary mechanisms (Burchell, Gordon, and Miller 1991). In other words, it is often through sets of documenting practices (Cambrosio, Limoges, and Pronovost 1990) that policy is constituted, as when accounting is elided with policing (Power 1999; Strathern 1997).

Documents do not merely serve as vehicles of control; they also enable the coordination of activities across space and time. They do so by serving as both accounts of and accounts for organizational activities. Dorothy Smith (2005, 174) calls this *double coordination*. Documents are not solely accounts *of* work and documentations *of* past activities. They are also accounts *for* work. Documents often offer models for practice, as the things to which people should pay attention, when and as they shape ongoing practices. The literature on double coordination illustrates how its quite different flavors depend on the position of the accountant. Yet when generated close to the unfolding work, the diverse coordinative and generative capabilities of documents are also worth focusing on. Documents allow some things to come into being (Frohmann 2008). Many computer-supported cooperative work (CSCW) studies (Bowker 1997; Schmidt and Bannon 1992; Suchman 1993) recount how accounts for work are modeled in documents. These allow communities to coordinate and synchronize their unfolding activities and help legitimize their role in and contribution to the community.

Accounts of work may also serve as a surveillance tool for managers (Suchman 1995). Accounts for work generated at a distance by managers, systems designers, or policy makers can carry prescriptive elements. Bechky and Østerlund (1994) studied a sales force information system developed by Xerox. They found that the documenting process built into the system influenced both the way that sales work is seen by other parts of the corporation (more specifically, higher management) and what the corporation wants the sales work to look like. At a broader political level, Shore and Wright (1999, 2003) studied British and Danish higher education. They illustrate how mechanisms to document teaching performance, research quality, and institutional effectiveness also introduce new disciplinary mechanisms that sometimes reduce professional relations to crude, quantifiable, and inspectable templates.

This of-and-for, double nature of accountability documents creates a tension that plays out in many practices. Systems designers must build into the documents decisions about how fine-grained the accounting should be. Such decisions have consequences for the producers who will have to fill out the forms and encode activities, as well as for consumers when assessing the validity of the account. Bowker and Star (2000) summarize these challenges as trade-offs between comparability, visibility, and control. The "trade-off" necessity stems from no account being able to specify all the wildness and complexity of the work it attempts to represent; choices have to be made. Yet the choices' implications are not the same for all and typically involve other negotiations. In such negotiations, documents often become the battleground for defining work and what constitutes its relevant contexts (Star and Strauss 1999). Accounting through documentation involves selecting indicators, whether finished artifacts, changes of affairs, or new relationships. The documents identify what work is made visible and invisible. Star and Strauss (1999) illustrate how the women's movement of the 1970s worked hard to make domestic work economically accountable via documenting it. Documented activities would no longer simply be treated as acts of love or expression of a natural role. Timmermans, Bowker, and Star (1998) detail attempts to make nurses' work visible by developing formal record systems with predefined categories for what accounts as legitimate activities. However, more visibility may entail more surveillance as well as more paperwork. Exactly what work is made visible and what invisible through the trace data left on, for instance, Amazon's Mechanical Turk or other crowdsourcing sites is still an open question (Irani and Silberman 2013).

The importance of documents' double accountability shines through in the STS literature addressing documents' reflexivity, that is, the ability of any account to explain its own formation (Ashmore 1989; Mulkay 1989). Moments of reflexivity are particularly important for STS scholars. They illuminate how sciences construct facts, including how scientific documents make some parts of researchers' work visible while leaving other parts invisible (Star and Strauss 1999). For instance, by changing or modifying the documents that embody conventions for how one accounts for scientific work, one can reflexively raise awareness about the consequences of particular documentation practices. Other STSers explore political reflexivity by studying the ways in which social groups reposition themselves in society. One way is to reappropriate bureaucratic documents and embody in them different accounts that serve the group's interests. For instance, Shore and Wright (1999) show how indigenous groups in Greenland,

Australia, and Brazil reflexively re-create traditions in ways that allow them to deal with government documentary practices more strategically. (For a more comprehensive review of different theoretical directions taken within the reflexivity debate, see Malcolm Ashmore [1989].)

In sum, how coordination and control are achieved in complex technoscience practices is now central to how STS scholars account for their social construction. Documents are often, maybe always, at the center of efforts to achieve coordination and control, so greater attention to the complexity of their role should result in better accounts of how this is accomplished.

Indexical Range and Background Knowledge

Conceptually, documents can interact with their contexts in two distinct but closely entwined ways. First, documents are often intended to be self-explicating devices (Brown and Duguid 1994; Harper 1998); that is, they denote the context of their use by asserting explicitly how they are to be understood. Producers of documents often invoke particular instructional genres (Bakhtin 2010; Bazerman 1994; Miller 1984) or they may demarcate a place with explicit limits for interactions over interpretation (Østerlund 2008). To put it differently, documents specify desired connections among people, objects, times, places, and events and thus constitute a structure of relevancies for discourses about organizational practices. Such documents and their content index the world, often directing people to which specific parts are to be brought into focus (Hanks 1996).

In contrast, a second kind of document is not intended to be self-explicating, pointing instead to the setting within which it is properly seen as existing. Prior experiences with the setting thus come to serve as resources whereby readers know what to expect and how to use the document. Rather than being structured rigidly, documents pointing largely at background knowledge are subject to a situation-by-situation reordering, according to priorities at hand (Hanks 1996). Background knowledge is incomplete and selective and often contains contradictions. This is why newcomers, even when able to read it repeatedly, often find it challenging to comprehend for what purposes a particular document is intended. What eludes them is a shared, practical, and entitled understanding of a common context between writer and reader (Garfinkel 1967). Consequently, presuming that a document of this second type "means what it says," looking at it in isolation and outside of unfolding social processes, makes little sense.

Thus, one needs to ask of a document, how much does it *actually* self-explicate, as well as how much and what background knowledge does it presume that writers and readers will have to bring to it? Some documents address a broad and heterogeneous audience and thus can rely very little on readers bringing background knowledge. A number of studies effectively demonstrate that bureaucratic documents are *self-contextualizing* and *self-analyzing* (Latour 1999) in this sense. Organizational members should be able to use them without placing them in direct relation to specific social settings and institutional goals outside the document itself (Reed 2006; Riles 1998, 2001). Other documents are written for a very narrow group or maybe even an individual and thus lean heavily on the writer's or reader's background knowledge. For example, until she enters it into the formal record, the notes a doctor scribbles in the margins of a patient's test results may only be decipherable by that doctor and make sense solely in the context of her immediate work practice (Berg and Bowker 1997). With such document forms, one must know important aspects of context in order to decide which presumptions are relevant.

Further, the relationship between documents' power to explicate, the relevant aspects of context, and background knowledge needed is dynamic. This might point analysis straight to the age-old dichotomies of general versus particular, local versus global, and implicit versus explicit. It may indeed be the case that documents tending to serve a general or global audience rely little on background knowledge, whereas documents with very particular content and form rely heavily on the writer's or reader's local background knowledge to discern the meaning of the document.

On reflection, however, it is difficult to treat these two types as airtight distinctions. A document may serve as general and explicit in some situations and become particular and implicit in others. Actual documents often contain elements of both strong and weak context dependence. All documents likely exist in relation to others. Documents' networks range from the highly organized to the informal and personal (e.g., the organization of personal documents and photos). In any case, there is likely to be a dynamic relationship between a document and its contexts, with some aspects seeming to presume no context relevance while other aspects presume heavy writer and reader reliance on background knowledge.

Moreover, documents do not simply force such distinctions upon the reader (Harper 1998). Instead writers and readers engage with documents in mutually informing, reciprocal relationships. Dorothy Smith (2005) refers to this as a document's *active text*. Through the engagement between text and user, both are altered. What a document stands for, the user's concern with that document, and what the document does are not stable. Depending on the context and the background knowledge writers and readers bring to bear, the meaning of a document and its power to explicate specific uses may change, whatever the self-explicating intentions of an author might be. Similarly, Bijker, Hughes, Pinch, and Douglas (2012) argue that artifacts hold *interpretive flexibility*, revealed through the different meanings attributed to them by different social groups. How much interpretive flexibility there is depends on the practical concerns and problems of each group and how the artifact allows them to address specific concerns or reach a solution to pending problems.

The metaphor of construction conjures up an image of something like a building site, of an artifact that eventually will be erected, "cast in concrete." When documents and documenting practices are ramshackle, poorly coordinated, and different in form, one might argue that documents are not constructed in this sense. Instead, if documenting practices are multiple, or a document simultaneously contains diverse guises, it makes more sense to see that different realities are being constructed, moment by moment (Law 2008). No longer a metaphorical building site, documentary practices are better understood as parts of larger performances or enactments (see Konrad et al., chapter 16 this volume). Mol illustrates this point beautifully in her book *The Body Multiple* (Mol 2002). Reading a medical textbook, one might assume that lower-limb atherosclerosis is one disease. Yet Mol finds that the practices for dealing with lower-limb atherosclerosis differ so greatly among different health services in the same Dutch town that they effectively enact different realities. The world is not merely epistemologically complex, associated with a plurality of interpretations; "It is ontologically multiple too" (Law 2008, 637). The realities of health-care work, or hotel services in Scott and Orlikowski's case (2012), mean the same document can be connected to simultaneous but very different performances. This also underlines the interpretive flexibility of documents, which are not generally representations of realities produced after the fact but are co-constituted in their enactment.

In sum, there are multiple ways in which a document's context can matter, and these multiple ways can matter simultaneously. These potentials of documents are directly relevant to the debate in STS over what it actually means to say a technoscience practice or artifact is socially constructed. There is great diversity among the characteristics and degree of documents' indexicality, as well as in the degree and type of background knowledge required to make sense of them. As with the three other issues on which this section has focused, a large part of the STS debate over social construction is being worked out on the terrain of the document.

Building an STS of Documents

The next section underlines how the study of documents has contributed to major debates in STS. It also identifies several document-related challenges currently facing

STSers. To help deal with these challenges while also affording a more mature STS of documents, in this section we focus on some things to do.

Foster an Ethnography of the Document/Documentary Practices

We argue that there is still a need for greater awareness of the similarities and differences among how STSers in different areas of the field deal with documentary practices. One way to increase such awareness would be to foster an explicit ethnography of document and documenting. Consider, for example, how acts of creating, managing, and using documents (reports, memos, legal records, etc.) serve as primary sources of socialization and learning. This is particularly true in organizations, the study of which has given rise to a literature around situated learning and legitimate peripheral participation (e.g., Lave and Wenger 1991). This literature offers numerous examples of how documentation practices help "shape" newcomers to become skillful participants in diverse communities of practice (Østerlund 2008; Riles 2006), such as science laboratories (Latour and Woolgar 1986; Shankar 2009), citizen science (Wiggins and Crowston 2011), environmental activism (Ottinger 2010), health care (Berg 1999; Berg and Bowker 1997), and police work (Van Maanen and Pentland 1994). To what extent is document making similar in various arenas, and to what extent is it different? How specifically, in each of these areas, does documentation work configure relationships, emphasize or create prevailing power structures, or empower the marginalized by making aspects of their work more visible through documented evidence?

Another example of how a comparative ethnography of documenting might be useful builds on the laboratory studies at the core of STS use of documents, such as laboratory notebooks (e.g., Latour and Woolgar 1986; Shankar 2004). For example, in his seminal study of the International Monetary Fund (IMF), Harper (1998) shows that an organization's inner workings stand out when we track the transformations taking place in a document's career through the organization, as does the changing relationship between the document and its writers and readers. It is these documenting processes that allow IMF employees to coalesce into the loosely coupled alliances that at its core constitute this organization (Harper 1998). Similarly, in the context of developmental aid in Southeast Asia, Jensen and Winthereik (2013) consider the flow of aid-related documents that help create transparency, improve accountability, and strengthen partnerships. Instead of taking as given that each document arose *de novo* or could be taken at face value, they track the processes, social practices, and infrastructures that emerge to monitor development aid. In Namibia as in other Global South contexts, researchers have helped to document formally indigenous knowledge (IK) so that it can be used against predatory agricultural or pharmaceutical companies (Okere, Njoku, and Devisch 2005; Winschiers-Theophilus et al. 2010).

Accounts of storytelling in oral traditions suggest the existence of a shared suite of practices. Taken together, these document studies suggest that documenting practices across arenas of technoscience are also alike enough to suggest the existence of something similar. Future comparative studies could build on these similarities by asking what other documents are already being used or circulated in the relevant domain, who narrates or writes them, which procedures are normally followed in communicating with other groups or entities, and how all these change (or don't change) with digitization. Ethnographic studies like those cited combined techniques like interviewing, analyzing sociotechnical systems (Biazzo 2002), and observing in field settings to learn more about the social contexts in which particular documents arise, and how they participate in constructing groups, delimiting boundaries, and structuring work practices.

Learn How to Deal with Collectivities of Documents

Another way forward for the STS of documents is working out how to deal not simply with individual documents but with document collections. Documents tend to be part of larger collections, either informal ones (boxes in an office) or formal ones (archives or repositories). The researcher often needs to interrogate not just the individual documents but also the collections and collectivities to which documents belong, as collections, too, have provenance. Questions should probably include: Where was this document found? What other documents are around it? How are they organized (or not)?

Both indexing and archiving choices have evolved over time and are imbued with political implications. Section II notes the tendency of early documentation and archival studies to approach documents as relatively stable artifacts (Cox and Wallace 2002). In contrast, the workplace studies cited in the previous section show how looking at documents as unchanging isolates separate from unfolding work processes makes little sense. More recently, archivists have developed tools to appraise documents for inclusion in collections based on institutional policies and mandates. Such tools work to maintain documents in the context in which they were found, a principle called *respect du fond* (or *respect des fonds*). However, it may not always be possible to do this, as when the original order was not "ordered." In any case, the researcher needs to develop skills to ascertain the relationships among documents and the choices made with respect to their assembly.

To tackle the challenges posed by collectivities, STS scholars can learn from archival studies (Zeitlyn 2012), but they can also use infrastructure and standards development studies within STS (Bowker and Star 2000; Ribes and Finholt 2009). Just as one can question the construction of individual documents, one can also study the construction practices around the collections in which they are found.

Develop Procedures to Analyze Digital Infrastructures

STS has long been concerned with the process of infrastructuring—how large-scale, complex technologies, such as telescopes, supercolliders, laboratories, or particular institutions, become part of the daily work of disciplines (Sands et al. 2012; Traweek 2009). Numerous studies have discussed specific assemblages of networks, artifacts, protocols, standards, and information technologies that are explicitly designed to become infrastructures for specific research communities, such as geologists and ecologists (Bowker et al. 2010; Karasti and Baker 2004). When digital, they become what Atkins (2003) calls *cyberinfrastructures*. These studies rely heavily on documents like contracts, e-mails, and grant proposals to understand how rhetorical strategies, political moves, and other practices are embedded in the technologies. For highly complex cyberinfrastructures, documents and their practices are often the most accessible evidence upon which to build such understandings. A reflective discourse that attends to how digital practices of annotating, mixing, recombining, reusing, and sharing would give STS researchers greater insight into how digitizing practices mediate the production of contemporary knowledge, groups, cultures, and social structures.

Navigate Ownership and Sharing of Documents

Several emergent issues that pose challenges for STS can be seen as being about ownership, broadly conceived. One is that STS researchers are going *in house*, becoming embedded in companies, which often means they do not own the research documents they rely on. Limitations on the use of such *born proprietary* documents, including materials purchased from private companies, can clash with the academic culture of openness. Similar concerns emerge around the use of social media data, as when the researcher scrapes data from Twitter, LinkedIn, or similar sites. Nondisclosure agreements also pose challenges for STS studies of high-tech innovation and related topics (Simakova 2010). Efforts to manage organization knowledge have elevated the strategic importance of documents. Documenting and leveraging indigenous knowledge, especially when codifying it for purposes of preservation and establishment of ownership and authority, can also engender conflict (Winschiers-Theophilus and Bidwell 2013). Even when STS researchers use, share, or author documents with the communities they study, they need to be sensitive to locally appropriate definitions and enactments of privacy, intellectual property, and ownership, especially when contested, as an intrinsic part of their studies (Ngulube 2002).

While many STS scholars have studied document sharing in disciplines, they now face the prospect of sharing their own data. Many funders and institutions are calling for researchers to make their publications and data (e.g., field notes) openly accessible. Proponents cite enhanced citation rates, increased collaborative and analytical possibilities, research transparency and accountability, reuse of expensive data, and other similar benefits. However, in addition to ownership issues, effective data sharing at scale, especially of digital documents, requires significant work and attention to data curation if the data are to be comprehensible and adhere to standards.

STS scholars may dismiss the utility of sharing their research data or question the usefulness of reuse of their data. The forms of some STS documents (e.g., field notes, documents gathered in the field, interview transcriptions, interview protocols, codebooks, and analytical memos) may simply mean they are too idiosyncratic to be useful. They may also compromise the privacy of study participants. Anonymization of large data sets is difficult, but STS's small samples, the rich context provided with interview transcripts, and the emphasis on the specific and local all make such data particularly difficult to anonymize (or even pseudonymize). Field notes usually contain reflections and observations particular to the researcher, and it is not surprising that scholars are reluctant to share their dead ends and personal and scholarly anxieties. Use of field notes also raises issues of indexical reach and background knowledge, making them resistant to the kinds of verifiability and reuse that are touted as advantages to data sharing.

In spite of these obstacles, STSers (with other qualitative and ethnographic researchers) have found ways to share their data and documents in structured ways to facilitate bigger projects. One fairly lightweight approach that some STS researchers have suggested is making interview protocols, analytical codebooks, and research designs "open" through institutional repositories, disciplinary repositories, or websites (Broom, Cheshire, and Emmison 2009; Corti 2011). This may be a way forward for those doing small-team research with relatively low amounts of funding and thus who are unable to fund bigger data-sharing initiatives.

Conclusion: Documents and Digitization

This chapter has explored a range of STS and cognate approaches to working with documents as objects of study. It has shown how STS has engaged documents as elements of ongoing practices in technoscience, as well as how this engagement has made substantial contributions to key issues in the field. By highlighting challenges facing those using documents in STS scholarship, it has made a case for fostering a more self-conscious approach to documents and documenting in STS. We intend for this chapter to lay the basis for a more developed theory of documenting in STS, as part of more general conversations about documents. We have highlighted several STS studies of documents, but more in-depth, comparison-informed studies into documenting in other environments (e.g., leisure or citizen) would reciprocally aid understanding of their place in other areas of technoscience practice. Ethnographies of documents, for example, would unearth new questions about the nature of organizational and group memory and practice, the purposes for creating particular documents, and the nature of personal, professional, and organizational accountability.

To conclude, we choose to discuss how the approach we have advocated is relevant to an obvious challenge that STSers share with many other scholars, the massive growth in the sheer number of documents. The so-called data deluge can legitimately be described as a *document deluge*—of paper as well as digital forms (Sellen and Harper 2002). A steady stream of articles discusses how a data deluge swamps not only the natural sciences but also the social sciences and humanities, and how large organizations are also grappling with this burden in what Buckland (2007) calls our "Document Age."

There is clearly a need to develop new strategies to deal with all these documentary traces of social practices. An increasing number of STS scholars are addressing the new, computation-based phenomena increasingly central to technoscience practices (Sayers 2013). Topics include big data, free/libre open source software (F/LOSS), open access/knowledge, citizen science, digital research ethics, and information and communication technology for development (ICT4D). In such areas, documents tend to become the field itself, but to encompass such fields means a more systematic approach. This is another reason for fostering an ethnography of documents. STS analyses of context increasingly address documents like algorithms and web pages. As digital documents are highly contingent, understanding documentary contexts means mastering digital tools. Digitization increases the malleability of documents, posing new questions about authenticity and maintenance over time. How to deal with the increasingly blurry line between documents and data will become an important issue.

All these issues are made more complex by the growing scale of digital documenting. The role of STS as document creators will be itself a prod to thinking more self-consciously about documents. Fortunately, we have the methods to do this work, and we have paragons who show us how to do it. We look forward to joining new STS scholars in growing the STS of documents.

References

Ashmore, Malcolm. 1989. *The Reflexive Thesis: Wrighting Sociology of Scientific Knowledge*. Chicago: University of Chicago Press.

Atkins, Daniel. 2003. "Revolutionizing Science and Engineering through Cyberinfrastructure: Report of the National Science Foundation Blue-Ribbon Advisory Panel on Cyberinfrastructure." Accessed at https://arizona.openrepository.com/arizona/handle/10150/106224.

Bakhtin, Mikhail Mikhaïlovich. 2010. *Speech Genres and Other Late Essays*. Austin: University of Texas Press.

Barad, Karen. 2003. "Posthumanist Performativity: Toward an Understanding of How Matter Comes to Matter." *Signs: Journal of Women in Culture and Society* 28 (3): 801–31.

___. 2007. *Meeting the Universe Halfway: Quantum Physics and the Entanglement of Matter and Meaning*. Durham, NC: Duke University Press.

Baudoin, Patsy. 2008. "The Principle of Digital Preservation." *The Serials Librarian: From the Printed Page to the Digital Age* 55 (4): 556–59.

Bazerman, Charles. 1994. "Systems of Genres and the Enactment of Social Intentions." In *Genre and the New Rhetoric*, edited by Aviva Freedman and Peter Medway, 79–104. London: Taylor & Francis.

Bechky, Beth, and Carsten Østerlund. 1994. "Qualifying the Customer: An Ethnographic Study of Sales." PARC Tech. Report. Palo Alto, CA: Xerox Palo Alto Research Center.

Berg, Marc. 1999. "Patient Care Information Systems and Health Care Work: A Sociotechnical Approach." *International Journal of Medical Informatics* 55 (2): 87–101.

Berg, Marc, and Geoffrey Bowker. 1997. "The Multiple Bodies of the Medical Record: Toward a Sociology of an Artifact." *The Sociological Quarterly* 38 (3): 513–37.

Biazzo, Stefano. 2002. "Process Mapping Techniques and Organisational Analysis: Lessons from Sociotechnical System Theory." *Business Process Management Journal* 8 (1): 42–52.

Bijker, Wiebe E., Thomas P. Hughes, Trevor Pinch, and Deborah G. Douglas, eds. 2012. *The Social Construction of Technological Systems: New Directions in the Sociology and History of Technology*. Cambridge, MA: MIT Press.

Bjørn, Pernille, and Carsten Østerlund. 2014. *Sociomaterial-Design: Bounding Technologies in Practice*. Switzerland: Springer.

Blanc-Brude, Tristan, and Dominique L. Scapin. 2007. "What Do People Recall about Their Documents? Implications for Desktop Search Tools." In *Proceedings of the 12th International Conference on Intelligent User Interfaces*, 102–11. New York: ACM.

Blomberg, Jeanette, Lucy Suchman, and Randall H. Trigg. 1996. "Reflections on a Work-Oriented Design Project." *Human-Computer Interaction* 11 (3): 237–65.

Bowker, Geoffrey C. 1997. "Lest We Remember: Organizational Forgetting and the Production of Knowledge." *Accounting, Management and Information Technologies* 7 (3): 113–38.

Bowker, Geoffrey C., Karen Baker, Florence Millerand, and David Ribes. 2010. "Toward Information Infrastructure Studies: Ways of Knowing in a Networked Environment." In *International Handbook of Internet Research*, edited by Jeremy Hunsinger, Lisbeth Klastrup, and Matthew Allen, 97–117. Dordrecht Springer. Accessed at http://link.springer.com/chapter/10.1007/978-1-4020 -9789-8_5.

Bowker, Geoffrey C., and Susan Leigh Star. 2000. *Sorting Things Out: Classification and Its Consequences*. Cambridge, MA: MIT Press.

Briet, Suzanne. 2006. *What Is Documentation*? Translated by Ronald E. Day, Laurent Martinet, with Hermania G. B. Anghelescu. Lanham, MD: Scarecrow.

Broom, Alex, Lynda Cheshire, and Michel Emmison. 2009. "Qualitative Researchers' Understandings of Their Practice and the Implications for Data Archiving and Sharing." *Sociology* 43 (6): 1163–80.

Brown, John Seely, and Paul Duguid. 1994. "Borderline Issues: Social and Material Aspects of Design." *Human-Computer Interaction* 9 (1): 3–36.

Buckland, Michael. 2007. "Northern Light: Fresh Insights into Enduring Concerns." In *Document (re)turn: Contributions from a Research Field in Transition*, edited by Roswitha Skare, Niels Windfeld, and Andreas Vaarheim, 315–22. Frankfurt am Main: Peter Lang.

Burchell, Graham, Colin Gordon, and Peter Miller, eds. 1991. *The Foucault Effect: Studies in Governmentality*. Chicago: University of Chicago Press.

Callon, Michel. 1986. "Some Elements of a Sociology of Translation: Domestication of the Scallops and the Fishermen of Saint Brieuc Bay." In *Power, Action and Belief: A New Sociology of Knowledge?*, edited by John Law, 196–233. London: Routledge and Kegan.

Cambrosio, Alberto, Camille Limoges, and Denyse Pronovost. 1990. "Representing Biotechnology: An Ethnography of Quebec Science Policy." *Social Studies of Science* 20 (2): 195–227.

Corti, Louise. 2011. "The European Landscape of Qualitative Social Research Archives: Methodological and Practical Issues." *Qualitative Social Research* 12 (3). Accessed at http://www .qualitative-research.net/index.php/fqs/article/view/1746/3247.

Cox, Richard J., and David A. Wallace. 2002. *Archives and the Public Good: Accountability and Records in Modern Society*. Westport, CT: Quorum Books.

Duranti, Luciana. 1997. "The Archival Bond." *Archives and Museum Informatics* 11 (3–4): 213–18.

Elyachar, Julia. 2006. "Best Practices: Research, Finance, and NGOs in Cairo." *American Ethnologist* 33 (3): 413–26.

Foucault, Michel. 1971. "The Discourse on Language." In *The Archaeology of Knowledge*, edited by Michel Foucault, 315–35. New York: Harper Colophon.

Frohmann, Bernd. 2008. "Documentary Ethics, Ontology, and Politics." *Archival Science* 8 (3): 165–80.

Fujimura, Joan H. 1992. "Crafting Science: Standardized Packages, Boundary Objects, and 'Translation.'" In *Science as Practice and Culture*, edited by Andrew Pickering, 168–211. Chicago: University of Chicago Press.

Garfinkel, Harold. 1967. *Studies in Ethnomethodology*. Englewood Cliffs, NJ: Prentice-Hall.

Goody, Jack. 1977. *The Domestication of the Savage Mind*. Cambridge: Cambridge University Press.

Hanks, William F. 1996. *Language and Communicative Practices*. Boulder, CO: Westview Press.

Haraway, Donna. 1987. Donna Haraway Reads "The National Geographic" on Primates. YouTube video. Accessed at http://www.youtube.com/watch?v=eLN2ToEIlwM.

Harper, Richard H. R. 1998. *Inside the IMF: An Ethnography of Documents, Technology and Organisational Action*. London: Routledge.

Hetherington, Kregg. 2008. "Populist Transparency: The Documentation of Reality in Rural Paraguay." *Journal of Legal Anthropology* 1 (1): 45–69.

Hodder, Ian. 1982. *Symbols in Action: Ethnoarchaeological Studies of Material Culture*. Cambridge: Cambridge University Press.

___. 1994. "The Interpretation of Documents and Material Culture. In *Handbook of Qualitative Research*, edited by Norman K. Denzin and Yvonna S. Lincoln, 393–402. Thousand Oaks, CA: Sage.

Hull, Matthew S. 2003. "The File: Agency, Authority, and Autography in an Islamabad Bureaucracy." *Language & Communication* 23 (3): 287–314.

___. 2012. "Documents and Bureaucracy." *Annual Review of Anthropology* 41: 251–67.

Ingold, Tim. 2012. "Toward an Ecology of Materials." *Annual Review of Anthropology* 41: 427–42.

Irani, Lilly C., and M. Six Silberman. 2013. "Turkopticon: Interrupting Worker Invisibility in Amazon Mechanical Turk." In *Proceedings of the SIGCHI Conference on Human Factors in Computing Systems*, 611–20. New York: ACM.

Jensen, Carlos, Heather Lonsdale, Eleanor Wynn, Jill Cao, Michael Slater, and Thomas G. Dietterich. 2010. "The Life and Times of Files and Information: A Study of Desktop Provenance." In *Proceedings of the SIGCHI Conference on Human Factors in Computing Systems*, 767–76. New York: ACM. Accessed at http://dl.acm.org/citation.cfm?id=1753439.

Jensen, Casper Bruun, and Brit Ross Winthereik. 2013. *Monitoring Movements in Development Aid: Recursive Partnerships and Infrastructures*. Cambridge, MA: MIT Press.

Jones, Matthew. 1998. "Information Systems and the Double Mangle: Steering a Course between the Scylla of Embedded Structure and the Charybdis of Strong Symmetry." In *Information Systems:*

Current Issues and Future Challenges, edited by Tor Larsen, Linda Levine, and Janice I. DeGross, 287–302. Laxenburg: International Federation for Information Processing.

Karasti, Helena, and Karen S. Baker. 2004. "Infrastructuring for the Long-Term: Ecological Information Management." In *Proceedings of the 37th Annual Hawaii International Conference on System Sciences*, January 5–8, 2004, IEEE. Accessed at http://ieeexplore.ieee.org/xpls/abs_all.jsp ?arnumber=1265077.

Knorr Cetina, Karin. 1997. "Sociality with Objects: Social Relations in Postsocial Knowledge Societies." *Theory, Culture & Society* 14 (4): 1–30.

Latour, Bruno. 1990. "Technology Is Society Made Durable." *The Sociological Review* 38 (S1): 103–31.

___. 1999. *Pandora's Hope: Essays on the Reality of Science Studies*. Cambridge, MA: Harvard University Press.

Latour, Bruno, and Steve Woolgar. 1986. *Laboratory of Life: The Construction of Scientific Facts*. Princeton, NJ: Princeton University Press.

Lave, Jean, and Etienne Wenger. 1991. *Situated Learning: Legitimate Peripheral Participation*. Cambridge: Cambridge University Press.

Law, John. 2004. *After Method: Mess in Social Science Research*. London: Routledge.

___. 2008. "On Sociology and STS." *The Sociological Review* 56 (4): 623–49.

Leonardi, Paul M. 2011. "When Flexible Routines Meet Flexible Technologies: Affordance, Constraint, and the Imbrication of Human and Material Agencies." *MIS Quarterly* 35 (1): 147–67.

Levy, David M. 1994. "Fixed or Fluid? Document Stability and New Media." In *Proceedings of the 1994 ACM European Conference on Hypermedia Technology*, 24–31. Edinburgh: ACM.

___. 2001. *Scrolling Forward: Making Sense of Documents in the Digital Age*. New York: Arcade Publishing.

Li, Fabiana. 2009. "Documenting Accountability; Environmental Impact Assessment in a Peruvian Mining Project." *PoLAR: Political and Legal Anthropology Review* 32 (2): 218–36.

Lund, Niels Windfeld. 2009. "Document Theory." *Annual Review of Information Science and Technology* 43 (1): 1–55.

Lynch, Clifford A. 2001. "When Documents Deceive: Trust and Provenance as New Factors for Information Retrieval in a Tangled Web." *Journal of the American Society for Information Science and Technology* 52 (1): 12–17.

Martinez-Comeche, J. A. 2000. "The Nature and Qualities of the Document in Archives, Libraries and Information Centres and Museums." *Journal of Spanish Research on Information Science* 1 (1): 5–10.

Miller, Carolyn R. 1984. "Genre as Social Action." *Quarterly Journal of Speech* 70 (2): 151–67.

Mol, Annemarie. 2002. *The Body Multiple: Ontology in Medical Practice*. Durham, NC: Duke University Press.

Mulkay, Michael. 1989. "Looking Backward." *Science, Technology, & Human Values* 14 (4): 441–59.

Ngulube, Patrick. 2002. "Managing and Preserving Indigenous Knowledge in the Knowledge Management Era: Challenges and Opportunities for Information Professionals." *Information Development* 18 (2): 95–102.

Okere, Theophilus, Chukwudi Anthony Njoku, and René Devisch. 2005. "All Knowledge Is First of All Local Knowledge: An Introduction." *Africa Development* 30 (3): 1–19. Accessed at http://www.ajol.info/index.php/ad/article/view/22226.

Østerlund, Carsten S. 2008. "Documents in Place: Demarcating Places for Collaboration in Healthcare Settings." *Computer Supported Cooperative Work (CSCW)* 17 (2–3): 195–225.

Otlet, Paul. 1903. "The Science of Bibliography and Documentation." In *International Organization and Dissemination of Knowledge: Selected Essays of Paul Otlet*, edited by W. Boyd Rayward, 71–86. Amsterdam: Elsevier.

___. 1907. "The Systematic Organization of Documentation and the Development of the International Institute of Bibliography." In *International Organization and Dissemination of Knowledge: Selected Essays of Paul Otlet*, translated and edited by W. B. Rayward, 105–11. Amsterdam: Elsevier.

Ottinger, Gwen. 2010. "Buckets of Resistance: Standards and the Effectiveness of Citizen Science." *Science, Technology, & Human Values* 35 (2): 244–70.

Pickering, Andrew. 1995. *The Mangle of Practice: Time, Agency and Science*. Chicago: University of Chicago Press.

Power, Michael. 1999. *The Audit Society: Rituals of Verification*. Oxford: Oxford University Press.

Reed, Adam. 2006. "Documents Unfolding." In *Documents: Artifacts of Modern Knowledge*, edited by Annelise Riles, 158–77. Ann Arbor: University of Michigan Press.

Ribes, David, and Thomas A. Finholt. 2009. "The Long Now of Technology Infrastructure: Articulating Tensions in Development." *Journal of the Association for Information Systems* 10 (5): 375–98.

Riles, Annelise. 1998. "Infinity within the Brackets." *American Ethnologist* 25 (3): 378–98.

___. 2001. *The Network Inside Out*. Ann Arbor: University of Michigan Press.

___. 2006. *Documents: Artifacts of Modern Knowledge*. Ann Arbor: University of Michigan Press.

Robichaud, Daniel, and François Cooren, eds. 2013. *Organization and Organizing: Materiality, Agency and Discourse*. New York: Routledge.

Sands, Ashley, Christine L. Borgman, Laura Wynholds, and Sharon Traweek. 2012. "Follow the Data: How Astronomers Use and Reuse Data." In *Proceedings of the American Society for Information Science and Technology* 49 (1): 1–3.

Sayers, Jentery. 2013. "Computation Cultures after the Cloud: A Special Issue of the Journal of e-Media Studies." *Journal of e-Media Studies* 3 (1). Accessed at http://journals.dartmouth.edu/cgi-bin/WebObjects/Journals.woa/1/xmlpage/4/issue/40.

Schatzki, Theodore. 2002. *The Site of the Social: A Philosophical Exploration of the Constitution of Social Life and Change.* University Park: Pennsylvania State University Press.

Schmidt, Kjeld, and Liam Bannon. 1992. "Taking CSCW Seriously." *Computer Supported Cooperative Work (CSCW)* 1 (1–2): 7–40.

Scott, Susan V., and Wanda J. Orlikowski. 2012. "Reconfiguring Relations of Accountability: Materialization of Social Media in the Travel Sector." *Accounting, Organizations and Society* 37 (1): 26–40.

Sellen, Abigail J., and Richard H. R. Harper. 2002. *The Myth of the Paperless Office.* Cambridge, MA: MIT Press.

Shankar, Kalpana. 2004. "Recordkeeping in the Production of Scientific Knowledge: An Ethnographic Study." *Archival Science* 4 (3–4): 367–82.

___. 2009. "Ambiguity and Legitimate Peripheral Participation in the Creation of Scientific Documents." *Journal of Documentation* 65 (1): 151–65.

Shen, Jianqiang, Erin Fitzhenry, and Thomas G. Dietterich. 2009. "Discovering Frequent Work Procedures from Resource Connections." In *Proceedings of the 14th International Conference on Intelligent User Interfaces,* 277–86. New York: ACM. Accessed at http://dl.acm.org/citation.cfm?id=1502690.

Shore, Cris, and Susan Wright. 1999. "Audit Culture and Anthropology: Neo-Liberalism in British Higher Education." *Journal of the Royal Anthropological Institute* 5 (4): 557–75.

___, eds. 2003. *Anthropology of Policy: Critical Perspectives on Governance and Power.* London: Routledge.

Simakova, Elena. 2010. "RFID 'Theatre of the Proof': Product Launch and Technology Demonstration as Corporate Practices." *Social Studies of Science* 40 (4): 549–76. Accessed at http://sss.sagepub.com/content/early/2010/06/11/0306312710365587.abstract.

Smith, Dorothy E. 2005. *Institutional Ethnography: A Sociology for People.* London: AltaMira Press.

Star, Susan Leigh, and Anselm Strauss. 1999. "Layers of Silence, Arenas of Voice: The Ecology of Visible and Invisible Work." *Computer Supported Cooperative Work (CSCW)* 8 (1–2): 9–30.

Strathern, Marilyn. 1997. "'Improving Ratings': Audit in the British University System." *European Review* 5 (3): 305–21.

___. 2006. "A Community of Critics? Thoughts on New Knowledge." *Journal of the Royal Anthropological Institute* 12 (1): 191–209.

Suchman, Lucy. 1993. "Do Categories Have Politics?" *Computer Supported Cooperative Work (CSCW)* 2 (3): 177–90.

___. 1995. "Making Work Visible." *Communications of the ACM* 38 (9): 56–64.

___. 1997. "Centers of Coordination: A Case and Some Themes." In *Discourse, Tools, and Reasoning*, edited by Lauren B. Resnick, Roger Säljö, Clotilde Pontecorvo, and Barbara Burge, 56–64. Berlin Heidelberg: Springer.

___. 2007. *Human-Machine Reconfigurations: Plans and Situated Actions*. Cambridge: Cambridge University Press.

Sweeney, Shelley. 2008. "The Ambiguous Origins of the Archival Principle of 'Provenance.'" *Libraries & the Cultural Record* 43 (2): 193–213.

Timmermans, Stefan, Geoffrey C. Bowker, and Susan Leigh Star. 1998. "The Architecture of Difference: Visibility, Control, and Comparability in Building a Nursing Interventions Classification." In *Differences in Medicine: Unraveling Practices, Techniques, and Bodies*, edited by Marc Berg and Annemarie Mol, 202–25. Durham, NC: Duke University Press.

Traweek, Sharon. 2009. *Beamtimes and Lifetimes: The World of High Energy Physicists*. Cambridge, MA: Harvard University Press. Accessed at https://books.google.com/books?hl=en&lr=&id=CdsgPG535C4C&oi=fnd&pg=PR7&dq=beamtimes+lifetimes&ots=zCaLAzADwA&sig=wPsWJXU61YdZs4lh6xgjUiMg2RQ.

Van Maanen, John, and Brian T. Pentland. 1994. "Cops and Auditors: The Rhetoric of Records." In *The Legalistic Organization*, edited by Sim B. Sitkin and Robert J. Bies, 53–90. Thousand Oaks, CA: Sage.

Weber, Max. 1978. *Economy and Society: An Outline of Interpretive Sociology*. Berkeley: University of California Press.

Wiggins, Andrea, and Kevin Crowston. 2011. "From Conservation to Crowdsourcing: A Typology of Citizen Science." In *Proceedings of the 2011 44th Hawaii International Conference on System Sciences (HICSS)*, 1–10. Washington, DC: IEEE. Accessed at http://ieeexplore.ieee.org/xpls/abs_all.jsp?arnumber=5718708.

Winschiers-Theophilus, Heike, and Nicola J. Bidwell. 2013. "Toward an Afro-Centric Indigenous HCI Paradigm." *International Journal of Human-Computer Interaction* 29 (4): 243–55.

Winschiers-Theophilus, Heike, Shilumbe Chivuno-Kuria, Gereon Koch Kapuire, Nicola J. Bidwell, and Edwin H. Blake. 2010. "Being Participated—A Community Approach." In *Proceedings of the 2010 Participatory Design Conference*, 1–10. New York: ACM.

Woolgar, Steve, Tarek Cheniti, Javier Lezaun, Daniel Neyland, Chris Sugden, and Christian Toennesen. 2008. "A Turn to Ontology in STS." A Turn to Ontology Workshop: University of Oxford.

Woolgar, Steve, and Javier Lezaun. 2013. "The Wrong Bin Bag: A Turn to Ontology in Science and Technology Studies?" *Social Studies of Science* 43 (3): 321–40.

Yates, JoAnne. 1993. *Control through Communication: The Rise of System in American Management,* vol. 6. Baltimore: Johns Hopkins University Press.

Zeitlyn, David. 2012. "Anthropology in and of the Archives: Possible Futures and Contingent Pasts. Archives as Anthropological Surrogates." *Annual Review of Anthropology* 41: 461–80.

3   Intellectual and Practical Contributions of Scientometrics to STS

Sally Wyatt, Staša Milojević, Han Woo Park, and Loet Leydesdorff

As an interdisciplinary specialty, science and technology studies (STS) deploys a wide variety of methods, including (participant) observation, interviews, focus groups, and close reading of textual and visual materials. These draw upon anthropology, sociology, and history, three of the core constituent disciplines of STS. Quantitative methods, based on numerical data and/or statistical analysis of large-scale surveys, experiments, national censuses and, more recently, data and information visualizations, are less visible within STS. The application of quantitative methods to study the formal and semantic aspects of the scientific literature as information and communication processes is known as scientometrics. Scientometrics and qualitative approaches within STS share a common origin, even if they have grown apart over the past decades in terms of research practices, norms, and standards. Different skills are needed, and the epistemological assumptions are also different. However, both quantitative and qualitative STS have always shared a deep commitment to the empirical study of science and technology, and practitioners of both can be reflexive about their own knowledge production practices, as will be seen below.

In this chapter, we argue for greater use of scientometric methods within STS. We suggest that widening the repertoire of methods available to STS scholars provides new insights into old questions and opens possibilities for raising new research questions. In addition, there are two more pragmatic reasons for greater attention to scientometrics within STS. The first is the growing attention for "big data," computational methods, and digital forms of representation of data and knowledge in the humanities and the social sciences (Borgman 2015; Mayer-Schönberger and Cukier 2013). The second reason, especially for those STS scholars based in the academy, is that a deeper understanding of scientometrics is necessary for making sense of and for formulating informed criticism about university rankings, evaluations, and the audit culture within which academics work (Dahler-Larsen 2011; Halffman and Radder 2015; Hicks et al. 2015 [also known as the Leiden Manifesto]; Strathern 2000).

In the next section, we provide an overview of the common origins of qualitative and quantitative forms of STS, offering a discursive account of this history. We then demonstrate how scientometric techniques can be used to address substantive research questions, and we provide examples relevant both to the origins of STS and its state of the art. Our purpose is not to provide an exhaustive review of either qualitative or quantitative methods, as there exist many methods textbooks for both (e.g., Franklin 2012; Moed et al. 2004), although STS has sometimes neglected methods.[1] The final section picks up the themes of "big data" and "reflexivity," and also reflects on the current use of indicators in the evaluation of research performance.

A Very Brief History of STS and Scientometrics

The origin of STS is often placed in the 1970s with the development of the "sociology of scientific knowledge" (SSK) and with political activism, especially against nuclear power (Bijker 1993). However, we suggest that a fuller understanding of both qualitative and quantitative STS needs to look back to the 1940s and the work of Robert K. Merton, one of the giants[2] of twentieth-century sociology. Merton was a remarkably prolific scholar and made major contributions not only to the sociology of science but also to functionalism and the sociology of deviance. In his œuvre, he combined elements from the philosophy and history of science with the sociological tradition (e.g., Émile Durkheim, 1858–1917; Karl Mannheim, 1893–1947; and Talcott Parsons, 1902–1979). At the same time, he insisted on an empirical approach and on the value of middle-range theorizing as most appropriate for developing STS as an interdisciplinary field (Merton 1973; Wyatt and Balmer 2007). In one of his most influential papers, Merton (1942) formulated the "CUDOS norms of science": Communalism, Universalism, Disinterestedness, and Organized Skepticism. He argued that together these norms distinguish science from other areas of activity.

Mertonian sociology of science, also known as the institutional sociology of science, focused mainly on topics such as the stratification of science and the development of specialties, using quantitative methods prevalent in the empirical American sociology of the 1960s (Mullins 1973). Stratification studies used measures of productivity for the evaluation of the sciences and also drew attention to formal and informal networks of researchers. These measures resonated with the work of Derek de Solla Price, historian of science and one of the founders of scientometrics. He published extensively in the 1960s and 1970s (e.g., Price 1965, 1970), laying the foundations for the newly emerging field of quantitative science studies. Price's contributions to measuring the productivity of scientists were of interest to the stratification-of-science group, while his operationalization of the notion of "invisible college" was of interest to researchers such as Diana Crane (1972), who was studying the formation and development of scientific specialties.

A major impetus for the development of institutional studies of science and scientometrics was provided by the development of the Science Citation Index (SCI) during the 1960s (Elkana et al. 1978). The SCI was first conceptualized and developed as a tool to ameliorate a well-known information science problem of retrieving relevant documents. (Information retrieval continues to guide much information and computer science research.) In an article published in *Science* in 1955, with the title "Citation Indexes for Science: A New Dimension in Documentation through Association of Ideas," Eugene Garfield, an information scientist, promoted the idea that indexing does not have to be done only in the form of keywords and descriptors; citations and references present in published journal articles can be used to bring related documents together. To materialize his approach, Garfield founded a company, the Institute of Scientific Information (ISI) in Philadelphia that started producing such indexes under the name Science Citation Index (Garfield 1964; Wouters 1999).[3]

Price (1965) made early and novel usage of the tool to measure networks of scientists, whereas others deployed it primarily to develop citation and co-citation measures to understand the emergence, growth, and decline of scientific specialties and the diffusion of ideas. Thus, early studies based on citation data were not so much focused on citation as a currency or as an indicator of quality and reputation as on the symbolic functions of citation (Small 1978).

Most studies within the empirical and institutional sociology of science could be described as scientometrics *avant la lettre*. However, quantitative studies of science soon began to attract researchers from the natural and life sciences and from information and library sciences, the former primarily concerned with measuring productivity of different groups, the latter primarily interested in information retrieval, but both concerned with tracing genealogies of concepts and theories. During the 1970s, scientometrics itself started to exhibit the properties of a specialty, with a new journal, *Scientometrics*, launched in 1978.

The 1970s also marked the beginning of the constructivist tradition in the studies of science and technology, which developed as a response not only to the analytic philosophy of science but also to the normative perspective of Mertonian sociology. In an article entitled "A Deviant Viewpoint," Barry Barnes and Alex Dolby (1970) proposed a constructivist alternative to Merton's "normative" orientation and argued for a focus on scientific practices. These authors were working within the emerging "sociology of scientific knowledge" (SSK), also known as the "Strong Program" or the Edinburgh

School (Bloor 1976). Adherents of the Strong Program consider science as community-based belief systems, and the units of analysis are actors or collectives driven by a blend of socio-epistemic interests.[4] From this perspective, Merton's norms can be considered *professed* norms, and citations as rhetorical instruments to be deployed in arguments (Gilbert 1977; Latour 1987). A constructivist interpretation of citation thus became juxtaposed to the Mertonian or normative interpretation in what Susan Cozzens (1989) has called "the citation debate."

Historical methods predominated during the further development of SSK. At the same time, an approach that came to be known as the "Bath School" focused on studying controversies in science using observational methods (Collins 2004). This marked the shift of attention away from structural or macro-level phenomena to micro-sociological processes that remained the focus of "laboratory studies" which observed the process of science in the making mostly in research laboratories (Latour and Woolgar 1979), or what Karin Knorr Cetina (1999) refers to as epistemic communities. Later work by Bruno Latour (1987, 2005) and others, in what came to be known as actor-network theory (ANT), rejected notions of scale implied by the use of "macro" and "micro" and highlighted the networks of associations between humans and nonhumans to create more or less stable assemblages.

In the 1980s, Michael Mulkay, Jonathan Potter, and Steven Yearley (1983) argued that priority should be given to the analysis of constructions in scientific discourse. Nigel Gilbert and Mulkay (1984) then distinguished between a contingent repertoire in scientific practices and the rationalized ("empiricist") discourse in formal publications. In their study of the scientific debate about oxidative phosphorylation, they concluded that scientists tend to use a "contingent repertoire" when explaining error versus an "empiricist repertoire" when hindsight proved them right. In other words, the epistemic contexts of discovery versus validation (Popper [1935] 1959) matter in social practices. Translations between these contexts are mediated and can be studied empirically using, among other techniques, co-word analysis (Callon et al. 1983).

In summary, the early disputes were around the nature of citations. From a Mertonian perspective, citation analysis had been considered primarily as a methodology for the historical and sociological analysis of the sciences (e.g., Cole and Cole 1973). However, David Edge, a member of the Edinburgh School, criticized citation analysis for ignoring the construction of knowledge claims in "the soft underbelly of science" (1979, 117). In his view, one should "give pre-eminence to the *account from the participant's perspective*, and it is the *citation analysis* which has to be 'corrected'" (1979, 111; italics in the original). Steve Woolgar (1991) reached "beyond the citation debate"—between normative and constructivist interpretations of citations—by emphasizing that scientometric measurement instruments are used in science policy. Thus, the processes of translations and codifications in and between scholarly and political discourses merit sociological attention from both normative and constructivist perspectives because of the use and performativity of scientometric indicators in science and technology policies.

Ever since, STS has moved beyond SSK to the postconstructivist, or "critical and cultural studies of science and technology" (Hess 1997, 85), drawing heavily on qualitative and interpretative theories and methods from queer, feminist, and postcolonial theories and from literary and cultural studies. During the same period, the scientometrics literature experienced an exponential growth with contributions from researchers from very diverse disciplinary backgrounds. This growth can in part be attributed to the increased reliance on quantitative metrics in research evaluation by funding agencies and academic administrators, as well as advances in computational capabilities, availability of data sets, and new analytic techniques. As a result, despite their common origins, as Daniele Rotolo, Diana Hicks, and Ben Martin (2015, 1838) demonstrate, "There has been relatively little interaction between scientometrics and STS since the late 1980s. Each ... has its own conferences and journals, and only a handful of researchers operate at the interface—most individuals would identify themselves as either 'scientometricians' or 'STS' scholars." We return to this later but first want to illustrate how scientometric methods can be used to address STS questions.

Using Scientometric Methods to Answer Research Questions

The account provided above about the development of STS, paying particular attention to the role of quantitative methods, is based on our multidisciplinary backgrounds and readings of the literature. It is not only brief but, like all histories, it is partial, identifying a few crucial moments relevant to our narrative. As yet, there is no canonical origin story of STS, perhaps as a result of the different histories of the contributing disciplines. Sergio Sismondo (2008) does provide an overview of the field, starting with Thomas Kuhn and emphasizing the construction of scientific and technical knowledge; however, he did not mention Merton. In the opening chapter of the 1995 *STS Handbook*, Edge offers his own perspective on the history of the field, in which he does acknowledge the influence of Merton, and in which he makes a plea for a "creative reconciliation" (1995, 15) between what he calls normative and reflexive approaches to STS and what we label here as quantitative and qualitative approaches. Elsewhere, Edge and Roy MacLeod (1986), then editors of *Social Studies of Science*, acknowledged that submissions using quantitative methods were routinely diverted to *Scientometrics*.

This practice has had long-lasting effects on both fields, as we shall see later in this chapter. In this section, we provide more input for a history of STS by providing some evidence based on scientometric methods. This illustration is also a means of providing guidance as to when such methods can be helpful in addressing research questions. Throughout the discussion, we pay particular attention to visualizations, how they are produced, and how they can be interpreted.

Our first example is based on co-word analysis, a technique that captures the frequency of pairs of words or phrases in and between documents. It is used to trace links between texts and to understand the development of scientific fields. We have applied this to bibliographic data from both *Scientometrics* and *Social Studies of Science* (*SSS*), key journals in, respectively, scientometrics and STS in order to show both topics studied and knowledge lineages. However, there is an important caveat to be made: because the subset of publications in *Scientometrics* is more than twice as large as the *SSS* set, one can expect the former to dominate the latter in the representations.[5]

We map the contents of the field in terms of the co-words used in the titles and abstracts of the articles in these two journals, providing an overview or "distant reading" (Moretti 2013). *Distant reading* is the term increasingly applied to analyzing large quantities of text in order to identify patterns and anomalies, and it is used in distinction to the "close reading" central to the hermeneutic tradition. Although computers cannot understand text, they can organize it in quantities that humans cannot read ("big data"), for example, into semantic maps. Semantic maps are based on clustering algorithms but translate the results into visualizations that can provide an orientation (Latour 1986). Richard Rogers and Noortje Marres (2000) map different repertoires and vocabularies confronting each other in controversy studies.

In the combined set of texts under study,[6] four main discourses were algorithmically distinguished (figure 3.1).* On the right, the red nodes indicate the vocabulary of qualitative studies of science and technology with an emphasis on words such as *theory*, *practice*, *debate*, *sociology*, *market*, and *controversy*. These are patterns of co-word relations, and the word *theory* itself occurs more frequently (41 times) in titles of *Scientometrics* articles than in those from *SSS* (30 times). On the left, the program distinguishes among a group of terms led by *journal* (in blue) representing studies of structural developments in the sciences as revealed by or measured in terms of the journal literature, and a third group (green) devoted to indicators of research output using words such as *institute*, *output*, and *China* (the latter reflecting the growth of attention to research evaluation in China in recent years) among the dominant terms. The fourth group (yellow)

*In the print edition, please see insert for color version.

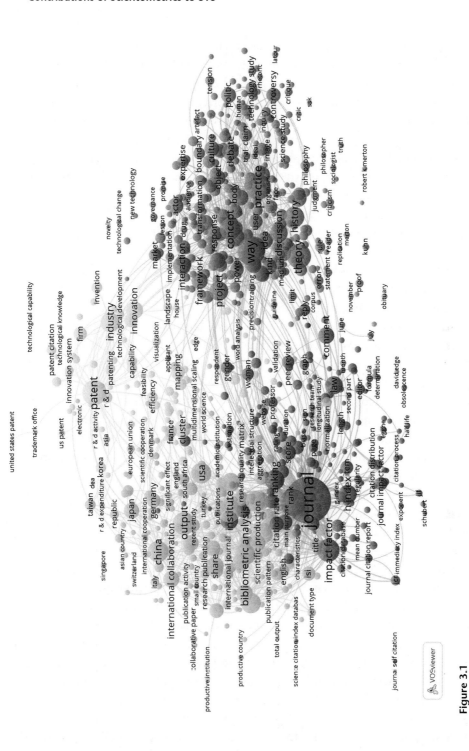

Figure 3.1

Semantic map of 896 most frequently used words in titles and abstracts of 5,677 publications in *Scientometrics* and *Social Studies of Science*. The map is available for web-starting at http://www.vosviewer.com/vosviewer.php?map=http://www.leydesdorff.net/sts_hbk/fig4map .txt&network=http://www.leydesdorff.net/sts_hbk/fig4net.txt&label_size_variation=0.5&n_lines=1000&zoom_level=1. (See Color Plate 3.)

is focused on *patents*, *industry*, and *innovation* and borders on the qualitative repertoire but on the opposite side from the structural cluster with blue nodes, using terms such as *technological change* (qualitative) versus *innovation* (quantitative), as innovations can be counted in terms of the number of patents involved.

Although the transitions among the repertoires are gradual and sometimes fuzzy, the analysis points to distinctions that are relevant for understanding the literature that cannot easily be demarcated otherwise because of the interdisciplinary character of the field. Such distinctions can be difficult for a single individual to capture through "close reading." Consequently, the indicated distinctions are not easy to validate; but they can be used as heuristics to guide further analysis and interpretation.

Figure 3.2 shows the results of using another common technique with these same data, namely co-citation analysis, based on the frequency with which two documents are cited together. This measure works from the assumption that if two authors are cited in the same studies, then they are providing a common knowledge base. In figure 3.2, Merton is represented as a node in the center among a group of yellow nodes at the interface between qualitatively oriented STS scholars (red nodes on the right) and three groups of authors adopting quantitative approaches (blue, green, and pink nodes on the left). Merton shares this position with, among others, Price (mentioned earlier) and Vasily Nalimov, founder of Russian scientometrics (*nauchometria*). Richard Whitley ([1984] 2000), known for his work about the intellectual and social organization of the sciences, is positioned next to Merton in the yellow-colored domain. Note that citation is more frequent and probably more disciplined on the quantitative side than on the qualitative one (Larivière et al. 2006), which also provides us with a different kind of evidence for the differences between epistemic cultures.

Among the "qualitative" authors on the right side of figure 3.2 (red nodes), the co-citation analysis highlights authors who have engaged in discussions with their "quantitative" colleagues. The bridging authors are Michel Callon, Harry Collins, Nigel Gilbert, and Bruno Latour. Kuhn is positioned among the "qualitative" authors in this set because his work is co-cited in a constructivist context—prevalent on the qualitative side of our field—more than in the structuralist (that is, Mertonian) context indicated in yellow. On the left side of figure 3.2, three clusters are distinguished: one with Eugene Garfield, Wolfgang Glaenzel, and Leo Egghe, prominent founding members of scientometrics (blue nodes). A second cluster indicated in green is led by Henry Small and Loet Leydesdorff, who share an interest in visualization, but this group also includes authors interested in university-industry relations and patenting, such as Martin Meyer and Robert Tijssen. Authors who focus on technology studies are also indicated with green dots (e.g., Nathan Rosenberg and Michael Gibbons). These authors combine scientometric methods with economic and sociological theorizing, more so than the methodologically oriented authors in the pink and blue clusters.

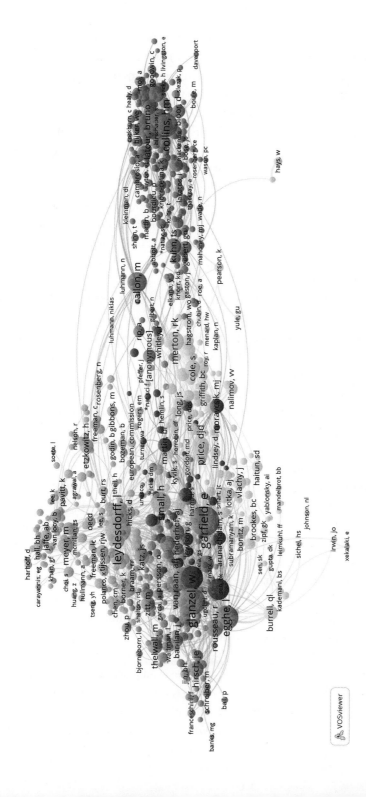

Figure 3.2

Co-citation analysis of 879 authors in *Scientometrics* and *Social Studies of Science* who are cited more than 20 times (6 October 2014). The map is available for web-starting at http://www.vosviewer.com/vosviewer.php?map=http://www.leydesdorff.net/sts_hbk/fig6map.txt&network=http:// www.leydesdorff.r.ret/sts_hbk/fig6net.txt&label_size_variation=0.4&n_lines=1000&visualization=1. (See Color Plate 4.)

Figures such as those above can be compared for different years so that one can visualize developments as animations, as Katy Börner (2010) and others have done (e.g., van Eck and Waltman 2014a). One can also combine the two maps because both words and references are attributes of documents and may thus co-occur or not. One can extend this to other characteristics (e.g., author name, address, or country information) and thus construct what can be considered a "heterogeneous network." Analytically, different possible perspectives are then combined into a more complex construct. Very different maps are possible using combinations of attributes, other algorithms or, more technically, different parameters. However, categories that are meaningful for humans cannot be expected to match algorithmic constructs. Thus, figures such as those above offer a platform for further discussion and more detailed reconstructions. Since there is no baseline of what is represented, as in a geographical map, other representations can be equally valid (Studer and Chubin 1980, 269 ff.).

Just as co-word analysis allows us to engage in distant reading, the technique of "algorithmic historiography" (Garfield, Sher, and Torpie 1964; Garfield, Pudovkin, and Istomin 2003; van Eck and Waltman 2014b)[7] facilitates a form of computational history by using the dates of publication and projecting and clustering the citations among them as arrows along the time axis. A historiogram of the set of documents we have used (not shown here) demonstrates that, with the exception of Kuhn (1962), authors in the qualitative tradition are cited only incidentally on the quantitative side. Furthermore, there is almost no citation traffic, or no flow of knowledge, *from* the quantitative *to* the qualitative tradition.

The central articles in the qualitative tradition are tightly knit together by citation relations (see figure 3.3 for the citation relations among 35 core papers), whereas the quantitative group in figure 3.2 is divided (by the clustering algorithm of the routine) into two groups: a core group who developed the specialty and a group with broader research interests who use scientometric methods. In summary, the historiographic analysis suggests that, in terms of citation relations, qualitative STS functions as a source of knowledge but is not itself informed by the scientometric analysis to the extent that qualitative STS scholars actually refer to this literature.

One observation that can be made from examining figure 3.3 is that the core documents in the set are relatively old. This is partly an effect of the accumulation of citations if there is no fixed citation window, but it is also typical for fields without a "research front" (Price 1970; Whitley [1984] 2000). However, sciences with research fronts tend to "obliterate [literature] by incorporation" (Cozzens 1989, 438; Garfield 1975). Older literature is incorporated in a process of periodic reviews (Price 1965) or in the production of textbooks. In the social sciences and the humanities, however, codification of a

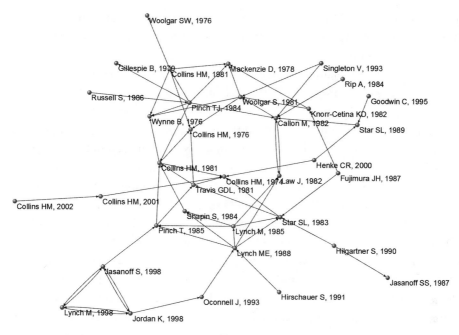

Figure 3.3

Citation relations among thirty-five core documents sorted into the qualitative set.

citation classic (e.g., "Merton 1942" or "Latour 1987") provides a different mechanism; a research front is then often absent. Another observation is that women are relatively absent from figure 3.3, as only eight of the thirty-five documents are (co-)authored by women, namely, Joan Fujimura, Sheila Jasanoff, Kathleen Jordan, Karin Knorr Cetina, Vicky Singleton, and Susan Leigh Star. Nonetheless, analyses such as those presented here can be used to contribute to discussions about the role of women in knowledge production. It is also relatively straightforward to capture and represent country differences using address information, but author characteristics such as ethnic, religious, and class backgrounds are difficult to analyze using documents as units of analysis.

Scientometrics: From Discourse Analysis to Policy Tool

In the previous section, we illustrated how scientometric instruments and techniques can be used to inform theoretical and historical debates at the core of STS, however defined. The scientometric map is performative because it seems to be objective. But the assumptions are hidden in parameter choices that may lead to interaction effects that are not easily accessible to readers. Thus, the status is that of a machine reading of the literature under study that should not be reified as "objective" or "intersubjective" because the scientometric constructs remain grounded in decisions about how best to represent latent structures in the data.

As shown above, the main units of analysis in scientometrics are documents, and these documents provide a wealth of data for analysis about the development of fields, the emergence of ideas and institutions, and the socio-epistemic organization of knowledge. However, the scientometric focus on the textual domain provides new perspectives for further developments of STS. In addition to social and epistemic dynamics, traditionally named "context of discovery" and "context of justification," communication networks and mediation increasingly co-structure the empirical advances of science and technology in society (Luhmann 1995). Codes of communication do part of the work in science as a symbolic system of signs that can reflexively be (re)invoked and discursively be modified (Latour 1987, 2005; Wouters 1998).

During the 1980s the analysis of scientific discourse was increasingly prioritized in STS (e.g., Amsterdamska and Leydesdorff 1989; Callon 1986; Myers 1985; Pinch 1985). Translations between discourses are mediated and can be studied empirically using, for example, co-word analysis. When studying scientific texts, however, one retrieves not only words but also references (Braam, Moed, and van Raan 1991). References enable authors to mobilize resources for making a persuasive argument (this is the constructivist interpretation) or to shortcut an argument reflexively. In other words, references code the communication, which no longer has to be elaborated in words (Bernstein 1971; Leydesdorff 2007), and thus enable authors to be more precise. References also enable authors to construct potential audiences and readers. This construction can be particularly important in interdisciplinary work, where mutual referencing can help to construct bridges between disparate epistemic communities (Fujigaki and Leydesdorff 2000). We have deliberately invoked authors from the qualitative STS canon in order to communicate with those readers of this chapter who are already familiar with this tradition.

The communication dynamics add another perspective to the sociological one focusing on agents (authors), communities, and institutions and to the philosophical one highlighting the epistemic dynamics of knowledge claims and validation. Citations can be a form of strategic behavior or can reflect a cognitive debt, and they may also be a reflection of a social hierarchy within the scientific community. The translations between the social and cognitive aspects of citations may be useful in the context of policy making and research management but tend to hide the analytically different functions of citations in different forms of knowledge production across the sciences (Whitley [1984] 2000; Wouters 1999; cf. Leydesdorff and Amsterdamska 1990). In our view, the three dimensions of texts, authors, and epistemics can be expected to shape a triple helix with *complex* dynamics. Densities in networks of documents and authors can be recognized as areas of shared interest. The epistemic contents, however, are only reflexively available to those participants and analysts enrolled in the discourses, who can move the discourse forward by contributing to the literature. Barnes and Dolby's (1970) "deviant viewpoint" restructured the meaning of the citation from a countable reference to a potentially strategic instrument in an argument (Latour 1983). Mulkay, Potter, and Yearley's (1983) emphasis on the analysis of discourse made it relevant to study the meaning of references in documents, as different from words and other textual marks. In other words, the challenge could be framed in terms of understanding how the sciences develop in a semiosis of texts and thus become amenable to measurement (Callon et al. 1983; Fujigaki 1998).

The shortcut of the translation between the cognitive and the social dimensions provides scientometric instruments and techniques with their performativity. By counting textual units as cognitive ones, it is possible to make evaluative inferences about social units of analysis (such as universities; e.g., Waltman et al. 2012). The resulting indicators have become increasingly important in the policy arena because of the gradual transition from industrial to knowledge-based economies in many parts of the world (Bell 1973; Castells 1996–98; David and Foray 2002; Godin 2006; OECD 1996). In addition to input indicators, such as the numbers of scientists and engineers (and their gender) and funding for research and development (R&D) (OECD 1963, 1976; OECD/Eurostat 1997), output indicators such as numbers of publications and patents became relevant in a policy context. Greater attention was given to both the measurement of science and technology and the statistical foundations of inferences based on the measurement (Martin and Irvine 1983; Moed et al. 1985; Narin 1976).

What were first conceptualized as relevant variables in STS have thus become indicators used in science policy and research management. Qualitative STS has informed and continues to inform science and technology policies via the development and circulation of concepts (Collins 1985; Martin 2012; Martin, Nightingale, and Yegros-Yegros 2012; Wagner et al. 2011). Quantitative STS also mediates between the theoretical contexts and the policy arena (Leydesdorff and van den Besselaar 1997). The Citation Indices first provided the sociology of science and scientometrics with new measurement instruments for understanding social and epistemic relationships between people, groups, and ideas across time and space. At the same time, however, the Citation Indices became the major source for measuring research performance.

In 1972 the U.S. National Science Board launched the biannual series of Science Indicators using, among other things, the SCI. Since then, funding agencies and research advisory bodies have increasingly taken up STS insights, both qualitative and quantitative, in their science and technology policies and R&D management. The construction and deconstruction of the indicators has become a literature in itself (e.g., Moed 2005; van Raan 1988) but at arm's length from STS (Martin 2012). Nevertheless, a leading producer of university rankings and purveyor of institutional and disciplinary evaluations continues to call itself the Centre for Science and Technology Studies (CWTS) at Leiden University, and similar spin-offs from STS departments can be found at other universities (e.g., in Montréal, Granada, and Atlanta, Georgia). At the institutional level it is common practice to integrate qualitative and quantitative STS, even though they may address different audiences and deploy different epistemic norms, standards, and codes of communication. In many instances, "multidisciplinarity" is probably more suited as a format for combining quantitative and qualitative STS than an ambition of interdisciplinary integration.

Research evaluation, or "evaluative bibliometrics" (Narin 1976; Pritchard 1969), has become a specialty in itself with sophisticated discussions about issues such as different normalizations. In the scientometric literature, however, the purpose of the analysis is mostly the further improvement of an indicator (e.g., Waltman et al. 2011). In the meantime, journal impact factors,[8] university rankings, and the *h*-index[9] have become an everyday part of academic life. In such policy contexts, these measures are used without scholarly articulation in terms of STS. In the next and final section, we turn to the question of how the development of scientometric methods can be expected to further contribute to the development of STS from the perspective of qualitative STS.

Conclusions and Future Directions

In this chapter, we have highlighted some of the common origins of (qualitative) STS and (quantitative) scientometrics and their common commitment to the empirical study of the sciences. Documents and texts broadly conceived are also an important, if not the most important, resource for STS researchers, whatever their theoretical and methodological commitments (see Shankar, Hakken, and Østerlund, chapter 2 this volume). Drawing on the scientometric tradition, we argued that scholarly publications can be analyzed both as constructs and as factual data amenable to text, data, and network analytics. For the latter, as illustrated above, agents are reconstructed as author names (*actants*), institutions as institutional addresses, and cognitions can be coded as references. The process of knowledge production can thus be considered one of argumentative purification and coding of texts at the network level.

In the long history of the dissemination of knowledge in the form of manuscripts, documents were first scarce and transcription was error-prone. The search was thus for "the original." The invention of the printing press in the latter half of the fifteenth century, analyzed by Elizabeth Eisenstein (1983), changed the textual dynamics, as successive versions could be corrected for errors apparent in previous editions (Luhmann 1981). By the end of the twentieth century, the selection mechanisms changed with the diffusion of the Internet. Text is everywhere, and texts are abundant as variation; word processing induces working in versions, where each new version is intended to improve the content and the formulation. Authors can envisage and engage with a variety of selection environments.

Scholarly manuscripts are processed and selected by journal editors before and after review. The composition of review boards remains contingent, and change in editorship may affect the selection mechanisms operating at the journal level. These selection processes can be studied empirically, using both qualitative and quantitative methods, as Michèle Lamont (2012) has done. However, the mechanisms develop at the supra-individual level, with their own dynamics. Yet, they are not closed off from agency. As Latour (1988) noted, authors have access to them "infra-reflexively," accessing the networks through their own activities such as publishing and reading. An author can read a journal's mission statement and author guidelines and use these instructions when preparing a manuscript; but individually or locally a single author cannot control the emerging network dynamics. In other words, only knowledge of selection mechanisms operating in the networks can sometimes inform authors' expectations or specify options for change.

From this perspective of uncertain selection environments, performativity and reflexivity are no longer localized and attributable to individuals but distributed in relations and network configurations, and are therefore potential objects of sociological analysis. Communications in network relations are units of analysis different from agency and provide another frame of reference for what the concepts may mean: while reflexivity and performativity have mainly been attributed to (individual or institutional) agency in the sociological tradition, the reflection of concepts in documents can be expected to lead to different meanings at other ends and can thus open or reinforce a discourse (Bourdieu 2004). A sociology of communication emerges alongside, in relation to but not isomorphic to the sociology of human agency and groups of individuals in social organizations (Luhmann 1995). Actor-network theory noted that this newly emerging sociology could be expected to include also nonhuman *actants*.

Our conclusions are of two types. The first relates to the nature of the scientometric enterprise itself. We have argued that scientometrics can be useful for understanding, among other things, the emergence and transformation of concepts and theories, the dynamics of the communication of science, and mapping the socio-epistemic organization of knowledge. However, we have warned against the unreflexive use of such techniques in STS and by policy makers and management, and thus the export of presumed theories in the dynamics of science and technology. The further refinement of theoretically compounded categories (e.g., citations, impact), especially if they are then used to produce highly consequential rankings of universities, departments, and individuals, can be resisted with scientometric arguments, as Diana Hicks and her colleagues have done (2015). The data, methods, and techniques remain as artifactual and constructed as any other kind of social scientific fact or method, and they need to be treated with due care for their theoretical backgrounds and foundations.

Let us repeat that the scientometric maps and tables are algorithmic constructs that remain heavily dependent on the parameters chosen for generating them, and this is true for all computational methods. Refining the parameter choices does not by itself validate the constructs. By presenting our own scientometric results in this chapter as algorithmic constructs, we have tried to re-open the discussion between quantitative and qualitative STS. Can the scientometric endeavor be made useful to the study of STS? Can the different approaches be mutually infused with one another's heuristics? Following Woolgar (1991), we have argued that one way of reconciling the epistemological contradictions between quantitative and qualitative methods and thus quantitative and qualitative epistemologies within STS is to recognize that all types of constructs require interpretation in contexts of use, not only by other researchers but also by social actors, including respondents, policy makers, and other audiences.

This brings us to the second type of conclusions, related to the future of STS more broadly, and the potential benefits of incorporating scientometric methods. As so-called big data become more prevalent in the social sciences and the humanities (leading to fields such as computational social science, network science, and digital humanities), we wonder whether the usage of massive data and associated statistical, mathematical, and computational methods of analysis could lead to significant changes within STS that would facilitate the exchange of knowledge between qualitative and quantitative STS.

Since the 1990s, with the widespread diffusion of the Internet and increased globalization of the research environment, publication and communication dynamics in the sciences have changed. "Open access" and the requirements to make data transparently available when submitting a manuscript, and at a later stage for the publication itself, add new dimensions to the selection environments. This more complex environment also generates expectations and standards that STS cannot afford to ignore. Inspired by the research program of scientometrics, we suggest that the socio-epistemic context has irrevocably been extended by a textual dimension that provides us with "big data" about, among other things, developments in science and technology. To address this complexity, STS needs to be able to build bridges and to engage in multiple processes of translation between qualitative and quantitative approaches.

The possibilities offered by digitally mediated communication have brought to the fore an awareness that the dynamics of communication in scholarly discourses are both specific and interacting with other forms of communication. The study of communication (e.g., in discourse and content analysis) invites questions about sample choices and statistical testing of significance (Krippendorff and Bock 2009). Communication in specific channels (e.g., scholarly or administrative) can always be improved and thus one can expect standards (i.e., quality measures) to emerge. From this perspective, scientometric indicators hitherto have mainly captured communication within the sciences, whereas future challenges include the measurement and analysis of communication across scientific domains and at the interfaces of science and society.

While the scientometric results presented in this chapter tend to confirm a lack of communication between quantitative and qualitative STS, the opening of a conversation between the two is essential for the successful utilization of novel data sources to advance STS. Monika Büscher and John Urry (2009) suggest that the problems in the exchange and transfer of knowledge within sociology are not surprising given the different epistemological foundations underlying the construction of the data and the very different skills researchers need in order to analyze and interpret the findings based on such different types of data. Our argument is different in that we emphasize the always-constructed nature of data, methods, and algorithms. But we do agree that the skill sets differ and that greater dialog could enrich the skills and methodological repertoires of both qualitative and quantitative researchers.

STS can contribute to "big data" debates by analyzing what big data means for other research fields, as Sabina Leonelli (2014) has done for biology and Clement Levallois and others for sociology and economics (Levallois et al. 2013). More general critiques of the uncritical use of big data have been made by danah boyd and Kate Crawford (2012) and Paul Edwards et al. (2011) and can be found in the journal *Big Data & Society*, launched in 2014. We also need to foster thinking about how STS itself can remain relevant to on-going debates, by making statistical, reflexive, and policy-relevant use of its own large archives of data, not only bibliographic databases and citation indices but also oral history archives and ethnographic field notes. STS could embrace data science, not only as an object of critique but also as a source of inspiration for future work. In this context, Tommaso Venturini, Pablo Jensen, and Bruno Latour (2015) have proposed to consider digital methods as "quali-quantitative methods." New *actants* such as blogs, Wikipedia, and mark-up languages are and will be generated in the digital domain, and their analysis requires the development of new methodologies (Light, Polley, and Börner 2014; Rogers 2013).

Our focus on the textual dimension of the sciences has theoretical implications, such as the attribution of performativity to networks of documents instead of to individuals, as would be common in the sociological tradition (e.g., Giddens 1979). We have criticized the ways in which performance indicators with time horizons of between two and five years are attributed as rankings to individual authors or institutes in the humanities and social sciences, with performative and exclusionary consequences (Leydesdorff and Milojević 2015). But we are aware that our own reconstruction of STS can also be considered performative and exclusionary: we did not fully address issues of gender and ethnicity in science, nor did we explicitly address publishing in national languages and in book or other formats. The history of scientometrics and the usage of indicators and rankings in science and technology policies and research management make a coupling to qualitative STS urgent because algorithmic reifications tend to prevail and to obscure the full stories. The stories of policy makers and indicator suppliers may be based on sophisticated and innovative statistics, but discussions and debates about science and technology have been intellectually indebted to STS in the past and can be expected to remain so in the future.

Acknowledgments

Diana Lucio-Arias and Selma Šabanović offered valuable support and input during early stages of preparation. We are also grateful to the editors of the *Handbook* for their patience and support and to the anonymous reviewers of an earlier draft for their challenging and constructive remarks. We thank Thomson Reuters for providing us with relevant data.

Notes

1. As Sergio Sismondo argued in the opening chapter of the 2008 *Handbook*, "STS looks to *how* the *things* it studies are constructed" (2008, 13; emphasis added). He goes on to show that those "things" have expanded in scope to include not only scientific and technical knowledge practices but also practices associated with policy, politics, and governance of science and technology. But the "how" is not elaborated, and indeed the 2008 *Handbook* does not have a section on method, nor even a mention of it in the index, though the editors lament this lack in their introduction (Hackett et al. 2008, 5).

2. "Stand on the shoulders of giants" is the motto adopted by Google Scholar. It invokes the letter that Isaac Newton (1642–1727) sent to Robert Hooke on February 5, 1676, in which he wrote: "If I have seen further, it is only by standing on the shoulders of giants." Merton borrowed the aphorism for the title of one of his books (1965).

3. From its beginnings, scientometrics has been characterized by the involvement of private and for-profit university-based groups in the provision of data, indices, rankings, and evaluations. The political economy of these activities is beyond the scope of this chapter but is nonetheless extremely important (Dahler-Larsen 2011).

4. There was a debate about "interests" in *Social Studies of Science* in the early 1980s, involving Donald MacKenzie, Steve Woolgar, Barry Barnes, and Michel Callon. It was agreed that the formation of interests is a slow historical process, that interests cannot always be read from subject positions, and that interests are both an actor's and an analyst's category.

5. The analyses are based on the full sets of publications in *Scientometrics* (since 1978) and *Social Studies of Science* (since 1971) downloaded from the installation of the Web of Science (WoS) on 6 October 2014.

6. The titles and abstracts contain 50,533 terms, of which 1,494 occur more than ten times. VOSviewer reduces this set using a relevance score (van Eck and Waltman 2010). Abstracts have been available in the database since 1991.

7. This was further made accessible for general use as a program HistCite™, developed by Garfield in collaboration with Alexander Pudovkin (Garfield, Pudovkin, and Istomin 2003).

8. The impact factor of a journal is the average number of times that articles published in the journal during the preceding two or five years are cited in the current year.

9. The *h*-index was proposed by the physicist Jorge Hirsch (2005). An author has an index of *h* where *h* is the number of papers that have each been cited at least *h* times. It has had a huge influence on scientometrics, but it is not so much reflected in the STS literature.

References

Amsterdamska, Olga, and Loet Leydesdorff. 1989. "Citations: Indicators of Significance?" *Scientometrics* 15 (5–6): 449–71.

Barnes, S. Barry, and R. G. Alex Dolby. 1970. "The Scientific Ethos: A Deviant Viewpoint." *European Journal of Sociology* 11 (1): 3–25.

Bell, Daniel. 1973. *The Coming of Post-Industrial Society*. New York: Basic Books.

Bernstein, Basil. 1971. *Class, Codes and Control*, vol. 1: *Theoretical Studies in the Sociology of Language.* London: Routledge and Kegan Paul.

Bijker, Wiebe E. 1993. "Do Not Despair. There Is Life after Constructivism." *Science, Technology, & Human Values* 18 (1): 113–38.

Bloor, David. 1976. *Knowledge and Social Imagery*. London: Routledge and Kegan Paul.

Borgman, Christine. 2015. *Big Data, Little Data, No Data: Scholarship in the Networked World*. Cambridge, MA: MIT Press.

Börner, Katy. 2010. *Atlas of Science: Visualizing What We Know*. Cambridge, MA: MIT Press.

Bourdieu, Pierre. 2004. *Science of Science and Reflexivity*. Chicago: University of Chicago Press.

boyd, danah, and Kate Crawford. 2012. "Critical Questions for Big Data: Provocations for a Cultural, Technological and Sociological Phenomenon." *Information, Communication & Society* 15 (5): 662–79.

Braam, Robert R., Henk F. Moed, and Anthony F. J. van Raan. 1991. "Mapping of Science by Combined Co-citation and Word Analysis. I. Structural Aspects." *Journal of the American Society for Information Science* 42 (4): 233–51.

Büscher, Monika, and John Urry, J. 2009. "Mobile Methods and the Empirical." *European Journal of Social Theory* 12 (1): 99–116.

Callon, Michel. 1986. "The Sociology of an Actor Network: The Case of the Electric Vehicle." In *Mapping the Dynamics of Science and Technology*, edited by Michel Callon, John Law, and Arie Rip, 19–34. London: Macmillan.

Callon, Michel, Jean-Pierre Courtial, William A. Turner, and Serge Bauin. 1983. "From Translations to Problematic Networks: An Introduction to Co-word Analysis." *Social Science Information* 22 (2): 191–235.

Castells, Manuel. 1996–98. *The Information Age: Economy, Society and Culture*. Oxford: Blackwell, vol. 1: *The Rise of the Network Society* (1996); vol. 2: *The Power of Identity* (1997); vol. 3: *End of Millennium* (1998).

Cole, Jonathan R., and Stephen Cole. 1973. *Social Stratification in Science*. Chicago: University of Chicago Press.

Collins, Harry. 1985. "The Possibilities of Science Policy." *Social Studies of Science* 15 (3): 554–58.

___. 2004. *Gravity's Shadow: The Search for Gravitational Waves*. Chicago: University of Chicago Press.

Cozzens, Susan E. 1989. "What Do Citations Count? The Rhetoric-First Model." *Scientometrics* 15 (5):437–47.

Crane, Diana. 1972. *Invisible Colleges*. Chicago: University of Chicago Press.

Dahler-Larsen, Peter. 2011. *The Evaluation Society*. Stanford, CA: Stanford University Press.

David, Paul A., and Dominique Foray. 2002. "An Introduction to the Economy of the Knowledge Society." *International Social Science Journal* 54 (171): 9–23.

Edge, David. 1979. "Quantitative Measures of Communication in Science: A Critical Overview." *History of Science* 17: 102–34.

___. 1995. "Reinventing the Wheel." In *Handbook of Science and Technology Studies*, edited by Sheila Jasanoff, Gerald E. Markle, James C. Petersen, and Trevor Pinch, 3–23. Thousand Oaks, CA: Sage.

Edge, David, and Roy MacLeod. 1986. "Editorial." *Social Studies of Science* 16 (1): 3–8.

Edwards, Paul N., Matthew S. Mayernik, Archer L. Batcheller, Geoffrey C. Bowker, and Christine L. Borgman. 2011. "Science Frictions: Data, Metadata, and Collaboration." *Social Studies of Science* 41 (5): 667–90.

Eisenstein, Elizabeth L. 1983. *The Printing Revolution in Early Modern Europe*. Cambridge: Cambridge University Press.

Franklin, Marianne I. 2012. *Understanding Research: Coping with the Quantitative-Qualitative Divide*. London: Routledge.

Fujigaki, Yuko. 1998. "Filling the Gap between Discussions on Science and Scientists' Everyday Activities: Applying the Autopoiesis System Theory to Scientific Knowledge." *Social Science Information* 37 (1): 5–22.

Fujigaki, Yuko, and Loet Leydesdorff. 2000. "Quality Control and Validation Boundaries in a Triple Helix of University-Industry-Government: 'Mode 2' and the Future of University Research." *Social Science Information* 39 (4): 635–55.

Garfield, Eugene. 1955. "Citation Indexes for Science: A New Dimension in Documentation through Association of Ideas." *Science* 122 (3159): 108–11.

___. 1964. "Science Citation Index—A New Dimension in Indexing." *Science* 144 (3619): 649–54.

___. 1975. "The 'Obliteration Phenomenon' in Science—and the Advantage of Being Obliterated." *Current Contents* 51–52: 396–98.

Garfield, Eugene, Alexander I. Pudovkin, and Vladimir S. Istomin. 2003. "Why Do We Need Algorithmic Historiography?" *Journal of the American Society for Information Science and Technology* 54 (5): 400–12.

Garfield, Eugene, Irving H. Sher, and Richard J. Torpie. 1964. *The Use of Citation Data in Writing the History of Science*. Philadelphia: Institute for Scientific Information.

Giddens, Anthony. 1979. *Central Problems in Social Theory*. London: Macmillan.

Gilbert, G. Nigel. 1977. "Referencing as Persuasion." *Social Studies of Science* 7: 113–22.

Gilbert, G. Nigel, and Mike J. Mulkay. 1984. *Opening Pandora's Box: A Sociological Analysis of Scientists' Discourse*. Cambridge: Cambridge University Press.

Godin, Benoît. 2006. "The Knowledge-Based Economy: Conceptual Framework or Buzzword?" *Journal of Technology Transfer* 31 (1): 17–30.

Hackett, Edward J., Olga Amsterdamska, Michael Lynch, and Judy Wajcman. 2008. "Introduction." In *Handbook of Science and Technology Studies*, edited by Edward J. Hackett, Olga Amsterdamska, Michael Lynch, and Judy Wajcman, 13–31. Cambridge, MA: MIT Press.

Halffman, Willem, and Hans Radder. 2015. "The Academic Manifesto: From an Occupied to a Public University." *Minerva* 53 (2): 165–87.

Hess, David J. 1997. *Science Studies: An Advanced Introduction*. New York: New York University Press.

Hicks, Diana, Paul Wouters, Ludo Waltman, Sarah de Rijcke, and Ismael Rafols. 2015. "The Leiden Manifesto for Research Metrics." *Nature* 520: 429–31.

Hirsch, Jorge E. 2005. "An Index to Quantify an Individual's Scientific Research Output." *Proceedings of the National Academy of Sciences of the USA* 102 (46): 16569–72.

Knorr Cetina, Karin. 1999. *Epistemic Cultures: How the Sciences Make Knowledge*. Cambridge, MA: Harvard University Press.

Krippendorff, Klaus, and Mary Angela Bock. 2009. *The Content Analysis Reader*. Los Angeles: Sage.

Kuhn, Thomas S. 1962. *The Structure of Scientific Revolutions*. Chicago: University of Chicago Press.

Lamont, Michèle. 2012. "Toward a Comparative Sociology of Valuation and Evaluation." *Sociology* 38 (1): 201–21.

Larivière, Vincent, Éric Archambault, Yves Gingras, and Étienne Vignola-Gagné. 2006. "The Place of Serials in Referencing Practices: Comparing Natural Sciences and Engineering with Social Sciences and Humanities." *Journal of the American Society for Information Science and Technology* 57 (8): 997–1004.

Latour, Bruno. 1983. "Give Me a Laboratory and I Will Raise the World." In *Science Observed*, edited by Karin Knorr Cetina and Michael Mulkay, 141–70. London: Sage.

___. 1986. "Visualisation and Cognition: Drawing Things Together." *Knowledge and Society: Studies in the Sociology of Culture Past and Present* 6: 1–40.

___. 1987. *Science in Action*. Milton Keynes: Open University Press.

___. 1988. "The Politics of Explanation: An Alternative." In *Knowledge and Reflexivity: New Frontiers in the Sociology of Knowledge*, edited by Steve Woolgar and Malcolm Ashmore, 155–77. London: Sage.

___. 2005. *Reassembling the Social: An Introduction to Actor-Network-Theory*. Oxford: Oxford University Press.

Latour, Bruno, and Steve Woolgar. 1979. *Laboratory Life: The Social Construction of Scientific Facts*. London: Sage.

Leonelli, Sabina. 2014. "What Difference Does Quantity Make? On the Epistemology of Big Data in Biology." *Big Data & Society* 1 (1): 1–11.

Levallois, Clement, Stephanie Steinmetz, and Paul Wouters. 2013. "Sloppy Data Floods or Precise Social Science Methodologies?" In *Virtual Knowledge, Experimenting in the Humanities and the Social Sciences*, edited by Paul Wouters, Anne Beaulieu, Andrea Scharnhorst, and Sally Wyatt, 151–82. Cambridge, MA: MIT Press.

Leydesdorff, Loet. 2007. "Scientific Communication and Cognitive Codification: Social Systems Theory and the Sociology of Scientific Knowledge." *European Journal of Social Theory* 10 (3): 375–88.

Leydesdorff, Loet, and Olga Amsterdamska. 1990. "Dimensions of Citation Analysis." *Science, Technology, & Human Values* 15 (3): 305–35.

Leydesdorff, Loet, and Staša Milojević. 2015. "The Citation Impact of German Sociology Journals: Some Problems with the Use of Scientometric Indicators in Journal and Research Evaluations." *Soziale Welt* 66 (2): 193–204.

Leydesdorff, Loet, and Peter van den Besselaar. 1997. "Scientometrics and Communication Theory: Towards Theoretically Informed Indicators." *Scientometrics* 38 (1): 155–74.

Light, Robert P., David E. Polley, and Katy Börner. 2014. "Open Data and Open Code for Big Science of Sciences Studies." *Scientometrics* 101 (2): 1535–51.

Luhmann, Niklas. 1981. "Gesellschaftsstrukturelle Bedingungen und Folgeprobleme der naturwissenschaftlich-technischen Fortschirtts." In *Fortschritt ohne Maß: Eine Ortsbestimmung der wissenschaftlich-technischen Zivilisation*, edited by Reinhard Loew, Peter Koslowski, and Philipp Kreuzer, 112–31. Munich: Piper.

___. 1995. *Social Systems*. Stanford, CA: Stanford University Press.

Martin, Ben R. 2012. "The Evolution of Science Policy and Innovation Studies." *Research Policy* 41 (2) (7): 1219–39.

Martin, Ben R., and John Irvine. 1983. "Assessing Basic Research: Some Partial Indicators of Scientific Progress in Radio Astronomy." *Research Policy* 12: 61–90.

Martin, Ben R., Paul Nightingale, and Alfredo Yegros-Yegros. 2012. "Science and Technology Studies: Exploring the Knowledge Base." *Research Policy* 41 (7): 1182–204.

Mayer-Schönberger, Viktor, and Kenneth Cukier. 2013. *Big Data: A Revolution That Will Transform How We Live, Work, and Think*. Boston: Houghton Mifflin Harcourt.

Merton, Robert K. 1942. "Science and Technology in a Democratic Order." *Journal of Legal and Political Sociology* 1: 115–26.

___. 1965. *On the Shoulders of Giants: A Shandean Postscript*. New York: Free Press.

___. 1973. *The Sociology of Science: Theoretical and Empirical Investigations.* Chicago: University of Chicago Press.

Moed, Henk F. 2005. *Citation Analysis in Research Evaluation.* Dordrecht: Springer.

Moed, Henk F., W. J. M. Burger, J. G. Frankfort, and Anthony F. J. van Raan. 1985. "The Use of Bibliometric Data for the Measurement of University Research Performance." *Research Policy* 14 (3): 131–49.

Moed, Henk F., Wolfgang Glaenzel, and Ulrich Schmoch. 2004. *Handbook of Quantitative Science and Technology Research: The Use of Publication and Patent Statistics in Studies of S & T Systems.* Dordrecht: Springer.

Moretti, Franco. 2013. *Distant Reading.* London: Verso.

Mulkay, Michael, Jonathan Potter, and Steven Yearley. 1983. "Why an Analysis of Scientific Discourse Is Needed." In *Science Observed: Perspectives on the Social Study of Science*, edited by Karin Knorr Cetina and Michael Mulkay, 171–204. London: Sage.

Mullins, Nicholas C. 1973. *Theories and Theory Groups in Contemporary American Sociology.* New York: Harper and Row.

Myers, Greg. 1985. "Texts as Knowledge Claims: The Social Construction of Two Biology Articles." *Social Studies of Science* 15 (4): 593–630.

Narin, Francis. 1976. *Evaluative Bibliometrics: The Use of Publication and Citation Analysis in the Evaluation of Scientific Activity.* Washington, DC: National Science Foundation.

OECD. 1963, 1976. *The Measurement of Scientific and Technical Activities: "Frascati Manual."* Paris: OECD.

___. 1996. *The Knowledge-Based Economy.* Paris: OECD, accessed at http://www.oecd.org/dataoecd/51/8/1913021.pdf.

OECD/Eurostat. 1997. *Proposed Guidelines for Collecting and Interpreting Innovation Data, "Oslo Manual."* Paris: OECD.

Pinch, Trevor J. 1985. "Towards an Analysis of Scientific Observation: The Externality and Evidential Significance of Observational Reports in Physics." *Social Studies of Science* 15 (1): 3–36.

Popper, Karl R. [1935] 1959. *The Logic of Scientific Discovery.* London: Hutchinson.

Price, Derek de Solla. 1965. "Networks of Scientific Papers." *Science* 149 (3683): 510–15.

___. 1970. "Citation Measures of Hard Science, Soft Science, Technology, and Nonscience." In *Communication among Scientists and Engineers*, edited by Carnot E. Nelson and Donald K. Pollock, 3–22. Lexington, MA: Heath.

Pritchard, Alan. 1969. "Statistical Bibliography or Bibliometrics?" *Journal of Documentation* 25 (4): 348–49.

Rogers, Richard. 2013. *Digital Methods.* Cambridge MA: MIT Press.

Rogers, Richard, and Noortje Marres. 2000. "Landscaping Climate Change: A Mapping Technique for Understanding Science and Technology Debates on the World Wide Web." *Public Understanding of Science* 9 (2): 141–63.

Rotolo, Daniele, Diana Hicks, and Ben Martin. 2015. "What Is an Emerging Technology?" *Research Policy* 44 (10): 1827–43.

Sismondo, Sergio. 2008. "Science and Technology Studies and an Engaged Program." In *The Handbook of Science and Technology Studies*, 3rd ed., edited by Edward J. Hackett, Olga Amsterdamska, Michael Lynch, and Judy Wajcman, 13–31. Cambridge, MA: MIT Press.

Small, Henry. 1978. "Cited Documents as Concept Symbols." *Social Studies of Science* 8 (3): 113–22.

Strathern, Marilyn. 2000. *Audit Cultures: Anthropological Studies in Accountability, Ethics and the Academy*. London: Routledge.

Studer, Kenneth E., and Daryl E. Chubin. 1980. *The Cancer Mission: Social Contexts of Biomedical Research*. Beverly Hills, CA: Sage.

van Eck, Nees J., and Ludo Waltman. 2010. "Software Survey: VOSviewer, a Computer Program for Bibliometric Mapping." *Scientometrics* 84 (2): 523–38.

___. 2014a. "Visualizing Bibliometric Networks." In *Measuring Scholarly Impact*, edited by Ying Ding, Ronald Rousseau, and Dietmar Wolfram, 285–320. Cham: Springer.

___. 2014b. "CitNetExplorer: A New Software Tool for Analyzing and Visualizing Citation Networks." *Journal of Informetrics* 8 (4): 802–23.

van Raan, Anthony F. J., ed. 1988. *Handbook of Quantitative Studies of Science and Technology*. Amsterdam: Elsevier.

Venturini, Tommaso, Pablo Jensen, and Bruno Latour. 2015. "Fill in the Gap: A New Alliance for Social and Natural Sciences." *Journal of Artificial Societies and Social Simulation* 18 (2): 11.

Waltman, Ludo, Clara Calero-Medina, Joost Kosten, Ed C. M. Noyons, Robert J. W. Tijssen, Nees Jan van Eck, Thed N. van Leeuwen, Anthony F. J. van Raan, Martijn S. Visser, and Paul Wouters. 2012. "The Leiden Ranking 2011/2012: Data Collection, Indicators, and Interpretation." *Journal of the American Society for Information Science and Technology* 63 (12): 2419–32.

Waltman, Ludo, Nees Jan van Eck, Thed N. van Leeuwen, Martijn S. Visser, and Anthony F. J. van Raan. 2011. "Towards a New Crown Indicator: Some Theoretical Considerations." *Journal of Informetrics* 5 (1): 37–47.

Wagner, Caroline S., J. David Roessner, Bob Kamau, Julie Thompson Klein, Kevin W. Boyack, Joann Keyton, Ismael Rafols, and Katy Börner. 2011. "Approaches to Understanding and Measuring Interdisciplinary Scientific Research (IDR): A Review of the Literature." *Journal of Informetrics* 5 (1): 14–26.

Whitley, Richard J. [1984] 2000. *The Intellectual and Social Organization of the Sciences*, 2nd ed. Oxford: Oxford University Press.

Woolgar, Steve. 1991. "Beyond the Citation Debate: Towards a Sociology of Measurement Technologies and Their Use in Science Policy." *Science and Public Policy* 18 (5): 319–26.

Wouters, Paul. 1998. "The Signs of Science." *Scientometrics* 41 (1): 225–41.

___. 1999. *The Citation Culture*. Unpublished Ph.D. Thesis. Amsterdam: University of Amsterdam.

Wyatt, Sally, and Brian Balmer. 2007. "Home on the Range: What and Where Is the Middle in Science and Technology Studies?" *Science, Technology, & Human Values* 32 (6): 619–26.

4 Ethnomethodology, Video Analysis, and STS

Philippe Sormani, Morana Alač, Alain Bovet, and Christian Greiffenhagen

Introduction

As Garfinkel has taught us: it's practice all the way down.
(Latour 2005, 135)

This chapter offers a synoptic overview of how ethnomethodology (EM) has developed in and contributed to science and technology studies (STS) with a particular emphasis on where, why, and how video-based methods fit in the arc of EM approaches in STS. Over the last fifty years, EM has diversified into a multiplicity of approaches, expanded worldwide, and consequently been variably, yet often tacitly, folded into STS discourse (Lynch 2011a, 2011b). The present chapter traces that multiplicity, expansion, and variability. In particular, the chapter charts the origins of video analysis outside STS, specifies its main applications inside the field, and hints at practice-based developments alongside its mainstream. In so doing, the chapter demonstrates the "diversity of ethnomethodology" (Maynard and Clayman 1991) as manifested, most recently, through the various ways of doing video analysis in and alongside STS. On this basis, the chapter discusses and reflects upon the multiple relationships between EM, video analysis, and STS.

EM's long-standing interest in the constitutive practices of social order, mundane realities, and technoscientific artifacts *anticipated* paradigmatic developments in social theory at large and later STS in particular, not least of which their "practice turn" (Pickering 1992a, 1992b; Soler et al. 2014a). Latour's allusion in the epigraph hints at this core contribution, as his allusion spells out one of the key lessons of Harold Garfinkel, the founder of EM. It is only and somewhat paradoxically because of STS having taken on board, and now largely taking for granted, many of EM's most basic assumptions, that EM might indeed appear "not much in evidence" (Lynch 2011b, 928). In particular, EM anticipated the practice turn in STS by turning to the situated accomplishment of social practices as an empirical topic (Garfinkel 1967), including the methodical pursuit of scientific practice and indeed *the very constitution of "evidence"* (cf. Benson and Hughes 1991; Lynch 1993). Recent characterizations of the practice turn in STS—itself a multifaceted endeavor in social theory more broadly (Schatzki, Knorr Cetina, and von Savigny 2001)—are in fact largely congruent with EM's long-standing empirical core interest:

> Beyond any divergent understandings of "practice," for both ethnographic and historical studies, the maxim "pay attention to scientific practice" conveys a crucial methodological ideal: to recover detailed actions and reasoning—including uncertainties, conflicting interpretations, and so on—*as they operate in the situation*, in contrast to retrospective reconstructions of actions and results provided by scientists and traditional philosophy of science. In other words, the ideal is to recover important aspects of actual scientific activity that are left out of scientific publications. (Soler et al. 2014b, 12–13)

In addition to ethnographic and historical studies in STS, ethnomethodological *video analysis* makes a distinctive contribution to the detailed study of the local accomplishment of practical activities, ranging from mundane conversation and embodied interaction to technical activities in the natural sciences and beyond. Therefore, the bulk of this chapter traces the origins, diversity, and heuristics of video analysis in an EM vein—that is, a video analysis that "recover[s] detailed actions and reasoning—including uncertainties, conflicting interpretations, and so on—*as they operate in the situation*" (Soler et al. 2014b, 13). To fully grasp the rich diversity and heuristic contribution of video analysis and its recent developments to STS, we start by spelling out EM's significance for the field, before taking a closer look at and reflecting upon ethnomethodological video analysis, STS, and their multiple relationships.

EM and STS: Practical Relevancies and Paradigmatic Significance

In 1967 Garfinkel published *Studies in Ethnomethodology*—the first collection of empirical studies of common, everyday methods of practical action and practical reasoning with a sociological rationale (e.g., Livingston 1987, 4). "Methods" of all kinds, including those of (professional) sociological inquiry, were turned into a prosaic phenomenon of *empirical* interest rather than (continued to be) drawn upon as special instruments of epistemic import (e.g., to improve, refine, or contribute to existing forms of social science). In particular, the tacit dependency of the methods of professional sociology on received views and vernacular formulations of society was to be made explicit (Garfinkel 1967; Rose 1960; Sacks 1963) rather than relied upon as an unacknowledged resource (Zimmerman and Pollner 1973). Accordingly, "'methodology' in 'ethnomethodology' stand[s] for a subject matter, rather than a scientific apparatus" (Turner 1974, 12). The idea for this reversal, turning methods and methodology into phenomena, occurred to Garfinkel (1974) in the course of a research project on jury deliberations in the United States, where jury members are drawn from civil society. "Ethnomethodology," then, provided a convenient neologism to characterize its lay participants' methods of reasoning, as it alluded to *their* "folk methods" in contrast to legal reasoning by trained professionals. An additional task, in turn, would soon be to make explicit how methods specific to professional occupations (e.g., lawyers' work) hinged themselves upon the mastery of all sorts of lay methods (professional sociology providing the first exemplary case of this more general interest).[1] In a nutshell, EM set out as a descriptive approach of *practical relevancies* in their methodical character and ordinary intelligibility, as achieved in particular situations, technically mediated and/or conversationally articulated. Accordingly, EM may be defined as a "sociology of the witnessable order" (Livingston 2008, 128).

The *paradigmatic significance* for STS of this analytic focus, in turn, holds in one technical expression: "respecification" (Lynch 1993, xi, note 1). The leading idea in respecification, which goes back to Garfinkel (1991), is not only to describe locally produced orders (if only to avoid their "misplaced abstraction" *ex cathedra*) but also to reflect upon previous theoretical accounts of social, sociotechnical, or socio-material order (e.g., Durkheim [1895] 1964; Parsons 1949; Callon and Latour 1981; or Law and Mol 1995) in light of that description. In this vein, Lynch (1993) worked out the paradigmatic significance of EM investigations for STS by introducing them as an "empirical approach to epistemology's traditional topics" (Lynch 1993, xviii):

Begin by taking up one or more of the epistopics [epistemological topics]. The epistopics have a prominent place in the large literatures in the history, philosophy, and sociology of science, but in this case our aim will be to break out of the academic literature by searching for what Garfinkel has called "perspicuous settings": Familiar language games in which one or another epistopic has a prominent vernacular role. So, for instance, although there are many interesting and erudite discussions of "observation" in the philosophy of science, "observation" has no less prominent a place in the practices, written and oral instructions, and reports in numerous other organized activities, some of which are quite humble and ordinary. The academic literature provides a relevant background for beginning such investigations, insofar as a long history of scholarly treatments and argumentative positions establishes the initial significance of the epistopic. Although for the program I am outlining, the literature cannot be disregarded—it does, after all, supply a current situation of inquiry—the academic conversation will be continued by other means than an explication of the classic literature. (Lynch 1993, 300)

The outlined program, casting EM as an "empirical epistemology" of a practice-interested kind, constituted a key move toward STS. Indeed, it allowed any of its practitioners, including Lynch (1985) himself, to examine in detail how particular concepts, distinctions, or "theories" gained their purchase in and through mundane scientific practices *and*, on that empirical basis, to reexamine classic scholarly topics, including "scientific discovery" (Garfinkel, Lynch, and Livingston 1981), "mathematical demonstration" (Livingston 1986), and "technical artifacts" (Lynch 1985). Each time the approach would be to describe how, when, and why such topics became practically relevant to the involved participants as (or as part of) their ongoing activities in their ordinary intelligibility, and then to reflect upon the new insights, theoretical or conceptual, gained from such description. The technical term *respecification* encapsulates this two-step procedure. Its precise form, in turn, would vary depending upon the examined case and the STS argument at stake. Ethnomethodological video analysis of situated practices remains of particular interest in this respect, as it has opened up new research avenues in and alongside STS, as the remainder of this chapter will demonstrate and discuss.[2]

Ethnomethodological Video Analysis: Heuristic Diversity, in and Alongside STS

Audiovisual documents are premier resources.
(Garfinkel 2002, 148)

Ethnomethodological video analysis, as developed over the last three decades, is characterized by heuristic diversity. Different kinds of approaches have made differing contributions, in and alongside STS. The present section of the chapter introduces three main strands of ethnomethodological video analysis. First, it spells out the conversation analytic strand of video analysis originating outside STS. Second, it charts video analysis's applications within the field, beginning from the "laboratory studies" tradition and leading to empirical respecifications of some of the field's key concepts. Finally, it hints at practice-based developments alongside STS's mainstream. The ethnomethodological analysis of audiovisual resources, as we shall see, constitutes a potent means to achieve not a single but a multifaceted practice turn in STS, not solely as a theoretical gesture but in empirical specifics.

Interaction: Video in Conversation Analysis and Its Contributions to STS
Interaction in everyday settings has constituted a primary site for ethnomethodological video analysis, insofar as its participants achieve mutual understanding not only in terms of a verbal exchange (via questions and answers, for example) but also by displaying and relying upon visual cues (including gestures, particular gazes, and body

postures). In tune with EM's policy of case-specific analysis, this section gives some "detailed examples" (Livingston 1987, 5) of video analysis of situated interaction. The section, more specifically, charts the origins of video analysis in conversation analysis (CA) while demonstrating that CA-based video analysis, even though it has been developed outside the field of STS, affords this field with a powerful "microscope of interaction" (Büscher 2005). When deftly used, this microscope of interaction can contribute to respecifying some of STS's central concerns, including the concrete operation of actor-networks in situ, as we shall see below. In sum, CA-based video analysis makes possible a thoroughly *interactionist* turn to practice in unprecedented empirical detail, by "address[ing] itself directly to actual and observable occurrences in and of the social [or socio-material] order" (Hutchinson, Rupert, and Sharrock 2008, 109).[3]

The enterprise known today as CA was invented by Harvey Sacks (1935–1975), one of the early research associates of Garfinkel, and then developed with Emanuel A. Schegloff and Gail Jefferson (e.g., Sacks, Schegloff, and Jefferson 1974). CA was based upon audio recordings of ordinary (telephone) conversations and their detailed transcription, in view of the descriptive analysis of interaction, social order, and their methodic production "in real time" (Sacks 1984; Schegloff and Sacks 1973). Examining the ethnomethods of talking together (when answering a question, for example), CA continues to contribute to EM's approach to real-time conversation as a social phenomenon, although it has evolved into a virtually autonomous discipline since its invention by Sacks in the 1960s. CA, in that sense, is the (ethno-)methodological basis of current video analysis of situated practices in various domains of knowledge and expertise, science and technology. In the early 1960s, the technical availability of audio recordings constituted an apt opportunity to capture actual courses of spoken activities *in vivo* and to make them available to a sociological analysis of their orderly production in situ:

• How do participants talk so that they understand each other?
• How do they take turns at talk for that purpose?
• Which institutional, professional, and/or technical tasks may their turn-taking procedures facilitate, in and as particular situations of interaction—how and why?

These were some of the key research questions that CA, as well as its early extensions to video analysis, made possible to ask, and answer. Over the last five decades, CA has played a prominent role not only within the diversified field of EM (Maynard and Clayman 1991; Wieder 1999) but also and especially for video analytic approaches. Pioneered by Charles Goodwin (1981) in the United States and Christian Heath (1986) in the United Kingdom, video analysis has become a rapidly growing trend, developing largely along the so-called institutional talk program (Drew and Heritage 1992) and examining the particular, often task-specific uses of talk in various institutional settings (e.g., medical and legal settings). Heath, Hindmarsh, and Luff (2010), Knoblauch et al. (2006), and Mondada (2008) provide a technical introduction to video analysis of situated practices; here we take closer look at Lorenza Mondada's (2007) video analysis of surgery as an instructive example of CA's extensions to video analysis and its contribution to STS.

Entitled "Operating Together through Videoconference," Mondada's (2007) video analysis examines the visual instruction to a particular kind of surgery—video-supported laparoscopy (or so-called keyhole surgery)—in a particular setting—a videoconference displaying the surgical intervention to a wider audience (including both students and experts). The paper describes how this minimally invasive kind of surgery, supported by small cameras introduced into the patient's body, is achieved so as to prove maximally instructive, involving not only a local team of surgeons but co-involving the mentioned mixed audience. First, the paper examines how the surgical procedure, to succeed safely, relies upon its *ongoing technically mediated visualization*. Laparoscopy requires its practitioners to make their successive incisions on the basis of the images displayed on a video screen rather than by having the operative field within direct eyesight. Consequently, Mondada's initial interest is in the "video image as a constitutive dimension of the ongoing activities" and more specifically, how keyhole surgeons are engaged in the "making of a visual space of action" (ibid., 52). Second, the paper describes the *occasional explicitly verbalized demonstration* of the technically mediated procedure, not only among the co-operating surgeons in situ but also to the remote audience in the conference room. Here, the "making (and unmaking) of participation spaces" (ibid., 61–66) becomes of particular interest, especially regarding when, why, and how off-site medical experts come into play in the on-site surgical operation (e.g., when a surgical procedure has to be corrected).[4]

Why analyze such "telesurgery"? Where does the STS interest in its video analysis lie? In answer to this question, we may note that Mondada's analysis not only homes in on surgery's visual instruction, relying on words, gestures, video images, and surgical tools, but it also demonstrates how such instruction is "creating wider networks" (Mondada 2007, 51). In a nutshell, Mondada's study provides for the *situated constitution* of a manifest actor-network (or sociomaterial assemblage), something that actor-network theory (ANT) arguably fails to do, at least as far as the audiovisual specifics of interaction are concerned. In Mondada's study, in turn, the "lived workings of the network" (Dant 2004, 81) were to be found in situ, in the tricky exposition of keyhole surgery to a remote yet co-involved and at times collaborating audience (which thus extended the working situation itself). For the "video image [to become] the major link between the various participants" (Mondada 2007, 53)—their "obligatory point of passage" (Latour 1987)—a "common space of action" had to be created among them. In examining surgeons' instructional "procedures for accomplishing a common space of action" (Mondada 2007, 51), the video analysis highlights the situated production of a "participation space" as the *"conditio sine qua non"* for networked collaboration (ibid., 61; emphasis in original). The detailed description of the mutually elaborating sense of verbal, bodily, and material aspects of the unfolding interaction (see also Deppermann 2013) turns out to be of STS interest, insofar as it demonstrates how a wider actor-network is progressively, yet locally, constituted. In that demonstration, Mondada's video analysis offers an elegant respecification of ANT, even though it was not intended as such. Let us now turn to explicit attempts at respecification.

Respecification: Laboratory Studies and Video Analysis in STS

Respecification has constituted an explicit move in EM to bring to bear its empirical insights on STS concerns. Its paradigmatic significance for STS was exposed above in terms of how detailed attention to practical relevancies can contribute to conceptual clarifications, if not theoretical "lessons," inside STS (cf. Lynch 1993). The turn to practice, in this vein of EM and ethnomethodological video analysis in particular, is not due to an empirical interest in interactional order per se (as in the case of CA-based video analysis), but rather from the outset is linked with an interest in conceptual matters. Indeed, conceptual interest has led to recent EM respecifications of central concerns in STS, sociology, and social theory more broadly, including materiality, the mind/body nexus, and ontological politics (Mol 1999). Before turning to exemplary studies in this vein, it is worthwhile to revisit briefly the constructivist tradition in laboratory studies (Doing 2008; Knorr Cetina 1995) and EM's contribution to its development, discussion, and respecification.

Science Observed (Knorr Cetina and Mulkay 1983), *Laboratory Life* (Latour and Woolgar 1979), *The Manufacture of Knowledge* (Knorr Cetina 1981)—these remain some of the seminal, now familiar titles associated with the ethnographic move into laboratories to study scientists at work and their practical "construction" of scientific facts. In contrast to the early "constructivist" lab studies influenced by the sociology of scientific knowledge (SSK), EM studies of laboratory work did not pursue an explanatory and critical agenda by seeking social or sociotechnical explanations of seemingly "messy" research against realist and rationalist pictures of "pure" science (Friedman 1998). Instead, they were devoted to the observational and descriptive tasks of capturing research on the fly, as it proved intelligible to its participants, in its ordinary pursuit and in the (partly) technical terms of their trade. On this basis, EM studies invited reflection on what could be learned from such description for STS. Two examples of this two-step approach may be given.

To begin with, the so-called pulsar paper published by Garfinkel and his colleagues (1981) provides an instructive case. This paper drew upon an audio recording of astronomers at work to describe how, as part of their observational work, the astronomical discovery of a (new) optical pulsar is progressively achieved and exhibited. The EM/STS crux of the paper was to demonstrate how the epistemic status of this cultural object, as an "astronomical discovery" in participants' own terms, hinges upon the embodied practices of its local production, including the running commentaries on the eventually emerging object (whose fallible and uncertain character may *at times* itself become a topic of conversation; ibid.). Lynch's (1985) *Art and Artifact* provides a similarly instructive case. Again, his key interest was not to document a new picture of "science as construction," but instead to describe how researchers themselves document (say) the "real," "constructed," or "artifactual" character of their phenomena. After a quasi-historical account of neurobiological objects (cast as an "archeology"), Lynch's study deals with researchers' "shop talk," and notably their ways of (tentatively) "achieving agreement" among themselves, through such talk, on microscopic slides of neurobiological relevance (ibid., see Part II). Lynch's study questions conventional sociological notions of common culture as a shared basis *for* situated interaction, describing instead how that very basis is tentatively achieved, and sometimes negotiated, *in* and *as* situated interaction. A pioneering EM respecification was the insightful result.

To further recover embodied, mutually available, visible specifics of doing science and enacting technology, the studies that followed added video analysis of situated practices. Those studies proved (and still prove) of strategic import. In line with EM's early interest (as outlined in the section "EM and STS" above), they made possible empirical demonstrations of how scientific practice(s) and technical activities (e.g., Heath 1986; Suchman 2000; Suchman and Trigg 1993) draw upon "embodied interaction" (Streeck, Goodwin, and LeBaron 2010), exchanged glances, and other *mundane* resources (e.g., Goodwin 1995; Lynch and Macbeth 1998; Nishizaka 2011; Rystedt et al. 2011; Sherman Heckler 2011). This strand of video analysis applies the core methods of CA (live recording, detailed transcription, and repeated inspection) to the examination of embodied, technical, and scientific action and interaction (e.g., Heath and Luff 2000), including the various uses of video itself (cf. Broth, Laurier, and Mondada 2014). In doing so, this strand of video analysis goes beyond the use of video in "straight" CA, which demonstrates how ordinary conversation is sustained by bodily expressions, including gestures, glances, and postures (Goodwin 1981; Sacks and Schegloff 2002). At the same time, it draws upon CA's "microscopic" power to examine seemingly mundane yet highly participant-relevant details in complex kinds of interaction (Luff and Heath 2010; Lynch 1987).

In this respect, Charles Goodwin's (1994) paper on "professional vision" occupies a special place. His paper is of STS interest insofar as it demonstrates, in audiovisual detail, *just how* (rather than *just that*) an intricate array of mundane resources, such as hints, timely glances, and pointing gestures, are involved in (and co-constitutive of) the core tasks of various, professional "communities of practice" (Wenger 1998). In particular, the paper demonstrates how elaborate and embodied "hints," in the sense of allusively displayed cues, are drawn upon (and partly spelled out) to conduct and teach a technical task in archeological field excavation. Similarly, the paper examines how such resources, in particular pointing gestures, were drawn upon in court to provide a documented interpretation of the video recording of the manifestly brutal beating of a black man, Rodney King, by several policemen in terms of their actually defensive, technically appropriate, professional intervention (which, we should add, was the defense expert's highly questionable reading). In so doing, the paper demonstrates how and which relevancies were highlighted in a situated practice (a particular court hearing), in and as its ordinary intelligibility, rather than as a practice-exogenous, abstract matter (recent shootings by U.S. police, some of which have been videotaped, bear witness to the continuing, social and political, relevance of the described case). Yet Goodwin's paper is also of more general STS interest, insofar as it analyzes an empirical case of interpretive flexibility (i.e., the possibility of various versions) in terms of an observed situated practice (similarly to Garfinkel, Lynch, and Livingston's 1981 paper), rather than glossing over the practical details constitutive of the intelligibility of the case, as a reconstructed historical account may be constrained to do (e.g., Pinch and Bijker 1984).[5]

The most recent video analytic studies in EM/STS draw upon EM (and in particular CA methods) to respecify topical conceptual concerns in STS (such as materiality, the mind/body nexus, and ontological politics). This new strand of research often combines ethnographic material with video recordings to make a distinctive contribution to STS, a contribution that describes technoscientific practice in its everyday pursuit. Morana Alač (2011), for example, describes how interacting participants in cognitive neuroscience, as they employ functional magnetic resonance imaging (fMRI) technology, use talk, gesture, gaze, prosody, facial expressions, body orientation, touch, material objects, and the dynamic aspects of space as constitutive parts of specific social scenes (similarly to Mondada 2007, discussed above in the section "Interaction"). On this basis, Alač also demonstrates how these situated actions and their manifest intelligence question dualist presuppositions of cognitive character, embedded not only in the fMRI brain imaging philosophy but also in the workings of its technological setup (as the latter, to make visible the "mind," has by and large to disembed its operations from bodily and interactional aspects, for example). By taking seriously the multimodal and multisensory aspects of practical encounters in fMRI to which the participants orient in interaction, she deals with such STS topics as the "body multiple" (Mol 2002), research artifacts in experimental data (Lynch 1985), and the numerical character of visual renderings (Beaulieu 2002). Each time, she traces these STS topics back to the methodical practices that give life to them in situ, in the technical surroundings of particular fMRI labs. Her detailed analysis thus affords the reader an EM respecification of conceptual interest, and that in the form of a video-enhanced lab study which questions dualist reasoning in cognitive neuroscience.

Analogously to Alač's book questioning dualist reasoning in cognitive neuroscience, Greiffenhagen and Sharrock's (2011) article "Does Mathematics Look Certain in the Front, but Fallible in the Back?" addresses the recurring contrast between doing and writing science and mathematics, a sociological dualism which has marked the field of STS (e.g., Schickore 2008). The paper is grounded in the authors' research on mathematicians' pedagogical practices, presents video recordings of graduate lectures and doctoral supervision meetings, and thereby offers a carefully constructed examination of the lived work of mathematics. By engaging the presumed contrast between how mathematics is practiced and how the results of those practices are reported, Greiffenhagen and Sharrock examine whether there is actually a concealed messiness in the practice of science and mathematics, and whether the certainties that enter the public discourse really aren't so certain. The argument is centered on Reuben Hersh's (1991) proposal which applies Goffman's (1956) dramaturgical metaphor of a "front" and "back" stage to mathematics. Greiffenhagen and Sharrock (2011) do not suggest that there aren't important differences between a front stage of public representations (through publications and lectures) and back-staged practices (how mathematics is really done). Instead, they aim at depolarizing the contrast and argue that, first, reporting on science and mathematics does not "misrepresent" how it is actually done and that, second, the two moments do not exhibit different standards of certainty or forms of reasoning. Sormani (2011) makes a similar case for depolarization on the basis of his video analysis of the local conduct of a physics experiment and the instant, yet qualified, appraisal of its manifest result (which, as it happened, anticipated its extended, yet similarly qualified, publication). In both studies, the video analysis of situated practice calls into question dualist modes of sociological reasoning.[6]

Practice: Self-Instruction and Video Analysis Alongside Mainstream STS

Practice is not only an empirical topic of video analysis but also a requirement for the (video) analyst to engage in the very activity that he or she sets out to study. This practical focus was one of Garfinkel's later leitmotifs (Garfinkel 1986, 2002; Lynch 1999). In his second major book, *Ethnomethodology's Program*, Garfinkel (2002) emphasizes repeatedly the self-instructive practice in any domain of technical activity that he would expect his students to engage in. In every case, the key point would be to make explicit the situated practice under scrutiny (be it jazz improvisation, legal reasoning, or mathematical demonstration) in terms of its "identifying details," requiring self-instruction by the analyst in that very practice (as these details, in their technical terms, could only be grasped, so Garfinkel's argument went, from within the practice's self-same conduct). In contrast to EM's prior interest in the mundane resources of conversational interaction and/or professional practice, this line of hybrid studies of work (ibid.) opened up a new avenue for EM investigation, now focused on the highly specialized character of an open range of technical activities, from alchemy (Eglin 1986) and truck tire repair (Baccus 1986) to origami (Livingston 2008) and coffee tasting (Liberman 2014). Scientific practices, relating to mathematics, medicine, or physics, are thus recast as one technical domain among others, without allocating it any special place in or for EM (or STS, for that matter).[7]

Before we turn to the STS interest of this hybrid practice of EM "all the way down" (Lynch 2015), we shall spell out one of its basic tricks, as well as its methodological implications for video analysis. Perhaps its most basic trick, to begin with, is to exploit "technical idiocy" rather than practical competency or linguistic fluency as a resource for descriptive analysis (see, e.g., Garfinkel 2002, 197–218; Lynch 2011b, 932–34; Lynch, Livingston, and Garfinkel 1983, 223–24). The leading idea is to engage oneself, as a novice practitioner, in the technical domain under scrutiny (surgery, mathematics, physics, or any other tricky domain). In turn, the likely difficulties of tentative self-instruction, including social, technical, and moral matters alike, provide unique insight into the practice's identifying details and constitutive procedures (its ethnomethods)—in short, an instructive "opportunity to see what experts frequently 'assume' and 'take for granted' as the most ordinary, pervasive, and characteristic features of their work practices" (Livingston 2008, 132). One way, then, to draw out the methodological implications of hybrid EM for video analysis is to design the latter as an explicitly practice-based mode of analysis, such that it makes explicit the analyst's own typically troubled practice in the activity that his or her video analysis will eventually examine. In his recent book, Sormani (2014a) develops such a hybrid form of video analysis, partly based on his own practice in a current domain of experimental physics, to tease out some of this domain's constitutive procedures. For this purpose, he follows a three-step procedure:

• First, video recordings of entitled practitioners at work (e.g., experimental physicists) provide a starting point.
• Second, the filmed practices and observed tasks are tentatively reenacted, with or without the technical assistance by expert practitioners.
• Third, this reenactment, given the difficulties it presents to the untrained analyst, provides technical insight into the competent achievement of the examined work practice.

One phenomenon that is often taken for granted, or at least counted upon, among competent practitioners, be it in experimental physics or other domains, is the achieved correspondence between technical tasks and technical argot, where the latter provides for the verbal identification of the former (as a "scanning tunneling spectroscopy" in physics, for example). In turn, the descriptive purpose of technical self-instruction, and practice-based video analysis in particular, is to make explicit how this correspondence is recognizably achieved, in and as a situated practice. The difficulties encountered in doing so provide as many occasions to figure out—or be taught—how it should be done (Garfinkel 2002, 264, note 2). Video analysis, as in the case of the outlined three-step procedure, may then become part and parcel of a "practice-immanent" pedagogy as much as the empirical basis for engaging STS concerns. The major STS interest of this hybrid mode of video analysis is thus "experiential" and heuristic rather than conceptual or interactional, as in the case of the two prior modes of ethnomethodological video analysis.[8]

Ethnomethodological Video Analysis and STS: Multiple Relationships

This third section of the chapter discusses and reflects upon the multiple relationships between ethnomethodological video analysis and STS, two fields marked by fuzzy fringes as well as internal diversity, as we have just seen. First, a reminder of the common ground of the presented modes of video analysis shall be offered: their empirical interest in the distinctively situated production of sociomaterial orders, anticipating the practice turn in STS and its similar focus on the situation (at least in Léna Soler and her colleagues' [2014b] understanding, quoted in the introduction). Second, this common ground shall allow us to clarify and reflect further upon the presented modes of video analysis and their respective contributions to STS. The reciprocally instructive character of these modes of video analysis—emphasizing interactional, conceptual, or practical matters—should then allow the interested STS scholar to chart his or her own course in EM-inspired video analysis.

Turning to Practices in Situ: Ethnomethodological Common Ground

EM's key notions of indexicality, reflexivity, and accountability, as introduced by Garfinkel (1967, 1–13), may help us to delineate what the three approaches presented in this chapter have in common. First of all, this modest conceptual equipment emphasizes the irremediably situated character of social action, be it the conduct of a laboratory experiment or of a dinner conversation. This understanding of indexicality has certainly inspired relativist and constructivist critiques of science (in denouncing the assumed inability of science to live up to its objective or trans-situational ambitions; e.g., Knorr Cetina 1981). However, indexicality's current relevance is to be found in the opposite direction: it is because indexical expressions (expressions such as "you," "I," "dinner at 8:30," or "yes!") can be reflexively related to familiar settings, types, rules, norms, and so forth, that we do not—or only exceptionally—encounter the world as underdetermined, experienced as one or multiple, whether in the sciences (e.g., Sormani 2011) or other domains (e.g., Mol 2002). Indexicality and reflexivity (in the sense of what elaborates the situation and is being elaborated by it) combine to ascribe a specific yet communicable meaning to settings and actions. This irremediably indexical and reflexively accountable character of sociomaterial practices, in turn, can be taken as a permanent invitation to observe and describe how scientists or laypersons deal with the unavoidable contingencies and specificities of each situation (see Goffman 1964 and the more recent Quéré 1998). This invitation has been taken seriously by the three video analytic approaches that were presented above. Let us take a second, yet closer look at how.

CA has from the very start devoted much attention to the indexical character of interaction (Sacks 1992a, 1992b). Transcripts of actual talk, for example, were expected to account for everything that was said "just how it was said" and not adapted, corrected, or simplified, according to some external criteria. The underlying idea of transcribing the recorded talk-in-interaction consists of treating it as a constitutive part of the situation, which in turn would be lost in any instance of correction, adaptation, or simplification. Somewhat paradoxically, this injunction frequently resulted in documenting silent courses of action, indicating thus that they "go without saying" for their participants. As Sacks noted, the observation that many aspects of social life are not spelled out should invite us to ask what a sociologist is doing when he or she characterizes (or elicits informants' formulations on) a silent situation (Sacks 1992a, 515–22). That a course of action goes without saying implies in no way that nothing happens (see also note 5). This action without talk probably makes the recourse to video the most relevant. By offering repeatable access to embodied action and interaction, as well as their spatial arrangement and sociomaterial environments, video recordings provide resources for resisting the social scientific tendency to decontextualize data. Whether words are exchanged or not, the verbal component is only a fragment of a meaningful situation, a situation that involves gestures, moves, gazes, spatial arrangements, and so on. In other words, video allows CA to "save the phenomena" instead of ignoring them or reducing them to preexisting analytical frameworks.

This indexical requirement also plays a central role in the respecification enterprise (e.g., Button 1991; Garfinkel 1991; Lynch 1993). Again, respecification does not begin with contesting, criticizing, or taking an ironic stance on social scientific concepts. Through the notion of "perspicuous setting," it proposes instead, as we have seen, to reexamine these concepts as concrete stakes that are addressed by practitioners in specific circumstances. In this perspective, video recordings provide a decisive resource for the re-contextualization of scientific or other concepts, thereby offering a fuller account of the language games and forms of life which make them intelligible to practitioners at work and, in a second step, to the video analyst of their unfolding practice. The explicitly practice-based approach in video analysis (e.g., Sormani 2014a, 2014b) can be said to "radicalize" this indexical injunction. Reenacting a given practice, in this case, allows one to gain technical access to its constitutive tasks, from within their tentative accomplishment in situ, an accomplishment that reflexively constitutes the working situation of which these tasks are part. This move involves "coming upon the phenomenon via the work ... of producing it" (Garfinkel 1996, 11), inviting "situated learning" (Lave and Wenger 1991) across any preestablished expert/novice dichotomy. In this vein, David Sudnow (1978) described his "ways of the hand" and autodidactic tribulations at the jazz piano, from one training session to another, having him and his fingers progressively "going for the jazz" (as Sormani [2014a] aims at "going for the physics").

The discussed indexical injunction—EM's central concern of turning to practices in situ—is shared by the three considered approaches in video analysis. As we have suggested, they have all made possible the detailed investigation of practical relevancies as pursued and displayed by interacting participants (Hester and Francis 2007). What for? The final section of this chapter deepens the reflection on the STS interest of the respective approaches.

Practicing Diversity: Distinctive Contributions to STS

The distinctive contribution of CA-inspired video analysis to STS may be briefly reflected upon in the light of the value and cost of CA's autonomy. For more than forty years, CA—which constitutes EM's interactionist approach to talk as a social phenomenon—has explored the unexpected complexity of talk-in-interaction. During the last two decades, the increasing recourse to video recordings has considerably enriched CA's observational sensibility. In this process, CA has progressively come to define specific procedures, concepts, and criteria, thereby achieving the status of an autonomous discipline and, to some extent, contributing to a cumulative "normal science." This unquestionable success is attested by a large corpus of systematic studies on a variety of phenomena of talk-in-interaction, including "repair" (the interactional organization of correction in conversation), "assessment" (the conversational practices by which evaluations are produced and received, agreed or disagreed with), and "participation" (the practices by which a variety of participatory statuses are dynamically established and possibly contested, based on such resources as age, gender, entitlement, and expertise). These interactional phenomena and the CA-inspired video analysis that have addressed them (e.g., Greiffenhagen and Watson 2009) provide a valuable resource for STS scholars willing to examine interaction taking place in (if not *as*) science- and technology-saturated settings; for an early example of continuing interest, see Suchman ([1987] 2007).

This invitation comes, however, with a caution. As previously mentioned, CA has gained its autonomy through the elaboration of a disciplinary apparatus that may constitute an important threshold for the newcomer, just as in any other discipline. Undoubtedly, CA has reached a level of technicality that is infrequent in social sciences. For example, the transcripts used for demonstrative purposes are at times difficult to read for the unacquainted reader, and, for all that video brought to CA, it often did not make the transcripts easier to read (at the exception of Goodwin's work, and, we hope, most of our own). Skipping transcripts, however, means missing out on the conversational phenomenon (respectively its textual rendition in transcribed form) on which any CA, video-enhanced or not, should be based. Another consequence of CA's disciplinary status lies in the establishment of an autonomous agenda, which may be remote from STS's primordial concerns. Initial interests in STS may have to be translated. While such a translation exercise is far from uncommon in STS, perhaps even constitutive of it, opportunity costs for engaging in CA (or drawing upon CA-inspired video analysis) may be judged too high, if the specific focus is not on situated interaction.

As we have seen, ethnomethodological respecification, even if it draws upon CA or CA-inspired video analysis, provides a way of *engaging with STS's topical agenda*, thus lowering said opportunity costs of "straight" CA. This enterprise consists in reconsidering concepts in the very settings that make them relevant and consequential to their participants. In this perspective, any STS concept can be submitted to a respecification process, a process through which its indexical use and reflexive determination will be meticulously observed and described. In a second step, the relevant literature in STS will be reconsidered in the light of that description. Construed as a complementary and sometimes critical endeavor, respecification can sustain a fruitful dialogue with STS, which in turn, due to its conceptual inventiveness, will provide respecification with an inexhaustible list of research topics.[9] In this perspective, the recourse to video recordings can only make STS scholars more alert to the situated, embodied, and sociomaterial character of the forms of life that make their concepts intelligible in the first place, if only to have them questioned upon a second look. However, the use of detailed video data for respecification purposes also runs the already mentioned risk of gaining in technicality what it loses in readability to nonspecialists. Whenever video-based respecification is taken on board by other fields such as STS, it must find a balance between EM's empirical rigor and these other fields' topical agenda. While the cost of empirical rigor might be loss of the addressed interlocutor, the cost of an exclusive focus on the interlocutor's concepts is the loss of EM's autonomous agenda. Additionally, there is a risk of "theoricism" in being confined to conceptual issues; this issue has been recently discussed by Hutchinson, Rupert, and Sharrock (2008), Livingston (2008), Lynch (2015), and Sormani (2015).

Faced with the challenge of theoricism, practice-based video analysis offers a distinctive alternative. Following Garfinkel's injunction to *engage oneself in technical practice* rather than to solely observe it, this approach examines (and reexamines) sociomaterial practices in situ not through scholarly concepts found in STS literature but in terms of their indexical properties as situated accomplishments. The video analytic interest, then, is not so much in what these practices may teach one on concepts of interest to STS (including fundamental concerns, regarding "materiality," for example) but rather in the tacit procedures which are drawn upon by participants to ensure the recognizable production of their practices, as ordinarily expressed in participants' own terms. As it requires self-instructive engagement, practice-based video analysis potentially recasts any logo- and visio-centric approach to interaction. It minimizes thus the risk of ascribing excessive weight to language and vision, at the expense of the configuration process by which a situated practice achieves a specific assemblage of language, vision, and many other of its constitutive aspects. Technical practices, rather than technical transcripts, are the core methodology of this approach; STS scholars are thus invited to grapple with the former rather than the latter.

The autodidactic attention to detail, enhanced by audiovisual means, is certainly attuned to Garfinkel's hybrid understanding of EM. Yet it also prevents any cumulative ambition and thus can hardly be expected to engender an autonomous discipline with its own procedures, concepts, and criteria, as in the case of CA-based video analysis. Practice-based video analysis resolutely assumes a form of idiosyncrasy that runs the risk of confining its audience to ethnomethodologists and/or STS scholars specifically interested in the examined practice. Arguably, this situation is akin to Ludwig Wittgenstein's (1953) invitation to study language games for what they are, if not for their own sake, instead of indulging in general theorizing about language for other purposes. Yet, and with such a precedent, the apparently inevitable "so what?" question may not be the end of the argument.

Conclusion

Because of EM's diversity, it has become almost as difficult to offer a synoptic overview of it as of STS (see, though, Lynch 2011a, 2011b; Psathas 2008; Wieder 1999). This chapter presented three distinctive articulations of ethnomethodological video analysis and STS, all of which contribute to renew the latter's "practice turn" in audiovisual specifics. *Interaction, respecification*, and *practice* provided the keywords to present the respective approaches. These approaches should be considered reciprocally instructive rather than mutually exclusive. Our subsequent clarification by contrast (Taylor 1985) spelled out and reflected upon their multiple relationships with (and within) STS, as well as their reciprocally instructive character. STS uptake has been varied, ranging from more applied versions of video analysis, either focusing upon interaction or conceptually relevant matters, to practice-based inquiry alongside STS's mainstream. Various configurations of "straight," "applied," or "hybrid" video analysis are thus to be observed, which—and this is important to note—in actual practice are often combined, reconfigured, and refined to suit the case at hand. We leave it up to the reader to engage in further, renewed, and exciting video analytic work.

Acknowledgments

We wish to acknowledge the comments, criticisms, and observations on previous drafts by Aug Nishizaka, the *Handbook* editors, as well as three anonymous reviewers. They all helped to improve the present chapter.

Notes

1. "Ethnomethodology, Garfinkel explains, was developed as a cognate with a set of perfectly standard anthropological terms—*ethnobotany*, *ethnomedecine*, etc. Just as *botany* in *ethnobotany* refers to a corpus to be treated as *data*, so does *methodology* in *ethnomethodology* stand for a subject matter rather than a scientific apparatus" (Turner 1974, 12). This longer quote anticipates later developments both in STS—its "laboratory studies" tradition (e.g., Knorr Cetina 1995)—and EM—its "hybrid studies of (scientific) work" (see the section "Ethnomethodological Video Analysis" in this chapter).

2. Not unlike STS, EM's development has been accompanied by edited collections and regular overviews of the field, including Button (1991), Heritage and Maynard (forthcoming), Lynch and Sharrock (2003, 2011), Maynard and Clayman (1991), Sharrock and Anderson (1987), Wieder (1999), and Zimmerman and Pollner (1973).

3. With respect to the ethnographic tradition in symbolic interactionism, "the issue is not simply one of detail, or as computer scientists sometimes say 'granularity' but rather that the emergent, practical, and contingent accomplishment of work and occupational life disappears from view— from analytic consideration—in these fine ethnographies. Social interaction is placed at the heart of the analytic agenda and yet the very concepts which pervade certain forms of ethnographic research, concepts such as 'negotiation,' 'bargaining,' 'career,' 'shared understanding,' 'trajectory,' even 'interpretive framework,' gloss the very phenomena that they are designed to reveal" (Heath and Hindmarsh 2002, 101).

4. Her own video data, as Mondada explains, "have to deal with various complex and intertwined activities, such as demonstrating and operating, where different kinds of participants are involved. We can therefore ask how those groups of participants, those corporate entities, incipient teams or sides are constituted through and for these particular actions" (Mondada 2007, 61).

5. On the key distinction between "practical activities" and their "discursive formulation," see Garfinkel and Sacks (1970). In their view, not only do discursive formulations (including historical accounts) trade upon the intelligible achievement of practical activities, but the latter can also be turned into an analyzable phenomenon of its own, such as the embodied methods of jazz improvisation (Sudnow 1978) or the technical practices typically closer to STS (see the section "Practice" below).

6. For recent video analytic work in related technical, scientific, and medical domains, see Greiffenhagen (2014), Koschmann and Zemel (2009), Lindwall (2008, 2014), and Lindwall and Lymer (2008). For a video-based respecification of "nonhuman agency," see Sormani, Bovet, and Strebel (2015). For a recent EM-inspired discussion of "ontological politics" in STS, see Lezaun and Woolgar (2013).

7. This line of EM research has led to "hybrid" studies of work that draw upon the technical skills gained by the analyst him- or herself (e.g., in experimental physics) in addition to his or her social science training (Bjelic 1996; Livingston 2008). Technical self-instruction would be cast by

the later Garfinkel as a "unique adequacy requirement of methods" for doing EM, see Garfinkel (2002, 175–76).

8. Hybrid video analysis treats audiovisual recordings not only as documents of the filmed scene but also as constitutive parts of it, parts that are reflexively related to the scene's endogenous organization. This reflexive potential of audiovisual recordings has been analyzed in two seminal papers: Macbeth (1999) and Relieu (1999), and in a more recent paper by Macbeth (2014).

9. The fruitfulness of such a dialogue is amply evidenced by the important position that Michael Lynch has come to occupy in STS, as both an author and an editor, as well as the momentum gained by video analysis in view of respecification.

References

Alač, Morana. 2011. *Handling Digital Brains: A Laboratory Study of Multimodal Semiotic Interaction in the Age of Computers*. Cambridge, MA: MIT Press.

Baccus, Melinda D. 1986. "Multipiece Truck Wheel Accidents and Their Regulations." In *Ethnomethodological Studies of Work*, edited by Harold Garfinkel, 20–59. London: Routledge and Kegan Paul.

Beaulieu, Anne. 2002. "Images Are Not the (Only) Truth: Brain Mapping, Visual Knowledge, and Iconoclasm." *Science, Technology, & Human Values* 27: 53–86.

Benson, Douglas, and John Hughes. 1991. "Method: Evidence and Inference—Evidence and Inference for Ethnomethodology." In *Ethnomethodology and the Human Sciences*, edited by Graham Button, 109–36. Cambridge: Cambridge University Press.

Bjelic, Dusan. 1996. "Lebenswelt Structures of Galilean Physics: The Case of Galileo's Pendulum." *Human Studies* 19 (4): 409–32.

Broth, Mathias, Eric Laurier, and Lorenza Mondada, eds. 2014. *Studies of Video Practices: Video at Work*. New York: Routledge.

Büscher, Monika. 2005. "Social Life under the Microscope?" *Social Research Online* 10 (1). Accessed at http://socresonline.org.uk/10/1/buscher.html.

Button, Graham, ed. 1991. *Ethnomethodology and the Human Sciences*. Cambridge: Cambridge University Press.

Callon, Michel, and Bruno Latour. 1981. "Unscrewing the Big Leviathan: How Actors Macro-Structure Reality and How Sociologists Help Them to Do So." In *Advances in Social Theory and Methodology: Toward an Integration of Micro- and Macro-Sociologies*, edited by Karin Knorr Cetina and Aaron V. Cicourel, 277–303. Boston: Routledge and Kegan Paul.

Dant, Tim. 2004. *Materiality and Society*. Maidenhead: Open University Press.

Deppermann, Arnulf. 2013. "Multimodal Interaction from a Conversation Analytic Perspective." *Journal of Pragmatics* 46 (1): 1–7.

Doing, Park. 2008. "Give Me a Laboratory and I Will Raise a Discipline: The Past, Present, and Future Politics of Laboratory Studies in STS." In *The Handbook of Science and Technology Studies*, 3rd ed., edited by Edward J. Hackett, Olga Amsterdamska, Michael Lynch, and Judy Wajcman, 279–318. Cambridge, MA: MIT Press.

Drew, Paul, and John Heritage, eds. 1992. *Talk at Work: Interaction in Institutional Settings*. Cambridge: Cambridge University Press.

Durkheim, Emile. [1895] 1964. *The Rules of Sociological Method*. New York: Free Press of Glencoe.

Eglin, Trent. 1986. "Introduction to a Hermeneutics of the Occult: Alchemy." In *Ethnomethodological Studies of Work*, edited by Harold Garfinkel, 121–56. London: Routledge and Kegan Paul.

Friedman, Michael. 1998. "On the Sociology of Scientific Knowledge and Its Philosophical Agenda." *Studies in History and Philosophy of Science* 29A (2): 239–71.

Garfinkel, Harold. 1967. *Studies in Ethnomethodology*. Englewood Cliffs, NJ: Prentice-Hall.

___. 1974. "The Origins of the Term Ethnomethodology." In *Ethnomethodology: Selected Readings*, edited by Roy Turner, 15–18. Harmondsworth: Penguin.

___, ed. 1986. *Ethnomethodological Studies of Work*. London: Routledge.

___. 1991. "Respecification: Evidence for Locally Produced, Naturally Accountable Phenomena of Order, Logic, Reason, Meaning, Method, etc. in and as of the Essential Haecceity of Immortal Ordinary Society (I)—an Announcement of Studies." In *Ethnomethodology and the Human Sciences*, edited by Graham Button, 10–19. Cambridge: Cambridge University Press.

___. 1996. "Ethnomethodology's Program." *Social Psychology Quarterly* 59 (1): 5–21.

___. 2002. *Ethnomethodology's Program: Working Out Durkheim's Aphorism*. Lanham, MD: Rowman and Littlefield.

Garfinkel, Harold, Michael Lynch, and Eric Livingston. 1981. "The Work of a Discovering Science Construed with Materials from the Optically Discovered Pulsar." *Philosophy of the Social Sciences* 11 (2): 131–58.

Garfinkel, Harold, and Harvey Sacks. 1970. "On Formal Structures of Practical Actions." In *Theoretical Sociology: Perspectives and Developments*, edited by John C. McKinney and Edward A. Tiryakian, 338–66. New York: Appleton-Century-Crofts.

Goffman, Erving. 1956. *The Presentation of Self in Everyday Life*. Edinburgh: University of Edinburgh, Social Sciences Research Centre.

___. 1964. "The Neglected Situation." *American Anthropologist* 66 (6): 133–36.

Goodwin, Charles. 1981. *Conversational Organization: Interaction between Speakers and Hearers*. New York: Academic Press.

___. 1994. "Professional Vision." *American Anthropologist* 96 (3): 606–33.

___. 1995. "Seeing in Depth." *Social Studies of Science* 25 (2): 237–74.

Greiffenhagen, Christian. 2014. "The Materiality of Mathematics." *British Journal of Sociology* 65 (3): 502–28.

Greiffenhagen, Christian, and Wes Sharrock. 2011. "Does Mathematics Look Certain in the Front, but Fallible in the Back?" *Social Studies of Science* 41 (6): 839–66.

Greiffenhagen, Christian, and Rod Watson. 2009. "Visual Repairables: Analysing the Work of Repair in Human-Computer Interaction." *Visual Communication* 8 (1): 65–90.

Heath, Christian. 1986. *Body Movement and Speech in Medical Interaction.* Cambridge: Cambridge University Press.

Heath, Christian, and Jon Hindmarsh. 2002. "Analysing Interaction: Video, Ethnography and Situated Conduct." In *Qualitative Research in Action*, edited by Tim May, 99–121. London: Sage.

Heath, Christian, Jon Hindmarsh, and Paul Luff. 2010. *Video in Qualitative Research: Analysing Social Interaction in Everyday Life.* London: Sage.

Heath, Christian, and Paul Luff. 2000. *Technology in Action.* Cambridge: Cambridge University Press.

Heritage, John, and Douglas Maynard, eds. Forthcoming. *Harold Garfinkel: Praxis, Social Order, and the Ethnomethodology Movement.* Oxford: Oxford University Press.

Hersh, Ruben. 1991. "Mathematics Has a Front and a Back." *Synthese* 88 (2): 127–33.

Hester, Stephen, and David Francis. 2007. "Analysing Orders of Ordinary Action." In *Orders of Ordinary Action: Respecifying Sociological Knowledge*, edited by Stephan Hester and David Francis, 3–12. Aldershot: Ashgate.

Hutchinson, Phil, Read Rupert, and Wes Sharrock. 2008. "Seeing Things for Themselves: Winch, Ethnography, Ethnomethodology and Social Studies." In *There Is No Such Thing as a Social Science: In Defence of Peter Winch*, 91–112. Aldershot: Ashgate.

Knoblauch, Hubert, Bernd Schnettler, Juergen Raab, and Hans-Georg Soeffner, eds. 2006. *Video Analysis: Methodology and Methods. Qualitative Audiovisual Data Analysis in Sociology.* Frankfurt am Main: Peter Lang.

Knorr Cetina, Karin. 1981. *The Manufacture of Knowledge: An Essay on the Constructivist and Contextual Nature of Science.* Oxford: Pergamon Press.

___. 1995. "Laboratory Studies: The Cultural Approach to the Study of Science." In *Handbook of Science and Technology Studies*, 2nd ed., edited by Sheila Jasanoff, Gerald E. Markl, James C. Peterson, and Trevor Pinch, 140–66. Thousand Oaks, CA: Sage.

Knorr Cetina, Karin, and Michael Mulkay, eds. 1983. *Science Observed: Perspectives on the Social Study of Science.* London: Sage

Koschmann, Timothy, and Alan Zemel. 2009. "Optical Pulsars and Black Arrows: Discoveries as Occasioned Productions." *Journal of the Learning Sciences* 18 (2): 200–46.

Latour, Bruno. 1987. *Science in Action: How to Follow Scientists and Engineers through Society*. Cambridge, MA: Harvard University Press.

___. 2005. *Reassembling the Social: An Introduction to Actor-Network Theory*. Oxford: Oxford University Press.

Latour, Bruno, and Steve Woolgar. 1979. *Laboratory Life: The Social Construction of Scientific Facts*. London: Sage.

Lave, Jean, and Étienne Wenger. 1991. *Situated Learning: Legitimate Peripheral Participation*. Cambridge: Cambridge University Press.

Law, John, and Annemarie Mol. 1995. "Notes on Materiality and Sociality." *The Sociological Review* 43 (2): 274–94.

Lezaun, Javier, and Steve Woolgar. 2013. "The Wrong Bin Bag: A Turn to Ontology in Science and Technology Studies?" *Social Studies of Science* 43 (3): 321–40.

Liberman, Ken. 2014. "The Phenomenology of Coffee Tasting: Lessons in Practical Objectivity." In *More Studies in Ethnomethodology*, 215–66. Albany: State University of New York Press.

Lindwall, Oskar. 2008. *Lab Work in Science Education: Instruction, Inscription, and the Practical Achievement of Understanding*. University of Linköping, Sweden: Studies in Arts and Science.

___. 2014. "The Body in Medical Work and Medical Practice: An Introduction." *Discourse Studies* 16 (2): 125–29.

Lindwall, Oskar, and Gustav Lymer. 2008. "The Dark Matter of Lab Work: Illuminating the Negotiation of Disciplined Perception in Mechanics." *Journal of the Learning Sciences* 17 (2): 180–224.

Livingston, Eric. 1986. *The Ethnomethodological Foundations of Mathematics*. London: Routledge and Kegan Paul.

___. 1987. *Making Sense of Ethnomethodology*. London: Routledge and Kegan Paul.

___. 2008. *Ethnographies of Reason*. Aldershot: Ashgate.

Luff, Paul, and Christian Heath. 2010. "Technologies in Collaboration: Video Analysis of Complex Activities and Material Artefacts." Paper presented at the EASST conference, Trento (I), 4 September.

Lynch, Michael. 1985. *Art and Artifact in Laboratory Science: A Study of Shop Work and Shop Talk in a Research Laboratory*. Boston: Routledge and Kegan Paul.

___. 1987. "Ethnométhodologie et pratique scientifique: La pertinence du detail." *Cahiers de Recherches Sociologiques* 5 (2): 47–64.

___. 1993. *Scientific Practice and Ordinary Action: Ethnomethodology and Social Studies of Science*. Cambridge: Cambridge University Press.

___. 1999. "Silence in Context: Ethnomethodology and Social Theory." *Human Studies* 22 (2/4): 211–33.

___. 2011a. "Ad Hoc Special Section on Ethnomethodological Studies of Science, Mathematics, and Technical Activity: Introduction." *Social Studies of Science* 41 (6): 835–37.

___. 2011b. "Harold Garfinkel (29 October 1917—21 April 2011): A Remembrance and Reminder." *Social Studies of Science* 41 (6): 927–42.

___. 2015. "Garfinkel's Legacy: Bastards All the Way Down." *Contemporary Sociology* 44 (5): 604–14.

Lynch, Michael, Eric Livingston, and Harold Garfinkel. 1983. "Temporal Order in Laboratory Work." In *Science Observed: Perspectives on the Social Study of Science*, edited by Karin Knorr Cetina and Michael Mulkay, 205–38. London: Sage.

Lynch, Michael, and Douglas Macbeth. 1998. "Demonstrating Physics Lessons." In *Thinking Practices in Mathematics and Science Learning*, edited by James Greeno and Shelley Goldman, 269–97. Mahwah, NJ: Lawrence Erlbaum Associates.

Lynch, Michael, and Wes Sharrock, eds. 2003. *Harold Garfinkel* [Four-volume set: SAGE Masters in Modern Social Thought series]. London: Sage.

___, eds. 2011. *Ethnomethodology* [Four-volume set: SAGE Benchmarks in Social Research Methods]. London: Sage.

Macbeth, Douglas. 1999. "Glances, Trances, and Their Relevance for a Visual Sociology." In *Media Studies: Ethnomethodological Approaches*, edited by Paul L. Jalbert, 135–70. Lanham, MD: University Press of America.

___. 2014. "Studies of Work, Instructed Action, and the Promise of Granularity: A Commentary." *Discourse Studies* 16 (2): 295–308.

Maynard, Douglas, and Steven Clayman. 1991. "The Diversity of Ethnomethodology." *Annual Review of Sociology* 17: 385–418.

Mol, Annemarie. 1999. "Ontological Politics: A Word and Some Questions." In *Actor Network Theory and After*, edited by John Law and John Hassard, 74–89. Oxford and Keele: Blackwell and the Sociological Review.

___. 2002. *The Body Multiple: Ontology in Medical Practice*. Durham, NC: Duke University Press.

Mondada, Lorenza. 2007. "Operating Together through Videoconference: Members' Procedures for Accomplishing a Common Space of Action." In *Orders of Ordinary Action: Respecifying Sociological Knowledge*, edited by Stephen Hester and David Francis, 51–68. Aldershot: Ashgate.

___. 2008. "Using Video for a Sequential and Multimodal Analysis of Social Interaction: Videotaping Institutional Telephone Calls." *Forum Qualitative Sozialforschung / Forum: Qualitative Social Research* 9 (3), accessed at http://www.qualitative-research.net/index.php/fqs/article/view/1161/2566.

Nishizaka, Aug. 2011. "The Embodied Organization of a Real-time Fetus: The Visible and the Invisible in Prenatal Ultrasound Examinations." *Social Studies of Science* 41 (3): 309–36.

Parsons, Talcott. 1949. *The Structure of Social Action: A Study in Social Theory with Special Reference to a Group of Recent European Writers*. New York: Free Press.

Pickering, Andrew, ed. 1992a. *Science as Practice and Culture*. Chicago: University of Chicago Press.

___. 1992b. "From Science as Knowledge to Science as Practice." In *Science as Practice and Culture*, edited by Andrew Pickering, 1–16. Chicago: University of Chicago Press.

Pinch, Trevor, and Wiebe E. Bijker. 1984. "The Social Construction of Facts and Artefacts: Or How the Sociology of Science and the Sociology of Technology Might Benefit Each Other." *Social Studies of Science* 14 (3): 399–441.

Psathas, George. 2008. "Reflections on the History of Ethnomethodology: The Boston and Manchester 'Schools'." *The American Sociologist* 39: 38–67.

Quéré, Louis. 1998. "The Still-Neglected Situation?" *Réseaux* 6 (2): 223–53.

Relieu, Marc. 1999. "Du tableau statistique à l'image audiovisuelle: Lieux et pratiques de la représentation en sciences sociales." *Réseaux* 17 (94): 49–86.

Rose, Edward. 1960. "The English Record of a Natural Sociology." *American Sociological Review* 25 (2): 193–208.

Rystedt, Hans, Jonas Ivarsson, Sara Asplund, Åsa Johnsson, and Magnus Båth. 2011. "Rediscovering Radiology: New Technologies and Remedial Action at the Worksite." *Social Studies of Science* 41 (6): 867–91.

Sacks, Harvey. 1963. "Sociological Description." *Berkeley Journal of Sociology* 8: 1–16.

___. 1984. "Notes on Methodology." *Structures of Social Action: Studies in Conversation Analysis*, edited by J. Maxwell Atkinson and John Heritage, 21–27. Cambridge: Cambridge University Press.

___. 1992a. *Lectures on Conversation*, vol. 1. London: Blackwell.

___. 1992b. *Lectures on Conversation*, vol. 2. London: Blackwell.

Sacks, Harvey, and Emanuel A. Schegloff. 2002. "Home Position." *Gesture* 2 (2): 133–46.

Sacks, Harvey, Emanuel Schegloff, and Gail Jefferson. 1974. "A Simplest Systematics for the Organization of Turn-Taking for Conversation." *Language* 50: 696–735.

Schatzki, Theodore, Karin Knorr Cetina, and Eike von Savigny, eds. 2001. *The Practice Turn in Contemporary Theory*. London: Routledge.

Schegloff, Emanuel A., and Harvey Sacks. 1973. "Opening Up Closings." *Semiotica* 7: 289–327.

Schickore, Jutta. 2008. "Doing Science, Writing Science." *Philosophy of Science* 75 (3): 323–43.

Sharrock, Wes, and Robert J. Anderson. 1987. *The Ethnomethodologists*. Chichester: Ellis Horwood.

Sherman Heckler, Wendy. 2011. "Discovering Work as 'Experimental Demonstration' in School Science Labs." *Ethnographic Studies* 12: 12–30.

Soler, Léna, Sjoerd Zwart, Michael Lynch, and Vincent Israel-Jost, eds. 2014a. *Science after the Practice Turn in the History, Philosophy, and Social Studies of Science*. London: Routledge.

___. 2014b. "Introduction." In *Science after the Practice Turn in the History, Philosophy, and Social Studies of Science*, 1–43. London: Routledge.

Sormani, Philippe. 2011. "The Jubilatory YES! On the Instant Appraisal of an Experimental Finding." *Ethnographic Studies* 12: 59–77.

___. 2014a. *Respecifying Lab Ethnography: An Ethnomethodological Study of Experimental Physics*. Aldershot: Ashgate.

___. 2014b. "Practice-Based Video Analysis: A DIY Tutorial in EM 3.0." Paper presented at the American Sociological Association Annual Meeting, Session on Topics and Methods in EMCA Studies of Work. San Francisco, 16–19 August.

___. 2015. "Fun in Go: The Timely Delivery of a Monkey Jump and Its Lingering Relevance to Science Studies." *Human Studies* 38 (2): 281–308.

Sormani, Philippe, Alain Bovet, and Ignaz Strebel. 2015. "Reassembling Repair: Of Botched Jobs, Maintenance Routine, and Situated Inquiry." *Tecnoscienza: Italian Journal of STS* 6 (2): 41–60.

Streeck, Jürgen, Charles Goodwin, and Curtis LeBaron, eds. 2010. *Embodied Interaction: Language and Body in the Material World*. Cambridge: Cambridge University Press.

Suchman, Lucy. 2000. "Embodied Practices of Engineering Work." *Mind, Culture and Activity* 7 (1–2): 4–18.

___. [1987] 2007. *Human-Machine Reconfigurations. Plans and Situated Actions*, 2nd ed. Cambridge: Cambridge University Press.

Suchman, Lucy, and Randy H. Trigg. 1993. "Artificial Intelligence as Craftwork." In *Understanding Practice*, edited by Set Chaiklin, and Jean Lave, 144–78. Cambridge: Cambridge University Press.

Sudnow, David. 1978. *Ways of the Hand: The Organization of Improvised Conduct*. Cambridge, MA: MIT Press.

Taylor, Charles. 1985. *Philosophy and the Human Sciences*. Cambridge: Cambridge University Press.

Turner, Roy, ed. 1974. *Ethnomethodology*. Harmondsworth: Penguin.

Wenger, Étienne. 1998. *Communities of Practice: Learning, Meaning, and Identity*. Cambridge: Cambridge University Press.

Wieder, D. Lawrence. 1999. "Ethnomethodology, Conversation Analysis, Microanalysis, and the Ethnography of Speaking (EM-CA-MA-ES): Resonances and Basic Issues." *Research on Language and Social Interaction* 32 (1–2): 163–71.

Wittgenstein, Ludwig. 1953. *Philosophical Investigations*. Oxford: Blackwell.

Zimmerman, Don H., and Melvin Pollner. 1973. "The Everyday World as a Phenomenon." In *Understanding Everyday Life: Towards the Reconstruction of Sociological Knowledge*, edited by Jack Douglas, 80–103. London: Routledge and Kegan Paul.

5   Art, Design, and Performance

Chris Salter, Regula Valérie Burri, and Joseph Dumit

Recent science and technology studies (STS) have been demonstrating an increasing interest in art and design as emerging fields of inquiry. While art practices have been the focus of a scattered but increasing number of individuals and projects in STS and related disciplines for decades, issues in design have also recently become a key concern of STS. Such studies draw on a long tradition in STS to explore how the social world is inscribed into technology in the processes of its making and use (Bijker, Hughes, and Pinch 1987; Latour and Woolgar [1979] 1986; MacKenzie and Wajcman 1999), including their design (Star and Ruhleder 1996; Suchman 2011). Other studies that have inspired the recent attraction of STS in art and design concentrate on the entanglements of art and design things and practices in heterogeneous and messy environments (Latour 2008; Law 2004). They explore how in such environments, actors and artifacts are continuously being (re-)arranged in changing hybrid formations by complex processes of translation, which stabilize and destabilize the sociomaterial order in a given space and time. The study of art has also been inspired by well-known works such as Caroline Jones and Peter Galison's (1998) *Picturing Science, Producing Art*, which analyzes the role of visual representations and cultures of vision in science and art. All these studies demonstrate that art and design are deeply entangled with sociotechnical worlds.

In this chapter we will develop an argument for art and design as new, growing, and challenging fields for future research in STS. Michael Fischer proposed as early as 2003 to bring together science and technology with "the social sciences, arts, and humanities, which constitute the analytic understanding and cultural commentaries about the societies of (post)modernity" (Fischer 2003, 305). Bringing these approaches into conversation might turn STS into a key player "which would be integrative, critical, technically competent, and culturally resonant" (ibid., 305). Following this view, we will explore how STS scholars can benefit from the ways artists and designers bring about new futures and work in speculative modes of inquiry that are not necessarily beholden to established epistemological frames, methods, conventions, and practices. We argue that STS could broaden its ways of investigating and intervening into technoscientific worlds through an engagement with art and design in four ways:

1. The involvement of art and design in exploring both the production and stabilization of scientific facts and the development and use of technological artifacts and sociomaterial networks can generate other enriched forms of knowledge, which includes aesthetics, across *all* of the senses (Benschop 2009).

2. STS (and science in general) could tap into enlarged methodological repertoires. Art and design works can thus counterbalance the more standard cognitive and social science approaches of STS by injecting ambiguity, complexity, speculation, and agonism when displaying and communicating science.

3. Art- and design-inspired ways of enacting and communicating research results can enable STS work to reach broader audiences than written scientific work, thus facilitating the inclusion of wider publics in the reflection on science and technology and contributing to its democratization.

4. Art and design interventions can be forms of radical political engagement with sociomaterial worlds and thus can participate in the shaping of technoworlds and the formation of technosocieties. Art and design methods might enable an alternative way for STS to get involved in sites where science and technology are constructed—experiments in codesign, participatory and adversarial design, and critical making are key in this respect.

Two questions are at the core of this enterprise. First, what can STS learn from art and design and apply to its broader scope of analysis?

The first half of this chapter, "STS Perspectives on Art and Design," asks how STS can—through an engagement with art and design—account differently for notions of performance, affect, multisensory embodiment, aesthetic play, invention, radical experiment and the forms of experience brought about by intervention, hacking, critical and disruptive design, "care," and critique that take place by way of art and design practices in science and technology. Throughout our review of the literature we point to a range of recent analytic subjects that have turned to new research directions such as improvisation and invention, embodiment, sensorial knowledge, and "haptic creativity" (Myers and Dumit 2011), "material engagement" (Malafouris 2008), contingency and material agency, performance and performativity, "adversarial design" (DiSalvo 2012), and "critical making" (Ratto and Hertz 2014).

Second, we ask if science has traditionally been seen as the premiere site of knowledge production (Knorr Cetina 1999), then what kind of knowledge does art and design produce and how does this knowledge challenge traditional paradigms of scientific knowledge (production)? What new research areas do art and design suggest for STS, from art dealing with living systems to questions of disruption and adversity in design practice? The second half of the chapter, "Art and Design as STS Methods," thus looks at ways to include art and design practices to explore issues in science, technology, and society. In this section, art and design and their performances are discussed as methods for STS analysis. This part will complement the future research agenda by suggesting issues for further examination such as genealogies of technoscientifically influenced art, "civic technoscience" (Wylie et al. 2014), emergent "biological sensibilities" (Fischer 2009), as well as "counterimages" (Burri and Dumit 2008).

STS Perspectives on Art and Design

Improvisation, Creativity, and Invention

In what ways do creativity and improvisation come into play in the (re-)configuration of sociomaterial arrangements such as the fabrication and appropriation of new works in art and design? How are unanticipated incidents handled in art and design practices? How can the subsumption of creativity and invention under neoliberal capitalist rhetorics of "creative industry" and "innovation" (Pang 2012) or of traditional gender, race, and culture distinctions carried forth in big science be critically interrogated (Fouché 2003; Lerman, Oldenziel, and Mohun 2003; Oldenziel 1999)? Such questions should be further explored by STS scholars, given STS's long-term exposure and examination of issues such as gendering in scientific practice, the questions of expertise in relationship to invention, and the role of material contingency in the production of knowledge.

In experimental practices in science, improvisation, creativity, and invention have always played important roles (Milburn 2015; Rheinberger 1997; Wylie 2015). STS may thus gain from carefully looking into the role and enactment of creativity, improvisation, and invention as crucial epistemic tools in art and design (Hallam and Ingold 2007; Montuori 2003). Theater and dance worlds, for example, explicitly theorize the *training* of improvisational skills, emphasizing that, "[i]mprovisation is by its very nature among the most rigorous of human endeavors ... Improvisers prepared themselves by making improvisation a regular practice, a daily practice even" (Gere 2003, xiv). Improvisers also emphasize the processes of developing collaborations, such as devising, again with an attention to training in the process.

Training in improvisation takes place through workshops and handbooks, and studies of these might provide STS with another level of attention to the making of scientists and engineers. Halpern (2010), for example, explores how theater-based role-playing can enable students to experience science, so-called theater in science education. At the same time, the new field of critical improvisation studies points to the long history in which creativity and improvisation have served as concepts to demarcate race and gender, reserving creativity as a label for privileged bodies, while nonetheless practically serving communities in political critique and survival (Fischlin, Heble, and Lipsitz 2013; Goldman 2010; Heble and Caines 2014; Moten 2003). These works point to an as-yet-understudied assumption within STS about the place and role of creativity and improvisation in science and technology–related practices.

Simultaneously, STS can combat the tendency to associate creativity and invention within Euro–North American capitalist, neo-liberal narratives of progress (Eglash 2004; Fouché 2006; Verran 2002). STS studies of non-Western design cultures can challenge predominant ways of understanding innovation as outcomes of mainly Western (creative) industries and countercultures. For instance, Lindtner and Li (2012) show how Chinese hackers appropriate Western countercultural approaches by drawing at the same time on China's tradition of open source manufacturing, thus redefining what "innovation" means for China.

Embodied Knowledge, Material Engagement, and the Senses

STS has long described scientific knowledge as being embodied, that is, tacit and implicit (Haraway 1991; Polanyi 1967), and has highlighted the role of emotions in the process of knowledge production. These discussions reflect long-standing debates in cognitive science, philosophy, and human-computer interaction (HCI) on the nature of knowing and the manner in which such embodied knowledge gets represented and enacted, particularly within systems, artifacts, and technical practice itself (Dreyfus 1979; Haraway 1988; Suchman 1987; Wenger 1998; Winograd and Flores 1986).

Paralleling Haraway's "situated knowledges," Suchman's (1987) notion of "situated action," defined as "simple actions taken in the context of particular, concrete circumstances," was originally directed at plan-based models within early AI and HCI-based systems but has recently been broadened by Suchman herself and others to reflect on media art, design, and digital gaming (Suchman 2004a, 2007).

Although emerging from different epistemic histories, these running debates on "situated knowledges" (Haraway), "situated actions" (Suchman), and more recently "enactive cognition" (a research program which argues that the experienced world is determined by mutual interactions of the physiology of the organism, its sensorimotor circuit, and the environment) (Alač, Movellan, and Tanaka 2011; Noë 2004; Varela, Thompson, and Rosch 1991), all revolve around a central critique of the reduction of human experience to formalized sets of rules or computational models at the price of neglecting embodied, sensorimotor, and concrete perceptual action in the world.

Polanyi's (1967) claim that "we know more than we can tell" has recently been extended by Cambridge archaeologist Lambros Malafouris. Investigating a potter working with clay, he claims that artistic and creative practice is not only a form of tacit knowledge but "material engagement": "brain, body, wheel and clay relate and interact with each other through different stages of an activity" (Malafouris 2008, 19) in complex ways.

Malafouris's argument that artistic practice is always a constellation of material, cognitive, and biological forces and that "verbal description, however detailed, can hardly capture the phenomenological perturbations of real activity" (Malafouris 2008, 20) also reconfirms the long-standing claim that certain forms of artistic knowledge can only be sited in what Csordas (1993) calls "somatic modes of attention." Such somatic modes are "culturally elaborated ways of attending to and with one's body in surroundings that include the embodied presence of others" (ibid., 135) that are difficult to render into signs or texts. Attending to and with one's own body and the bodies of others exemplifies a core STS argument that knowing is always framed by the unknown acts and becomings of bodily existence.[1]

What "embodied knowledge," "material engagement," and "somatic modes of attention" ultimately suggest is that knowledge itself is forever entangled with bodily senses. While the visual sense has received much attention in STS when looking at the ways scientific images are interpreted and perceived (e.g., Coopmans et al. 2014; Dumit 2004; Fischer 2003; Jones and Galison 1998; Lynch and Woolgar 1990), more recent studies draw attention to bodily senses other than vision that are involved in scientific and artistic knowledge production, such as listening (Helmreich 2015; Mody 2005; Pinch and Bijsterveld 2004; Roosth 2009; Supper 2014, 2015), touching (Fischer 2003; Myers and Dumit 2011), and tasting (Burri, Schubert, and Strübing 2011; Paxson 2013; Roosth 2013b; van de Port and Mol 2015). Such body-centric studies framed through the more general concept of "affect," the process by which emotions and feelings affect bodies without being able to be described in language or representations (Angerer, Bösel, and Ott 2014; Gregg and Seigworth 2010), deserve much further exploration in STS. Concepts like "good hands" involving embodied skills used in biology (Fischer 2003) or "haptic creativity" in which scientists and artists' "data, instruments, and bodies and identities are continually reconfigured within their apparatuses" (Myers and Dumit 2011, 241) suggest new avenues for exploring how researchers' "thinking with eyes and hands" (Latour 1986) generate new forms of knowledge.

One area that STS already benefits from but could help enlarge its sensorial and affective scope in the future is the burgeoning cross-disciplinary field of sensory studies, which brings together a range of disciplines interested in "a cultural approach to the study of the senses and a sensory approach to the study of culture" (Howes 2013). Debates about artistic research or research-creation that understand art practices as forms of tacit knowledge production are another field of reference for STS (e.g., Borgdorff 2012; Riley and Hunter 2009).

The use of bodily senses in interacting with the world is part of what makes up an aesthetic experience. The study of embodied knowledge in art and design and the manner in which practice is understood through its own "endogenous" theoretical and methodological claims may thus enrich STS insights into processes of knowledge production.

Performance and Performativity

The focus on embodied knowledge always is related to two key words that have gained much credence in STS over the past twenty-five years: *performance* and *performativity*. As Herzig (2004, 130) writes, "given the heightened recognition of contingency, temporality, and reflexivity made possible by performative analyses, it is perhaps not surprising that a number of recent studies of science reveal a quiet but steady turn toward this useful analytical tool."

The appeal of performance and performativity in a study of STS and art and design is that both terms do boundary work by embodying core concerns that link STS and artistic and design practices: the tension between representations and objects/processes in the world; the sociotechnical construction of things and phenomena; questions of material agency, contingency, and the power of actants beyond humans; and finally, the tension between observation and participation.

The word *performance* was originally used to describe actions, happenings, and time-based events (particularly involving the body of the artist) emerging out of the visual arts during the 1950s through the 1980s and referring to human activity involving a physically sited, co-present live and social event with the *human body* and *human agency* at the center of action (Goldberg 1976; Phelan 1993; Schneider 1997).

In art, the body has always been an important subject of inquiry and representation. Artists have used and experimented with their bodies forever—by dancing, playing theater, and doing artistic performances that have involved manipulating the body through prosthetics, surgery, and technical adornment (Fischer 2003), forming social plastics, or simply inquiring into their own corporeal existence in space and time. Performance studies' conceptions of corporealization, or "body techniques" (Mauss 1973), as a performative act (Butler 1988) thus align specifically with STS's focus on the body as a subject (Berg and Mol 1998; Cussins 1996; Mol 2002) and as a scientific instrument within research processes and material practices (Knorr Cetina 1999; Latour 1986,

2004; Shapin 1994), as well as on the role of gestures in the construction of expertise, which play an important role in the ways research is carried out and results are interpreted and experienced (Alač 2008; Burri 2008).

One area that could benefit from further STS studies, however, is a focus on the *moving* body as an important instrument in the manufacturing of science. While there has been substantial work from historians of science like Dagognet (1992) on Étienne-Jules Marey's chronophotography and Landecker (2006) on the use of microcinematography by tissue culture pioneer Alexis Carrel, both have focused more on the role of scientific instruments in creating an analytic of time and motion beyond what the eye can see than on how the performing body itself becomes an essential part of the research process. Here, concepts of exploring how researchers "lean into the data" (Hustak and Myers 2012; Myers and Dumit 2011) and perform their bodies (Herzig 2004; Myers 2015) when producing and displaying science as well as the body as scientific instrument in research processes (Fischer 2003) would also benefit from an understanding of performance in the artistic realm.

The term *performance* was quickly recontextualized in the first "performative turn" emerging from anthropology, sociology, and linguistics in the mid to late 1960s, particularly influenced by the work of Cambridge linguist J. L. Austin, who coined the term *performative* to describe speech as enacting rather than simply describing a situation (e.g., the words "I do" spoken at a wedding, for example, do a specific kind of work (Austin [1955] 1975). This performative turn, which eventually birthed the anthropologically centered discipline performance studies in the 1970s, sought to shift focus away from the stage to phenomena such as everyday social interactions, rituals, festivals, sports, games, and urban practices (Conquergood 1989; Garfinkel 1967; Goffman 1959; Turner and Schechner 1983).

Principally, anthropologist Victor Turner's notions of "culture as performance" or "cultural performance" both stem from a deeper anthropological project to critique a concept of culture that Dwight Conquergood (1989, 83) described as filled with "static structures and stable systems with variables that can be measured, manipulated and managed." Thus, by highlighting the tensions between both observing and acting, direct experience and the representation of an experience in textual form, performance thus came to be seen as an anthropological method troubling these dichotomies by examining the frenetic sites of conflict and human action and their potential for transformation.

In the 1980s and 1990s, however, a substantial shift took place with the introduction of the term *performativity*. Reflecting back on Austin's notion of the performative as a linguistic act rather than description, the increased use of the term is chiefly attributed to philosopher Judith Butler, who sought to describe gender not as an ontologically "stable identity or locus of agency [...] but rather an identity tenuously constituted in time—an identity instituted through a stylised repetition of acts" (Butler 1988, 519).

One influential STS-derived response to Butler's linguistic and human body-centered focus is physicist and feminist STS scholar Karen Barad's notion of a "post-human performativity"—a concept that brings together a range of natural-cultural material, social, scientific, human, and nonhuman factors in order to question "the givenness of the differential categories of 'human' and 'nonhuman,' examining the practices through which these differential boundaries are stabilized and destabilized" (Barad 2003, 808). In broad strokes, scholars like Barad (ibid.) and Andrew Pickering (1995) have used the notion of performance and performativity to critique what Pickering has labeled the representational idiom of science—the idea that scientists are "disembodied intellects making knowledge in a field of facts and observations" (ibid. 1995, 6). Instead, performative forms of knowledge production focus on "doing" in scientific practice (Barad 2003; Biagioli 1999; Callon 2007; Latour 1986; Law and Singleton 2000; Mol and Law 2004; Shapin and Lawrence 1998).

A closer reading of the literature, however, demonstrates more nuanced and complementary ways in which the terms *performance* and *performativity* are utilized and interchanged in STS parlance (Salter 2010). First, performance describes the actions of human scientists in their experimental work within the confines of the laboratory, aligning with early microstudies of laboratories in the 1970s and 1980s from scholars like Latour and Woolgar, Knorr Cetina, Fujimura, Traweek, and others who focused on the shift to science as a cultural practice. Scientists "perform" experiments—that is, they directly *manipulate* and *transform* materials through instruments and apparatuses in real time. Hacking (1983) argued that experimental practice is not about merely shaping existing phenomena but instead radically constituting new phenomena by manipulating, altering, abusing, or subjecting it to "interventions," "interferences," and transient processes (Knorr Cetina 1992, 127).

Second, performance is used for the attribution of acts to nonhuman things, substances, and processes; the acts of material itself or what matter *does* rather than what it is. As Herzig (2004, 130) writes, matter thus "appears not as some passive, fixed physicality chastely awaiting discovery but as contingent, temporal and active. The genes, fruit flies, computers and atoms that populate scientific practice emerge only as nodes of interminable, multivalent, contested processes."

Pickering (1995) makes a distinction between representations and performances, assigning performance to acts and forces beyond human bodies. "Performances are what agents do, whether human or nonhuman. My conviction is that we need to move to a performative (rather than representational) idiom for studying and reflecting on science (and on being in general)" (Pickering 2010, 195).

Yet, according to Barad, Pickering's post-humanist idea of performativity leaves aside the political import of performativity, particularly its "arguably inherently queer— genealogy" and performativity's connection to post-structural, feminist, and queer studies theory and activism (Barad 2003, 807). Instead, Barad's notion of "intra-action" suggests that entities do not preexist the world in which they operate but are relationally constituted through a continual process of material-discursive possibilities. In other words, knowers and things known (what Barad labels *phenomena*) mutually arise through their intra-actions based on concrete, ever-evolving material circumstances (for example, scientific instruments and apparatuses) and post-human performativity describes the manner in which such phenomena act and become embodied with social-cultural-technical meaning, forms of power, and significance (Barad 2003, 815).

Finally, performance signifies the manner in which scientists "stage" or "direct" (to use theatrical language) their findings and knowledge—or, the acts of matter, before a public. Scholars such as Latour (1987) and Schmidgen (2011), who have described the theatrical acts inherent in laboratory practices in the biological sciences such as the evolution of anatomical theaters or "spectatoriums" in the nineteenth century, have long used theatrical language to characterize the techniques by which scientists "stage" for the public (the audience) the results of things "rehearsed" in laboratory settings in order to create levels of belief. Stephen Hilgartner (2000), as another example, analyzed science advice as a performance of experts toward their audiences while engaged in impression management.

That performance is so freely utilized within STS is not without its critics. Herzig argues that even though performance and performativity have taken deep root in STS parlance, these notions still rely on an ultimate conception of performance as endlessly producing something—knowledge, bodies, or relations. As she provocatively asks, "must all intra-actions be generative?" (Herzig 2004, 138). Instead, Herzig cites Zoë Sofia's (1993) criticism of STS's "use value" and "practicality" perspective on science, asking whether there is room for an approach to performativity that is not based on the generative power of human-nonhuman agency but rather from unproductive acts like "squander," "excess," "desire," "unproductive expenditure" (Bataille 1991), and waste. Thus, by depicting the irrational, the aberrant, and the queer, performance and performativity could also be useful in questioning the political underpinnings of human and nonhuman doing and agency inherent to scientific (and artistic) practices.

Contingency and Material Agency

Another arena where questions of human and material doing and agency have made their mark is design. In fact, design practices have become an inspiring field of scrutiny for STS in recent years (Coletta et al. 2014; Ehn 2011; Irani, Dourish, and Mazmanian 2010; Latour 2008; Shove 2014; Storni 2012; Volonté 2014; Woodhouse and Patton 2004; for ethics in design see Le Dantec and DiSalvo 2013; Shilton 2012; Steen 2015). In a broad understanding of embodiments and enactments, and adopting an understanding of the performance of nonhuman things, STS studies on architecture, in particular, have demonstrated how material objects like buildings perform the "scripts" and (sociotechnical) worlds inscribed into their design (Farías and Bender 2010; Galison and Thompson 1999; Moore and Karvonen 2008; Yaneva 2005, 2009, 2012).

Using actor-network theory (ANT) methodologies such as following the actors and their networks combined with controversy studies (Pinch and Leuenbeger 2006; Venturini 2010), analysts aim at understanding the social-material context of architectural *practice*. In *Mapping Controversies in Architecture*, for example, Yaneva (2012) asks the question, "Who designed the Sydney Opera House?" Was it the Danish architect called Jørn Utzon, who eventually withdrew from the project but whose name still adorns the sail-like building? Was it the engineering firm Ove Arup, without whose expertise in structural engineering "Utzon's" building could never have been? Or, is design spread among the various constituencies (the Labor government at the time, the Australian public), instruments, materials, computational models, and systems of calculations?

What forms such questions is not how the fixed and finished architectural object can reveal the social forces that made it possible but, rather, the exploration of specific "shapes, fabrics and material arrangements" demonstrates how "the built environment either reflects or produces social life" (Yaneva 2012, 106). The concept that the materiality of a building (or any art or design object or environment) is contingent, created by different forces, events, and trajectories and "shaped more by external conditions than by the internal processes of the architect" (ibid., 105) suggests a different approach to previous theories of material agency like Pickering's (1995, 2010), in which agency is situated within artifacts.

Disruption and Critical Design

While design and architecture—just like art—can be understood as unstable material practices, they can also serve to disrupt established methods and ideologies. In what DiSalvo labels "adversarial design," both the design process itself as well as its objects give form to problematic situations, providing a method of inquiry to grapple with a diversity of actors and agendas "which often seem incongruent" (DiSalvo 2012,

290). Adversarial design is only one of a number of contemporary forms of design inquiry that go by names such as "critical design" (Ward and Wilkie 2009), "participatory design" (Ehn 2008), "contestational design" (Hirsch 2008) and "reflective design" (Sengers et al. 2005) and that are influenced by or serve as subjects for STS. The common element across all of these different forms of design inquiry is that they seek to challenge the normative frameworks of what design is and its assumed constituencies, as well as focus on how design practices operate *critically* in relationship to technology and society, producing new forms of (speculative) objects (Dunne and Raby 2013; Ehn 2008; Loukissas 2012) and publics (Callon, Lascoumes, and Barthe 2009; Latour and Weibel 2005; Marres 2007).

A case in point is Latour and Weibel's 2005 *Zentrum für Kunst und Medientechnologie* (ZKM) exhibition *Making Things Public*. The exhibition inquired into the political representation of things in democracies by precisely assembling a heterogeneous array of projects, ranging from texts, images, installations, and interactive interventions, staging encounters between science, art, and politics, and thus using design to produce and debate knowledge and things.[2] Design is critical not just in terms of its resultant objects but also in terms of the manner in which such objects and products operate in the formation of new democratic forms of participation that, similar to the objects of science, are equally contingent, pluralistic, and dynamic (Bjoergvinsson, Ehn, and Hillgren 2010; Le Dantec 2010).

Extending earlier STS notions of infrastructure (Bowker and Star 1999; Star and Ruhleder 1996) as sociotechnical structures that organize human doing and the technologies that enable such doing, the concept of *infrastructuring* (Le Dantec and DiSalvo 2013; Pipek and Wulf 2009) describes the process of bringing together the full range of design activities, from development, deployment, and enactment (so-called a priori activities such as interpreting and articulating) to appropriation, redesign, and maintenance (so-called design in use) (Karasti and Baker 2008). With infrastructuring as a framework, the notion of innovation thus arises from both emerging and established communities of practice and their actions rather than through strictly technocratic systems already in place.

Already Suchman's (2007, 2011) work at Xerox PARC in the 1990s began to think about infrastructuring in her arguments that "local improvisation," the appropriation of specific work practices within local sites, resulted in the reception of an existing technology and its integration into new sociotechnical contexts. Such local improvisational activities function as "the generative practices out of which new technologies are made" (Suchman 2004b, 170). Following Suchman's lead that design is always a contextually and geopolitically situated activity and that much design literature focuses on work paradigms within the "hyperdeveloped world," recent work from Irani and Silberman (2013) on Amazon's Mechanical Turk (AMT) "human computation resource" system describes the author's tactical media CSCW intervention "Turkopticon," a software extension which acts as an "ethically motivated response to workers' invisibility on the AMT system" (Irani and Silberman 2013, 614). In this sense the Turkopticon project opens up the question of how design is always political and can both operate "in the world" as an activist intervention in critical crowdsourcing and simultaneously be tied to and dependent on a large-scale, functioning sociotechnical infrastructure.

Similarly, the paradigm of "critical making" describes the role of creative, critical activity within academic and design contexts (Maeda 2013; Ratto and Hertz 2014) as well as a material-theoretical research program (Ratto 2009, 2011). By using material practices to extend, and unpack concepts such as "materiality," "affordances," and "objects," critical making acts both as a series of methodological commitments and as a metaframe for generalizing across a range of diverse sociomaterial-technical contexts and sites (Ratto and Boler 2014; Whitson 2013).

Art and Design as STS Methods

In addition to being subjects of STS analysis, art and design may also be used as (experimental) methods in STS inquiry—thus contributing to unconventional ways to explore the social (Lury and Wakeford 2012). The inclusion of artistic and design practices contributes to deeper and more multifaceted insights into science and technology and their entanglements with complex sociomaterial worlds.

Genealogies of Technoscientific Art

Experimenting with science and innovative technologies has been a long-standing interest and persistent engagement of artists. They have explored early painting and graphic technologies through to digital plotting, image apparatuses, and sensors, while at the same time making these very technologies the subject of inquiry. As early as 1923, the inaugural exhibition of the Bauhaus in Weimar was entitled "Art and Technology: A New Unity." In the 1960s cybernetics, what Edward Shanken has called a "cultural mindset," began influencing a wide range of artists/thinkers, from John Cage and Nam June Paik to Roy Ascott, Gordon Pask, and James Tenney, all of whom became interested in the application of feedback-based systems to art (Kahn 2007; Shanken 2003).

STS scholars like Turkle (2005) and Suchman (2007) have long explored the performativity inherent in cybernetic and interactive systems from within the sites of research laboratories and everyday use of technologies. While increasing numbers of (media) art historians and media theorists have examined the successes and failures of the complex of "art and technology" (Grau 2003; Kwastek 2013), STS has mainly ignored the historical genealogies of such work—an area that would be worthy of future study given art and technology's long-time entanglement with questions of infrastructure, invention, innovation, and experimental practice.

This complex has been made even more prescient given the range of cultural institutions internationally dedicated to explorations in art and technoscience, ranging from the ZKM in Karlsruhe, Germany, V2 in Rotterdam, the NTT-ICC in Tokyo, and the annual Ars Electronica Festival in Linz, Austria, to organizations outside the European/North American/Japanese axis such as the HONF (House of Natural Fibers) in Indonesia, the ArtScience Museum in Singapore, ArtCenterNabi in Korea, Arte Alameda in Mexico, or a range of centers in Brazil, Peru, Colombia, Senegal, Tunis, and Chile. Such institutions have been dedicated to providing both infrastructure and know-how to artists and designers to realize artistic works focused on new technologies while supporting critical discussion and debate with their broader publics on the sociocultural implications of such technologies. This has been particularly important in settings outside of North America and Europe where access to such knowledge and technologies has been limited. While these institutions strive to reflect on the role and aesthetic and ethical implications of artistic inquiry into our technoscientific worlds within their very specific cultural milieus, STS insights, particularly around the role of infrastructuring, intervention, and the role of material knowledge in the formation of new sociotechnical imaginaries discussed in our chapter, have not been a primary focus of their attention.

Art, Innovation, and Policy: The Studio Lab

One site in which art, science, and technology have been strongly linked is studies of research settings (government- and industry-sponsored research labs) within science policy studies examining interdisciplinarity (Gibbons et al. 1994). In a white paper for the Rockefeller Foundation, Century (1999) focused on the relationship between science policy studies and innovation applied to art science projects. In particular, Century's coining of the term *studio laboratory*, which described "a new class of hybrid innovative institution (...) where new media technologies are designed and developed in co-evolution with their creative application" (historical examples included Bell Labs, Xerox PARC, EVL, and the CAVS-MIT) was influential in U.S.-based private foundations like Rockefeller and Ford and in U.S., Canadian, and UK research policy circles in articulating the relationship between new media artistic practices and the sociology and economics of innovation.

More recent work from Barry and Born (2013) building on their collaboration with Weszkalnys (Barry, Born, and Weszkalnys 2009) has further explored the relationship between art and science as part of a larger trend toward new forms of interdisciplinary knowledge production remaking humanities and social science research while sociology of knowledge studies also have examined artistic practice and the intersection of research policy, institutions, and innovation agendas (Nelson 2015). For example, Hennion (1989) and Born (1995) have applied methodological formulations from STS to the sociology of music and larger issues of cultural production. Fourmentraux (2007, 2010, 2011) has also extensively studied the organizational dynamics of innovation within the interface of art, science, and industry, focusing on large-scale, university-cultural-industry new media consortiums like Hexagram in Montreal.

Intervention, Engagement, and Civic Technoscience

One of the major arguments for the inclusion of art and design in STS is that both practices engage broader publics beyond the strictly peer-based ones found in science. In his study of Dunne and Raby's notion of speculative design, Michael (2012) argues for "designerly public engagement" over "public understanding" with artists, the former making use of ambiguity, inventive problem-making, and explorations of complexity. Indeed, there has been much STS literature dedicated to the argument that art can work in public engagement or art methods can function as "out-of-the-box" informal science education (de Ridder-Vignone 2012).

Yet, as Barry and Born (2013) point out in their ethnographic study of interdisciplinarity, there are different "logics" at play in the manner in which art and science configurations operate in their agendas, intentions, and goals. Indeed, part of artists' interventionary interest to STS is the boundary work such projects do, collapsing distinctions between art and science and art and society. In particular, some art science constellations involving engagement and the "public understanding of science" tend to function under a "logic of accountability," which reduces or mitigates artists' critical work in scientific contexts to instrumental, "decorative," "celebratory," or "superficial."

Instead, another logic, that of ontology, operates in many art science projects in which a more radical form of "interdisciplinary production" takes place in which both the object and subject of research is questioned, reworked, and reinvented through sustained, deep, and long-term mutual collaboration and where new forms of material and social objects are invented (Barry and Born 2013). Outside of institutional contexts, such practices as hacking involving DIY biology (Meyer 2014), hacking as an intervention into political economy (Söderberg and Delfanti 2015) and the larger field of "amateur" and "citizen science" (Rogers 2011) act artistically and technologically to change worlds. For example, Wylie et al. have provocatively called for a "civic technoscience," where "new material technologies in combination with new social and literary technologies can sustain a civic research space external to the academy and where nonacademics can credibly question the state of things" (Wylie et al. 2014, 118).

Emergent Biosensibilities

Artists' involvement with biotechnologies has received concentrated attention from STS scholars. Indeed, this larger, very heterogeneous category of artistic work with living systems called *bioart(s)* engages with the ethical, aesthetic, and social implications of (engineered) living matter, and is carried out in both science labs and art studios, taught in art schools, and increasingly institutionalized in special programs and academic centers.[3] In what Fischer (2009) called emergent "biological sensibilities," artists working with biotechnology propose new ways of living with (and caring for) emerging ecological forces of biology and biotechnology as well as "forging modes of representation and intervention that synthesize practices from science and engineering" (da Costa and Philip 2008, xix–xx).

One of the most important institutional catalysts for this work has been SymbioticA, the Centre for Excellence in Biological Arts located within the Biology, Physiology, and Human Anatomy Department of the University of Western Australia in Perth. Founded in 2000 by designer Oron Catts, cell biologist Miranda Grounds, and neurologist Stewart Bunt with the support of Western Australian state lottery funds, SymbioticA has attracted international artists, designers, scientists, anthropologists, sociologists, STS scholars, and others to research and experiment with living matter within a university research context.

SymbioticA and the bioarts in general have been the subject for a number of recent STS related studies (Dixon 2009, 2011; Rogers 2011; Roosth 2013a; Salter 2015). These works have focused particularly on the manner in which bioarts destabilize assumed ontological categories of living and nonliving, explore care and regeneration in relationship to the use of biomatter, and detail new art forms based on the use of genetic engineering techniques to transfer synthetic genes to an organism or to transfer natural genetic material from one species into another, to create unique living beings.

One of the central foci for STS in studies of the bioarts has been articulating the entanglements of institutional settings, technical work in the science lab, and the differing intentions and agendas of artists within such settings (Dixon 2009; Dixon, Straughan, and Hawkins 2011; Kac 1998). For example, in their early tissue culture work under the moniker TCA (Tissue Culture and Art), Catts and partner Ionat Zurr (currently the academic director of SymbioticA) developed tissue-culture-based art works that not only brought laboratory procedures into cultural settings but also worked toward an "aesthetics of care"—attending to and being mindful of the use of living material in aesthetic practice (Catts and Zurr 2002).

Other fields in the life sciences are also addressed by artistic inquiry, including synthetic biology, nanotechnology's obsessions with scale (de Ridder-Vignone and Lynch 2012), and biomedical subjects such as reproductive medicine and the neurosciences (e.g., Scott and Stoeckli 2012). On the one hand, biomedical and brain imaging used in these fields are aesthetically very appealing for artists. On the other hand, artistic engagement with biomedicine and the life sciences has been fostered by a variety of artist-in-residence programs, for example, at the Wellcome Trust in the UK (http:// wellcome.ac.uk).

Moving between Art, Design, and STS

The inclusion of art and design in STS raises methodological challenges: How can artistic and design methods be involved and used in STS to reflect on science and technology? In other words, how to do (social) science through art and design? When exploring this issue, STS itself becomes experimental; it not only observes people "thinking with eyes and hands" but uses eyes and hands to intervene and interfere in spaces and sites where science and technology are constructed, distributed, used, incorporated, and enacted. Such a performative approach to the exploration of science and technology can be understood as practices of "witnessing," which produce "knowledge without contemplation" (Dewsbury 2003). Rather than just passively observing and reflecting, knowledge emerges through an active engagement and interplay of senses, bodies, implicit knowledge, and material objects.

Researchers moving between STS and art have been undertaking first steps to linking artistic inquiry into science and technology to STS. By drawing on both artistic and scientific knowledge when doing research, they perform STS-informed art practice, and, more often, collaborate with artists and designers in their research projects on science and technology.

Pioneering scholars in this emerging field were mostly trained as both scientists and artists or have developed a special interest in STS while having a background in art.[4] The outcomes of their projects are mostly art works, which only in a few cases are complemented with written work. For example, Natalie Jeremijenko's environmental health clinic located at New York University (environmentalhealthclinic.net) approaches the topic of health from a contextual environmental perspective. Instead of diagnosing individual illnesses and providing patients with pharmaceutical prescriptions, the clinic "treats" peoples' environmental concerns with knowledge and specific data on environmental aspects, and with prescriptions for local actions (Schaffer 2008). This art project relates scientific and STS knowledge to political intervention.

Some STS researchers strive for a transdisciplinary mode of inquiry into science and technology when looking for project specific collaborations with artists and designers. The outcomes of such projects have been written rather than art and design works (Halpern et al. 2013). Collaborations between scientists and artists have been practiced in a variety of projects in the past (e.g., Gabrys and Yusoff 2012; see also Hannah 2013). For example, a joint research project run by Stanford University and the University of Edinburgh called Synthetic Aesthetics sought to bring together artists, designers, synthetic biologists, and STS scholars to examine the aesthetic-sociotechnical issues surrounding the burgeoning field of synthetic biology (Ginsberg et al. 2014). Synthetic Aesthetics seeks the possibility of radical experimental collaboration, "opening up" synthetic biology by making possible emergent forms of critique that might not occur within traditional disciplinary boundaries (Frow and Calvert 2013).

It is a rare STS project that includes art in its outline. One example of such a collaboration is the ModLab, a project carried out at the University of California–Davis. STS served as a bridging language between humanities, performance studies, sciences, and social sciences to theorize the co-construction of art practices (drawing, painting, theater, and dance), game design, and interdisciplinary tool development in the sciences, in this case a 3-D virtual environment and modeling language. Collaboratively designing performances and art installations in turn are also generating concepts for STS scholars.[5]

Counterimages and Future Work

These examples have to be seen as only first steps in an endeavor to include art and design in STS thinking about science and technology more systematically. Building on John Law's (2004) quest to invent new methods in order to grasp the multiple social realities and shape the social worlds in intended ways, the inclusion of art and design practices in STS contributes to both explore *and* intervene in sociomaterial arrangements.

STS can benefit from such an approach in both analytical and normative terms. Analytically, the involvement of art and design in exploring both the production and stabilization of scientific facts, and the development and use of technological artifacts and sociomaterial networks, may generate another, enriched form of knowledge that includes aesthetics, such as visual, sonic, and haptic dimensions. The whole repertoire of sensual and performative practices and resources may be used as methods in such inquiry and become part of the research process, thus contributing to a broader perspective on and understanding of the research subject. STS (and science in general) may profit from such enlarged methodological repertoires also in the communication of research results. In normative terms, artistic and design methods may facilitate STS interventions in sites where science and technology are constructed—experiments in codesign and participatory design are important in that respect—and become part of our everyday lives and practices.

We suggest seeing the growing conversations and hybridization between STS and art as a site where *counterimages* are being produced; creative responses to the rhetorical power of dominant cultural scientific images and visualizations also contribute to the social studies of scientific imaging and visualization (Burri and Dumit 2008). STS not only may profit from a critical appraisal of these works but also may be inspired to develop and extend its own methods by including art and design in its analysis. By doing science through art and design, STS may extend its capability of reflection, reach new publics, and, most important, get new insights into the cultural workings and implications of science and technology.

Notes

1. A case in point is the field of dance, which has increasingly turned to interactive new media technologies (navigable CD-ROMs/DVDs as well as online databases and interactive websites) that combine still and moving images and sounds to articulate different forms of knowledge systems that are difficult to describe in textual form. For example, the "Nagarika" interactive DVD by the Bangalore-based dance company Attakkalari, described as an "integrated information system on Indian physical traditions" is an extensive research project that attempts to communicate the complex movement vocabulary and taxonomies of gesture in the Southern Indian dance form of Bharatanatyam through interactive video demonstration of expert practitioners (nagarika. attakkalari.org). Similar approaches have been developed with the American choreographer William Forsythe, who has long utilized interactive new media to communicate complex movement knowledge or what he calls "movement research."

2. An earlier exhibition from Latour and Weibel, *Iconoclash* in 2002, may be seen as an iconic precursor to the approach of *Making Things Public*, in which heterogeneous projects of artists and theorists were displayed in order to question the issue of representation in science, art, and religion (Latour and Weibel 2002).

3. Bioart is taught and institutionalized, for example, at the Bio Art Lab at the School of Visual Arts, New York—SVA (bioart.sva.edu), SUNY Buffalo, and Rensselaer Polytechnic Institute, New York. The Arts & Genomics Center at University of Leiden (NL) also engages in this field (artsgenomics.org). Examples of projects in bioart are discussed in Reichle (2009).

4. Examples are Natalie Jeremijenko (nataliejeremijenko.com), Chris Csikszentmihályi (edgyproduct.org), Phoebe Sengers (www.cs.cornell.edu/people/sengers), Hanna Rose Shell (web .mit.edu/~hrshell/www), Tad Hirsch (publicpractice.org), Yanni Loukissas (yloukissas.com), and Simon Penny (simonpenny.net).

5. Projects so far include Dream Vortex, Play the Knave, Datamining Beckett, Glitch, and Frack: The Game. See http://modlab.ucdavis.edu and https://ucsota.wordpress.com/2011/11/30/joe -dumit-expressing-the-caves/.

References

Alač, Morana. 2008. "Working with Brain Scans: Digital Images and Gestural Interaction in fMRI Laboratory." *Social Studies of Science* 38 (4): 483–508.

Alač, Morana, Javier Movellan, and Fumihide Tanaka. 2011. "When a Robot Is Social: Spatial Arrangements and Multimodal Semiotic Engagement in the Practice of Social Robotics." *Social Studies of Science* 41 (6): 893–926.

Angerer, Marie-Luise, Bernd Bösel, and Michaela Ott, eds. 2014. *Timing of Affect*. Berlin and Chicago: Diaphanes and University of Chicago Press.

Austin, John L. [1955] 1975. *How to Do Things with Words*. Cambridge, MA: Harvard University Press.

Barad, Karen. 2003. "Posthuman Performativity: Towards an Understanding of How Matter Comes to Matter." *Signs: Journal of Women in Culture and Society* 28 (3): 801–31.

Barry, Andrew, and Georgina Born, eds. 2013. *Interdisciplinarity: Reconfigurations of the Social and Natural Sciences*. London: Routledge.

Barry, Andrew, Georgina Born, and Gisa Weszkalnys. 2009. "Logics of Interdisciplinarity." *Economy and Society* 37 (1): 20–49.

Bataille, Georges. 1991. *The Accursed Share: An Essay on General Economy*. Vol. 1, *Consumption*, translated by Robert Hurley. New York: Zone Books.

Benschop, Ruth. 2009. "STS on Art and the Art of STS: An Introduction." *Krisis: Journal for Contemporary Philosophy* 1: 1–4.

Berg, Marc, and Annemarie Mol, eds. 1998. *Differences in Medicine: Unraveling Practices, Techniques, and Bodies*. Durham, NC: Duke University Press.

Biagioli, Mario, ed. 1999. *The Science Studies Reader*. London: Routledge.

Bijker, Wiebe, Thomas Hughes, and Trevor Pinch, eds. 1987. *The Social Construction of Technological Systems: New Directions in the Sociology and History of Technology*. Cambridge, MA: MIT Press.

Bjoergvinsson, Erling, Pelle Ehn, and Per-Anders Hillgren. 2010. "Participatory Design and Democratizing Innovation." In *Proceedings of the 11th Biennial Participatory Design Conference*, 41–50. New York: ACM.

Borgdorff, Henk. 2012. *The Conflict of the Faculties: Perspectives on Artistic Research and Academia.* Leiden: Leiden University Press.

Born, Georgina. 1995. *Rationalizing Culture: IRCAM, Boulez, and the Institutionalization of the Musical Avant-Garde.* Berkeley: University of California Press.

Bowker, Geoffrey C., and Susan L. Star. 1999. *Sorting Things Out: Classification and Its Consequences.* Cambridge, MA: MIT Press.

Burri, Regula Valérie. 2008. *Doing Images: Zur Praxis medizinischer Bilder.* Bielefeld: transcript.

Burri, Regula Valérie, and Joseph Dumit. 2008. "Social Studies of Scientific Imaging and Visualization." In *Handbook of Science and Technology Studies*, 3rd ed., edited by Edward J. Hackett, Olga Amsterdamska, Michael E. Lynch, and Judy Wajcman, 297–317. Cambridge, MA: MIT Press.

Burri, Regula Valérie, Cornelius Schubert, and Jörg Strübing. 2011. "The Five Senses of Science." *Science, Technology, & Innovation Studies* 7 (1): 3–7.

Butler, Judith. 1988. "Performative Acts and Gender Constitution: An Essay in Phenomenology and Feminist Theory." *Theatre Journal* 40 (4): 519–31.

Callon, Michel. 2007. "Performative Economics." In *Do Economists Make Markets? On the Performativity of Economics*, edited by Donald A. MacKenzie, Fabian Muniesa, and Lucia Siu, 311–57. Princeton, NJ: Princeton University Press.

Callon, Michel, Pierre Lascoumes, and Yannick Barthe. 2009. *Acting in an Uncertain World.* Cambridge, MA: MIT Press.

Catts, Oron, and Ionat Zurr. 2002. "Growing Semi-Living Sculptures: The Tissue Culture Art Project." *Leonardo* 35 (4): 365–70.

Century, Michael. 1999. *Pathways to Innovation in Digital Culture.* Centre for Research on Canadian Cultural Industries and Institutions / Next Century Consultants. Accessed April 18, 2015, at http://www.nextcentury.ca/PI/PImain.html.

Coletta, Claudio, Sara Colombo, Paolo Magaudda, Alvise Mattozzi, Laura Lucia Parolin, and Lucia Rampino, eds. 2014. *A Matter of Design: Making Society through Science and Technology.* Milano: STS Italia, Open Access Digital Publication (stsItalia.org).

Conquergood, Dwight. 1989. "Poetics, Play, Process, and Power: The Performative Turn in Anthropology." *Text and Performance Quarterly* 9 (1): 82–95.

Coopmans, Catelijne, Janet Vertesi, Michael Lynch, and Steve Woolgar, eds. 2014. *Representation in Scientific Practice Revisited.* Cambridge, MA: MIT Press.

Csordas, Thomas J. 1993. "Somatic Modes of Attention." *Cultural Anthropology* 8 (2): 135–56.

Cussins, Charis. 1996. "Ontological Choreography: Agency through Objectification in Infertility Clinics." *Social Studies of Science* 26 (3): 575–610.

da Costa, Beatriz, and Kavita Philip, eds. 2008. *Tactical Biopolitics: Art, Activism, and Technoscience.* Cambridge, MA: MIT Press.

Dagognet, François. 1992. *Étienne-Jules Marey: A Passion for the Trace.* New York: Zone Books.

de Ridder-Vignone, Kathryn. 2012. "Public Engagement and the Art of Nanotechnology." *Leonardo* 45 (5): 433–38.

de Ridder-Vignone, Kathryn, and Michael Lynch. 2012. "Images and Imaginations: An Exploration of Nanotechnology Image Galleries." *Leonardo* 45 (5): 447–54.

Dewsbury, John David. 2003. "Witnessing Space: 'Knowledge without Contemplation'." *Environment and Planning A* 35 (11): 1907–32.

DiSalvo, Carl. 2012. *Adversarial Design.* Cambridge, MA: MIT Press.

Dixon, Deborah P. 2009. "Creating the Semi-Living: On Politics, Aesthetics and the More than Human." *Transactions of the Institute of British Geographers* 34 (4): 411–25.

Dixon, Deborah, Elizabeth Straughan, and Harriet Hawkins. 2011. "When Artists Enter the Laboratory." *Science* 331 (6019): 860.

Dreyfus, Hubert L. 1979. *What Computers Can't Do: The Limits of Artificial Intelligence*, vol. 1972. New York: Harper and Row.

Dumit, Joseph. 2004. *Picturing Personhood: Brain Scans and Biomedical Identity.* Princeton, NJ: Princeton University Press.

Dunne, Anthony, and Fiona Raby. 2013. *Speculative Everything: Design, Fiction, and Social Dreaming.* Cambridge, MA: MIT Press.

Eglash, Ron, ed. 2004. *Appropriating Technology: Vernacular Science and Social Power.* Minneapolis: University of Minnesota Press.

Ehn, Pelle. 2008. "Participation in Design Things." In *Proceedings of the Tenth Anniversary Conference on Participatory Design 2008*, 92–101. Indianapolis: Indiana University.

___. 2011. "Design Things: Drawing Things Together and Making Things Public." *Tecnoscienza* 2 (1): 31–52.

Farías, Ignacio, and Thomas Bender, eds. 2010. *Urban Assemblages: How Actor-Network Theory Changes Urban Studies.* New York: Routledge.

Fischer, Michael M. J. 2003. *Emergent Forms of Life and the Anthropological Voice.* Durham, NC: Duke University Press.

___. 2009. *Anthropological Futures.* Durham, NC: Duke University Press.

Fischlin, Daniel, Ajay Heble, and George Lipsitz. 2013. *The Fierce Urgency of Now: Improvisation, Rights, and the Ethics of Cocreation.* Durham, NC: Duke University Press.

Fouché, Rayvon. 2003. *Black Inventors in the Age of Segregation: Granville T. Woods, Lewis H. Latimer, and Shelby J. Davidson*. Baltimore: Johns Hopkins University Press.

___. 2006. "Say It Loud, I'm Black and I'm Proud: African Americans, American Artifactual Culture, and Black Vernacular Technological Creativity." *American Quarterly* 58 (3): 639–61.

Fourmentraux, Jean-Paul. 2007. "Governing Artistic Innovation: An Interface among Art, Science and Industry." *Leonardo* 40 (5): 489–92.

___. 2010 "Linking Art and Sciences, an Organizational Dilemma. About Hexagram Consortium (Montreal, Canada)." *Creative Industries Journal* 3 (2): 137–50.

___. 2011. *Artistes de laboratoire: Recherche et création à l'ère numérique*. Paris: Hermann.

Frow, Emma, and Jane Calvert. 2013. "Opening Up the Future(s) of Synthetic Biology." *Futures* 48: 32–43.

Gabrys, Jennifer, and Kathryn Yusoff. 2012. "Arts, Sciences and Climate Change: Practices and Politics at the Threshold." *Science as Culture* 21 (1): 1–24.

Galison, Peter, and Emily Thompson. eds. 1999. *The Architecture of Science*. Cambridge, MA: MIT Press.

Garfinkel, Harold. 1967. *Studies in Ethnomethodology*. Englewood Cliffs, NJ: Prentice-Hall.

Gere, David. 2003. "Introduction." In *Taken by Surprise: A Dance Improvisation Reader*, edited by Ann Cooper Albright and David Gere, xiii–xvii. Middletown, CT: Wesleyan University Press.

Gibbons, Michael, Camille Limoges, Helga Nowonty, and Simon Schwarzman. 1994. *The New Production of Knowledge: The Dynamics of Science and Research in Contemporary Societies*. London: Sage.

Ginsberg, Alexandra Daisy, Jane Calvert, Pablo Schyfter, Alistair Elfick, and Drew Endy. 2014. *Synthetic Aesthetics: Investigating Synthetic Biology's Designs on Nature*. Cambridge, MA: MIT Press.

Goffman, Erving. 1959. *The Presentation of Self in Everyday Life*. Garden City, NY: Doubleday.

Goldberg, Rose Lee. 1976. *Performance Art: From Futurism to the Present*. New York: Harry Abrams.

Goldman, Danielle. 2010. *I Want to Be Ready: Improvised Dance as a Practice of Freedom*. Ann Arbor: University of Michigan Press.

Grau, Oliver. 2003. *Virtual Art: From Illusion to Immersion*. Cambridge, MA: MIT Press.

Gregg, Melissa, and Gregory J. Seigworth, eds. 2010. *The Affect Theory Reader*. Durham, NC: Duke University Press.

Hacking, Ian. 1983. *Representing and Intervening: Introductory Topics in the Philosophy of Natural Science*. Cambridge: Cambridge University Press.

Hallam, Elizabeth, and Tim Ingold, eds. 2007. *Creativity and Cultural Improvisation* (vol. 44 of Association of Social Anthropologists Monographs). Oxford: Berg.

Halpern, Megan. 2010. "Science Theater." In *Encyclopedia of Science and Technology Communication*, edited by Susanna Hornig Priest, 742–46. Thousand Oaks, CA: Sage.

Halpern, Orit, Jesse LeCavalier, Nerea Calvillo, and Wolfgang Pietsch. 2013. "Test-Bed Urbanism." *Public Culture* 25 (2): 272–306.

Hannah, Dehlia. 2013. "Art and Climate Change References." In *Climate Change: An Encyclopedia of Science and History*, edited by Brian C. Black, David M. Hassenzahl, Jennie C. Stephens, Gary Weisel, and Nancy Gift, 90–96. Santa Barbara, CA: ABC-CLIO.

Haraway, Donna J. 1988. "Situated Knowledges: The Science Question in Feminism and the Privilege of Partial Perspective." *Feminist Studies* 14 (3): 575–99.

___. 1991. *Simians, Cyborgs, and Women: The Reinvention of Nature*. London: Free Association Books.

Heble, Ajay, and Rebecca Caines, eds. 2014. *The Improvisation Studies Reader: Spontaneous Acts*. London: Routledge.

Helmreich, Stefan. 2015. *Sounding the Limits of Life: Essays in the Anthropology of Biology and Beyond*. Princeton, NJ: Princeton University Press.

Hennion, Antoine. 1989. "An Intermediary between Production and Consumption: The Producer of Popular Music." *Science, Technology, & Human Values* 14 (4): 400–424.

Herzig, Rebecca. 2004. "On Performance, Productivity, and Vocabularies of Motive in Recent Studies of Science." *Feminist Theory* 5 (2): 127–47.

Hilgartner, Stephen. 2000. *Science on Stage: Expert Advice as Public Drama*. Stanford, CA: Stanford University Press.

Hirsch, Tad. 2008. *Contestational Design: Innovation for Political Activism*. Unpublished dissertation. MIT, Department of Architecture, Program in Media Arts and Sciences.

Howes, David. 2013. "The Expanding Field of Sensory Studies." Accessed April 18, 2015, at http://www.sensorystudies.org/sensorial-investigations/the-expanding-field-of-sensory-studies/.

Hustak, Carla, and Natasha Myers. 2012. "Involutionary Momentum: Affective Ecologies and the Sciences of Plant/Insect Encounters." *differences: A Journal of Feminist Cultural Studies* 23 (3): 74–118.

Irani, Lilly, Paul Dourish, and Melissa Mazmanian. 2010. "Shopping for Sharpies in Seattle: Mundane Infrastructures of Transnational Design." In *Proceedings of the 3rd International Conference on Intercultural Collaboration*, 39–48. New York: ACM.

Irani, Lilly C., and M. Six Silberman. 2013. "Turkopticon: Interrupting Worker Invisibility in Amazon Mechanical Turk." In *Proceedings of the SIGCHI Conference on Human Factors in Computing Systems*, 611–20. New York: ACM.

Jones, Caroline A., and Peter Galison, eds. 1998. *Picturing Science, Producing Art*. New York: Routledge.

Kac, Eduardo. 1998. "Transgenic Art." *Leonardo Electronic Almanac* 6 (11).

Kahn, Douglas. 2007. "Between a Bach and a Bard Place: Productive Constraint in Early Computer Arts." In *MediaArtHistories*, edited by Oliver Grau, 423–51. Cambridge, MA: MIT Press.

Karasti, Helena, and Karen S. Baker. 2008. "Community Design: Growing One's Own Information Infrastructure." In *Proceedings of the Tenth Anniversary Conference on Participatory Design 2008*, 217–20. New York: ACM.

Knorr Cetina, Karin. 1992. "The Couch, the Cathedral, and the Laboratory: On the Relationship between Experiment and Laboratory in Science." In *Science as Practice and Culture*, edited by Andrew Pickering, 113–38. Chicago: University of Chicago Press.

___. 1999. *Epistemic Cultures: How the Sciences Make Knowledge*. Cambridge, MA: Harvard University Press.

Kwastek, Katja. 2013. *Aesthetics of Interaction in Digital Art*. Cambridge, MA: MIT Press.

Landecker, Hannah. 2006. "Microcinematography and the History of Science and Film." *Isis* 97 (1): 121–32.

Latour, Bruno. 1986. "Visualization and Cognition: Thinking with Eyes and Hands." *Knowledge and Society: Studies in the Sociology of Culture Past and Present* 6: 1–40.

___. 1987. *Science in Action: How to Follow Scientists and Engineers through Society*. Cambridge, MA: Harvard University Press.

___. 2008. "A Cautious Prometheus? A Few Steps toward a Philosophy of Design (with Special Attention to Peter Sloterdijk)." Keynote lecture for the *Networks of Design* Meeting of the Design History Society. Falmouth, Cornwall, 3 September 2008.

Latour, Bruno, and Peter Weibel, eds. 2002. *ICONOCLASH: Beyond the Image Wars in Science, Religion, and Art*. Cambridge, MA: MIT Press.

___, eds. 2005. *Making Things Public: Atmospheres of Democracy*. Cambridge, MA: MIT Press.

Latour, Bruno, and Steve Woolgar. [1979] 1986. *Laboratory Life: The Construction of Scientific Facts*. Princeton, NJ: Princeton University Press.

Law, John. 2004. *After Method: Mess in Social Science Research*. London: Routledge.

Law, John, and Vicky Singleton. 2000. "Performing Technology's Stories." *Technology and Culture* 41 (4): 765–75.

Le Dantec, Christopher A. 2010. "Situating Design as Social Creation and Cultural Cognition." *CoDesign* 6 (4): 207–24.

Le Dantec, Christopher A., and Carl DiSalvo. 2013. "Infrastructuring and the Formation of Publics in Participatory Design." *Social Studies of Science* 43 (2): 241–64.

Lerman, Nina, Ruth Oldenziel, and Arwen P. Mohun. 2003. *Gender and Technology: A Reader*. Baltimore: Johns Hopkins University Press.

Lindtner, Silvia, and David Li. 2012. "Created in China: The Makings of China's Hackerspace Community." *Interactions* 19 (6): 18–22.

Loukissas, Yanni A. 2012. *Co-Designers: Cultures of Computer Simulation in Architecture*. New York: Routledge.

Lury, Celia, and Nina Wakeford, eds. 2012. *Inventive Methods: The Happening of the Social*. Abingdon: Routledge.

Lynch, Michael, and Steve Woolgar, eds. 1990. *Representation in Scientific Practice*. Cambridge, MA: MIT Press.

MacKenzie, Donald, and Judy Wajcman, eds. 1999. *The Social Shaping of Technology*. 2nd ed. Buckingham, PA: Open University Press.

Maeda, John. 2013. "Foreword." In *The Art of Critical Making: Rhode Island School of Design on Creative Practice*, edited by Rosanne Somerson and Mara L. Hermano, 5–10. Hoboken, NJ: Wiley.

Malafouris, Lambros. 2008. "At the Potter's Wheel: An Argument for Material Agency." In *Material Agency: Towards a Non-Anthropocentric Approach*, edited by Carl Knappett and Lambros Malafouris, 19–36. New York: Springer.

Marres, Noortje. 2007. "The Issues Deserve More Credit: Pragmatist Contributions to the Study of Public Involvement in Controversy." *Social Studies of Science* 37 (5): 759–80.

Mauss, Marcel. 1973. "Techniques of the Body." *Economy and Society* 2 (1): 70–88.

Meyer, Morgan. 2014. "Hacking Life? The Politics and Poetics of DIY Biology." In *META-Life: Biotechnologies, Synthetic Biology, ALife and the Arts*, edited by Annik Bureaud, Roger F. Malina, and Louise Whiteley. Cambridge, MA: MIT Press, Leonardo e-Book Series.

Michael, Mike. 2012. "'What Are We Busy Doing?' Engaging the Idiot." *Science, Technology, & Human Values* 37 (5): 528–54.

Milburn, Colin. 2015. *Mondo Nano: Fun and Games in the World of Digital Matter*. Durham, NC: Duke University Press.

Mody, Cyrus C. M. 2005. "The Sounds of Science: Listening to Laboratory Practice." *Science, Technology, & Human Values* 30 (2): 175–98.

Mol, Annemarie. 2002. *The Body Multiple: Ontology in Medical Practice*. Durham, NC: Duke University Press.

Mol, Annemarie, and John Law. 2004. "Embodied Action, Enacted Bodies: The Example of Hypoglycaemia." *Body & Society* 10 (2–3): 43–62.

Montuori, Alfonso. 2003. "The Complexity of Improvisation and the Improvisation of Complexity: Social Science, Art and Creativity." *Human Relations* 56 (2): 237–55.

Moore, Stevan A., and Andrew Karvonen. 2008. "Sustainable Architecture in Context: STS and Design Thinking." *Science Studies* 21 (1): 29–46.

Moten, Fred. 2003. *In the Break: The Aesthetics of the Black Radical Tradition.* Minneapolis: University of Minnesota Press.

Myers, Natasha. 2015. *Rendering Life Molecular: Models, Modelers, and Excitable Matter.* Durham, NC: Duke University Press.

Myers, Natasha, and Joe Dumit. 2011. "Haptic Creativity and the Mid-Embodiments of Experimental Life." In *A Companion to the Anthropology of the Body and Embodiments*, edited by Frances E. Mascia-Lees, 239–61. Chichester: Blackwell.

Nelson, Andrew J. 2015. *The Sound of Innovation: Stanford and the Computer Music Revolution.* Inside Technology Series. Cambridge, MA: MIT Press.

Noë, Alva. 2004. *Action in Perception.* Cambridge, MA: MIT Press.

Oldenziel, Ruth. 1999. *Making Technology Masculine: Men, Women and Modern Machines in America, 1870–1945.* Amsterdam: Amsterdam University Press.

Pang, Laikwan. 2012. *Creativity and Its Discontents: China's Creative Industries and Intellectual Property Rights Offenses.* Durham, NC: Duke University Press.

Paxson, Heather. 2013. *The Life of Cheese: Crafting Food and Value in America*, vol. 41. Berkeley: University of California Press.

Phelan, Peggy. 1993. *Unmarked: The Politics of Performance.* New York: Routledge.

Pickering, Andrew. 1995. *The Mangle of Practice: Time, Agency, and Science.* Chicago: University of Chicago Press.

___. 2010. "Material Culture and the Dance of Agency." In *The Oxford Handbook of Material Culture Studies*, edited by Dan Hicks and Mary C. Beaudry, 191–208. Oxford: Oxford University Press.

Pinch, Trevor, and Karin Bijsterveld. 2004. "Sound Studies: New Technologies and Music." *Social Studies of Science* 34 (5): 635–48.

Pinch, Trevor, and Christine Leuenberger. 2006. "Studying Scientific Controversy from the STS Perspective." Accessed November 20, 2015, at http://stspo.ym.edu.tw/easts/concluding%20 remarks.doc.

Pipek, Volkmar, and Volker Wulf. 2009. "Infrastructuring: Toward an Integrated Perspective on the Design and Use of Information Technology." *Journal of the Association for Information Systems* 10 (5): 1.

Polanyi, Michael. 1967. *The Tacit Dimension.* London: Routledge and Kegan Paul.

Ratto, Matt. 2009. "Critical Making: Conceptual and Material Studies in Technology and Social Life." Hybrid Design Practices workshop, Ubicomp, Orlando, Florida, September 30–October 3.

___. 2011. "Critical Making: Conceptual and Material Studies in Technology and Social Life." *The Information Society* 27 (4): 252–60.

Ratto, Matt, and Megan Boler, eds. 2014. *DIY Citizenship: Critical Making and Social Media*. Cambridge, MA: MIT Press.

Ratto, Matt, and Garnet Hertz. 2014. "Critical Making." Special Issue on The Culture of Digital Education: Innovation in Art, Design, Science and Technology Practices. *Leonardo Electronic Almanac*.

Reichle, Ingeborg. 2009. *Art in the Age of Technoscience: Genetic Engineering, Robotics, and Artificial Life in Contemporary Art*. Vienna: Springer.

Rheinberger, Hans-Jörg. 1997. *Toward a History of Epistemic Things: Synthesizing Proteins in the Test Tube*. Stanford, CA: Stanford University Press.

Riley, Shannon Rose, and Lynette Hunter, eds. 2009. *Mapping Landscapes for Performance as Research: Scholarly Acts and Creative Cartographies*. Basingstoke: Palgrave Macmillan.

Rogers, Hannah Star. 2011. "Amateur Knowledge: Public Art and Citizen Science." *Configurations* 19 (1): 101–15.

Roosth, Sophia. 2009. "Screaming Yeast: Sonocytology, Cytoplasmic Milieus, and Cellular Subjectivities." *Critical Inquiry* 35 (2): 332–50.

___. 2013a. "Biobricks and Crocheted Coral: Dispatches from the Life Sciences in the Age of Fabrication." *Science in Context* 26 (1): 153–71.

___. 2013b. "Of Foams and Formalisms: Scientific Expertise and Craft Practice in Molecular Gastronomy." *American Anthropologist* 115 (1): 4–16.

Salter, Chris. 2010. *Entangled: Technology and the Transformation of Performance*. Cambridge, MA: MIT Press.

___. 2015. *Alien Agency: Experimental Encounters with Art in the Making*. Cambridge, MA: MIT Press.

Schaffer, Amanda. 2008. "Prescriptions for Health, the Environmental Kind." *New York Times* (online publication). Accessed November 20, 2015, at http://www.nytimes.com/2008/08/12/health/12clin.html.

Schmidgen, Henning. 2011. "1900—The Spectatorium. On Biology's Audio-Visual Archive." *Grey Room* 43: 42–65

Schneider, Rebecca. 1997. *The Explicit Body in Performance*. New York: Routledge.

Scott, Jill, and Esther Stoeckli, eds. 2012. *Neuromedia: Art and Neuroscience Research*. Heidelberg: Springer.

Sengers, Phoebe, Kirsten Boehner, Shay David, and Joseph Kaye. 2005. "Reflective Design." In *Proceedings of the 4th Decennial Conference on Critical Computing: Between Sense and Sensibility*, 49–58. New York: ACM.

Shanken, Edward A., ed. 2003. "From Cybernetics to Telematics: The Art, Pedagogy, and Theory of Roy Ascott." In *Telematic Embrace: Visionary Theories of Art, Technology, and Consciousness*, 1–96. Berkeley: University of California Press.

Shapin, Steven. 1994. *A Social History of Truth: Civility and Science in Seventeenth-Century England.* Chicago: University of Chicago Press.

Shapin, Steven, and Christopher Lawrence. 1998. "Introduction: The Body of Knowledge." In *Science Incarnate: Historical Embodiments of Natural Knowledge*, edited by Steven Shapin and Christopher Lawrence, 1–19. Chicago: University of Chicago Press.

Shilton, Katie. 2012. "Values Levers: Building Ethics into Design." *Science, Technology, & Human Values* 38 (3): 374–97.

Shove, Elizabeth. 2014. "On the Design of Everyday Life." *Tecnoscienza* 5 (2): 33–42.

Söderberg, Johan, and Alessandro Delfanti. 2015. "Hacking Hacked! The Life Cycles of Digital Innovation." *Science, Technology, & Human Values* 40 (5): 793–98.

Sofia, Zoë. 1993. *Whose Second Self? Gender and (Ir)rationality in Computer Culture.* Geelong, Victoria: Deakin University Press.

Star, Susan Leigh, and Karen Ruhleder. 1996. "Steps toward an Ecology of Infrastructure: Design and Access for Large Information Spaces." *Information Systems Research* 7 (1): 111–34.

Steen, Marc. 2015. "Upon Opening the Black Box and Finding It Full: Exploring the Ethics in Design Practices." *Science, Technology, & Human Values* 40 (3): 389–420.

Storni, Cristiano. 2012. "Unpacking Design Practices: The Notion of Thing in the Making of Artifacts." *Science, Technology, & Human Values* 37 (1): 88–123.

Suchman, Lucy. 1987. *Plans and Situated Actions: The Problem of Human-Machine Communication.* New York: Cambridge University Press.

____. 2004a. "Talking Things." In *First Person: New Media as Story, Performance, and Game*, edited by Noah Waldrip-Fruin, 262–65. Cambridge, MA: MIT Press.

____. 2004b. "Decentering the Manager/Designer." In *Managing as Designing*, edited by Richard J. Boland Jr. and Fred Collopy, 169–73. Stanford, CA: Stanford Business Books.

____. 2007. *Human-Machine Reconfigurations.* New York: Cambridge University Press.

____. 2011. "Anthropological Relocations and the Limits of Design." *Annual Review of Anthropology* 40: 1–18.

Supper, Alexandra. 2014. "Sublime Frequencies: The Construction of Sublime Listening Experiences in the Sonification of Scientific Data." *Social Studies of Science* 44 (1): 34–58.

____. 2015. "Data Karaoke: Sensory and Bodily Skills in Conference Presentations." *Science as Culture* 24 (4): 436–57.

Turkle, Sherry. 2005. *The Second Self: Computers and the Human Spirit.* Cambridge, MA: MIT Press.

Turner, Victor, and Richard Schechner. 1983. *Between Anthropology and Performance*. New York: Routledge.

van de Port, Mattijs, and Annemarie Mol. 2015. "Chupar Frutas in Salvador da Bahia: A Case of Practice-Specific Alterities." *Journal of the Royal Anthropological Institute* 21 (1): 165–80.

Varela, Francisco, Evan Thompson, and Eleanor Rosch. 1991. *The Embodied Mind: Cognitive Science and Human Experience*. Cambridge, MA: MIT Press.

Venturini, Tommaso. 2010. "Diving in Magma: How to Explore Controversies with Actor-Network Theory." *Public Understanding of Science* 19 (3): 258–73.

Verran, Helen. 2002. "A Postcolonial Moment in Science Studies: Alternative Firing Regimes of Environmental Scientists and Aboriginal Landowners." *Social Studies of Science* 32 (5–6): 729–62.

Volonté, Paolo. 2014. "Design Worlds and Science and Technology Studies." *Tecnoscienza* 5 (2): 5–14.

Ward, Matt, and Alex Wilkie. 2009 "Made in Criticalland: Designing Matters of Concern." *Networks of Design: Proceedings of the 2008 Annual International Conference of the Design History Society (UK)*. Conference paper. Goldsmiths Research Online.

Wenger, Étienne. 1998. *Communities of Practice: Learning, Meaning, and Identity*. Cambridge: Cambridge University Press.

Whitson, Roger. 2013. *Critical Making in Digital Humanities: A MLA 2014 Special Session Proposal*. Accessed February 10, 2015, at http://www.rogerwhitson.net/?p=2026.

Winograd, Terry, and Fernando Flores. 1986. *Understanding Computers and Cognition: A New Foundation for Design*. Norwood, NJ: Ablex.

Woodhouse, Edward, and Jason W. Patton. 2004. "Design by Society: Science and Technology Studies and the Social Shaping of Design." *Design Issues* 20 (3): 1–12.

Wylie, Caitlin Donahue. 2015. "'The Artist's Piece Is Already in the Stone': Constructing Creativity in Paleontology Laboratories." *Social Studies of Science* 45 (1): 31–55.

Wylie, Sara A., Kirk Jalbert, Shannon Dosemagen, and Matt Ratto. 2014. "Institutions for Civic Technoscience: How Critical Making Is Transforming Environmental Research." *The Information Society* 30 (2): 116–26.

Yaneva, Albena. 2005. "Scaling Up and Down: Extraction Trials in Architectural Design." *Social Studies of Science* 35 (6): 867–94.

___. 2009. "Making the Social Hold: Towards an Actor-Network Theory of Design." *Design and Culture* 1 (3): 273–88.

___. 2012. *Mapping Controversies in Architecture*. Farnham: Ashgate.

6 Engaging, Designing, and Making Digital Systems

Janet Vertesi, David Ribes, Laura Forlano, Yanni Loukissas, and Marisa Leavitt Cohn

Introduction

Science and technology studies (STS) has a long history of studying information technologies in social contexts (Boczkowski 2005; Edwards 1997; Knorr Cetina and Bruegger 2000; Woolgar 1990). However, like scholarly interventions in the arenas of law (Lynch and Cole 2005), policy (Jasanoff 1990), or science communication (Oreskes and Conway 2010), scholars in STS have also gone beyond analysis to impact the *design* of information technologies. This chapter describes four modes of engagement (Ribes and Baker 2007) and associated sites of scholarly intersection that constitute rich trading zones (Galison 1997) between STS and design. We identify these engagements as *corporate*, *critical*, *inventive*, or focused on *inquiry*, depending on the nature of the exchange between STS principles and design practice. We use examples of classic and contemporary instantiations of these interactions to describe current debates about the role of STS research in the design of new technologies.

STS is an evolving interdisciplinary field with porous boundaries. An increasing number of scholars who identify with STS hold positions today in institutions that specialize in the design of new technologies, while many technology designers find themselves working in STS-aligned humanities and social sciences programs. We describe these trading zones in North America, the United Kingdom, and Europe as they were guided and shaped by constellations of people, institutions, publication venues, conferences, and research practices. In some cases, STS practitioners contribute either directly or indirectly to building new technologies; in others, they critique the products or the practices of design; in still others they build digital systems or suggest novel processes for technical work; and across the board they confront hybrid identities and novel arrangements of expertise alongside their STS commitments. While we do not offer a definitive taxonomy, or a pathway for scholars seeking to bridge these fields, the modes of interaction that characterize these existing spaces encourage continued cross-fertilization, and opportunities for future generations of designers and STS practitioners.

Corporate Engagement: Anthropologists in the Laboratory

Just as STS work has informed legal institutions (Jasanoff 1990), STS concepts and theories are drawn upon and co-articulated in the development of information technologies in industry centers such as Silicon Valley. Here, STS approaches offered more nuance in describing the relationships between the social and the technical, or the human and the machine, than narratives from engineering disciplines. In what we term *corporate* relationships between STS scholars and their technical research sites, STS concepts are somewhat uniquely "made durable" (Latour 1991) as researchers' contributions are incorporated into information systems with broad distribution and reach, presenting implications for downstream usage and the development of future technologies. At the same time, elements from information design also entered into STS.

An example is the case of Xerox's Palo Alto Research Center (PARC), the location from which Lucy Suchman conducted her seminal work in human-machine interaction in the early 1980s. Within STS, Suchman's early work is known for exploring the micropractices of sensemaking with machines, respecifying cognition as a situated, practical, and negotiated achievement. However, Suchman's writing was also influential among a community of technical researchers and designers working on collaboration tools and environments, human-computer interaction, and artificial intelligence. Her empirical study was, after all, a story of users struggling to use a photocopier conducted under the auspices of the company that made those very machines (Suchman 1987).

Suchman's work challenged a naive rationalist understanding of how people engage with artifacts through stepwise plans, offering an alternative model in which users are simultaneously interacting with machines, making sense of their interactions, and reworking their plans. To STS audiences, this notion of continual sensemaking became a formative description of human-machine (mis-)communications; to human-computer interaction (HCI) audiences, this was read as a substantive critique of dominant artificial intelligence models that envisioned the accomplishment of a goal (in this case, making a photocopy) as acting out a fully articulated plan to be followed stepwise. This latter audience grasped hold of concepts such as situated action, the mutual understandings of humans and machines, and local practices of sensemaking in a way that continues to suffuse publications in these subdisciplines in computer science (e.g.,

Rooksby 2013). When we completed this chapter, Suchman's *Plans and Situated Actions* had over 800 citations in the Association of Computing Machinery's digital library.

Suchman's book was published in the same year as *The Social Construction of Technological Systems,* edited by Wiebe Bijker, Thomas Hughes, and Trevor Pinch (1987), and just as an emerging community of science studies scholars began to turn their attention to technology and to machines. However, her work was also produced in conversation with the growing fields of human-computer interaction, artificial intelligence, and the design of everyday objects. The site of these encounters matter. By the early 1980s Xerox PARC was an important node in the dense sociotechnical networks of Silicon Valley, then transitioning from semiconductor hub, to biotech hub, to software hub (Saxenian 1996). Having produced the technologies such as the graphical user interface and the mouse (later appropriated by Apple Computers), PARC boasted tight linkages to Stanford University and to technology companies in the region. Its proponents— individuals like John Seely Brown—moved easily between research and building institutions. Suchman was one of several ethnographers that PARC hired in the 1980s to assist with product development and workforce management issues.

It is in this context that a review of Suchman's book appeared in the journal *Artificial Intelligence.* Its author, computer scientist Phil Agre, was an interlocutor between cognitive science, system design, and social studies of cooperative work (Agre 1990). Within parallel university-industry laboratories and Ph.D. training programs on the East Coast of the United States centered around MIT, *Plans and Situated Actions* became a staple reference in the growing fields of AI and human-computer interaction (Wania, Atwood, and McCain 2006). The situated action approach is evident in the work of Rodney Brooks, former director of the MIT Artificial Intelligence Lab and founder of iRobot, whose insect-like robot prototypes, Attila and Hannibal, and ultimately the Roomba espouse "intelligence without representation" (Brooks, 1991; Brooks's Cog robot is critiqued in Suchman 2007). Suchman's insights not only circulated within communities dedicated to artificial intelligence and information technology design, they were ultimately embedded in the design of novel technical artifacts.

Qualitative insights provided to the company by this community of researchers were also called upon to "improve" the company itself. At Xerox, Julian Orr documented how photocopy technicians would meet for breakfast, lunch, dinner, or drinks to share "war stories"—the informative narratives characterizing their successes and challenges working with the machines that technicians shared with one another rather than referring to lengthy manuals (Orr 1996). Orr's work is considered a classic for STS scholars interested in tacit knowledge, labor relations, and maintenance. But Orr's analysis was also appreciated by management (Brown 1998; Contu and Willmott 2003). In their popular press book *The Social Life of Information*, Paul Duguid and John Seely Brown, head of PARC during Orr's appointment, recount how the company responded to Orr's work by giving repair technicians walkie-talkies so that they could exchange war stories over radio and eliminate long lunch breaks. Xerox also created a knowledge management database where technicians could contribute their expertise, ensuring it would be "captured" by the company in case a technician took a new job, retired, or was let go—a system that saved Xerox $100 million in yearly service costs (Brown and Daguid 2000).

Companies like Yahoo, Google, Intel, and Microsoft continue to host qualitative researchers trained in anthropology or sociology who draw upon the insights of STS analysis in the construction of novel artifacts. There are tensions inherent to crossover work in this style of engagement with industrial partners (Suchman and Bishop 2000). Situated as they are at the intersection of academic and corporate interests, researchers may struggle to make their work legible to both the field of STS and a company's priorities or bottom line. Those who work in this vein of corporate engagement may yet level critiques at the functioning or the intellectual commitments inherent to the sociotechnical systems at hand as part of their published scholarship from within the institutional frame of the company. For instance, scholars affiliated with Microsoft's and Intel's research units have critiqued "ICT for development" using postcolonial frameworks (Taylor 2011), the politics of big data and algorithms (boyd and Crawford 2012; Gillespie 2014), and computer system deployment in global contexts (Dourish and Bell 2011). STS crossings in this trading zone site—where the action is, in terms of technological production—lend access to resources with which to conduct international research and produce impact within computer science and corporate policy, even as it can provoke questions in the discipline as to the outcomes of STS work.

Critical Engagement: Cyberinfrastructures and Values in Design

While corporate engagements involve researchers embedded in technology companies who deploy critiques that influence design, we also identify an additional form of "critical" engagement, one in which the STS scholar analyzes the politics or practices of information technology design while embedded at a site of use, influencing production but without contributing to a "bottom line." Like Lynch and Cole's unpacking of biological technologies in judicial contexts (2005), such scholarship intersects with STS work on the politics of knowledge construction by situating the design of information technologies as political work, using the tools of the sociology of knowledge. These scholars are typically in conversation with designers of IT artifacts and may offer "implications for design" but operate principally from within an academic or otherwise nonprofit context.

Susan Leigh Star's work and that of her students and collaborators is a touchstone example. Star's influential study of infrastructure was published first at the Computer Supported Cooperative Work (CSCW) conference (Star and Ruhleder 1994)—a conference of ethnographers, organizational analysts, sociologists, computer scientists, information systems developers, and other digital tool builders sponsored by the Association of Computing Machinery—only later publishing the version familiar to STS (Star 1999). Star argued that infrastructure could be approached as a relation rather than a thing, that it need not be large or even necessarily "technical" (though it is always material and practical), and that the work of librarians, archivists, technicians, nurses, and homemakers could be considered infrastructural in the sense that they provide support or do articulation work needed for the accomplishment of interdependent tasks. For STS, this opened a new approach to the study of infrastructure, following the discussion of large technical systems that had developed from the work of Thomas Hughes. For computer science, this provided new tools for conceptualizing and designing for sociotechnical relations.

Star's perspectives on infrastructures and on boundary objects (Star and Greisemer 1989) remain deeply influential for a community of "researchers of the sociotechnical" in information schools, who maintain a Consortium for the Study of Sociotechnical Systems and associated summer school (www.sociotech.net). Star's collaborations with computer and information scientists, along with the "domain" scientists that such infrastructure serves, also inspired a lineage of on-the-ground, ethnographic investigations of the development, use, and sustainability of research computing conducted in close collaboration with developers (Ribes and Baker 2007). An important trading zone in this network is the CSCW conference itself (and its European sister conference, eCSCW), a conference of record for the scholarly and professional organization for computer engineering, the Association of Computing Machinery. CSCW papers deploy ethnographic and interview methods to produce findings about the current and prospective role for technologies in social and cultural life, and its attendees also participate in the Society for Social Studies of Science meetings and publish in *Science, Technology, & Human Values* (Jackson and Buyuktur 2014; Jackson et al. 2011; Ribes and Lee 2010; Vertesi and Dourish 2011; Zimmerman 2008).

Scholars of information systems and infrastructures in this critical mode frequently engage in close collaborations with their field sites through providing feedback and requirements analysis or by taking on governance roles in system-building endeavors. They take seriously the claim that the organizations and people they study are inseparable from the networks and infrastructures that connect them, and consider how to build systems that not only reflect local modes and practices but also can supplant inequalities and silences as well. Their guidelines and critiques for system development draw on classic STS themes such as residual categories, visibility and invisibility, boundary objects, and heterogeneous engineering (Bowker et al. 2010; Jirotka, Lee, and Olson 2013). Thus, their findings are not merely descriptive, but they aim to influence the production of new sociotechnical systems even as they provide critique.

Like STS studies of architectural practices and studios (Latour and Yaneva 2008), such studies may also suggest critical alternatives for design futures. For instance, over the past five years, a series of U.S. government–funded conferences on "values in design" led by Geof Bowker and Helen Nissenbaum further cemented a network of scholar-practitioners across these sites of STS and design practice, inspiring individuals to evaluate the values and ethical properties of digital devices and systems and to build alternatives (Knobel and Bowker 2011; Nissenbaum 2001). In a pedagogical exercise called "values at play," participants design a game to satisfy multiple conflicting sets of values, serving as a site for critical thinking about how privacy, democracy, or equality are embedded or abandoned in technology design (Flanagan and Nissenbaum 2014). In a related vein, design exercises by Ron Eglash develop "culturally situated design tools" to teach computer science and mathematical principles using non-Western frameworks (Eglash 1999).

If corporate work engages from within a technology company, *critical* engagement relies upon trading zone sites that link domain scientists, computer scientists, and engineers. Star's own academic appointment during the early phase of her career was at the University of California–Irvine's Department of Information and Computer Science alongside Jonathan Grudin, Rob Kling, Walt Scacchi, and others who formed a nascent field of social informatics. This forged enduring links across STS, sociology, organization studies, and computer science (Kling 1991; Orlikowski 2000; Sawyer and Tapia 2007), resulting in concepts such as "webs of computation" (Kling and Scacchi 1982) and the relationship between information categories and power (Pine and Liboiron 2015; Suchman 1993). To facilitate this continued exchange, interdisciplinary centers for computer science, the social sciences, and the humanities known as Information Schools (abbreviated as iSchools) spread rapidly in the early 2000s, frequently emerging from library schools and located near powerful centers of corporate computing such as the University of California–Berkeley, the University of Washington, and the University of Michigan. These sites provide an institutional home in the United States for the conversation between STS-inflected studies of technology and information technology design by providing critiques of the underlying politics and practices of design directed at both scholarship and practice.[1] University affiliations do not preclude corporate sponsorship. Individual grants and working relationships are common, including an experimental consortium funded by Intel Corporation, which linked researchers in this domain from 2011 to 2014.[2]

Contemporary work in this trading zone examines, for instance, mobile "constant connectivity" (Mazmanian and Erickson 2014), GPS tracking (Troshynski, Lee, and Dourish 2008), repair work (Jackson 2014), scientific collaboration (Jackson et al. 2011; Wynholds et al. 2011), sustainable design (Brynjarsdóttir et al. 2012; DiSalvo, Sengers, and Brynjarsdóttir 2010; Dourish 2010), and notions of innovation in hacker spaces and incubators (Lindtner, Hertz, and Dourish 2014). Such studies leverage critiques on modes of production and accounts of economic value using language from STS alongside organizational studies, criminology, and critical anthropology. Others import the language of multisitedness, transnationalism, or postcolonialism to illuminate the power structures reified in global IT circulation (Burrell 2012; Irani et al. 2010; Shklovski et al. 2014; Williams et al. 2014). In such studies, the implications are not necessarily system requirements generated from ethnographic observation. Instead STS, anthropology, and sociology provide perspectives on the design process itself, with implications for the development and implementation of new projects.

Critical engagements typically embrace intersections between IT research and corporations yet eschew immediate pay-offs for companies or designers. For example, Paul Dourish's oft-cited paper at the ACM flagship conference for Human-Computer Interaction argues against providing "implications for design" in papers that deploy STS, ethnographic, or anthropological tools and techniques (Dourish 2006). The author, himself once a Xerox PARC employee and now professor at University of California–Irvine's Informatics Department, argues that implications for theory or method are more appropriate than distilling complex social analysis into bullet points for engineering. In other cases, researchers draw on science studies and laboratory ethnography to support scientific work in nonprofit sectors or in academia. For instance, a generation of scholars trained with Bowker and Star took this perspective to the development and implementation of cyberinfrastructure, a major funding framework at the National Science Foundation at the turn of the twenty-first century, which is equivalent to the e-science initiative in Europe (Hine 2006). Results were published at trading zone venues like CSCW, *Science, Technology, & Human Values,* and MIT Press series (Bietz, Lee, and Baumer 2010; Bowker et al. 2010; Cohn 2016; Edwards et al. 2013; Lee, Dourish, and Mark 2006; Millerand et al. 2013; Ribes and Bowker 2008; Ribes 2014; Ribes and Lee 2010; Vertesi 2014). Still, while they speak to computer scientists through publication in relevant conferences and journals and provide insight into crafting information systems, the majority of studies in this vein are not expected to directly impact digital production.

Inventive Engagement: Reflective Design and Critical Engineering

"Everyone designs who devises courses of action aimed at changing existing situations into preferred ones," states design theorist Herbert Simon in what has become an influential definition of design (Simon [1969] 1996, 111). For many STS practitioners, the follow-up question is, preferred by whom? We characterize a third facet of work at the intersection of STS and information technology design as a form of inventive engagement that builds STS principles into digital artifacts. We use the term *inventive* to signal the interventionist character of their work. Like Oreskes and Conway's provocative book and film on climate change science, *Merchants of Doubt* (2010), these activities are outcome-focused. They produce artifacts that embody or put forward new politics of making, design, and use and that implicitly challenge dominant technical conceptions of those practices.

These scholars draw on foundational STS work, beginning with the concept of artifacts "having politics" (Winner 1986) and including Bowker and Star's (1999) sensibility to inequalities and residual categories in networked systems. They take the social construction of technology seriously as both a program of analysis and process of invention. They do not simply ask, "Could it have been otherwise?" by imagining variants of artifacts, interpretations, and social groups (as in Pinch and Bijker 1984); instead they explore alternative technological solutions and allow different actors to define the design space. Thus, they operationalize STS theories to accommodate excluded user communities, embody alternate value frameworks, or inspire reflective critique.

One example is the program of reflective design. Its goal is to inspire "critical reflection," such that users can make conscious choices about technologies instead of "unthinkingly adopt[ing] attitudes, practices, values, and identities we might not consciously espouse" (Sengers et al. 2005, 50). Reflective design therefore aims to inspire a form of lay STS work in users who encounter such systems, making visible the politics of design and inspiring alternatives to the increasingly consumption-oriented space of information technology. Reflective design is itself a product of many interinstitutional and disciplinary trading zones. The theory was developed during a period of considerable cross-over between Information Science and Science & Technology Studies at Cornell University. Moreover, lead author Sengers holds a dual Ph.D. in critical theory and artificial intelligence. The approach also draws explicitly on critical technical practice

(CTP), a sensibility described by Phil Agre, an early adopter of Suchman's work mentioned above. According to Agre, design in a critical vein should itself be a form of inquiry into the systematic failures, limitations, and confusions that arise in the design of digital systems (Agre 1997). Guided by critical theory, designers should see obstacles as opportunities not only for innovation but also for questioning the underlying values in technology design.

Scholars in this mode of engagement deploy STS concepts to build artifacts that reconfigure the intertwining of humans and machines, and to suggest alternative visions and distributions of power and agency. Their work is interlaced with STS theories and concepts, with outcomes encoded into software, hardware, or other interactive artifacts. For instance, Yanni Loukissas and David Mindell's *Visual Apollo* (2014)[3] is an interactive display of sociotechnical data from the lunar landing that invites new interpretations of mediated relations between humans and machines, even as it reveals the otherwise subtended labor of technicians working far below in ground control or behind the scenes of software interfaces. Lilly Irani and Michael Silberman's *Turkopticon* (Irani and Silberman 2013)[4] make visible the problems of contingent, invisible, and underpaid labor inherent in crowdsourcing platforms such as Amazon Mechanical Turk through a website that enables "Turkers" (i.e., Mechanical Turk workers) to speak back to their employers, thus inverting global flows of power and knowledge inherent in "postcolonial computing" (Irani et al. 2010; Philip, Irani, and Dourish 2012). Natalie Jeremijenko's feral robotic dogs, outfitted with sensors for toxicity and deployed by school children in poor urban centers, materially engage with both rhetoric and practices of participatory technology as educational project and critical urban geography (Jeremijenko 2009). Carl DiSalvo's program of adversarial design explicitly carves out space for designers interested in doing the work of social justice. Outcomes of this practice are explicitly meant to "represent and enact the political conditions of contemporary society and function as contestational objects that challenge and offer alternatives" (DiSalvo 2012, 115).

True to bidirectional exchange in this trading zone, affiliated design labs are also inflected with the critical stance offered by STS. This is visible in hybrid spaces of design such as at Goldsmiths College in London. There, Celia Lury and Nina Wakeford (2012) explore intersections of design, the social, and critical theory (ibid.) and Alex Wilkie and Jennifer Gabrys inhabit spaces of critical overlap between STS, sociology, and design (Farías and Wilkie 2015; Gabrys 2011; Wilkie and Michael 2009). In Paris, Bruno Latour's *Inquiry into the Modes of Existence* (2013) features interactive online text and commentary that repositions the reader as coauthor of the text, thereby troubling the ontological work and authority of authorship. In Berlin, the "Critical Engineering Manifesto" draws on actor-network theory to suggest that "the Critical Engineer expands 'machine' to describe interrelationships encompassing devices, bodies, agents, forces and networks" (Oliver, Savičić, and Vasiliev 2011–2014, 1), producing objects such as *Transparency Grenade* or *Border Bumping* that "subvert communication networks and other technopolitical infrastructure in order to begin productive, critical conversations regarding their impact and governance" (Oliver and Hart 2014). Such projects are digitally akin to those of artist-activist scholars such as Beatriz da Costa (2008), Eben Kirksey (Kirksey and Dragomar 2012), and Eugene Thacker (Thacker et al. 2005), whose work engages STS through the mutual entanglement of political geographies, cosmopolitics, and biotechnological futures.

Rather than eschewing technical modes of production, scholars in the inventive vein seek to transform IT design by folding the humanist critique of technology that STS offers into the practice. For instance, Phil Agre's intention in critical technical practice was not to deny the foundations of artificial intelligence but rather to infuse the field with critical theory. The authors of "Reflective Design" similarly suggest that critical reflection "can and should be a core principle of technology design for identifying blind spots and opening new design spaces" and that user and designer reflection offers "a crucial element of a socially responsible technology design practice" (Sengers et al. 2005, 49). While critical engineers are exhorted to "deconstruc[t] and incit[e] suspicion of rich user experiences" (i.e., user experiences to which scholars in the collaborative engagement mode contribute), they propose "the exploit" or "digital excavation" alongside other computational techniques to produce technological artifacts and art projects (Oliver, Savičić, and Vasiliev 2011–2014, 1).

Agre once posited that the critical designer would require a new infrastructure of praxis: "forms of language, career strategies, and social networks that support the exploration of alternative work practices that will inevitably seem strange to insiders and outsiders alike" (Agre 1997, 155). The scholars cited have sought to produce the infrastructure and networks to support this rich trading zone, integrating critical and constructive modes in a single professional persona. Their publication venues gather a like-minded audience across STS, computer science, and design at conferences like Research Through Design or Designing Interactive Systems (DIS). They teach at schools of design, information, or media alongside colleagues who have very different attitudes toward technology, critical theory, or both. Thus, these scholars deploy technology and design practice themselves as a way of putting STS into the world.

Design as Inquiry: Participatory Design and Speculative Fictions

Even as they attempt to instill design pedagogy and practice with the sensibility that STS analyses of technologies can offer, scholars in the critical mode do not aim to unseat the praxis or role of design in the manufacture of technological artifacts. In yet other cases, we see scholars offer alternative formulations of design practice itself, not simply its outcomes. We describe this as an intersection focused on *the problem of inquiry*. Scholars in this trading zone between design and STS ask how design practices might be reconfigured to embody different knowledge, models of expertise, and politics as inspired by STS analyses. Designs or artifacts may or may not result from these practices and the focus is rarely placed on the outcome. Like ACT-UP (AIDS Coalition to Unleash Power) activists' reformulation of pharmaceutical research as described by Epstein (1995), the result is a reformulation of both design and designer to reflect alternative associations and commitments.

A formative example is the tradition of participatory design (PD), a technique born in Scandinavia which brings disparate groups together to design for their own conditions of interaction instead of relying on an external designer or a managerial perspective on the organization to do so (e.g., Greenbaum and Kyng 1991; Schuler and Namioka 1993; Spinuzzi 2005). Participatory design is also linked with the human-computer interaction and CSCW communities, but its power lies in its reformulation of design practice as the requirement to bring multiple stakeholders together around complex issues. For instance, bringing workers or users into the design of their own tools and systems that embed and celebrate their tacit knowledge. Thus, users engage directly in shaping their technology rather than accepting technologies implemented by management for the purposes of control or efficiency. Such participation on the ground in the bottom-up building of technical systems and infrastructures establishes a "co-design" relationship (Sanders and Stappers 2008) that has effects on scientific and technical practice (Binder, Brandt, and Gregory 2008), shifts the role of the designer from expert to facilitator, and reflects a commitment to situated expertise.

Marc Berg (1998) draws specific associations between STS and participatory design, tracing the foundation of the CSCW community itself as due to both a "discovery of the user" within engineering companies and management theory on the one hand and Scandinavian collectivist labor politics on the other. Like STS, participatory design holds that "knowledge is situated in a complex of artifacts, practices, and interactions; it is essentially interpretive, and therefore it cannot be decontextualized and broken into discrete tasks, nor totally described and optimized" (Spinuzzi 2005, 3). As Berg puts it, "CSCW cannot be disentangled from the political impetus of the Scandinavian approach and from the philosophical critique of 'rationalism'" (1998, 463). Berg identified that the rational and instrumental technical world that fueled most CSCW work was linked to a politics that aligned IT requirements with managerial concerns and a rationalized view of the organization, making those who use and operate IT systems into unruly or unpredictable elements of the system. Thus he argued that designers of new technologies must seek to keep technology modest and open in order to keep instrumental rationality in check, explore putting theories to work as direct materialized alternatives for the design of IT, and resist subverting the human to the logic of technology.

In this framing of participatory design, technology is an emergent and undetermined actor drawing from Latour and Haraway's conception of subject-object relations. The researcher-designer is not an "overseer" over the design process (another version of Haraway's "God-trick," 1988) but an actor that can be enrolled into the politics of the IT design process. Thus Bannon and Ehn revisit Suchman and Star to claim that participatory designers must consider *infrastructuring* as an ongoing process of "continuous appropriation and redesign," involving many stakeholders and aligning conflicting interests (Bannon and Ehn 2012, 57). Meanwhile, DiSalvo (2009) draws on Dewey to argue that design should move beyond the design of objects and toward the construction of publics. Along with Le Dantec he argues that "infrastructuring ... [moves] past participation as a framing for design, toward participation as an ongoing act of articulating and responding to dynamic attachments. The public, however it might be constituted, is a sociomaterial response to these dynamics" (Le Dantec and DiSalvo 2013, 260).

Contemporary PD draws upon actor-networks and ontologies, understandings of design as a practice of producing "interventionist assemblages" of heterogeneous entities that resist "purify[ing] categories of 'design' or 'research'" (Andersen, Halse, and Moll 2011, 7). Work in this vein therefore displays a strong bent toward hybridity and the formation of publics, viewing design as "an ontological practice" that "quer[ies] what kinds of objects, actors, or constructed publics emerge" (Lenskjold 2015, 7; citing Bredies et al. 2011). For instance, focusing on the use of design prototypes, scholars explore how examining prototypes reveals "unsettled opportunities" and shifting ontological status throughout the "process of creative becoming" that comprises an artifact's design (Andersen, Halse, and Moll 2011; see also Bødker and Grønbæk 1991; Danholt 2005). The "design things" concept developed by the Atelier (Architecture and Technology for Inspirational Living) deploys *interessement* and alliance building to design "not just a thing (an object, an 'entity of matter') but also a thing (a sociomaterial assembly that deals with matters of concern)" (Binder et al. 2011, 1; see also Ehn 2011; Latour 2008).

Within this trading zone that addresses design processes, other scholars aim to reframe the practice by situating it as one of "design fiction" or "speculative design" that produces technoscientific imaginaries (Haraway 2011; Jasanoff and Kim 2015; Marcus 1994). Such scholars produce reflections, provocations, projections, extrapolations, and anticipations of multiple possible near-future worlds. For instance, at the Near Future Laboratory, scholars argue that the purpose of design fiction is to disrupt simple stories about scientific objects, their disinterestedness, and their provenance, adding that designers must "speculate about new, different, distinctive social practices that assemble around and through these objects" (Bleecker 2009, 8). Laura Watts, Lucy Suchman, and Pelle Ehn's keynote address at the 2012 Society for Social Studies of Science annual meeting, under the conference theme "Design and Displacement: Social Studies of Science and Technology," used the reading of fictional messages in a bottle to describe the travels, circumspections, and circumlocutions of design, designers, and design thinking to unseat our notion of the hegemony of certain forms of design (Watts, Ehn, and Suchman 2014). Design fiction challenges or otherwise exposes dominant visions of social media, ubiquitous computing, and the contemporary economic landscape of IT and consumerism (Wilkie and Michael 2009). Such studies not only draw from but also contribute to STS because design fiction and speculative design open new ways of participating in discussions around scientific issues that are not limited by technical expertise. Mike Michael (2012) therefore suggests that STS look to speculative design for a new model of public engagement with science and technology.

In this mode of inquiry, design is also investigated as presenting a problem of knowledge. If design is a new type of knowledge (Buchanan 2001), then what kind of knowledge does it produce, how does it do so, and who might benefit from and/or co-create this knowledge (Bardzell, Bardzell, and Hansen 2015)? The Virtual Knowledge Studio in the Netherlands (2005–2010) brought together designers, computer scientists, and social scientists under the support of the Royal Netherlands Academy of Arts and Sciences to consider how design and information technologies might contribute to knowledge production in STS and elsewhere. This work in turn has brought perspectives on these crossings home to STS to consider our own forms of inquiry, evidenced in copious published studies of ethnographic methods, computational simulations, and virtual worlds (Beaulieu 2010; Wouters et al. 2008, 2012). While the studio has disbanded, scholars once affiliated with VKS occupy visible roles in the global community of STS practitioners.

Engagement and the Researcher

Engaging, making, and designing digital technologies from an STS perspective relies upon trading zones that bring together STS, computation, and design at specific times, places, and configurations: from corporate research labs to exhibition galleries, from iSchools to computer science conferences, from private to publicly funded research centers. Wherever STS scholars participate in the construction of digital systems, they bring insights on situated knowledge, institutional logistics of funding, regulation, and collaboration, the importance of worker skill sets, or ethnographically inspired design methods. There are important implications for continuing research—and research-ers—in this intersectional space. After all, this type of work requires new forms of interactional expertise (Collins and Evans 2007). Corporate and critical work requires expertise in social organization, management, work practices, hardware, and software systems alongside domain knowledge, STS training, and technical skills. STS schol-ars at this intersection contribute to a body of interdisciplinary knowledge that may not be recognizable to the expert communities in which their designs intervene. And they often find themselves to be a repository of social knowledge for those scientists and engineers that they work with and study. Further, this design work often engages explicitly with the politics of expertise, breaking down the barriers between scholar and research subject or between designer and user. This sensibility contributes to an agenda for a new model of public understanding of science and technology, evident in projects like the Public Lab which encourages open, public, critical practice (Dosema-gen, Warren, and Wylie 2011). It also contributes to STS. Much like the material turn in the history of science (Sibum 1995; Smith, Meyers, and Cook 2014), design as a form of participant observation in material culture can "bridge the gap between creative physical and conceptual exploration" to enrich the analyst's own social and cultural understanding of technology (Ratto 2011, 11; see also Ratto, Wylie, and Jalbert 2014). Thus, the politics of expertise is not merely an analytical theme but a lived experience for scholars who occupy the hybrid spaces between STS and design.

Second, the work we describe requires forging new hybrid identities for the STS practitioner, in addition to existing interdisciplinary commitments. The scholars cited above nurture ties with anthropology and sociology, computer science, and art. Their sites of work bridge academia and industry, the gallery and the street, the scientific conference of record and the laboratory. Hybridity and expertise are not new topics for the STS scholar. But in these spaces of interaction that encircle the design of digital arti-facts, professionalization can be an uncertain process. Within STS, the scholarly valid-ity of building a laboratory, creating gallery work, or crafting software—which play a role in tenure cases in the sciences, computer sciences, or visual arts—pose challenges for evaluation to those who typically review the written scholarship of practitioners. Within design, alternative forms of engagement and intentionally subversive goals can be antagonistic to normative values in the field. Recently such practices of design have become more commonplace and visible at venues such as the Society for Social Studies of Science (4S). The exhibit at a 4S meeting in Paris in 2005, the design workshop in Copenhagen (Forlano et al. 2012), digitalSTS workshops (www.digitalsts.net) and the Making and Doing section and award initiated at the 4S conference in Denver in 2015, all speak to field-sanctioned spaces to explore such intersections, to discuss their relative merits, and to expand relevant trading zones critical to the advancement of the field.

Finally, it is worth noting that these trading zones sit at the intersection of powerful transnational networks and enroll centers of calculation such as resource-rich universities and technology companies located in urban centers in the global North. These locations stand to gain from employing STS researchers and insights to create better products and to benefit from academic-industrial ties, while STS researchers through their connections to these networks gain access to data sets, resources, tools, and relationships that expand the capabilities and reach of their work. Yet the networks of IT scholarship and practice are themselves political. Because these associations produce specific artifacts, labor relations, use cases, and design practices, scholars who work in the intersection of STS, design, and information technology participate in the very world-making practices that they might otherwise critique. Researchers in this space are well aware of these tensions and often give it explicit voice in their work as they critique powerful concepts like "innovation," "design thinking," and "entrepreneurial development" (Irani 2015; Lindtner 2015; Suchman and Bishop 2000). In an effort to expand the definition of which users matter and whose making counts, the increasing intermingling of STS scholars and the design of computational and informational systems does not require abandoning critical positioning but returning to it with vigor.

Conclusions

From infrastructure studies to reflective and participatory design, the scholarship we have described in this chapter draws upon theoretical foundations established by core STS literature. At sites of interdisciplinary intersections like PARC, CSCW, design studios, and iSchools, STS continues to impact the development of information technologies. Its insights inspire designerly attention to the social and political relations between individuals and machines, both the human and nonhuman as well as the hybridities that emerge at boundaries. It has given voice to an ethnographic and an historical understanding of technology that contributes to system design and provides a vocabulary for sociotechnical construction and maintenance as well as for situated knowledge production. Scholars who work at the intersection of STS and design imagine different categorical understandings of the world, commit to a praxiological stance on the development of new systems, and participate in assembling them along alternative lines of power and knowledge. Whether directly involved in the construction and evaluation of sociotechnical systems or through critique, STS has for the past thirty years exerted an influence in the development of our information infrastructure.

Examining these trading zones can guide scholars from both design and STS toward productive existing intersections, with their relevant guiding scholarship, publication venues, conferences, and critical questions. Yet even as we offer a review of past and current work, this chapter also looks prospectively to expanding the STS tool kit to include participation and design as important and established avenues for research and practice. The coming years offer an opportunity for new expressions of STS that merge interactional expertise and a critical stance on the sociology of knowledge and technology on the one hand with the *techne* of building, designing, and implementing new technological systems and concomitant "matters of concern" (Latour 2008) on the other. This will require addressing anew the difficulties of constructing intersectional identities and interdisciplinary personas even as novel opportunities for engagement arise. Parallel changes in the digital humanities and media studies may yet serve as models or cautionary tales along the way. Regardless of the specific path forward, the continued, successful integration of design and STS scholarship will require sustaining those sensibilities essential to STS honed over the early decades of the field, while seeking out new ways to put them into practice.

Notes

1. The "studio" model is more prevalent across European sites. Several examples will be discussed below.

2. The Intel Science & Technology Center for Social Computing linked researchers at the University of California–Irvine Informatics Department, Cornell University's Information School, the Information School at Indiana University, the program for Literature, Culture and Communication at Georgia Institute of Technology, and the Media, Culture and Communication department at New York University.

3. See also http://mindell.scripts.mit.edu/homepage/.

4. See http://turkopticon.differenceengines.com and browser extensions at http://addons .mozilla.org/en-US/firefox/addon/turkopticon/.

References

Agre, Phil. 1990. "Book Review: Lucy A. Suchman, Plans and Situated Actions: The Problem of Human-Machine Communication." *Artificial Intelligence* 42 (3): 369–84.

___. 1997. "Toward a Critical Technical Practice: Lessons Learned in Trying to Reform AI." In *Bridging the Great Divide: Social Science, Technical Systems, and Cooperative Work*, edited by Geof Bowker, Susan Leigh Star, William Turner, and Les Gasser, 131–58. New York: Erlbaum Press.

Andersen, Tariq, Joachim Halse, and Jonas Moll. 2011. "Design Interventions as Multiple Becomings of Healthcare." *Nordes* 4. Accessed April 20, 2016, at http://www.nordes.org/opj/index.php/n13/article/view/106/0.

Bannon, Liam, and Pelle Ehn. 2012. "Design Matters in Participatory Design." In *Routledge International Handbook of Participatory Design*, edited by Toni Robertson and Jesper Simonsen, 37–63. New York: Routledge.

Bardzell, Jeffrey, Shaowen Bardzell, and Lone Koefoed Hansen. 2015. "Immodest Proposals: Research through Design and Knowledge." In *Proceedings of the 33rd Annual ACM Conference on Human Factors in Computing Systems* (CHI '15), 2093–102. New York: ACM.

Beaulieu, Anne. 2010. "Research Note: From Co-Location to Co-Presence: Shifts in the Use of Ethnography for the Study of Knowledge." *Social Studies of Science* 40 (3): 453–70.

Berg, Marc. 1998. "The Politics of Technology: On Bringing Social Theory into Technological Design." *Science, Technology, & Human Values* 23 (4): 456–90.

Bietz, Matthew, Charlotte Lee, and Eric Baumer. 2010. "Synergizing in Cyberinfrastructure Development." *Computer Supported Cooperative Work* 19 (3–4): 245–81.

Bijker, Wiebe E., Thomas Hughes, and Trevor J. Pinch, eds. 1987. *The Social Construction of Technological Systems*. Cambridge, MA: MIT Press.

Binder, Thomas, Eva Brandt, and Judith Gregory. 2008. "Design Participation(s)." *CoDesign* 4 (1): 1–3.

Binder, Thomas, Giorgio De Michelis, Pelle Ehn, Giulio Jacucci, Per Linde, and Ina Wagner. 2011. *Design Things*. Cambridge, MA: MIT Press.

Bleecker, Julian. 2009. "Design Fiction: A Short Essay on Design, Science, Fact and Fiction." *Near Future Laboratory*. Accessed January 28, 2016, at http://www.nearfuturelaboratory.com/2009/03/17/design-fiction-a-short-essay-on-design-science-fact-and-fiction/.

Boczkowski, Pablo. 2005. *Digitizing the News: Innovation in Online Newspapers*. Cambridge, MA: MIT Press.

Bødker, Susanne, and Kaj Grønbæk. 1991. "Cooperative Prototyping: Users and Designers in Mutual Activity." *International Journal of Man-Machine Studies* 34 (3): 453–78.

Bowker, Geoffrey, Karen Baker, Florence Millerand, and David Ribes. 2010. "Toward Information Infrastructure Studies: Ways of Knowing in a Networked Environment." In *International Handbook of Internet Research*, edited by Jeremy Hunsinger, Lisbeth Klastrup, and Matthew Allen, 97–117. New York: Springer.

Bowker, Geoffrey, and Susan Leigh Star. 1999. *Sorting Things Out: Classification and Its Consequences*. Cambridge, MA: MIT Press.

boyd, danah, and Kate Crawford. 2012. "Critical Questions for Big Data: Provocations for a Cultural, Technological, and Scholarly Phenomenon." *Information, Communication & Society* 15 (5): 662–79.

Bredies, Katharina, Simon Bowen, Carl DiSalvo, Li Joensson, Tobie Kerridge, Tau Ulv Lenskjold, Ramia Mazé, Regina Peldzus, and Alex Wilkie. 2011. "Before and After Critical Design." Design Research Network Feature Discussion, August 31, 2011. Accessed January 30, 2016, at www.designresearchnetwork.org/drn/content/feature-discussion%3A-and-after-critical-design.

Brooks, Rodney A. 1991. "Intelligence without Representation." *Artificial Intelligence* 47 (1–3): 139–59.

Brown, John Seely. 1998. "Internet Technology in Support of the Concept of 'Communities-of-Practice': The Case of Xerox." *Accounting Management and Information Technologies* 8 (4): 227–36.

Brown, John Seely, and Peter Duguid. 2000. *The Social Life of Information*. Boston: Harvard Business School Press.

Brynjarsdóttir, Hrönn, Maria Håkansson, James Pierce, Eric Baumer, Carl DiSalvo, and Phoebe Sengers. 2012. "Sustainably Unpersuaded: How Persuasion Narrows Our Vision of Sustainability." In *Proceedings of the SIGCHI Conference on Human Factors in Computing Systems* (CHI '12), 947–56. New York: ACM.

Buchanan, Richard. 2001. "Design Research and the New Learning." *Design Issues* 17 (4): 3–23.

Burrell, Jenna. 2012. *Invisible Users: Youth in the Internet Cafés of Urban Ghana*. Cambridge, MA: MIT Press.

Cohn, Marisa. 2016. "Convivial Decay: Entangled Lifetimes in a Geriatric Infrastructure." In *Proceedings of 2016 ACM Conference on Computer Supported Collaborative Work* (CSCW '16), 1511–23. New York: ACM.

Collins, Harry, and Robert Evans. 2007. *Rethinking Expertise*. Chicago: University of Chicago Press.

Contu, Alessia, and Hugh Willmott. 2003. "Re-Embedding Situatedness: The Importance of Power Relations in Learning Theory." *Organization Science* 14 (3): 283–96.

da Costa, Beatriz. 2008. "Reaching the Limit. When Art Becomes Science." In *Tactical Biopolitics: Art, Activism, and Technoscience*, edited by Beatriz da Costa, Kavita Philip, and Joseph Dumit, 365–86. Cambridge, MA: MIT Press.

Danholt, Peter. 2005. "Prototypes as Performative." In *Proceedings of the 4th Decennial Conference on Critical Computing: Between Sense and Sensibility*, 1–8. New York: ACM.

DiSalvo, Carl. 2009. "Design and the Construction of Publics." *Design Issues* 25 (1): 48–63.

___. 2012. *Adversarial Design*. Cambridge, MA: MIT Press.

DiSalvo, Carl, Phoebe Sengers, and Hrönn Brynjarsdóttir. 2010. "Mapping the Landscape of Sustainable HCI." In *Proceedings of the SIGCHI Conference on Human Factors in Computing Systems* (CHI'10), 1975–84. New York: ACM.

Dosemagen, Shannon, Jeff Warren, and Sara Wylie. 2011. "Grassroots Mapping: Creating a Participatory Map-Making Process Centered on Discourse." *Journal of Aesthetics and Protest* 8. Accessed at http://www.joaap.org/issue8/GrassrootsMapping.htm.

Dourish, Paul. 2006. "Implications for Design." In *Proceedings of the SIGCHI Conference on Human Factors in Computing Systems* (CHI'06), 541–50. New York: ACM.

___. 2010. "HCI and Environmental Sustainability: The Politics of Design and the Design of Politics." In *Proceedings of the 8th ACM Conference on Designing Interactive Systems* (DIS'10), 1–10. New York: ACM.

Dourish, Paul, and Genevieve Bell. 2011. *Designing a Digital Future: Mess and Mythology in Ubiquitous Computing*. Cambridge, MA: MIT Press.

Edwards, Paul. 1997. *The Closed World: Computers and the Politics of Discourse in Cold War America*. Cambridge, MA: MIT Press.

Edwards, Paul, Steven Jackson, Melissa Chalmers, Geoffrey Bowker, Christine Borgman, David Ribes, Matt Burton, and Scout Calvert. 2013. *Knowledge Infrastructures: Intellectual Frameworks and Research Challenges*. Ann Arbor, MI: Deep Blue.

Eglash, Ron. 1999. *African Fractals: Modern Computing and Indigenous Design*. New Brunswick, NJ: Rutgers University Press.

Ehn, Pelle. 2011. "Design Things: Drawing Things Together and Making Things Public." *Tecnoscienza: Italian Journal of Science & Technology Studies* 2 (1): 31–52.

Epstein, Steven. 1995. "The Construction of Lay Expertise: AIDS Activism and the Forging of Credibility in the Reform of Clinical Trials." *Science, Technology, & Human Values* 20 (4): 408–37.

Farías, Ignacio, and Alex Wilkie. 2015. *Studio Studies: Operations, Topologies & Displacements*. CRESC Series. London: Routledge.

Flanagan, Mary, and Helen Nissenbaum. 2014. *Values at Play in Digital Games*. Cambridge, MA: MIT Press.

Forlano, Laura, Dehlia Hannah, Kat Jungnickel, Julian McHardy, and Hannah Star Rogers. 2012. "Experiments in (and Out of) the Studio: Art and Design Methods for Science and Technology

Studies." Accessed January 30, 2016, at http://www.academia.edu/2527061/Experiments _in_and_out_of_the_studio_Art_and_design_methods_for_Science_and_Technology_Studies _2012.

Gabrys, Jennifer. 2011. *Digital Rubbish: A Natural History of Electronics*. Ann Arbor: University of Michigan Press.

Galison, Peter. 1997. *Image and Logic*. Chicago: University of Chicago Press.

Gillespie, Tarleton. 2014. "The Relevance of Algorithms." In *Media Technologies: Essays on Communication, Materiality, and Society*, edited by Tarleton Gillespie, Pablo Boczkowski, and Kirsten Foot, 167–94. Cambridge, MA: MIT Press.

Greenbaum, Joan, and Martin Kyng. 1991. *Design at Work*. New York: Erlbaum.

Haraway, Donna J. 1988. "Situated Knowledges: The Science Question in Feminism and the Privilege of Partial Perspective." *Feminist Studies* 14 (3): 575–99.

___. 2011. "SF: Science Fiction, Speculative Fabulation, String Figures, So Far." Acceptance speech for Pilgrim Award, July, 7, 2011. Accessed January 30, 2016, at http://people.ucsc.edu/~haraway/ Files/PilgrimAcceptanceHaraway.pdf.

Hine, Christine, ed. 2006. *New Infrastructures for Knowledge Production: Understanding E-science*. London: Information Science Publishing.

Irani, Lilly. 2015. "Hackathons and the Making of Entrepreneurial Citizenship." *Science, Technology, & Human Values* 40 (5): 799–824.

Irani, Lilly, and M. Six Silberman. 2013. "Turkopticon: Interrupting Worker Invisibility in Amazon Mechanical Turk." In *Proceedings of the SIGCHI Conference on Human Factors in Computing Systems* (CHI'13), 611–20. New York: ACM.

Irani, Lilly, Janet Vertesi, Paul Dourish, Kavita Philip, and Rebecca E. Grinter. 2010. "Postcolonial Computing: A Lens on Design and Development." In *Proceedings of the SIGCHI Conference on Human Factors in Computing Systems* (CHI'10), 1311–20. New York: ACM.

Jackson, Steven. 2014. "Rethinking Repair." In *Media Technologies: Essays on Communication, Materiality and Society*, edited by Tarleton Gillespie, Pablo Boczkowski, and Kirsten Foot, 221–39. Cambridge, MA: MIT Press.

Jackson, Steven, and Ayse Buyuktur. 2014. "Who Killed WATERS? Mess, Method, and Forensic Explanation in the Making and Unmaking of Large-Scale Science Networks." *Science, Technology, & Human Values* 39 (2): 285–308.

Jackson, Steven, Davi Ribes, Ayse Buyuktur, and Geoffrey Bowker. 2011. "Collaborative Rhythm: Temporal Dissonance and Alignment in Collaborative Scientific Work." In *Proceedings of the ACM 2011 Conference on Computer Supported Cooperative Work*, 245–54. New York: ACM.

Jasanoff, Sheila. 1990. *The Fifth Branch: Science Advisers as Policymakers*. Cambridge, MA: Harvard University Press.

Jasanoff, Sheila, and Sang-Hyun Kim, eds. 2015. *Dreamscapes of Modernity: Sociotechnical Imaginaries and the Fabrication of Power*. Chicago: University of Chicago Press.

Jeremijenko, Natalie. 2009. *Feral Robot Dogs*. Accessed January 30, 2016, at http://www.nyu.edu/projects/xdesign/feralrobots/.

Jirotka, Marina, Charlotte Lee, and Gary M. Olson. 2013. "Supporting Scientific Collaboration: Methods, Tools and Concepts." *Computer Supported Cooperative Work* 22 (4): 667–715.

Kirksey, Evan, and Krista Dragomer. 2012. "The Multispecies Salon." Exhibited at Proteus Gowanus, Brooklyn, NY. Accessed April 21, 2016, at http://www.multispecies-salon.org/migrations.

Kling, Rob. 1991. "Computerization and Social Transformations," *Science, Technology, & Human Values* 16 (3): 342–67.

Kling, Rob, and Walter Scacchi. 1982. "The Web of Computing: Computer Technology as Social Organization." *Advances in Computers* 21: 1–46.

Knobel, Cory, and Geoffrey Bowker. 2011. "Values in Design." *Communications of the ACM* 54 (7): 26–28.

Knorr Cetina, Karin, and Urs Bruegger. 2000. "The Market as an Object of Attachment: Exploring Postsocial Relations in Financial Markets." *Canadian Journal of Sociology* 25 (2): 141–68.

Latour, Bruno. 1991. "Technology Is Society Made Durable." In *A Sociology of Monsters: Essays on Power, Technology and Domination*, edited by John Law, 103–32. London: Routledge.

___. 2008. "A Cautious Prometheus? A Few Steps toward a Philosophy of Design." In *Networks of Design: Proceedings of the 2008 Annual International Conference of the Design History Society* (Falmouth, 3–6 September 2009), edited by Fiona Hackne, Jonathan Glynne, and Viv Minto, 2–10. Boca Raton, FL: BrownWalker Press.

___. 2013. *An Inquiry into the Modes of Existence: An Anthropology of the Moderns*. Cambridge, MA: Harvard University Press.

Latour, Bruno, and Albena Yaneva. 2008. "Give Me a Gun and I Will Make All Buildings Move: An ANT's View of Architecture." In *Explorations in Architecture: Teaching, Design, Research*, edited by Reto Geiser, 80–89. Basel: Birkhauser.

Le Dantec, Christopher, and Carl DiSalvo. 2013. "Infrastructuring and the Formation of Publics in Participatory Design." *Social Studies of Science* 43 (2): 241–64.

Lee, Charlotte, Paul Dourish, and Gloria Mark. 2006. "The Human Infrastructure of Cyberinfrastructure." In *Proceedings of the 2006 20th Anniversary Conference on Computer Supported Cooperative Work*, 483–92. New York: ACM.

Lenskjold, Tau Ulv. 2015. "Objects of Entanglement and Allure: Steps Towards an Anthro-De-Centric Position in Speculative Design." Ph.D. Dissertation. The Royal Danish Academy of Fine Arts, Schools of Architecture, Design and Conservation, Copenhagen.

Lindtner, Silvia. 2015. "Hacking with Chinese Characteristics: The Promises of the Maker Movement against China's Manufacturing Culture." *Science, Technology, & Human Values* 40 (5): 854–79.

Lindtner, Silvia, Garnet D. Hertz, and Paul Dourish. 2014. "Emerging Sites of HCI Innovation: Hackerspaces, Hardware Startups and Incubators." In *Proceedings of the SIGCHI Conference on Human Factors in Computing Systems* (CHI'11), 439–48. New York: ACM.

Loukissas, Yanni, and David Mindell. 2014. "Visual Apollo: A Graphical Exploration of Computer-Human Relationships." *Design Issues* 30 (2): 4–16.

Lury, Celia, and Nina Wakeford, eds. 2012. *The Happening of the Social*. Abingdon: Routledge.

Lynch, Michael, and Simon Cole. 2005. "Science and Technology Studies on Trial: Dilemmas of Expertise." *Social Studies of Science* 35 (2): 269–311.

Marcus, George. 1994. *Technoscientific Imaginaries: Conversations, Profiles, and Memoirs*. Chicago: University of Chicago Press.

Mazmanian, Melissa, and Ingrid Erickson. 2014. "The Product of Availability: Understanding the Economic Underpinnings of Constant Connectivity." In *Proceedings of the SIGCHI Conference on Human Factors in Computing Systems* (CHI'14), 763–72. New York: ACM.

Michael, Mike. 2012. "'What Are We Busy Doing?' Engaging the Idiot." *Science, Technology, & Human Values* 37 (5): 528–54.

Millerand, Florence, David Ribes, Karen Baker, and Geoffrey Bowker. 2013. "Making an Issue Out of a Standard: Storytelling Practices in a Scientific Community." *Science, Technology, & Human Values* 38 (1): 7–43.

Nissenbaum, Helen. 2001. "How Computer Systems Embody Values." *Computer* 34 (3): 117–19.

Oliver, Julian, and Sam Hart. 2014. "Engineered Ecologies: An Interview with Julian Oliver." Accessed November 1, 2015, at http://avant.org/media/julian-oliver.

Oliver, Julian, Gordan Savičić, and Danja Vasiliev. 2011–2014. "The Critical Engineering Manifesto." Accessed November 1, 2015, at http://criticalengineering.org.

Oreskes, Naomi, and Erik Conway. 2010. *Merchants of Doubt*. London: Bloomsbury Press.

Orlikowski, Wanda. 2000. "Using Technology and Constituting Structures: A Practice Lens for Studying Technology in Organizations." *Organization Science* 11 (4): 404–28.

Orr, Julian E. 1996. *Talking about Machines: An Ethnography of a Modern Job*. Ithaca, NY: Cornell University Press.

Philip, Kavita, Lilly Irani, and Paul Dourish. 2012. "Postcolonial Computing: A Tactical Survey." *Science, Technology, & Human Values* 37 (1): 13–29.

Pinch, Trevor J., and Wiebe E. Bijker. 1984. "The Social Construction of Facts and Artifacts: Or How the Sociology of Science and the Sociology of Technology Might Benefit Each Other." *Social Studies of Science* 14 (3): 399–441.

Pine, Kathleen, and Max Liboiron. 2015. "The Politics of Measurement and Action." In *Proceedings of the 33rd Annual ACM Conference on Human-Computer Interaction* (Chi'15), 3147–56. New York: ACM.

Ratto, Matt. 2011. "Critical Making: Conceptual and Material Studies in Technology and Social Life." *The Information Society* 27 (4): 252–60.

Ratto, Matt, Sara Wylie, and Kirk Jalbert. 2014. "Introduction to the Special Forum on Critical Making as Research Program." *The Information Society* 30 (2): 85–95.

Ribes, David. 2014. "Ethnography of Scaling, or, How to a Fit a National Research Infrastructure in the Room." In *Proceedings of the 17th ACM Conference on Computer Supported Cooperative Work*, 158–70. New York: ACM.

Ribes, David, and Karen Baker. 2007. "Modes of Social Science Engagement in Community Infrastructure Design." In *Communities and Technologies 2007: Proceedings of the Third Communities and Technologies Conference*, edited by Charles Steinfield, Brian T. Pentland, Mark Ackerman, and Noshir Contractor, 107–30. London: Springer.

Ribes, David, and Geoffrey Bowker. 2008. "Organizing for Multidisciplinary Collaboration: The Case of the Geosciences Network." In *Scientific Collaboration on the Internet*, edited by Gary Olson, Ann Zimmerman, and Nathan Bos, 311–30. Cambridge, MA: MIT Press.

Ribes, David, and Charlotte Lee. 2010. "Sociotechnical Studies of Cyberinfrastructure and E-Research: Current Themes and Future Trajectories." *Journal of Computer Supported Cooperative Work* 19 (3–4): 231–44.

Rooksby, John. 2013. "Wild in the Laboratory: A Discussion of Plans and Situated Actions." *ACM Transactions on Computer-Human Interaction (TOCHI)* 20 (3): 17.

Sanders, Elizabeth, and Pieter Jan Stappers. 2008. "Co-creation and the New Landscapes of Design." *CoDesign* 4 (1): 5–18.

Sawyer, Steve, and Andrea Tapia. 2007. "From Findings to Theories: Institutionalizing Social Informatics." *The Information Society* 23 (4): 263–75.

Saxenian, AnnaLee. 1996. *Regional Advantage: Culture and Competition in Silicon Valley and Route 128*. Cambridge, MA: Harvard University Press.

Schuler, Douglas, and Aki Namioka, eds. 1993. *Participatory Design: Principles and Practices*. Aarhus: Lawrence Erlbaum Associates.

Sengers, Phoebe, Kirsten Boehner, Shay David, and Joseph Kaye. 2005. "Reflective Design." In *Proceedings of the 4th Decennial Conference on Critical Computing*, 49–58. New York: ACM.

Shklovski, Irina, Janet Vertesi, Silvia Lindtner, and Lucy Suchman. 2014. "Introduction to This Special Issue on Transnational HCI." *Journal of Human-Computer Interaction* 29 (1): 1–21.

Sibum, Otto. 1995. "Reworking the Mechanical Value of Heat: Instruments of Precision and Gestures of Accuracy in Early Victorian England." *Studies in History and Philosophy of Science* 26 (1): 73–106.

Simon, Herbert A. [1969] 1996. *The Sciences of the Artificial.* 3rd ed. Cambridge, MA: MIT Press.

Smith, Pamela, Amy Meyers, and Harold Cook, eds. 2014. *Ways of Making and Knowing: The Material Culture of Empirical Knowledge.* Ann Arbor: University of Michigan Press.

Spinuzzi, Clay. 2005. "The Methodology of Participatory Design." *Technical Communication* 52 (2): 163–74.

Star, Susan Leigh. 1999. "The Ethnography of Infrastructure." *American Behavioral Scientist* 43 (3): 377–91.

Star, Susan Leigh, and James Griesemer. 1989. "Institutional Ecology, 'Translations,' and Boundary Objects: Amateurs and Professionals in Berkeley's Museum of Vertebrate Zoology, 1907–1939." *Social Studies of Science* 19 (3): 387–420.

Star, Susan Leigh, and Karen Ruhleder. 1994. "Steps towards an Ecology of Infrastructure: Complex Problems in Design and Access for Large-Scale Collaborative Systems." In *Proceedings of the ACM Conference on Computer Supported Cooperative Work*, 253–64. New York: ACM.

Suchman, Lucy. 1993. "Do Categories Have Politics? The Language/Action Perspective Reconsidered." In *Proceedings of the Third European Conference on Computer-Supported Cooperative Work*, 1–14. Norwell, MA: Kluwer Academic Publishers.

___. 1987. *Plans and Situated Actions: The Problem of Human-Machine Communication.* Cambridge: Cambridge University Press.

___. 2007. *Human-Machine Reconfigurations: Plans and Situated Actions.* 2nd ed. Cambridge: Cambridge University Press.

Suchman, Lucy, and Libby Bishop. 2000. "Problematizing 'Innovation' as a Critical Project." *Technology Analysis & Strategic Management* 12 (3): 327–33.

Taylor, Alex. 2011. "Out There." In *Proceedings of the SIGCHI Conference on Human Factors in Computing Systems* (CHI'11), 685–94. New York: ACM.

Thacker, Eugene, Denna Jones, Heath Bunting, and Natalie Jeremijenko. 2005. *Creative Biotechnology: A User's Manual.* Newcastle upon Tyne: Locus+ Publishing.

Troshynski, Emily, Charlotte Lee, and Paul Dourish. 2008. "Accountabilities of Presence: Reframing Location-Based Systems." In *Proceedings of the SIGCHI Conference on Human Factors in Computing Systems* (CHI'08), 487–96. New York: ACM.

Vertesi, Janet. 2014. "Seamful Spaces: Heterogeneous Infrastructures in Interaction." *Science, Technology, & Human Values* 39 (2): 264–84.

Vertesi, Janet, and Paul Dourish. 2011. "The Value of Data: Considering the Context of Production in Data Economies." In *Proceedings of the ACM 2011 Conference on Computer Supported Cooperative Work*, 533–42. New York: ACM.

Wania, Christine, Michael Atwood, and Katherine McCain. 2006. "How Do Design and Evaluation Interrelate in HCI Research?" In *DIS '06: Proceedings of the 6th Conference on Designing Interactive Systems*, 90–98. New York: ACM.

Watts, Laura, Pelle Ehn, and Lucy Suchman. 2014. "Prologue." In *Making Futures: Marginal Notes on Innovation, Design, and Democracy*, edited by Pelle Ehn, Elisabet M. Nilsson, and Richard Topgaard, x–xxxix. Cambridge, MA: MIT Press.

Wilkie, Alex, and Mike Michael. 2009. "Expectation and Mobilisation: Enacting Future Users." *Science, Technology, & Human Values* 34 (4): 502–22.

Williams, Amanda, Silvia Lindtner, Ken Anderson, and Paul Dourish. 2014. "Multisited Design: An Analytical Lens for Transnational HCI." *Human-Computer Interaction* 29 (1): 78–108.

Winner, Langdon. 1986. *The Whale and the Reactor: A Search for Limits in an Age of High Technology*. Chicago: University of Chicago Press.

Woolgar, Steve. 1990. "Configuring the User: The Case of Usability Trials." *The Sociological Review* 38 (1): 58–99.

Wouters, Paul, Katie Vann, Andrea Scharnhorst, Matt Ratto, Iina Hellsten, Jenny Fry, and Anne Beaulieu. 2008. "Messy Shapes of Knowledge: STS Explores Informatization, New Media and Academic Work." In *The Handbook of Science and Technology Studies*. 3rd ed., edited by Edward J. Hackett, Olga Amsterdamska, Michael Lynch, and Judy Wajcman, 319–52. Cambridge, MA: MIT Press.

Wouters, Paul, Anne Beaulieu, Andrea Scharnhorst, and Sally Wyatt. 2012. *Virtual Knowledge: Experimenting in the Humanities and the Social Sciences*. Cambridge, MA: MIT Press.

Wynholds, Laura, David Fearon, Christine L. Borgman, and Sharon Traweek. 2011. "Awash in Stardust: Data Practices in Astronomy." In *Proceedings of the 2011 iConference* (iConference'11), 802–4. New York: ACM.

Zimmerman, Ann. 2008. "New Knowledge from Old Data: The Role of Standards in the Sharing and Reuse of Ecological Data." *Science, Technology, & Human Values* 33 (5): 631–52.

7   Experiments in Participation

Javier Lezaun, Noortje Marres, and Manuel Tironi

Introduction

The idea that experiments represent key settings and instruments for participation in public affairs has acquired a distinctly contemporary flavor. Social and political experiments feature prominently in our daily lives, whether it is in the form of new entertainment genres, such as reality television, in the many media outlets that regularly report experimental findings from the social sciences, or in the multitude of recent government- and business-led experiments in behavioral change. When the social media company Facebook introduced a new button in its interface that allowed users to tell their "friends" that they had voted in recent U.S. and UK elections, this small technical intervention was presented as "an experiment" that pursued two inextricable goals: to foster greater political participation and to gain new insights into the behavior of the platform's users and the ability of new interface features to alter their conduct (Healy 2015). Participatory initiatives that adopt an explicitly experimental orientation are now commonplace in urban planning, architecture, art, service design, environmental management, and public health, to name a few prominent domains. In all these cases, lay or amateur audiences are invited to engage with technical, scientific, or aesthetic matters that used to be the preserve of experts, and to do so in an explicitly creative or innovative fashion that pushes the boundaries of traditional ways of enacting public affairs and performing democratic governance. The use in all these contexts of experimental formats, such as collaborative mapping tools or interactive exhibitions, is presented as a means of intensifying the generative potential of these participatory experiences, while in the process producing new evidence and documentation about social and political life (Horst and Michael 2011). Deploying settings, devices, and/or things experimentally makes it possible to curate novel forms of participation, eliciting expressions or accounts of public issues that would otherwise remain underarticulated or exist only in potentia (Lury and Wakeford 2012; Marres 2012).

The proliferation of experimental forms of participation has attracted the interest of researchers in science and technology studies (STS)—not surprisingly, given the central place that experimentation occupies in the history of our field. STS not only has studied the role that experimental practices and apparatuses have played in the evolution of the modern sciences but has also drawn attention to experiments as crucial occasions for the articulation of the relationship between science and society. Work in STS has long argued that experimental settings and situations not only play an important role in the acquisition of new knowledge about the natural or the social world but also offer exceptional opportunities for intervening in and changing those realities. Yet as experiments are today explicitly designed and defined as a privileged format of participation in public life, an important question arises: can the proliferation of experimental formats facilitate meaningful engagement with public affairs, or does it threaten to impoverish or even undermine political democracy?

As "experiments in participation" are today undertaken across a range of settings and spheres, core concerns of STS become newly relevant to social and public life: preoccupations with the authority of experts and the distribution of agency in the design and interpretation of experimental interventions; the rhetorical power of public demonstrations and their capacity to elicit engagement, consent, and "lock-in"; or the role of experimental situations in recasting the relationship between natural and social domains. At the same time, by highlighting novel configurations of the relationship between experimentation and participation, recent work in this area puts to the test many received wisdoms in STS. For one, the proliferation of experimental formats in social, economic, and public life—such as "living experiments" conducted in domestic settings, or collective experiments in "sustainable transitions" sponsored by industry or government—implies that scientific registers of validity and value lose some of their hold over the deployment and interpretation of these interventions. Epistemological considerations must contend with alternative repertoires of evaluation, as experiments bring into relation diverse knowledge cultures, innovation paradigms, and material practices, opening up new possibilities for encounters, exchanges, and conflicts among different constituencies. Partly as a consequence of this, experiments acquire a formative ambiguity in relation to the nature and purposes of public life in technological societies: ostensibly meant to enable new or enriched forms of participation, they also configure participation as an object of research, innovation, valuation, and manipulation (Chilvers and Kearnes 2016). This ambivalence—or rather "multivalence" (Marres 2012)—is key to elucidating the specific roles that experiments in participation can play in contemporary democratic politics.

This chapter discusses experiments in participation as an emerging nexus of research, theory, and practice in STS. It addresses a growing body of work that examines participation in public affairs—and the reconfiguration of situations, actors, and issues through participatory processes—as possessing a crucial experimental quality. To map out this field, the chapter draws together sometimes divergent strands of work in STS and cognate fields. We begin by briefly discussing the constitutive role of experiments as a research topic in the history of STS, including the significant body of work on experimentation in the social and political sciences. We then turn to an experimental practice of long standing in our field, namely, work in the field of public engagement with science (PES) that attends explicitly to the experimental and performative dimensions of participatory mechanisms. Finally, we review some current work on participatory experiments, organizing our discussion around three broad areas of concern: object-centered practices and forms of engagement; design, digital, and "inventive" methods; and public experimentation as prototyping. Bringing these different strands of work into dialogue with one another is crucial if we are to fully grasp the distinctive features, possibilities, and challenges of experiments in participation as an STS method. We will conclude by making the case that current work in this area, while still very much under development, offers our field an opportunity to expand the range of its engagements with science and technology, and with research and innovation more broadly conceived. This expansion in the repertoire of STS intervention practices (see also chapter 8 this volume) should strengthen the ability of our field to participate in wider experimental cultures and contribute to the activation of new forms of collective imagination.

Experimentation in STS: Redefining Relations among Science, Technology, and Society

Experimentation was a key theme in the emergence of STS as an original field of research and scholarship. Paying close attention to the quandaries of experimental practices, describing in detail what scientists did in their laboratories and field sites, served as a corrective to the traditional focus of the philosophy of science on already formalized knowledge and helped reformulate representation as a form of instrumentalized intervention (Hacking 1983).

This turn to experimentation advanced on a number of fronts. Historical studies of the rise of experimental cultures in the early modern period redefined the meaning of the Scientific Revolution and the Enlightenment, showing that the invention of modern experimental knowledge went hand in hand with the production of new kinds of audiences and publics. The organization of demonstrations in the newly established "houses of experiment" or the founding of journals and other "literary technologies" helped create the forms of public witnessing congruent with the sort of evidence that the new experimental ethos sought to produce (Dear 1985; Shapin 1988; Shapin and Schaffer 1985). In parallel to this historical work on modern experimental cultures, sociologists of scientific knowledge began to reevaluate the role of experiments in the establishment or refutation of scientific facts, focusing on the contingencies and paradoxes implied by any attempt to create universally valid and publicly legitimate knowledge through the staging of unique events held in closely guarded spaces (Collins and Pinch 1982). Last but not least, the physical presence of STS researchers in experimental settings, primarily as ethnographers of laboratory practices, gave us the first close look at science in action, revealing a world of artifacts, equipment, and inscription devices at odds with the sanitized version of fact-making conveyed by traditional epistemology (Latour and Woolgar 1979; Lynch 1985).

Why was experimentation such a productive focus for the development of the analytical sensibilities that would eventually coalesce into STS? The reasons are too many to list here, but it is clear that attending to experimental settings and situations was a powerful way of circumventing the traditional framework of epistemology. Classic distinctions in the philosophy of science, such as that between the "context of discovery" and the "context of justification" (Popper 1963; Reichenbach 1938) collapsed in the face of detailed investigations of experimental work that showed an inextricable combination of empirical and normative elements.

Experiments also confronted STS researchers with the unavoidable technical and material mediations of scientific knowledge production. Experimental settings were full of machinery, devices, and materials. Philosophers of science had recognized the importance of these technical infrastructures as "conditions of possibility" for scientific knowledge production, but STS demonstrated that artifactual elements played a far more active and formative role in the making of scientific knowledge, leaving their traces in the very claims advanced by the experimental sciences. If one wanted to understand the ability of experiments to settle controversies or establish new facts, it was imperative to come up with a better account of how nonhuman entities contributed to the production of scientific and social realities.

Furthermore, it became apparent that in the course of experimentation, theoretical constructions were often overtaken by the sheer productivity of research apparatuses. Experiments were best understood as highly choreographed practices whose performance, if successful, resulted in the production of surprises (Rheinberger 1997). The notion of "method" inherited from the philosophy of science, with its connotations of ordered procedures, predictable transitions, and replicable outcomes, had to be severely qualified, if not discarded altogether.

Finally, experiments drew attention to practices of demonstration, the curation of controlled displays of evidence or instrumental action designed to persuade audiences of the existence of experimentally generated entities and phenomena (Collins 1988). The intimate connection between the validation of experimental knowledge and the creation of specific forms of public witnessing and testimony highlighted a crucial and highly productive ambiguity in the organization of public culture in scientific and technological democracies: the fact that *public* refers at once to genres, procedures, and apparatuses of knowledge-making (as in transparent or accountable), and to a distinctive kind of political collective, the gathering of strangers around a common object of interest (as in stakeholder or audience). The study of experimental demonstrations and public displays of technical competence thus became a strategic site for working out the evidentiary underpinnings of different political cultures, whether it was the gentlemanly polity of Restoration England (Shapin 1994; Shapin and Schaffer 1985), or the liberal-democratic aspirations to transparency of the American republic and other contemporary Euro-American polities (Ezrahi 1990; Hilgartner 2000; Jasanoff 1998, 2005).

STS work on these issues has been primarily concerned with experimentation in the natural sciences and engineering, but the field has also harbored an expanding body of research on experiments in the social and political sciences. This literature includes studies of the rise of experimental settings and procedures in the human sciences (Danziger 1994; Lemov 2005; Mayer 2013) and of the use of these settings and procedures for the articulation of social and political issues (Gillespie 1993; Gross and Krohn 2005). Indeed, the role of the social sciences in the development of new techniques for *representing the public to the public* has been a central tenet of much of the recent historical scholarship (Igo 2007; Osborne and Rose 1999; Porter 1996). Work informed by STS sensibilities often adds to these discussions a closer examination of the technical apparatuses through which social-scientific knowledge is produced (Derksen, Vikkelsø, and Beaulieu 2012; Haffner 2013; Lezaun and Calvillo 2014). It has shown, for instance, the distinctly situated understanding of the experimental "truth spot" that characterized classic sociological approaches to the "city as laboratory" (Gieryn 2006; see also Guggenheim 2012), or how research technologies such as new survey designs made possible the expression of societal phenomena and their formatting for public and political intervention (Didier 2002).

These studies of the social and political sciences have particular relevance for the emergent STS approach to experiments in participation, since they concern situations in which the public is mobilized in at least three distinct ways: (1) as the subject matter of a research apparatus, (2) as an audience for the evidence produced by that apparatus, and (3) as an active source (or agent) of knowledge about social and political matters. In the first case, participation is instrumental to the production of experimental knowledge in an immediate, material way, as individuals, now recast as "research participants," must engage personally and directly with the experimental apparatus. In the second, the organization of a certain public is the goal of experimentation, in the sense of seeking the validation of knowledge propositions through specific forms of public witnessing. And in the third, participation is the object of scientific experimentation in a more indirect but no less pertinent sense: in their experimental interventions the social and political sciences articulate a particular vision of society and the polity, whether this relates to the capacities of social actors to know and act upon the world or to the possibility of envisioning, managing, or contesting social change. In a wider sense, then, the nature of citizenship in democratic societies is inextricably linked to experimental performances, whether those involve the representation of the political will of the nation in an electoral contest (Miller 2004), unfold in the relative privacy of a focus group discussion designed to surface the preferences of the population (Lezaun 2007), or precipitate the assembling of diverse actors to articulate issues in which they are jointly and antagonistically implicated, thus giving these issues a new public form (Callon, Lascoumes, and Barthe 2009; Marres 2007).

By expanding the range of matters at stake in experimental situations, STS has thus managed to revive the multiple connotations that have historically been attached to the notion of "social experiment." A scientific demonstration or proof—the ability of an experiment to validate or refute a scientific hypothesis—is only one of several evidentiary registers available to assess the purpose of an experimental intervention. The category of "social experiment" is best understood as a format or genre that can circulate across scientific, professional, political, public, and everyday settings—not simply as a procedure for testing scientific claims. Related literatures on the "enactment" or "happening" of "the social" (Law and Urry 2004; Lury and Wakeford 2012), studies of the role of experimental technologies and provocations in the constitution of markets and economies (Callon 2009; Muniesa 2014), or the developing research agenda on "the social life of methods" (Law and Ruppert 2013) all speak to this growing interest in the ability of social scientific experimentation to *perform* new collectives.

In sum, STS has long advanced the idea that experiments constitute a crucial site for bringing science, technology, and the public into intimate relation. In doing so, the field has offered an expansive account of experimentation as entailing not just a distinctive method of scientific inquiry but also a genre, an apparatus, and a particular form of publicity or sociality. The relation between experimental practices and their publics, in other words, is not that between an inside and an outside, or between a scientific activity and its social or political context. The *public* of an experiment is not an ingredient added to the production of technoscience after the fact, so to speak, but a form of relationality that emerges—is invoked, put to the test, validated, or discarded— as part of the progress of the experiment itself (Marres 2009).

In advancing these arguments STS initiated a broad reconceptualization of publics and participation. As we will see, current STS scholarship makes a double move in rela- tion to experiments and publics. By scrutinizing the role of experimentation in social and public life, it unsettles the question of how science, technology, and public relate or should relate to one another in contemporary societies. At the same time, STS research- ers adopt experiments as a *resource or instrument for social and public inquiry*, developing their own experimental techniques to probe and perhaps even alter the very meaning of democracy in technological societies. We will next revisit a significant tradition of experimental practice in STS: the creation of experimental situations designed to foster the public understanding of, or public engagement with, science and technology.

PUS/PES Experiments: Redistributing Expertise, Creating Public Situations

STS has long encompassed a set of experimental practices aimed at involving citizens in debates about science, technology, and society. A significant portion of the work that emerged in the 1980s under the rubric of the public understanding of science (PUS), and much of what nowadays is described as public engagement with science (PES), is informed by STS sensibilities and has a direct experimental dimension, even if this dimension has not always been articulated, or even acknowledged, in an overt fash- ion. Teasing out the implications of this tradition of experimentation leads to a more explicit consideration of how STS-inspired technical practices of participation can give form and help curate particular publics (Irwin 2001, 2006).

PUS/PES work in the STS tradition advocates the creation of opportunities for the public to engage with scientific research and technological innovation, and has typi- cally understood engagement as participation in forums of deliberative exchange. This strand of STS thus endeavors to create *situations of publicity*, formally designated and stage-managed occasions where members of the public are invited to discuss techno- scientific topics, deliberate with experts, or question policy makers on controversial issues in science and innovation policy. In the pursuit of this agenda, the field has developed or used a series of semi-standardized formats of public participation, such as the consensus conference (Blok 2007; Grundahl 1995), citizen juries (Crosby, Kelly, and

Schaefer 1986; Stilgoe 2007), multiple deliberative methodologies (Burgess et al. 2007; Rogers-Hayden and Pidgeon 2007), constructive technology assessment (Schot and Rip 1997), or hybrid forums (Callon, Lascoumes, and Barthe 2009).

The instrumental value of these methods to disclose hidden or tacit public opinions has progressively been overtaken, however, by a growing interest in their quality as experimental interventions in their own right (Felt and Fochler 2008, 2010). Public engagement events, in other words, can be approached as situations in which the expressions or accounts elicited by a participatory mechanism potentially disrupt any preformatting of issues, actors, or the participatory event itself. The question, as Michael (2012, 534) puts it, is, "What sort of events might our PUS/PES events precipitate that are not necessarily graspable within the frameworks informing the design of those events?" (See also Michael 2009.)

Coming to terms with the fact that STS conducts its own public experiments opens up new research questions. For instance, formats of public engagement can be subjected to an analysis inflected by STS sensibilities. Historical accounts of the origins and evolution of some of these techniques, such as Soneryd's (2016) work on scenarios workshops, or Voß and Amelung's (2016) study of citizen panels, show the tortuous career of the deliberation tools adopted by STS. Indeed, multiple studies have recently examined the complicated transportation of participatory methods into new contexts and issues. For instance, Laurent (2009) has described the attempted application of a participatory device—the consensus conference pioneered by the Danish Board of Technology—to a novel technoscientific area—nanotechnology—and in two different countries—the United States and France. The difficulties encountered in preserving a seemingly ready-made format across political or scientific domains allow Laurent to make visible "the investments and works that are required to replicate and stabilize forms of public participation" (Laurent 2009, 2). At the same time, the "cracks and gaps" that emerge as a participatory device is stretched to meet a new topic or operate in a new environment provide opportunities to explore the "ambivalence" inherent in participation procedures—an ambivalence that, Laurent argues, ought to be part of our definition of successful public participation.

Describing a similar sort of travel and a similarly vexing process of experimental replication, Ureta (2015) has explored the use of the consensus conference format in Chile to encourage further public engagement on the management of patient health records. In Ureta's account, the format travels well to the new environment as far as its ability to generate a discrete moment of deliberation and consensus is concerned, but it fails to live up to its implicit promise to revamp the role of citizens in the oversight of patient records. The experiment, in other words, did not contribute to an intensification of public engagement with the issue at hand, serving only to realize, as Ureta puts it, "a small and secluded version of Danish democratic deliberation in the midst of the Chilean wilds" (Ureta 2015, 11; see also Bogner 2012). In a similar vein, Tironi (2015) has described the deployment of the "hybrid forum" model in the context of postdisaster reconstruction in Chile. Originally introduced as an apparatus to radicalize public engagement, in Chile the model encountered publics that did not behave as predicted by the assemblage of theories, principles, and methodological protocols articulated in this experimental formula. By exploring the assumptions about democracy, politics, and participation that were brought along with the model as it traveled from Europe to Chile, Tironi challenges the expectation of transportability that is often attached to experimental political forms.

Exploring public engagement as an experimental practice reveals some obvious but long-neglected empirical realities. Much of the PUS/PES literature had initially defined public participation rather narrowly, characterizing engagement as a discursive phenomenon involving primarily talk and the expression of opinions, and specifying its features in terms of procedural rules and roles. In contrast, approaching public engagement events as experimental interventions immediately redirects our attention to the fact that these events are saturated with things, machines, and other stuff, that they unfold in settings and under material conditions specifically tailored to the requirements of participatory action (Davies et al. 2012; Marres and Lezaun 2011). In their work on "competency groups," for instance, Whatmore and Landström (2011) explore how civic involvement with controversies—in this case flood defense schemes in rural Yorkshire (UK)—is enabled through the deployment of seemingly mundane artefacts, as when participants were invited to bring along a relevant object, and a piece of carpet salvaged during a recent flood came to instantiate the issues at hand. Drawing on the work of Isabelle Stengers and Karen Barad, Whatmore and Landström argue for the invention of research apparatuses able to "slow down" expert reasoning and redistribute agency among specialists, lay people, and nonhuman entities. In a similar vein, Waterton and Tsouvalis (2015) show how public engagement with lively materials—this time cyanobacteria in the Lake District—led to envisioning a new form of relationality of people and things, what they describe as an "intra-active collective politics."

While this emerging body of work develops a broad normative argument in favor of an experimental approach to participation, it is also increasingly alert to the crucial question of whether—and how—discrete settings and moments of experimentation can index wider political constellations (Barry 2001, 2013; Lezaun 2011). The question of the experimental performance of democracy and the role of knowledge and technology in public life is thus posed anew, this time around the connectivity of discrete experimental interventions (Laurent, in press). In a recent volume entitled *Remaking Participation*, Chilvers and Kearnes (2016, 52) offer an "ecological" perspective on this question, arguing that "it is not possible to properly understand any one collective of participation without understanding its relational interdependence with other participatory practices, technologies of participation, spaces of negotiation, and the cultural-political settings in which they become established."

This ecological perspective has direct implications for STS, as our field can be one of the actors that contributes to the "relational interdependence" of situated experiments in participation. Exploring the potential role of STS as a mediator or connector requires, however, reflexive attention to the formation and deployment of STS participatory expertise (Chilvers 2008a, 2008b), a critical examination of the discursive and instrumental dimensions of STS' own experimental practices (Felt 2016; Lezaun and Soneryd 2007; Tironi 2015; Voß, 2016), and a continuing exploration of how the production of epistemic orders, including those of our own making, relates to the stabilization or disruption of institutional dynamics (Ezrahi 2012; Jasanoff 2004). The confluence of much of this work around concepts such as "technologies of humility" (Jasanoff 2003) or "technologies of democracy" (Laurent 2011) and the willingness to consider our own participatory experiments as part and parcel of the contested emergence of new technoscience (Bellamy and Lezaun 2015; Irwin, Jensen, and Jones 2013; Stilgoe 2015) express a commitment to develop an approach that recognizes the confluences, asymmetries, and tensions among science, social science, political democracy, and social democracy, and the modest but significant role that our field can play in modulating those relationships.

New Themes: Experiments in Participation Unbound

Alongside the well-established tradition of PUS/PES interventions in STS, several emerging strands of scholarship are shifting the emphasis of work in this area from treating experiments as objects of STS study to approaching them as devices of STS research, and from considering experimentation as a procedural activity in which actors take part to exploring how the condition of experimentality enables the enactment of actors and their relationships in specific ways. This shift takes advantage of the aforementioned ambiguity of experiments as both objects of and resources for research on science, technology, and society. In this section we will organize our discussion by foregrounding three distinctive themes in this emerging literature: object-centered engagement, inventive methods, and prototyping. Each of these themes brings with it different empirical and conceptual definitions of experimentation, showcasing the diversity of approaches that characterizes current STS work in this area.

Object-Centered Engagement: Expanding the Settings of Participation

Renewed attention to the role of material objects in public participation processes has helped reorient the study of science, technology, and democracy toward contemporary challenges. Specifically, it has broadened the concept of experimentation to denote not just the methods and techniques used to curate particular forms of public participation but also the specific capacities of the often mundane objects and devices used to this end. Everyday things such as thermostats and wristbands acquire the capacity to mediate involvement with issues such as climate change and public health (Hawkins 2011; Wilkie 2014). Marres's (2012) work on "material participation," for instance, observes that governments, corporations, and civil society organizations configure everyday material practices as significant sites of participation in problems such as climate change. The fact that people are materially implicated in this issue by way of everyday practices—such as cooking, heating, or gardening—provides an opening for object-centered and technological strategies for societal change, including "ethical consumption" and "behavioral change." But the material implication of actors also provides opportunities for the experimentalization of political participation and the development of alternative formats of ecological involvement. Marres develops this argument through an analysis of everyday practices of engagement with climate change and environmental sustainability, including so-called living experiments. These experiments serve to thematize—that is, to make public—the implicit normative powers of material objects and their capacity to implicate us in matters of common concern and to put our ontological commitments to the test (see also Murphy 2006).

In reframing participation as something done with things, however, this strand of work also highlights that participation is not contained or overdetermined by its location. In his analysis of the formation of a new collective around the issue of consumer debt, for instance, Joe Deville (2015) pays particular attention to how individuals deploy the letters they receive from debt collection agencies to organize into a consumer debt public. They do so by uploading the documents onto online discussion forums dedicated to the topic of consumer debt, thus employing private communications to stage public demonstrations of the issues of consumer indebtedness, using the letters as a sort of lure to enable wider political engagement with the issue. This and other examples of how mundane, everyday objects can feature in the formation of publics suggest that the efficacy of participation initiatives derives to some extent from the experimental qualities that these objects acquire when they are deployed *in* and *as* the apparatus of participation.

Experiments in participation are in this sense object-dependent, insofar as everyday things such as thermostats or debt collection letters can bear an explicit normative or political charge that enables new forms of participation. This object dependency, however, does not imply that these experiments are dependent on a specific physical setting (e.g., the laboratory) or a specific procedure of participation (e.g., the debate). These "political things" are circulating objects. Furthermore, these objects not only include discrete or concrete entities—such as household appliances —but much more fluid and complex material realities—such as the (green) electricity grid (Schick and Winthereik 2013), the (polluted) atmosphere (Tironi and Calvillo 2016), or the Internet of Things (Gabrys 2014). When we consider these scattered techno-environmental arrangements it becomes clear that objects do not just play a role as props—rhetorical devices that dramatize the issue or demand in question—but operate also as diffuse mediators with specific powers of engagement and the material element in which engagement may find its practical justification (Marres 2009). By the same token, referring to participation initiatives as experiments does not just highlight their intrinsic potential to generate surprises or unexpected results but also implies that these initiatives often serve to test the capacity of objects, as well as subjects, to render wider issues relevant, above and beyond already-established problem definitions. This entails a further reframing of publicity and participation in relation to science and technology; rather than being simply objects worthy of public participation, distributed material practices of research and innovation become a distinctive register of participation in public problems.

Furthermore, work on the role of material objects in participation multiplies the traditions of experimentation relevant to the enactment of participation, loosening the hold of scientific understandings of experimentation in liberal democracy. In her previously mentioned work on sustainable living experiments and demonstrational ecohomes, Marres (2012) shows how these devices and practices draw on a variety of experimental formats, originating, for instance, in the monitoring of building performance in construction research, in ecological movements committed to living "in tune with nature," or in formats of marketing research designed to assess people's willingness to engage. Similarly, feminist-informed work in STS by scholars like Murphy (2006), Roberts (2006) and Puig de la Bellacasa (2014) demonstrates the intersection of multiple experimental forms in lay and scientific practices of environmental monitoring, as moral traditions focused on the care for the self are brought into relation with a technoscientific and/or ecological preoccupation with the monitoring of chemicals in the environment. The body as a site of experimentation and an incarnation of public evidence has also been front-staged in recent STS-inspired work on atmospheric contamination and "chemical attunement" that similarly highlight the precarious existence of technologically mediated collectives (Shapiro 2015; see also Choy 2011; Tironi 2014b).

In sum, a variety of genealogies and understandings of experimentation intersect in object-oriented approaches to participation, and STS researchers pay particular attention to how this cross-fertilization of different traditions, knowledges, and skills shapes contemporary politics. For one, the multiplicity of relevant traditions signals that experiments in participation are often unstable in terms of the political agenda they further—they are highly malleable and appropriable by a multitude of constituencies. For example, while the aforementioned scenario workshop methodology was invented to further the goals of the ecology movement and extend awareness to distant futures, it was subsequently adopted and absorbed by oil companies to organize debates among their own stakeholders. Partly for this reason, STS work in this area is not particularly interested in fixing the meaning of any given participatory experiment—for example, by anchoring it in a singular experimental tradition, scientific, political, or artistic. Instead, greater sensitivity to the variability of participatory forms has reactivated a commitment to what we might call a politics of underdeterminacy. Seen from this perspective, the multivalence of participatory experiments does not denote a lack of consistency or dependability; it rather points to their ability to circulate across multiple domains, facilitate encounters between different traditions and sensibilities, and enable ways of articulating public concerns that cannot be fully anticipated or contained by any given design.

Inventive Methods in STS: Experiments between What Is and What Might Be

As we noted earlier, the broadening of the STS perspective on participation to include material and technological practices has led STS researchers to reconsider their own role in experiments in participation. One reason for this is fairly simple: as a growing number of actors in policy, activism, social research, art, and design use a wider range of devices to foster participation, STS researchers have started to wonder how they may productively take up such instruments themselves. But there is also a more complex reason: as STS researchers account for participation in performative terms—as something accomplished through the deployment of settings, devices, and objects–participation as a *topic* is to some extent destabilized. Devices of participation may of course still be approached as an object of study, but they also represent a possible resource to be deployed, and often (a bit of) both at the same time. How can we deploy this ambiguity in STS research? And what politics of knowledge would it enable?

Before further discussing this methodological challenge, we want to emphasize that these shifts in STS approaches to participation are partly predicated on empirical developments. The dissemination of digital technologies is perhaps the most salient among these. For example, Plantin (2015) and Petersen (2014) have described the role of online mapping technologies in allowing new forms of public engagement in the context of emergencies—the 2007 San Diego wildfires and the Fukushima Daiichi nuclear disaster in 2011. In these cases, digital technologies are configured as instruments of participation—citizens can use online cartographic tools to assemble data and in the process constitute themselves as a new public. At the same time, online mapping tools can be used to conduct research *on* participation, producing for instance new evidence on which lay communities collected and uploaded radiation data in the aftermath of the Fukushima disaster. Digital technologies thus invite STS researchers to reflect on the increasing continuities between the technical apparatuses of participatory research that we seek to analyze and the technologies that we may wish to deploy in our own investigative work.

This multifaceted nature of technologies of participation—as both object and device of social research—has led some STS researchers to take the next step and get involved in the design and development of experimental devices (Jalbert and Kinchy 2016). This work is sometimes framed as a contribution to the development of inventive methods for social and cultural research (Lury and Wakeford 2012). The characteristic STS orientation toward the performative capacities of devices, objects, and settings—their ability to bring new phenomena into being—is translated here into efforts to deploy those entities experimentally with the aim of eliciting participation as a phenomenon to both cultivate and investigate.

Thus, the Austrian-Swiss collective Xperiment! has used creative drawing techniques to allow elderly patients to record and visualize their everyday lives, a project that resulted not only in a series of written research articles but also in gallery installations where the drawings are displayed (Kräftner and Kröll 2005; see also Guggenheim 2011). In another example of inventive methods at work, Wilkie, Michael, and Plummer-Fernandez (2014) explore the topic of "energy demand reduction and community engagement" by creating an experimental device of participation, the Energy Babble Box. This device combines a radio function—broadcasting content from the web and social media that deals with energy demand—with the interactivity offered by a microphone—allowing users to input their own "energy talk" into the device and to circulate this talk to other users of the Energy Babble Box. The experimental device facilitates at least three distinct operations upon participation in energy demand reduction: it renders visible current enactments of energy publics, it facilitates a playful engagement with the issue, and it offers speculative proposals for the reorganization of participation in this area. By deploying creative devices to organize experiments in engagement, this and similar STS-informed projects embrace the formative ambiguity of devices to open up an interstitial space between research and development, working around this traditional distinction to enable a movement or oscillation between the observation of what is given and the cultivation of new entities and relations.

Finally, efforts to render the performative capacities of methods and technologies productive for social and cultural research are being pursued extensively in digital media studies and digital social research. Some work in this area is inspired by STS research on the politics of technology and the politics of method and seeks to deploy dominant digital devices, such as search engines and social media analytics, for the study of public controversies and issue formation around science, technology, and society. This is, for instance, the case with the Issue Crawler, a web-based research tool for the location and analysis of "issue networks" online (Rogers 2010; Rogers and Marres 2000), or the Twitter Capture and Analysis Toolkit (T-CAT) developed at the University of Amsterdam (Borra and Rieder 2014). These instruments adapt tools of online data capture, analysis, and visualization to enable research on issue formation by academics, activists, advocates, journalists, and so on. Arguably, these interventions can be qualified as experiments in participation in themselves: by taking up and repurposing research instruments and infrastructures that were developed and are largely owned by private and for-profit organizations, they test the feasibility of a more public-oriented form of inquiry by digital means.

In sum, the various projects discussed in this section draw on and engage with very different traditions of practice-based work—product design, data visualization, installation art, and software development—but they have in common an orientation toward the performative capacities of devices and settings of participation and a commitment to move from the description of such settings and devices to their design and deployment. They attempt to open up a space between knowledge and invention, conjuring up forms of participation that would otherwise remain unavailable and that, in many respects, are yet to be fully imagined. In doing so, STS researchers adopt the role of participants in wider research and innovation cultures, seeking to enrich and radicalize traditional ways of conceiving and doing participatory research and design.

Experimentation as Prototyping: Participation in Times of Environmental, Technological, and Social Change

While much of the work discussed so far focuses on specific devices of public engagement, a growing literature highlights the broader ontological, epistemic, and political contexts in which experiments in participation become salient. Current research in STS focuses, for instance, on the forms of political experimentation that are deployed in post-disaster situations. In his study of participation initiatives in the wake of the 2010 earthquake in Talca, Chile, Tironi (2014a) argues that the disaster, as a social phenomenon located in the space between radical ontological uncertainty and the need for immediate action, gave rise to an iterative and all-encompassing form of experimentation (see also Tironi and Calvillo 2016). Here, the experiment in participation does not refer simply to a discrete apparatus used to elicit solutions to preestablished problems but points to a broader experimental atmosphere in which questions, solutions, and their context of application are speculated into being in the face of complete uncertainty. In other words, the concept of experiment does not just pertain to the methods and technologies deployed to engage people in current affairs. Disasters produce deep disturbances in ingrained ways of being and doing, forcing a radical experimentalization of the questions of how to live, and how to live together. They offer a most vital demonstration of the motto "No issue, no public" (cf. Marres 2005).

Using the concept of the experiment as a heuristic for the study of engagement practices also enables a different understanding of what it means to change the settings of participation. By approaching participation not as a procedure or mechanism but as an experimental practice, it becomes possible to loosen the association between participation and stabilization—to unsettle the assumption that participation primarily serves to steady or "fix" a certain situation by providing legitimacy or ensuring consent. Participatory practices can serve as a source of more disruptive kinds of social and political experimentalism, novel ways of equipping actors to deal with change in the face of pressing issues (e.g., climate change), extreme settings (e.g., disasters), and/or recalcitrant objects (e.g., digital infrastructures). The understanding of experiments in participation developed in social and political theories of scientific and technological democracy, which foregrounds how experimental arrangements serve to enact publics, encounters here a different notion of experimentation, one that traces its genealogy to progressive social movements, technological cultures, and the arts. Experiments provide settings for collective tinkering in vivo with objects and environments whose status and value are called into question by the emergence of a new political situation (Estalella and Sánchez Criado 2015).

Attention to collaborative experimentation has been particularly productive in the intersections of STS with the fields of architecture, urbanism, environmentalism, and the "maker culture" (Corsín-Jiménez, Estalella, Zoohaus Collective 2014; Farías 2015; Guggenheim 2011; Jungnickel 2013; Nold 2015; Papadopoulos 2015; Yaneva 2013). In this work, the city, the region, the neighborhood, or the project—architectural, urban, environmental, or otherwise—emerges as a concrete site in which innovation and development may be politicized, becoming a focal point of creative and material experimentation. Crucially, these experiments serve multiple purposes all at once: they mediate between institutions and communities, bring diverse actors together (sometimes to dramatize their differences), produce hands-on solutions, pilot unorthodox technologies, and, last but not least, test new ways of articulating issues. The experiment is a way of shifting the initiative, of demonstrating that people possess greater capacities to transform the conditions of their everyday life than they had previously assumed.

It should by now be apparent that to study experiments in participation does not imply that we scale down our perspective on democracy—or on science and technology, for that matter—and consider only the most immediate environments in people's daily lives. For one, much of the work on experiments in participation discussed so far stresses the public mediation of these experiments by media technologies. Experiments enable the assembling of new collectives around contentious objects that are, at the same time, the political, epistemological, and environmental media through which these collectives seek to act. Any modification of those objects of engagement is simultaneously an intervention into the conditions of publicity of those collectives— and vice versa. This is the idea conveyed by Kelty's concept of recursive publics in his work on the free software movement. These are publics that operate directly, both discursively and materially, on the infrastructures that allow their coming into being. These collectives "argue about technology, but they also argue through it. They express ideas, but they also express infrastructures through which ideas can be expressed (and circulated) in new ways" (Kelty 2008, 29; see also Coleman 2014).

The normative effect of this recursive logic is what could be labeled, after the work of Corsín-Jiménez and Adolfo Estalella on open source urbanism (Corsín-Jiménez 2014a), a politics in beta: an experimental mode of inquiry in which constantly changing conditions, materials, and spaces invoke an equally mutable and transient public sphere. Corsín-Jiménez uses the figure of the prototype to bring into theoretical relief the political implications of such collective experiments. Prototyping, in this context, is not just a particular way of configuring and staging a technical device—as in the release of beta, nonstable, or work-in-progress versions in software or architectural development—but a process characterized by the "mutual prefiguration of objects and sociality" (Corsín-Jiménez 2014b, 383). Prototyping, as Corsín-Jiménez puts it, should be investigated "as something that happens to social relationships when one approaches the craft and agency of objects in particular ways" (ibid., 383). In their work with disability and independent-living activists in Barcelona, Sánchez Criado,

Rodríguez Giralt, and Mencaroni (2015) present a collaborative process of open source prototyping aimed at tactically altering the urban structures that constrain movement. Through targeted material interventions, such as the design and deployment of portable wheelchair ramps, these interventions do not simply produce a public statement about the need for more inclusive cities, but also exemplify a practical way of doing this inclusiveness—they make manifest the process of "taking part in the definition of the technical and material aspects defining independent living" (Sánchez Criado, Rodriguez-Giralt, and Mencaroni 2015, 14). In this particular context, STS is explicitly cast as a practical resource in the quest for new collaborative methods and solutions (see also Gabrys and Pritchard 2015).

Conclusion: Linking Up the Experimental Dimensions of STS

It should by now be clear why we believe that the theme of experiments in participation represents a productive nexus of theory and practice for STS. Not only does it help us think about the technoscientific dimension of politics, or about the politics of technoscience, but it also broadens the domains of science, technology, and innovation and allows us to attend to a much wider range of practices of research and invention across social, cultural, and political life. This broadening is certainly not meant to undo the commitment to specificity, granular description, and empirical situatedness that has been distinctive of much of the best work in STS. It is rather a way of following through on the original commitment of our field to a symmetrical treatment of science and its publics, of technical expertise and other ways of knowing and acting in the world.

We have approached the theme of experiments in participation as a channel to include the creative and generative practices of design, art, computing, digital media, and architecture in the fields of STS—a way of recognizing the critical role these practices play today in bringing science, innovation, and society into new sorts of relationships with one another. A focus on experiments in participation allows us to recognize the expansion of participatory registers in technological societies. Participation in public affairs is performed today in a multitude of everyday, workplace, cultural, environmental, and digital settings and media. Multiple traditions of experimentation are relevant to these performances—not only those of institutional technoscience but also many that emanate from social movements, computing, creative practices, and the arts (Born and Barry 2010; Gabrys and Yusoff 2012; Kelty 2016).

Participatory experiments thus represent trading zones between different traditions of experimentation, and demand that we develop modes of analysis and intervention that distribute the initiative more evenly across diverse and heterogeneous forms of practice. While STS shares this overarching objective with other approaches and agendas, such as those of digital culture or ecological politics, it is our view that the conceptual, empirical, and normative sensibilities of STS have a crucial role to play in this task. STS offers critical intellectual resources that we simply cannot do without if we are to address key political challenges in contemporary technological societies: the lasting appeal of technological determinism, renewed assertions of the sovereign power of expert authority, and the proliferation of narrow framings of "evidence-based" policy—framings that sometimes infuse digital culture, the creative economy, or sustainability transitions almost as much as they do technoscience.

Finally, we have emphasized that experiments in participation are a productive field for the further development of STS scholarship because their formative ambiguity makes it possible to reframe and elaborate key insights and approaches of the field. Experiments in participation represent important phenomena to be described and analyzed as well as instruments to be deployed in intervention-oriented strands of STS. This dual character offers an opportunity to elaborate the reflexive capacities of our field. In experiments in participation, the characteristic STS orientation toward the constructed, performed, and technologically mediated nature of our world becomes deployable as part of the conduct of social and political inquiry; our analytical and critical sensibilities can be put to the test in the process of curating new public situations. Elaborating the experimental dimensions of STS will bring our field into a more productive dialogue with broader contemporary efforts to redefine democratic culture in technological and knowledge-intensive societies. The larger aim of STS research and intervention, however, remains the same as it has been for the last few decades, namely, to activate new collective imaginations of what an epistemically, technically, environmentally and materially engaged polity might be.

References

Barry, Andrew. 2001. *Political Machines: Governing a Technological Society*. London: Athlone Press.

___. 2013. *Material Politics: Disputes along the Pipeline*. West Sussex: Wiley-Blackwell.

Bellamy, Rob, and Javier Lezaun. 2015. "Crafting a Public for Geoengineering." *Public Understanding of Science*, Published online before print August 27, 2015. doi:10.1177/0963662515600965.

Blok, Anders. 2007. "Experts on Public Trial: On Democratizing Expertise through a Danish Consensus Conference." *Public Understanding of Science* 16 (2): 163–82.

Bogner, A. 2012. "The Paradox of Participation Experiments." *Science, Technology, & Human Values* 37 (5): 506–27.

Born, Georgina, and Andrew Barry. 2010. "Art-Science: From Public Understanding to Public Experiment." *Journal of Cultural Economy* 3 (1): 103–19.

Borra, Erik, and Bernhard Rieder. 2014. "Programmed Method: Developing a Toolset for Capturing and Analyzing Tweets." *Aslib Journal of Information Management* 66 (3): 262–78.

Burgess, Jacquelin, Andy Stirling, Judy Clark, Gail Davies, Malcolm Eames, Kristina Staley, and Suzanne Williamson. 2007. "Deliberative Mapping: A Novel Analytic-Deliberative Methodology to Support Contested Science-Policy Decisions." *Public Understanding of Science* 16 (3): 299–322.

Callon, Michel. 2009. "Civilizing Markets: Carbon Trading between in Vitro and in Vivo Experiments." *Accounting, Organizations and Society* 34 (3): 535–48.

Callon, Michel, Pierre Lascoumes, and Yannick Barthe. 2009. *Acting in an Uncertain World*. Cambridge, MA: MIT Press.

Chilvers, Jason. 2008a. "Deliberating Competence: Theoretical and Practitioner Perspectives on Effective Participatory Appraisal Practice." *Science, Technology, & Human Values* 33 (3): 421–51.

___. 2008b. "Environmental Risk, Uncertainty, and Participation: Mapping an Emergent Epistemic Community." *Environment and Planning A* 40 (12): 2990–3008.

Chilvers, Jason, and Matthew Kearnes, eds. 2016. *Remaking Participation: Science, Environment and Emergent Publics*. London: Routledge.

Choy, Timothy K. 2011. *Ecologies of Comparison: An Ethnography of Endangerment in Hong Kong*. Durham, NC: Duke University Press.

Coleman, Gabriella. 2014. *Hacker, Hoaxer, Whistleblower, Spy: The Many Faces of Anonymous*. London: Verso Books.

Collins, Harry. 1988. "Public Experiments and Displays of Virtuosity: The Core-set Revisited. *Social Studies of Science* 18 (4): 725–48.

Collins, Harry, and Trevor Pinch. 1982. *Frames of Meaning: The Social Construction of Extraordinary Science*. London: Routledge and Kegan Paul.

Corsín-Jiménez, Alberto. 2014a. "The Right to Infrastructure: A Prototype for Open Source Urbanism." *Environment and Planning D: Society and Space* 32 (2): 342–62.

___. 2014b. "Introduction: The Prototype: More Than Many and Less Than One." *Journal of Cultural Economy* 7 (4): 381–98.

Corsín-Jiménez, Alberto, Adolfo Estalella, and Zoohaus Collective. 2014. "The Interior Design of (Free) Knowledge." *Journal of Cultural Economy* 7 (4): 493–515.

Crosby, Ned, Janet Kelly, and Paul Schaefer. 1986. "Citizens Panels: A New Approach to Citizen Participation." *Public Administration Review* 46 (2): 170–78.

Danziger, Kurt. 1994. *Constructing the Subject: Historical Origins of Psychological Research*. Cambridge: Cambridge University Press.

Davies, Sarah R., Cynthia L. Selin, Gretchen Gano, and Ângela G. Pereira. 2012. "Citizen Engagement and Urban Change: Three Case Studies of Material Deliberation." *Cities* 29 (6): 351–57.

Dear, Peter. 1985. "Totius in Verba: Rhetoric and Authority in the Early Royal Society." *Isis* 76 (2): 145–61.

Derksen, Martin, Signe Vikkelsø, and Anne Beaulieu. 2012. "Social Technologies: Cross-Disciplinary Reflections on Technologies in and from the Social Sciences." *Theory & Psychology* 22 (2): 139–47.

Deville, Joe. 2015. *Lived Economies of Default: Consumer Credit, Debt Collection and the Capture of Affect*. London: Routledge.

Didier, Emmanuel. 2002. "Sampling and Democracy: Representativeness in the First United States Surveys." *Science in Context* 15 (3): 427–45.

Estalella, Adolfo, and Tomás Sánchez Criado. 2015. "Experimental Collaborations: An Invocation for the Redistribution of Social Research." *Convergence: The International Journal of Research into New Media Technologies* 21 (3): 301–5.

Ezrahi, Yaron. 1990. *The Descent of Icarus: Science and the Transformation of Contemporary Democracy*. Cambridge, MA: Harvard University Press.

___. 2012. *Imagined Democracies: Necessary Political Fictions*. Cambridge: Cambridge University Press.

Farías, Ignacio. 2015. "Epistemic Dissonance: Reconfiguring Valuation in Architectural Practice." In *Moments of Valuation: Exploring Sites of Dissonance*, edited by Ariane Berthoin Antal, Michael Hutter, and David Stark, 271–89. Oxford: Oxford University Press.

Felt, Ulrike. 2016. "The Temporal Choreographies of Participation: Thinking Innovation and Society from a Time-Sensitive Perspective." In *Remaking Participation: Science, Environment and Emergent Publics*, edited by Jason Chilvers and Matthew Kearnes, 178–98. London: Routledge.

Felt, Ulrike, and Maximilian Fochler. 2008. "The Bottom-up Meanings of the Concept of Public Participation in Science and Technology." *Science and Public Policy* 35 (7): 489–99.

___. 2010. "Machineries for Making Publics: Inscribing and De-scribing Publics in Public Engagement." *Minerva* 48 (3): 219–38.

Gabrys, Jennifer. 2014. "Programming Environments: Environmentality and Citizen Sensing in the Smart City." *Environment and Planning D: Society and Space* 32 (1): 30–48.

Gabrys, Jennifer, and Helen Pritchard. 2015. "Just Good Enough Data and Environmental Sensing: Moving beyond Regulatory Benchmarks toward Citizen Action." In "Environmental Infrastructures and Platforms 2015—Infrastructures and Platforms for Environmental Crowd Sensing and Big Data Proceedings of the Workshop," Proceedings of the European Citizen Science Association General Assembly, Barcelona, Spain, 28–30 October, eds. Arne J. Berre, Sven Schade, and Jaume Piera.

Gabrys, Jennifer, and Kathryn Yusoff. 2012. "Arts, Sciences and Climate Change: Practices and Politics at the Threshold." *Science as Culture* 21 (1): 1–24.

Gieryn, Thomas. 2006. "City as Truth-Spot Laboratories and Field-Sites in Urban Studies." *Social Studies of Science* 36 (1): 5–38.

Gillespie, Richard. 1993. *Manufacturing Knowledge: A History of the Hawthorne Experiments*. Cambridge: Cambridge University Press.

Gross, Matthias, and Wolfgang Krohn. 2005. "Society as Experiment: Sociological Foundations for a Self-Experimental Society." *History of the Human Sciences* 18 (2): 63–86.

Grundahl, Johs. 1995. "The Danish Consensus Conference Model." In *Public Participation in Science: The Role of Consensus Conferences in Europe*, edited by Simon Joss and John Durant, 31–40. London: Science Museum.

Guggenheim, Michael. 2011. "The Proof Is in the Pudding: On 'Truth to Materials' in STS, Followed by an Attempt to Improve It." *Science, Technology, and Industry Studies* 7 (1): 65–86.

___. 2012. "Laboratizing and De-laboratizing the World: Changing Sociological Concepts for Places of Knowledge Production." *History of the Human Sciences* 25 (1): 99–118.

Hacking, Ian. 1983. *Representing and Intervening: Introductory Topics in the Philosophy of Natural Science*. Cambridge: Cambridge University Press.

Haffner, Jeanne. 2013. *The View from Above: The Science of Social Space*. Cambridge, MA: MIT Press.

Hawkins, Gay. 2011. "Packaging Water: Plastic Bottles as Market and Public Devices." *Economy and Society* 40 (4): 534–52.

Healy, B. 2015. "Facebook Button Lets Users Tell Friends They Voted in UK Election." *Mashable* (May 5).

Hilgartner, Stephen, 2000. *Science on Stage: Expert Advice as Public Drama*. Stanford, CA: Stanford University Press.

Horst, Maja, and Mike Michael. 2011. "On the Shoulders of Idiots: Re-thinking Science Communication as 'Event'." *Science as Culture* 20 (3): 283–306.

Igo, Sarah Elizabeth. 2007. *The Averaged American: Surveys, Citizens, and the Making of a Mass Public*. Cambridge, MA: Harvard University Press.

Irwin, Alan. 2001. "Constructing the Scientific Citizen: Science and Democracy in the Biosciences." *Public Understanding of Science* 10 (1): 1–18.

___. 2006. "The Politics of Talk Coming to Terms with the 'New' Scientific Governance." *Social Studies of Science* 36 (2): 299–320.

Irwin, Alan, Torben Elgaard Jensen, and Kevin Jones. 2013. "The Good, the Bad and the Perfect: Criticizing Engagement Practice." *Social Studies of Science* 43 (1): 118–35.

Jalbert, Kirk, and Abby J. Kinchy. 2015. "Sense and Influence: Environmental Monitoring Tools and the Power of Citizen Science." *Journal of Environmental Policy & Planning* 18 (3): 1–19.

Jasanoff, Sheila. 1998. "The Eye of Everyman Witnessing DNA in the Simpson Trial." *Social Studies of Science* 28 (5–6): 713–40.

___. 2003. "Technologies of Humility: Citizen Participation in Governing Science." *Minerva* 41 (3): 223–44.

___, ed. 2004. States of Knowledge: The Co-production of Science and the Social Order. New York: Routledge.

___. 2005. *Designs on Nature: Science and Democracy in Europe and the United States.* Princeton, NJ: Princeton University Press.

Jungnickel, Katrina. 2013. *DiY WiFi: Re-imagining Connectivity.* Basingstoke: Palgrave Macmillan.

Kelty, Christopher. 2008. *Two Bits: The Cultural Significance of Free Software.* Durham, NC: Duke University Press.

___. 2016. "Too Much Democracy in All the Wrong Places: Towards a Grammar of Participation." *Current Anthropology*, S13.

Kräftner, Bernd, and Judith Kröll. 2005. "What Is a Body / a Person? Topography of the Possible." In *Making Things Public: Atmospheres of Democracy*, edited by Bruno Latour and Peter Weibel, 906–10. Cambridge, MA: MIT Press.

Latour, Bruno, and Steve Woolgar. 1979. *Laboratory Life: The Social Construction of Scientific Facts.* Beverly Hills, CA: Sage.

Laurent, Brice. 2009. *Replicating Participatory Devices: The Consensus Conference Confronts Nanotechnology.* CSI Working Papers Series 018.

___. 2011. "Technologies of Democracy: Experiments and Demonstrations." *Science and Engineering Ethics* 17 (4): 649–66.

___. In press. "Political Experiments That Matter. Ordering Democracy from Experimental Sites." *Social Studies of Science.*

Law, John, and John Urry. 2004. "Enacting the Social." *Economy and Society* 33 (3): 390–410.

Law, John, and Evelyn Ruppert. 2013. "The Social Life of Methods: Devices." *Journal of Cultural Economy* 6 (3): 229–40.

Lemov, Rebecca Maura. 2005. *World as Laboratory: Experiments with Mice, Mazes, and Men.* New York: Macmillan.

Lezaun, Javier. 2007. "A Market of Opinions: The Political Epistemology of Focus Groups." *Sociological Review* 55 (s2): 130–51.

___. 2011. "Offshore Democracy: Launch and Landfall of a Socio-Technical Experiment." *Economy and Society* 40 (4): 553–81.

Lezaun, Javier, and Nerea Calvillo. 2014. "In the Political Laboratory: Kurt Lewin's Atmospheres." *Journal of Cultural Economy* 7 (4): 434–57.

Lezaun, Javier, and Linda Soneryd. 2007. "Consulting Citizens: Technologies of Elicitation and the Mobility of Publics." *Public Understanding of Science* 16 (3): 279–97.

Lury, Celia, and Nina Wakeford, eds. 2012. *Inventive Methods: The Happening of the Social.* London: Routledge.

Lynch, Michael. 1985. *Art and Artifact in Laboratory Science: A Study of Shop Work and Shop Talk in a Research Laboratory.* Boston: Routledge and Kegan Paul.

Marres, Noortje. 2005. "No Issue, No Public: Democratic Deficits after the Displacement of Politics." Doctoral Dissertation, University of Amsterdam.

___. 2007. "The Issues Deserve More Credit: Pragmatist Contributions to the Study of Public Involvement in Controversy." *Social Studies of Science* 37 (5): 759–80.

___. 2009. "Testing Powers of Engagement: Green Living Experiments, the Ontological Turn and the Undoability of Involvement" in "What Is the Empirical?," edited by Lisa Adkins and Celia Lury, special issue, *European Journal of Social Theory* 12 (1): 117–33.

___. 2012. *Material Participation: Technology, the Environment and Everyday Publics.* Basingstoke: Palgrave Macmillan.

Marres, Noortje, and Javier Lezaun. 2011. "Materials and Devices of the Public: An Introduction." *Economy and Society* 40 (4): 489–509.

Mayer, Andreas. 2013. *Sites of the Unconscious: Hypnosis and the Emergence of the Psychoanalytic Setting.* Chicago: University of Chicago Press.

Michael, Mike. 2009. "Publics Performing Publics: Of PiGs, PiPs and Politics." *Public Understanding of Science* 18 (5): 617–31.

___. 2012. "'What Are We Busy Doing?' Engaging the Idiot." *Science, Technology, & Human Values* 37 (5): 528–54.

Miller, Clark. 2004. "Interrogating the Civic Epistemology of American Democracy Stability and Instability in the 2000 US Presidential Election." *Social Studies of Science* 34 (4): 501–30.

Muniesa, Fabian. 2014. *The Provoked Economy: Economic Reality and the Performative Turn.* London: Routledge.

Murphy, Michelle. 2006. *Sick Building Syndrome and the Problem of Uncertainty: Environmental Politics, Technoscience, and Women Workers.* Durham, NC: Duke University Press.

Nold, Christian. 2015. "Micro/Macro Prototyping." *International Journal of Human-Computer Studies* 81 (September): 72–80.

Osborne, Thomas, and Nikolas Rose. 1999. "Do the Social Sciences Create Phenomena? The Example of Public Opinion Research." *British Journal of Sociology* 50 (3): 367–96.

Papadopoulos, Dimitris. 2015. "From Publics to Practitioners: Invention Power and Open Technoscience." *Science as Culture* 24 (1): 108–21.

Petersen, Katrina. 2014. "Producing Space, Tracing Authority: Mapping the 2007 San Diego Wildfires." *Sociological Review* 62 (s1): 91–113.

Plantin, Jean-Christophe. 2015. "The Politics of Mapping Platforms: Participatory Radiation Mapping after the Fukushima Daiichi Disaster." *Media, Culture & Society* 37 (6): 904–21.

Popper, Karl. [1963] 2002. *Conjectures and Refutations: The Growth of Scientific Knowledge*. London: Routledge.

Porter, Theodore. 1996. *Trust in Numbers: The Pursuit of Objectivity in Science and Public Life*. Princeton, NJ: Princeton University Press.

Puig de la Bellacasa, Maria. 2014. "Encountering Bioinfrastructure: Ecological Struggles and the Sciences of Soil." *Social Epistemology* 28 (1): 26–40.

Reichenbach, Hans. 1938. *Experience and Prediction: An Analysis of the Foundations and the Structure of Knowledge*. Chicago: University of Chicago Press.

Rheinberger, Hans-Jörg. 1997. *Toward a History of Epistemic Things: Synthesizing Proteins in the Test Tube*. Stanford, CA: Stanford University Press.

Roberts, Celia. 2006. "'What Can I Do to Help Myself?': Somatic Individuality and Contemporary Hormonal Bodies." *Science Studies* 19 (2): 54–76.

Rogers, Richard. 2010. "Mapping Public Web Space with the Issuecrawler." In *Digital Cognitive Technologies: Epistemology and the Knowledge Economy*, edited by Bernard Reber and Claire Brossaud, 89–99. West Sussex: Wiley-Blackwell.

Rogers, Richard, and Noortje Marres. 2000. "Landscaping Climate Change: A Mapping Technique for Understanding Science and Technology Debates on the World Wide Web." *Public Understanding of Science* 9 (2): 141–63.

Rogers-Hayden, Tee, and Nick Pidgeon. 2007. "Moving Engagement 'Upstream'? Nanotechnologies and the Royal Society and Royal Academy of Engineering Inquiry." *Public Understanding of Science* 16 (3): 346–64.

Rosental, Claude. 2013. "Toward a Sociology of Public Demonstrations." *Sociological Theory* 31 (4): 343–65.

Sánchez Criado, Tomás, Israel Rodríguez-Giralt, and Arianna Mencaroni. 2015. "Care in (Critical) Making: Prototyping as a Radicalisation of Independent-Living." *ALTER: European Journal of Disability Research*. Online first 11 September 2015, http://dx.doi.org/10.1016/j.alter.2015.07.002.

Schick, Lea, and Brit Ross Winthereik. 2013. "Innovating Relations—or Why Smart Grid Is Not Too Complex for the Public." *Science & Technology Studies* 26 (3): 82–102.

Schot, Johan, and Arie Rip. 1997. "The Past and Future of Constructive Technology Assessment." *Technological Forecasting and Social Change* 54 (2): 251–68.

Shapin, Steven. 1988. "The House of Experiment in Seventeenth-Century England." *Isis* 79 (3): 373–404.

___. 1994. *A Social History of Truth: Civility and Science in Seventeenth-Century England*. Chicago: University of Chicago Press.

Shapin, Steven, and Simon Schaffer. 1985. *Leviathan and the Air-Pump: Hobbes, Boyle, and the Experimental Life*. Princeton, NJ: Princeton University Press.

Shapiro, Nicholas. 2015. "Attuning to the Chemosphere: Domestic Formaldehyde, Bodily Reasoning, and the Chemical Sublime." *Cultural Anthropology* 30 (3): 368–93.

Soneryd, Linda. 2016. "Technologies of Participation and the Making of Technologized Futures." In *Remaking Participation: Science, Environment and Emergent Publics*, edited by Jason Chilvers and Matthew Kearnes, 144–61. London: Routledge.

Stilgoe, Jack. 2007. "The (Co-)Production of Public Uncertainty: UK Scientific Advice on Mobile Phone Health Risks." *Public Understanding of Science* 16 (1): 45–61.

___. 2015. *Experiment Earth: Responsible Innovation in Geoengineering*. London: Routledge.

Tironi, Manuel. 2014a. "Atmospheres of Indagation: Disasters and Politics of Excessiveness." *Sociological Review* 62 (s1): 114–34.

___. 2014b. "Hacia una política atmosférica: Químicos, afectos y cuidado en Puchuncaví." [Towards Atmospheric Politics: Chemicals, Affects and Care in Puchuncaví] *Pléyade* 14: 165–89.

___. 2015. "Disastrous Publics: Counter-Enactments in Participatory Experiments." *Science, Technology, & Human Values* 40 (4): 564–87.

Tironi, Manuel, and Nerea Calvillo. 2016. "Water and Air: Excess, Planning and the Elemental Textility of Urban Cosmopolitics. In *Urban Cosmopolitics: Agencements, Assemblies, Atmospheres*, edited by Anders Blok and Ignacio Farías. London: Routledge.

Ureta, Sebastián. 2015. "A Failed Platform: The Citizen Consensus Conference Travels to Chile." *Public Understanding of Science*, published online January. doi:10.1177/0963662514561940.

Voß, Jan-Peter. 2016. "Reflexively Engaging with Technologies of Participation: Constructive Assessment for Public Participation Methods." In *Remaking Participation: Science, Environment and Emergent Publics*, edited by Jason Chilvers and Matthew Kearnes, 238–60. London: Routledge.

Voß, Jan-Peter, and Nina Amelung. 2016. "Innovating Public Participation Methods: Technoscientization and Reflexive Engagement." *Social Studies of Science* (April 25).

Waterton, Claire, and Judith Tsouvalis. 2015. "On the Political Nature of Cyanobacteria: Intraactive Collective Politics in Loweswater, the English Lake District." *Environment and Planning D: Society and Space* 33 (3): 477–93.

Whatmore, Sarah, and Catharina Landström. 2011. "Flood Apprentices: An Exercise in Making Things Public." *Economy and Society* 40 (4): 582–610.

Wilkie, Alex. 2014. "Prototyping as Event: Designing the Future of Obesity." *Journal of Cultural Economy* 7 (4): 476–92.

Wilkie, Alex, Mike Michael, and Matthew Plummer-Fernandez. 2014. "Speculative Method and Twitter: Bots, Energy and Three Conceptual Characters." *Sociological Review* 63 (1): 79–101.

Yaneva, Albena. 2013. *Mapping Controversies in Architecture*. London: Ashgate.

8 Making and Doing: Engagement and Reflexive Learning in STS

Gary Lee Downey and Teun Zuiderent-Jerak

Introduction: Turning STS Lessons onto STS Work

A collection of faculty and graduate student scholars integrate science and technology studies (STS) with design practices to "frame" the educational experiences of undergraduate majors in engineering, computer science, business management, and communication (Nieusma et al. 2015). Seven STS scholars strive to persuade the European Science Foundation to frame policies for science in society, such as making space-time for scientists to engage in "reflexive work" (Felt et al. 2013). An STS scholar teaches a collection of nurses how to produce video ethnographies of themselves to help them "problematize" their own deficit model of safety and learn from them how they accomplish safety in practice (Mesman 2015). STS scholars "stand with" women living at the borderline of personality disorders to engage their lived experiences "diffractively" (Whynacht and Westby 2015). STS scholars mobilize an online "platform" to pose questions to federal scientists and ministers on matters of public and environmental health and safety, inspiring more than 4,000 letter writers to raise their voices (Myers et al. 2015). An STS scholar produces a "manual" for participants in controversies over science and technology, to foster more fair and open debate (Martin 2014). A collection of STS scholars prototypes a "feedback website" to reflexively rate such feedback websites as TripAdvisor, Yelp, and Amazon reviews (Ziewitz, Woolgar, and Sugden 2015). STS scholars developed interactive exhibits, hands-on activities, and research projects that helped more than eleven million people in 2015 teach themselves about the "social dimensions" of nanotechnology (Ostman, Bennett, and Wetmore 2015).

Science and technology studies has long provided intellectual and institutional space for projects that extend beyond the academic paper or book, both to make STS knowledge and expertise travel as quickly and widely as possible and to produce and express STS knowledge and expertise in novel ways. The contents, forms, and scope of such work are expanding rapidly, challenging and redrawing boundaries around the notion and practices of STS scholarship. The main purpose of this chapter is to call your attention to these projects, whether as a prospective or an existing STS researcher or STS practitioner.[1] We hope to persuade you to consider developing, enhancing, and reflecting critically upon your own versions.

To that end, we highlight in this review those examples through which STS researchers have been turning STS lessons about the production, expression, and travels of knowledge and technologies back onto their own initiatives and experiences in their fields of study. We examine a distributed collection of projects, identifying constituent elements, highlighting key dimensions of their ecologies, and diving down into three distinct clusters. We call this distinctively performative array of scholarly practices "STS making and doing."

Building on Engagement and Reflexivity

STS work has frequently built on critiques of the linear model of knowledge creation, diffusion, and utilization. This widely held image pictures sciences and technologies as produced in a social vacuum by individual human creators who diffuse their creations into the world, where they become facts and technologies that other people use for various purposes. Developing alternative accounts, STS researchers have shown that technologies and forms of knowledge always develop in specific social settings and that myriad agents and agencies participate in their production, expression, and travel.

STS accounts have ventured into a multitude of empirical domains for studying the work required to produce knowledge (e.g., Knorr Cetina and Mulkay 1983), transform it into facts (e.g., Latour and Woolgar [1979] 1986), and make technologies function and act (e.g., Bijker, Hughes, and Pinch 1987). Following both the shaping of technoscience and the way it brings specific worlds into being also opened up sciences and technologies to empirical analyses of in-built assumptions about what those worlds should become, including who benefits from particular versions and who bears the costs (Star 1991; Traweek 1988). STS work has also found material devices to possess potentially powerful and often surprising agencies, whether by dissolving the human/nonhuman dichotomy (Latour 1988) or articulating the active, often tricky, visions of technologies and their knowledges (Haraway 1991).

Since repeated studies have shown how the creation, materialization, travel, and utilization of the facts and artifacts under examination were intertwined right from the start, it should come as no surprise that many STS researchers are trying to cultivate scholarly self-understandings that recognize the same in their own work. How can STS scholars, including both researchers and practitioners, avoid separating the work of conceptualizing objects of study from that of diffusing research outcomes to knowledge recipients where it could have, hopefully laudable, effects and implications? Via findings about the nonlinearity of knowledge and the materiality of scholarly practices, students of science and techonology become symmetrically obliged to consider their relations with their own fields of study and action.

STS making and doing is, therefore, a mode of scholarship that involves attending not only to what the scholar makes and does but also to how the scholar and the scholarship get made and done in the process. On the one hand, this entails examining how STS scholars and scholarship actively engage the settings they study or otherwise enter, including the agents that occupy them (Hackett et al. 2008; Sismondo 2008), asking such questions as what consequences does STS scholarship have in these settings, and might STS notions help STS scholars determine whether and how their work may or may not bring value to those settings? On the other hand, it involves reflecting critically on how the work and identities of STS scholars are constructed in the process. Such reflection implies mobilizing STS notions—whether one prefers such offerings as "(social) construction" (Latour and Woolgar [1979] 1986), "translation" (Callon 1986), "boundary objects" (Star and Griesemer 1989), "socially robust knowledge" (Gibbons et al. 1994), "co-production" (Jasanoff 1996, 2004), or "undone science" (Frickel et al. 2010)—as relevant not only for the fields we study but also for understanding STS scholarly practices.

Practices of STS making and doing build upon those of engagement and reflexivity. In the examples offered at the outset, STS scholars venture into undergraduate education, science policy, nursing practice, medical diagnostics, environmental policy, science controversy, crowdsourced judgment, and informal science education. Each project engages audiences in the field, beyond the boundaries of academic STS. And arguably each project reflexively mobilizes and adjusts STS notions in its formulation.

What warrants the label "STS making and doing" is therefore simultaneous attention to the engagement of actors and practices in STS fields of work and to reflexive learning from those actors and practices. Even as you, the aspiring or established STS researcher or practitioner, offer your interlocutors something new to inflect their understandings of themselves and possible future actions, accepting the challenge to theorize in situated, localized, and material terms also challenges you to learn from the interlocutors and settings in which you work. Practices of STS making and doing thus both draw upon and extend academic critiques of the linear model by enacting two-way, or multiple-way, travels of knowledge production and expression.

Practices of STS Making and Doing

We draw on two sets of sources in this review. The first is a selection of contributions to the inaugural STS Making and Doing Program that took place during the 2015 annual conference of the Society for Social Studies of Science (4S).[2] The program included but was not limited to projects that resulted in (a) policy papers, recommendations, regulations, devices, decision-making practices, or other policy outcomes; (b) design and creation of products, graphics, spaces, and landscapes; (c) artistic creations, including those in audiovisual format or in public installations, exhibits, and performances; (d) practices for education and training; and (e) informational or material infrastructures for the construction, operation, and travel of STS as a field or discipline. The second set of sources for the analyses below consists of publications by STS researchers that reflect upon or serve as making and doing practices.

The gatherings and distinctions we outline below attend first to what we call the "elements" of STS making and doing practices that establish their directionalities, or valenced pathways of travel, recognizing that many of the consequences of STS scholarship are beyond the scholar's control.[3] We then turn to dimensions of the "ecologies," or dynamic relations in the fields of study, that they envision, encounter, and produce.[4] This dual focus on elements and ecological dimensions calls attention to how scholarly projects in STS making and doing build on practices of engagement and reflexive learning to enact their two-way or multiple-way flows of knowledge production and expression. Accordingly, this chapter repeatedly poses the following questions: What sorts of elements constitute distinguishable practices of STS making and doing? How do practices of STS making and doing envision and learn from the ecologies of knowledge-making and world-making that they join or help create?

After spelling out elements and ecological dimensions, we investigate how these may appear in, and sometimes infuse, distinguishable clusters of STS making and doing. We identify three such clusters: (1) boundary-crossing STS claims, (2) meta-activism, and (3) experiments in participation. By organizing this array of work into three clusters, we seek to be generative rather than exhaustive in content.

Elements

In order for STS claims to become relevant for others beyond the field, STS scholars must build the elements necessary for those claims to travel into new settings and gain position and status within them. Such elements pertain to what STS scholars judge to be the key analytical issues at stake, as well as to the concrete activities they undertake to address those issues within the settings.

One: Frictions and Alternate Images Like much STS research for scholarly publication, practices of STS making and doing position themselves by identifying some dominant or otherwise problematic images of science and/or technology and their analogs in medicine, engineering, and so on. The term *dominant images* refers here to ideas or meanings whose acceptance has traveled sufficiently across some population and terrain to become given, or true—for that population and across that space (Downey 2009, 60). Dominant images of science in the singular, for example, tend to highlight creative discovery that produces Truth. Dominant images of technology in the singular tend to posit autonomous developments that become external forces. STS scholars who "get involved" with science and technology encounter and try to learn from versions of dominant images that are localized and, hence, diverse.

A common first step is to identify "frictions" within the field (Kember 2003)—places where dominant images lose their smoothness and become multiple. All images make some things visible while hiding others. Much STS scholarship published for academic audiences makes visible for those readers the frictions that dominant images of science and technology hide. A key element of STS making and doing involves making frictions visible for audiences within the field.

Two: Techniques, Devices, and Infrastructures Projects of STS making and doing draw on learning to develop and situate specific, localized techniques and devices to contribute to the field. These may be such discursive forms as policy analyses, op-ed articles, (participation in) public debates, focus-group reports, and so on. They may include such materializations as modified information communication technology (ICT) systems, art installations, or market devices. Techniques, devices, and, sometimes, infrastructures for activating them tend to be tangible and carry both the pleasures of construction and demands of maintenance. To identify them as elements of STS making and doing, it is important to question their directionalities. Where are they headed—for what or for whom, and with what expected bandwidths of influence or effect? In which unexpected places do they end up?

Three: STS Expertise and Identities Who an STS scholar is sometimes figures in specifying the pathways across which making and doing practices can or cannot travel. It can matter, for example, whether one's scholarly formation included education in the sciences, engineering, medicine, and so on, in addition to STS. Also, STS scholars may adopt, resist, or even attempt to transform the expert positions they encounter in the field in order to involve themselves deeply. In what ways do STS scholars adopt, resist, or transform expert positions ascribed by actors in the field? What dimensions of the STS scholar *qua* person and agent are relevant in how her or his scholarship relates to its fields? In what ways does subject-matter expertise inflect the positioning of the work? How do such positions shift over time, and what possibilities do different stances open up or close down?

Ecological Dimensions

STS scholars must take account of the ecologies (Star 1995) within which they situate the work of making and doing. The important notion of ecologies highlights the agencies of learning, webs of influence, and hierarchical and dynamic orderings that the scholarship and scholars both encounter and enact across the locations and settings of their work.

Four: Audiences, Partners, and Engaged Practitioners Instances of STS making and doing are performed for specific audiences and often include partners or engaged practitioners as the work aims to produce learners who benefit and necessarily accept the risks of producing victims. Which audiences or collaborators do scholars build into their techniques and devices, and who become actual audiences or implied practitioners? How do practices of making and doing learn from and handle differences encountered? The interest here includes the identities and statuses of audiences, partners, and others as active agents in the STS project at hand. How and in which directions do STS scholars expect their practices of making and doing to travel? What are the geographies of their consequences? Who learns, with what benefits and what costs?

Five: STS Sensibilities Out There STS sensibilities about the making of sciences and technologies, as well as their travels and lives, are by no means the exclusive domain of STS scholars. Encountering such sensibilities among so-called research subjects is rather common. After all, STS scholars learn, or acquire, sensibilities from closely studying the work of actors in empirical fields. Furthermore, some educational programs for engineers, health care professionals, and business managers are heavily infused with STS work. This leads to interesting empirical puzzles: how do STS sensibilities in the wild relate to STS making and doing practices?

Six: Feedback and Reframing Practices of STS making and doing deal quite differently from one another with the return flows of claims, techniques, and devices—the feedback—that their activities generate. Those who position the work as pedagogical pronouncements by informed STS scholars may be interested mostly in strategic alliances with like-minded actors. Those who explicitly search for generative instantiations of messy knowledge production and expression may find themselves compelled to learn by exploring mutual feedback that changes both the empirical domain and STS understandings of it. How do different approaches to STS making and doing deal with feedback, controversies, and, sometimes, success? Do these become ethnographic moments for reframing the goals of scholarship, strategic moments for finding out who one's allies are or are not, or what?

Clusters of Making and Doing

Having outlined elements and ecologies, we now turn to introduce three clusters of STS making and doing by exploring scholarly publications as well as nineteen projects that were part of the STS Making and Doing Program at the 2015 4S annual meeting in Denver, Colorado. All expand knowledge production, expression, and travel in STS. We offer "cluster" as a heuristic for identifying activities that display similarities in the collections of elements they mobilize and ecological dimensions they encounter in two-way or multiple-way flows of learning. At the same time, each activity within an assigned cluster includes distinctive features, and many projects have activities that span more than one cluster. At this writing, most have websites. We invite you to explore them. After briefly introducing each cluster, we examine more closely its constituent elements and ecological dimensions.

Cluster One: Boundary-Crossing STS Claims

Concerns have long persisted about how the unique dimensions of STS facilitate novel contributions in the arenas that STS researchers study. For example, in his final presentation at the 4S annual meeting in 2002, David Edge cautioned that popular debates and discussions about science and technology tended to "make ... no reference whatsoever to the STS literature[s] ... on [those] topics ..." (Edge 2003, 162). "It is hard to start any sort of conversation," Edge pointedly continued, "when your conversation partner believes, in all sincerity, that your aim is to silence them" (ibid., 167).

One way in which the scholarship of making and doing in STS seeks involvement in shaping the worlds it enters is by building practices that aim to help STS claims travel across the boundaries of the field into those worlds. What counts as the relevant boundary around the field depends on how the specific making and doing project positions itself in relation to its empirical area of work. Boundary-crossing STS claims appear *inter alia* in blogs, op-ed articles and columns, expert testimony, and policy reports, with both individual and organizational authors frequently adopting the figure of the public intellectual (Bijker 2003; Society for Social Studies of Science 2014).

Boundary-crossing STS practices infuse pedagogies, artistic creations, and a range of experiences designed for learners beyond the field.

One: Frictions and Alternate Images   Making and doing scholarship in this cluster tends to portray the areas of technoscience it encounters as instrumental and in need of reflective awareness and reflexive practices. Social dimensions and implications of these technical fields appear to their practitioners as marginalized, subordinated, or otherwise backgrounded in relation to narrowly defined, yet privileged, knowledge or technical contents.

Boundary-crossing STS claims and practices tend to highlight the benefits of apprehending deep interrelations among science, technology, and society, typically evidenced in a specific case or cases at hand. Such can mean making science and technology "subject to political debate" or "[expanding] the information needed to make sound policy," including "intertwined expert knowledge about the biological, material and social worlds" (Jasanoff 2011, 622). It can also mean developing boundary-crossing pedagogies. An STS department runs an interdisciplinary set of degree programs in design, innovation, and society that "bring ... STS to the worlds of engineering, computer science, business management, and communications—all via design" (Nieusma et al. 2015). Developers of these degree programs found the technical fields to perform narrow conceptions of problem solving and design that are out of step with the complex, collaborative design problems graduates will encounter on the job, producing an interesting friction within engineering practice to which STS sensibilities could speak. In this case, an extensive set of design practices spans a four-year curriculum. The goal is to "*frame* our students' entire educational experience ... within a more expansive vision of technology-in-society" than what those students routinely encounter (Nieusma et al. 2015). Included in the alternate vision it enacts are skill sets to prepare graduates for leadership positions in worlds with multiple "design cultures" (Department of Science and Technology Studies, Rensselaer Polytechnic Institute 2015).

Two: Techniques, Devices, and Infrastructures   Some discursive techniques and devices that STS scholars have used to transport STS questions and insights to learners beyond the field include writing opinion pieces in journalistic publications (Shapin 2006), public lectures for diverse audiences (e.g., Harvard's Science and Democracy Lecture Series 2015), broadcasting provocative short takes on issues of public concern (STS blogs), sharing maps of STS practices and practitioners for public consumption (Pfaffenberger and Hunsinger 2015) (http://www.stswiki.org), organizing public debates to democratize technology development (Sclove 1995), and training STS scholars to "participate more effectively in decision processes and public affairs" (Jasanoff, Wellerstein, and Rabinowich 2015).

The Critical Futures Lab (http://criticalfutureslab.org/) uses a game to enter the worlds of labor activists and entice them to collaborate with scholars, designers, and technologies to imagine the future of work. The classic scholarly insight that "it could be different" (Boas 1940; Hughes 1970; Lévi-Strauss 1966) is materialized in the infrastructural move of a limited, one-day participatory design workshop. The game challenges participating activists to "explore historical and present technologies, socio-economic conditions and labor realities in order to open up discussions around the way in which technologies shaped and were shaped by social, economic, political and cultural contexts" (Forlano and Halpern 2015). After opening up these discussions, the game helps participants create "counterfactual histories that might allow for alternate relationships, outcomes and possibilities that might benefit workers" (ibid.). They come to play, imagine, and, it is hoped, learn ways of repositioning themselves (ibid.).

Three: STS Expertise and Identities   In accepting the challenges of transporting STS insights to the publics of technoscience, STS scholars must develop and display unique expertise in producing the claims and practices they offer. Expertise on blurring the social and the technical claims relevance, sophistication, and, sometimes, superiority by providing intellectually distinct, revealing, and socially significant angles on topics at hand.

Cleverly accepting the widespread claim that the "innovation process" stands "at the heart of reimagining Europe," seven STS scholars explicitly activate lessons from STS scholarship in a Science Policy Briefing designed to persuade the European Science Foundation (ESF) that both science and society are fluid entities whose innovations are in "continuous co-evolution" (Felt et al. 2013, 3). Their thirty-page report forcefully asserts that the "often narrow evaluation criteria used in research, innovation and education policy" are sorely at odds with the "broader value systems employed by [diverse] societal actors to assess science as a public good." The strategy is to nominate a "science *in* society" model to replace the reactionary, yet dominant, "science *and* society" model in formal ESF policy making. Such authorizes what they call a "logic of care" in what should be seen as the "governance" of science in society. It grounds ambitious initiatives to grant scientists "more time and space for reflexive work within research" (Felt et al. 2013, 3–4).

In such instances, while claiming to be of central importance to debates and deliberations over science and technology, the arrival of STS expertise in boundary-crossing moves may be seen by others as coming out of left field. While challenging the notions embodied in established practices of technoscience, the risks can be high that such expertise will be misunderstood, will be judged to be threatening, or will not resonate sufficiently with the concerns of actors involved. Those risks are clearly high, so the necessary STS expertise must be enacted strongly.

Four: Audiences, Partners, and Engaged Practitioners The STS work in producing, expressing, and transporting boundary-crossing claims typically aims to expand audiences and multiply pathways for reaching them. The expectation is that the learning that results will help expand distributions of benefits and reduce distributions of costs. Bijker's (2003) call for public-intellectual work, for example, seeks not only to reach "politicians, engineers, scientists, and the general public" with the STS insight that "science and technology are value laden." It also aspires to elevate the public profile of technoscience by making clear that "all aspects of modern culture are infused with science and technology, that science and technology do play key roles in keeping society together, and that they are equally central in all events that threaten its stability" (ibid., 444). Accordingly, the transport of STS findings and claims in this cluster typically construes audiences as external to the field and the pathways for reaching them as multiple and overlapping. The cluster tends to define success as transforming external audiences into apprentices for STS learning and STS-grounded action.

Seeking "public learning" of "STS ideas," for example, the Nano and Society project reaches millions of people across the United States each year (Ostman, Bennett, and Wetmore 2015). Through science museums, exhibits for rent or purchase, free videos, print media, public forums, theater and stage presentations, and programming for more than a week of NanoDays, the seven regional hubs of Nano and Society create mazes of pathways for transporting STS-informed questions and accounts to non-STS audiences (NISE Network 2015). Built around the core question in one exhibit, "Nanotechnology: What's the big deal?," this self-styled initiative in "informal science education" connects research to museums to wrestle with the "complexity and abstract nature of nano, the ubiquity of its applications in society, and the relative lack of knowledge about the ethics and the impacts on health, environment and society of these applications" (Lundh, Stanford, and Shear 2014, 5). Expanding attention to the social dimensions of nano beyond the narrow early conception of "ethical, legal, and social implications," Nano and Society endeavors to build a public geography for this technological emergence that is informed about its social and ethical dimensions and hungry for further insights.

Five: STS Sensibilities Out There   Audiences of STS claims may have strong appreciation for frictions that their particular positions produce, for example, for the politics of knowledge or for self-limiting practices of technical fields. Such does not, however, equal or necessarily lead to developing and displaying an STS understanding of those frictions. Within this cluster, boundary-crossing STS claims are ascribed something of a privileged status in terms of conceptualizations, understandings, and contributions, tending to portray others as more the recipients than the bearers of STS insights. To the extent practitioners in the fields of study become agents of STS sensibilities or analysis, they typically enact it implicitly. Anything more explicit becomes a successful travel of STS insights.

Crafting Digital Selves is one such project. It asks if preservice science teachers can become agents of STS sensibilities and scholarship. Through the infrastructure of a required course for preservice science teachers (an accomplishment in itself), STS scholars help the teachers acquire "sustainability literacy" so they can, in turn, become agents of its pedagogy (Warren et al. 2015). The prior knowledge that teachers tend to bring to the course predisposes them to expect, for example, that solutions to the problem of sustainable water supply depend primarily, if not solely, upon large technical infrastructures. But then they encounter a ten-minute "digital story" that highlights how infrastructural solutions in the past have helped to produce the problems of the present. Technical infrastructures gain social dimensions and futures become multiple, subject to the choices of the present. Instructors seeking to achieve a "more inclusive form of STS" produce such stories and then use them in homework assignments as well as in both face-to-face and asynchronous online discussions (Warren et al. 2015). To persuade teachers to help produce "globally-minded and knowledgeable citizens who are able to analyze sustainability challenges and work toward solutions," the STS-informed instructors must find acceptable ways to interpolate sustainability literacy into the identities and directionalities—the selves—that the preservice learners are already building as prospective science teachers (Biodesign Institute 2015).

Six: Feedback and Reframing   Given the status of an STS insight as frequently drawing on experiences of a given setting, the expectation of feedback, pushback, and other responses can certainly affect or reframe the STS scholar's understanding of that setting. At the same time, however, given that the boundary-crossing claims also draw on insights from what is now an established academic field, they may be resistant to substantial reformulation once offered or implemented. STS learning and reframing may hone in on better ways to transport risky messages to their audiences. Moments of learning can also congregate around reformulating alternate images, to increase their chances of traveling sufficiently to become new realities.

The Energy Walk produces learning about one's experiences of renewable wave energy in the sea. Wave, wind, geothermal, solar, and other renewable energy technologies can feel like alien objects. Energy is supposed to come powerfully, but invisibly, through wires or pipelines in the ground, from distant sources. Renewable energy objects can be big, in one's face, and, hence difficult to understand or accept (Winthereik et al. 2015). The Energy Walk puts in people's hands a digital walking stick with an audio player activated by radio frequency ID tags, similar to those used to tag pets. It invites the walker to experience the landscapes of a wave energy center and its environs. In an energy center, where the invisibility of power generation and distribution is given, the experience produced by the walk could well generate resistance, pushback, or rejection. But The Energy Walk happens at the "energy edge," especially island settings whose residents cannot assume the existence of reliable, invisible energy supply (Watts and Winthereik 2015). Perhaps places where "people see the end" are particularly fertile sites for boundary-crossing learnings to generate alternate images and experiences (Ford 2015). The walking stick is designed to help researchers learn how "green energy creates and reconfigures relations between humans, technologies, and nature at particular project sites" (Winthereik et al. 2015).

In summary, this cluster of making and doing activities specializes in rendering visible STS insights to new audiences, providing them with fresh, more sensitive perspectives on or experience of the connections among science, technology, and society. As a consequence, elements of making and doing in this cluster tend to be strongly developed, whereas the ecological dimensions may prove to be less formative or consequential. To the extent these activities remain occasional for STS scholars and position their audiences mainly as recipients of STS knowledge, this cluster of STS scholarship and its practitioners may not become integral parts of the ecologies in which the audiences operate. Indeed, the ecology may become first and foremost an object of study. When the boundary-crossing project is successful, it may afford new opportunities for strategic alliances with those who are receptive to STS insights. To the extent activities of boundary crossing expand or become persistent, features more characteristic of the second or third cluster could emerge and come to predominate.

Cluster Two: Meta-Activism

We call the second, related, cluster of practices through which STS gets involved in making and doing the worlds it inhabits "meta-activism."[5] With this term, we refer to the activities deployed by STS scholars to assist or support actors who may already be resisting, challenging, or seeking alternatives to dominant images in their fields. While such scholarly practices draw upon insights from STS research to enhance repertoires of resistance, challenge, or the formulation of alternates, they are "meta" in the sense that they still largely delegate the contents and work of advocacy itself to actors in the field.

One: Frictions and Alternate Images Meta-activist projects of making and doing tend to highlight the political contents and power dimensions of the frictions that dominant images of science and technology produce and perform. They frequently make the case that subordinated or marginalized positions and silenced voices deserve both visibility and strength. They find actors who are part of or represent the subaltern or the silenced and contribute to their initiatives by formulating, expressing, and transporting alternate images aimed at facilitating empowerment.

The Scientific Legislation project (Comisso, Maciel, and Roberts 2015b), for example, aims to "empower several disciplinary undergrad students with law creation techniques, knowledge about science & technology in contemporary controversies and policy analysis in the frontier of STS to elaborate innovative and transdisciplinary proposals in scientific law for Chile." Through a curriculum for undergrad students, including law-development workshops with participatory methods, and a session presenting the results to senators and members of congress, the aim is to "construct legal projects that are consistent and pertinent for Chilean legislation concerning matters of science and technology, in a collective manner, using suitable learning elements for the appropriate legal, scientific/technological, and social interpretation of the phenomenon" (Comisso, Maciel, and Roberts 2015a, 2).[6] In this manner, students of specific technoscientific issues mobilize frictions to provide alternate images of innovative "civilian-academic" and citizen-led laws, without providing subject-matter expertise on the specific phenomena for which those laws are made (Comisso, Maciel, and Roberts 2015b).

Drawing attention to power relations can help contest the subordination of selves as well as that of positions or groups. The Department of Play project creates temporary play zones through which the initiators "explore the potential of play" to tease out experiences of public, urban spaces that "neat, authoritative urban visions" tend to hide (Balug and Vidart-Delgado, 2015; Vidart-Delgado and Balug 2015a). Play supports resistance to frictions as it invites participants to "envision alternate futures, share life experiences and different kinds of knowledge, collaboratively create artifacts, and negotiate different ways of relating to each other beyond established norms" (Vidart-Delgado and Balug 2015a). Without offering or imposing alternate images themselves, the initiators explore how diversely designed playdates can contribute to "forging a common vocabulary among diverse actors" and potentially "facilitate ... the pursuit of pluralistic visioning" (Vidart-Delgado and Balug 2015b).

Two: Techniques, Devices, and Infrastructures   Meta-activism mobilizes diverse arrays of techniques, devices, and infrastructures. A common technique is to provide strategy advice to activists, informed by academic STS scholarship. An exceptional example, Martin's *The Controversy Manual* (2014, back cover) informs its readers that it does "not tak[e] sides on individual controversies." Rather, as a meta-activist device, it "offer[s] practical advice for campaigners ... provid[ing] information for understanding controversies, arguing against opponents, getting your message out, and defending against attack." Made available by Irene Publishing, it contributes to an infrastructure designed to help effect "peace through peaceful means." The Civic Laboratory: Plastics project (Liboiron 2015), by contrast, addresses a specific issue. Microplastics "are ingested by marine life, their associated chemicals bioaccumulate in animals and biomagnify up the food chain," making them "an environmental justice issue." Civic Laboratory tools seek to make the problem "visible and actionable" through citizen science projects that create feminist technologies to monitor" even microscopic plastics. The project makes available "open source, affordable, hackable, do-it-with-others (DIWO) devices" such as the Plastic Eating Device for Rocky Ocean Coasts (P.E.D.R.O.C.). With fairly detailed building instructions, this wooden/metal device, anchored on rocky coastlines, can collect plastic over time. With two sieves and a flag, it mobilizes a heterogeneous infrastructure that includes both rocky coastlines and passers-by.

Three: STS Expertise and Identities   Similar to the first cluster, meta-activist projects tend to position the STS researcher as expert, in this case on political patterns within technoscience. The relevant expertise can consist of both analytical understandings of technopolitics and power relations and strategic sensibilities about how to act in the midst of technopolitical complexities. Whether by enacting advice to the subaltern, building new organizations, or producing devices to expand participation, meta-activist making and doing tends to proactively position STS expertise as strategy that others can bring to bear to transform the experience of frictions into more overt action. In this sense there is a division of labor, in which STS scholars serve as experts on certain sensibilities so others can rethink and revise technopolitical positions and strategies.

As the STS+ With Practitioners initiative puts it (de la Torre et al. 2015), participants aim to help medical students in post-Fukushima Japan and both K–12 and engineering students in the United States animate environmental thinking within formal pedagogies that all-too-often exclude them. Rather than straightforwardly teaching STS findings, the project "cultivate[s] the 'thought styles' of STS" in students being trained otherwise. Its practices involve building "scaffolding," or "light structure," to facilitate reading, listening, and observing "with STS sensibilities," to enhance students' capacities to analyze, evaluate, and creatively intervene in key exclusions.

Four: Audiences, Partners, and Engaged Practitioners   Audiences of meta-activism consist, in the first instance, of those bearing the costs of dominant images of science and technology. Such audiences can therefore be quite large and heterogeneously populated, and the pathways for reaching them quite diverse. In addition to work that seeks to empower marginalized positions and perspectives, as both the *Controversy Manual* and the Scientific Legislation project expect, some meta-activist making and doing focuses its attention on the means through which dominant images of sciences and technologies produce subordination and silences in the first place. Favorite sites for such work are the performative operations of technoscience institutions, which can produce diverse, even surprising, arrays of potential audiences and partners for meta-activist scholarship.

The "War on Science" by the Harper administration in Canada included "cancellation of over 100 federal research programs; the closure and destruction of libraries and archives; the firing of thousands of federal scientists; and … government policies that constrain federal scientists' freedom to speak to the public." The Write2Know project challenges both explicit and implicit silencing of evidence-based decision making via large-scale letter campaigns (Myers et al. 2015). Demanding that undone science be both done and revealed, Write2Know offers a "platform" that those who have stakes in such science can use to "pose questions to federal scientists and ministers on matters of public and environmental health and safety." Addressing research on topics including "endocrine disruption, uranium mining, and lead toxicity" (Myers and Liboiron 2015), it provides STS-informed templates that highlight "gaps between research and government policy" and make visible issues of social and environmental justice, including the "impacts of resource extraction, oil sands pollution, [and] marine plastics." The project also "foreground[s] colonial contexts in Canada, which render aboriginal communities especially vulnerable" (Myers et al. 2015). In addition to facilitating the generation of more than 4,000 individually signed letters on eleven topics, Write2Know also offers the option to ask a new question via a stepped approach, a sample email, and an overview of federal ministers and their critics in the political opposition. While not necessarily problematizing the politics of evidence, the project directly problematizes the power relations in its production. Its audiences and partners therefore evolve with the performative operations of the institutions that command its attention.

Five: STS Sensibilities Out There   Rather than starting from more definitive STS notions about a certain topic, meta-activist projects typically begin with problem statements advanced by marginalized actors in the field. The STS work then involves reconceptualizing those problem statements in STS terms to open up new repertoires of action. STS sensibilities about the creation, materialization, travel, and utilization of facts and artifacts may be largely latent until activated by meta-activist scholarship, and need not even be identified as such, or at least not prominently. The key point is that marginalized actors in the field are likely experiencing problems to which STS sensibilities can speak or relate.

Although many meta-activist projects of making and doing position others to be, or become, bearers or agents of STS knowledge and expertise, some projects do not. In the *Controversy Manual*, STS scholarship on controversies appears in an appendix, while the Write2Know project links to an STS scholarly endeavor in the "Who we are" section of the project website. Yet other initiatives, such as the STS+ With Practitioners project, expect practitioners precisely to become well versed in those sensibilities as STS sensibilities.

Six: Feedback and Reframing   Feedback within meta-activism is likely to appear as overt troubles. Instances of pushback could prove to be confirming signs of the political conservatism and power hierarchies performed and effected by dominant technoscientific practices. As support for activists seeking to expand frictions into opportunities to reframe or replace the performers of dominant images, meta-activism can prompt rebuttal, reassertion, retribution, and reprisal.

Such feedback does promote learning. For meta-activist scholars advising participants in controversies, pushback could increase the sophistication of *The Controversy Manual* and other similarly positioned techniques and devices. For policy-oriented meta-activism, feedback and critique could lead to better understandings of how to advise effectively. The possibility of obstruction or impediments also raises the question of which dominant images of sciences and technologies, or which aspects of locally dominant images, meta-activist projects leave intact while contesting another image or aspect of an image, and why. For example, in fighting a technopolitics designed to create ignorance, the Politics of Evidence Working Group in the Write2Know project aims at supporting a "right to know about the health and wellbeing of our bodies, communities, and environments" (2015). This is a laudable aim and courageous attempt to address undone science that affects the lives of many. And such a focus on undone science probably stands the best chance if it brackets—in some settings and for some audiences—how the politics of evidence is at other times understood in STS, namely, as the politics *of* evidence rather than of its *absence* (see Ottinger, Barandarián, and Kimura, chapter 35 this volume).

Collaborative learning within meta-activist scholarship may thus involve finding out what sorts of outcomes are achievable or potentially workable and selecting among diverse potential pathways, with distinct gains and losses. No project contests all images. Indeed, reflexively learning from initiatives to identify and promote undone science may challenge STS scholars to interrogate and postpone, or even undo, some of their own knowledges in order to facilitate specific meta-activist projects.

In summary, the second cluster of STS making and doing projects specializes in finding strategic routes into dominant settings and practices. When successful, scholars articulate STS insights for audiences that have experiential sensibilities of the politics and power implications of dominant images and practices but may be locked into the frames and hierarchies those images and practices enact. Meta-activist scholarship travels through interactions with those implicated in scientific and technical controversies, as well as by engaging professionals or those who may not even realize that their experiences are relevant and subordinated. It can also travel through new publication formats, such as the strategy guide or foundation report, as well as through procedural innovations in legislative practices and even the creation of temporary play zones. Collaborative learning develops further understanding about both the technopolitics and the power relations in the setting and the feasibility of additional meta-activist initiatives.

Cluster Three: Experiments in Participation

Drawing on our linked interests in "sociological experiments" (Zuiderent-Jerak 2015b) and "critical participation" (Downey 2009), we call this cluster of projects "experiments in participation" (see Lezaun, Marres, and Tironi, chapter 7 this volume). Projects in this cluster formulate, enact, and reflexively learn from novel, STS-inspired practices within their fields of study. Both posing and performing alternates to dominant images of sciences and technologies, participatory experiments actively blur boundaries between the project and the field. Some projects seek to open up new possibilities for action and interpretation by emphasizing and mobilizing the "improvised, surprising, generative side[s]" of existing settings (Zuiderent-Jerak 2015, 20). Some seek to facilitate the travels of STS scholarship within the field by materializing new pathways and settings for its formulation and expression (Downey 2009, 63–66). All generate guideposts, or "fingerposts" (Hacking 1983, 249), to point subsequent learning in both or multiple directions.

One: Frictions and Alternate Images Experiments in participation involve STS scholars deeply with both the existing frictions they identify and the new frictions they produce. Approaching technoscience through an understanding of the "'increasable complication' of practice" (Strathern 1991, xiv), participatory experiments actively mobilize the localized complexities in technoscientific practices that dominant images and practices hide or ignore. Whereas scholarship in the first cluster mainly envisioned boundaries between empirical fields and STS scholarship, work within this cluster rather emphasizes internal inconsistencies *within* empirical fields. Also, large-scale issues involving the status, power, and authority of technoscience become localized and complicated in specific moments of formulation, enactment, and learning.

The Now(here) project explores "an inverse approach to science education, where [participants] are invited to learn about BPD [Borderline Personality Disorder] diffractively and with attention to the implications in the lives of affected women" (Whynacht and Westby 2015). The project consists of "a participatory, digital installation created with a group of women living with the diagnosis of [BPD]" (ibid.). It explores "materialist feminisms and controversy mapping approaches to science engagement [that] allow for new techniques with which to address 'marks on bodies' (Barad 2007) in ethical ways" (ibid.). Rather than simply resisting, critiquing, or rejecting biomedical and legal experts, experimental involvement involves both "education and engagement with [such] experts" (ibid.).

Two: Techniques, Devices, and Infrastructures The techniques, devices, and infrastructures in participatory experiments typically seek to inflect the settings under study in specific directions, drawing upon prior learning. Their specific contents are crucial to their effects, both expected and unexpected. The array already in use in STS projects is dizzying.

In an experimental collaboration between an STS scholar and care professionals in a high-risk medical ward, the care professionals video themselves engaged in clinical practices that they judge to be significant to patient safety. The outcome is "exnovation" in patient care—challenging a locally dominant image of safety by building new practices upon "overlooked or forgotten" competencies, while also informing STS understandings of patient safety in the process (Mesman 2015). Values in Design, another initiative, mobilizes a wealth of techniques, from design workshops to card games, to challenge the reactionary view of critical theory as tackling technologies once they are in place. Its sociotechnical design experiments both reflexively "build values into design" and seek to learn from the valuations that such design processes produce

(Knobel and Bowker 2011, 2). Further examples include developing new organizational formats in the management of patient trajectories as techniques and devices for learning about standardization in health care (Zuiderent-Jerak 2007). Also, the Engineering as Problem Definition and Solution (PDS) project advances friction-based pedagogical techniques, such as multiple images of engineers and role-playing exercises, as part of an experiment to displace a dominant image of engineering problem solving by juxtaposing alongside it an image of collaborative problem definition (Downey 2008, 2015; Han and Downey 2014).

Three: STS Expertise and Identities   A challenge for scholars contributing to this cluster involves accepting both the constraints and opportunities of joining the field under study. With scholars informed by STS sensibilities, introducing techniques and devices into the fields of study becomes a way to learn more, and more collaboratively, about the field, while also inflecting it. Enacting STS expertise may consist of explicit efforts to "keep open" (Jerak-Zuiderent 2015) acquired understandings of practice and "hold the tension" (Star 1991) that such opening affords. Developing attachments within the field also becomes part of the experiment, for, in addition to producing commitments and responsibilities, they may also produce moments of surprise (Jensen 2007). Also, acquiring subject matter expertise can be essential to "becom[ing] interesting enough for practices [in the field] to care about" (Jensen and Lauritsen 2005, 72).

"Interventions," the Experimental Methods project reminds us, "can function as sites where divergent investigative approaches can become visible, actionable, manipulable, and theoretically viable to each other" (Klein and Gluzman 2015). The Experimental Methods project searches for ways of "including" scientists. "While scientists and their daily practices are the objects of our research," its advocates acknowledge, they also ask, "[H]ow might we do the difficult work of including them as interlocutors?" (ibid.). Offering an alternative to the notion of the STS expert, the project deploys "empirical structures (experiments) that involve scientists' participation, as co-designers, subjects, and/or co-interpreters" (ibid.) in order to explore the potential for having scientists as scholarly interlocutors.

Four: Audiences, Partners, and Engaged Practitioners   Experiments in participation foreground local audiences in the ecologies they join because the material, organizational outcomes of experiments tend to be highly localized, with travel inherent in the experimental practices themselves. The scholar may also build attachments to still other audiences through published accounts, workshops, demonstrations, and other locally appropriate pathways of travel. These latter audiences may prove important especially when the audience implied in the initial experiment inhabits only one of many possible sites for the lessons learned. Sometimes, these extensions turn back to fellow STS scholars as well.[7]

Five: STS Sensibilities Out There Although STS scholars acquire sensibilities by studying interesting practices of technoscience, this does not mean that practitioners in the field are able to examine and assess what they do in terms other than those established by dominant images and practices. Taking seriously the pragmatist insight that knowing and acting are deeply intertwined (Dewey [1951] 1998), agents in the field may possess both authority and authoritative knowledge but do not necessarily occupy a more privileged knowledge position. Experiments in participation share with other clusters the expectation that examining and rethinking the boundaries between the social and technical/scientific dimensions of technoscience has performative value. The particular challenges in experimental work lie in simultaneously collaborating with people and things in the field to produce such performances and learning from the often-surprising ways that collaborators formulate, or reformulate, the contents of the experiment itself.

In Mesman's video ethnography, neither the STS analysts nor the care professionals knew prior to the experiments how they might reposition "safe care" or even the problem space of safety. It was actually "one of the NICU-nurses on the video team [who] ... proposed to use the footage as a visual aid to the protocols" (2015, 189). While Mesman was looking for ways to widen the analytical focus from "eliminating causes of error" to "includ[ing] causes of strength" (Mesman 2009, 1705), this nurse taught her what could be seen as an STS lesson about the importance of materializing connections between conceptualization and practice. Bringing STS understandings of risk and safety together with the knowledges and techniques of specialized nurses thereby achieved "artful contamination" (Zuiderent-Jerak 2010) of both simultaneously.

Six: Feedback and Reframing Because this cluster of projects treats frictions as invitations, experiences of pushback, critique, and anger, including the stress these may generate for the STS scholar (Mascarenhas-Keyes 1987), can become ethnographic moments for further learning about the setting and the possible directions of future experiments. Even the potentially evolving politics of the experiment and the scholar can become positioned "as one dimension of theorizing rather than as its foundation" (Downey and Rogers 1995, 272). Keeping the setting open for further experimentation and reframing requires "accepting the risk of greater ambivalence" (ibid., 277)—but also comes with the attendant benefits that may emerge from the outcomes of the experimental work.

In summary, this cluster of projects in STS making and doing highlights the intertwinement of experimentation, knowledge production, and world making. Contributions often place both their concepts and their politics at risk in embracing participatory practices of collaborative, multiple-way learning. Frictions appear magnetic to scholars who join practitioners in the field to identify with and respond to them. Drawing on critical STS work demonstrating the frictions in knowledge production, expression, and travel in technoscience despite the ecological authority of dominant images and practices, STS experiments in participation make "keeping open" a tool for both STS-informed alternates and ongoing collaborative learning.

Conclusion: Enacting Multiple-Way Travels of Knowledge Production, and Expression

Performing knowledge production, expression, and travel in STS through practices of making and doing offers important opportunities for expanding the contents, the scope, and, at times, the influence of STS scholarship. As we have seen, many STS scholars are turning STS lessons back onto STS work. We identify and outline three distinct clusters of STS making and doing to call greater attention to the relations between STS findings and insights, on the one hand, and the myriad of settings into which these travel, on the other—with reflexive and collaborative learning as the planned, expected, or hoped-for outcome. Also, by detailing the elements and ecological dimensions of individual projects and clusters, we have elaborated how the directionalities of STS making and doing initiatives vary with their precise scholarly contents. A willingness to get involved is essential but provides only a prefatory condition for the high-quality scholarship required.

This survey of scholarly activities in STS making and doing suggests that it is less a new direction for STS as a field or discipline and more an existing array of projects that, when made more visible and prominent, can serve as attractive touchstones for further work. At the same time, analyzing such projects in relation to one another may have something to offer to longtime debates about the directionality of STS as a field. An oft-told story about STS is that it developed through two distinct, contrasting threads of scholarship. The theoretical, academic thread took on dominant images of science and technology through the metaphor of construction. The activist, social movement thread took on power inequalities produced by enacting those very images. Each thread tended to judge the other to be lacking in significant respects. This way of cutting up the field has appeared, for example, in debates over whether the acronym

Table 8.1
Elements and ecologies of clusters of making and doing.

Clusters of Making and Doing	Elements			Ecological Dimensions		
	Frictions and Alternate Images	Techniques, Devices, and Infrastructures	STS Expertise and Identities	Audiences, Partners, and Engaged Practitioners	STS Sensibilities Out There	Feedback and Reframing
Boundary-Crossing STS Claims	Social dimensions are marginalized. Frames deep inter-relations among S, T, and society.	Op-ed articles, statements, expert witnesses, documentaries, blogs, design workshops. Can include pedagogies, games, etc.	STS scholars produce unique insights on blurring social/technical-scientific divide. Expertise must be enacted strongly.	STS scholars share insights with external audiences: politicians, engineers, scientists, general public.	Others are usually recipients. Learners can become practitioners.	Learning from resistance may lead to better ways to reach audiences or better alternates. Realities made through getting insights across.
Meta-Activism	Highlights political contents and power dimensions that dominant images perform. Aims at more dynamic, open, and innovative S&T. Details of alternate images left to other actors.	Strategy support, advice for participation in controversies. Developing infrastructures for distributed activism through devices and STS thought styles.	Specialized analytical understandings of technopolitics and power relations. Subject-matter expertise mostly left to others.	Those facing problems, either as marginalized knowledge producers, general scientists being stuck in a conservative system, or those suffering from undone science.	Others usually not expected to become agents of STS knowledge, but can benefit from STS sensibilities.	Can prompt pushback, confirming power hierarchies. Learning can lead to better understandings of how to advise. Re-framing mostly left to the field. Re-framing STS is risky.
Experiments in Participation	S&T have 'increasable complication.' STS scholars "get involved" with frictions found with dominant images to open up action repertoires.	Building devices, changing organizations, video-reflexivity, art installations, novel curricula, etc.	Accept constraints and opportunities of participation. Focus on "keeping open" acquired understandings. Focused learning from collaborators in the field.	Audiences and pathways implied in the local ecologies of experiments. STS is equally an audience of the experiment.	Agents in the field may possess authority and authoritative knowledge, but not necessarily a more privileged knowledge position. Can be sources of surprise.	Feedback and re-framing as core elements of learning, for opening up the field and for developing STS understandings. Realities are enacted in the experiments.

STS stands for the academic enterprise called Science and Technology Studies or the societally involved project of Science, Technology and Society studies (Rip 1999). It has also appeared in tendencies to treat engagement and reflexivity as mutually exclusive alternatives to theorizing and enacting involvements in STS fields of study.

By contrast, the scholarship of STS making and doing treats practices of engagement and reflexivity as intimately linked and builds on both by calling attention to the two-way, or multiple-way, travels of knowledge production and expression. The scholarship of STS making and doing is about both care and learning. As the examples above repeatedly demonstrate, STS scholars learn by using involvement to problematize our own intentions and actions, even as we justify the value of our work to others.

Acknowledgments

We want to thank the participants, including both students and speakers, in a summer school of the Netherlands Graduate Research School of Science, Technology, and Modern Culture (WTMC), that we organized and ran in 2014 (together with Geert Somsen, Maastricht University), for indulging in our early efforts to develop and elaborate the notion of STS making and doing. Downey also acknowledges students and interlocutors in the experimental 2012 course "What is STS for? What are STS scholars for?" Kristen Koopman provided research assistance, and Samantha Fried and Jennifer Henderson helpfully commented on a previous draft, as did the *Handbook* reviewers and editors. Thanks especially to Clark Miller and Ulrike Felt for their close, critical reading, and to Victoria Neumann for meticulous editorial assistance. Finally, we thank all those involved with the inaugural STS Making and Doing Program, held in Denver at the annual 4S meeting, for their inspiring work.

Notes

1. Gaining the PhD typically qualifies a learner as a researcher. We take STS practitioners to be those learners and performers of STS knowledge and expertise who neither have sought the PhD nor claim career identities as researchers. STS scholars are now giving more overt attention to the figure of the STS practitioner (Downey et al. 2015). The boundary between STS researcher and STS practitioner is porous and, hopefully, increasingly contested.

2. We acknowledge other members of the 4S Making and Doing Committee—Sulfikar Amir (chair), Joseph Dumit, Nina Wakeford, Chia-Ling Wu, and Sara Wylie, and all contributors to the STS Making and Doing Program.

3. See Downey (2016) for more on directionalities as vectors or valenced pathways of travel for knowledge and people.

4. See Zuiderent-Jerak (2015), particularly the conclusions, for elaboration of the related notion of ecologies of intervention.

5. We thank Ernst Thoutenhoofd for co-developing this notion.

6. Translated by the authors.

7. To name some dedicated collections: Cohen and Galusky 2010; Caswill and Shove 2000; Downey and Dumit 1997; Jespersen et al. 2012; Pors et al. 2002; Woolgar, Coopmans, and Neyland 2009; Zuiderent-Jerak and Jensen 2007.

References

Balug, Katarzyna, and Maria L. Vidart-Delgado. 2015. "Imagine! You Have Nothing to Lose: Collaboration and Play in Urban Development." *Critical Sociology* 41 (7–8): 1027–44.

Barad, Karen. 2007. *Meeting the Universe Halfway: Quantum Physics and the Entanglement of Matter and Meaning*. Durham, NC: Duke University Press.

Bijker, Wiebe E. 2003. "The Need for Public Intellectuals: A Space for STS." *Science, Technology, & Human Values* 28 (4): 443–50.

Bijker, Wiebe E., Thomas Hughes, and Trevor J. Pinch, eds. 1987. *The Social Construction of Technological Systems: New Directions in the Sociology and History of Technology*. Cambridge, MA: MIT Press.

Biodesign Institute, Arizona State University. 2015. "Sustainability Science Education." Accessed October 24, 2015, at http://sse.asu.edu/.

Boas, Franz. 1940. *Race, Language, and Culture*. New York: The Free Press.

Callon, Michel. 1986. "Some Elements of a Sociology of Translation: Domestication of the Scallops and the Fishermen of St. Brieuc Bay." In *Power, Action and Belief: A New Sociology of Knowledge*, edited by John Law, 196–233. London: Routledge and Kegan Paul.

Caswill, Chris, and Elizabeth Shove, eds. 2000. "Interactive Social Science." Special issue, *Science and Public Policy* 27 (3).

Cohen, Benjamin R., and Wyatt Galusky, eds. 2010. "Embodying STS: Identity, Narrative, and the Interdisciplinary Body." Special issue, *Science as Culture* 19 (1).

Comisso, Martín Pérez, Cynthia Maciel, and Raimundo Roberts. 2015a. *Sciences and Technology: Law and Policy for Chile*. Universidad de Chile, Biblioteca del Congreso Nacional (BCN).

___. 2015b. "Scientific Legislation." Accessed October 25, 2015, at http://www.4sonline.org/md/post/scientific_legislation.

de la Torre, Pedro, Kim Fortun, Scott Kellog, Alli Morgan, Brandon Costelloe-Kuehn, Amir H. Hirsa, Aalok Khandekar, and Rethy Chhem. 2015. "STS+ with Practitioners." Accessed October 25, 2015, at http://www.4sonline.org/md/post/sts_with_practitioners.

Department of Science and Technology Studies, Rensselaer Polytechnic Institute. 2015. "B.S. in Design, Innovation, and Society." Accessed October 23, 2015, at http://www.sts.rpi.edu/pl/design-innovation-society-dis.

Dewey, John. [1951] 1998. *Experience and Education*. Indianapolis, IN: Kappa Delta Pi.

Downey, Gary Lee. 2008. "The Engineering Cultures Syllabus as Formation Narrative: Critical Participation in Engineering Education through Problem Definition." *St. Thomas Law Journal: Special Symposium Issue on Professional Identity in Law, Medicine, and Engineering* 5 (2): 101–30.

___. 2009. "What Is Engineering Studies For? Dominant Practices and Scalable Scholarship." *Engineering Studies: Journal of the International Network for Engineering Studies* 1 (1): 55–76.

___. 2015. "PDS: Engineering as Problem Definition and Solution." In *International Perspectives on Engineering Education: Engineering Education and Practice in Context*, vol. 1, edited by Steen Hyldgaard Christensen, Christelle Didier, Andrew Jamison, Martin Meganck, Carl Mitcham, and Byron Newberry, 435–55. Dordrecht: Springer International.

___. 2016. "Directionalities in Engineering and Engineers." Ms. available from author.

Downey, Gary Lee, and Joseph Dumit. 1997. *Cyborgs and Citadels: Anthropological Interventions in Emerging Sciences and Technologies*. Santa Fe, NM: School of American Research Press.

Downey, Gary Lee, Ulrike Felt, Kim Fortun, Yuko Fujigaki, Bruce Lewenstein, Noortje Marres, Hernán Thomas, John Willinsky, and Chia-Ling Wu. 2015. Opening Presidential Plenary: Making and Doing II: The Formation and Ecologies of STS Practitioners, Annual Meeting of the Society for Social Studies of Science, Denver, CO. Accessed January 27, 2016, at http://www.4sonline.org/meeting/15.

Downey, Gary Lee, and Juan D. Rogers. 1995. "On the Politics of Theorizing in a Postmodern Academy." *American Anthropologist* 97 (2): 269–81.

Edge, David. 2003. "Celebration and Strategy: The 4S after 25 Years, and STS after 9-11." *Social Studies of Science* 33 (2): 161–69.

Felt, Ulrike, Daniel Barben, Alan Irwin, Pierre-Benoit Joly, Arie Rip, Andy Stirling, and Tereza Stöckelová. 2013. *Science in Society: Caring for Our Future in Turbulent Times, Science Policy Briefing 50*. Strasbourg: European Science Foundation.

Ford, Rebecca. 2015. *Invisible Work at EMEC: European Marine Energy Centre, Orkney, Scotland*. Copenhagen: IT University of Copenhagen.

Forlano, Laura, and Megan Halpern. 2015. "Reimagining Work." Accessed October 23, 2015, at http://www.4sonline.org/md/post/reimagining_work.

Frickel, Scott, Sahra Gibbon, Jeff Howard, Joanna Kempner, Gwen Ottinger, and David J. Hess. 2010. "Undone Science: Charting Social Movement and Civil Society Challenges to Research Agenda Setting." *Science, Technology, & Human Values* 35 (4): 444–73.

Gibbons, Michael, Peter Scott, Helga Nowotny, Camille Limoges, Simon Schwartzmann, and Martin Trow. 1994. *The New Production of Knowledge: The Dynamics of Science and Research in Contemporary Societies*. London: Sage.

Hackett, Edward J., Olga Amsterdamska, Michael Lynch, and Judy Wajcman. 2008. "Introduction." In *The Handbook of Science and Technology Studies*. 3rd ed., edited by Edward J. Hackett, Olga Amsterdamska, Michael Lynch, and Judy Wajcman, 1–7. Cambridge, MA: MIT Press.

Hacking, Ian. 1983. *Representing and Intervening: Introductory Topics in the Philosophy of Natural Science*. Cambridge: Cambridge University Press.

Han, Kyonghee, and Gary Lee Downey. 2014. *Engineers for Korea*. San Rafael, CA: Morgan & Claypool.

Haraway, Donna J. 1991. *Simians, Cyborgs and Women: The Reinvention of Nature*. New York: Routledge.

Hughes, Everett. 1970. *The Sociological Eye*. Chicago: Aldine.

Jasanoff, Sheila. 1996. "Beyond Epistemology: Relativism and Engagement in the Politics of Science." *Social Studies of Science* 26 (2): 393–418.

___, ed. 2004. *States of Knowledge: The Co-production of Science and Social Order*. London: Routledge.

___. 2011. "Constitutional Moments in Governing Science and Technology." *Science and Engineering Ethics* 17 (4): 621–38.

Jasanoff, Sheila, Alex Wellerstein, and Shana Rabinowich. 2015. "Science and Democracy Network." Accessed January 27, 2016, at http://www.hks.harvard.edu/sdn/.

Jensen, Casper Bruun. 2007. "Sorting Attachments: On Intervention and Usefulness in STS and Health Policy." *Science as Culture* 16 (3): 237–51.

Jensen, Casper Bruun, and Peter Lauritsen. 2005. "Qualitative Research as Partial Connection: Bypassing the Power-Knowledge Nexus." *Qualitative Research* 5 (1): 59–77.

Jerak-Zuiderent, Sonja. 2015. "Keeping Open by Re-imagining Laughter and Fear." *Sociological Review* 63 (4): 897–921.

Jespersen, Astrid Pernille, Morten Krogh Petersen, Carina Ren, and Marie Sandberg, eds. 2012. "Cultural Analysis as Intervention." Special issue, *Science & Technology Studies* 25 (1).

Kember, Sarah. 2003. *Cyberfeminism and Artificial Life*. London: Routledge.

Klein, Sarah, and Yelena Gluzman. 2015. "Experimental Methods." Accessed October 25, 2015, at http://www.4sonline.org/md/post/experimental_methods.

Knobel, Cory, and Geoffrey C. Bowker. 2011. "Values in Design: Focusing on Socio-technical Design with Values as a Critical Component in the Design Process." *Communications of the ACM* 54 (7): 1–3.

Knorr Cetina, Karin, and Michael Mulkay, eds. 1983. *Science Observed: Perspectives on the Social Study of Science*. London: Sage.

Latour, Bruno. 1988. "Mixing Humans and Nonhumans Together: The Sociology of a Door-Closer." *Social Problems* 35 (3): 298–310.

Latour, Bruno, and Steve Woolgar. [1979] 1986. *Laboratory Life: The Construction of Scientific Facts*. Princeton, NJ: Princeton University Press.

Lévi-Strauss, Claude. 1966. *The Savage Mind*. Chicago: University of Chicago Press.

Liboiron, Max. 2015. "Civic Laboratory: Plastics." Accessed October 25, 2015, at http://www.4sonline.org/md/post/civic_laboratory_plastics.

Lundh, Patrik, Tina Stanford, and Linda Shear. 2014. *Nano and Society: Case Study of a Research-to-Practice Partnership between University Scientists and Museum Professionals*. Menlo Park, CA: SRI International.

Martin, Brian. 2014. *The Controversy Manual*. Sparsnaes: Irene Publishing.

Mascarenhas-Keyes, Stella. 1987. "The Native Anthropologist: Constraints and Strategies in Research." In *Anthropology at Home*, edited by Anthony Jackson. London: Tavistock Publications.

Mesman, Jessica. 2009. "The Geography of Patient Safety: A Topical Analysis of Sterility." *Social Science & Medicine* 69 (12): 1705–12.

___. 2015. "Boundary-Spanning Engagements on a Neonatal Ward: A Collaborative Entanglement between Clinicians and Researchers." In *Collaboration across Health Research and Care*, edited by Bart Penders, Niki Vermeulen, and John Parker, 171–94. Aldershot: Ashgate.

Myers, Natasha, and Max Liboiron. 2015. "Write2Know." Accessed October 24, 2015, at http://write2know.ca.

Myers, Natasha, Max Liboiron, Peter Hobbs, Kelly Ladd, and Sabrina Scott. 2015. "The Write2Know Project." Accessed October 25, 2015, at http://www.4sonline.org/md/post/the_write2know_project.

Nieusma, Dean, James W. Malazita, Michael Lachney, David A. Banks, and Brandon Costelloe-Kuehn. 2015. "STS Design Program." Accessed October 23, 2015, at http://www.4sonline.org/md/post/sts_design_program.

NISE Network. 2015. "Nanoscale Informal Science Education." Accessed October 24, 2015, at http://nisenet.org.

Ostman, Rae, Ira Bennett, and Jameson Wetmore. 2015. "Nano and Society." Accessed October 24, 2015, at http://www.4sonline.org/md/post/nano_society.

Pfaffenberger, Bryan, and Jeremy Hunsinger. 2015. "STSwiki." Accessed April 27, 2015, at http://www.stswiki.org/index.php?title=Main_Page.

Politics of Evidence Working Group. 2015. "The Politics of Evidence: Where Science and Technology Intersect with Social and Environmental Justice." Accessed October 25, 2015, at https://politicsofevidence.wordpress.com.

Pors, Jens Kaaber, Dixi Henriksen, Brit Ross Winthereik, and Marc Berg, eds. 2002. "Ethnography and Intervention." Special issue, *Scandinavian Journal of Information Systems* 14 (2).

Rip, Arie. 1999. "STS in Europe." *Science Technology & Society* 4 (1): 73–80.

Science and Democracy Lecture Series. 2015. Harvard University. Accessed October 24, 2015, at http://sts.hks.harvard.edu/events/lectures/.

Sclove, Richard E. 1995. *Democracy and Technology*. New York: Guilford Press.

Shapin, Steve. 2006. "Eat and Run: Why We're So Fat." *The New Yorker*, January 16, 76–82.

Sismondo, Sergio. 2008. "Science and Technology Studies and an Engaged Program." In *The Handbook of Science and Technology Studies*. 3rd ed., edited by Edward J. Hackett, Olga Amsterdamska, Michael Lynch, and Judy Wajcman, 13–32. Cambridge, MA: MIT Press.

Society for Social Studies of Science. 2014. "Guidelines for 4S Resolutions and Reports." Accessed April 27, 2015, at http://www.4sonline.org/resolutions_reports/2327.

Star, Susan Leigh. 1991. "Power, Technology and the Phenomenology of Conventions: On Being Allergic to Onions." In *A Sociology of Monsters: Essays on Power, Technology and Domination*, edited by John Law, 25–56. London: Routledge.

___, ed. 1995. *Ecologies of Knowledge: Work and Politics in Science and Technology*. New York: State University of New York Press.

Star, Susan Leigh, and James R. Griesemer. 1989. "Institutional Ecology, 'Translation' and Boundary Objects: Amateurs and Professionals in Berkeley's Museum of Vertebrate Zoology, 1907–39." *Social Studies of Science* 19: 387–420.

Strathern, Marilyn. 1991. *Partial Connections*. Savage, MD: Rowman & Littlefield.

Traweek, Sharon. 1988. *Beamtimes and Lifetimes: The World of High Energy Physicists*. Cambridge, MA: Harvard University Press.

Vidart-Delgado, Maria L., and Katarzyna Balug. 2015a. "Department of Play." Accessed October 25, 2015, at http://www.4sonline.org/md/post/department_of_play.

___. 2015b. "Department of Play-Research." Accessed October 25, 2015, at http://www.deptofplay.com/research.

Warren, Annie, Rider Foley, Jen Richter, Leanna Archambault, Omaya Ahmad, and John Harlow. 2015. "Crafting Digital Stories." Accessed October 23, 2015, at http://www.4sonline.org/md/post/crafting_digital_stories.

Watts, Laura, and Brit Ross Winthereik. 2015. "The Energy Walk." Accessed October 24, 2015, at http://www.4sonline.org/md/post/the_energy_walk.

Whynacht, Ardath, and Margaret Jean Westby. 2015. "The Now(here) Project." Accessed October 25, 2015, at http://www.4sonline.org/md/post/the_nowhere_project.

Winthereik, Brit Ross, Laura Watts, Louise Torntoft Jensen, James Maguire, Simon Carstensen, and Rebecca Ford. 2015. "Alien Energy: Social Studies of an Emerging Industry." Accessed October 24, 2015, at http://alienenergy.dk/the-energy-walk/.

Woolgar, Steve, Catelijne Coopmans, and Daniel Neyland, eds. 2009. "Does STS Mean Business?" Special issue, *Organization* 16 (1).

Ziewitz, Malte, Steve Woolgar, and Chris Sugden. 2015. "How's My Feedback?" Accessed December 15, 2015, at http://www.4sonline.org/md/post/hows_my_feedback.

Zuiderent-Jerak, Teun. 2007. "Preventing Implementation: Experimental Interventions with Standardization in Healthcare." *Science as Culture* 16 (3): 311–29.

___. 2010. "Embodied Interventions—Interventions on Bodies: Experiments in Practices of Science and Technology Studies and Hemophilia Care." *Science, Technology, & Human Values* 35 (5): 677–710.

___. 2015. *Situated Intervention: Sociological Experiments in Health Care*. Cambridge, MA: MIT Press.

Zuiderent-Jerak, Teun, and Casper Bruun Jensen, eds. 2007. "Unpacking 'Intervention' in Science and Technology Studies." Special issue, *Science as Culture* 16 (3).

II   Making Knowledge, People, and Societies

Ulrike Felt

That knowledge and social orders are co-produced remains one of the centerpieces of STS thought, and the chapters assembled in this section bear witness to the rich and powerful insights a close attention to practices of co-production can deliver. Investigating technoscience in laboratories and beyond through this lens highlights how the development of knowledge about the world is always entangled with both how people inhabit it and how they imagine the futures to which they aspire. Co-production invites the analyst to move beyond unidirectional, often deterministic thinking and to reflect on the complex global and local choreographies through which science, technology, and society are brought into being. Making knowledge is thus never an "innocent" activity; nothing can be regarded as "natural" or "simply given." Rather, knowledge plays a central role in the way we understand ourselves, people around us, and our societies. At the same time, people and societies as well as the values, aspirations, and imaginaries they develop, circulate, and share, also shape the making of knowledge.

This understanding explains why, across many of the chapters gathered in this section, the central questions are: Who has a voice to make legitimate knowledge claims, who defines what matters, or who participates in imagining and shaping the future? Asking these "who questions" points to concerns not only about actors and identities but also about exclusion, oppression, inequality, and social justice. It means proactively looking for resistance to conventional knowledge orders, for dissident voices, for neglected knowledge and experiences, and for alternative conceptualizations of progress and futures. These are long-standing questions in science and technology studies (STS), posed in new ways in the following chapters, updating previous debates and offering new areas of focus.

In multiple ways the authors address the relations of knowledge and power, of politics of and through science and technology. They share a concern that while in the past STS has been very good at studying the microdynamics at work in conflicts over specific knowledge claims, the field should pay closer attention to more systemic perspectives on the problems at stake and to the more durable (infra)structures that stabilize sociotechnical orders. Examples of such (infra)structures are systems and procedures of classification and standardization (around, e.g., race, gender, sexuality, ethnicity) that tend to be taken for granted in contemporary societies; specific regimes of valuation which foreground some values and neglect or reject others; established practices of interaction within and between institutions; (sociotechnical) imaginaries which shape collective visions of technoscientific futures; and institutionalized ways of public sense-making, which lack spaces where so-called nonexperts are given voice—to mention but a few. In doing so, several of the chapters underline that this thinking has been deeply rooted in and inspired by feminist scholarship and postcolonial studies.

The chapters also open up—more or less explicitly—questions regarding the epistemic and institutional geographies of science and technology. While STS has incorporated comparative work in the past, an explicit reflection of what it means to do comparative work and the insights it generates is stronger today. A comparative lens allows for opening up and questioning taken-for-granted practices, institutions, and structures and for asking how, despite making reference to ostensibly the same scientific and technological rationality and to shared (democratic) values, different nation-states or regions sometimes draw fundamentally different conclusions. Comparson raises awareness of technoscientific divides across the globe, of the diverse technopolitical cultures which develop within different places and embrace or reject knowledge and innovations. And it highlights the hierarchical knowledge orders at work, which create different impacts in different parts of the world, in different societies. This also means paying careful attention to the places we speak from and to avoiding the reproduction of classical divides such as "West/non-West" and "North/South."

The section starts with several chapters that present recent STS scholarship on the complex articulations and entanglements of science and democracy. Sheila Jasanoff (chapter 9) manages to cover in an impressive breadth the ways in which STS has generated valuable insights on the relations of knowledge, power, and politics. The reader is skillfully guided through historical and contemporary debates on science and democracy, showing how STS analysts have conceptualized the relationship between good knowledge and good governance and highlighting the novel theoretical and methodological contributions to interpreting key terms such as *citizenship*, *states*, and *democracy*.

Chapters 10 and 11 share an interest in social movements and their relation to STS, as well as a deep concern for how science and technology relate to inequalities. Steve Breyman, Nancy Campbell, Virginia Eubanks, and Abby Kinchy (chapter 10) offer a subtle and dense reflection on how STS scholars have researched, theorized, and analyzed the role of social movements in contemporary societies, as well as how forms of social movements evolve within changing technoscientific worlds. At the same time they remind the reader how deeply STS as a field was and still is entangled with concerns we share with these groups. David J. Hess, Sulfikar Amir, Scott Frickel, Daniel Lee Kleinman, Kelly Moore, and Logan D. Williams (chapter 11) approach the question of inequality through close reflections on the role of more enduring structures (e.g., social disparities of class, race, and gender) and the conditions that lead to ever-new reconfigurations of science, technology, and politics. Pointing to a growing interest in thinking about inequality together with race and gender, with the uneven geography of technoscientific developments, as well as with the ways states and industries relate to knowledge generation and technology making, the authors stress the hope that through a conscious diversification of the STS community itself a richer picture might emerge.

The next two chapters examine the powerful ways in which biomedicine defines human identities and creates or reinforces specific orders in society. Ramya M. Rajagopalan, Alondra Nelson, and Joan H. Fujimura (chapter 12) give a lively account of the cultural, political, and social practices of classifying humans, a topic that has a long and contentious history of social debate and a rich history of STS analysis and critique. Against the backdrop of more recent developments of new molecular technologies, the authors call for new attention to the ways in which the relations of race and biomedicine are imagined and practiced in the twenty-first century. The chapter illustrates nicely how technological innovation not only reinvigorates long-standing biomedical efforts to explain differential life outcomes across individuals and groups but also serves as a site for "making up people," their identities, and the forms of life they inhabit.

Continuing some lines of reflection from previous chapters, Jennifer R. Fishman, Laura Mamo, and Patrick R. Grzanka (chapter 13) engage with a broad body of STS work on the scientific production of sexed bodies, identities, and sexual desire, always carefully pointing to the ways in which these are co-produced with social and cultural conceptualizations of sex, gender, and sexuality. They also discuss the implications of this for both bodies and the lives that can be lived. The authors reflect on how the making of categories and the classificatory work of science creates pathologies and inequalities and show, in turn, how social movements, activist groups, and others have responded by resisting scientific claims and the ordering work they do. They invite the reader to develop sensibilities toward the multiple coexisting axes of difference, power, and inequality within contemporary biomedicine.

These two perspectives on ordering work done through biomedical knowledge are then complemented and enlarged by the essay of Banu Subramaniam, Laura Foster,

Sandra Harding, Deboleena Roy, and Kim TallBear (chapter 14). The authors bring in a strong postcolonial perspective, while simultaneously critically reflecting on the term *postcolonial* itself. The chapter aptly guides the reader through a broad field of theoretical thinking and empirical observations drawn from feminist and postcolonial STS. At the same time, the chapter opens up a critical perspective on some of the assumptions embedded in these analyses, for example, by looking through the lenses of indigenous knowledge-making or theories of decolonization prevalent in Latin American scholarship. The authors convincingly argue for the need to bring different strands of theorizing and analysis, that is, STS, feminist studies, and postcolonial studies, into closer conversation in order to provide richer accounts of how different formations of knowledge travel. This in turn would not only contribute to a better understanding of how knowledge, people, and societies are co-produced but open up space for alternative imaginations and knowledges.

The last two chapters in this section add a reflection on the concept of imaginaries in STS as well as on the ways in which futures come to matter in the making of knowledge, people, and societies. Maureen McNeil, Michael Arribas-Ayllon, Joan Haran, Adrian Mackenzie, and Richard Tutton (chapter 15) start the penultimate chapter of this section by remarking that over the last few decades an increasing number of STS analysts have embarked on investigations of imaginaries associated with science and technologies. The concept of imaginaries of science and technology, which is defined as culturally stabilized sets of shared beliefs about how a world worth inhabiting can be realized through science and technology, offers new ways to capture the normative dimensions of the complex relationships among science, technology, and society. The chapter introduces the reader to the different traditions of thought that nourished the concept's development and reflects on the explanatory power it offers within different strands of STS research. Imaginaries as discussed in this chapter, as well as in others in this section (e.g., chapter 9) may be read as an example of a new kind of infrastructure, one that has been largely absent in the explicit STS reflections of previous handbooks.

The final chapter in this section complements the reflections on imaginaries through a focus on the construction and governance of (representations of) futures, the dynamics of expectations that emerge around them, and how they become a key resource for shaping sociotechnical developments in the present. Kornelia Konrad, Harro van Lente, Christopher Groves, and Cynthia Selin (chapter 16) offer a detailed picture of the rapidly growing body of scholarly work that addresses the future as an object of representation while simultaneously investigating the practices of futuring that have emerged in different domains, taking stock of key insights, drawing together different strands of thought and empirical experimentation, and pointing out directions for further research. Konrad et al. elaborate both on STS analyses of how futures are shaped, narrated, and traded as well as on STS engagements in developing methods and experiments to make space for alternative visions of future possibilities that could be used to critically appraise and challenge existing normative assumptions about how innovation can and should drive societal developments.

Not only analyzing but trying *to make a difference in the world* is strongly present in all of the chapters in this section. Sheila Jasanoff asks, for example, how STS scholars have turned their knowledge about science and democracy into actual interventions into public policy. Steve Breyman and coauthors remind us of the field's common history with social movements, despite more recent efforts to turn STS into a respected discipline. Or Jennifer Fishman and coauthors invite STS not only to contribute to the understandings of sex, gender, and sexualities but also to cultivate new ways of actively destabilizing the power and authority of certain technoscientific orders in this domain. Together the chapters thus call for innovative analyses of how knowledge, people, and societies get made together and for STS scholars to proactively use the field's research to intervene in the processes of ordering of the world.

9 Science and Democracy

Sheila Jasanoff

The relationship between science and democracy occupies a central place in the canon of science and technology studies (STS), although the topic interested philosophers and social scientists long before STS emerged as a distinct discipline. The heading encompasses a set of theoretical questions about the ways in which scientific knowledge and technical expertise intersect with people's ability to hold state power accountable to democratic values: how does power know; does politics shape science or does science govern political power; and can the politically powerless use their knowledge to influence decisions rendered at the seats of power? Put differently, if democratic government demands that the *demos* should have a role in building the technical basis for framing and resolving public problems, then what analytic resources does STS provide to facilitate such participation?

These questions gained urgency against a backdrop of broad institutional changes that distanced people from government in the modern period: the emergence of the nation-state as the most significant repository of political power; increasing technical specialization and the exclusion of laypeople from expert judgments; the spread of industry and growing mechanization of life. States (and, earlier, empires—see, e.g., Cohn 1996) deployed knowledge and expertise inaccessible to ordinary people, creating pockets of ungovernable power that undermined the principle of popular self-rule (Foucault 2007; Scott 1988). Asymmetries between rulers and ruled became more apparent as waves of emancipation brought new claims and claimants into the public sphere: abolition of slavery, women's suffrage, decolonization, civil rights activism, new social movements, and, recently, unprecedented access to information and opportunities for expression and communication through the digital media. Freshly arrived on the political scene, once disenfranchised actors asked why the old orders had neglected their concerns, indeed sometimes their very existence; in the process, they also challenged rulers' claims of superior expertise.

By the end of the twentieth century, worries about who governs expanded from local and national scales to include the global. Increasing mobility, pressure on resources, and conflicting ideologies gave rise to potentially catastrophic risks, from pandemics to terrorism, that national authorities seemed ill equipped to handle. Climate change, flagged as a planetary threat by international scientific experts (Edwards 2010), summoned people everywhere to assume responsibility for their common future. Yet, the potential for political action remained frustratingly sequestered within nation-states, answerable to domestic economic interests and national cultures of lawmaking and public reason (Jasanoff 2005, 2010). Norms of global citizenship began to form, but global institutions remained for the most part resistant to direct public participation (Jasanoff 2011c). Subnational communities, too, often found it difficult to win respect for their knowledge when pitted against countervailing state claims of superior expertise (Espeland 1998; Fairhead and Leach 1996; Leach, Scoones, and Wynne 2005; Mathews 2011).

By the mid-twentieth century, "big science" projects (Price 1963) consumed substantial slices of public revenue, ostensibly geared toward solving big social problems. Though scientists still enjoyed considerable freedom to define and pursue their own research agendas (Kitcher 2001), politicians felt compelled to justify expenditures and demonstrate results. Scientists had to balance freedom of inquiry against growing demands for goal-oriented research (Böhme, van den Daele, and Krohn 1976); scientific fraud, misconduct, and ethics emerged as significant concerns (Guston 2000); and the private sector, especially in the United States, contested the adequacy of the science used in regulatory decisions (Jasanoff 1990). State authority eroded as the political transformations of the post-Soviet era concentrated knowledge and power within multinational corporations. With intensifying global competition, not only states but corporate interests forged closer ties to science and technology, now seen as engines of innovation and economic growth (Gibbons et al. 1994; Nowotny, Scott, and Gibbons 2001).

As the configurations of science, technology, and politics evolved, questions about their interrelationship came into sharper focus within STS. In this literature, as in political theory, inquiry began with a foundational question about representation: how are the few authorized to speak for the many (Brown 2009)? To be sure, scientists claim to speak for nature, whereas democratic governments speak on behalf of public wants, needs, interests, and visions of the good. Yet, since both forms of representation are implicated in the exercise of power, scientists, political rulers, and ordinary people all have an interest in the legitimacy of both, and hence also in the practices by which both scientists and politicians gain and exercise authority.

STS scholarship, however, deviates from other disciplinary approaches in insisting that science-based representations of the world, conceptual and material, are products of social work, and therefore open to social analysis. The realities we inhabit are painstakingly constructed, even though many aspects of these "imagined and invented worlds" (Jasanoff and Kim 2015) may become so taken for granted that it seems pointless to question their foundations. In the past decade or so, STS scholarship has embraced constructivism not simply as a descriptive framework that displays the processes of world-making, but as a springboard for normative analysis, illuminating how science and values—or, more generally, knowledge and norms—are co-produced (Hilgartner, Miller, and Hagendijk 2015; Jasanoff 2004). Put differently, accounts of how the world is and how its inhabitants choose to live in it are now seen as mutually constitutive. Studying the processes of co-production clarifies relationships between social norms and what is taken to be good knowledge or good technology, allowing the analyst to critique both.

It follows, then, that STS, unlike classical political theory, refuses to take either epistemic or political legitimacy as given by nature or fixed in principle. Instead, recognizing that scientific and technological practices matter in the making of democracy, and vice versa, STS scholars attend to the ways legitimacy comes into being (is *constructed*) in scientific thought, in technological choices, and in political action. This constructivist orientation demands particular attentiveness to the bases for acceptance or rejection of public knowledge claims. It becomes important to ask not only how assertions of expertise incorporate or sideline public values but also how expert authority gains, retains, or fails to retain authority in politically charged environments.

Political theory undergoes a tectonic shift if who knows and how things are known become issues that power must take seriously in order to govern well, and that citizens, too, must question in order to rule themselves responsibly. To take just one example, worries about *technocracy* originally revolved around the premise that expert "technocrats" might seize the reins of power without respect for public preferences. The issues for democracy become infinitely more complex if one must consider how an expert class comes into being in a specific culture, how political institutions delegate authority to technical experts, why technocratic claims are or are not contested, and under what conditions expert pronouncements achieve credibility or win deference. By posing these questions among others, STS studies have reconfigured some of the most basic concepts of political theory: citizenship, the state, political culture, public reason, constitutionalism, and, most foundationally, democracy itself.

This chapter aims to give readers, regardless of their home disciplines, a usable (if tightly compressed) map of STS's distinctive contributions to political thought, a map that not only locates the major monuments but also offers sufficient orientation to enable future research. Any such effort necessarily involves multiple kinds of line drawing: temporally, where does one begin; conceptually, what does one take to be the boundaries of the two terms (*science, democracy*) that define the topic; and, disciplinarily, what does one include under the umbrella of science and technology studies? The choices made here may strike some as incomplete, even arbitrary, but the hope is less for total coverage than to open doors to promising lines of inquiry. Most basically, the chapter explores how STS analysts have conceived the relationship between good knowledge and good governance. How, in other words, are epistemic and political virtue connected, and what potential exists for positive synergies among science, technology, and democratic politics?

The chapter begins with concerns about science and democracy from the turn of the twentieth century. Authors from that period remain highly relevant to STS thinking today, although it would be misleading to bundle them under a disciplinary rubric that was not yet in use when they were writing. Pragmatically, it makes sense to speak of an STS literature as such only after the field acquired institutional coherence from about 1970 onward,[1] but the historical context also brings into relief the novelty of the STS moves. Next, the chapter reviews the major theoretical and methodological trajectories traced by scholars who have interrogated the mutually constitutive legitimation practices of science, technology, and politics. The concluding section assesses the resources that STS offers to meet society's demands for usable knowledge along with a just and inclusive politics.

A Genealogy of Concern

Origin stories are as suspect as they are appealing. They construct the lineages they purport to be merely describing; yet, they help explain how new modes of thought diverge from those that went before. The topic of science and democracy particularly benefits from a genealogical approach because it illuminates the distinctive ways that STS scholarship approaches problems with a longer history in social thought. Earlier thinkers reflected on the limits of democracy in expert-dominated political systems, and the tendency of science and technology to produce not only expertise but also docile and governable subjects. STS, we will see, continues those concerns but opens up dimensions that classical political theory left unexamined.

It is almost a truism that the expansion of natural knowledge (science) and its instrumental uses (technology) affected possibilities for political self-rule (democracy). The challenge has been to specify precisely how. Immanuel Kant famously described the

Enlightenment as the emancipation of the human intellect from subservience to others, a coming of age, or becoming *mündig*. Kant's notion of "public" reason was predicated on the freedom, even the duty, of informed minds to assert opinions without constraint; yet social order, in his view, demanded that persons acting in a "private" professional capacity should comply with the rules applicable to that position (Kant 1784). Thus, Kant asserted, a citizen who is duty-bound to pay taxes to the state may nevertheless freely speak out "as a scholar" (*als Gelehrter*) on the inappropriateness or injustice of taxation; similarly, soldiers and clerics may criticize the rules that bind them while obeying those rules in practice. The Kantian public sphere was a space for the free exchange of informed beliefs, open in principle to anyone with the capacity to engage in reasoned debate. Science would serve as a liberating force because more knowledge would enable people to reason better for themselves. Scientists themselves initially bought into this idea of universal access (Broman 1998), even though growing specialization meant, in practice, that only knowledgeable and resourceful persons could participate in most technical debates.

By the beginning of the twentieth century, belief in an open public sphere weakened in the face of rising social complexity and reliance on expertise. Some saw the delegation of political power to technocrats as essential and inevitable (Veblen 1906); others as pernicious and excessive (Laski 1930). Calls arose on both sides of the Atlantic to limit expert influence in order to safeguard people's right to design and deliberate on their own futures. Skeptics countered by asking whether the idea of an informed public, the sine qua non of functioning democracies, any longer made sense. In the United States, a well-known debate between the philosopher and social reformer John Dewey (1927), firm believer in the power of education to create informed publics, and the journalist Walter Lippmann (1927), deeply pessimistic about the public's capacity to reason, crystallized Progressive Era anxieties about improving society. Dewey maintained that publics could form around issues of shared concern and acquire the knowledge needed for rational self-governance. Lippmann, by contrast, saw the public as a fiction in a world where common people are alienated from politics by money and power, including the power of the mass media to mislead and manipulate opinion. Contradicting Dewey's presumption of an educable public, Lippmann (1927, 4) said of the private citizen: "He reigns in theory, but he does not govern."

If laypeople cannot provide the knowledge needed for governance, then how should governments go about producing good public knowledge? Political scientists and policy analysts asked whose knowledge is needed to govern well, and how scientific knowledge, the territory of experts, should stack up beside other forms of knowing. Here again alternative theories of democratic capacity came into play. Charles

Lindblom (1959) at Yale University critiqued the overreliance on science—what might today be called scientism—in his work on "muddling through." "Usable knowledge," Lindblom argued, should encompass the knowledge of ordinary people—that is, "common sense, casual empiricism, or thoughtful speculation and analysis" (Lindblom and Cohen 1979, 12). At Harvard, Don K. Price, founding dean of the school of public policy, thought, by contrast, that technical decision making calls for experts trained to marry judgment with values. For Price (1965), policy education would bridge the "spectrum from truth to power." Whereas Lindblom put his faith in mining the wisdom of the crowd, Price offered a new brand of expertise as the right means for striking the balance between science and democracy.

The challenges for democracy grew more complicated as the scientific achievements of the Enlightenment became ever more tightly coupled to the life-changing technologies of modernity. A second world war, enforced population migration and genocide, the atomic bomb, slums and alienation, and the specter of environmental decay made it impossible to ignore the runaway force of a technological civilization that seemed to upend cherished human values and elude efforts at management and control. Mid-century shocks and upheavals gave birth to a global order in which concerns about the antidemocratic power of expertise, and the public's incapacity to reason, were inflected with new worries about technological risk and human survival (Beck 1992, 2007). Antinuclear and antiwar scientists and engineers joined concerned citizens to form organizations such as the Pugwash Conference, the British Society for Social Responsibility in Science, and the Union of Concerned Scientists. The perception of a world at risk spawned a range of writing, from despairing predictions by scientists and modelers (Ashley 1983; Carson 1962) to work by philosophers and social scientists meditating on the fate of human agency in an increasingly machine-dominated and technologically standardized world.

A major strand of theorizing focused on the implications of technological advances for the freedom of political thought and action. Hannah Arendt (1958) began her influential essay on the human condition with a nod toward Sputnik, which she thought had radically reshaped understandings of how humans should lead active lives upon the Earth. Though not intended as a reflection on science and democracy, Arendt's essay mourned the loss of the human ability to think meaningfully about a future in which scientists, "the least practical and the least political members of society," had become the most concerted and powerful actors (Arendt 1958, 323–24).

Arendt was hardly alone in perceiving technology and science as undermining people's political autonomy. A generation earlier, Max Weber had deplored the advent of a made and measured world, disenchanted by technology, that traps humans into the "iron cage" (*stahlhartes Gehäuse*) of bureaucratic efficiency (Weber 1930). Jacques Ellul in France articulated another bleak vision of society, in which humans are reduced to being the objects of machines and technique: "Today technique has taken over the whole of civilization" (Ellul [1954] 1964, 128). From the interwar through the postwar years, the Frankfurt School developed a powerful critique of technology and culture, blaming technology and the mass media for Germany's descent into barbarism and for producing the "one-dimensional man" (Marcuse 1964), an object of control and domination (Horkheimer and Adorno 2002; Marcuse 1941). American writing echoed similar concerns about the deleterious effects of the machine on human freedom, although Deweyan faith in citizen empowerment through participation persisted, for example, in Lewis Mumford's works on community involvement in regional scale planning and urban development (Mumford 1934, 1938).

Postwar German and French social thinkers, from Arendt's teacher Martin Heidegger ([1954] 1977) to Jürgen Habermas and Michel Foucault, continued to associate technology with forces of rationalization, standardization, mental and physical disciplining, and the hollowing out of human lifeworlds. In their disillusioned minds, the threat to self-rule lay not so much in the control of political agendas by experts—still the focus of much deliberative democracy theory (see, e.g., Dryzek 2014)—as in the curtailment of liberal democracy's promises of an open life (Bauman 1991; Foucault 1979; Popper 1945). Reacting to the student unrest of the 1960s, Habermas (1970), the foremost democracy theorist of his era, charged the alliance between elites and technical experts with depoliticizing the public sphere, replacing collective deliberation on the public good with morally empty, technical problem solving.

Although not a self-described theorist of democracy, Foucault pervasively influenced critical thinking on the alliance between power and knowledge and its implications for the human subject. His primary preserve was subjugation of the human body to ruling institutions through the practices of "governmentality" (Foucault 2007). In pathbreaking lectures and publications, he explored how the human sciences enabled modern states to exercise a mode of "biopower" that renders subjects governable as individuals and as members of populations. Many of Foucault's ideas are now part of the lexicon of social theory, essential for any attempt to understand the capacities of the modern political subject: the panoptic gaze and the apparatus of bureaucratic power that discipline bodies and selves (Foucault 1979); the role of the clinic and its trained ways of seeing that make bodies normal or abnormal; and technologies of the self that enable individuals to retain some control over their persons and even to refashion themselves at need.

A New Politics of Knowledge

Preoccupied with the impact of science and technology on people's civic capacities, classical political thought elided the part that politics plays in making authoritative facts and working technologies. STS scholars in the 1960s began to question that omission. Starting with the discovery, widely if too generously attributed to Thomas Kuhn (1962),[2] that "normal science" is embedded in social understandings, researchers began investigating how science and its technological applications relate to particular constellations and ideologies of power. An emerging body of work, encompassing a diversity of theoretical commitments, methods, and practices, revealed unsuspected connections between power and knowledge: that states and other governing bodies construct the very sciences they claim to rely on, while invoking objective science to legitimize their actions; that rationality is multiple, and it takes work, both normative and epistemic, to generate univocal reason; and that the practices of politics, science, and technology work together to produce effects of naturalness, neutrality, facticity, objectivity, and inevitability—as modes of depoliticization.

STS in the 1970s responded to a significant degree to external events—most notably the Vietnam War and the ongoing threat of the nuclear arms race, along with rising social concerns such as feminism, environmentalism, and postcolonialism. The much cited "military-industrial complex" that President Dwight D. Eisenhower warned against in his 1961 farewell address pointed to secrecy and ungovernability at the heart of democratic political systems. Capitalism continued to manifest its discontents. Despite political emancipation, the global south still saw the north as maintaining dominance through resource exploitation and control of markets. For analysts of science and technology, there was no question that the modern world was a profoundly unequal and dangerous place, in which large groups, indeed entire regions, remained unrepresented. In that world, technology, together with its economic and institutional underpinnings, functioned predominantly as a ratifier of hegemonic, antidemocratic tendencies. The analyst's task was to hold those dynamics up for critique and resistance.

"Left" critics of the period included well-known historians of technology as well as scholars explicitly identified with STS. Their analytic repertoire stressed the ways in which established structures of money and political power influence technological design and applications. These analyses framed the issue for democracy as one of regaining control over institutions of governance, particularly those entrusted with making war and peace. Examples include David Noble (1977) on *corporate capitalism*, documenting the simultaneous rise of technology and corporate power in the United States; Langdon Winner (1986) on *political artifacts*, demonstrating how political assumptions become embedded in technological systems; David Edgerton (2006) showing that interwar Britain was as much a *warfare state* as a welfare state; and Brian Martin (2007) on *backfire* and *blowback*, illustrating how covert government policies in wartime trigger unforeseen political consequences. All these writings could be read to some extent as accounts of cooptation, of science and technology falling into the wrong hands of antidemocratic and nonaccountable power, and giving rise to new forms of postmodern imperialism (Hardt and Negri 2000). The institutional dynamics of both state making and knowledge-making, however, remained largely black-boxed and out of view.

A more radical critique of the relations among power, knowledge, and technology emerged from a school of development critics in India who asked whether the root causes of technologically inflicted violence and injustice lie deeper than the intentional choices of militarists and marketeers. Is there something intrinsic in the very nature of the scientific enterprise that leads to its abuse, especially when it is harnessed to "development," the process by which less advanced nations and regions are brought up to speed with those imagined to be further along on the arc of progress (Nandy 1989)? These scholars looked upon science and technology as modes of governance with powerful, yet hidden, normative presumptions. Shiv Visvanathan's (1997) polemic on the *laboratory state* eloquently argued that genocidal tendencies are built into a developmental project committed to rooting out what is scientifically defined as obsolescent. Violence, on this view, arises from the moment that a backward "other" is designated as incurable in order to save what, in that culture, is seen as not dead, dying, or outmoded.

Feminist studies of science and technology agreed with development critics that the exercise of power without accountability is built into the scientific method and cannot be ascribed simply to breakdowns in democratic control. Evelyn Fox Keller (1985) found in the writings of Robert Boyle an early construction of a masculine science investigating a female nature. Keller's (1983) controversial biography of Barbara McClintock offered an account of a female scientific outsider who contravened mainstream, male scientific reductionism by studying plant genetics with "a feeling for the organism." Other prominent feminist historians and philosophers traced gendering into scientific worldviews and practices, especially in the biological and environmental sciences, reinforcing conventional hierarchies of power between the sexes (Haraway 1989, 1991; Merchant 1980; Schiebinger 1989). Haraway in particular laid bare the myriad ways in which gender-imbued ideas of nature shape representations of natural phenomena in museums and in cyborg bodies. Work by feminist historians, philosophers, and sociologists problematized universal notions of objectivity in science, not merely by historicizing the concept (see, by contrast, Daston and Galison 2007), but by calling attention to the "situated" (in Haraway's terms) and political nature of objectivity within regimes of experimentation that simultaneously validate male forms of knowledge and gendered orderings of power.

Gendering, feminists argued, also infiltrates the politics of technology. Innovation, even in domestic technologies, may reinforce rather than cure gender inequality (Cowan 1983). But feminist technology studies were particularly drawn to the representation and treatment of women's bodies by a male-dominated hierarchy in the life sciences and technologies. *Our Bodies, Ourselves*, compiled by the Boston Women's Health Book Collective (1973), marked and catalyzed a moment of awakening with regard to gender and biomedicine. In succeeding decades, feminist writing reclaimed a territory in which women's perspectives had long been underrepresented (Clarke and Olesen 1998; Fausto-Sterling 2000; Oudshoorn 2003; see also Fishman, Mamo, and Grzanka, chapter 13 this volume; Subramaniam et al., chapter 14 this volume). Human reproduction provided a particularly rich site for critique (Hartouni 1997; Thompson 2005). This work underscored the *ontological politics* (Mol 2002) that accompanies biomedical science's recognition and naming of new natural entities. Affairs of state more rarely came under the feminist lens, but in a celebrated exception Carol Cohn (1987) used participant observation to reveal how largely male, defense-policy experts employed gendered language to make a future of nuclear devastation seem rational and acceptable.

The Epistemic Turn

By the late 1970s, STS research to some extent turned away from following already defined lines of political cleavage—militarism, colonialism, gender—into the internal workings of science and technology. The field adopted a more explicitly constructivist stance toward knowledge-making as a phenomenon worth studying in its own right. Major schools of thought emerged, with programmatic ideas for how to challenge earlier realist and positivist orthodoxies about how facts and artifacts are made: sociology of scientific knowledge (SSK) (Bloor 1976), social construction of technological systems (SCOT) (Bijker, Pinch, and Hughes 1987), and actor-network theory (ANT) (Callon 1986; Latour 1990). With the production of knowledge and materiality as its prime subjects, this work generated a new discourse for talking about the practices of scientists, for example, social construction, negotiation, boundary work, enrollment, immutable mobiles, black-boxing, and closure.

A unifying feature of constructivist work in STS is a commitment to the principle of methodological symmetry.[3] This requires the researcher to set aside given in advance categorical boundaries, even notions of truth or falsity, when investigating facts or things in the making. Symmetry means evenhandedness: the analyst cannot presume to know in advance how problems of order will or should be settled, regardless of whether the matter under investigation involves factual disputes (e.g., claims vs. counterclaims, good vs. bad science, science vs. nonscience), divisions between nature and culture (e.g., what counts as human or nonhuman in politics or law), or relations between social and material technologies (e.g., technology as law vs. law as technology). Extrapolated to the political realm, methodological symmetry influences studies of science and democracy in two ways: first, it means that both must be seen as historically situated, culturally inflected practices rather than as claims or norms that transcend time and place; second, it means that neither should be subordinated to the other but should be treated instead as mutually reinforcing, sometimes competing, forms of institutionalized authority.

A range of qualitative methods to study the practices of knowledge-making proved useful in investigating questions of science and democracy. Foremost among these, indeed a staple of constructivist STS analysis, is the widely used method of *controversy studies*, but participant observation, focus group research, and comparison (especially cross-national) also generated new understandings. These methods helped establish that public epistemic conflicts are often symptomatic of deeper normative disagreements: who should define the good; whose experiences deserve public acknowledgment; and which costs and benefits should be deemed commensurable in the official calculus of decision making?

Controversy studies can be grouped for convenience under three headings: internalist, interactional, and institutional. Internalist controversy studies take disputes about right knowledge as their focal subject matter. Often, though not always, such conflicts arise among experts inside well-demarcated scientific workplaces or technological systems, but even highly technical disputes may reflect background political concerns. For SSK scholars, controversies between experimental traditions or schools of thought illuminate the stakes involved in making definitive knowledge. Analysts treat the disputed claims symmetrically, probing their underlying assumptions, without introducing explanatory factors that the disputants themselves did not control during the heat of the controversy. Arguably the most influential product of the SSK school, Steven Shapin and Simon Schaffer's (1985) account of the fight for epistemic authority between Thomas Hobbes and Robert Boyle interpreted the protagonists' rivalry against a wider struggle for political authority in Restoration England. The conflict between Hobbes and Boyle about ways of doing science reflected, in the authors' view, shifting understandings about how social order should be achieved. The political scientist

Yaron Ezrahi (1990) saw this as a defining moment for democracy: in winning ground, Boyle's experimental science enabled a new mode of critique, as if the political arena was like a laboratory in which states were accountable to witnessing subjects who could demand justification for the sovereign's claimed superiority.

For SCOT analysts, key controversies are those among adherents of disparate technological frames and user groups that, together, influence design. Values and judgment, as SCOT studies demonstrate, affect decisions to adopt particular technological systems or trajectories. Important for democracy, black boxing technology renders those value conflicts invisible—and therefore apolitical—behind expert assertions of accuracy, impartiality, and objectivity (Barry 2001; MacKenzie 1990; Porter 1995).

Both SSK and SCOT attend principally to controversies among human actors. ANT analyses, by contrast, conceive of science and technology as heterogeneous actor networks, in which human actors struggle with nonhuman "actants," such as microbes or inert materials, to produce stability and closure. ANT's version of methodological symmetry refuses a priori distinctions between the social and the nonsocial—for example, in quarrels between scientists and their study subjects about how well claimants for each side represent their respective groups (Callon 1986). That approach calls attention to the myriad ways in which humans both act and are acted upon: the material components of technoscience constrain and direct human choices, including political ones.

Interactional controversy studies pay greater attention to science's external relations with society than to epistemic struggles within science. Dorothy Nelkin (1979) pioneered a genre of public controversy studies that continues into the present (e.g., Kleinman, Kinchy, and Handelsman 2005). Conflicts over technological developments, such as nuclear power (Nelkin and Pollak 1981), show that, in areas of high polarization, organized interests line up behind competing knowledge claims. Fact making, in short, aligns with political interests and commitments that reflect background structures of power. Interactional controversy studies are sometimes seen as intrinsically democratizing because, in deconstructing expert claims and displaying the interpretive flexibility of evidence, study authors inevitably help the "underdogs" (Scott, Richards, and Martin 1990). Others agree that disclosing the values latent in epistemic conflicts can never be politically neutral, but that the more important aim of such studies is to display the complex lifeworlds at stake in technological controversies, and to explain why some views rise to dominance while others remain marginal.

Institutional controversy studies similarly elevate the importance of contexts beyond the scientific workplace but emphasize how the logics and practices of ruling institutions affect the framing and resolution of epistemic conflicts. A seminal contribution, Brian Wynne's ([1982] 2011) *Rationality and Ritual* demonstrated how an official judicial inquiry limited participation by excluding knowledge claims that were not compatible with the logic of the proceedings. Bodies such as courts and regulatory agencies routinely engage in boundary drawing to certify evidentiary claims as science or not science, objective or subjective, reliable or unreliable (Gieryn 1999). Such work, in turn, safeguards the authority of what is ultimately deemed to be scientific, even though—as in the production of "regulatory science" and "litigation science" (Jasanoff 1990, 1995)—the bright lines drawn around reliable knowledge may reflect unexamined institutional ideologies and purposes.

Another popular method, *participant observation*, probes science in the making by situating the observer within communities conducting research or affected by it. Participant observation permits investigators to study in real time the interplay of diverse epistemic and social assumptions in the production of authoritative knowledge, even in the absence of controversy. Here, the symmetry principle entails equally attentive treatment of lay and expert positions, or dominant and marginal communities. Medical and environmental sciences offer especially fruitful sites for participant observation, as illustrated in case studies of radioactive fallout from Chernobyl (Wynne 1989), HIV-AIDS treatment (Epstein 1996), and dam building (Espeland 1998), among many other examples. Such cases have documented the emergence of "citizen science" (Corburn 2005)—that is, experiential knowledge and proactive fact making by laypeople to contest or amplify expert claims.

Focus groups and other methods of creating mini-forums for debate and exchange have been used to probe people's attitudes toward developments in science and technology. This method allows STS researchers to uncover culturally held, commonsense beliefs about risks and benefits that tend to be suppressed or excluded from consideration in formal decision making. Such studies helped to establish the constitutive role of identity and memory in assessments of technology by lay publics (Felt 2015; Felt and Fochler 2011). Contradicting presumptions of apathy and ignorance, publics consulted in such studies frequently display a capacity to shape the aims, meanings, and responsibilities associated with technological futures (Centre for Study of Environmental Change 2001).

Cross-national *comparative studies* have played a significant role in STS by elevating the scale of analysis to the macro level of nation-states and shifting the unit of analysis from disputes over facts to debates about collective action. Comparison is well established as a method in political science. STS, however, broadens the scope and aims of comparison by looking beyond the structural variables that form the staples of political inquiry and by symmetrically scrutinizing national knowledge practices, not privileging any single approach as rational. Why, researchers ask, do modern states equally committed to scientific rationality, technological advancement, and democratic values reach different conclusions on what is hazardous, who is at risk, and what protection is warranted? Answers have led to a renewed engagement with political culture, a neglected topic in traditional political science, and to the identification of public ways of knowing and reasoning as core elements of political culture (Brickman, Jasanoff, and Ilgen 1985; Jasanoff 2005; Parthasarathy 2007). This work underscores the definitive role of organizations such as political parties (Gottweis 1998), courts (Jasanoff 1995), and bioethics bodies (Hurlbut forthcoming) in framing public knowledge debates. It also calls attention to durable *civic epistemologies*, or citizens' culturally conditioned expectations about the legitimate forms of public reason (Jasanoff 2012).

Overall, the epistemic turn in STS not only resurrected lost and submerged voices in the politics of knowledge but directed attention toward formerly neglected sites that provide rich insights into the mutually constitutive interplay of science and democracy. Analysts uncovered a largely invisible politics at micro scales, often centering on episodes of fact making in such forums as advisory committees (Brown 2009; Jasanoff 1990), public hearings (Lynch and Bogen 1996; Wynne [1982] 2011), legal proceedings (Aronson 2007; Cole 2001; Jasanoff 1995), patent offices (Parthasarathy 2017), social movements and public interest activism (Callon, Lascoumes, and Barthe 2009; Epstein 1996; Reardon 2004). Scholars writing in the framework of co-production (Hilgartner, Miller, and Hagendijk 2015; Jasanoff 2004; Latour 1993) showed that controversies function not only as struggles between institutionalized forces of power and knowledge, but—more important—as agonistic fields where demarcations between right and wrong, legitimate and illegitimate, are constituted or transgressed through the mundane but ideologically shaped practices of both science and politics.

Theoretical Advances

Moves to study knowledge-making in and especially outside the lab redefined key terms in the discourse on power and knowledge, offering novel interpretations of classical concepts such as state and market, citizenship and expertise, lawfulness, and democracy to supplement more traditional definitions from political theory, history, and sociology.

States of Knowledge

In the early 1980s Benedict Anderson (1983) revolutionized thinking about nation-states by linking top-down practices of circulating social knowledge (print capitalism) to the self-identification of citizens with their nations (*imagined communities*). STS scholarship adds to Anderson's account that modern states are also quintessentially "states of knowledge" (Jasanoff 2004), whose practices of collective sense-making include the production and deployment of scientific knowledge. Knowledge circulates, in Bruno Latour's (1990) terms, through *centers of calculation* that powerfully aggregate and disseminate facts, as well as through shared modes of reasoning that are taken to be legitimate (Jasanoff 2005, 2012). In a parallel move, Ezrahi (1990) underscored how the state's instrumental and celebratory uses of technology, such as bombs and satellites, generate assent from *attestive citizens*—in much the same way that Shapin and Schaffer (1985) discussed how the legitimacy of experimental science depends on the testimony of willing witnesses. If states hold national polities together, it is not only because of their territorial claims or status in positive international law, but also because they command the epistemic and material resources to propagate powerful *sociotechnical imaginaries* (Felt 2015; Hecht 1998; Jasanoff and Kim 2015). The capacity to enroll and enact collective imaginations through shared knowledge is increasingly a feature of global governance, overtaking to some degree the power and capacity of nation-states (Miller 2015; Miller and Edwards 2001).

No theoretical account of states today would be complete without a complementary account of markets. Not only is governance occurring in an era of market ascendancy (some say market fundamentalism, cf. Stiglitz 2002), but imaginaries of the state, and indeed of democracy more generally, depend upon a "significant other"—the non-state, the private, the market—spaces governed by radically different understandings of human agency and the role of knowledge. STS perspectives have made distinctive contributions in this area, through work on how markets reaffirm sociotechnical imaginaries, produce governing technologies, form consumer-citizens, and legitimize public reason. Perhaps predictably, the earliest contributions to studying markets focused on the technical instruments that markets use. One theoretical insight from such work is that these instruments are *performative* in the sense that they enable transactions that they claim only to describe (MacKenzie, Muniesa, and Shiu 2008; also see Law 2009). An accumulating body of work delineates the kinds of agency that make markets work (Callon 1998; MacKenzie 2009; see Preda, chapter 21 this volume), supplemented by a critical focus on *biocapital* and the *bioeconomy* (Dumit 2012; Sunder Rajan 2006). Arguably however, fields such as critical legal studies and economic sociology have gone further in a fully symmetrical, co-productionist direction, by problematizing how processes of economization contribute to varieties of political inclusion and exclusion. STS has still to learn from those endeavors.

Citizens and Experts

If states exert power through authoritative knowledge-making, then citizenship must include the rights and obligations of members of a polity to contribute to and act upon those collective ways of knowing. STS's attentiveness to multiple, coexisting knowledge regimes and rationalities complicates earlier understandings of citizens as either permanently clueless or eternally educable. It opens up a third possibility: the *knowledgeable* public that can process information, learn, and produce or enroll expertise when the situation demands (Jasanoff 2011b). The concept of the *epistemic citizen* (Leach, Scoones, and Wynne 2005) draws attention to the crucial role of lay knowledge in good government. Indeed, citizens as lay experts are entitled to *epistemic justice* (Visvanathan 2005), that is, a measure of respect for the experiential knowledge they bring to politics.[4] Beside Foucault's clinical subjects, disciplined by governmentality, recent engagements with the politics of the life sciences and technologies have given meaning to the concept of *biological citizenship*, which regards citizens as capable of forming new collectives and using biological knowledge for informed self-governance (Petryna 2002; Rabinow 1992; Rose 2007).

Despite repeated demonstrations of citizen competence, however, governing institutions tend to operate with an asymmetrical *deficit model* of the public (Wynne 1994), regarding dissenters against the official epistemic consensus as cognitively impaired. Though the idea of public engagement has gained ground, largely as a result of STS research (Chilvers and Kearnes 2016), disagreements remain over what constitutes "good" public engagement in technical decisions, reflecting competing understandings of citizens as knowledge-able or in a state of deficiency. Challenges to the idea of intellectually able citizens come not only from natural scientists and expert bodies but from other branches of the social and human sciences. One such trend still awaiting full-blown critical commentary derives from cognitive psychology and behavioral economics, with their findings of mental biases that disable rational economic behavior (Kahneman, Slovic, and Tversky 1982). Building on those findings, and on the neoliberal turn assimilating states to markets, "nudge theory" proposes that it is the policy maker's duty to build "choice architectures" that will induce citizen-consumers to make only those choices that rulers designate as rational (Thaler and Sunstein 2008).

STS's symmetrical gaze, however, questions the institutional logics and social interactions behind such forms of rationality, including the varied means by which expertise is certified as deserving of public trust. Constructivist analyses portray experts as active agents in building and protecting their privileged status: through skillful *boundary work* (Gieryn 1999; Jasanoff 1990), through membership in cohesive *moral economies* (Shapin 1994), and through elaborate public *performance* (Ezrahi 1990; Hilgartner

2000), all under the pervasive influence of *political culture* (Jasanoff 2005). Democracy, following these analyses, demands watchfulness not merely downstream, in the uses or applications of science and technology, but also upstream at sites in which expertise, know-how, and power mingle with hard materiality to depoliticize the role of experts. These insights, as discussed below, have shaped the prescriptive dimensions of STS work on the politics of knowledge.

In a turn away from the field's orthodoxies, Harry Collins (2014) and his coauthors (Collins and Evans 2002, 2007; see also Collins, Evans, and Weinel, chapter 26 this volume) stress the crucial importance of knowledge and skills in constituting both good expertise and good politics. Democracy, in their view, demands that citizens should not be treated the same as experts in technical deliberations unless they can contribute valid knowledge to a propositional field rightfully controlled by experts. That analysis, however, fails to attend to a massive critical literature showing that a narrow focus on citizens' technical competence misses the politics that frames the debate, determines which skills are or are not relevant (Jasanoff 2003; Wynne 2003), and narrows the imagination of expertise needed to address complex social concerns (Hurlbut 2015).

Law and Constitutionalism

In conventional political theory, values, upheld by law, are thought to kick in where facts end. STS questions that linear representation of facts and norms, showing instead that law is a site where both the rightness of knowledge claims and the rightful place of science are constantly under construction (Jasanoff 1995, 2012). In this sense, STS work is of increasing relevance to constitutional theory. In the simplest cases, legal proceedings validate novel technoscientific claims and practices (e.g., DNA typing, quantitative indicators) that have consequences for social order (Aronson 2007; Cole 2001). But the law is also a site where political subjectivities are made and unmade. *Bioconstitutionalism* (Hurlbut forthcoming; Jasanoff 2011a), in particular, points to changing ideas of personhood in a constitutional order as life itself becomes known, commodified, and handled in new ways (Sunder Rajan 2006).

Globalization and multilevel governance raise constitutional questions of *epistemic subsidiarity* (Jasanoff 2013), focusing on the conditions under which political communities can be allowed to stand by their own knowledge. More foundationally, law reaffirms and reinforces epistemic norms that are constitutive of political culture, for example, who gets to define when knowledge is reliable enough, how closure should be achieved when facts are insufficient, what forms of explanation and proof conform to legally supported civic epistemologies, and (reflexively) when legal institutions should defer to scientific ones.

Democracy

Scholarship in the idiom of co-production investigates not only the construction of robust public facts but the simultaneous construction of the right publics for purposes of deliberation. How do representative assemblies form around technical issues, and who properly belongs in them (Chilvers 2013; Horlick-Jones et al. 2005; Laurent 2013)? Following from ANT's symmetrical treatment of humans and nonhumans, Latour (1993) advocated the inclusion of inanimate and nonhuman entities in *parliaments of things*. Though the inclusion of artifacts in political analysis is not new with him (see, e.g., Winner 1986), Latour's insistence on the hybridity of networks played an important role in highlighting materiality as an element in politics, whether or not one ascribes agency to nonhumans. Widely influential within STS (Law 2002; Marres 2012), Latour's politicization of the material has also reached into adjacent fields such as geography (Barry 2001) and political economy (Mitchell 2002, 2011).

The composition of the *demos* is not merely a point of academic interest but an active concern of governing institutions, which have strong interests in excluding segments of the public that may destabilize existing orders. Work on the public understanding of science calls attention to the myriad ways in which modern states construct the baseline for democratic participation by defining what publics ought to know (see, e.g., Wynne 1994). These observations have led some STS scholars to see democratization as a form of necessarily oppositional activity, conducted by *concerned groups* "from the bottom up" (Callon, Lascoumes, and Barthe 2009; Irwin and Wynne 1996). Others, however, view the concept of democracy as always/already contested, and the nature of the *demos*, as well as its purposes and needs, to be therefore in continual flux. From that standpoint, regimes of democratization demand ongoing critique by scholars and actors to discern which modes of participation are being espoused or disavowed, by whom and for what purposes, under the rubric of being "democratic." More specifically, an important horizon for future scholarship lies in interrogating the place of science and technology in alternative imaginaries of democratization, where contestation over knowledge and technique reveals not merely the internal dynamics of expert disagreement but also their constitutive role in envisioning democracy (Jasanoff 2012).

Implications for Policy: STS in Practice

Has critical scholarship on science and technology prompted new thinking on how to govern technoscience in modern democracies? Have the findings of such research affected policy? And have STS studies of science and democracy translated into actual policy interventions? The answer to all three questions is a cautious yes, although the conclusions are far from definitive.

In Ian Hacking's (1999) famous formulation, social science knowledge, producing "social kinds," differs from natural science and its "natural kinds" because the former loops back and refashions the objects of its study. One can observe similar interaction between STS scholarship and the objects it studies, including the practices of democracy. Interaction began with growing recognition for the field. Governmental bodies around the world have proved willing to call upon (if not abide by) STS expertise, although with significant differences across nations and regions (Fujigaki 2009). Findings from the field helped reshape thinking on communicating risks to publics, engaging them in policy, and exploring the ethical dimensions of technology. An example is Europe's adoption of ELSA for its bioethics programs rather than the American rubric ELSI— substituting "ethical, legal and social *aspects*" for the more linear "ethical, legal and social *implications*" (see Hilgartner, Prainsack, and Hurlbut, chapter 28 this volume).

STS has generated if not a coherent discourse then at least a cluster of concepts that are closely associated with policy making and have been taken up to some degree by ruling institutions. Included are terms of broad theoretical relevance, such as _boundary_ _work_, _boundary organization_ (Guston 2000), and _deficit model_, as well as more instrumental concepts that help policy makers demonstrate their commitment to the democratic governance of science and technology. In this toolkit are process-oriented concepts such as *constructive technology assessment* (Schot and Rip 1996), *public engagement, anticipatory governance* (Guston 2014), and *responsible innovation* (Owen, Bessant, and Heintz 2013). Generally directed toward broadening the inputs to science and technology policy, this conceptual repertoire reassures decision makers that means exist to secure public buy-in for state-sponsored research and development. Such reassurance, however, does little if anything to dethrone experts' power to frame issues or exclude supposedly ignorant publics.

STS scholars have sought to democratize knowledge, not only through their writings but by directly injecting their disciplinary expertise into advisory bodies, media organizations, legal processes (Lynch and Cole 2005; Winickoff et al. 2005), museum displays, activist organizations, and the like. In turn, these involvements have generated publications ranging from technical reports on how to be more inclusive (Felt et al. 2007) to scholarly papers reflecting on the risks and limits of wider public involvement (Jasanoff 2011c). This record, though extensive, should not be read in triumphalist terms. Offsetting the tangible contributions to governance are continuing echoes of the "science wars" of the 1990s, as well as charges that STS—through its commitment to methodological symmetry—unduly complicates issues, offers few clear alternatives, remains overly relativist with respect to good and bad knowledge (cf., Collins and Evans 2007; Latour 2004), and thus cannot unambiguously advocate for justice or the public good. The fact that some of this critique is echoed by major players within the field suggests that, as Hacking (1999) long ago noted, the scope and aims of constructivist analysis remain a source of unresolved tension inside STS.

The confluence of the "old" STS of the 1960s, rooted in political activism and discoveries of the social within science and technology, and the "new" constructivist STS of the epistemic turn has produced a powerful synthesis that is poised to transform thinking about science and democracy. While regarding both science and democracy as deeply human achievements, STS scholarship calls attention to the fact that practices of collective knowledge-making shape not only the ground rules for political engagement but our very understanding of the *demos* to be served by democracy. What are the rights, responsibilities, and entitlements of knowledge-able citizens in today's states of knowledge, and what legitimate expectations for self-rule follow from these? The literature reviewed here indicates that the field has developed substantial resources for addressing those questions but that a too insistent focus on scientific knowledge-making may at times obscure the need for symmetrical attentiveness to democracy's practices for defining the collective good. Remedying this imbalance will put STS scholarship in a better position to reinvigorate the basic institutions of democracy so as to become more responsive to the demands and capabilities of mature and enlightened citizens.

Notes

1. The journal *Social Studies of Science* was launched in 1971. The Society for Social Studies of Science was formed in 1975. The oldest U.S. and European training programs in STS go back to the late 1960s.

2. This is not the place to discuss in detail the field's origins in diverse national and disciplinary contexts, but running through all the separate emergences was a common perception that science and technology are social and political all the way through—from the definition of aims and objects to choices of method and eventual uptake in society. Kuhn's *Structure of Scientific Revolutions* stands as a convenient marker for this shift in thought, but the Polish physician Ludwik Fleck, in his *Genesis and Development of a Scientific Fact*, had already charted a deeply sociological path to studying scientific change (Fleck [1935] 1979). Moreover, the "STS revolution" encompasses strands as diverse as German ideas of "finalization" (Böhme, van den Daele, and Krohn 1976); the debate between Kuhn and Karl Popper on method and error in science; Foucault's poststructuralist studies; and the strands of Marxist, feminist, and postcolonial critique discussed above.

3. The word *methodological* is important here. Methodological symmetry does not mean that the STS researcher eschews normative judgment. It simply calls for analytic attentiveness to the ways in which categorical distinctions come into being, are dissolved, or are sustained, without presuming to know ahead of time which distinctions are the right ones, epistemically or normatively.

4. It may be tempting to see these discussions simply as contributions to deliberative democracy in political science, but that large body of literature mostly fails to engage meaningfully with scientific knowledge and institutions. Thus, there is no mention of science's agency in a recent work on democratic agents of justice by a major deliberative democracy theorist (Dryzek 2015).

References

Anderson, Benedict. 1983. *Imagined Communities*. New York: Verso.

Arendt, Hannah. 1958. *The Human Condition*. Chicago: University of Chicago Press.

Aronson, Jay D. 2007. *Genetic Witness: Science, Law, and Controversy in the Development of DNA Profiling*. New Brunswick, NJ: Rutgers University Press.

Ashley, Richard K. 1983. "The Eye of Power: The Politics of World Modeling." *International Organization* 37 (3): 495–535.

Barry, Andrew. 2001. *Political Machines: Governing in a Technological Society*. London: Athlone Press.

Bauman, Zygmunt. 1991. *Modernity and Ambivalence*. Ithaca, NY: Cornell University Press.

Beck, Ulrich. 1992. *Risk Society: Towards a New Modernity*. London: Sage.

___. 2007. *World at Risk*. Cambridge: Polity.

Bijker, Wiebe, Trevor Pinch, and Thomas Hughes, eds. 1987. *The Social Construction of Technological Systems*. Cambridge, MA: MIT Press.

Bloor, David. 1976. *Knowledge and Social Imagery*. Chicago: University of Chicago Press.

Böhme, Gernot, Wolfgang van den Daele, and Wolfgang Krohn. 1976. "Finalization in Science." *Social Science Information* 15 (2/3): 307.

Boston Women's Health Collective. 1973. *Our Bodies, Ourselves*. New York: Simon and Schuster.

Brickman, Ronald, Sheila Jasanoff, and Thomas Ilgen. 1985. *Controlling Chemicals: The Politics of Regulation in Europe and the U.S.* Ithaca, NY: Cornell University Press.

Broman, Thomas. 1998. "The Habermasian Public Sphere and 'Science *in* the Enlightenment'." *History of Science* 36: 123–49.

Brown, Mark B. 2009. *Science in Democracy: Expertise, Institutions, and Representation*. Cambridge, MA: MIT Press.

Callon, Michel. 1986. "Some Elements of a Sociology of Translation: Domestication of the Scallops and the Fishermen of St. Brieuc Bay." In *Power, Action, and Belief: A New Sociology of Knowledge?*, edited by John Law, 196–223. London: Routledge and Kegan Paul.

___. 1998. *The Laws of the Markets*. Oxford: Blackwell.

Callon, Michel, Pierre Lascoumes, and Yannick Barthe. 2009. *Acting in an Uncertain World: An Essay on Technical Democracy*. Cambridge, MA: MIT Press.

Carson, Rachel. 1962. *Silent Spring*. Boston: Houghton Mifflin.

Centre for Study of Environmental Change. 2001. "Public Attitudes to Agricultural Biotechnologies in Europe: Final Report of PABE Project." Lancaster: Centre for Study of Environmental Change, Lancaster University.

Chilvers, Jason. 2013. "Reflexive Engagement? Actors, Learning, and Reflexivity in Public Dialogue on Science and Technology." *Science Communication* 35 (3): 283–310.

Chilvers, Jason, and Matthew Kearnes, eds. 2016. *Remaking Participation: Science, Environment and Emergent Publics*. New York: Routledge.

Clarke, Adele E., and Virginia Olesen. 1998. *Revisioning Women, Health and Healing: Feminist, Cultural and Technoscience Perspectives*. New York: Routledge.

Cohn, Bernard S. 1996. *Colonialism and Its Forms of Knowledge*. Princeton, NJ: Princeton University Press.

Cohn, Carol. 1987. "Sex and Death in the Rational World of Defense Intellectuals." *Signs: Journal of Women in Culture and Society* 12 (4): 687–718.

Cole, Simon. 2001. *Suspect Identities: A History of Fingerprinting and Criminal Identification*. Cambridge, MA: Harvard University Press.

Collins, Harry. 2014. *Are We All Scientific Experts Now?* Cambridge: Polity.

Collins, Harry, and Robert Evans. 2002. "The Third Wave of Science Studies: Studies of Expertise and Experience." *Social Studies of Science* 32 (2): 235–96.

___. 2007. *Rethinking Expertise*. Chicago: University of Chicago Press.

Corburn, Jason. 2005. *Street Science: Community Knowledge and Environmental Health Justice*. Cambridge, MA: MIT Press.

Cowan, Ruth Schwartz. 1983. *More Work for Mother: The Ironies of Household Technology from the Open Hearth to the Microwave*. New York: Basic Books.

Daston, Lorraine, and Peter Galison. 2007. *Objectivity*. New York: Zone Books.

Dewey, John. 1927. *The Public and Its Problems*. New York: Holt.

Dryzek, John S. 2014. "Institutions of the Anthropocene: Governance in a Changing Earth System." *British Journal of Political Science*. Available on CJO 2014 doi:10.1017/S0007123414000453.

___. 2015. "Democratic Agents of Justice." *Journal of Political Philosophy* 23 (4): 361–84.

Dumit, Joseph. 2012. *Drugs for Life: Managing Health and Happiness through Facts and Pharmaceuticals*. Durham, NC: Duke University Press.

Edgerton, David. 2006. *Warfare State: Britain, 1920–1970*. Cambridge: Cambridge University Press.

Edwards, Paul N. 2010. *A Vast Machine: Computer Models, Climate Data, and the Politics of Global Warming*. Cambridge, MA: MIT Press.

Ellul, Jacques. [1954] 1964. *The Technological Society*. New York: Alfred A. Knopf.

Epstein, Steven. 1996. *Impure Science: AIDS, Activism, and the Politics of Knowledge*. Berkeley: University of California Press.

Espeland, Wendy N. 1998. *The Struggle for Water: Politics, Rationality, and Identity in the American Southwest*. Chicago: University of Chicago Press.

Ezrahi, Yaron. 1990. *The Descent of Icarus: Science and the Transformation of Contemporary Democracy*. Cambridge, MA: Harvard University Press.

Fairhead, James, and Melissa Leach. 1996. *Misreading the African Landscape: Society and Ecology in a Forest-Savanna Mosaic*. Cambridge: Cambridge University Press.

• Fausto-Sterling, Anne. 2000. *Sexing the Body: Gender Politics and the Construction of Sexuality*. New York: Basic Books.

Felt, Ulrike. 2015. "Keeping Technologies Out: Sociotechnical Imaginaries and the Formation of Austria's Technopolitical Identity." In *Dreamscapes of Modernity: Sociotechnical Imaginaries and the Fabrication of Power*, edited by Sheila Jasanoff and Sang-Hyun Kim, 103–25. Chicago: University of Chicago Press.

Felt, Ulrike, and Maximilian Fochler. 2011. "Slim Futures and the Fat Pill: Civic Imaginations of Innovation and Governance in an Engagement Setting." *Science as Culture* 20 (3): 307–28.

Felt, Ulrike, Brian Wynne, Michel Callon, Maria Eduarda Gonçalves, Sheila Jasanoff, Maria Jepsen, Pierre-Benoît Joly, Zdenek Konopasek, Stefan May, Claudia Neubauer, Arie Rip, Karen Siune, Andy Stirling, and Mariachiara Tallacchini. 2007. *Taking European Knowledge Society Seriously: Report of the Expert Group on Science and Governance to the Science, Economy, and Society Directorate*. Brussels: European Commission.

• Fleck, Ludwik. [1935] 1979. *Genesis and Development of a Scientific Fact*, translated by Thaddeus J. Trenn. Chicago: University of Chicago Press.

• Foucault, Michel. 1979. *Discipline and Punish*. New York: Vintage.

• ___. 2007. *Security, Territory, Population: Lectures at the Collège de France 1977–1978*. Basingstoke: Palgrave Macmillan.

Fujigaki, Yuko. 2009. "STS in Japan and East Asia: Governance of Science and Technology and Public Engagement." *East Asian Science, Technology and Society* 3 (4): 511–58.

Gibbons, Michael, Camille Limoges, Helga Nowotny, Simon Schwartzman, Peter Scott, and Martin Trow. 1994. *The New Production of Knowledge: The Dynamics of Science and Research in Contemporary Societies*. London: Sage.

Gieryn, Thomas. 1999. *Cultural Boundaries of Science: Credibility on the Line*. Chicago: University of Chicago Press.

Gottweis, Herbert. 1998. *Governing Molecules: The Discursive Politics of Genetic Engineering in Europe and the United States*. Cambridge, MA: MIT Press.

Guston, David. 2000. *Between Politics and Science: Assuring the Integrity and Productivity of Research*. New York: Cambridge University Press.

___. 2014. "Understanding 'Anticipatory Governance'." *Social Studies of Science* 44 (2): 218–42.

Habermas, Jürgen. 1970. *Toward a Rational Society: Student Protest, Science, and Politics*. Boston: Beacon Press.

Hacking, Ian. 1999. *The Social Construction of What?* Cambridge, MA: Harvard University Press.

Haraway, Donna. 1989. *Primate Visions: Gender, Race, and Nature in the World of Modern Science*. New York: Routledge.

___. 1991. *Simians, Cyborgs and Women: The Reinvention of Nature*. London: Routledge.

Hardt, Michael, and Antonio Negri. 2000. *Empire*. Cambridge, MA: Harvard University Press.

Hartouni, Valerie. 1997. *Cultural Conceptions: On Reproductive Technologies and the Remaking of Life*. Minneapolis: University of Minnesota Press.

Hecht, Gabrielle. 1998. *The Radiance of France: Nuclear Power and National Identity after World War II*. Cambridge, MA: MIT Press.

Heidegger, Martin. [1954] 1977. "The Question Concerning Technology." In *Basic Writings*, edited by David Farrell Krell, 308–41. New York: Harper and Row.

Hilgartner, Stephen. 2000. *Science on Stage: Expert Advice as Public Drama*. Stanford, CA: Stanford University Press.

Hilgartner, Stephen, Clark A. Miller, and Rob Hagendijk. 2015. *Science and Democracy: Making Knowledge and Making Power in the Biosciences and Beyond*. New York: Routledge.

Horkheimer, Max, and Theodor W. Adorno. 2002. *Dialectic of Enlightenment*. Stanford, CA: Stanford University Press.

Horlick-Jones, Tom, Nick Pidgeon, Gene Rowe, and John Walls. 2005. "Difficulties in Evaluating Public Engagement Initiatives: Reflections on an Evaluation of the UK GM Nation? Public Debate about Transgenic Crops." *Public Understanding of Science* 14 (4): 331–52.

Hurlbut, J. Benjamin. 2015. "Remembering the Future: Science, Law, and the Legacy of Asilomar." In *Dreamscapes of Modernity: Sociotechnical Imaginaries and the Fabrication of Power,* edited by Sheila Jasanoff and Sang-Hyun Kim, 126–51. Chicago: University of Chicago Press.

___. Forthcoming. *Experiments in Democracy: Human Embryo Research and the Politics of Public Bioethics.* New York: Columbia University Press.

Irwin, Alan, and Brian Wynne, eds. 1996. *Misunderstanding Science? The Public Reconstruction of Science and Technology.* Cambridge: Cambridge University Press.

Jasanoff, Sheila. 1990. *The Fifth Branch: Science Advisers as Policymakers.* Cambridge, MA: Harvard University Press.

___. 1995. *Science at the Bar: Law, Science and Technology in America.* Cambridge, MA: Harvard University Press.

___. 2003. "Breaking the Waves in Science Studies: Comment on H. M. Collins and Robert Evans, 'The Third Wave of Science Studies'." *Social Studies of Science* 33 (3): 389–400.

___, ed. 2004. *States of Knowledge: The Co-production of Science and Social Order.* London: Routledge.

___. 2005. *Designs on Nature: Science and Democracy in Europe and the United States.* Princeton, NJ: Princeton University Press.

___. 2010. "A New Climate for Society." *Theory, Culture & Society* 27 (2–3): 233–53.

___, ed. 2011a. *Reframing Rights: Bioconstitutionalism in the Genetic Age.* Cambridge, MA: MIT Press.

___. 2011b. "The Politics of Public Reason." In *The Politics of Knowledge,* edited by Patrick Baert and Fernando D. Rubio, 11–32. Abingdon: Routledge.

___. 2011c. "The Practices of Objectivity in Regulatory Science." In *Social Knowledge in the Making,* edited by Charles Camic, Neil Gross, and Michèle Lamont, 307–37. Chicago: University of Chicago Press.

___. 2012. *Science and Public Reason.* Abingdon: Routledge-Earthscan.

___. 2013. "Epistemic Subsidiarity: Coexistence, Cosmopolitanism, Constitutionalism." *European Journal of Risk Regulation* 2: 133–41.

Jasanoff, Sheila, and Sang-Hyun Kim, eds. 2015. *Dreamscapes of Modernity: Sociotechnical Imaginaries and the Fabrication of Power.* Chicago: University of Chicago Press.

Kahneman, Daniel, Paul Slovic, and Amos Tversky, eds. 1982. *Judgment under Uncertainty: Heuristics and Biases.* Cambridge: Cambridge University Press.

Kant, Immanuel. 1784. "Was Ist Aufklärung?" ["What Is Enlightenment?"]. *Berlinische Monatsschrift* (December): 481–94.

Keller, Evelyn Fox. 1983. *A Feeling for the Organism: The Life and Work of Barbara McClintock.* New York: W. H. Freeman.

•   ___. 1985. *Reflections on Gender in Science*. New Haven, CT: Yale University Press.

Kitcher, Philip. 2001. *Science, Truth and Democracy*. New York: Oxford University Press.

Kleinman, Daniel L., Abby J. Kinchy, and Jo Handelsman. 2005. *Controversies in Science and Technology, Volume 1: From Maize to Menopause*. Madison: University of Wisconsin Press.

•   Kuhn, Thomas. 1962. *The Structure of Scientific Revolutions*. Chicago: University of Chicago Press.

Laski, Harold J. 1930. "The Limitations of the Expert." *Harper's Monthly Magazine* 162: 101–10.

Latour, Bruno. 1990. "Drawing Things Together." In *Representation in Scientific Practice*, edited by Michael Lynch and Steve Woolgar, 19–68. Cambridge, MA: MIT Press.

___. 1993. *We Have Never Been Modern*. Cambridge, MA: Harvard University Press.

___. 2004. "Why Has Critique Run Out of Steam? From Matters of Fact to Matters of Concern." *Critical Inquiry* 30 (2): 225–48.

Laurent, Brice. 2013. "Nanomaterials in Political Life: In the Democracies of Nanotechnology." In *Nanomaterials: A Danger or a Promise? A Chemical and Biological Perspective*, edited by Roberta Brayner, Fernand Fiévet, and Thibaud Coradin, 379–99. London: Springer.

Law, John. 2002. *Aircraft Stories: Decentering the Object in Technoscience*. Durham, NC: Duke University Press.

___. 2009. "Assembling the World by Survey: Performativity and Politics." *Cultural Sociology* 3 (2): 239–56.

Leach, Melissa, Ian Scoones, and Brian Wynne, eds. 2005. *Science and Citizens*. London: Zed Books.

Lindblom, Charles E. 1959. "The Science of 'Muddling Through'." *Public Administration Review* 19 (2): 79–88.

Lindblom, Charles E., and David K. Cohen. 1979. *Usable Knowledge: Social Science and Social Problem Solving*. New Haven, CT: Yale University Press.

Lippmann, Walter. 1927. *The Phantom Public*. New York: Macmillan.

Lynch, Michael, and David Bogen. 1996. *The Spectacle of History: Speech, Text, and Memory at the Iran-Contra Hearings*. Durham, NC: Duke University Press.

Lynch, Michael, and Simon Cole. 2005. "Science and Technology Studies on Trial: Dilemmas of Expertise." *Social Studies of Science* 35 (2): 269–311.

MacKenzie, Donald. 1990. *Inventing Accuracy: A Historical Sociology of Nuclear Missile Guidance*. Cambridge, MA: MIT Press.

___. 2009. *Material Markets: How Economic Agents Are Constructed*. Oxford: Oxford University Press.

MacKenzie, Donald, Fabian Muniesa, and Lucia Shiu. 2008. *Do Economists Make Markets? On the Performativity of Economics*. Princeton, NJ: Princeton University Press.

Marcuse, Herbert. 1941. "Some Social Implications of Modern Technology." *Studies in Philosophy and Social Sciences* 9 (3): 414–39.

___. 1964. *One-Dimensional Man*. Boston: Beacon Press.

Marres, Noortje. 2012. *Material Participation: Technology, the Environment and Everyday Publics*. London: Palgrave Macmillan.

Martin, Brian. 2007. *Justice Ignited: The Dynamics of Backfire*. Lanham, MD: Rowman and Littlefield.

Mathews, Andrew. 2011. *Instituting Nature: Authority, Expertise, and Power in Mexican Forests*. Cambridge, MA: MIT Press.

Merchant, Carolyn. 1980. *The Death of Nature: Women, Ecology and the Scientific Revolution*. New York: Harper and Row.

Miller, Clark. 2015. "Globalizing Security: Science and the Transformation of Contemporary Political Imagination." In *Dreamscapes of Modernity: Sociotechnical Imaginaries and the Fabrication of Power*, edited by Sheila Jasanoff and Sang-Hyun Kim, 277–99. Chicago: University of Chicago Press.

Miller, Clark, and Paul Edwards, eds. 2001. *Changing the Atmosphere: Expert Knowledge and Environmental Governance*. Cambridge, MA: MIT Press.

Mitchell, Timothy. 2002. *Rule of Experts: Egypt, Techno-Politics, Modernity*. Berkeley: University of California Press.

___. 2011. *Carbon Democracy: Political Power in the Age of Oil*. London: Verso.

• Mol, Annemarie. 2002. *The Body Multiple: Ontology in Medical Practice*. Durham, NC: Duke University Press.

Mumford, Lewis. 1934. *Technics and Civilization*. New York: Harcourt.

___. 1938. *The Culture of Cities*. New York: Harcourt Brace.

Nandy, Ashis, ed. 1989. *Science, Hegemony and Violence: A Requiem for Modernity*. New Delhi: Oxford University Press.

• Nelkin, Dorothy. 1979. *Controversy: Politics of Technical Decisions*. Beverly Hills, CA: Sage.

Nelkin, Dorothy, and Michael Pollak. 1981. *The Atom Besieged: Extraparliamentary Dissent in France and Germany*. Cambridge, MA: MIT Press.

Noble, David. 1977. *America by Design: Science, Technology, and the Rise of Corporate Capitalism*. New York: Alfred A. Knopf.

Nowotny, Helga, Peter Scott, and Michael T. Gibbons. 2001. *Re-Thinking Science: Knowledge and the Public in an Age of Uncertainty*. Cambridge: Polity.

Oudshoorn, Nelly. 2003. *The Male Pill: A Biography of a Technology in the Making*. Durham, NC: Duke University Press.

Owen, Richard, John Bessant, and Maggy Heintz, eds. 2013. *Responsible Innovation: Managing the Responsible Emergence of Science and Innovation in Society*. Chichester: Wiley.

Parthasarathy, Shobita. 2007. *Building Genetic Medicine: Breast Cancer, Technology, and the Comparative Politics of Health Care*. Cambridge, MA: MIT Press.

___. 2017. *Patent Politics: Life Forms, Markets, and Public Interest in the United States and Europe*. Chicago: University of Chicago Press.

Petryna, Adriana. 2002. *Life Exposed: Biological Citizens after Chernobyl*. Princeton, NJ: Princeton University Press.

Popper, Karl R. 1945. *The Open Society and Its Enemies*. Princeton, NJ: Princeton University Press.

Porter, Theodore M. 1995. *Trust in Numbers: The Pursuit of Objectivity in Science and Public Life*. Princeton, NJ: Princeton University Press.

Price, Derek de Solla. 1963. *Little Science, Big Science . . . and Beyond*. New York: Columbia University Press.

Price, Don K. 1965. *The Scientific Estate*. Cambridge, MA: Harvard University Press.

Rabinow, Paul. 1992. "Artificiality and Enlightenment: From Sociobiology to Biosociality." In *Zone 6: Incorporations*, edited by Jonathan Crary, 234–52. Cambridge, MA: MIT Press.

Reardon, Jennifer. 2004. *Race to the Finish: Identity and Governance in an Age of Genomics*. Princeton, NJ: Princeton University Press.

Rose, Nikolas. 2007. *The Politics of Life Itself: Biomedicine, Power, and Subjectivity in the Twenty-First Century*. Princeton, NJ: Princeton University Press.

Scott, James C. 1988. *Seeing Like a State: How Certain Schemes to Improve the Human Condition Have Failed*. New Haven, CT: Yale University Press.

Scott, Pam, Evelleen Richards, and Brian Martin. 1990. "Captives of Controversy: The Myth of the Neutral Social Researcher in Contemporary Scientific Controversies." *Science, Technology, & Human Values* 15 (4): 474–94.

Schiebinger, Londa. 1989. *The Mind Has No Sex? Women in the Origins of Modern Science*. Cambridge, MA: Harvard University Press.

Schot, Johan, and Arie Rip. 1996. "The Past and Future of Constructive Technology Assessment." *Technological Forecasting and Social Change* 54 (2): 251–68.

Shapin, Steven. 1994. *A Social History of Truth*. Chicago: University of Chicago Press.

Shapin, Steven, and Simon Schaffer. 1985. *Leviathan and the Air-Pump*. Princeton, NJ: Princeton University Press.

Stiglitz, Joseph E. 2002. *Globalization and Its Discontents*. New York: Norton.

Sunder Rajan, Kaushik. 2006. *Biocapital: The Constitution of Post-Genomic Life*. Durham, NC: Duke University Press.

Thaler, Richard H., and Cass S. Sunstein. 2008. *Nudge: Improving Decisions about Health, Wealth, and Happiness*. New Haven, CT: Yale University Press.

Thompson, Charis. 2005. *Making Parents: The Ontological Choreography of Reproductive Technologies*. Cambridge, MA: MIT Press.

Veblen, Thorstein. 1906. "The Place of Science in Modern Civilization." *American Journal of Sociology* 11 (5): 585–609.

Visvanathan, Shiv. 1997. *A Carnival for Science: Essays on Science, Technology and Development*. New Delhi: Oxford University Press.

___. 2005. "Knowledge, Justice and Democracy." In *Science and Citizens*, edited by Melissa Leach, Ian Scoones, and Brian Wynne, 83–94. London: Zed Books.

Weber, Max. [1905] 1930. *The Protestant Ethic and the Spirit of Capitalism*. London: Allen and Unwin.

Winickoff, David, Sheila Jasanoff, Lawrence Busch, Robin Grove-White, and Brian Wynne. 2005. "Adjudicating the GM Food Wars: Science, Risk, and Democracy in World Trade Law." *Yale Journal of International Law* 30: 81–123.

Winner, Langdon. 1986. *The Whale and the Reactor: A Search for Limits in an Age of High Technology*. Chicago: University of Chicago Press.

Wynne, Brian. 1989. "Sheep Farming after Chernobyl: A Case Study in Communicating Scientific Information." *Environment* 31 (2): 10–39.

___. 1994. "Public Understanding of Science." In *Handbook of Science and Technology Studies*, edited by Sheila Jasanoff, Gerald Markle, James Peterson, and Trevor Pinch, 361–88. Thousand Oaks, CA: Sage.

___. 2003. "Seasick on the Third Wave? Subverting the Hegemony of Propositionalism." *Social Studies of Science* 33 (3): 401–17.

___. [1982] 2011. *Rationality and Ritual: Participation and Exclusion in Nuclear Decision-Making*. 2nd ed. Abingdon: Routledge Earthscan.

10 STS and Social Movements: Pasts and Futures

Steve Breyman, Nancy Campbell, Virginia Eubanks, and Abby Kinchy

This chapter surveys how growing communities of science and technology studies (STS) scholars engage with social movements in contemporary societies. The central—though often unrecognized—role of social movements in phenomena of interest to STS suggests that studying them is not merely one facet of a highly diverse field but rather a core idea. This chapter traces interconnections among social movements and intellectual currents in STS, reform efforts in scientific and technical fields, the study of scientific controversies, and the shaping of technology via lay and expert knowledge. We aim to stimulate fuller recognition of the multiple and changing ways in which social movements connect with science and technology, and how the study of social movements can advance knowledge of science, technology, and society when conducted with reflexive awareness in a field whose own history is deeply intertwined with societal change.

We begin by tracing the marked influence of social movements on the formation and development of the field of STS, which has demonstrated the effects of social movements on formation of scientific specialties, intellectual development, and research agendas of established fields (Epstein 1996; Frickel 2004; Hess 2005, 2015; Jamison 2003; Moore 2008; Tesh 1988, 2000). Here, we turn this insight upon our own field. While many origins stories might be told about STS, the genealogy that resonates most strongly with us includes many activists, thinkers, and writers whose critiques of science and technology emerged in contexts of social struggle and conflict. With particular attention to feminist and postcolonial movements, we make the case that social movements have intersected with our field, and its participants, in profound ways.

The second part of the chapter turns to the question of how STS scholars have theorized social movements in key areas of research such as controversy studies, the social shaping of technology, embodied health movements, and governance of risk. We observe increasing theoretical sophistication and broad public relevance of these

STS approaches, using health social movements to highlight some unique theoretical perspectives elaborated by STS scholars.

In the third and final part of the chapter, we examine emerging movements and directions for future research with a focus on the underexamined use of digital media technologies by movements. We seek to inspire reflection about how interaction with new movements might generate new ideas in STS.

Intersections with Social Movements: From the Luddites to STS

Social Movement Forerunners of STS

Organized STS groupings first emerged—often through the efforts of concerned natural and life scientists—in North America and Western Europe amid the tumult of rapid social change characteristic of postindustrial society, postmaterialist culture, and the extra-parliamentary, extra-institutional New Politics of the 1960s and 1970s. As Leonard J. Waks (1993, 406) argues, "[A]ctivists see STS as a part, a component, of a larger social movement associated with the 'sixties,' with personal autonomy, decentralization of power over social systems, and hence with 'antitechnology' when this refers not to particular artifacts but the sociotechnical requirements of scientific-industrial society itself." A heightened awareness of how rapid technological change can create specific kinds of political order and carry significant risk of harm to people and the environment was a growing focus of technology-centered activism in Western Europe and North America after 1960.

While the social upheaval of the 1960s clearly shaped the emerging field of STS, a much longer social movement history also informs theories and practices in our field. In 1811, during a period of unemployment and rapid industrialization, police in Nottingham broke up a protest for more work and better wages; in retaliation, angry workers smashed textile machinery in a neighboring town (Bailey 1998; Coniff 2011; Sale 1996). The machine-smashing protests caught on and spread across northern England. Despite retaliation from police and factory owners, the Luddite rebellions continued for five years. In some ways, science- and technology-focused social movements and scholars have been pushing back against the charge that they are "antiprogress" ever since.

Insisting that they were not antitechnology but rather against the "worship of technology," social reformers argued that a more equitable, peaceful society required more thoughtful relationships with machines. In India in the 1920s, Gandhi's *charkha* and the All-India Spinners' Association provided an icon of self-reliance and illustrated an economy devoted to "inefficient" labor that would employ thousands of weavers impoverished by British textile manufacturing (Brown 2010). After the devastation of

World War II, and at the dawn of the *green revolution*, Japanese farmers began experimenting with nontechnological, no-till, "one-straw" agriculture. In the United States, the 1950s saw labor movement negotiation of the increased role of automation by creating "Automation Fund Committees" to support workers displaced by new technology (Ullman 1969, 35). In the same era, African American activists and intellectuals protested nuclear weapons, viewing their development and use in terms of racial inequality and oppression (Kinchy 2009).

By the late 1950s and early 1960s, Minimata disease (mercury poisoning) in Japan, C. Wright Mills's (1956) *The Power Elite* (whence the term *academic-military-industrial complex*), Rudi Dutschke's flight to West Berlin (1961), and Rachel Carson's *Silent Spring* (1962)—a trigger for environmental activism in the United States—signaled growing concern for power imbalances among people, corporations, and state institutions. Various New Left formations amplified this concern across the industrialized and industrializing worlds. A significant portion of early STS scholars took part in and had their work shaped in diverse ways by movements of the era.

These students and faculty rejected middle-class conformism and worried about Big Science (a term coined by Derek de Solla Price in 1963). They wrestled with Marxism and evinced strong antiauthoritarian tendencies. They read E. P. Thompson, Raymond Williams, and Stuart Hall, Simone de Beauvoir and Betty Friedan, Jacques Ellul and Lewis Mumford, Herbert Marcuse and Paul Goodman. They took part in resistance to the Vietnam War, and allied with the new waves of feminist, environmental, anticolonial ferment of the era. The blend of scholarly purpose and moral mission led some—especially those from physical and life science backgrounds—to form Science for the People in 1969, an organization that served as a thorn in the side of the American Association for the Advancement of Science for nearly a decade. Science for the People's magazine focused critical attention on several controversial areas of applied and emerging science, including genetics, sociobiology, and the participation of scientists and engineers in military research (Garvey 2014; Moore 2008).

These new movements—both intellectual and political—were generally united in opposition to specific, untested, risky, or potentially harmful technologies. Women Strike for Peace—part of the "Ban the Bomb" movement—organized a daylong strike of 50,000 women worldwide to protest nuclear weapons testing in 1961. They declared "End the Arms Race, Not the Human Race," and raised awareness that high levels of strontium-90 were found in breast milk, cow's milk, and children's teeth. "Appropriate technologists" critiqued the Western development model based on reductionist science and complex, expensive technology manufactured in industrialized countries and exported to former colonies hungry for jobs and growth. They believed an alternative, human-centered development was possible in the global south were it able to wean itself of dependence on Western technology, investment, and concomitant debt. "Appropriate technology" (AT) was to be labor intensive (where labor was abundant and capital scarce), decentralized, human scale, energy efficient, and environmentally sound (Ghosh and Morrison 1984). AT initiatives—including low-cost computers and telephony—across the globe survive and even thrive today, building cross-national technoscientific cooperation that empowers communities and improves the lives of a growing number of the world's poor.

While many of the interventions suggested by rich-country social movements could be described as "technologically cautious," Latin American movements took a more technologically optimistic approach. Project Cybersyn was an ambitious attempt to combine cybernetic systems thinking, new digital computers, and socialist state building in Allende's Chile (Medina 2011). In the 1980s, Brazilian company Unitron reverse engineered the *Mac de periforia* (Mac of the periphery) as a response to state incentives to design and manufacture products to lessen economic inequality and encourage Brazilian innovation (Marques 2005; Irani et al. 2010). The "barbarian" pirate figure—who uses existing resources but repurposes legal and social regimes—has long been a part of postcolonial resistance narratives and practices (Philip 2005).

Social Movements in the Formation of the Field

Some early STSers considered the 1970s citizen initiatives and single-issue movements signals for their partial flight from the disciplines, and for their self-conscious constitution of complex interdisciplinary hybrids. Early STS scholars did not wholly give up their disciplinary identities or approaches—they remain central to this day—but believed the traditional academic disciplines on their own lacked the questions, theoretical diversity, and breadth to address the problems and concerns of the age.

Not all the pioneers were unabashed movement partisans. But for a significant number, participation in and sympathy for grassroots movements helped shape the questions asked, the conceptual and methodological approaches taken, and the presentation of ideas to diverse audiences. The cultural prevalence of social Darwinist ideas moved Hilary Rose and Steven Rose to criticize the separation between science and society and to argue that science was socially determined and could not be analytically separated from society (1969). The explicitly critical and political project within STS (Martin 1979) opened the field to charges that it was "politicizing science," or "antiscience," commonly heard later during the 1990s science wars in the United States (a series of hostile exchanges among scientific realists and postmodernists; Sokal 1996).

A division developed within academic STS by the 1990s. Steve Fuller (1993) diagnosed a discipline-centric, scholarly High Church STS and a Low Church, activist-oriented

STS rooted in social movements. Juan Ilerbaig (1992) described "two subcultures" in the growing interdiscipline. David Hess (1997) identified a split between the sociology of scientific knowledge (SSK) and "critical STS." Brian Martin (1993) lamented the "academization" of the critique of science. Narrow and insulated research projects, he believed, pushed the politicized analysis characteristic of some early STS to the margins. Bruno Latour feared division into what he considered "an applied but soft branch—STS—and a basic but isolated one—science studies" (1993, 384).

Numerous suggestions arose to bridge the gap (Breyman 1997a). Langdon Winner (1993, 374) suggested we consider "which ends, principles, and conditions deserve ... our commitment." Fuller recommended reconstituting STS departments (especially their graduate programs) as "social movements" (1992, 3). Hess proposed paths for peaceful coexistence (1997). Martin suggested STSers work on technoscientific problems relevant to social movements and intervene as "open partisans" in scientific controversies (1996). Others recommended "weakly asymmetrical third positions" that are "situated, partial and committed in a knowledge political sense" (Pels 1996, 304) or contextualized and policy-relevant (Richards 1991), and shift from SSK's restrictive controversy framework to one that explains the "co-production" of science and society (Jasanoff 1996).

Despite the divide, STS scholars of both tendencies contributed to several important cultural changes, perhaps best reflected in the conceptual move from "science and society" to "science in society." Science and technology were knocked off their cultural pedestal (Chubin and Chu 1988). Daniel Sarewitz (1996) criticized science policy for its foundation on myths about the relationship between science and progress. STS students and scholars, many openly partisan, called for participatory and deliberative decision-making processes around science and technology policy (Fischer 2000).

"Given the insular and culturally exalted status of science," reported Richard Worthington et al. (2012, v), "a significant trend of the past several decades has centered on broader access to and participation in technological policies and practices." While hyperbolic to speak of the "democratization of science," specific features of the still unfolding cultural trend include (1) diversification of the science and technology workforce; (2) participation by ordinary people in citizen science interventions and community-based research endeavors; (3) challenges by lay citizens to the authority of experts; and (4) the appearance of dissident scientists, whistleblowers, and counter-experts of all descriptions who directly challenge corporation, state, and university—what David Hess (2007) terms "epistemic modernization."

An increasing number of scholars use STS frameworks to study science and social justice with the aim to democratize science and technology (Eubanks 2011; Fisher

2009; Hilgartner, Miller, and Hagendijk 2015; Sclove 1995). This occurs not only in academia but also in centers like the Loka Institute, which works for increased citizen participation in technology assessment, and a "science and technology of, by, and for the people." New materialist modes of knowing, making, and doing have emerged to counter large-scale technological systems that appear self-governing and self-reproducing (Winner 1978; see chapter 8 this volume).

The U.S. National Institutes of Health established a low-percentage set-aside for ethical, legal, and social scientific studies in an attempt to anticipate and forestall adverse effects of the Human Genome Project. A similar model was used for publicly funded nanotechnology research (Fisher, Selin, and Wetmore 2008). Although limited by the type of implications considered, these programs are one outcome of social movement pressure on state-sponsored research and innovation. Social movements that seek to renegotiate the social contract for science and technology in response to the pervasive presence of new technologies themselves shape science, medicine, and technology (Wajcman 2004). New forms of agency and patterns of social solidarity emerge in response to new sciences and technologies, and with them, the idea, central to STS, that science and technology should be responsive to the needs of a diverse swath of humanity.

Feminist, Antiracist, and Postcolonial Movements

Feminist, antiracist, and postcolonial movements have had particularly porous boundaries with STS. The concept of objectivity came under attack in anticolonial movements, which linked colonialism, internalized racism, and dehumanization (Fanon [1952] 1967, [1961] 1963). Fanon argued that formerly colonized people had to unlearn intellectual colonialism. Edward Said (1978) critiqued the "will to power" expressed in the Western appetite for producing knowledge about "the Orient." Orientalism was both a representation and an instrument of domination working through colonialism. These two streams of critique of the colonial and imperial tendencies of modern Western science strongly inflect the exciting, but still underdeveloped, field of postcolonial STS (Goonatilake 1999; Harding 2011; Medina, Marques, and Holmes 2014; Reardon, Kowal, and Radin 2013; Shiva 1997; Visvanathan 1997).

Continued critique of colonial and postcolonial deployment of science and technology for resource extraction in the global south (Hecht 2012) led to important overlaps between social and academic movements. Development of medical humanitarianism and proliferation of professional nongovernmental organizations (NGOs) such as *Médecins Sans Frontières* compelled STS scholars to scrutinize the role of science and technology in development, immigration, social welfare, and well-being (Redfield 2013).

The postcolonial critique helped give rise to advocacy for and scholarship on sustainable development and sustainability. Growing unease about the effects of the *green revolution* (Shiva 1992) and economic policies that contributed to inequality spurred STS interest in movements seeking to contain downstream effects of some *products* of technoscience—from genetically modified organisms to weaponry—and the nondemocratic *processes* of technoscience (dominated by giant corporations and nonresponsive governments) upon human health and social well-being.

National liberation struggles addressed both material social relations and the politics of "objective" knowledge about non-Western peoples. The postcolonial critique of Western appropriation of indigenous knowledge systems evidences much overlap with feminist STS (Figueroa and Harding 2003; Harding 1991, 1993, 1998, 2006, 2015). Both approaches illuminate extractive relationships between Western science and indigenous knowledge systems (Breyman 1994; Hayden 2003; Reardon 2005; Reardon and TallBear 2012). Casting science and technology as techniques of domination that reproduced both patriarchy and colonialism, during the 1980s and 1990s North American and European women's movements showed how intersections of difference among women (class, race, ethnicity, sexual identity, and ability) rendered the very category "woman" open to question. The rich confluence of liberal, radical, socialist, and deeply cultural roots at the nexus of feminist theory and STS led to the gradual emergence of feminist STS.

Sparked by responses to doctors' failure to inform women of adverse effects of oral contraception, the movement to wrest knowledge about and control over women's bodies and minds from the medical and pharmaceutical establishment relied on raising consciousness as the basis for creation and circulation of new forms of scientific and medical knowledge. This included the landmark *Our Bodies, Ourselves* (Boston Women's Health Collective and Norsigian 1979; Davis 2007). Women's health activism, in conjunction with the HIV/AIDS movement (Epstein 1996) promoted new forms of public participation on "contested illnesses" (Brown, Morello-Frosch, and Zavestoski 2011), toxic exposure (Brown 2007; Gibbs 2002), chronic fatigue syndrome (Murphy 2004, 2006), and environmental health (Daniels 1993). Di Chiro (2008) used intersectionality theory (Crenshaw 1991) to argue that "all environmental issues are reproductive issues." The reproductive rights movement attended to conditions of "stratified reproduction" (Colen 1995), differential vulnerability of women to forced sterilization and compulsory birth control (Briggs 2003), and gendered and sexualized subjection within social movements themselves.

As the 1970s women's liberation movement focused on the neglect of women's health and the objectification of women's bodies, the feminist critique of masculinist bias in the conduct of science, medicine, and technological innovation became influential for many STS scholars (Wylie, Potter, and Bauchspies 2015). Scholars asked during the 1980s, what might a "feminist science" look like (Bleier 1986; Harding 1986, 1991; Hubbard 1990; Keller 1978, 1985; Longino 1987; Rose 1983; Rosser 1990)? The confluence of women's consciousness raising, activism, and rapid entry into higher education contributed to the focus on goals ranging from production of new forms of knowledge, concepts, epistemologies, and methodologies to workforce diversification and quality of life for women in workplace cultures of science, medicine, and technology.

Feminist critique of the sciences emerged first in biology, physics, and history and philosophy of science. Strong connections formed between the projects to diversify the sciences and medicine, activism around reproductive politics and technologies, global women's health, and environmental movements. Hilary Rose's (1994) "Love, Power, and Knowledge" negotiated new terms for feminist engagement with the sciences. Feminist knowledge transformation extended beyond science studies and into the postcolonial project (Harding 1993, 1998, 2006).

Donna Haraway's (1991) "Cyborg Manifesto" reckoned with the limits of socialist feminism for the politics of women's liberation, ecofeminism, and an emerging informatics of domination. The manifesto catalyzed debate between feminists who identified positive potentials in the cyborgian figure and those who rejected it on grounds that technology represented little more than a patriarchal means to wrest the power of procreation from women. As the women's movement became institutionalized in women's studies programs, attention to science and technology remained a niche interest until feminists elaborated new methodologies for examining topics such as objectification of the body (Martin 1987, 1991) and science and technology as "conceptual practices of power" that sustain "ruling relations" (Smith 1987, 1990).

Judy Wajcman (1991, 25) theorized the "gendered character of technology" as adopting neither an "essentialist position that sees technology as inherently patriarchal [nor] losing sight of the structure of gender relations through an emphasis on the historical variability of the categories of 'women' and 'technology'." Feminist STS challenged not only the research agenda and composition of science but also commitments to "universal" and supposedly value neutral forms of objectivity that suppress diversity (Harding 2015). Advancing the claim that all knowledge systems and the logics of inquiry upon which they rest should be valued according to the extent to which they promote social justice, Harding (2015) emphasizes that emancipatory movements raise research questions that should be consulted in the setting of research agendas. As knowledge production and circulation became central activities for social movements, study of the role of science and technology in feminist, antiracist, and postcolonial movements became central to STS.

Theorizing Social Movements

STS scholars examined social movements with increasing theoretical sophistication. The dominant research approach to social movement interactions with science and technology in the seventies was "controversy studies." This mostly descriptive research field—pioneered by Dorothy Nelkin—provided accounts of citizen groups' grappling with unwanted technologies and development projects. Nelkin studied citizen mobilizations surrounding nuclear power, airport siting, creation science, methadone maintenance, and myriad other topics in which movements were potent actors (Nelkin 1992). Science itself was understood to be controversial in nature (Brante, Fuller, and Lynch 1993). Transnational environmental movements spurred comparative social studies of science (Yearley 1989, 1995). Nelkin's and Yearley's analyses contrasted with early studies of public understanding of science (PUS) based on large-scale surveys of "indicators" of "science literacy" that adopted a deficit model (people reject science because they do not understand it). Lacking a theory of social movements, early PUS studies relied on individualist framings of "mental models" and "cognitive approaches" and failed to grasp the significance of organized activism and collective struggles for social change.

Brian Wynne's study of sheep farmers dealing with radioactive fallout from the Chernobyl disaster inspired an important shift in STS scholarship on public perceptions and understanding of science and technology (Wynne 1992). Wynne observed that local lay knowledges permitted farmers to renegotiate expert claims. Urging scientific experts to examine their own scientific cultures and institutions in the process of meeting citizen demands for candor and openness, Wynne and collaborators helped shift STS beyond the limitations of PUS and toward more expansive comparative and transnational studies of citizen engagement in controversies (Leach, Scoones, and Wynne 2005).

STS and Social Movement Theory

As STS scholars paid more attention to social movement engagements with science and expertise, Steve Epstein (1996, 19) observed, "[F]ew sociologists of scientific knowledge have engaged significantly with the sociological literature on social movements." STS scholars found little use for theories in sociology and political science developed to grapple with collective action and contentious politics. Generally speaking, the same is true today; although social movement theories have a greater presence in STS, their application is eclectic. A key reason for the gap between STS and sociology of social movements is that until fairly recently, sociological research did not focus on social movements that challenge institutions of science, medicine, and engineering, so insights relevant to STS were limited. Brown et al. (2004a) considered these movements "uncharted territory" in the broader context of social movement studies. When Brown et al. (2004b) examined embodied health movements, they discarded the dominant social movement theories as inadequate due to lack of consideration for class, experiences of illness, and social networks. However, when STS scholars adapted social movement theories to diverse institutional contexts, they made theoretical contributions that have influenced new theory-building efforts in sociology (Breyman 1997c). As a result, the gap between STS and sociology of social movements is slowly shrinking. Before discussing this recent shift, however, a brief history of the sociological study of social movements, and its influence in STS, is in order.

Social movement theory underwent a series of theoretical leaps, reflecting major shifts in understanding how and why people organize for social change. Prior to the 1970s, scholars treated social movements as social psychological phenomena (Gurr 1970; Hoffer 1951), an approach strongly criticized by those who observed and participated in the civil rights, antiwar, and labor struggles of their generation. In the turn away from social psychological understandings of protest, theories of social movements centered on state power, emphasizing how people excluded from formal channels of political power engage in contentious collective action to express grievances and seek change. Viewing social movements as political processes rather than psychological aberrations, the political process approach produced key concepts still in wide use, notably political opportunity structure (McAdam 1982; McAdam, McCarthy, and Zald 1996; Tilly 1978), resource mobilization (McCarthy and Zald 1977), and collective action framing (Benford and Snow 2000). STS scholars adopted some of these concepts, for example, in studies of how cancer activists formed coalitions and mobilized elite allies (Petersen and Markle 1981), how scientist-activists established the field of genetic toxicology (Frickel 2004), and how citizens mobilized scientific data when organizing against dams in Brazil or causes of environmental breast cancer (Ley 2009; McCormick 2009).

European social theorists formulated an alternative theoretical approach focused on "new social movements" (Laraña, Johnston, and Gusfield 1994; Melucci 1980; Touraine 1981). Central to this perspective was the idea that social movement struggles are no longer primarily rooted in class conflict, rather, they challenge cultural codes and ways of life, particularly surrounding identity, sexuality, and the body (Melucci 1989). Debate about the novelty of these movements (Breyman 1997b) led researchers to point out that even the "old" class-based movements engaged in identity formation and cultural work. Nevertheless, the ideas of the new social movement theorists offer conceptual tools to interpret movements that seek not just political change but also changes in knowledge, technical practices, ideas about health and the body, and activist engagement with scientific knowledge production (Epstein 1996).

An early attempt to connect the sociology of social movements with STS can be found in Ron Eyerman and Andrew Jamison's (1991) cognitive approach. They sought to integrate the prevalent American and European approaches and address their respective shortcomings. Eyerman and Jamison (1991, 52) focused on how movements mediate "both in the transformation of everyday knowledge into professional knowledge, and, perhaps even more importantly, in providing new contexts for the reinterpretation of professional knowledge." This insight was subsequently deployed in studies of the knowledge work of the antibiotechnology movement (Schurman and Munro 2006) and antitoxics activists (Tesh 2000).

A recent turn in the sociology of social movements responded to economic and cultural globalization. The antiglobalization movement prompted theoretical work on the "new" transnational activism (Tarrow 2005). Sociologists borrowed key ideas such as transnational advocacy networks (Keck and Sikkink 1998) and venue shopping from political science, and refined earlier frameworks to create concepts such as nested political opportunity structures—the idea that local opportunity structures are nested in national and global opportunity structures (Rothman and Oliver 1999). Many of these ideas proved useful to STS scholars writing about movements that transcended national borders, including collective opposition to transgenic crops (Kinchy 2012; Scoones 2008), ecofeminism (Leach 2007), and mobilizations against nuclear weapons (Breyman 2001).

STS research on movements that seek change in science and medicine (Epstein 1996, 2007; Moore 2008; Waidzunas 2013), education (Rojas 2007), agriculture (Kinchy 2012), and industry (Hess 2010; Seidman 2008) demonstrated that social movements often spur reform in institutions beyond the state. Some social movements arise within scientific fields themselves. Frickel and Gross (2005, 204) developed a theory of "science and intellectual movements" (SIMs), which are "collective efforts to pursue research programs or projects for thought in the face of resistance from others in the scientific or intellectual community." Kelly Moore (2008) examined how scientists opposed to U.S. militarism transformed science as an institution. She showed that science has vulnerabilities that include schisms among "insiders" that become the basis for movement-like organizing within scientific fields.

This research has made a mark on the sociology of social movements, where scholars now ask why challenges take the forms that they do in different contexts (Walker,

Martin, and McCarthy 2008), what interactions between challengers and targets tell us about domination in society (Armstrong and Bernstein 2008), and how struggles for power in one field ripple across other fields (Fligstein and McAdam 2011). In sum, STS and the sociology of social movements benefit from exchanges of ideas.

Health Social Movements

Research on health social movements is an area where STS has posed productive challenges to the sociology of social movements. Brown et al. (2004b) divide health social movements into health access movements, constituency-based health movements addressing inequality and health disparities, and embodied health movements responsive to the experience of disease, illness, and disability (Brown et al. 2004a, 2004b). Organized collectivities assert alternative means of self-help, support, alternative therapies, or ways to live with chronic conditions other than those held out in public health education and policy. Such movements enable members to redefine themselves by overcoming prevailing "epistemologies of ignorance" (Proctor and Schiebinger 2008; Tuana 2006). Examples include reproductive rights activism around technologies such as abortion and birth control (Murphy 2012); self-help health access such as the now global *Our Bodies, Ourselves* (Davis 2007); breast cancer activism (Klawiter 2008; McCormick 2009); and movements that form around specific environmental injustices entailing health consequences (Ottinger 2009, 2013).

Environmental health and justice movements exploded into public consciousness with the controversy over toxins buried beneath Love Canal in a working-class neighborhood of Niagara Falls, New York, and with the organization of exploited Brazilian rubber tappers by Chico Mendes. These struggles framed the tensions—scientific expertise versus local knowledge, public participation versus technocratic decision making, environmental equity versus racism and classism—that underlay innumerable disputes in the United States throughout the eighties and nineties. Several pivotal studies of the centrality of social and racial justice concerns to local struggles over contamination and waste facilities followed (e.g., Bullard [1990] 2000), as did tangible victories for the movement, including closing many notorious polluters and forcing government agencies and the environmental movement as a whole to take environmental justice and workers' rights seriously (including the establishment of pollution remediation funds in the global north and extractive reserves in the Amazon).

STS scholars have since added important layers to our understanding of environmental justice mobilizations, with a focus on knowledge-making practices of movement groups in their entanglements with authorities. Barbara Allen looks at citizen-expert alliances in environmental justice campaigns along Louisiana's Cancer Alley. Allen develops the concept of "toxic narratives," situated, collective ways of understanding that become knowledge claims and "help negotiate the fissure dividing science from everyday experience" (2003, 49). Jason Corburn (2005) illustrates how local knowledge can join professional techniques to address environmental health problems in Brooklyn. Melissa Checker (2007) shows how environmental risk assessments can exclude the experience of the poor and people of color, and proposes mandating meaningful public participation. Gwen Ottinger (2009) examines the effectiveness of the famed "bucket brigades" employed by fenceline communities to generate scientific proof of their exposures to airborne hazards.

Forms of embodied experience like environmental justice campaigns enable people to respond to health conditions by forming online and virtual communities now central to patients' rights movements around a variety of chronic conditions (Callon and Rabeharisoa 2008; Rapp 2011). Governments typically function at a distance from lived experience of risky realities, relying on population-level public health approaches designed for containing contagious or infectious disease. The rubric of "biomedicalization" (Clarke et al. 2010) introduced a sophisticated framework for the globalized commodification and pharmaceuticalization of health (Dumit 2012; Petryna 2009). These processes draw connections between social stratification and the globalization of clinical trials (Fisher 2009; Petryna, Lakoff, and Kleinman 2006), producing new concerns about social justice, human rights, labor, and the body on which health social movements focus.

Health movements challenge governmentality in myriad ways. Epstein argued that a tacit coalition called the "anti-standardization resistance movement" composed of actors from the disability rights, women's, and HIV/AIDS movements advanced frames that succeeded at the national level "because the expert counter-frames lacked punch in the face of the strength and diversity of the reform coalition" (2007, 109). Diversity was a source of tactical strength in this movement, which successfully changed governmental resource allocation and regulatory oversight.

Harm-reduction movements are an interesting illustration of the diversity of state and nonstate targets of social movements. European "Health for All" movements revolve around ideas concerning community empowerment, harm reduction, and health promotion (Ashton and Seymour 1988). Harm reduction—around HIV/AIDS, Hepatitis C, opiate overdose, and other preventable conditions—works via community-based organizations that adopt rapid response, low-threshold services to their communities. These movements advocate responses that emphasize dignity, access, and shift of the locus of change to production of "enabling environments" and away from "individuals alone to the social situations and structures in which individuals find themselves"

(Rhodes 2002, 91). Such movements self-organize, typically forming underground distribution sites and networks, encouraging constituents to pressure for changes such as Good Samaritan laws and laws governing access to prescription or experimental medicines, clean syringes, and naloxone. These movements articulate localized harm reduction with the need for social change while challenging mainstream notions of illness and disease (Krieger 2011) and identity-based inequalities and health disparities. Harm-reduction movements both challenge and collaborate with elites (including police and EMTs), appealing to overarching commitments to human rights, social responsibility, and environmental justice while undertaking practical steps to improve near-term experiences.

Emerging Movements and Directions for Research

Social movements have been a core part of STS since its inception. STS students and scholars, committed both to advancing and studying collective action, have pushed back against depoliticized currents within the academy. Exemplary STS work advances social movements by examining issues that arise when science and technology collide with inequalities based on race, class, gender, sexuality, ability, mental health status, and immigrant or refugee status (Benjamin 2013; Bliss 2012; Braun 2014; Duster 1990; Fausto-Sterling 1990; Fouché 2003; Gray 2009; Mamo 2007; Montoya 2011; Nelson 2011; Roberts 2012; Waidzunas 2013). In particular, STS critically examined the role of science and technology in constructing race, ethnicity, and racism (Barkan 1992; Efron 1994; Fernández Kelly 1987; Gould 1981). STS scholarship would be further strengthened by directly engaging conceptual frameworks for critical study of race, racism, and racial-ethnic formation (Omi and Winant 2015), intersectionality (Combahee River Collective 1977; Crenshaw 1991), and whiteness (Harris 1993; Hartigan 1999). Integrating critical race theory or critical whiteness studies would be a productive way to build on Sandra Harding's edited collections on science and race (Figueroa and Harding 2003; Harding 1993).

STS research about social movements has been theoretically innovative and topically diverse, but it has tended to pay more attention to some movements than to others. STS scholars regularly attend to movements that mobilize scientific knowledge claims, challenge dominant modes of knowledge production, oppose risky technologies, and call for more democratic governance. However, few studies have analyzed the technoscientific entanglements of movements for racial justice (as distinct from environmental justice), lesbian, gay, bisexual, trans, and queer/questioning (LGBTQ) liberation, prison abolition, indigenous sovereignty, immigrant rights, and welfare rights. Significant sectors of the global social movement universe remain underexamined.

To begin to remedy this gap, this final section explores the role of digital technology in activism as an important and evolving new study area. The advent of the World Wide Web, GPS, the smartphone, and other new tools is changing how social movements exploit technology's liberatory potential. Though it has hardly lived up to the hype as the revolutionary "flattener" of hierarchies in power, resources, and status, or the great democratizer of early predictions, the Internet nevertheless fundamentally altered the terrain, tactics, targets, and reach of social movements. From open source/ free software in the 1980s (Kelty 2008) to the birth of the independent media movement (Castells 2012; Gerbaudo 2012; Wolfson 2014), from the use of social media by ISIS to Wikileaks and the post-Snowdon hacktivism of Anonymous (Coleman 2014), the proliferation of new tools for movement mobilization provides scholars ample opportunities for study and theorizing.

Some Internet-based progressive organizing focuses on "making power" (Smith 2008), creating autonomous zones within organizations, movements, and communities "that model the world we are trying to create" rather than focusing on "taking power" from the state. Indymedia, Makerspaces, and hacktivism around the world may be seen as creating alternative structures and anarchic publics outside of traditional targets of movement activism (Anderson 2014; Kelty 2005; Turner 2008). These generally anti-authoritarian campaigns—including critical making and hacktivism—eschew charismatic leadership and celebrate temporary, strategic, radically democratic and often satirical interactions with structures of power (Coleman 2014; Pickard 2006). Similarly, the "civic" in civic media is more often related to global civil society and distributed decision making than it is to interventions in formal governance (Zuckerman 2014), despite the work of government-facing organizations such as Code for America and *Laboratorio para la Ciudad* in Mexico City. What might STS contribute to our understanding of the development of these new technology-infused cultural and social spaces and actors?

Internet-based organizing in North American communities of color celebrates broad-based participatory engagement and storytelling through new technology tools without abandoning critical engagement with the state. The Black Lives Matter movement directly challenges the extrajudicial killings of black people in the United States while affirming "black folks' contributions to this society, our humanity, and our resilience in the face of deadly oppression" (Garza 2015). Organizations such as ColorOfChange .org, the Center for Media Justice, and Allied Media Projects explicitly identify with the high-tech tools they use but also directly engage the state in struggles over network neutrality, media ownership and fairness, broadband accessibility, racial domination, and police violence.

Legacy civil rights organizations such as the NAACP, the Council of La Raza, and Asian Americans Advancing Justice translate historical concerns such as real estate redlining into the age of big data, arguing for preservation of constitutional principles and procedural fairness in government and corporate decision making (Eubanks 2014b). Immigrant rights groups turned to new technology tools like Voz Mob to tell their stories and articulate their concerns—as in the Undocumented and Unafraid campaign—but remain active in legislative battles (Costanza-Chock 2014). How might new or enduring STS concepts advance our understanding of these diverse challengers to the technosocial status quo?

Whatever degrees of freedom offered by open source, hacktivism, and making, it is hard not to see the "Anonymous turn" in technology and social movements as also a rearguard reaction to the successful (and ongoing) challenges posed to white male–dominated progressive movements by women, men of color, poor and working-class people, and the LGBTQ community. It is perhaps not surprising that the success of identity politics should result in a philosophical backlash celebrating flexible and collective identities cut free from demographic markers. The most extreme variety of this thinking has been described as techno- or cyber-libertarian (Boorsook 2001), a brand of anti-statist, free-market, antiregulation individualism that revels in the hypercapitalist "sharing economy" of Uber, Airbnb, and eBay and the "creative destruction" of Silicon Valley start-ups. Unlike Black Lives Matter, technolibertarian interventions generally fail to ground themselves in the material, embodied experience of oppressed people existing in physical/networked community. It can be said that the focus is on "coding freedom" rather than "creating justice."

Current underexamined social movements present STS with opportunities for growth and critical reflection. How do Black Lives Matter to STS? There is also room for growth in studying scientific and technological social movements outside of the United States and Europe. The BRICS cable—a 21,000 mile, 12.8 terabyte per second fiber system connecting Brazil, Russia, India, China, South Africa, and Miami—is creating an alternative data pipeline to lower the cost of communication among major economies of the global south and provide non-U.S. routes for world communications. Brazil recently completed the Marco Civil da Internet, an "Internet Constitution," that connects digital communication to democratic values (Eubanks 2014a).

Accelerated scientific discovery and rapid technological innovation present significant challenges to the lives, livelihoods, and identities of people everywhere. At the same time, science and technology color new and enduring conflicts over nation, state, ethnicity, religion, and international relations. What further insights might STS scholars bring to bear on movements probing states' use of technology to surveil (Monahan 2006), incarcerate, police (Brucato 2015), and dispossess (Eubanks 2014a)?

Some movements take what Virginia Eubanks (2011) calls a "popular technology" approach that recognizes technology not just as a tool for mobilization but as a social justice issue in its own right. Technology touches every aspect of movement work from economic equity and jobs with justice (thus the concern from left movements and parties like Syriza in Greece and Podemos in Spain) to LGBTQ liberation and feminism, to immigrant and asylum seekers' rights, to military violence and criminal justice reform. This approach asks new questions in collaboration with progressive social movements: was the Arab Spring an unprecedented and networked series of "Facebook revolutions" or more traditional mobilizations using social media for communication and coordination (Breyman 2011)? What role do new technologies like body cameras, drones, and predictive algorithms play in police brutality and global militarization? How is the "new punitiveness" in social services linked to automated decision-making systems and administrative databases?

Scholarly recognition of the myriad entanglements of science, technology, and social movements in recent decades transformed the latter into a core idea of STS. As movements shaped STS ideas in the past, how might new movements generate new ideas in STS? We suggest that rich opportunities for research and civic participation await students of science and technology studies who investigate and participate in contemporary social movements. We invite our colleagues to study and engage new and emerging movements in innovative ways.

References

Allen, Barbara L. 2003. *Uneasy Alchemy: Citizens and Experts in Louisiana's Chemical Corridor Disputes*. Cambridge, MA: MIT Press.

Anderson, Chris. 2014. *Makers: The New Industrial Revolution*. New York: Crown Business.

Armstrong, Elizabeth A., and Mary Bernstein. 2008. "Culture, Power, and Institutions: A Multi-Institutional Politics Approach to Social Movements." *Sociological Theory* 26 (1): 74–99.

Ashton, John, and Howard Seymour. 1988. *The New Public Health: The Liverpool Experience*. Philadelphia: Open University Press.

Bailey, Brian. 1998. *The Luddite Rebellion*. Gloucestershire: Sutton Publishing.

Barkan, Elazar. 1992. *The Retreat of Scientific Racism*. Cambridge: Cambridge University Press.

Benford, Robert D., and David A. Snow. 2000. "Framing Processes and Social Movements: An Overview and Assessment." *Annual Review of Sociology* 26: 611–39.

Benjamin, Ruha. 2013. *People's Science: Reconstituting Bodies and Rights on the Stem Cell Frontier.* Palo Alto, CA: Stanford University Press.

Bleier, Ruth, ed. 1986. *Feminist Approaches to Science.* New York: Routledge.

Bliss, Catherine. 2012. *Race Decoded: The Genomic Fight for Social Justice.* Palo Alto, CA: Stanford University Press.

Boorsook, Paulina. 2001. *Cyberselfish: A Critical Romp through the Terribly Libertarian Culture of High-Tech.* New York: Public Affairs.

Boston Women's Health Book Collective. 1979. *Our Bodies, Ourselves.* Boston: Boston Women's Health Book Collective.

Brante, Thomas, Steve Fuller, and William Lynch, eds. 1993. *Controversial Science: From Content to Contention.* Albany: State University of New York Press.

Braun, Lundy. 2014. *Breathing Race into the Machine: The Surprising Career of the Spirometer from Plantation to Genetics.* Minneapolis: University of Minnesota Press.

Breyman, Steve. 1994. "Local Lore and Science: Toward a Sociology of Ecology Movement Knowledge." In *Green Politics Three*, edited by Wolfgang Ruedig, 184–215. Edinburgh: Edinburgh University Press.

___. 1997a. "Social Studies of Science & Activism: STS as a Campus Greening Movement." *Philosophy and Social Action: Philosophy, Science & Society* 23 (1): 11–22.

___. 1997b. "Were the 1980s' Anti-Nuclear Weapons Movements New Social Movements?" *Peace & Change: A Journal of Peace Research* 22 (3): 79–94.

___. 1997c. *Movement Genesis: Social Movement Theory and the West German Peace Movement.* Boulder, CO: Westview Press.

___. 2001. *Why Movements Matter: The West German Peace Movement and US Arms Control Policy.* Albany: State University of New York Press.

___. 2011. "The Tao of Media." *OurTown* 24: 26–31.

Briggs, Laura. 2003. *Reproducing Empire: Race, Sex, Science and U.S. Imperialism in Puerto Rico.* Berkeley: University of California Press.

Brown, Phil. 2007. *Toxic Exposures: Contested Illnesses and the Environmental Health Movement.* New York: Columbia University Press.

Brown, Phil, Rachel Morello-Frosch, and Stephen Zavestoski, eds. 2011. *Contested Illnesses: Ethnographic Explorations.* Berkeley: University of California Press.

Brown, Phil, Stephen Zavestoski, Sabrina McCormick, Brian Mayer, Rachel Morello-Frosch, and Rebecca Gasior Altman. 2004a. "Embodied Health Movements: Uncharted Territory in Social Movement Research." *Sociology of Health and Illness* 26 (6): 1–31.

Brown, Phil, Stephen Zavestoski, Sabrina McCormick, Brian Mayer, Rachel Morello-Frosch, and Rebecca Gasior Altman. 2004b. "Embodied Health Movements: New Approaches to Social Movements in Health." *Sociology of Health and Illness* 26 (1): 50–80.

Brown, Rebecca. 2010. *Gandhi's Spinning Wheel and the Making of India*. New York: Routledge.

Brucato, Ben. 2015. "Watching Police Violence: Negotiating the Politics of Visibility." Unpublished dissertation. Rensselaer Polytechnic Institute.

Bullard, Robert D. [1990] 2000. *Dumping in Dixie: Race, Class, and Environmental Quality*. Boulder, CO: Westview Press.

Callon, Michel, and Vololona Rabeharisoa. 2008. "The Growing Engagement of Emergent Concerned Groups in Political and Economic Life: Lessons from the French Association of Neuromuscular Disease Patients." *Science, Technology, & Human Values* 33 (2): 230–61.

Carson, Rachel. 1962. *Silent Spring*. New York: Houghton Mifflin.

Castells, Manuel. 2012. *Networks of Outrage and Hope: Social Movements in the Internet Age*. New York: Polity.

Checker, Melissa. 2007. "'But I Know It's True': Environmental Risk Assessment, Justice and Anthropology." *Human Organization* 66 (2): 112–24.

Chubin, Daryl, and Ellen Chu, eds. 1988. *Science Off the Pedestal: Social Perspectives on Science and Technology*. Belmont, CA: Wadsworth.

Clarke, Adele E., Janet Shim, Laura Mamo, Jennifer Fosket, and Jennifer Fishman. 2010. *Biomedicalization: Technoscience, Health and Illness in the U.S.* Durham, NC: Duke University Press.

Coleman, Gabriella. 2014. *Hacker, Hoaxer, Whistleblower, Spy: The Many Faces of Anonymous*. New York: Verso.

Colen, Shellee. 1995. "'Like a Mother to Them': West Indian Childcare Workers and Employers in New York." In *Conceiving the New World Order*, edited by Faye Ginsburg and Rayna Rapp, 78–102. Berkeley: University of California Press.

Combahee River Collective. 1977. "The Combahee River Collective Statement." A widely circulated pamphlet, later published in Zillah Eisenstein (1978), *Capitalist Patriarchy and the Case for Socialist Feminism*. New York: Monthly Review Press.

Coniff, Richard. 2011. "What the Luddites Really Fought Against." *Smithsonian Magazine* (March).

Corburn, Jason. 2005. *Street Science: Community Knowledge and Environmental Health Justice*. Cambridge, MA: MIT Press.

Costanza-Chock, Sasha. 2014. *Out of the Shadows, in the Streets! Transmedia Organizing and the Immigrant Rights Movement.* Cambridge, MA: MIT Press.

Crenshaw, Kimberle. 1991. "Mapping the Margins: Intersectionality, Identity Politics, and Violence against Women of Color." *Stanford Law Review* 43 (6): 1241–99.

Daniels, Cynthia. 1993. *At Women's Expense: State Power and the Politics of Fetal Rights.* Cambridge, MA: Harvard University Press.

Davis, Kathy. 2007. *The Making of Our Bodies, Ourselves: How Feminism Travels across Borders.* Durham, NC: Duke University Press.

Di Chiro, Giovanna. 2008. "Living Environmentalisms: Coalition Politics, Social Reproduction, and Environmental Justice." *Environmental Politics* 17 (2): 276–98.

Dumit, Joseph. 2012. *Drugs for Life: How Pharmaceutical Companies Define Our Health.* Durham, NC: Duke University Press.

Duster, Troy. 1990. *Backdoor to Eugenics.* New York: Routledge.

Efron, John M. 1994. *Defenders of the Race: Jewish Doctors and Race Science in Fin-de-Siècle Europe.* New Haven, CT: Yale University Press.

Epstein, Steven. 1996. *Impure Science: AIDS, Activism, and the Politics of Knowledge.* Berkeley: University of California Press.

___. 2007. *Inclusion: The Politics of Difference in Medical Research.* Chicago: University of Chicago Press.

Eubanks, Virginia. 2011. *Digital Dead End: Fighting for Social Justice in an Information Age.* Cambridge, MA: MIT Press.

___. 2014a. "Want to Predict the Future of Surveillance? Ask Poor Communities." *The American Prospect* online, January 15, 2014, Accessed April 1, 2015, at http://prospect.org/article/want-predict-future-surveillance-ask-poor-communities.

___. 2014b. "How Big Data Could Undo Our Civil Rights Laws." *The American Prospect* online, accessed April 1, 2015, at http://prospect.org/article/how-big-data-could-undo-our-civil-rights-laws.

Eyerman, Ron, and Andrew Jamison. 1991. *Social Movements: A Cognitive Approach.* Cambridge: Polity Press.

Fanon, Frantz. [1952] 1967. *Black Skin, White Masks.* New York: Grove Press.

___. [1961] 1963. *The Wretched of the Earth.* New York: Grove Press.

Fausto-Sterling, Anne. 1990. *Myths of Gender: Biological Theories about Women and Men.* New York: Basic Books.

Fernández Kelly, Patricia. 1987. "Technology and Employment along the US-Mexico Border." In *The United States and Mexico: Face to Face with New Technology*, edited by Cathryn Thorup, 152–68. New Brunswick, NJ: Transaction Books.

Figueroa, Stephen, and Sandra G. Harding, eds. 2003. *Science and Other Cultures: Issues in Philosophies of Science and Technology*. New York: Routledge.

Fischer, Frank. 2000. *Citizens, Experts, and the Environment: The Politics of Local Knowledge*. Durham, NC: Duke University Press.

Fisher, Erik, Cynthia Selin, and Jameson M. Wetmore, 2008. *Yearbook of Nanotechnology in Society*. vol. 1, *Presenting Futures*. New York: Springer.

Fisher, Jill A. 2009. *Medical Research for Hire: The Political Economy of Pharmaceutical Trials*. New Brunswick, NJ: Rutgers University Press.

Fligstein, Neil, and Doug McAdam. 2011. "Toward a General Theory of Strategic Action Fields." *Sociological Theory* 29 (1): 1–26.

Fouché, Rayvon. 2003. *Black Inventors in the Age of Segregation: Granville T. Woods, Lewis H. Latimer, and Shelby J. Davidson*. Baltimore: Johns Hopkins University Press.

Frickel, Scott. 2004. *Chemical Consequences: Environmental Mutagens, Scientist Activism, and the Rise of Genetic Toxicology*. New Brunswick, NJ: Rutgers University Press.

Frickel, Scott, and Neil Gross. 2005. "A General Theory of Scientific/Intellectual Movements." *American Sociological Review* 70 (2): 204–32.

Fuller, Steve. 1992. "STS as Social Movement: On the Purpose of Graduate Programs." *Science, Technology & Society* 91: 1–5.

___. 1993. *Philosophy, Rhetoric, and the End of Knowledge: The Coming of Science and Technology Studies*. Madison: University of Wisconsin Press.

Garvey, Colin. 2014. "Tactical Evolution of an Intellectual Insurgency: Science for the People and the AAAS, 1969–78." Department of Science and Technology Studies. Rensselaer Polytechnic Institute. Unpublished manuscript.

Garza, Alicia. 2014. "A Herstory of the #BlackLivesMatter Movement." *The Feminist Wire*, accessed April 18, 2015, at http://www.thefeministwire.com/2014/10/blacklivesmatter-2/.

Gerbaudo, Paolo. 2012. *Tweets and the Streets: Social Media and Contemporary Activism*. London: Pluto Press.

Ghosh, Pradip K., and Denton Morrison, eds. 1984. *Appropriate Technology in Third World Development*. Westport, CT: Greenwood Press.

Gibbs, Lois. 2002. "Citizen Activism for Environmental Health: The Growth of a Powerful New Grassroots Health Movement." *Annals of the American Academy of Political & Social Science* 584 (1): 97–109.

Goonatilake, Susantha. 1999. "A Post-European Century in Science." *Futures* 31 (9): 923–27.

Gould, Stephen J. 1981. *The Mismeasure of Man*. New York: W. W. Norton.

Gray, Mary. 2009. *Out in the Country: Youth, Media, and Queer Visibility in Rural America*. New York: New York University Press.

Gurr, Ted Robert. 1970. *Why Men Rebel*. Princeton, NJ: Princeton University Press.

Haraway, Donna J. 1991. "The Cyborg Manifesto: Science, Technology, and Socialist Feminism in the Late Twentieth Century." In *Simians, Cyborgs and Women: The Reinvention of Nature*, edited by Donna J. Haraway, 149–81. New York: Routledge.

Harding, Sandra. 1986. *The Science Question in Feminism*. Ithaca, NY: Cornell University Press.

___. 1991. *Whose Science? Whose Knowledge? Thinking from Women's Lives*. Ithaca, NY: Cornell University Press.

___, ed. 1993. *The "Racial" Economy of Science: Toward a Democratic Future*. Bloomington: Indiana University Press.

___. 1998. *Is Science Multicultural? Postcolonialisms, Feminisms, and Epistemologies*. Bloomington: Indiana University Press.

___. 2006. *Science and Social Inequality: Feminist and Postcolonial Issues*. Urbana: University of Illinois Press.

___, ed. 2011. *The Postcolonial Science and Technology Studies Reader*. Durham, NC: Duke University Press.

___. 2015. *Objectivity & Diversity: Another Logic of Scientific Research*. Chicago: University of Chicago Press.

Harris, Cheryl I. 1993. "Whiteness as Property." *Harvard Law Review* 106 (8): 1707–91.

Hartigan, John Jr. 1999. *Racial Situations: Class Predicaments of Whiteness in Detroit*. Princeton, NJ: Princeton University Press.

Hayden, Cori. 2003. *When Nature Goes Public: The Making and Unmaking of Bioprospecting in Mexico*. Princeton, NJ: Princeton University Press.

Hecht, Gabrielle. 2012. *Being Nuclear: Africans and the Global Uranium Trade*. Cambridge, MA: MIT Press.

Hess, David J. 1997. *Science Studies: An Advanced Introduction*. New York: New York University Press.

___. 2005. "Technology and Product-Oriented Movements: Approximating Social Movement Studies and STS." *Science, Technology, & Human Values* 30 (4): 515–35.

___. 2007. *Alternative Pathways in Science and Technology: Activism, Innovation, and the Environment in an Era of Globalization*. Cambridge, MA: MIT Press.

___. 2010. "Environmental Reform Organizations and Undone Science in the United States: Exploring the Environmental, Health, and Safety Implications of Nanotechnology." *Science as Culture* 19 (2): 181–214.

___. 2015. "Undone Science and Social Movements: A Review and Typology." In *The International Handbook of Ignorance*, edited by Matthias Gross, 141–54. New York: Routledge.

Hilgartner, Stephen, Clark Miller, and Rob Hagendijk, eds. 2015. *Science and Democracy: Making Knowledge and Making Power in the Biosciences and Beyond.* New York: Routledge.

Hoffer, Eric. 1951. *True Believers: Thoughts on the Nature of Mass Movements.* New York: Harper & Brothers.

Hubbard, Ruth. 1990. *The Politics of Women's Biology.* New Brunswick, NJ: Rutgers University Press.

Ilerbaig, Juan. 1992. "The Two STS Subcultures and the Sociological Revolution." *Science, Technology & Society* 90: 1–6.

Irani, Lilly, Janet Vertesi, Paul Dourish, Kavita Philip, and Rebecca E. Grinter. 2010. "Postcolonial Computing: A Lens of Design and Development." *CHI 2010*, April 10–15. Atlanta, GA.

Jamison, Andrew. 2003. "The Making of Green Knowledge: The Contribution from Activism." *Futures* 35 (7): 703–16.

Jasanoff, Sheila. 1996. "Beyond Epistemology: Relativism and Engagement in the Politics of Science." *Social Studies of Science* 26 (2): 393–418.

Keck, Margaret E., and Kathryn Sikkink. 1998. *Activists beyond Borders: Advocacy Networks in International Politics.* Ithaca, NY: Cornell University Press.

Keller, Evelyn Fox. 1978. "Gender and Science." *Psychoanalysis & Contemporary Thought* 1: 409–33.

___. 1985. *Reflections on Gender and Science.* New Haven, CT: Yale University Press.

Kelty, Chris. 2005. "Geeks, Social Imaginaries, and Recursive Publics." *Cultural Anthropology* 20 (2): 185–214

___. 2008. *Two Bits: The Cultural Significance of Free Software.* Durham, NC: Duke University Press.

Kinchy, Abby J. 2009. "African Americans in the Atomic Age: Post-War Perspectives on Science, Technology and the Bomb, 1945–1960." *Technology & Culture* 50 (2): 291–315.

___. 2012. *Seeds, Science, and Struggle: The Global Politics of Transgenic Crops.* Cambridge, MA: MIT Press.

Klawiter, Maren. 2008. *The Biopolitics of Breast Cancer: Changing Cultures of Disease and Activism.* Minneapolis: University of Minnesota Press.

Krieger, Nancy. 2011. *Epidemiology and the People's Health.* Oxford: Oxford University Press.

Laraña, Enrique, Hank Johnston, and Joseph R. Gusfield, eds. 1994. *New Social Movements: From Ideology to Identity*. Philadelphia: Temple University Press.

Latour, Bruno. 1993. "Acceptance." *Science, Technology, & Human Values* 18 (3): 384–88.

Leach, Melissa. 2007. "Earth Mother Myths and Other Ecofeminist Fables: How a Strategic Notion Rose and Fell." *Development and Change* 38 (1): 67–85.

Leach, Melissa, Ian Scoones, and Brian Wynne. 2005. *Science and Citizens: Globalization and the Challenge of Engagement*. London: Zed Books.

Ley, Barbara. 2009. *From Pink to Green: Disease Prevention and the Environmental Breast Cancer Movement*. New Brunswick, NJ: Rutgers University Press.

Longino, Helen. 1987. "Can There Be a Feminist Science?" *Hypatia* 2 (3): 51–64.

Mamo, Laura. 2007. *Queering Reproduction: Achieving Pregnancy in the Age of Technoscience*. Durham, NC: Duke University Press.

Marques, Ivan da Costa. 2005. "Cloning Computers: From Rights of Possession to Rights of Creation." *Science as Culture* 14 (2): 139–60.

Martin, Brian. 1979. *The Bias of Science*. Canberra: Society for Social Responsibility in Science.

___. 1993. "The Critique of Science Becomes Academic." *Science, Technology, & Human Values* 18 (2): 247–59.

___, ed. 1996. *Confronting the Experts*. Albany: State University of New York Press.

Martin, Emily. 1987. *The Woman in the Body*. Baltimore: Johns Hopkins University Press.

___. 1991. "The Egg and the Sperm: How Science Has Constructed a Romance Based on Stereotypical Male-Female Roles." *Signs* 16 (3): 485–501.

McAdam, Doug. 1982. *Political Process and the Development of Black Insurgency, 1930–70*. Chicago: University of Chicago Press.

McAdam, Doug, John D. McCarthy, and Mayer N. Zald. 1996. *Comparative Perspectives on Social Movements: Political Opportunities, Mobilizing Structures, and Cultural Framings*. Cambridge: Cambridge University Press.

McCarthy, John D., and Mayer N. Zald. 1977. "Resource Mobilization and Social Movements: A Partial Theory." *American Journal of Sociology* 82 (6): 1212–41.

McCormick, Sabrina. 2009. *Mobilizing Science: Movements, Participation, and the Remaking of Knowledge*. Philadelphia: Temple University Press.

Medina, Eden. 2011. *Cybernetic Revolutionaries: Technology and Politics on Allende's Chile*. Cambridge, MA: MIT Press.

Medina, Eden, Ivan da Costa Marques, and Christina Holmes. 2014. *Beyond Imported Magic: Essays on Science, Technology and Society in Latin America*. Cambridge, MA: MIT Press.

Melucci, Alberto. 1980. "The New Social Movements: A Theoretical Approach." *Social Science Information* 19 (2): 199–226.

___. 1989. *Nomads of the Present: Social Movements and Individual Needs in Contemporary Society.* Philadelphia: Temple University Press.

Mills, C. Wright. 1956. *The Power Elite.* Oxford: Oxford University Press.

Monahan, Torin. 2006. *Surveillance and Security: Technological Politics and Power in Everyday Life.* New York: Routledge.

Montoya, Michael. 2011. *Making the Mexican Diabetic: Race, Science, and the Genetics of Inequality.* Berkeley: University of California Press.

Moore, Kelly. 2008. *Disrupting Science: Social Movements, American Scientists, and the Politics of the Military, 1945–1975.* Princeton, NJ: Princeton University Press.

Murphy, Michelle. 2004. "Uncertain Exposures and the Privilege of Imperception: Activist Scientists and Race at the US Environmental Protection Agency." *Osiris* 19: 266–82.

___. 2006. *Sick Building Syndrome and the Problem of Uncertainty.* Durham, NC: Duke University Press.

___. 2012. *Seizing the Means of Reproduction: Entanglements of Feminism, Health, and Technoscience.* Durham, NC: Duke University Press.

Nelkin, Dorothy, ed. 1992. *Controversy: Politics of Technical Decisions.* 3rd ed. Newbury Park, CA: Sage.

Nelson, Alondra. 2011. *Body and Soul: The Black Panther Party and the Fight Against Medical Discrimination.* Minneapolis: University of Minnesota Press.

Omi, Michael, and Howard Winant. 2015. *Racial Formation in the United States.* 3rd ed. New York: Routledge.

Ottinger, Gwen. 2009. "Buckets of Resistance: Standards and the Effectiveness of Citizen Science." *Science, Technology, & Human Values* 35 (2): 244–70.

___. 2013. "Changing Knowledge, Local Knowledge, and Knowledge Gaps: STS Insights into Procedural Justice." *Science, Technology, & Human Values* 38 (2): 250–70.

Pels, Dick. 1996. "The Politics of Symmetry." *Social Studies of Science* 26 (2): 277–304.

Petersen, James C., and Gerald E. Markle. 1981."Expansion of Conflict in Cancer Controversies." *Research in Social Movements, Conflict and Change* 4: 151–69.

Petryna, Adriana. 2009. *When Experiments Travel: Clinical Trials and the Global Search for Human Subjects.* Princeton, NJ: Princeton University Press.

Petryna, Adriana, Andrew Lakoff, and Arthur Kleinman. 2006. *Global Pharmaceuticals: Ethics, Markets, Practices.* Durham, NC: Duke University Press.

Philip, Kavita. 2005. "What Is a Technological Author? The Pirate Function and Intellectual Property." *Postcolonial Studies* 8 (2): 199–218.

Pickard, Victor W. 2006. "Assessing the Radical Democracy of Indymedia: Discursive, Technical, and Institutional Constructions." *Critical Studies in Media Communication* 23 (1): 19–38.

Price, Derek J. de Solla. 1963. *Little Science, Big Science.* New York: Columbia University Press.

Proctor, Robert N., and Londa Schiebinger, eds. 2008. *Agnotology: The Making and Unmaking of Ignorance.* Palo Alto, CA; Stanford University Press.

Rapp, Rayna. 2011. "Chasing Science: Children's Brains, Scientific Inquiries, and Family Labors." *Science, Technology, & Human Values* 36 (5): 662–84.

Reardon, Jenny. 2005. *Race to the Finish: Identity and Governance in an Age of Genomics.* Princeton, NJ: Princeton University.

Reardon, Jenny, Emma Kowal, and Joanna Radin. 2013. "Indigenous Body Parts, Mutating Temporalities, and the Half-Lives of Postcolonial Technoscience." *Social Studies of Science* 43 (4): 465–83.

Reardon, Jenny, and Kim TallBear. 2012. "Your DNA Is Our History: Genomics, Anthropology, and the Construction of Whiteness as Property." *Current Anthropology* 53 (S5): S233–45.

Redfield, Peter. 2013. *Life in Crisis: The Ethical Journey of Doctors Without Borders.* Berkeley: University of California Press.

Rhodes, Tim. 2002. "The 'Risk Environment': A Framework for Understanding and Reducing Drug-Related Harm." *International Journal of Drug Policy* 13 (2): 85–94.

Richards, Evelleen. 1991. *Vitamin C and Cancer: Medicine or Politics?* London: Macmillan.

Roberts, Dorothy. 2012. *Fatal Invention: How Science, Politics, and Big Business Re-create Race in the 21st Century.* New York: New Press.

Rojas, Fabio. 2007. *From Black Power to Black Studies.* Baltimore: Johns Hopkins University Press.

Rose, Hilary. 1983. "Hand, Brain, and Heart: A Feminist Epistemology for the Natural Sciences." *Signs: Journal of Women in Culture and Society* 9 (1): 73–90.

___. 1994. *Love, Power, and Knowledge: Towards a Feminist Transformation of the Sciences.* Bloomington: Indiana University Press.

Rose, Hilary, and Steven Rose. 1969. *Science and Society.* London: Penguin Press.

Rosser, Sue. 1990. *Female-Friendly Science.* Elmsford, NY: Pergamon Press.

Rothman, Franklin Daniel, and Pamela E. Oliver. 1999. "From Local to Global: The Anti-Dam Movement in Southern Brazil, 1979–1992." *Mobilization* 4 (1): 41–57.

Said, Edward. 1978. *Orientalism*. London: Routledge and Kegan Paul.

Sale, Kirkpatrick. 1996. *Rebels against the Future: The Luddites and Their War on the Industrial Revolution: Lessons for the Computer Age*. New York: Basic Books.

Sarewitz, Daniel. 1996. *Frontiers of Illusion: Science, Technology and the Politics of Progress*. Philadelphia: Temple University Press.

Schurman, Rachel, and William Munro. 2006. "Ideas, Thinkers, and Social Networks: The Process of Grievance Construction in the Anti-Genetic Engineering Movement." *Theory and Society* 35 (1): 1–38.

Sclove, Richard. 1995. *Democracy and Technology*. New York: Guilford Press.

Scoones, Ian. 2008. "Mobilizing against GM Crops in India, South Africa and Brazil." *Journal of Agrarian Change* 8 (2–3): 315–44.

Seidman, Gay. 2008. "Transnational Labour Campaigns: Can the Logic of the Market Be Turned against Itself?" *Development and Change* 39 (6): 991–1003.

Shiva, Vandana, 1992. *The Violence of the Green Revolution: Ecological Degradation and Political Conflict in Punjab*. New Delhi: Zed Press.

___. 1997. *Biopiracy: The Plunder of Nature and Knowledge*. Boston: South End Press.

Smith, Andrea. 2008. "American Studies without America: Native Feminisms and the Nation-State." *American Quarterly* 60 (2): 309–15.

Smith, Dorothy E. 1987. *The Everyday World as Problematic: A Feminist Sociology*. Boston: University Press of New England.

___. 1990. *The Conceptual Practices of Power: A Feminist Sociology of Knowledge*. Toronto: University of Toronto Press.

Sokal, Alan. 1996. "Transgressing the Boundaries: Towards a Transformative Hermeneutics of Quantum Gravity." *Social Text* (46–47): 217–52.

Tarrow, Sidney. 2005. *The New Transnational Activism*. Cambridge: Cambridge University Press.

Tesh, Sylvia Noble. 1988. *Hidden Arguments: Political Ideology and Disease Prevention Policy*. Camden, NJ: Rutgers University Press.

___. 2000. *Uncertain Hazards: Environmental Activists and Scientific Proof*. Ithaca, NY: Cornell University Press.

Tilly, Charles. 1978. *From Mobilization to Revolution*. New York: McGraw-Hill.

Touraine, Alain. 1981. *The Voice and the Eye: An Analysis of Social Movements*. Cambridge: Cambridge University Press.

Tuana, Nancy. 2006. "The Speculum of Ignorance: The Women's Health Movement and Epistemologies of Ignorance." *Hypatia* 21 (3): 1–19.

Turner, Fred. 2008. *From Counterculture to Cyberculture: Stewart Brand, the Whole Earth Network, and the Rise of Digital Utopianism.* Chicago: University of Chicago Press.

Ullman, Joseph C. 1969. "Helping Workers Locate Jobs Following a Plant Shutdown." *Monthly Labor Review* 92: 35–40.

Visvanathan, Shiv. 1997. *A Carnival for Science: Essays on Science, Technology, and Development.* Oxford: Oxford University Press.

Waidzunas, Tom. 2013. "Intellectual Opportunity Structures and Science-Targeted Activism: Influence of the Ex-Gay Movement on the Science of Sexual Orientation." *Mobilization* 18 (1): 1–18.

Wajcman, Judy. 1991. *Feminism Confronts Technology.* State College: Pennsylvania State University Press.

___. 2004. *Technofeminism.* Cambridge: Polity Press.

Waks, Leonard J. 1993. "STS as an Academic Field and a Social Movement." *Technology and Society* 15 (4): 399–408.

Walker, Edward T., Andrew W. Martin, and John D. McCarthy. 2008. "Confronting the State, the Corporation, and the Academy: The Influence of Institutional Targets on Social Movement Repertoires." *American Journal of Sociology* 114 (1): 35–76.

Winner, Langdon. 1978. *Autonomous Technology: Technics Out of Control as a Theme in Political Thought.* Cambridge, MA: MIT Press.

___. 1993. "Upon Opening the Black Box and Finding It Empty: Social Constructivism and the Philosophy of Technology." *Science, Technology, & Human Values* 13 (3): 362–78.

Wolfson, Todd. 2014. *Digital Rebellion: The Birth of the Cyber Left.* Urbana: University of Illinois Press.

Worthington, Richard, Darlene Cavalier, Mahmud Farooque, Gretchen Gano, Henry Geddes, Steven Sander, David Sittenfeld, and David Tomblin. 2012. "Technology Assessment and Public Participation: From TA to pTA." Report for Expert and Citizen Assessment of Science and Technology (ECAST).

Wylie, Alison, Elizabeth Potter, and Wenda K. Bauchspies. 2015. "Feminist Perspectives on Science." In *The Stanford Encyclopedia of Philosophy*, edited by Edward N. Zalta. (Archived Summer 2015 at http://plato.stanford.edu/cgi-bin/encyclopedia/archinfo.cgi?entry=feminist-science.)

Wynne, Brian. 1992. "Misunderstood Misunderstanding: Social Identities and Public Uptake of Science." *Public Understanding of Science* 1 (3): 281–304.

Yearley, Steven. 1989. "Environmentalism: Science and a Social Movement." *Social Studies of Science* 19 (2): 343–55.

___. 1995. "The Environmental Challenge to Science Studies." *Handbook of Science and Technology Studies*, edited by Sheila Jasanoff, Gerald E. Markle, James C. Petersen, and Trevor Pinch, 457–79. Thousand Oaks, CA: Sage.

Zuckerman, Ethan. 2014. *Digital Cosmopolitans: Why We Think the Internet Connects Us, Why It Doesn't, and How to Rewire It.* New York: W. W. Norton.

11   Structural Inequality and the Politics of Science and Technology

David J. Hess, Sulfikar Amir, Scott Frickel, Daniel Lee Kleinman, Kelly Moore, and Logan D. A. Williams

Although the politics of science and technology is a topic of long-standing interest in science and technology studies (STS), the ways in which STS scholars have conceptualized and studied politics have changed considerably over the past half century. During the 1960s and 1970s researchers frequently focused on Cold War–era geopolitics, and they also reacted to the growing vulnerability of scientific research to the influence of powerful state and market actors. Various essays in the first edition of the STS handbook (Spiegel-Rösing and Price 1977) reflected these concerns and approached the problem of politics largely as a matter of policy and government. During the 1980s and 1990s—when studies of laboratories, networks, and controversies were especially prominent—a different set of ideas about the politics of science and technology gained traction. Constructivist research tended to frame the problem of politics in epistemic terms, and it focused on conflicts over knowledge claims and technological designs. To the extent that issues of power and politics were addressed, they were studied mostly at the microscale of local settings and interactions among networked experts.

"Structural" accounts of the politics of science and technology attracted far less attention during this period (with some exceptions, e.g., Mukerji 1989), but they have gained greater salience since 2000. By "structural" we mean approaches that focus on the "durable inequalities" (Tilly 1999) of power and resources as features of scientific and social life that can be challenged and rechanneled but are not reducible to the outcomes of microsocial processes. This approach to STS is nourished by scholars' growing awareness of global inequality associated with neoliberalization, which we define as the political shift toward a reduced role for government regulation and redistribution, often coincident with the liberalization of global trade and finance. This more recent body of STS research investigates how science and technology contribute to or enable resistance to enduring inequalities structured around race, gender, and global and class position, especially as these are associated with industrial and state power.

This chapter reviews the different streams of STS research that investigate the relationship among science, technology, and structural inequalities. We consider some of the ways these different pathways converge and diverge, and we focus on a growing body of recent research distinguished by its integration of political sociological theory and related methods.[1] This chapter reviews three aspects of this important and enduring topic of STS research: structural inequality and the politics of science and technology as they were developed in previous generations of STS research; work since 2000 illustrating the reemergence of attention to this problem area; and current STS research that advances a new agenda for investigating the causes, consequences, and possible interventions that challenge existing structures of enduring inequality.

There are two important clarifications for our approach. First, the analysis of structural inequality does not imply advocacy of structural determinism. To the contrary, attention to structural inequality provides the basis for building a relatively balanced and interactive approach to the structure-agency relationship. We see a need for a deeper appreciation of the ways in which the politics of fact construction and of technological design both shape and are shaped by the more enduring structures of local, national, and global inequality. Second, although the STS field has become increasingly internationalized, the study of race, class, gender, and colonial position has not been evenly distributed. For example, several of the most prominent North American STS researchers have focused on gender, sexuality, and to a lesser degree race (e.g., Haraway 1991; Harding 1998), and issues of colonialism and dependent development often appear in work from less developed countries (Medina, Marques, and Holmes 2014; Quet and Noel 2014). Thus, the literature is potentially quite large; our focus here is on research and concepts that have significantly advanced the STS literature on the politics of structural inequality. Because it is impossible to cover the history of the entire STS field, in the first section we rely heavily on previous editions of STS handbooks as a way to sample and characterize changes in the STS literature.

Structural Inequality in the STS Literature

Structural Inequality as Gender and Race Positions

In the 1970s and early 1980s, there was a paucity of STS scholarship on the politics of gender, ethnicity, and race as reflected in the first edition of the STS handbook (Spiegel-Rösing and Price 1977). Indeed, such analyses were only being developed by a small community of scholars who were turning attention away from the study of women's science careers to the gendered assumptions about and content of scientific knowledge itself. Importantly, many of these writers had connections with the health, feminist, civil rights, and antiwar movements of the 1960s and 1970s, in which scientists themselves were deeply involved (Hubbard 1990; Moore 2008).

Gender received more attention in the second edition of the STS handbook than in the first, and discussions reflected two major concerns about gender and science: women's exclusions from the centers of knowledge production and the gendered assumptions built into scientific knowledge. Keller (1995) drew on earlier and contemporaneous work such as Haraway (1991), Harding (1991), Martin (1991), and Tuana (1989)—whose foundational scholarship rejected dichotomous scientific thinking about gender and race—and illuminated the way that *gender ideology* shapes ideas about whose knowledge matters and how gender and sex are constituted by science. Like Keller, Wajcman's (1995) chapter on feminist technology studies showed that gendered interests shaped the way technologies are made and used (see Layne, Vostral, and Boyer 2010 for subsequent developments). Less interested in origins and impacts than in processes, these early approaches sharpened the focus on how political assumptions are built into knowledge and material design.

In the 2008 edition of the STS handbook, feminist science studies was represented in work that explored in greater detail the role of the taken-for-granted in shaping knowledge. Suchman (2008) advanced the debate about the origins and functions of dichotomies on which science rests—such as human/animal, culture/nature, and male/female—and she noted that the first referent is assumed to be most important and the second subordinate. Consistent with arguments by Star (1989) and Haraway (1991), Suchman drew attention to "the politics of ordering" (2008, 140). She also engaged questions taken up by scholars such as Schiebinger (1993) about how the absence of women and Africans from science shaped the content of knowledge. By showing that somatic position was an important part of scientific knowledge-making, Suchman turned on its head the old dichotomy between body/woman and reason/man. Other scholarship in the 2008 edition raised questions about who was missing in studies of scientific knowledge production and use. Spearheaded by Clarke and Star (2008), "social worlds" scholarship illuminated the importance of considering actors, including women, who were often hidden but implicated in the development and use of new sciences and technologies. In doing so, Clarke (1990) presaged later analyses of who is left out of debates and decisions about sociotechnical issues, and she offered causal analysis to explain absences.

The relationship between racism and science had almost no presence in these earlier versions of the STS handbook. Indeed race had not yet appeared as a focal area of research in STS, although there were scholarly analyses of science and racism being published in the 1980s and 1990s. For example, Duster (1990) had been writing about race and medical interventions since the 1980s, and Pearson (1985) analyzed race and careers in science. These studies complemented analyses by biological scientists and anthropologists, who showed how racial logics were built into scientific research (Segerstråle 2000; Teslow 2014). In an analysis of apartheid in South Africa, Bowker and Star (2000) argued that systems of classification impose a hierarchical social order that requires an intense regulatory apparatus to create and maintain. But the dearth of structural analyses of racism during this earlier period is undoubtedly related to another kind of absence: the underrepresentation of people of color in academic positions generally and in STS positions more specifically.

Structural Inequality as Global Position

In the 1970s science and technology entered the discourse of international development after political leaders realized that scientific and technical superiority affected political power among nations (Schroeder-Gudehus 1977). This concern prompted an analysis in the first edition of the STS handbook of the impact of science and technology on the international system and its political dynamics (Skolnikoff 1977). Sardar and Rosser-Owen (1977) then advanced a critical analysis of development to argue that power and structural inequality emerged from the Western-biased processes in developmental policies and programs. Approaches to development covered a broad spectrum of theories from modernization (Rostow 1960) and dependency (Cardoso and Faletto 1979) to appropriate technology (Eckaus 1977).

Divergent perspectives emerged from the late 1980s onwards, and the focus shifted to more pragmatic concerns. Shrum and Shenhav (1995) observed a diversity of scholarly approaches and practices aiming to solve perennial problems that hinder the developing world from achieving socioeconomic progress through scientific and technological change (Cozzens et al. 2008). Development practices also received accumulated criticisms due to disappointing outcomes of modernization projects in the global south, and new approaches emerged in response (Escobar 1995; Evans 1995). STS researchers located in the global south also contributed analyses of the relationships among science, universities, and lagging economic development in the region (Vessuri 1988).

By the late 1990s global studies of science and technology were developing complementary perspectives to the critical literature on science, technology, and development. The growth of this research field coincided with interest in the rise of the networked global society (Castells 1996) and in the impact of globalization on the multipolarities of modernity (Appadurai 1996). Modernity was no longer seen as a unifying process of cultural and sociotechnical transformation; instead, researchers unpacked the historical and geographical complexity of the movement of technoscience from one place to another (Anderson 2002; Feenberg 1995). This trend was reinforced by recognition of the "situatedness" of knowledge (Haraway 1991) and the relationship between diversity and strong objectivity (Harding 1998, 2015). Some of the work examined the potentially liberating role of participatory science (Nelson and Wright 1995), but researchers also noted how colonial and postcolonial science promoted Western hegemony while also providing opportunities for contestation (Prakash 1999). Visvanathan (1997) drew on Indian independence and nationalist movements to develop a critical analysis of Western science.

Taken together, these earlier approaches contributed a great deal to understanding how politics and science are tightly intertwined in the globalization of technoscience and the structuring of a hierarchical world system. However, many of the studies lacked a more integrated framework that combined attention to the role of systems of race, gender, and economic inequality in reproducing and altering structural inequality.

Structural Inequality as Position with Respect to State and Industry

With some exceptions, such as work that brings together class and gender (Cockburn 1981) or work on labor and technology (e.g., Noble 1977), the concept of "class" is rarely used as a central analytical framework in the STS literature, and even notions of business and political elites have not always found favor in STS scholarship. In the sociology of scientific knowledge, methodological questions were raised about attempts to utilize class interests as a factor that could partially explain technical positions in scientific controversies (MacKenzie 1981). But given Cozzens and Woodhouse's (1995, 535) assertion that business is perhaps the most powerful actor in the "current structure of influence around science," there was surprisingly limited attention to the intersection of industry, the state, and science among STS scholars during this period.

Previous generations did show some concern with the related problem of technocracy, that is, the domination of political processes by scientific and technical experts. For example, Lakoff (1977) concluded that although there was substantial evidence that scientists influenced policy outcomes, politics as we know it was not giving way to rule by experts. Winner (1986) noted particular circumstances and types of policy arenas where technocratic decision making was especially prominent, and he noted that reliance on risk assessment tended to restrict decision-making criteria rather than to open it up to democratic deliberation.

Arguing against control by technical elites, Nelkin (1977) suggested that in addition to the "reactive controls" that regulatory agencies and the courts offer, both "participatory controls" (such as citizen group involvement and public interest science) and "anticipatory controls" (such as technology assessment) could strengthen the potential for a more democratic politics of the governance of technology. However, Nelkin also recognized the potential for such mechanisms of democratic deliberation to be coopted and channeled in directions that do not pose significant obstructions to technological progress. This concern has been borne out by a subsequent generation of STS research that has shown the potential for public consultation exercises to fail to embody public input and instead to legitimate or at best to modify decisions that have already been made by elites (e.g., Irwin 2008). To address these concerns, Nelkin suggested the need to design mechanisms of participation and anticipation that are more decentralized, adversarial, and autonomous. Increasingly, STS research on publics, democracy, and participation has turned to the role of mobilized publics such as social movements as an important ingredient in maintaining the integrity of public participation (Hess 2011; Thorpe 2007).

In addition to the emerging critique of state power and the governance of science and technology, scholars interested in industrial power also developed research on the influence of industry on academic science (Croissant and Smith-Doerr 2008). Some pointed to a threat to the goals of openness and information sharing posed by the spread of relations between academic science and industry; these changes are associated especially with the emergence of the biotechnology industry in the late 1970s and 1980s (e.g., Kenney 1985). Implicit in this scholarship is the claim that business has overwhelming power to restructure academic science to make it more oriented toward the innovation economy. In subsequent work, the analysis of the commercialization (and thus business domination) of science became clearer. For example, Kleinman (1998, 2003) and Mirowski and Sent (2008) argued that the microscale of most STS work during the 1980s and 1990s led to the failure to consider fully and to understand the commercialization of science.

Three strands of work address the entanglement of academic science and business: the triple helix model (Etzkowitz and Leydesdorff 2000), the mode 1/mode 2 debate (Gibbons, Limoges, and Nowotny 1994), and research on academic capitalism (Slaughter and Rhoades 2004). Effectively ignoring the dangers of business domination, triple-helix and mode 2 theorists advocate understanding the transformation of late twentieth-century science in largely optimistic terms. In contrast, those who work on academic capitalism tend to stress structural inequality in the form of increasing power of the market and businesses over academic science and education.

Political Sociological Perspectives in STS and Structural Inequality

Political Sociology of Science and Technology

The political sociology of science and technology is an important area of the interdisciplinary STS field that has emerged to examine systematically the issue of structural inequality. In *Toward a Political Sociology of Science*, Blume (1974) grappled with the increasingly visible and troubling political uses and implications of scientific research in contemporary society, an area ignored at the time by mainstream sociology of science. These sociologists asserted that science was a meritocratic realm in which power and politics played little or no part, and consequently they treated science as largely autonomous from other social institutions (Cole and Cole 1973; Merton 1973). In contrast to understanding science as a largely self-contained and self-governing system, Blume (1974, 279) argued that "sociologists of science must discard their assumption of the autonomy of the social system of science."[2] He showed that national systems of science in Britain and the United States were highly dependent on and integrated with economic and political institutions. At the top, interconnections with political and economic agents reinforced elite scientists' managerial power to control access to funding and to shape the direction of research. Further down the structural hierarchy, nonelite scientists followed different political and economic logics in their collective efforts to improve working conditions and to disseminate scientific information to the broader public. Indeed, concerned scientists who engaged in active social critique and political protest were another important source of influence (Moore 1996). In addition to challenging assumptions about race and gender as discussed above, scientists were active in a range of issues, including weapons development and arms control, environmental degradation, and surveillance associated with new information technologies (Egan 2012; Wisnioski 2012).

Three decades after Blume's book, in *The New Political Sociology of Science* (*NPSS*) Frickel and Moore (2006) drew explicit attention to the role of science in concentrating power and reproducing inequality within existing institutions, and they highlighted the political relations between science and other major social institutions. The project called for the investigation of structured interactions across institutional spheres or social fields, especially science, the state, industry, and social movements. The NPSS also encouraged scholars to use diverse methodological strategies for studying these interconnections at broader scales of analysis, and the book challenged the field's still-tight focus on microlevel interactions among networked experts. Focusing less than Blume on national-level systems, the NPSS drew attention to streams of research in STS that offered analyses of science as an essentially political institution and helped to rearticulate the problem of enduring inequalities in science, technology, and society as an organizing principle for research. At its core, this work took social (and sociotechnical) structure seriously—not as a determining set of factors but as dynamically influencing the shape and consequences of social action. It also strove to explain processes and outcomes and to emphasize the value of multiscalar analytic strategies that consciously attend to broader historical and institutional patterns, which in turn affect and are affected by local sources of action and power.

Seen from this angle, political sociological approaches stand in productive tension with certain other prominent approaches to politics and inequality in STS. They are broadly consonant with Foucault-inspired research on biopolitics that take structural inequalities as a background assumption; however, political sociological approaches also examine how durable systems, such as capitalism and racism, produce variegated but predictable results and why some discourses and practices are able to proliferate, whereas others are not. Such concerns often fall outside the scope of biopolitical analysis.

Theoretical Innovations

In the decade since *NPSS* was published, the nascent field has grown in productive ways. Here we highlight three distinctive theoretical contributions that this body of work offers to STS.

Responding to the perceived need for broader sociopolitical analysis, Moore and colleagues (2011, 506) developed a historical and comparative framework for studying change in the "institutional and extra-institutional matrix of the scientific field" during the contemporary period. This framework builds on *NPSS*'s original four-part optic of science, industry, state, and social movements but sets these interrelationships more explicitly within the dynamic context of neoliberal globalization in order to identify important tensions and processes that today are shaping science's winners and losers. This approach argues that an important dimension of the analysis of science and technology from the perspective of structural inequality requires attention to the neoliberalization of the political field and its broader effects on academic science. Thus, one goal of this area of theorization is to put STS into conversation with broader interdisciplinary research on neoliberalism and politics (Harvey 2005; Mirowski 2011).

Another area of concentrated theoretical activity has been critical engagement with various forms of field theory (Albert and Kleinman 2011; Hess 2009; Panofsky 2014). A field is a structured social space where actors have different capacities to affect outcomes based on their levels of capital and on their habitus (Bourdieu 2001). The edited collection *Fields of Knowledge* (Frickel and Hess 2014) offered two complementary directions of field-level empirical analyses of science and technology under contemporary neoliberal arrangements: relations among fields and the institutional logics or systems of meaning operating within fields. Collectively, this body of work interrogated the autonomy assumption common in various strands of older sociology of science scholarship (Bourdieu 1975, 2001; Merton 1973) and in the microsociological constructivist studies of the 1980s and 1990s described above. It also questioned assumptions built into field theory itself, for example, that fields are stable and coherent social spaces (Moore and Hoffman 2014; Schweber 2014). In addition to developing and extending field-theoretic frameworks (Albert and McGuire 2014; Berman 2014; Lave 2014), NPSS work found theoretical sustenance in other diverse quarters, including Foucault's governmentality approach (Schweber 2014), political sociological theories of the state and social movements (Hess 2014; Suryanarayanan and Kleinman 2014), and institutional theory (Berman 2014; Kinchy, Jalbert, and Lyons 2014; Moore and Hoffman 2014). These studies developed social theories of science by focusing on the conditions under which field-specific social structures become institutionalized or are disrupted. The studies also examined how (and how well) endogenous field-level pressures pattern knowledge practices and the distribution of scientific outcomes.

A third area of theoretical innovation in political sociology of science is the study of undone science and knowledge gaps (Frickel 2008; Frickel et al. 2010; Hess 2007, 2015; Kleinman and Suryanarayanan 2013). The study of ignorance is not new, and some of the research described earlier in the essay raised the question of the absence of certain kinds of knowledge related to structural inequality. (For a review of the now burgeoning research field, see Gross and McGoey 2015.) The specific contribution of NPSS is to develop a framework for analysis of the structural causes and consequences of ignorance. One example is research on a particular type of ignorance, undone science, or conditions characterized by the absence of *socially desirable* knowledge. Accumulating evidence from diverse case studies of epistemic inequality suggests that such conditions are socially produced by uneven research agenda setting and by the politicization of absent knowledge (Frickel 2014; Frickel et al. 2010; Hess 2015). The resulting epistemic conflict leads social groups (often, social movements and their scientist allies) to advocate for research to produce missing knowledge aimed at, for example, low-cost medical treatments for certain diseases (Cleary 2012) or at environmentally risky technologies (Brown 2007; Kinchy and Perry 2012). These goals are often in conflict with industrialists and their political and scientific allies, whose interests lie in continued not-knowing (Ottinger 2013).

Methodological Innovations

Both laboratory ethnographies and controversy studies showed how the micropolitics of social networks shape choices among interpretations of data, theoretical frameworks, and technological designs. They also showed that the tools that scientists and inventors use—including "raw" materials—are imbued with meaning and that the construction of scientific facts and artifacts relies heavily on language, instruments, and social conventions. However, some of the laboratory studies were limited by their focus on microsocial processes. Latour and Woolgar ([1979] 1986) noted that things went in and out of the laboratory, but they and some of the other ethnographers of science were often less concerned about where those things came from or where they went. Even when reaching outside the laboratory (Latour 1988), the emphasis was on the social effects of scientific knowledge rather than on the interaction of structural inequality and choices for scientific knowledge (for an exception, see Traweek 1992). Studies of scientific controversies tended to have a broader focus than the laboratory ethnographies, such as the work of Collins and Pinch (1982) on the parapsychology controversy and Nelkin (1992) and Pinch and Bijker (1987) on technological controversies. Nevertheless, there was considerable room to build on the traditions of micro- and mesosocial analyses of laboratories and controversies by resituating them in broader interfield and institutional dynamics.

In general, researchers who focused on laboratories and controversies did not attend to the problem of structural inequality. There was some recognition of the need to study the problem (e.g., in the third phase of the empirical program of relativism; Collins 1983), but typically the issue was not salient. In contrast, emergent NPSS work during the 1990s grappled with issues of structural inequality and tended to examine science on a broader scale. For example, Kleinman (1998, 2003) offered an alternative method for the laboratory ethnography that linked the microsocial to the subtle and indirect ways in which industrial and state power shape academic science. In an analysis of nuclear submarine development, Frickel engaged actor-network theory by arguing that a sociological explanation of the success of heterogeneous networks requires reconsidering "the social contexts in which actor networks are embedded" (1996, 48). At roughly the same time, Hess (1997) examined the construction of science and technology at the field level of dominant and subordinate networks of cancer research and treatment, and Moore (1996) analyzed scientists who moved into the broader political field through the creation of public interest science organizations. Likewise, Epstein (1996) analyzed the important role of social movements and their effects on research agendas and regulatory policy.

These studies were part of a trend in STS to resituate the analysis of science and expertise in a broader context than the laboratory, the core set of scientists in a controversy, or even the networks that extend out from the laboratory. The trend was evident in a range of STS subfields, including the anthropology of science and technology (Downey and Dumit 1997), public understanding of science studies (Irwin and Wynne 1996), and technology policy studies (Jasanoff 1994). In turn, the shift of attention to wider social scales was linked to a methodological transition away from the case study and toward comparative, global, and transnational perspectives in STS. This transition built on the field's tradition of comparative research, but it often differed from predecessors by including social movements as central actors (Allen 2014; Clarke 1998; Kinchy 2012; McCormick 2009; Suryanarayanan and Kleinman 2014). Furthermore, STS research that is concerned with the issue of structural inequality sometimes adopts a transnational and global perspective. Some work challenges traditional stories of North-South technology diffusion by examining South-South trajectories. For example, Williams (2013) explored the circulation of scientific knowledge and technological innovation among countries in the global south. Likewise, Prasad (2014) argued that the investigation of technology transfer requires that we track the multisited and global dimensions of emerging technologies, where such "technoscientific trails" do not depend upon a particular national innovation system or epistemic community (Anderson and Adams 2007). Yet other work examined the transnational, collaborative relations between scientists from northern and southern countries. For example, the World Science Project used longitudinal surveys and social network analysis to explore the utility of the Internet to facilitate such collaborations (Shrum, Genuth, and Chompalov 2007; Ynalvez and Shrum 2011).

New Directions in the Analysis of Structural Inequality, Science, and Technology

Structural Inequality as Race and Gender

As the relationship between structural inequality and science and technology has gained the attention of STS researchers, a growing body of work has examined how the use of new technologies and research methods reinvent old concepts such as race and gender. Several analyses of race and racism in science have focused on the promise of genomics to offer "personalized" medicine that purports to provide health and economic benefits but also can undermine biological conceptions of race. Reardon (2005) showed that genomic scientists remained mired in the idea of race as a biological property, even though they often used phrases such as "population" instead of race. She also showed that "participation" in scientific research—touted by many as a panacea for the problems of race and gender exclusion from knowledge production—was a poor solution to the problem, as long as scientists failed to understand how their own assumptions about race remained unexamined. Other analysts examined similar dynamics and captured how racism continued to be reproduced in scientists' thinking and practice (Duster 2012; Fujimura and Rajagopalan 2011; Hatch 2014; Morning 2011; Nelson 2008; Shim 2005).

Like Reardon, Epstein (2007) traced how race and gender came to be reproduced as sociobiological categories. He showed that women's and African Americans' health groups pursued "inclusion" in medical studies as a means to generate better health for their constituents. But by presuming that health inequalities were "raced" and "gendered," new rules for inclusion in research underlined the very differences that were supposed to be eliminated. Similarly, Fisher (2009) showed that a complex organizational system of institutional review boards, compliance companies, and recruitment specialists work to recruit and keep African Americans in drug safety trials, despite the fact that the trials rarely help any of the enrollees, and few have access to the drugs that might result from the trials.

Another strand of work on race in STS has also examined the promise of social movements to undermine scientific race categories (Bliss 2012). Nelson (2011) demonstrated that the Black Panther Party played a critical role in challenging the biomedical experiences of African Americans. Through demonstrations, community organizations, and alliances with other groups, the Black Panthers were able to acquire some new health benefits for their communities and to raise key questions about studies of "criminal violence" that considered only people of color. At the same time, recent work has questioned the effectiveness of social movements with respect to goals of racial justice. Benjamin (2013) studied the California Stem Cell Research Referendum, which was supposed to benefit all Californians, and showed that the ultimate beneficiaries of this initiative were not people with disabilities, people of color, and the poor. Taken together, this body of research suggests that structures of racism embedded in rules, profit making, and the logics of political movements continue to shape the ways that racial science and racial experience are perpetuated and undermined.

Structural Inequality as Global Position

Since 2000 STS research has increasingly paid attention to science, technology, and power in the global south. Examples include nationalist rhetoric and the design of Indonesian aircraft technology (Amir 2007), Brazilian scientists as both colonizers and colonized (Cukierman 2014), the industrial-chemical disaster in Bhopal (Fortun 2001), intellectual property and local biological knowledge (Delgado and Rodríguez-Giralt

2014; Hayden 2003), the relationship between nuclear energy and decolonization (Hecht 2002), the project of developing decolonized knowledges (Mignolo 2009), and the links between genomic research and global capital in India and the United States (Sunder Rajan 2006). We focus on one aspect of this substantial literature: new directions in the study of science, technology, and development.

Much of this work involves a critical appraisal of high-technology development projects but without returning to the models of the appropriate technology movement. Amir (2012, 2014) showed how states in industrializing countries have intentionally selected specific technological trajectories inspired by nationalist identities and structurally empowered by authoritarian politics. He argued that in the case of Indonesia technological development was deeply embedded in the power structure that underpinned New Order authoritarianism. Such a structure gave rise to the "technological state," a distinct type of state that was organized around an ambition of socioeconomic transformations and with a strong emphasis on state-led, high-technology development. In this political structure, government goals of economic development were aligned with large-scale, capital-intensive, high-tech projects.

Other researchers have developed a critical analysis of models of innovation that focus on industrialized countries as the primary sources of innovation. In a study of drug discovery in South Africa, Pollock (2014) challenged the bifurcation between the global centers of knowledge production in the pharmaceutical industry, generally located in the global north, and the global south as a place for bioprospecting and the consumption of drugs and production of generics. The company iThemba attempted to break down this division by developing new drugs that also addressed the needs of the South African poor. Although one vehicle for this new model of technology innovation and development is the private-sector firm, Williams and Woodson (2012) argued that the fields of innovation and technology studies have not been attentive enough to the roles of civil society organizations. Using the examples of science-based agricultural innovation and the invention of low-cost intraocular lenses and surgical techniques for avoidable blindness, they showed the importance of innovation by civil society organizations that is oriented toward the rural poor in less developed countries (Williams 2013). In contrast with the older division in technology studies between innovation clusters focused on national industrial development and low-tech appropriate technology for the rural poor, this work demonstrated that high-technology innovation can also be reshaped to address the pervasive problems of poverty in less developed countries. Thus, it represented "sciences from below" (Harding 2008), where the perspectives of the local rural poor are drivers of technological innovation and scientific research that is emerging and spreading from countries in the global south.

Some of the literature on technology and development has also critically examined the emergence of projects and programs that are in some ways heirs to the appropriate technology movement. Among the criticisms are the one-way technology transfer model and the use of low-technology solutions for the global poor (Eglash 2004). Research on the One Laptop per Child project suggested connections with the history of racism in the United States (Fouché 2011) and the reproduction of center-periphery relations (Chan 2014). Likewise, Fressoli, Dias, and Thomas (2014) developed a critical, comparative analysis of various efforts to develop "inclusive innovation" in Latin America and India. In Latin America the reform efforts include collaborations among universities, civil society organizations, social movements, and government agencies, whereas in India there is a greater focus on private-sector actors and the creation of business opportunities. Their comparative analysis drew attention to both the shortcomings of these projects and the potential for them to be linked to a democratic politics of technology. In a similar way, Amir and Nugroho (2013) argued for an approach that takes into account structural, cultural, and epistemological changes by which technoscience can be produced with broad, democratic participation in response to local conditions.

STS researchers have developed similar analyses for science and development, examining both the problems and the potential that have emerged in the international collaborations among scientists from wealthier and poorer countries. As with the analyses of inclusive innovation, recent research suggests that there is still potential for these new emerging forms of scientific and technological innovation to reproduce relations of structural inequality. A case in point is the new "global health sciences" paradigm that many American universities have adopted. Crane argued that despite attempting to invoke a northern-southern "partnership" as an ethical bulwark, researchers in northern academic positions "benefit from the opportunities afforded by global inequalities" (2013, 169). The opportunities for northern researchers in the global south may have unsettling parallels with the economic prospects for European and American merchants as they pursued imperial conquests and trade in Africa and Asia (Adas 2006; Harris 2011). Although northern medical scientists had good intentions when collecting data for "pure science," this goal led to testing antiviral drugs, developed for the HIV strain prevalent in North America, on African bodies exposed to a different HIV strain (Crane 2013).

In another example of the problem of inequality in international research partnerships, Nieusma and Riley (2010) examined collaboration among engineers from two universities in Nicaragua and two in the United States for a project that involved product design for local economic development. In the collaboration, differences in perspectives emerged over what constituted meaningful community involvement and what economic models were best for launching products in the marketplace. In summary, STS researchers can provide insights into the international teams of well-intentioned engineers, physicians, and scientists as they think through the implications of appropriate technology, knowledge transfer, and research partnerships with communities that they intend to help.

Structural Inequality as Position with Respect to State and Industry

As societies have become increasingly dependent on complex technologies, scientific research has become both more important to the state and industry and more politicized. From the perspective of structural inequality, we might think of the transformation of the relationship between the scientific field and other social fields as having two main dimensions: asymmetrical convergence and epistemic modernization.

Kleinman and Vallas (2001) coined the term *asymmetrical convergence* to capture the idea that although codes and practices are being traded between academia and industry in a new hybrid environment, it is also the case that such exchange is fundamentally shaped by industry interest in profits (Vallas and Kleinman 2007). In developing this and related lines of analysis, Kleinman and Vallas (2001), Smith-Doerr (2004), and other STS scholars (Berman 2012a, 2012b; Hoffman 2011; Owen-Smith 2003) have been influenced by the new institutionalism in organizational studies. At the center of discussion in much of this work is how certain codes and practices—often those associated with business—become legitimate and taken for granted in academia. Attention to when and why cracks develop in legitimacy can help to explain the factors that make change (toward or away from industry domination) more or less likely. According to Kleinman and Osley-Thomas, "Proposed innovations that fit more easily into existing institutionalized organizational narratives and structures are more likely to persist than those that do not" (2014, 3). While this change might advantage commercial codes and practices, academic scientists nevertheless do not always accept business-oriented practices and have developed strategies of resistance to them (Colyvas and Jonsson 2011; Smith-Doerr and Vardi 2015; for a comparative perspective, see Bak 2014).

Another strand of research oriented toward state and industrial power involves the countervailing trend of "epistemic modernization," or the increasing openness of the scientific field to research agendas that are responsive to the needs of those who have been historically excluded from the corridors of power and governance: women, ethnic minority groups, gays and lesbians, workers, indigenous people, and persons located in less privileged geographical regions of the world (Hess 2007; Moore et al. 2011). These changes emerge from the increased diversification of the social composition of science as well as from interactions between mobilized publics and scientists. Activists, advocates, and community groups sometimes approach researchers for help, often in response to perceptions of industrial risk and the perceived need for the regulation of new technologies. These groups identify underline science, and they place a spotlight on the politics of choices among research agendas. At the same time, the increasing politicization of research claims and agendas results in what Moore (2008) describes as the unbinding of scientific authority: the authority to speak credibly in the name of science can become separated from scientists who are expert in a field.

The epistemic modernization of the scientific field and its relations to other social fields creates openings for social movements that have mounted challenges to political and economic elites. STS research on social movements and mobilized publics has resulted in a range of new concepts and insights of use to STS scholars interested in inequality. Scientists may respond to mobilized publics by developing partnerships with community groups and advocacy organizations in the form of citizen-scientist alliances (Brown 2007) and shadow mobilizations (the informal networks of scientists who work with social movements; Frickel, Torcasso, and Anderson 2015). These mobilizations can also result in the development of new technologies and products in hybrid organizations that often link entrepreneurship and activism (Hess 2007). In some cases, scientists become public figures involved in policy advocacy (Moore 1996, 2008), and they may also support efforts of activists in less developed countries by joining transnational networks (Kinchy 2012).

This body of STS scholarship shifts the analysis of the public as the repository of lay knowledge to a more mobilized and segmented public of groups that claim to speak in the public interest as a counterpublic (Hess 2011). Frequently, these groups push for a more precautionary approach to technology regulation, and they are able to have traction in the political field in part due to the credibility of partnerships with scientists. Thus, dilemmas emerge in the political coalitions of social movements and scientists over how to participate in a policy field that may be dominated by "scientized" decision making, which Kinchy (2012, 2) defines as "the transformation of social conflict into a debate, ostensibly separated from its social context, among scientific experts." Furthermore, industry may also mobilize its experts and assist in the circulation of their views through the media (e.g., Oreskes and Conway 2010), and sometimes the attacks result in the intellectual suppression of scientists (Delborne 2008; Martin 2007). These attacks can lead to backfire, which occurs when the public reacts with outrage to an action by established authorities (Hess and Martin 2006).

Conclusion

The reinvigoration of attention to structural inequality in STS has led to the emergence of a suite of new concepts and methods. These sensitizing concepts—among them undone science, citizen-scientist alliances, shadow mobilizations, asymmetrical convergence, and epistemic modernization—constitute the elements of an emergent conceptual framework in STS. Again, we are not advocating that STS researchers leave behind the world of boundary objects, interpretive flexibility, and enrollment. STS researchers with a range of theoretical perspectives have introduced many useful concepts, and they have helped us to conduct research with more clarity. But we also suggest that there is room for additional concepts especially attuned to the problem of structural inequality.

We expect that as the STS field becomes more diverse and internationalized, there will be greater attention to the problem of structural inequality worldwide. A complete review of STS literatures throughout the world was outside the scope of this chapter, but at several points we have suggested the potential for engaging the unique perspectives of STS literatures that have developed in Latin America, Asia, and other areas outside the North Atlantic (Dagnino 2010; Medina, Marques, and Holmes 2014; Quet and Noel 2014). Historically STS researchers located in or from the global south have brought important perspectives to the field by drawing attention to colonialism, poverty, and dependent development. Some of this scholarship has been critical of the university systems and research funding priorities of these countries, building on the foundational work of Herrera (1972). Other research has drawn inspiration from the social movements of the global south to develop critical appraisals of cosmopolitan science and development projects in the traditions of Visvanathan (1997) and Escobar (1995). In either case, a fundamental contribution of STS researchers from the global south has been to develop theoretical frameworks and to develop research projects focused on poverty and inequality. As we have indicated, some STS researchers have continued with this tradition by developing critical appraisals of state-directed, high-technology projects of development without romanticizing appropriate technology and similar movements today.

One of the next steps in the study of structural inequality is a better integrated investigation of race, gender, sexuality, and related studies with work on global poverty, inequality, and underdevelopment. The field could also benefit from reflexive studies of how different national and continental traditions of STS have developed different emphases in the study of structural inequality. In addition to the various research programs outlined here, both steps would represent significant new directions in the development of the STS field, and these steps would contribute to research that is especially relevant to the problems of the twenty-first century.

Notes

1. Coverage is necessarily selective. We do not, for example, attend to substantive similarities and differences in conceptualizations of power as the term has been employed across different theoretical traditions. We also do not deal specifically with the problem of science and democracy, a topic covered by Jasanoff (chapter 9 this volume). Whereas she focuses on politics and the state, our lens is broader and includes social movements and the private sector.

2. For another early nod to the political sociology of science, see Sklair (1970).

References

Adas, Michael. 2006. *Dominance by Design: Technological Imperatives and America's Civilizing Mission*. Cambridge, MA: Harvard University Press.

Albert, Mathieu, and Daniel Lee Kleinman, eds. 2011. "Beyond the Canon: Pierre Bourdieu and Science and Technology Studies." *Minerva* 49 (3): 263–348.

Albert, Mathieu, and Wendy McGuire. 2014. "Understanding Change in Academic Knowledge Production in a Neoliberal Era." *Political Power and Social Theory* 27: 33–57.

Allen, Barbara. 2014. "From Suspicious Illness to Policy Change in Petrochemical Regions." In *Powerless Science? Science and Politics in a Toxic World*, edited by Soraya Boudia and Nathalie Jas, 152–69. New York: Berghahn Books.

Amir, Sulfikar. 2007. "Nationalist Rhetoric and Technological Development: The Indonesian Aircraft Industry in the New Order Regime." *Technology in Society* 29 (3): 283–93.

___. 2012. *The Technological State in Indonesia: The Co-constitution of High Technology and Authoritarian Politics*. London: Routledge.

___. 2014. "Risk State: Nuclear Politics in an Age of Ignorance." In *Routledge Handbook of Science, Technology, and Society*, edited by Daniel Lee Kleinman and Kelly Moore, 292–306. London: Routledge.

Amir, Sulfikar, and Yanuar Nugroho. 2013. "Beyond the Triple Helix: Framing STS in the Developmental Context." *Bulletin of Science, Technology & Society* 33 (3–4): 115–26.

Anderson, Warwick. 2002. "Introduction: Postcolonial Technoscience." *Social Studies of Science* 32 (5–6): 643–58.

Anderson, Warwick, and Vincanne Adams. 2007. "Pramoedya's Chicken: Postcolonial Studies of Technoscience." In *The Handbook of Science and Technology Studies*, 3rd ed., edited by Edward J. Hackett, Olga Amsterdamska, Michael Lynch, and Judy Wajcman, 181–204. Cambridge, MA: MIT Press.

Appadurai, Arjun. 1996. *Modernity at Large: Cultural Dimensions of Globalization*. Minneapolis: University of Minnesota Press.

Bak, Hee-Je. 2014. "The Utilitarian View of Science and Norms and Practices of Korean Scientists." In *Routledge Handbook of Science, Technology, and Society*, edited by Daniel Lee Kleinman and Kelly Moore, 406–18. London: Routledge.

Benjamin, Ruha. 2013. *People's Science: Bodies and Rights on the Stem Cell Frontier*. Stanford, CA: Stanford University Press.

Berman, Elizabeth Popp. 2012a. *Creating the Market University: How Academic Science Became an Economic Development Engine*. Princeton, NJ: Princeton University Press.

___. 2012b. "Explaining the Move toward the Market in U.S. Academic Science: How Institutional Logics Can Change without Institutional Entrepreneurs." *Theory and Society* 41 (3): 261–99.

___. 2014. "Field Theories and the Move toward the Market in U.S. Academic Science." *Political Power and Social Theory* 27: 193–221.

Bliss, Catherine. 2012. *Race Decoded: The Genomic Fight for Social Justice*. Stanford, CA: Stanford University Press.

Blume, Stuart. 1974. *Toward a Political Sociology of Science*. New York: Free Press.

Bourdieu, Pierre. 1975. "The Specificity of the Scientific Field and the Social Conditions of the Progress of Reason." *Social Sciences Information. (Information Sur les Sciences Sociales)* 14 (6): 19–47.

___. 2001. *Science of Science and Reflexivity*. Chicago: University of Chicago Press.

Bowker, Geoffrey, and Susan Leigh Star. 2000. *Sorting Things Out: Classification and Its Consequences*. Cambridge, MA: MIT Press.

Brown, Phil. 2007. *Toxic Exposures: Contested Illnesses and the Environmental Health Movement*. New York: Columbia University Press.

Cardoso, Fernando Henrique, and Enzo Faletto. 1979. *Dependency and Development in Latin America*. Berkeley: University of California Press.

Castells, Manuel. 1996. *The Rise of the Network Society*. Cambridge, MA: Blackwell Publishers.

Chan, Anita Say. 2014. "Balancing Design: OLPC Engineers and ICT Translations at the Periphery." In *Beyond Imported Magic: Essays on Science, Technology, and Society in Latin America*, edited by Eden Medina, Ivan da Costa Marques, and Christina Holmes, 181–205. Cambridge, MA: MIT Press.

Clarke, Adele. 1990. "A Social Worlds Adventure." In *Theories of Science in Society*, edited by Susan Cozzens and Thomas Gieryn, 15–42. Bloomington: Indiana University Press.

___. 1998. *Disciplining Reproduction: Modernity, American Life Sciences, and "the Problems of Sex."* Berkeley: University of California Press.

Clarke, Adele, and Susan Leigh Star. 2008. "The Social Worlds Framework: A Theory/Methods Package." In *The Handbook of Science and Technology Studies*, 3rd ed., edited by Edward J. Hackett, Olga Amsterdamska, Michael Lynch, and Judy Wajcman, 113–38. Cambridge, MA: MIT Press.

Cleary, Tom. 2012. "Undone Science and Blind Spots in Medical Treatment Research." *Social Medicine* (Social Medicine Publication Group) 6 (4): 234–39.

Cockburn, Cynthia. 1981. "The Material of Male Power." *Feminist Review* 9: 41–58.

Cole, Jonathan R., and Stephen Cole. 1973. *Social Stratification in Science*. Chicago: University of Chicago Press.

Collins, Harry. 1983. "An Empirical Relativist Programme in the Sociology of Scientific Knowledge." In *Science Observed*, edited by Karin Knorr Cetina and Michael Mulkay, 85–113. Thousand Oaks, CA: Sage.

Collins, Harry, and Trevor Pinch. 1982. *Frames of Meaning: The Social Construction of Extraordinary Science*. London: Routledge.

Colyvas, Jeannette, and Stephan Jonsson. 2011. "Ubiquity and Legitimacy: Disentangling Diffusion and Institutionalization." *Sociological Theory* 29 (2): 27–53.

Cozzens, Susan, Sonia Gatchair, Kyung-Sup Kim, González Ordóñez, and Anupit Supnithadporn. 2008. "Knowledge and Development." In *The Handbook of Science and Technology Studies*, 3rd ed., edited by Edward J. Hackett, Olga Amsterdamska, Michael Lynch, and Judy Wajcman, 787–812. Cambridge, MA: MIT Press.

Cozzens, Susan, and Edward Woodhouse. 1995. "Science, Government, and the Politics of Knowledge." In *The Handbook of Science and Technology Studies*, 2nd ed., edited by Sheila Jasanoff, Gerald E. Markle, James C. Petersen, and Trevor J. Pinch, 533–53. Thousand Oaks, CA: Sage.

Crane, Johanna. 2013. *Scrambling for Africa : AIDS, Expertise, and the Rise of American Global Health Science*. Ithaca, NY: Cornell University Press.

Croissant, Jennifer, and Laurel Smith-Doerr. 2008. "Organizational Contexts of Science: Boundaries and Relationships between University and Industry." In *The Handbook of Science and Technology Studies*, 3rd ed., edited by Edward J. Hackett, Olga Amsterdamska, Michael Lynch, and Judy Wajcman, 691–718. Cambridge, MA: MIT Press.

Cukierman, Hernique. 2014. "Who Invented Brazil?" In *Beyond Imported Magic: Essays on Science, Technology, and Society in Latin America*, edited by Eden Medina, Ivan da Costa Marques, and Christina Holmes, 27–46. Cambridge, MA: MIT Press.

Dagnino, Renato, ed. 2010. *Estudos Socias de Ciência e Tecnología & Política de Ciência e Tecnologia*. Campina Grande, Brazil: Eduepb.

Delborne, Jason. 2008. "Transgenes and Transgressions: Scientific Dissent as Heterogeneous Practice." *Social Studies of Science* 38 (4): 509–41.

Delgado, Ana, and Israel Rodríguez-Giralt. 2014. "Creole Interferences: A Conflict over Biodiversity and Ownership in the South of Brazil." In *Beyond Imported Magic: Essays on Science, Technology, and Society in Latin America*, edited by Eden Medina, Ivan da Costa Marques, and Christina Holmes, 331–48. Cambridge, MA: MIT Press.

Downey, Gary Lee, and Joseph Dumit, eds. 1997. *Cyborgs and Citadels: Anthropological Interventions in Emerging Sciences and Technologies*. Santa Fe, NM: School of American Research Press.

Duster, Troy. 1990. *Backdoor to Eugenics*. New York: Routledge.

___. 2012. "The Combustible Intersection: Genomics, Forensics, and Race." In *Race after the Internet*, edited by Lisa Nakamura and Peter Chow-White, 310–27. London: Routledge.

Eckaus, Richard. 1977. *Appropriate Technologies for Developing Countries*. Washington, DC: National Academy of Sciences.

Egan, Michael. 2012. *Barry Commoner and the Science of Survival: The Remaking of Modern Environmentalism*. Cambridge, MA: MIT Press.

Eglash, Ron. 2004. "Appropriating Technology: An Introduction." In *Appropriating Technology: Vernacular Science and Social Power*, edited by Ron Eglash, Jennifer Croissant, Giovanna DiChiro, and Rayvon Fouché, vii–xxi. Minneapolis: University of Minnesota Press.

Epstein, Steven. 1996. *Impure Science: AIDS, Activism, and the Politics of Knowledge*. Berkeley: University of California Press.

___. 2007. *Inclusion: The Politics of Difference in Medical Research*. Chicago: University of Chicago Press.

Escobar, Árturo. 1995. *Encountering Development: The Making and Unmaking of the Third World*. Princeton, NJ: Princeton University Press.

Etzkowitz, Henry, and Loet Leydesdorff. 2000. "The Dynamics of Innovation: From National Systems and 'Mode 2' to a Triple Helix of University-Industry-Government Relations." *Research Policy* 29 (2): 109–23.

Evans, Peter. 1995. *Embedded Autonomy: States and Industrial Transformation*. Princeton, NJ: Princeton University Press.

Feenberg, Andrew. 1995. *Alternative Modernity: The Technical Turn in Philosophy and Social Theory*. Berkeley: University of California Press.

Fisher, Jill. 2009. *Medical Research for Hire: The Political Economy of Pharmaceutical Clinical Trials*. New Brunswick, NJ: Rutgers University Press.

Fortun, Kim. 2001. *Advocacy after Bhopal: Environmentalism, Disaster, New Global Orders*. Chicago: University of Chicago Press.

Fouché, Rayvon. 2011. "From Black Inventors to One Laptop Per Child: Exporting a Racial Politics of Technology." In *Race after the Internet*, edited by Lisa Nakamura and Peter Chow-White, 661–83. New York: Routledge.

Fressoli, Mariano, Rafael Dias, and Hernán Thomas. 2014. "Innovation and Inclusive Development in the South: A Critical Perspective." In *Beyond Imported Magic: Essays on Science, Technology, and Society in Latin America*, edited by Eden Medina, Ivan da Costa Marques, and Christina Holmes, 47–66. Cambridge, MA: MIT Press.

Frickel, Scott. 1996. "Engineering Heterogeneous Accounts: The Case of Submarine Thermal Reactor Mark-1." *Science, Technology, & Human Values* 21 (1): 28–53.

___. 2008. "On Missing New Orleans: Lost Knowledge and Knowledge Gaps in an Urban Landscape." *Environmental History* 13 (4): 643–50.

___. 2014. "Not Here and Everywhere: The Non-Production of Knowledge." In *Routledge Handbook of Science, Technology, and Society*, edited by Daniel Lee Kleinman and Kelly Moore, 256–69. London: Routledge.

Frickel, Scott, Sahra Gibbon, Jeff Howard, Joanna Kempner, Gwen Ottinger, and David Hess. 2010. "Undone Science: Charting Social Movement and Civil Society Challenges to Research Agenda Setting." *Science, Technology, & Human Values* 35 (4): 444–73.

Frickel, Scott, and David J. Hess. 2014. "Fields of Knowledge: Science, Politics, and Publics in the Neoliberal Age." Special issue, *Political Power and Social Theory*, vol. 27. London: Emerald.

Frickel, Scott, and Kelly Moore, eds. 2006. *The New Political Sociology of Science: Institutions, Networks, and Power*. Madison: University of Wisconsin Press.

Frickel, Scott, Rebakah Torcasso, and Annika Anderson. 2015. "The Organization of Expert Activism: Shadow Mobilization in Two Social Movements." *Mobilization: An International Quarterly* 21 (3): 305–33.

Fujimura, Joan H., and Ramya Rajagopalan. 2011. "Different Differences: The Use of 'Genetic Ancestry' versus Race in Biomedical Human Genetic Research." *Social Studies of Science* 41 (1): 5–30.

Gibbons, Michael, Camille Limoges, and Helga Nowotny. 1994. *The New Production of Knowledge: The Dynamics of Science and Research in Contemporary Societies*. London: Sage.

Gross, Matthias, and Linsey McGoey, eds. 2015. *International Handbook of Ignorance Studies*. London: Routledge.

Haraway, Donna J. 1991. *Simians, Cyborgs, and Women: The Reinvention of Nature*. New York: Routledge.

Harding, Sandra G. 1991. *Whose Science? Whose Knowledge? Thinking from Women's Lives*. Ithaca, NY: Cornell University Press.

___. 1998. *Is Science Multicultural? Postcolonialisms, Feminisms, and Epistemologies*. Bloomington: Indiana University Press.

___. 2008. *Sciences from Below: Feminisms, Postcolonialities, and Modernities*. Durham, NC: Duke University Press.

___. 2015. *Objectivity and Diversity*. Chicago: University of Chicago Press.

Harris, Steven J. 2011. "Long-Distance Corporations, Big Sciences, and the Geography of Knowledge." In *The Postcolonial Science and Technology Studies Reader*, edited by Sandra Harding, 61–83. Durham, NC: Duke University Press.

Harvey, David. 2005. *A Brief History of Neoliberalism*. Oxford: Oxford University Press.

Hatch, Anthony. 2014. "Technoscience, Racism, and the Metabolic Syndrome." In *Routledge Handbook of Science, Technology, and Society*, edited by Daniel L. Kleinman and Kelly Moore, 38–55. New York: Routledge.

Hayden, Cori. 2003. *When Nature Goes Public: The Making and Unmaking of Bioprospecting in Mexico*. Princeton, NJ: Princeton University Press.

Hecht, Gabrielle. 2002. "Rupture-talk in the Nuclear Age: Conjugating Colonial Power in Africa." *Social Studies of Science* 32 (5–6): 691–727.

Herrera, Amílcar. 1972. "Social Determinants of Science Policy in Latin America." *Journal of Development Studies* 9 (1): 19–37.

Hess, David J. 1997. *Can Bacteria Cause Cancer? Alternative Medicine Confronts Big Science*. New York: New York University Press.

___. 2007. *Alternative Pathways in Science and Industry: Activism, Innovation, and the Environment in an Era of Globalization*. Cambridge, MA: MIT Press.

___. 2009. "Bourdieu and Science and Technology Studies: Toward a Reflexive Sociology." *Minerva* 49 (3): 333–48.

___. 2011. "To Tell the Truth: On Scientific Counterpublics." *Public Understanding of Science* 20 (5): 627–41.

___. 2014. "When Green Became Blue: Epistemic Rift and the Corralling of Climate Science." *Political Power and Social Theory* 27: 123–53.

___. 2015. "Undone Science, Industrial Innovation, and Social Movements." In *Routledge International Handbook of Ignorance Studies*, edited by Matthias Gross and Linsey McGoey, 141–54. London: Routledge.

Hess, David J., and Brian Martin. 2006. "Backfire, Repression, and the Theory of Transformative Events." *Mobilization: An International Quarterly* 11 (2): 249–67.

Hoffman, Steve G. 2011. "The New Tools of the Science Trade: Contested Knowledge Production and the Conceptual Vocabularies of Academic Capitalism." *Social Anthropology* 19 (4): 439–62.

Hubbard, Ruth. 1990. *The Politics of Women's Biology*. New Brunswick, NJ: Rutgers University Press.

Irwin, Alan. 2008. "STS Perspectives on Scientific Governance." In *The Handbook of Science and Technology Studies*, 3rd ed., edited by Edward J. Hackett, Olga Amsterdamska, Michael Lynch, and Judy Wajcman, 583–607. Cambridge, MA: MIT Press.

Irwin, Alan, and Brian Wynne, eds. 1996. *Misunderstanding Science? The Public Reconstruction of Science and Technology*. Cambridge: Cambridge University Press.

Jasanoff, Sheila. 1994. *The Fifth Branch: Science Advisors as Policymakers*. Cambridge, MA: Harvard University Press.

Keller, Evelyn F. 1995. "The Origin, History, and Politics of the Subject Called 'Gender and Science': A First Person Account." In *Handbook of Science and Technology Studies*, 2nd ed., edited by Sheila Jasanoff, Gerald Markle, James Petersen, and Trevor Pinch, 80–94. Thousand Oaks, CA: Sage.

Kenney, Martin. 1985. *Biotechnology: The University–Industrial Complex*. New Haven, CT: Yale University Press.

Kinchy, Abby. 2012. *Seeds, Science, and Struggle: The Global Politics of Transgenic Crops*. Cambridge, MA: MIT Press.

Kinchy, Abby, Kirk Jalbert, and Jessica Lyons. 2014. "What Is Volunteer Water Monitoring Good For? Fracking and the Plural Logics of Participatory Science." *Political Power and Social Theory* 27: 259–89.

Kinchy, Abby, and Simona L. Perry. 2012. "Can Volunteers Pick Up the Slack? Efforts to Remedy Knowledge Gaps about the Watershed Impacts of Marcellus Shale Gas Development." *Duke Environmental Law & Policy Forum* 22 (2): 303–39.

Kleinman, Daniel Lee. 1998. "Untangling Context: Understanding a University Laboratory in the Commercial World." *Science, Technology, & Human Values* 23 (3): 285–314.

___. 2003. *Impure Cultures: University Biology and the World of Commerce*. Madison: University of Wisconsin Press.

Kleinman, Daniel Lee, and Robert Osley-Thomas. 2014. "Uneven Commercialization: Contradiction and Conflict in the Identity and Practices of American Universities." *Minerva* 52 (1): 1–26.

Kleinman, Daniel Lee, and Sainath Suryanarayanan. 2013. "Dying Bees and the Social Production of Ignorance." *Science, Technology, & Human Values* 38 (4): 492–517.

Kleinman, Daniel Lee, and Steven Vallas. 2001. "Science, Capitalism, and the Rise of the 'Knowledge Worker': The Changing Structure of Knowledge Production in the United States." *Theory and Society* 30 (4): 451–92.

Lakoff, Sanford. 1977. "Scientists, Technologists, and Political Power." In *Science, Technology, and Society: A Cross-Disciplinary Perspective*, edited by Ina Spiegel-Rösing and Derek J. de Solla Price, 335–91. Beverly Hills, CA: Sage.

Latour, Bruno. 1988. *The Pasteurization of France*. Cambridge, MA: Harvard University Press.

Latour, Bruno, and Steve Woolgar. [1979] 1986. *Laboratory Life: The Social Construction of Scientific Facts*. 2nd ed. Princeton, NJ: Princeton University Press.

Lave, Rebecca. 2014. "Neoliberal Confluences: The Turbulent Evolution of Stream Mitigation Banking in the U.S." *Political Power and Social Theory* 27: 59–88.

Layne, Linda, Sharra Vostral, and Kate Boyer, eds. 2010. *Feminist Technology*. Champaign: University of Illinois Press.

MacKenzie, Donald. 1981. "Interests, Positivism, and History." *Social Studies of Science* 11 (4): 498–501.

Martin, Brian. 2007. *Justice Ignited: The Dynamics of Backfire*. Lanham, MD: Rowman & Littlefield.

Martin, Emily. 1991. "The Egg and the Sperm: How Science Has Constructed a Romance Based on Stereotypical Male-Female Roles." *Signs: Journal of Women in Culture and Society* 16 (3): 485–501.

McCormick, Sabrina. 2009. *Mobilizing Science: Movements, Participation, and the Remaking of Knowledge*. Philadelphia: Temple University Press.

Medina, Eden, Ivan da Costa Marques, and Christina Holmes, eds. 2014. *Beyond Imported Magic: Essays on Science, Technology, and Society in Latin America*. Cambridge, MA: MIT Press.

Merton, Robert. 1973. *The Sociology of Science*. Chicago: University of Chicago Press.

Mignolo, Walter. 2009. "Epistemic Disobedience, Independent Thought, and Decolonial Freedom." *Theory, Culture & Society* 26 (7–8): 1–23.

Mirowski, Philip. 2011. *Science-Mart: Privatizing American Science*. Cambridge, MA: Harvard University Press.

Mirowski, Philip, and Esther-Mirham Sent. 2008. "The Commercialization of Science and the Response of STS." In *The Handbook of Science and Technology Studies*, 3rd ed., edited by Edward J. Hackett, Olga Amsterdamska, Michael Lynch, and Judy Wajcman, 635–89. Cambridge, MA: MIT Press.

Moore, Kelly. 1996. "Organizing Integrity: American Science and the Creation of Public Interest Science Organizations, 1955–1975." *American Journal of Sociology* 101 (6): 1592–627.

___. 2008. *Disrupting Science: Social Movements, American Science, and the Politics of the Military, 1945–1975*. Princeton, NJ: Princeton University Press.

Moore, Kelly, and Matthew Hoffmann. 2014. "'The Tip of the Day': Field Theory and Alternative Nutrition in the U.S." *Political Power and Social Theory* 27: 223–58.

Moore, Kelly, Daniel L. Kleinman, David J. Hess, and Scott Frickel. 2011. "Science and Neoliberal Globalization: A Political Sociological Approach." *Theory and Society* 40 (5): 505–32.

Morning, Ann. 2011. *The Nature of Race: How Scientists Think and Teach about Human Difference*. Berkeley: University of California Press.

Mukerji, Chandra. 1989. *A Fragile Power: Scientists and the State*. Princeton, NJ: Princeton University Press.

Nelkin, Dorothy. 1977. "Technology and Public Policy." In *Science, Technology, and Society: A Cross-Disciplinary Perspective*, edited by Ina Spiegel-Rösing and Derek J. de Solla Price, 393–441. Beverly Hills, CA: Sage.

___, ed. 1992. *Controversy: Politics of Technical Decisions*. Newbury Park, CA: Sage.

Nelson, Alondra. 2008. "Bio Science: Genetic Genealogy Testing and the Pursuit of African Ancestry." *Social Studies of Science* 38 (5): 759–83.

___. 2011. *Body and Soul: The Black Panther Party and the Fight against Medical Discrimination*. Minneapolis: University of Minnesota Press.

Nelson, Nici, and Susan Wright, eds. 1995. *Power and Participatory Development: Theory and Practice*. London: Intermediate Technology Publications.

Nieusma, Dean, and Donna Riley. 2010. "Designs on Development: Engineering, Globalization, and Social Justice." *Engineering Studies* 2 (1): 29–59.

Noble, David. 1977. *American by Design: Science, Technology, and the Rise of Corporate Capitalism*. Oxford: Oxford University Press.

Oreskes, Naomi, and Erik Conway. 2010. *Merchants of Doubt: How a Handful of Scientists Obscured the Truth on Issues from Tobacco Smoke to Global Warming*. London: Bloomsbury Press.

Ottinger, Gwen. 2013. *Refining Expertise*. New York: New York University Press.

Owen-Smith, Jason. 2003. "From Separate Systems to a Hybrid Order: Accumulative Advantage across Public and Private Science at Research One Universities." *Research Policy* 32 (6): 1081–104.

Panofsky, Aaron. 2014. *Misbehaving Science: Controversy and the Development of Behavior Genetics*. Chicago: University of Chicago Press.

Pearson, Willie. 1985. *Black Scientists, White Scientists, and Colorless Science*. Millwood, NY: Associate Faculty Press.

Pinch, Trevor J., and Wiebe E. Bijker. 1987. "The Social Construction of Facts and Artifacts." In *The Social Construction of Technological Systems*, edited by Wiebe E. Bijker, Thomas Hughes, and Trevor J. Pinch, 17–50. Cambridge, MA: MIT Press.

Pollock, Anne. 2014. "Places of Pharmaceutical Knowledge-Making: Global Health, Postcolonial Science, and Hope in South African Drug Discovery." *Social Studies of Science* 44 (6): 848–73.

Prakash, Gyan. 1999. *Another Reason: Science and the Imagination of Modern India*. Princeton, NJ: Princeton University Press.

Prasad, Amit. 2014. *Imperial Technoscience : Transnational Histories of MRI in the United States, Britain, and India*. Cambridge, MA: MIT Press.

Quet, Mathieu, and Marianne Noel. 2014. "From Politics to Academics: Political Activism and the Emergence of Science and Technology Studies in South Korea." *EASTS* 8 (2): 175–93.

Reardon, Jenny. 2005. *Race to the Finish: Identity and Governance in an Age of Genomics*. Princeton, NJ: Princeton University Press.

Rostow, Walt. 1960. *The Stages of Economic Growth: A Non-communist Manifesto*. Cambridge: Cambridge University Press.

Sardar, Ziauddin, and Dauwd Rosser-Owen. 1977. "Science Policy and Developing Countries." In *Science, Technology, and Society: A Cross-Disciplinary Perspective*, edited by Ina Spiegel-Rösing and Derek J. de Solla Price, 535–75. Beverly Hills, CA: Sage.

Schiebinger, Londa. 1993. *Nature's Body: Gender in the Making of Modern Science*. New Brunswick, NJ: Rutgers University Press.

Schroeder-Gudehus, Brigitte. 1977. "Science, Technology, and Foreign Policy." In *Science, Technology, and Society: A Cross-Disciplinary Perspective*, edited by Ina Spiegel-Rösing and Derek J. de Solla Price, 473–506. Beverly Hills, CA: Sage.

Schweber, Libby. 2014. "The Cultural Role of Science in Policy Implementation: Voluntary Self-Regulation in the U.K. Building Sector." *Political Power and Social Theory* 27: 157–91.

Segerstråle, Ullica. 2000. *Defenders of the Truth: The Sociobiology Debate*. New York: Oxford University Press.

Shim, Janet K. 2005. "Constructing 'Race' across the Science-Lay Divide: Racial Formation in the Epidemiology and Experience of Cardiovascular Disease." *Social Studies of Science* 35 (3): 405–36.

Shrum, Wesley, Joel Genuth, and Ivan Chompalov. 2007. *Structures of Scientific Collaboration*. Cambridge, MA: MIT Press.

Shrum, Wesley, and Yehouda Shenhav. 1995. "Science and Technology in Less Developed Countries." In *Handbook of Science and Technology Studies*, edited by Sheila Jasanoff, Gerald E. Markle, James C. Peterson, and Trevor J. Pinch, 627–51. Beverly Hills, CA: Sage.

Sklair, Leslie. 1970. "The Political Sociology of Science: A Critique of Current Orthodoxies." *Sociological Review* 18 (S1): 43–59.

Skolnikoff, Eugene B. 1977. "Science, Technology, and the International System." In *Science, Technology, and Society: A Cross-Disciplinary Perspective*, edited by Ina Spiegel-Rösing and Derek J. de Solla Price, 507–35. Beverly Hills, CA: Sage.

Slaughter, Sheila, and Gary Rhoades. 2004. *Academic Capitalism and the New Economy: Markets, State, and Higher Education*. Baltimore: Johns Hopkins University Press.

Smith-Doerr, Laurel. 2004. "Flexibility and Fairness: Effects of the Network Form of Organization on Gender Equity in Life Science Careers." *Sociological Perspectives* 47 (1): 25–54.

Smith-Doerr, Laurel, and Itai Vardi. 2015. "Mind the Gap: Formal Ethics Politics and Chemical Scientists' Everyday Practices in Academia and Industry." *Science, Technology, & Human Values* 40 (2): 176–98.

Spiegel-Rösing, Ina, and Derek J. de Solla Price, eds. 1977. *Science, Technology, and Society: A Cross-Disciplinary Perspective*. Beverly Hills, CA: Sage.

Star, Susan Leigh. 1989. *Regions of the Mind: Brain Research and the Quest for Scientific Certainty*. Stanford, CA: Stanford University Press.

Suchman, Lucy. 2008. "Feminist STS and the Sciences of the Artificial." In *The Handbook of Science and Technology Studies*, 3rd ed., edited by Edward J. Hackett, Olga Amsterdamska, Michael Lynch, and Judy Wajcman, 140–53. Cambridge, MA: MIT Press.

Sunder Rajan, Kaushik. 2006. *Biocapital: The Constitution of Postgenomic Life*. Durham, NC: Duke University Press.

Suryanarayanan, Sainath, and Daniel Lee Kleinman. 2014. "Beekeepers' Collective Resistance and the Politics of Pesticide Regulation in France and the United States." *Political Power and Social Theory* 27: 89–122.

Teslow, Tracy. 2014. *Constructing Race: The Science of Bodies and Race in American Anthropology*. Cambridge: Cambridge University Press.

Thorpe, Charles. 2007. "Political Theory in Science and Technology Studies." In *The Handbook of Science and Technology Studies*, 3rd ed., edited by Edward J. Hackett, Olga Amsterdamska, Michael Lynch, and Judy Wajcman, 62–82. Cambridge, MA: MIT Press.

Tilly, Charles. 1999. *Durable Inequality*. Berkeley: University of California Press.

Traweek, Sharon. 1992. "Border Crossing: Narrative Strategies in Science Studies and among Physicists in Tsukuba Science City, Japan" In *Science as Practice and Culture*, edited by Andrew Pickering, 429–65. Chicago: University of Chicago Press.

Tuana, Nancy, ed. 1989. *Feminism and Science*. Bloomington: Indiana University Press.

Vallas, Steven, and Daniel Lee Kleinman. 2007. "Contradiction, Convergence and the Knowledge Economy: The Confluence of Academic and Commercial Biotechnology." *Socio-economic Review* 6 (2): 283–311.

Vessuri, Hebe. 1988. "The Universities, Scientific Research, and the National Interest in Latin America." *Minerva* 24 (1): 1–38.

Visvanathan, Shiv. 1997. *A Carnival for Science: Essays on Science, Technology, and Development*. Delhi: Oxford University Press.

Wajcman, Judy. 1995. "Feminist Theories of Technology." In *Handbook of Science and Technology Studies*. 2nd ed., edited by Sheila Jasanoff, Gerald Markle, James Petersen, and Trevor Pinch, 189–204. Thousand Oaks, CA: Sage.

Williams, Logan D. A. 2013. "Three Models of Development: Community Ophthalmology NGOs and the Appropriate Technology Movement." *Perspectives on Global Development and Technology* 12 (4): 449–75.

Williams, Logan D. A., and Thomas Woodson. 2012. "The Future of Innovation Studies in Less Economically Developed Countries." *Minerva* 50 (2): 221–37.

Winner, Langdon. 1986. *The Whale and the Reactor: A Search for Limits in an Age of High Technology*. Chicago: University of Chicago Press.

Wisnioski, Matthew. 2012. *Engineers for Change: Competing Visions of Technology in the 1960s*. Cambridge, MA: MIT Press.

Ynalvez, Marcus Antonius, and Wesley M. Shrum. 2011. "Professional Networks, Scientific Collaboration, and Publication Productivity in Resource-Constrained Research Institutions in a Developing Country." *Research Policy* 40 (2): 204–16.

12   Race and Science in the Twenty-First Century

Ramya M. Rajagopalan, Alondra Nelson, and Joan H. Fujimura

The production of knowledge and its limits and implications comprise a central set of concerns within science and technology studies (STS). Scholars in the field have long been concerned with how knowledge is made and used, how it is authorized and legitimated, contested and displaced, as well as applied and appropriated in different contexts. Within this set of concerns are embedded questions about how politics and culture are complicit in and shape scientific activity. Contemporary knowledge practices produce scientific and biomedical understandings of human bodies that proliferate in a multiplicity of technoscientific arenas. This chapter explores a highly contested and highly politicized site of knowledge production: the scientific practices that have produced knowledge about human difference, and in particular, racialized understandings of human bodies. Drawing on developments in genetics over the past decade primarily in the United States, this chapter focuses on how, where, and when scientific claims about race are produced and with what consequences. Studies of race and science bring together two important lines of inquiry within STS: analysis of the production of knowledge through examinations of scientific practice and analysis of how these knowledge forms are mobilized and enacted in larger social and institutional terrains.

For centuries, sociologists, biologists, anthropologists, historians, and philosophers of science have sought to pin down what race "actually is." Meanwhile, through these and other actors (including institutions), varied conceptions of race have proliferated across social and political domains. Invested actors have argued for varied notions of race: as socially constructed; as natural, innate, and biologically determined; or as something in between.

This chapter provides a historically situated overview of the last decade of debates and scholarship concerning the nature of race and of race classification in the context of contemporary biology and genetics. Although many twenty-first century social scientists view race formation and race categories as processual, historical, and dynamic, debates over the "nature" of race have been invigorated with renewed intensity in recent years with the rise of genomic science. Examination of these debates permits exploration of and sheds light on core theoretical concerns within STS, for example, classification and categorization; the contextual dimensions of the production and management of scientific knowledge and authority; and the artificial binaries of nature and culture, of science and society, and their mutual co-construction. The arena of race and genetics is a complex and fruitful site in which to examine the ways in which knowledge is produced in the domains of the sciences and how it is then used and applied in other sociopolitical arenas, sometimes to authorize and legitimize, and sometimes to destabilize or replace, prevailing sociocultural belief systems and practices.

Understanding these debates is important because struggles over the definitions and uses of concepts of race have historically indexed larger struggles for power and resources, political legitimacy, and health and human rights. In recent years genetics has risen to prominence within biomedicine and the life sciences, allowing this field to wield increasing scientific authority. Findings from genetics are being employed to transform the categorization of human diversity, with significant stakes across realms of social and political life, including medicine, forensics and the law, and individual, group, and national identity.

Early Race Concepts in the Sciences and Scientific Racism

Social, cultural, and political practices of classifying humans have a long and contentious history. Frequently, these practices have pivoted around conceptions of difference anchored in ideas of "race," and related concepts of ethnicity, ancestry, and national origin. These concepts have been used for centuries as the basis for lumping and splitting humans into various groups to achieve particular political, economic, and sociocultural purposes. Invariably, these practices have arisen from and resulted in systems of oppression, subversion, and domination built on social hierarchies in which some groups (usually those wielding power) were seen to be superior to other, usually economically and politically subordinated, groups.

The ideology of race emerged during the heyday of European expansion around the world, as colonial empires attempted to justify the socioeconomic privilege of white colonists, the plundering of "their" colonies, and the subjugation of indigenous groups and involuntary migrants in far-flung colonies. This race ideology was based on the belief that white Europeans were culturally, mentally, and physically more evolved than and thus superior to colonized peoples (Smedley 2007).

In the seventeenth and eighteenth centuries, some practitioners in the natural sciences began creating human classification schemes that reflected these sociopolitical racial hierarchies. In the seventeenth century, French physician and voyager François Bernier was among the first to try to classify peoples around the world into distinct races using morphological traits (Smith 2015). Partly influenced by Bernier, eighteenth-century Swedish naturalist Carolus Linnaeus and his student, Johannes Blumenbach, formalized the first detailed Euro-American taxonomy of races as natural kinds, one that has changed over time but whose values have persisted. Linnaeus and Blumenbach proposed hierarchies of four and five major races, respectively, each originating on a different continent, implicitly ranked along visible phenotypic attributes such as skin and hair color. Their schemes attempted to extend Aristotle's "great chain of being," a ranking of the animal, vegetable, and mineral world, to humans. Through Linnaeus and Blumenbach, the "great chain of being" was reinvigorated to subclassify humans by continental region, placing "evolved" Europeans at the top of a ranking of beauty and worth, and ranking others lower, closer to "savage" animals (Gould 1996; Marks 1995).

In the following three centuries, a handful of white Euro-American scientists continued to build the theory that darker-skinned people were biologically inferior in their skeletal features, brain size, and temperament (Gould 1996; Schiebinger 1993). This "scientific racism" quantified aspects of nonwhite bodies, attempting to demonstrate that they were physiologically and mentally deficient and thus deserving of their lower socioeconomic status. In the process, racial differences were essentialized—that is, cast as natural, innate, and immutable. For example, during the height of slavery in the United States, proslavery physicians like Samuel A. Cartwright argued that black and white bodies exhibited significant anatomical and skeletal differences, as well as differences in bile and blood composition. They concluded that black physical features were perfectly biologically suited to, and indeed made for, slavery (Hammonds and Herzig 2009; Jackson 2014). Braun (2014) similarly demonstrates how U.S. Civil War–era scientists turned the spirometer into an instrument of racism, generating questionable data about the supposedly reduced lung capacity of black soldiers compared to white soldiers. Even into the twentieth century, such racist sciences have purported to identify and measure racial differences in physiology, such as in bone density (Fausto-Sterling 2008), between groups defined as white, black, Asian, and Native American. Racist sciences constructed biological research outcomes that underscored the very differences they were looking for, and these alleged biological differences were in turn mobilized as "explanations" for still other alleged differences across race, for example, in ability, intellect, and health. Racist sciences attached value and meaning to any differences they claimed to have found, building conceptual hierarchies of race with the imprimatur of "objective" science.

Actors in positions of political, economic, and social power used these conceptual hierarchies to legitimize existing systems of inequality in society. Hierarchies founded on these so-called "scientific" definitions of race formed the basis of early twentieth-century misuses of scientific authority to legitimize racist political agendas and systems of political and economic oppression against subjugated groups. STS scholars have analyzed the foundational assumptions of these regimes and the ways in which they perpetuated systemic inequalities using findings from "science" as justification. For example, Nazi scientists in mid-twentieth-century Germany proposed that people of Jewish ancestry, as well as groups defined as non-Aryan (including those with African and Asian ancestry), were biologically inferior. These interpretations supported Nazi doctrines of racial hygiene and purity and culminated in the persecution and mass genocide of over six million Jews and members of other racial and ethnic minorities. From 1948 to 1994, the all-white Afrikaner National Party legislated the apartheid regime in South Africa, mobilizing racist science, racial classifications, and doctrines of racial purity to segregate blacks to the poverty of shantytowns and ban interracial contact (Bowker and Star 1999).

As racially motivated atrocities and systems of white domination and oppression proliferated around the world, many biological and social scientists in the United States and Europe began by the mid-twentieth century to disavow "pseudoscientific" investigations of race. Anxieties arose among many Western intellectuals, that the atrocities of racial domination and racial hygiene had in many cases been committed with the blessing of science, or alternatively in the name of science (such as Nazi experiments of torture to test the physical limits of racially devalued bodies). American and European anthropologists, sociologists, and biological scientists began for the first time to systematically intervene to try to "settle" the question of race and biology.

Through the UNESCO Statements (Montagu 1972)—originally in 1950 and subsequently in 1951, 1964, 1967, and 1978—scientists asserted that race and race classifications were produced and given meaning through social, political, and cultural processes in different times and places. On the surface, the UNESCO statements appeared to reflect an anti-essentialist, anti-racist consensus within the human sciences, but recently STS scholars and historians have exposed some of the cracks in this seemingly unified position. They note that some of the scientists who drafted and signed on to the UNESCO statements, for instance, argued that the use of race concepts to study evolutionary processes should be allowed to continue but that these should not be used to justify social or political ends. Race, these biologists contended, was a natural concept and had no inherent social meaning (Brattain 2007; Reardon 2004). Thus, although human biologists and geneticists had supported the UNESCO statements that decried racism and the use of science to support racist doctrines, many did not relinquish the idea that race and related concepts had some biological basis. Some continued to mobilize race concepts in their scientific research.

As Donna Haraway has shown, throughout the twentieth century a complex web of anxieties, fears, tensions, and technoscientific commitments surrounded biological investigations of race (Haraway 1997). In the postwar period, a range of race concepts were used in biology, including typological races based on Blumenbach's scheme. However, the most popular notion of human group differences was "population." Through the notion of population, geneticists like Theodosius Dobzhansky (an adviser to the UNESCO statements) and Leslie Dunn advanced the idea that human groups are differentiated not by the absolute presence or absence of different traits but by the statistical frequencies of different versions of genes (known as variants or alleles). The different frequencies, they argued, were unique to each population and reflected evolutionary processes, resulting partly from geographical localization on different continents in earlier eras of human history, as well as breeding laws and cultural customs of reproduction. This notion of population reflected a particular conception of race, one that was meant to be nonideological and nonhierarchical and to signify genetically open, fluid, and internally heterogeneous groups rather than discrete or typological groups. Nevertheless, it was a concept of race grounded in biology and theories of evolution. Although Dobzhansky was instrumental in shifting discussions in the biological sciences from earlier "typological" views of race to "populationist" notions, the latter nevertheless laid the groundwork for a "genetic race concept," retaining the core idea of bounded, genetically differentiated groups (Gannett 2001, 2013; Reardon 2007).

Historians, philosophers, and sociologists of science have shown that notions of race as biological continued to be a specter in scientific work throughout the twentieth century (Brattain 2007; Gissis 2008; Lipphardt 2014; Morning 2011; Müller-Wille 2007). They argue that the notion of population permitted ideas of race as biology to persist and sanitized the study of human genetic differences in an attempt to diffuse some of the post–World War II anxieties around race. Some have examined how scientists engaged in this race-making work and pedagogy themselves think and write about human difference (Bliss 2012; C. Lee 2009; Morning 2011). Thus, despite claims to the contrary, the biological race concept remained in play, even in populationist thinking.

Meanwhile, and in contrast, some scholars in sociology and anthropology, as well as evolutionary genetics, began to argue in the twentieth century that race and race categories were not durable, natural, biological, or innate groupings (Boas 1912; Du

Bois 1940; Gould 1996; Lewontin 1972; Montagu 1942). For example, Lewontin (1972) examined fifteen genes and demonstrated that variation within the same race group was actually *much larger* than variation between race groups. These scholars saw race and race categories as sociohistorical, dynamic, and specific to particular times and places, and produced relationally and processually (Blumer 1958; Duster 1990; Hacking 2005; Haraway 1997; Stepan 1986). Race and ethnicity were enacted as cultural idioms and political processes, discursive frames, and political projects (López 1996; Omi and Winant 1986). By the end of the twentieth century, there was a clear consensus, at least among social scientists and some biologists, that race should be studied as political, social, and cultural constructs.

Race-Based Experimentation and Medical Racism

Despite broad consensus in the social sciences that race and race categories are delineated and produced by social forces for different political ends, practitioners in the biological and allied medical sciences repeatedly proposed the existence of racial bodily differences throughout the twentieth century and continued to mobilize biological conceptions of human difference coded in terms of race. Echoing attempts to define anatomical, physiological, and blood-based inferiority of blacks relative to whites in colonial contexts, these studies were often ethically fraught, and not widely known in public circles, coming to light years or decades later through the work of sociologists, anthropologists, historians of science, and science journalists. Studies of racialized bodies exploited racial and ethnic minorities, particularly those at considerable political and socioeconomic disadvantage.

Medical experiments on racialized bodies involved dangerous and often fatal procedures, often without the consent of participants. For example, physicians in the United States Public Health Service in the 1930s were interested in racial differences in the long-term progression of the deadly syphilis disease. In the now infamous Tuskegee Syphilis Study, conducted between 1932 and 1972 in the Tuskegee area of Macon County, Alabama, 399 black men who had contracted the disease were medically tracked for any changes in cardiovascular or mental health over time, most without their knowledge or consent. When the study finally reached the attention of the media in 1972, it precipitated widespread public outcry and was shut down by the United States government. Ethicists called for stricter research protections for vulnerable medical research subjects (Reverby 2009). Within African American communities, skepticism and mistrust of the medical establishment and public health authorities predates the Tuskegee experiment but was certainly exacerbated by it.

During World War II, the Pentagon attempted to study so-called racial differences in responses to chemical weapons, conducting clandestine experiments with over 60,000 American troops. African American, Puerto Rican, and Japanese American soldiers were included as test subjects, as well as white men, whose responses to chemical weapons exposure were used as a baseline against which to compare the physiologic responses of other "races" to mustard gas chambers. Soldiers in the study did not receive follow-up treatment for painful blisters and burns. Later in life, they suffered debilitating illnesses and psychological traumas due to the chemical tests to which they had been subjected (Smith 2008).

As these examples demonstrate, the medical sciences continued to propose and test the existence of racial bodily differences throughout the twentieth century. Such was the faith in innate racial differences that racialized groups were exploited repeatedly for these and other experiments. Nevertheless, even in these examples, race was not a stable concept. Rather, conceptualizations, operationalizations, and constructions of race through these medical experiments were ongoing, though often self-reinforcing.

Race and Human Genetics after World War II

Many prominent early twentieth-century human geneticists in the United States and Europe were also eugenicists, believing that certain debilitating mental and physical diseases ran in particular families, which suggested to them that these diseases were hereditary. They advocated that affected individuals and families practice voluntary eugenics by abstaining from reproducing. Some of the more extreme eugenics movements advocated forced sterilization of imprisoned criminals, patients in psychiatric or mental institutions who were deemed to be mentally deficient, and other groups— including racial and ethnic minorities—deemed to be of "inferior stock" and therefore unfit to reproduce. Eugenics advocates argued that this was the safest course of action to protect the human population from further suffering due to the persistence of mental and other diseases (Kevles 1995; Paul 1995). From the 1930s to the 1970s in American states like North Carolina and California, eugenics through sterilization was de rigueur though little known.

By the 1970s, with the advent of new molecular tools for analyzing chromosomes and manipulating DNA, geneticists began to look for the molecular causes of some of the diseases that were highly hereditary and appeared in families or in groups that tended to intermarry. In time, for many of these diseases, researchers uncovered a single associated gene and determined that symptoms were largely caused by variations in the gene sequence, which compromised the function of the gene's protein product. But

in the process of studying and treating these diseases, including Tay-Sachs, phenylke-
tonuria (PKU), and sickle cell anemia, medicine and medical research constructed these
diseases as if they were exclusively associated with specific racialized and ethnicized
groups. Researchers and doctors viewed people of Jewish descent to be at much higher
risk for Tay-Sachs and PKU (Paul and Brosco 2013), and Tay-Sachs and PKU were soon
widely believed to be exclusively "Jewish diseases." Similarly, science and medicine
constructed sickle cell anemia as an "African American disease" (Wailoo and Pemberton 2006) even though mounting evidence has demonstrated that the variant gene,
the disease, and associated blood afflictions are prevalent across the Middle East and
parts of Asia and prevalent only in certain parts of Africa.

Nevertheless, the latter half of the twentieth century also saw a more complex set
of public reactions to the relationship between race and disease (Duster 1990), with
the rise of health activism around particular diseases. Jewish communities supported
genetic research on Tay-Sachs and other diseases that appeared frequently in their
communities and even took measures to eradicate these diseases through voluntary
family planning. In contrast, reactions to research on sickle cell disease among African American communities were more complex. Duster argues that the differential
interest in genetic research has been in direct proportion to the ability of these communities to exert control over policy outcomes of research. Nelson (2011) notes that
the Black Panther Party built free health care clinics in the 1960s and 1970s and created educational programs to encourage early detection, monitoring, and treatment
of sickle cell disease. Party members were also vocal critics of the lack of attention
given to "black" diseases like sickle cell disease in federal health research programs. At
the same time, the activists urged blacks to avail themselves of the Party's free genetic
testing and counseling and encouraged active research into therapies and possible
cures for sickle cell anemia.

Race and Genomics in Contemporary Biosciences

The twenty-first-century biosciences have amplified the complexity of both genetic
research and debates about race, medicine, health, and ethics. STS researchers have
taken different approaches to understand how race as biology has continued as an
object of study and a tool in contemporary science, technology, and medicine. Scholars
of the contemporary era of genomics have examined how scientific theories linking
biology and race are being newly constructed by some geneticists, using new genomics
technologies. This work has explored how race continues to be used as a shorthand for
bodily (and biological) difference in biomedicine, and the issues of concern that these

practices raise, particularly around our understandings of what race is, what it means for identity and health, and how such uses do or do not address ongoing structural racial inequalities and persistent discrimination.

Over the past twenty-five years, coalescing around the decade-long Human Genome Project (HGP), the study of DNA and genomics has quickly accelerated to prominence within biomedicine. The HGP, lasting from 1990 to 2003, was a multinational effort to generate a reference sequence of all three billion base pairs in a human genome, along with physical and genetic "maps" of the chromosomes. The HGP marked an era during and after which DNA and genetics came to be seen by many as the ultimate source of disease. As STS scholars have shown, this view is problematic for a number of reasons, not least of which are the conceptual connections that continue to drive popular perceptions of the association between disease incidence, genetics, and certain racialized groups. Duster's (1990) concept of the "prism of heritability" and Lippman's (1991) concept of "geneticization" captured what became in the early 1990s a dominant turn within biomedicine to genetic explanations for health and disease. Both warned during the HGP's early phases that the social determinants and environmental causes of disease were being overshadowed by a linear discourse that placed genetics at the center of discussions of disease etiology. Duster, for example, drew attention to the implications for at-risk racial minority communities for whom interventions to improve health care access, education, sanitation, and other necessary social services would have a much more pronounced and enduring impact than genetic intervention. Patient activists and other laypeople also became engaged in the processes of privileging genetic explanations for disease, often in an effort to curb health disparities and other inequalities, but not always with the full implications of this genetics turn in mind.

STS studies since have examined how contemporary scientific practices and technologies in biomedical and population genetics construct scientific theories, technologies, practices, and evidence about race or genetic groups. Since the completion of the HGP, genomics studies have become increasingly concerned with examining human genetic variation, focusing on tiny points of difference among human genomes. The assumptions built into some of these studies and their conclusions have reinvigorated debates about the relationship between biological difference and race, the social underpinnings of race, and racial classification schemes that continue to be used in the biosciences.

Contemporary genomics has been guided by theories of human evolution emerging out of twentieth-century evolutionary and population genetics and biological anthropology, specifically, hypotheses about the distribution of genetic variation in human groups (Fujimura et al. 2014). These theories postulated that morphological and genetic differences accumulated over thousands of years, during and after the early migrations of anatomically modern humans out of Africa, and during successive waves of expansion spread to different continents. They argued that on separate continents, the ancestors of anatomically modern humans spent some part of their history reproductively isolated from each other; within geographical locales, laws and cultural customs over millennia specified accepted reproductive norms and interbreeding relations among tribes and nations; and along with processes of natural selection, processes of selective mating shaped human genomes. Geneticists use different interpretations of this basic understanding of human evolution as they conduct different types of genomics projects.

Concurrent with the HGP in the early 1990s, Stanford University geneticist Luca Cavalli-Sforza, University of California, Berkeley geneticist Allan Wilson, and their colleagues proposed the Human Genome Diversity Project (HGDP), an international effort to collect and study the DNA of peoples from around the world, which these researchers believed would shed light on human migration and variation. After Wilson's death in 1991, Cavalli-Sforza's framing of the project took precedence over Wilson's idea of evenly spaced human DNA sampling using a "grid" thrown around the globe. In contrast, Cavalli-Sforza's frame used predesignated social and linguistic groups as the basis for collecting biological samples. He further proposed to study and preserve the "genetic heritage" of humanity by beginning with isolated and "vanishing" tribes whose language, culture, and livelihood, he argued, were increasingly being threatened to extinction or hybridization by the encroaching forces of globalization. This approach to the HGDP sparked critical responses. Activist groups argued there was no clear indication that participant groups would have a voice in the framing or execution of the project. They dubbed it the "Vampire Project" and criticized it as a new form of colonization, biopiracy, and scientific racism posing as thinly veiled international cooperative science. STS scholars showed how the project intensified debates about the nature of race by attempting to bring biology and genetics into the construction of distinctions between groups in far-flung corners of the globe (M'charek 2005; Reardon 2004). Indeed, the HGDP as conceptualized by Cavalli-Sforza resembled mid-twentieth-century efforts to map genetic diversity among populations through a "genogeographic atlas" in the Soviet Union (Bauer 2014).

Studies of human variation also emerged from other quarters. In 2003 the National Human Genome Research Institute (NHGRI), a unit of the United States National Institutes of Health, launched an international study to catalog human DNA variation, the International Human Haplotype Mapping Project (or HapMap). More specifically, the

HapMap was interested in documenting single points of DNA difference known as single nucleotide polymorphisms (SNPs), in people around the world.

Partly in response to criticisms of the HGDP, the HapMap organizers explicitly and from the outset highlighted the importance of attending to issues of donor consent and data privacy. They implemented community engagements and consultations with participating groups prior to collecting DNA and follow-up engagement activities after collecting DNA. Nevertheless, the slate of initially sampled groups—Yoruba in Ibadan, Nigeria; Japanese in Tokyo, Japan; Han Chinese in Beijing, China; and Utah residents with ancestry from Northern and Western Europe—engendered some criticism. Scholars worried that their strong resemblance to the continental race classifications "European," "Asian," and "African" might be misread by the media and publics, as well as by scientists using the data, as evidence that races were genetically defined categories (Hamilton 2008; S. Lee 2015; Reardon 2007).

As the HapMap was ongoing, genome scientists also began to prepare to use genetic variation as a tool to study disease. Many recognized that racial and continental groupings are internally very heterogeneous, socially, culturally, and genetically. But as early as 1998, a few biological anthropologists and population geneticists had begun to develop SNP markers that they claimed could be used to distinguish different continental ancestries, including European, African, Asian, or Native American. They called these markers "population-specific alleles" and later "ancestry informative markers" (AIMs). STS scholars have deconstructed the AIMs technology, showing how human populations and genetic markers were carefully designed and selected in order to construct AIMs that would seem to correspond with minority race groups in the United States. Although their intent was to facilitate studies of disease in different racial and ethnic populations, AIMs researchers conflated genetics and race in explicit, often totalizing ways, claiming to be able to use DNA to quantify the different proportions of "continental ancestries" and even "European" versus "African" chromosomes in an individual. Equally important, AIMs researchers used these markers to design "admixture mapping" methods to study disease risk in these groups. Their conclusions have circulated widely in the media and reinforced the perception that certain race groups are more susceptible to certain diseases, even though these diseases are prevalent throughout the human population (Fullwiley 2008; Rajagopalan and Fujimura 2012a).

However, some researchers involved in admixture mapping themselves identified with minority groups and argued that their genomic studies would provide a measure of social justice and health equity to marginalized communities historically overlooked or exploited as research subjects in biomedicine (Bliss 2012; Fullwiley 2008). That is, by appealing to efforts at "inclusion" of historically underrepresented groups

(Epstein 2007), admixture mapping studies have attempted to legitimize the use of race in genetics through an appeal to the eradication of health disparities. Some scholars have described these as deliberate strategies to distance current genome sciences from the taints that plagued the HGDP and earlier projects on race and science, and to render contemporary genomics efforts as democratic and "antiracist" work intended to advance public health (Bliss 2012; Reardon 2012).

More recently, biomedical genetics researchers have employed the genome-wide association study (GWAS) method to study genetic associations to disease. Unlike admixture mapping, GWAS did not overtly rely on notions of race (Fujimura and Rajagopalan 2011). Instead of using race as a surrogate for population, some GWAS researchers devised statistical genomics tools and software that allowed them to control for confounding genetic similarities among cases or controls without naming or ascribing these similarities to any particular population or race group. These GWAS researchers were interested in understanding whether any regions of the genome were associated with common complex diseases but were less interested in tracing disease through populations or race groups. They viewed race categories as too imprecise for scientific work, artificially homogenizing what were actually quite genetically heterogeneous groups. Unlike admixture mapping, which assumed different continental race groups were distinguishable through genetics, the practitioners of GWAS conceptualized populations on finer geographic scales. Nevertheless, some members of the media and some in the scientific community read early GWAS studies as an affirmation that race had a biological component.

Meanwhile, race was also being used to construct profit-driven pharmaceuticals and to reimagine racial and ethnic groups as having presumptive and differential genetic responses to prescription medications via "personalized medicine," often without sufficient evidence (Kahn 2013; Pollack 2012; Prainsack 2015; Rajagopalan and Fujimura 2012b; Roberts 2011). For example, in 2005 the heart failure drug BiDil became the first race-specific pharmaceutical approved by the FDA. BiDil was a combination of two generic drugs for which there was no data to suggest differential efficacy across race groups. Nevertheless, BiDil was patented and marketed as a "race-specific drug" exclusively for African American heart patients (Kahn 2013). Though BiDil ended up a commercial failure, it serves as a troubling reminder of the extent to which social constructs and concepts of race are deeply embedded in, and built on by, activities in the biomedical sciences.

Classification, Categorization, and Social Implications of Research on Race, Genetics, and Disease

As investigations of the sociomaterial production of knowledge, empirical analyses of the role of race in the contemporary biosciences illustrate how our representations and understandings of nature and human difference are co-constituted. They also demonstrate the many ways in which STS scholars can attend to the varied practices at play, and their consequences, in a complex field where actors and institutions pursuing different political projects continually negotiate, split, and consolidate concepts and categories of difference.

The ways in which human groups have historically been identified, labeled, and valued are intimately tied to the ways in which these groups are constituted both from the outside, at the level of institutions and governance, and from within, at the level of individual and group identities. Classification and categorization have long been core theoretical concerns within STS, and classification schemes are closely tied to language, meaning, and interpretation, as many scholars in STS and beyond have shown (Bowker and Star 1999; Douglas and Hull 1992; Haraway 1989; Stepan 1986). They also establish boundaries and divisions. Conceptual meanings of race in different times and places have sedimented in labels, terms, and practices that became political ways of counting, and accounting, for populations. Braun and Hammonds (2008) illustrate how the concept of population was put into practice by mid-twentieth-century anthropologists in Africa, who named and cataloged tribes as discrete population units to render them more amenable to scientific study and classification. Similarly, Bowker and Star (1999) examine the production of official race classifications in apartheid South Africa, using STS approaches to expose the unstable and ambiguous nature of race amid competing interpretations of color and class. In nineteenth-century British-occupied South India, colonial and indigenous ethnographers used anthropometric science in efforts to name, count, and hierarchically classify local tribes and castes, rendering prevailing social divisions as seemingly natural (Philip 2004).

As Nancy Leys Stepan (1986) has argued, the language, metaphors, analogies, and terms that people use all play a significant role in constructing the objects that they are meant to signify in the world around us. Within the contested field of race and genomics, many scholars have explored the dangers of using race categories and labels as variables in biomedical research, including the imprecise nature of such terms for capturing genetic diversity in large heterogeneous race groups, and the lack of control about how they are viewed by the media or readers unfamiliar with the debates (Ellison et al. 2008; Fujimura and Rajagopalan 2011; C. Lee 2009; Sankar and Cho 2002; Smart et al. 2008).

Social scientists have critiqued studies that link race and genetics because they threaten to reinvigorate the belief that human social classification systems like race have a biological, "natural" basis. Following on early critiques during the HGP, scholars have argued that contemporary genomics has ushered in what has been called the "molecular" or "genetic reinscription of race" (Abu El-Haj 2012; Duster 2006; Ossorio and Duster 2005) or the "molecularization of race" (Fullwiley 2007). An overemphasis on biological and genetic views of disease causation eclipses the significance of more proximal causal factors such as unequal living conditions. Mexican Americans at the Texas border, for example, have been enrolled through "bioethnic conscription" into genetic studies of diseases in their communities (Montoya 2007), which reads race, and disease, as innate properties of the body and DNA rather than as a product of sociocultural forces that produce inequalities across race and class.

Importantly, as Duster (1990) has pointed out, though race is a product of human societies and prevailing social values rather than a product of our biology or our genes, it is no less "real." The practices, structures, and institutions that enact race inflict "real" biological consequences. The literature on race, health, and health disparities demonstrate the very material effects of the negative experiences of race and racial discrimination. Poor environments and socioeconomic conditions, including limited access to adequate education, employment opportunities, and health care, likely play a much larger role than genetics in the differential disease burdens experienced by different communities. For example, many scholars have shown how racial minorities are disproportionately subjected to resource-poor living conditions and environments and lower socioeconomic status in the United States. They demonstrate how the negative experiences of being identified as a "minority race" in a racialized society can induce physical stress or exacerbate early symptoms of disease. In this way the negative experiences of race can manifest as disease and "become" biology, inscribed on and in the racialized body (Duster 2015; Geronimus 1992; Gravlee 2009; Holmes 2013; James 1994; Krieger and Bassett 1993; Montoya 2007; Pearson 2008; Viruell-Fuentes 2007).

When biomedical researchers do pay attention to socioeconomic or environmental factors, genetic explanations are often seen to trump other accounts, even when there is little evidence. For example, scholars have shown that researchers who study whether interactions between specific genes and environments might lead to disease, nevertheless often privilege genes as authoritative measurements of race and as sources of information about both race and disease (Shim et al. 2014; Whitmarsh 2008). As scholars have noted since the earliest days of the HGP, this genetic reductionism and reinscription of race as genetic could have profound social implications and consequences for certain groups. If genomic research is conducted on groups who already experience stigma and discrimination, for example, such studies could impose greater harm on these groups (de Vries et al. 2012). For another example, the use of race as a proxy for genetics could result in individuals from certain sociocultural groups being uncritically categorized or "lumped" together, assumed to share biomedical and genetic similarities in susceptibility and drug tolerance, and potentially leading to incorrect diagnoses and treatments. As Duster argues, the authority of genetic and genomic explanations requires political and legal systems to take extra cautions to guard against the most extreme abuses and avoid ascribing genetic meanings to race that might open a "backdoor to eugenics" whereby certain groups are subjected to policies that limit or restrict their life chances and reproductive choices (Duster 1990). Without sensitivity to these concerns and potential pitfalls, applications of genomics could exacerbate rather than alleviate poor health, stigma, and discrimination.

Race and Genetics in Identity, Governance, Law, and Forensics

Genome scientists have emphasized that associations between DNA and disease must be further characterized to determine if they have any clinical value for disease prediction or treatment. Despite these cautions, an entire industry in direct-to-consumer (DTC) genetic testing sprung up as companies in the United Kingdom, United States, and Iceland began to commercialize the genetic markers through fee-based tests that analyze GWAS-identified SNPs, as well as AIMs, mitochondrial DNA, and Y-chromosome DNA. Some companies claimed to predict consumers' personal disease susceptibilities, while others claimed they could infer and quantify individual genetic ancestry. Geneticists and their professional associations have critiqued these tests in both the United States and the United Kingdom, arguing that such "recreational genomics" are providing inaccurate and misleading information to consumers. Social science scholars have also intervened, warning consumers about the technical limitations of the science underlying the tests, which makes them highly unreliable for predicting individual medical susceptibilities or ancestry (Bolnick et al. 2007; S. Lee 2013). They have voiced concerns that the juxtaposition of DNA, ancestry, and race in the direct-to-consumer marketplace may position the genetic sciences as the ultimate arbiters of identity, buttressing scientistic claims to "objective truth" about human origins. However, DTC genetic ancestry testing holds particular appeal for individuals who may not have other ways to access information about their ancestry, such as adoptees and the descendants of enslaved persons (Nelson 2016). Some companies marketed the technology by appealing to individuals whose family ancestry information was missing or absent in the archives due to the legacies of slavery or genocide. In this way, genetics, genomics, and race became implicated in "reconciliation projects" (Abu El-Haj 2012; Nelson 2012, 2016; Schramm, Skinner, and Rottenburg 2012).

STS scholars have explored the allure of genetic information within this contested terrain of commercial testing and tried to understand if and how social identities are being remade or transformed by personal genetic readouts. Nelson (2008a, 2008b, 2016), TallBear (2013), and Tamarkin (2014) show that some consumers react in much more complicated ways, using new genetic information as just one of many dimensions along which they navigate and negotiate among potentially conflicting narratives of selfhood, biographical history, and group belonging. For example, African American "root seekers" have employed genetic ancestry testing to forge community, solidarity, and citizenship with groups living on the African continent. In many instances, they do not take the genetics results as fact but instead use this information to develop sociopolitical projects, self-consciously employing the methods for their own ends. They combine DNA ancestry results with other genealogical stories to craft new genealogies and histories for themselves and their families. Genetic testing is also being used by some native and tribal communities in the United States to both enroll and disenroll members. Thus, through courts of law, historical institutions, and activist campaigns, communities and individual test takers are using new genetic technologies to develop community cohesion and collective memory, to achieve social justice, to reunite family members, and simultaneously to rebuild histories of community, nation-state, and diaspora. Their projects concern the adjudication and resolution of contentious issues in reconciliation efforts, and in this way, they act as cultural and political entrepreneurs toward particular political ends. As Nelson (2008a) argues, DNA analysis is sometimes being used to redress—rather than create—inequality.

Scholars in STS have also been interested in how scientific arguments and evidence about race and identity are used in political debates and governance processes. Increasingly, regional and national governments, private enterprises, and genome scientists have turned to the authority of genomics to build narratives of national, tribal, and ethnic origins and genetic uniqueness. Some genomics researchers, for example, began to use GWAS methods and tools for the purpose of tracing differences among different ethnic and regional groups and for constructing historical patterns of migration and colonization in different nations. STS scholars have pointed out the ways in which these attempts to map race onto place through genetics have been mobilized to reconstruct and reify race and identity as molecular, with implications for understandings of citizenship and rights.

In 2005 the National Geographic Society launched the Genographic Project, a worldwide survey of the DNA of indigenous communities and regional ethnic groups. It posited, like the HGDP, that collecting and analyzing DNA from donors around the world, but especially from tribal groups, could help construct the stories of how humans moved, migrated, and evolved from ancient to modern times. However, these knowledge-making regimes exclude and often clash with and endanger local histories and sensibilities about ancestry and place and have the potential to weaken indigenous claims to land, resources, and sovereignty (TallBear 2013). For example, the appropriation of tribal DNA in the Genographic Project threatens to destabilize, displace, and colonize indigenous narratives of identity. Reardon and TallBear (2012) seek to disrupt the biocolonial interactions whereby researchers appropriate indigenous DNA to further their science, by proposing more democratic, mutually beneficial, and ethical modes of engagement between genomics researchers and the indigenous groups whom they study.

Although this chapter focuses on Euro-American constructions of race, we acknowledge that race is understood and practiced differently elsewhere, and many within STS have been attentive to local practices that imbricate biology, genetics, and understandings of human difference. Several countries have built national DNA biobanks to investigate relationships between diseases and genetics among their citizens, but also to reconstruct their demographic history and to map ethnicity and indigeneity. These projects rely on and construct "genome geographies," topologies of difference that flatten rich local histories into linear stories tied to molecular genetic understandings of diversity (Fujimura and Rajagopalan 2011). When researchers use DNA to trace the historical roots of different peoples within their borders, they lend scientific credence to particular, often politicized, narratives about citizenship, indigeneity, and belonging. STS scholars have examined how these state-based projects are used to craft narratives of regional, racial, and ethnic identity, for example, through national biobank projects in Iceland (Fortun 2008), Taiwan (Tsai 2012), and the United Kingdom (Fortier 2012; Nash 2013; Tutton, Kaye, and Hoeyer 2004). Elsewhere, state-based genomics projects have attempted to narrate population origins and a biology of difference tied to specific ethnicized groups within national territories, including French-Canadians in Québec (Hinterberger 2012), census race groups in Brazil (Santos et al. 2009), regional and caste groups in India (Egorova 2009; Title VI Committee 2006), Taiwanese peoples as distinct from Chinese (Liu 2010), ethnic groups in China (Sung 2010), census populations in Latin America (Wade et al. 2014), and the male Jewish line known as the Cohanim (Abu El-Haj 2012). Some of these national projects are framed through the lens of "genomic sovereignty," in which genomics is leveraged as a political instrument to assert scientific progress and economic independence in response to increasing capital accumulation in the global north (Benjamin 2009). STS scholars have also examined how DNA and race have become implicated in immigration policies in the United States (C. Lee 2013), Germany (Heinemann and Lemke 2014), and Europe (M'charek, Schramm, and Skinner 2014). When genomics becomes part of nation-building tools, states align themselves with particular ways of circumscribing the limits of citizenship and belonging. In this way, genetics practices become embedded within the politics of ethnicity, indigeneity, and nationalism.

STS scholars have also studied how race and science have become increasingly conflated in the domain of law enforcement in the genome era, at a time when racial and ethnic minorities are more than ever disproportionately punished within the American judicial system. Genotyping crime scene DNA and comparing it to a suspect's DNA has become a core part of the judicial system's evidence-building tool kit. Although currently routine in law enforcement and forensics arenas, historically the introduction of DNA as evidence was a contentious affair, and its admissibility when left at crime scenes was a matter of debate for the courts (Lynch et al. 2010). In many cases, DNA markers have been used to make probabilistic predictions about a suspect's race, using statistical estimates of allele frequencies in particular groups (M'charek 2008). However, the technology is prone to error, with significant consequences for innocent individuals at risk of becoming suspects through the technology (Ossorio 2006; Ossorio and Duster 2005). The use of DNA as evidence has raised the concern that those already disproportionately targeted by the criminal justice system, particularly black men, may be subjected to increased scrutiny and racial targeting, mandatory surrender of DNA on demand, and heightened rates of incarceration by law enforcement (Ossorio and Duster 2005).

More recently, proponents of a new generation of DNA genotyping tools claim they can computationally reconstruct suspects' faces, based on predictions of a suspect's phenotype from DNA left at crime scenes. Among the predicted features are hair, eye, and skin color (M'charek 2008; Sankar 2010) even though the few genes researchers have identified as being associated with these traits have a very low prediction accuracy. Such technologies highlight the concern that using DNA to predict not only the race of suspects but the contours of physical features generally could make genetics complicit in intensifying racial profiling and stereotyping.

The Future of Studies of Race and Science: Theoretical and Methodological Implications for STS

The scholarly work described above offers tools for understanding the intersections of scientific research and political projects with ideas about human difference. In analyzing how biology and difference are constructed, STS scholars have provided key analytic frameworks for thinking about how artificial binaries are shaped, maintained, reinforced, and entrenched between, for example, nature and culture, genetics and the environment, and the biological and the social (Duster 2015; Fausto-Sterling 2008; Fujimura 2006; Haraway 1997). They also offer tools for considering how to fruitfully move beyond these dichotomies. Pamela Sankar (2008) notes that genomics renders biological views of race as "statistical" rather than typological. She argues that constructionist views of race continue to argue against typological race and miss this crucial distinction, which weakens the constructionist view and opens it to attack. Rather, against these new claims from "science" for a statistical and genomic view of race, social scientists must shore up the constructionist argument by attending to the reasoning, often circular, that underlies how the new statistical genomic arguments are made. In attending to the production of scientific knowledge, STS scholars can reveal the multidirectional traffic of ideas and practices that undergird artificial dichotomies such as the biological and the social, and illuminate more nuanced paths forward in the debates. For example, Fujimura and Rajagopalan (2011) and Fujimura et al. (2014) deconstruct new genomics claims, and show how "populations" in statistical genomics are fashioned as synonymous with racialized groups, not already out there in "nature" for scientists to discover, but emerging out of the ways geneticists construct their tools and representations.

When choosing which problems to study, what questions to ask, and what frames to use, a key scholarly consideration is one's personal positionality with respect to the issues at stake. Some STS scholars choose problems based on their ability to intervene in the "real world," which often constitutes an ethical position. A related methodological and ethical question for future scholars of race and science to consider is whether and how STS scholars should attempt to intervene in debates about race and science. Together, both of these questions are central to existing debates within STS about the place of positionality in the discipline. Does the way that scholars position themselves in these debates shape what kind of STS is performed? Do differing approaches have differing consequences, for science, society, and STS?

These debates about positionality have particular relevance with respect to race and science. Although the aspirations of the new genetics share little in common with those of coercive eugenics movements and other racially motivated atrocities of the nineteenth and twentieth centuries, they do lend themselves to the interpretation that race might be innate and natural, sparking debates with renewed vigor in both the biological sciences and social sciences. Some scholars assume that STS research should not take positions about what constitutes "good" versus "bad" science. In contrast, some scholars who study race and genetics decline to be associated with STS because they disagree with this position. Many, including STS scholars of race and science, argue that STS research should take positions, in part because debates about whether race is genetic have profound consequences. As a small number of geneticists, social scientists, biological scientists, and members of the public media continue to use genomics to argue that racial groupings are genetically determined, inciting intensified debate, many STS scholars hold that it is necessary to find ways to intervene. How especially should constructivist STSers respond to this re-opening of the debate? How do STS scholars deal with and address these new arguments, in light of a strong consensus across STS that race is a set of processual, social, and historically constructed concepts and categories?

STS has long been concerned with how scientific controversies begin, are negotiated, and become settled, but most of STS has not been interventionist. Should STS scholars attempt to take a step back and analyze the debates around race and genetics as one would any other controversy? Or, is it the job of STS to attempt to intervene in and settle such debates (Fujimura 1998; Lynch and Cole 2005)? What kinds of interventions would be considered "STS enough"?

Recently, a small number of social scientists have made attempts to bring genetics into the social sciences, which has raised serious concerns about the potential conflation of race, genetics, and social behavior. These developments hearken back to earlier eras in the latter half of the twentieth century, when a few social scientists attempted to resurrect essentialist views of race as biology. Arthur Jensen, a psychologist and hereditarian, argued in the 1960s that genetics played a role in behavioral traits like intelligence and personality and that these traits differed by race, concluding that the political correctness of the times had left these differences largely unacknowledged (Kevles 1995). In 1994 political scientist Charles Murray and psychologist Richard Herrnstein coauthored the contentious book *The Bell Curve: Intelligence and Class Structure in American Life*. They argued that differences in IQ scores were race-based and likely inherited, which they believed accounted for lower test scores in African Americans as well as their lower levels of education, employment, and socioeconomic success. In a firestorm of critique, social scientists shredded their conclusions, citing hundreds of studies which showed that IQ was determined by educational opportunity, not heredity, and that socioeconomic inequalities were due to structural and institutional forces, not innate differences in IQ.

Articles published in reputable sociology journals have argued that racial groups in the United States can be described as genetic groups. This position is supported by only a small number of sociologists and has prompted critiques from other sociologists and STS scholars (Frank 2014; Fujimura et al. 2014; HoSang 2014; Morning 2014). These critiques show *how* these articles misinterpret a large body of genetic data which clearly illustrates that race categories do not represent genetic groups (Fujimura et al. 2014).This response could be characterized as a *strong interventionist* critique, which mobilizes findings from the sciences to critique misrepresentations of those findings. This approach is in line with theoretical streams within STS which view science and society as so tightly imbricated that any separation is arbitrary and hampers studies of how complex assemblages of people and things together constitute the world (Latour 2012). We argue that STS scholars too can take positions and stakes in the worlds they study, even as they actively try to remove themselves from the story. Interventionist approaches in race and genetics are especially important in light of Donna Haraway's (1988) call to all scholars who construct knowledge to "situate" these knowledges, to reject "universal objectivity" as unattainable and instead acknowledge that all observers and participants in world-making have stakes in the issues at hand and only a "partial perspective" on their resolution. According to Haraway, STS scholars must account for their positionality as much as any of the science makers that they study. What this means for those studying race and genetics is that not only can scholars take and defend political positions about the social construction of race, but to do so, they may need to mobilize resources from within the "science" itself to counter scientistic arguments about biology and race. Armed with this STS tool kit, race scholars can effectively balance the lingering tensions and seeming contradictions of promoting the idea that race is socially constructed, while maintaining a commitment to attend to the real biological effects of the experiences of race.

Conclusion

This chapter has contextualized a recent and emerging set of debates and scholarship concerning the nature of race in the context of contemporary biology and genetics. STS scholarship on human genetic variation research, emerging over the last decade, has grown rapidly, helping to sharpen discussion on matters of social, political, and cultural significance, including questions of human identity and human difference.

STS approaches have illuminated the historical struggles over the role of science in adjudicating questions of race, and the role of race in the practice of science. STS studies have enhanced our understanding of how race has been produced historically, including the situational and contextual dimensions of the production of knowledge and the ways in which knowledge produced by the sciences is used and applied in sociopolitical domains, sometimes to authorize and legitimize, and sometimes to destabilize or replace, prevailing sociocultural belief systems and practices. Studies of race and genetics have tied together the core parts of STS: an analysis of the production of knowledge through micro and macro examinations of bioscientific work and developments, as well as their enactments on larger institutional terrains. They reveal how far conversations about race have come in the field of STS, and they emphasize that, in the early years of the twenty-first century, the stakes remain just as high.

STS research on race and genetics is becoming an increasingly visible site of investigation and attention, as illustrated by several volumes which have been devoted to its study. They offer a variety of perspectives across the disciplines (Koenig, Lee, and Richardson, 2008; Wade et al. 2014; Wailoo, Nelson, and Lee 2012; Whitmarsh and Jones 2010), as do several journal special issues on race and genetics (Duster 2015; Fujimura, Duster, and Rajagopalan 2008; M'charek, Schramm, and Skinner 2014; Wade et al. 2015; Winther 2015).

Contemporary genomics studies have sometimes been viewed as coming perilously close to recapitulating the disavowed doctrines of earlier eras, wherein race was considered to be innate, biological, genetic, and inscribed on bodies prior to and outside of enculturation or socialization. These debates are far from settled and continue to escalate in genomics, as researchers pursue the collection of ever more massive repositories of DNA samples and databases of human DNA sequence information from individuals and groups around the world. As these projects develop, scholars will continue to ask and explore why race continues to be such a durable and salient lens through which to view genetic difference.

Acknowledgments

This work was supported in part by funding to A.N. from the Office of the Executive Vice President of the Faculty of Arts and Sciences, Columbia University, and the Office of the Provost, Columbia University.

References

Abu El-Haj, Nadia. 2012. *The Genealogical Science: The Search for Jewish Origins and the Politics of Epistemology*. Chicago: University of Chicago Press.

Bauer, Susanne. 2014. "Virtual Geographies of Belonging: The Case of Soviet and Post-Soviet Human Genetic Diversity Research." *Science, Technology, & Human Values* 39 (4): 511–37.

Benjamin, Ruha. 2009. "A Lab of Their Own: Genomic Sovereignty as Postcolonial Science Policy." *Policy and Society* 28 (4): 341–55.

Bliss, Catherine. 2012. *Race Decoded: The Genomic Fight for Social Justice*. Stanford, CA: Stanford University Press.

Blumer, Herbert. 1958. "Race Prejudice as a Sense of Group Position." *Pacific Sociological Review* 1 (1): 3–7.

Boas, Franz. 1912. "Changes in the Bodily Form of Descendants of Immigrants." *American Anthropologist* 14 (3): 530–62.

Bolnick, Deborah A., Duana Fullwiley, Troy Duster, Richard S. Cooper, Joan H. Fujimura, Jonathan Kahn, Jay S. Kaufman, Jonathan Marks, Ann Morning, Alondra Nelson, Pilar Ossorio, Jenny Reardon, Susan M. Reverby, and Kimberly TallBear. 2007. "The Science and Business of Genetic Ancestry Testing." *Science* 318 (5849): 399–400.

Bowker, Geoffrey, and Susan Leigh Star. 1999. *Sorting Things Out: Classification and Its Consequences*. Cambridge, MA: MIT Press.

Brattain, Michelle. 2007. "Race, Racism, and Antiracism: UNESCO and the Politics of Presenting Science to the Postwar Public." *American Historical Review* 112 (5): 1386–413.

Braun, Lundy. 2014. *Breathing Race into the Machine: The Surprising Career of the Spirometer from Plantation to Genetics*. Minneapolis: University of Minnesota Press.

Braun, Lundy, and Evelynn Hammonds. 2008. "Race, Populations, and Genomics: Africa as Laboratory." *Social Science & Medicine* 67 (10): 1580–88.

de Vries, Jantina, Muminatou Jallow, Thomas N. Williams, Dominic Kwiatkowski, Michael Parker, and Raymond Fitzpatrick. 2012. "Investigating the Potential for Ethnic Group Harm in Collaborative Genomics Research in Africa: Is Ethnic Stigmatisation Likely?" *Social Science & Medicine* 75 (8): 1400–1407.

Douglas, Mary, and David Hull, eds. 1992. *How Classification Works: Nelson Goodman among the Social Sciences*. Edinburgh: University of Edinburgh Press.

Du Bois, William E. B. 1940. *Dusk of Dawn: An Essay toward an Autobiography of a Race Concept*. New York: Harcourt, Brace.

Duster, Troy. 1990. *Backdoor to Eugenics*. New York: Routledge.

___. 2006. "The Molecular Reinscription of Race: Unanticipated Issues in Biotechnology and Forensic Science." *Patterns of Prejudice* 40 (4–5): 427–41.

___. 2015. "A Post-Genomic Surprise: The Molecular Reinscription of Race in Science, Law and Medicine." *British Journal of Sociology* 66 (1): 1–27.

Egorova, Yulia. 2009. "De/geneticizing Caste: Population Genetic Research in South Asia." *Science as Culture* 18 (4): 417–34.

Ellison, George T. H., Richard Tutton, Simon M. Outram, Paul Martin, Richard Ashcroft, and Andrew Smart. 2008. "An Interdisciplinary Perspective on the Impact of Genomics on the Meaning of 'Race' and the Future Role of Racial Categories in Biomedical Research." *NTM Journal of History of Sciences, Technology, and Medicine* 16 (3): 378–86.

Epstein, Steven. 2007. *Inclusion: The Politics of Difference in Medical Research.* Chicago: University of Chicago Press.

Fausto-Sterling, Anne. 2008. "The Bare Bones of Race." *Social Studies of Science* 38 (5): 657–94.

Fortier, Anne-Marie. 2012. "Genetic Indigenisation in 'The People of the British Isles.'" *Science as Culture* 21 (2): 153–75.

Fortun, Michael. 2008. *Promising Genomics: Iceland and deCODE Genetics in a World of Speculation.* Berkeley: University of California Press.

Frank, Reanne. 2014. "The Molecular Reinscription of Race: A Comment on 'Genetic Bio-Ancestry and Social Construction of Racial Classification in Social Surveys in the Contemporary United States.'" *Demography* 51 (6): 2333–36.

Fujimura, Joan H. 1998. "Authorizing Knowledge in Science and Anthropology." *American Anthropologist* 100 (2): 347–60.

___. 2006. "Sex Genes: A Critical Sociomaterial Approach to the Politics and Molecular Genetics of Sex Determination." *Signs: Journal of Women in Culture and Society* 32 (1): 49–82.

Fujimura, Joan H., Deborah A. Bolnick, Ramya Rajagopalan, Jay S. Kaufman, Richard C. Lewontin, Troy Duster, Pilar Ossorio, and Jonathan Marks. 2014. "Clines without Classes: How to Make Sense of Human Variation." *Sociological Theory* 32 (3): 208–27.

Fujimura, Joan H., Troy Duster, and Ramya Rajagopalan. 2008. "Introduction: Race, Genetics, and Disease: Questions of Evidence, Matters of Consequence." *Social Studies of Science* 38 (5): 643–56.

Fujimura, Joan H., and Ramya Rajagopalan. 2011. "Different Differences: The Use of 'Genetic Ancestry' versus Race in Biomedical Human Genetic Research." *Social Studies of Science* 41 (1): 5–30.

Fullwiley, Duana. 2007. "The Molecularization of Race: Institutionalizing Human Difference in Pharmacogenetics Practice." *Science as Culture* 16 (1): 1–30.

___. 2008. "The Biologistical Construction of Race: 'Admixture' Technology and the New Genetic Medicine." *Social Studies of Science* 38 (5): 695–735.

Gannett, Lisa. 2001. "Racism and Human Genome Diversity Research: The Ethical Limits of 'Population Thinking.'" *Philosophy of Science* 68 (3): S479–92.

___. 2013. "Theodosius Dobzhansky and the Genetic Race Concept." *Studies in History and Philosophy of Science Part C: Studies in History and Philosophy of Biological and Biomedical Sciences* 44 (3): 250–61.

Geronimus, Arline T. 1992. "The Weathering Hypothesis and the Health of African-American Women and Infants: Evidence and Speculations." *Ethnicity & Disease* 2 (3): 207–21.

Gissis, Snait B. 2008. "When Is 'Race' a Race? 1946–2003." *Studies in History and Philosophy of Science Part C: Studies in History and Philosophy of Biological and Biomedical Sciences* 39 (4): 437–50.

Gould, Stephen Jay. 1996. *The Mismeasure of Man*. New York: W. W. Norton.

Gravlee, Clarence C. 2009. "How Race Becomes Biology: Embodiment of Social Inequality." *American Journal of Physical Anthropology* 139 (1): 47–57.

Hacking, Ian. 2005. "Why Race Still Matters." *Daedalus* 134 (1): 102–16.

Hamilton, Jennifer A. 2008. "Revitalizing Difference in the HapMap: Race and Contemporary Human Genetic Variation Research." *Journal of Law, Medicine & Ethics* 36 (3): 471–77.

Hammonds, Evelynn M., and Rebecca M. Herzig. 2009. *The Nature of Difference: Sciences of Race in the United States from Jefferson to Genomics*. Cambridge, MA: The MIT Press.

Haraway, Donna J. 1988. "Situated Knowledges: The Science Question in Feminism and the Privilege of Partial Perspective." *Feminist Studies* 14 (3): 575–99.

___. 1989. *Primate Visions: Gender, Race, and Nature in the World of Modern Science*. Hove, UK: Psychology Press.

___. 1997. *Modest_Witness@Second_Millennium.FemaleMan_Meets_OncoMouse: Feminism and Technoscience*. Hove, UK: Psychology Press.

Heinemann, Torsten, and Thomas Lemke. 2014. "Biological Citizenship Reconsidered: The Use of DNA Analysis by Immigration Authorities in Germany." *Science, Technology, & Human Values* 39 (4): 488–510.

Hinterberger, Amy. 2012. "Publics and Populations: The Politics of Ancestry and Exchange in Genome Science." *Science as Culture* 21 (4): 528–49.

Holmes, Seth. 2013. *Fresh Fruit, Broken Bodies: Migrant Farmworkers in the United States*. Berkeley: University of California Press.

HoSang, Daniel Martinez. 2014. "On Racial Speculation and Racial Science: A Response to Shiao et al." *Sociological Theory* 32 (3): 228–43.

Jackson, Myles W. 2014. "The Biology of Race: Searching for No Overlap." *Perspectives in Biology and Medicine* 57 (1): 87–104.

James, Sherman A. 1994. "John Henryism and the Health of African-Americans." *Culture, Medicine and Psychiatry* 18 (2): 163–82.

Kahn, Jonathan. 2013. *Race in a Bottle: The Story of BiDil and Racialized Medicine in a Post-Genomic Age*. New York: Columbia University Press.

Kevles, Daniel J. 1995. *In the Name of Eugenics: Genetics and the Uses of Human Heredity*. Cambridge, MA: Harvard University Press.

Koenig, Barbara A., Sandra Soo-Jin Lee, and Sarah S. Richardson, eds. 2008. *Revisiting Race in a Genomic Age*. New Brunswick, NJ: Rutgers University Press.

Krieger, Nancy, and Mary Bassett. 1993. "The Health of Black Folk." In *The "Racial" Economy of Science: Toward a Democratic Future*, edited by Sandra Harding, 161–69. Bloomington: Indiana University Press.

Latour, Bruno. 2012. *We Have Never Been Modern*. Cambridge, MA: Harvard University Press.

• Lee, Catherine. 2009. "'Race' and 'Ethnicity' in Biomedical Research: How Do Scientists Construct and Explain Differences in Health?" *Social Science & Medicine* 68 (6): 1183–90.

___. 2013. *Fictive Kinship: Family Reunification and the Meaning of Race and Nation in American Immigration*. New York: Russell Sage Foundation.

Lee, Sandra Soo-Jin. 2013. "Race, Risk, and Recreation in Personal Genomics: The Limits of Play." *Medical Anthropology Quarterly* 27 (4): 550–69.

___. 2015. "The Biobank as Political Artifact: The Struggle over Race in Categorizing Genetic Difference." *Annals of the American Academy of Political and Social Science* 661 (1): 143–59.

• Lewontin, Richard C. 1972. "The Apportionment of Human Diversity." *Evolutionary Biology* 6: 381–98.

Lipphardt, Veronika. 2014. "'Geographical Distribution Patterns of Various Genes': Genetic Studies of Human Variation after 1945." *Studies in History and Philosophy of Science Part C: Studies in History and Philosophy of Biological and Biomedical Sciences* 47: 50–61.

Lippman, Abby. 1991. "Prenatal Genetic Testing and Screening: Constructing Needs and Reinforcing Inequities." *American Journal of Law & Medicine*. 17 (1): 15–50.

Liu, Jennifer A. 2010. "Making Taiwanese (Stem Cells): Identity, Genetics, and Hybridity." In *Asian Biotech: Ethics and Communities of Fate*, edited by Aihwa Ong and Nancy Chen, 239–62. Durham, NC: Duke University Press.

López, Ian F. Haney. 1996. *White by Law: The Legal Construction of Race*. New York: New York University Press.

Lynch, Michael, and Simon Cole. 2005. "Science and Technology Studies on Trial: Dilemmas of Expertise." *Social Studies of Science* 35 (2): 269–311.

Lynch, Michael, Simon A. Cole, Ruth McNally, and Kathleen Jordan. 2010. *Truth Machine: The Contentious History of DNA Fingerprinting*. Chicago: University of Chicago Press.

• Marks, Jonathan. 1995. *Human Biodiversity: Genes, Race, and History*. Hawthorne, NY: Aldine de Gruyter.

• M'charek, Amade. 2005. *The Human Genome Diversity Project: An Ethnography of Scientific Practice*. Cambridge: Cambridge University Press.

___. 2008. "Silent Witness, Articulate Collective: DNA Evidence and the Inference of Visible Traits." *Bioethics* 22 (9): 519–28.

M'charek, Amade, Katharina Schramm, and David Skinner. 2014. "Technologies of Belonging: The Absent Presence of Race in Europe." *Science, Technology, & Human Values* 39 (4): 459–67.

Montagu, Ashley. 1942. *Man's Most Dangerous Myth: The Fallacy of Race*. New York: Harper.

___. 1972. *Statement on Race: An Annotated Elaboration and Exposition of the Four Statements on Race Issued by the United Nations Educational, Scientific, and Cultural Organization*. 3rd ed. Westport, CT: Greenwood Press.

Montoya, Michael J. 2007. "Bioethnic Conscription: Genes, Race, and Mexicana/o Ethnicity in Diabetes Research." *Cultural Anthropology* 22 (1): 94–128.

• Morning, Ann. 2011. *The Nature of Race: How Scientists Think and Teach about Human Difference*. Berkeley: University of California Press.

___. 2014. "Does Genomics Challenge the Social Construction of Race?" *Sociological Theory* 32 (3): 189–207.

Müller-Wille, Staffan. 2007. "Hybrids, Pure Cultures, and Pure Lines: From Nineteenth-Century Biology to Twentieth-Century Genetics." *Studies in History and Philosophy of Science Part C: Studies in History and Philosophy of Biological and Biomedical Sciences* 38 (4): 796–806.

Nash, Catherine. 2013. "Genome Geographies: Mapping National Ancestry and Diversity in Human Population Genetics." *Transactions of the Institute of British Geographers* 38 (2): 193–206.

Nelson, Alondra. 2008a. "Bio Science: Genetic Genealogy Testing and the Pursuit of African Ancestry." *Social Studies of Science* 38 (5): 759–83.

___. 2008b. "The Factness of Diaspora: The Social Sources of Genetic Genealogy." In *Revisiting Race in a Genomic Age*, edited by Barbara Koenig, Sandra Lee, and Sarah Richardson, 253–70. New Brunswick, NJ: Rutgers University Press

___. 2011. *Body and Soul: The Black Panther Party and the Fight against Medical Discrimination*. Minneapolis: University of Minnesota Press.

___. 2012. "Reconciliation Projects: From Kinship to Justice." In *Genetics and the Unsettled Past: The Collision of DNA, Race, and History*, edited by Keith Wailoo, Alondra Nelson, and Catherine Lee, 20–31. New Brunswick, NJ: Rutgers University Press.

___. 2016. *The Social Life of DNA: Race, Reparations, and Reconciliation after the Genome*. Boston: Beacon Press.

Omi, Michael, and Howard Winant. 1986. *Racial Formation in the United States*. New York: Routledge.

Ossorio, Pilar N. 2006. "About Face: Forensic Genetic Testing for Race and Visible Traits." *The Journal of Law, Medicine & Ethics* 34 (2): 277–92.

Ossorio, Pilar N., and Troy Duster. 2005. "Race and Genetics: Controversies in Biomedical, Behavioral, and Forensic Sciences." *American Psychologist* 60 (1): 115–28.

Paul, Diane B. 1995. *Controlling Human Heredity: 1865 to the Present*. Atlantic Highlands, NJ: Humanities Press.

Paul, Diane B., and Jeffrey P. Brosco. 2013. *The PKU Paradox: A Short History of a Genetic Disease*. Baltimore: Johns Hopkins University Press.

Pearson, Jay A. 2008. "Can't Buy Me Whiteness." *Du Bois Review: Social Science Research on Race* 5 (01): 27–47.

Philip, Kavita. 2004. *Civilizing Natures: Race, Resources and Modernity in Colonial South India*. New Brunswick, NJ: Rutgers University Press.

Pollack, Anne. 2012. *Medicating Race: Heart Disease and Durable Preoccupations with Difference*. Durham, NC: Duke University Press.

Prainsack, Barbara. 2015. "Is Personalized Medicine Different? (Reinscription: The Sequel): A Response to Troy Duster." *The British Journal of Sociology* 66 (1): 28–35.

Rajagopalan, Ramya, and Joan H. Fujimura. 2012a. "Making History via DNA, Making DNA from History: Deconstructing the Race-Disease Connection in Admixture Mapping." In *Genetics and the Unsettled Past: The Collision between DNA, Race, and History*, edited by Keith Wailoo, Catherine Lee, and Alondra Nelson, 143–63. New Brunswick, NJ: Rutgers University Press.

___. 2012b. "Will Personalized Medicine Challenge or Reify Categories of Race and Ethnicity?" *American Medical Association Journal of Ethics* 14 (8): 657–63.

Reardon, Jenny. 2004. *Race to the Finish: Identity and Governance in an Age of Genomics*. Princeton, NJ: Princeton University Press.

___. 2007. "Democratic Mis-Haps: The Problem of Democratization in a Time of Biopolitics." *Bio-Societies* 2 (2): 239–56.

___. 2012. "The Democratic, Anti-Racist Genome? Technoscience at the Limits of Liberalism." *Science as Culture* 21 (1): 25–47.

Reardon, Jenny, and Kim TallBear. 2012. "Your DNA Is Our History." *Current Anthropology* 53 (S5): S233–45.

Reverby, Susan M. 2009. *Examining Tuskegee: The Infamous Syphilis Study and Its Legacy.* Chapel Hill: University of North Carolina Press.

Roberts, Dorothy. 2011. *Fatal Invention: How Science, Politics, and Big Business Re-Create Race in the Twenty-First Century.* New York: New Press.

Sankar, Pamela. 2008. "Moving beyond the Two-Race Mantra." In *Revisiting Race in a Genomic Age*, edited by Barbara Koenig, Sandra Soo-Jin Lee, and Sarah Richardson, 271–84. New Brunswick, NJ: Rutgers University Press.

___. 2010. "Forensic DNA Phenotyping: Reinforcing Race in Law Enforcement." In *What's the Use of Race?*, edited by David Jones and Ian Whitmarsh, 49–62. Cambridge, MA: MIT Press.

Sankar, Pamela, and Mildred K. Cho. 2002. "Toward a New Vocabulary of Human Genetic Variation." *Science* 298 (5597): 1337.

Santos, Ricardo Ventura, Peter H. Fry, Simone Monteiro, Marcos Chor Maio, José Carlos Rodrigues, Luciana Bastos-Rodrigues, and Sérgio D. J. Pena. 2009. "Color, Race, and Genomic Ancestry in Brazil: Dialogues between Anthropology and Genetics." *Current Anthropology* 50 (6): 787–819.

Schiebinger, Londa L. 1993. *Nature's Body: Gender in the Making of Modern Science.* New Brunswick, NJ: Rutgers University Press.

Schramm, Katharina, David Skinner, and Richard Rottenburg. 2012. *Identity Politics and the New Genetics: Re/creating Categories of Difference and Belonging.* Oxford: Berghahn Books.

Shim, Janet K., Katherine Weatherford Darling, Martine D. Lappe, L. Katherine Thomson, Sandra Soo-Jin Lee, Robert A. Hiatt, and Sara L. Ackerman. 2014. "Homogeneity and Heterogeneity as Situational Properties: Producing—and Moving Beyond?—Race in Post-Genomic Science." *Social Studies of Science* 44 (4): 579–99.

Smart, Andrew, Richard Tutton, Paul Martin, George T. H. Ellison, and Richard Ashcroft. 2008. "The Standardization of Race and Ethnicity in Biomedical Science Editorials and UK Biobanks." *Social Studies of Science* 38 (3): 407–23.

Smedley, Audrey. 2007. *Race in North America: Origin and Evolution of a Worldview.* Boulder, CO: Westview Press.

Smith, Justin E. H. 2015. *Nature, Human Nature, and Human Difference: Race in Early Modern Philosophy.* Princeton, NJ: Princeton University Press.

Smith, Susan L. 2008. "Mustard Gas and American Race-Based Human Experimentation in World War II." *Journal of Law, Medicine & Ethics* 36 (3): 517–21.

Stepan, Nancy Leys. 1986. "Race and Gender: The Role of Analogy in Science." *Isis* 77 (2): 261–77.

Sung, Wen-Ching. 2010. "Chinese DNA: Genomics and Bionations." In *Asian Biotech: Ethics and Communities of Fate*, edited by Aihwa Ong and Nancy Chen, 263–92. Durham, NC: Duke University Press.

TallBear, Kim. 2013. *Native American DNA: Tribal Belonging and the False Promise of Genetic Science*. Minneapolis: University of Minnesota Press.

Tamarkin, Noah. 2014. "Genetic Diaspora: Producing Knowledge of Genes and Jews in Rural South Africa." *Cultural Anthropology* 29 (3): 552–74.

Title VI Committee. 2006. "Biogenetic Data and Historical Scholarship: Sources of Evidence Regarding 'Aryan Migration.'" Report Presented to the California State Board of Education, Sacramento.

Tsai, Yu-Yueh. 2012. "The Geneticization of Ethnicity and Ethnicization of Biomedicine: On the 'Taiwan Bio-Bank.'" In *Biomapping Indigenous Peoples: Towards an Understanding of the Issues*, edited by Susanne Berthier-Foglar, Sheila Collingwood-Whittick, and Sandrine Tolazzi, 183–220. Amsterdam: Rodopi.

Tutton, Richard, Jane Kaye, and Klaus Hoeyer. 2004. "Governing UK Biobank: The Importance of Ensuring Public Trust." *Trends in Biotechnology* 22 (6): 284–85.

Viruell-Fuentes, Edna A. 2007. "Beyond Acculturation: Immigration, Discrimination, and Health Research among Mexicans in the United States." *Social Science & Medicine* 65 (7): 1524–35.

Wade, Peter, Carlos López Beltrán, Eduardo Restrepo, and Ricardo Ventura Santos, eds. 2014. *Mestizo Genomics: Race Mixture, Nation, and Science in Latin America*. Durham, NC: Duke University Press.

———. 2015. "Genomic Research, Publics and Experts in Latin America: Nation, Race and Body." *Social Studies of Science* 45 (6): 775–96.

Wailoo, Keith, Alondra Nelson, and Catherine Lee, eds. 2012. *Genetics and the Unsettled Past: The Collision of DNA, Race, and History*. New Brunswick, NJ: Rutgers University Press.

Wailoo, Keith, and Stephen Gregory Pemberton. 2006. *The Troubled Dream of Genetic Medicine: Ethnicity and Innovation in Tay-Sachs, Cystic Fibrosis, and Sickle Cell Disease*. Baltimore: Johns Hopkins University Press.

Whitmarsh, Ian. 2008. *Biomedical Ambiguity: Race, Asthma, and the Contested Meaning of Genetic Research in the Caribbean*. Ithaca, NY: Cornell University Press.

Whitmarsh, Ian, and David S. Jones, eds. 2010. *What's the Use of Race? Modern Governance and the Biology of Difference*. Cambridge, MA: MIT Press.

Winther, Rasmus Gronfeldt. 2015. "Special Issue: Genomics and Philosophy of Race." *Studies in History and Philosophy of Science Part C: Studies in History and Philosophy of Biological and Biomedical Sciences* 52: 1–4.

13　Sex, Gender, and Sexuality in Biomedicine

Jennifer R. Fishman, Laura Mamo, and Patrick R. Grzanka

Much of the social sciences and humanities, including the interdisciplinary field of science and technology studies (STS), have treated the three concepts of sex, gender, and sexuality (SG&S) as distinct. One of feminist STS's major contributions has been to show the interrelatedness of SG&S as material, embodied, *and* discursive sites in and through which power and power relations coalesce. Due in large part to this body of scholarship, these categories are no longer understood through either essentialist or constructivist frames; they are no longer debated as the results of either nature or nurture. Instead, feminist STS has encouraged an understanding of SG&S as complex entanglements among what we know as nature/biology and culture/technology (e.g., Braidotti 1994; Grosz 1994; Haraway 1989; Keller 1985). We use the acronym SG&S here to indicate this nexus of sex, gender, and sexualities as objects of academic analysis.

In the 1990s a proliferation of STS scholarship, largely driven by feminist analyses, began to explore the ways biomedicine discursively and materially constructs SG&S, analyzing the technoscientific production of sexed bodies, identities, and sexual desire, the ways these are co-produced by social and cultural ideas about SG&S, and the implications for bodies, lives, and experiences of health and illness (Clarke et al. 2003). Beginning with the life sciences and expanding to other knowledge domains and practices, STS scholars examined how technologies and scientific knowledges configure and reconfigure the meanings, embodiments, and technologies of sex, sexed bodies, genders, and sexualities. Many argued that cultural ideas and norms about SG&S filtered into and shaped scientific claims about these categories. Yet the seeming unidirectionality of culture seeping into what we know as nature was quickly replaced with a more nuanced analysis of these categories—nature and culture as intertwined and co-produced.

"The laboratory" provided a space to analyze the ways scientific practitioners organize and engage in their work activities, engage in claims or "fact making," endow their claims with credibility, and defend or extend their boundaries of practice based on these claims. Studies gradually moved beyond the lab to include broader dimensions of public engagements, politics, and practices across and within sciences and technologies. One area of scholarship significant to SG&S was how claims of *difference* were understood as part of classification practices that reflect prevailing hierarchies of power, serve to shape social order, and sometimes provide rationales for exclusion (Berg and Mol 1998; Keller and Longino 1996). As Bowker and Star (1999) revealed, such classifications became standardized practices and taken-for-granted infrastructures. Many examined the emergent varied and complex ways sexed bodies and sexuality are mutually produced through sociopolitical controversies, social movement resistance, and practices and interactions with biomedicine itself. Prominent in this literature is exploration of the making of normative and naturalized scientific categories of SG&S and, moreover, how these have been contested and defied by those individuals and social groups defined—often marginally—by these categories (Hacking 1986). This literature has consistently suggested that the very categories on which ontologies of SG&S rest need to be interrogated and their production uncovered.

Despite a robust research program on SG&S in feminist STS (e.g., Wyer et al. 2013), this literature often remains under-cited in "mainstream" STS. Evidence is thus equivocal regarding the degree to which critical work on SG&S has made a substantial impact on STS broadly or is considered part of foundational STS scholarship. On the one hand, high-profile feminist STS scholars such as Adele Clarke, Donna Haraway, Sandra Harding, and Nelly Oudshoorn have had a marked influence on STS methods and theories beyond the study of sex/gender and/or women's position in the sciences (e.g., grounded theory, situational analysis, studies of technologies' users). In this sense, certain strands of feminist STS have experienced sustained uptake across the social sciences (cf. Susan Leigh Starr [1990], Anne Fausto-Sterling [1992], Judy Wajcman [1991], and others). On the other hand, at the 2012 annual meeting of the Society for Social Studies of Science (4S), Voss and Lock (2012) argued that mainstream STS may be "bolstering scientific heteronormativity" by ignoring the mutual construction of science and sexualities, a hallmark of critical feminist STS. Based on research conducted from reviewing four major STS journals and a literature review of STS books, they found little scholarship exploring the interactions, in particular, of queer sexualities, scientific knowledge, and technologies.

That scholarship on sex and gender may be incorporated into mainstream STS while sexuality remains confined to explicitly feminist STS inquiry is disconcerting for many reasons, not the least of which is that there is a great deal of feminist STS work on SG&S that reveals the deep interconnectedness of these social categories and their co-constitution. Though gender and sexuality (including what is often referred to as

"sexual orientation") are not equivalent constructs, <u>eliding the significance of heter-</u>
<u>onormativity on the production of scientific knowledge</u> about gender and sex (and
perhaps other areas of inquiry, as well) is to impose an artificial, ideological divide
between things that are already empirically intertwined. To investigate gender and sex
in science and technology at the expense of sexuality—including the production of
normative heterosexuality—is to efface the co-constitution of the body, gender, and
sexuality in historical and contemporary technoscience, which is both contrary to the
goals of STS and effectively undermines the potential for critical STS work to uncover
the unexpected (i.e., to see what science is actually doing, as opposed to what we presume it is doing) (Haraway 1989).

I'D ADD RACE

    If sexuality is a kind of "forbidden knowledge" (Kempner, Perlis, and Merz 2005),
informally excluded from research by norms that are highly resistant to change—even
within STS—then we should be especially wary of how <u>sexuality and queer studies'</u>
<u>relative marginality in our field</u> may be affecting the form and content of STS, even
across sites of investigation and critical inquiry that may not immediately appear to
have anything to do with sexuality (Stein and Plummer 1994). For example, in their
analysis of cisgenderism in psychological research on gender-nonconforming children,
Ansara and Hegarty (2012) found that an "invisible college" exists in the psychological
literature such that a relatively small group of frequent collaborators conduct similar
research and frequently cite each other. This reproduces ideologies and practices that
pathologize what are today categorized as gender "atypical" or "nonconforming" children, reinforces heteronormativity, and legitimizes biased science about transgender
identities and experiences (Ansara and Hegarty 2012). This informal knowledge production process has the effect of creating both inclusion and exclusion: an elite group
of scientists cite one another's work and produce consensus around a topic while other
researchers offer different perspectives that are dismissed or ignored. If, as Voss and
Lock (2012) suggest, a kind of "invisible college" may exist in the STS mainstream to
the exclusion of queer and sexualities-focused concerns, then heteronormativity may
affect and constrain STS as a field in ways we have yet to fully understand. Nevertheless, we argue <u>SG&S work has been vital to STS,</u> and when we focus specifically on
feminist and other critical STS scholarship, a rich body of work emerges that not only
<u>pushes against epistemic norms in STS</u> and beyond but also is already influential and
poised to influence the fields of STS and STS-inflected scholarship further.

    Our goal for this chapter is to examine the robust feminist STS scholarship on
SG&S as a way into critical reflection on STS's current engagement with this set of
concerns and to examine some promising directions and conceptual challenges facing scholars committed to this work. While we focus here on SG&S, we do so with full

understanding that ignoring or separating these social categories from those of race, class, and nation fails to acknowledge how these not only intersect but are co-constitutive. We choose, here, to tell a narrative of STS and SG&S that begins with what we regard as central STS approaches that interrogate the ways SG&S is not only produced within the (bio)sciences and technosciences more generally but also entangled in their production. We then selectively review scholarship that reveals and challenges scientific and biomedical productions of inequalities and "pathologies" especially as these produce and maintain social hierarchies through scientific claims making. We specifically examine how social movements/activists and scholar-activists challenge power in historical and empirical work revealing scientific claims as sociopolitical knowledges, often value laden (Keller 1985; Keller and Longino 1996) and, at times, deeply flawed (Jordan-Young 2010; Stein 2001). We then turn toward inquiries into bioscientific category making to examine how STS scholars from across the social sciences and humanities foreground the simultaneous and "intersectional" production of various axes of difference, power, and inequality.

Collectively, these three parts engage the following questions, pertinent to the past and future of STS research on SG&S and biomedicine: (1) In what ways do biomedicine and its concomitant technologies create and respond to SG&S as key categories of difference? Given that classification and categorization processes are a key analytic concern for much of STS, biomedicine's persistent "obsession" (Terry 1999) with sex and sexuality proffers the exigency for continued attention to these important sites of knowledge and policy making. (2) How do biomedical phenomena and, more specifically, the study of them, create pathologies and inequalities that are both pervasively influential and differentially resisted? STS offers important answers and innovative methods for exploring the constitution of SG&S and other categories and the scientific controversies and health social movements that have coalesced around them. Finally, (3) how do STS scholars conceptualize the complex relationships among sciences, SG&S, and other axes of difference through which biomedical knowledge about bodies is produced and organized? SG&S, once a central component of some STS scholarship, has shifted to less of a matter of concern even as race has "returned" as a primary site of inquiry in STS. While categories are co-constitutive, analytic projects need not necessarily attend to their mutuality as their focus of analysis. Yet, understanding the ways categories are mutually productive in ways that always and already overlay meaning on one another is essential to STS. The work we engage in this chapter does not conceptualize SG&S and other categories, such as race and disability, as neutral, equivalent, or discrete categories; rather, this work examines how these dimensions of difference are crafted, deployed, and transformed by scientists, clinicians, corporations, and policy makers—as well as by STS scholars themselves.

Making Sex: Science, Bodies, and Sexualities

An early contribution of STS on SG&S is the thorough examination of the ways in which the life sciences and the practices of biomedicine, often in "the laboratory," have contributed to our fundamental understandings about the nature of sex, gender, and sex *difference* (e.g., Dreger 1998; Fausto-Sterling 2000; Oudshoorn 1994; and more recently, Ha 2011; Richardson 2013). This includes work on the scientization of sex via disciplinary fields such as primatology (Haraway 1989), sexology (Irvine 1990), endocrinology (Oudshoorn 1994), plant biology (Schiebinger 1989), reproductive sciences (Clarke 1998), and anatomy (Fausto-Sterling 2000; Moore and Clarke 1995) among others (Waldby 1996) that interrogated the normalization and naturalization of the binary categories of sex.

A collection of key work in STS emerged that interrogated these scientific projects by studying scientists, their knowledge claims, and their work practices to uncover the gendered and cultural assumptions embedded within, and how they have become part of pervasive cultural understandings about, the "nature" of difference. Some of this work can be best characterized by examining the ways in which our assumptions about *gender* (i.e., masculinity and femininity) influenced our studies of sex, sexuality, and reproduction. Martin (1991), for example, examines the ways in which scientific descriptions and conceptual understandings of sexed body parts and organs mapped onto traditional gendered norms, adjectives, and anthropomorphic language. Most notably, she details the ways in which sperm is described by metaphors of action and competition while eggs are described as passive and "damsels in distress." Not only did this affect our cultural imagination of these gametes, but these assumptions also influenced what reproductive scientists might "see" in the laboratory. For example, alternative narratives of the reproductive process instead noted the "active" role that eggs and female organs play in the reproductive process (also Moore 2007). This latter contribution is what carries through this strain of scholarship: assumptions about gender influence the production of medical knowledge about sex and sexed bodies in ways that affect subsequent research, and indeed, lived experiences. Oudshoorn (1994) shows how the development of endocrinology and the "making" of sex hormones were based on already-held beliefs about sex differentiation and a binary model of male and female, which gave rise to the notion of "male" and "female" hormones. Clarke (1998) also demonstrates how the rise of the reproductive sciences as a discipline was predicated on, and deeply intertwined with, similar premises from fields such as animal husbandry, human embryology, and elsewhere.

Irvine (1990) similarly charts the rise of the discipline of sexology as it scientized sexuality over the course of the twentieth century, through the political and ideological dimensions of its most well-known researchers, Alfred Kinsey and Masters and Johnson. She explains how a dedication to the ideal of egalitarianism (at least biologically if not politically) gave rise to the search for *similarities* between men and women in terms of physiological sexual response and sexual desire (Jackson 1996; Tiefer 1995). Their decisions to emphasize the similarities between men and women stemmed from an ideological commitment to the maintenance of heterosexual relationships. Realizing that the sexual ideology of the Victorian era, which emphasized inherent differences between the sexes, had been seriously eroded by the middle of the twentieth century, Masters and Johnson instead chose to focus on the similarities between men and women. It was thought that this emphasis could help to achieve sexual parity by touting women's capacities for sexual behavior as *equal* to that of men's. This could then be used to bolster heterosexual marriages (Segal 1994). This was achieved primarily through an emphasis on the "naturalness" of male *and* female sexual desire and sexual expression. Irvine (1990) and Tiefer (1995) effectively convey how the legacy of this thinking influences the study of sex to this day (Marshall and Katz 2002).

One of the most influential STS scholars in the study of sex difference is the feminist biologist Anne Fausto-Sterling. Like other feminist STS scholars studying SG&S, Fausto-Sterling (1992, 2000) questions the underlying premises of developmental biology and other disciplines regarding the dual system of male-female—that one is "naturally" born male or female. Through an interrogation of the science itself, she meticulously shows us how these assumptions were born and, moreover, how things could have been otherwise conceived, classified, and explained, effectively illustrating the ways a dual system of sex is not supported by the science itself. This also serves to undermine the deterministic mapping of sex (i.e., male/female) onto gender (i.e., masculine/feminine), positing that we should not think about nature *versus* nurture, but rather nature *and* nurture (see also Schiebinger 1989 on sex classification in plant biology).

Other feminist scholars continue to work in this tradition, studying SG&S by examining the medical production of sexed bodies through the clinical deployment of medico-scientific knowledge. Terry (1999), for example, reveals how scientific and medical ideas about homosexuality are co-produced by cultural ideas about bodies, sex, race, and difference through her sociohistorical analysis of sexology. Somerville (1994) likewise traces the racialized constitution of homosexuality in medicine, as well as the sexual underpinnings of scientific racism. As Meyerowitz (2002) and Hirschauer (1998) show, the medical creation and management of the transsexual involves the practical transformation of the meaning of sex. Jordan-Young (2010) examines the "brain" sciences' (psychology and neuroscience) study of sex difference to show how assumptions about the "hardwiring" of the brain for these differences was an assumption bred into the making of the science itself (Fine 2010; Tavris 1993). Richardson's (2013) research focuses on genetic theories of sex—about the search for "male" and "female" in the genome and in our chromosomes. For example, the idea that sex is binary played an important role in the research project seeking the "master gene" for sex differentiation; the idea that females are unpredictable and capricious played a role in theorizing the significance of the X chromosome; the idea of a "war between the sexes" continues to play a role in theorizing the role of the Y chromosome; and an investment in the significance of sex difference (as opposed to a more fundamental human similarity) plays a role in a broad range of sex-based biology (Haslanger 2015).

Some of the most influential work understanding the "making" of the categories of SG&S analyzes the co-constitution of medico-technologies with the bodies they purport to diagnose, manipulate, or "fix" because of how they transgress normative classifications. Karkazis and colleagues (2012), for example, question the knowledge claims around testosterone and performance, examining the ways bodies are "sex tested" and objects of social controversy around the very definition of "the" female sex. Others explore how scientific assumptions shape lay beliefs about the nature of sexual orientation, including whether or not lesbian, gay, and bisexual (LGB) individuals are "born this way" (Arseneau et al. 2013); whether the body reveals "truths" about their desires (Waidzunas and Epstein 2015); or if LGB individuals can be "converted" to heterosexuality (Brian and Grzanka 2014; Waidzunas 2015). Others explore the development of "sexed" illnesses and disorders from "ailments" assumed to be gendered—read as women's problems, such as premenstrual syndrome (Figert 1996; Greenslit 2005; Markens 1996), fertility and (in)fertility disorders (Ikemoto 2009; Mamo 2007; Thompson 2005), menopause and aging (Bell 1987; Fishman, Flatt, and Settersten 2015; Tavris 1993), pregnancy and childbirth (Casper 1998; Layne 2002), and menstruation (Mamo and Fosket 2009).

Feminist STS scholars also focus on the gendered assumptions in biomedical knowledge beyond hormonal and reproductive conditions such as anemia (Hirschauer and Mol 1995) and depression (Blum and Starcuzzi 2004; Metzl 2003). Kempner (2014), for example, studies the politics of gender and health in the biomedical understanding of migraines, with the most common type associated with women and the specialized "cluster headaches" understood as a male disorder. The relative absence of inquiry into men/maleness/masculinities in early STS work on health and illness is taken up in studies that uncover the ways men's health is gendered and relational (Oudshoorn 2003), revealing biomedicine's interest in masculinity only when characterized as enhancing and/or as a potential emerging market (Fishman and Mamo 2002; Mamo 2013) or as a presumed pathology and object of intervention (such as disorders of "sexual development," excess sexual desires, or sexual object choices). As shown more fully in the next section, these technoscientific knowledges, as well as the identities they produce, are not only scripted into what we know and do not know about SG&S but are always embedded in practices of power (Wajcman 1991).

Power, Pathologies, and Resistance

Despite this large and varied contribution to the study of SG&S in STS that specifically reveals the ways scientific knowledge is itself social knowledge, tension exists regarding the degree to which STS scholarship is political in its assessment of power and power relations as well as its treatment of scientists and scientific knowledge. Yet, since at least the 1960s, Marxist and/or socialist-influenced social and humanistic scholars, including feminists, have examined knowledge-making practices for the ways these shape, reproduce, and provide possibilities for intervening in power, attentive to the uneven and simultaneous manifestations of these processes. Feminist STS and other critical STS scholarships are intertwined with challenges to the ideologies and practices that produce, justify, and maintain social inequalities (Duster 1991; Fausto-Sterling 1992; Haraway 1991; Harding 1986; Waldby 1996). Harding (1986), for example, explicitly argues that what is taken as "expert" knowledge and the focus of these knowledge practices, including STS, can be improved by the contributions and, importantly, redirections of those who have been made marginal to the knowledge-production enterprise.

Examining the social nature of scientific knowledges and the ways these produce and reflect inequalities through epistemological frames such as binary logics of normal/pathological and sociopolitical enterprises has long been at the heart of STS projects. SG&S literature often questions the ways science, medicine, and professional practices produce, maintain, and resist the production of sexed and gendered pathologies and the inequalities they manifest, uncovering, for example, the ways biomedical phenomena create pathologies and inequalities that are both pervasively influential and differentially resisted. This is also how many scholars approached SG&S within STS. In fact, much of women's health activism challenged the scientific ideal of men's bodies as normal and women's as pathological (Clarke and Olesen 1999; Murphy 2012; Ruzek 1978). While men's health was medicalized, especially vis-à-vis syphilis and other so-called venereal diseases of the nineteenth and early twentieth centuries, their diseases were largely framed as coming from the outside (e.g., from sexual excess) and framed in collectivist terms associated with the degeneration of the race or family (Bland and Doan 1998; Rohden 2009). Women's bodies, in contrast, were biologized as inherently weaker than men's and, in turn, their functioning pathologized, ultimately giving rise to the women's health movements which countered these claims. Similar to women's health activists, gay and lesbian activists of the 1970s contested the classification of homosexuality as a disease in the American Psychiatric Association's *Diagnostic and Statistical Manual of Mental Disorders* (Kirk and Kutchins 1992; Martin and Lyon 1972). Disability rights activists responded to the social construction of people with (dis)abilities as permanent patients in need of biomedical and technological fixes; seeking to be "patient no more" (Paul K. Longmore Institute on Disabilities 2015), advocating for neuro- and body diversity and against medicalization of and in their lives.

STS-trained scholars have continued to inquire into the various and multidimensional ways policies, practices, and facts emerge not out of unidirectional pressures but also out of conflicts and compromise (Frickel and Moore 2006), and how health and illness are recast through political matters (what Clarke et al. [2003] term "biomedicalization" and Rose [1997] calls "vital politics"). In recent decades, anthropologists and sociologists of science and technology have contributed the concept of biosociality to indicate the ways the "biological" can serve as a meaningful ground for membership and, therefore, as a response to resource and political exclusions (Heinemann and Lemke 2014; Petryna 2002; Shah 2001).

Inclusion in scientific research based on sex-, gender-, sexuality-, or even race-based claims have been widely taken up in the United States context. As sociologist and STS scholar Epstein (2007) argued, by the 1980s group identities and group differences established via biomedical claims-making had become commonplace and part of biomedical research. Driven by the assumptions that these SG&S and racial identities correspond to distinct kinds of bodies, they were codified into biomedical research, laws, policies, and guidelines, while stratified research—often gender-based—came to dominate over "one size fits all" approaches (Epstein 2007; Shim 2014).

Epstein (1996) detailed the ways AIDS treatment activists emerged in response to the AIDS crisis as a movement not only advocating for a place at the scientific table as nonscientists but also as a challenge to the pathologization of gay men and especially their presumed sexual "excess" as an etiology of disease and justification for lack of health care response to the needs of gay men. Epstein's research showed how nonscientists could gain the credibility necessary to revise the questions, goals, and processes of the medical enterprise. In another example, the grassroots campaign known as *A New View of Women's Sexual Problems* advocates against defining a "normal" or "natural" female sexuality, especially those that rely on heteronormative models of male and female sexuality (Tiefer and Kaschak 2002). These, they argue, have ossified into normative expectations of men's and women's sexual functioning, such as men's sexual virility and women's sexual availability that are encouraged through new sexuo-pharmaceutical drugs (ibid.).

The rise of "modern medicine" in the nineteenth and early twentieth centuries is replete with many instances of the co-constitution of biological claims with sociopolit-ical power. Uncovering the varied ways biological categories have been linked to such things as presumed moral character, psychological attributes, physical features, and social behaviors has long been a commitment of feminist STS scholarship. Such work has included studies of how sciences measured the clitorises of American prostitutes (Groneman 2001) and the buttocks and genitalia of African women (Fausto-Sterling 1992; Somerville 1994) and what Terry (1999) described as the "scientific search for homosexual bodies" by locating proof of "deviancy" in what were claimed to be innate constitutional deficiencies. These studies together demonstrate that the classification and categorization of people by virtue of their "natural differences" had become the project of "modern" science (Foucault 1978; Hacking 1986). Social inequalities and medical pathologies operate in tandem: the "congenital" diseases of disability, homo-sexuality, nonwhiteness, Jewishness, and womanhood all created through the life sciences justified racist, homophobic, and other discriminatory practices. Behavioral distinctions also became the object of biomedicine as another way to distinguish the assumed differences among groups and, more so, to justify power hierarchies. Gen-dered conditions also named "excesses" of sexuality, such as nymphomania (Grone-man 2001), sexual addiction (Irvine 1995), and even adolescents' desire for sex (Patton 1995) also reveal how sciences are constitutive of power.

Here, constructions of SG&S justify gender and other social hierarchies, through sci-entifically establishing a presumed natural hierarchical order of dominant "races" and "sexes." The methods, theories, and occupations of STS scholarship have revealed the ways scientific epistemologies, technologies, and their many organizational entities are never *not* racialized, gendered, and naturalizing. Further, the projects of denaturalizing the scientific assumptions and claims of bodies followed scientific practices to show how the idea of the hormonal body, for example, came to be aligned both with wom-en's bodies and with "natural" bodies. Once "sex" or "race" is denaturalized and the constructed binary opposition of male/female and black/white is revealed, responses and challenges are made possible. That is, once sex was denaturalized, first gender and later intersex activists and advocates were able to draw on these assertions to chal-lenge the universal status of this constructed ideal to further dismantle the injustices of these sex/gender/sexuality frames (Dreger 1998; Fausto-Sterling 2000). Defining sex, for example, has required negotiation and provoked controversy about binary frameworks heretofore held as truths by various communities, including sex/gender, nature/nurture, and male/female (Dreger 1998; Laqueur 1992; Schiebinger 1989). Ettorre and Riska (1995) foreground the multifarious intersections of masculinity, sexuality, and biomedical science by critiquing scientifically legitimated answers to culturally contingent questions about what it means to be an appropriately masculine man. In her ethnographic study of controversy around intersex, for example, Karkazis (2008) examines methods used for infant sex assignment and controversy among clinicians, patients, and intersex people. She examined the ways disciplinary knowledge differentiates what are referred to as "sex" (biological and anatomical traits that are used to label a person as female or male) from "gender" (psychological and behavioral traits that are designated masculine or feminine). The breadth of human physical variance is more complex than the categories suggest, and boundaries demarcating what is meant by and constitutes the social are under debate.

Returning to the centrality of technologies to STS, this analytic move is particularly salient for revealing mutual processes of inscription and description. Akrich (1992) argued that technical designers build in and, therefore, materialize assumptions about their users: the interests, motives, skills, and other attributes of "imagined" technological futures, using the term *script* as a way to capture the materialization of anticipated uses and users. Many SG&S scholars draw heavily on this approach to explore the gendered and sexual dimensions of technological design and use (Balsamo 1996; Fishman and Mamo 2002; Mackenzie and Wajcman 1999). Technologies are viewed here as identity projects and through study of their inscription reveal how hegemonic ideals of SG&S as well as other categories of difference emerge and become stable. Studies of assisted reproduction, for example, have shown how "technosemen" used in assisted reproduction, for example, script normative ideals of masculinity into the technology itself (Almeling 2011; Daniels 2006; Mamo 2007; Moore 2007). Metzl (2003) showed how gendering and racialization are co-constitutive social processes embedded in pharmaceuticals such as Prozac. Waidzunas and Epstein (2015), for example, examine sciences of sexual desire and, specifically, phallometric testing, demonstrating how assemblages of this technoscience of arousal produce "truths" about and therefore inscribe bodily and subjective desires. The public health projects of AIDS prevention technologies and the construct of the category "MSM" (for men who have sex with men) similarly have been shown to produce sexual formations and sexual subjects (Boellstorff 2011; Carrillo 2002).

Many SG&S scholars, noting the significance of this conceptual leap, followed Akrich (1992) and examined the ways technologies not only are built with scripts for users but also the ways users might *de-script* their intended meaning. In sites as varied as cosmetic surgery (Balsamo 1996), pharmaceuticals for erectile dysfunction (Fishman and Mamo 2002), oral contraception (Mamo and Fosket 2009), "safe sex" technologies (Moore 1997), procreative technologies such as IVF and infertility biomedicine (Mamo 2007; Thompson 2005), and "conversion therapies" (Brian and Grzanka 2014), feminist STS scholars interrogate how gender and sexual norms and technosciences are deeply co-constitutive phenomena, particularly in late twentieth- and early twenty-first-century biomedicine (Clarke et al. 2003). Oudshoorn's (2003) work on the male contraceptive pill, for example, unravels the alignment of "femininity" with contraception technologies, thereby creating challenges to contraceptive innovations for men due to a conflict with hegemonic masculinity. While Mamo's (2007) analysis of the infertility industry and procreative technologies showed that heteronormative ideals are built into the technologies of assisted reproduction and thus materialize the ideal user as a partnered, heterosexual woman, she demonstrates how users are able to descript technologies for their own purposes. Materialities of users' needs and desires provide the basis of resistance efforts.

The processes of inscribing certain bodies and bodily processes as "normal" or "natural" and then scripting these ideas into the technologies and practices of biomedicine has provided activists and advocates a challenge to the ways health and health problems are constructed, as well as the application of certain pharmaceutical, medical, and technological treatments for them. Concepts such as pharmaceuticalization and biomedicalization provide a macrostructural framework for thinking through and critiquing the co-constitution of SG&S in biomedicine and reveal the ways these are pliant to power (Bell and Figert 2012). As a critical conceptual framework, biomedicalization (Clarke et al. 2003) offers applications in specific social worlds and arenas where science, technology, health/illness, and bodies collide, as well as sites in which biomedicalization may or may not work in counterintuitive ways (e.g., Grzanka and Mann 2014; Mackenzie 2013). Biomedical economies of such things as stem cell research (Benjamin 2013; Thompson 2013) and the "clinical labor" of tissue donors and human subjects (Cooper and Waldby 2014), as well as reproductive labor (Winddance Twine 2011) among others, reveal the ways these rely on and produce gendered bodily and global economies.

In some ways this work follows scholarship that challenged medicalization, interrogating how, when, and through what knowledge claims are gender and sexuality—in particular, women's bodies—placed under the medical gaze. While the categorical construction of "disorders of sexual and gender identity" in the various versions of the DSM (Bryant 2006; Russo 2004), or the ways plastic surgery (Leem 2015), hormone replacement therapies (Fishman, Flatt, and Settersten 2015), and other techniques inscribe and reproduce ideals of gender and sexuality, these are also indicative of bioeconomies, politics, and technosciences that together inscribe difference and produce practices that often sustain power and inequities. With the emergence of Viagra and the creation of categories of sexual dysfunction, for example, many STS scholars argued that Viagra scripted a particular heteronormative masculinity onto its users (Fishman and Mamo 2002; Giami 2004; Marshall and Katz 2002; Rohden 2009). Fishman (2004) continued this work to examine the production of "female sexual dysfunction" as well as anti-aging medicine to reveal the ways gender is scripted into ideals of aging (Faro et al. 2013; Fishman, Flatt, and Settersten 2015).

The queer and sexualities scholarship on biomedicalization has been particularly robust and increasingly concerned with how these processes intersect with other axes or dimensions of difference and inequality, namely race, nation, ethnicity, class, and ability. Aizura (2010), for example, examines the scientific understandings of gender/ sexuality and explores how queer and transgender bodies shape and are shaped by technologies of race, gender, transnationality, medicalization, and political economy. Their research takes up gender reassignment surgery tourism in Thailand and elsewhere, looking at the circuitry of nationalism and biopolitics and value as they relate to gender-variant life. Drawing on ethnographic research in gender clinics in Thailand and with trans women and men who obtained gender reassignment surgery there, they consider how understandings of Orientalized Thai femininity structured non-Thai patients' experiences of care, community, and transition in the space of the clinic and in tourist encounters with Thailand. Examining maternal and reproductive medicine in the United States, ethnographic research by Bridges (2011) reveals the over-medicalization of poor and Black women's pregnant bodies and the ways health care practices participate in the co-construction of racial, class, and gendered hierarchies.

Decades of critical STS scholarship on SG&S has illustrated that "sex," "gender," and "sexuality" are historically contingent and meaningful distinct constructs. Simultaneously, a primary contribution of this work has been to expand our understanding of the interconnectedness and co-constitution of SG&S, which are always embedded in power relations that are variously asymmetrical and unequal. Much of this scholarship is expressly political and concerned with challenging how sciences and technologies contribute to various forms of social oppression, including heterosexism, sexism, and transphobia, leading us to the question of broader social justice concerns in STS, which we engage below.

Reengaging Sex, Gender, and Sexualities in STS: Emergent Approaches and Epistemologies

Our review of the literature thus far has elaborated a robust program of feminist and queer theory-informed STS work on sex, gender, and sexuality that has contributed transformational knowledge and offered potent methodological frameworks for investigating how scientific knowledge, technologies, social categories, and inequalities co-construct one another. STS scholars of SG&S have long taken scientific technoscience as social knowledge and shown the ways this can be scripted into practices in ways that produce and buttress inequalities. STS scholarship here has been essential to SG&S for its engagement with the life sciences and the social world as traffic and entanglements, what Haraway (2000, 105) termed "naturecultures." The assertion that there is no "culture" as separate from some available material "nature" per se but that instead both material and discursive practices constitute the social world continues to go unheard in many disciplinary engagements with SG&S.

Thinking about the future of SG&S research in STS means taking seriously Mamo and Fishman's (2013) provocation about the question of "justice" in STS. To Mamo and Fishman, the question of justice means doing STS work over and against the narrow confines of ethics, which are predicated on socially accepted norms about procedures and best practices: "principles for creating rules of conduct and moral courses of action" (2013, 161). Justice, as their work suggests, can concern what is just and how the law may distribute or execute justice; however, the study of justice in STS specifically may also draw our attention to the production of *injustice* in scientific knowledge production and practices. In this sense, injustice is not simply posited as the unfair and asymmetrical distributions of life chances, which are the target of (neo)liberal social justice projects that seek reparations or inclusion within civil rights frameworks (Duggan 2003; Spade 2015). Instead, justice and therefore injustice are integral and co-implicated in the production and practice of science itself. Justice, then, is neither antecedent nor consequent but actually co-constitutive with processes of science and scientization. In surveying work on genomic, environmental, and reproductive justice, they illuminate how contemporary STS scholarship has heterogeneously engaged (in)justice(s) while also moving beyond reductionist debates between so-called normative versus descriptive STS approaches. Mamo and Fishman stress that this emergent justice work in STS does not represent a singular or unified approach or framework, but they also do not represent justice-focused scholarship as a radical break from the established STS tradition; to the contrary, they emphasize how much of the canon of critical social theory in STS has attended to power, politics, and inequality in ways that could be framed as "justice" concerns (Reardon et al. 2015; Thompson 2015). Grzanka (2014, 260) likewise asserts that critical STS is often not conducted simply to understand how science happens but in "identifying how scientific practices manage and order bodies, influence the distribution of life-saving and life-enhancing resources, control and manipulate natural resources, determine public policy, and reflect broader social norms." Together, we argue that a labile focus on justice as both object of study and/or research outcome facilitates innovative STS work on SG&S and intersecting forms of social inequality that envisions STS as a socially engaged field with the potential to investigate injustice and promote equity in science, technology, and society. Finally, this work—much of which is being forged by leaders in STS and allied fields—highlights ways through which STS scholarship on SG&S may interface with other scholarly and activist projects that seek to catalyze meaningful social change and interrupt scientific practices, programs, and systems that subordinate and harm vulnerable populations and communities.

A focus on justice and inequalities in STS may also resist the arbitrary compartmentalization of sex-, gender-, and sexualities-focused inquiry that results in work that imagines these dimensions of social life as fundamentally distinct from other forms of social inequality and objects of technoscientific inquiry. Stratified biomedicalization's concerns—sometimes implicit and increasingly explicit—with multidimensional analyses of SG&S, race, nation, and other axes of social difference raises the question of how STS has, can, and will engage with problems of what scholars in other fields refer to as "intersectionality," a Black feminist framework for studying complex inequalities (see Grzanka 2014 for an overview). Similar to the ways biomedicalization (Clarke et al. 2003) functions as an historical, inductive analytic for interrogating the ways that power works in contemporary technoscientific systems, practices, and controversies, intersectionality research directs attention to the fundamental relationality of dimensions of inequality and the heterogeneous ways in which individuals, groups, and social movements resist their subordination in the face of interlocking structural oppressions. Kennedy (2005, 472) argued that "feminist STS needs to acknowledge that techno-experiences cannot be understood by reference to only one aspect of identity, like gender, and to engage with debates about intersectionality, an engagement which is comparatively absent from studies of gender and technology." STS studies of gender, according to Kennedy, are often conceptualized and represented as being primarily about gender—or sex, or sexuality—despite the empirical inextricability of gender from other forms of difference and other dimensions of social inequality, such as age, race, ethnicity, nation, and ability.

"Debates about intersectionality" to which Kennedy (2005) refers are, we assert, actively ongoing in critical STS and represent a fruitful area in which to explore Mamo and Fishman's (2013) question of justice in STS. Epstein's (2007) landmark work on social differences in biomedical research, for example, has investigated the polyvalent consequences of incorporating "diversity" into technoscientific research and development. According to Epstein, the inclusion imperative offers opportunities for challenging dominant scientific knowledges about minority groups *and* the reification of social inequalities themselves. Notably, Epstein does not find that incorporating racial minorities, women, and sexual minorities in biomedical research represents discrete or parallel processes; instead, he argues that diversification and inclusion reflect a larger institutional imperative, often with uneven and contradictory consequences for groups being studied. Likewise, Shim's (2014) work on cardiovascular disease (CVD) asks how race, ethnicity, class, gender, and other dimensions of difference influence epidemiology (i.e., disciplinary norms and the operationalization of key constructs, such as risk) and the lived experiences of CVD treatment. She finds that the concept of "culture" in CVD discourse is not outside or exterior to science, nor a byproduct of biomedicine. Rather, Shim illustrates how culture becomes a malleable and dynamic scientific-medical construct in which ideas about racial, ethnic, and class differences are taken up by research scientists, deployed in epidemiological studies, re-inscribed in CVD treatment, and negotiated, modified, and rejected by CVD patients. The "prism of culture" she describes is an intersectional logic in which multiple dimensions of human cultural diversity become scientific techniques by which members of marginalized groups become further stigmatized by biomedicine.

Roberts (1997, 2011), Holloway (2011), Shim (2014), Takeshita (2012), Benjamin (2013) and others offer empirically rigorous and theoretically diverse contributions to our thinking about the "problem" of intersectionality in feminist STS scholarship: how questions of SG&S in science and technology are never divorced from other ideological, political, and material dimensions of inequality and are indeed co-constituted by other axes of difference. The issues facing contemporary STS scholars are not strictly contemporary problems per se. In other words, it is not that the landscape of social inequalities has necessarily become more complex in recent decades (Dill 2014); rather, STS projects need to attend to the interlocking relationships among systems of social injustice that have always been present while always changing across social and historical contexts. For example, Somerville's (1994) history of scientific racism and the science of homosexuality illuminates how race and sexual orientation were co-constructed by Western biologists, psychologists, physicians, and anthropologists. Africans' (and African Americans') racial inferiority emerged with the nascent science of the homosexuality as scientists drew upon ideas about racial difference to construct sexual deviance and vice versa. Though the scientists themselves may have imagined their projects to be about sexuality *or* about race, Somerville's work shows that these projects only appear unrelated in historical retrospective (Stoler 1995).

One of STS's key analytic strengths is its empirical approach to studying the construction and material effects of social-scientific problems; in this sense, STS has much to contribute and much to learn from ongoing interdisciplinary work on intersectionality that explores SG&S in relation to other dimensions of difference and inequality. Because STS does not presume that scientific, technological, and social categories function statically or in isolation, it may complement intersectional perspectives that stress the cultural and historical contingency of "identity" categories and demographic "variables" deployed in scientific research and practice, as well as different communities' complex relationships with health systems and technologies. Furthermore, new work in feminist STS that addresses long-standing calls for greater attention to intersectionality (Kennedy 2005) can broaden the scope of the triple-trope of "sex, sexuality, and gender" to consider the intersectional dynamics immanent to stratified biomedicalization and other systems of inequality, such as the racialization of deviant sexualities (Bridges 2011; Mackenzie 2013), transnational deployments of reproductive technologies (Takeshita 2012) and sex-selection practices (Bhatia 2010), the class and racial dynamics of pharmaceuticals (Carpenter and Casper 2009; Mamo, Nelson, and Clark 2010), as well as mutual processes of sexualization and pharmaceuticalization (Mamo and Epstein 2014). Finally, these projects collectively aim to imagine otherwise and reconfigure social relations in ways that offer rubrics for how STS might do "justice work" in the world, including influencing policy, harmful practices, and centering knowledge-making and technoscientific practices among "marginalized" populations in the United States and worldwide.

Conclusion

The critique of the foundational heteronormative binaries of male/female, masculine/feminine, homosexual/heterosexual, and ultimately sex/gender is a key contribution of feminist STS. Though technoscientific and biomedical projects continue to recast and reconstitute bio-determinist renderings of SG&S, feminist STS scholarship has generated a robust and growing archive of critical inquiry that undermines the presumed sanctity of a priori sex and its relationship to flexible-but-still-biological framings of gender and sexual desires. While driven by a fundamental commitment to social constructionism, feminist STS has *rejected* simplistic constructionisms that position themselves in dichotomous opposition with essentialist or determinist epistemologies that downplay or negate the significance of the social and cultural. Much of the feminist STS work we have elaborated here actually facilitates the destabilization of the reductionist social constructionism/essentialism binary by taking seriously the interplay of what Haraway (1997) refers to as the material and the semiotic, or "material-semiotics." By interrogating the materiality of bodies alongside the ideological, philosophical, and immaterial forces that co-create those bodies and make them matter (Butler 1993), feminist STS remains a crucial lens for thinking through and advancing some of the important medical, environmental, technological, and social policy issues of our time.

In conclusion, we assert that feminist STS offers an indispensable perspective in contemporary bioethical and environmental debates, projects, and controversies that will determine the present and futures of SG&S in science, technology, industry, policy, and everyday life. In particular, we find it disconcerting that so much contemporary debate in both public (i.e., "lay") and medical-scientific (i.e., "expert") spheres about the origins, immutability, and discreteness of sexual orientation categories and desires lacks an STS perspective. Likewise, debates about hormone "blockers" and other therapies for gender nonconforming, nonbinary, and trans children and adolescents are organized by ethical debates rather than meaningful STS-informed consideration of what gender, hormones, and the body actually are (cf. Bryant 2006). We recognize that such conversations are not easy to have and that STS insights often get lost in translation or are incomprehensible when communicating across diverse discourses and publics. Nonetheless, if STS does conceptualize itself as possessing a capacity for justice and as a potential engine of social transformation—and we think the field should!—STS scholars should commit to thinking through ways to have constructive and meaningful conversations with dissimilar audiences, including intellectual-political movements such as contemporary feminist and postcolonial studies, so that critical STS perspectives may influence consequential decisions about all kinds of lives and social worlds, but especially on the lives of those who are most vulnerable and voiceless in major policy debates.

Ultimately, this chapter has also sought to demonstrate how inextricable SG&S issues are from race, nation, ethnicity, class, and other salient dimensions of social inequality that are co-constituted by historical and contemporary technoscientific knowledge production and practices. Accordingly, we conclude by affirming that our aim has been to illuminate and *open up* the connections between SG&S and other consequential domains of social life—those dimensions on which opportunity, privilege, and oppression are asymmetrically distributed across populations, environments, and even species. We should not use contemporary rubrics of social inequality to foreclose the possibilities of other unpredictable or heretofore unseen entanglements between SG&S and other domains of material-semiotic relations of power. In other words, SG&S should not be displaced in our attempts to engage new sites of technoscientific power relations, nor should our extant understandings of social inequality (e.g., racism, sexism, xenophobia) prematurely dictate how we conceptualize intersecting systems of domination in critical STS scholarship. Energizing work in disability studies and "crip theory" (Kafer 2013; McRuer and Mollow 2012) and trans-species scholarship (Chen 2012; Hayward 2010) that engages concepts of virality, capacity, and debility—largely emanating from queer, gender, postcolonial, and critical race studies—reminds us that STS has a place in conversations that are radically emergent and that are still very much uncharted (Brown 2015; Clough and Puar 2012; Puar 2009; Shildrick 2009). As an interdisciplinary field, STS is adept at resisting the boundary policing that we observe in other disciplines and should embrace rather than retreat from the undetermined futures of critical, feminist STS.

References

Aizura, Aren Zachary. 2010. "Feminine Transformations: Gender Reassignment Surgical Tourism in Thailand." *Medical Anthropology: Cross-Cultural Studies in Health and Illness* 29 (4): 424–43.

Akrich, Madeleine. 1992. "The De-Scription of Technical Objects." In *Shaping Technology/Building Society: Studies in Sociotechnical Change*, edited by Wiebe E. Bijker and John Law, 205–24. Cambridge, MA: MIT Press.

Almeling, Rene. 2011. *Sex Cells: The Medical Market for Eggs and Sperm*. Berkeley: University of California Press.

Ansara, Y. Gavriel, and Peter Hegarty. 2012. "Cisgenderism in Psychology: Pathologising and Misgendering Children from 1999 to 2008." *Psychology & Sexuality* 3 (2): 137–60.

Arseneau, Julie R., Patrick R. Grzanka, Joe R. Miles, and Ruth E. Fassinger. 2013. "Development and Initial Validation of the Sexual Orientation Beliefs Scale (SOBS)." *Journal of Counseling Psychology* 60 (3): 407–20.

Balsamo, Anne Marie. 1996. *Technologies of the Gendered Body: Reading Cyborg Women*. Durham, NC: Duke University Press.

Bell, Susan E. 1987. "Changing Ideas: The Medicalization of Menopause." *Social Science and Medicine* 24 (6): 535–42.

Bell, Susan E., and Anne E. Figert. 2012. "Medicalization and Pharmaceuticalization at the Intersections: Looking Backward, Sideways, and Forward." *Social Science and Medicine* 75 (5): 775–83.

Benjamin, Ruha. 2013. *People's Science: Bodies and Rights on the Stem Cell Frontier*. Stanford, CA: Stanford University Press.

Berg, Marc, and Annemarie Mol. 1998. *Differences in Medicine*. Durham, NC: Duke University Press.

Bhatia, Rajani. 2010. "Constructing Gender from the Inside Out: Sex-Selection Practices in the United States." *Feminist Studies* 36 (2): 260–91.

Bland, Lucy, and Laura Doan. 1998. *Sexology Uncensored: The Documents of Sexual Science*. Cambridge: Polity Press.

Blum, Linda M., and Nena F. Starcuzzi. 2004. "Gender in the Prozac Nation Popular Discourse and Productive Femininity." *Gender and Society* 18 (3): 269–86.

Boellstorff, Tom. 2011. "But I Do Not Identify as Gay: A Proleptic Genealogy of the MSM Category." *Cultural Anthropology* 26 (2): 287–312.

Bowker, Geoffrey C., and Susan Leigh Star. 1999. *Sorting Things Out: Classification and Its Consequences*. Cambridge, MA: MIT Press.

Braidotti, Rosi. 1994. *Nomadic Subjects: Embodiment and Sexual Difference in Contemporary Feminist Theory, Gender, and Culture*. New York: Columbia University Press.

Brian, Jennifer D., and Patrick R. Grzanka. 2014. "The Machine in the Garden of Desire." *American Journal of Bioethics: Neuroscience* 5 (1): 17–18.

Bridges, Khiara M. 2011. *Reproducing Race: An Ethnography of Pregnancy as a Site of Racialization*. Berkeley: University of California Press.

Brown, Jayna. 2015. "Being Cellular: Race, the Inhuman, and the Plasticity of Life." *GLQ: A Journal of Lesbian and Gay Studies* 21 (2–3): 321–41.

Bryant, Karl. 2006. "Making Gender Identity Disorder of Childhood: Historical Lessons for Contemporary Debates." *Sexuality Research & Social Policy* 3 (3): 23–39.

Butler, Judith. 1993. *Bodies That Matter: On the Discursive Limits of "Sex."* New York: Routledge.

Carpenter, Laura M., and Monica J. Casper. 2009. "A Tale of Two Technologies: HPV Vaccination, Male Circumcision, and Sexual Health." *Gender and Society* 23 (6): 790–816.

Carrillo, Héctor. 2002. *The Night Is Young: Sexuality in Mexico in the Time of AIDS*. Chicago: University of Chicago Press.

Casper, Monica J. 1998. *The Making of the Unborn Patient: A Social Anatomy of Fetal Surgery*. New Brunswick, NJ: Rutgers University Press.

Chen, Mel Y. 2012. *Animacies: Biopolitics, Racial Mattering, and Queer Affect*. Durham, NC: Duke University Press.

Clarke, Adele E. 1998. *Disciplining Reproduction: Modernity, American Life Sciences, and "The Problems of Sex."* Berkeley: University of California Press.

Clarke, Adele E., and Virginia Olesen. 1999. "Revising, Diffracting, Acting." In *Revisioning Women, Health, and Healing: Feminist, Cultural, and Technoscience Perspectives*, edited by Adele. E. Clarke and Virginia. L. Olesen, 3–48. New York: Routledge.

Clarke, Adele E., Janet Shim, Laura Mamo, Jennifer R. Fosket, and Jennifer R. Fishman. 2003. "Biomedicalization: Theorizing Technoscientific Transformations of Health, Illness, and U.S. Biomedicine." *American Sociological Review* 68 (2): 161–94.

Clough, Patricia T., and Jasbir K. Puar, eds. 2012. Special issue, "*WSQ: Viral*." *Women's Studies Quarterly* 40 (1–2).

Cooper, Melinda, and Catherine Waldby. 2014. *Clinical Labor: Tissue Donors and Research Subjects in the Global Bioeconomy, Experimental Futures*. Durham, NC: Duke University Press.

Daniels, Cynthia R. 2006. *Exposing Men: The Science and Politics of Male Reproduction*. Oxford: Oxford University Press.

Dill, Bonnie Thornton. 2014. "Epilogue: Frontiers." In *Intersectionality: A Foundations and Frontiers Reader*, edited by Patrick R. Grzanka, 341–43. Boulder, CO: Westview Press.

Dreger, Alice Domurat. 1998. *Hermaphrodites and the Medical Invention of Sex*. Cambridge, MA: Harvard University Press.

Duggan, Lisa. 2003. *The Twilight of Equality? Neoliberalism, Cultural Politics, and the Attack on Democracy*. Boston: Beacon Press.

Duster, Troy. 1991. *Backdoor to Eugenics*. New York: Routledge.

Epstein, Steven. 1996. *Impure Science: AIDS, Activism, and the Politics of Knowledge, Medicine and Society*. Berkeley: University of California Press.

___. 2007. *Inclusion: The Politics of Difference in Medical Research*. Chicago: University of Chicago Press.

Ettorre, Elizabeth M., and Elianne Riska. 1995. *Gendered Moods: Psychotropics and Society*. New York: Routledge.

Faro, Livi, Lilian Chazan, Fabiola Rohden, and Jane Russo. 2013. "Man with Capital 'M': Masculinity Ideals Reconstructed in Pharmaceutical Marketing." *Cadernos Pagu* 40: 287–321.

Fausto-Sterling, Anne. 1992. *Myths of Gender: Biological Theories about Women and Men*. 2nd ed. New York: Basic Books.

___. 2000. *Sexing the Body: Gender Politics and the Construction of Sexuality*. New York: Basic Books.

Figert, Anne E. 1996. *Women and the Ownership of PMS: Structuring a Psychiatric Disorder*. Hawthorne, NY: Aldine de Gruyter.

Fine, Cordelia. 2010. *Delusions of Gender: How Our Minds, Society, and Neurosexism Create Difference*. New York: W. W. Norton.

Fishman, Jennifer R. 2004. "Manufacturing Desire: The Commodification of Female Sexual Dysfunction." *Social Studies of Science* 34 (2): 187–218.

Fishman, Jennifer R., and Laura Mamo. 2002. "What's in a Disorder: A Cultural Analysis of Medical and Pharmaceutical Constructions of Male and Female Sexual Dysfunction." *Women & Therapy* 24 (1–2): 179–93.

Fishman, Jennifer R., Michael A. Flatt, and Richard A. Settersten. 2015. "Bioidentical Hormones, Menopausal Women, and the Lure of the 'Natural' in U.S. Anti-aging Medicine." *Social Science and Medicine* 132: 79–87.

Foucault, Michel. 1978. *The History of Sexuality, Volume 1: An Introduction.* New York: Pantheon Books.

Frickel, Scott, and Kelly Moore. 2006. "Prospects and Challenges for a New Political Sociology of Science." In *The New Political Sociology of Science: Organizations, Networks, and Institutions,* edited by Scott Frickel and Kelly Moore, 3–31. Madison: University of Wisconsin Press.

Giami, Alain. 2004. "De l'impuissance à la dysfonction éretile: Destins de la médicalization de la sexualité." In *Le gouvernement des corps,* edited by Didier Fassin and Dominique Memmi, 77–108. Paris: Éditions EHESS.

Greenslit, Nathan. 2005. "Dep®ession and Consum♀tion: Psychopharmaceuticals, Branding, and New Identity Practices." *Culture, Medicine and Psychiatry* 29 (4): 477–502.

Groneman, Carol. 2001. *Nymphomania: A History.* New York: W. W. Norton.

Grosz, Elizabeth A. 1994. *Volatile Bodies: Toward a Corporeal Feminism, Theories of Representation and Difference.* Bloomington: Indiana University Press.

Grzanka, Patrick R. 2014. "Science and Technology Studies as Tools for Social Justice." In *Intersectionality: A Foundations and Frontiers Reader,* edited by Patrick. R. Grzanka, 259–66. Boulder, CO: Westview Press.

Grzanka, Patrick R., and Emily S. Mann. 2014. "Queer Youth Suicide and the Psychopolitics of 'It Gets Better.'" *Sexualities* 17 (4): 369–93.

Ha, Nathan Q. 2011. "The Riddle of Sex: Biological Theories of Sexual Difference in the Early Twentieth Century." *Journal of the History of Biology* 44 (3): 505–46.

Hacking, Ian. 1986. "Making Up People." In *Reconstructing Individualism: Autonomy, Individuality, and the Self in Western Thought,* edited by Thomas C. Heller, Morton Sosna, and David E. Wellbery, 222–36. Stanford, CA: Stanford University Press.

Haraway, Donna J. 1989. *Primate Visions: Gender, Race, and Nature in the World of Modern Science.* New York: Routledge.

___. 1991. *Simians, Cyborgs, and Women: The Reinvention of Nature.* New York: Routledge.

___ 1997. *Modest_Witness@Second_Millennium.FemaleMan_Meets_OncoMouse™: Feminism and Technoscience.* New York: Routledge.

___. 2000. *How Like a Leaf.* New York: Routledge.

Harding, Sandra G. 1986. *The Science Question in Feminism.* Ithaca, NY: Cornell University Press.

Haslanger, Sally. 2015. "Sarah S. Richardson. Sex Itself: The Search for Male and Female in the Human Genome." Review. *Hypatia Reviews Online.* Accessed April 12, 2016 at http://hypatiaphilosophy.org/HRO/reviews/content/228.

Hayward, Eva. 2010. "FINGERYEYES: Impressions of Cup Corals." *Cultural Anthropology* 25 (4): 577–99.

Heinemann, Torsten, and Thomas Lemke. 2014. "Biological Citizenship Reconsidered: The Use of DNA Analysis by Immigration Authorities in Germany." *Science, Technology, & Human Values* 39 (4): 488–510.

Hirschauer, Stefan. 1998. "Performing Sexes and Genders in Medical Practices." In *Differences in Medicine: Unraveling Practices, Techniques, and Bodies,* edited by Marc Berg and Annemarie Mol, 13–27. Durham, NC: Duke University Press.

Hirschauer, Stefan, and Annemarie Mol. 1995. "Shifting Sexes, Moving Stories: Feminist/Constructivist Dialogues." *Science, Technology, & Human Values* 20 (3): 368–85.

Holloway, Karla F. C. 2011. *Private Bodies, Public Texts: Race, Gender, and a Cultural Bioethics.* Durham, NC: Duke University Press.

Ikemoto, Lisa C. 2009. "Eggs as Capital: Human Egg Procurement in the Fertility Industry and the Stem Cell Research Enterprise." *Signs: Journal of Women in Culture and Society* 34 (4): 763–81.

Irvine, Janice M. 1990. *Disorders of Desire: Sexuality and Gender in Modern American Sexology.* Revised and expanded edition. Philadelphia: Temple University Press.

___. 1995. "Reinventing Perversion: Sex Addiction and Cultural Anxieties." *Journal of the History of Sexuality* 5 (3): 429–50.

Jackson, Stevi. 1996. *Theorising Heterosexuality.* Buckingham: Open University Press.

Jordan-Young, Rebecca M. 2010. *Brain Storm: The Flaws in the Science of Sex Differences.* Cambridge, MA: Harvard University Press.

Kafer, Alison. 2013. *Feminist, Queer, Crip.* Bloomington: Indiana University Press.

Karkazis, Katrina. 2008. *Fixing Sex: Intersex, Medical Authority, and Lived Experience.* Durham, NC: Duke University Press.

Karkazis, Katrina, Rebecca Jordan-Young, Georgiann Davis, and Sylvia Camporesi. 2012. "Out of Bounds: A Critique of the New Policies of Hyperandrogenism in Elite Female Athletes." *American Journal of Bioethics* 12 (7): 3–16.

Keller, Evelyn Fox. 1985. *Reflections on Gender and Science.* New Haven, CT: Yale University Press.

Keller, Evelyn Fox, and Helen E. Longino. 1996. *Feminism and Science.* Oxford: Oxford University Press.

Kempner, Joanna. 2014. *Not Tonight: Migraine and the Politics of Gender and Health*. Chicago: University of Chicago Press.

Kempner, Joanna, Clifford S. Perlis, and Jon F. Merz. 2005. "Forbidden Knowledge." *Science* 307: 854.

Kennedy, Helen. 2005. "Subjective Intersections in the Face of the Machine." *European Journal of Women's Studies* 12: 471–87.

Kirk, Stuart A., and Herb Kutchins, eds. 1992. *The Selling of the DSM: The Rhetoric of Science in Psychiatry*. New York: Aldine de Gruyter.

Laqueur, Thomas. 1992. *Making Sex: Body and Gender from the Greeks to Freud*. Cambridge, MA: Harvard University Press.

Layne, Linda L. 2002. *Motherhood Lost: A Feminist Account of Pregnancy Loss in America*. New York: Routledge.

Leem, So Yeon. 2015. "The Dubious Enhancement: Making South Korea a Plastic Surgery Nation." *East Asian Science, Technology, and Society: An International Journal* 10: 1–21.

MacKenzie, Donald A., and Judy Wajcman. 1999. *The Social Shaping of Technology*. 2nd ed. Philadelphia: Open University Press.

Mackenzie, Sonja. 2013. *Structural Intimacies: Sexual Stories in the Black AIDS Epidemic (Critical Issues in Health and Medicine)*. New Brunswick, NJ: Rutgers University Press.

Mamo, Laura. 2007. *Queering Reproduction: Achieving Pregnancy in the Age of Technoscience*. Durham, NC: Duke University Press.

___. 2013. "Queering the Fertility Clinic." *Journal of Medical Humanities* 34 (2): 227–39.

Mamo, Laura, and Steven Epstein. 2014. "The Pharmaceuticalization of Sexual Risk: Vaccine Development and the New Politics of Cancer Prevention." *Social Science and Medicine* 101: 155–65.

Mamo, Laura, and Jennifer R. Fishman. 2013. "Why Justice? Introduction to the Special Issue on Entanglements of Science, Ethics, and Justice." *Science, Technology, & Human Values* 38: 159–75.

Mamo, Laura, and Jennifer Ruth Fosket. 2009. "Scripting the Body: Pharmaceuticals and the (Re) Making of Menstruation." *Signs: Journal of Women in Culture and Society* 34 (4): 925–49.

Mamo, Laura, Amber Nelson, and Aleia Clark. 2010. "Producing and Protecting Risky Girlhoods: How the HPV Vaccine Became the Right Tool to Prevent Cervical Cancer." In *Three Shots at Prevention: The HPV Vaccine and the Politics of Medicine's Simple Solutions*, edited by Julie Livingston, Keith Wailoo, Steven Epstein, and Robert Aronowitz. Baltimore: Johns Hopkins University Press.

Markens, Susan. 1996. "The Problematic of 'Experience': A Political and Cultural Critique of PMS." *Gender and Society* 10 (1): 42–58.

Marshall, Barbara L., and Stephen Katz. 2002. "Forever Functional: Sexual Fitness and the Ageing Male Body." *Body & Society* 8 (4): 43–70.

Martin, Del, and Phyllis Lyon. 1972. *Lesbian/Woman*. San Francisco: Glide Publications.

Martin, Emily. 1991. "The Egg and the Sperm: How Science Has Constructed a Romance Based on Stereotypical Male-Female Roles." *Signs: Journal of Women in Culture and Society* 16 (3): 485–501.

McRuer, Robert, and Anna Mollow. 2012. *Sex and Disability*. Durham, NC: Duke University Press.

Metzl, Jonathan. 2003. *Prozac on the Couch: Prescribing Gender in the Era of Wonder Drugs*. Durham, NC: Duke University Press.

Meyerowitz, Joanne J. 2002. *How Sex Changed: A History of Transsexuality in the United States*. Cambridge, MA: Harvard University Press.

Moore, Lisa Jean. 1997. "'It's Like You Use Pots and Pans to Cook. It's the Tool': The Technologies of Safer Sex." *Science, Technology, & Human Values* 22 (4): 434–71.

___. 2007. *Sperm Counts: Overcome by Man's Most Precious Fluid*. New York: New York University Press.

Moore, Lisa Jean, and Adele E. Clarke. 1995. "Clitoral Conventions and Transgressions: Graphic Representations in Anatomy Texts, c 1900–1991." *Feminist Studies* 21 (2): 255–301.

Murphy, Michelle. 2012. *Seizing the Means of Reproduction: Entanglements of Feminism, Health, and Technoscience*. Durham, NC: Duke University Press.

Oudshoorn, Nelly. 1994. *Beyond the Natural Body: An Archaeology of Sex Hormones*. New York: Routledge.

___. 2003. *The Male Pill: A Biography of a Technology in the Making*. Durham, NC: Duke University Press.

Patton, Cindy. 1995. "Between Innocence and Safety: Epidemiological and Popular Constructions of Young People's Need for Safe Sex." In *Deviant Bodies*, edited by Jennifer Terry and Jacqueline Urla. Bloomington: Indiana University Press.

Paul K. Longmore Institute on Disability. Accessed November 12, 2015, at https://sites7.sfsu.edu/longmoreinstitute/patient-no-more.

Petryna, Adriana. 2002. *Life Exposed: Biological Citizens after Chernobyl*. Princeton, NJ: Princeton University Press.

Puar, Jasbir K. 2009. "Prognosis Time: Towards a Geopolitics of Affect, Debility and Capacity." *Women & Performance: A Journal of Feminist Theory* 19 (2): 161–72.

Reardon, Jenny, Jacob Metcalf, Martha Kenney, and Karen Barad. 2015. "Science & Justice: The Trouble and the Promise." *Catalyst: Feminism, Theory, Technoscience* 1 (1): 1–48.

Richardson, Sarah S. 2013. *Sex Itself: The Search for Male and Female in the Human Genome*. Chicago: University of Chicago Press.

Roberts, Dorothy E. 1997. *Killing the Black Body: Race, Reproduction, and the Meaning of Liberty*. New York: Pantheon Books.

___. 2011. *Fatal Invention: How Science, Politics, and Big Business Re-create Race in the Twenty-First Century*. New York: New Press.

Rohden, Fabíola. 2009. "Gender Differences and the Medicalization of Sexuality in the Creation of Sexual Dysfunctions Diagnosis." *Revista Estudos Feministas, Florianópolis* 17 (1): 89–109.

Rose, Nikolas. 1997. *The Politics of Life Itself: Biomedicine, Power, and Subjectivity in the Twenty-First Century*. Princeton, NJ: Princeton University Press.

Russo, Jane. A. 2004. "Do desvio ao transtorno: A medicalização da sexualidade na nosografia psiquiátrica contemporânea." In *Sexualidade e saberes: Convenções e fronteiras*, edited by Adriana Piscitelli, Maria Filomena Gregori, and Sérgio Carrara, 95–114. Rio de Janeiro: Garamond.

Ruzek, Sheryl Burt. 1978. *The Women's Health Movement: Feminist Alternatives to Medical Control*. New York: Praeger.

Schiebinger, Londa L. 1989. *The Mind Has No Sex: Women in the Origins of Modern Science*. Cambridge, MA: Harvard University Press.

Segal, Lynn. 1994. *Straight Sex: Rethinking the Politics of Pleasure*. Berkeley: University of California Press.

Shah, Nayan. 2001. *Contagious Divides: Epidemics and Race in San Francisco's Chinatown*. Berkeley: University of California Press.

Shildrick, Margrit. 2009. *Dangerous Discourses of Disability, Subjectivity and Sexuality*. New York: Palgrave Macmillan.

Shim, Janet K. 2014. *Heart-sick: The Politics of Risk, Inequality, and Heart Disease*. New York: New York University Press.

Somerville, Siobhan. 1994. "Race and the Construction of the Homosexual Body." *Journal of the History of Sexuality* 5 (2): 243–66.

Spade, Dean. 2015. *Normal Life: Administrative Violence, Critical Trans Politics, and the Limits of the Law*. Revised edition. Durham, NC: Duke University Press.

Star, Susan Leigh. 1990. "This Is Not a Boundary Object: Reflections on the Origins of a Concept." *Science, Technology, & Human Values* 35 (5): 601–17.

Stein, Arlene, and Ken Plummer. 1994. "I Can't Even Think Straight: 'Queer' Theory and the Missing Sexual Revolution in Sociology." *Sociological Theory* 12 (2): 178–87.

Stein, Edward. 2001. *The Mismeasure of Desire*. London: Oxford University Press.

Stoler, Ann Laura. 1995. *Race and the Education of Desire: Foucault's 'History of Sexuality' and the Colonial Order*. Durham, NC: Duke University Press.

Takeshita, Chikako. 2012. *The Global Biopolitics of the IUD: How Science Constructs Contraceptive Users and Women's Bodies*. Cambridge, MA: MIT Press.

Tavris, Carole. 1993. "The Mismeasure of Woman." *Feminism & Psychology* 3 (2): 149–68.

Terry, Jennifer. 1999. *An American Obsession: Science, Medicine, and Homosexuality in Modern Society*. Chicago: University of Chicago Press.

Thompson, Charis. 2005. *Making Parents: The Ontological Choreography of Reproductive Technologies*. Cambridge, MA: MIT Press.

___. 2013. *Good Science: The Ethical Choreography of Stem Cell Research*. Cambridge, MA: MIT Press.

___. 2015. "Move beyond Difference." *Nature* 522: 415.

Tiefer, Leonore. 1995. *Sex Is Not a Natural Act and Other Essays*. Chicago: Westview Press.

Tiefer, Leonore, and Ellyn Kaschak, eds. 2002. *A New View of Women's Sexual Problems*. New York: Routledge.

Voss, Georgina, and Simon J. Lock. 2012. "STS and Sexuality: Why the Silence?" Conference paper. 4S Annual Meeting, Copenhagen Business School, Denmark, October 17.

Waidzunas, Tom. 2015. *The Straight Line: How the Fringe Science of Ex-Gay Therapy Reoriented Sexuality*. Minneapolis: University of Minnesota Press.

Waidzunas, Tom, and Steven Epstein. 2015. "'For Men Arousal Is Orientation': Bodily Truthing, Technosexual Scripts, and the Materialization of Sexualities through the Phallometric Test." *Social Studies of Science* 45 (2): 187–213.

Wajcman, Judy. 1991. *Feminism Confronts Technology*. University Park: Pennsylvania State University Press.

Waldby, Catherine. 1996. *AIDS and the Body Politic: Biomedicine and Sexual Difference*. London: Routledge.

Winddance Twine, France. 2011. *Outsourcing the Womb: Race, Class and Gestational Surrogacy in a Global Market*. New York: Routledge.

Wyer, Mary, Mary Barbercheck, Donna Cookmeyer, Hatice Örün Öztürk, and Marta Wayne. 2013. *Women, Science, and Technology: A Reader in Feminist Science Studies*. 3rd ed. New York: Routledge.

14   Feminism, Postcolonialism, Technoscience

Banu Subramaniam, Laura Foster, Sandra Harding, Deboleena Roy, and Kim TallBear

This chapter identifies an emerging cluster of work that brings together the intersecting concerns of science and technology studies (STS), feminist STS, and postcolonial STS.

We begin by identifying a few of the central themes in each field and then introduce an emerging cluster of scholarship that works across all three. We then discuss three recent themes that highlight the key issues for STS: (1) critiques of colonial science and its hierarchies of gender/race/class, (2) Latin American decolonial theory and its feminist insights, and (3) how indigenous peoples' knowledge challenges all three of the above mentioned fields. We end with some reflections for the future of STS by arguing for more scholarly work that engages with all three of these intersecting fields. We believe that this is important for the field of STS because a singular focus on gender, race, coloniality, or indigeneity alone leaves numerous gaps in our understanding of the co-constitution of science and society.

Studying Science, Studying Society: The Importance of Feminist Postcolonial STS

Feminist STS, postcolonial STS, and STS itself are all recognized research fields with powerful theoretical and methodological projects. Lively literatures in postcolonial and feminist STS have produced valuable insights. Yet, work that draws upon the fundamental insights of all three remains underexplored and theorized. In this essay, we examine emerging scholarship that brings these three fields together, exploring some of their common themes and tensions. The emerging work is heterogeneous (as is work in the three constituent fields), fluid, porous, and polyvocal and defies simplified definitions or categories. Bringing these fields together is important for developing more robust accounts of the co-constitution of science and society.

STS asks how societies, on the one hand, and their technosciences, on the other, shape each other in any particular historical era. A central approach in STS is to examine these relationships through practices of co-production; that is, we cannot study science or society in isolation; we must always understand them as mutually producing each other and thus as constitutive of each other. In doing so, scholars ask how historically situated relationships of science and society co-produce new forms of knowledge production, as well as understandings of the social order (Jasanoff 2004; Reardon 2005; Shapin and Schaffer 1985). Thus influential STS projects have focused on sociologies of scientific knowledge (Collins 1983; Gieryn 1982), networks of human and nonhuman actors (Callon 1986; Latour 1987, 2004), technological artifacts (MacKenzie and Wajcman 1985; Pinch and Bijker 1984), boundary objects (Clarke and Casper 1996; Fujimura 1992; Star and Griesemer 1989), genetic technologies (Duster 2003; Franklin, Lury, and Stacey 2000; Rabinow 1999; Reardon 2005; Sunder Rajan 2006) and more. Feminist scholars contribute much to this work, bringing attention to how gender, inequality, and power figure centrally within relationships of science and society. Similarly, postcolonial STS demonstrates that science is best understood as co-constituted with colonialism (Anderson 2002; Anderson and Adams 2008; McNeil 2005; Seth 2009). Science and technology, it argues, is best understood as "sciences of empire" (Schiebinger 2004), and indeed almost all modern science should be understood as "science in a colonial context" (Seth 2009). A singular focus on gender, race, indigeneity, or coloniality however leaves gaps in our understanding of the co-constitution of science and society; therefore it becomes important to draw upon insights from all three fields of feminist STS, postcolonial STS, and STS.

Feminist STS has examined how scientific knowledge does not equally benefit everyone; it produces even more unequal social relations in significant respects, thus emphasizing the need for new ways of producing science. The work that emerged from the women's movements of the 1970s focused mainly on five questions. Where were the women in the history and present practices of technosciences? How did applications and technologies of technosciences affect women and our conceptions of gender, race, and sexuality? What should be feminist priorities for scientific and technical education for a more socially just world? What should be the feminist research priorities in natural and social sciences? What should more adequate epistemologies, ontologies, and philosophies of science look like? Early and more recent review essays provide valuable analyses of how such questions have been pursued in different research disciplines and political contexts (Harding 1986; Schiebinger 1987). Today many of these questions have actually reached top levels of science policy in the United States as well as in many other countries around the globe.[1]

As a product of the women's movements, feminist STS engaged more directly with social justice issues than did STS. Carolyn Merchant's influential 1980 work *The Death of Nature* in fact begins from the social concerns of women's liberation and ecology movements (Merchant 1980). Organizations such as the Boston Women's Health Collective and SisterSong Women of Color Reproductive Justice Collective also take different strategies to increase the participation, access, and advocacy of patients historically marginalized within the health care system.

Yet feminist STS is by no means a unified, coherent, or complete area of study. Questions about women in science and about the gendering of science, although related, often do not overlap in their foci. The former questions are accused of failing to challenge historical and epistemological assumptions of science, while the latter are charged with implementing simplistic understandings of scientific practice. Additionally many "northern" feminist STS projects do not address colonialism and imperialism, while leaving indigenous peoples and their knowledge achievements and needs unrecognized (Harding 2008, 103).

Postcolonial STS focuses on colonialisms, their aftermaths, and how these are co-produced with a diversity of technoscience projects. Warwick Anderson writes that the term *postcolonial* "refers both to new configurations of technoscience and to the critical modes of analysis that identify them" (Anderson 2002, 643). Yet activists and scholars have been questioning the usefulness of the term for understanding, for instance, Latin America with its differing colonial histories (Medina, Marques, and Holmes 2014; Moraña, Dussel, Cuato, and Jáuregai 2008), and the diverse situations of indigenous/aboriginal groups around the globe. Recent special journal issues and edited collections have called for understandings of technosciences that account for colonial histories and postcolonial conditions (Anderson 2002; Harding 2011; McNeil 2005; Medina, Marques, and Holmes 2014; Phalkey 2013; Schiebinger 2005a; Seth 2009). This field, too, has rapidly expanded. Its distinctive focuses include contradictory spaces of postcolonial technoscience (Abraham 2006; Anderson and Adams 2008; Hecht 2012), histories of colonial expansion (Brockway 1979; Cañizares-Esguerra 2006; Schiebinger 2004; Tilley 2011), indigenous knowledge systems (Verran 2002; Watson-Verran and Turnbull 1995), and the circulation of technology within postcolonial technoscience projects (Crow and Carney 2013; de Laet and Mol 2000; Marques 2012). Scholars have explored the production of counter histories of Western science, the residues and reinventions of colonial science, reevaluations of traditional knowledge, and the development of alternative technoscience projects (Harding 2011). In later sections we identify some of the important additional insights that have emerged from anticolonial tendencies that focus on Spanish and Portuguese colonialism in the Americas, and those that focus on indigenous/aboriginal situations around the globe.

Similar to feminist STS, postcolonial STS also emerged from social justice movements of the 1960s and 1970s. Postcolonial STS took shape as newly independent states began establishing their own science research and education programs. Both argue that science is not value free or, insofar as that is a requirement, objective, and that Western modern technosciences tend to distribute their benefits primarily to already well-resourced groups and their costs to economically and politically vulnerable groups. Western epistemologies of science have been used to justify the oppression of both kinds of marginalized groups (Harding 2009, 403). Both fields also call for recognition of multiple scientific traditions and for engagement with how those historically characterized as "other" produce valuably different ways of knowing (ibid., 403). Postcolonial STS differs however in its primary attention to colonialism and imperialism, leaving gender often unacknowledged (ibid., 406). Feminist STS shares similar limitations in its lack of attention to relations of colonialism and indigeneity (ibid., 406).

To address these limitations, scholars have begun exploring the conjunctions of STS, feminist STS, and postcolonial STS. Theorists such as Donna Haraway (1989), Sandra Harding (1986), Mary Louise Pratt (1992), Londa Schiebinger (1989), Vandana Shiva (1997), and Sharon Traweek (1988) defined important projects at these intersections. Contributions have also been made through feminist work on Third World development, which critique patriarchal and colonial science projects as well as basic Enlightenment assumptions undergirding modern Western sciences (Braidotti 1994; Shiva and Moser 1995; Visvanathan et al. 2011).

Drawing upon these legacies, recent scholarship continues to interrogate and develop new directions in these three conjoined fields (Benjamin 2009; Carney 2001; Foster 2016; Harding 2015; Hayden 2003; Philip 2004; Pollock 2014; Reardon 2005; Reardon and TallBear 2012; Roy and Subramaniam 2016; Subramaniam 2013, 2014; TallBear 2013b). It examines issues of science and technology of political concern to women in the global south and those who are considered "other" in the global north, including issues of environment, development, corporatization, and militarism. Additionally, it analyzes the science and technology traditions of non-European cultures and how colonial histories have and continue to shape European and American science and technology in new ways. It considers gendered social relations as always implicated within colonialism and imperialism, avoiding privileging gender as the sole site of analysis. Relations of power such as race, indigeneity, and settler colonialism are always deemed important and equally relevant. It also interrogates the term *postcolonial* more broadly as signifying a temporal moment, political condition, theoretical critique, subjectivity, and counter politic.

While such emerging scholarship has produced valuable key insights, this chapter focuses on more recent scholarship to show how issues of gender, race, coloniality, and indigeneity simultaneously come to the forefront, thus demonstrating the importance of drawing upon feminist STS, postcolonial STS, and STS to provide new ways of understanding how the co-constitution of science and society shapes and is shaped by relations of power and inequality.

Colonial Legacies: Colonization, Racism, Sexism, and Science

Although many feminist analyses of science might have begun with a focus on questions related to gender alone, many feminist scholars quickly learned to appreciate the mutual imbrication of the category of gender with those of race, class, age, sexual orientation, ability, and other categories of classification. Feminists of color and postcolonial feminist scholars have been quick to point out that it is *specific bodies* that have not only served as the objects of scientific inquiry but also as the *raw materials* needed for the "manufacture" of modern Western scientific theories and knowledge claims. Together these disciplines have brought to light evidence that many modern Western scientific theories, particularly the creation of scientific classification and taxonomical systems, have been built upon perceived differences identified in the bodies and lives of gendered, raced, classed, and colonized subjects. Drawing upon key examples, this section highlights the entangled histories of gender/race/class hierarchies and coloniality.

Modern Sciences and Their Colonial Encounters

Postcolonial scholars have suggested that naturalists, anthropologists, and ethnographers of the colonial era used colonized spaces as their labs, exploiting so-called tribals along with the flora and fauna of their newly occupied lands as sources for data. The colonies, often viewed as being less civilized, served as the raw materials for the development of scientific theories of race and resources. It is not surprising, therefore, that scientific racism and scientific sexism flourished alongside colonial expansion during the late seventeenth and eighteenth centuries.[2] The bodies of individuals living in premodern conditions represented for European naturalists, such as the comparative anatomist George Cuvier, various points of evolutionary development along what was taken as a teleological view of human existence. The treatment of Saartjie Baartman for example, the "Hottentot Venus" who was taken from South Africa in 1810 and put on display as a "savage female" in London and Paris, remains a perfect case in point.[3] Following her death in 1815, Baartman's remains were dissected by Cuvier and her brain, skeleton, and genitalia remained on display at the Musée de l'Homme in Paris until 1974. The question that scientists such as Cuvier at the time were interested in pursuing was whether or not some populations of humans they encountered should be classified as being human at all. Ideas of hierarchies of being were reinforced by a

scientific vision with the white Western heterosexual couple as the pinnacle of evolution (Markowitz 2001), lending credence to the logics of slavery and colonial domination. Indeed, Darwin and Malthus, two key figures that shaped the logics of evolution
and population growth, developed their ideas during the height of colonial expansion
and times when there would have been an influx of slaves and indentured laborers
not only inhabiting newly colonized lands but also arriving on the shores of England.

Claims of modernity and its contrast with barbarianism and savagery also played an
integral part in distinguishing Western scientific ideas and practices from local knowledge systems that were encountered in the colonies. We should remember that Darwin
was writing at the beginning of seventy years of British colonialism in Africa and in the
middle of two hundred years of colonialism in India. As Kavita Philip notes, "One of
the effects of nineteenth-century scientific theories of non-Western nature and natives
is the belief, persistent to this day, in an epistemological divide between universal science and local knowledge" (Philip 2004, 6). Through the extension of ideals from the
Enlightenment era, the practices of modern Western science and scientific rationality were viewed as indications of civilization and social progress. By articulating what
constituted the tools, technical aspects, and practices of "proper" modern scientific
inquiry, scientists also simultaneously assigned local knowledges and practices in colonized nations to the status of backward, uncivilized, and premodern.

This legacy of scientific inquiry, which emerged as a result of colonial encounters,
informs present-day struggles over indigenous knowledge, patent ownership, and benefit sharing. Laura Foster's work, for instance, examines how South African Indigenous
San peoples, scientists, and growers animate multiple modalities of the Hoodia plant as
natural, molecular, or cultivated to assert unequal modes of belonging in South Africa.
Indigenous San, for instance, made strategic claims for benefit sharing with South African scientists in 2003 by articulating an affective kinship with Hoodia as a "natural"
plant through gendered histories of its traditional use by both San men and women,
thus asserting themselves as both modern and nonmodern subjects and simultaneously
contesting and reinforcing colonial and colonial settler histories (Foster 2012, 2016). In
doing so, San reconfigured themselves as producers of Hoodia knowledge in ways that
changed the relationships between scientists and indigenous peoples, but at the same
time they became stakeholders within Hoodia commercialization and its attachments
to Hoodia as used only by San male hunters. Such relationships shifted though when
the plant's patented molecular properties interacted with human bodies during clinical
trials, foiling scientists' efforts to transform Hoodia into a weight-loss treatment and
contributing to the termination of its commercialization in late 2008, thus making
San claims for benefit sharing more difficult. Foster's work produces critical insights

for STS by accounting for how multiple modalities of human–nonhuman relations are entangled within gendered, indigenous, and colonial histories.

Biological Differences and the Eugenic Project

Beginning with a central concept in evolutionary biology, variation, I have argued that this biological concept has been deeply intertwined with cultural ideas about diversity and difference since its very inception. Put starkly, evolutionary theories and models of variation owe their formulation to cultural debates around diversity and difference, culminating in their *eugenic scripts* that have haunted us ever since. (Subramaniam 2014, 224)

Many historians of science have traced the implementation of hierarchical political and racist measures of human worth to the emergence of scientific, and particularly, taxonomical classification systems, beginning with Carl Linnaeus's *Systema Naturae*, first published in 1735. For instance, Linnaeus's taxonomy drew on ideological beliefs about sex differences to scientifically categorize "male" and "female" parts in plants and flowers in traditional modes of human sexuality and the subordinate status of female gender roles (Schiebinger 1995). Early modern naturalists also used comparisons to female human anatomy to delineate distinctions between apes and humans. As Londa Schiebinger has suggested however, Linnaeus's classification of animals into six classes was dependent not only on notions of gender and female inferiority but also on ideas of racial inferiority. It is precisely within this "interplay between racial and sexual science" (1995, 116) that "[t]he body—stripped clean of history and culture as it was of clothes and often skin—became the touchstone of political right and social privileges" (1995, 116).

The entwined histories of racism and sexism can also be found in scientific theories of human intelligence. Differences in intellect, measured by questionable experimental methods, have made clear references to the undeveloped and premodern knowledge systems of "savages" (Gould 1996; Philip 2004) and human females historically understood as closely aligned with apes (Schiebinger 1995). These theories shaped the development of scientific research programs that were based on racist as well as eugenic principles and practices. For example, the "father of eugenics" Francis Galton, Darwin's half-cousin and contemporary, devoted his career to studying human differences and developing scientific measures for analyzing the inheritance of intellect.

This legacy of biological difference informs contemporary practices of a "new" eugenics, as articulated by Nobel Laureate James Watson (who has also shown interest in the biological underpinnings of intellect), which is based on the idea that humans should in fact desire better outcomes for the propagation of the human race. In his own words, Watson (2004, 401) describes his vision:

My view is that, despite the risks, we should give serious consideration to germ-line gene therapy. I only hope that the many biologists who share my opinion will stand tall in the debates to come and not be intimidated by the inevitable criticism. Some of us already know the pain of being tarred with the brush once reserved for eugenicists. But that is ultimately a small price to pay to redress genetic injustice. If such work be called eugenics, then I am a eugenicist.

From Francis Galton to our current atmosphere of scientific progress in genetics and molecular biology through innovations in directed evolution, we can witness the imprint that colonial encounters and legacies of racism and sexism have left behind on our scientific imaginations and practices. Such legacies inform contemporary debates over the reemergence of race as bio-genetic and its implications for racial hierarchies (Duster 2003; Koenig, Lee, and Richardson 2008; Roberts 2011). By conducting our analyses along interdisciplinary frameworks informed by feminist STS, postcolonial STS, and STS, we can begin to understand the history of our cultural and scientific obsessions with human differences.

A Focus on Latin America

Spanish and Portuguese colonialism in the Americas and its co-production with sciences and technologies have been almost entirely neglected in postcolonial STS, with its focus typically restricted to British and French colonialism in Asia and Africa (Medina, Marques, and Holmes 2014; Mignolo 2000; Rajão, Duque, and De' 2014). There are important similarities between the two literatures but also significant differences.[4] Feminist concerns deeply permeate the Latin American issues, though their relevance is not always fully addressed by these authors.

Thus one cannot today just "disseminate" to Latin America forms of social studies of science and technology and their feminist issues that were designed a few decades ago for other historical and social contexts. Such a practice will not access the richness and radicalness of Latin American projects. Moreover, it risks being experienced as one more example of residual coloniality. Colonial assumptions are by no means only a feature of the past. The experiences and practices of knowledge production in Latin America today continue to be shaped by their distinctive histories of colonialism and its aftermaths, including gender commitments.

This new focus is especially relevant now when institutional relations have been rapidly developing between STS organizations in North America and Europe, on the one hand, and in Latin America and other parts of the world, on the other. Essays in two publications were prepared to coincide with the 2014 Buenos Aires conference cosponsored by the Society for Social Studies of Science (4S), and the Sociedad Latino-americana de Estudios Sociales de la Ciencia y la Technología (ESOCITE). These provide innovative analyses of how modern Western sciences never, in fact, were disseminated to Latin America culturally intact. Rather they were always repositioned institution-ally and in practice. Elements of them were "sutured" into local cultural, political, and material environments, which required innovations by the locals. Moreover, the Latin American innovations sometimes traveled to Europe and North America, in turn reshaping science and technology projects there (Medina, Marques, and Holmes 2014; Rajão, Duque, and De' 2014). One of these publications, "Voices from Within and Out-side the South: Defying STS Epistemologies, Boundaries, and Theories" signals in its subtitle the question of how self-conscious of its own local character, of its parochiality, Northern STS itself is willing to become. That is, is Northern STS willing to give up its universalizing assumptions about its own endeavors when encountering the different and sometimes conflicting projects of Latin American STS? More generally, what could and should STS pluralism look like?

Pursuing such an issue is important also because these Latin American projects join a new world of STS analyses that start off from the different ways that science and technology projects and their societies co-produce each other around the globe. *East Asian Science, Technology and Society: An International Journal* began publication in 2007. In 2013, STS-Africa (sts-africa@lists.uni-halle.de) began to publish announcements of conferences, publications, jobs, and research initiatives around the globe focused on social studies of technosciences in or about Africa. Moreover, indigenous groups around the globe have been networking with one another. Their science and technol-ogy concerns are beginning to be recognized and institutionalized in United Nations and regional contexts. Several programs at the 2015 4S meetings were focused on such issues.[5] Clearly perspectives from diverse global colonialities, and resistances to them, are already beginning to transform what count as social and philosophic studies of sciences and technologies, as well as the choices, conceptions, and practices of diverse scientific and technology projects themselves. The distinctive history of state control of science and technology has led much of Latin America STS to focus on projects that increase social equality.[6] What are the central themes that appear in this Latin Ameri-can literature?[7]

The modernity/coloniality research network that formed in the 1990s was influ-enced by many distinctively Latin American political impulses, most notably libera-tion theology, dependencia theory, the rise of indigenous social movements, and then of the World Social Forums.[8] These Decolonial[9] theorists argue that the connection between the virtually simultaneous emergence of Spanish and Portuguese colonialism and early aspects of modernity was not merely the temporal accident implied by standard histories. Rather, Europeans invented modernity as an explicit response to the discovery of the "new world" of the Americas, populated with its savages and barbarians, "noble" or not.[10]

Modernity appears when Europe organizes the initial world-system and places itself at the center of world history over against a periphery equally constitutive of modernity . . . When one conceives modernity as part of a center-periphery system instead of an independent European phenomenon, the meanings of modernity, its origin, development, present crisis, and its postmodern antithesis change. (Dussel 1995, 9–10, 11)

In the case of both modernity and Iberian colonialism, scientific rationality and technical expertise were conceived as the motors of modernity and its progress.[11]

So modern Western sciences and technologies are implicated in the violent, oppressive, and destructive consequences of colonialism from 1492 on, not just beginning with the British occupation of India. European colonialism and its persistent residues and reinventions ("coloniality," in these writings) constitute the "darker side of modernity" (Marques 2014; Mignolo 2011; Santos 2014).

Elite European women certainly benefitted in significant ways from colonialisms, and they were complicit in exercising its prerogatives. Yet, they did not participate in colonial decision making, nor did they receive the kinds of benefits of wealth and power that their brothers did (McClintock 1995; Pratt 2008). They suffered the imposition of new, restrictive standards of "proper" womanhood, namely, confinement to the newly invented sphere of private domesticity, which were constituted to contrast not only with the public sphere of men but also with the perceived degraded conditions of colonized women—indigenes, slaves, peasants.

As a number of feminist scholars have argued, violence and exploitation of colonized women's sexuality and labor were crucial to the establishment of colonial control (Lugones 2010; McClintock 1995; Mendoza 2016; Pratt 2008). Colonialism could not succeed without the constant miscegenation that was Iberian colonial policy.[12] Thus gender and sexuality considerations should not be regarded as an optional addition to analyses of modernity and colonialism; they are an intrinsic element of such phenomena. Moreover, the Iberians legitimated and managed this constant miscegenation through the introduction of new, complex, and rigid pre-Darwinian racial categories about blood purity that persist in Latin America to this day. Thus new sciences establishing the natural foundations of social order—ones that were always simultaneously about race, gender, sexuality and class—were co-produced in both the colonized and the colonizing societies.

Meanwhile, getting from Europe to the Americas, traveling and living there, and getting back to Europe required new kinds of sciences and technologies. The voyagers needed astronomy of the Southern Hemisphere and better principles of cartography to enable them to chart their locations and routes across the Atlantic and in the Americas.[13] They needed better climatology and oceanography and better nautical engineering to enable the voyagers and their precious cargoes to survive the journeys. They needed knowledge of the dangerous or valuable new flora and fauna that they encountered, as well as social knowledge of the societies that they intended to conquer (Todorov 1984). Of course, they extracted virtually all of their knowledge of "natural history" from the indigenes: "discovery" consisted primarily of asking indigenes to share their knowledge. They needed the geology to enable them to extract the gold and silver that they found in Central America. It was three kinds of long-distance corporations that created these particular sciences: the Jesuits, the European trading companies, and the European empires (Harris 1998). These sciences were far more important to the creation of modern European science and technology than is usually acknowledged in the standard histories of science. Moreover, these sciences tend only rarely to advance through Kuhnian paradigm shifts. The classed and racialized gender relations between men in the development of colonial sciences have also begun to be examined (Cañizares-Esguerra 2006; Harris 2005; Schiebinger 2004, 2005a).

Both the Europeans and the indigenes they encountered had different economic, political, and social worlds in 1492 than did the British and the indigenes they encountered in the eighteenth century. Moreover, Spain and Portugal were Catholic monarchies. The conquistadors conceptualized the purpose of their colonization of the Americas and its "infidels" as economic, and importantly in the same religious terms that had directed the expulsion of the Moors (and Jews) from Granada in 1492; they were "unifying Jerusalem." The virtually incomparable bravery and heroism of the conquistadors, hideous as its practices and consequences were, is comprehensible only as a religious mission.

Formal emancipation was virtually complete in the Americas by 1830, more than a century before it began to occur in Asia and Africa. But the newly emancipated countries had no models of what "independence" could or should be. They went through more than six decades of exhausting political uprising, coups, and shifting forms of government: an era of bold "social experiment" (Pratt 2008). French Comteian positivism, which had begun to appear in the early 1800s, appealed for different reasons to both conservatives and liberals, and eventually "saved" Mexico and Brazil from this chaos. Yet its adoption by governments faced with increasing resistance to its democratizing tendencies led to its transformation into a recipe for dictatorial technocracies that vastly increased inequality. It then was completely rejected by 1910—decades before the German version emerged in the Vienna Circle (Gilson and Levinson 2013). Yet this history of (eventual) resistance to positivism can seem to haunt Latin American stances toward science and technology today—a topic for another time.

Finally, Andean highland indigenous movements, Buen Vivir in particular, managed to get their social and environmental agendas into the new national constitutions of Bolivia and Ecuador in 2005 and 2006. Here "nature" is a legal entity with rights that can be defended by advocates, much as children and the disabled have litigable rights. Debates over if, when, and how such epistemological, ontological, and political rights will prevail are currently at a high intensity (de la Cadena 2010; Gudynas 2011; Walsh 2010).

Finally, colonial practices of "just observing" and "just reporting what was seen" were often framed as having no consequences at all for what was subsequently done by others (such as militaries and corporations) with those observations and reports of them. Yet in fact colonial scientists were always also commenting on the value of local plants and indigenous practices for Europeans, and on how "nature" (including the indigenes) could be improved, as they collected samples and renamed indigenous plants and animals. Scientific exploration inherently makes use of "imperial eyes" (Pratt 2008). The positivist epoch in Mexico and Brazil was another form of this persistent attempt to define social progress in "innocent" terms that could be perceived as socially neutral with respect to conservative and liberal political impulses. Today's conventional Western historians of science fully recognize that the single most effective generator of scientific achievements has been militarism. Yet there is a general disconnect between this acknowledged fact and appreciation of the role of colonial violence in advancing modern Western sciences. Establishing this disconnect has taken a lot of epistemic and ontological work on the part of Western scientific communities.

How can the social studies of science and technology acknowledge and work against this tendency? Latin Americans have been pondering such issues.[14]

Nonmodern and Anticolonial Knowledge: Multiple Ways of Knowing, Being, and Doing

As we have seen, the notion of *post*colonial knowledge production has its limitations and does not capture indigenous knowledge, both pre- and post-contact. Of course, we have only a very partial view on what constitutes *indigenous* knowledge from peoples who inhabited, say the Americas, pre-European arrival. However, in this section we address indigenous knowledge as a set of dynamic, articulated knowledge practices that living indigenous peoples engage in. *Articulation* refers to borrowing, reinterpretation, and reconfiguration—the unhooking and recombining of different knowledges and practices into new cultural and social formations (Clifford 2001, 2003, 2007; Hall 1986a, 1986b; TallBear 2013a; Tsing 2007; Yeh 2007).

Thus indigenous peoples draw on both contemporary and more ancient traditions—in order to resist further decimation of their bodies, lands, and traditions. Beyond survival, indigenous peoples also engage with knowledge production in order to thrive in today's world. Therefore, while indigenous knowledge draws on "traditional" knowledges, it is also in flux today as it no doubt always was, as the world in which indigenous peoples must live has shifted. The term *postcolonial* implies, of course, an *after* colonialism. At odds with this idea, however, is the way in which *indigenous* is defined. Indigenous peoples in the very fact of differentiating themselves from the settler state as first nations, tribal nations, first peoples, and the like, assert an explicitly political definition of indigeneity. *Indigenous* in this definition does not imply simply firstness in place (i.e., autochthony), cultural difference/isolation, or economic marginalization, although these are common components of how the state defines indigeneity that resonate in part with indigenous peoples. But for indigenous groups, adding the umbrella term to their people-specific identities (e.g., Maori, Cree, Dayak) helps organize their ongoing resistance to the authority of the genocidal and/or assimilative (settler) colonial state (TallBear 2013a). Because indigenous resistance to state hegemony is ongoing, indigenous peoples in the United States and Canada, for example, do not generally speak in terms that imply an *after* colonialism. Rather, settler state colonial practice continues in the very existence of those states.

Central to the production of the settler state and its ongoing projects has been knowledge production, or science. The project of the settler state has been organized around a trope of modernity versus savagery, including bringing Christianity, capitalism, agriculture, nuclear family, and so-called reasoned inquiry to civilize the savage lands and peoples of the Americas. Part of the right to civilize had been the right to inquire (Reardon and TallBear 2012; TallBear 2013b). The entire civilizational project has included not only rights of access to "nature" or "resources" that include indigenous lands and sacred sites but also the bodies, blood, and bones of indigenous and other marginalized subjects. Modernity also requires and in turn supports civilizational hierarchies of bodies and their knowledges with indigenous peoples being viewed as primitive, albeit sometimes the noble savage, in both regards.

Instead, we want to explore the idea of indigenous knowledges—in the past and in the present—as in part "nonmodern." That is, indigenous knowledges can be seen as organized around not a uniquely human right to inquire but within systems in which many diverse humans and nonhumans together constitute ways of knowing the world. In one example, Oceti Sakowiŋ (Dakota, Lakota, or Nakota) people have referred to plants revealing their names and properties to patient or ready humans in dreams or visions (Howe and TallBear 2006). Similarly, the Cherokee, in addition to "the powers of observation" over long periods of time, sought "guidance through dreams or divination." The "medicines of the new country were [after forced U.S. removal from previous lands] shown to the people through communication with animal spirits, the Little People, or even *Unehlanvhi*, the Creator" (Carroll 2015, 61). Whether or not the reader finds such ideas salient, the underlying principle is that such indigenous thinkers—both in the past and in the present—know that it is not simply the prerogative of humans alone to go about the world (re)naming and classifying all beings, including entities settler society deems to be inanimate. There are instead more multidirectional relationships of knowing and action at play between humans and nonhuman persons. On the other hand, modernity compels a more unidirectional human ordering and right to study and know other supposedly more primitive humans and nonhumans. The idea that there is a right to know the bodies of others cannot but involve nonconsent, exploitation, and violence (Deloria [1969] 1988; Mosby 2013; Reverby 2009, 2011; Skloot 2010).

Another Dakota thinker provides a more detailed example that can help illustrate how indigenous peoples might constitute knowledge differently. Travis Erickson, a quarrier and expert carver at the Pipestone Quarries in southwestern Minnesota, United States, possesses expert knowledge about the red stone that is a key source for ceremonial pipes used by indigenous peoples throughout North America. The pipestone site is an important site at which tribes are thought to have gathered for hundreds of years—a place where custom dictates peaceful relations among humans. The stone was unsurprisingly renamed by whites for science. They called it *catlinite* after the nineteenth-century American artist George Catlin, who painted the site in 1836–1837. The assumed authority to rename is part of the colonial claim of ownership, a move to control through knowing. During that era human "explorers" moved into territories new to them, intending to name, classify, and report on the vast spaces of the continent. Erickson, who is approximately 50 years old, has in contrast spent his entire life living near to and working in the Pipestone quarries. Over decades he has engaged in diverse ways of interacting with stone and earth. When he describes how he acquires knowledge of the site, he is not the only agent in knowledge production. Erickson lives near the site year round and has one of the deepest quarries there. With such sustained exposure to the stone and after being taught by elders who also quarried for decades, he has acquired techniques for cutting through granite to get the soft pipestone beneath.

He also speaks of reading stories in the earth, histories of earth's movements and the glaciers that brought rocks and bones and other items south to lodge in the soil there now. His "stories" seem to be in part co-constituted with geologists and other scientists own narratives of the geological history of Pipestone. The Pipestone National Monument is a U.S. Park Service site and so environmental and other scientists regularly also produce knowledge there.

Travis Erickson's multiple ways of working demonstrate that important differences in indigenous ways of knowing, both in the past and the present, are not all classified completely or easily as nonmodern, however. Indigenous peoples also use so-called modern or Western knowledges to live meaningful lives even within settler society and to further contemporary indigenous political goals. Again, since indigeneity itself is an umbrella term added by indigenous peoples to their people-specific understandings of themselves and used to organize politically in a global context (TallBear 2013a), one can understand that "indigenous knowledge" can then also be understood as forms of knowledge that serve indigenous peoples' survival, indeed their thriving, in (settler) colonial societies. These articulated indigenous knowledges might include older forms of knowledge acquisition such as knowledge through visions or dreams or intuitions combined with observations, learning-through-doing, and knowledge gleaned in conversations between, say, a quarrier and a geologist, both of whom interact closely with stone and earth. As Maggie Walter and Chris Andersen (2013, 17) explain in *Indigenous Statistics: A Qualitative Research Methodology*, "We argue for an accounting of modernity within first world Indigenous methodologies; we also argue against positioning Indigenous methodology dichotomously in opposition to Western frames and against grounding it in a concept of traditional knowledge and culture 'outside of modernity'."

Finally, while Erickson is one of the few indigenous thinkers who minimize their use of the words *spiritual* and *sacred*, his listening and waiting sometimes for knowledge to come to him and his recognition that humans are not the only agents brings to mind a final problem with so-called modern knowledge: that is, its inability to engage with immaterial forces such as "spirits." While recent critical approaches to animal studies seek to dismantle hierarchies in the relationships of "Westerners" with their nonhuman others, they have largely restricted themselves to beings that live—defining beings organismically, for example, dogs, bears, mushrooms, microorganisms (e.g., Kirksey 2014). Such work is methodologically and ethically innovative in that it highlights how organisms' livelihoods are co-constituted with cultural, political, and economic forces. But these approaches cannot fully contain indigenous standpoints. Indigenous peoples have not forgotten that nonhumans are agential beings engaged in social relations in which human lives are intimately and intricately shaped. To that end, recent work in the "new materialisms" is complementary as it attempts sometimes to capture the vibrancy and forces in human lives of nonliving things (a chair, a mineral), but in ways that, similarly to multi- or interspecies thinkers, divides the world into life and not life organismically defined (e.g., Bennett 2010; Chen 2012). But for many indigenous peoples, the other-than-human beings that shape their lives might also include objects and force" such as stones, thunder, stars, or spirits (for lack of a better word) (Shorter 2012; TallBear 2015). Recognition of the ability of nonhumans to actively shape human lives serves as an epistemological and ethical challenge to settler colonialism and its ordering (say into human versus animal), management of, and violence toward nonhuman others.

Toward the Future

This brief account highlights the mutual importance of STS, feminist STS, and postcolonial STS, as well as the fault lines and limitations of each. Feminist STS often marginalizes questions of race and postcoloniality. Postcolonial STS pays little attention to the complexities of gender and indigeneity. We have also stressed that colonial histories in different parts of the world and their subsequent resistance have taken different shapes, leading to related but distinct theories of postcoloniality and decoloniality. Issues of gender, race, postcoloniality, and indigeneity remain at the peripheries of STS. One of the central arguments we have been making is that questions of gender, race, coloniality, and indigeneity are not optional variables or analytics that each field can choose whether to consider. In bringing together central insights of feminism, postcolonialism, and the social studies of technosciences, we can begin to appreciate the inextricable interconnections of the three.

It is worth noting two other social movements that are shifting the narratives of technosciences and their practitioners, albeit in different ways to indigenous peoples' projects. As patient groups and consumers have laid claim to science through political mobilization as well as DIY science movements, science has moved from the laboratory and the hallways of power into an era of citizen-driven science (Benjamin 2013; Epstein 1996; Hess 2011). We have also seen shifts in geopolitics, as emerging economies have thrown their national economies into state-supported national genomic projects, resulting in new configurations of "genomic nationalism" (Benjamin 2009). Both projects depend on entangling race, gender, and narratives of technoscience. As colonial and postcolonial STS remind us, science is not a Western invention that has slowly diffused to the other societies around the globe, thereby ushering them into the glorious world of modernity (Basalla 1967; Prasad 2008; Raina 1999). Rather, from its beginnings, science and the European modernity of which it was a central part were intimately involved with coloniality and empire. Thus modern Western technosciences were not developed outside of politics; rather they were at the heart of colonialism's political ideologies and institutional structures (Baber 1996; Prakash 1999).

Elaborating the logic of difference—of sex, gender, race, sexuality—was central to the operations of colonial government. Science played a critical role in naturalizing such logics of difference. In such a reframing, we cannot tell the story of colonialism without attention to the roles of science and of gender hierarchies, or narrate a history of Western science without attention to coloniality and gender hierarchies, or a history of gender without attention to science and colonialism. Yet contemporary histories also remind us that colonized nations and their indigenes have never been the passive victims of colonial rule but rather vibrant actors with agency, active engagement, and resistance, albeit in situations with grossly unequal power.

What emerges in these new studies is an entangled history of gender hierarchies and coloniality, in which modern Western technosciences have all too often provided powerful resources. Today we need to understand the unequal, varied, and complex practices of technosciences across the globe as they encounter the "sticky materiality of practical encounters" (Tsing 2005, 1). We need to move to more mobile, fluid, and "entangled" (Prasad 2014) models of feminist STS, postcolonial STS, and STS whether they are through "contact zones" (Pratt 1992), "a moving metropolis" (MacLeod 2000), or "science as circulation" (Raj 2013). In the case of the European technosciences, these alternative metaphors shift our understanding from the old linear diffusion model from west to east, to a more vibrant and dynamic account of the long and varied travels and interlocutions of technosciences, gender, and colonialities. The same is the case for our research disciplines that study such phenomena.

STS, feminist STS, and postcolonial/decolonial STS need one another to provide a richer account of the extensive travels of colonialism and colonial knowledge formations. It is in these shifts that one can find possibilities of new histories and new accounts of how the world, its peoples, our disciplines, and knowledges have *become* what they are. And from such accounts of becoming we can find ways and modes of *unbecoming* and becoming anew.

Acknowledgments

Many thanks to the Institute for Research on Women and Gender at the University of Michigan for sponsoring a feminist research seminar during Fall 2014 on "Feminist Postcolonial Science and Technology Studies: Instigations, Interrogations and New

Developments." The many spirited and engaging conversations about feminism, post-colonialism, and technoscience during this seminar helped shape this essay. We are grateful to our participating colleagues: Irna Aristarkhova, Ruha Benjamin, May-Britt Öhman, Anne Pollock, Lindsey Smith, Sari van Anders, and Angela Willey.

Notes

1. Londa Schiebinger's website on "Gendered Innovations in Science, Health & Medicine, Engineering, and Environment" (genderedinnovations.stanford.edu) documents policy initiatives and conferences around the globe that address such issues.

2. This standard chronology of the origins of scientific racism is challenged (or at least expanded) by the Latin American decolonial claim that it was the arrival of the distinctive policies of Spanish and Portuguese coloniality in the Americas in 1492, two centuries earlier, that introduced systematic racial classifications to the world and that inextricably entangled racial and gender hierarchies in Latin America to this day, as will be discussed later.

3. The case of Saartjie Baartman also gives us pause and is indicative of the contradictory terrain in which a decolonial feminist STS is situated. Feminist scholars have recently questioned the continual use of Baartman to discuss colonial violence, arguing that to do so re-inscribes such violence. At the same time, contemporary indigenous Khoi and San peoples active in political mobilization to return the body of Baartman to South Africa in 2002 note the importance of making such histories known as they are critical for supporting San and Khoi efforts at self-determination. Yet, even the return of Baartman raised new questions as to who could speak for her and how. Emerging decolonial feminist STS work provides insights that bring such contradictory and complex tensions to the forefront.

4. See, for example, Rodriguez (2001).

5. For example, # 036 "Indigenous Knowledge Sovereignties and Scientific Research."

6. See Saldana (2006) and Vessuri (2006) for histories of science in Latin America, and Vessuri (1987) for the earliest survey of the social study of science in Latin America.

7. A huge proportion of the Latin American decolonial literature and its STS components is now in English, including many primary sources. This is the literature addressed here.

8. See Escobar (2010) for a review of the history and distinctive concerns of the modernity/coloniality network, including an extensive analysis of gender issues and the innovative treatment of nature.

9. This term is capitalized to distinguish this theoretical group from the more general use of the term.

10. But see Brotherston (2008) and Mann (2002) for recent evaluations of what the indigenes knew in 1491. Of course it is only in the "high modernity" of the eighteenth and nineteenth centuries that the sociological markers of modernity are fully established, such as public and

private spheres, and the disaggregation from the family of economic, political, education, and religious/moral institutions.

11. This is so even though Todorov (1984) and others have pointed out that Columbus was very much a medieval man. Cortes, on the other hand, already possessed distinctively modern characteristics. Of course, as a navigator, Columbus highly valued empirical observation.

12. Whether this was a practice of rape or consent may not be a meaningful distinction in the context of colonial inequalities.

13. See the discussion in Mignolo (2011, 78) of the first appearance of "globalization": a "god's eye" map of the known world.

14. In addition to earlier citations, see for example, Isasi-Diaz and Mendieta (2012); Maldonalda-Torres (2012); Morana et al. (2008); and Santos (2014).

References

Abraham, Itty. 2006. "The Contradictory Spaces of Postcolonial Techno-Science." *Economic and Political Weekly* 41 (3): 210–17.

Anderson, Warwick. 2002. "Introduction: Postcolonial Technoscience." *Social Studies of Science* 32 (5–6): 643–58.

Anderson, Warwick, and Vincanne Adams. 2008. "Pramoedya's Chickens: Postcolonial Studies of Technoscience." In *Handbook of Science and Technology Studies*. 3rd ed., edited by Edward J. Hackett, Olga Amsterdamska, Michael E. Lynch, and Judy Wajcman, 181–204. Cambridge, MA: MIT Press.

Baber, Zaheer. 1996. *The Science of Empire: Scientific Knowledge, Civilization, and Colonial Rule in India*. Albany: State University of New York Press.

Basalla, George. 1967. "The Spread of Western Science." *Science* 156: 611–22.

Benjamin, Ruha. 2009. "A Lab of Their Own: Genomic Sovereignty as Postcolonial Science Policy." *Policy and Society* 28 (4): 341–55.

___. 2013. *People's Science: Bodies and Rights on the Stem Cell Frontier*. Stanford, CA: Stanford University Press.

Bennett, Jane. 2010. *Vibrant Matter: A Political Ecology of Things*. Durham, NC: Duke University Press.

Braidotti, Rosi. 1994. *Women, the Environment and Sustainable Development: Towards a Theoretical Synthesis*. London: Zed Books.

Brockway, Lucile H. 1979. "Science and Colonial Expansion: The Role of the British Royal Botanic Gardens." *American Ethnologist* 6 (3): 449–65.

Brotherston, Gordon. 2008. "America and the Colonizer Question: Two Formative Statements from Early Mexico." In *Coloniality at Large: Latin America and the Postcolonial Debate*, edited by Mabel Moraña, Enrique Dussel, Marcos Cuanto, and Carlos A. Jáuregui, 23–42. Durham, NC: Duke University Press.

Callon, Michel. 1986. "Some Elements of a Sociology of Translation: Domestication of the Scallops and the Fishermen of St. Brieuc Bay." In *Power, Action, and Belief: A New Sociology of Knowledge?*, edited by John Law, 196–223. London: Routledge.

Cañizares-Esguerra, Jorge. 2006. "Chivalric Epistemology and Patriotic Narratives: Iberian Colonial Science." In *Nature, Empire, and Nation: Explorations of the History of Science in the Iberian World*, 7–14. Stanford, CA: Stanford University Press.

Carney, Judith Ann. 2001. *Black Rice: The African Origins of Rice Cultivation in the Americas*. Cambridge, MA: Harvard University Press.

Carroll, Clint. 2015. *Roots of Our Renewal: Ethnobotany and Cherokee Environmental Governance*. Minneapolis: University of Minnesota Press.

Chen, Mel. 2012. *Animacies: Biopolitics, Racial Mattering, and Queer Affect*. Durham, NC: Duke University Press.

Clarke, Adele A., and Monica J. Casper. 1996. "From Simple Technology to Complex Arena: Classification of Pap Smears, 1917–1990." *Medical Anthropology Quarterly* 10 (4): 601–23.

Clifford, James. 2001. "Indigenous Articulations." *Contemporary Pacific* 13 (2): 468–90.

___. 2003. *On the Edges of Anthropology: Interviews*. Chicago: Prickly Paradigm Press and University of Chicago Press.

___. 2007. "Varieties of Indigenous Experience: Diasporas, Homelands, Sovereignties." In *Indigenous Experience Today*, edited by Marisol de la Cadena and Orin Starn, 197–223. Oxford: Berg.

• Collins, Harry M. 1983. "The Sociology of Scientific Knowledge: Studies of Contemporary Science." *Annual Review of Sociology* 9: 265–85.

Crow, Britt, and Judith Carney. 2013. "Commercializing Nature: Mangrove Conservation and Female Oyster Collectors in the Gambia." *Antipode* 45 (2): 275–93.

Darwin, Charles. [1876] 1993. *The Autobiography of Charles Darwin*, edited by Nora Barlow. New York: W. W. Norton.

de la Cadena, Marisol. 2010. "Indigenous Cosmopolitics in the Andes," *Cultural Anthropology* 25 (2): 334–70.

de Laet, Marianne, and Annemarie Mol. 2000. "The Zimbabwe Bush Pump: Mechanics of a Fluid Technology." *Social Studies of Science* 30 (2): 225–63.

• Deloria Jr., Vine. [1969] 1988. *Custer Died for Your Sins: An Indian Manifesto*. Reprint. Norman: University of Oklahoma Press.

Dussel, Enrique. 1995. *The Invention of the Americas: Eclipse of "the Other" and the Myth of Modernity*, translated by Michael D. Barber. New York: Continuum.

Duster, Troy. 2003. *Backdoor to Eugenics*. 2nd ed. New York: Routledge.

Epstein, Steve. 1996. *Impure Science: AIDS, Activism, and the Politics of Knowledge*. Berkeley: University of California Press.

Erickson, Travis. 2014. Conversations with Kim TallBear at Pipestone, Minnesota.

Escobar, Árturo. 2010. "Worlds and Knowledges Otherwise: The Latin American Modernity/ Coloniality Research Program." In *Globalization and the Decolonial Option*, edited by Walter D. Mignolo and Árturo Escobar, 33–64. New York: Routledge.

Foster, Laura. 2012. "Patents, Biopolitics, and Feminisms: Locating Patent Law Struggles over Breast Cancer Genes and the *Hoodia* Plant." *International Journal of Cultural Property* 19 (3): 371–400.

___. 2016. "The Making and Unmaking of Patent Ownership: Technicalities, Materialities, and Subjectivities." *PoLAR: Political and Legal Anthropology Review* 39 (1): 1–17.

Franklin, Sarah, Celia Lury, and Jackie Stacey, eds. 2000. *Life Itself: Global Nature and the Genetic Imaginary*. London: Sage.

Fujimura, Joan H. 1992. "Crafting Science: Standardized Packages, Boundary Objects, and 'Translation'." In *Science as Practice and Culture*, edited by Andrew Pickering, 168–211. Chicago: University of Chicago Press.

Gieryn, Thomas F. 1982. "Relativist/Constructivist Programmes in the Sociology of Science: Redundance and Retreat." *Social Studies of Science* 12 (2): 279–97.

Gilson, Gregory D., and Irving W. Levinson. 2013. *Latin American Positivism: New Historical and Philosophical Essays*. Lanham, MD: Lexington Books.

Gould, Stephen J. 1996. *The Mismeasure of Man*. New York: W. W. Norton.

Gudynas, Eduardo. 2011. "Buen Vivir: Today's Tomorrow." *Development* 54 (4): 441–47.

Hall, Stuart. 1986a. "Gramsci's Relevance for the Study of Race and Ethnicity." *Journal of Communication Inquiry* 10 (2): 5–27.

___. 1986b. "On Postmodernism and Articulation: An Interview with Stuart Hall." *Journal of Communication Inquiry* 10 (2): 45–60.

Haraway, Donna. 1989. *Primate Visions: Gender, Race, and Nature in the World of Modern Science*. New York: Routledge.

Harding, Sandra. 1986. *The Science Question in Feminism*. Ithaca, NY: Cornell University Press.

___. 2008. *Sciences from Below: Feminisms, Postcolonialities, and Modernities*. Durham, NC: Duke University Press.

• ___. 2009. "Postcolonial and Feminist Philosophies of Science and Technology: Convergences and Dissonances." *Postcolonial Studies* 12 (4): 401–21.

• ___. 2011. *The Postcolonial Science and Technology Studies Reader.* Durham, NC: Duke University Press.

• ___. 2015. *Objectivity and Diversity: Another Logic of Scientific Research.* Chicago: University of Chicago Press.

Harris, Steve. 1998. "Long-Distance Corporations and the Geography of Natural Knowledge." *Configurations* 6 (2): 269–304.

___. 2005. "Jesuit Scientific Activity in the Overseas Missions, 1540–1773." *Isis* 96 (1): 71–79.

Hayden, Cori. 2003. *When Nature Goes Public: The Making and Unmaking of Bioprospecting in Mexico.* Princeton, NJ: Princeton University Press.

Hecht, Gabrielle. 2012. *Being Nuclear: Africans and the Global Uranium Trade.* Cambridge, MA: MIT Press.

Hess, David J. 2011. "Science in an Era of Globalization: Alternative Pathways." In *The Postcolonial Science and Technology Studies Reader,* edited by Sandra Harding, 419–38. Durham, NC: Duke University Press.

Howe, Craig, and Kim TallBear, eds. 2006. *This Stretch of the River: Lakota, Dakota, and Nakota Responses to the Lewis and Clark Expedition and Bicentennial.* Sioux Falls, SD: Oak Lake Writers Society and Pine Hill Press.

• Isasi-Diaz, Ada Maria, and Eduardo Mendieta, eds. 2012. *Decolonizing Epistemologies.* New York: Fordham University Press.

Jasanoff, Sheila, ed. 2004. *States of Knowledge: The Co-production of Science and Social Order.* New York: Routledge.

Kirksey, Eben. 2014. *The Multispecies Salon.* Durham, NC: Duke University Press.

Koenig, Barbara A., Sandra Soo-Jin Lee, and Sarah Richardson, eds. 2008. *Revisiting Race in a Genomic Age.* New Brunswick, NJ: Rutgers University Press.

• Latour, Bruno. 1987. *Science in Action: How to Follow Scientists and Engineers through Society.* Cambridge, MA: Harvard University Press.

___. 2004. "Whose Cosmos, Which Cosmopolitics?" *Common Knowledge* 10 (3): 450–62.

• Lugones, Maria. 2010. "The Coloniality of Gender." In *Globalization and the Decolonial Option,* edited by Walter D. Mignolo and Árturo Escobar, 369–90. New York: Routledge.

MacKenzie, Donald A., and Judy Wajcman. 1985. *The Social Shaping of Technology: How the Refrigerator Got Its Hum.* Philadelphia: Open University Press.

• MacLeod, Roy, ed. 2000. *Nature and Empire: Science and the Colonial Enterprise. Osiris,* vol. 15. Chicago: University of Chicago Press.

Maldonada-Torres, Nelson. 2011. "Epistemology, Ethics, and the Time/Space of Decolonization: Perspectives from the Caribbean and the Latina/o Americas." In *Decolonizing Epistemologies: Latina/o Theology and Philosophy*, edited by Ada Maria Isasi-Diaz and Eduardo Mendieta, 193–206. New York: Fordham University Press.

Mann, Charles C. 2002. "1491." *Atlantic Monthly* 289 (3): 41–53.

Markowitz, Sally. 2001. "Pelvic Politics: Sexual Dimorphism and Racial Difference." *Signs: Journal of Women and Culture in Society* 26 (2): 389–414.

Marques, Ivan da Costa. 2012. "Ontological Politics and Situated Public Policies." *Science and Public Policy* 39 (5): 570–78.

___. 2014. "Ontological Politics and Latin American Local Knowledges." In *Beyond Imported Magic: Essays on Science, Technology, and Society in Latin America*, edited by Eden Medina, Ivan da Costa Marques, and Christina Holmes, 85–110. Cambridge, MA: MIT Press.

McClintock, Anne. 1995. *Imperial Leather: Race, Gender and Sexuality in the Colonial Contest*. New York: Routledge.

McNeil, Maureen. 2005. "Introduction: Postcolonial Technoscience." *Science as Culture* 14 (2): 105–12.

Medina, Eden, Ivan da Costa Marques, and Christina Holmes, eds. 2014. *Beyond Imported Magic: Essays on Science, Technology, and Society in Latin America*. Cambridge, MA: MIT Press.

Mendoza, Breny. 2016. "Coloniality of Gender and Power: From Postcoloniality to Decoloniality." In *The Oxford Handbook of Feminist Theory*, edited by Lisa Disch and Mary Hawkesworth, 100–121. Oxford: Oxford University Press.

Merchant, Carolyn. 1980. *The Death of Nature: Women, Ecology, and the Scientific Revolution*. San Francisco: Harper and Row.

Mignolo, Walter D. 2000. *Local Histories/Global Designs: Coloniality, Subaltern Knowledges, and Border Thinking*. Princeton, NJ: Princeton University Press.

___. 2011. *The Darker Side of Western Modernity: Global Futures, Decolonial Options*. Durham, NC: Duke University Press.

Moraña, Mabel, Enrique Dussel, Marcos Cueto, and Carlos A. Jáuregui, eds. 2008. *Coloniality at Large: Latin America and the Postcolonial Debate*. Durham, NC: Duke University Press.

Mosby, Ian. 2013. "Administering Colonial Science: Nutrition Research and Human Biomedical Experimentation in Aboriginal Communities and Residential Schools, 1942–1952." *Histoire Sociale (Social History)* 46 (91): 145–72.

Phalkey, Jahnavi. 2013. "Introduction: Focus: Science, History, and Modern India." *Isis* 104 (2): 330–36.

Philip, Kavita. 2004. *Civilizing Natures: Race, Resources, and Modernity in Colonial South India*. New Brunswick, NJ: Rutgers University Press.

Pinch, Trevor J., and Wiebe E. Bijker. 1984. "The Social Construction of Facts and Artefacts: Or How the Sociology of Science and the Sociology of Technology Might Benefit Each Other." *Social Studies of Science* 14 (3): 399–441.

Pollock, Anne. 2014. "Places of Pharmaceutical Knowledge-Making: Global Health, Postcolonial Science, and Hope in South African Drug Discovery." *Social Studies of Science* 44 (6): 848–73.

Prakash, Gyan. 1999. *Another Reason: Science and the Imagination of Modern India*. Princeton, NJ: Princeton University Press.

Prasad, Amit. 2008. "Science in Motion: What Postcolonial Science Studies Can Offer." *Electronic Journal of Communication Information & Innovation in Health (RECIIS)* 2 (2): 35–47.

___. 2014. "Entangled Histories and Imaginative Geographies of Technoscientific Innovations." *Science as Culture* 23 (3): 432–39.

Pratt, Mary Louise. 1992. *Imperial Eyes: Travel Writing and Transculturation*. New York: Routledge.

___. 2008. *Imperial Eyes: Travel Writing and Transculturation*. 2nd ed. New York: Routledge.

Rabinow, Paul. 1999. *French DNA: Trouble in Purgatory*. Chicago: University of Chicago Press.

Raina, Dhruv. 1999. "From West to Non-West? Basalla's Three-Stage Model Revisited." *Science as Culture* 8: 497–516.

Raj, Kapil. 2013. "Beyond Postcolonialism ... and Postpostivism: Circulation and the Global History of Science." *Isis* 104 (2): 337–47.

Rajão, Raoni, Ricardo B. Duque, and Rahul De', eds. 2014. "Voices from Within and Outside the South—Defying STS Epistemologies, Boundaries, and Theories." Special issue, *Science, Technology, & Human Values* 39 (6): 767–874.

Reardon, Jenny. 2005. *Race to the Finish: Identity and Governance in an Age of Genomics*. Princeton, NJ: Princeton University Press.

Reardon, Jenny, and Kim TallBear. 2012. "'Your DNA Is Our History': Genomics, Anthropology, and the Construction of Whiteness as Property." *Current Anthropology* 53 (S5): 233–45.

Reverby, Susan M. 2009. *Examining Tuskegee: The Infamous Syphilis Study and Its Legacy*. Chapel Hill: University of North Carolina Press.

___. 2011. "'Normal Exposure' and Inoculation Syphilis: A PHS 'Tuskegee' Doctor in Guatemala, 1946–1948." *Journal of Policy History* 23 (1): 6–28.

Rodriguez, Ileana, ed. 2001. *The Latin American Subaltern Studies Reader*. Durham, NC: Duke University Press.

Roberts, Dorothy E. 2011. *Fatal Invention: How Science, Politics, and Big Business Re-create Race in the Twenty-First Century*. New York: The New Press.

Roy, Deboleena, and Banu Subramaniam. 2016. "Matter in the Shadows: Feminist New Materialism and the Practices of Colonialism." In *Mattering: Feminism, Science and Materialism*, edited by Victoria Pitts Taylor. New York: New York University Press.

Saldana, Juan José. 2006. *Science in Latin America: A History*. Austin: University of Texas Press.

Santos, Boaventura de Sousa. 2014. *Epistemologies of the South: Justice against Epistemicide*. Boulder, CO: Paradigm Publishers.

Schiebinger, Londa. 1987. "The History and Philosophy of Women in Science: A Review Essay." *Signs: Journal of Women and Culture in Society* 12 (2): 305–32.

___. 1989. *The Mind Has No Sex? Women in the Origins of Modern Science*. Cambridge, MA: Harvard University Press.

___. 1995. *Nature's Body: Gender in the Making of Modern Science*. Boston: Beacon Press.

___. 2004. *Plants and Empire: Colonial Bioprospecting in the Atlantic World*. Cambridge, MA: Harvard University Press.

___. 2005a. "Introduction: Forum: Colonial Science." *Isis* 96 (1): 52–55.

___. 2005b. "Prospecting for Drugs: European Naturalists in the West Indies." In *Colonial Botany: Science, Commerce and Politics in the Early Modern World*, edited by Londa Schiebinger and Claudia Swan, 119–33. Philadelphia: University of Pennsylvania Press.

Seth, Suman. 2009 "Putting Knowledge in Its Place: Science, Colonialism, and the Postcolonial." *Postcolonial Studies* 12 (4): 373–88.

Shapin, Steven, and Simon Schaffer. 1985. *Leviathan and the Air-Pump: Hobbes, Boyle, and the Experimental Life*. Princeton, NJ: Princeton University Press.

Shiva, Vandana. 1997. *Biopiracy: The Plunder of Nature and Knowledge*. Boston: South End Press.

Shiva, Vandana, and Ingunn Moser. 1995. *Biopolitics: A Feminist and Ecological Reader on Biotechnology*. London: Zed Books and Third World Network.

Shorter, David. 2012. *Indian, Frequencies: A Collaborative Genealogy of Spirituality Blog*. Accessed June 6, 2015, at http://frequencies.ssrc.org/2012/01/03/indian/.

Skloot, Rebecca. 2010. *The Immortal Life of Henrietta Lacks*. New York: Broadway Books.

Star, Susan Leigh, and James R. Griesemer. 1989. "Institutional Ecology, 'Translations' and Boundary Objects: Amateurs and Professionals in Berkeley's Museum of Vertebrate Zoology, 1907–39." *Social Studies of Science* 19: 387–420.

Subramaniam, Banu. 2013. "Re-owning the Past: DNA and the Politics of Belonging." In *Negotiating Culture: Heritage, Ownership, and Intellectual Property*, edited by Laetitia Amella La Follette, 147–69. Amherst: University of Massachusetts Press.

___. 2014. *Ghost Stories for Darwin: The Science of Variation and the Politics of Diversity*. Champaign: University of Illinois Press.

• Sunder Rajan, Kaushik. 2006. *Biocapital: The Constitution of Postgenomic Life*. Durham, NC: Duke University Press.

• TallBear, Kim. 2013a. "Genomic Articulations of Indigeneity." *Social Studies of Science* 43 (4): 509–34.

___. 2013b. *Native American DNA: Tribal Belonging and the False Promise of Genetic Science*. Minneapolis: University of Minnesota Press.

___. 2015. "Dossier: Theorizing Queer Inhumanisms: An Indigenous Reflection on Working beyond the Human/Not Human." *GLQ: A Journal of Lesbian and Gay Studies* 21 (2–3): 230–35.

Tilley, Helen. 2011. *Africa as a Living Laboratory: Empire, Development, and the Problem of Scientific Knowledge, 1870–1950*. Chicago: University of Chicago Press.

Todorov, Tzvetan. 1982. *The Conquest of America: The Question of the Other*, translated by Richard Howard. New York: Harper and Row.

Traweek, Sharon. 1988. *Beamtimes and Lifetimes: The World of High Energy Physicists*. Cambridge, MA: Harvard University Press.

Tsing, Anna Lowenhaupt. 2005. *Friction: An Ethnography of Global Connection*. Princeton, NJ: Princeton University Press.

___. 2007. "Indigenous Voice." In *Indigenous Experience Today*, edited by Marisol de la Cadena and Orin Starn, 33–67. Oxford: Berg.

Verran, Helen. 2002. "A Postcolonial Moment in Science Studies: Alternative Firing Regimes of Environmental Scientists and Aboriginal Landowners." *Social Studies of Science* 32 (5–6): 729–62.

Vessuri, Hebe M. C. 1987. "The Social Study of Science in Latin America." *Social Studies of Science* 17 (3): 519–54.

___. 2006. "Academic Science in Twentieth-Century Latin America." In *Science in Latin America*, edited by Juan José Saldana, 197–230. Austin: University of Texas Press.

Visvanathan, Nalini, Lynn Duggan, Nan Wiegersma, and Laurie Nisonoff, eds. 2011. *Women, Gender, and Development Reader*. 2nd ed. New York: Zed Books.

Walsh, Catherine. 2010. "Development as *Buen Vivir:* Institutional Arrangements and (De)colonial Entanglements." *Development* 53 (1): 15–21.

Walsh, Catherine. Forthcoming. "On Gender and Its Otherwise." In *The Palgrave Handbook on Gender and Development: Critical Engagements in Feminist Theory and Practice*, edited by Wendy Harcourt. London: Palgrave.

Walter, Maggie, and Chris Andersen. 2013. *Indigenous Statistics: A Quantitative Research Methodology*. Walnut Creek, CA: Left Coast Press.

Watson, James. 2004. *DNA: The Secret of Life*. New York: Knopf Books and Random House.

Watson-Verran, Helen, and David Turnbull. 1995. "Science and Other Indigenous Knowledge Systems." In *The Handbook of Science and Technology Studies*, edited by Sheila Jasanoff, Gerald E. Markle, James Petersen, and Trevor Pinch, 115–30. Thousand Oaks, CA: Sage.

Weatherford, Jack MacIver. 1988. *Indian Givers: What the Native Americans Gave to the World*. New York: Crown.

Yeh, Emily T. 2007. "Tibetan Indigeneity: Translations, Resemblances, and Uptake." In *Indigenous Experience Today*, edited by Marisol de la Cadena and Orin Starn, 69–97. New York: Berg.

15 Conceptualizing Imaginaries of Science, Technology, and Society

Maureen McNeil, Michael Arribas-Ayllon, Joan Haran, Adrian Mackenzie, and Richard Tutton

Introduction

During the last few decades, an increasing number of science and technology studies (STS) researchers have embarked on investigations of *imaginaries* associated with science and technologies (see figure 15.1). As the use of this concept seems to offer new ways to investigate the relationships among science, technology, and society, it merits attention. Observing the proliferation of work on imaginaries, we became aware of two patterns: the first is the *plurality* in STS approaches. While there were explicit connections between some STS research on imaginaries, there were also some strikingly distinct clusters of work. Second, examining the clusters of STS research on the imaginary, we found limited reference to the theoretical hinterland of the concept. Authors frequently used it with little or no reference to a theoretical or a methodological repertoire but with, nonetheless, a strong sense of its relevance. This lacuna dissipated somewhat during the drafting of this chapter (Jasanoff 2015a; Nerlich and Morris 2015). Nevertheless, Sheila Jasanoff's (2015a) recent exposition of the "theoretical precursors" and "major methodological approaches" in a new collection of research on the imaginary remains exceptional.

Our curiosity about the plural yet disparate imaginaries in STS led us to a genealogical approach in this chapter. This enables acknowledgment of the diversity of STS research undertaken with reference to this concept, without assuming a unified understanding of the concept or a single program or trajectory for its development. This approach informs the structure of the chapter: we begin from the etymology and features of the term itself as used in STS, and then trace a genealogy, highlighting the main lineages of imaginaries. We identify three key clusters of STS research stemming from the genealogy.

Our tracking, genealogy, and reflections address the emergence of the concept with reference primarily to Euro-American sciences and technologies. Figure 15.2 maps the

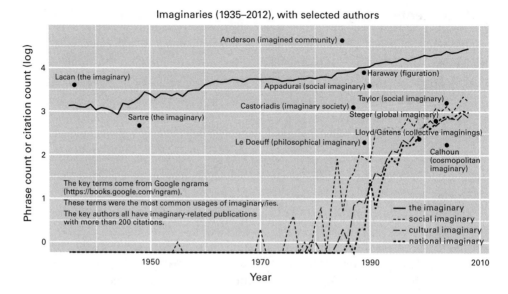

Figure 15.1
Chart of use of the concept of imaginaries (1935–2012) by selected authors. (See Color Plate 5.)

complex terrain covered in this study. Its complexity indicates the proliferation and diversity of the term, which we could not consider in detail in our examination of prominent clusters of STS research. Nevertheless, we draw attention to notable STS research deriving from the study of other cultures, including of Australian aboriginal (Verran 1998), South Korean (Jasanoff and Kim 2009), Japanese (Fujimura 2003; Mikami 2014), and Indian (Prasad 2014) contexts.

Etymology and STS Deployments

The etymology of the term *imaginary* yields a fascinating conceptual interplay of notions of reality, thought, and images. The *Oxford English Dictionary* (2009) gives this contemporary definition: "existing only in the imagination or fancy; having no real existence; not real or actual," which is remarkably consistent with its original meaning (classical Latin *imāginārius*) as "a mere semblance, unreal, fictitious, pretended." The semantic consistency of imaginary extends to its adjectival use as relating to *imagination*, implying "having no real existence." As these definitions suggest, imaginary properties are often deemed irrational or arbitrary and therefore not to offer reliable accounts of reality.

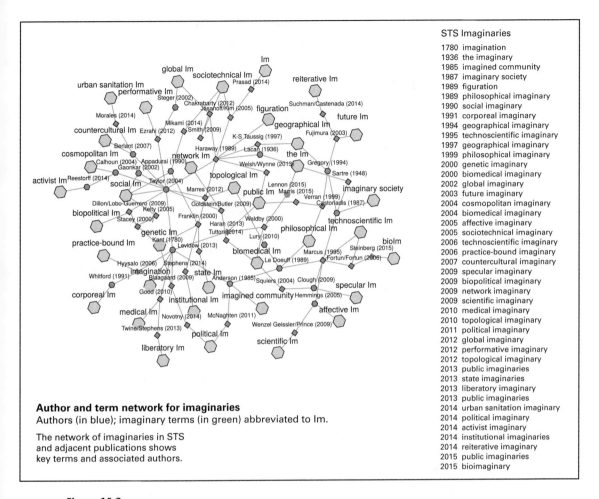

Author and term network for imaginaries
Authors (in blue); imaginary terms (in green) abbreviated to Im.

The network of imaginaries in STS
and adjacent publications shows
key terms and associated authors.

STS Imaginaries

1780	imagination
1936	the imaginary
1985	imagined community
1987	imaginary society
1989	figuration
1989	philosophical imaginary
1990	social imaginary
1991	corporeal imaginary
1994	geographical imaginary
1995	technoscientific imaginary
1997	geographical imaginary
1999	philosophical imaginary
2000	genetic imaginary
2000	biomedical imaginary
2002	global imaginary
2003	future imaginary
2004	cosmopolitan imaginary
2004	biomedical imaginary
2005	affective imaginary
2005	sociotechnical imaginary
2006	technoscientific imaginary
2006	practice-bound imaginary
2007	countercultural imaginary
2009	specular imaginary
2009	biopolitical imaginary
2009	network imaginary
2009	scientific imaginary
2010	medical imaginary
2010	topological imaginary
2011	political imaginary
2012	global imaginary
2012	performative imaginary
2012	topological imaginary
2013	public imaginaries
2013	state imaginaries
2013	liberatory imaginary
2013	public imaginaries
2014	urban sanitation imaginary
2014	political imaginary
2014	activist imaginary
2014	institutional imaginaries
2014	reiterative imaginary
2015	public imaginaries
2015	bioimaginary

Figure 15.2
Author and term network for imaginaries. (See Color Plate 6.)

However, there are two significant developments involving more positive meanings of imaginary. From the twelfth century, various usages among British sources begin to denote the faculty of imagination as a specific kind of thought (e.g., "imaginable, thinkable"), which is a direct antecedent of contemporary accounts of cognition and creativity. Appearing at the end of the sixteenth century, *imaginaries* is frequently used in the plural to denote "an imagination; a fancy; something imagined." The plural form also appears in the eighteenth century to designate "an imaginary quantity or expression" in mathematics. Hence, early definitions pertained to the interface between thought and reality, where *imaginary* designates a frivolous or inconsistent mapping between the two domains. The term is also frequently used in poetry and personal narratives associated with introspection, reflection, or affective writing. Nearly all examples of usage refer to processes of or about individuals; there is no indication that imaginaries refer to collective processes until the twentieth century.

The pluralization and nominalization of the term follows an historical pattern characterized by the increasing conceptualization of abstract processes. That is to say, *imaginary* is transformed from a word that designates processes pertaining to reality to one associated with interpretations of reality. To illustrate this, the *OED* cites a recent (1999) example of the noun form: "Such 'imaginaries' are crucial because they shape urban development patterns." The source for this is a publication of *Friends of the Earth*, an environmental campaigning organization in the United Kingdom. The term appears in quotes, indicating technical terminology. As the grammatical agent in the clause, it is registered as acting upon "urban development patterns," which is itself a complex noun. In this example, *imaginaries* seems to denote patterns of thought that have an influence on wider social processes. It is this sense of the term—conceptualizing significant and sometimes abstract social processes—which resonates with STS usage.

Recent iterations of imaginaries in academic work often appear in conjunction with a modifier. For example, Le Doeuff (1989) writes of "the philosophical imaginary." Some political and social theorists (including Charles Taylor) have added the label "social." STS scholars have offered their own specifications, most notably Jasanoff and Kim's (2009) rendering of "sociotechnical imaginaries." Franklin (2000) and Stacey (2010) specify "the genetic imaginary" and Waldby's (2000) coinage is "biomedical imaginary," while Steinberg (2015) invokes "bioimaginary," and some anthropologists favor "medical imaginary" (Good 2010).

We might query the usefulness of some of these modifiers. First, as many of those using the term within STS indicate, all imaginaries are necessarily social in some way. While the addition of "social" may seem rather superfluous,[1] through such labeling STS researchers emphasize both that they are held collectively and draw attention to embedded visions of the social in operation, in particular technological and scientific developments or regimes.

There are similar issues around the attachment of the adjectives *scientific, technical, sociotechnical,* or *technoscientific* to imaginaries in STS work. Since the focus of such research is scientific and technological, this qualifier sometimes seems redundant. However, the framing of imaginaries as technoscientific may be a way of asserting that such imaginaries are embedded in the science and technology being investigated and a crucial part of the social. This also challenges assumptions about technoscience as exclusively the realm of facts and, hence, disturbs assumptions about science as "a system of pure logic" (Waldby 2000, 137). Moving from these features of STS terminology, the next section surveys some sources which have inspired and sustained the field's research on imaginaries.

Genealogy of Imaginaries

A number of streams of theory can be traced in the genealogy of the concept of the imaginary. Our review registers the complexity of the concept's genealogy and derives from references and affiliations cited in STS research. However, it also tracks work not explicitly deploying the term but significant for its usage in STS. The purpose of this genealogy is both to track the emergence of the imaginary concept over time and to register the latent baggage of the way STS scholars use it. We contend that working through the concept's past lives could enrich work in the field, given that STS has only relatively recently come to it. Such genealogical work may help to provide a history of the contemporary concept and its usages. We have identified the following key categories of genealogical resources: (1) Western philosophy, (2) psychoanalysis, (3) late twentieth-century sociopolitical theory, and (4) science fiction. All but the third category designates broad fields relevant to STS conceptualizations of imaginaries. The third title is more specific and refers to the body of work that has been most influential in the deployment of the term in STS.

Western Philosophy

Reflections on imaginaries can be traced as a narrow, but significant thread in Western philosophy: the works of Kant, Sartre, Le Doeuff. This leads back to Kant's (1787) demarcation of "the real" from "the imaginary," which became a touchstone for modern Western science and reemerges with notable contributions from existential and feminist philosophy in the twentieth century.

Jean Paul Sartre's *The Imaginary: A Phenomenological Psychology of the Imagination* (1940) was the first work to explore systematically the psychological domain of the imagination. For Sartre, the imaginary was the sphere and set of psychological and mental operations associated with imagination. Although not cited in the STS research we have consulted, Sartre's text is important because of its designation of the imaginary as a distinctive and powerful domain. By delineating how imagination differed from perception, Sartre was identifying its key features but also insisting on its importance as a parallel mode of knowing. Hence, he brought the term *imaginary* into critical political discourse, proposing it as a vital sphere of mental operations and knowledge production. Sartre's exploration proved to be an important resource for both philosophical and psychoanalytic investigations.

Another much later philosophical text that has been influential in STS research is Michèle Le Doeuff's *The Philosophical Imaginary* ([1980] 1989). Le Doeuff's critical rereading of Western philosophy, including her account of Kant's paradoxical use of imagery and metaphor to characterize modern reason as requiring the exclusion of the imaginary, has been particularly important for feminist STS, as we show below.

Psychoanalysis

Jacques Lacan's conceptualization of the imaginary as a psychic realm of subjective identification with images that exist prior to identification through language emerged almost concurrently with Sartre's. Lacan accounted for subjective misrecognition of the self's unity and coherence through his positing of the mirror stage of psychic development in which the young child views him- or herself in the mirror and identifies with the specular image. He maintained that this takes place before language acquisition so that the ego is formed "prior to its social determination, in a fictional direction" (Lacan 2006, 76). Lacan asserted that by understanding this experience, psychoanalysis is "at odds with any philosophy directly stemming from the *cogito*" (ibid., 75)." In his 1936 essay—having suggested that the imaginary can only be brought fully into consciousness through psychotherapy—Lacan (2006, 70) claimed that "physical science, as purified as it may seem in its modern progress from any intuitive category, nevertheless betrays, indeed all the more strikingly, the structure of the intelligence that constructed it."

STS has had limited engagement with the psychoanalytic tradition in its use of the concept of imaginaries. The Harvard Imaginaries (Harvard University 2016) project website does reference Lacan and a number of feminist researchers have drawn on this tradition, but otherwise, it has not been a main resource for STS. Nevertheless, Lacanian psychoanalysis's identification of the imaginary with the fragmentary character of subjectivity draws attention to the mediated nature of knowledge production (attending particularly to images) and, as some STS scholars (Stacey 2010; Waldby 2000) have noted, psychoanalysis has made a significant contribution to the conceptualization of subjectively experienced collective imaginaries.

Late Twentieth-Century Sociopolitical Theory

STS scholars have relied mainly on a fairly fixed body of sociopolitical theory in their investigations of the imaginary. Primary reference points have been the work of Benedict Anderson, Cornelius Castoriadis, Arjun Appadurai, and Charles Taylor. This section probes the contributions of these theorists to the conceptualization of imaginaries and assesses the significance of Althusser's theories in such conceptualization.

Benedict Anderson's (1991) *Imagined Communities: Reflections on the Origin and Spread of Nationalism* has inspired a number of STS researchers, as well as myriad other political theory and cultural studies analysts. His key insight was that nations were "*imagined* because the members of even the smallest nation will never know most of their fellow-members, meet them, or even hear of them, yet in the minds of each lives the image of their communion" (Anderson 1991, 15).

Although he does not explicitly use the term *imaginary*, Anderson's historical reflections brought collective imaginative capacities to the fore as vital elements in the making of nations, through literary and media technologies. Anderson (1991, 49) contended that "the convergence of capitalism and print technology on the fatal diversity of human language created the possibility of a new form of imagined community, which in its basic morphology set the stage for the modern state." Anderson's general conceptualization of how collective imaginary capacities are generated and sustained has had considerable resonance for STS research. His work suggests that in analyzing *scientific* imaginaries, it may be important to explore both the communication processes and the media that enable/instantiate these imaginaries, such as maps and census taking. Moreover, Anderson's influential investigation of nation formation has been linked to technoscientific imaginaries in another way. As Jasanoff and Kim (2009) and other STS researchers have demonstrated, sociotechnical imaginaries are often (although not exclusively [Jasanoff 2015a]) cast with reference to nations.[2]

While Anderson's project was a Marxist theorization of nationhood, Castoriadis's approach to the imaginary emerged from his dissatisfactions with late twentieth-century Marxism. He articulated new ways of thinking about the "revolutionary project" with reference to the "institution of society" (Castoriadis 1987) and alienation. He was gripped by the failure to acknowledge the subjective dimensions of "the idea of another society" (Castoriadis 1987, 90). He arrived at the concept of imaginary through reflections both on the limitations of Marxism and on how societies function:

The social world is, in every instance, constituted and articulated as a function of such a system of significations, and these significations exist, once they have been constituted, in the mode of what we called the *actual imaginary* (or the *imagined*).... . Every society up to now has attempted to give an answer to a few fundamental questions: The role of imaginary significations is to provide an answer to these questions, an answer that, obviously, neither "reality," nor "rationality" can provide. (Castoriadis 1987, 146–47; original emphasis)

The social imaginary became for Castoriadis not only the register of the character of a specific society; it provided the basis for his "schemata" for distinguishing societies.

Arjun Appadurai crafted a distinctive way into imaginaries by probing the features of the contemporary world. He declared that "the image, the imagined, the imaginary ... are all terms that direct us to something critical and new in global cultural processes: *the imagination as a social practice*" (Appadurai 1996, 31). For us to understand this "new role for the imagination in social life," he argues that it will be necessary to bring together

the old idea of images, especially mechanically produced images (in the Frankfurt School sense); the idea of the imagined community (in Anderson's sense) and the French idea of the imaginary (*imaginaire*) as a constructed landscape of collective aspirations, which is no more and no less real than the collective representations of Emile Durkheim, now mediated through the complex prism of modern media. (Appadurai 1996, 31)

Appadurai explicitly articulates sociopolitical theory with attention to images, mediation, and electronic media, which is methodologically suggestive for investigating imaginaries. Extending Anderson's ideas, Appadurai (1996, 33) addresses "*imagined worlds*, that is, the multiple worlds that are constituted by the historically situated imaginations of persons and groups spread around the globe," and he suggests that the building blocks of these imagined worlds are "five dimensions of global cultural flows that can be termed (a) *ethnoscapes*, (b) *mediascapes*, (c) *technoscapes*, (d) *financescapes*, and (e) *ideoscapes*."

Charles Taylor's *Modern Social Imaginaries* (2004) was both a product of collaboration with Appadurai and others and a response to Castoriadis's theorizing of the social imaginary. It emerged from the interdisciplinary Center for Transcultural Studies (CTS) and a smaller subgroup convened there in 1999 comprising Charles Taylor, Arjun Appadurai, Benjamin Lee, Michael Warner, and Dilip Parameshwar Gaonkar, which produced a position statement on "new imaginaries"[3] and a 2002 special issue of *Public Culture*. This issue situated the project within a larger set of collective preoccupations about modernity, civil society, and the public sphere.

Taylor's (2002) essay is cited as providing the "conceptual frame" (Gaonkar 2002, 10) for the special issue of *Public Culture*. This article also staked the terrain for Taylor's (2004) subsequent book, addressing what the CTS group regarded as weaknesses in Castoriadis's theorization of the social imaginary, as well as key global political developments of the early twenty-first century. In this essay Taylor (2002, 92) offered "an account of the forms of social imaginary that have underpinned the rise of Western modernity." Whereas Castoriadis had propounded a dualistic framework for characterizing social imaginaries, Taylor insisted that the Western social imaginary sustained a specific version of modernity. Taylor's narrative traces the development of "the modern theory of moral order" which later "infiltrates" (ibid., 110) the Western social imaginary.

Taylor (2002, 108) reflected that "the social imaginary" is that "which enables us to carry out the collective practices that make up our social life." Clarifying the distinction between social theory and the social imaginary, Taylor proposed that "the social imaginary is that common understanding that makes possible common practices and a widely shared sense of legitimacy" and was "the way we are able think or imagine the whole of society" (Taylor 2004, 23, 63).

While considerations of scientific imaginaries may be more specific than the social imaginaries Taylor explored, he does offer perspectives relevant to STS. This includes his handling of charges that the concept of social imaginary "smacks" of idealism. He insisted that this is "based on a false dichotomy"

between ideas and material factors as rival causal agencies. In fact, what we see in human history is ranges of human practices that are both at once, that is, material practices carried out by human beings in space and time, and very often coercively maintained, and at the same time, self-conceptions, modes of understanding. These are often inseparable, in the way described in the discussion of social imaginaries, just because the self-understandings are the essential condition of the practice making the sense that it does to the participants. (Taylor 2004, 31–32)

It is impossible to review the sociopolitical theory which has influenced STS work on imaginaries without considering the return of the repressed—the contribution of Louis Althusser. Although, as our conclusion will suggest, turns to the imaginary may be displacing the concept of ideology, Althusser's work on ideological state apparatuses and his conceptualization of interpellation (hailing) remain influential in analyzing imaginaries (Warner 2002). Notably, Althusser claimed, "*Ideology represents the imaginary relationship of individuals to their real conditions of existence*" (Althusser 1971, 109; original emphasis). Formulated in the shadow of Lacan's account of subjectivity, he suggests a mechanism by which the relations of production are reproduced through institutions, rituals, and practices. Moreover, given its implicit dependence on ideas of performativity, the concept of interpellation can be mobilized to think about the ways in which imaginaries engage subjects (Haraway 1997).

In sum, late twentieth-century sociopolitical theory has provided important resources for STS of imaginaries. However, there has been relatively little critical commentary about the adaptations and translations of this tradition into STS, prior to Jasanoff's (2015a) recent review.

Science Fiction

Donna Haraway's work has been crucial to STS explorations of technoscientific imaginaries. Refusing reductive distinctions between science fiction and science fact, her creative engagement with and redeployment of science fiction has extended the resources for investigating the imaginaries of technoscience. Until fairly recently, there was scarcely any mention of science fiction in STS. However, on the margins of the field, its speculative fiction mode has long been understood as a medium for diagnosing the history of the present.[4] Furthermore, some feminist STS researchers have been particularly sensitized to the visual aspects of imaginaries through their encounters with science fiction film and with key research in this domain. As Lisa Yaszek (2008, 385) observes: "Science fiction enables authors to dramatize widespread cultural hopes and fears about new technoscientific formations as they emerge at specific historical moments."

Yaszek is one of many feminist scholars inspired by Haraway's lead, who, in her "Cyborg Manifesto" (1991), hailed feminist science fiction writers as storytellers who muse on what it means to live in and with modern technoscience. In *Primate Visions* (1989), science fiction is utilized forcefully in Haraway's representation of the twentieth-century science of primatology:

science fiction has provided one of the lenses for reading primatological texts. Mixing, juxtaposing, and reversing reading conventions appropriate to each genre can yield fruitful ways of understanding the production of origin narratives in a society that privileges science and technology in its construction of what may count as nature and for regulating the traffic between what it identifies as nature and culture (Haraway 1989, 373).

Drawing on the tropes of science fiction, Haraway (1989, 376) challenged her readers to consider how the core narratives of primatology might have been or could be different, by taking "the next logical step" of moving "from reading primatology as science fiction" to "reading science fiction as primatology." Haraway's experimentation with science fiction raised questions about the imaginaries of modern sciences and about alternative versions and visions of technosciences. Recently, Sheila Jasanoff (2015a, 2015b, 337) has explicitly acknowledged science fiction as "a repository of sociotechnical imaginaries."

In this section, we have outlined four intellectual traditions from which STS scholars have drawn, exposition of which helps to situate and enrich understandings of how such scholars conceive of the imaginary. We now analyze how STS researchers have engaged with and related to the notion of imaginaries in a striking variety of ways.

Clusters of STS Work with Imaginaries

In this section, we map several clusters of STS research around the concept of imaginary with reference to the nature of their objects and modes of investigation, their conceptual frameworks, and their ethico-political relations to sciences and technologies. As figure 15.2 shows, these clusters are poly-vocal and plural. The chronology of several dozen STS imaginaries shown on the right of the figure illustrates a proliferation of imaginaries. The network diagram shown on the left centers on key terms and maps its key authors connected through citation links. The expanding periphery of terms (such as "reiterative" and "biological" imaginaries) deployed here suggests both ongoing development and plurality. Amid the many patterns of citation, influence, and dialogue running through these links and lists, we identify exemplary contributions to STS under the following orientations: (1) cultures, communities and practices; (2) nations, institutions, and policies; and (3) bodies, subjects and differences. These headings point to the multiple orientations of the concept and an inherently labile composition, which opens the imaginaries concept to ongoing development and a wide range of STS uses.

Cultures, Communities, and Practices

This orientation of the concept of imaginaries is best represented by research in the anthropological STS literature, beginning with George Marcus's edited collection *Technoscientific Imaginaries: Conversations, Profiles, and Memoirs* (1995b). Marcus (1995a, 3) explains that "the term *imaginary* emerged effortlessly and just seemed to fit the topic" of the volume, which was, "an optimistic assemblage" of studies of the conditions of work in science and technology at the end of the twentieth century. Marcus (1995a, 4) assessed that the investigators contributing to the collection were primarily interested in "the imaginaries of scientists tied more closely to their current positionings, practices, and ambiguous locations in which the varied kinds of science they do are possible at all." This led him to reflect on the concept that informed their studies: "This is a socially and culturally embedded sense of the imaginary that indeed looks to the future and future possibility through technoscientific innovation but is equally constrained by the very present conditions of scientific work" (Marcus 1995a, 4).

Marcus (1995a, 4) notes that if future visioning figures in these conditions, it is a "cautiously imagined emergent future, filled with volatility, and uncertainty, but in which faith in practices of technoscience become even more complex and interestingly constructed." Hence, the collection revolves around technoscientific imaginaries perceived to derive from tensions between practices and discourses within the work of scientists. It is this gap which Marcus envisages his contributing authors and other STS researchers exploring, thereby generating "a completely transformed and vast field of inquiry on which a distinctly *cultural* studies of science might establish itself" (Marcus 1995a, 7; original emphasis).

Joan Fujimura (2003) offers an anthropological STS research project focused on the crafting of future imaginaries as a constitutive part of the work of scientists. She considers the research of two leading Japanese scientists in the fields of genomics and computer science who she regards as having constructed two different imaginaries that link investment in innovative science and technology with discourses of cultural and religious distinctiveness. She uses the term *technosocial imaginaries* since she regards them as conjuring both alternative futures for scientific practice and re-visioned versions of Japanese culture in the context of transnational economies in biology, genomics, and computing.

Citing Appadurai, Fujimura (2003, 192) insists that scientists' future imaginaries are not "mere fantasy" but are implicated in the formation and practices of scientific communities, involving "enterprises that have enrolled and engaged many people, funds, and government agencies, and much public and consumer interest." For her, imaginaries are enabling visions that involve persuasive rhetoric and possibly hyperbole, but that facilitate community formation and the marshaling of resources. Her emphasis on technosocial imaginaries as implicated in national formations anticipates Jasanoff and Kim's (2009) later comparative study of sociotechnical imaginaries discussed below.

Kim Fortun and Mike Fortun (2005) address notions of the imaginary in their research on the recent state of toxicology, advocating the idea of toxicology becoming a "civic science" that would protect public health rather than serving the interests of industry or the state. They suggest that anthropologists and other STS researchers could help in facilitating this vision. A study of imaginaries, they contend, might—in the spirit of "friendship" rather than criticism, as is usually the case—help scientists negotiate change in their field by enabling them to engage more fully with the social, ethical and legal implications of their practices. Referencing Marcus's (1995a) discussion of "technoscientific imaginaries," they propose the study of imaginaries as a way of looking at large-scale changes over time and at how these are understood locally. Moreover, they contend that a focus on imaginaries may enable analysts to study the factors constitutive of subjectivity. They signal their interest in extending Sharon Traweek's (1988) exploration of subject formations within and through scientific practices.

In contrast to anthropological studies of contemporary scientists, Karen-Sue Taussig (1997) deploys the concept in her situated study of preimplantation genetic diagnosis (PGD) in the Netherlands to look at parents' decision making with reference to understandings of difference(s)—geographical, social, and religious. Coining the term *geographical imaginary* to characterize the way that "deeply embedded understandings of geographically specific social practices ... play [out] in daily interactions that simultaneously produce people and their social worlds" (Taussig 1997, 497), her exploration concerns the potential users of this technology rather than scientists or doctors. Drawing on Edward Said's study of *Orientalism* (1977), Taussig shows how such imaginaries, behaviors, and identities are often ascribed to others in ways that serve the interest of the majority. Thus, Taussig uses the notion of imaginary to draw attention to the way that social and cultural conceptions of others are often aligned with geographical locality and gain widespread currency.

While Taussig's ethnography pertains to a specific location, Helen Verran (1998) extends the use of the term *imaginary* in a comparative study of competing claims to land ownership between Cape York pastoralists and Australian aboriginals. Verran's study juxtaposes the openness about the picturing, storytelling, and the working up of metaphors in the knowledge-making and negotiating among aboriginal peoples with the denial of equivalent practices in Western knowledge making. Observing the differences between the knowledge practices of these two cultures, Verran (1998, 238) comments:

Looking at some of their puzzles [faced by participants in negotiations over native title and pastoral leases] allows us to see an element almost entirely ignored by modern practices and accounts of knowledge. I call this element "the imaginary" and point to its necessary involvement in knowing and knowledge making. I show the imaginary as something constitutive of, and constituted by, ontic and epistemic commitments.

Verran (1998, 243) holds out the challenge of acknowledging and addressing the imaginaries in Western science and knowledge production that have generally been denied or obscured because, as she sees it, "Modernity circumscribes its imaginary as of aesthetic, but not ontic or epistemic interest." She states that "by restoring imaginaries to modern theories of knowledge, we [Westerners] will rediscover the capacity to re-imagine ourselves, and devise ways they can work with other communities— human and non-human" (Verran 1998, 249). So imaginaries are not only about possible futures involving visions and speculation but also about knowledge production itself, and commitments to certain forms of reality.

Verran's take on imaginaries is linked to her reading of Kant through feminist philosopher Michèle Le Doeuff ([1980] 1989) and his exclusion of the imaginary as the defining property of reason. Le Doeuff contends that Kant could only represent this exclusion through the use of imagery and extended metaphors (picturing and storytelling). Moreover, imaginaries are very much associated with practices, not minds, for Verran. It is in the everyday messing with mucky, obdurate stuff, and in conversations and texts that imaginaries are enacted and enact. Hence, imaginaries interpellate objects/subjects that/who are implicated in and by practices, thereby constituting them as objects/subjects (Verran 1998, 252). Verran insists that imaginaries must be acknowledged and recognized with their operations in what she calls a "logic"—in performative modes of knowledge production.

The preceding review provides a sample of the range and diversity of recent anthropological studies of technoscience that have used the concept of imaginaries in relation to scientific cultures, communities, and practices. It is perhaps not surprising to find anthropologists, who focus on narrative and storytelling in communities, generally through ethnographic methods,[5] drawn to the study of imaginaries. Nevertheless, these studies vary, with much anthropological STS focused on particular scientific communities or specific technoscientific sites (e.g., laboratories or clinics), while some investigate the imaginaries of nonprofessionals (Taussig's case of parents using PGD testing) or, in Verran's case, offering a comparative analysis of Western scientific and Australian aboriginal knowledge systems. For some of these scholars, researching imaginaries becomes more than a theoretical or methodological pivot for their own research, as they recommend it as possibly a potential reorienting tool for STS more generally.

Nations, Institutions, and Policies

A second strand of STS research has drawn heavily on political theory in its use of the concept of imaginaries. Oriented toward analyzing and characterizing the policies and practices of states and large institutions, this work has generally not relied on ethnographic modes. An influential example of such research is Jasanoff and Kim's (2009) comparative study of U.S. and South Korean orientations toward civil nuclear power technology. Jasanoff and Kim advocated the use of the notion of "sociotechnical imaginaries" as a way of encouraging STS research on national and state technoscientific policies and politics. This has proven to be a generative line of inquiry (Jasanoff and Kim 2013, 2015; Kim 2013; Mikami 2014). Across this body of work, Anderson's (1991) study of the origins of national "imagined communities," as well as Castoriadis's (1987) and Taylor's (2007) studies of modern political formations constructed in and through "social imaginaries," have been vital reference points.

Jasanoff and Kim (2009) contend that the relationship between science and technology and political institutions has been relatively neglected by STS. They pose the question: "How do national science and technology projects encode and reinforce particular conceptions of what a nation stands for?" (Jasanoff and Kim 2009, 120). Setting out to explore how national political orders and technoscientific projects co-produce each other, they offer a definition of national sociotechnical imaginaries as "collectively imagined forms of social life and social order reflected in the design and fulfillment of nation-specific scientific and/or technological projects" (ibid., 120). They add that such imaginaries "at once describe attainable futures and prescribe futures that states believe ought to be obtained" and that they operate "in the understudied regions between imagination and action, between discourse and decision, and between inchoate public opinion and instrumental state policy" (ibid., 120, 123). Jasanoff and Kim (2009, 123) argue that despite globalization, sociotechnical imaginaries are intertwined with the production and reproduction of nations, insisting that the national is not simply given or immutable, but continuously "reimagined, or re-performed, in the projection, production, implementation, and uptake of sociotechnical imaginaries."

Jasanoff and Kim draw extensively on sociopolitical theory in framing their study. They also discuss a cluster of STS work that they regard as either implicitly or explicitly sharing their interest in sociotechnical imaginaries. They demarcate between STS and histories of science and technology claiming the latter field tends to regard the imagination as an individualized mental capacity (Jasanoff and Kim 2009, 122).[6] They distinguish between their coinage—"sociotechnical imaginaries"—and Marcus's term—"technoscientific imaginaries"—on grounds that imaginaries of technoscience pertain to "the social world writ large" [and not just scientists] since "'social imaginaries' encode collective visions of the good society" (Jasanoff and Kim 2009, 123). They clarify the conceptual scope of imaginary by contrasting it with other concepts used in exploring the cultural, social, or political dimensions of technoscience, such as "policy agendas," "master narratives," or "media packages" around "discursive frames." They note that, while they project hopes and promises, imaginaries may also entail fears and risks around innovation.

Jasanoff and Kim (2009, 123) explain that, "sociotechnical imaginaries as we define them are associated with active exercises of state power" and that while multiple discursive framings may circulate in any society, some become filtered and selected, emerging as dominant, embedded in the goals and priorities of state and public action. Koichi Mikami (2014) takes up this observation in his case study of regenerative medicine research in Japan. He argues that the state's early material commitments to certain technologies, with reference to its visions for the nation's future, can lead to certain imaginaries becoming "locked-in."

Elta Smith (2009) elaborates on Jasanoff and Kim's notion of the "sociotechnical imaginary" in her study of the Rockefeller Foundation as an actor that is beyond any national or state-bounded political sphere. Smith investigates the Rockefeller

Foundation's fifty-year funding of rice research and uses "the term *imaginaries* to characterize the Foundation's conceptions of 'development' and its changing role in rice experimentation over time" (Smith 2009, 462; original emphasis). She contends that there are "always multiple imaginaries at play in a society and within institutions" and explains that her study explores how "particular imaginaries" emerged and prevailed through the Rockefeller Foundation to become the "best, most appropriate, or even inevitable—and how they became hegemonic while seeming apolitical or value-neutral" (Smith 2009, 462). Smith (2009, 479) concludes by noting that "imaginaries of development have history and politics," observing that the "the imaginaries projected and actualized by the Rockefeller Foundation" have altered in relation to changing political imperatives and rationalities.

For Smith (2009, 462), an imaginary is a "particular, often complex view of the world that comes to shape agendas, research trajectories, projects, and policies," noting here the work of Taylor (2004) and Anderson (1991). She uses the term to denote "normatively loaded visions not only of what should be done 'in the world' but also how it should be undertaken and why" (Smith 2009, 462), adding that "imaginary also refers to a larger constellation of ideologies, and social factors that enables or constrains discourse in certain ways," quoting Appadurai's (1996) evaluation that it is an "organized field of social practices" (Smith 2009, 462). Smith (2009, 463) explains that "the imaginaries concept suggests that the world has been consequentially envisioned in certain ways, at certain moments in time, by actors who have the capacity to materialize these abstractions."

A recent collection, edited by Jasanoff and Kim (2015), has developed and extended the study of sociotechnical imaginaries by clarifying the scope of the concept and demonstrating its usefulness for STS. Jasanoff's (2015a, 2015b) introductory and concluding chapters (cited previously) set out its theoretical underpinning, methodological parameters, phases, and analytical benefits. The collection includes a range of studies deriving from diverse national (for example, Felt 2015; Kim 2015), institutional (Smith 2015), and global (Miller 2015) investigations.

Other researchers have deployed the concept of the imaginary in this national and institutional thread of STS work to consider how publics are positioned and perceived by national institutions or policy makers. For example, Neil Stephens, Paul Atkinson, and Peter Glasner (2013) adapt Jasanoff and Kim's approach, shifting the focus to particular state institutions—United Kingdom and Spanish stem cell banks—tracking how these bodies conjure and enact "institutional imaginaries of publics." Ian Welsh and Brian Wynne (2013) investigate the imaginaries of publics that have animated recent U.K. technoscientific policy making. While they do use the term *technoscientific imaginary*, their primary reference is Charles Taylor's concept of the "social imaginary." Influenced by Welsh and Wynne's research, Claire Marris (2015) explores the imaginaries of publics emerging in synthetic biology, while Stevienna de Saille (2015), considers how United Kingdom policy imaginaries of publics impact environmental activists. David Hess also cites Welsh and Wynne (2013) but considers how the imaginary resonates with a range of other concepts in anthropology and sociology, including cultural logics, cultural code, discourse, ideology, and frame (Hess 2014, 76). He suggests that the use of the concept of imaginary fits well with Clifford Geertz's call for an "interpretative science that attends to webs of meaning" (Hess 2014, 76). While Jasanoff and Kim (2009), and Smith (2009) also acknowledge multiple imaginaries, they conclude that over time certain imaginaries prevail. Hess, however, stresses the importance of paying greater attention to times of contestation and to the social positions and power of actors who articulate "counter-imaginaries."

The stream of research focused on national imaginaries of science and technology has predominantly been either explicitly or implicitly comparative (Jasanoff 2015a, 24). Its analytical gaze has been mainly Euro-American, but some studies investigate other contexts (for example, Jasanoff and Kim 2009). However, one of the more challenging deployments of imaginaries proposes movement beyond "the Western versus non-Western technological divide" and "relativistic comparisons across nations" toward accounts of the "hierarchically entangled histories of technoscience practices" (Prasad 2014, 7). Combining empirical and deconstructivist methods and drawing on ethnographic research, Prasad examines the relational and hierarchical imaginaries that have emerged around MRI technology in the United States, Britain, and India.

This cluster of work on national and institutional imaginaries is distinguished by its foregrounding of issues of governance and, to some extent, policy. Methodologically, much of the research in this mode has been realized through textual analysis (of various sorts) and there has often been an historical and comparative (Jasanoff 2015a, 2015b) dimension to these investigations. Once again, there is a reformist agenda emerging here, particularly with Jasanoff and Kim's recommendations for more attention to state policies on technoscience and national comparisons in STS research. Some researchers (Smith 2009) and, subsequently Jasanoff and Kim (2015) themselves, have extended their focus to include the study of other institutions and publics in explorations of technoscience imaginaries. In addition, through ethnographic research and deconstructive methods, Prasad (2014) has raised critical questions about the reproduction of binaries and the neglect of hierarchies in the study of technoscience practices.

Bodies, Subjects, and Differences

Since the close of the twentieth century, some feminist STS scholars have worked intensely with the notion of imaginaries in relation to the life sciences, influenced particularly by the work of Le Doeuff (Squier 2004; Verran 1998; Waldby 2000). However, it is important to situate the deployment of the concept within a wider feminist STS context. This brings us back again to Haraway—a crucial figure beckoning her readers to decipher the imaginative dimensions of technoscience. Perhaps more than any other researcher, she has dispersed the STS gaze, demonstrating the need to investigate the makings and remakings of science and technology in a range of diverse sites, drawing attention to locations that would previously have been regarded as ephemera in relation to modern science (e.g., cinema, advertising, etc.).

Haraway's foregrounding of storytelling as a key mode of modern science has been another important vehicle for her exploration of the imaginaries of technoscience, marking a significant shift in STS practice. *Primate Visions* examined the strands of storytelling that were woven into the twentieth-century science of primatology. Moreover, as outlined above, a further reflexive spin was added through Haraway's use of science fiction to raise questions about her own storytelling about the making of this science and to encourage her readers to conjure alternative stories about it.

Haraway's attention to the imagery of modern technoscience is another crucial mode in her explorations of the imaginaries of science. In *Modest Witness* (1997), for example, she provides a reading of a set of visual texts (advertisements and cartoons) which represented modern genetics during the period of the Human Genome Project. This is accompanied by an examination of the trope of mapping, which was crucial to the Human Genome Project. Haraway (1997, 172) considers the making and interpellation of "technoscientific subjects" through the imaginary of genomics, during the completion of the Human Genome Project. She shows that interpellation works through the mobilizing of a repertoire of cultural resources and references (e.g., high art, Christian iconography), involving complex psychosocial processes, including investment and attachment. Humor emerges as an important vehicle in such interpellation, indicating that engagement with scientific imaginaries is not exclusively cognitive.

Haraway (1997, 151) discerns a "technoscientific unconscious" in operation in "the processes of formation of the technoscientific subject" and she sets out to identify the "structures of pleasure and anxiety" contributing to the formation and reproduction of this subject. Her accounts highlight various moments in the continuous making and remaking of technoscience in diverse, often mundane but complex, processes which she sees as enabling the formation of technoscientific subjects, involving pleasure, anxiety, and other emotions. Narrative and figuration are presented as the modes through which the technoscientific unconscious operates that can be traced through a multitude of media. Haraway maintains that we are constituted as technoscientific subjects through diverse practices and encounters. Moreover, she insists that STS analysts are themselves not exempt from such interpellation.

Having foregrounded Haraway's significance in STS research on the imaginary, we now turn to the work of other scholars, beginning with Catherine Waldby's examination of the "biomedical imagination" and "biomedical imaginary." Waldby provides a detailed explication of Le Doeuff's ([1980] 1989) notion of the "philosophical imaginary." Bringing this together with Derrida's discussion of metaphor, Waldby (1996, 29) considers figurations in science, insisting on "the absolute indissociability of figure from technical language, the impossibility of controlling its connotative force, the irreducible operation of the metaphor in scientific textual practice." She introduces her own term—*biomedical imagination*—"to emphasize the speculative, 'fictional' dimensions of the medical enterprise" (Waldby 1996, 5, 16).

Waldby's (2000) study of the Visible Human Project uses the term *imaginary*. Taking her cue from Le Doeuff, she registers the importance of imagery as a marker of points of tension in a system of logic or knowledge—in this case—biomedical knowledge. She explains that

[t]he biomedical imaginary refers to the speculative, propositional fabric of medical thought, the generally disavowed dream work performed by medical theory and innovation. It is a kind of speculative thought which supplements the more strictly systematic, properly scientific thought of medicine, its deductive strategies and empirical epistemologies. (Waldby 2000, 136)

Hence, Waldby also follows Le Doeuff in associating the imaginary with the excessive—that which "supplements" the bare bones of logic. Accordingly, she notes that the imaginary includes fantasy, myth, and so on. Despite her explicit discussion of "biomedical imaginary," Waldby continues to use the term *biomedical imagination* at various points, as if the terms were interchangeable. Moreover, she posits that in the contemporary era the maintenance of canonical scientific meanings may be difficult to enforce.

Alongside Waldby's biomedical explorations there has also been a set of feminist STS research projects analyzing the imaginary of genetics and genomics. A relatively early investigation was José van Dijck's *Imagenation: Popular Images of Genetics*, which examined "the role of images and imagination in popular representations of the new genetics since the late 1950s" (van Dijck 1998, 3) and coined the term *imagenation*. Adapting Katherine Hayles's characterization of science as a "theatre of representation" (ibid., 3), van Dijck (1998, 3) offers pictures of the four stages of "imagenation," which she considered constituted the phases of popular imagery associated with new human genetics from the 1950s to the Human Genome Project. Van Dijck deciphered specific popular images, but her analysis concerns "imagenations" which range from "biofears" and "biofantasies" to notions of the gene as "master controller" (van Dijck 1998, 91).

Following van Dijck's project, Sarah Franklin (2000) and Jackie Stacey both offer different versions of what they call the "genetic imaginary" (see also Steinberg 2015).[7] Franklin provides a case study of the popular Hollywood film *Jurassic Park* (1993) and of its cultural offshoots. She unpacks the many layers of the film's representation of "life itself" and indicates its cultural reverberations—not just in a plethora of tie-in products but in high-cultural manifestations, including an exhibition at the American Natural History Museum in New York. Franklin undertakes an examination of how the cultural phenomena of *Jurassic Park* instantiate the new genetic imaginary associated with late-twentieth-century genomics. Introducing her terminology, she explains:

> If part of the way life itself, as a discursive condition, or as historical epistemology, calibrates its syntax is at the level of politics, truth or liberation, another level of this syntax can be defined as an *imaginary*. Not in the technical sense of a psychoanalytic pre-symbolic realm of undifferentiated toti-potency, but in the more quotidian sense of a realm of imagining the future, and re-imagining the borders of the real, life itself is dense with the possibility of both salvation and catastrophe. (Franklin 2000, 198)

One crucial feature of Franklin's perception of the "genetic imaginary" is that it involves the breaking down of established distinctions: "[I]n its blend of sober scientific prediction, speculative commercial ventures, virtual cinematic effects and popular narrative forms, *Jurassic Park* is a film that collapses distinctions between fact and fiction, life and art, science and entertainment" (Franklin 2000, 215). Franklin (2000, 216) highlights the "public witnessing" of the making and remaking of life as it is being "manufactured and marketed" in and through the new practices of the genomic biosciences, epitomized by the media attention given to the cloning of Dolly the sheep. She proposes that "in tracing the work of the genetic imaginary ... an essential critical dimension can be added to the analysis of global culture, global nature" (Franklin 2000, 224). Her argument and deployment of the concept of "genetic imaginaries" is "concerned not only with how we imagine genes, genetics or genealogy, but with a much wider set of orienting devices through which the world is both imagined and reproduced" (ibid., 222). Hence, Franklin regards "genetic imaginaries" as a critical tool that could generate awareness of shifts in key social/political categories and orientations realized in and through technoscientific change.

Jackie Stacey's *The Cinematic Life of the Gene* (2010) is also centrally concerned with the genetic imaginary. The book offers an exploration of "the changing relationship between biological and cultural forms at the current conjuncture of science, feminism, and the cinema" (Stacey 2010, x) and consists of a collection of readings of cultural theories and six films, released between 1995 and 2005, which revolve around genomics and cloning. Offering her own definition of the genetic imaginary, she relates it to the contemporary era:

I define the genetic imaginary as the mise-en-scène of those anxieties [concerning the reconfiguration of the boundaries of the human body, the transferability of its informational components, and the imitative potentialities of geneticized modes of embodiment], a fantasy landscape inhabited by artificial bodies that disturb the conventional teleologies of gender, reproduction, racialization, and heterosexual kinship. (Stacey 2010, 8)

Stacey both acknowledges the genealogical resources informing her use of the term and distinguishes her own from other deployments. She registers, in particular, Haraway's (1997) use of notions of the "gene fetish" and the "genetic unconscious" as in tune with her own psychoanalytic take on the genetic imaginary. Stacey is more explicit than any other researcher we have discussed about her psychoanalytically informed uses of imaginary.[8]

In summary, the STS researchers whose research we have reviewed in this section all register that images and imagery are highly significant in all instantiations of technoscience, which they regard not as extraneous or merely illustrative but as integral to technoscience. Feminist STS researchers have been particularly concerned with the visual forms and aspects of imaginaries, and some investigate links to subjectivity and subject formation. Feminist researchers have explored the fantasies, hopes, and fears engendered by recent developments in life sciences, particularly by genomics. To date, however, in comparison with other approaches discussed in this chapter, feminist work on imaginaries has garnered limited attention.

Conclusion

What is at stake in the investigations of imaginaries in STS? Why this turn in STS research? The first two sections of this chapter explored the rich hinterland of the concept of imaginaries. As such, they should help both those using it and those observing its deployment in STS to have a better sense of its framings and orientations. Genealogical mapping highlighted the plurality of trajectories and the diversity in the resourcing of this STS work. It also demonstrates that with the exception of some feminist research, STS has been much more open to the traditions of sociopolitical theory than to psychoanalysis and science fiction as resources for the investigation of scientific imaginaries. Our extended mapping is thus a reminder that there has been and continues to be other ways into imaginaries. As such, our exploration constitutes something of an invitation to STS scholars to sample the breadth of this field.

We identified clusters of research foregrounding their characteristic terrains of investigation (scientists, clinics, and scientific communities; national and institutional policies; popular culture and technoscientific imagery), their research methods (ethnographic and textual; comparative historical analysis, visual and cultural studies), and their registers (scientific communities and scientific practice; national, institutional and global identities; corporeality, subjects and differences). By no means rigid and fixed, these groupings could perhaps be thought of as fluid research assemblages.

The clusters foreground some differences in research methods. Jasanoff and Kim emphasize the importance of comparative methods (latterly, particularly historically based comparisons) (Jasanoff 2015a, 24–28; Jasanoff and Kim 2009) in STS research on imaginaries. However, Prasad (2014, 6) has warned that some comparative STS research reproduces rather than challenges, "the West versus non-West technocultural divide." As noted previously, he opts instead for a focus on "hierarchically entangled histories of technoscientific practices" (ibid., 6) and deconstructive methods to avoid such binaries. More generally, alternative ways of exploring the performativity of imaginaries may be required. There are signs of methodological experimentation and adaptation, including the use of focus groups and memory work in Felt et al.'s (2015) study of the evolution of a distinctive sociotechnical imaginary in Austria from the 1970s to the present. Given that imaginaries are far reaching, it may take something other than the conventional techniques of exposition and argument to conjure their features. Thus, it is not surprising that Haraway experiments with the form of her texts—dabbling with humor and shock, as well as playing with science fiction.

Research into imaginaries is often presented as a vehicle for reorienting STS. For example, Marcus (1995a, 3) regarded the notion of imaginaries as a tool for moving toward "a distinctly cultural study of science"—encouraging explorations of the tensions between scientific discourses and practices. Jasanoff and Kim (2009) called for STS to cast its investigative gaze beyond professional scientific actors and communities to analyze national cultures of technoscience, facilitated by their notion of *sociotechnical imaginaries*. More recently, Jasanoff (2015a, 5) has contended that such imaginaries are "not limited to nation-states but can [also] be articulated by other organized groups, such as corporations, social movements, and professional societies." The concept has also sometimes become the lynchpin for researchers' ambitions for STS. Hence, Fujimura (2003) advocated the use of "imaginaries" in forging "sociologies of the future." Invoking imaginaries, Kim Fortun and Mike Fortun (2005) had aspirations for a new "civic science" of toxicology and an STS "ethics and friendship with the sciences."

Recent research on imaginaries has also been part of a more general shift within the field. STS's earlier preoccupations with logic and epistemology have been supplemented or, indeed, replaced with a much broader agenda that includes research on aesthetics, values, and emotions. The sociology of expectations (Borup et al. 2006; Brown and Michael 2003; van Lente 1993) and concern with hope, promise, and hype (Brown 2003; Hedgecoe and Martin 2003, 2008; Michael 2000; Pollock and Williams 2010; Wyatt 2000) have opened STS to the study of social and psychological investments and future visions linked to specific technoscientific developments.[9] Whereas science and technology were formerly generally regarded as the domains of facts and artifacts, they are now also associated with storytelling, imaging, and imagining.

Beyond strategic, ethical, and methodological reorientations of the field, imaginaries matter in normative work on technoscience. While investigation of normative aspects of science and technology is not new, feminist and postcolonial scholarship has intensified this concern. In demonstrating how modern Western science has been implicated in gendered and postcolonial power relations, these movements have opened the field to studies of imaginaries (see esp. Prasad 2014). Imaginaries research sometimes seems to bring a new humanist inflection to STS—concerned as it is with human vision, values, aesthetics, and power. Indeed, Jasanoff characterizes imaginaries research as "a fundamentally humanistic inquiry" (Jasanoff 2015b, 322), as a reaction against "the flatness of networks" (Jasanoff 2015a, 15). More generally, engagement with imaginaries may also constitute a critical response to some exclusively materialist dispositions within STS, opening the field to psychosocial perspectives on science and technology and/or to investigations of the interplay between the imaginary and the material (Taylor 2004, 31–32).

The flourishing of the concept of imaginaries also registers a more specific theoretical shift. Until recently, discussions of values within technoscience were generally handled through notions of "interests" and/or "ideologies." These have proven to be limited theoretical tools for pursuing the normative dimensions of science as they operate primarily in a cognitive register—neglecting affective dimensions. Moreover, both concepts are linked to distortion, misrepresentation, and manipulation, whereas invoking the imaginary allows for consideration of the productive—of expectations, hopes, and dreams, as well as fears—which are increasingly attracting attention.

From this perspective, the circulation of the concept of imaginaries marks the relative decline in the deployment of the notion of ideology in STS research.[10] Taylor (2004, 183) has noted that, while the concept of social imaginary could designate elements traditionally associated with "distorted or false consciousness" that he associates with ideology, it may also entail "what we imagine can be something new, constructive, opening new possibilities." Haraway (2000, 77–78) also cautions that there is a need for precision in the use of the term *ideology*, distinguishing the "fantastic, the mythological and the ideological as three different registers of an imaginary relationship."

Here it may be appropriate to return to Verran's (1998), Waldby's (2000), and Squier's (2004) contentions that the denial of imaginaries has been a crucial feature of Western science and to assumptions about there being clear demarcations between fact and fiction or fantasy that may still linger around STS. Likewise, subjectivity is another domain that may be uncomfortable territory for STS, even if Steven Shapin (2012) has nominated it as *the* new challenge for the field. While there may be unease about bringing subjectivity and fantasy into STS research, the concept of imaginaries may provide an avenue onto that terrain. If this is to occur, there may need to be more awareness of and recourse to the diverse repertoires through which the concept has emerged.

Finally, we must return to note again that the range and diversity of STS scholarship that pivots on notions of imaginaries is impressive. Our investigation has generated a sense of the many flowers blooming in this rich field. Our concern has been to broaden awareness of this complex development and to encourage further experimentation in STS investigations of imaginaries.

Acknowledgments

We are grateful for the support provided by the ESRC Centre for Economic and Social Aspects of Genomics (Cesagen), a collaboration between Cardiff University and Lancaster University, UK (2002–2012), where this research project began. Joan Haran is also grateful for the support provided by the Hanse-Wissenschaftskolleg, Institute of Advanced Studies, Delmenhorst, Germany.

Notes

1. See Latour and Woolgar's ([1979] 1986) deletion of "Social" from the original title of *Laboratory Life*.

2. Jasanoff (2015a, 8) is critical of Anderson, Taylor, and Appadurai for not offering detailed investigations of science and technology in their accounts of modernity. Nevertheless, Anderson and Appadurai do acknowledge the importance of printing and communication technologies and Anderson considered technoscientific developments associated with mapping, census, and museum constructions in the revised edition of his book.

3. This position statement presented "five key ideas" associated with the concept: "social imaginaries were ways of understanding the social that become social entities themselves and mediate collective life"; "modernity has multiple forms that rely on forms of social imaginary based on

relations amongst strangers and … stranger sociability was made possible through mass mediation"; "the national people is a paradigmatic case of modern social imaginary"; "a national people lives amid many other social imaginaries"; "the agency of modern social imaginaries comes into being in a number of secular temporalities" (Gaonkar 2002, 5).

4. Nevertheless, there are other links between STS and science fiction that could be traced but that are beyond the remit of this article. For example, before establishing his reputation as a science fiction novelist, the British author Brian Stableford produced an STS Ph.D. thesis on the sociology of science fiction.

5. Hyysalo (2006) also uses ethnography within a symbolic-interactionist framework to analyze practice-based imaginaries in design.

6. Although not included in our study, Kay's (2000) history of the emergence of molecular biology as an informational technoscience uses, but does not directly discuss, the term *imaginaries* and is a counterexample to this assessment.

7. For other versions of the genetic imaginary following Franklin, see Gerlach (2004), Blaagard (2009), and Steinberg's (2015) related study of the bioimaginary.

8. Haraway distinguishes clearly what she calls "the philosophical/aesthetic" tradition from her favored psychoanalytic version of the concept. The former is particularly evident in Waldby's (1996, 2000) and Verran's (1998) studies. Nevertheless, Le Doeuff ([1980] 1989) and Waldby certainly draw on the psychoanalytical tradition in formulating their takes on imaginaries.

9. There are obvious connections between the sociology of expectations and the conceptualization of imaginaries in STS. However, the disciplinary specificity, the focus on particular technoscientific developments, and exclusively on orientations toward the future distinguish the former from the explorations of imaginaries considered in this chapter.

10. Nevertheless, some STS researchers (as noted above) do use the concept together with the notion of imaginaries.

References

Althusser, Louis. 1971. "Ideology and Ideological State Apparatuses." In *Lenin and Philosophy and Other Essays*, edited by Louis Althusser. London: Verso.

Anderson, Benedict. 1991. *Imagined Communities: Reflections on the Origin and Spread of Nationalism*. New York: Verso.

Appadurai, Ajun. 1996. *Modernity at Large: Cultural Dimensions of Globalization*. Minneapolis: University of Minnesota Press.

Blaagard, Bolette B. 2009. "The Flipside of My Passport. Myths of Origin and Genealogy of White Supremacy in the Mediated Genetic Imaginary." In *Complying with Colonialism: Gender, Race and Ethnicity in the Nordic Region*, edited by Suvi Keskinen, Salla Tuori, Sari Irni, and Diana Mulinari, 51–66. Farnham: Ashgate.

Borup, Mads, Nik Brown, Kornelia Konrad, and Harro van Lente. 2006. "The Sociology of Expectations in Science and Technology Analysis." *Technology Analysis and Strategic Management* 18 (3–4): 285–98.

Brown, Nik. 2003. "Hope against Hype: Accountability in Biopasts, Presents, and Futures." *Science Studies* 16 (2): 3–21.

Brown, Nik, and Mike Michael. 2003. "A Sociology of Expectations: Retrospecting Prospects and Prospecting Retrospects." *Technology Analysis and Strategic Management* 15 (1): 3–18.

Castoriadis, Cornelius. 1987. *The Imaginary Institution of Society*, translated by Kathleen Blamey. Cambridge, MA: MIT Press.

de Saille, Stevienna. 2015. "Dis-inviting the Unruly Public." *Science as Culture* 24 (1): 99–107.

Felt, Ulrike. 2015. "Keeping Technologies Out: Sociotechnical Imaginaries and the Formation of Austria's Technopolitical Identity." In *Dreamscapes of Modernity: Sociological Imaginaries and the Fabrications of Power*, edited by Sheila Jasanoff and Sang-Hyun Kim, 103–25. Chicago: University of Chicago Press.

Fortun, Kim, and Mike Fortun. 2005. "Scientific Imaginaries and Ethical Plateaus in Contemporary U.S. Toxicology." *American Anthropologist* 107 (1): 43–54.

Franklin, Sarah. 2000. "Life Itself: Global Nature and the Genetic Imaginary." In *Global Nature, Global Culture*, edited by Sarah Franklin, Celia Lury, and Jackie Stacey 188–227. London: Routledge.

Fujimura, Joan. 2003. "Future Imaginaries: Genome Scientists as Sociological Entrepreneurs." In *Genetic Nature/Culture: Anthropology and Science beyond the Two-Culture Divide*, edited by Alan H. Goodman, Deborah Heath, and M. Susan Lindee, 176–99. Berkeley: University of California Press.

Gaonkar, Dilip Parameshwar. 2002. "Toward New Imaginaries: An Introduction." *Public Culture* 14 (1): 1–19.

Gerlach, Neil. 2004. *The Genetic Imaginary: DNA in the Canadian Justice System*. Toronto: University of Toronto Press.

Good, Mary-Jo DelVecchio. 2010. "The Medical Imaginary and the Biotechnical Embrace: Subjective Experiences of Clinical Scientists and Patients." In *A Reader in Medical Anthropology: Theoretical Trajectories, Emergent Realities*, edited by Byron J. Good, Michael M. J. Fischer, Sarah S. Willen, and Mary-Jo DelVecchio Good 272–83. Oxford: Wiley-Blackwell.

Haraway, Donna J. 1989. *Primate Visions: Gender, Race and Nature in the World of Modern Science*. London: Routledge.

___. 1991. "A Cyborg Manifesto: Science, Technology, and Socialist-Feminism in the Late Twentieth Century." In *Simians, Cyborgs and Women: The Reinvention of Nature*, edited by Donna Haraway. London: Free Association Books.

___. 1997. *Modest_Witness@Second_Millennium.Femaleman©_Meets_Oncomouse™: Feminism and Technoscience*. London: Routledge.

Harvard University. 2016. "STS Research Platform: Sociotechnical Imaginaries." Accessed January 28, 2016, at http://harvard.edu/research/platforms/imaginaries.

Hedgecoe, Adam, and Paul Martin. 2003. "The Drugs Didn't Work: Expectations and the Shaping of Pharmacogenetics." *Social Studies of Science* 33 (3): 327–64.

___. 2008. "Genomics, STS, and the Making of Sociotechnical Futures." In *The Handbook of Science and Technology Studies*. 3rd ed., edited by Edward J. Hackett, Olga Amsterdamska, Michael Lynch, and Judy Wajcman, 817–40. Cambridge, MA: MIT Press.

Hess, David J. 2014. "Publics as Threats? Integrating Science and Technology Studies and Social Movement Studies." *Science as Culture* 24 (1): 69–82.

Hyysalo, Sampsa. 2006. "Representations of Use and Practice-Bound Imaginaries in Automating the Safety of the Elderly." *Social Studies of Science* 36 (4): 599–626.

Jasanoff, Sheila. 2015a. "Future Imperfect: Science, Technology and the Imaginations of Modernity." In *Dreamscapes of Modernity: Sociological Imaginaries and the Fabrications of Power*, edited by Sheila Jasanoff and Sang-Hyun Kim, 1–33. Chicago: University of Chicago Press.

___. 2015b. "Imagined and Invented Worlds." In *Dreamscapes of Modernity: Sociological Imaginaries and the Fabrications of Power*, edited by Sheila Jasanoff and Sang-Hyun Kim, 321–42. Chicago: University of Chicago Press.

Jasanoff, Sheila, and Sang-Hyun Kim. 2009. "Containing the Atom: Sociotechnical Imaginaries and Nuclear Power in the United States and South Korea." *Minerva* 47 (2): 119–46.

___. 2013. "Sociotechnical Imaginaries and National Energy Policies." *Science as Culture* 22 (2): 189–96.

___, eds. 2015. *Dreamscapes of Modernity: Sociological Imaginaries and the Fabrications of Power*. Chicago: University of Chicago Press.

Kant, Immanuel. [1787] 2007. *Critique of Pure Reason*, translated by Marcus Weigelt. London: Penguin Books.

Kay, Lily E. 2000. *Who Wrote the Book of Life? A History of the Genetic Code*. Stanford, CA: Stanford University Press.

Kim, Sang-Hyun. 2013. "The Politics of Human Embryonic Stem Cell Research in South Korea: Contesting National Sociotechnical Imaginaries." *Science as Culture* 23 (3): 293–319.

___. 2015. "Social Movements and Contested Sociotechnical Imaginaries in South Korea." In *Dreamscapes of Modernity: Sociological Imaginaries and the Fabrications of Power*, edited by Sheila Jasanoff and Sang-Hyun Kim. Chicago: University of Chicago Press.

Lacan, Jacques. 2006. *Écrits: The First Complete Edition in English*, translated by Bruce Fink. New York: W. W. Norton.

Latour, Bruno, and Steve Woolgar. [1979] 1986. *Laboratory Life: The Social Construction of Scientific Facts.* 2nd ed. Princeton, NJ: Princeton University Press.

Le Doeuff, Michele. [1980] 1989. *The Philosophical Imaginary,* translated by Colin Gordon. Stanford, CA: Stanford University Press.

Marcus, George. 1995a. "Introduction." In *Technoscientific Imaginaries: Conversations, Profiles and Memoirs,* edited by George Marcus. Chicago: University of Chicago Press.

___, ed. 1995b. *Technoscientific Imaginaries: Conversations, Profiles and Memoirs.* Chicago: University of Chicago Press.

Marris, Claire. 2015. "The Construction of Imaginaries of the Public as a Threat to Synthetic Biology." *Science as Culture* 24 (1): 83–98.

Michael, Mike. 2000. "Futures of the Present: From Performativity to Prehension." In *Contested Futures: A Sociology of Prospective Techno-Science,* edited by Nik Brown, Brian Rappert, and Andrew Webster, 21–42. Aldershot: Ashgate.

Mikami, Koichi. 2014. "State-Supported Science and Imaginary Lock-In: The Case of Regenerative Medicine in Japan." *Science as Culture* 24 (2): 183–204.

Miller, Clark A. 2015. "Globalizing Security: Science and the Transformation of Contemporary Political Imagination." In *Dreamscapes of Modernity: Sociological Imaginaries and the Fabrications of Power,* edited by Sheila Jasanoff and Sang-Hyun Kim. Chicago: University of Chicago Press.

Nerlich, Brigitte, and Carol Morris. "Imagining Imaginaries." 23 April 2015. Accessed January 28, 2016, at http://blogs.nottingham.ac.uk.

Pollock, Neil, and Robin Williams. 2010. "The Business of Expectations: How Promissory Organisations Shape Technology and Innovation." *Social Studies of Science* 40 (4): 525–48.

Prasad, Amit. 2014. *Imperial Technoscience: Transnational Histories of MRI in the United States, Britain, and India.* Cambridge, MA: MIT Press.

Said, Edward. 1977. *Orientalism.* Harmondsworth: Penguin Books.

Sartre, Jean-Paul. [1940] 2004. *The Imaginary: A Phenomenological Psychology of Imagination,* translated by Jonathan Webber. London: Routledge.

Shapin, Steven. 2012. "The Sciences of Subjectivity." *Social Studies of Science* 42 (2): 170–84.

Smith, Elta. 2009. "Imaginaries of Development: The Rockefeller Foundation and Rice Research." *Science as Culture* 18 (4): 461–82.

___. 2015. "Corporate Imaginaries of Biotechnology and Global Governance: Sygenta, Golden Rice and Corporate Social Responsibility." In *Dreamscapes of Modernity: Sociological Imaginaries and the Fabrications of Power,* edited by Sheila Jasanoff and Sang-Hyun Kim. Chicago: University of Chicago Press.

Squier, Susan. 2004. *Liminal Lives: Imaging the Human at the Frontiers of Biomedicine*. Durham, NC: Duke University Press.

Stacey, Jackie. 2010. *The Cinematic Life of Genes*. Durham, NC: Duke University Press.

Steinberg, Deborah Lynn. 2015. *Genes and the Bioimaginary: Science, Spectacle, Culture*. London: Routledge.

Stephens, Neil, Paul Atkinson, and Peter Glasner. 2013. "Institutional Imaginaries of Publics in Stem Cell Banking: The Cases of the UK and Spain." *Science as Culture* 22 (4): 497–515.

Taussig, Karen-Sue. 1997. "Calvinism and Chromosomes: Religion, the Geographical Imaginary, and Medical Genetics in the Netherlands." *Science as Culture* 6 (4): 495–524.

Taylor, Charles. 2002. "Modern Social Imaginaries." *Public Culture* 14 (1): 95–124.

___. 2004. *Modern Social Imaginaries*. Durham, NC: Duke University Press.

Traweek, Sharon. 1988. *Beamtimes and Lifetimes: The World of High Energy Physics*. Cambridge, MA: MIT Press.

van Dijck, José. 1998. *Imagenation: Popular Images of Genetics*. London: Macmillan.

van Lente, Harro. 1993. "Promising Technology: The Dynamics of Expectations in Technological Developments." Ph.D. dissertation, Twente University, Enschede, Netherlands.

Verran, Helen. 1998. "Re-imagining Land Ownership in Australia." *Postcolonial Studies: Culture, Politics, Economy* 1 (2): 237–54.

Waldby, Catherine. 1996. *Aids and the Body Politic: Biomedicine and Sexual Differences*. London: Routledge.

___. 2000. *The Visible Human Project: Informatic Bodies and Posthuman Medicine*. London: Routledge.

Warner, Michael. 2002. "Publics and Counterpublics." *Public Culture* 14 (1): 49–90.

Welsh, Ian, and Brian Wynne. 2013. "Science, Scientism and Imaginaries of Publics in the UK: Passive Objects, Incipient Threats." *Science as Culture* 22 (4): 540–66.

Wyatt, Sally. 2000. "Talking about the Future: Metaphors of the Internet." In *Contested Futures: A Sociology of Prospective Techno-Science*, edited by Nik Brown, Brian Rappert, and Andrew Webster, 109–28. Aldershot: Ashgate.

Yaszek, Lisa. 2008. *Galactic Suburbia: Recovering Women's Science Fiction*. Columbus: Ohio State University Press.

16 Performing and Governing the Future in Science and Technology

Kornelia Konrad, Harro van Lente, Christopher Groves, and Cynthia Selin

Introduction

The practices of science and technology are saturated with expectations, promises, and prospective claims. Scientists "have the future in their bones" as C. P. Snow (1964, 17) phrased it fifty years ago in his famous essay about the *Two Cultures*. The performative force of such claims and promises has been studied in various traditions, in science and technology studies (STS) and elsewhere. Rather than being reducible to merely descriptive statements of what may or may not happen in the future, expectations, visions, scenarios, and other forms of anticipation affect what may actually happen. They are performative. Promises and concerns around technologies in the making mobilize and legitimate the activities of scientists, innovators, policy makers, and regulators, along with NGOs and other societal actors. In broader societal discourses, expectations of scientific progress are enduring components of "grand challenges" narratives, buttressing arguments about how science and technology are supposed to address the wicked problems of the twenty-first century.

In this chapter we present a review of STS (and closely related) research on future-oriented representations and anticipatory practices. While STS as a field has looked at the future through a number of different lenses, in this chapter we hone in on concepts that circulate in inquiries related to the study of expectations, sociotechnical imaginaries, and the sociology of time, and relay how these concepts are operationalized in relation to the governance of emerging technologies.

In doing so, we will link and synthesize research addressing these themes at different degrees of resolution. We start with studies which are predominantly concerned with understanding the performativity and shaping of expectations in science and technology building on concrete empirical cases. Then we review how such dynamics are embedded in broader, historically contingent modes of future orientation that characterize cultures and societies. We next move to investigate cases of anticipatory practices where the future is intentionally utilized by STS scholars who seek to intervene in the governance of science and technology. Through this threefold exercise, we will bring together various strands of literature in and close to STS that have emerged over the last ten to fifteen years, in particular work related to the sociology of expectations, socio-technical imaginaries, the role of the future in social theories of time, and anticipatory governance. The future in STS, we conclude, is a fertile ground for theoretical, empirical, and practical work on performativity, temporality, and anticipation.

Expectations in Current Science and Technology

In this section, we consider work that traces the role of performative future-oriented statements, representations, and practices in concrete, empirical contexts. First, we review work that illuminates the performative effects of expectations. In a second step, we turn to literature which studies the dynamic patterns of how expectations evolve and change over time, as well as the different practices and modalities of how futures are constructed, promoted, and referred to. Before examining the role played by the future in current science and technology, we reflect on the particular connotations of terms such as *expectations*, *visions*, and *imaginaries*, which have been used for denoting future-oriented statements and representations in the STS literature.

As a key feature, the sociology of expectations stresses the collective character of expectations and focuses on statements that are more or less publicly available as part of a social repertoire of smaller or larger communities or an element of particular dis-courses, also expressed by the term of "collective expectations" (Konrad 2006). The interest is in expectations as social facts, in the basic sense that Durkheim introduced to grasp the dimensions of collective conscience that constrain and guide behavior (Durkheim 1988). This focus on the collective can be distinguished from a psychological understanding of expectations which instead highlights the specific expectations an individual may hold. This is not to deny that expectations and visions of influential individuals may be important for the development of particular technologies, but the main performative roles of expectations in mobilizing, guiding, and coordinating diverse sets of actors involved in technoscientific fields require expectations which are to some degree common, shared reference points. Expectations, then, can be defined as statements about future conditions or developments that imply assumptions about how likely these are supposed to be and that travel in a community or public space. This definition is at the same time broader and narrower than the frequently cited definition by Borup et al.'s (2006, 286) of "real-time representations of future techno-logical situations and capabilities" and emphasizes both the collective character and the implied likelihood. At the same time, it takes into account that expectations are typically heterogeneous in that they refer (both implicitly and explicitly) to technical, economic, or cultural developments and are not the exclusive domain of straight technical trajectories. Such collective expectations emerge as the result of strategic voicing and dedicated promotional efforts of actors, and as the aggregated effects of discursive dynamics or the outcome of collective anticipatory practices, such as foresight, road mapping, or other forms of joint, systematic exploration of future possibilities and developments.

Expectations may also circulate as indirect, materialized assessments of the potential of an emerging technology, as in when, for instance, a government invests in a technology. The very act of investment, together with the accompanying promotional material, can be unpacked as statements revealing a commitment to a particular future. In a similar vein, scholarship suggests that the future is not only evident in written or spoken claims. Representations can circulate as graphs, as in the case of Moore's law or the "hockey stick" of climate change; as iconic images, like the ubiquitous nanolouse that repairs body cells (Ruivenkamp and Rip 2011); or as embodied in artifacts, like the "car of the future" displayed in car shows and magazines (Bakker, van Lente, and Engels 2012) or as early prototypes imagined by scientists and engineers (Selin and Boradkar 2010).

Expectations are rarely presented as neutral, value-free statements, innocently referring to a range of possible developments, but instead can be read as promises or concerns and warnings, implying a positive or negative valuation. Promises refer to optimistic expectations, sketching the potential and assumed benefits which may follow from a technology, and usually require additional work, alliances, or investments to reach the implied outcomes (van Lente 1993). Concerns or warnings, in contrast, refer to expectations about possible problems and risks related to the development and application of a technology (te Kulve et al. 2013). Promissory rhetoric has received more attention in the sociology of expectations, but in recent years a number of authors have treated negative expectations and concerns (Nerlich and Halliday 2007; Tutton 2011) or the relation between both (Kitzinger and Williams 2005; te Kulve et al. 2013; McGrail 2010).

Another common genre of anticipatory concepts is visions. While expectations may be confined to particular technological developments or future states, the concept visions often refers to more or less coherent packages of potential future states (Berkhout 2006; Eames et al. 2006). That is, visions often relay a fuller portrait of an alternative world that includes revised social orders, governance structures, and societal values. Visions usually imply normative connotations, often being statements of desirable or preferable futures, while not necessarily including assessments of likelihood or plausibility.

In recent years, the concept of sociotechnical imaginaries as collectively imagined forms of social life and social order reflected in the design and fulfillment of scientific or technological projects has informed a growing body of work (Felt 2015; Jasanoff and Kim 2009, 2015; Levidow and Papaioannou 2013; Taylor-Alexander 2014). Sociotechnical imaginaries stress the politics and power of deeply rooted collective ideas shaping expectations and visions, often, though not exclusively, explored at a national level, and studies have largely focused on the ways in which broad future-oriented narratives legitimate certain pathways, while foreclosing others (for a more thorough discussion of this concept of imaginaries, see McNeil et al., chapter 15 this volume).

The Performativity of Expectations

A key concern of STS treatments of expectations, visions, and imaginaries has been to investigate the performativity of future representations. These studies show how future-oriented discourses, practices, and materialities shape the way society makes sense of science and technology, adjust how actors create strategies, and contribute to the shaping of technologies, as well as the development of entire technology fields.

Studies of prospective technoscientific discourses have largely focused on the critical examination of future representations related to new and emerging scientific and technological fields such as nano- or biotechnology (Chiles 2013; Kitzinger and Williams 2005; Selin 2007; te Kulve et al. 2013; Väliverronen 2004). These studies are based on the key insight that "futures occupy a contested terrain" carrying a wide range of different interests (Brown, Rappert, and Webster 2000, 6). Many studies analyze rhetoric and metaphors and show how, through discourses, meaning is constructed and interpretive social repertoires are formed, be they media, policy, or scientific. This shapes conceptions of what a technology or what a technoscientific field is supposed to be or to become, its public image, the images of the public which accompany it (Horst 2007), and what are considered to be plausible and legitimate promises and concerns relating to it (Konrad 2006).

While this research emphasizes discursive effects and implications thereof, others have traced how collective expectations mobilize, legitimate, and coordinate concrete activities involved in the real-world construction of science and technology (Borup et al. 2006; van Lente 1993). In this way, expectations are seen to contribute to what can be called the de facto governance of science and technology, that is, the patterns and structures of coordination which emerge from the interaction of many actors, intentionally or not (Konrad and Alvial Palavicino, forthcoming; Rip 2010). STS scholars have revealed how actors have to operate in a "sea of expectations," or a "prospective structure," which guides and constrains their room to maneuver (Brown, Rappert, and Webster 2000; van Lente 2012). Expectations can even turn into taken-for-granted assumptions of what is going to happen, thus falling into a deterministic lock-in, where future states become not just a promise but an inevitability (Konrad 2006). In many cases, expectations tend to appear in the imperative mode, where expectations as statements about what might occur are transformed into requirements for what should occur: the "promise-requirement cycle" (van Lente and Rip 1998). Expectations and visions serve to mobilize actors but also to legitimate and justify their actions and plans, for instance, when technical departments report to high-level decision makers in a firm or in policy making (Bakker and Budde 2012; Berti and Levidow 2014; Konrad et al. 2012).

Expectations and other prospective claims also have an important role in coordinating the often quite diverse sets of actors in research and innovation. The specific modes of how coordination takes place may take a range of forms, exhibiting different degrees of bindingness. There is, for example, the comparatively open mode of coordination of actor strategies and actor constellations at times when expectations are rather fluid, for example, in emerging technological fields, when actors reciprocally position themselves by way of discursively exchanging and mutually adapting expectations, which may result in the emergence of patterns and paths (van Merkerk and van Lente 2008). Expectations become more binding at later stages, when institutionalized, collective expectations coalesce, creating a pressure to respond to them. Such future-oriented representations may become further institutionalized and solidified if they are taken up in formal institutional or legal arrangements, for instance, when integrated as requirements and yardsticks into funding schemes (Konrad et al. 2012) or as highly organized road maps in industry which tend to serve as quasi contracts between actors. A well-known example for the latter is Moore's law, referring to the expected continuous increase of performance of computer chips, which turned from a retrospective observation to a broadly shared expectation and requirement (van Lente and Rip 1998) that disciplines a shared sense of the "shape" of the future (Watts 2008), leading to the well-organized industry-wide road-mapping process it is now (Le Masson et al. 2012; Schubert, Sydow, and Windeler 2013).

These understandings underlie an increasing body of empirical studies that investigate across a broad range of scientific and technological fields how expectations and visions of sociotechnical change shape concrete action and strategies, policies, and eventually research directions and the development of sociotechnical fields. It has for instance been shown how the creation of a rhetorical entity such as "membrane technology" or "nanotechnology" (McGrail 2010; van Lente and Rip 1998) or claims about a "breakthrough" help to define and create momentum for a new field of research (Nowotny and Felt 1997). Selin (2007) studied how speculative claims have created legitimacy for nanotechnology, while others followed the role of expectations for the evolvement of technology fields such as solar or fuel cells through field-configuring events and arenas (Hultman and Nordlund 2013; Nissilae, Lempiae, and Lovio 2014; Rip 2010; van Merkerk and Robinson 2006). Paul Martin and colleagues revealed how distinct visions of pharmacogenetics have been inscribed within research agendas, experimental technologies, and emerging industrial structures, thus guiding this emerging field in different directions (Hedgecoe and Martin 2003; Smart and Martin 2006). The concept of sociotechnical imaginaries has been informative in revealing how historically rooted visions shape policy practices, notably in the energy sector in relation to policy formation (Jasanoff and Kim 2009, 2013, 2015; Levidow and Papaioannou 2013), or how they shape up to lock in certain research trajectories (Mikami 2015).

For the main part, the focus in investigating performativity has been on representations of the future in various types of documents and speech, distilled through a range of methods from ethnography, interviews, and discourse analysis to visual semiotics. At the same time, expectations and imaginaries have been traced at work in material embodiments, such as images, film, technological objects, or strategic practices engaged in by science and technology actors and, increasingly, the performativity of such alternative forms has been investigated (Alvial Palavicino and Konrad 2015; Michael 2000; Selin 2015). Ethnographic investigations of future imaginaries demonstrate how futures are constructed through material orderings of space, thus supporting and reproducing particular expectations (Watts 2008). Recently, more quantitative studies have emerged (Bakker 2010; Budde, Alkemade, and Hekkert 2015).

The types of activities and actors probed by STS scholars have also varied. Scholars have focused on the concrete strategies and specific interactions of and between heterogeneous actors embedded in different societal settings, technology fields, and industrial sectors. In earlier studies, the emphasis was often on scientists and scientific promises. More recently, the scope of interest has been widened to actors and phenomena relating to innovation more broadly, to the strategies of industry actors (Budde, Alkemade, and Weber 2012; Garud, Schildt, and Lant 2014; Konrad et al. 2012; Parandian, Rip, and te Kulve 2012), to the role of intermediaries such as consultants or specialized news services shaping expectations (Morrison and Cornips 2012; Pollock and Williams 2010; Wüstenhagen et al. 2009), or to policy discourses and policy formation, and to regulation (Bakker, van Lente, and Meeus 2012; Berti and Levidow 2014;

Beynon-Jones and Brown 2011; Budde and Konrad 2015; Groves and Tutton 2013; Melton, Axsen, and Sperling 2016; Nerlich 2012). Contexts for research have also been geographical or jurisdictional, relating to practices and representations encountered in distinct national institutional settings, such as the United States or Japan (Fujimura 2003), South Korea (Jasanoff and Kim 2009; Kim 2013), Indonesia (Afiff 2014), the United Kingdom (Berti and Levidow 2014), or Austria (Felt 2015). It is quite obvious though that the bulk of research has so far concentrated on Europe and to some extent the United States, whereas a broader international perspective has only been emerging in recent years.

The Dynamics and Shaping of Expectations

A further prominent concern in the sociology of expectations (but also beyond) has been how expectations change and evolve over time, triggered by dynamics of rising optimism and attention, often followed by disappointment, which expectations exhibit in many fields. This dynamic temporal pattern within collective expectations has been known as a hype cycle or hype-disappointment cycle (Borup et al. 2006). The notion of hype is also widely used to denote promises which are strategically exaggerated in order to gain attention and interest (Brown 2003). From early on, studies in the field embarked on a critical analysis of hype and its effects (Bakker 2010; Brown 2003; Petersen 2009). Pollock and Williams (2010) have followed how the Gartner Group and other consultancies employ hype: they refer to these as "promissory organizations" because they operate by shaping and modifying expectations within procurement and innovation markets. Recently, the patterns of expectation dynamics have been scrutinized more closely and for different technology fields (Alkemade and Suurs 2012; Budde, Alkemade, and Hekkert 2015; Melton, Axsen, and Sperling 2016; van Lente, Spitters, and Peine 2013), as well as the conditions and processes which contribute to hype and its effects (Bakker 2010; Hedgecoe 2010; Konrad 2006; Konrad et al. 2012; Morrison and Cornips 2012). The ups and downs of expectations have furthermore been investigated as the result of contestation, competition, and also supportive relations between expectations, mainly in relation to recent developments in energy technologies such as fuel cells, hydrogen and electric cars (Alkemade and Suurs 2012; Bakker, van Lente, and Engels 2012; Budde and Konrad 2015).

Hypes and hype cycles are not primarily a concept of STS studies but have developed into a rather common idea—a "folk theory" (Rip 2006)—among those involved in research and innovation and many will easily point out examples of technologies which went through such hype and often also disappointment phases. This raises the question if and how research and innovation actors reflect on earlier, often nonrealized expectations, or how they retrospect prospects. Also of concern is how they may prospect retrospects, that is, which conclusions they may draw from former disappointments for current and upcoming promising technologies. The reasons for expectations to fail are often rationalized so as not to threaten current promises (Brown and Michael 2003); or the performative effect is valued highly enough to justify exaggerated claims (Oerlemans et al. 2014). Brown and Michael (2003) have furthermore pointed out that it depends on the position of actors which conclusions actors may draw from retrospecting prospects, in particular how close they are to actual knowledge production and thus aware of the involved uncertainties, just as for the general trust put into prospective claims.

In recent years, we have also seen more and more studies that investigate not only future representations and their dynamics and performative effects but also the strategies which underlie the contributions of various actors to expectation building—their expectation work (Bakker, van Lente, and Meeus 2011; Budde, Alkemade, and Hekkert 2015; Konrad et al. 2012; Nissilae, Lempiae, and Lovio 2014), and the different anticipatory practices which contribute to the shaping and spreading of expectations, that is different types of formal and informal activities and their material outcomes. These practices and their specific performative effects have been studied for instance for the work of consultants (Pollock and Williams 2010), online news providers (Morrison and Cornips 2012), venture capitalists (Wüstenhagen et al. 2009), online advertisement for health treatments (Petersen and Seear 2011), press releases (Budde, Alkemade, and Hekkert 2015), road mapping (McDowall 2012), the work of bioethicists (Hedgecoe 2010), forms of calculation, scenario building, performances, and simulations (Anderson 2010), as well as the way sets or assemblages of practices interact (Alvial Palavicino 2016). One of the questions tackled is not only how and to what extent particular practices contribute to the hyping of expectations and to promoting rather linear accounts of the future of a technology field but also to the concrete shaping of technologies, the formation of fields, or the support of transition processes.

Joly and colleagues have suggested a perspective on the performativity and dynamics of expectations, which goes beyond tracing particular manifestations of performativity, but rather points to different modalities of constructing and relating to collective futures varying with different types of technology fields, respectively, and types of innovation processes. The recurring pattern of supposed breakthrough technologies, which promise a vast potential of market prospects and solving societal problems and create a sense of urgency in the context of international competition, is conceptualized as a "regime of economics of technoscientific promises," characteristic in particular for innovation in new and emerging technologies. This is posed against a "regime of collective experimentation," which can be observed, for instance, for open source–based innovation which does not seem to relate in the same way to "breakthrough" promises (Felt et al. 2007; Joly 2010). In a similar vein, the concept of a "regime of hope" contrasting with a "regime of truth" has been introduced by a number of scholars researching biomedical technologies to capture the distinctive way that certain medical offers, treatments, and trials create value mainly based on potential future therapeutic and economic value rather than on present use and evidence (Martin, Brown, and Turner 2008; Moreira and Palladino 2005).

Taking a more fine-grained, empirical approach, Konrad and Alvial Palavicino (forthcoming) have traced how the modalities of constructing and coordinating expectations and their role in the governance of the emerging technology field of graphene differed among actor groups and how these modalities evolved over time.

Such a conception of different modalities of constructing and relating to futures within and among technology fields leads us further to the question of how future orientation may differ across and among societies beyond the institutions of technoscience. In the next section we turn to examine scholarship on this topic and the insights which it can contribute to the study of expectations. The insights outlined above about the diverse forms, practices, contexts, and performativity of future representations and their importance for the de facto governance of science and technology have also informed many studies that look for ways to mobilize these insights for *intentional* anticipatory governance—which we will discuss in the last section in this chapter on anticipation as forms of intervention.

Temporality and Modes of Future Orientation

The question of the performativity of expectations is also inseparable from broader questions regarding the constitutive social role of future orientation. While the sociology of expectations and closely related research has examined the role of anticipatory and performative practices and representations within science and technology with a focus on current and mostly quite concrete empirical manifestations, scholarship which explores the constitutive role of future orientation within social life extends further back into the sociological tradition, as well as beyond the discipline.

Like the STS work inventoried above, these broader streams of work explore how the future, as a dimension of the present, is constructed through practices as well as through discourse and thus contributes to the production and reproduction of social reality. This diverse body of work includes scholarship by a number of specialists in the sociology of time, along with conceptual historians and writers dealing with risk and uncertainty. One of its key contributions is to examine how and why societies, cultures, and historical periods manifest different ways of constructing and relating to the future. This has importance for STS scholars insofar as differences in modes of future orientation within and among societies may influence the dynamics of expectations, as sketched above, and underline the strategic and political significance of expectations.

Future Orientation as a Social and Historical Phenomenon

Theories concerning the constitutive function of future orientation are foreshadowed within classical social theory. Durkheim proposed that there exists a "social time" just as there existed other "social facts," a dimension of the collective conscience that constrained and guided what people do (Durkheim 1988). This differed from Kant's earlier understanding of time as a transcendental condition of experience, as it theorized the temporal organization of experience as socially determined. Later, George Herbert Mead and Alfred Schutz enriched our understanding of temporality as a basic category of social and personal life. Mead (1938), for instance, viewed orientation toward an open future, toward novelty, in human consciousness, thought, and action as pointing toward what is distinctive about human beings. But if consciousness and action hold open the future and resist determination by the past, they also orient themselves toward particular visions of what the future might turn out to be. Alfred Schutz (1962, 1964) emphasized that anticipating the future draws on a shared 'stock of knowledge' about the past, but also shows how novelty may emerge from this. These features of the future-orientedness of human action make experience possible in the present and enable historical consciousness of past influences on the present.

Within this tradition of thinking about the future as constitutive of social reality, the ways in which individuals view, imagine, and act upon their own futures are seen as shaped by "social facts," which, as scholars such as Giddens (1991) and Beck (1992) affirm, may change between societies and come to characterize historical periods. For example, Giddens (1991) argues that modernity is characterized by the emergence of concepts such as risk for organizing social practice. In this way, an explicit concern with the future enters social practice and shapes correlated forms of "modern" consciousness. Others make distinctions among the ways of constructing futures more cautiously, suggesting that it goes too far to state that different sociohistorical formations can be identified with a greater or lesser orientation to the future. Instead, a concern with an uncertain future can be seen as universal, though this future is posited or constructed intersubjectively in a range of ways (Marris 1991).

Surfacing and analyzing the differences among ways of constructing futures has also been pursued in terms of distinctions among social repertoires, or modes, which depend on the kinds of performative representations of the future discussed in the previous section. Such work complements scholarship on future imaginaries in technoscience discussed earlier (Fujimura 2003). Images of the future (Bell and Mau 1971) are, for example, identified as shared assumptions through which the future is made concrete in discourse and representation, enabling actors in the present to anticipate, act, and prepare. Others have explored how the future is constructed across society through repertoires of practices (Thrift 2008), as well as through artifacts (Michael 2000). The performativity of a promise, which might be thought of as binding how language shall be used and thus shaping practices (in a way similar to how a contract establishes rules which then constrain practices), might therefore be regarded as a special case of how future orientation is realized and futures posited or constructed through language, images, and practices.

The historical dimension of such repertoires, which sees them evolving in ways which cannot easily be constrained within clear periodizations (e.g., of modernity and premodernity), has been addressed by conceptual historians. For example, Lucian Hölscher (1999) explores how the future in Western antiquity, up to the medieval period, was constructed as the future or fate of individuals, shaped primarily by the future tense in grammar. He explores how the growth of mechanistic views of the universe helped to reshape how societies oriented themselves toward the future via conceptual frameworks and calculative practices (such as the mathematics of probability), orienting representation and practice toward the future as a totality of possible events shaped by the past. But rather than simply marking a clear distinction between medieval and modern, this evolution, he argued, made possible the "mechanization" [*Technisierung*] of the future. Mechanizing the future involves a broad repertoire of images and practices, including the construction of uncertainty as risk and probability through knowledge practices of quantification and ideas of progress as quantitative material betterment. In a related way, Reinhart Koselleck (2002) explores how the eighteenth century extended "horizons of expectation" beyond the present, reflecting a growing conviction that scientific expertise could lead to a degree of control over the future. Such reflections are complemented by studies which show how the future became an object of knowledge through this period, thanks first to Newtonian conceptualizations of the predictive vocation of scientific knowledge (Adam and Groves 2007) and then through the new sciences of probability (Hacking 1990).

Changing Practices of Future Orientation
A more in-depth cultural and historical approach taken by analysts of the representational and practical repertoires through which futures are constructed also extends

to the study of contemporary future making. If it is too easy to say that modernity is uniquely characterized by consciousness of the future, then it is nonetheless the case that forms of expertise directly concerned with projecting and forecasting have become increasingly central to governance in recent decades. Work examining future studies as a distinct discipline that emerged in the decades since World War II has shown how ways of constructing the future evolve and change over time in conjunction with the development and valorization of distinct forms of expertise. Andersson, in particular, has argued that such changes bring forth a politics of future knowledge, a struggle over how "certain predetermined traces" of futures could be known, and what futures as "objects of the human imagination, creativity and will" should be treated as desirable (Andersson 2012, 1413; see also Andersson and Keizer 2014).

In a similar way, Seefried (2013) has studied different conceptual and metaphorical repertoires or modes that are employed within future studies as a discipline and a practice. The first mode, forecasting, is based on forms of quantitative and modeling expertise stemming from the uptake of risk-cost–benefit analysis, game theory, and other approaches developed by U.S. government and military agencies before and during World War II—see also the studies of Edwards (1996) and Porter (1996) on the political force of numbers and calculation. The positivistic conception of expertise of this mode is often seen as exemplified by the RAND Corporation's approach to framing the future. Here, the future is framed as an object of management, in ways which reflect the broader preoccupations of modernization theory (Gilman 2007). Other forms of expertise were leveraged by futurists who, like Bertrand de Jouvenel, saw in the production of future scenarios not a predictive but a prognostic activity based on individual skill and experience, which could lead to preferable possibilities being identified and then used to organize action to bring them about (Jouvenel 1967). Finally, Seefried (2013) identifies an "emancipatory" form of anticipation, promoted by Robert Jungk (1952) and Johan Galtung (Jungk and Galtung 1969), which sought to open up the construction of possible and preferable futures to a wider range of social actors, including lay citizens (a development which links to the theme of anticipation as intervention discussed in the next section).

What united these modes and styles of anticipation was a common optimism that the future could be managed, or at least steered. Where they differed was on where legitimacy rested when it came to claiming knowledge about and the authority to realize potential futures. In the 1970s this optimism was replaced by pessimism, relating to increasing concern over the unintended ecological and social consequences of efforts to manage the future—famously diagnosed by Ulrich Beck (1992) as "the risk society." Since then, the repertoires and modes represented in future studies have become embodied in new forms of governance, or "styles of anticipation": prevention, precaution, and preemption, in which rather than material or social progress, security and safety become primary concerns (Anderson 2007). In these ways, the study of repertoires and modes of constructing futures points to the different ways in which expectations and promises may be institutionalized across history and within diverse societies, broadening insights discussed earlier into the different modalities that coexist in contemporary Western innovation regimes.

Theorizing Future Construction

The diverse approaches to the empirical study of repertoires or modes of futures construction discussed above are also complemented by more theoretical reflections on the relationship between these studies and broader sociological traditions. Developing a broadly Foucauldian approach to mapping social patterns in how futures are constructed, Adam and Groves (2007) draw on the sociology of time, futures studies, and continental philosophy to explore a variety of ways of "telling" and "taming" futures from distinct cultures and across different historical periods of European and North American history. They distinguish multiple geometries or "horizons" of the future: cycles (as in expectations about the return of agricultural seasons), circles (in which the future is symbolically constrained to repeat the past, as in collective rituals), and lines (the idea of the open future). They also explore further nuances within these categories, finding within modern linear horizons specimens of distinct future horizons, which they title "abstract," "open," and "empty," respectively. The first, abstract future horizon is shaped by mechanistic visions of the universe, in which the future is seen as naturalistically determined by the past, and which are represented in rhetoric around expectations that "naturalize" particular anticipated technologies or paths (as in Moore's law, for example). An open future horizon, by contrast, is a historicist and often utopian vision in which shared values motivate collective action in the name of a more or less concretely envisaged sociopolitical goal, as with patient groups who campaign for scientific research into overlooked conditions or into new technologies to make living with such conditions easier. Against an empty future horizon, however, the future appears as a totality of more or less quantifiable possibilities, from which the use of mathematical tools like modeling and cost/benefit analysis can select preferable outcomes and help plan their realization.

Each of these future horizons, Adam and Groves (2007) suggest, is conditioned and reproduced through specific ways of knowing, practices (along with the technologies which support them), and normative frameworks. Abstract and empty future horizons, in particular, are oriented toward exerting control over the future, rendering it predictable, at least for certain actors. A different mode constitutes what Adam and Groves call the "living future," which points toward the "futures-in-the-making," which are latent within the interactions between plans and efforts at intentional control on the one hand and sociotechnical or natural processes on the other, which are outside the forms of control projected through abstract or empty future horizons. It underlines the fact that the future may be socially constructed, but it is not entirely untenanted. Unintended, emergent effects are always on the way to being realized beyond the horizons of expertise and willed outcomes (Adam and Groves 2007).

The repertoires or modes ("future horizons" or Anderson's "styles of anticipation") through which social actors orient themselves toward and construct futures therefore constitute different ways of living the future in the present. In a similar way to Deleuze's and Guattari's notion of "assemblage" or Foucault's notion of *dispositif*, a mode of future orientation can be defined as a heterogeneous set of practices, patterns of desire, forms of knowledge and ethical imperatives which hang together with a degree of consistency in producing and reproducing social forms in the present, and which also help to shape the future itself in intended and unintended ways (Deleuze and Guattari 1980; Foucault 1980). Modes of future orientation may be heterogeneous with and even antagonistic to each other (cf. Rammstedt 1975), but may also often coexist. They condition expectations, images, and imaginaries, which in their turn, reinforce the modes of future orientation. Examples of this dual dynamic are images and narratives of inevitable technical progress (Lopez 2004); the extension of control over matter and life into the nanoscale and into the genome (Kearnes 2008); future worlds governed through big data and self-quantification (Whitson 2013); and financial speculation constructing radically uncertain futures, as a process central to neoliberalism (Dillon 2008).

If modes of future orientation help to produce the present, they also undoubtedly materially shape the future, and this raises questions relating to the politics not only of how particular future expectations and images become hegemonic but also of how we live the future in the present and how different ways of constructing the future have significant political implications in the present. Questions about how the future is collectively produced and governed therefore raise additional questions about directions and desirable forms of governance. Analytical work on modes of future orientation, then, also raises questions about the future orientation of STS scholarship itself, considered as a form of futures-related expertise, set of practices, and mode of anticipation. STS work which examines scholarship as a species of intervention in the governance of technoscience has brought such issues into the foreground.

Anticipation as Intervention

It follows from studies of expectations, visions, and imaginaries in technoscience (and also from the broader literature on how the future is socially constructed and performed in the present) that different modes of constructing the future can be resources for producing expectations in a more intentional mode, staging interventions in the evolution of science and technology. Acknowledging the performativity of expectations and other ways of constructing the future has led scholars to ask to what extent these can also be mobilized productively by STS researchers to contribute intentionally to the governance of science and technology. Further, normative questions about how to best manage, influence, and intervene in emerging developments within science and technology have been enduring, though not unproblematic, concerns for scholars in the field who recognize that technology and society are co-produced and could evolve in radically different directions.

Anticipatory Practices for Intervention

STS researchers have been working pragmatically in concert with scientists and engineers, as well as with policy makers and publics, to intervene in the governance of science and technology through a variety of bespoke anticipatory practices. Whether the focus falls upon technology assessment (TA), ethical, legal, and social implications (ELSI), environmental health and safety (EHS) research, anticipatory governance, or responsible research and innovation (RRI), future-oriented inquiry and engagement is considered a means to build more flexibility, resilience, and responsiveness to social priorities into institutions, laboratories, and communities. Each of these areas of practice takes up anticipation, temporality, and performativity using different methods, with different intentions, and with different theoretical departure points that are worth investigating further (see also Hilgartner, Prainsack, and Hurlbut, and Stilgoe and Guston, chapters 28 and 29 this volume).

To pursue the nature of this more interventionist work, we draw attention to the ways in which scholars of technoscience deploy the future not as an analytical object, tracing the role of expectations, or as a conceptual object, charting how the future is constitutive of social realities, but as a *resource* to modulate the directions and outcomes of science and technology. In covering this work, we keep the notion of governance broad, noting that work on governance covers a sprawling terrain dotted with work that delves into power and politics, law and regulation, and the different ways that science and technology relate to social ordering. Yet amid this spread, commonalities are apparent. Much STS interventionist work related to anticipation looks at the role of scientific authority, at ways to extend the democratization of science and technology, and at different mechanisms for public and stakeholder engagement. There are key debates around these interventionist efforts about the extent of steering possible, about the rationales for intervening, and about the appropriate timing for interventions, whether they be upstream, midstream, or downstream (Fisher, Mahajan, and Mitcham 2006; Parandian, Rip, and te Kulve 2012).

Recent work related to emerging nanotechnologies provides key examples of how STS scholars have designed and implemented a constellation of anticipatory practices as interventions within the governance of science and technology. At the Center for Nanotechnology in Society at Arizona State University (CNS-ASU), for example, such efforts are pursued in the form of scenario development within university laboratories or within multi-stakeholder communities, through design fiction studios, prototyping emerging technologies, and traditional workshops. This more interventionist work at CNS-ASU is informed by theoretical considerations about the historical importance of plausibility in constructing futures (Selin and Pereira 2013), and about the distinction between plausibility and probability in thinking about alternative futures (Ramírez and Selin 2014). Another approach, making use of scenarios informed by STS insights on sociotechnical dynamics (including the influence of expectations), and targeted mainly at innovation actors and stakeholders, is constructive technology assessment (te Kulve 2011; Parandian and Rip 2013).

While in both traditions scenarios are developed carefully and grounded in empirical analysis tempered by expert knowledge (often to circumvent facile speculation), a third approach explicitly avoids the use of prefabricated, expert-driven scenarios. Here, imagination and debate among citizens about how emerging technologies may (or should) develop in the future is pursued. Novel methods for engaging the public in future-oriented inquiry are burgeoning and vary to include card games (Felt et al. 2014), museum exhibitions (Selin 2015), science fiction (Burnam-Fink 2015), storytelling (Davies and Macnaghten 2010), and urban walking tours (Altamirano-Allende and Selin 2015). Working beyond a traditional workshop setting or consensus conference, these diverse modalities of public engagement resonate with the notion of "material deliberation" (Davies et al. 2012), connecting with the attention given to material embodiments of anticipation as reported earlier in this chapter.

Another example that bridges both traditional scenario development and novel approaches of engagement is the work of Frow and Calvert (2013), which engaged scientists working in synthetic biology. They designed a lively intervention during an international scientific conference that aimed to draw "attention to the often implicit framing conditions and assumptions that underlie discussions of the future" to "enable new questions to be asked of the future, neglected issues to be addressed and alternative pathways to be explored" (ibid., 33).

STS Approaches to Anticipatory Governance

In the various STS efforts under the umbrella of responsible research and innovation (RRI), one finds similar examples and urges to be critical and reflexive about the modes of future orientation. The general aim of RRI is to increase the societal value of science and technology, not only by striving to anticipate unforeseen consequences but also by ensuring that technoscience becomes more responsive to societal needs. The focus of RRI therefore falls on motivating positive developments in line with public needs (see Stilgoe and Guston, chapter 29 this volume). As Stilgoe, Owen, and Macnaghten (2013, 1570) characterize RRI, a turn to anticipation is derived from "critiques of the limitations of top-down risk-based models of governance to encapsulate the social, ethical and political stakes associated with technoscientific advances." Crucially, they juxtapose prediction with participation, gesturing to the political weight often granted to experts making projections, and note how the former tends to reify particular futures while the latter opens up futures and allows more diversity.

Such frameworks for anticipation as intervention explicitly attempt to create a difference in the world, engaging in future-oriented practices in order to unearth and challenge often taken-for-granted assumptions through which futures are produced. Here again, as we discussed earlier, the future is not treated as an object that can be observed, studied, or predicted. Instead, it is mobilized in the present as promises or expectations in policy documents or in legislation, as the object of public imagination, or as embodied in artifacts and systems primed with an intended use. Working intentionally with futures is, similarly, not about making prognoses of different possible future states but rather seeking to activate particular anticipatory practices here and now. It is key for anticipatory governance, as well as work under the banner of responsible research and innovation, to be reflexive about the way the future is represented and enacted, or, in the terms discussed above, to be reflexive about the mode of future orientation. The broader task of such modes of governance is to increase reflexivity throughout society, encouraging in the process critical thought about alternative trajectories of change without falling prey to the impossible task of predicting either the direction of technological advance or the complex interactions between technology and society.

While work on anticipation in STS recognizes the difficulty and even impossibility of prediction, this is not to deny the de facto role and import of prediction in society as one of the dominant modes of managing and taming the future—the rise of big data and IT-driven modeling underpinned by a deterministic approach to technological change and a framing of technology as extrapolatable trajectories are the norm—against which reflexive approaches to anticipation and governance are small and modest interventions. Yet in their approach of anticipation, many STS scholars increasingly focus on what is not captured in linear models of innovation: surprise, radical contingency, or simply the messy interplay between humans and their technological culture.

While the future is used as a resource for critical thinking, the purpose of such interventions is firmly rooted in the present. Manufacturing, negotiating, and debating alternative futures should therefore be seen as related primarily to current concerns. We can therefore examine visions, expectations, promises, and scenarios for what they can reveal about current agendas, motivations, and existing social orders. For instance, when Shell publicly releases their annual scenarios, they should not be assessed for accuracy but instead used to decode the company's worldview, thus providing an opportunity to better understand the desires and anxieties that inform its commercial endeavors and research directions. STS scholars, however, can also intentionally intervene in the production of such visions, calling attention to trade-offs, winners and losers, or unforeseen alternatives.

A common thread that underpins future-oriented STS approaches to governance is a concern with greater participation—supporting more diverse imaginaries and experimenting with methods that explore alternative futures in a participatory fashion. Participatory practices involve a sort of "extended peer review," as proposed by proponents of postnormal science (Funtowicz and Ravetz 1993) or Mode 2 science (Gibbons et al. 1994) that seek to broaden the knowledge base that underpins policy, often through deliberative processes that work to ensure more democratic governance mechanisms. What draws many of these future-oriented deliberative activities together is the effort to productively mobilize diverse epistemologies, professional experiences, and personal reflections into the decision-making or policy-making process. Clearly, what STS approaches to anticipation as intervention share is, in the terminology Seefried (2013) finds in the work of Galtung and Junck, an "emancipatory" form of anticipation.

An emancipatory form of anticipation requires practices that unearth and interrogate metanarratives which buttress dominant visions of technological progress. A number of experiments in public engagement with emerging technologies indicate that while seemingly esoteric technologies like nanotechnology or synthetic biology may not yet be part of a widely shared public consciousness, the issues raised connect to deeply rooted, archetypal cultural narratives, which act as resources for lay public responses to novel, emerging technologies (Davies and Macnaghten 2010). Similarly, the importance of national and cultural repertoires linked to experiences with technologies in the past has been highlighted (Burri 2009; Felt, Schumann, and Schwarz 2015).

The approaches we have highlighted here therefore call attention to the ways in which anticipatory practices are fundamentally about reflexivity and collective learning. Within STS scholarship, reflexivity has been embraced as "the process of identifying, and critically examining (and thus rendering open to change), the basic, pre-analytic assumptions that frame knowledge-commitments" (Wynne 1993, 324). Put another way, a focus on the constitutive role of futurity within the present opens up the possibility of an alternative perspective on present conditions. Following Mazé (2014), "'the future' is not a destination that might be defined and reached with the right methods, but a 'supervalence' (Grosz 1999), an outside to an experienced present." As such, futurity represents a resource through which critical distance may be attained.

Conclusion

We would like to close our *tour d'horizon* of three strands of scholarship engaged with the role of futures in science, technology, and society by highlighting a number of implications of the insights we have assembled above for STS. We saw that interventionist approaches to anticipatory governance are already reflexive toward the sources of knowledge which help to create spaces in which futures are deliberated, and in this sense reflect important insights from critical analysis of expectations and other future-oriented practices and discourses. There are further avenues along which insights on the constitutive role of futurity within present practice may be developed in order to inform and tailor interventionist STS approaches to the governance of technoscience. Considering the dynamics of expectations in a technology field, such as phases of hype or uncritical acceptance of promises, or, in a similar vein, determining the degree of fluidity or entrenchment of expectations, could be helpful in preparing interventions and specifying their goals, including particular desired impacts of governance. While in some cases opening up entrenchment and supposed inevitable outcomes of innovation may deserve particular attention, supporting forms of coordination among actors may be more appropriate in other cases. This may also mitigate, even if not completely avoid, the possible unintended consequence that some anticipatory practices may actually contribute to reinforcing the very expectations they aim to critically investigate (Hedgecoe 2010; Nordmann 2007). On the other hand, acknowledging the forcefulness of common anticipatory practices can also help in considering how practices and spaces for deliberation may be best connected to ongoing debates in order to make a difference most effectively.

We also have seen that the studies of dynamics of expectations (such as promise-requirement and hype-disappointment cycles) tend to focus on specific contemporary developments and patterns, often not taking explicit account of how these are embedded in broader, long-term cultural patterns. Such studies are only starting to pay attention to long-established practices and modalities of constructing futures, or national, cultural, or other socially distinct forms of anticipation, concerns which are much better represented in sociological and historical studies of the constitutive role of futurity. Drawing on insights from these traditions can surely be informative for charting relationships among contemporary practices. They may furthermore highlight how the promises attached to particular technologies and specific technoscientific fields are embedded in the broader cultural promises associated with technology and its contribution to solving social problems, issues to which sociologists like Giddens and Beck have drawn our attention with their emphasis on the central constitutive role played by technology in modernity and its effect on how societies orient themselves to the future. Historically, the rise of the modern notion of progress is paralleled by the cultural prominence of a generalized belief in technology as the key to solving social problems. A question, thus, is how particular promises draw from the generic cultural promise of science and technology and, by doing so, reinforce or change this central feature of modernity and its characteristic future horizons.

The various scholarly approaches within STS of the future reviewed here all point to the future as constitutive of the present, be it in an analytical or engaged, normative way. They reveal a generalized concern in STS with the future that focuses on how to steer and shape technoscience in positive directions before undesired path dependencies and lock-in occur. Through this concern and the various scholarly resources for articulating it that we have reviewed, STS scholars have an important contribution to make to futures thinking in light of their critical attention to the ethical, political, cultural, and social dimensions of science and technology and a more nuanced approach to temporality (Felt 2015; Selin and Sadowski 2015). Scrutinizing, designing, and adapting future-oriented methods presents an opportunity for STS scholars to critically appraise and challenge existing assumptions and visions, infusing them with the critical understandings of innovation that animate STS scholarship.

Acknowledgments

Dr. Selin's work on this chapter is supported by a Marie Curie International Incoming Fellowship within the 7th European Community Framework Programme. In addition, Dr. Selin's contribution to this chapter is in part based upon work supported by the U.S.

National Science Foundation cooperative agreement #0937591 and grant #1257246. Any findings, conclusions, or opinions are those of the author and do not necessarily represent NSF.

References

Adam, Barbara, and Christopher Groves. 2007. *Future Matters: Action, Knowledge, Ethics*. Leiden: Brill.

Afiff, Suraya. 2014. "Engineering the Jatropha Hype in Indonesia." *Sustainability* 6 (4): 1686–704.

Alkemade, Floortje, and Roald Suurs. 2012. "Patterns of Expectations for Emerging Sustainable Technologies." *Technological Forecasting and Social Change* 79 (3): 448–56.

Altamirano-Allende, Carlo, and Cynthia Selin. 2015. "Seeing the City: Photography as a Place of Work." *Journal of Environmental Studies and Sciences*, 1–10.

Alvial Palavicino, Carla. 2016. *Mindful Anticipation: A Practice Approach to the Study of Emergent Technologies*. Ph.D. thesis, University of Twente.

Alvial Palavicino, Carla, and Kornelia Konrad. 2015. "Doing Is Believing: How Material Practices Shape the Future in 3D Printing." In *Practices of Innovation, Governance and Action—Insights from Methods, Governance and Action*, edited by Diana Bowman, Anne Dijkstra, Camilo Fautz, Julia Guivant, Kornelia Konrad, Harro van Lente, and Silvia Woll, 141–63. Berlin: AKA / IOS Press.

Anderson, Ben. 2007. "Hope for Nanotechnology: Anticipatory Knowledge and the Governance of Affect." *Area* 39 (2): 156–65.

___. 2010. "Preemption, Precaution, Preparedness: Anticipatory Action and Future Geographies." *Progress in Human Geography* 34 (6): 777–98.

Andersson, Jenny. 2012. "The Great Future Debate and the Struggle for the World." *American Historical Review* 117 (5): 1411–30.

Andersson, Jenny, and Anne-Greet Keizer. 2014. "Governing the Future: Science, Policy and Public Participation in the Construction of the Long Term in the Netherlands and Sweden." *History and Technology: An International Journal* 30 (2): 104–22.

Bakker, Sjoerd. 2010. "The Car Industry and the Blow-out of the Hydrogen Hype." *Energy Policy* 38 (11): 6540–44.

Bakker, Sjoerd, and Björn Budde. 2012. "Technological Hype and Disappointment: Lessons from the Hydrogen and Fuel Cell Case." *Technology Analysis and Strategic Management* 24 (6): 549–63.

Bakker, Sjoerd, Harro van Lente, and Remko Engels. 2012. "Competition in a Technological Niche: The Cars of the Future." *Technology Analysis & Strategic Management* 24 (5): 421–34.

Bakker, Sjoerd, Harro van Lente, and Marius Meeus. 2011. "Arenas of Expectations for Hydrogen Technologies." *Technological Forecasting and Social Change* 78 (1): 152–62.

___. 2012. "Credible Expectations—The US Department of Energy's Hydrogen Program as Enactor and Selector of Hydrogen Technologies." *Technological Forecasting and Social Change* 79 (6): 1059–71.

Beck, Ulrich. 1992. *Risk Society: Towards a New Modernity*. London: Sage.

Bell, Wendell, and James A. Mau. 1971. "Images of the Future: Theory and Research Strategies." In *The Sociology of the Future*, edited by Wendell Bell and James A. Mau, 6–44. New York: Russell Sage Foundation.

Berkhout, Frans. 2006. "Normative Expectations in Systems Innovation." *Technology Analysis and Strategic Management* 18 (3–4): 299–311.

Berti, Pietro, and Les Levidow. 2014. "Fuelling Expectations: A Policy-Promise Lock-in of UK Biofuel Policy." *Energy Policy* 66: 135–43.

Beynon-Jones, Sian M., and Nik Brown. 2011. "Time, Timing and Narrative at the Interface between UK Techno-science and Policy." *Science and Public Policy* 38 (8): 639–48.

Borup, Mads, Nik Brown, Kornelia Konrad, and Harro van Lente. 2006. "The Sociology of Expectations in Science and Technology." *Technology Analysis and Strategic Management* 18 (3–4): 285–98.

Brown, Nik. 2003. "Hope against Hype—Accountability in Biopasts, Presents and Futures." *Science Studies* 16 (2): 3–21.

Brown, Nik, and Mike Michael. 2003. "A Sociology of Expectations: Retrospecting Prospects and Prospecting Retrospects." *Technology Analysis & Strategic Management* 15 (1): 4–18.

Brown, Nik, Brian Rappert, and Andrew Webster. 2000. "Introducing Contested Futures: From Looking into the Future to Looking at the Future." In *Contested Futures: A Sociology of Prospective Techno-science,* edited by Nik Brown, Brian Rappert, and Andrew Webster, 3–20. Aldershot: Ashgate.

Budde, Björn, Floortje Alkemade, and Markko Hekkert. 2015. "On the Relation between Communication and Innovation Activities: A Comparison of Hybrid Electric and Fuel Cell Vehicles." *Environmental Innovation and Societal Transitions* 14: 45–59.

Budde, Björn, Floortje Alkemade, and K. Matthias Weber. 2012. "Expectations as a Key to Understanding Actor Strategies in the Field of Fuel Cell and Hydrogen Vehicles." *Technological Forecasting and Social Change* 79 (6): 1072–83.

Budde, Björn, and Kornelia Konrad. 2015. "Governing Fuel Cell Innovation in a Dynamic Network of Expectations." In *Hopes, Hypes and Disappointments: On the Rise of Expectations for Sustainability Transitions: A Case Study on Hydrogen and Fuel Cell Technology for Transport*, edited by Björn Budde, 143–71. Ph.D. dissertation, Twente University, Enschede, Netherlands.

Burnam-Fink, Michael. 2015. "Creating Narrative Scenarios: Science Fiction Prototyping at Emerge." *Futures* 70: 48–55.

Burri, Regula. 2009. "Coping with Uncertainty: Assessing Nanotechnologies in a Citizen Panel in Switzerland." *Public Understanding of Science* 18 (5): 498–511.

Chiles, Robert M. 2013. "If They Come, We Will Build It: In Vitro Meat and the Discursive Struggle over Future Agrofood Expectations." *Agriculture and Human Values* 30 (4): 511–23.

Davies, Sarah R., and Phil Macnaghten. 2010. "Narratives of Mastery and Resistance: Lay Ethics of Nanotechnology." *NanoEthics* 4 (2): 141–51.

Davies, Sarah R., Cynthia Selin, Gretchen Gano, and Angela Pereira. 2012. "Citizen Engagement and Urban Change: Three Case Studies of Material Deliberation." *Cities* 29 (6): 351–57.

Deleuze, Gilles, and Félix Guattari. 1980. *A Thousand Plateaus*. London: Athlone.

Dillon, Michael. 2008. "Underwriting Security." *Security Dialogue* 39 (2–3): 309–32.

Durkheim, Emile. [1895] 1988. *Les Règles de la Méthode Sociologique*. Paris: Flammarion.

Eames, Malcolm, William McDowall, Mike Hodson, and Simon Marvin. 2006. "Negotiating Contested Visions and Place-Specific Expectations of the Hydrogen Economy." *Technology Analysis and Strategic Management* 18 (3–4): 361–74.

Edwards, Paul N. 1996. *The Closed World: Computers and the Politics of Discourse in Cold War America*. Cambridge, MA: MIT Press.

Felt, Ulrike. 2015. "Keeping Technologies Out: Sociotechnical Imaginaries and the Formation of a National Technopolitical Identity." In *Dreamscapes of Modernity: Sociotechnical Imaginaries and the Fabrication of Power*, edited by Sheila Jasanoff and Sang-Hyun Kim, 103–25. Chicago: University of Chicago Press.

Felt, Ulrike, Simone Schumann, and Claudia Schwarz. 2015. "(Re)assembling Natures, Cultures and (Nano)technologies in Public Engagement." *Science as Culture* 24 (4): 458–83.

Felt, Ulrike, Simone Schumann, Claudia Schwarz, and Michael Strassnig. 2014. "Technology of Imagination: A Card-Based Public Engagement Method for Debating Emerging Technologies." *Qualitative Research* 14 (2): 233–51.

Felt, Ulrike, Brian Wynne, Michel Callon, Maria Eduarda Gonçalves, Sheila Jasanoff, Maria Jepsen, Pierre-Benoît Joly, Zdenek Konopasek, Stefan May, Claudia Neubauer, Arie Rip, Karen Siune, Andy Stirling, and Mariachiara Tallacchini. 2007. "Taking European Knowledge Society Seriously." Report of the Expert Group on Science and Governance to the Science, Economy and Society Directorate, Directorate-General for Research. Brussels: European Commission.

Fisher, Erik, Roop L. Mahajan, and Carl Mitcham. 2006. "Midstream Modulation of Technology: Governance from Within." *Bulletin of Science, Technology & Society* 26 (6): 485–96.

Foucault, Michel. 1980. "The Confession of the Flesh." In *Power/Knowledge: Selected Interviews and Other Writings*, edited by Colin Gordon, 194–228. New York: Pantheon.

Frow, Emma, and Jane Calvert. 2013. "Opening Up the Futures of Synthetic Biology." *Futures* 48: 32–43.

Fujimura, Joan. 2003. "Future Imaginaries: Genome Scientists as Cultural Entrepreneurs." In *Genetic Nature/Culture*, edited by Alan H. Goodman, Deborah Heath, and Susan M. Lindee, 176–99. Los Angeles: University of California Press.

Funtowicz, Silvio, and Jerome R. Ravetz. 1993. "Science for the Post-normal Age." *Futures* 25 (7): 739–59.

Garud, Raghu, Henri A. Schildt, and Theresa K. Lant. 2014. "Entrepreneurial Storytelling, Future Expectations, and the Paradox of Legitimacy." *Organization Science* 25 (5): 1479–92.

Gibbons, Michael, Camille Limoges, Helga Nowotny, Simon Schwartzman, Peter Scott, and Martin Trow. 1994. *The New Production of Knowledge: The Dynamics of Science and Research in Contemporary Societies*. London: Sage.

Giddens, Anthony. 1991. *Modernity and Self-Identity: Self and Society in the Late Modern Age*. Stanford, CA: Stanford University Press.

Gilman, Nils. 2007. *Mandarins of the Future: Modernization Theory in Cold War America*. Baltimore: Johns Hopkins University Press.

Grosz, Elizabeth, ed. 1999. *Becomings: Explorations in Time, Memory, and Futures*. Ithaca, NY: Cornell University Press.

Groves, Christopher, and Richard Tutton. 2013. "Walking the Tightrope: Expectations and Standards in Personal Genomics." *BioSocieties* 8 (2): 181–204.

Hacking, Ian. 1990. *The Taming of Chance*. Cambridge: Cambridge University Press.

Hedgecoe, Adam. 2010. "Bioethics and the Reinforcement of Socio-technical Expectations." *Social Studies of Science* 40 (2): 163–86.

Hedgecoe, Adam, and Paul Martin. 2003. "The Drugs Don't Work: Expectations and the Shaping of Pharmacogenetics." *Social Studies of Science* 33 (3): 327–64.

Hölscher, Lucian. 1999. *Die Entdeckung der Zukunft*. Frankfurt: Fischer.

Horst, Maja. 2007. "Public Expectations of Gene Therapy Scientific Futures and Their Performative Effects on Scientific Citizenship." *Science, Technology, & Human Values* 32 (2): 150–71.

Hultman, Martin, and Christer Nordlund. 2013. "Energizing Technology: Expectations of Fuel Cells and the Hydrogen Economy, 1990–2005." *History and Technology* 29 (1): 33–53.

Jasanoff, Sheila, and Sang-Hyun Kim. 2009. "Containing the Atom: Sociotechnical Imaginaries and Nuclear Power in the United States and South Korea." *Minerva* 47 (2): 119–46.

___. 2013. "Sociotechnical Imaginaries and National Energy Policies." *Science as Culture* 22 (2): 189–96.

Jasanoff, Sheila, and Sang-Hyun Kim, eds. 2015. *Dreamscapes of Modernity: Sociotechnical Imaginaries and the Fabrication of Power*. Chicago: University of Chicago Press.

Joly, Pierre-Benoît. 2010. "On the Economics of Techno-scientific Promises." In *Débordements: Mélanges offerts à Michel Callon*, edited by Madeleine Akrich, Yannick Barthe, Fabian Muniesa, and Philippe Mustar, 203–22. Paris: Presses des Mines.

Jouvenel, Bertrand. 1967. *The Art of Conjecture*. New York: Basic Books.

Jungk, Robert. 1952. *Die Zukunft hat schon begonnen. Amerikas Allmacht und Ohnmacht*. Stuttgart: Scherz & Goverts.

Jungk, Robert, and Johann Galtung, eds. 1969. *Mankind 2000*. London: Allen & Unwin.

Kearnes, Matthew. 2008. "Informationalising Matter: Systems Understandings of the Nanoscale." *Spontaneous Generations* 2 (1): 99–111.

Kim, Sang-Hyun. 2013. "The Politics of Human Embryonic Stem Cell Research in South Korea: Contesting National Sociotechnical Imaginaries." *Science as Culture* 23 (3): 293–319.

Kitzinger, Jenny, and Clare Williams. 2005. "Forecasting Science Futures: Legitimising Hope and Calming Fears in the Embryo Stem Cell Debate." *Social Science and Medicine* 61 (3): 731–40.

Konrad, Kornelia. 2006. "The Social Dynamics of Expectations: The Interaction of Collective and Actor-Specific Expectations on Electronic Commerce and Interactive Television." *Technology Analysis & Strategic Management* 18 (3–4): 429–44.

Konrad, Kornelia, and Carla Alvial Palavicino. Forthcoming. "Evolving Patterns of Governance of and by Expectations—the Graphene Hype Wave." In *Embedding and Governing New Technologies: A Regulatory, Ethical and Societal Perspective*, edited by Diana Bowman, Elen Stokes, and Arie Rip. Singapore: Pan Stanford.

Konrad, Kornelia, Jochen Markard, Annette Ruef, and Bernhard Truffer. 2012. "Strategic Responses to Fuel Cell Hype and Disappointment." *Technological Forecasting and Social Change* 79 (6): 1084–98.

Koselleck, Reinhart. 2002. *The Practice of Conceptual History: Timing History, Spacing Concepts*. Stanford, CA: Stanford University Press.

te Kulve, Haico. 2011. *Anticipatory Interventions and the Co-evolution of Nanotechnology and Society*. Ph.D. dissertation, University of Twente.

te Kulve, Haico, Kornelia Konrad, Carla Alvial Palavicino, and Bart Walhout. 2013. "Context Matters: Promises and Concerns Regarding Nanotechnologies for Water and Food Applications." *NanoEthics* 7 (1): 17–27.

Le Masson, Pascal, Benoit Weil, Armand Hatchuel, and Patrick Cogez. 2012. "Why Are They Not Locked in Waiting Games? Unlocking Rules and the Ecology of Concepts in the Semiconductor Industry." *Technology Analysis and Strategic Management* 24 (6): 617–30.

Levidow, Les, and Theo Papaioannou. 2013. "State Imaginaries of the Public Good: Shaping UK Innovation Priorities for Bioenergy." *Environmental Science and Policy* 30 (1): 36–49.

Lopez, José. 2004. "Bridging the Gaps: Science Fiction in Nanotechnology." *Hylé* 10 (2): 129–52.

Marris, Peter. 1991. "The Social Construction of Uncertainty." In *Attachment across the Life Cycle*, edited by Colin Murray Parkes, Joan Stevenson-Hinde, and Peter Marris, 77–90. London: Routledge.

Martin, Paul, Nik Brown, and Andrew Turner. 2008. "Capitalizing Hope: The Commercial Development of Umbilical Cord Blood Stem Cell Banking." *New Genetics and Society* 27 (2): 127–43.

Mazé, Ramia. 2014. "The Future Is Not Empty: Design Imaginaries and Design Determinisms." In *Proceedings of the Oxford Futures Forum*. https://www.sbs.ox.ac.uk/school/events-0/oxford-futures-forum-2014.

McDowall, Will. 2012. "Technology Roadmaps for Transition Management: The Case of Hydrogen Energy." *Technological Forecasting and Social Change* 79 (3): 530–42.

McGrail, Stephen. 2010. "Nano Dreams and Nightmares: Emerging Technoscience and the Framing and (Re)interpreting of the Future, Present and Past." *Journal of Futures Studies* 14 (4): 23–48.

Mead, George Herbert. 1938. *The Philosophy of the Act*. Chicago: University of Chicago Press.

Melton, Noel, Jonn Axsen, and Daniel Sperling. 2016. "Moving beyond Alternative Fuel Hype to Decarbonize Transportation." *Nature Energy* 1 (3): 16013.

Michael, Mike. 2000. "Futures of the Present—from Performativity to Prehension." In *Contested Futures: A Sociology of Prospective Techno-science*, edited by Nik Brown, Brian Rappert, and Andrew Webster, 21–42. London: Ashgate.

Mikami, Koichi. 2015. "State-Supported Science and Imaginary Lock-in: The Case of Regenerative Medicine in Japan." *Science as Culture* 24 (2): 183–204.

Moreira, Tiago, and Paolo Palladino. 2005. "Between Truth and Hope: On Parkinson's Disease, Neurotransplantation and the Production of the 'Self'." *History of the Human Sciences* 18 (3): 55–82.

Morrison, Michael, and Lucas Cornips. 2012. "Exploring the Role of Dedicated Online Biotechnology News Providers in the Innovation Economy." *Science, Technology, & Human Values* 37 (3): 262–85.

Nerlich, Brigitte. 2012. "'Low Carbon' Metals, Markets and Metaphors: The Creation of Economic Expectations about Climate Change Mitigation." *Climatic Change* 110 (1–2): 31–51.

Nerlich, Brigitte, and Christopher Halliday. 2007. "Avian Flu: The Creation of Expectations in the Interplay between Science and the Media." *Sociology of Health & Illness* 29 (1): 46–65.

Nissilae, Heli, Tea Lempiaelae, and Reimo Lovio. 2014. "Constructing Expectations for Solar Technology over Multiple Field-Configuring Events: A Narrative Perspective." *Science and Technology Studies* 27 (1): 54–75.

Nordmann, Alfred. 2007. "If and Then: A Critique of Speculative NanoEthics." *NanoEthics* 1 (1): 31–46.

Nowotny, Helga, and Ulrike Felt. 1997. *After the Breakthrough: The Emergence of High-Temperature Superconductivity as a Research Field*. Cambridge: Cambridge University Press.

Oerlemans, Anke, Maria van Hoek, Evert van Leeuwen, and Wim Dekkers. 2014. "Hype and Expectations in Tissue Engineering." *Regenerative Medicine* 9 (1): 113–22.

Parandian, Alireza, and Arie Rip. 2013. "Scenarios to Explore the Futures of the Emerging Technology of Organic and Large Area Electronics." *European Journal of Futures Research* 1 (1): 1–18.

Parandian, Alireza, Arie Rip, and Haico te Kulve. 2012. "Dual Dynamics of Promises, and Waiting Games around Emerging Nanotechnologies." *Technology Analysis and Strategic Management* 24 (6): 565–82.

Petersen, Alan. 2009. "The Ethics of Expectations: Biobanks and the Promise of Personalised Medicine." *Monash Bioethics Review* 28 (1): 05.1–05.12.

Petersen, Alan, and Kate Seear. 2011. "Technologies of Hope: Techniques of the Online Advertising of Stem Cell Treatments." *New Genetics and Society* 30 (4): 329–46.

Pollock, Neil, and Robin Williams. 2010. "The Business of Expectations: How Promissory Organizations Shape Technology and Innovation." *Social Studies of Science* 40 (4): 525–48.

Porter, Theodore M. 1996. *Trust in Numbers: The Pursuit of Objectivity in Science and Public Life*. Princeton, NJ: Princeton University Press.

Ramírez, Rafael, and Cynthia Selin. 2014. "Plausibility and Probability in Scenario Planning." *Foresight* 16 (1): 54–74.

Rammstedt, Otthein. 1975. "Alltagsbewußtsein von Zeit." *Koelner Zeitschrift für Soziologie und Sozialpsychologie* 27 (1): 47–63.

Rip, Arie. 2006. "Folk Theories of Nanotechnologies." *Science as Culture* 15 (4): 349–65.

___. 2010. "De Facto Governance of Nanotechnologies." In *Dimensions of Technology Regulation*, edited by Morag Goodwin, Bert-Jaap Koops, and Ronald Leenes, 285–308. Nijmegen: Wolf Legal Publishers.

Ruivenkamp, Martin, and Arie Rip. 2011. "Entanglement of Imaging and Imagining of Nanotechnology." *NanoEthics* 5 (2): 185–93.

Schubert, Cornelius, Jörg Sydow, and Arnold Windeler. 2013. "The Means of Managing Momentum: Bridging Technological Paths and Organisational Fields." *Research Policy* 42 (8): 1389–405.

Schutz, Alfred. 1962. "Common-sense and Scientific Interpretation of Human Action." In *Collected Papers I—The Problem of Social Reality*, 3–47. The Hague: Martinus Nijhoff.

___. 1964. "Tiresias, or Our Knowledge of Future Events." In *Collected Papers II, Studies in Social Theory*, 277–93. The Hague: Martinus Nijhoff.

Seefried, Elke. 2013. "Steering the Future: The Emergence of 'Western' Futures Research and Its Production of Expertise, 1950s to Early 1970s." *European Journal of Futures Research* 2 (1): 1–12.

Selin, Cynthia. 2007. "Expectations and the Emergence of Nanotechnology." *Science, Technology, & Human Values* 32 (2): 196–220.

___. 2015. "Merging Art and Design in Foresight: Making Sense of Emerge." *Futures* 70: 24–35.

Selin, Cynthia, and Prasad Boradkar. 2010. "Prototyping Nanotechnology: A Transdisciplinary Approach to Responsible Innovation." *Journal of Nano Education* 2 (1–2): 1–12.

Selin, Cynthia, and Angela Pereira. 2013. "Pursuing Plausibility." *International Journal of Foresight and Innovation Policy* 9 (2/3/4): 93–109.

Selin, Cynthia, and Jathan Sadowski. 2015. "Against Blank Slate Futuring: Noticing Obduracy in the City through Experiential Methods of Public Engagement." In *Remaking Participation: Science, Environment and Emerging Publics,* edited by Matthew Kearnes and Jason Chilvers, 218–37. London: Routledge.

Smart, Andrew, and Paul Martin. 2006. "The Promise of Pharmacogenetics: Assessing the Prospects for Disease and Patient Stratification." *Studies in History and Philosophy of Science Part C: Studies in History and Philosophy of Biological and Biomedical Sciences* 37 (3): 583–601.

Snow, Charles Percy. 1964. *The Two Cultures and a Second Look.* New York: Mentor Books.

Stilgoe, Jack, Richard Owen, and Phil Macnaghten. 2013. "Developing a Framework for Responsible Innovation." *Research Policy* 42 (9): 1568–80.

Taylor-Alexander, Samuel. 2014. "Bioethics in the Making: 'Ideal Patients' and the Beginnings of Face Transplant Surgery in Mexico." *Science as Culture* 23 (1): 27–50.

Thrift, Nigel J. 2008. *Non-representational Theory: Space, Politics, Affect.* Abingdon: Routledge.

Tutton, Richard. 2011. "Promising Pessimism: Reading the Futures to Be Avoided in Biotech." *Social Studies of Science* 41 (3): 411–29.

Väliverronen, Esa. 2004. "Stories of the 'Medicine Cow': Representations of Future Promises in Media Discourse." *Public Understanding of Science* 13 (4): 363–77.

van Lente, Harro. 1993. *Promising Technology: The Dynamics of Expectations in Technological Developments.* Ph.D. dissertation, University of Twente.

___. 2012. "Navigating Foresight in a Sea of Expectations: Lessons from the Sociology of Expectations." *Technology Analysis & Strategic Management* 24 (8): 769–82.

van Lente, Harro, and Arie Rip. 1998. "Expectations in Technological Developments: An Example of Prospective Structures to Be Filled in by Agency." In *Getting New Technologies Together: Studies in Making Sociotechnical Order,* edited by Cornelis Disco and Barend van der Meulen, 203–30. New York: Walter de Gruyter.

van Lente, Harro, Charlotte Spitters, and Alexander Peine. 2013. "Comparing Technological Hype Cycles: Towards a Theory." *Technological Forecasting and Social Change* 80 (8): 1615–28.

van Merkerk, Rutger O., and Douglas Robinson. 2006. "Characterizing the Emergence of a Technological Field: Expectations, Agendas and Networks in Lab-on-a-Chip Technologies." *Technology Analysis & Strategic Management* 18 (3–4): 411–28.

van Merkerk, Rutger O., and Harro van Lente. 2008. "Asymmetric Positioning and Emerging Paths: The Case of Point-of-Care." *Futures* 40 (7): 643–52.

Watts, Laura. 2008. "The Future Is Boring: Stories from the Landscapes of the Mobile Telecoms Industry." *Twenty-First Century Society* 3 (2): 187–98.

Whitson, Jennifer R. 2013. "Gaming the Quantified Self." *Surveillance & Society* 11 (1–2): 163–76.

Wüstenhagen, Rolf, Robert Wuebker, Mary Jean Bürer, and Dale Goddard. 2009. "Financing Fuel Cell Market Development: Exploring the Role of Expectation Dynamics in Venture Capital Investment." In *Innovation, Markets, and Sustainable Energy: The Challenge of Hydrogen and Fuel Cells*, edited by Stefano Pogutz, Angeloantonio Russo, and Paolo Migliavacca, 118–37. Cheltenham: Edward Elgar.

Wynne, Brian. 1993. "Public Uptake of Science: A Case for Institutional Reflexivity." *Public Understanding of Science* 2 (4): 321–37.

III Sociotechnological (Re-)configurations

Rayvon Fouché

Technology, in its multiple forms, and as a mediator of a multitude of social, cultural, and political interactions, influences the shape of the modern world. From individual devices to vast infrastructural systems, humanity continues to pour massive amounts of physical and mental energy into the design, creation, construction, and distribution of technologies. STS, from its origins, has focused on understanding two key technological questions: (1) How do people shape the development of new technologies? (2) How, in turn, does the development of technological networks, systems, or infrastructures shape, impact, and (re-)configure the human condition? With this critical agenda, the chapters in this section explore how human efforts to create technology—materially and conceptually—continue to force societies to rethink, reassess, and reexamine the morphing relationships among themselves and with intertwining complex sociotechnical arrangements.

The ways in which the authors write about technology in the world clearly illustrate the opening up of technology as a process in the making. The authors' interchangeable and overlapping usages of the terms *technology*, *sociotechnical*, and *technoscience* allude to the complexity of fully grasping technology as an object of inquiry, site of investigation, or mode of political action. In the past few decades, the expansion of STS understandings of technology has led to transformative analyses embracing the comingling and potential merger of body and machine. This expanding work has pushed the study of human and technology relationships into diverse contexts like the environment, cultural movements, political apparatuses, and nation-state infrastructures. The chapters in this section, in addressing different moments of technological activity and experience, all contend that an STS approach to technology is deeply context dependent. Whatever their subject—ethnographies of users, infrastructures, or built environments, technology's role in sustaining or unraveling forms of difference, mediations of human interaction on- and off-line, configurations of financial markets, assemblages of urban spaces, or technoscientific development in remote locations—these chapters are invested in tracing and mapping the ways STS can provide effective, critical, and illuminating methods, approaches, and techniques through which we might understand the evolving place of technology in the world.

Studies of users as agents of technological consumption, modification, and resistance broaden our understanding of the interaction between humans and machines. In the first chapter of the section, Lisa-Jo van den Scott et al. delve into human and machine relationships and illustrate how richer ethnographic accounts of technologies in use can substantively add to existing ethnographic STS work. Their contribution also highlights the merits of ethnography as a valuable method to reveal new ways to understand users. They contend that deeper and broader ethnographic accounts will more fully explicate "(a) the relationship of technology to social control; (b) the distribution of agency and structure in human-technological relations; (c) the use of technology at the organizational, institutional, and macrosociological levels; and (d) the everyday experiences of technological use in cross-cultural, global, and non-Western contexts." Though ethnography has been a key method informing much STS scholarship, they assert that more deeply rooted ethnographic research will provide more finely grained analyses of how technology mediates everyday human interactions.

While van den Scott et al. emphasize heightening the focus on users, the chapter by Slota and Bowker argues for a renewed emphasis on infrastructure. They define infrastructure as "those systems, technologies, organizations, and built artifacts that do not need to be reconsidered at the start of a new venture." It is their preexisting and embedded qualities that often makes them disappear into the background of everyday human existence, yet they take on enormous significance. The sheer constructed opacity of infrastructures allows the social, cultural, and political impacts of infrastructure to hide in plain sight. But, this does not of course mean that infrastructure is any less potent than other technologies in shaping human experience. Stephen Slota and Geoffrey Bowker supply a genealogy of the multiple ways STS scholars, directly or tacitly, deploy an infrastructural approach to the world governed by science and technology. From this foundation, they provide a compelling argument for why infrastructure, and the ways cognate disciplines and fields use the concept of infrastructure, should be brought to the foreground in future STS-centered research.

Ignacio Farías and Anders Blok effectively show how urban studies, and specifically studies of cities, can be productive sites for STS scholarship to engage. Similar to Slota and Bowker, Farías and Blok illustrate how a closer STS-grounded analysis of cities not only can reveal the ignored technoscientific infrastructure of cities but also compel scholars to see the city itself as a technoscientific organism worthy of STS study. They prod scholars invested in the interlinked study of urban spaces and their relations to technoscience to move beyond the important historical foundations represented in the work of early analysts of technology and cities like Lewis Mumford to approaches and methods seen in actor-network theory and the scholarship of Bruno Latour. This refocusing creates new sets of questions—dealing with contemporary concerns—to discern how cities function as technoscientific artifacts. By considering the built environment of buildings, roads, and the network of material objects that are the infrastructural elements of cities, STS can participate in the continual reassemblage of cities themselves. In this regard, Farías and Blok argue that it is time to embrace the potential of actor-network theory to reassemble the city, where "urban politics becomes a version of cosmopolitics, the politics of searching for and building the shared common cosmos, an urban common world."

Continuing the theme of infrastructure, Hector Postigo and Casey O'Donnell select networked information communication technologies (ICTs) to explore the digital interstitial tissues connecting myriad forms of social interaction, from simple communication to play. The platform—whether it be the Internet, YouTube, a PC, or a video gaming system—is just that, a platform from which to express one's innovative creativity. Platforms come in many forms, but what makes ICT platforms unique is their built-in flexibility and malleability that allows individuals with access to tinker and reconfigure the platform's function through user-generated content. They argue that familiar STS concepts that stabilize and structurally bind the parameters of a moment of technological change, like closure, lose their explanatory relevance in realms where those that can participate in technological innovation are seemingly limitless. Certain digital architectures can inspire forms of freedom that the world outside of a global network of connected computing machines cannot match. Using user-created content in video games and on YouTube, Postigo and O'Donnell show how the massive technoscientific indeterminacy, or as they call it "interactive plasticity," of digital environments demands that STS integrate new methods and techniques to understand these worlds.

The previous two chapters illustrate how cities and digital platforms supply infrastructures for living and communicating. Chapter 21 by Alex Preda continues the development of the theme of technological infrastructure by outlining the importance of STS analysis for the study of financial markets and, especially, for exploring the role of technology in shaping their organization and dynamics. Faster computational practices—whether in the form of adding machines or of contemporary super computers—have always been a key element of financial markets. Today, financial markets are a complex mix of humans and machines executing a host of different tasks, many of which have enabled the creation of new forms of financial institutions and exchanges. Preda takes on three aspects of this world: "first, the transformation of market institutions in relationship to technology; second, the new formats of social interactions emerging in electronic markets; and third, new types of (academic) knowledge and disciplines accompanying these transformations." Altogether, Preda shows, financial markets can be an extremely productive site to explore the growing nexus of human interactions within a society's institutions as they rapidly change and demand new sociotechnical organizations and structures.

Andrew Feenberg in chapter 22, "A Critical Theory of Technology," conceptualizes the role of technology in the shape of political and social action. Situated within Frankfurt School critical theory, Feenberg presents a critical theory of technology as a methodological approach "concerned with the threat to such critical potential posed by the pretensions of the technocracy." This approach emphasizes how racial disparity, ethnic difference, geographical distance, political confrontation, and class struggle are critical to broadening our understanding of the place of technology in the world. Critical theory of technology reveals certain opportunities to think through the ways capitalist structures, rooted within technological modes of production, maintain socially, culturally, and politically manufactured forms of marginality. Thus, Feenberg in part motivates us to take on the responsibility of critically engaging the agential force of technology. By doing so, we might create a world where technology is more responsive to human needs and not seen narrowly as a tool for capitalists or authoritarian ends.

In thinking about the response to human needs, the final chapter of this section, by Aalok Khandekar et al., is concerned with the development of formerly colonized countries of the global south and what it means to speak of technoscientific development in regions replete with histories and traditions of inequitable distributions of wealth. Building upon histories of the interconnections of STS, development, and postcolonial studies, the chapter clearly articulates how and why technoscience played a key role in "(neo)colonial expansion, postcolonial nation-building, and on the interplay between traditional and scientific knowledges." Khandekar et al. suggest that by engaging development studies, policy, governance, and practice, STS can live up to its transformative aims. They impel STS scholarship to not only question the uneven distribution of resources between the global north and the global south but also reflexively question itself—as an area of scholarly inquiry—and the commitments it makes to the place of the worlds, the peoples, and the technoscientific interactions it chooses to study and not study.

By extending boundaries and engaging new intellectual areas, STS can continue to develop scholarship to critically study, examine, and discuss the multifaceted social, cultural, and political relationships manifested by and through technology. Technology, today, is omnipresent, making it crucial to continue investigating the ways that technological projects have shaped humanity's pasts and presents and will shape its futures. The perspectives presented in the following chapters forge important interdisciplinary new ground to analyze the roles that social and cultural dynamics have and will play in the production and consumption of technological artifacts, practices, and knowledge. Though these chapters can only provide an incomplete glimpse into the variety and diversity of STS scholarship on technology, they aim to inspire and motivate creative and imaginative work explicating the relationships between technology and human existence.

17   Reconceptualizing Users through Enriching Ethnography

Lisa-Jo K. van den Scott, Carrie B. Sanders, and Antony J. Puddephatt

Science and technology studies (STS) has long been interested in users and how they "consume, modify, domesticate, design, reconfigure, and resist" (Oudshoorn and Pinch 2005, 1) technologies. Since human groups perceive, define, and value things differently, designers do not know how technologies will be adopted or rejected until they are distributed into the everyday lives and workflows of sociocultural groupings. It is here where users react, imaginatively reconfigure, and provide feedback, such that designs are reshaped by the unanticipated meanings people develop through practice.

In this chapter we reimagine the power of ethnography for exploring and enriching the study of users. We move beyond the traditional question of how "users matter" and explore the ways in which ethnographic methods provide a means to explore new and compelling contemporary issues. Such themes concern (1) the relationship of technology to social control; (2) the distribution of agency and structure in human-technological relations; (3) the use of technology at the organizational, institutional, and macrosociological levels; and (4) the everyday experiences of technological use in cross-cultural, global, and non-Western contexts. We begin by providing a brief review of the key concepts and contributions derived from user studies as they relate to our present argument.[1] We then describe the strengths and advantages of ethnography for studying user-technology relations. Finally, we consider the most promising conceptual directions for studying users of technology by utilizing the benefits of ethnographic methods.

Studying Users in Science and Technology Studies

Technologies hold no one inherent meaning or objective, but instead "it is the use made of [them] that brings [them] into the life of the society" (Wagner-Pacifici and Schwartz 1991, 416). Previous research has treated technologies and users as separate objects of analyses, often perceiving technological development as occurring "outside society, independently of social, economic and political forces" (Wyatt 2008, 168). Yet, the social and technological are co-constituted—with one making up the other (Clarke 2005). Scholarship on user studies, however, has drawn attention to the embedded nature of technologies in their contexts of use, showing how technology is both emergent from and generative of our social worlds (Wyatt 2008). As such, research on user-technology relations has moved beyond technological determinism to illuminate the importance of users in technological design and development, and how processes of adoption and resistance reflect wider cultural and social issues. This has been accomplished using a range of methods, such as case studies, document analysis, interviews, focus groups, weblogs, and ethnographic accounts. In this chapter we explore how ethnography can help extend the theorizing of user technology relationships in new and conceptually innovative ways.

The Turn to Users: The Consumption Junction

Early studies on technology largely focused on the design of an innovation and, by default, the designer. Feminist scholars drew attention to the neglect of women in the development of technology and urged researchers to move beyond the designer and to pay more analytical attention to the user. Ruth Schwartz Cowan (1987, 263) was a prominent figure in the turn to users with her work on the "consumption junction," which is the "place and time at which the consumer makes choices between competing technologies." Cowan brought her analysis of users to the structural level by regarding consumers as those "embedded in a network of social relations that limits and controls the technological choices that she or he is capable of making" (1987, 262). By recognizing the social structural networks embedded within technologies, researchers were able to see the way in which users were both active and passive members of technological innovation.

Household technologies are a useful site for studying users and technologies. Cowan (1983) examined household inventions that were intended by the designers to be time-savers for women accomplishing work in the home. She found that these technologies actually increased the workload of women, even while increasing the standard of living (see also Cockburn and Ormrod 1993). By attending to the intersection of user and technological artifact, we learn that time-saving technologies often produce increased, rather than decreased, workloads. We use the consumption junction as a place to begin our discussion of users because it draws attention to the analytical complexity of users and the social structural networks in which they engage and resist technologies. By being attentive to the social structural "network of utilization" (Hyysalo 2004), research on users has illuminated the multiple sociotechnical configurations that are required to keep networks functioning.

Conceptualizing, Configuring, and Scripting the User

At the initial phase of design, designers "project," or imagine, the preferences of various user groups (Akrich 1992; Berg 1997). In this regard, designers inscribe visions onto a technology which "depend on hypotheses about the situations in which the device will be used" (Samuelsson and Berner 2013, 726), by anticipating the skills, behaviors, motives, and interests of future users. Woolgar's (1991, 59) research on the design of personal computers, for example, illuminates the "configuring process" where future interactions are constrained by the designer's projection of a particular user through technological designs. Indeed, Bardini and Horvath (1995) have found that designers' conceptualizations of users is more often a reflection of the social qualities of the designers themselves rather than the actual users the designs are intended for. As such, the images, prejudices, and assumptions of designers may be vastly different from the actual needs and desires of the users (Hyysalo 2004).

Fairhurst (1999), for example, reveals how the stereotypical ideas around the needs of older people not only become manifest in the designs of nursing homes by young architects but constrain the ways in which older people must then move through and in these spaces. Assumptions around "privacy, security and ... maintainable living space, in the context of declining health" are "informed by abstract knowledge" (Fairhurst 1999, 107). As such, the architecture of these spaces works against the actual needs and desires of the older people housed therein. Transitions from designers to users is a particularly crucial question in regards to the care of the elderly because the "convergence" of design and use often creates problems when the technology introduces unfit procedures (Berg 1997; Hasu 2000; Hyysalo 2004).

Designers not only configure users but can also script use (Akrich 1992). While a designer may imagine a user and proceed to the design, their anticipated ideals also become scripted and materialized into a piece of technology, providing the user with directives on how to act. Scripting the perceived needs and competencies of users into objects serves to transform, delegate, attribute, or reinforce existing "geographies of responsibilities" (Akrich 1992, 207). When users engage with a scripted piece of technology they engage in "reading" its script—"which essentially involves adapting the new product to user environments—the meanings, uses, or even the products themselves can be changed and adapted" (van Oost, Verhaegh, and Oudshoorn 2009, 186).

Central to Akrich's (1992) concept of script is the symmetrical analysis of the interaction between user and artifact. Both can be "attributed with (inscribed) agency and meaning that enable and constrain user practices and users' agency. Both user and artifact shape and at the same time are shaped by the practice of usage" (van Oost, Verhaegh, and Oudshoorn 2009, 187). Both forms of agency are exposed in van den Scott's (forthcoming) study of government housing plans for the Inuit of Nunavut, Canada. She found that systems of knowledge and ways of being were scripted into the technology of the houses themselves. For example, many residents of the new houses had trouble fathoming how sex was accomplished previous to the internal division of space in the new structures. They had "learned" from the houses that a certain division of individuals in their sleeping arrangements was more appropriate.

Users, however, can resist programmed scripts, exercising their own agency through antiprograms. The Inuit resisted the novel scripts inherent to the newly imposed technologies of walls by "spatially grafting" their cultural practices indoors as much as possible (Dawson 2008). They refused to simply discard their traditional heritage for the more mainstream Western ideals that were scripted within the houses. One might find, for example, seal being butchered in a bathtub or on a kitchen floor (Dawson 2008), despite the cultural scripts of the artifacts that suggest they ought to act otherwise. Technologies impart scripts to users, but as an incomplete project. Users balance between what is prefaced in intended scripts and in their resistance to or unexpected interpretations of these scripts.

Problematic Distributions of Power: Gender Scripts and Implicated Actors

Feminist researchers have extended Akrich's "script" concept to explore designers' representations of gender in technology (Oudshoorn 1996; Rommes 2002; Rommes, van Oost, and Oudshoorn 1999). Gender scripts "function on an individual and a symbolic level, reflecting and constructing gender identities, and on a structural level, reflecting and constructing gender differences in the division of labour" (van Oost 2003, 199). For example, van Oost's (2003) study on the design of shavers illustrates how artifacts can reflect dominant gender symbols and identities. The intentional gender scripting of the women's Ladyshave, Oost argues, inhibits "the ability of women to see themselves as interested in technology and as technologically competent," while the men's Philishave invites men to see themselves as technologically savvy (2003, 207).

Collins-Nelsen (2010) has shown how existing gender scripts can be resisted with new technologies that encourage women to pursue practices in traditionally male domains. In her study of home-repair tools designed for women, she found that the tools were interpreted both as empowering and oppressive, since their designs often resorted to clichéd feminine cues such as cute, pink imagery. In this sense, research on gender scripts has identified the ways in which technologies can construct, reinforce, and shape gender and gender identities—though it is not always a simple process, nor is it free of contradictions (Oudshoorn 2005).

Research has also found users to be "implicated actors," those who are "silent or not present but affected by the action" (Clarke 1998, 267). For example, Clarke and Montini (1993) identified how "the exclusion of women as patients and users/consumers from participation at design stages has constituted millions of women as implicated rather than agentic actors in the contraceptive arena for almost a century" (Clarke and Star 2008, 125). Implicated actors have also been identified in computer design (Bishop et al. 2000) where the assumption is that "I am the world," where the designers "needs" take precedence (Forsythe 2001). These categories of users are not necessarily mutually exclusive. Active users of technology may become implicated in the web of technocultural relations in ways they had not previously anticipated. On the other hand, implicated actors may become reflexive enough about their unanticipated role in technologies and decide to try to bring about change (Collyer 2011).

Another type of implicated actor is the non-user. Non-use encompasses both "voluntary and involuntary aspects of non-use" (Oudshoorn and Pinch 2008, 555). Kline and Pinch (1996), for example, showed how non-users actively resisted automobiles in rural America. Oudshoorn and Pinch (2008) define four categories of non-users: resisters, those who choose not to use; rejecters, those who have tried but no longer use a technology; the excluded, those who are excluded due to cost or access; and the expelled, those who used a technology and no longer can (Oudshoorn and Pinch 2008, citing Wyatt 2003; Wyatt, Thomas, and Terranova 2003). Actors are implicated in various ways in each of these groups.

Relevant Social Groups and Interpretive Flexibility

Research on users has also furthered our understanding of human agency and the meaning-making processes embedded in sociotechnical relations (Manning 2013; Mukerji 1994; Neff 2005; Wagner-Pacifici and Schwartz 1991). Users and non-users, as "relevant social groups," generate different meanings around an artifact, known as "interpretive flexibility." As such, users play important feedback roles, deeply influencing subsequent technological designs (e.g., Bijker 1993; MacKenzie and Wajcman 1985; Pinch and Trocco 2002). This concept is well illustrated in Pinch and Bijker's (1984) classic account of the development of the bicycle where relevant social groups (e.g., older men, women, young men) defined the high-wheeled bicycle in vastly different ways (e.g., "unsafe machine" and "macho machine"), which reshaped designs, leading to the bicycle we know today. Eventually "closure" happens as one form of the technology is adopted by the most influential and relevant social groups, establishing it as the most dominant.

Innovation studies also demonstrate the active role of users. For example, von Hippel's (2005) research has explored the role "lead users" play in designing and altering material artifacts. Lead users are the first to "face needs that will be general in a marketplace—but face them months or years before the bulk of that marketplace encounters them" so they are positioned to "benefit significantly by obtaining a solution to those needs" (von Hippel 1986, 796). Von Hippel's (1976) research has demonstrated how lead users can, but do not always, invent, build, and design technological solutions. Much of this research has been market oriented, alerting companies to the rich potential users play in technological development. This research has identified the importance of creating a forum—"nexus" (Schot 1992) or "mediation junction" (Schot and de la Bruheze 2003)—where producers and users meet to facilitate social learning processes "in which alignments in articulation processes between various actors from both contexts can be established" (van Oost, Verhaegh and Oudshoorn 2009, 186).

The literature on innovation studies has illuminated the broader economic, normative, and political contexts that shape design and use. For example, it has illustrated how user-initiated feedback has led to innovations in the design of the mountain bike (Luthje, Herstatt, and von Hippel 2005). This debunks the notion that innovation results from purely economic interests, since the ethical and political convictions of users can also drive innovation (see, for example, Morton and Podolny 2002; Shah and Tripsas 2007). For example, hackers often share code freely for the betterment of their craft and for wider political beliefs in an open society (Söderberg 2011; see also Coleman 2004; Opel 2004). Research has also uncovered how a technological artifact can result from the innovative energy of a community of users, as well as give rise to an entire user community (van Oost, Verhaegh and Oudshoorn 2009).

The perceptions and meanings users hold toward technology are often situated in the context of particular "technological frames" (Bijker 1995). Where technological frames of key groups "are significantly different, difficulties and conflict around the development, use, and change of technology may result" (Orlikowski and Gash 1994, 174; see also Allen and Wilson 2005; Chan 2003). By focusing on technological frames, researchers have discovered two types of disconnects between design and use. "Functional disconnects" refer to how an artifact was designed to function versus the way it is used in practice (Sanders and Henderson 2013). "Ideological disconnects" refer to how designers perceive the need for and value of a technology differently than users (Sanders 2014). These disconnects demonstrate how designers cannot entirely capture or "plan" for the situated actions of users prior to the technology in question being introduced to the market.

Research on computer-supported cooperative work has also drawn attention to the challenges of capturing work processes due to the "visible-invisible matrix" (Star and Strauss 1999, 23) because "what counts as work is a matter of definition" (ibid., 15). Indeed, these studies "have cautioned that forced representation of work (especially that which results in computer support) may kill the very processes that are the target of support, by destroying naturally occurring information exchange, stories, and networks" (ibid., 24).[2] Research that explores how technologies capture work processes has provided important insights into the way visibility is connected to bigger questions concerning discretion, power, and autonomy (Berg 1997; Bowker, Timmermans, and Star 1995; Star 1991, 1995).

Literature on the implementation and appropriation of technologies has shown the creativity and artistry of users as they develop work-arounds to fit technology into their everyday activities (Suchman 2002; Suchman et al. 1999).This includes the "articulation work" used to keep things on track (Star and Strauss 1999; see also Bowker and Star 1999; Clarke and Star 2008) and the "reformulation of work practice, including the shaping of technology in question" (Hyysalo 2004, 25; see also Hasu and Engestrom 2000). Such research has identified processes of "localization" (Berg 1997) and "taming" (Lie and Sørensen 1996) to capture the negotiated and uncertain character of technologies. For example, Berg's (1997) ethnographic research on the integration of a computer system into a hospital setting illustrates how hospital staff adapt technology in accordance with their local circumstances (space), limiting its use (scope) and clearly defining the needs and purposes it serves (rationale). As such, Berg demonstrates how, through users' ongoing adjustments, the full potential of the computer system was not always realized. A piece of technology is "tamed" when users have manipulated it to make it relevant and workable for them. However, processes of localization and taming do not result in technology being fully "domesticated" and therefore never fully achieve a "recognizable repertoire of uses" (Lie and Sørensen 1996; Pols and Willems 2011). In the following section we discuss ethnography and consider how it enables an in-depth analysis of technology-user relations.

Methodological Leverage: Ethnography Unleashed

On the ground, experiences of everyday life involve interactions with technologies as they accommodate us and we accommodate them (Giddens 1979; Kennedy 2010; Miller 2002). One way to study this "mutual accommodation" is through ethnographic

methods. There are myriad approaches to ethnography—but the common denominator involves going out into the world, living among cultural groups, and observing
users in action—moving beyond previously imagined constructs in ways that are not
harmful or intrusive (van den Hoonaard 2014). The ethnographer attempts to record
the experiences of those being observed in detailed field notes while remaining committed to defining "the world from the perspective of those studied" (Shaffir and Stebbins 1991, 5). Data collection for ethnographers often begins in an unstructured way
and does not follow a "fixed and detailed research design specified at the start" (Hammersley and Atkinson 2007, 3). Results are emergent as the researcher is sensitized by
participants to local meanings, actions, relationships, and processes. Issues that are
salient and relevant are inductively arrived at over a period of time during which the
researcher discovers recurring patterns of meaning and use, or non-use.

Ethnographers often rely on three forms of data: observations, participants' observations, and interviews, each of which enables the researcher to become more intimately
familiar with the lived experiences of others while being far enough outside to analyze
emerging patterns (Becker 1986; Emerson, Fretz, and Shaw 1995; Prus 1996). Ethnography, we argue, provides researchers with an opportunity to see the ways in which
users of technology are rooted in richly defined social contexts with embedded and
hierarchical constraints of time, space, and power (Hall 1983; Lewis and Weigert 1981).

An ethnographic approach provides for a cyclical data collection and analysis process. As insights arise they present new opportunities and directions for data collection.
Further, the categories, classifications, and codes used to make sense of what is happening are not predefined or used to structure data collection, but instead arise out of
the data analysis itself (see Charmaz 2014; Glaser and Strauss 1967). One method of
analysis is Clarke's (2005) situational analysis which identifies and maps the relationships among human actors, nonhuman actants, implicated actors, and the discourses
embedded within arenas of action as they play out across social worlds. In summary,
ethnography enables us to analyze and observe passive resistance as well as agentic
incorporation in the creative reimagining of objects throughout various contexts of
use. Yet it can take us further in our understanding of users, technology, and social life
by pursuing new questions about social control, structure and agency, organizations,
and global issues.

Exploring Technology and Social Control

As noted earlier, user-studies research has done an excellent job of studying the distribution of power and social inequality. Ethnographies of users can press further by providing insights into the more subtle experiences, practices, and mechanisms of social

control and marginalization. Ethnographers can ask how technologies are embedded, evolving, and impactful in our personal and social lives, and how these tie into issues of social control.

Let us consider technologies of surveillance and big data[3] in contemporary society. Closed circuit television (CCTV) cameras have been constructed as "neutral technologies of surveillance," put in place to enhance public safety (Hier 2004, 542). Ethnographic research on CCTV cameras, however, draws attention to both their intended and unintended consequences. Hier's study on the sociotechnical work of CCTV camera operators illuminates how they participate in "selective social monitoring" that reproduces and reinforces categories of suspicion, marginalization, stigmatization, and criminalization. Camera operators focus their "surveillance gaze overwhelmingly ... upon individuals occupying categories of suspicion—youth, homeless persons, street traders and black men" (2004, 543). Hier's study is illustrative of the value of ethnographic methods for providing insights into the construction of categories such as risk and threat, and the practices of social control and marginalization in the use of technology.

Everyday life is filled with "countless interactions with massive technological systems that support our communication, our transport, our retail activities, and much more. Through using our own technologies as private citizens, we unwittingly generate increasing volumes of big data, "documenting our collective behaviour at an unprecedented scale" (Moat et al. 2014, 92). The advent of big data has made our lives ever more "transparent" and visible to other individuals, and private and public organizations (Bennett et al. 2014). Intelligence and security services are turning to big data as a means to identify and track potential terrorists. Using tweets, blog and Facebook posts, and geocodes from cell phones, policing agencies are identifying potential security threats and developing preemptive strategies. Yet, injustices can be perpetuated when we do not understand how big data are being collected, analyzed, and used (boyd and Crawford 2012; Ess 2002; Moses and Chan 2014). The inferences and predictions made from big data "provide fertile ground for speculation, innuendo, and the exercise of pre-existing biases for, and particularly against, racial, ethnic, religious, and socioeconomic stereotypes" (Wigan and Clarke 2013, 49). Thus, important questions arise concerning social control and marginalization as we consider ethnographic inquiries into the people and agencies who make use of big data on a broad scale.

Ethnography provides an excellent means for exploring such questions and allows us to understand how technologies are used and resisted, while also providing critical insight into the processes of social control and marginalization. For example, how do technologies originally developed for legitimate surveillance purposes transform into exploitative practices? How do these technologies undergo "function creep," where the technology designed for one purpose takes on a different function within the social world for which it was designed and in various, unrelated social worlds (Curry, Phillips, and Regan 2004)? By employing an ethnographic approach, one places analytic attention on users' activities and definitions, as well as the structural contexts and material realities that influence, shape, and guide the sociomaterial interactions (Clarke 1991). This invites an analysis of social difference and a means to uncover how social control is enacted and resisted.

The Distribution of Agency and Structure

In the field of science and technology studies, scholars have long been interested in understanding how technology impacts society. Interpretive scholars argue that overly structural approaches unfairly treat users as "cultural dopes" (Garfinkel 1967). The late 1970s saw theorists beginning to reconcile such structural and interpretive tensions, also known as the structure-agency divide, by developing models that allow for spontaneous and creative human action while still providing an important place for the impact of structural forces (Bourdieu 1990; Giddens 1979). The "social construction of technology" approach (Bijker and Law 1992; Pinch and Bijker 1984) inspired many of the new technology studies we have mentioned thus far. Following this, scholars began to explore how artifacts take on responsibilities, act relatively independently, and socialize human actors through scripts in a reflexive process of "co-construction" (Jasanoff 2006; Jerolmack & Tavery 2014; Puddephatt 2005; Verbeek 2006). Yet in doing so, STS scholars still tend to ignore the agency-structure tension entirely, in favor of lopsided, purely agency-based accounts ignoring or downplaying structural analysis (a point made by Hess et al., chapter 11 this volume).

This tendency to ignore structural explanations is a problem. Ethnographers are well situated to study micro- and macrostructural realities in user-technology relations by focusing on the intimate contexts where the imposition of structure is experienced firsthand. The question is not only how agency is distributed through the technologies we use but also what types of structural patterns begin to emerge and consolidate over time. Attending to structural issues in our ethnographic work may, ironically, be the key to innovating and expanding our explanatory tools in accounting for user-technology relationships.

There are a number of studies that examine how agency is distributed at the intersection of users and technologies. Scholars began to take seriously the question of "what things do" (Verbeek 2006) by inquiring about the agentic roles of, for example, computer applications (McEneaney 2013), medical technologies (Fleischman 2006; Joyce

2005; Moreira 2004), equipment in sports and leisure activities (Barratt 2011), technologies in financial markets (Preda 2006), and other objects in everyday encounters (Jerolmack and Tavory 2014). While ethnographers have excelled in demonstrating the agency of things, artifacts, and technologies, their structural aspects often remain under-theorized. One exception is Freeman's (2001) study of how the implementation of a *National Vocational Education and Training System* (NVETS) in Australia was a source of organizational chance that structured the training process in highly mechanized and bureaucratized ways. Freeman demonstrates that as objects and technologies first emerge, they demonstrate agency through their direct impact on human communities and organizations, often bringing with them widespread changes. However, once distributed and extended, these same technologies can impose highly rigid structures that force users to adjust their learning styles, evaluation procedures, and administrative practices in ways that seem unnatural.

For Bruno Latour (1992), once technologies are distributed through the collective, they become materially sedimented, and are "society made durable." His example of the seat-belt car alarm may be illustrative. He found that he was unable to break the law and drive without his seat belt, as a result of the annoying beeping that would emanate from the car. Latour argued that responsibility and morality was thus taken out of his hands, now delegated within the workings of the alarm. Yet nowadays seat-belt car alarms are ubiquitous among all new production vehicles. Given this, is the seat-belt car alarm best seen as holding agency over a would-be lawbreaker within a specific microsociological interaction or as a structural condition that applies across society and improves automobile safety? At what stage do we reach a "tipping point" where technological agency is so widely distributed that it is better thought of as a more constant structural condition? STS scholars might look to ethnography for aid in finding more nuanced interpretations of the relationship, and tipping point between, agency and structure, ensuring that both sides are accounted for in understanding the other.

Fuchs (2001), for example, conceives of agency not as a given but as a dependent variable that is influenced by the size and scope of social networks. Small, familiar, and intimate networks tend to view actors operating with free will, while longer, abstract networks tend to result in more structural explanations for behavior. Knorr Cetina's (1999) ethnography, as an exemplar, observes the tendency for scientists to anthropomorphize instruments in the science laboratory as a result of a loose-working and intimate web of relationships. This results in what Pickering (2005) called the "dance of agency" between human and nonhuman actors. When networks are spread out and extended, however, structural understandings of technology are more likely (Latour 2005). Rather than reaching for philosophical speculations of how people are intrinsically different from machines, we can analyze change at the level of dynamic interaction.

Randall Collins (2004) laments that ethnographers too often ignore the microstructures that shape everyday interaction. Sociotechnical microstructures are surely no less important. For example, Gary Alan Fine's study of competitive chess shows the importance of how chess clocks structure and enforce the "temporal tapestry" of the game (Fine 2015), shortening the time required for games to play out. Now, games could be played in hours rather than days. Then, "digital clocks provided the technology necessary to create a swift game, as time could be measured in small increments. Games shrank: fifteen minutes, ten minutes, less. How little can minds and hands accept?" (Fine 2015, 97). Yet more than simply contracting time, recent innovations in digital chess clocks now enable innovative formats for play (e.g., "Fischer delay time," where players can get time instantly added back onto their side with each move). Again, what might begin as a study of how technology carries agency by impacting human behavior might shift to consider the microstructural conditions that place constraints and affordances on social interaction (e.g., Puddephatt 2008). Ethnographers would do well to pay attention to how technologies are experienced both as agentic and structural and to how these concepts may be mutually related across a range of local settings.

Users as Acting Units: Organization- and Institutional-Level Analysis

Ethnography can provide the methodological leverage to better understand the relationship of technology to user groups (Orlikowski 2000). By studying "organizations" themselves as "users" of technology, we provide insight into the way organizations shape technological functioning and, subsequently, how technology alters organizational practices, cultures, and structural arrangements (Orr 1996; Suchman 1987). It is important to unpack how users are embedded in contextual webs and networks. These influence how they move through the world, make technological decisions, and interact with technologies and each other (Michael 2000). By focusing on "technologies-in-practice," the ethnographer analyzes "how people act as competent practitioners, and how they organize their action practically as methods of seeing, listening, reasoning, and responding to human and nonhuman elements" (Samuelsson and Berner 2013, 726). It requires the ethnographer to study practice from the inside in order to explore how the structures inscribed in information technologies shape "action by facilitating certain outcomes and constraining others" (Orlikowski and Robey 1991, 148; see also Orlikowski 2000).

For example, Sanders's (2014) ethnography on emergency technologies in police, fire, and emergency medical services (EMS) uncovered how the functionality of the technologies were shaped by the structural, cultural, and operational contexts of the various organizations. Her study demonstrates how these contexts provide interpretive flexibility to workers, as members of different organizations define and give meaning to emergency technologies differently, sometimes impeding their functionality. Information technologies in emergency response have been implemented to overcome communication and information barriers, to facilitate collaborative action on the ground. Yet, hierarchical relations and organizational conflicts among police, fire, and EMS are played out and reinforced through the use of emergency technologies, harming communication and collaborative action. This study reaffirms how organizational contexts shape both technological artifacts and users' actions, which have an "irreducible and emergent effect on the way complex information is transmitted, communicated, processed and stored" (Vaughan 1999, 916).

Berg's (2001) ethnographic study on the implementation of patient care information systems (PCIS) in health care settings also illustrates a "process of mutual transformation" and "mutual learning" where organizational practices transform technologies, and the technologies, in turn, transform organizational processes and structures. His research illustrates the importance of studying organizations as users for extending our theoretical understanding of technological diffusion, resistance, and acceptance, as well as organizational reform and change (see also Ericson and Haggerty 1997; Meehan 1998; Preda 2006). By studying the in situ use of technologies, ethnographers can uncover the way organizational users "enact structures which shape their emergent and situated use of that technology" (Orlikowski 2000, 404).

Ethnographies that perceive and analyze the organization as a user of technology can also provide insight into organizational practices and cultures. For example, the integration of information technologies into animal shelters provides an opportunity to explore knowledge production in organizations. Irvine's (2003, 561) study on unwanted pets illustrates how the introduction of standardized forms in the animal shelter removed the complexity of clients' experiences and, in turn, assisted in reinforcing and strengthening the institutional model by producing "particular characterizations of social problems and solutions" (see also Fox 1999).

The implementation of technologies within organizations may also contribute to organizational change, reform, and restructuring. Fleischmann (2006) considers how the institutional worlds of gross anatomy instruction, educational administration, and simulation design all played a part in the construction of the "cyber-cadaver" technology. Once this technology was introduced into the practices of anatomical education, it had unintended effects, reshaping each of these social worlds in unforeseen ways. For example, educational administrators could now shift costs away from gathering traditional cadavers, and these digital technologies could be used in mobile environments and the developing world, where traditional cadavers are harder to come by. Thus, users' social worlds are often reshaped in marked ways by the introduction of new technologies, which can create rapid changes in how educational and administrative practices are organized.

Including the concept of "organization-as-user" into the STS vocabulary of user studies enables a meso- and macro-level analysis that can expand our theoretical understanding of how sociotechnical relations and technologies shape knowledge production (Vaughan 1999). Ethnographers can attend to the way organizational structures shape technical use/functioning, and how sociomaterial interactions shape organizational structures and processes. As Vaughan (1999) demonstrates in her study of the *Challenger* launch decision, organizations have powerful and long-lasting impacts on how information is constructed, gathered, exchanged, stored, and used. We argue that ethnographic studies on organizations as users enables meso-level analyses that can provide critical insight into occupational cultures, organizational processes, and change, and how organizations shape knowledge production.

Cross-Cultural Users and Global Ethnography

STS scholarship intersects with comparative-historical work in its approach to cross-cultural and global users, exploring the relationship of technology to globalization, colonization, and modernization (e.g., Taylor 2012). Kennedy (2010) argues that many of these studies often default to extrapolating concepts to the global scene from institutional ethnographies conducted in Western contexts. Charles Taylor (1999) agrees that how modernity affects different non-Western groups is poorly understood. Cross-cultural groups are too often regarded as helpless victims who become increasingly constrained by imported technologies. Ethnographers should consider how to balance sweeping, global, often colonizing, forces and the unique resistance and contributions of local lived experience. In this section, we explore how global forces interact with local communities across different global settings. Through this, we explore how interpretive flexibility, identity, and distributions of power are key concepts in the study of cross-cultural users. We argue that ethnography is an important tool to understand how mundane technologies are received, interpreted, and resisted across the globe by users from within their local, lived experiences.

In considering users across networks of culture, space, and power networks, ethnography provides access to everyday uses of technology in global spaces often far removed from more familiar Western contexts. These studies show that globalization does not simply mean the imposition of Western culture onto non-Western groups. Not only can one find the global in the local, but one must turn to localities in order to achieve an understanding of the global and the actual realities of globalization in the lives of ordinary people around the world. Conducting "global ethnography" (Burawoy 2010) enables scholars to focus on localities and relevant social groups across cultures, avoiding abstract theorizing (Robertson 2001), while allowing for greater insight into the specific experiences of technologies in terms of global identities, power, and resistance.

Globalization manifests as a web of connectivity, interdependence, and "global scripts" (Lechner and Boli 2005). We are, indeed, globally interconnected as never before. In this age of modernization, globalization is not a "project" (Albrow 1996), but a market-driven juggernaut in which "the local is a personally lived experience" (Kien 2009, 89) with multiple and varied meanings (Schudson 2002). New forms of governance, markets, technologies, and ideas interact with local cultures, values, and structures (Arnold and DeWald 2012). When finding the global in the local, one discovers not only homogenizing effects but also "increasing diversifications of local singularities and particularisms" (Entrena 2003, xv). These singularities are discovered on the ground, through methods of ethnographic research.

When a technological object is taken up or introduced to a non-Western group, local contingencies, beliefs, and the structures of daily life remain at the forefront of local experience. These contexts provide the basis for interpretive flexibility, resulting in unique practices as artifacts "are made meaningful in localized settings" (Kien 2009, 7). For example, Arnold and Dewald show how the uptake of the bicycle in India and Vietnam "became implicated in the lifestyle and work regimes of a significant section of the population, and was caught up in issues of race, class, and gender, and of national identity and colonial state power" (2011, 995). Most prominently, the bicycle became an important source of middle-class mobility by opening up economic opportunities, enabling the promise of change for many of the lower-class groups. How the bicycles were used and who they privileged cannot be understood apart from local contexts. Ethnographic approaches have the potential to move beyond the limits of historical data, and to more deeply explore how interpretive flexibility plays out in local communities, by analyzing the relevant social processes live and in action.

Globalization creates an implicit tension in the face of encroaching cultural scripts embedded in newly introduced technologies. Finding the global in the local means attending to identities created through the hybridization of global interdependence and localized identities (Cvetkovich and Kellner 1997). Tunc (2009), for example, explores the intersection of global and local identities in American-style malls in Turkey. Consumers in these malls "combine the messages embedded in signs with their own interpretive stream ... which is a direct product of their" (Tunc 2009, 133) social location and experiences. These malls, as terrains full of meanings and as liminal spaces where a "collage of traditionalism and modernity" are expressed (Tunc 2009, 135) allow for the construction of a hyper-real and more global Turkish hybrid identity, which Turks perform through their consumption practices.

Hybrid-identity practices may also, however, cause a pushback where some relevant social groups perform "nation-work" (Surak 2013) by exuberantly performing the identity of their nation or marginalized group. Van den Scott (2009), for example, finds that the Inuit constantly strive to practice Inuitness through performative preference for the cold and aversion to warmer climates, despite the introduction of heated structures and other technologies which help to mitigate the cold.

Ethnography also allows us to examine problematic distributions of power in global contexts. The local built environment may represent power dynamics stemming from historical and/or colonial contingencies, or from capitalist endeavors impinging on life-as-it-was, used by dominant groups as a symbol and tool for power (King 1989; Lawrence and Low 1990; Scott 1998). People relate to their built environment symbolically, and power "is experienced by the local in everyday situations which reveal the global" (Kien 2009, 6). In addressing the question of power across cultures, ethnographies would do well to address both "epochal" technologies, such as the printing press, ICTs, and airplanes (Arnold and DeWald 2012), as well as "invisible," mundane technologies, such as walking boots (Michael 2000) or patio heaters (Hitchings 2007). Arnold and DeWald (2012) argue that powerful groups have less control over mundane, invisible technology than over epochal technology because epochal technologies are either regulated or have high barriers to access. We must move away from diffusionist models which privilege Euro-American innovation and entrepreneurship in favor of the "social life of things" (Appadurai 1986) within the local group (Arnold and DeWald 2011, 971).

Scott (1998) describes how states organize people to control, account for, and help them. The technology of housing is often an important part of this process and embodies the cultural values and assumptions of the designers. Users are imagined to be either like Westerners or in such a state that they should become more like Westerners. Thus, designs embody normative assumptions. Dawson (2008) terms these kinds of structures "unfriendly architecture." Buildings are symbolic technologies which influence and stabilize local, lived routines, creating a "normative landscape" (Gieryn 2000). The Inuit of Arviat, for example, were only implanted into their houses by the Canadian government in the last fifty years (Damas 2002; Duffy 1988; van den Scott 2009). Thomas and Thompson (1972) document the "crash" program of housing and aver that it was aimed to modify the behavior of the Inuit into Euro-Canadian patterns.

Indigenous groups are often overlooked in global ethnography (van den Scott, forthcoming) but have a particular interest in asserting and protecting the "particularities of their affiliations, territories and cultural meanings" (Kennedy, 2010, 193).

Traditional social constructivist studies of users have argued that technologies do not carry essential meanings, even if they carry scripts from the designers. Instead, such meanings can only arise through the social process of everyday use and interaction. It is a relevant next step for the ethnographer to explicitly look for users with a different cultural makeup to analyze this question in a more radical way. Hence, studying the reception of new technologies in the consumption junction of the remote and developing world pays dividends in appreciating interpretive flexibility at work. By paying close attention to the local, we deepen our understanding of how ordinary people navigate technologies within wider processes of globalization and networks of power.

Conclusion

Throughout this chapter we have argued that ethnography provides an avenue for those researching users of technology to advance and extend important theoretical debates, concepts, and understandings. We argue that ethnography facilitates such a study by capturing the "messy complexities" of everyday life (Clarke 2005). By studying situated actions and knowledges, ethnographers are able to study the unsettled and dynamic aspects of technology in relation to user groups, often in the midst of wider power struggles. Ethnography allows us to be attentive to the content of technology and its relation to the social world in which it gets shaped. Ethnography gathers discourse about the material agency of technology while being attentive to human-nonhuman interactions. It moves us away from the study of static objects toward a sociology of objects that move, but sometimes in ordered ways.

We began the chapter by reviewing how previous research on user studies has made evident the multiple and conflicting identities of users, non-users, and designers that have been articulated, scripted, performed, and transformed in both the design and use of technologies (Oudshoorn and Pinch 2005). We argue that ethnography provides a means for revitalizing research into users, but this requires asking new conceptual questions. This also means moving ethnography out of its old habits as well; ethnographers need to consider the structuring effects of technologies, move beyond methodological individualism, consider organizations as users, and look at how global technological expansion is in creative dialogue with the cultural particularities of non-Western groups. By expanding the conceptual repertoire in these ways, while maintaining the methodological advantage of studying users' actions, meanings, and interactions in situ, we believe that researchers can ask deeper questions about the human-technology relationship and contribute to broader debates on some of the rich conceptual themes we have raised here.

Acknowledgments

We would like to thank Fiona Miller for her involvement in the genesis of this chapter, and to acknowledge Sanders SSHRC Insight Development Grant #210201.

Notes

1. For in-depth reviews of "user-studies" literature, see Oudshoorn and Pinch 2005, 2008.

2. For a fascinating illustration of this problem, see Orr's (1996) research on the social worlds of photocopy repair technicians.

3. "Big data" refers to large data sets, including those that "consolidate many datasets from multiple sources" (Wigan and Clarke 2013, 46) and the tools and techniques used to analyze them.

References

Akrich, Madeleine. 1992. "The De-scription of Technical Objects" In *Shaping Technology, Building Society*, edited by Wiebe E. Bijker and John Law, 205–24. Cambridge, MA: MIT Press.

Albrow, Martin. 1996. *The Global Age: State and Society beyond Modernity*. Stanford, CA: Stanford University Press.

Allen, David K., and Tom D. Wilson. 2005. "Action, Interaction and the Role of Ambiguity in the Introduction of Mobile Information Systems in a UK Police Force." *Mobile Information Systems IFIP TC8 Working Conference on Mobile Information Systems (MOBIS)*, 15–17 September 2004. Oslo: Springer, 15–36.

Appadurai, Arjun. 1986. *The Social Life of Things: Commodities in Cultural Perspective*. Cambridge: Cambridge University Press.

Arnold, David, and Erich DeWald. 2011. "Cycles of Empowerment? The Bicycle and Everyday Technology in Colonial India and Vietnam." *Comparative Studies in Society and History* 53 (4): 971–96.

___. 2012. "Everyday Technology in South and Southeast Asia: An Introduction." *Modern Asian Studies* 46 (1): 1–17.

Bardini, Thierry, and August Horvath. 1995. "The Social Construction of the Personal Computer User." *Journal of Communication* 45 (3): 40–66.

Barratt, Paul. 2011. "Vertical Worlds: Technology, Hybridity, and the Climbing Body." *Social and Cultural Geography* 12 (4): 397–412.

Becker, Howard S. 1986. *Doing Things Together: Selected Papers*. Evanston, IL: Northwestern University Press.

Bennett, Colin, Kevin Haggerty, David Lyon, and Valerie Steeves. 2014. *Transparent Lives: Surveillance in Canada*. Edmonton: Athabasca University Press.

Berg, Marc. 1997. *Rationalizing Medical Work: Decision Support Techniques and Medical Practices*. Cambridge, MA: MIT Press.

___. 2001. "Implementing Information Systems in Health Care Organizations: Myths and Challenges." *International Journal of Medical Informatics* 64: 143–56.

Bijker, Wiebe E. 1993. "Do Not Despair: There Is Life after Constructivism." *Science, Technology, & Human Values* 18 (1): 113–38.

___. 1995. *Of Bicycles, Bakelites, and Bulbs: Toward a Theory of Sociotechnical Change*. Cambridge, MA: MIT Press.

Bijker, Wiebe E., and John Law. 1992. *Shaping Technology/Building Society: Studies in Sociotechnical Change*. Cambridge, MA: MIT Press.

Bishop, Ann, Laura Neumann, Susan Leigh Star, Cecelia Merkel, Emily Ignacio, and Robert Sandusky. 2000. "Digital Libraries: Situating Use in Changing Information Infrastructure" *Journal of American Society for Information Science* 51 (4): 394–413.

Bourdieu, Pierre. 1990. *The Logic of Practice*. Stanford, CA: Stanford University Press.

Bowker, Geoffrey, and Susan Leigh Star. 1999. *Sorting Things Out: Classification and Its Consequences*. Cambridge, MA: MIT Press.

Bowker, Geoffrey, Stefan Timmermans, and Susan Leigh Star. 1995. "Infrastructure and Organizational Transformation: Classifying Nurses' Work." In *Information Technology and Changes in Organizational Work*, edited by Wanda Orlikowski, Geoff Walsham, Matthew Jones, and Janice DeGross, 344–70. London: Chapman and Hall.

boyd, danah, and Kate Crawford. 2012. "Critical Questions for Big Data." *Information, Communication & Society* 15 (5): 662–79.

Burawoy, Michael. 2010. *Global Ethnography: Forces, Connections, and Imaginations in a Postmodern World*. Berkeley: University of California Press.

Chan, Janet. 2003. "Police and New Technologies" In *Handbook of Policing*, edited by Tim Newburn, 665–69. Cullompton, UK: Willan Publishing.

Charmaz, Kathy. 2014. *Constructing Grounded Theory*. Thousand Oaks, CA: Sage.

Clarke, Adele. 1991. "Social Worlds/Arenas Theory as Organizational Theory." In *Social Organization and Social Process: Essays in Honor of Anselm Straus,* edited by David Maines, 119–58. New York: Aldine de Gruyter.

___. 1998. *Disciplining Reproduction: Modernity, American Life Sciences, and "The Problems of Sex."* Berkeley: University of California Press.

___. 2005. *Situational Analysis: Grounded Theory after the Postmodern Turn.* Thousand Oaks, CA: Sage.

Clarke, Adele, and Theresa Montini. 1993. "The Many Faces of RU486: Tales of Situated Knowledges and Technological Contestations." *Science, Technology, & Human Values* 18 (1): 42–78.

Clarke, Adele, and Susan Leigh Star. 2008. "The Social Worlds Framework: A Theory/Methods Package." In *The Handbook of Science and Technology Studies.* 3rd ed., edited by Edward J. Hackett, Olga Amsterdamska, Michael Lynch, and Judy Wajcman, 113–37. Cambridge, MA: MIT Press.

Cockburn, Cynthia, and Susan Ormrod. 1993. *Gender and Technology in the Making.* Thousand Oaks, CA: Sage.

Coleman, Gabriella. 2004. "The Political Agnosticism of Free and Open Source Software and the Inadvertent Politics of Contrast." *Anthropological Quarterly* 77: 507–19.

Collins, Randall. 2004. *Interaction Ritual Chains.* Princeton, NJ: Princeton University Press.

Collins-Nelsen, Rebecca. 2010. *Retooling Gender? A Constructivist Analysis of Tomboy Tools.* Master's thesis, Lakehead University, Thunder Bay, Ontario.

Collyer, Fran. 2011. "Reflexivity and the Sociology of Science and Technology: The Invention of 'Eryc' the Antibiotic." *Qualitative Report* 16 (2): 316–40.

Cowan, Ruth Schwartz. 1983. *More Work for Mother: The Ironies of Household Technology from the Open Hearth to the Microwave.* New York: Basic Books.

___. 1987. "The Consumption Junction: A Proposal for Research Strategies in the Sociology of Technology." In *The Social Construction of Technological Systems: New Directions in the Sociology and History of Technology,* edited by Wiebe E. Bijker, Thomas Parke Hughes, and Trevor J. Pinch, 261–80. Cambridge, MA: MIT Press.

Curry, Michael, David Phillips, and Priscilla Regan. 2004. "Emergency Response Systems and the Creeping Legibility of People and Places." *The Information Society* 20: 357–69.

Cvetkovich, Ann, and Douglas Kellner. 1997. "Thinking Global and Local." In *Articulating the Global and the Local: Globalization and Cultural Studies,* edited by Ann Cvetkovich and Douglas Kellner, 1–32. Boulder, CO: Westview Press.

Damas, David. 2002. *Arctic Migrants/Arctic Villagers: The Transformation of Inuit Settlement in the Central Arctic.* Montreal: McGill-Queen's University Press.

Dawson, Peter C. 2008. "Unfriendly Architecture: Using Observations of Inuit Spatial Behavior to Design Culturally Sustaining Houses in Arctic Canada." *Housing Studies* 23 (1): 111–28.

Duffy, Ronald Q. 1988. *The Road to Nunavut: The Progress of the Eastern Arctic Inuit since the Second World War*. Kingston: McGill-Queen's University Press.

Emerson, Robert M., Rachel I. Fretz, and Linda L. Shaw. 1995. *Writing Ethnographic Fieldnotes*. Chicago: University of Chicago Press.

Entrena, Francisco. 2003. "Facing Globalization from the Local." In *Local Reactions to Globalization Processes*, edited by Francisco Entrena, ix–xix. New York: Nova Science Publishers.

Ericson, Richard V., and Kevin D. Haggerty. 1997. *Policing the Risk Society*. Toronto: University of Toronto Press.

Ess, Charles. 2002. "Ethical Decision-Making and Internet Research." *Association of Internet Researchers*. Available at http://aoir.org/reports/ethics.pdf.

Fairhurst, Eileen. 1999. "Fitting a Quart into a Pint Pot: Making Space for Older People in Sheltered Housing." In *Ideal Homes? Social Change and Domestic Life*, edited by Tony Chapman and Jenny Hockey, 96–107. London: Routledge.

Fine, Gary A. 2015. *Players and Pawns: How Chess Builds Community and Culture*. Chicago: University of Chicago Press.

Fleischmann, Kenneth R. 2006. "Boundary Objects with Agency: A Method for Studying the Design-User Interface." *Information Society* 22: 77–87.

Forsythe, Diana E. 2001. *Studying Those Who Study Us: An Anthropologist in the World of Artificial Intelligence*. Stanford, CA: Stanford University Press.

Fox, Kathryn. 1999. "Reproducing Criminal Types: Cognitive Treatment for Violent Offenders in Prison." *Sociological Quarterly* 40 (3): 435–53.

Freeman, David. 2001. "Antinomies of Agency, Structure, and Technology." *Melbourne Journal of Politics* 28: 30–52.

Fuchs, Stephan. 2001. "Beyond Agency." *Sociological Theory* 19 (1): 24–40.

Garfinkel, Harold. 1967. *Studies in Ethnomethodology*. Englewood Cliffs, NJ: Prentice Hall.

Giddens, Anthony.1979. *Central Problems in Social Theory*. Berkeley: University of California Press.

Gieryn, Thomas F. 2000. "A Space for Place in Sociology." *Annual Review of Sociology* 26: 463–96.

Glaser, Barney G., and Anselm L. Strauss. 1967. *The Discovery of Grounded Theory: Strategies for Qualitative Research*. Chicago: Aldine.

Hall, Edward T. 1983. *The Dance of Life: The Other Dimension of Time*. Garden City, NY: Anchor Press/Doubleday.

Hammersley, Martyn, and Paul Atkinson. 2007. *Ethnography Principles in Practice*. 3rd ed. Hoboken, NJ: Taylor & Francis.

Hasu, Mervi. 2000. "Constructing Clinical Use: An Activity-Theoretical Perspective on Implementing New Technology." *Technology Analysis & Strategic Management* 12: 369–82.

Hasu, Mervi, and Yrjoe Engestrom. 2000. "Measurement in Action: An Activity-Theoretical Perspective on Producer-User Interaction." *International Journal of Human-Computer Studies* 53 (1): 61–89.

Hier, Sean P. 2004. "Risky Spaces and Dangerous Faces: Urban Surveillance, Social Disorder and CCTV." *Social and Legal Studies* 13 (4): 541–54.

Hitchings, Russell. 2007. "Geographies of Embodied Outdoor Experience and the Arrival of the Patio Heater." *Area* 39 (3): 340–48.

Hyysalo, Sampsa. 2004. "Technology Nurtured—Collectives in Maintaining and Implementing Technology for Elderly." *Science Studies* 12 (2): 23–43.

Irvine, Leslie. 2003. "The Problem of Unwanted Pets: A Case Study of How Institutions 'Think' about Clients' 'Needs'." *Social Problems* 50 (4): 550–66.

Jasanoff, Sheila. 2006. *States of Knowledge: The Co-production of Science and the Social Order*. London: Routledge.

Jerolmack, Colin, and Iddo Tavory. 2014. "Molds and Totems: Nonhumans and the Constitution of the Social Self." *Sociological Theory* 32 (1): 64–77.

Joyce, Kelly. 2005. "Appealing Images: Magnetic Resonance Imaging and the Production of Authoritative Knowledge." *Social Studies of Science* 35 (3): 437–62.

Kennedy, Paul T. 2010. *Local Lives and Global Transformations: Towards World Society*. New York: Palgrave Macmillan.

Kien, Grant. 2009. *Global Technography: Ethnography in the Age of Mobility*. New York: Peter Lang.

King, Anthony D. 1989. "Colonialism, Urbanism and the Capitalist World Economy." *International Journal of Urban and Regional Research* 13: 1–18.

Kline, Ronald, and Trevor Pinch. 1996. "Users as Agents of Technological Change: The Social Construction of the Automobile in the Rural United States." *Technology and Culture* 37 (4): 763–95.

Knorr Cetina, Karin. 1999. *Epistemic Cultures: How the Sciences Make Knowledge*. Cambridge, MA: Harvard University Press.

Latour, Bruno. 1992. "Where Are the Missing Masses? The Sociology of a Few Mundane Artefacts." In *Shaping Technology/Building Society: Studies in Sociotechnical Change*, edited by Wiebe Bijker and John Law, 225–58. Cambridge, MA: MIT Press.

___. 2005. *Reassembling the Social: An Introduction to Actor-Network Theory*. Oxford: Oxford University Press.

Lawrence, Denise L., and Setha M. Low. 1990. "The Built Environment and Spatial Form." *Annual Review of Anthropology* 19: 453–505.

Lechner, Frank J., and John Boli. 2005. *World Culture: Origins and Consequences*. Malden, MA: Blackwell.

Lewis, J. David, and Andrew J. Weigert. 1981. "The Structures and Meanings of Social Time." *Social Forces* 60 (2): 432–62.

Lie, Merete, and Knut H. Sørensen, eds. 1996. *Making Technology Our Own? Domesticating Technology into Everyday Life*. Oslo: Scandinavian University Press.

Luthje, Christian, Cornelius Herstatt, and Eric von Hippel. 2005. "User-Innovators and 'Local' Information: The Case of Mountain Biking." *Research Policy* 34: 951–56.

MacKenzie, Donald, and Judy Wajcman. 1985. *The Social Shaping of Technology*. Milton Keynes: Open University Press.

Manning, Peter K. 2013. "Information Technology and Police Work." In *The Springer Encyclopedia of Criminology and Criminal Justice*, edited by Gerben Bruinsema and David Weisburd, 2501–13. New York: Springer.

McEneaney, John. 2013. "Agency Effects in Human-Computer Interaction." *International Journal of Human-Computer Interaction* 29 (12): 798–813.

Meehan, Albert. 1998. "The Impact of Mobile Data Terminal (MDT) Information Technology on Communication and Recordkeeping in Patrol Work." *Qualitative Sociology* 21 (3): 225–53.

Michael, Mike. 2000. "These Boots Are Made for Walking ... : Mundane Technology, the Body and Human-Environment Relations." *Body & Society* 6 (3–4): 107–26.

Miller, Daniel. 2002. "Accommodating." In *Contemporary Art and the Home*, edited by Colin Painter, 115–30. Berg: Oxford.

Moat, Helen, Tobias Preis, Christopher Olivola, Chengwei Liu, and Nick Chater. 2014. "Using Big Data to Predict Collective Behavior in the Real World." *Behavioral and Brain Sciences* 37 (1): 92–93.

Moreira, Tiago. 2004. "Self, Agency, and the Surgical Collective: Detachment." *Sociology of Health and Illness* 26 (1): 32–49.

Morton, Fiona M. Scott, and Joel Podolny. 2002. "Love or Money? The Effects of Owner Motivation in the California Wine Industry." *Journal of Industrial Economics* 50: 431–56.

Moses, Lyria, and Janet Chan. 2014. "Using Big Data for Legal and Law Enforcement Decisions: Testing the New Tools." *UNSW Law Journal* 37 (2): 642–78.

Mukerji, Chandra. 1994. "Toward a Sociology of Material Culture: Science Studies, Cultural Studies and the Meanings of Things." In *The Sociology of Culture: Emerging Theoretical Perspectives*, edited by Diana Crane, 143–62. Oxford: Blackwell.

Neff, Gina. 2005. "The Changing Place of Cultural Production: The Location of Social Networks in a Digital Media Industry." *Annals of the American Academy of Political and Social Science* 597: 134–52.

Opel, Andy. 2004. *Micro Radio and the FCC: Media Activism and the Struggle over Broadcast Policy.* Westport, CT: Praeger.

Orlikowski, Wanda J. 2000. "Using Technology and Constituting Structures: A Practice Lens for Studying Technology in Organizations." *Organization Science* 11 (4): 404–28.

Orlikowski, Wanda J., and Debra Gash. 1994. "Technological Frames: Making Sense of Technology in Organizations." *ACM Transactions on Information Systems* 12 (2): 174–207.

Orlikowski, Wanda J., and Daniel Robey. 1991. *Information Technology and the Structuring of Organizations.* Cambridge, MA: MIT Press.

Orr, Julian E. 1996. *Talking about Machines: An Ethnography of a Modern Job.* Ithaca, NY: ILR Press.

Oudshoorn, Nelly. 1996. "Gender Scripts in Technologie: Noodlot of Uitdaging?" *Tijdschrift voor Vrouwenstudies* 17 (4): 350–68.

___. 2005. "Clinical Trials as a Cultural Niche in Which to Configure the Gender Identities of Users: The Case of Male Contraceptive Development." In *How Users Matter: The Co-Construction of Users and Technology*, edited by Nelly Oudshoorn and Trevor Pinch, 209–27. Cambridge, MA: MIT Press.

Oudshoorn, Nelly, and Trevor Pinch, eds. 2005. *How Users Matter: The Co-Construction of Users and Technology.* Cambridge, MA: MIT Press.

___. 2008. "User-Technology Relationships: Some Recent Developments" In *The Handbook of Science and Technology Studies.* 3rd ed., edited by Edward J. Hackett, Olga Amsterdamska, Michael Lynch, and Judy Wajcman, 541–66. Cambridge, MA: MIT Press.

Pickering, Andrew. 2005. "Decentering Sociology: Synthetic Dyes and Social Theory." *Perspectives on Science* 13 (3): 352–405.

Pinch, Trevor, and Wiebe E. Bijker. 1984. "The Social Construction of Facts and Artefacts: Or How the Sociology of Science and the Sociology of Technology Might Benefit Each Other." *Social Studies of Science* 14 (3): 399–441.

Pinch, Trevor, and Frank Trocco. 2002. *Analog Days: The Invention and Impact of the Moog Synthesizer.* Cambridge, MA: Harvard University Press.

Pols, Jeannette, and Dick Willems. 2011. "Innovation and Evaluation: Taming and Unleashing Telecare Technology." *Sociology of Health and Illness* 33 (2): 484–94.

Preda, Alex. 2006. "Socio-Technical Agency in Financial Markets: The Case of the Stock Ticker." *Social Studies of Science* 36 (5): 753–82.

Prus, Robert. 1996. *Symbolic Interaction and Ethnographic Research: Intersubjectivity and the Study of Human Lived Experience.* Albany: State University of New York Press.

Puddephatt, Antony. 2005. "Mead Has Never Been Modern: Using Meadian Theory to Extend the Constructionist Study of Technology." *Social Epistemology* 19 (4): 357–80.

___. 2008. "Incorporating Ritual into Greedy Institution Theory: The Case of Devotion in Organized Chess." *Sociological Quarterly* 49 (1): 155–80.

Robertson, Roland. 2001. "Globalization Theory 2001+: Major Problematics." In *Handbook of Social Theory*, edited by George Ritzer and Barry Smart, 458–71. London: Sage.

Rommes, Els. 2002. *Gender Scripts and the Internet: The Design and Use of Amsterdam's Digital City.* Ph.D. thesis. Twente University Press, Enschede, The Netherlands.

Rommes, Els, Ellen van Oost, and Nelly Oudshoorn. 1999. "Gender and the Design of a Digital City." *Information Technology, Communication and Society* 2 (4): 476–95.

Samuelsson, Tobias, and Boel Berner. 2013. "Swift Transport versus Information Gathering: Telemedicine and New Tensions in the Ambulance Service." *Journal of Contemporary Ethnography* 42 (6): 722–44.

Sanders, Carrie B. 2014. "Need to Know vs. Need to Share: The Intersecting Work of Police, Fire and Paramedics." *Information, Communication and Society* 17 (4): 463–75.

Sanders, Carrie B., and Samantha Henderson. 2013. "Police 'Empires' and Information Technologies: Uncovering Material and Organisational Barriers to Information Sharing in Canadian Police Services." *Policing and Society: An International Journal of Research and Policy* 23 (2): 243–60.

Schot, Johan. 1992. "Constructive Technology Assessment and Technology Dynamics: The Case of Clean Technologies." *Science, Technology, & Human Values* 17 (1): 36–56.

Schot, Johan, and Adri de la Bruheze. 2003. "The Mediated Design of Products, Consumption, and Consumers in the Twentieth Century." In *How Users Matter: The Co-Construction of Users and Technologies*, edited by Nelly Oudshoorn and Trevor Pinch. Cambridge, MA: MIT Press.

Schudson, Michael. 2002. "How Culture Works: Perspectives from Media Studies on the Efficacy of Symbols." In *Cultural Sociology*, edited by Lyn Spillman, 141–48. Malden, MA: Blackwell.

Scott, James C. 1998. *Seeing Like a State: How Certain Schemes to Improve the Human Condition Have Failed.* New Haven, CT: Yale University Press.

Shaffir, William, and Robert A. Stebbins. 1991. *Experiencing Fieldwork: An Inside View of Qualitative Research.* Newbury Park, CA: Sage.

Shah, Sonali K., and Mary Tripsas. 2007. "The Accidental Entrepreneur: The Emergent and Collective Process of User Entrepreneurship." *Strategic Entrepreneurship Journal* 1 (1–2): 123–40.

Söderberg, Jan. 2011. "Free Space Optics in the Czech Wireless Community: Shedding Some Light on the Role of Normativity for User-Initiated Innovations." *Science, Technology, & Human Values* 36 (4): 423–50.

Star, Susan Leigh. 1991. "Invisible Work and Silenced Dialogues in Representing Knowledge." In *Women, Work and Computerization: Understanding and Overcoming Bias in Work and Education,* edited by Inger Eriksson, Barbara A. Kitchenham, and Kea G. Tijdens, 81–92. Amsterdam: North Holland.

___. 1995. "The Politics of Formal Representations: Wizards, Gurus, and Organizational Complexity." In *Ecologies of Knowledge: Work and Politics in Science and Technology*, edited by Susan L. Star, 88–118. Albany: State University of New York Press.

Star, Susan Leigh, and Anselm Strauss. 1999. "Layers of Silence, Arenas of Voice: The Ecology of Visible and Invisible Work." *Computer Supported Cooperative Work* 8: 9–30.

Suchman, Lucy A. 1987. *Plans and Situated Actions: The Problem of Human-Machine Communication*. New York: Cambridge University Press.

___. 2002. "Located Accountabilities in Technology Production." *Scandinavian Journal of Information Systems* 14 (2): Article 7.

Suchman, Lucy A., Jeannette Blomber, Julian E. Orr, and Randall Trigg. 1999. "Reconstructing Technologies as Social Practice." *American Behavioral Scientist* 43 (3): 392–408.

Surak, Kristin. 2013. *Making Tea, Making Japan: Cultural Nationalism in Practice*. Stanford, CA: Stanford University Press.

Taylor, Charles. 1999. "Two Theories of Modernity." *Public Culture* 11 (1): 153–74.

Taylor, Jean Gelman. 2012. "The Sewing-Machine in Colonial-Era Photographs: A Record from Dutch Indonesia." *Modern Asian Studies* 46: 71–95.

Thomas, David K., and Charles Thomas Thompson. 1972. *Eskimo Housing as Planned Culture Change*. Ottawa: Northern Science Research Group, Department of Indian Affairs and Northern Development.

Tunc, Tanfer Emin. 2009. "Technologies of Consumption: The Social Semiotics of Turkish Shopping Malls." In *Material Culture and Technology in Everyday Life*, edited by Phillip Vannini, 131–44. New York: Peter Lang.

van den Hoonaard, Deborah K. 2014. *Qualitative Research in Action: A Canadian Primer*. 2nd ed. Don Mills, Ontario: Oxford University Press Canada.

van den Scott, Lisa-Jo K. 2009. "Cancelled, Aborted, Late, Mechanical: The Vagaries of Air Travel in Arviat, Nunavut, Canada." In *The Cultures of Alternative Mobilities: Routes Less Travelled*, edited by Phillip Vannini, 211–26. Surrey: Ashgate.

___. Forthcoming. "Mundane Technology in Non-Western Contexts: Wall-as-Tool." In *Sociology of Home: Belonging, Community and Place in the Canadian Context*, edited by Laura Suski, Joey Moore, and Gillian Anderson. Toronto: Canadian Scholars Press International.

van Oost, Ellen. 2003. "Materialized Gender: How Shavers Configure the Users' Femininity and Masculinity." In *How Users Matter: The Co-Construction of Users and Technology*, edited by Nelly Oudshoorn and Trevor Pinch, 193–208. Cambridge, MA: MIT Press.

van Oost, Ellen, Stefan Verhaegh, and Nelly Oudshoorn. 2009. "From Innovation Community to Community Innovation: User-Initiated Innovation in Wireless Leiden." *Science, Technology, & Human Values* 34 (2): 182–205.

Vaughan, Diane. 1999. "The Role of the Organization in the Production of Techno-Scientific Knowledge." *Social Studies of Science* 29 (6): 913–43.

Verbeek, Peter-Paul. 2006. *What Things Do: Philosophical Reflections on Technology, Agency, and Design*. University Park: Pennsylvania State University Press.

von Hippel, Eric. 1976. "The Dominant Role of Users in the Scientific Instrument Innovation Process." *Research Policy* 5 (3): 212–39.

___. 1986. "Lead Users: A Source of Novel Product Concepts." *Management Science* 32 (7): 791–805.

___. 2005. *Democratizing Innovation*. Cambridge, MA: MIT Press.

Wagner-Pacifici, Robin, and Barry Schwartz. 1991. "The Vietnam Veterans Memorial: Commemorating a Difficult Past." *American Journal of Sociology* 97 (2): 376–420.

Wigan, Marcus, and Roger Clarke. 2013. "Big Data's Big Unintended Consequences." *Computer Society* 46 (6): 46–53.

Woolgar, Steve. 1991. "Configuring the User: The Case of Usability Trials." In *The Sociology of Monsters*, edited by John Law. London: Routledge.

Wyatt, Sally. 2003. "Non-Users Also Matter: The Construction of Users and Non-Users of the Internet." In *How Users Matter: The Co-Construction of Users and Technologies*, edited by Nelly Oudshoorn and Trevor J. Pinch. Cambridge, MA: MIT Press.

___. 2008. "Technological Determinism Is Dead; Long Live Technological Determinism." In *The Handbook of Science and Technology Studies*. 3rd ed., edited by Edward J. Hackett, Olga Amsterdamska, Michael Lynch, and Judy Wajcman, 165–80. Cambridge, MA: MIT Press.

Wyatt, Sally, Graham Thomas, and Tiziana Terranova. 2003. "They Came, They Surfed, They Went Back to the Beach: Conceptualizing Use and Non-Use of the Internet." In *Virtual Society? Technology, Cyberhole, Reality*, edited by Steve Woolgar, 23–41. Oxford: Oxford University Press.

18   How Infrastructures Matter

Stephen C. Slota and Geoffrey C. Bowker

Introduction

In this chapter we discuss the sociotechnical concept of infrastructure with a particular focus on those theories and discussions of infrastructure relevant to various forms of knowledge and scientific work. Infrastructure, in a simple (though somewhat flawed) formulation, refers to the prior work (be it building, organization, agreement on standards, and so forth) that supports and enables the activity we are really engaged in doing. More particularly, infrastructure refers to those systems, technologies, organizations, and built artifacts that do not need to be reconsidered at the start of a new venture. A chef does not need to know everything about the infrastructural network of pumps, sewers, reservoirs, filters, and regulations to fill a stockpot with water for soup—she turns the handle and clean, potable water comes out of the faucet. In much the same way, a librarian does not need to rebuild his classification scheme each time he wants to add a new book to the collection. Electricians and contractors, with their infrastructure of regulations and formal standards, do not worry that their building projects will fail to connect to the relevant grids or be accessible by vehicle when connected to local roads—the work of interoperability and access has already been done. Uber and Lyft did not need to rebuild the entire network of roads to establish their transportation system. Infrastructure is pervasive and ubiquitous, and many otherwise distant fields of scholarly work have seen (or are discovering) the need to negotiate infrastructure in order to support fundamentally new kinds of work. While it is difficult to point to a stable and consistent definition of *infrastructure*, or even to find a singular point of origin for what we think of as modern infrastructural theory, there is a cogency to the emergent field which we will endeavor to set forth. Throughout this chapter, we will describe several seemingly disparate historical trajectories (or relatively isolated instances) of thinking and research that have worked together to inform how we understand infrastructure theory and infrastructural work.

Infrastructure subtends our lives in so many ways. Without built infrastructure, such as roads and railways, we can reach neither work nor the seaside. Without an electricity grid or gas and oil pipelines, we would be miserable once we got there. Without an information infrastructure (the Internet, mobile communications), our work and play would be vastly different. However, often, in the way we tell stories about history or society, all of this infrastructure is left out or made invisible. It fades into the background so that the real story can be told: the development of a social movement, the toppling of a power structure, the invention of the standard model of particle physics. Yet, as we will argue, understanding infrastructure—it's affordances and constraints—will be core to understanding any of these. At first blush, it would seem that the highway system has very little in common with the Worldwide Large Hadron Collider Computing Grid, but both are clearly forms of infrastructure despite their substantial differences. In what sense, then, is it useful to assert that roads, pipes, electrical networks, the Internet, and cyberinfrastructures (CI) fall under a single rubric? It is one of the goals of infrastructure studies to address infrastructure as infrastructure despite the material heterogeneity of its physical appearances. This only works if the concept provides a sharing ground for both theoretical and practical developments based on commonalities among otherwise disparate entities such that insights from roads really do inform how scientists cooperate over shared information resources and technology. The central task of this chapter is to demonstrate that this is the case and to show the importance of thinking about both technoscientific and societal change in terms of infrastructure.

One of the most important developments in science and technology studies (STS) has been to refocus attention away from the spectacle of the pageant of history toward the formation and operation of infrastructure—an approach that has been called infrastructure inversion (Bowker 1994). The core reason for making this move is the postulate that infrastructure matters. Infrastructure is not a neutral background that enables an infinite set of activities. Infrastructure holds values, permits certain kinds of human and nonhuman relations while blocking others, and shapes the very ways in which we think about the world. This is evident in Veyne's (2013) proposition that it is impossible to trace the development of the concept of democracy over time because the performance of democracy changes fundamentally with new infrastructural developments. Meeting in an agora or town square to determine matters of concern is fundamentally different from holding discussions through print media or following a 24-hour news cycle on electronic media and voting by mail. As Richard John (2009) has pointed out, the United States of America could not coalesce as a nation without the cheap circulation of newspapers across the territories through the infrastructure of the Post Office: this permitted the engagement of a national debate among residents of the otherwise remote places. As U.S. news and deliberations about them have increasingly shifted online, not only is the viability of newspapers increasingly threatened but the conceptual and organizational foundations of U.S. democracy are shifting.

For this chapter we take a relaxed definition of science and technology studies. The field has developed without the brick and mortar instantiations of the traditional discipline. It itself is a traveling field that has become infrastructural, dare we say, to a number of other fields, including computer-supported cooperative work, media studies, and organization theory. As it becomes more evident that infrastructure matters, scholars and theorists from a wide range of disciplines are finding it necessary to grapple with infrastructure from their own perspectives, introducing significant variety and a multitude of historical vectors. In this chapter we take STS infrastructural theory to include work from any field producing analyses about infrastructure that makes use of and develops concepts we think of as being integral to the field of STS because of how they draw on and contribute to the canon of STS literature.

We argue that one of the key insights of STS has been to treat infrastructure relationally: it is not so much a single thing as a bundle of heterogeneous things (standards, technological objects, administrative procedures—in Foucault's term, a *dispositif technique* [Foucault 1979])—which involves both organizational work as well as technology. An infrastructural relationship is one among goals such as getting to work, being able to see in order to read, and sharing data in order to produce transformative science and means such as roads, light bulbs connected to the electricity network, and cyberinfrastructure. Infrastructure does not exist in a vacuum: much as in Engeström's (1990) argument about tools, the question is not whether a given thing *is in essence* an infrastructure but *when* it is an infrastructure. There is no system that is inherently infrastructural; there are only observed infrastructural relationships. As Star and Ruhleder (1996) have argued so eloquently, one person's invisible infrastructure is another person's job, to be faced materially and directly every day. Infrastructure, as they argue, is inherently relational—a given system, technology, or organization is *infrastructural to* a particular activity at a particular time.

Looking Back to Understand the Present: Infrastructure as Physical Substrate

Some key moments in the study of built infrastructure presage the STS study of infrastructure. Writing in the nineteenth century, the golden age of modern historiography, Karl Marx made a distinction between base and superstructure (1970), which can be read as an early infrastructural turn. The base—that which underpinned all of social action—was the economic mode of production, be this slavery or feudalism, bourgeois or revolutionary. The superstructure comprised the content that was produced—culture, ideology, social institutions, and so forth. In a telling phrase, he argued that human history was determined *in the last instance* by the base (1970). Louis Althusser (2006) later developed a highly sophisticated version of the relationship between the two as mutually constitutive—though preserving the last instance—partly in order to explain the Russian revolution occurring before what Marxists saw as the logically prior revolution (because the contradictions were further developed) in Western Europe. However, while the specter of Marx (Sartre 2001) haunted much of early STS research, it has not been manifest in most, sadly idealist, accounts of the development of our field. One of the testaments to his influence is Lewis Mumford's magisterial *City in History* (1961), which developed Marxist attention to "substructure" to draw attention to the role of civic infrastructure (sewers, roads, aquifers) in the development of urban life. (Mumford, J. D. Bernal, and Joseph Needham—three founders of the influential journal *Science and Public Policy*—were central to bringing this perspective into the nascent field of science studies.)

A more recent starting point for infrastructure studies is Thomas Hughes's *Networks of Power* (1993), a study of the development of electricity networks in the late nineteenth and early twentieth centuries in Germany, Great Britain, and the United States. Hughes, coming from engineering and systems thinking, laid the basis for a general history of physical infrastructure, noting the continuity in personnel and techniques between the canal builders of the late eighteenth century and the architects of nineteenth- and twentieth-century railway, telephone, and electrical networks. It is simple (by way, for example, of Shannon and Weaver [1948]) to extend this to the rise of the Internet. The core insight that Hughes developed was that of the reverse salient. A reverse salient can be defined as an obstruction that prevents an infrastructural system from being developed. This might be a technical matter (could you find the right filament to make incandescent lights which would last?) or a social one (fear of direct current because it was tied to the electric chair, so it was seen as a dangerous housemate). From a systems perspective, it made no difference where the reverse salients were occurring: they had to be addressed for the infrastructure itself to develop. This systems view found ready acceptance in Bruno Latour's *Science in Action* (1987). For Hughes and Latour, infrastructure was not just technology: it was always already braided with social, cultural, and political actors and their values.[1]

These threads bring up a couple of important methodological points about how the overall field of STS approaches infrastructural concerns in its own scholarly work. One is the curious reluctance of the field of STS to address issues of systems, even though much of the work in our field has been directly or indirectly affected by cybernetics (concepts of the black box, the analytic interchangeability of human and nonhuman actors in actor-network theory (ANT), for example—although the latter is more often attributed to semiotics [Propp 2010]). The allergy to "systems" is due to the founding methodological commitments of much science studies. Thus, a number of canonical early works (Knorr Cetina and Mulkay 1983; Latour 1987; Lynch 1985) were written from a nascent ANT or ethnomethodological perspective, each of which eschewed systemic factors for the active work of building networks or sustaining truths. Indeed, Latour (1987) argued that the ANT and ethnomethodological traditions were largely one. The systemic was largely seen as too determinative, partly as a fallout from the supposed problems of (structural) functionalism. Callon's (1990) work on technoeconomic systems and irreversibility is one of the few to address the issue directly from a science studies perspective—in general, networks were seen as more actively constructed than systems.

The second is about the methodology of this chapter and is rather more complex. As for many fields of intellectual history, within STS accounts—paradoxically—we pay little attention to our own sociomaterial infrastructure. Were we to tell the tale in fully consistent fashion, we would need to tell the story of the set of institutional configurations which permitted the rise of the field—how science studies formed separately in England, France, the United States, and the Netherlands, say, with different sets of commitments which have continued to mesh awkwardly. However, in the interests of time and diplomacy, we will follow the conceptual threads rather than their infrastructural roots—even if this is at the price of some inconsistency, since we often call attention to the role of infrastructure in the ideas of others.[2]

Alongside Hughes's work as foundational to infrastructure studies must figure Langdon Winner's influential paper "Do Artefacts Have Politics?" (1980). The argument most picked up from this work deployed the case of New York city planner Robert Moses to demonstrate that built infrastructure was itself politically active: if you build bridges that prevent buses from passing under expressways, then you can exclude the great unwashed from public beaches in ritzy neighborhoods. Although the case itself has been comprehensively disproven (Joerges 1999; Woolgar 1993), its message has continued to resonate, as indicated by Latour's dictum, adapting Clausewitz, that technology is politics by other means (1987 cf. Callon 1986, and Strum and Latour 1987). Callon (1986) wrote a canonical article in this area, which characterized a debate between two plans for the use of electric cars in France as being actually a contest between two incompatible theories of political sociology. In this time there is also a growing trend, as noted by Joerges, toward conceiving of the social effects of infrastructure—in a "heroic simplification" (Joerges 1999, 422)—as characterized by control or contingency. In a discourse of control, according to Joerges, social change is the result of intentional action, with all other effects being classed as something like unintended side effects. A contingent discourse (for example, cf. Woolgar 1986), alternatively, sees social order and disorder as a result of the confluence of a variety of consequences of many small actions. Each action is intentional, but the result, for all intents, is blind—coming from many independent actors making sense as best they can. Both these discourses inform later discussions of infrastructure, particularly when attempting to address the broader effects of new information infrastructure and algorithmic ways of knowing.

A Note on the Return to the Material

In recent years there have been moves, particularly within anthropological STS, to study aspects of physical infrastructure in a new light. Within STS, Dourish and Mazmanian (2013) propose a returned focus on the material in analyzing digital representations and information/organizational practice. From a more anthropological perspective, Ashley Carse's (2012) examination of the Panama Canal, Klose and Macrum's (2015) work on the history of containers, and Penny Harvey and Hannah Knox's rich analysis of road building in Peru each have drawn on studies of infrastructure through the lens of technopolitics, which works from the understanding that technology is primarily a political actor (Larkin 2013). In similar fashion, STS analyses of information and knowledge infrastructures have developed a sensitivity to the importance of material conditions for knowledge production. There has recently begun a corresponding move back to the materiality of fixed infrastructure as a means of understanding sociopolitical discourse. "Roads and railways are not just technical objects then but also operate on the level of fantasy and desire. They encode the dreams of individuals and societies and are the vehicles whereby those fantasies are transmitted and made emotionally real" (Larkin 2013, 333). Infrastructure in this mode of inquiry is seen as integrally material and political: it promises a certain kind of utopian future, in the form of a world subtended by and supported through new infrastructure. A new infrastructural project then is fundamentally about the future and possibility. The material of failed or abandoned infrastructures is a record of the dreams and imaginations of the past.

As prospective statements about the future, infrastructures can be discussed in terms of their poetics. In the poetic mode of infrastructure, the technical capability of the system is treated separately from the future that infrastructure is represented to enable. "Infrastructures are the means by which a state proffers these representations to its citizens and asks them to take those representations as social facts" (Larkin 2013, 335). The centrality of the material in anthropological infrastructure studies engenders a discussion of "embodied experience governed by the ways infrastructures produce the ambient conditions of everyday life: our sense of temperature, speed, florescence, and the ideas we have associated with these conditions" (Larkin 2013, 335). Interactions with infrastructure govern not just the aesthetic experience of the world, they define imaginaries of what is possible and potentially possible and are presented politically as a pathway to those potentials.

Responding to Changing Technology: Information Infrastructures

As the newly forming field of STS took a closer look at the development of new infrastructure rather than study extant infrastructures, a new phrase entered the lexicon: *information infrastructure*. We now need to use the retronym of *built infrastructure* to designate what *infrastructure* used to mean (cf. Lakoff 1993), much as we use *snail mail* to designate the postal service. Of course, information infrastructures are built, too, (Starosielski 2015) and earlier ideas about how to think about infrastructure could be ported into the new space. However, the new formations did cause a change in analytic and methodological focus toward studying "assemblages" of human and non-human actors (or, in simpler language, people and technology taken together). The underlying reason for this is in the way that new technology made the distribution of agency between humans and nonhumans more immediately obvious. More and more often people were interacting primarily with a relatively automated piece of technology where they used to interact with a person. In a sense, the world has become more overtly amenable to science studies: the discussion within communities building the Internet of Things bears striking similarity to STS infrastructural understandings of the distribution of agency between humans and nonhumans (Latour 1992).

What is now called information infrastructure was once the domain of the library scientist (managing of books, questions of information retrieval) and the organizational theorist (file structures and so forth). The nature of scholarly production and the profession of librarian have changed greatly over the past thirty years. The very infrastructure that had grown throughout the nineteenth and twentieth centuries (file folders, typewriters, carbon paper, hanging files, Xerox machines) has been largely displaced (Yates 1993), though its terms linger. In step, the dominance of computer science across multiple disciplines has increased (echoing the earlier dominance of statistics with the development of governmentality [Foucault 1991]).

The work of economist Paul David (1990) provides a double transition here. First, David argued classically that computers were like electric dynamos, in the sense that when both were introduced into the workplace, there was a cultural lag between innovation and usability. One needed not just the infrastructural change (from gas to electric, from typewriters to personal computers), one needed to develop the associated infrastructural imaginary. In both cases, as he pointed out, there was a "productivity gap" of about twenty years before the use of new technology started to make a fundamental change. Hannemyr (2003) has made the related argument that the Internet did not "explode," as many assume, but rather followed a trajectory of deployment very similar to other communication technologies such as the telephone. Second, David (1990) developed the theory of network externalities. If I am a member of a phone network with two others, it costs me nothing if three more users are added, but it affords me value. Where Hughes had argued the importance of load bearing for the economics of built infrastructure, David argued the importance of network effects for the new economy.

In many ways, the emergent focus on information infrastructures over built infrastructure was a recognition of the presence, importance, and social effects of existing information infrastructures. A possible narrative of the development of modern infrastructure studies could trace its origin to the broad successes of distributed scientific collaboration during World War II and the recognition of the potential and actual impacts of basic research (especially in physics). The period of time following World War II is generally characterized as the point in time where *big science* began to emerge. Following the war, the scale of inquiry among theoretical and especially experimental physicists expanded substantially, and other areas of scientific study followed suit (Galison and Hevly 1992). Vannevar Bush's 1945 proposal, "Science: The Endless Frontier"—a proposal that contributed in part to the 1950 passage of the National Science Foundation Act—takes on the appearance, in hindsight, of a proto-cyberinfrastructural effort with the stated goal of capitalizing on existing scientific capital through investment in basic research and policy support (Bush 1945). Bush's vision of interdisciplinarity and distant collaboration has become a clarion cry for cyberinfrastructural development. Core pathways for achieving this in the 1980s and 1990s were the funding of collaboratories (Bos et al. 2007; Kouzes, Myers, and Wulf 1996; Olson and Olson 2000) and digital libraries (Bishop et al. 2000; Van House, Bishop, and Buttenfield 2003) as infrastructural efforts attempting to capitalize on new technology.

Collaboratories were initially conceived as a "center without walls, in which the nation's researchers can perform their research without regard to physical location, interacting with colleagues, accessing instrumentation, sharing data and computational resources, [and] accessing information in digital libraries" (Wulf 1989). These were shown to be not just "an elaborate collection of information and communications technologies," they were "a new networked organizational form that also include[d] social processes; collaboration techniques; formal and informal communication; and agreement on norms, principles, values, and rules" (Cogburn 2003).

It was in the era of collaboratories that Star and Ruhleder published a seminal work in infrastructure studies, one that recognized and formalized many of the preceding concepts of infrastructure and heavily influenced future methodological work in infrastructure studies. "Steps towards an Ecology of Infrastructure" (1996) was an ethnographic study of a collaboratory of biologists and computer scientists working with the Worm Community System (WCS), a digitized library of *C. elegans* flatworm specimens and technologically mediated paths for collaboration among the biologists working with them. In many ways the WCS was an ideal site for the implementation of new computing infrastructure: there was already a social expectation of collaboration and a well-established network of biologists sharing specimens (one of the traits making *C. elegans* an excellent laboratory subject was its amenability to being transported by post). However, by most metrics the WCS was a failed project with little uptake among the studied communities. It is in their accounting for that failure that Star and Ruhleder propose infrastructural issues as major factors influencing that outcome.

Infrastructure, as defined by Star and Ruhleder (1996), has a variety of different dimensions. It is embedded and transparent; infrastructure exists (metaphorically) within or underneath other social, technological, and built worlds and does not need to be reconsidered at the moment of each task it enables. Infrastructure is learned as a part of membership in a given community and linked with the conventions of practice therein and embodies some set of standards. It is built above an installed base, becomes visible upon breakdown, and is of a scale or scope that exceeds a single "site"—however that might be conceived. The heterogeneous nature of this list was quite deliberate: it was less Borges's Chinese encyclopedia[3] cited by Foucault (2002) than a recognition that infrastructure was integrally a social, organizational, and physical phenomenon.

The shift from collaboratories and digital libraries to cyberinfrastructure (CI), or e-science, is characterized by the uptake of concepts and methods from science studies, which added sociotechnical understanding to developments previously conceived as purely technical (Atkins 1996; Jackson and Buyuktur 2014; Lee, Dourish, and Mark 2006; Ribes and Finholt 2009). Latour and Woolgar's work in *Laboratory Life* (1979) had already drawn attention to the substrate of intellectual production—the document as "immutable mobile" that could circulate along scientific networks. Star and Ruhleder's research helped to point out something that was becoming increasingly evident: infrastructure to support large-scale, distributed, scientific collaboration responds to factors other than the availability and ease of use of new communicative and sharing resources. That the WCS was underused was substantively less interesting than why and how it was.

As the collaboratory and the digital library gave way, in the wake of National Science Foundation (NSF) programs, to the concept of cyberinfrastructure in the United States (eScience/eResearch elsewhere), there were two corresponding intellectual developments. The uptake of research infrastructure as a significant topic among scholars in STS saw a growing agreement on methods (the influence of ethnomethodology and grounded theory are core: both pay close attention to analysis of social discourse in material settings) and a broadening of the theoretical space of infrastructure into issues of policy, temporality, design, and values. Scholars studied the sociology of knowledge production in the presence of particular funding regimes (Gibbons et al. 1994) at a time where emergent roles in knowledge production, such as developing simulations and computer models of observed phenomena, were deeply entangled in the development of information infrastructure (Galison 1996).

In an influential report for the NSF, Edwards et al. (2013) argued that one could learn a lot about CI from studying the lessons of built infrastructure. Thus, "path dependence" (the concept that sociotechnical trajectories can get locked in, making switching to another path progressively more difficult) is key to the development of transportation or information networks; it is also crucial for CI. Similarly, the need to study standards (as brought out by Egyedi [2014]) works across both domains. It is possible to reconsider the information qualities of built infrastructure for insights about both. Brand's (1995) writing about how buildings learn shows one trajectory of discussing the built world as an informational good embodying a given record of human activity. Busch's (2011) *Standards: Recipes for Reality* and Lampland and Star's *Standards and Their Stories* (2009) (which includes an appendix on how to teach about standards) develop these issues further by providing a rich array of examples of the centrality of standards like the "standard size" of a human rear end for determining the minimum width of an airline seat, or the definition of the "standard human" in life insurance.

Thinking Algorithmically: Knowledge Infrastructures

Edwards et al. (2013) have argued that we need to move beyond the concept of information infrastructure to that of knowledge infrastructure in order to explore the ways in which knowledge work is changing in the twenty-first century: that is to say that publishing and exchanging knowledge requires different configurations from publishing and exchanging information and data—a core concern here is that new kinds and configurations of actors are needed to support knowledge infrastructures. Studies of infrastructure at work seek to unearth marginalized work, recognize the invisibles and occlusions as they influence action at a variety of levels, and propose ways in which these marginalia might be accounted for in future work. Knowledge infrastructures expand beyond instrumentation and work practice to account for the presence of political considerations, values and other invisibles that work to allow the exchange and proliferation of knowledge throughout a group. Instead of an either/or distinction between group membership and social isolation, researchers can bring to bear in their analysis a set of structural variables, such as the density of a network, how tightly it is bounded, and whether it is diversified or constricted in its size and heterogeneity, how narrowly specialized or broadly multiplex are its relationships, and how indirect connections and positions in social networks affect behavior.

On February 24, 2014, *Nature* author Van Noorden reported: "The publishers Springer and IEEE are removing more than 120 papers from their subscription services after a French researcher discovered that the works were computer-generated nonsense." The proximate culprit was a program called SciGen, developed at MIT to prove that conferences and journals would accept gobbledegook. Cyril Labbé (Labbé and Labbé 2013), who detected the fakes (not through reading but through an algorithm) had increased his own *h*-factor in Google Scholar by seeding the academic world with his own automatic writing. On April 3, 2015, *Science* author John Bohannon reported on the serious use of these papers: rather than demonstrate a flaw in our academic system, they exploit it.

What both together suggested was that the problem lay not with the disciplines per se but with the current state of the work of reading. We piece apart our papers into least publishable units so that rather than produce carefully reasoned works containing rich arguments, we publish hither and yon snippets of findings which in some notional database could be assembled into something resembling a coherent text. The work for new knowledge infrastructures here is to develop new modes of publication and expression (which are core to all academic work). Deploying the knowledge infrastructure inherited from the Enlightenment, in many disciplines, we write algorithmically: that is to say, we conjure our research into a format prescribed by a given journal or field (for example, sections literature review, methods, findings, discussion, and conclusion). Though there are far too few studies of reading practices in science studies, there are a number of fields in which methods sections are "plug and play" such that nobody reads in depth, just scans to see if the references are there—this is similarly evident in literature reviews. The goods that circulate, the academic papers, increasingly take the form of pure commodities, circulating in an ever-more rapid space and time. Put this way, it is relatively clear to see that our current knowledge infrastructure partakes of the same complex sociotemporality as the world around it—there is a resonance between our economic and knowledge infrastructures.

Indeed, Alfred Sohn-Rethel (1987) in a classic article suggests that seemingly pure scientific concepts such as the spatiotemporal framework of Galilean absolute space and time emerged from a reflection on the state of the economy—emergent capitalism created the commodity form, and within that form the ideal was that capital and commodities should flow in a frictionless time and space. The move here is the argument that many of our apparently most abstract thoughts can be seen as meditations on our infrastructures. David Deutsch (1997) is one of a number who have maintained that our theories of the brain have been modeled after the height of technology in their time. Brains have been metaphorically characterized as hydraulic engineering structures (grand dams being written by Freud into the fabric of our brains) to the telephone switchboard in the 1920s to the computer today, with our binary neurons, firing ones or zeros into the substrate of the brain. When we take our academic productions (journal articles, monographs) as independent entities which are removed from the complexity of socioeconomic life, we make precisely the mistake of assuming that our institutions could still/did ever exist as ivory towers. The knowledge infrastructure aligns with and meditates on the set of infrastructures that undergird it—just as the others align with and are influenced by it.

We use a metaphor from botany—subtension—to express a particular attribute of the infrastructural relationship. In botany, a bract is a structure that supports and grows underneath a flower and serves to connect that flower to the stem of the plant. It is said, then, that the bract subtends the flower—they grow together, but the bract supports the continued growth of the flower and maintains its connection to the central plant. Infrastructure, then, subtends different fields and forms of work. Many knowledge infrastructures can be observed to even be in reciprocal subtension, where change bears consequence not just for the infrastructure in question but also for each subtending infrastructure. In these cases the design characteristics of each mutually constitute the characteristics of both the subtending and subtended infrastructure to the point that it is difficult to determine where each underlies the other. Changing a data standard to encourage interoperability as part of a CI project may also require a substantive set of changes in the standards employed by data centers, in the organizational practices of scientists making use of that data and working with those data centers, and so on (Hepsø, Montiero, and Rolland 2009). Infrastructures in reciprocal subtension cascade change, or introduce, moments of systemic failure when the subtending infrastructure does not respond or counterintuitively responds to that change. This is why changing a knowledge infrastructure is about a lot more than developing new digital libraries and databases to allow us to deploy the methods that we know and trust; it is fundamentally about engaging with and understanding the social and the political.

A major new infrastructural force in our lives in the past ten years has been the development of the field of big data analytics. With the rise of algorithmic reasoning through big data, it is no surprise that we get equally from the computer evangelist (Anderson 2008) and the social theorist (Latour 2007) a call for an end of theory, where theory is understood as classificatory reasoning. For Latour, the argument goes that we are moving beyond the world of Durkheim (1884), whose sociology has reified categories such as "society," "gender," "class," and "race." Rather than see these as fixed categories in the world, we can examine the variable collectivities that operate at any one instance in any one place. We only needed the categories as theoretical units when we did not have access to all the data. Anderson's argument is much the same, though dressed in different garb. He was drawing on algorithmic analysis of behavior through big data as superseding categories: we can just get information about individuals without passing through, say, census classifications. Marketing firms no longer need to know what "middle-class women" want if they have access to each of us individually. The map is indeed coextensive with the territory. Nick Seaver (2012), studying music recommender systems, has shown that this is how the algorithms that help shape our taste work. Natasha Dow-Schüll (2012), through the example of gambling in Las Vegas, has shown how the new form of social engineering can be precisely a Skinnerian black box. In the latter, human and animal behavior was seen as a conditioned response to external stimulus: there was no analytic necessity to study the physiology or thought process of the person since only the "outputs," or conditioned responses, mattered. (Thus, for example, Pavlov's dogs could be trained to respond by salivating to given stimuli). For the social engineers, we really do not need to know what is going on inside the brain of the gambler, we only need to be able to predict the behavior of the individual gambler.

There are two figures who find the speed of light too slow. The first is the cosmologist, the archetypal pure scientist, who would like to be able to see further back into the putative origin of the universe. The other is the trader, who moves her offices ever closer to Wall Street to gain femtosecond advantages over her competitors, who rely on electrons traveling over wires at achingly slow rates. At first blush, this is a pretty paradox, since it indicates that only at two extremities of our activity (the intensely social world of business—see Preda, chapter 21 this volume—and the intensely abstract world of cosmology) butt up against this restriction. Viewed infrastructurally, it is of the essence of our enterprise. Algorithms inherently act in time (first do this, then that, then you get the result) and without codified knowledge laid out in a table. As we move into the world of the algorithm, we necessarily temporalize our knowledge in new ways.

There is a strong argument, then, that our deeply embedded knowledge infrastructures align well with our information and economic infrastructures. A corollary is that as our epochal shift in knowledge and information infrastructures is taking place across many levels simultaneously, we are not yet locked in through generative entrenchment (Wimsatt 2001) to the perhaps troubling forms discussed thus far. There is certainly a move to develop lockstep relationships among economy, management, and knowledge. The ideology of inevitability is prosecuted most strongly at inflection points that might lead to new kinds of social, organizational, and infrastructural arrangement. However, new forms of knowledge infrastructure, involving a new cognitive division of labor and new knowledge objects, are proliferating in this rich space. Dominique Boullier (2014) has explored this theoretically as the creation of a new kind of social fact. New social theory does away with the reification (cf. Latour 2002 on Gabrielle Tarde) alluded to above, arguing that this reification was based on the instrument of the social survey with its fixed categories. Rather, it deals in new social facts—vibrations—caused by the ever more rapid feedback loops between knowledge, information, and data generation. Thus, for the spread of ideas we might consider memes going viral; for social movements, the use of Twitter in the so-called Arab Spring; for the relationship between producer and consumer, the development of customized advertising by Google. Crucially, as we develop new kinds of social fact, the infrastructural tools for social action also change. As discussed above, when a fundamentally new information and communication infrastructure comes into being, society as a whole changes its very nature. While the process of moving from one infrastructure to another is a long one, once the transition is made, there is an epistemic break—just as we argued above that democracy is a different entity depending on its infrastructure.

A number of science studies scholars have begun to explore the force of these changes. Gillespie (2011) and Sandvig et al. (2014) describe the "politics of algorithms": the ways in which proprietary algorithms developed by Facebook, Google, or Twitter shape our access to information and view of the world evoke a need for the field of science and technology studies to develop tools to explore these issues. The question of the Internet of Things—a panoply of intelligent devices communicating with each other as much as with humans and forms of social tracking[4] raises a series of related issues. If we are increasingly driven with intelligent devices and have most aspects of our lives being constantly monitored (one might call this, following Gregory [1998], an "incomplete dystopian project"), then we are different kinds of beings in the world. This developing area of concern marks a shift in experimentation from university and industrial laboratories to the world at large—thus Facebook's A/B experiment attempted somewhat successfully (Kramer, Guillory, and Hancock 2014) to affect users' emotional states by manipulating their news feeds. Our field is particularly well poised to analyze these issues.

Infrastructure and the Role of the Social Scientist

A contentious problem in science and technology studies has been the role of the scholar in infrastructural development. The first wave of modern STS researchers stood resolutely apart from the scientists we were studying: our role was to observe, not to engage (Collins 1983). Within the infrastructure community, this stance lost its relevance—we were drawn into CI projects to carry out formative evaluations, to perform codesign, and to provide organizational advice (Ribes 2007)—many early CI projects failed due to social rather than technical reverse salients.

As infrastructure project funders began to recognize the role of the social in infrastructure development, there began an uneasy alliance of social scientists with the technical experts already working to develop infrastructure. In multidisciplinary work, social scientists are often in a "response mode," where they support and study research work initiated by others rather than define the trajectory of the project itself (Strathern 2004). Engagement with infrastructure projects, then, generally saw those from science and technology studies informing, advising, and observing/reporting rather than guiding, initiating, or leading. Social scientists in this role found themselves rehearsing a debate endemic to anthropology and ethnography—to what extent ought the researcher engage with the studied community? The classic (and flawed) image of the anthropologist as a disengaged, objective reporter who seeks to avoid going native seems inappropriate to the roles social researchers often found themselves filling in infrastructure work. And while there is not a straight trajectory where objectivity is lost in favor of participation, level and mode of engagement within an infrastructure project is still a significant methodological issue. Ribes and Baker identify four elements that work to inform the role of the social scientist as a part of an infrastructure project: the development timeline, the state of the project at the beginning of social science engagement, the type of participation, and the details of how the social scientists are involved (Ribes and Baker 2007).

The role of the STS scholar in information systems design was explored in a series of cohort-building workshops on Values in the Design of Information Systems and Technology (VID) led by Geoffrey C. Bowker, Helen Nissenbaum, and Susan Leigh Star.

The argument that we should actively engage in design has been developed recently in a classic paper by Steve Jackson, Tarleton Gillespie, and Sandy Payette (2014) on the policy knot: here they argue for the need to integrate policy, practice, and design in the development of, say, large-scale social media. Dan Atkins (2005)—a founder in the United States of the field of cyberinfrastructure through his work first at the University of Michigan and then at the National Science Foundation—used to speak similarly of the Borromean ring consisting of social, organizational, and technical strands at the core of CI (Atkins 1996), saying that none of these by themselves could be separated from either of the other if one wanted to understand the field. This itself was an echo of Hughes's reverse salients.

Infrastructure and Policy

As it relates to infrastructure studies, policy is something of a chimera. Even though policy concerns are presented as significant factors in the major infrastructure reports (Edwards et al. 2013) there is a conflation—or at the least a blurred distinction—between policy as law; policy as organizational practice and rules; policy as embodied in standards; policy as systems of classification and work practice; and policy as issues relating to the governance of particular infrastructure projects. While this might make the concept of policy as related to infrastructure appear at first blush to be something of a misplaced concretism, the apparent heterogeneity of what might be termed "policy issues" can be viewed as more a product of Latour's (among others) claim that "science is politics by other means" (1987). For Latour, there are no simple distinctions of science, values, society, and political power: they all happen together, or at least, at the same time (cf. Latour and Callon 1992; Jackson et al. 2014). The apparent chimerism of specific policy topics in infrastructure studies reflects more the embodiment of values, practices, and preferences in the material and organizational substrates from which infrastructure emerges.

Policy is most brazenly present in infrastructure in the general need for financial support and upward accountability, and that has a powerful effect on design decisions for and restrictions on the introduction of new infrastructural elements (Sahay, Monteiro, and Aanestad 2009). Increasing amounts of resources needed to expand infrastructural capacity or provide the basis for new infrastructural networks moves decision making for large-scale research projects to ever larger groups and recognizes its impacts on more diverse social, academic, and industrial institutions (Galison and Hevly 1992). As such, representations of infrastructural results to policy makers are both attended to and avoided in equal measure. Rip and Voß describe an entity called an "umbrella term" that mediates between the work of science and the political and social understandings of that science (Rip and Voß 2013). Umbrella terms (like nanotechnology) provide a basis for innovation by providing a *de facto* understanding of emergent scientific work that does not need to be referenced back to individual research efforts. Like a snowball, these terms collect nuance and context as they move, only to be emptied in a moment as a new term gains traction, leaving only the nuance and context, the organizational realities and human connections.

It might be argued that the larger political effects of infrastructure, much like the infrastructure itself, are easily relegated to the marginalia, rendered invisible by a focus on the subject discipline of a given infrastructure problem. Support work such as maintenance and repair that is not readily susceptible to a decontextualized measurement is marginalized, while still having profound effects on organizational realities and local practice (Bowker and Star 2000). Agreement on protocols, standards, and even effective measures is a political process: the preferences, values, and practices of one group are adopted and supported above others.

The concept of "the standard" here is a strong one. A standard conceptually is able to represent a wide variety of design considerations, from physical attributes to classification systems (cf. the discussion of the International Classification of Diseases in Bowker and Star 2000) to technical protocols, common pieces of software, and file formats. Standards serve as a gateway between disparate sociotechnical systems, and their equalization of the design space is a vital component in how these systems interlink into networks (Jackson et al. 2007) and into the material world (Busch 2011). Though standards have an "intuitive tension" with system flexibility, the establishment of standards in one area of a system tends to increase flexibility in others (Egyedi 2014; Mulgan 1991). As standards simultaneously support the linking of systems into networks and the catalyzing of systemic change by stabilizing a set of design considerations, standards "achieve some small or large transformation of an existing social order" (Timmermans and Epstein 2010, 83).

When one builds a standard, it is built in such a way that one's own interests are promulgated (Latour 2007). From this perspective, adoption of certain standards is a sort of contest with the "victor" achieving formalization of their own values and practices as agreed-upon or default policy (Jasanoff 2007; Latour 1987). In other words, standards, protocols, and systems of classification expose embedded sociotechnical values as they make changes to regimes of decision making (Edwards and Hecht 2010; Kranzberg 1986; Lampland and Star 2009). Sensitive to this, DeNardis (2009) traces the negotiations and agreements that were infrastructural to the interoperability characteristic of the modern Internet with a special focus on the political effects of the standardization of protocols. Edwards (2010), also, accounts for the political effects of legacy code, data regimes, and the formal adoption and use of specific models in climate science. The standard can be pictured as an actor in its own right, catalyzing change and being changed through use; following the standard in use provides a basis for understanding the social effects of the network itself (Latour 2007).

Policy, then, is not a distinct and separate feature of infrastructure. Instead, infrastructural work is fundamentally and pervasively political. Successful CI projects are those that result in the creation of a social reality for its participants where their individual work practice is shaped in relation to the attendant infrastructure. Organizational practice imports a set of values, ethics, and implicit knowledge all its own. Even before work on standardization begins, the negotiations and agreements necessary to allow heterogeneous groups to work together inevitably ensconces some agendas while marginalizing others. As participants and researchers are sensitized to the social effects of design decisions at the infrastructural scale, a space for a more nuanced discussion of those decisions sensitive to their effects at the moment of design becomes possible, perhaps even inevitable.

Conclusion

It is difficult to put one's arms around a set of texts and declare that they constitute the nature of science and technology studies. One reading of our field—among many others—is that by our constant attention to the mundane work of doing science and building technology, we have consistently drawn attention to the infrastructural. It is possible that the success of the field in influencing numerous cognate fields (computer-supported cooperative work, organization theory, management theory, and human-computer interaction, to name a few) has been the development of this angle of vision. Accordingly, we have taken a capacious view in this chapter in order to follow the actants. An infrastructural account of this would include the observation that because we have not managed in general to create a discipline of our own with institutional roots (Bowker and Latour 1987), we have had to lodge ourselves in a number of disciplinary homes, with infrastructure being one thematic which permits this move.

While we as STS scholars as a group have not in any sense developed a grand unified theory of infrastructure—indeed we have done much to show that this very coinage is oxymoronic—we have been central to the emergence of infrastructural concerns in a number of domains. As we have endeavored to show in this chapter, the resulting skein of intertwined threads has its own richness and continuity. The role of our own scholarship in designing infrastructure is an exciting frontier. The ontological commitment that the apparently ideal worlds of form and consciousness are subtended by an active infrastructure, which we need to understand still, has many roads to travel.

Notes

1. Hughes's work is somewhat infrastructural to infrastructure studies; it can also be seen as part of the sociotechnical systems literature. STS was first developed through the work of the Tavistock Institute—initially through an analysis of coal mining work in England (Trist and Bamforth 1951). While this work did not have a large direct influence on science studies (Kaghan and Bowker 2001), it did follow a parallel track, leading to some interesting entanglement between the two over the decades. Thus Rose's 1990 classic *Governing the Soul* both tied to the radical psychoanalytic tradition of the Tavistock Institute and brought Foucault's analysis of power into play—the latter itself becoming variably central to science studies and infrastructure. A more direct filiation of Hughes's work is the rich tradition of large-scale sociotechnical systems which drew heavily on the systems literature and on STS (in our acceptation of the term). A third, more recent tradition on the sociotechnical has been pioneered by Steve Sawyer (2001) and Wayne Lutters et al. (2000) among others.

2. Here we are taking more the approach of Kuhn (2012) than Lakatos (1970).

3. Borges Chinese Encyclopedia, or *The Celestial Emporium of Benevolent Knowledge*, is a fictional taxonomy of animals intended to point out the culturally relative and arbitrary nature of classification systems, cf. his 1942 essay *The Analytical Language of John Wilkins* (Borges 1964).

4. The Quantified Self movement (cf. Nafus 2016) is significant here—for example, in the way one's scale might interact with one's Fitbit and one's electronic medical records.

References

Anderson, Chris. 2008. "The End of Theory: The Data Deluge Makes the Scientific Method Obsolete." *Wired Magazine* June 24. Accessed at http://www.wired.com/2008/06/pb-theory/.

Althusser, Louis. 2006. "Ideology and Ideological State Apparatuses (Notes towards an Investigation)." In *The Anthropology of the State: A Reader*, edited by Aradhana Sharma and Akhil Gupta, 86–111. Hoboken, NJ: Blackwell.

Atkins, Daniel. 1996. "Electronic Collaboratories and Digital Libraries." *Neuroimage* 4 (3): S55–58.

Atkins, Daniel E. 2005. "Cyberinfrastructure and the Next Wave of Collaboration." Report for EDUCAUSE Australia.

Bishop, Ann Peterson, Laura J. Neumann, Susan Leigh Star, Cecelia Merkel, Emily Ignacio, and Robert J. Sandusky. 2000. "Digital Libraries: Situating Use in Changing Information Infrastructure." *Journal of the American Society for Information Science* 51 (4): 394–413.

Bohannon, John. 2015. "Hoax-Detecting Software Spots Fake Papers." *Science* 348 (6230): 18–19.

Borges, Jorge Luis. 1964. *Other Inquisitions: Essays 1937–1952*, translated by Ruth L. C. Simms. Austin: University of Texas Press.

Bos, Nathan, Ann Zimmerman, Judith Olson, Jude Yew, Jason Yerkie, Erik Dahl, and Gary Olson. 2007. "From Shared Databases to Communities of Practice: A Taxonomy of Collaboratories." *Journal of Computer-Mediated Communication* 12 (2): 652–72.

Boullier, Dominique. 2014. "Habitele: Mobile Technologies Reshaping Urban Life." *URBE* 6 (1): 13–16.

Bowker, Geoffrey C. 1994. *Science on the Run: Information Management and Industrial Geophysics at Schlumberger, 1920–1940*. Cambridge, MA: MIT Press.

Bowker, Geoffrey C., Karen Baker, Florence Millerand, and David Ribes. 2010. "Toward Information Infrastructure Studies: Ways of Knowing in a Networked Environment." In *International Handbook of Internet Research*, edited by Jeremy Hunsinger, Lisbeth Klastrup, and Matthew M. Allen, 97–117. Dordrecht, Netherlands: Springer.

Bowker, Geoffrey C., and Bruno Latour. 1987. "A Booming Discipline Short of Discipline: (Social) Studies of Science in France." *Social Studies of Science* 17 (4): 715–48.

Bowker, Geoffrey C., and Susan Leigh Star. 2000. *Sorting Things Out: Classification and Its Consequences*. Cambridge, MA: MIT Press.

Brand, Stewart. 1995. *How Buildings Learn: What Happens after They're Built*. New York: Penguin.

Busch, Lawrence. 2011. *Standards: Recipes for Reality*. Cambridge, MA: MIT Press.

Bush, Vannevar. 1945. "Science: The Endless Frontier." *Transactions of the Kansas Academy of Science (1903–)* 48 (3): 231–64.

Callon, Michel. 1986. "The Sociology of an Actor Network: The Case of the Electric Vehicle." In *Mapping the Dynamics of Science and Technology*, edited by Arie Rip, Michel Callon, and John Law, 19–36. London: Macmillan.

___. 1990. "Techno-Economic Networks and Irreversibility." *Sociological Review* 38 (S1): 132–61.

Carse, Ashley. 2012. "Nature as Infrastructure: Making and Managing the Panama Canal Watershed." *Social Studies of Science* 42 (2): 539–63.

Cogburn, Derrick L. 2003. "HCI in the So-called Developing World: What's in It for Everyone." *Interactions* 10 (2): 80–87.

Collins, Harry. 1983. "An Empirical Relativist Programme in the Sociology of Scientific Knowledge." In *Science Observed: Perspectives on the Social Study of Science*, edited by Karin Knorr Cetina and Michael Mulkay, 85–113. London: Sage.

David, Paul A. 1990. "The Dynamo and the Computer." *American Economic Review* 80 (2): 355–61.

DeNardis, Laura. 2009. *Protocol Politics: The Globalization of Internet Governance*. Cambridge, MA: MIT Press.

Deutsch, David. 1997. *The Fabric of Reality*. London: Penguin.

Dourish, Paul, and Melissa Mazmanian. 2013."Media as Material: Information Representations as Material Foundations for Organizational Practice." In *How Matter Matters: Objects, Artifacts, and Materiality in Organization Studies*, edited by Paul R. Carlile, Davide Nicolini, Ann Langley, and Haridimos Tsoukas, 92–118. Oxford: Oxford University Press.

Dow-Schüll, Natasha. 2012. *Addiction by Design: Machine Gambling in Las Vegas*. Princeton, NJ: Princeton University Press.

Durkheim, Emile. 1884."The Division of Labor in Society." *Journal des Economistes*, 211.

Edwards, Paul N. 2010. *A Vast Machine: Computer Models, Climate Data, and the Politics of Global Warming*. Cambridge, MA: MIT Press.

Edwards, Paul N., and Gabrielle Hecht. 2010. "History and the Technopolitics of Identity: The Case of Apartheid South Africa." *Journal of Southern African Studies* 36 (3): 619–39.

Edwards, Paul N., Steven J. Jackson, Melissa K. Chalmers, Geoffrey C. Bowker, Christine L. Borgman, David Ribes, Matt Burton, and Scout Calvert. 2013."Knowledge Infrastructures: Intellectual Frameworks and Research Challenges." Accessed at http://knowledgeinfrastructures.org/.

Egyedi, Tineke. 2014. "Standards and Sustainable Infrastructures: Matching Compatibility Strategies with System Flexibility Objectives." In *Papers on Open Innovation! A Collection of Papers on Open Innovation from Leading Researchers in the Field*, edited by Shane Coughlan, 55–80. OpenForum Europe LTD.

Engeström, Yrjö. 1990."When Is a Tool? Multiple Meanings of Artifacts in Human Activity." In *Learning, Working and Imagining*, 171–95. Helsinki: Orienta-Konsultit Oy.

Foucault, Michel. 1979. *Discipline and Punish: The Birth of the Prison*. London: Penguin.

___. 1991. "Governmentality." In *The Foucault Effect: Studies in Governmentality*, edited by Michel Foucault, Graham Burchell, Colin Gordon, and Peter Miller. Chicago: University of Chicago Press.

___. 2002. *The Order of Things: An Archaeology of the Human Sciences*. New York: Routledge.

Galison, Peter. 1996. "Computer Simulations and the Trading Zone." In *The Disunity of Science: Boundaries, Contexts, and Power*, edited by Peter Galison and David J. Stump, 118–57. Stanford, CA: Stanford University Press.

Galison, Peter, and Bruce Hevly. 1992. *Big Science: The Growth of Large-Scale Research*. Stanford, CA. Stanford University Press.

Gibbons, Michael, Camille Limoges, Helga Nowotny, Simon Schwartzman, Peter Scott, and Martin Trow. 1994. *The New Production of Knowledge: The Dynamics of Science and Research in Contemporary Societies*. London: Sage.

Gillespie, Tarleton. 2011. "Can an Algorithm Be Wrong? Twitter Trends, the Specter of Censorship, and Our Faith in the Algorithms around Us." *Culture Digitally*. Accessed at http://culturedigitally.org/2011/10/can-an-algorithm-be-wrong/.

Gregory, Judith. 1998. "Envisioning and Historicizing: Incomplete Utopian Projects." Unpublished Paper. 1SCRAT Conference, Aarhus, Denmark.

Hannemyr, Gisle. 2003."The Internet as Hyperbole: A Critical Examination of Adoption Rates." *The Information Society* 19 (2): 111–21.

Harvey, Penny, and Hannah Knox. 2008. "'Otherwise Engaged': Culture, Deviance and the Quest for Connectivity through Road Construction." *Journal of Cultural Economy* 1 (1): 79–92.

Hepsø, Vidar, Eric Monteiro, and Knut H. Rolland. 2009. "Ecologies of e-Infrastructures." *Journal of the Association for Information Systems* 10 (5): 430–46.

Hughes, Thomas P. 1993. *Networks of Power: Electrification in Western Society, 1880–1930*. Baltimore: Johns Hopkins University Press.

Jackson, Steven J., and Ayse Buyuktur. 2014."Who Killed WATERS? Mess, Method, and Forensic Explanation in the Making and Unmaking of Large-Scale Science Networks." *Science, Technology, & Human Values* 39 (2): 285–308.

Jackson, Steven J., Paul N. Edwards, Geoffrey C. Bowker, and Cory P. Knobel. 2007. "Understanding Infrastructure: History, Heuristics and Cyberinfrastructure Policy." *First Monday* 12 (6). Accessed at http://dx.doi.org/10.5210/fm.v12i6.1904.

Jackson, Steven J., Tarleton Gillespie, and Sandy Payette. 2014. "The Policy Knot: Re-integrating Policy, Practice and Design in CSCW Studies of Social Computing." In *Proceedings of the 17th ACM Conference on Computer Supported Cooperative Work and Social Computing*, 588–602. Baltimore: AMC.

Jasanoff, Sheila. 2007. "Technologies of Humility." *Nature* 450 (7166): 33.

Joerges, Bernward. 1999. "Do Politics Have Artefacts?" *Social Studies of Science* 29 (3): 411–31.

John, Richard R. 2009. *Spreading the News: The American Postal System from Franklin to Morse*. Cambridge, MA: Harvard University Press.

Kaghan, William N., and Geoffrey C. Bowker. 2001. "Out of Machine Age? Complexity, Sociotechnical Systems and Actor Network Theory." *Journal of Engineering and Technology Management* 18 (3): 253–69.

Klose, Alexander, and Charles Marcrum. 2015. *The Container Principle: How a Box Changes the Way We Think*. Cambridge, MA: MIT Press.

Knorr Cetina, Karin, and Michael Mulkay. 1983. *Science Observed: Perspectives on the Social Study of Science*. London: Sage.

Kouzes, Richard T., James D. Myers, and William Wulf. 1996. "Collaboratories: Doing Science on the Internet." *Computer* 29 (8): 40–46.

Kramer, Adam D. I., Jamie E. Guillory, and Jeffrey T. Hancock. 2014. "Experimental Evidence of Massive-Scale Emotional Contagion through Social Networks." *Proceedings of the National Academy of Sciences* 111 (24): 8788–90.

Kranzberg, Melvin. 1986. "Technology and History: 'Kranzberg's Laws'" *Technology and Culture* 27 (3): 544–60.

Kuhn, Thomas S. [1962] 2012. *The Structure of Scientific Revolutions*. 50th anniversary ed. Chicago: University of Chicago Press.

Labbé, Cyril, and Dominique Labbé. 2013. "Duplicate and Fake Publications in the Scientific Literature: How Many SCIgen Papers in Computer Science?" *Scientometrics* 94 (1): 379–96.

Lampland, Martha, and Susan Leigh Star. 2009. *Standards and Their Stories: How Quantifying, Classifying, and Formalizing Practices Shape Everyday Life*. Ithaca, NY: Cornell University Press.

Lakatos, Imre. 1970. "Falsification and the Methodology of Scientific Research Programmes." In *Criticism and the Growth of Knowledge*, edited by Imre Lakatos and Alan Musgrave, 91–196. Cambridge: Cambridge University Press.

Lakoff, George. 1993. "The Contemporary Theory of Metaphor." *Metaphor and Thought* 2: 202–51.

Larkin, Brian. 2013. "The Politics and Poetics of Infrastructure." *Annual Review of Anthropology* 42: 327–43.

Latour, Bruno. 1983. "Give Me a Laboratory and I Will Raise the World." In *Science Observed: Perspectives on the Social Study of Science*, edited by Karin Knorr Cetina and Michael Mulkay, 258–75. London: Sage.

___. 1987. *Science in Action: How to Follow Scientists and Engineers through Society*. Cambridge, MA: Harvard University Press.

___. 1992. "Where Are the Missing Masses? The Sociology of a Few Mundane Artifacts." *Shaping Technology/Building Society: Studies in Sociotechnical Change*, edited by Wiebe E. Bijker and John Law. 225–58. Cambridge, MA: MIT Press.

___. 1993. *The Pasteurization of France*. Cambridge, MA: Harvard University Press.

___. 1996. *Aramis, or, the Love of Technology*. Cambridge, MA: Harvard University Press.

___. 2002. "Gabriel Tarde and the End of the Social." In *The Social in Question: New Bearings in History and the Social Sciences*, edited by Patrick Joyce, 117–32. East Sussex: Psychology Press.

___. 2007. *Reassembling the Social: An Introduction to Actor-Network Theory*. New York: Oxford University Press.

Latour, Bruno, and Michel Callon. 1992. "Don't Throw the Baby Out with the Bath School." In *Science as Practice and Culture*, edited by James Pickering, 343–68. Chicago: University of Chicago Press.

Latour, Bruno, and Steve Woolgar. 1979. *Laboratory Life: The Construction of Scientific Facts*. Thousand Oaks, CA: Sage.

Lee, Charlotte P., Paul Dourish, and Gloria Mark. 2006. "The Human Infrastructure of Cyberinfrastructure." In *Proceedings of the 2006 20th Anniversary Conference on Computer Supported Cooperative Work*, 483–92.

Lutters, Wayne G., Mark S. Ackerman, James Boster, and David W. McDonald. 2000. "Mapping Knowledge Networks in Organizations: Creating a Knowledge Mapping Instrument." *AMCIS 2000 Proceedings*, 315.

Lynch, Michael. 1985. *Art and Artefact in Laboratory Science*. London: Routledge.

Marx, Karl, and Friedrich Engels. 1970. *The German Ideology*, edited by Christopher John Arthur. New York: International Press.

Mulgan, Geoffrey J. 1991. *Communication and Control: Networks and the New Economies of Communication*. New York: Guilford Press.

Mumford, Lewis, and George Copeland. 1961. *The City in History: Its Origins, Its Transformations, and Its Prospects*. New York: Harcourt, Brace & World.

Nafus, Dawn. 2016. *Quantified: Biosensing Technologies in Everyday Life*. Cambridge, MA: MIT Press.

Olson, Gary M., and Judith S. Olson. 2000. "Distance Matters." *Human-Computer Interaction* 15 (2): 139–78.

Propp, Vladimir. 2010. *Morphology of the Folktale*, vol. 9. Austin: University of Texas Press.

Ribes, David, and Karen Baker. 2007. "Modes of Social Science Engagement in Community Infrastructure Design." In *Communities and Technologies: Proceedings of the Third Communities and Technologies Conference, Michigan State University 2007*, edited by Charles Steinfield, Brian T. Pentland, Mark Ackerman, and Noshir Contractor, 107–30. London: Springer.

Ribes, David, and Thomas A. Finholt. 2009. "The Long Now of Technology Infrastructure: Articulating Tensions in Development." *Journal of the Association for Information Systems* 10 (5): 375–98.

Rip, Arie, and Jan-Peter Voß. 2013. "Umbrella Terms as a Conduit in the Governance of Emerging Science and Technology." *Science, Technology & Innovation Studies* 9 (2): 39–59.

Rose, Nikolas. 1990. *Governing the Soul: The Shaping of the Private Self*. London: Routledge.

Sahay, Sundeep, Eric Monteiro, and Margunn Aanestad. 2009. "Configurable Politics and Asymmetric Integration: Health e-Infrastructures in India." *Journal of the Association for Information Systems* 10 (5): 399–414.

Sandvig, Christian, Kevin Hamilton, Karrie Karahalios, and Cedric Langbort. 2014. "Auditing Algorithms: Research Methods for Detecting Discrimination on Internet Platforms." Paper presented at the 64th Annual Meeting of the International Communication Association, Seattle, WA, USA, May 22.

Sartre, Jean-Paul. [1964] 2001. *Colonialism and Neocolonialism,* translated by Azzedine Haddour, Steve Brewer, and Terry McWilliams. New York: Routledge.

Sawyer, Steve. 2001. "A Market-Based Perspective on Information Systems Development." *Communications of the ACM* 44 (11): 97–102.

Seaver, Nick. 2012. "Algorithmic Recommendations and Synaptic Functions." *Limn* 2, "Crowds and Clouds." Retrieved from http://limn.it/algorithmic-recommendations-and-synaptic -functions/.

Shannon, Claude E., and Warren Weaver. [1948] 2015. *The Mathematical Theory of Communication.* Chicago: University of Illinois Press.

Sohn-Rethel, Alfred. 1987. *The Economy and Class Structure of German Fascism.* London: Free Association.

Star, Susan Leigh, and Karen Ruhleder. 1996. "Steps toward an Ecology of Infrastructure: Design and Access for Large Information Spaces." *Information Systems Research* 7 (1): 111–34.

Starosielski, Nicole. 2015. *The Undersea Network.* Durham, NC: Duke University Press.

Strathern, Marilyn. 2004. *Commons and Borderlands: Working Papers on Interdisciplinarity, Accountability and the Flow of Knowledge.* Herefordshire: Sean Kingston.

Strum, Shirley S., and Bruno Latour. 1987. "Redefining the Social Link: From Baboons to Humans." *Social Science Information* 26 (4): 783–802.

Timmermans, Stefan, and Steven Epstein. 2010. "A World of Standards but Not a Standard World: Toward a Sociology of Standards and Standardization." *Annual Review of Sociology* 36: 69–89.

Trist, Eric L., and Ken W. Bamforth. 1951. "Some Social and Psychological Consequences of the Longwall Method." *Human Relations* 4 (3): 3–38.

Van House, Nancy A., Ann P. Bishop, and Barbara P. Buttenfield. 2003. "Introduction: Digital Libraries as Sociotechnical Systems." In *Digital Library Use: Social Practice in Design and Evaluation,* edited by Nancy A. Van House, Ann P. Bishop, and Barbara P. Buttenfield, 1–21. Cambridge, MA: MIT Press.

Van Noorden, Richard. 2014. "Publishers Withdraw More Than 120 Gibberish Papers." *Nature* 24.

Veyne, Paul. 2013. *Foucault: His Thought, His Character.* Hoboken, NJ: John Wiley & Sons.

Wimsatt, William C. 2001. "Generative Entrenchment and the Developmental Systems Approach to Evolutionary Processes." In *Cycles of Contingency: Developmental Systems and Evolution,* edited by Paul E. Griffiths and Russell D. Gray, 219–37. Cambridge, MA: MIT Press.

Winner, Langdon. 1980. "Do Artifacts Have Politics?" *Daedalus* 109 (1): 121–36.

Woolgar, Steve. 1986. "On the Alleged Distinction between Discourse and Praxis." *Social Studies of Science* 16 (2): 309–17.

___. 1993. "What's at Stake in the Sociology of Technology? A Reply to Pinch and to Winner." *Science, Technology, & Human Values* 18 (4): 523–29.

Wulf, William A. 1989. "Towards a National Collaboratory." In *Towards a National Collaboratory: Report of an Invitational Workshop at the Rockefeller University*, edited by Joshua Lederberg and Keith Uncaphar. March 17–18 (appendix A). Washington, DC: National Science Foundation, Directorate for Computer and Information Science Engineering.

Yates, JoAnne. 1993. *Control through Communication: The Rise of System in American Management*, vol. 6. Baltimore: Johns Hopkins University Press.

19 STS in the City

Ignacio Farías and Anders Blok

Introduction: STS and Urban Studies

In the last decade, science and technology studies (STS) is increasingly becoming a highly influential source of inspiration for other disciplinary and interdisciplinary fields. It is important to trace such intellectual trajectories because they function as test sites for our conceptual repertoires. In this chapter we explore the borrowings and travelings between STS and urban studies, two highly interdisciplinary and heterogeneous fields. We are interested in two types of displacement. On the one hand, the city provides us with a fascinating empirical site (or sites) to explore the workings of science and technology. On the other hand, STS conceptual repertoires are increasingly being used to explore urban phenomena beyond science and technology. In this way, the city and urban studies potentially also challenge STS to renew its own capacities, a point to which we return in the conclusion.

Bringing STS into the city remains urgent, we contend, considering how even in the last decades new technologies have been praised and introduced as a powerful force reshaping urban settings worldwide. Since the 1980s, new information and communication technologies (ICTs) have been seen as promising (or threatening) to overcome the need for physical proximity and thus dissolve what we know as cities. More recently, versions of citizen science and algorithmic regulation of infrastructures have become the new panacea for "smart urbanism" in order for cities to solve demographic, economic, and ecological challenges. Needless to say, such high-modernist hopes for technological fixes are nothing new to the city, as witnessed for instance by long-standing themes of development and mobility. In most of these discourses and actual policies and practices of city administrations all over the planet, we encounter the myth of a "technological essence" (Graham 1997)—the idea that technologies possess intrinsic logics or qualities, which they inevitably bring along in causing specific and necessary forms of urban change.

Within urban research, key traditions have generally tended to neglect the workings of science and technology and more broadly the type of sociomaterial processes
underscored in STS. In a nutshell, urban studies has approached cities as spatial forms
and ecological niches, as capitalist politico-economic actors, or as involving specific
cultures of practice (see also Farías 2009). First, the still influential innovation of the
Chicago School of sociology of the 1920s was to conceive of cities as ecological niches
within which human communities settle down in discernible sociospatial patterns
resulting from competition for location, as well as invasion and succession processes.
This perspective contributed crucial insights into the relationships, for instance, among
neighborhoods, socioeconomic structure and segregation, the dynamics of real estate
markets, and gentrification. Second, and at least since Max Weber, cities have been
studied as key politico-economic formations shaped and shaping capitalist dynamics. The large influence exerted by French Marxist philosopher and sociologist Henri
Lefebvre has been crucial to reimagining cities as "the major actors in the new global
economy" (Sassen 1991, 14) and as key sites of political struggle. Third, following on
Georg Simmel's groundbreaking essay on the mental life of big city dwellers, cities have
been studied as involving specific cultures of urban practice. Since the 1960s and 70s,
the tradition of everyday urbanism associated with such authors as activist and planner Jane Jacobs or historian and philosopher Michel de Certeau has explored urbanity as involving spontaneous choreographies of multiple subtle orders of practice and
contestation. Urban culture is recognized here in residual and transient spaces, seen as
opposed to "the" city made of bounded places, fixed meanings, and big history.

Given this context, it is not surprising that when STS scholars approach the city, they
tend to look for antecedents outside the field of urban studies. One key reference here
has been the work of urban historian Lewis Mumford (1937, 1961), who is often cited
as one of the few to have taken seriously how technological innovation, especially with
regards to construction materials and techniques, influenced the built environments
and cultures of cities (e.g., Aibar and Bijker 1997). The work of Mumford is, however,
in many important respects antithetical to the STS project. In his view, the city is ultimately the technomaterial expression of superior and persistent symbolic functions,
including military control, sovereignty, worship, and social integration. Despite his
detailed analyses of how new technologies shaped towns, these remain in the end incidental. Mumford is thus far from constituting a direct precedent for an approach that
takes seriously the mutually shaping roles of science and technology in cities.

In sum, the field of urban studies may thus be said to have been historically moving
among three untenable positions: one in which urban technologies serve the eternal
and essential social functions of the city; another in which urban technologies are

not problematized as constitutive of the definition of the city as a research object; and yet another in which urban technologies possess intrinsic qualities, which when implemented in cities necessarily change their functioning, thereby condemning or redeeming the city. It is against this general backdrop that we can begin to assess the fundamental intervention that STS can make, and indeed has increasingly been making, in urban studies.

Our review and exploration in the following revolves around an identification of three distinct avenues of STS in the city, each shaped by specific academic traditions, empirical sensibilities, and political concerns. The first section will review STS approaches that explicitly address the workings of science and technology in the city as a whole and unpack the technoscientific objects and practices overlooked by urban studies. The second section will be dedicated to STS work on the built environment, focusing on conceptual challenges associated with understanding what buildings do and how architects work on their realization. The third and final section will review what has been recently discussed as assemblage urbanism and which involves various attempts at studying cities and urban life via perspectives and concepts provided by actor-network theory (ANT) and related intellectual projects. We speak of avenues to highlight that each encounter enables two-way circulations of theories, concepts, and methods between STS and urban studies and to avoid a simple historical periodization. Indeed, each of these ways of bringing STS into the city maintains its actuality, within and across the increasingly heterogeneous fields of both STS and urban studies.

Science and Technology in the City

It probably all began with an "urban legend" (Woolgar and Cooper 1999), the one about the Long Island Parkway underpasses in New York, built under the direction of influential urban planner Robert Moses in pursuit of a subtle form of racial discrimination. Constructed with low heights, these two hundred or so underpasses impeded the circulation of public buses and thus the access of poor, black urban populations to the white, middle-class resort areas of Long Island. In his classic STS piece "Do Artifacts Have Politics?," Langdon Winner (1980) tells this story to argue that urban and infrastructural artifacts should be seen as technical devices through which powerful individuals or social groups pursue their strategic interests and settle contested political issues.

The particular interest of this case for STS in cities is related to the various answers and debates triggered by Winner's piece, which generally contested his intentionalist reading of urban artifacts. As Joerges (1999) in particular elaborates, a more detailed empirical analysis of Moses's underpasses demonstrates that, rather than by individual intentionality and strategy, these artifacts were shaped by the conjunction of various engineering, economic, and legal commitments of a wider planning culture. Woolgar and Cooper (1999), in turn, warned that Joerges's critique would imply that one could get to the heart of the matter, discover the true intentions of Moses, instead of tracing the effects of the bridge story, itself a shifting urban legend. What such positions on the bridge story share, however, is an understanding of the effects of sociotechnical artifacts in cities (as elsewhere) as contingent and subject to change, rather than (over-) determined by specific political strategies.

The City as a Technological Artifact

The contingency of urban change came first to be emphasized in studies using insights from the social construction of technology (SCOT) to study city-planning initiatives. Different from Winner, the analytical departure point here is that the city, understood as "a 'seamless web' of material and social elements" (Hommels 2005, 15), is a giant sociotechnical artifact whose change could be understood "with the same conceptual tools that are applied to other technologies such as bicycles, transport systems and refrigerators" (ibid., 21). The classical study by Aibar and Bijker (1997) on the urban controversy about the extension of Barcelona in the mid-nineteenth century inaugurates this approach. In line with general SCOT commitments, Aibar and Bijker adopt the methodological strategy of studying historical controversies in order to bring out the "interpretative flexibility" whereby different "relevant social groups" attribute different meanings to and valorize emerging technological artifacts differently. Accordingly, in the case of Barcelona's extension plans, Aibar and Bijker show how these resulted from negotiating different "technological frames" shaping the perspectives of architects, engineers, and industrial workers.

Whereas, in most SCOT controversy analyses, one technological frame becomes dominant and stabilizes the artifact (Pinch and Bijker 1984), the interest of the Barcelona planning case was that closure involved a compromise between the technological frames of architects and engineers at the expense of industrial workers' way of problematizing the plans. Subsequent work at the intersections of urban history, the history of technology, and SCOT-informed STS has added further empirical and conceptual sophistication to the study of how urban technologies, infrastructures, planning expertise, and democracy relate (e.g., Hård and Misa 2008). The study by Bijker and Bijsterveld (2000) provides a particularly striking case of the role of nonexperts in shaping urban technologies, by showing how citizen groups of Dutch women came to strategically influence the shaping of public housing and city planning in the years following World War II.

Aibar and Bijker (1997, 23) explicitly presented their analysis as an attempt to "draw the city into the limelight of social studies of technology." At the same time, however, one problem with their SCOT-inspired perspective is that the specificity of the city as a difficult and messy object is conceptually backgrounded. The assumption that "the city as a kind of artifact" (ibid., 6) may legitimately be equated to other technological artifacts simultaneously facilitates *and* constrains dialogue between STS and urban studies. In this context, perhaps the most promising attempt at engaging with the specificity of the city from within the SCOT tradition is Hommels's work (2005) on the obduracy of urban sociotechnical change. By looking at a highway, a commercial center, and a high-rise housing project in different Dutch cities—all of which persistently resist the broad social consensus about the need for their redesign and reconfiguration— Hommels depicts cities as key sites to understand how technological artifacts are made obdurate in specific sets of practices and discourses. Obduracy, in this sense, represents one way in which the city has challenged STS analysts to expand their conceptual repertoires.

Sociotechnical Systems and Splintering Urban Infrastructures

Even though the STS study of large technical systems (LTS) was never primarily concerned with *urban* technical systems, it has had an important influence for the current study of urban infrastructural transitions. Its particular usefulness for urban studies resides in the very change of focus it proposed, from singular technological artifacts to large technical systems, that is, spatially extended and functionally integrated sociotechnical networks. Notably, the explicit systems perspective put forward by people such as historian of technology Thomas P. Hughes (1983) did not just involve linking technical artifacts to encompassing engineering systems but also the latter to organizational, economic, and political actors, institutions, and processes. The complexity of technical systems was thus seen from early on as involving various nontechnical contexts, and their very intermingling as giving shape to industrial modern life and, one should add, the modern industrial and networked city (cf. Coutard 1999).

Three types of systems in particular were seen by LTS analysts as crucial: "the modern transportation, communication and supply systems, which one might subsume under the heading infrastructural systems, since their primary function consists in enabling a multitude of specific activities to take place" (Mayntz and Hughes 1988, 233). Alongside Susan Leigh Star's work on information infrastructures, LTS should be seen as a key antecedent for the subsequent development of relational sociotechnical understandings of urban infrastructures. The key insight is that infrastructures are not an ontologically fixed substrate but a relational arrangement that can simultaneously enable specific types of activity and function as a technological barrier for other activities or social groups. Accordingly, the key question is not so much *what* an infrastructure is but *when* and *for whom* an urban sociotechnical arrangement becomes infrastructural in its effects (Star and Ruhleder 1996).

These insights have played a major role in the development of a prolific research tradition at the boundary between STS and urban studies, focusing specifically on late capitalist collapse of the modern networked city. The fundamental contribution to this debate is geographer Stephen Graham and urban planner Simon Marvin's (2001) *Splintering Urbanism*, a book that summarizes a highly ambitious project of describing contemporary trends of urban infrastructural change. Notably, the book starts by reconstructing the modern ideal of the networked city, said to have prevailed in Western cities from mid-nineteenth century until the 1960s and centered on the drive to integrate existing fragmentary pockets of urban infrastructure into centralized and standardized technical systems of sewage, electricity, transport, water, and so on. Two operations were key to this modern urban ideal. First, issues of social cohesion, order, and justice came to be seen as infrastructural problems, problems to be addressed by perfecting the operation of the city as a machine or an organism. Second, city space underwent a sharp bifurcation. The new underground city was seen as a purely technoscientific space, whereas the surfaces of boulevards, streets, and parks became realms where new purified forms of social life could emerge (Domínguez Rubio and Fogué 2013).

Relational concepts of infrastructure developed in STS emerged here as particularly well suited not just to unveil the hybridity of modern infrastructural arrangements but especially to elucidate the collapse of the modern integrated ideal since the 1960s. As Coutard (2008) summarizes, this collapse is generally connected with the global expansion of neoliberalism, withdrawal of the state, and various related phenomena, including notably the unbundling of urban infrastructures, that is, the process of "segmenting integrated infrastructure into different network elements and service packages" (Graham and Marvin 2001, 141). Vertical forms of unbundling, and what the authors call "bypass strategies," lead to the emergence of premium infrastructural spaces for "valued" or "powerful" users and places. A new urban landscape emerges in which interlinked premium spaces come to be disconnected from their immediate urban contexts. Infrastructural unbundling leads to a wider process of splintering urbanism, where new conflicts and struggles over infrastructural privatization and democratization begin to take place across both "developed" and "developing" cities in North and South America, Asia, Europe, and beyond.

Graham and Marvin's global narrative has generated fruitful critical debates among STS scholars (Coutard and Guy 2007), who see here yet another story of universal alarmism, so common in (post-Marxian) urban studies, yet based this time not on economic but rather on a soft form of technological-cum-infrastructural determinism. In the view of Coutard and Guy (2007, 713), an STS perspective on these matters is helpful exactly to "move beyond this 'universal alarmism' by emphasizing the ambivalence inherent to all technologies." Two elements of this response are indeed critical. First, whereas Graham and Marvin claim to study the city as a sociotechnical process, Coutard and Guy note that their empirical analysis of splintering urbanism ends up contradicting processual, contingent, and relational approaches. Second, Coutard and Guy suggest that STS scholars should adopt an explicit ethical and political commitment to producing more hopeful accounts of urban change, in the understanding that such accounts will have performative effects and are thus always-already part of political projects. Here again, we see how more general debates, this time on reflexivity and positionality in STS analysis, are played out and diffracted through the city as a challenging test site.

Metabolisms and the Urban Politics of Nature(s)

STS scholarship has also played a role in rethinking the entanglement of natural and urban processes. Hence, whereas the human ecology of the Chicago School aimed at understanding how humans adapt to their environment, the city as such was always conceived as a purified social phenomenon. Louis Wirth (1938, 1–2), for example, would write that "nowhere has mankind been farther removed from organic nature than under the conditions of life characteristic of great cities." It is only in the 1960s that industrial ecologists begin to develop models to think through the intermingling of natural and urban processes. Yet, these models propose a cybernetic separation between city systems and natural environments: imagining the city as a machine converting natural resources into waste, nature remains a realm outside the city and urban operations.

While noteworthy exceptions may be cited—such as environmental historian William Cronon's (1991) famous study into the natural causation of Chicago's development path—urban studies has thus tended to uphold, rather than to challenge, the society-nature binary. This is the backdrop against which the work of STS authors such as Donna Haraway (1991) and Bruno Latour (1993) has become widely influential in the more recent development of so-called urban political ecology (UPE)—by now a widely influential urban research tradition that proposes to think of cities and their development as processes of "cyborg urbanization" (Swyngedouw 1996). What distinguishes the city as a complex infrastructural and sociotechnical apparatus, according to the UPE approach, is that all of its components, human and nonhuman, are constantly in flux, interacting with each other, and exchanging their properties. The city involves thus "a perpetual passing through deterritorialized materials" (Kaika 2005, 27), a socio-natural-technical process to be grasped by means of an old Marxian concept: metabolism. Indeed, as geographer Erik Swyngedouw (1996) stresses, the notion of metabolism understood as involving the human transformation of nature through labor was the building stone of Marx's early version of historical materialism. Labor, as a hybrid but asymmetrical process, is thus the key to understanding how social relations and regulations channel the process of environmental production in cities.

On these grounds, UPE scholars put forward their own distinct theoretical articulation between post-Marxian traditions of critical urban studies and STS insights to study how social histories, power structures, and capital accumulation dynamics shape urban socionatural environments, often in ways that reproduce ruling class privilege (Swyngedouw and Heynen 2003). Accordingly, UPE's political challenge involves the struggle for environmental justice, that is, the work of unveiling, resisting, and ultimately changing the differentiated environmental impacts of capitalist or neoliberal urbanization on urban populations worldwide such as, for instance, the unequal exposure to toxic waste of the urban poor.

As the latter point suggests, UPE's attention to natural entities and processes amounts largely to a methodological enhancement of Marxian-Lefebvrian urbanization theory rather than to any decentering of human (or other) agencies, let alone the economy. This arguably makes the approach sit awkwardly vis-à-vis widely held STS commitments. As Holifield has noted (2009, 646), "the significance of nonhuman agents here lies in their 'social mobilization.'" Even in empirical studies that more strongly acknowledge the capacity of nonhuman entities to, for example, resist commodification and thus change capitalist accumulation patterns, there is a tendency to imagine the economy "as an already constituted structural unity *that only consequently comes into contact with a recalcitrant non-human nature*" (Braun 2008, 669, emphasis in original). It becomes apparent that the Marxian definition of metabolism is still anchored in a purified understanding of nature as that which has not (yet) been transformed by labor.

There are, however, some key contributions to urban political ecology that adopt a more symmetrical look at the capacities of human and nonhuman actors, while also breaking with the idea of imagining the city as one single, overarching metabolic process. One important example here is Paul Robbins and Julie Sharp's (2006) analysis of the capacities of urban lawns to interpellate homeowners as subjects, thus contributing to the reproduction of these ubiquitous urban mono-cultures. Another key contribution is the work of geographers Sarah Whatmore (2002) and Steve Hinchliffe et al.

(2005) on what they dub, following Isabelle Stengers, the "politics of conviviality" at play in civic practices of caring for biodiversity within urban brownfields and otherwise wild spaces of Birmingham. Conversely, anthropologists Ann Kelly and Javier Lezaun (2014) have studied municipal programs of mosquito surveillance and larval elimination in Dar es Salaam as processes of multispecies disentanglement, thus raising the vexed question of the role of separation practices in a politics of nature. Here, STS commitments to civic practices of knowledge-making and world-making come to articulate with emerging concerns for specific and hybrid urban natures.

Similarly, related STS sensibilities toward the co-production (Jasanoff 2004) of science, politics, and urban nature(s) has been brought to bear on more historically oriented studies into urban change. Sociologist Jens Lachmund (2013), for instance, tells the fascinating history of the birth of urban ecology as a scientific discipline in divided West Berlin after World War II and traces the radical reshaping of the city's greenery effectuated since the 1970s via its articulation into new planning regimes of biotope protection. In a similar vein, geographer Andrew Karvonen (2011) explores the various technonatural paradigms through which cities in North America and Europe have traditionally dealt with the problem of urban water flows—including the rise in recent years of more ecological and relational approaches to landscape architecture. What becomes clear in these studies is the strong sense in which the politics of urban nature(s), including in the realm of urban climate mitigation and adaptation, is always also a politics of shifting and competing expert knowledge regimes and practices (Blok 2013). Here again, urban studies is made to resonate with long-standing STS preoccupations, including questions as to how expert worlds of city planning may be further democratized through new forms of civic engagement in urban knowledge-making.

The Discovery of the Built Environment (in the City)

For a long time, STS did not really take into account the shaping capacities of buildings in the production of science and technology. This is particularly apparent in laboratory studies, where a whole subfield of research is defined by reference to a building typology, the laboratory. And yet, the architectural mediation of laboratory science has hardly been discussed, except perhaps for the fact that the lab enacts a carefully policed epistemic space vis-à-vis "society." This is, for example, the point made by Karin Knorr Cetina (1989, 129): "if the laboratory has come of age as a continuous and bounded unit that encapsulates internal environments, it has also become a link between internal and external environments, a border in a wider traffic of objects and observations."

However, apart from such general observations, one finds little research into how the actual design of laboratories shapes knowledge production.

An instructive case is Latour and Woolgar's (1986) purposely naïve description of the activities occurring in two main areas of a biological laboratory: the bench, where technological equipment is located and technicians, in particular, pursue activities such as cutting, mixing, shaking, and marking; and the office, where scientists, in particular, engage in activities such as reading, writing, and typing. According to the authors, this spatial division offers a suitable entry point to understanding knowledge production as a process of circulating inscriptions. Such analysis, however, assumes the built environment of the laboratory to be an unproblematic expression of the activities "contained" within it, implying that the architectural layout of the laboratory fully corresponds with certain sets of functions and practices, thereby tracing no distinction between the design of the building and its practical appropriation.

Over the past fifteen years, however, STS scholars have come to actively address architecture, buildings, and the built environment from different perspectives and across divergent empirical sites, although almost always within urban settings. There are many ways of bringing together these inventive STS accounts; yet, one fruitful route is precisely in terms of the problem of the "gap" between the design and use of buildings. This gap is well known in urban studies, where it was made famous not least by Stewart Brand's (1994) classic (proto-STS) work on *How Buildings Learn* after they are built. Here, we deploy the gap rather as a device for distinguishing different strands of STS work.

Laboratories of Architectural Design

In two papers written in the mid-1990s, ANT theorist Michel Callon proposes a radical redescription of the processes of architectural conception (Callon 1996, 1997). Whereas this native term usually denotes an individual mental process, Callon reads it as a collective process involving not just different voices (Cuff 1992), but mediated by various material supports such as plans and models. Paying attention to such material mediators, Callon demonstrates that each of them settles in specific ways the epistemic and evaluative differences among the multiple actors involved in the design process. In fact, he goes even further to argue—in ways reminiscent of Antoine Hennion's (2015) study of music—that the entire architectural design process cannot be understood as incrementally advancing toward the realization of an object but rather toward the creation of a multiplicity of mediators. "There exists no equivalence between what it [the building] is on paper, what it is in the scale model, and the final construction that finds its place in a social space which gives it its measure" (Callon 1996, 29; our translation).

Architectural design thus would be a classic example of what John Law (1987) calls heterogeneous engineering, but with one difference: what holds together such multiplicity is not just its black-boxing toward the end of the process but the maintenance of a certain *style* throughout the process.

Whereas the distributed production and maintenance of an architectural style has remained an unexplored hypothesis, Callon's work has been influential in the growing field of STS-inspired studies into architectural design. Much work, by now, has focused on design as an epistemic practice shaped by the problem of producing knowledge about a not-yet existing object. In a key contribution, ANT scholar Albena Yaneva (2005) has shown how architects gain knowledge of their emerging buildings through processes of scaling their physical Styrofoam models up and down, thereby enabling jumps between otherwise irreconcilable visual perspectives. In a similar vein, but invoking historian of science Hans-Jörg Rheinberger (1997), organization studies scholars Boris Ewenstein and Jennifer Whyte (2009) propose to understand architectural plans as epistemic objects, that is, as well-defined but abstract and incomplete objects calling for completion and thus inspiring architects to contribute to its realization. They also show how plans function as boundary objects (Star and Griesemer 1998) facilitating collaboration among different professional disciplines without the need for strong coordination. In related fashion, Ignacio Farías (2015) has described the organized occurrence of epistemic dissonance in architectural design processes, paying particular attention to how the work with visual mediators, such as photorealistic renderings and video animations, enable the production of uncertainty.

Apart from such epistemic problems, key contributions by Yaneva and anthropologist Sophie Houdart have highlighted the nonlinear, multiple, and oftentimes controversial character of architectural design processes. In documenting the nonrealization of a highly contested extension of the Whitney Museum of American Art in New York, Yaneva (2009) lays the foundation for a more general approach to understanding architectural design via the public controversies it generates (Yaneva 2011). In detailed ethnographic accounts of a well-known Tokyo-based architectural firm, Houdart and Chihiro (2009) show that what matters in architecture is not the specific sequence of the versions of a building (as in scientific inscriptions), but rather how their simultaneous presence defines a space of architectural conception and alternatives.

While analytically generative, however, the fact that laboratory studies has been taken as a role model for these new "studio studies" has also posed problems (cf. Farías and Wilkie 2015). In other words, and in spite of programmatic statements (e.g., Latour and Yaneva 2008), taking the equivalents to the lab (the office) and the experiment (building project) as key research objects has, for example, happened at the expense of a more precise understanding of architecture as a complex and power-laden expert-client assemblage (Cuff 1992). Be that as it may, part of what these studies demonstrate is the sheer extent to which buildings are imagined, within processes of architectural design, mostly as technological artifacts unproblematically enabling a certain type of social uses. As such, it has taken different kinds of STS approaches to detect the possible gap between the design and use of buildings.

What Do Buildings Actually Do?

The critical importance of the question of what buildings actually do, addressed by sociologist Thomas Gieryn in the STS field in the early 2000s, becomes particularly evident when noting the ubiquity of certain modernist understandings of buildings as technologies tailored for highly specific human activities. In the famous words of Le Corbusier, homes would be "machines for living in." Gieryn's (2002) discovery of buildings as somehow "difficult" technologies opened a whole set of reflections on such modernist conceptualizations.

As his starting point, Gieryn points out that popular social theories of the time (the late 1990s), such as those of Bourdieu and Giddens, dilute the question of what buildings do in the old problem of structure and agency. As such, they miss the varying capacities of buildings as unfolding material objects. In a move similar to that of SCOT readings of cities, Gieryn (2002, 41) proposes to understand buildings as technological artifacts: "Buildings, as any other machine or tool, are simultaneously the consequence and structural cause of social practice." Accordingly, his study of how a new building for biotechnology research comes into being at Cornell University mobilizes three concepts originally coined for grasping technological artifacts: heterogeneous design, blackboxing, and interpretative flexibility. With their help, Gieryn shows how design of the building's material form is tied into processes of enrolling and articulating multiple human and nonhuman actors and how in this process it reaches a point where it attains stability and gets built. This blackboxing, however, remains open to interpretation and practical reuse and retrofitting; in Gieryn's apt terms (2002, 35; emphasis in original) "buildings stabilize social life [...] yet, buildings stabilize *imperfectly*."

This understanding of buildings as always contestable technological black boxes has gained some traction in STS-inspired urban geography and been deployed to, for example, understand the modernist mass high-rise building, such as the Red Road development in Glasgow, which within only forty years went from enthusing inauguration to being earmarked for demolition (Jacobs, Cairns, and Strebel 2007). As any other technological artifact, buildings attain the uncontroversial status of a black box only provisionally. Not just their past is plagued by controversy; so too is their future, lead-ing to interpretative contestation (as shown by Gieryn) or, in other cases, to downright demolition. Indeed, as with any other technology, the key to a building's continued existence is maintenance and repair (Strebel 2011)—a point that has also been made forcefully about cities as such (Graham and Thrift 2007). There is now a growing STS-informed literature on the myriad roles of repair and maintenance work in the city, showing, for instance, how constant work is needed to maintain the dwelling ecology of high-rise housing in Singapore (Cairns and Jacobs 2011). Here, maintenance and repair not only keeps in check the recalcitrant agencies of water, mold, and mosquitoes but also serves to manage the agency of residents through forms of cyclical checking and surveillance.

So far, what buildings do and how they fit into their urban surroundings resemble other material objects; interpretive flexibility, as noted, is how SCOT describes any technological artifact, not something specific to a building. Other STS scholars, how-ever, have taken the opportunity to deploy the specificity of buildings as a challenge to such uniform ideas of (socio)materiality. Michael Guggenheim (2009), in particular, has pointed to three characteristics of buildings as objects that make them stand apart from such objects as artworks, technological artifacts, and scientific inscriptions. Build-ings, he argues, occupy a stable location; they are singulars with distinct biographies; and they are used by different people at the same time for different purposes. This implies that buildings are defined by their environment in stronger ways than other objects. Hence, Guggenheim inverts the classic ANT analysis of technological artifacts as immutable mobiles: buildings, he argues (2009), are mutable immobiles and, as such, only qualify as quasi-technologies. The mutability of buildings may pertain either to their material configuration or to the usages to which they are put, and changes often generate controversies mediated by building codes and zoning laws that regulate pos-sible sociomaterial covariations: at what point are members of the Muslim minority in a Swiss city, as new users of a former factory, allowed to erect a minaret onto the build-ing? (Guggenheim 2010). Here, the law—another expert practice—becomes a key site of building conversion, and hence of negotiating the stability and mutability of the urban built environment.

Among other things, Guggenheim's work clearly pinpoints the gap between the design and use of buildings; a gap centrally mediated, in his reading, by legal means. Yet, this reading arguably still falls short of fully engaging the question of how built environments are used and experienced as material settings for everyday urban life, sometimes in ways that diverge from any encoding by design.

The Sociomaterial Mediation of Urban Built Environments

The built environment of cities, as involving more than the sum of single buildings, is a long-standing topic of urban studies. At least since American urban planner Kevin Lynch's (1960) inaugural work on *The Image of the City*, the urban built environment has been cast as a text that is written and read by different urban actors in different ways. Lynch studied how the built environment is mentally read and mapped by different types of city users. Other analyses of the built environment focus rather on the processes of writing and, especially, on the production of urban symbolic landscapes of power by certain institutions and social groups. Such processes may serve variously to naturalize the privileged positions of elites (Zukin 1996) or the politics of national identity (Jones 2006), as well as to attract corporate investments by sending star-architectural messages regarding the position of the city in the global system of capital circulation. In this context, STS approaches to the built environment are making significant contributions that take seriously the material capacities of urban environments to mediate such social practices and processes (e.g., Göebel 2015).

Building on Bruno Latour's description of how human bodies are formatted, for instance, cultural sociologists Degen, Rose, and Basdas (2010, 62) have pointed to the urgent need of understanding "how the design of the material environment and people's embodiment co-constitute the experience of [...] places" in cities. Their study of shopping practices in two English commercial streets is an important contribution, as it shows the extent to which the different formatting offers made by the various elements of the urban environment are variably effective in affecting urban dwellers. In paying close attention to the various bodily practices in which people engage, whether task-oriented shopping, waiting, caring for others, hanging out, and so forth, they observe how the affordances of the built environment gain or lose their capacity to actually accommodate bodies. The built environment, they conclude, is "not only multiple in the sense of many, but multiple in the sense of ambivalent" (ibid., 73).

A perhaps extreme example of a similar approach is geographers Ralf Brand and Sarah Fregonese's (2013) study of the role played by the built environment in the political polarization and radicalization of urban conflicts in Belfast, Beirut, Berlin, and Amsterdam. Apart from pointing to features of the urban environment that act as mirrors of preexisting conflicts, such as fences, peace lines, and graffiti, they are interested in subtler mediations of polarization and conflict. This analysis entails showing, for instance, how a new footbridge to cross the Westlink carriageway in Belfast played a crucial role in the escalation of violence in 2007, as it provided access "to a launching spot for missile attacks," while giving "youths an easy escape route" (2013, 16).

Analyses such as these valuably demonstrate how the built environment is a constitutive mediator of urban practices, in the sense of opening up both foreseen and unforeseen spaces for activities. Moreover, by engaging both forms of embodiment and the capacities of materials, such as brick stones (Edensor 2013), to enable sensual and imaginative experiences, this line of work serves to open up new exchanges between STS and urban studies on the question of how to retheorize the specific "affective atmospheres" (Latham and McCormack 2009) of city settings in nonreductive and materially sensitive ways. This may, in turn, invite analysts to ask questions about those affective energies of specific urban milieus, such as the Biopolis research center of Singapore (Ong 2013), where sciences are made to thrive. So far, however, while studies of the built environment are almost invariably placed in the city, the STS approaches reviewed here cannot be said to have conclusively shown how buildings contribute to the making, remaking, and unmaking of entire cities. On this point as well, there is potential for further cross-fertilization of STS and urban studies.

Reassembling the City

Yet a third type of STS accounts of cities has relied on the conceptual repertoires of ANT and related intellectual projects. In this third avenue, which has come to be known as assemblage urbanism (e.g., Farías and Bender 2009; McFarlane 2011a), the city is cast not primarily as a novel site in which to study science and technology or as confronting STS scholars with new difficult artifacts, such as buildings. Rather, what is at stake is the extent to which it is necessary to recast and reassemble the very object of urban studies: the city.

Making the Invisible City Visible

Whereas strongly relational and postrepresentational approaches to the city began to emerge in the fields of urban studies and urban geography in the mid-1990s, probably the first dedicated work by an STS scholar to address the city on such terms was the book-website *Paris: Ville Invisible* by Bruno Latour and photographer Emilie Hermant (1998). This work focuses on different urban sites of material practice, embodied circulation, and infrastructural maintenance and coordination. While the deployment of the medium of the web mimics the dystopian imagination of the death of physical urban space with the rise of ICT, nothing could be further from this work's main tenets. The book-website serves rather to demonstrate that visual and textual representations of cities are always locally assembled and that the urban experience involves a constant passage through a proliferating array of interconnected locales.

One key focus of the book is the control rooms, in which urban technical systems and urban natures are made visible, coordinated, and organized. Within these confined sites, visual, textual, and numeric inscriptions of urban processes are accumulated, aligned, and used to inform practices of knowledge-making and intervention in urban realities. In Latour's ANT vocabulary, these sites are urban oligoptica, that is, places in which very little can be seen at any one time but in which everything that enters appears with great precision. The oligopticon stands in contrast to Foucault's panopticon; indeed, the notion aims to counteract the fantasy of totalizing overviews associated with the latter. At the same time, it shifts attention toward those crisscrossing networks of urban actors, practices, and material devices that are needed for any inscriptions to enter these often hidden places. In this sense, oligoptica not just interpret the city according to the different functions they address; rather, they involve different and overlapping ways of visualizing, constructing, and practicing a city.

Building on Latour's early work on visualization and cognition in scientific laboratory work, Swiss geographer Ola Söderström (1996) studied the role played by different visualization techniques in the history of urban planning, paying special attention to their varying capacities to make a complex object such as the city visible. Retracing the invention of the geometrical plan of the city, Söderström shows how the historical transition from an oblique to a zenithal, bird's eye gaze led to a naturalization of the city as a measurable object that could be classified in zones according to indicators such as socioeconomic profiles, criminality rate, and life quality. More recently, and along similar lines, STS-informed scholars have explored the current decentering of city visualizations resulting from the proliferation of digital interfaces articulating new relationships between citizens and urban infrastructures. Anthropologist Jennifer Gabrys (2014), for instance, has shown how smart city infrastructures perform the city as data sets to be managed and how they redefine citizenship as segmented practices of producing, managing, and monitoring data.

These developments support in different ways Latour's key claim that there is not one Paris, but multiple Parises; that is, that the city needs to be understood as a multiplicity that is simply impossible to totalize or to fix. Anticipating what later becomes his generalized social ontology, Latour (2005) extends this point to every urban agency: persons, institutions, social movements, tourists, political parties, and so on. In urban spaces one does not encounter stable subjects but rather flexible and fluid agencies being co-defined by different regimes of materiality, affectivity, and intelligence. "In front of the bank automat I had to act as a generic individual endowed only with an individual pin code; pressed against the barrier on the pavement I was a mechanical force weighing against another mechanical force; in front of the traffic light I became a reader of signs, capable of understanding a prohibition; by swearing at a reckless driver I am transformed into an indignant moral citizen [...]" (Latour and Hermant 1998, plan 33).

The greatest challenge posed by the city to the conceptual repertoires of ANT, arguably, is precisely how to think of this multiplicity. Whereas it might be evident that different urban technical systems, institutions, and actors build up different networks, the city entails a complex multiplicity, "folded perhaps, and folded again like an origami, but flat everywhere" (ibid., plan 31). Such "flat" multiplicity remains however invisible; it is a virtual plane of potential associations. The city appears thus as a *terra incognita*, a plasma waiting to take shape (Latour 2005). In this sense, *Paris: Ville Invisible* is certainly the most overtly Deleuzian book ever written by Latour. And as such, it contains many of the key propositions that have gone into current discussions of urban assemblages and assemblage urbanism, while perhaps not yet being quite recognized for it.

Reimagining the Urban as Assemblage

While hard to pin down conceptually, the notion of assemblage has gradually come to reshape urban studies in terms of ANT-consonant principles of symmetry, flatness, and multiplicity. This reshaping, arguably, was crucially facilitated via the book *Reimaging the Urban* by geographers Ash Amin and Nigel Thrift (2002). Urban everyday life, urban politics, urban economies, urban technical systems are all recast in this book as sets of constantly evolving assemblages that collectively form what these authors (echoing Deleuze) call a mechanosphere, a virtual plane of abstract machines informing the constitution and operation of cities. Relying on theoretical inputs from ANT and technoscience studies, this is perhaps the first book to propose a radical decentering of urban actors and spaces. Tools, machines, and technical systems are to be refigured as integral parts of human actors, just as bacteria, plants, animals, and humans enter relationships of co-production in the same urban symbiotic sphere. The city appears thus as a site of intensive encounters of humans, technology, and nature. These encounters, which escape the dynamics of metabolic organization, are grasped better with the language of chemistry in terms of compositions, reactions, emergences, and intensities (cf. Stengers 2005).

In one sense, what the concept of urban assemblages does is to make explicit the key theoretical displacements in the understanding of cities put forward in such contributions: the human-nonhuman hybridity of urban associations (e.g., Hinchliffe et al. 2005); the flattening of scalar and nested models of urban space (e.g., Latham and McCormack 2009); and the redefinition of the city as a multiplicity of intensities and ordering practices. This redefinition is indeed how ANT might be said to change urban studies (Farías and Bender 2009). Yet, at the same time, urban assemblages has also come to denote a more complicated set of two-way exchanges with urban studies, challenging ANT in particular to move in novel directions (cf. Blok 2012, 2013; Blok and Farías 2016).

Perhaps one of the more far-reaching routes opened up by assemblage thinking is the radical redefinition of urban economies and politics. Paying attention to the more-than-human passions, attachments, and entanglements occurring in urban spaces, for instance, Amin and Thrift (2002) suggest reconsidering the economic role of cities, away from the traditional macroeconomic focus on regional clusters and other urban geographies as assets for production. Instead, they underline the way cities shape the economy in terms of how urban intensive encounters are capable of constantly generating new affects and passions (e.g., Tironi 2009), thereby eventually constituting new types of demands for goods and services, demands which are however not strictly economic (e.g., Färber 2014). As such, they invite new reflections at the intersection of urban studies and ANT insights into economization processes (Callon 2007).

The city as an intensive, affective, and passionate site also defines and recasts the urban political. In one language, rediscovered recently by ANT theorists (e.g., Latour 2007; Marres 2007), it reconstitutes urban politics by way of what American pragmatist John Dewey (1927) described as publics of variable geometry and duration, constituted around emerging issues of shared concern. Following this Latourian (and Stengerian) recasting of political philosophy, urban politics becomes a version of cosmopolitics, the politics of searching for and building the shared common cosmos, an urban common world (Blok and Farías 2016; Farías 2011; Tironi and Sánchez Criado 2015). More than anything, this is a politics of urban knowledge-making, one committed to new forms of collective experimentation and learning in the city by way of constituting and strengthening urban democratic publics (McFarlane 2011a).

The Assemblage Urbanism Debate

Since 2011, lively exchanges have unfolded, primarily in the pages of the journal *City*, fueled by theoretical critiques of assemblage urbanism coming from critical urban scholars. This debate is interesting because it gives us clues to how ANT, and also more general STS insights and analytical tensions, are currently traveling across academic fields of inquiry, being taken into account, transformed, and contested.

A major critique has focused on the empirical commitments of ANT analyses of the city, and often more generally STS approaches to technoscience, as involving a form of naïve realism. Accordingly, post-Marxian critical scholars have attempted to adjudicate different ways of using the notion of urban assemblages and to argue for "a narrower, primarily methodological application" (Brenner, Madden, and Wachsmuth 2011, 230) that could serve as add-on to more substantial forms of theoretical and critical engagement. Such a position is perhaps unsurprising when considering that the ethical and political consequences of thinking with assemblage seek to debunk, or at least seriously deflate, the very premises upon which classical critiques of ideology rest.

The first of such premises is that urban politics results from struggles among well-defined classes of humans over the appropriation of urban space (Brenner, Madden, and Wachsmuth 2011, 236). In this context, the city is conceived as a "point of collision" between the mobilizations of the deprived, the discontented, and the dispossessed on the one side and, on the other, ruling class strategies to instrumentalize, control, and colonize social and natural resources" (Brenner, Marcuse, and Mayer 2010, 182). ANT complicates this picture, not least by pointing to the importance of objects and sociomaterial devices for equipping humans with agency in the first place. At the same time, ANT also entails a shift from a conflict-based model of politics, rooted in structural (capitalist) contradictions, toward a controversy-based model of urban politics based on the eruption of uncertainty and critique (Farías 2011).

The second main challenge involves the task of the critical scholar, which in critical urban studies is usually described as deciphering the hidden structural contradictions and injustices, unveiling the ideologies of the ruling class, and enlightening people about the structural forces lurking behind their apparent matters of concern. ANT's empirical stance, and arguably also that of much other STS work, is fully incompatible with this position, as it implies modest, careful, and analytically respectful engagements with the various actors involved in urban politics—including financial capitalists and neoliberal technocrats—in order to "not impose 'ready-made explanations' upon the cartographies of actors and networks" (Puig de la Bellacasa 2011, 88).

This ANT position, however, is not without tensions of its own. Indeed, one important route for strengthening the urban assemblages approach will be to cross-fertilize it with the long feminist STS tradition and its strong focus on questions of asymmetry, invisibilization, and exclusion. This might involve following philosopher Maria Puig de la Bellacasa (2011) when she proposes to treat sociotechnical assemblages as "matters of care" rather than through the Latourian language of "matters of concern." A focus on caring entails posing the question of who actually does the devalued doings necessary to sustain urban assemblages, including sustainable or "smart" infrastructures. Thereby, the point is not just to make urban caring practices visible, but to actively generate care by way of maintaining a commitment to the possible and alternative becomings of things. Indeed, the key ethical and political question resulting from treating urban assemblages as matters of care is perhaps not *whose* assemblage or *for whom* to care but rather *how* to care, *how* to carefully (re)assemble urban life.

Concluding Remarks: New STS Avenues, New Cities?

In this chapter we have pointed to three important avenues by which STS concepts and approaches have engaged in studies of the city and urban life, thereby entering into dialogues with the similarly heterogeneous field of urban studies. In the first avenue, well-established STS frameworks were mobilized to rethink the city as a technological artifact. In the second avenue, STS scholars discovered the urban built environment as a challenging new technical object of study. In the third avenue, ANT and Deleuzian intersections seek to reconstitute the ontology and politics of cities. While we have narrated these various encounters mainly from the point of view of important and innovative STS scholars, we have stressed throughout that intellectual borrowings and travelings are indeed two-way streets, with urban phenomena and analysts at times calling on STS scholars to tinker with and rework their conceptual tools.

Indeed, arguably the most fascinating feature of the city for STS is that it confronts us with ethicopolitical questions associated with the articulation and composition of common worlds of sociotechnical cohabitation. Multiplicity in cities is not just an analytical insight from a sophisticated STS-cum-ethnographic reflexivity but a fundamental urban experience not the least channeled into political mobilizations and controversies. Thus, whether we study urban technologies, infrastructures, socio-natures or buildings, questions concerning the local articulation of multiple, often incommensurable ways of enacting the city keep emerging and challenging our concepts and methods. Notably, after decades of conceptual work to explore network and fluid topologies of technoscience assemblages (cf. Law and Mol 2001), the city arguably challenges STS to once again rethink notions of place, of regional topology, of localized heterogeneities. In terms of methods, it invites us to not just follow the objects throughout translocal networks but also learn to stay put and study how urban sites are made and unmade through their multiple sociotechnical enactments, exploring the urban politics of coexistence and copresence. The city, in short, emerges as a crucial site in which to explore all the key political problematiques of a hybrid, technoscientific world (Blok and Farías 2016).

By way of concluding this tour of the urban test site, we want to briefly point to some of the routes less traveled so far—as an invitation also for future STS engagements in the city. One striking white spot in this respect is the relative lack of engagement from STS scholars with long-standing questions of global urban hierarchies and, more generally, the problem of urban difference. No two cities are quite the same. This simple fact has long been reckoned with in urban studies, mainly through vocabularies of Northern and Southern urbanisms (Roy 2011), global versus ordinary cities (Robinson 2006; Sassen 1991), and various postcolonial urban legacies outside of the West (Yeoh 2005). Yet, STS scholars have had little to say on these issues, in part reflecting the—perhaps paradoxical—fact that many have preferred to stay at home, in a Euro-American metropolis, turning this into their truth spot (Gieryn 2006). With attention to other-than-Western contexts of science and technology picking up across STS these years, this picture may be expected to change—thus opening up new challenges of how to adapt and decenter STS concepts into those globally traveling knowledges through which cities are shaped and reshaped (McCann and Ward 2011; McFarlane 2011b).

A second and related lacuna concerns what might be called the performativity of STS vis-à-vis the formation of urban knowledges, including those more long-standing urban professions of architecture, design, and planning. While these worlds of urban knowledge and practice are by now the object of STS inquiry, as noted, it remains to be seen what might emerge once STS concepts and approaches start leaving their mark more strongly on the very socialization of urban professionals and, more generally, the formation of urban policy–related claims. In this respect, one hopeful projection is to imagine the formation of an engaged program (Sismondo 2007) of urban STS—one willing and capable of posing critical questions and providing constructive input at the moving boundaries of science, technology, and democratic politics in the city. Here, addressing the prospects and limitations for democratizing urban expertise, and for inventing new forms of technical democracy in the city, constitutes one important route ahead for STS in its situated search for the future of cities.

References

Aibar, Eduard, and Wiebe E. Bijker. 1997. "Constructing a City: The Cerda Plan for the Extension of Barcelona." *Science, Technology, & Human Values* 22 (1): 3–30.

Amin, Ash, and Nigel Thrift. 2002. *Cities. Reimagining the Urban.* Cambridge, Oxford: Polity.

Bijker, Wiebe, and Karin Bijsterveld. 2000. "Women Walking through Plans: Technology, Democracy, and Gender Identity." *Technology and Culture* 41 (3): 485–515

Blok, Anders. 2012. "Wandering around Cities with ANTs." *Science as Culture* 21 (2): 283–87.

___. 2013. "Urban Green Assemblages: An ANT View on Sustainable City Building Projects." *Science & Technology Studies* 26 (1): 5–24.

Blok, Anders, and Ignacio Farías, eds. 2016. *Urban Cosmopolitics: Agencements, Assemblies, Atmospheres*. London: Routledge.

Brand, Ralf, and Sara Fregonese. 2013. *The Radicals' City: Urban Environment, Polarisation, Cohesion*. Aldershot: Ashgate.

Brand, Stewart. 1994. *How Buildings Learn: What Happens after They're Built*. New York: Viking Press.

Braun, Bruce. 2008. "Environmental Issues: Inventive Life." *Progress in Human Geography* 32: 667–79.

Brenner, Neil, David J. Madden, and David Wachsmuth. 2011. "Assemblage Urbanism and the Challenges of Critical Urban Theory." *CITY* 15 (2): 225–40.

Brenner, Neil, Peter Marcuse, and Margit Mayer. 2010. "Cities for People, Not for Profit." *CITY* 13 (2–3): 176–84.

Cairns, Stephen, and Jane Jacobs. 2011. "Ecologies of Dwelling: Maintaining High-Rise Housing in Singapore." In *Companion to the City*, edited by Gary Bridge and Sophie Watson, 79–94. London: Wiley-Blackwell.

Callon, Michel. 1996. "Le Travail de la Conception en Architecture." *Situations: Les Cahiers de la Recherche Architecturale* 37 (1): 25–35.

___ . 1997. "Concevoir: Modèle Hiérarchique et Modèle Négocié." In *L'élaboration des Projets Architecturaux et Urbains en Europe*, edited by Michel Bonnet, 169–74. Paris: Plan Construction et Architecture.

___. 2007. "What Does It Mean to Say That Economics Is Performative?" In *Do Economists Make Markets? On the Performativity of* Economics, edited by D. MacKenzie, F. Muniesa, and L. Siu, 311–57. Princeton, NJ: Princeton University Press.

Coutard, Olivier. 1999. *The Governance of Large Technical Systems*. London: Routledge.

___. 2008. "Placing Splintering Urbanism: Introduction." *Geoforum* 39 (6): 1815–20.

Coutard, Olivier, and Simon Guy. 2007. "STS and the City: Politics and Practices of Hope." *Science, Technology, & Human Values* 32 (6): 713–34.

Cronon, William. 1991. *Nature's Metropolis: Chicago and the Great West*. New York: W. W. Norton.

Cuff, Dana. 1992. *Architecture: The Story of Practice*. Cambridge, MA: MIT Press.

Degen, Monica, Gillian Rose, and Begum Basdas. 2010. "Bodies and Everyday Practices in Designed Urban Environments." *Science Studies* 23 (2): 60–76.

Dewey, John. 1927. *The Public and Its Problems*. New York: Holt.

Domínguez Rubio, Fernando, and Uriel Fogué. 2013. "Technifying Public Space and Publicizing Infrastructures: Exploring New Urban Political Ecologies through the Square of General vara del Rey." *International Journal of Urban and Regional Research* 37 (3): 1035–52.

Edensor, Tim. 2013. "Vital Urban Materiality and Its Multiple Absences: The Building Stone of Central Manchester." *Cultural Geographies* 20 (4): 447–65.

Ewenstein, Boris, and Jennifer Whyte. 2009. "Knowledge Practices in Design: The Role of Visual Representations as 'Epistemic Objects'." *Organization Studies* 30 (1): 7–30.

Färber, Alexa. 2014. "Low-Budget Berlin: Towards an Understanding of Low-Budget Urbanity as Assemblage." *Cambridge Journal of Regions, Economy and Society* 7 (1): 119–36.

Farías, Ignacio. 2009. "Introduction: Decentering the Object of Urban Studies." In *Urban Assemblages: How Actor-Network Theory Changes Urban Studies*, edited by Ignacio Farías and Thomas Bender, 1–24. London: Routledge.

___. 2011. "The Politics of Urban Assemblages." *CITY* 15 (3–4): 365–74.

___. 2015. "Epistemic Dissonance: Reconfiguring Valuation in Architectural Practice." In *Moments of Valuation. Exploring Sites of Dissonance*, edited by Ariane Berthoin Antal, Michael Hutter, and David Stark, 271–89. Oxford: Oxford University Press.

Farías, Ignacio, and Thomas Bender, eds. 2009. *Urban Assemblages: How Actor-Network Theory Changes Urban Studies*. London: Routledge.

Farías, Ignacio, and Alex Wilkie, eds. 2015. *Studio Studies: Operations, Topologies & Displacements*. London: Routledge.

Gabrys, Jennifer. 2014. "Programming Environments: Environmentality and Citizen Sensing in the Smart City." *Environment and Planning D: Society and Space* 32 (1): 30–48.

Gieryn, Thomas F. 2002. "What Buildings Do." *Theory and Society* 31 (1): 35–74.

___. 2006. "City as Truth-Spot." *Social Studies of Science* 36 (1): 5–38.

Göbel, Hanna. 2015. *The Re-use of Urban Ruin: Atmospheric Inquiries of the City*. London: Routledge.

Graham, Stephen. 1997. "Telecommunications and the Future of Cities: Debunking the Myths." *Cities* 14 (1): 21–29.

Graham, Stephen, and Simon Marvin. 2001. *Splintering Urbanism: Networked Infrastructures, Technological Mobilities, and the Urban Condition*. London: Routledge.

Graham, Stephen, and Nigel Thrift. 2007. "Out of Order: Understanding Repair and Maintenance." *Theory, Culture & Society* 24 (3): 1–25.

Guggenheim, Michael. 2009. "Mutable Immobiles: Building Conversion as a Problem of Quasi-Technologies." In *Urban Assemblages: How Actor-Network Theory Changes Urban Studies*, edited by Ignacio Farías and Thomas Bender, 161–78. London: Routledge.

___. 2010. "The Law of Foreign Buildings. Flat Roofs and Minarets." *Social and Legal Studies* 19 (4): 441–60.

Hård, Mikael, and Thomas J. Misa, eds. 2008. *Urban Machinery: Inside Modern European Cities.* Cambridge, MA: MIT Press.

Haraway, Donna J. 1991. "A Cyborg Manifesto: Science, Technology, and Socialist-Feminism in the Late Twentieth Century." In *Simians, Cyborgs and Women: The Reinvention of Nature*, 149–81. New York: Routledge.

Hennion, Antoine. 2015. *The Passion for Music: A Sociology of Mediation.* Adlershot: Ashgate.

Hinchliffe, Steve, Matthew Kearnes, Monica Degen, and Sarah Whatmore. 2005. "Urban Wild Things: A Cosmopolitical Experiment." *Environment and Planning D: Society and Space* 23 (5): 643–58.

Holifield, Ryan. 2009. "Actor-Network Theory as a Critical Approach to Environmental Justice: A Case against Synthesis with Urban Political Ecology." *Antipode* 41 (4): 637–58.

Hommels, Anique. 2005. *Unbuilding Cities: Obduracy in Urban Sociotechnical Change.* Cambridge, MA: MIT Press.

Houdart, Sophie, and Minato Chihiro. 2009. *Kuma Kengo: An Unconventional Monograph.* Paris: Éditions Donner Lieu.

Hughes, Thomas P. 1983. *Networks of Power: Electrification in Western Society, 1880–1930.* Baltimore: Johns Hopkins University Press.

Jacobs, Jane M., Stephen Cairns, and Ignaz Strebel. 2007. "'A Tall Storey ... But, a Fact Just the Same': The Red Road High-Rise as a Black Box." *Urban Studies* 44 (3): 609–29.

Jasanoff, Sheila. 2004. "The Idiom of Co-production." In *States of Knowledge: The Co-production of Science and Social Order*, edited by Sheila Jasanoff, 1–12. London: Routledge.

Joerges, Bernward. 1999. "Do Politics Have Artefacts?" *Social Studies of Science* 29 (3): 411–31.

Jones, Paul R. 2006. "The Sociology of Architecture and the Politics of Building: The Discursive Construction of Ground Zero." *Sociology* 40 (3): 549–65.

Kaika, Maria. 2005. *City of Flows: Modernity, Nature, and the City.* New York: Routledge.

Karvonen, Andrew. 2011. *Politics of Urban Runoff: Nature, Technology, and the Sustainable City.* Cambridge, MA: MIT Press.

Kelly, Ann H, and Javier Lezaun. 2014. "Urban Mosquitoes, Situational Publics, and the Pursuit of Interspecies Separation in Dar es Salaam." *American Ethnologist* 41 (2): 368–83.

Knorr Cetina, Karin. 1989. "The Couch, the Cathedral, and the Laboratory: On the Relationship between Experiment and Laboratory in Science." In *Science as Practice and Culture*, edited by Andrew Pickering, 113–38. Chicago: University of Chicago Press.

Lachmund, Jens. 2013. *Greening Berlin: The Co-production of Science, Politics, and Urban Nature.* Cambridge, MA: MIT Press.

Latham, Alan, and Derek P. McCormack. 2009. "Globalizations Big and Small: Notes on Urban Studies, Actor-Network Theory, and Geographical Scale." In *Urban Assemblages: How Actor-Network Theory Changes Urban Studies*, edited by Ignacio Farías and Thomas Bender, 53–72. London: Routledge.

Latour, Bruno. 1993. *We Have Never Been Modern*. Cambridge, MA: Harvard University Press.

___. 2005. *Reassembling the Social: An Introduction to Actor-Network-Theory*. Oxford: Oxford University Press.

___. 2007. "Turning around Politics: A Note on Gerard de Vries' Paper." *Social Studies of Science* 37 (5): 811–20.

Latour, Bruno, and Emilie Hermant. 1998. *Paris: Ville Invisible*. Paris: La Découverte.

Latour, Bruno, and Steve Woolgar. 1986. *Laboratory Life: The Construction of Scientific Facts*. Princeton, NJ: Princeton University Press.

Latour, Bruno, and Albena Yaneva. 2008. "'Give Me a Gun and I Will Make All Buildings Move': An ANT's View of Architecture." In *Explorations in Architecture: Teaching, Design, Research*, edited by Reto Geiser, 80–89. Basel: Birkhäuser.

Law, John. 1987. "Technology and Heterogeneous Engineering: The Case of Portuguese Expansion." In *The Social Construction of Technological Systems: New Directions in the Sociology and History of Technology*, edited by Wiebe E. Bijker, Thomas P. Hughes, and Trevor J. Pinch, 111–34. Cambridge, MA: MIT Press.

Law, John, and Annemarie Mol. 2001. "Situating Technoscience: An Inquiry into Spatialities." *Environment and Planning D: Society and Space* 19 (5): 609–21.

Lynch, Kevin. 1960. *The Image of the City*. Cambridge, MA: MIT Press.

Marres, Noortje. 2007. "The Issues Deserve More Credit: Pragmatist Contributions to the Study of Public Involvement in Controversy." *Social Studies of Science* 37 (5): 759–80.

Mayntz, Renate, and Thomas Parke Hughes. 1988. *The Development of Large Technical Systems*. Frankfurt am Main: Campus Verlag.

McCann, Eugene, and Kevin Ward. 2011. *Mobile Urbanism: Cities and Policymaking in the Global Age*. Minneapolis: University of Minnesota Press.

McFarlane, Colin. 2011a. "Assemblage and Critical Urbanism." *CITY* 15 (2): 204–24.

___. 2011b. *Learning the City: Knowledge and Translocal Assemblage*. Oxford: Wiley-Blackwell.

Mumford, Lewis. 1937. "What Is a City?" *Architectural Record* 82 (5): 59–62.

___. 1961. The City in History: Its Origins, Its Transformations, and Its Prospects. New York: Harcourt, Brace.

Ong, Aihwa. 2013. "A Milieu of Mutations: The Pluripotency and Fungibility of Life in Asia." *East Asian Science, Technology and Society* 7: 69–85.

Pinch, Trevor J., and Wiebe E. Bijker. 1984. "The Social Construction of Facts and Artefacts: or How the Sociology of Science and the Sociology of Technology Might Benefit Each Other." *Social Studies of Science* 14 (3): 399.

Puig de la Bellacasa, Maria. 2011. "Matters of Care in Technoscience: Assembling Neglected Things." *Social Studies of Science* 41 (1): 85–106.

Rheinberger, Hans-Jörg. 1997. *Toward a History of Epistemic Things: Synthesizing Proteins in the Test Tube*. Stanford, CA: Stanford University Press.

Robbins, Paul, and Julie Sharp. 2006. "Turfgrass Subjects: The Political Economy of Urban Mono-culture." In *In the Nature of Cities: Urban Political Ecology and the Politics of Urban Metabolism*, edited by Nikolas Heynen, Maria Kaika, and Erik Swyngedouw, 110–28. London: Routledge.

Robinson, Jennifer. 2006. *Ordinary Cities: Between Modernity and Development*. London: Routledge.

Roy, Ananya. 2011. "Slumdog Cities: Rethinking Subaltern Urbanism." *International Journal of Urban and Regional Research* 35 (2): 223–38.

Sassen, Saskia. 1991. *The Global City: New York, London, Tokyo*. Princeton, NJ: Princeton University Press.

Sismondo, Sergio. 2007. "Science and Technology Studies and an Engaged Program." In *The Handbook of Science and Technology Studies*. 3rd ed., edited by Edward J. Hackett, Olga Amsterdamska, Michael Lynch, and Judy Wajcman, 13–32. Cambridge, MA: MIT Press.

Söderström, Ola. 1996. "Paper Cities: Visual Thinking in Urban Planning." *Cultural Geographies* 3 (3): 249–81.

Star, Susan Leigh, and James R. Griesemer. 1998. "Institutional Ecology, 'Translations,' and Boundary Objects: Amateurs and Professionals in Berkeley's Museum of Vertebrate Zoology, 1907–39." *Social Studies of Science* 19 (3): 387–420.

Star, Susan Leigh, and Karen Ruhleder. 1996. "Steps toward an Ecology of Infrastructure: Design and Access for Large Information Spaces." *Information Systems Research* 7 (1): 111–34.

Stengers, Isabelle. 2005. "A Cosmopolitical Proposal." In *Making Things Public: Atmospheres of Democracy*, edited by Bruno Latour and Peter Weibel, 994–1003. Cambridge, MA, Karlsruhe: MIT Press; ZKM / Center for Art and Media in Karlsruhe.

Strebel, Ignaz. 2011. "The Living Building: Towards a Geography of Maintenance Work." *Social & Cultural Geography* 12 (3): 243–62.

Swyngedouw, Erik. 1996. "The City as a Hybrid: On Nature, Society and Cyborg Urbanization." *Capitalism, Nature, Socialism* 7: 65–80.

Swyngedouw, Erik, and Nikolas C. Heynen. 2003. "Urban Political Ecology, Justice and the Politics of Scale." *Antipode* 35 (5): 898–918.

Tironi, Manuel. 2009. "Gelleable Spaces, Eventful Geographies: The Case of Santiago's Experimental Music Scene." In *Urban Assemblages. How Actor-Network Theory Changes Urban Studies*, edited by Ignacio Farías and Thomas Bender, 27–52. London: Routledge.

Tironi, Martin, and Tomás Sánchez Criado. 2015. "Of Sensors & Sensitivities: Towards a Cosmopolitics of 'Smart Cities'?" *Tecnoscienza* 6 (1): 89–108.

Whatmore, Sarah. 2002. *Hybrid Geographies: Natures, Cultures, Spaces*. London: Sage.

Winner, Langdon. 1980. "Do Artifacts Have Politics?" *Daedalus* 109 (1): 121–36.

Wirth, Louis. 1938. "Urbanism as a Way of Life." *American Journal of Sociology* 44 (1): 1.

Woolgar, Steve, and Geoff Cooper. 1999. "Do Artefacts Have Ambivalence? Moses' Bridges, Winner's Bridges and Other Urban Legends in S&TS." *Social Studies of Science* 29 (3): 433–49.

Yaneva, Albena. 2005. "Scaling Up and Down: Extraction Trials in Architectural Design." *Social Studies of Science* 35 (6): 867–94.

___. 2009. *The Making of a Building: A Pragmatist Approach to Architecture*. Bern: Peter Lang.

___. 2011. *Mapping Controversies in Architecture*. Aldershot: Ashgate.

Yeoh, Brenda S. A. 2005. "The Global Cultural City? Spatial Imagineering and Politics in the (Multi-)Cultural Marketplaces of South-East Asia." *Urban Studies* 42 (5–6): 945–58.

Zukin, Sharon. 1996. "Space and Symbols in an Age of Decline." In *Re-presenting the City: Ethnicity, Capital and Culture in the Twenty-First Century Metropolis*, edited by Anthony D. King, 43–59. Hampshire, London: Macmillan.

20   The Sociotechnical Architecture of Information Networks

Hector Postigo and Casey O'Donnell

Prologue

Before the first wave of dot-coms bubbled and then burst in 1999, many investors wondered if they would ever see their investments bear fruit. The burgeoning World Wide Web was gathering users, but the business models still were premised on distribution of consumer goods and the price differentials were not enough to cause shoppers to leave their brick and mortar stores for the web. Those untested business models appeared to many observers as "isomorphs," as Lewis Mumford might dub them: an old organizational and distribution structure imported to a categorically different sociotechnical milieu (Mumford 1963). We were not quite in the grips of Castell's "information society," but we certainly were heading in that direction (Castells 1996). Dubbed "irrational exuberance" (Shiller 2006), market capitalization for many Internet companies and businesses defied what had been traditionally understood as the standard for making wise investments: the price to earnings ratio. With many start-ups making little money at the time of their initial public offering, investments appeared inconsistent with behavior attributed to rational actors.

In 1999, when the bubble did burst, it did so in ways both expected and not. Many Silicon Valley web companies found themselves in a slump, but others showed promise. AOL, for example, continued to make inroads into defining web services, acquiring Time Warner, an "old media" company. While AOL never realized its potential, at the time the bubble burst the idea that an Internet service provider could serve as curating platform over the otherwise chaotic web was in many ways ahead of its time.[1] AOL's failure in the long run to enter the mass media business effectively may not be one of vision but rather one of timing. On the other hand, Amazon.com, working under old "isomorphic" business models without brick and mortar outlets remains one of the only consistently successful retailers on the web by focusing initially on selling books priced significantly below traditional retailers. Over time it has moved to serve as a retailer for just about any consumer good.

Our historical perspective notwithstanding, to those living through the first years of a rapidly growing World Wide Web, the Internet was moving fast with many retail and entertainment venues popping up. Surely it would achieve some form of closure in the coming decade, given that it was so earnestly attempting to replicate existing business models. Investors and users thought the Internet, as a ubiquitous information communication technology (ICT), could not escape the inevitability of "domestication" (Berker 2005). Surely it would achieve some form of "closure" in the coming decade given that business models therein were so earnestly attempting to replicate existing business models.

For those making their first foray into science and technology studies (STS), closure is a well-known element of the generative SCOT (social construction of technology) theory for elucidating the processes by which social dynamics shape technology. Specifically, closure is the process by which a technology and its meaning are settled into a stable form after periods of change and contention. Closure occurs when groups invested in a technology as a commercial venture or for its use value believe the problem the technology was meant to solve is settled or when a new more appropriate problem for the technological design is found (Pinch and Bijker 1984).

We began researching information technology at a time when AOL was the premier Internet service provider (ISP) and cell phones were not quite yet so ubiquitous. It did not take us long to realize that if anything crystallized so many of the core concepts in technology studies, it was the Internet and its other associated ICTs. Their deep connection to the technocratic desire to build redundant fail-safe systems to resist nuclear war and the influence of the "hacker ethic" made the Internet an unquestionably political artifact. Many had invested their ideologies about Cold War America, computers, and decentralization into the very protocols that made networked ICTs work at the most fundamental levels. (Abbate 1999; S. Levy 1984; Winner 1985). Observing the myriad of social groups that adopted networked ICT use and redesigned it on the fly caused us to wonder if this collage of increasingly networked communication technologies would ever find a stable single design, meaning, and use.

Nearly twenty years later, what began as objects of study not yet widespread in their adoption has morphed into a nearly ubiquitous, complex collection of data communication networks. Networked ICTs comprise a host of interconnected software and hardware technologies that afford users a number of communication and consumption options. Collectively they are conduits for novel business models, civil society, hate groups, news, entertainment, surveillance, and social networking. They solve all manner of problems from finding one's way on the roadways to an unknown location via global positioning systems to pirating copyright works and distributing them. The transmission points for all that data and services can be fixed on a terminal via a PC or video game console or can be mobile via a cell phone or tablet. Transmission, while facilitated by a few media companies serving as ISPs or mobile data companies, can also be entirely decentralized, as in the case of mesh networks (De Filippi 2014).

We opened our chapter with a bit of Internet business history[2] as a way of historically framing our perspective on *networked* ICT platforms.[3] We use it to help our readers imagine the initial indeterminacy in networked ICT penetration, design, use, and meaning we observed. As we had done when we began our research into networked ICT platforms, we imagined entrepreneurs wondering if the burgeoning Internet and mobile networks would congeal into being another form of mass media enterprise, retailer, or both. With the benefit of hindsight, we can reasonably conclude that it is far more than both and may never experience an STS kind of closure.

In subsequent sections we engage with case studies of networked ICT use to illustrate the ongoing indeterminacy of networked ICTs. We focus on the practice of play and production of user-generated content (UGC) in two case studies to illustrate how networked ICTs exhibit indeterminacy in their use and meaning and thus resist closure. The first case focuses on YouTube video game commentary video production and the second on video game design and play. In this chapter we also distinguish *networked* ICTs from what are generally referred to as simply ICTs because we recognize that the latter can apply to any artifact that serves as a means to ensure a message is delivered to a receiver. The printed word on a book or newspaper, material symbols of status or lack thereof, the telegraph, the television, and so on all qualify as forms of ICT in our opinion and the opinions of communication scholars before us (Carey 1989; Hebdige 1979). Networked ICTs, on the other hand, can be distinguished from other ICTs because they not only connect legions of users but also are themselves connected, communicating information and creating a *capture matrix* for incidental and uploaded user data (Postigo and Novack 2015).[4] *Networked ICTs* connect through application programming interfaces (APIs),[5] operating system level permissions or geolocation correlation algorithms. Together they create a nearly inescapable user data footprint and afford a level of plasticity in use that can lead to a level of indeterminacy in use and design that doesn't show signs of resolving. Our understanding of networked ICTs herein is based on that distinction.

Networked ICTs' plasticity[6] leads to variable uses and structures by way of their connectedness and interactive features. It allows users to resist and invent practices beyond what designers intended. Often, they orient their use to purposefully counter what designers intended as, for example, users did with video game platforms described below. Briefly, video games were designed to be entertaining to play but not designed for being entertaining to *watch someone else play as the prime source of* consumption. However, because users decided to record, share, comment about, and modify game play, video games are now also an entirely novel entertainment genre (Postigo 2014). With such persistent and dynamic permeability between networked ICTs' design and the always emerging user practices, it becomes increasingly difficult to predict what ICTs will ultimately become as fixed technological systems (Suchman 2007).

Because of that difficulty, it appears that ICTs, once bathed in the glow of novelty, are in theoretical tension as the idea of closure is rendered problematic when confronting how users and designers have come to understand them, use them, and wonder what else they can do. Others have also commented on the idea that whatever closure is for one technological system, along with the processes that achieve it, it is not a fixed recipe for how all technological systems will shape use and what users will ultimately do about it (Akrich 1992). We embrace Akrich's more dynamic understanding of user behavior in light of designer intentions. We agree that what a technology settles into after closure is ultimately a liminal configuration between the user in his or her social context and the imagined user inscribed in the technological design; or as Akrich calls it, the "script." However, we propose that in the case of networked ICTs, where interactivity and cross-platform functionality are part of being networked, an essential nature of the technological system is its interactivity and perpetually plasticity, open to whatever behavior and form users adopt and, therefore, indeterminate and beyond absolute closure as formally defined. History will tell if this conclusion bears out but so far enough diversity has emerged in the networked ICT ecology that this conjecture has been valid. We acknowledge technology studies has been deeply influenced by, but also struggled with, the idea of closure. As we noted previously, we believe that ICTs are another case of technological systems that render the idea of closure problematic. The continued indeterminacy we document in our case studies goes beyond cases where closure is reassessed via continual reopening of the problems originally meant to be solved by a technological system. Indeterminacy in networked ICT technological stability, the problems defined, and user creativity all cause us to imagine networked ICTs via the playful metaphor of the celestial Swiss army knife, capable of being nearly infinite things for nearly infinite life-world contexts.

Existing work has shown the above to be the case in our opinion. Some users resist what ICTs are becoming, others embrace them, some see them as new canvases for creative expression, and still others reinvent them (boyd 2014b; Coleman 2013; Earl and Kimport 2011; Gillespie 2006; Jenkins 2006; Lessig 1999; Postigo 2012). One cannot say that what they are will ever be settled because, like all things that are the work of human hands, they are in motion through history—dynamic and always in a state of metamorphosis. Closure is neither a given nor absolutely tenable because networked ICT indeterminacy consistently straddles a clear tension between the determinism of designed systems and the plasticity of user behavior, literally playing and creating novel communication modalities with them.

What are we to make of closure in light of networked ICT indeterminacy and unresolved tensions between theories foundational for understanding technology and society? Taking into account Winner's and others' critique of the social construction of technology (SCOT) (Winner 1993), and embracing Akrich's understating of ongoing liminality between designer visions and user-centered contexts, we propose that closure is relative to those that have the power to decide when and if a problem is solved or goal attained by a technology. We suggest that in the case of networked ICTs, even those actors that presumably have power may not be able to exercise it definitively since networked ICT platforms tend to be deeply interactive technologies, where user practices are fluid, creative, and emergent. STS scholars (Gillespie 2006; Langlois 2012; Mager 2009) have recognized this situation and most recently posited that networked ICTs' fluidity is as much a social construct as is its fixity:

... what may be remarkable about technology as a social achievement, and of media technologies in particular, is that they must be maintained, that their contested meanings persist and thrive, that they are the fragile residue of constant activity, and that they must be made and remade in every instance. Their seeming stability is itself a social accomplishment and an important myth to preserve in the face of a reality in which they require constant handling, ongoing repair, and regular upkeep of their public legitimacy. (Gillespie, Boczkowski, and Foot 2014, 12–13)

We couldn't agree more with the caveat that we separate the relative nature of historical and social context from the absolute nature of material reality. In other words, we agree that in the context of social histories artifacts do require social labor for "upkeep," even beneath apparent fixity. As an absolute claim about the role of a material artifact in social dynamics, however, the assertion seems to reduce the whole of a material construct—be it information architecture, hardware, or processes defined by algorithm—to ultimately *only* a social construct. When considering that reductionism, we disagree with it as a proposition that would resolve the liminality between technological determinism and social construction. Thus, we feel social construction and technological/material determinism are not mutually exclusive, nor is any form of closure ever assured in the contexts of networked ICTs' plasticity.

Where should we locate ourselves as researchers digging through artifacts, society, and culture to find how it is that they are coinstantiated? Where can human agency be found in the Ouroboros-like cause and effect relationship of society, culture, and technology? We retreat from extremes and, accepting the limits of human epistemology, adopt the view that the structure of society and artifact meet at indiscernible boundaries. If we, like others before us, contemplate the permeability between the socially and technologically determined, we have to consider that the dynamic tension between them can encumber closure. From there we have to accept that to theoretically belabor the tension between social construction and technological determinism in an attempt to settle into one or the other is a Sisyphean task. If we, like Sisyphus labor to roll our theories into one form of determinisms, we will find through the long arc of history that material and social conditions always change and unravel our conclusions. So we will find ourselves once again rolling our theories in one direction or another.[7] Networked ICTs are so imbued with plasticity that they will never be settled. Long-standing discussions and debates about whether technology is absolutely deterministic or socially constructed are conceptual loops and the debate continues to return to starting premises without ever absolutely settling the question at hand. We argue that once upon a time recognizing and attempting to resolve that tension was a convenient tool for compartmentalizing units of analysis and theories for description and prediction in the closed circle of technology and society. But today, given the empirical difficulty of producing such narrow work, it seems hardly enough to hold the unbearable lightness of networked ICT platforms that are organic even as they are computational.[8]

Examining Video Games and Playing with YouTube

As noted previously, the systems in our case studies, which explore the indeterminacy of networked ICTs and how they may elude closure, fall in the category of user-generated content (UGC) platforms and the playfulness that occurs therein. We focus on video games and YouTube as platforms that welcome and trade on emergent user behavior as users engage in the affordances therein. The unpredictable nature of user engagement with any interactive platform is understood as a form of emergence, a concept we will flesh out before explicating our case studies.

When thinking about video game technologies during our research, we understand video games as computational systems first and foremost. However, a video game cannot be considered to be solely an isolated system with variables of play that can be predicted or controlled. A game space is connected to a whole set of other systems (the console, the web, the living room, etc.) and, though the software-based system is a function of precise programming, play therein can be *emergent*. Emergence means that behavior and outputs are unpredictable and sometimes entirely in violation of computational rules set in place by designers and programmers (exploits, cheating, mods, viewership, etc.). At times emergence is hoped for and designers afford random- ness in play by computational means like random number generator (RNG) algorithms. RNGs can afford computationally indeterminate possibilities. That strategy is used by designers and programmers who recognize that the myriad of feedback possibilities in a particular input/output system would be too cumbersome to program. They let the RNG "roll the dice" and define possible outputs a priori (before play) or ad hoc during play.[9]

As a result, when studying emergence in game platforms one must confront two possibilities for emergence's origins: (1) designers, when they build a video game, set in motion a Corliss engine with so many parts in motion that designers themselves do not fully understand or (2) once in play, the computational system's input/output feedback loop, engaged by multitudes of online players, returns that which could never have been programmed by a singular or group of programmers/designers, thus the net- work and the game constitute their own RNG algorithm.[10]

The first possibility above implies a level of planned indeterminacy or acceptable chaos in the system. It yields an interactive space that affords unpredictability in play beyond game rules and logics of victory or success as determined by the design- ers. While some games are labeled "sandbox" environments because they purpose- fully afford players the ability to explore and define their purpose in play, not all are intended to be such. Nonetheless, we argue that many games, though not designed to be sandbox environments are such because of the interactive nature of their systems. How and if that will be discovered or explored during use (or play) and what it will look like is not something that can be planned exactly by designers.

Closure in such a system would elude both players and designers. Random emer- gent properties would necessarily define new goals and means of solving them. Thus, a technological system apparently meant for one thing could change iteratively so long as the system exists. Take the example of *Twitch Plays Pokemon*. Twitch is an online platform where video gamers can broadcast their game play live on the web (a practice called live streaming). The Twitch platform allows a player's audience to interact with the player through chat. In February of 2013 one player modified his Nintendo inter- face to accept text commands from the Twitch audience's chat stream, thus allowing the audience to execute play commands along with the player. He called his Twitch channel *Twitch Plays Pokemon*. As a result, on some days during play, this initiative attracted close to eighty thousand audience members, thousands of which were issu- ing play commands and massively playing what had been originally designed to be a single player game ("TPP Victory! [...]" 2014). Furthermore, the playing members of the audience changed the rules of the game by differently valuing in-game items denoting one item to be the winning token. The form the game took was not planned by designers (at Twitch or Nintendo) but was made possible by the networked nature of Twitch, Nintendo, and the web communication protocols that can be hacked, modified, and mashed up to yield an entirely user-generated game process.

We would suggest that video games are not alone in their emergent properties and that those properties limit closure. Social media platforms and swarm behavior present in web search can exhibit emergence when we note that what eventually yields their function are neither machine nor wetware. Take the example of a search algorithm basing its returns on tracked user behavior when selecting returns to input search terms. Over time this hypothetical algorithm will aggregate user behavior, noting whether users accept or reject the returns, follow links in the return, or choose other sites after the fact. It would then triangulate user behavior with other metrics of validity such as link-based rank for the return most preferred or keywords ranking. The algorithm may sometimes tailor its returns for a single user based on that user's search history and his or her membership in some demographic.

However, social circumstances may change over time and user behavior will diverge from validity metrics, showing that the social context has made a new return the best return for a single or many users. Designers will then have to contend with the emergent and socially dependent validity of a return. In that sense an algorithm designed to return the most objectively "valid" return, by incorporating user behavior as a validity metric, is subject to whatever social or individual user experiences change users' disposition about what is or is not a useful return to a given search query.[11] This is an element of randomness that will make closure, regarding the search return algorithm's exact function as a finder of valid returns, impossible because it returns neither "objectively valid" nor longitudinally stable outputs. What makes this example troubling for those of us who consider the social construction of facts is that if an algorithm or search interface can be manipulated, it can be used to corral users into a particular affect even if a return is not "valid" in an objective way. Recent research on search returns presentation, for example, has shown that voter disposition can change after a particular presentation of search returns for candidate queries (Epstein and Robertson 2015; Shultz 2015).

YouTube as a Slippery Boundary Object

Recent research has shown a confluence of social and technical affordances manifest in social media, broadcast media, leisure, and communication. Work on social networking platforms, blogs, algorithms, digital copyright, and civic engagement suggests that networked ICTs are a field of contention, collaboration, and shifting power that at some level cannot be controlled (boyd 2014a; Deuze 2007; Jordan 2015; Marwick 2015). If we take Star and Griesemer (1989) as correct about both the immutability and plasticity of boundary objects, then there may be room for modifying their conclusions in the face of emergence in today's networked ICT platforms which rely on constant user input, making the most dynamic and immutable forms of material and meaning. We suggest that for networked ICTs, neither the builders nor the users will ever anchor them to a set of affordances, social or technical.

YouTube can be understood as a collection of interactive features that are fixed but then change either by design or user appropriation. Understanding a platform as a boundary object may be useful as a theoretical perspective because it allows for framing how technology users and the platform administrators utilize and understand differently, depending on the context. Understanding networked ICT platforms as boundary objects change what we as analysts or users come to expect from them. That understanding shifts from seeing a platform as a space where users are invited to participate to a space where wrangling over meaning and possibilities can happen.

The more relatively fixed elements of the platform are copyright monitoring algorithms and licensing arrangements with media companies. The terms of service agreement between YouTube and its users has always been couched as being user-centric, yet it hosts a tremendous amount of content from media industries. It should appear both discursively contradictory and financially logical that a business like YouTube would spend a significant amount of time and resources courting content from established media businesses, while at the same time telling users it's about them as producers.[12] The logical part of the strategy is that without having vetted UGC in traditional media markets, it makes sense to afford commercial media content a lot of purchase in defining what YouTube is doing. Ad revenue sharing and promotional models are some of the ways those arrangements yield profitability for both YouTube and the institutional copyright owners who partner with them.

Capitalizing on UGC is another matter altogether, since much of the content uploaded by users is homemade and often a remix of proprietary content. While YouTube secured licensing from institutional mass media copyright owners, it nonetheless invites users to contribute so it is in the awkward position of having to mind copyright boundaries for UGC content. Copyright and private ordering constraints notwithstanding, the platform has to be plastic enough to afford a great deal of invention and experimentation with video, audio, and distribution models. That means that YouTube cannot police the copyright boundaries too stringently lest it take the *You*, which is user participatory culture,[13] out of the *Tube*, which is the platform. The flexible policies remain open but also cannot predict what users will do or understand YouTube to mean and how existing affordances will be used to accomplish goals entirely user-defined.

For example, recently YouTube has seen an increase in uploads of music playlists drawn from video game musical scores or a compilation of different film soundtracks. These videos, while posted in a video format like .mp4, don't actually have any video content save for a still image. What is being posted is music. One could easily download the "video clip" and in a matter of seconds extract the sound from the video and listen to it as a stand-alone sound track. YouTube did not get into the web video business to become a music-sharing site, but it can become that with a little tweaking by users. As a response, YouTube can execute its copyright infringement, take down algorithms, and remove videos it deems violate copyright, or place digital copyright protection measures on some posted music content. But circumventing those measures is a trivial matter. Once the video is streaming on users' computers, pulling the audio is fairly easy, so closure eludes and YouTube becomes exactly what its designers didn't want it to be: an updated version of Napster circa 1999.

Because of their imaginative engagement with the platform, users find themselves at play with YouTube's affordances, plugging various elements of their lives into the platform and presenting it as entertainment. One such life-element is recorded video game play. Video game researchers have pointed out that the video game industry is big business, and recent research points out that video game play is becoming more than a casual participatory pastime. It is quickly becoming a spectator sport, and play is being professionalized (Taylor 2012). One need not be a professional video game player to have impact in the growing niche of video game spectatorship as entertainment, however. Recent research on YouTube's affordance infrastructure has demonstrated that average gamers can produce game play commentary (a novel genre on YouTube) that garners not only millions of views but also ad revenue dollars for the video game commentators and for YouTube (Postigo 2014).

YouTube's plasticity is confluent with remix culture and technologies already present in game consoles, video capture cards, game peripherals, and the video game industry. Neither YouTube nor the video game industry anticipated the synergy and revenues that could be created by the interactions of video game culture, Internet culture, and remix culture. Various platforms were pieced together by users to afford video game commentary. The bricolage of platforms affording game play commentary were not implemented by disparate designers with that sort of integration in mind, but it was made possible by users who first started recording game play with portable cameras pointed at their television sets as they played. They then began recording game play using "pass-through" analog connections between game consoles, televisions, VCRs, and computers. These technology remixes ultimately evolved into digital capture configurations using PCs as personal DVRs or stand-alone external capture devices designed specifically to record digital content from consoles onto PCs.[14] In this particular case, video game commentary is clearly an example of emergence in play, platform configurations, and uses afforded by platforms of disparate networked ICT systems. This practice is quite different from other public commentary practices in media consumption such as commentary on music or art because game play is a novel instantiation of the models and processes in games. The commentary is aimed not at the proprietary game elements only but also at the game play, which is wholly produced by the player. Some copyright scholarship has argued that game play is transformative enough to constitute a transformative work with fair use privileges (Drachtman 2014).

Its censorship and "take-down" policies notwithstanding, one could conjecture that YouTube's plasticity invited whatever video genre a user saw fit to conjure and therein would lay the crux of the point about emergence and slippery boundary objects. The participatory architectures across platforms at some level welcomed a creative chaos that only a collective intelligence like users on the World Wide Web could provide (Lévy 1997). One could argue for absolute determinacy as a counter to our point about plasticity and suggest that all of the uses were somehow predetermined by designers' meticulous attention to designing affordances and all the possible uses a person would make of all the various networked ICT systems to make and distribute video game play commentary.

In such a hypothetical situation an algorithm could find correlation among every possible permutation of video content afforded by the feature set at play on YouTube or an Xbox. From that correlation, it may be possible to discern a predictive pattern from the chaos of nearly infinite ways a user can engage with YouTube. That sort of algorithm would present the randomness that emergence may take as an illusion to the uninitiated observer, when in reality it was a planned outcome that could be purposefully managed. But this is a hypothetical. Without access to Google's YouTube platform data, the proposition that somehow all that user creativity and randomness was absolutely planned by purposeful design is conjecture at best. Therefore, we cannot say with absolute certainty what were the causative agents that led to a combination of technology/media penetration (video games, Internet video content, etc.), cultural shifts (Internet cultures, DIY culture, etc.), and a desire to share one's hobby as game play commentary. We can just say that the platforms in place afforded it insomuch as they created a Lego-like framework for just about anything that users could imagine they wanted to record from their lives and share on networked ICTs and digital technologies.

The level of indeterminacy described in the previous paragraph would lead one to conclude that YouTube and its associated platforms, cobbled together for game play commentary, are functioning as a truly social-constructivist architecture, a veritable sandbox for user creativity. But research would also show the power of determinism. There are various features across platforms and YouTube in particular that orient users toward a particular model for production, distribution, and networking. First are the various elements that afford UGC contributors information about their audience, which of their videos gets the most views, demographics based on viewers' Google or YouTube profiles, and so on. The data isn't specific enough to compromise user privacy but it does give YouTube channel "owners" (the term used by YouTube users who have regular content uploads on dedicated pages) enough information to determine which types of videos within their chosen genres are likely to succeed in the future and how preferences might change. Also, as much as it serves as a social networking platform, YouTube doesn't exclude competing social networks from its interface. Facebook and Twitter links are common among popular channels and audience maintenance (recruitment and involvement) happens across social network platforms.

The data collected across networked ICT platforms from users' profiles, channels, and viewership are a treasure trove for triangulated measures of users' practices. Whatever remains indeterminate about production in UGC-dedicated platforms is afforded by platform owners so long as it serves the underlying business logics. The moment user practices run counter to a platform business's logic, that practice is reined in via warning systems or excluded via practices like takedowns, as channel owners call it when YouTube deletes their channel or blocks it from public view.

Thus, the technical affordances that create a menu of user practices simultaneously frame social affordances that shape how users create structures that govern their interpersonal communication and outlook about what is possible on the platform.[15] Creating a menu of possibilities based on affordances is not a case of absolute determinism, however. Those of us familiar with jazz history might think of it as the key signatures and general tempo parameters set by Miles Davis on otherwise empty sheet music before he and his famous quintet sat down and recorded *Kind of Blue*. Except, in the case of YouTube, a user base is not a quintet and the number of features and networked ICTs that can come together are legion, compared with a limited number of notes and keys. So, while framed by affordances, networked ICT combinations across platforms are so diverse that what the output ultimately looks like and means to users and designers eludes closure.

If YouTube game play commentary tells us anything about the boundaries between fixity and plasticity or the user subject position and platform, it is that both are murky during emergence. If STS can tell us anything about the processes that shape social structure and technological architecture, it is that the networked ICT platform is never closed. The platform then is a noncommittal boundary object, very slippery and therefore difficult to fix in one place or another.[16] Watching designers trying to corral user creativity in one way or another, we are reminded of Charlie Chaplin hopelessly chasing his hat along the floor as he kicks it just before he is about to pick it up.

The Update as the Tug of War between Emergence and Fixity

It is somewhat surprising that games, and video games in particular, have largely avoided the attention of STS scholars interested in ICTs, especially given the pervasiveness and reach of digital games.[17] Indeed, we would argue that often games have pioneered innovations that we find operating in sociotechnical architectures across life worlds. Put another way, it is quite possible that the plasticity of sociotechnical architecture in networked ICTs appeared first in the realm of entertainment technology and has changed user expectations in other technopractices.

While the field of game studies (GS) has explicitly claimed the domain of games and interactive media technologies as a specific realm of inquiry, STS scholars have been more reluctant to engage with the medium. There are of course notable exceptions, exploring the online cultural spaces and worlds (Malaby 2009; Nardi 2010; O'Donnell 2014; Taylor 2006). However, the conceptual frameworks and methods of STS lend themselves to studying the worlds of games and networked ICTs. Perhaps more interestingly, scholars of game studies have turned to other fields and frameworks outside of, though frequently connected to, STS. For example, the "object-oriented ontology" approaches to understanding games has gained a great deal of relevance in GS, which while connected to concepts within STS like actor networks and messes, instead become insular references within GS (Bogost 2012). In part this is due to STS's relative reticence to engage with more media-specific technologies. As these have become more closely intertwined with networked ICTs, such reluctance makes less analytic sense.

Networking technologies came first to personal-computer (PC) games and influenced console game technologies later. Networked ICTs and their ability to afford persistent synchronous and asynchronous culture and practice emerged in the area of online multiplayer games. Studies exploring online multi-user dungeons (MUDs) (Turkle 2011), massively multiplayer online roleplaying games (MMORPGs), and the companies creating these technologies show a trajectory that was made possible by accidents of history and technology designed for other purposes (Lenoir and Lowood 2003; Malaby 2009; Nardi 2010; O'Donnell 2014; Taylor 2006). If the Internet had not been already in place thanks to the Cold War and ARPA's investment in networked ICT systems (Abbate 1999), then networked digital games might have been confined to modem connections between a few users who knew each other's telephone numbers. Thus, MMORPG platforms and networked game consoles compare quite similarly to YouTube, Facebook, Twitter, Instagram, and so on.

The technologies of "game spaces" provided a very specific social and technological infrastructure that shaped the kinds of activities that could occur. T. L. Taylor's account of World of Warcraft raiding activity (large, coordinated multiplayer missions in the game) notes the sociotechnical architecture of the play world as bricolage.[18] The bricolage made up of structured play, emergent phenomena, technology, modifications, and social arrangements were all critical elements of the system. User-developed interventions that modified the game interface specifically yielded social affordances for coordination and team hierarchies that would have been impossible without the emergent qualities of raiding that required affordances more attuned and malleable to player desires (Taylor 2006).

Perhaps most interesting, networked game console manufacturers, as they leveraged console properties as sociotechnical architecture capable of emergence, also designed them to take control. Beginning in the early 2000s, game consoles in addition to being able to connect users to networks to play began using networks to execute periodic updates through downloadable "patches," automatic updates, or "flash" console hardware ROMs with updates known as "firmware" updates. Through upgradable "firmware" (rather than "software" or "hardware"), developers can change device capabilities over the lifetime of its use. While these features are often touted as a means by which the capacities of a device can be expanded and upgraded, so too can capabilities be removed. For example, Sony's PlayStation 3 (PS3) game console originally shipped with the ability to load a version of the open source software operating system Linux. While many companies leveraged this to explore new uses for the device, eventually some users exploited that feature to circumvent the console's copy-protection features. Sony, realizing that this level of plasticity could run afoul of its business model and copyright interests, removed that capability with a "pushed" firmware update to the console. Users who didn't want to install the firmware update were denied access to Sony's proprietary network, which in many cases meant it no longer functioned as a working game console. Similar practices have been used to shut off entire communities of users devoted to hobbyist work on these devices (O'Donnell 2014).

If our argument that platforms affording communication through digital information networks straddle a boundary between fixity and emergence holds, then the example above shows that the tools of contention and control available to platform designers and users increasingly take the form of code. Sony users were left with little choice if they wanted to take part in networked play. Those wishing to continue using the PS3 had to either accept the update or build an alternative network capable of connecting users on the PS3 while remaining outside Sony's control. While such a user-generated implementation of gaming networks has not been observed on other platforms, other user bases are already circumventing such management over their user experience. Apple Corporation, for example, is finding that some users who wish to monitor firmware and software "updates" are using "hackintosh" machines to take granular-level control over their hardware and networked ICT experience. Hackintosh machines are Intel-based PCs typically running the Microsoft Windows OS that have been converted to Apple OS machines through use of open source or third-party installation protocols like Clover or Unibeast.

If there is an historical analog to the emergent quality present in networked ICT platforms that allows tinkering and a tug-o-war over platform, it is "car culture" in the United States. American consumers have always tinkered with their automobiles, looking under the hood for maintenance or to improve on one element or another. The categorical difference is that unlike car culture, the car's hardware (carburetors, radiators, exhaust systems, etc.) was difficult to modify after it had been sold (Greenberg 2005; Volti 2006). Updates came in the form of the next model car, which the consumer would have to purchase. In networked ICTs, however, integral operating parts are either software installed on a hard disk or coded onto CPUs or memory sticks that can subsequently be changed via online flash updates. Upgrades and modifications are more available and easier to make for both users and manufacturers alike. Flash updates, however, are difficult and prone to causing hardware-level failures if undertaken without some amount of research before the user implements the process. Therefore, it serves platform owners who wish to keep users out from under the "hood" to use these types of updates to reconfigure essential platform functionality.

As game companies seek to fine-tune the user experience, designers' impetus to control what goes on with a games functionality has increased. Video games companies began tracking a wide variety of data in order to diagnose problems ("bugs") in games or understand the computer systems that hosted them. However, this information rapidly began to prove productive for understanding how players were playing a game. The initiative was meant to create what video game developers call "balance," a check on game elements that might prove too weak or powerful or might somehow circumvent how a game was "supposed to be" played.

The rise of "analytics" in games proved to be a siren call for many developers (Whitson 2014). Increasingly, games have become rapidly enmeshed in a wide array of other platforms, from Facebook games to games that will happily Tweet your progress (given the permission) as you advance further or play a game longer, and the data keeps getting bigger. The bricolage of platforms only adds to the indeterminacy of what a video game can do as an entertainment enterprise, as social media venue, and as a source of big data. Those who chose not to plug their gameplay into the bricolage of networked ICTs may find no cost to their gaming experience or may find themselves excluded from the social nature of many video game titles.

Interactive Plasticity: A Theory on Continuously Emergent Sociotechnical Structures

Networked ICTs pose a particularly difficult empirical domain for researchers in STS because they serve as technological platforms, sites for subculture/resistance, as well as sites for producing, remixing, modifying, and distributing mass media.[19] Thus, much of the existing research takes some range of analysis, attempting to disentangle the bricolage that composes networked ICTs. These technologies shift over time and prove difficult to analytically "pin down" because they themselves react to shifting fields of production and use. Yet, at the same time, they provide a kind of analytic purchase as architecture. They have impact in ways that make them look and feel like "durable" technological systems subject to the predictive power of theory and description.

What makes networked ICTs compelling as objects of study for indeterminacy in closure is their empirically proven unpredictable and iterative modus operandi. They are not just technologies for play or interpersonal/group communication, but they are also distribution/production platforms and more. Generally speaking, users do not just use a networked ICT for its intended use but discover uses as they discover desires and needs in their own lives and incorporate them into networked ICT use. The intrinsic complexity of the platform, hardware, and software bricolage goes beyond appropriation of technology. The roles that users/players can take are significantly shaped by the set of tools provided to them, allowing differing levels of access to the underlying sociotechnical architecture of a given platform.

The tools provided for users whisper clues and insights about a broader set of logics, ones that may or may not change over time. Tools and platforms "encourage a particular kind of use," they "direct the user to follow a particular design and implementation path that encodes a level of knowledge about [...] practice" (O'Donnell 2013, 173). They invite users and designers to engage in practice, but the receiver of the original "invitation" differentially binds how the response to the invitation is returned and articulated. Users then are in part socially constructing what they have been technologically determined to build.

Thus, to one degree or another, all networked ICTs exhibit a variety of elements that necessitate a kind of dual-analytic framework. What appears, on the one hand, an aspect of "the platform," may prove to be part of a broader set of sociotechnical logics; they may also be arbitrary, accidental, or emergent. Networked ICTs, more than any other pervasive technological system, are Janus-faced, with one face oriented toward one designed outcome and the other face oriented toward a multiplicity of futures. Closure then, may continuously elude them.

Networked ICTs remain mutable in ways that necessitate a kind of sociotechnical archeology. Sociotechnical architecture shifts and moves over time, sometimes responding to users, sometimes to designers, and in other times in response to corporate shareholders. These interests, values, and logics become encoded into the very systems and devices. Even hardware, once thought of as immutable, is now flashed, updated, or modded whereby new sociotechnical logics can be inserted into a system.

Architecture proves a productive metaphor because it does shift and change, but the physical construct of architecture proves stickier than simply being "socially constructed." The very physicality of a building is subject to its materiality. Concrete, iron, wood, and glass all "behave" differently. Yet, architecture can change and should change and may need to respond to external pressures from zoning laws to earthquakes. It is not above or beyond governance in ways that technology often expects to be (Gieryn 2002).

As we have illustrated, networked ICT platforms deploy deliberately designed features, carrying with them a very specific set of social and technical affordances. At the same time, however, when users and designers create a complex, deeply interactive system, they can use and change the meaning of the system and create indeterminacy eluding closure. What makes YouTube and video game platforms appropriate examples is that they demonstrate precisely the tension at play between technological determinism and social construction that is foundational in indeterminacy as driver for emergence. On the one hand, clearly, both platforms shape, enable, and constrain user activities. Yet, users continue to leverage and play within that space, requiring a more complex framing of data collection and analysis, design strategies, and sometimes changing platform business models altogether. The emergent aspect of how these systems are *played* by users is interwoven with how networked ICTs have *interactive plasticity*.

As a first element of our theory for continuously emergent sociotechnical systems, we define interactive plasticity as the capacity for some networked ICT platform features to be repurposed or reconfigured through purposeful *use hacking*. Users in video game platforms or social media venues often use a feature or affordance for ends entirely unpredicted by designers. This process-driven hack of the intended use requires little or no coding knowledge. It can be accomplished by knowledge of how various platform elements can creatively be cobbled together or with another platform's feature set. The new use might often still serve the ends designers intended but may also serve ends entirely valuable to users alone. Platforms exhibiting strong interactive plasticity will elude closure in both the short and the long term.

Interactive plasticity is more than what STS scholars might recognize as "appropriation." In our estimation, appropriation requires a user-directed reorienting of the whole of the use function for a technology being appropriated. Interactive plasticity preserves the original intent, still serving original designers' intent but also serving ends that may be entirely unexpected and user centered. In other words, in appropriation tool use stops serving the original intent and serves a new purpose for which the appropriation took place. In plasticity the tool never stops serving the original designer even as it serves users' novel purposes. The technology, then, quite literally simultaneously serves two masters.

We find it useful to think of any single, networked ICT platform as really two, the simulation and the simulacra (Baudrillard 1995): the platform we interact with, see, and present to others and ourselves and the one designers see as they see us represented by a collection of analytics. It's a two-way mirror through which designers and platform owners see us. They collect data, deploy analytics, and tweak the platform through obvious modifications like updates and not-so-obvious tweaks, like those Facebook made to test user responses to positive or negative news feed presentations. In so doing, Facebook attempted to manipulate our emotional affect, even as users intended the same platform to build their own disposition by the way they curate friends' news and how they present their response to it (Hill 2014).

A platform's ultimate structure and what stands in the way of strict closure then is a trick of the subject position's perception about how the platform mirrors. Closure eludes because the tug-o-war across the looking glass is not often so contentious to cause one participant to let go altogether their vision of what a networked ICT means or is meant to do nor pull the other completely through and have designers, for example, just surrender the machine to the ghost in the shell. In the context of video games, this tug-o-war takes the form of play with game content and its emergent qualities. On game consoles it takes the form of applicable uses and the nature of networks as means for communication versus user or corporate control.

Beginning students in general chemistry often confuse the idea of equilibrium with equity, believing that the concentrations of product and reactant are equal at equilibrium. But that is not the case. At equilibrium the rate of a chemical reaction toward or from product or reactant has reached a stable pace that establishes a consistent product output rate. We take that as our model for interactive plasticity and continuous emergence. As users and designers change their use practices and create new ways of understanding the value of that use, they shift the equilibrium back and forth, but it is never settled because users always change and designers either change their goals or attempt to corral users.

Any analysis that hopes to at least provide us with a heuristic for explaining this ever-dynamic process has to contend with this equilibrium. To reach optimum equilibrium would not be closure but rather to find the rate balance between emergence and fixity (afforded by plasticity) that serves the greater good. An ethics of this theory would require us to ask how platforms serve human flourishing for the one, the few, and the many, and welcome more voices, all while turning a profit.

Notes

1. The term *platform* has of late suffered from semantic indeterminacy because of its application to a myriad of structures that straddle or undergird the Internet. Therefore, herein we stick with the definition from the *Merriam-Webster English Dictionary*: "a flat surface that is raised higher than the floor or ground and that people stand on when performing or speaking." Taken as a metaphor then, the idea of a platform means any human-computer interaction software/hardware implementation that sits on top of existing standardized Internet and mobile data communication protocols. If by means of application programming interfaces (APIs), or other interplatform protocols, a series of platforms forms an infrastructure for data collection or communication, it does so as a platform of platforms but is not infrastructure the way we understand standardized Internet protocols like TPC/IP (which itself doesn't hold as the only Internet protocol).

2. For recent work on Internet history, see Megan Ankerson (2009).

3. For a nearly prescient and extensive deliberation of what networked ICTs would mean for societies, see Manuel Castells's three-volume series *The Rise of the Network Society: The Information Age: Economy, Society, and Culture*.

4. The capture matrix is a theoretical concept meant to illustrate the complex arrangement of uses and functions for a host of networked ICTs that collectively capture (or are willingly offered by users) information about our daily life activities. It illustrates the structure of metadata (i.e., data about data) but goes beyond thinking of it as the Internet of things by adding to it the idea of the Internet of people using the Internet of things.

5. APIs are a set of protocols for graphical user interfaces (GUIs) or other software. By having APIs for a social networking platform user interface, for example, it is technically easier to add new components and features to the original application. In the context of networked ICTs, APIs

allow third parties to create features for proprietary platforms, adding third-party functionality and connecting the proprietary platform databases to the third party.

6. Plasticity in the context of our analysis refers to the characteristics of ICTs that allow a reconfiguration of possible uses. Plasticity can be achieved purposefully by design or as a consequence of unpredictable use contexts. We will elaborate on this key concept in the sections that follow our case studies. We argue that plasticity is a key feature of ICTs that inoculates against an absolute moment of closure.

7. In the context of recent scientific discoveries in physics, for example, classical mechanics, which frayed in the context of quantum mechanics and relativity, were still believed reconcilable to some form of unified theory by Einstein and his collaborators when they rejected the "uncertainty principle" and proposed local hidden variables as causative agents in quantum entanglement behavior (one quanta causing the behavior of another quanta not co-located and at superluminal speed). They called it "spooky action at a distance"(Einstein, Podolsky, and Rosen 1935) and posited quantum mechanics incomplete for not accounting for hidden variables that classical mechanics demands be local and real. However, subsequent experiments in quantum mechanics suggested that indeed there are causative connections between entangled quanta that are nonlocal and not "real" (i.e., capable of traversing classical space at speeds faster than light) in classical and relativistic mechanics (Bell 1964; Hensen et al. 2015). Thus, those hoping to settle on one determining natural law for motion and matter across quantum and classical space-time have had to reexamine their conclusions and commitments to unified theories. If research in physics in elementary particles, classical mechanics, and the laws discovered therein must continuously be revisited because absolute epistemology eludes us, then, we would argue, so does research on technology and social theories about what determines the structure of human life. To belabor settling into one completely is, therefore, Sisyphean and an act of hubris; for that reason, we retreat from one absolute and espouse a more dynamic model.

8. By "computational" we mean a software-generated environment whose inputs are defined so that any possible permutation of inputs/outputs can be known so that the outcome is predictable or predetermined. As an example, imagine chess as it progresses from opening to end game. To say a platform is both computational and organic is to acknowledge that in the case of some networked ICTs, there are unknowable outcomes that resist computational prediction or control.

9. Natasha Schüll (2012) explores the "really new god" of the RNG in particularly compelling ways, though less focused on video games and more on the role of the RNG in gambling machines.

10. It may be the case that emergent outputs in games are predictable in chaos models that reflect the structure of the game's programming. While that may be the case, we would suggest that so long as a separate computational assessment is needed to discern the pattern in emergent output, *randomness is real and exists relative to the designers and programmers who built and set the system in motion.*

11. Rick Santorum not long ago saw the impact of a manipulated swarm behavior on search returns for Google. See "Will Rick Santorum's "Frothy" Google Problem Return?" at http://www

.motherjones.com/politics/2015/05/rick-santorum-2016-dan-savage-google for history and current speculation on the case.

12. See Tarleton Gillespie (Gillespie 2010) for a generative discussion about the trope "platform" in defining the meaning of networked ICTs.

13. For a discussion of participatory culture, see Henry Jenkins's generative work (Jenkins 1992).

14. The Hauppauge HD DVR, a high-definition video recorder is an example of that. It can record game play as it happens on the game console, process it to HD format, and store the output on a PC's hard drive.

15. The interplay between technical and social affordances is the backbone of our argument for emergence in a fixed system. Social affordances are social structures like norms, mores, social hierarchies, and "best practices" that take place because of or in spite of the technical structure. See Ian Hutchby's (2001) *Conversation and Technology: From the Telephone to the Internet* for a generative discussion on affordances in communication technology.

16. We remind our readers that platforms are themselves networked ICTs. We chose to mention them as a special category because the whole of the networked ICT infrastructure is actually an Internet of Internets. Platforms are the specific networked ICTs straddling and intermingled with more general networked ICTs and vice versa.

17. There are some exceptions; see other work from Sarah Grimes (2006) and T. L. Taylor (2006).

18. In our use of the term *bricolage*, we note that it is most closely related to the idea of "assemblage" in Law and Callon's (1994) "The Life and Death of an Aircraft: A Network Analysis of Technical Change."

19. Some readers may think that this is a process of domestication as expounded by Roger Silverstone (1994). It is not. While domestication may intimate how a technology becomes part of everyday life, it cannot say what everyday life will be tomorrow and how ultimately it will domesticate technology differently.

References

Abbate, Janet. 1999. *Inventing the Internet*. Cambridge, MA: MIT Press.

Akrich, Madeleine. 1992. "The De-scription of Technical Objects." In *Shaping Technology/Building Society: Studies in Sociotechnical Change*, edited by Wiebe E. Bijker and John Law, 205–24. Cambridge, MA: MIT Press.

Ankerson, Megan. 2009. "Historicizing Web Design." In *Convergence Media History*, edited by Janet Staiger, Sabine Hake, and Megan Ankerson, 192–203. New York: Routledge.

Baudrillard, Jean. 1995. *Simulacra and Simulation*, translated by Sheila Faria Glaser. Ann Arbor: University of Michigan Press.

Bell, John S. 1964. "On the Einstein-Podolsky-Rosen Paradox." *Physics* 1 (3): 195–200.

Berker, Thomas. 2005. *Domestication of Media and Technology*. Berkshire, UK: McGraw-Hill Education.

Bogost, Ian. 2012. *Alien Phenomenology, or What It's Like to Be a Thing*. Twin Cities: University of Minnesota Press.

boyd, danah. 2014a. *It's Complicated: The Social Lives of Networked Teens*. New Haven, CT: Yale University Press.

___. 2014b. "What Does the Facebook Experiment Teach Us?" *Social Media Collective*. July 3, 2014. Accessed at http://socialmediacollective.org/2014/07/01/facebook-experiment/.

Carey, James W. 1989. *Communication as Culture : Essays on Media and Society*. Boston: Unwin Hyman.

Castells, Manuel. 1996. *The Rise of the Network Society: The Information Age: Economy, Society, and Culture*. Cambridge, MA: Blackwell. 3 vols.

Coleman, E. Gabriella. 2013. *Coding Freedom: The Ethics and Aesthetics of Hacking*. Princeton, NJ: Princeton University Press.

Drachtman, Craig. 2014. "Do 'Let's Play' Videos Constitute Fair Use?" *Intellectual Property Law Society*. Accessed January 18, 2016, at https://iplsrutgers.wordpress.com/2014/01/26/do-lets -play-videos-constitute-fair-use/.

De Filippi, Primavera. 2014. "It's Time to Take Mesh Networks Seriously (and Not Just for the Reasons You Think)." *Wired* magazine, January 2, 2014, accessed January 16, 2016, at http:// www.wired.com/2014/01/its-time-to-take-mesh-networks-seriously-and-not-just-for-the -reasons-you-think/.

Deuze, Mark. 2007. *Media Work*. Cambridge: Polity Press.

Earl, Jennifer, and Katrina Kimport. 2011. *Digitally Enabled Social Change: Activism in the Internet Age*. Cambridge, MA: MIT Press.

Einstein, Albert, Boris Podolsky, and Nathan Rosen. 1935. "Can Quantum-Mechanical Description of Physical Reality Be Considered Complete?" *Physical Review* 47 (10): 777–80.

Epstein, Robert, and Ronald E. Robertson. 2015. "The Search Engine Manipulation Effect (SEME) and Its Possible Impact on the Outcomes of Elections." *Proceedings of the National Academy of Sciences* 112 (33): E4512–21.

Gieryn, Thomas F. 2002. "What Buildings Do." *Theory and Society* 31 (1): 35–74.

Gillespie, Tarleton. 2006. "'Designed to Effectively Frustrate': Copyright, Technology, and the Agency of Users." *New Media and Society* 8 (4): 651–69.

___. 2010. "The Politics of 'Platforms.'" *New Media & Society* 12 (3): 347–64.

Gillespie, Tarleton, Pablo J. Boczkowski, and Kirsten A. Foot. 2014. "Introduction." In *Media Technologies: Essays on Communication, Materiality, and Society*, 12–13. Cambridge, MA: MIT Press.

Greenberg, Joshua M. 2005. "Between Expert and Lay." *IEEE Annals of the History of Computing* 27 (2): 96.

Grimes, Sara M. 2006. "Online Multiplayer Games: A Virtual Space for Intellectual Property Debates?" *New Media & Society* 8 (December): 969–90.

Hebdige, Dick. 1979. *Subculture, the Meaning of Style*. London: Methuen.

Hensen, Bas, Hannes Bernien, Anaïs E. Dréau, Andreas Reiserer, Norbert Kalb, Machiel S. Blok, Just Ruitenberg, R. F. L. Vermeulen, R. N. Schouten, Carlos Abellán, Waldimar Amaya, Valerio Pruneri, Morgan W. Mitchell, M. Markham, Daniel J. Twitchen, D. Elkouss, S. Wehner, Tim Hugo Taminiau, and Roland Hanson. 2015. "Experimental Loophole-Free Violation of a Bell Inequality Using Entangled Electron Spins Separated by 1.3 Km." *Nature* 526 (7575): 682–86.

Hill, Kashmir. 2014. "Facebook Manipulated 689,003 Users' Emotions for Science." *Forbes*. June 28, 2014, accessed February 5, 2016, at http://www.forbes.com/sites/kashmirhill/2014/06/28/facebook-manipulated-689003-users-emotions-for-science/.

Hutchby, Ian. 2001. *Conversation and Technology: From the Telephone to the Internet*. Cambridge: Polity Press.

Jenkins, Henry. 1992. "Strangers No More, We Sing: Filking and the Social Construction of the Science Fiction Fan Community." In *The Adoring Audience : Fan Culture and Popular Media*, edited by Lisa A. Lewis, 208–36. London: Routledge.

___. 2006. *Fans, Bloggers, and Gamers: Exploring Participatory Culture*, edited by Henry Jenkins. New York: New York University Press.

Jordan, Tim. 2015. *Information Politics: Liberation and Exploitation in the Digital Society*. London: Pluto Press.

Langlois, Ganaele. 2012. "Participatory Culture and the New Governance of Communication: The Paradox of Participatory Media." *Television & New Media* 14 (2): 91–105.

Law, John, and Michel Callon. 1994. "The Life and Death of an Aircraft: A Network Analysis of Technical Change." In *Shaping Technology/Building Society: Studies in Sociotechnical Change*, edited by Wiebe E. Bijker and John Law, 21–52. Cambridge, MA: MIT Press.

Lenoir, Timothy, and Henry Lowood. 2003. "All But War Is Simulation: The Military Entertainment Complex—Lenoir-Lowood_TheatersOfWar.pdf." In *Collection, Laboratory, Theater*, edited by Jan Lazardzig, Helmar Schramm, and Ludger Schwarte, 427–56. Berlin: Walter de Gruyter Publishers. Accessed at http://web.stanford.edu/dept/HPS/TimLenoir/Publications/Lenoir-Lowood_TheatersOfWar.pdf.

Lessig, Lawrence. 1999. *Code and Other Laws of Cyberspace*. New York: Basic Books.

Lévy, Pierre. 1997. *Collective Intelligence: Mankind's Emerging World in Cyberspace*. Cambridge, MA: Perseus Books.

Levy, Steven. 1984. *Hackers: Heroes of the Computer Revolution*. New York: Penguin Books.

Mager, Astrid. 2009. "Mediated Health: Sociotechnical Practices of Providing and Using Online Health Information." *New Media & Society* 11 (7): 1123–42.

Malaby, Thomas M. 2009. *Making Virtual Worlds: Linden Lab and Second Life*. Ithaca, NY: Cornell University Press.

Marwick, Alice E. 2015. *Status Update: Celebrity, Publicity, and Branding in the Social Media Age*. New Haven, CT: Yale University Press.

Mumford, Lewis. 1963. *Technics and Civilization*. New York: Harcourt, Brace.

Nardi, Bonnie. 2010. *My Life as a Night Elf Priest: An Anthropological Account of World of Warcraft*. Ann Arbor: University of Michigan Press.

O'Donnell, Casey. 2013. "Wither Mario Factory? The Role of Tools in Constructing (Co)Creative Possibilities on Videogame Consoles." *Games and Culture* 8 (3): 161–80.

___. 2014. *Developer's Dilemma: The Secret World of Videogame Creators*. Cambridge, MA: MIT Press.

Pinch, Trevor J., and Wiebe E. Bijker. 1984. "The Social Construction of Facts and Artefacts: Or How the Sociology of Science and the Sociology of Technology Might Benefit Each Other." *Social Studies of Science* 14 (3): 399–441.

Postigo, Hector. 2012. "Podcast: Hector Postigo, 'Cultural Production and Social Media as Capture Platforms: How the Matrix Has You.'" *MIT Comparative Media Studies/Writing*. Podcast, September 16. Accessed at http://cmsw.mit.edu/hector-postigo-cultural-production-social-media/.

___. 2014. "The Socio-Technical Architecture of Digital Labor: Converting Play into YouTube Money." *New Media & Society* 18 (2): 332–49.

Postigo, Hector, and Alison Novack. 2015. "A Modest Proposal: Maybe the Matrix Isn't So Bad If We Can Watch the Watchmen—Culture Digitally." *Culture Digitally*, June 11, 2015, accessed at http://culturedigitally.org/2015/06/a-modest-proposal-maybe-the-matrix-isnt-so-bad-if-we-can-watch-the-watchmen/.

Schüll, Natasha Dow. 2012. *Addiction by Design: Machine Gambling in Las Vegas*. Princeton, NJ: Princeton University Press.

Shiller, Robert J. 2006. *Irrational Exuberance*. 2nd ed. New York: Crown Business.

Shultz, David. 2015. "Internet Search Engines May Be Influencing Elections." *ScienceMag*, August 7, 2015. Accessed January 18, 2016, at http://www.sciencemag.org/news/2015/08/internet-search-engines-may-be-influencing-elections.

Silverstone, Roger. 1994. *Television and Everyday Life*. New York: Routledge.

Star, Susan Leigh, and James Griesemer. 1989. "Institutional Ecology, Translations and Boundary Objects: Amateurs and Professionals in Berkeley's Museum of Vertebrate Zoology, 1907–1939." *Social Studies of Science* 19 (3): 387–420.

Suchman, Lucy. 2007. *Human-Machine Reconfigurations*. Cambridge: Cambridge University Press.

Taylor, T. L. 2006. *Play between Worlds: Exploring Online Game Culture*. Cambridge, MA: MIT Press.

___. 2012. *Raising the Stakes: E-Sports and the Professionalization of Computer Gaming*. Cambridge, MA: MIT Press.

"TPP Victory! The Thundershock Heard around the World." 2014. Commercial. *The Official Twitch Blog*. Accessed at http://blog.twitch.tv/2014/03/twitch-prevails-at-pokemon/.

Turkle, Sherry. 2011. *Life on the Screen*. New York: Simon and Schuster.

Volti, Rudi. 2006. *Cars and Culture: The Life Story of a Technology*. Baltimore: Johns Hopkins University Press.

Whitson, Jennifer. 2014. "Technologies of Control and Domination? Foucault, Governance and Gamification." In *The Gameful World: Approaches, Issues, Applications*, edited by Steffen P. Walz and Sebastian Deterding, 339–57. Cambridge, MA: MIT Press.

Winner, Langdon. 1985. "Do Artifacts Have Politics?" In *The Social Shaping of Technology*, edited by Donald McKenzie and Judy Wajcman. London: Open University Press.

___. 1993. "Upon Opening the Black Box and Finding It Empty: Social Constructivism and the Philosophy of Technology." *Science, Technology, and Human Values* 18 (3): 362–78.

21 Machineries of Finance: Technologies and Sciences of Markets

Alex Preda

Introduction

One winter afternoon, not so long ago, I took my students on a visit to the data center of the London Stock Exchange. They had been eagerly anticipating this and, because only thirty-five visitors were allowed in at any given time, we had to organize a lottery. At the time mutually agreed with our hosts, we were waiting in the City of London in front of a nondescript, nineteenth-century building lacking any of the customary brass plates and Corinthian columns usually associated with the temples of trade of past centuries. We were whisked inside, and soon the excitement of the students turned into surprise followed by boredom. There was complete silence; no cries, no shouts, no traders wearing colorful jackets and elbowing each other out of the way. All that was there were seven floors of servers neatly arranged in metal cages, with some empty spaces reserved for lease to firms wishing to bring their own servers as close as possible to those of the exchange. All servers were double mirrored—the same data were stored in another two different locations at once.

We were taken to the core of the building: a windowless control room with a semicircular desk in the middle and a couple of gigantic computer screens on the wall facing the desk. A handful of people were staring at the screens, on which a few dials similar to the speedometers seen in cars were displayed. As the hosts explained to us, the dials displayed the latencies (i.e., the speed with which orders arrive) on various European exchanges. We also learned that a number of these exchanges had their servers in London, while maintaining offices in their official locations. London has its own specific latency, which needs to be kept as constant as possible. Even slight variations could be noticed and exploited by arbitrageurs (i.e., traders taking advantage of mismatches across trading venues).

There is probably no greater contrast than between this setting, apparently stripped down to a bare minimum of human interactions, and the stock exchanges of the twentieth century, where human interactions were amplified to a constant frenzy. Yet, the very notion of market implies interaction (in the form of exchanges, haggling, negotiating, etc.), so that the question immediately arising is, how can finance work without interactions? And the complementary question is, how, if at all, can technology replace interactions?

This question has broader implications and has been recently hotly debated. Arguments have been formulated on both sides, namely, that market automation will either make everything better or will make everything worse (Baron, Brogaard, and Kirilenko 2014; Carrion 2013; Foresight 2012; Hasbrouck and Saar 2013).

In the present context, however, the question about the impact of technology on trading concerns less liquidity and market depth. It brings into spotlight the institutional, cognitive, and cultural changes which have impacted the ways in which financial markets work. Admittedly, it is not the only question one can raise about science and technology in relationship to finance (see also Callon and Muniesa 2005; Kliger and Gilad 2012; Lave, Mirowski, and Randalls 2010; MacKenzie 2012b; Mirowski 1999, 2003; Poon 2007; Schull and Zaloom 2011). The latter encompasses an array of institutions and practices which go well beyond the domain of markets. Such an array includes, but is not limited to, corporations, banks, and other financial firms (e.g., Hardie 2004; Hardie and MacKenzie 2007); economic institutions situated in the private and public spheres, as well as at the boundaries between them (e.g., Davis and Kim 2015; French, Leyshon, and Wainwright 2011; Knorr Cetina 2007; Linhardt and Muniesa 2011); practices of valuation distributed across various kinds of institutions (e.g., Lilley and Papadopoulos 2014; MacKenzie and Spears 2014a, 2014b; Patel 2007; Zajac and Westphal 2004; Zeckenhauser 2006; Zuckerman 2004; 2012); regulatory practices (e.g., Davies 2013; Ford 2013; Ortiz 2011; Power 2009; Reichman and Setiha 2013; Riles 2011; Vlcek 2012); processes of standardization, measurement, and ranking. Their investigation is embodied in a rich and complex body of work on STS and finance in the broader sense (e.g., Lenglet 2011; Millo, Muniesa, Panourgias, and Scott 2005; Muniesa 2014; Poon 2012; Power 1994), a body which partly intersects with and partly complements the sociology of finance and of markets generally speaking (see also Carruthers and Kim 2011; Chen 2011; Healy 2015; Knorr Cetina and Bruegger 2002b; Mirowski 2003; Montagne and Ortiz 2013).

The relevance of finance (and, more specifically, of financial markets) for science and technology studies is manifold. First, financial institutions and markets have become technologized to a considerable degree; information and communication technologies are not mere auxiliaries of financial transactions but shape the character and features of said transactions, as well as the organization of financial institutions and their relationship with regulatory institutions. Second, financial technologies change the nature of participating actors and of the relationships among them; market actors, for instance, are nowadays humans as well as robots, and the majority of transactions is done with the participation of robots (or algorithms). Third, technology plays a crucial role in the emergence of new types of financial exchanges, in the replication of financial institutions across the world, and in the networking of these institutions on a global scale. If finance is intrinsic to globalization, then technology is intrinsic to a globalized finance. Fourth, science permeates the organization of financial institutions and markets. It does so not only in that these institutions employ large numbers of graduates from the natural sciences and engineering but also in that (sometimes competing) bodies of scientific knowledge lay at the core of the procedures for developing financial instruments and contracts and for devising various types of financial transactions. Fifth, the presence of scientific expertise changes the professional composition of the financial services industry and with it the character of financial institutions. For all these reasons, understanding contemporary, globalized finance cannot occur without science and technology studies bringing a significant and diverse contribution.

For reasons of space and focus, though, I will not discuss here this contribution in its entirety. Instead, I will shed light on one aspect: the impact of automation upon financial markets. Markets have come to symbolize the central place finance takes not only in advanced, but in developing economies as well (e.g., Ervine 2013; Knorr Cetina 2006; MacKenzie 2007; Miyazaki 2013; Ortiz 2013; Randalls 2010). They have come to symbolize globalization. They are the object of intense debates about the morality and desirability of the capitalist mode of social organization. It is not only the symbolic power and status of financial markets in capitalist societies which make me adopt here a focused approach (see also Fridman 2010; Hessling and Pahl 2006; Hope 2006; Kristal 2013; Lepinay 2007; Martin 2010; Roberts and Joseph 2015). It is the real power, the impact markets have on economic and social life as well as which warrant this approach (e.g., Arup 2010; Beunza and Stark 2003a; Carruthers 2010; Crain 2014; Hardie 2011; Krippner 2005; MacKenzie 2011, 2012a; Perez and Rutherford 2009; Walby 2010). Since the developments I discuss here come without exception from developed, Western societies, I will necessarily focus the analysis on this part of the world.

I structure the analysis around three aspects: first, the transformation of market institutions in relationship to technology; second, the new formats of social interactions emerging in electronic markets; and third, new types of (academic) knowledge and disciplines accompanying these transformations.

Automation and Market Institutions

The question of market automation can be approached on several levels: first, there are the institutional shifts in the structure of markets, which have promoted an increasing degree of automation. This automation has generated vast amounts of market data, probably much vaster than anything produced by social media. Data require analysis and are crucial in developing quantitative models, which are then translated into market machines. The second level concerns the scientific disciplines that play a key role not only in the analysis of market data but also in developing machines, which are used to conduct market transactions. These relevant disciplines include not only financial economics but also information and computer science, as well as subdisciplines of physics and mathematics. Finally, the third level concerns the social interactions (or lack thereof). Does market automation, broadly understood, mean that social interactions all but disappear? This is more than just a philosophical question. Where, and in what format can we find social interactions in electronic markets, and to what extent are they still relevant?

I will start here with the shifts in the institutional setup of financial markets and the ways in which they have affected automation. This aspect is directly related to the question, what exactly do we mean by automation? Do we mean the automatic recording and storage of transaction data (such as price and volume)? Do we mean the development of analytical tools which compile these data in real time into visual representations? Do we mean a new mode of conducting financial transactions, where orders are not open cry (that is, conducted through face-to-face interactions) but sent to a server by clicking a button on a computer screen? Or do we mean the development of software tools which analyze price and volume data in real time and, as this happens, send buy and sell orders to a server? This latter aspect also includes the use of quantitative models incorporated into such software tools.

Market automation implies thus at least (a) real-time data recording, (b) real-time data analysis and visualization, (c) decision and execution tools grounded in real-time data analysis and operating according to a set of theoretical principles, (d) communication procedures for the execution of trading decisions, (e) aggregation of trading decisions and execution procedures and rules at an aggregated level, and (f) monitoring of aggregated trading decisions. All these are crucial aspects of trading. Market automation means that the ways in which trading is conducted changes fundamentally, and such changes cannot occur without significant modification in the institutional setup of markets (e.g., MacKenzie and Pardo Guerra 2014; Muniesa 2005, 2007; Wansleben 2012).

On the regulatory level, the beginnings of market automation in the United States can be traced back to 1963, when the Securities and Exchange Commission (SEC) sent to the U.S. Congress the *Report of Special Study of Securities Markets* (Preda 2009), recommending automation in relationship to keeping securities markets as public institutions and to creating a national market system. Automation was seen as a means of ensuring the broad access of various types of investors to financial transactions on all market venues, regardless of geographical distance. At that time, computerized systems for recording price and volume data had been already implemented on the New York Stock Exchange (NYSE). Major European Stock Exchanges started automation efforts in the late 1960s and early 1970s, though it was only the wave of financial deregulation in the 1980s which gave automation a major impetus (Pardo Guerra 2012).

Concrete efforts at automation started not in the glamorous front office (that is, on trading desks) but in the back office (Markham and Harty 2012), where clerks working on clearing and settlement had less influence than the floor and pit traders, who held expensive seats. Various types of markets automated at various speeds and with various degrees of internal resistance. Futures markets, for instance, automated later, not only because their regulatory body in the United States (the Commodities and Futures Trading Commission) was created only in 1974 but also because influential figures in the industry initially bitterly opposed automation (Zaloom 2006). Automation of the front office started in the 1980s, later than back office automation. Software engineers were actively involved and had to confront the resistance of traders and managers of the exchanges, who at first did not fully realize the implications of introducing computers in the front office (Pardo Guerra 2010). Over the ensuing three decades, the professional structure of trading desks was much changed: software engineers, mathematicians, and physicists took a prominent role in the emerging electronic markets.

Overall, market automation in finance neither followed a linear path of "progress" nor was a conflict- and resistance-free process. What rather happened was a process of "fragmented innovation" (Pardo Guerra 2012, 581), in which technological change was implemented in "niches" from where it spread at various speeds, meeting with different degrees of resistance. At least as important though is that market automation intervenes in jurisdictional processes and transforms both the nature of market actors and the relationships among them.

The traditional stock exchanges were quasi-monopolistic partnerships emphasizing status and exclusivity and going to great lengths to protect their privileges. The expansion of automated trading at first challenged trading monopolies; however, as automation consolidated, it also helped set in place new quasi-monopolistic arrangements. Automation has changed not only the nature of trading venues but also their variety and the nature of trading firms. Up to the 1970s there were, broadly speaking, two types of trading venues: the established exchanges such as the New York Stock Exchange, the Chicago Board of Trade, and the Chicago Mercantile Exchange, on the one hand, and over-the-counter (OTC) trading, on the other hand. An instance of OTC trading was the Nasdaq market, which was the first exchange to trade stocks electronically. In the 1990s, however, the variety of trading venues began to multiply. Electronic communication networks (ECNs) emerged. ECNs are systems which initially catered to individual traders but then expanded their services to institutional traders as well. They automatically match orders, bypassing traditional exchanges (e.g., Mizrach 2006). They were launched in the late 1960s but first fully developed in the 1990s. In addition to ECNs, the early 2000s saw the emergence of dark pools (Banks 2010), which are private, members-only exchanges, where transactions are conducted exclusively among participating members.

Thus, it is difficult to argue that automation has been a straightforward process of "democratization," that is, easier access to markets. First, automation has been linked to increased technology costs because both infrastructure and data transmission are expensive. Second, automation has created the possibility of setting up private exchanges, reemphasizing status and prestige.

On the side of firms participating in financial transactions, new types of firms have emerged. Traditionally, institutions trading financial securities were brokerages (some of them with market making roles), funds (mutual or pension), and banks. At least two new types of firms deserve to be mentioned here: high-frequency trading (HFT) firms and quantitative trading firms (some of which are hedge funds). HFT firms mainly exploit the advantages provided by low latency—that is, by sending orders to the market within the interval of microseconds. This requires a technological setup which minimizes the distance traveled by the trading signal—that is, it requires placing a firm's servers as close as possible to those of the trading venue, together with using special transmission networks (usually microwave networks) that increase the speed at which trading signals travel. As such orders cannot be sent manually, HFT firms rely on software to execute transactions, to monitor price and volume movements, and to intervene in markets. Quantitative trading firms rely on proprietary mathematical models, which identify and exploit particular market discrepancies, models which are implemented through coding into a piece of software. Trading algorithms based on quantitative models are usually accompanied by execution algorithms, which define how particular transactions should be put into practice, not only on which venues, and when, but also how. In other words, execution algorithms define how a particular order should be split or sliced into smaller orders of different sizes (making it thus more difficult to recognize by competitors), which are then sent for previously timed execution to particular venues.

An additional aspect of algorithmic trading that has recently emerged in the retail sector is provided by social trading platforms (STPs), which are a combination between social media and trading platforms (e.g., Chen et al. 2014; Neumann, Paul, and Doering 2014). STPs use specific metrics to rank and make public the trading performance of their members (which can go into the hundreds of thousands). Members who believe they are less skilled can choose to automatically mirror, through a replication algorithm, the transactions of those whom they think have better skills (so-called trade leaders). The latter receive monetary incentives for attracting followers who instantaneously copy their transactions. As STPs are completely unregulated, the relationships between trade leaders and followers are entirely informal (there are no legally binding contracts), yet grounded in a regime of permanent reciprocal observation (Knorr Cetina 2003, 2004, 2009; also Beunza and Stark 2012; Pryke 2010). In a manner specific to social media, all actions and communication of the participants are disclosed in real time to each other. In this case, technology directly intervenes in creating and reproducing different classes of market participants: leaders and copiers, as well as a regime of reciprocal observation, which directly contributes to reproducing social positions in the market.

Market Automation and Social Interactions

One question is whether the technological evolution of markets has pushed out or minimized the role of human interactions. The trading pits of old were full of verbal and nonverbal communication, which was intrinsic to transactions. Nowadays, when transactions happen within nanoseconds, communication seems to have vanished. Indeed, are there any human actors left? Electronic finance seems to conjure dark visions of worlds where robots interact, with little if any human intervention.

The reality, which I hinted at in the above lines, is more complex. In traditional financial markets, human interactions were colocated with transactions. Nowadays, colocation, in the parlance of markets, means placing one's servers as closely as possible to the servers of the exchange platform one uses. Yet, social interactions do not disappear at all. While colocated interactions have been partly eliminated or transformed, pre- and postlocated interactions remain in place, even if their character has changed. That means social interactions unfold as the indispensable, if hidden, work of building trading machines (including trading algorithms as well as quantitative models), of creating data transmission networks, and of positioning one's machines as close as possible to those of the exchange.

These interactions necessarily have a collaborative dimension, which has been often investigated in the past in relationship to scientific work (see also Hackett et al., chapter 25 this volume). Trading machines are so complex that it is not possible to build them in isolation. Teams of machine builders are set up, both in the formal context of financial institutions and outside it. In a manner akin to scientists, though to a much lesser extent openly, these teams exchange ideas and meet regularly in formal or quasi-formal contexts such as professional conferences. This exchange undermines one cornerstone idea of earlier economic theories, namely, that market transactions are (exclusively) individual and competitive. (Although I hasten to add here that economists have revised this view more recently—see for instance Duflo and Saez 2003.) Transactions in electronic finance are as competitive as they are collaborative. Market machines could not be built without collaborative efforts in teams, and without (even rival) teams exchanging ideas.

What I call prelocated sociality is only one of the aspects I want to highlight here. Another important aspect is postlocation sociality. Trading robots, once built, will have to be uploaded and put to work. As we know from previous STS studies, machine "work" is never flawless, nor entirely autonomous (e.g., Orr 1996; Pinch and Bijker 2012; Vaughan 1996). Accidents can happen, and such accidents can have significant consequences. As with any machine, working life is limited. Existing research on trading algorithms suggests that the average lifespan of a trading robot is three months (e.g., Stafford 2013) and that some of them will be used for only one week. An additional problem related to the behavioral predictability of an algorithm is their testing. Usually, there are three kinds of data one can test an algorithm on: historical market data, simulated data, and real-time market data (Preda 2016). A trading machine will not deliver the same kind of results and reliability with real-time market data as achieved with historical data. Ethnographic observations (i.e., MacKenzie 2016; Preda 2016) suggest that many algorithm ("algo") builders prefer to test their machines on all three kinds of data. Some builders will enlist lay users and form groups within which a hierarchy of skills is established: builders will stay in constant communication with the users and will gather information about how their algorithm behaves in markets. Machine trading requires constant supervision and correction, as well as constant communication within the group operating an algorithm. In this sense, sociality is not only prelocated but also co- and postlocated with trading, albeit in forms different from those of the traditional trading floors and pits.

A further aspect of market sociality that needs to be highlighted here is that technological hybridization combines features across domains of actions, as illustrated by the growing and very popular STPs, where social media meets trading. While the mainstream social media were geared toward what one could call a general and autonomous form of sociality (social relationships for their own sake), STPs as technological hybrids promote a differentiated, specialized form of sociality, subordinated to the aim of financial trading. Indeed, repeated ethnographic observations suggest that participants on STPs will be shunned if they engage in the kind of general conversations and commentary encountered in the mainstream social media (Preda 2016). STPs prize collaborative formats and communication geared toward one specific goal: trading. This focus indicates that electronic finance not only pre-, co-, and post-locates some forms of sociality (i.e., collaborative work geared toward building trading machines, collaboration between builders and users during the use of a trading machine, supervision of trading machines), it also embraces, adapts, and hybridizes forms of "virtual interactions" initiated in the sphere of autonomous sociality.

These hybridized technologies have social effects, in the sense that user communities are reconfigured around technologies to create new forms of social stratification in markets. In the case of STPs, they create two classes of traders: leaders and copiers. This division is difficult to break because all the participants' actions are there for everybody to see. These technologies also differ significantly from the mainstream institutional regime of financial disclosure. Mutual and pension funds for instance disclose information only quarterly. Hedge funds disclose information on an entirely discretionary basis: to whom they want, when they want, and what they want. STPs disclose information to their members in real time, as it happens. The effects of this scopic regime are not yet fully understood. Nevertheless, available evidence suggests that STPs tend to have strong stratifying tendencies, in the sense that they create categories of traders (leaders and followers), which become very stable and with very little mobility across them.

While the introduction and evolution of market automation has been studied more extensively, there are few studies about the political arenas in which uses of market automation unfold (but see Kunz and Martin 2015). These uses have not been without controversy among financial institutions, in part because algorithmic trading has been accompanied by the emergence of new categories of market players (e.g., new providers of technological infrastructures, new types of market actors) who have challenged the positions of more established players. Oftentimes, the purely financial benefits of algorithmic trading have been questioned by regulators as well as by some industry players, and we have seen academic expertise being mobilized on both sides. Concomitantly, conflicts can be observed between providers and users of infrastructural technology (e.g., microwave transmission networks), in relationship to the quasi-monopoly position of a few technology providers and the costs of using it.

Overall, it would be a mistake to think that electronic finance is devoid of human interactions, of political conflicts, or purely competitive. One can make a solid case that electronic trading has more complex and varied forms of sociality than face-to-face trading. At the same time, one can make the case that the introduction and uses of automated trading hasn't been a purely utilitarian process of increasing the efficiency of data transmission but rather is accompanied by conflicts and political fights which are not dissimilar to what STS has documented for other technologies (e.g., Callon and Law 1982; MacKenzie 1990).

What Concept of Technology?

Based on the above considerations, and taking a step back to reassess the field, what concept of technology emerges from finance (see also Beunza and Stark 2003b; Larkin 2013; MacKenzie and Pardo Guerra 2014; McMahon 2004; Muniesa 2005, 2007; Pollock and Williams 2010; Reith 2013)? It would be tempting to argue that trading technologies simply standardize markets—and indeed, in other domains of activity we have seen this standardization at work (e.g., Timmermans and Epstein 2010). With respect to data transmission infrastructures, (microwave) technology has been indeed a step toward standardization of transmission times. However, in other domains automation has led to diversification and fragmentation and to the emergence of new exchange venues. It has led to hybridization of trading technologies with media-specific technologies. It has led to trading machines which have short lifespans and have to be replaced all the time. It has led to intensified group work and to the need to constantly monitor machines in the market. It has also changed the timescale of action, from seconds to microseconds (see Foresight 2012 for an analysis of these processes).

At least with respect to finance, technology doesn't simply standardize data or transmission channels, or storage and processing (Preda 2006). Technology generates rhythms of activity to which human actors have to respond and adapt. This responsive adaptation has institutional consequences (e.g., the emergence of new trading venues, of new institutional actors within electronic finance), but also social ones (e.g., stratifying processes on STPs). Responses to technological rhythms have a collaborative as well as a strategic dimension: groups of algorithm builders collaborate but also compete. Yet, they try to guess how each other's machines work; they try to disguise this work; they replace their machines often.

Having been asked a while ago by a technology journalist whether in finance machines are going to take over, I answered that I still have yet to see a trading algorithm building another trading algorithm. Financial technologies cannot be seen as autonomous action, however tempting that may be from a science fiction point of view. Still, action in markets becomes hybridized not only in the sense that groups of human actors respond to machines, but also in the sense that social action itself is divided into autonomous and non-autonomous parts (copying algorithms are a case in point here), which then mingle. This hybridization of social action can be seen as a direct consequence of the rhythms of activity generated by financial technologies. Overall, it is perhaps more productive to regard technology as a generator rather than a standardizer of data, transmission, products, processing, and so on.

The Sciences of the Market

Up to this point, I have discussed mainly how new trading technologies have impacted market institutions and behavior. Yet, one question remains: what kinds of science(s) of the market have crystallized around electronic finance? What types of systematic knowledge are not merely involved in trading processes but also claim to represent such processes (and their consequences) in a rigorous fashion? Of course, practical activities in electronic finance require a variety of complex types of knowledge which was not seen in this intensity in floor trading: mathematical modeling, data processing, programming, or communications engineering, to name but a few.

What I mean here in the first place is the kinds of systematic knowledge claiming to provide a comprehensive yet economical account of electronic markets: comprehensive in the sense that this account explains (and in some instances tries to forecast) most, if not all market phenomena, and economical in the sense that it is organized around a set of interrelated concepts. Following Franck Jovanovic's account (2012), one can distinguish two different sets of approaches. The first set emerged in relationship to financial economics and belongs, broadly speaking, to the discipline of economics (although it is not necessarily central to the discipline itself). The second set emerged outside the discipline of economics and is often at odds with some basic conceptual assumptions (but not all) which can be found within the first set. The approaches belonging to the first set can often be at odds with each other and they do not necessarily share the same conceptual assumptions. Not all these approaches emerged in direct relationship to electronic finance. Some of them had developed before and simply adapted to the new empirical field (or became more prominent because of it); some though emerged in the 1990s, in direct relationship to the rise of electronic markets.

Before proceeding with a more systematic discussion of their basic tenets, a listing of these approaches is required. First, the approaches belonging to the economics discipline (not necessarily to the core) are financial economics, market microstructure, and behavioral finance. Second, the approaches from outside economics are social studies of finance and econophysics. Institutionally speaking, a majority of these approaches, but not all, are located in the social sciences. Econophysics, however, is located in the natural sciences (in mathematics or physics departments). Many but not all of these approaches underwent a rapid institutionalization process, in the sense of offering master's and doctoral programs (this is the case for financial economics, econophysics, and, to some extent, behavioral finance).

As I mentioned before, the conceptual tenets are not always in harmony with one another. There are oppositions and tensions, which are not necessarily arranged along methodological lines such as quantitative vs. qualitative approaches. Somewhat surprisingly perhaps, two of the most heavily quantified approaches, financial economics and econophysics, are at odds in their basic assumptions. Of interest here is how their conceptual and methodological apparatuses relate to market (more specifically, trading) technologies and how the latter impacts the basic assumptions of the former.

Historically speaking, financial economics, the oldest of these approaches, is not that old at all. Although there were forerunners in the nineteenth century, financial economics as an academic discipline crystallized in the 1950s and 1960s around two major schools, the University of Chicago and MIT (Jovanovic 2008). Financial economics was made possible by the formulation of the stochastic probability theory in the 1930s, which allowed the quantitative analysis of data such as stock prices and returns. The conceptual foundations were laid out by Eugene Fama and Harry Markowitz in relationship to the notions of market efficiency and portfolio construction, respectively (Jovanovic 2008). They enabled financial economics to link into the basic economics notion of equilibrium and thus integrate into the discipline of economics, something which had proven difficult previously.

The much-discussed hypothesis of market efficiency has price behavior and information at its core and, indirectly, a set of assumptions about human behavior in economic exchanges: individualism, utility maximization, and calculative action (or decisions) based on information available to all participants. This latter assumption has been later softened by distinguishing among various versions (strong, soft, and semi-strong) of the market efficiency hypothesis (Fama 1970). Since individual human actors make decisions based on the available information (mostly stock prices, but other data as well, such as publicly available accounting data from corporations), the outcomes of these decisions (i.e., post-trading stock prices) will incorporate this information.

Market prices become thus a self-sustaining mechanism grounded in a simple model of decision making which doesn't need any psychological or sociological assumptions about human beings (this has led to the emergence of competing approaches).

The kind of data this (benchmark) model of market efficiency rely on consist mainly in publicly available data (especially accounting information provided by public corporations at regular, mostly quarterly intervals). Information is also assumed to be time-scale insensitive, in the sense that information which changes every few seconds (or milliseconds) does not have properties different from information which changes quarterly. This assumption makes trading decisions timescale insensitive, too, in the sense that decision making on a millisecond time frame should not have properties different from decision making on an annual time frame.

The model of asocial market actors processing publicly available, timescale-insensitive information did not have universal acceptance from the start. Finance scholars at MIT were less enthusiastic about the idea of market efficiency than their colleagues at the University of Chicago, but over time they lost ground to the latter (Jovanovic 2008). Market efficiency became the standard approach in financial economics for the next decades.

Starting from the late 1970s, the academic notion of market efficiency provided the benchmark against which alternative approaches emerged. This notion is related to, but does not entirely overlap with, the notion of efficiency as used by finance practitioners (Ortiz 2014). The probably best known of these, behavioral finance, combines insights from cognitive psychology with finance and endows market actors with a series of psychological attributes such as emotions, or attention, or aversions, which have consequences for their decision making. These psychological attributes are often called "biases," not in the sense of human "imperfections" but in the sense of departures from the a-psychological model of market actors. (A-psychological means here that in the benchmark model, market actors are not affected in their judgments by inattention, emotions, or other psychological factors). Markets with psychological actors (i.e., with fading attention, or with specific aversions) cannot be efficient (see also Miyazaki 2006, 2007). Prices cannot be a self-sustaining mechanism reproducing invariably in the same ways by incorporating (publicly) available information. There will always be distortions, or "effects," in the incorporation of this information. Indeed, a large corpus of behavioral research in finance has focused on documenting the existence of such effects, caused by fading attention, by emotional reactions to events, or simply by the tendency of market actors to copy each other. Some of this research has also acknowledged the role of institutional and social factors in market "inefficiencies." For instance, with respect to herding—one of the most widely analyzed behavioral processes—scholars have acknowledged the role of incentive structures or of status and reputation in triggering copying processes.

From the viewpoint of the data they make use of, behavioral finance studies do not differ substantially from mainstream financial economics. Information is mostly treated as publicly available, as timescale invariant, and as consisting in mandatory, periodical disclosures made by corporations (e.g., earnings, dividends, and the like). At the same time, some behavioral topics assume that psychological factors influence stock price movements, making these movements inefficient. Thus, the characteristics of market actors have an impact on the characteristics of prices. Although more recently behavioral finance research has forayed into the effects of technology, and especially the new media's impact on market behavior (such as Twitter—see Blankespoor, Miller, and White 2014; Sprenger et al. 2014), we do not see a tight coupling between technology and this approach.

One such coupling is between studies of market microstructure and technology. Mainstream financial economics has traditionally shown little interest, if any, in how a market order is executed and the impact of this execution upon price movements. In trading pits, order execution meant shouting out loud and making eye contact with a potential counterpart—types of interaction that have been the domain of sociological investigations (e.g., Zaloom 2006). The move to electronic trading in the early 1990s has changed the style of execution substantially: market orders became electronic inscriptions arriving on the servers of the exchange and being ordered, or queued up, there. The rules of queuing orders up and executing them, of matching price and volume of execution has a significant impact upon prices, which was felt within a very short period of time. The relevant data became minute price and volume data, and the tenet of timescale insensitivity was put into question. The publication of an influential volume (O'Hara 1995) in mid-1995 raised issues such as the role of order inventory in price volatility, the strategic behavior of traders, and the impact of timescales on trading (see also Miyazaki 2003). Technology, and the reconfiguration of markets around it, thus made its way into a new domain of inquiry which over the last twenty years has become part and parcel of the academic approach to markets (Madhavan 2000, 206). While issues related to market microstructure had already been debated in the mid-1980s, it is since the mid-1990s that it became prominent as a *sui generis* domain of inquiry.

While there are several differences between mainstream financial economics and market microstructure approaches (e.g., the first uses a frequentist probabilistic approach while the second uses a Bayesian approach), one of primary significance is the treatment of information within the framework of specific behavioral assumptions.

The main assumption is that market actors behave strategically and that they will not (fully) disclose what they know. Information is thus asymmetrical and, by definition, will not be incorporated into prices efficiently.

The notion that strategies are shaped in relation to technological contexts brings market microstructure studies close to social studies of finance. Sociological studies of financial (SSF) transactions had already been initiated in the mid-1980s (Adler and Adler 1984; Baker 1984), at about the same time with behavioral finance. SSF can be seen as comprising mainly historical and ethnographic studies of financial markets and institutions, which share a number of assumptions: institutional logics shape the ways in which particular procedures, technologies, and products become adopted in markets or not (e.g., Callon 1998; Lepinay 2011; MacKenzie 2011; MacKenzie and Millo 2003); group-shared assumptions, views, and routines shape market behavior, with the consequence that there is not a single dominant behavioral format (e.g., Knorr Cetina and Bruegger 2002a); market transactions are social interactions and therefore the unit of analysis is the (small) group level rather than the individual; market actors (or better, market groups) respond to technological constraints and to opportunities in ways that cannot be (entirely) predetermined.

A significant number of studies have investigated the institutional paths and the group processes which lead to the adoption of particular market technologies; trading media as mechanisms of coordination in electronic finance; group communication and its role in generating shared views in electronic finance; the adoption of diverse valuation tools and models of financial analysis at group level, and the social processes which shape this adoption. Overall, SSF is less concerned with price behavior (i.e., variations) in markets in relationship to returns and much more concerned with the social behavior that underlies the formation of market institutions and their day-to-day workings.

While this preoccupation may seem far away from those of financial economics and other approaches, surprisingly or not, SSF is very close in its assumptions to at least parts of behavioral finance, to market microstructure approaches, as well as to econophysics (discussed below). The approaches share an empirical interest in observable market behavior, although there are differences with respect to the form and nature of observation. They share a skepticism (and sometimes outright indifference) to a benchmark model of rationality, which has often attracted accusations of being a-theoretical (e.g., DeBondt et al. 2009). They share an interest in the dynamic interplay between market behavior and technology, acknowledging a diversity of behavioral formats, as well as adaptability to various technological environments.

Econophysics, together with SSF, is the only influential conceptual approach of finance situated outside the discipline of economics. Econophysics is an analytical approach to price variations, derived from physics (e.g., Bouchaud and Muzy 2003; Bouchaud and Potters 2000). While SSF is primarily concerned with explaining trading practices, econophysics is concerned with explaining patterns of price variations, including practical trading applications as well. It is on an ascendant path and substantially more influential than SSF. It is institutionally supported by master's and doctoral programs situated in natural science departments, as well as by significant demand on the side of financial institutions for graduates of such programs. It has been institutionalized within a short period of time, the approach itself having emerged in the mid-1990s. One question arising immediately here is why econophysics hasn't been integrated into financial economics and has continued a parallel, yet very successful, existence. Oftentimes, when walking on North American or U.K. campuses, one passes by the buildings of business schools (where financial economics is usually located) and a few steps away (but institutionally completely separated) there is the building of financial mathematics, where econophysicists are active. Econophysicists publish the majority of their papers in natural science journals, not in finance journals (Gingras and Schinckus 2012).

Econophysics emerged in the mid-1990s as a direct consequence of, among others, changes in the nature of financial data recordings triggered by computerization. The shift to electronic trading—supported by computer technologies introduced since the 1980s—made possible the recording of every market order in a database, which could then be accessed for analysis. The shift to electronic trading also increased substantially the amount of orders compared with face-to-face trading, so that the frequency of orders coming to trading venues was much higher. All this was supported by better computing technology, which was also available to natural scientists, primarily mathematicians and physicists. (Many universities locate high-performance computing in natural science departments.) At the same time, mathematicians and physicists possessed the skills (e.g., programming) required to process large amounts of data, such as those being newly generated in finance. Some interviews also suggest that cuts in physics research budgets (see Collins 2004) in the mid-1990s have also prompted scientists to look for new fields of investigation.

The discipline of physics had a long-standing interest in economic and social processes (Jovanovic and Schinckus 2013). Yet, this interest, which applied to the study of various social phenomena, cannot on its own explain why in the mid-1990s physicists turned to study financial data, in a very quick institutionalization process (e.g., regular conferences, papers published in physics journals, establishing groups within university departments, and later launching master's programs).

The statistical toolbox favored by econophysicists doesn't completely overlap with that of financial economists: the former prefer to work with Lévy distributions while the latter prefer Gaussian distributions (Jovanovic and Schinckus 2013). This is more than a technical preference: Lévy distributions allow for infinite variance. This assumption, applied to financial markets, would translate into infinite price variance, something which is difficult to justify conceptually. (This variance could also mean sudden and dramatic, quasi-catastrophic price drops, of the kind observed during the Flash Crash of May 2010.) The growing fit between observable phenomena and methodological assumptions may also contribute to the popularity of econophysics.

Besides this popularity, however, there is another significant element to consider: econophysicists have been (and continue to be) trained in the natural sciences traditions, not in the economics tradition. Basic assumptions about market equilibrium and efficiency are not part of this training. Therefore, there is no adherence (or opposition) to these benchmark tenets and no striving to prove or disprove them. The approach of econophysics is an empirical one, characterized by an almost complete lack of interest in market equilibrium models (Jovanovic and Schinckus 2013). Oftentimes, at conferences, I have witnessed conflicts of opinion in which financial economists reproached econophysicists for their lack of "model," while the latter retorted by saying that regression analysis is unfit for the new kind of financial data.

This mutual disinterest (accompanied by tensions) helps explain why econophysics has continued to thrive outside financial economics and why the institutionalization paths have been parallel. At the same time, the absence of a benchmark model has made econophysics receptive to social behavioral research, of the kind practiced by SSF. There is acknowledgment in the econophysics community that trading technologies impact behavior in nondeterministic ways and that studying social interactions in electronic finance can provide explanatory templates which help understand data patterns.

Overall, seen in historical perspective, finance's relationship with (social) science has been a complex one. The period of financial economics' almost sole dominance has covered only about three decades, followed by an institutionalization of multiple, partly overlapping and partly diverging approaches. This divergence can also be seen as a sign of the growing complexity and importance of electronic finance, where a sole approach cannot cover all angles of financial phenomena. The degree to which these approaches place technology at their core is different. Financial economics and behavioral finance do it to a lesser extent than market microstructure, econophysics, or SSF. For the latter, technology and the effects thereof are a central intellectual challenge.

Conclusion

It is nowadays a truism to say that financial markets have been profoundly changed by their adoption of new technologies. As I have argued in this paper, electronic markets have both expanded and diversified during the past two decades. Technological changes have led to a diversification of trading venues and actors, while being accompanied by an institutionalization of multiple explanatory approaches. While some of these approaches are concerned primarily with the properties of price movements, others are concerned with how behavioral phenomena impact these properties, or with new interaction formats and social forms brought about by electronic finance. Why this sustained interest? If we are to go beyond a strictly utilitarian point of view (which reduces finance to a liquidity supplier), we can well argue that electronic finance has become a kind of social laboratory in which new interactions and new forms of sociality emerge, change, and adapt very rapidly, a laboratory in which institutions are reconfigured through technology. In a sense, electronic finance has become one of the social laboratories where forms of social life are produced and tested, forms which are neither limited to human-human interactions nor devoid of interactions, neither limited to routines nor devoid of stability. They are, however, new and, to some extent, nonforeseeable forms, the (moral) implications of which haven't been fully reflected upon. This, perhaps, is the next step that could be undertaken in the social science reflection upon electronic finance.

References

Adler, Patricia, and Peter Adler, eds. 1984. *The Social Dynamics of Financial Markets*. Greenwich, CT: JAI Press.

Arup, Christopher. 2010. "The Global Financial Crisis: Learning from Regulatory and Governance Studies." *Law & Policy* 32 (3): 363–81.

Baker, Wayne. 1984. "The Social Structure of a National Securities Market." *American Journal of Sociology* 89 (4): 775–811.

Banks, Eric. 2010. *Dark Pools: The Structure and Future of Off-Exchange Trading and Liquidity*. New York: St. Martin's Press.

Baron, Matthew, Jonathan Brogaard, and Andrei Kirilenko. 2014. "Risk and Return in High Frequency Trading." Accessed at http://dx.doi.org/10.2139/ssrn.2433118.

Beunza, Daniel, and David Stark. 2003a. "The Organization of Responsiveness: Innovation and Recovery in the Trading Rooms of Lower Manhattan." *Socio-Economic Review* 1 (2): 135–64.

___. 2003b. "Tools of the Stock Market: The Sociotechnology of Arbitrage in a Wall Street Trading Room." *Reseaux* 21 (122): 63–109.

___. 2012. "From Dissonance to Resonance: Cognitive Interdependence in Quantitative Finance." *Economy and Society* 41 (3): 383–417.

Blankespoor, Elizabeth, Gregory S. Miller, and Hal D. White. 2014. "The Role of Dissemination in Market Liquidity: Evidence from Firms' Use of Twitter." *Accounting Review* 89 (1): 79–112.

Bouchaud, Jean-Phillippe, and Jean-Francois Muzy. 2003. "Financial Time Series: From Bachelier's Random Walks to Multifractal Cascades." In *The Kolmogorov Legacy in Physics*, edited by Roberto Livi and Angelo Vulpiani, 229–47. New York: Springer.

Bouchaud, Jean-Phillippe, and Mark Potters. 2000. *Theory of Financial Risks: From Statistical Physics to Risk Management*. Cambridge: Cambridge University Press.

Callon, Michel, ed. 1998. *The Laws of Markets*. London: Blackwell.

Callon, Michel, and John Law. 1982. "On Interests and Their Transformation: Enrolment and Counter-Enrolment." *Social Studies of Science* 12 (4): 615–25.

Callon, Michel, and Fabian Muniesa. 2005. "Peripheral Vision: Economic Markets as Calculative Collective Devices." *Organization Studies* 26 (8): 1229–50.

Carrion, Allen. 2013. "Very Fast Money: High Frequency Trading on the NASDAQ." *Journal of Financial Markets* 16 (4): 680–711.

Carruthers, Bruce G. 2010. "Knowledge and Liquidity: Institutional and Cognitive Foundations of the Subprime Crisis." In *Markets on Trial: The Economic Sociology of the U.S. Financial Crisis: Part A (Research in the Sociology of Organizations, Volume 30 Part A)*, edited by Michael Lounsbury and Paul M. Hirsch, 157–82. Bingley: Emerald Group Publishing Ltd.

Carruthers, Bruce G., and Jeong-Chul Kim. 2011. "The Sociology of Finance." *Annual Review of Sociology* 37: 239–59.

Chen, Chuan. 2011. "Beyond the Embeddedness Model: The Origin, Development, and New Issues of Financial Sociology." *Shehui/Society: Chinese Journal of Sociology* 31 (5): 207–25.

Chen, Hailiang, Prabuddha De, Yu Hu, and Byoung-Hyoung Hwang. 2014. "Wisdom of Crowds. The Value of Stock Opinions Transmitted through Social Media." *Review of Financial Studies* 27 (5): 1367–403.

Collins, Harry. 2004. *Gravity's Shadow: The Search for Gravitational Waves*. Chicago: University of Chicago Press.

Crain, Matthew. 2014. "Financial Markets and Online Advertising: Reevaluating the Dotcom Investment Bubble." *Information, Communication & Society* 17 (3): 371–84.

Davies, Will. 2013. "When Is a Market Not a Market? 'Exemption,' 'Externality' and 'Exception' in the Case of European State Aid Rules." *Theory, Culture & Society* 30 (2): 32–59.

Davis, Gerald F., and Suntae Kim. 2015. "Financialization of the Economy." *Annual Review of Sociology* 41: 203–21.

DeBondt, Werner, Gulnur Muradoglu, Hersh Shefrin, and Sotiris K. Staikouras. 2009. "Behavioural Finance: Quo Vadis?" *Journal of Applied Finance* 18 (2): 7–21.

Duflo, Esther, and Emmanuel Saez. 2003. "The Role of Information and Social Interactions in Retirement Plan Decisions: Evidence from a Randomized Experiment." *Quarterly Journal of Economics* 118 (3): 815–42.

Ervine, Kate. 2013. "Carbon Markets, Debt and Uneven Development." *Third World Quarterly* 34 (4): 653–70.

Fama, Eugene. 1970. "Efficient Capital Markets: A Review of Empirical and Theoretical Work." *Journal of Finance* 25 (2): 383–417.

Ford, Cristie. 2013. "Innovation-Framing Regulation." *The Annals of the American Academy of Political and Social Science* 649 (1): 76–97.

Foresight. 2012. *Foresight: The Future of Computer Trading in Financial Markets: Final Project Report.* London: The Government Office for Science.

French, Shaun, Andrew Leyshon, and Thomas Wainwright. 2011. "Financializing Space, Spacing Financialization." *Progress in Human Geography* 35 (6): 798–819.

Fridman, Daniel. 2010. "From Rats to Riches: Game Playing and the Production of the Capitalist Self." *Qualitative Sociology* 33 (4): 423–46.

Gingras, Yves, and Christophe Schinckus. 2012. "The Institutionalization of Econophysics in the Shadow of Physics." *Journal of the History of Economic Thought* 34 (1): 109–30.

Hardie, Iain. 2004. "'The Sociology of Arbitrage': A Comment on MacKenzie." *Economy and Society* 33 (2): 239–54.

Hardie, Iain, and Donald MacKenzie. 2007. "Assembling an Economic Actor: The Agencement of a Hedge Fund." *Sociological Review* 55 (1): 57–80.

___. 2011. "How Much Can Governments Borrow? Financialization and Emerging Markets Government Borrowing Capacity." *Review of International Political Economy* 18 (2): 141–67.

Hasbrouck, Joel, and Gideon Saar. 2013. "Low Latency Trading." *Journal of Financial Markets* 16 (4): 646–79.

Healy, Kieran. 2015. "The Performativity of Networks." *Archives Européennes de Sociologie (European Journal of Sociology)* 56 (2): 175–205.

Hessling, Alexandra, and Hanno Pahl. 2006. "The Global System of Finance: Scanning Talcott Parsons and Niklas Luhmann for Theoretical Keystones." *American Journal of Economics and Sociology* 65 (1): 189–218.

Hope, Wayne. 2006. "Global Capitalism and the Critique of Real Time." *Time & Society* 15 (2–3): 275–302.

Jovanovic, Franck. 2008. "The Construction of the Canonical History of Financial Economics." *History of Political Economy* 40 (2): 213–42.

___. 2012. "Finance in Modern Economic Thought." In *The Oxford Handbook of the Sociology of Finance*, edited by Karin Knorr Cetina and Alex Preda, 546–66. Oxford: Oxford University Press.

Jovanovic, Franck, and Christophe Schinckus. 2013. "The Emergence of Econophysics: A New Approach in Modern Financial Theory." *History of Political Economy* 45 (3): 443–74.

Kliger, Doron, and Dori Gilad. 2012. "Red Light, Green Light: Color Priming in Financial Decisions." *Journal of Socio-Economics* 41 (5): 738–45.

Knorr Cetina, Karin. 2003. "From Pipes to Scopes: The Flow Architecture of Financial Markets." *Distinktion* 4 (2): 7–23.

___. 2004. "Capturing Markets? A Review Essay on Harrison White on Producer Markets." *Socio-Economic Review* 2 (1): 137–47.

___. 2006. "The Market." *Theory, Culture & Society* 23 (2–3): 551–56.

___. 2007. "Global Markets as Global Conversations." *Text & Talk* 27 (5–6): 705–34.

___. 2009. "What Is a Pipe?" *Theory, Culture & Society* 26 (5): 129–40.

Knorr Cetina, Karin, and Urs Bruegger. 2002a. "Global Microstructures: The Virtual Societies of Financial Markets." *American Journal of Sociology* 107 (4): 905–50.

___. 2002b. "Traders' Engagement with Markets: A Postsocial Relationship." *Theory, Culture & Society* 19 (5–6): 161–85.

Krippner, Greta. 2005. "The Financialization of the American Economy." *Socio-Economic Review* 3 (2): 173–208.

Kristal, Tali. 2013. "The Capitalist Machine: Computerization, Workers' Power, and the Decline in Labor's Share within U.S. Industries." *American Sociological Review* 78 (3): 361–89.

Kunz, Karin, and Jena Martin. 2015. "Into the Breach: The Increasing Gap between Algorithmic Trading and Securities Regulation." *Journal of Financial Services Research* 47 (1): 135–52.

Larkin, Brian. 2013. "The Politics and Poetics of Infrastructure." *Annual Review of Anthropology* 42: 327–43.

Lave, Rebecca, Philip Mirowski, and Samuel Randalls. 2010. "Introduction: STS and Neoliberal Science." *Social Studies of Science* 40 (5): 659–75.

Lenglet, Marc. 2011. "Conflicting Codes and Coding: How Algorithmic Trading Is Reshaping Financial Regulation." *Theory, Culture & Society* 28 (6): 44–66.

Lepinay, Vincent. 2007. "Economy of the Germ: Capital, Accumulation and Vibration." *Economy and Society* 36 (4): 526–48.

___. 2011. *Codes of Finance: Engineering Derivatives in a Global Bank*. Princeton, NJ: Princeton University Press.

Lilley, Simon, and Dimitris Papadopoulos. 2014. "Material Returns: Cultures of Valuation, Biofinancialisation and the Autonomy of Politics." *Sociology* 48 (5): 972–88.

Linhardt, Dominique, and Fabian Muniesa. 2011. "Take the Place of Politics: The Paradox of 'Economization Policies'." *Politix* 95 (3): 7–21.

MacKenzie, Donald. 1990. *Inventing Accuracy: A Historical Sociology of Nuclear Missile Guidance*. Cambridge, MA: MIT Press.

___. 2007. "The Material Production of Virtuality: Innovation, Cultural Geography and Facticity in Derivatives Markets." *Economy and Society* 36 (3): 355–76.

___. 2011. "The Credit Crisis as a Problem in the Sociology of Knowledge." *American Journal of Sociology* 116 (6): 1778–1841.

___. 2012a. "Knowledge Production in Financial Markets: Credit Default Swaps, the ABX and the Subprime Crisis." *Economy and Society* 41 (3): 335–59.

___. 2012b. "Material Markets: How Economic Agents Are Constructed." *Theory, Culture & Society* 29 (2): 150–53.

___. 2016. "How Algorithms Interact. Goffman's 'Interaction Order' in Automated Trading." *Theory, Culture & Society*, forthcoming.

MacKenzie, Donald, and Yuval Millo. 2003. "Constructing a Market, Performing a Theory: The Historical Sociology of a Financial Derivatives Exchange." *American Journal of Sociology* 109 (1): 107–45.

MacKenzie, Donald, and Juan Pablo Pardo Guerra. 2014. "Insurgent Capitalism: Island, Bricolage and the Re-making of Finance." *Economy & Society* 43 (2): 153–82.

MacKenzie, Donald, and Taylor Spears. 2014a. "'A Device for Being Able to Book P&L': The Organizational Embedding of the Gaussian Copula." *Social Studies of Science* 44 (3): 418–40.

___. 2014b. "'The Formula that Killed Wall Street': The Gaussian Copula and Modelling Practices in Investment Banking." *Social Studies of Science* 44 (3): 393–417.

Madhavan, Ananth. 2000. "Market Microstructure: A Survey." *Journal of Financial Markets* 3 (3): 205–58.

Markham, Jerry W., and Daniel H. Harty. 2012. "The Impact of Electronic Communication Networks on Exchange Trading Floors and Derivatives Regulation." In *Handbook of Research on Stock Market Globalization*, edited by Geoffrey Poitras, 244–95. Cheltenham: Edward Elgar.

Martin, Randy. 2010. "The Good, the Bad and the Ugly: Economies of Parable." *Cultural Studies* 24 (3): 418–30.

McMahon, Peter. 2004. "Money, Markets and Microelectronics: Building the Infrastructure for the Global Finance Sector." *Prometheus* 22 (1): 71–82.

Millo, Yuval, Fabian Muniesa, Nikiforos S. Panourgias, and Susan V. Scott. 2005. "Organized Detachment: Clearinghouse Mechanisms in Financial Markets." *Information and Organization* 15 (3): 229–46.

Mirowski, Philip. 1999. "Cyborg Agonistes: Economics Meets Operations Research in Mid-Century." *Social Studies of Science* 29 (5): 685–718.

___. 2003. "The Sociology of Science and the 'New Information Economy.'" *Reseaux* 21 (122): 167–87.

Miyazaki, Hirokazu. 2003. "The Temporalities of the Market." *American Anthropologist* 105 (2): 255–65.

___. 2006. "Economy of Dreams: Hope in Global Capitalism and Its Critiques." *Cultural Anthropology* 21 (2): 147–72.

___. 2007. "Between Arbitrage and Speculation: An Economy of Belief and Doubt." *Economy and Society* 36 (3): 396–415.

___. 2013. *Arbitraging Japan: Dreams of Capitalism at the End of Finance*. Berkeley: University of California Press.

Mizrach, Bruce. 2006. "Does SIZE Matter? Liquidity Provision by the NASDAQ Anonymous Trading Facility." Rutgers University, Department of Economics, Departmental Working Papers. Accessed at ftp://snde.rutgers.edu/Rutgers/wp/2006-02.pdf.

Montagne, Sabine, and Horacio Ortiz. 2013. "Sociology of the Financial Branch: Issues and Perspectives." *Sociétés Contemporaines* 92 (4): 7–33.

Muniesa, Fabian. 2005. "Containing the Market: The Transition from Open Outcry to Electronic Trading at the Paris Bourse." *Sociologie du Travail* 47 (4): 485–501.

___. 2007. "Market Technologies and the Pragmatics of Prices." *Economy and Society* 36 (3): 377–95.

___. 2014. *The Provoked Economy: Economic Reality and the Performative Turn*. London: Routledge.

Neumann, Sascha, Stephan Paul, and Philipp Doering. 2014 "A Primer on Social Trading—Remuneration Schemes, Trading Strategies and Return Characteristics." Accessed at http://dx.doi.org/10.2139/ssrn.2291421.

O'Hara, Maureen. 1995. *Market Microstructure Theory*. Malden, MA: Blackwell.

Orr, Julian E. 1996. *Talking about Machines: An Ethnography of a Modern Job*. Ithaca, NY: Cornell University Press.

Ortiz, Horacio. 2011. "Efficient Markets, Investors, and Guarantor States Guarantees: Political Threads of Professional Financial Practice." *Politix* 95 (3): 155–80.

___. 2013. "Investing: A Disseminated Decision. Fieldwork-Based Research on Credit Derivatives." *Sociétés Contemporaines* 92 (4): 35–57.

___. 2014. "The Limits of Financial Imagination: Free Investors, Efficient Markets, and Crisis." *American Anthropologist* 116 (1): 38–50.

Pardo Guerra, Juan Pablo. 2010. "Creating Flows of Interpersonal Bits. The Automation of the London Stock Exchange c. 1955–1990." *Economy and Society* 39 (1): 84–109.

___. 2012. "Financial Automation: Past, Present, and Future." In *The Oxford Handbook of the Sociology of Finance*, edited by Karin Knorr Cetina and Alex Preda, 567–86. Oxford: Oxford University Press.

Patel, Geeta. 2007. "Imagining Risk, Care and Security: Insurance and Fantasy." *Anthropological Theory* 7 (1): 99–118.

Perez, Carlota, and Jonathan Rutherford. 2009. "Financial Bubbles and Economic Crises." *Soundings: A Journal of Politics and Culture* 41 (1): 30–44.

Pinch, Trevor, and Wiebe E. Bijker. 2012. "The Social Construction of Facts and Artifacts. Or How the Sociology of Science and the Sociology of Artifacts Might Benefit Each Other." In *The Social Construction of Technological Systems: New Direction in the Sociology and History of Technology*, anniversary edition, edited by Wiebe E. Bijker, Thomas P. Hughes, and Trevor Pinch, 11–44. Cambridge, MA: MIT Press.

Pollock, Neil, and Robin Williams. 2010. "The Business of Expectations: How Promissory Organizations Shape Technology and Innovation." *Social Studies of Science* 40 (4): 525–48.

Poon, Martha. 2007. "Scorecards as Devices for Consumer Credit: The Case of Fair, Isaac & Company Incorporated." In *Market Devices*, edited by Michel Callon, Yuval Millo, and Fabian Muniesa, 284–306. London: Blackwell.

___. 2012. "Rating Agencies." In *The Oxford Handbook of the Sociology of Finance*, edited by Karin Knorr Cetina and Alex Preda, 272–92. Oxford: Oxford University Press.

Power, Michael, ed. 1994. *Accounting and Science: Natural Inquiry and Commercial Reason*. Cambridge: Cambridge University Press.

___. 2009. "Bankrupt: Global Lawmaking and Systemic Financial Crisis." *Canadian Journal of Sociology* (*Cahiers Canadiens de Sociologie*) 34 (4): 1124–27.

Preda, Alex. 2006. "Socio-Technical Agency in Financial Markets: The Case of the Stock Ticker." *Social Studies of Science* 36 (5): 753–82.

___. 2009. *Framing Finance: The Boundaries of Markets and Modern Capitalism*. Chicago: University of Chicago Press.

___. 2016. *Noise: Living and Trading in Electronic Finance*. Chicago: University of Chicago Press.

Pryke, Mike. 2010. "Money's Eyes: The Visual Preparation of Financial Markets." *Economy and Society* 39 (4): 427–59.

Randalls, Samuel. 2010. "Weather Profits: Weather Derivatives and the Commercialization of Meteorology." *Social Studies of Science* 40 (5): 705–30.

Reichman, Nancy, and Ophir Sefiha. 2013. "Regulating Performance-Enhancing Technologies: A Comparison of Professional Cycling and Derivatives Trading." *The Annals of the American Academy of Political and Social Science* 649 (1): 98–119.

Reith, Gerda. 2013. "Techno-Economic Systems and Excessive Consumption: A Political Economy of 'Pathological' Gambling." *British Journal of Sociology* 64 (4): 717–38.

Riles, Annelise. 2011. *Collateral Knowledge: Legal Reasoning in the Global Financial Markets*. Chicago: University of Chicago Press.

Roberts, John Michael, and Jonathan Joseph. 2015. "Beyond Flows, Fluids and Networks: Social Theory and the Fetishism of the Global Informational Economy." *New Political Economy* 20 (1): 1–20.

Stafford, Philipp. 2013. "Q&A: Algorithms." FT Trading Room, *Financial Times* August 22.

Schüll, Natasha Dow, and Caitlin Zaloom. 2011. "The Shortsighted Brain: Neuroeconomics and the Governance of Choice in Time." *Social Studies of Science* 41 (4): 515–38.

Sprenger, Timm O., Philipp G. Sandner, Andranik Tumasjan, and Isabell M. Welpe. 2014. "News or Noise? Using Twitter to Identify Company-Specific News Flow." *Journal of Business Finance and Accounting* 41 (7–8): 791–830.

Timmermans, Stefan, and Steven Epstein. 2010. "A World of Standards but Not a Standard World. Toward a Sociology of Standards and Standardization." *Annual Review of Sociology* 36: 69–89.

Vaughan, Diane. 1996. *The Challenger Launch Decision: Risk, Technology, Culture, and Deviance at the NASA*. Chicago: University of Chicago Press.

Vlcek, William. 2012. "Power and the Practice of Security to Govern Global Finance." *Review of International Political Economy* 19 (4): 639–62.

Walby, Sylvia. 2010. "A Social Science Research Agenda on the Financial Crisis." *Twenty-First Century Society: Journal of the Academy of Social Sciences* 5 (1): 19–31.

Wansleben, Leon. 2012. "Heterarchies, Codes, and Calculi: Contribution to a Sociology of Algo Trading." *Soziale Systeme* 18 (1–2): 225–59.

Zajac, Edward J., and James D. Westphal. 2004. "The Social Construction of Market Value: Institutionalization and Learning Perspectives on Stock Market Reactions." *American Sociological Review* 69 (3): 433–57.

Zaloom, Caitlin. 2006. *Out of the Pits: Traders and Technology from Chicago to London*. Chicago: University of Chicago Press.

Zeckenhauser, Richard. 2006. "Investing in the Unknown and Unknowable." *Capitalism and Society* 1 (2): 1–39.

Zuckerman, Ezra W. 2004. "Structural Incoherence and Stock Market Activity." *American Sociological Review* 69 (3): 405–32.

___. 2012. "Construction, Concentration, and (Dis)continuities in Social Valuations." *Annual Review of Sociology* 38: 223–45.

22   A Critical Theory of Technology

Andrew Feenberg

Introduction

This chapter summarizes the main ideas of critical theory of technology and shows how it relates to its two sources, Frankfurt School Critical Theory and early work in science and technology studies (STS).[1] Critical theory of technology is concerned with the threat to human agency posed by the technocratic system that dominates modern societies. Two early trends in STS, various versions of social constructivism and actor-network theory (ANT), addressed this threat implicitly, through challenging positivist and determinist ideologies that left little place for democratic control of technology. Critical theory of technology agrees with STS that technology is neither value neutral nor universal while proposing an explicit theory of democratic interventions into technology.

As STS has responded in recent years to the emergence of public participation in determining technology policy, it has moved closer to the concerns of critical theory of technology (Chilvers and Kearnes 2016). Critical theory of technology is still distinguished from most contributions to STS by its emphasis on certain themes derived from the Frankfurt School, especially the critique of rationality in modern culture. It thus puts STS in communication with traditions of social critique often overlooked. In this respect it is not so much an alternative to STS as an invitation to open STS to a wider range of philosophical and social theories of modernity.

The first sections of this chapter will map the relation between critical theory of technology and some of the major scholars and methodological innovations of STS. Next, the chapter explains relevant reservations concerning the concept of symmetry which historically is a central concern of STS scholarship. The succeeding sections will explain the principal concepts and methods of critical theory of technology and discuss its political implications. The concluding sections will interpret an interesting STS case study and discuss the methodological implications of the combined theory.

Mapping Critical Theory of Technology

Before the formalization of STS into a scholarly field of inquiry, the social study of technology was associated with Marxism, pragmatism, Heideggerian phenomenology, and various theories of modernity. These broad and often speculative theories explored the relation of technology to society. They attempted to understand the specificity of modernity in terms of the scientific and technological revolutions and on that basis to account for the many ills of modernity, especially the decline of human agency in a technologized society. Their themes are now familiar: technocracy, the tyranny of expertise, the substitution of knowledge for wisdom and information for knowledge, a vision of the human being and of society as a complex of functional systems, the meaninglessness of modern life, the obsolescence of man, and so on. Lost amid these vast concerns is technology itself.

STS has been largely successful in supplanting these competing approaches with empirical case studies of actual technologies. Today few look to Mumford or Dewey, Heidegger or Marcuse for insight into technology. However, when STS turned to case studies, with methods rooted in other intellectual traditions, it focused less frequently and ambitiously on wider social and political concerns. Of particular relevance to the argument of this chapter was the decreased emphasis on the modernity theorists' concern with the contradiction between political agency and technocratic rationality.

This abstention from politics was due to what Wiebe Bijker called the "detour into the academy," deemed necessary to establish STS as a social science (Bijker 1995, 5). Of course not everyone made the famous detour, but the field was sufficiently marked by abstention from political controversy to trouble some who belonged to the earlier critical tradition. Langdon Winner spoke for them in an article significantly entitled "Upon Opening the Black Box and Finding It Empty: Social Constructivism and the Philosophy of Technology" (Winner 1993). I responded differently by revising Critical Theory to accommodate the methodological innovations of STS (Feenberg 1991). Rather than calling for STS to adopt the critical spirit, I adopted the antideterminism and antipositivism of STS to support a critical theory of technology.

The concerns of STS broadened as the widespread controversies over medical care, the Internet, and the environment directly implicated technology in so many different aspects of contemporary political life. In response STS has become political, although sometimes with an unconvincing concept of politics (Brown 2015; Soneryd 2016). Actor-network theory (ANT) and the work of Sheila Jasanoff, Brian Wynne, and many others have had a broad influence on attempts in STS to understand the politics of technology (Jasanoff 2004; Latour and Weibel 2005;[2] Wynne 2011). Studies of hybrid forums and co-production challenge narrow understandings of democracy prevalent in philosophy and political theory (Callon, Lascoumes, and Barthe 2011; Chilvers and Kearnes 2016; Jasanoff 2004). Some STS researchers have now also become aware of the highly politicized approaches favored in the developing world, especially Latin America (Dagnino 2008; Kreimer et al. 2004; Rajão, Duque, and De' 2014). But how can the prior achievements of STS, exemplified in so many brilliant case studies, be preserved in the context of politically charged investigations of controversial issues? For reasons rooted in the origins of STS, this poses problems. This chapter proposes one way of addressing those problems.

The critical theory of technology draws on STS while placing the issues in the context of the Frankfurt School's critique of modernity. Critical Theory was developed by German Marxists in the 1920s and 1930s. Its most famous members were Max Horkheimer, Theodor Adorno, Herbert Marcuse, and Walter Benjamin. They were influenced by Georg Lukács, whose concept of "reification" described the reduction of complex and dynamic social relations to apparently law-governed interactions of (social) things (Lukács 1971). Lukács argued that reification constitutes the members of society as isolated individuals. In that condition they cannot change the laws of social life, only use them as the basis of technical manipulations. The Frankfurt School continued this line of criticism, demystifying reified institutions and opening up possibilities of critique foreclosed by the tendentious appeal to social and economic laws. Increasingly, from the 1940s on, the members of the Frankfurt School focused on the collapse of both bourgeois culture and the proletarian movement in the face of mass culture and fascism.

The dominant liberal ideology of the post–World War II era brought technocratic claims into the center of public discourse. Social arrangements were justified by reference to their rational character and opposition dismissed as sentimental nonsense. In the 1960s Marcuse distinguished himself by the popular success of his attack on liberalism.[3] His 1964 book *One-Dimensional Man* had a significant influence on the New Left. His critique of American society as a highly integrated system governed by "technological rationality" resonated with the concerns of youth in the advanced capitalist world. The technical details of Marcuse's theory of technology were not widely studied or understood at the time, but today it has surprising relevance. He not only claimed that technology has been shaped by the capitalist social forces that presided over its creation, but he argued for the possibility of progressive technological change under the influence of more humane social forces.

Marcuse's concept of "technological rationality" as a legitimating ideology updated earlier Marxist notions of market rationality. Social life in our time appears increasingly not only to depend on science and technology but also to mirror scientific and technical procedures. Efficiency is said to be rational and commands respect in every area of social life. Rationality thus serves as the justification and alibi for many kinds of social change. The "mantra of efficiency" draws strength from this connection even as it has disastrous consequences for some of those affected (Alexander 2008). Critique is disarmed before it can get off the ground by a blanket accusation of irrationality. Who dares question the universality, the neutrality, the progressive contribution of science? Luddites and other "romantic irrationalists" are easily dismissed with a reference to the overwhelming success of modern science and technology.

Marcuse's version of Critical Theory recapitulates the essential content of Lukács's concept of reification, the notion that capitalism imposes a rational culture that privileges technical manipulation over all other relations to reality. That culture narrows human understanding and lives to conform with the requirements of the economic system. Capitalism thus determines the form of social interaction and experience. Marcuse contends, "When technics becomes the universal form of material production, it circumscribes an entire culture; it projects an historical totality—a 'world'" (Marcuse 1964, 154).

Following this approach, critical theory of technology identifies the intrinsic bias of "efficient" solutions to social and technical problems. But it differs substantially from the earlier works of Lukács and Marcuse. The major shift is methodological. Critical theory of technology draws upon fundamental assumptions of STS in elaborating the themes treated by earlier traditions.

Like the modernity theorists, STS reacted against technocratic ideology but did not embrace sweeping philosophical critique. The case study method employed empirical evidence to subvert the positivistic and determinist assumptions underlying the liberal celebration of the rational society. The key alternative assumptions proposed in the social constructivist tradition are the notions of actors, underdetermination, interpretive flexibility, and closure. Although introduced to account for particular cases, these assumptions can lend support to the critique of the ideological deployment of the concept of rationality. Critical theory of technology incorporates these assumptions along with the concepts of program, delegation, and co-production drawn from ANT. However, critical theory of technology questions the most radical conclusions of STS theorists, such as the symmetrical treatment of disputants in technological controversies and the symmetry of humans and nonhumans. The following sections will elaborate where and how critical theory of technology draws on and diverges from certain STS theories and methods.

Contributions of Social Constructivism and ANT

The application of social constructivist methods to particular technologies is fruitful. It blocks the ideological recourse to pseudorational justification by showing that social factors intervene in the decisions that lead to "closure," or the success of a particular design. Social constructivism argues that the perception of technical problems depends on the interpretations of social groups or "actors." The early stages of the development of an artifact often involve a multiplicity of actors with conflicting interpretations of the nature of the problem to be solved. Different social groups may assign different purposes to devices that are basically similar from a technical standpoint. Design decisions flow from these assignments.

In one of the most cited instances of this approach, Trevor Pinch and Wiebe Bijker offer the example of the early history of the bicycle (Pinch and Bijker 1987). They present two competing types of bicycles in the early days of bicycle design: a fast bicycle with a large front wheel and a small rear wheel and a safer but slower bicycle with two wheels the same size. Today the large front wheelers appear to be primitive predecessors of the bicycles we ride, but in its own day the design suited a specific group of users. Pinch and Bijker propose a "symmetrical" treatment of the two main designs that takes account of their contemporary social meaning rather than viewing them in terms of an imaginary chronology.

This constructivist "principle of symmetry" was initially introduced to guide the study of scientific controversy toward an even-handed treatment of both winners and losers. The commonplace attribution of superior rationality to the winner was to be resisted in favor of an appreciation of the mixed motives and questionable assumptions on all sides of the controversy. In its application to technology, the constructivist principle of symmetry required a balanced view of the various designs competing at the outset, no one of which was obviously superior in the eyes of contemporaries.

Each of the bicycle designs Pinch and Bijker studied appealed to different actors: the high front-wheelers to young men who liked to race and the more stable design to ordinary people using a bicycle for transportation. Most of the parts were similar and both versions looked like a bicycle, but they were actually two different technologies understood in different ways by different social groups. Eventually, through innovation, the safer model prevailed. Its triumph was not due to absolute technical superiority but to contingent historical developments.

The outcome was underdetermined by purely technical considerations and can only be understood by taking into account the actors' struggles for control of the design process. The technical underdetermination of artifacts leaves room for social choice between different designs that have overlapping functions but better serve one or another social interest. This "interpretive flexibility" of artifacts concerns a hermeneutic dimension overlooked in standard instrumentalist accounts.

As Pinch and Bijker write, "The different interpretations by social groups of the content of artifacts lead by means of different chains of problems and solutions to different further developments ..." (1987, 42). Their key point is the influence of the social on "the content of the artifact" itself and not merely on such external factors as the pace of development, the packaging, or the usages. This means that context is not merely external to technology but actually penetrates its rationality, carrying social requirements into the very workings of the device. Thus, the "rational society" is not the "one best way" but contingent on values and interests.

This argument could lend support to the Marxist account of the development of a specifically capitalist technology in opposition to the deterministic arguments of the postwar technocracy. In fact, anticipations of Pinch and Bijker's approach can be seen in the work of Marxist historians of technology, Harry Braverman (1974) and David Noble (1977). The Frankfurt School affirmed the capitalist nature of technology on the basis of the same sources in Marx's work that influenced these scholars. Adorno writes, for example,

It is not technology which is calamitous, but its entanglement with societal conditions in which it is fettered ... . Considerations of the interests of profit and dominance have channeled technical development: by now it coincides fatally with the needs of control. Not by accident has the invention of means of destruction become the prototype of the new quality of technology. By contrast, those of its potentials which diverge from dominance, centralism and violence against nature, and which might well allow much of the damage done literally and figuratively by technology to be healed, have withered. (Adorno 2000, 161–62, note 15)

These arguments offer a possible bridge between Marxism and constructivism. Adorno, like the constructivists, attributes the design of technology to the actors who dominate the design process. However, other scholars, like those represented by the work of Bruno Latour, object to this approach because it absolutizes society at the same time as it relativizes technology. Latour's formulation of actor-network theory attempts to disengage constructivism from what he considers an overemphasis on human intention. In order to bring the material layers of the network into focus, ANT therefore extends the constructivist argument to the things incorporated into technical networks. It argues for a conceptual and functional "symmetry of humans and nonhumans" different from the social constructivist version of symmetry.[4] Symmetry in ANT is achieved by blurring the distinction between interpretive and intentional acts of humans and the causal powers of things, signifying both with the neutral term *agency*. People and things link together in networks and have effects on the networks to which they belong. The concept of "program" in ANT does the work of the constructivist notion of interpretation with the proviso that things too may have programs in the sense that their actions play a role in the life of the network.

This approach avoids the subjectivism and relativism sometimes attributed to social constructivism, but it does so in a strange way—not by reintroducing the objective properties of things identified by scientific research but rather by describing their roles in the networks to which they belong. A similar reduction strips human beings of inwardness and initiative. People and things are to be understood as *essentially* actors in networks, not as subjects and objects. The division between subject and object, meaning and causality is then explained as an illegitimate epistemological operation specific to modernity, which Latour calls "purification" (Latour 1993). This after-the-fact operation obscures the foundational significance of the hybrid sociotechnical realities of the networks.

ANT thus posits the hybrids prior to their components. Its concept of "co-construction," or "co-production," calls attention to the interdependence of the human actors and the technical world in which they find themselves. Actors are not constituted by purely social bonds but form around the technologies that support the interactions of their members. Human agency must not be privileged over the agency of the things that support the sociotechnical networks in which society consists. This argument, like that of social constructivism, is subversive of a naïve confidence in the purely "rational" character of the technical world, which is now shown to be a scene on which many types of agents are active in terms of a variety of programs.

Critical theory of technology draws on social constructivism for an alternative to technological determinism and on ANT for an understanding of networks of persons and things. The constructivist approach emphasizes the role of interpretation in the development of technologies. Actor-network theory explores the implications of technical networks for identities and worlds. These notions are congruent with the critique of context-free rationality in the early Frankfurt School which provides a background to the concept of the bias of technology in critical theory of technology (Feenberg 2014; Horkheimer 1995). Critical theory of technology thus concretizes the Frankfurt School approach through the application of STS methods.

The Limits of Symmetry

Constructivist STS research methods have been fruitful. They have introduced new ideas about technological design and the relation of publics to the technical mediations that bind their members together. This is an important advance over standard social and political theories that abstract from the technological altogether or fetishize it deterministically. However, less persuasive are the moves beyond critique and methodology to found a relativistic epistemology and a new network ontology. The problems show up in attempts to generalize STS as a full-fledged political theory. As we have seen, the two principles of symmetry require that the same methods and terms be used in the treatment of participants in controversies and of humans and nonhumans. These two principles have contradictory political implications. On the one hand, they weaken the hegemony of the technocracy and carve out a place for democratic initiatives in the technical sphere. But on the other, they make it difficult to understand the nature of social conflict in a heterogeneous environment such as a modern capitalist society. In the remainder of this section I will explain these limitations.

The constructivist principle of symmetry proves particularly effective in valorizing the contributions of ordinary people to the redesign of flawed or unnecessarily limited technologies. Experts bound by interests and traditions sometimes overlook problems and potentials revealed only once their products circulate widely (Oudshoorn and Pinch 2003). Environmentalism is based in large part on the intolerance of users and victims for levels of pollution deemed acceptable at first by business and its expert cadres. With the Internet, users have made an undeniable contribution to the evolution of a major technological system.

Some of these lay interventions involve significant conflict with established institutions. Conflict in society was of course a central concern of Marxists such as Lukács and Marcuse. Early social constructivism modeled its discussion of technology on a different type of conflict, scientific controversy. This poses a problem for the generalization of STS methods to society at large. Many later attempts in STS to understand social conflict have broken with this early model, but constructivist symmetry remains an important concept to which reference is often made. This appears to me to be an inconsistency.

Although there are exceptions, scientists typically act in good faith and on the basis of evidence, even when they disagree over its interpretation. The social aspect of science is not primarily a matter of motives. The constructivist principle of symmetry was introduced in acknowledgment of this fact. Its application in social studies of science was intended to ensure that the same methods would be applied to all parties to the dispute, avoiding a one-sided treatment of scientific controversies. As noted above, the methodological relativism imposed by symmetrical treatment counteracts the tendency to idealize the winner and undervalue the intelligence and rationality of the loser (or, vice versa, to demonize the winner and overvalue the rights and justice of the loser).

For example, Priestley's rejection of Lavoisier's discovery of the mechanism of combustion cannot be laid to mere dogmatism, self-interest, or stubbornness; his point of view too must be considered as a rational, if unsuccessful, attempt at understanding.[5] Unfortunately, many technical controversies are quite different from this model. One or both sides are often biased by economic interests, dishonest claims, irrational panic, racial or gender prejudice, and the corruption of scientific and public actors. The principle of symmetry can mislead in this context, if applied injudiciously. Its application to particular cases risks providing alibis for the machinations of unscrupulous actors or systematic discrimination. A relativistic method is of no use where dishonesty or prejudice prevails.[6] Not only is symmetry ill-suited to the rough and tumble of technological controversy, but it risks canceling the normal attributions of responsibility on which we rely in public life. An even-handed treatment of bad decisions can slip over into a justification of those responsible.

Consider the case of the *Challenger* accident as explained by Trevor Pinch and Harry Collins (Collins and Pinch 1998). The common view of the accident attributes blame to NASA managers' impatience. This asymmetrical explanation conforms with our usual notions of responsibility, but is it right? The danger of a cold weather launch had not yet been proven experimentally on the fatal day. But there was cause for concern: the informed observations of the engineer assigned to investigate the problem. His observations were ignored, Pinch and Collins write, not because management rejected reasonable caution but because they did not meet "prevailing technical standards" (Collins and Pinch 1998, 55). Symmetry prevails, but responsibility is defeated. The question remains of why "technical standards" were preferred in this instance, why expert observations were ignored in favor of more rigorous proof that was unavailable. Could it be that symmetry is broken at the level of epistemology? All too often, scientism trumps all other evidence when it serves the interests of the dominant social actors, but only then. This is a good illustration of Marcuse's notion of "one-dimensional thought," which privileges quantitative precision over experiential knowledge.[7]

ANT has other problems with politics. The network approach led to the widely adopted concept of co-production of society and technology. This concept is well suited to understanding political controversies over technology. It has the potential to revolutionize political theory by focusing attention on the technical mediation of social organization. But Latour's ambitious theoretical program is not as successful as the case histories in which the concept of co-production is applied. The principle of symmetry of humans and nonhumans was intended to orient research toward the structure of the networks uniting them. These networks were said to explain all macro entities such as "state," "ideology," "class," "culture," "nature," and "economic interest." But critics accused Latour of bias in favor of the victors in the struggle to define nature since he argues, in accordance with good STS practice, that nature in the only meaningful sense is established by the network. But what if the nature so defined is discriminatory? To what can the losers in struggles over race or gender discrimination appeal if not to a "natural" equality grounded on a different definition of nature (Radder 1996)? Latour eventually came to agree that he had gone too far toward a Machiavellian affirmation of success (Latour 2013).

He responded to his critics in his writings on ecology, arguing that actors can introduce new objects into the taken-into-account world, for example, such objects as toxic wastes and smog. Freedom of discussion in the constitution of the "collective" would ensure against economic or technocratic domination. This is not a bad start toward understanding environmental issues, such as climate change, but it is not very useful as an account of the actual struggle between affirmers and deniers and the gaps in national uptake of the policy recommendations of the UN panel on climate change. Latour's rejection of "critique" and of macro concepts in social theory deprives him of the means to address the role of interests and ideologies in determining positions on the issues. But without access to these categories, research cannot address the principle insight of the Frankfurt School, namely, the role of capitalism in the cultural generalization of rationality. Indeed, good old-fashioned Marxist notions of interest and ideology are obviously in play when energy bosses like the Koch brothers mobilize a billion dollars or more to sponsor climate denial and support political candidates whose policies protect their purse (Rich 2010).[8]

In an early work, Latour introduced the term "anti-program" to signify the conflictual aspect of his networks (Latour 1992, 251–52). Critical theory of technology introduces a new principle of symmetry based on this notion. I have proposed what I call the "symmetry of program and antiprogram" in order to avoid any bias in favor of the dominant network actor (Feenberg 1999). Programs corresponding to actors' intentions carve out subsets of the interconnected elements brought together in the network. Where actors are in conflict, different programs may highlight different elements. For example, the same factory that appears to its managers as an economic entity may appear to its neighbors as a source of pollution. Both managers and neighbors belong to the factory network, but their different relations to it are manifested in different programs, for example, a business plan and a lawsuit.

Critical theory of technology argues for a discriminating application of the two symmetry principles of social constructivism and ANT and rejects empiricism and methodological individualism. This does not imply a return to pre-constructivist realism and humanism, but it does open a bridge to the recovery of key insights of the tradition of social thought, insights that help to understand the tensions between individuals and a rationalized society.

As STS has evolved over the years it has engaged increasingly with politically sensitive issues. The problems with symmetry are rarely addressed directly. But the austere exigencies of the early methodological struggles are left behind and the useful methods and insights applied in combination with ideas drawn from many fields. The result is promising. STS has a great deal to contribute to an understanding of contemporary politics. Critical theory of technology is an attempt to provide one possible theoretical framework for such methodological *bricolage*.

Technical Citizenship

Critical theory agrees with ANT that individuality cannot be conceived in independence of other people and things. It is nonsense to speak of pure consciousness in abstraction from all material support. The individual emerges from the "network" constituted by the family and its material and cultural environment and is always thereafter conditioned by its roles in the networks to which it belongs. But once constituted the individual retains its identity as it switches from network to network. It cannot be dissolved into its roles. The relative stability of individuality is the basis of the reflective capacities that enable it to distance itself from and criticize the networks in which it participates.

Critical theory of technology is concerned with the threat to such critical potential posed by the pretensions of the technocracy. Reflective rationality is disarmed by the claims of technical rationality. In the tradition of the Frankfurt School individuality is regarded as an historical achievement. Bourgeois culture generalized the capacity for independent thought to an unprecedented degree. This is the basis of personal and political agency, the power to define one's identity and to further one's interests. In principle, socialism would extend this capacity to every human being, but in the twentieth century it appeared that individuality belonged to an all-too-brief interregnum between societies in which independent thought is overwhelmed, either by custom and religious conformity or by the technocratic legitimation of a mass society.

The Frankfurt School's notion of individuality grounded a critique of the reified rationality regulating more and more of social life in advanced societies. The destruction of the capacity for individuality testified to the rise of technocratic domination. The mass of the population was condemned to passive conformity while a minority preserved its mental independence through theoretical and artistic critique (Marcuse 2001). Critical theory of technology takes up this critical perspective in a different historical period in which resistance to technocratic domination has appeared in new forms.

In recent years we have seen the sphere of public debate and activity expand to take in technological issues that were formerly considered beyond the bounds of discussion. With the expansion of the public sphere, new forms of technical agency have emerged. This has given rise to what David Hess calls "object conflicts," conflicts over how to configure technologies so as to serve various interests and conceptions of the good life (Hess 2007, 80–84). The nature of these conflicts lies at the heart of this chapter. Their proliferation raises new questions about technology and democracy. Have we become technical citizens? More precisely, is there something we could call political agency in the technical sphere? And if there is, what is its relation to technical expertise and traditional political agency?[9]

Agency in the sense in which I use the term is not a free-floating matter of arbitrary preferences but is rooted in the experience made possible by specific social situations. Technical systems evoke what I call participant interests. The systems enroll individuals in networks which associate them in various roles, for example, as users of the technology or workers building it, or even as victims of its unanticipated side effects. Interests flow from these roles where the individuals have the capacity to recognize them.

The drivers of automobiles discover an interest in better roads they would have had no reason to feel before joining the automotive network. Similarly, the victims of pollution discover an interest in clean air that would never have occurred to them had they or their children not suffered from respiratory problems caused by those drivers. Drivers, sufferers, and cars co-produce a network to which all belong and it is this which makes certain interests salient that might otherwise have remained dormant or had no occasion to exist at all.

Once enrolled in a network, individuals not only acquire new interests, in some cases they also acquire a situated knowledge of the network and potential power over its development. This knowledge from below and insider power is different from that of individuals who have no connection to the network. Even without expert knowledge, insiders can identify problems and vulnerabilities. They have a platform for changing the design codes that shape the artifacts incorporated into the network. This is conscious co-production: the reciprocal interactions of members of the network and the codes that define roles and designs.

In critical theory of technology, the actions of citizens involved in conflicts over technology are called "democratic interventions." Most of these are "a posteriori," occurring downstream after the release of technologies into the public domain. There are many contemporary examples such as controversies over pollution or medical treatments, leading to hearings, lawsuits, and boycotts. Such controversies often result in changed regulations and practices. A second mode of intervention, the creative appropriation of technology, involves hacking or the reinvention of devices by their users to meet unanticipated demands. This mode has played an important role in the evolution of the Internet (Abbate 1999). A third mode of intervention can be called "a priori" because it involves action prior to the release of technologies. This mode takes two main forms, public participation in "citizen juries" or "hybrid forums" to evaluate proposed innovations and collaboration in the design process. In these cases, the individuals are solicited to participate by the authorities rather than entering into an a posteriori conflictual dialogue (Callon, Lascoumes, and Barthe 2011; Chilvers and Kearnes 2016; Feenberg 1999; Salter, Burri, and Dumit, chapter 5 this volume).

Differentiation and Translation

The notion of technical citizenship raises questions concerning the role of expertise. Ordinary people intervene in technical decisions on the basis of everyday experience rather than through mastery of a technical discipline. Experts possess such mastery and are qualified to implement technical decisions as most lay people are not. Somehow the claims of experience and those of technical disciplines must be reconciled in the design process. The conundrum only seems insoluble from a narrow and dogmatic perspective. In the real world of technology, a largely unacknowledged dialogue between lay and expert is a normal feature of technical decision making and should be further developed (Collins and Evans 2002). Examples of the challenges of design at the interface between engineers and users can be found in Joyce et al. (chapter 31 this volume) and van den Scott, Sanders, and Puddephatt (chapter 17 this volume).

If we have a different impression, and fear both arrogant expertise and irrational experience, that is a function of a unique historical situation that arose in the nineteenth century. Before that time negotiation between craftsmen and communities under the control of judicial authorities regulated the harmful externalities of production. The case law embodied the accumulated wisdom of experience as it applied to technical activity (Fressoz 2012). In the nineteenth century the path was cleared for rapid technical advance at the expense of workers, communities, and users of technology. Central administrative controls buttressed by expert authority replaced the traditional judicial restraints on technology. This change accompanied a much-enhanced differentiation of society under the impact of industrial capitalism.

The separation of technical work from everyday life is an important aspect of the differentiating process of modernization. Medieval craft guilds were social as well as professional organizations. In addition to regulating prices, training, and quality, they had many other functions. The crafts were not based on specialized technical disciplines in the modern sense but on traditional knowledge of materials and practices, rules of thumb, and what the French call *tours de main*. Their "secrets" needed to be kept secret precisely because they were communicable to experienced consumers. In fact the final stages of production often required consumers to finish the artifact in a process called "breaking in."

Modern technical work depends on specialized scientific knowledge. The language of the technical disciplines can only be understood by initiates, those trained in the profession. The social and religious concerns of the guilds are stripped away along with the independence of the technical worker. Today, most technical work goes on in business enterprises, which significantly changes its character and goals.

The property system on which business is based is also affected by the process of differentiation. Ownership in precapitalist societies involved broad responsibilities. Landowners had political, judicial, and religious functions. Capitalism strips away all these obligations and powers and focuses the owner on making a profit. Other goals such as providing employment and protecting the community are gradually abandoned (Simmel 1978). This new form of property explains the destructive logic of the industrial revolution. Indifference to nature and human beings contributed to the shaping of modern technology. Throughout the development process, scientific and technical knowledge was applied in the pursuit of profit without regard for the social and natural context of enterprise. Narrow specializations and narrow economic goals complemented each other. The resulting simplifications accelerated technical progress but also led to problems we are only beginning to address today.

For generations the victims of progress were too weak, ignorant, or marginalized to protest effectively. But conditions gradually changed, especially after World War II. The side effects of more powerful technologies became visible and provoked a public response. Unions and social movements gained influence and demanded the regulation of industry. As a result, a new stage of "reflexive modernization" engaged a slow corrective process that still continues (Beck 1992).

In this context everyday experience takes on renewed importance. Where formerly cognitive success required breaking all dependence of technical knowledge on everyday experience, Bacon's famous "idols," experience now measures the consequences of technical knowledge and designs (Wynne 2011). Those consequences can no longer be ignored and are traced back to their origin in the blind spots of technical disciplines and the limitations of the business perspective. Users and victims now defend themselves against narrowly conceived technology on the basis of their understanding of their experience.[10] These democratic interventions constitute the social background to the broad success of new interdisciplinary initiatives such as STS that attempt to understand the emerging forms of technical citizenship.

These postwar trends constitute original forms of de-differentiation that are progressive rather than regressive in nature. On the one hand, the technosciences bring science and technology together in powerful combinations, crossing well-established boundaries between the true and the useful. On the other hand, corresponding to the emergence of technoscience and its increasingly dangerous side effects, government regulation crosses the lines between state and economy, forcing capitalist enterprise to work under a widening range of constraints. The new relationship must develop its own institutions for translating social knowledge about technology's harmful effects or overlooked potentialities into technical specifications for better designs. Such translation processes will become routine in the long run as public involvement increases, closing the circle in which technology modifies society while itself being modified by society.

Revising Rationality

The early Frankfurt School addressed a cultural environment characterized by an unprecedented faith in technical rationality. It attributed the decline of agency to the rational culture of modernity. This is not merely a subjective disposition but is reflected in the multiplication of bureaucracies and technologies that effectively organize and control most of social life on the basis of technical disciplines. The knowledge of ordinary people is increasingly devalued and human agency is reduced to technical manipulations of the rational systems.

This is still the situation in which democratic interventions challenge the technocracy today. But the original all-encompassing formulations of the Frankfurt School left no room for the return of agency. To account for struggles over technology, critical theory of technology reconstructs the critique of rationality in a more empirically oriented form.

Where the Frankfurt School proposed a very general critique of "reified" or "instrumental rationality," critical theory of technology looks to a more concrete critique of the bias of social institutions and technologies. Rational culture is shown to depend on the imitation of methods and concepts modeled on mathematics and natural science, generalized as a framework for thought and action in every sphere. For example, market relations rely on quantification in the form of prices. Similarly, bureaucracy subsumes specific cases under precisely formulated rules that resemble laws of nature in their formalism and pretention to universality. Technology is implicated in scientific development. The identification of biases in these domains employs methods explored in STS and yields a cultural and political critique of modern institutions.

The elaboration of this critique poses a challenge because we normally identify bias where prejudices, emotions, and pseudofacts influence judgments that ought to be based on objective standards. I call this "substantive bias" because it rests on a content of belief such as, for example, the idea that some races possess inferior intelligence. The Enlightenment showed us how to criticize this type of bias. The philosophers of that era appealed to rational foundations, facts and theories unbiased by prejudice and on that basis refuted the narrative legitimations of feudal and religious institutions. There is no doubt that Enlightenment critique played and still plays an important role in emancipatory politics. However, it has a significant limitation since it implies the neutrality and universality of institutions that claim a rational foundation. This is the case, for example, with the market, which is justified not by myths, stories, or emotional appeals but by the dry logic of the equivalence of money for goods.

Rationality in this social form did not go unchallenged. Romantic critique implausibly attributed substantive bias to rational systems and thereby denied the rationality of rationality as such. The choice of reason over passion is supposedly biased by a preference for a bourgeois life style or, in some recent formulations, by patriarchal ideology. This stance is associated with artistic and political sub-cultures but has little impact on the organization of modern societies (Löwy and Sayre 2001, 35). The effective critique of a rational system such as markets, technology, and administrative procedure requires a different approach. A subtler analysis must find the bias in the concrete realization of a rational form. Marx initiated this new type of critique with his analysis of the bias of the market. Today STS shows that technically rational design is underdetermined by purely technical considerations and thus is biased under the influence of social criteria (Pinch and Bijker 1987).[11] The materialization of interests and ideologies in technical disciplines and designs I call "formal bias."

Formal bias has political implications. Some benefit more than others from the technologies that surround us. The sidewalk ramp is a case in point (Winner 1989). The ordinary high curb is fine for pedestrians but blocks the free circulation of wheelchairs. Sidewalk ramps were introduced in response to the demands of the disabled. A suppressed interest was incorporated into the system. The outcome is not an unbiased technology but, more precisely, a technology that translates a wider range of interests.

The familiar opposition of irrational society and rational technology invoked by technocratic ideology has no place in this context. The biased design that eventually prevails in the development of each technology is the framework within which that technology is rational and efficient. Efficiency is not an absolute standard since it cannot be calculated in the abstract but only relative to specific contingent demands which bias design. Technology is value-laden like other social realities that frame our everyday existence. But after technologies are well established, their particular bias seems obvious and inevitable. We cease to conceive it as a bias at all and assume that the technology had to be as we find it. Uncovering the implicit bias constitutes the "rational critique of reason" promised by the Frankfurt School (Adorno 1973, 85).

Layers and Codes

The emergence of technical citizenship highlights the inherent contingency and complexity of technical artifacts masked by the coherence of technical explanations. In this context, I suggest that the concept of a palimpsest can serve as a useful analogy. A palimpsest has diverse layers or aspects apparent beneath the surface. Technological design is similar: multiple layers of influence coming from very different regions of society and responding to different, even opposed logics converge on a shared object. Critical theory of technology is a "palimpsestology."

Social history has long treated artifacts as palimpsests. In the course of his comments on the history of money, Marx sketched the rationale for such an approach. He writes that "[t]he concrete is concrete, because it is a combination of many objects with different destinations, i.e., a unity of diverse elements. In our thought, it therefore appears as a process of synthesis, as a result, and not as a starting point, although it is the real starting point and, therefore, also the starting point of observation and conception" (Marx [1857] 1904, 293). Marx rejects an Aristotelian notion of thinghood in which an essence endures through accidental changes. Analysis must identify the ontological differences in the construction and meaning of objects at each stage in their development. This is a de-reifying approach that treats social "things," such as artifacts, institutions, and laws, as assemblages of functional components held together by their social roles. The components disaggregate and recombine as society changes.

In the case of technology and technical systems, these constructions reflect the relative power of the actors engaged with design. The outcome of their struggles and collaborations is a "technical code." The code identifies the larger social meaning of technical choices embodied in the stabilized intersection of social choice with technical specification. Technical codes translate the one into the other through what ANT calls "delegation." So, the social demand for more navigable sidewalks became a speci-fication for construction projects. The rights of the disabled were translated into a specific slope. Taken in isolation the slope appears merely technical, but in its context it has a political significance captured in the code.[12] Such codes are incorporated not only into designs but also penetrate technical disciplines.

Critical theory of technology distinguishes two types of technical codes, the codes of particular artifacts and the codes of whole technical domains. The sidewalk example illustrates the artifact code. Codes relevant to whole technical domains are involved in the definition of progress. The domain code under which industrial progress was pur-sued in the nineteenth century required the replacement of skilled labor by machines. This code is still influential to this day. Where it is contested we see the continuing role of public action in determining the technical future.[13] Domain codes in modern capitalist societies are translated into higher-level meanings, such as ideologies and worldviews. For example, the technocratic concept of efficiency translates particular interests into technical arrangements conducive to the exercise of managerial authority (Alexander 2008). Reification may be considered the ultimate domain code of capital-ism, describing the core principles to which all the lesser domains conform.

Critical theory of technology expresses these complexities through the analysis of design in terms of functional layers.[14] Design is a terrain on which social groups advance their interests. It proceeds through bringing together layers of function cor-responding to the various meanings actors attribute to the artifact. The study of tech-nology must identify the layers and explain their relations. This yields a "concrete" account in Marx's sense. It reveals the co-production of the social groups formed around the technology and the design of the technology that forms them.

Adding layers corresponds to accepting more social inputs. This takes several forms. Often apparently conflicting interests are reconciled to some extent in the final design. The metaphor of the palimpsest is illuminating in such cases since each relevant social group contributes a layer to the final design. Artifacts are not coherent individuals so much as they are concatenations, assemblages of more or less integrated parts. Like a palimpsest, their parts embody levels of meaning that reflect a variety of social and technical influences.

The result may involve trade-offs, compromises resulting in a less than perfect design for all parties. More interesting are those cases in which elegant innovations make it possible to satisfy all the different demands without loss in efficiency. Such innovations are called "concretizations" by Gilbert Simondon (1958, chapter 1). This term is decep-tive since Simondon does not mean to contrast the concrete with the conceptually abstract. His terminology, like Marx's, is loosely Hegelian. He defines concretization as the merging of several functions in a single structure. This can be seen in the Pinch and Bijker (1987) bicycle case where inflatable tires satisfied both the racers' desire for speed and the ordinary users' transportation needs.

This concretizing innovation reconciled all the relevant actors in a single perfected design. Concretizations construct alliances among the actors whose various demands are materialized in a single object. That object operates across the boundaries of different social groups, each interpreting it in accordance with its own conception of its needs, each incorporating it into its own world. Such materialized "boundary objects" are increasingly sought in the struggle between environmentalists and the representatives of industry (Dusyk 2013; Star and Griesemer 1989). Concretizations make it possible for industry to find a new trajectory of development that satisfies a range of demands that were formerly ignored. Concretizing advances refute by example the supposed opposition of facts and values, rational achievements and ideological opposition that justifies the technocracy in its resistance to change. Identifying such advances validates a democratic politics of technology.

An Exemplary Case

In this section I propose to apply the concepts of bias, technical codes, layers, and rationality to an exemplary case. These key elements of critical theory of technology bridge the gap between particular cases and the wider cultural world of modernity. The case I have chosen illustrates a common type of technoscientific controversy in which the same artifact plays very different roles in the different worlds of the actors. In such cases, conflicting interpretations of the artifact resulting from different goals and epistemic tests may eventually lead to design changes or the displacement of one design by another more powerfully supported design. From the standpoint of critical theory of technology, this case illustrates the legitimating role of rational criteria introduced to suppress subordinate actors, and countered by those actors on the basis of their own epistemic resources.

Medicine is an especially rich field for the application of STS. For example, Tiago Moreira poses the problem of the relation between supposedly universal "rational" standards and personal experience through the example of the denial of a medication for treatment of Alzheimer's disease by the British National Health Service (NHS) (Moreira 2012).

The NHS evaluates medications and decides on their cost/benefit ratio in terms of quasi-scientific measures. In the case of Alzheimer's, the evaluation did not include issues of quality of life but was based on measures of cognition and hospitalization.

When an existing treatment was found not to be cost-effective on this basis, it was withdrawn. From the critical theory perspective, this represents a reified basis for decision making, aligning scientific and economic regimes but ignoring other aspects of the network in which the treatment is embedded. The occlusion of those aspects is the formal bias of the reified NHS approach.

The program under which the NHS interprets treatments has curing as its goal. It did not measure or concern itself with the network of palliative care constructed around the Alzheimer's medication but isolated it as a more or less effective technical device. But patients and caregivers were upset by the agency's decision. Even though the medication in question did not do much to slow cognitive decline or prevent many hospitalizations, it did have a significant impact on quality of life. Here we have an alternative program focused on caring, an aspect of the network the scientistic bias of the NHS ignored. A typical clash of interpretations ensued, calling into question the definition of the artifact.

Moreira's article concerns how the victims of the decision made their point through an epistemological shift: telling stories about how the medicine had changed and improved the patients' quality of life and their own experience of caring. These stories evoked emotional responses in the form of anger, disappointment, and depression. They operated as allegories, much like human-interest stories in newspapers. Everyone can identify with the subject of a human-interest story through sharing imaginatively the affect it evokes. Similarly, the stories told by the patients' caregivers solicited identification and formed community on a different basis from the quasi-scientific "rational" standards applied by the NHS, with different results for the definition of the medication and indeed of the medical system itself. Eventually the NHS agreed to supply the medication to some patients at a certain stage in the progress of the disease.

In this example, the actors' programs highlight two layers of the medical network constructed around the medication, a curing and a caring layer. The chemistry of the medication was not changed by the caregivers' intervention, but its meaning and usage *qua* medication was decided by the controversy. Meaning and usage, too, are aspects of the reality of technical artifacts. Only a partial alignment between the layers was achieved by the clumsy compromise that settled the controversy.

This example shows the role of democratic interventions in resisting the imposition of a biased rationality representing the interests of dominant actors. Naturally, the effectiveness of the caregivers' campaign was due in part to the existence of widespread concern about both the NHS and Alzheimer's disease. This is all about politics, but politics in one of several unfamiliar domains in which we must get accustomed to seeing ever more frequent public challenges to the dominant methods and decisions.

Technocratic technical codes are called into question in struggles such as this. As "rational," technology takes on an apparent inevitability. It is assumed that devices and systems do what they do because of what they "are." This is the dangerous tautology of the illusion of technology. To create a place for agency, technical citizens must struggle to overcome this illusion and restore consciousness of the contingency of the technical domain. The very definition of progress is at stake in this struggle.

Democratic interventions are translated into new regulations, new designs, even in some cases the abandonment of technologies. They give rise to new technical codes both for particular types of artifacts and for whole technological domains, as in the case of energy production and computing. This is a special and irreplaceable form of activism in a technological society. It limits the autonomy of experts and forces them to redesign the worlds they create to represent a wider range of interests. Insofar as STS contributes to the understanding of these movements, it plays a progressive political role.

Methodological Suggestions

I call this concluding section "suggestions" because that is what a formal methodology can provide. In the end there is no substitute for insight, which cannot be formalized. But methods do suggest perspectives on cases and that can be helpful in pursuing and organizing research. Here then are a few methodological suggestions that follow from the approach of critical theory of technology.

A Dialectical Approach

Political and economic power are legitimated in modern societies by reference to rational criteria such as efficiency, and powerful institutions rely on technical disciplines to satisfy those criteria in practice. This sets the stage for the confrontation of expert and lay actors in many domains, as in the medical example described above. These confrontations are not challenges to rationality as such, but to the particular form it takes in specific cases. Critical theory of technology focuses on such conflicts and argues that they be understood as a unifying theme in the study of the role of technology in modern society.

The different concerns of expert and lay actors engaged with a technical artifact or system often place conflicting demands on design. This is illustrated in the example, where an object conflict arose over a medication. Such conflicts reflect conflicting programs which carve out different subsystems from the total network. In the example, that network extended to pharmaceutical companies, the NHS, hospitals, doctors, patients, caregivers, and all the artifacts they employ, including the medication in question. The curing and caring programs foregrounded different aspects of the network, the one emphasizing economic costs of treatment, the other the quality of life of patients and their caregivers. Similar conflicts are found throughout modern life, in relation to environmental issues, transportation, urban design, worker health and safety, food safety, and many other issues. Research can be structured around the unfolding of such conflicts.

Symmetry and Asymmetry

Any method which fails to recognize the widespread existence of deception and corruption is fatally naïve. (Volkswagen has a car for those who dismiss the critique of hidden motives as outdated.) STS must be able to distinguish these cases from authentic disagreements and employ appropriate methods to study each. Asymmetrical methods such as old-fashioned muckraking and ideology critique are useful in cases such as the financing of climate change denial by the energy industry and research on the harms of tobacco by cigarette companies. There is no symmetry between the painstaking work of real science and the manufacture of propaganda.

Symmetrical methods are appropriate for other cases where the actors are engaged in a real controversy. In such cases actual knowledge is invoked with reason and conviction on all sides, other motives notwithstanding. Claims are not offered simply to create artificial controversy or to hide costly failures from public view. The question is how to understand these conflicts of rationality. Maintaining symmetry is problematic where one side in the argument can stand on material achievements while the other is able to mobilize only words. This is typically the case where dominant actors meet public protests. This is where the symmetry of program and antiprogram plays an essential methodological role. It overcomes the apparent asymmetry between dominant actors with their established technologies and subordinate challengers with only a discursively formulated antiprogram.

Double Aspects of Rationality

Critical theory of technology argues that modernity is characterized by a dominant rational culture. Rationality in this sense is not universal but is context-bound like other aspects of culture. Critical theory of technology claims further that the rationality of actors' positions has both strictly technical and intrinsic normative aspects. It is not necessary to invoke extrinsic values to get at the normative aspect because it is implied in the nature of the technology as understood by the actors. Both sides of the argument between the actors rest on rational principles of technical construction, which are also the basis of normative claims. The double aspects of technical rationality thus transcend the supposed gap between "ought" and "is." Furthermore, technical rationality in different forms is equally available to expert and lay actors. Basic technical categories such as "efficiency" and "compatible" are refined versions of categories of everyday experience and so communicate across the boundary between expert and lay. The methodological suggestion that follows from this approach is "Follow the actors' *reasons*" (Bensaude-Vincent 2013). I will again use the medical example to illustrate this approach.

Technologies depend most fundamentally on an act of abstraction in which the useful aspects of a natural entity are isolated and privileged for incorporation into a device. Abstraction is a rational procedure. It supplies technical elements that combined constitute a thing with a purpose.

The chemists who devised the Alzheimer's medication were not simply abstracting useful aspects from nature; they were also creating a meaningful object with a purpose from which obligations would flow. Those obligations were interpreted differently by the actors because they understood the product differently. For the NHS, the medication had to "work" by curing or at least slowing the progress of the disease. For the caregivers, "working" had a different meaning: the medication had to relieve the burden of care. These normative aspects of the relation to the medication flow directly from its nature as interpreted by the actors.

A second example can further clarify the double aspects of rationality. Consistency is another basic rational principle invoked and employed in the production and evaluation of technologies. A technical device must be consistent with its material environment to function. This is a matter of adaptation: voltage requirements of a hair dryer must be compatible with the voltage delivered by the electric outlet, and so on. But the device must also be consistent with a social environment of meanings and values. The demand for consistency at this level is just as essential as material adaptation. In the medical example, a medication consistent with a neoliberal economic environment is maladapted to the ethical environment of family caregivers and their sympathizers and vice versa.

Note that these examples show the role of rationality not only in the dominant program but also in the antiprogram of the lay advocates of an alternative understanding of the technology. Of course there is an important difference between program and antiprogram: experts must translate the lay position into technically rational specifications for it to achieve success on a par with the already realized program of the dominant actors. In the case in point, physicians would understand the function of the medication differently in the two opposing cases. While this does not change its chemical composition, it does define it differently within the medical system. In other cases the actual design of the technology might have to be changed to adapt it to lay demands. Interpretative flexibility understood as a conflict of rationalities reflects the claims of differently situated actors.

Layers

The multiple demands on design are reflected in discursive forms, practices, and specifications. Technologies can thus be analyzed as layered phenomena, reaching from the heights of full-blown ideology down to the details of technical design. At each level, further layers appear, reflecting different degrees of abstraction. For example, the neoliberal ideology of the British state inspired the quantitative evaluation methods of the NHS, which in turn reflected the institutional preference for curing typical of modern medical practice. The program of the NHS achieved congruence at all levels except the crucial one, the level of the medication itself, which did not perform effectively on the terms of the NHS. The caregivers' antiprogram introduced a different epistemology, one appropriate to an ethic of compassion and oriented toward the practices of caring for the chronically ill. An appeal to empathy through personal case narratives corresponded in their program to the quantitative methods of the NHS. On these terms, the medication performed effectively.

The Internet illustrates layering at the level of design. Technical features of the Internet serve a wide variety of functions, reflecting the demands of different actors. For example, a feature such as anonymous online presence is rooted in the technical protocols of the Internet. The Internet has no gatekeeper, such as a telephone company, that can demand real name identities. Anonymity, while imperfect, serves political dissenters, individuals pursuing online partners, criminals engaged in buying and selling drugs, and those engaged in illegal downloading of music and films, among many others.

The different functions that cluster around this feature create tensions over the most appropriate design of the Internet. Music and film companies find it difficult to defend their intellectual property on the Internet. They would like to make it easier to identify users who appropriate their contents without payment. This would require changes in the Internet protocol or the regulation of Internet service providers, and it would have an impact on all other uses of anonymity. At the ideological level, advocates of freedom of speech oppose advocates of the free market in the contest for control of this aspect of design. Case studies of issues such as this can be organized from top to bottom, from ideological formulations of the desiderata to their realization in usages, technical features, and functions.[15]

Many more methodological suggestions flow from the focus on the role of biased realizations of rationality in modern culture. Critical theory of technology examines these realizations in particular cases in the context of a theory of modernity. In so doing, it aims to create a bridge between the two "layers" I have been discussing throughout this chapter, empirical research and general social theory.

Notes

1. In the mid-1980s, when I first worked on the theme of this chapter, the phrase "Critical Theory" was associated with the critique of positivist and technocratic ideology in Marcuse, Habermas, and other members of the Frankfurt School. Today the phrase has no very specific referent, unless capitalized, in which case it still refers to the Frankfurt School. In miniscule it is loosely associated with the critique of these same ideologies and might refer to work derived from the writings of Deleuze and Foucault among others.

2. This text is a massive collection of articles and graphics reconceptualizing the political in the light of STS. My contribution to this effort is found on pages 976–77.

3. The first edition of *One-Dimensional Man* sold 300,000 copies (Aronson 2014).

4. For the early debate, see Pickering (1992, chapters 10–12).

5. For a brief, clear illustration of the significance of the symmetry principle in this case, see Mauskopf (2006, 76).

6. Feminists within STS were among the first to see the problem with symmetry. See Wajcman (2004, 126). See also Michaels (2008) and Oreskes and Conway (2010).

7. I engaged in a debate over this claim with Kochan (2006), Feenberg (2006), and Collins and Pinch (2007).

8. Is social constructivism better equipped than ANT to deal with such cases? Indeed, it would seem so since it emphasizes the role of interests and resources in decision making. But in this case, symmetry in the strong sense in which it was initially proposed might play an ambiguous role, equating such "resources" as real scientific knowledge with the well-rewarded pseudoscience of climate change denial. Or, more sensibly, symmetry would play no role at all since there are no serious rational resources on one side of the argument, just propaganda. In that case, it is difficult to see how the constructivist analysis would differ from a conventional political analysis.

9. Because the concept of agency has been applied to things under the influence of ANT, a preliminary clarification of my use of the term is necessary. I do not intend by "agency" any and every activity, whether of persons or things, that has an impact on a network. I will stick to the usual meaning of political agency as the ability to perform intentional acts of public consequence.

10. This new public involvement is not an unmixed blessing. The public makes mistakes too, for example, in the case of the rejection of vaccinations for childhood diseases. But every advance of

democracy grants new powers to the "unqualified." Only after the individuals have acquired citizenship are they in a position to engage the learning process that qualifies them to exercise it.

11. This is a point also made in the field of design. See Friedman, Kahn Jr., and Borning (2008) and Friedman (1996).

12. I have developed what I call the "instrumentalization theory" to explain these "double aspects" of technology. See Grimes and Feenberg (2013) for a brief account and examples.

13. For a contemporary example drawn from the field of education, see Hamilton and Feenberg (2012).

14. Two studies exemplifying this method are Feenberg (2010, chapter 6) and Cressman (2016).

15. For an example from my work, see Feenberg (2010, chapter 5).

References

Abbate, Janet. 1999. *Inventing the Internet*. Cambridge, MA: MIT Press.

Adorno, Theodor. 1973. *Negative Dialectics*, translated by Ernst Basch Ashton. New York: Seabury.

___. 2000. *Introduction to Sociology*, translated by Edmund Jephcott. Cambridge: Polity Press.

Alexander, Jennifer. 2008. *The Mantra of Efficiency: From Waterwheel to Social Control*. Baltimore: Johns Hopkins University Press.

Aronson, Ronald. 2014. "Marcuse Today." Retrieved January 27, 2016, http://www.bostonreview .net/books-ideas/ronald-aronson-herbert-marcuse-one-dimensional-man-today.

Beck, Ulrich. 1992. *Risk Society*, translated by Mark Ritter. London: Sage.

Bensaude-Vincent, Bernadette. 2013. *L'opinion publique et la science*. Paris: La Découverte.

Bijker, Wiebe E. 1995. *Of Bicycles, Bakelites, and Bulbs: Toward a Theory of Sociotechnical Change*. Cambridge, MA: MIT Press.

Braverman, Harry. 1974. *Labor and Monopoly Capital*. New York: Monthly Review.

Brown, Mark. 2015. "Politicizing Science: Conceptions of Politics in Science and Technology Studies." *Social Studies of Science* 45 (1): 3–30.

Callon, Michel, Pierre Lascoumes, and Yannick Barthe. 2011. *Acting in an Uncertain World*, translated by Graham Burchell. Cambridge, MA: MIT Press.

Chilvers, Jason, and Matthew Kearnes, eds. 2016. *Remaking Participation: Science, Environment and Emergent Publics*. London: Routledge.

Collins, Harry, and Robert Evans. 2002. "The Third Wave of Science Studies: Studies of Expertise and Experience." *Social Studies of Science* 32 (2): 235–96.

Collins, Harry, and Trevor Pinch. 1998. *The Golem at Large: What You Should Know about Technology*. Cambridge: Cambridge University Press.

___. 2007. "Who Is to Blame for the *Challenger* Explosion?" *Studies in History and Philosophy of Science Part A* 38 (1): 254–55.

Cressman, Daryl. 2016. *Building Musical Culture in Nineteenth-Century Amsterdam: The Concertgebouw*. Amsterdam: Amsterdam University Press.

Dagnino, Renato. 2008. *Neutralidade da Ciência e Determinismo Tecnológico*. Campinas: Editora Unicamp.

Dusyk, Nicole. 2013. *The Transformative Potential of Participatory Politics: Energy Planning and Emergent Sustainability in British Columbia, Canada*. Ph.D. dissertation, University of British Columbia.

Feenberg, Andrew. 1991. *Critical Theory of Technology*. Oxford: Oxford University Press.

___. 1999. *Questioning Technology*. New York: Routledge.

___. 2006. "Symmetry, Asymmetry and the Real Possibility of Radical Change: Reply to Kochan." *Studies in the History and Philosophy of Science* 37 (4): 721–27.

___. 2010. *Between Reason and Experience: Essays in Technology and Modernity*. Cambridge, MA: MIT Press.

___. 2014. *The Philosophy of Praxis*. London: Verso.

Fressoz, Jean-Baptiste. 2012. *L'Apocalypse Joyeuse: Une Histoire du Risque Technologique*. Paris: Le Seuil.

Friedman, Batya. 1996. "Value-Sensitive Design." Interactions, November-December. Accessed at https://cseweb.ucsd.edu/~goguen/courses/271/friedman96.pdf.

Friedman, Batya, Peter H. Kahn Jr., and Alan Borning. 2008. "Value Sensitive Design and Information Systems." In *The Handbook of Information and Computer Ethics*, edited by Kenneth Einar Himma and Herman T. Tavani, 69–101. New York: John Wiley & Sons.

Grimes, Sara, and Andrew Feenberg. 2013. "Critical Theory of Technology." In *The Sage Handbook of Digital Technology Research*, edited by Sara Price, Carey Jewitt, and Barry Brown, 121–29. London: Sage.

Hamilton, Edward, and Andrew Feenberg. 2012. "Alternative Rationalisations and Ambivalent Futures: A Critical History of Online Education." In *(Re)Inventing the Internet*, edited by Andrew Feenberg and Norm Friesen, 43–70. Rotterdam: Sense Publishers.

Hess, David. 2007. *Alternative Pathways in Science and Industry: Activism, Innovation and the Environment in an Era of Globalization*. Cambridge, MA: MIT Press.

Horkheimer, Max. 1995. "On the Problem of Truth." In *Between Philosophy and Social Science: Selected Early Writings*, translated by G. Frederick Hunter, Matthew S. Kramer, and John Torpey, 177–215. Cambridge, MA: MIT Press.

Jasanoff, Sheila. 2004. "The Idiom of Co-production." In *States of Knowledge: The Co-production of Science and Social Order*, edited by Sheila Jasanoff, 1–12. London: Routledge.

Kochan, Jeff. 2006. "Feenberg and STS: Counter-reflections on Bridging the Gap." *Studies in History and Philosophy of Science Part A* 37 (4): 702–20.

Kreimer, Pablo, Hernán Thomas, Patricia Rossini, and Alberto Lalouf, eds. 2004. *Producción y Uso Social e Conocimientos: Estudios Sociales de la Ciencia y la Tecnología en América Latina*. Quilmès: UNQ, Bernal.

Latour, Bruno. 1992. "Where Are the Missing Masses? The Sociology of a Few Mundane Artifacts." In *Shaping Technology/Building Society: Studies in Sociotechnical Change*, edited by Wiebe E. Bijker and John Law, 225–58. Cambridge, MA: MIT Press.

___. 1993. *We Have Never Been Modern*, translated by Catherine Porter. Cambridge, MA: Harvard University Press.

___. 2013. *An Inquiry into the Modes of Existence*, translated by Catherine Porter. Cambridge, MA: Harvard University Press.

Latour, Bruno, and Peter Weibel, eds. 2005. *Making Things Public: Atmospheres of Democracy*. Cambridge, MA: MIT Press.

Löwy, Michael, and Robert Sayre. 2001. *Romanticism against the Tide of Modernity*. Durham, NC: Duke University Press.

Lukács, Georg. 1971. *History and Class Consciousness*, translated by Rodney Livingstone. Cambridge, MA: MIT Press.

Marcuse, Herbert. 1964. *One-Dimensional Man*. Boston: Beacon Press.

___. 2001. "The Individual in the Great Society." In *Towards a Critical Theory of Society*, edited by Douglas Kellner. New York: Routledge.

Marx, Karl. [1857] 1904. *A Contribution to the Critique of Political Economy*, translated by Nahum Isaac Stone. Chicago: Charles H. Kerr.

Mauskopf, Seymour. 2006. "A Tale of Two Chemists." *American Scientist* 94 (1), accessed at https://www.americanscientist.org/bookshelf/bookshelf.aspx?name=a-tale-of-two-chemists&content=true.

Michaels, David. 2008. *Doubt Is Their Product: How Industry's Assault on Science Threatens Your Health*. Oxford: Oxford University Press.

Moreira, Tiago. 2012. "Health Care Standards and the Politics of Singularities: Shifting in and out of Context." *Science, Technology, & Human Values* 37 (4): 307–31.

Noble, David. 1977. *America by Design: Science, Technology, and the Rise of Corporate Capitalism*. New York: Knopf.

Oreskes, Naomi, and Erik M. Conway. 2010. *Merchants of Doubt: How a Handful of Scientists Obscured the Truth on Issues from Tobacco Smoke to Global Warming*. New York: Bloomsbury Press.

Oudshoorn, Nelly, and Trevor Pinch, eds. 2003. *How Users Matter: The Co-Construction of Users and Technology*. Cambridge, MA: MIT Press.

Pickering, Andrew, ed. 1992. *Science as Practice and Culture*. Chicago: University of Chicago Press.

Pinch, Trevor, and Wiebe E. Bijker. 1987. "The Social Construction of Facts and Artefacts." In *The Social Construction of Technological Systems*, edited by Wiebe E. Bijker, Thomas Hughes, and Trevor Pinch, 17–50. Cambridge, MA: MIT Press.

Radder, Hans. 1996. *In and about the World: Philosophical Studies of Science and Technology*. Albany: State University of New York Press.

Rajão, Raoni, Ricardo B. Duque, and Rahul De', eds. 2014. "Voices from within and outside the South—Defying STS Epistemologies: Boundaries, and Theories." Special issue of *Science, Technology, & Human Values* 39 (6): 767–72.

Rich, Frank. 2007. "The Billionaires Bankrolling the Tea Party." *New York Times*, August 28.

Simmel, Georg. 1978. *The Philosophy of Money*, translated by Tom Bottomore and David Frisby. Boston: Routledge & Kegan Paul.

Simondon, Gilbert. 1958. *Du Mode d'Existence des Objets Techniques*. Paris: Aubier.

Soneryd, Linda. 2016. "Technologies of Participation and the Making of Technologized Futures." In *Remaking Participation: Science, Environment and Emergent Publics*, edited by Jason Chilvers and Matthew Kearnes, 144–61. London: Routledge.

Star, Susan Leigh, and James Griesemer. 1989. "Institutional Ecology, 'Translations' and Boundary Objects: Amateurs and Professionals in Berkeley's Museum of Vertebrate Zoology, 1907–1939." *Social Studies of Science* 19 (3): 387–420.

Wajcman, Judy. 2004. *Technofeminism*. Cambridge: Polity Press.

Winner, Langdon. 1989. *The Whale and the Reactor: A Search for Limits in an Age of High Technology*. Chicago: University of Chicago Press.

___. 1993. "Upon Opening the Black Box and Finding It Empty: Social Constructivism and the Philosophy of Technology." *Science, Technology, & Human Values* 18 (3): 365–68.

Wynne, Brian. 2011. *Rationality and Ritual: Participation and Exclusion in Nuclear Decision-Making*. London: Earthscan.

23 STS for Development

Aalok Khandekar, Koen Beumer, Annapurna Mamidipudi, Pankaj Sekhsaria, and
Wiebe E. Bijker

Introduction: Development in Technological Cultures

We live in technological cultures. In saying that, we want to suggest that we live in
societies that are thoroughly constituted by science and technology, shaping not only
their social structures and stratifications but also their political processes, everyday
interactions, and dominant values (Hommels, Mesman, and Bijker 2014). Disaster,
innovation, war, solidarity, poverty, sustainability—whichever manifestation of the
entanglements of science, technology, and society we analyze, the global interconnectedness of technological cultures intensifies and shapes these processes. No place or
event remains isolated in time or space.

The statement that globalization plays out differently now because of technologically mediated interconnectedness is partly a claim about the world in which we live
and partly a heuristic advice about studying it. The world today is radically different
from fifty years ago: Climate change, international migration, and the rise of new economies with their associated changes in international relations—all these emergences
index a newer global order that is fundamentally constituted by the particular forms
of today's technosciences. Science and technology studies (STS) scholarship, with its
rich set of tools to analyze science-technology-society interactions, is thus particularly
well positioned to help make sense of such transformations. In this essay, we extend
this observation specifically to the realms of development: Our overarching argument
is that an engagement between the domains of STS and development scholarship and
practice can be mutually beneficial.

By "development," we refer broadly to the ensemble of actors, ideas, and institutions
through which aspirations for societal change are articulated and enacted. Historically,
as we outline in the following section, the idea of development, and its institutionalization through various United Nations and other international organizations, evolved in
the early post–World War II period. Development then often referred to bringing about

social transformation and material prosperity in newly decolonized nations, often located in those parts of the world that are today broadly identified as comprising the global south. We explore the connections between development and colonialism and of both to technoscience in the following sections.

But, as we argue in this essay, both the empirical and conceptual underpinnings of development today are shifting. Taking seriously the densely interconnected nature of contemporary technological cultures has implied that development itself be understood differently: as a set of shared concerns, opportunities, and responsibilities across countries in the global north as well as the global south, rather than merely as a concern with social transformation and poverty alleviation in formerly colonized nations. At the same time, practitioners themselves now also acknowledge the key role of technoscience in development and are factoring this recognition into their policies and practices. STS scholars, too, are beginning to advocate for taking on pressing societal challenges more directly, be they in the form of the EU's "grand societal challenges" or the UN's "sustainable development goals" (Bijker 2003; Collins and Evans 2002; Fortun 2012; Hackett 2008). Previous editions of the STS handbook have featured essays on the connections between STS and development (Cozzens et al. 2008; Shrum and Shenhav 1995). Building on those analyses, in this essay, we contend that at this historical moment, newer vocabularies and conceptual frameworks are needed to more productively connect STS to development scholarship and practice.

We make our argument in the following manner: In the next section, we offer an historical overview of the shifts in development thinking and practice vis-à-vis the role of technoscience in development. We then offer an overview of literature at the intersection of STS and development studies. Here, drawing largely on the tradition of postcolonial studies, we identify three strands of scholarship focused broadly on the role of technoscience in (neo)colonial expansion, in postcolonial nation building, and on the interplay between traditional and scientific knowledges. Next, building on this literature review, we explore possible engagements between STS and development by addressing issues of governance, vulnerability, and practice. The aim of this section is to demonstrate, with a broad array of examples, different ways that STS provides fresh insights into development.

Last, to conclude the essay, we suggest that engaging with questions of development can benefit STS scholarship in crucial ways—by expanding not only the conceptual and methodological toolkits available to STS scholars but also the very modes of intellectual production and political engagement that have characterized the field thus far.

Historicizing the Role of Technoscience in Development

Most histories trace the origins of modern-day development theory and practice to the early post–World War II years. Political realignments in the postwar period, Escobar (1995) argues, shaped institutions and practices of development as we know them today. The period from 1945 to 1955, which was also the prime period of decolonization, witnessed dramatic shifts in the organization of world power with the United States manifesting itself as the dominant economic and political powerhouse. The consolidation of U.S. hegemony in the world capitalist system, in turn, expanded global markets for U.S. products, launched a search for newer sites for investing surplus capital, and relied on access to cheap raw materials to sustain industrial growth. At the same time, the emerging Cold War and its tripartite division of the globe into the first, second, and third worlds—comprising rich industrialized, communist industrialized, and poor nonindustrialized nations, respectively—meant that expanding their respective spheres of cultural and political influence became a primary goal for both capitalist and communist nations. Development aid to third-world countries, channeled through institutions such as the International Bank for Reconstruction and Development (World Bank) and the International Monetary Fund (IMF), became an important mechanism in the pursuit of geopolitics. However, as we note in the following section, development wasn't merely an imposition of Western will on its erstwhile colonies either: it was often actively desired by many in the third world. Through development, poorer countries sought to liberate themselves from the painful legacies of protracted colonial rule and, in so doing, to become fully modern on an equal footing with Western societies. And, as we discuss at length below, development inevitably encountered resistance and subversion that transformed programs toward greater accommodation of local sensibilities, imaginaries, and ends.

Science and technology were at the very core of the development project from the outset. Systematic knowledge about the third world, produced through sciences such as economics and demography, was crucial to get a scientifically accurate picture of a country's social and economic problems and resources (Ehrlich 1971; Halfon 2007; Rostow 1960). Rapid technological advancement—through aggressive industrialization, agricultural modernization, and infrastructural expansion, for example—became the very cornerstone of development and hence also the primary index of its success (Adas 1989). Hence, large-scale technological projects such as the commissioning of big dams and heavy industries and interventions such as the mechanized and chemical and water-intensive technologies of the agricultural green revolution came to be prominently championed by development agencies such as the World Bank. Providing technical assistance and technology transfer to replicate the successes of the West under the watchful eyes of its experts, through bodies such as the United Nations Development Programme (UNDP) and other UN agencies, became a central *modus operandi* of development interventions during this period (Cherlet 2014).

Beginning in the 1960s and 1970s, however, it became increasingly apparent that approaches to development based around technology transfer and technical assistance were proving counterproductive and perpetuated dependencies among third-world countries on their counterparts in the developed world. Oppositional movements championing "appropriate technology" also prominently came to the fore during this period (Schumacher 1973), gathering momentum especially from various environmental and civil rights movements. As more third-world countries sought equitable access to science and technology rather than technology transfer (Shah 2009), dominant approaches to development also shifted their emphasis toward fostering local capacities and innovation systems, thus drawing attention also to the societal contexts within which technological advancement was pursued (Cherlet 2014). The neoliberal turn in global capitalism from the 1980s onwards further shifted the nature of the developmental project, making science and technology into commodities to be traded just like any other. Concomitantly, the locus of development also shifted from the state to markets and individuals. While formerly the state bore the responsibility for the welfare of its citizens, development now was to be pursued, on the one hand, through markets regulated via international trade treaties (e.g., General Agreement on Tariffs and Trade [GATT]) and institutions (e.g., World Trade Organization [WTO]), and on the other hand, by empowered individuals who were now expected to prioritize their own economic betterment (Sharma 2008).

Recent trends in development hence display greater corporatization overall: ideas such as corporate social responsibility (CSR), for instance, have acquired greater prominence in the realms of development along with concomitant efforts to convert the poor into consumers (Prahalad 2005). Increased corporatization has also resulted in greater attention to the formulation of voluntary and involuntary compliance standards for ethical (e.g., fair trade) and sustainable (e.g., organic) consumption, a topic of significant import within STS scholarship (Höhler and Ziegler 2010; Loconto 2014). Two additional dynamics stand to influence the shape of development in the years to come. First is the set of U.N. Sustainable Development Goals—which succeeded the Millennium Development Goals in September 2015 (United Nations Development Programme 2015)—that identify key developmental priorities agreed upon by U.N. member states for targeted intervention. These development goals codify a sensibility whereby various nations recognize shared stakes and responsibilities, albeit in a differentiated manner, toward ensuring our planetary future (Marnie, Millennium Development Goals Task Force, and United Nations Development Group 2013; United Nations Development Group 2003, 2014) in an inclusive and equitable fashion (Vaughan 2015). Second, the advent of newer donor nations such as China, India, and South Korea and institutions such as the New Development Bank, financed by leading developing countries, significantly alter the dynamics of development aid since these countries no longer share a formerly colonizer-colonized relationship with each other (Mosse 2013). Development projects sponsored by these countries, consequently, are not driven by long-standing animosities or claims to entitlement but rather primarily by economic and geopolitical considerations.

We draw three provisional conclusions from this brief overview. First, development has been a dynamic and evolving assemblage of actors, ideas, and institutions since its earliest days. While one set of understandings of development tends to highlight it as yet another form of oppression of formerly colonized nations by their erstwhile colonial masters, other interpretations of development emphasize that it is the predominant register on which societies and nations across the globe articulate their aspirations. Second, an emphasis on science and technology as agents of development continues to remain at the core of contemporary imaginaries of development, albeit in ways different from before. Last, it follows then that even as STS scholars continue to pay critical attention to the uneven historical legacies of development, newer idioms and modes of engagement with development will have to be figured out if STS is to contribute toward producing better and fairer developmental outcomes. There is, by now, a rich body of work that can inform a robust STS engagement with development policy and practice; we now turn to outlining the contours of this scholarship.

STS and Development: An Overview

Postcolonial studies of technoscience in STS have prominently focused on the unequal relations of exchange between the West and East that characterize technoscientific knowledge production and distribution (Abraham 2006; Anderson 2002; Harding 1998; McNeil 2005; Seth 2009). Such scholarship, as Anderson (2002, 644) suggests, destabilizes received dichotomies such as "global/local, first world/third world, Western/Indigenous, modern/traditional, developed/underdeveloped, big-science/small-science, nuclear/nonnuclear, and [...] theory/practice," that effectively work to maintain Western dominance. Below, we identify three broad themes within this literature that inform STS engagements with studies and policies of development.

Development as Neocolonialism

An overarching critique of development has focused on its implicit continuation of colonial relations of domination between a developed West and a developing East (broadly comprising countries in Africa, Asia, and Latin America). Dominant theories of development adopted a teleological model that assumed that all societies, irrespective of their histories and cultures, would pass through similar stages of economic growth (Rostow 1960). In this scheme of thinking, the modern West represented the apex of human development; its model of society would eventually be replicated in other parts of the world as well. The East was understood as simply being in an earlier stage of development than the West. Its traditions, modes of social organization, and philosophies were deemed anachronistic and in need of being scrapped in the quest for rapid economic progress (United Nations and Department of Economic Affairs 1951). Consequently, the West was supposedly in a position to provide expertise and resources toward replicating its material prosperity and economic growth in the East as well. Inspired by Foucauldian approaches to discourse analysis, scholars hence have been extremely critical of development as an inherently orientalist discourse that was legitimized by the very same dichotomies that constituted colonial reason (Bhabha 1993; Escobar 1995; Said 1978).

STS scholars following this lead investigate the role of technoscience in furthering (neo)colonial relationships between the West and East. They highlight the role of science and technology in enabling colonialism in the first place, underscoring not only the technological prowess of the West but also its complementary ideology of colonialism as a civilizing mission that materialized through the spread of such gifts as science, technology, and modern statecraft (Adas 1989; Conklin 1997; Seth 2009).

According to Seth (2009), this scholarship emphasizes two key aspects of colonial technoscience. The first is the creation of taxonomic "sciences of empire" under various colonial regimes. Sciences of classification, such as anthropology, botany, and cartography, flourished during the colonial era precisely because of the need to understand colonial subjects better. Indeed, as Seth points out, it is scarcely possible to imagine Darwin's postulation of his theory of evolution without access to the vast variety of flora and fauna gathered from several European empires. The second focus is on the role of technoscience in shaping colonial subjectivities. In this regard, the building of various large-scale sociotechnological infrastructures, such as railways, irrigation networks, and public health systems, served to further colonial expansion by helping gain control over colonial landscapes and populations. Work on colonial medicine in particular has highlighted the co-production of scientific knowledge, subjectivity, and colonial order (Anderson 2006; Arnold 1993; Rogaski 2004). Frantz Fanon's (1963,

1967) work on the French occupation of Algeria is an early inspiration here, leading scholars to label medicine as among the most important "tool[s] of empire" alongside the power of steamships and machine guns (Headrick 1981).

However, scholars are also quick to caution against reductionist readings of science and technology as purely oppressive (Storey 1997, 2008). Scientific knowledge produced under colonial regimes, scholars argue, has had ambiguous legacies (Schiebinger 2005). Conklin (1997), for instance, argues that anthropological knowledge produced under French colonialism has had the contradictory effects of simultaneously enabling modern racism and antiracism. Moon (2007) similarly shows that technical projects in Indonesia, while maintaining the ultimate goal of extracting value for the Dutch empire, were often tailored to fit local social, ecological, and economic imperatives. Thus, Moon (2007, 6) argues, in contrast to the metropolitan high modernism embodied by such projects as the construction of large dams and irrigation networks, colonial modernization in Indonesia proceeded through "small-scale, locally sensitive projects of technological development [that] emerged not despite colonial rule, but because of it." In a similar fashion, Tilley (2010, 2011) also demonstrates that Western scientists did not merely advance imperial agendas but were autonomous agents who sometimes did not deliver what metropolitan governments demanded of them.

Development as Postcolonial Nation Building

Shifting attention from postcolonial relations between the East and West to the newly independent nations themselves, scholars have also traced how technoscience is enrolled in projects of postcolonial nation building. Such nation building was articulated through a fundamental paradox: even as various anticolonial nationalisms contested the sovereign authority of their colonial masters, they nonetheless looked toward the fruits of modern science, technology, and industry to emancipate them from colonial underdevelopment as well as such social dispositions as religion and superstition, which they deemed antithetical to modern nationhood (Abraham 2006; Chatterjee 1993; Prakash 1999). In this scheme of thinking, secular science was to displace prior tradition: large-scale investments in projects such as the building of big dams and industrial manufacturing units hence became ubiquitous in many newly independent nations, with science and technology as key signifiers of social and economic progress (Abraham 2006; Prakash 1999), and, in some instances, the very ends of state power (Abraham 1998; Amir 2012; Phalkey 2013).

In turn, science and technology also became powerful sources for shaping postcolonial subjectivities: Akrich (1992), for instance, demonstrates how electricity networks in the Ivory Coast allowed for the emergence of the idea of national citizens by, among other things, fixing land allocation that was formerly done on a temporary basis by village elders and by allowing the government to collect taxes (Moon 2009). Gupta's (1998) argument about development as a form of identity that teaches its subjects to think of themselves as perpetually underdeveloped similarly underscores the profound ways in which technoscience is enrolled in projects of national development and progress. Moreover, as several scholars point out, these nation-building projects are nonetheless caught up in a web of transnational relations: For example, only in the context of international isolation did nuclear energy and computers become key sites for nation building in apartheid South Africa (Edwards and Hecht 2010; Hecht 2006; see also Gupta 1992).

Technoscience and Traditional Knowledge

Early work in STS highlighted the marginalization of traditional knowledge systems under colonial and later newly independent regimes. More recent work, however, problematizes the categorical opposition between scientific and traditional knowledge. Instead, scholars now emphasize "the reciprocal influences and multidirectional flows of materials, knowledge, and people between multiple centers (some European, some not) and diverse peripheries" (Schiebinger 2005, 53). Prominent here is the work of David Turnbull (2000), who demonstrated the situatedness of all knowledges (Haraway 1988), be they traditional or scientific. In contrast, de Laet and Mol (2000) show how the Zimbabwean bush pump could travel widely precisely because of its "fluidity" in adapting to different social and environmental contexts, exemplifying how traditional knowledge can also move across disparate contexts with relative ease. By demonstrating, on the one hand, so-called local knowledge to be highly mobile, and on the other hand, modern scientific knowledge as less than universal, such scholarship effectively "provincializes Europe" (Chakrabarty 2000) and delocalizes traditional knowledge (Saidi 2016; Sekhsaria 2013; Watson-Verran and Turnbull 1995).

But of course, both scientific and traditional knowledges are produced and travel within already existing fields of power, even as these relations of power are remade during the processes of knowledge production and dissemination (Hayden 2003). A recurring example here is the case of various alternative systems of medicine, which repeatedly have to legitimize themselves within the terms and categories of modern medicine in order to be accepted. Such legitimation work risks reifying the various dichotomies between Western and traditional knowledge but may also sometimes enable local practitioners to gain from the ensuing encounters (Kim 2007). The task for critical scholars, then, is to explicate what kind of work traditional knowledge has to do in order to travel, and to what effect (Hayden 2005).

Directly challenging the spatial correlations of the global north as a center of science and the global south as a site of tradition, recent scholarship also reveals the global south as a space for technoscientific knowledge production in itself, rather than merely a space in which such knowledge is licensed, used, or distributed (Medina, Marques, and Holmes 2014; Pollock 2014). Such a sensibility is also already evident in historical studies of colonial technoscience, highlighting for instance the centrality of tropical island ecologies for modern environmentalism (Grove 1995) or the multidirectional traffic between imperial metropoles and colonies since the earliest days of colonialism (Raj 2007; Storey 1997, 2008).

Scholars interested in the knowledge politics of development highlight the irresolvable paradox that is at the heart of colonial discourse: it simultaneously posits Western science and colonial tradition as fundamentally opposed while also depending on the ability of colonized populations to be reasonable enough so as to acknowledge themselves as less knowledgeable (Prakash 1999). Precisely this paradoxical condition, scholars argue, opens up colonial discourse to resistance and subversion. Such work draws inspiration from Bhabha's (1993) theorization of the failures of colonial "mimicry" (the idea that colonized populations could achieve the successes of their imperial masters by wholly replicating their ideas, institutions, and modes of social organization) and the hybrid subjectivities that result from such failures.

For example, Gupta's (1998, 5) study of the agricultural green revolution in rural India shows farmers strategically drawing on both traditional and modern scientific agricultural knowledge, depending on an array of considerations that include the availability of finances, local understandings of soil health, prevailing conceptions of development and the role of the state, as well as the politics of class, caste, and gender. Eglash et al. (2004, vii) similarly show how users outside centers of scientific power often actively reinvent and appropriate technologies subversively "in ways that embody critique, resistance, or outright revolt." In such accounts, development appears also as a form of identity, whereby its subjects contest their position as lagging behind the West in the "waiting room of history" (Chakrabarty 2000, 8).

In hybridity, scholars sense political opportunity: by making visible the many ways in which modernity and tradition co-constitute each other, scholars lay the ground for a more serious engagement with a variety of nonscientific knowledges as a matter of epistemic (Harding 1998) or cognitive (Visvanathan 2009) justice. Visvanathan (2009), for example, reflects on the tension between Indian national strategies and tribal knowledges from remote regions of the country that effectively serve to challenge the political sovereignty of the Indian state.

These analyses of hybrid knowledge systems for development also foreground the many different brokers and mediators that translate among disparate social domains (Lewis and Mosse 2006) so as to enable a "loose coupling" between very different spaces (donors, aid recipients, international organizations, development agencies, NGOs), such that it becomes possible to maintain "development's contradictory commitment[s] to difference and similarity, progress and emancipation, efficiency and local ownership, by allocating incongruous principles to separated contexts" (Mosse 2013, 233; Rottenburg 2009). From such a perspective, then, development is not merely imposed through the top-down mechanisms of international organizations such as the World Bank but rather is thoroughly indigenized and remade through locally salient categories as it traverses messy social and political fields (Pigg 1992).

Postcolonial studies of technoscience thus serve to problematize the uneven legacies of development projects and practices and the role that science and technology have played in both enabling and contesting these. Such studies also challenge dichotomous distinctions that have characterized dominant articulations of development and instead highlight the historical processes through which such binarisms have been co-produced. We now turn to outlining how such scholarship can contribute to development, in both its conception and practice.

STS for Development: Three Themes

In this section we highlight three themes—governance, vulnerability, and practice—that represent a broad spectrum of engagements between STS and development policy and praxis. This list is by no means exhaustive but documents a wide range of possible interfaces—from micro to macro, from politics to intervention, and from rethinking governance to rethinking sustainability and livelihoods—between these two domains.

Governing Development and Technoscience
Governance has been one important focus for STS scholarship on development. The notion of governance goes beyond state institutions and includes a broad array of formal and informal arrangements that coordinate actions between and among various governmental and nongovernmental entities. Thus, while remaining attentive to policy making and state-based legal and regulatory mechanisms, scholarship on governance explicates the wide range of historically and culturally specific factors that shape how technoscience furthers or inhibits certain developmental goals.

The understanding that technoscience is governed differently in different places is axiomatic for such scholarship: There is, by now, a rich body of work that documents how past experiences, locally specific value systems, and entrenched styles of regulation all inform which technologies are deemed appropriate for particular social contexts and under what conditions (Beumer 2016; Cozzens 2012; Jasanoff 2005). More generally, such work is informed by earlier STS scholarship on topics such as social constructivism, social movements, public participation, and expertise that argues against an understanding that assumes that the public needs greater education in order to more willingly embrace technoscience (Bijker, Bal, and Hendriks 2009; Bijker, Hughes, and Pinch 2012; Bucchi and Neresini 2008; Hess et al. 2008; Marques 2014; Marres 2007; Tironi and Barandiarán 2014). Instead, this scholarship emphasizes the different kinds of knowledge that publics bring to technoscientific decision making and thus the different ways in which publics and technoscientific development co-produce each other (Jasanoff 2005; Wynne 2013).

From this perspective, development is not a straightforward application of technoscience to solve a societal problem. Indeed, critical scholars of development have documented how various interventions have often failed and even exacerbated the very conditions that they were meant to address. Bauchspies (2014), for example, documents the complete disruption of otherwise successful practices of statistical recordkeeping in Togo because the previous practice of recordkeeping with typewriters by secretaries came to be replaced by more prestigious "Western" computers, which were then moved into the director's office. Deploying Western models of technoscientific education without adequate consideration of the particular contexts of implementation, similarly, has had limited success (Leslie and Kargon 2006) and has historically been an important facilitator for the movement of technoscientifically skilled experts from developing to developed countries (Bassett 2009; Khandekar 2013). This can be contrasted to the successful provision of clean water in Guinea through complicated local systems, thereby challenging the common assumption of the superiority of modern technologies over traditional knowledge systems (Bauchspies 2012; Lansing 1991; Pfaffenberger 1992). Similarly, Adams et al. (2007) demonstrate that in the context of global health, informed consent protocols cannot be simply transferred across cultural, national, and ethnic groups. Neither can responses to key challenges in public health, such as HIV/AIDS, be formulated without incorporating how local populations already understand these challenges, informed by locally specific ways of knowing disease and well-being (Pigg and Pike 2001). From the vantage point of such scholarship, even well-intentioned developmental interventions fail because they do not recognize the complex and interconnected nature of contemporary technological cultures: they depend on "rendering technical" an otherwise messy social landscape (Ferguson 1990; Li 2007; Rottenburg 2009), reducing social complexity to a set of rules and procedures that, for the most part, are unable to accommodate historical and cultural specificities of particular locales (Bauchspies 2014; Scott 1998).

Moreover, large-scale technocratic interventions such as the commissioning of large dams necessarily displace local populations in big numbers; even when "successfully" relocated, those displaced find themselves in culturally and socially alien landscapes with often greatly diminished means of livelihood (Roy 1999). Some critical scholars thus see violence, physical and otherwise, as inherent to Western conceptions of modernity (Nandy 1990), going so far as to label development as genocidal and calling for abandoning the notion of development altogether (Visvanathan 1997).

Transnational comparisons have yielded important insights for STS scholarship on governance (Beumer 2016; Jasanoff 2005). As a conceptual and methodological principle, transnationalism moves beyond the nation-state as the default unit of analysis: even when focusing on issues of national governance, transnational studies remain alert to ways in which subnational or international dynamics influence eventual outcomes in important ways (Basch, Schiller, and Szanton Blanc 1994). Applied to governance practices, such scholarship provides empirical grounding to the constructivist principle that realities could have been otherwise. Thus Beumer (2016), for instance, demonstrates that India and South Africa have taken very different approaches toward addressing the risks of nanotechnology. India started from the expectation that explicitly regulating risk would stifle innovation and adopted a wait-and-watch approach until an international consensus on risk governance had fully crystallized. South African governance of nanotechnology, in contrast, builds on the assumption that governing risks is a precondition for reaping the benefits of nanotechnology. South Africa hence proactively contributes to international efforts to conduct risk analyses while focusing on those domains where it expects to benefit most.

Studies of governance in STS thus highlight the different trajectories that technoscientific development can trace within particular historical and cultural contexts. Importantly, remaining cognizant of postcolonial critiques, these studies create a foundation for articulating development agendas that privilege particular societal aspirations instead of received ideals such as technoscientific advancement as progress that have historically characterized development ideology. Important examples of efforts to govern science and technology for development in this manner include the Indian and African manifestos for science and technology in which Indian, African, and European STS scholars have worked with activists and policy makers to formulate agendas for science and technology with Indian and African developmental priorities at their core, rather than following the trends set by Western technoscience (SET-DEV and ATPS 2011; SET-DEV and KICS 2011).

Development and Vulnerability

A recognition of the densely interconnected nature of contemporary technological cultures, powerfully highlighted through such events as the Fukushima nuclear accident or the ongoing and anticipated impacts of global climate change, has brought ideas such as managing vulnerability and enhancing resilience to the center stage of contemporary development agendas. Reducing vulnerability of people, communities, livelihoods, ecologies, economies, and nations has become a key goal for many development endeavors (e.g., United Nations Development Programme 2015).

An analysis of vulnerability and resilience as characteristics of contemporary technological cultures allows for a broadening of the issues beyond hazard and risk. STS researchers have sought to expand our existing vocabulary of risk: By adding the cultural to the societal, *Gemeinschaft* to *Gesellschaft*, solidarity to security, precaution to prevention, and justice to legality, they argue that the emphasis on probabilistic analysis of problems and decision making (which have characterized the classic risk analysis approach), must be complemented with a more open-ended, qualitative, and discourse-based attention to ethics and justice (Hommels, Mesman, and Bijker 2014). Such a broadening of the risk discourse to vulnerability aligns with Jasanoff's (2007) plea for humility in recognizing the limitations of human capacity to harness the world, and as such emphasizes the need for greater attention to the contexts of technoscientific development and dissemination.

Exemplary here is Delgado and Rodríguez-Giralt's (2014, 336) description of deploying "creole seeds" in Brazil as a development strategy. Creole seeds are "developed, adapted or produced by [...] local farmers, landless people and indigenous people" and recognized as such under Brazilian law. In comparison to the strategy of importing standardized, often genetically modified seeds, creole seeds present an alternative pathway to development as "local traditional communities [usually] diversify their seeds, ensuring that they will behave differently, thereby producing a seed stock that is supposed to be resilient to changing local environmental conditions: If one variety fails, another will succeed" (ibid., 336). This approach to "the production of safety (and risk)" (ibid., 336) does not fit the classic risk management style but rather draws on a broader vocabulary of vulnerability, plurality, and adaptability.

Another consequence of the focus on vulnerability is to recognize that in vulnerable technological cultures, constant maintenance and repair, of social as well as technical infrastructures, is a *conditio sine qua non* for all stable societies. Jackson (2014, 226) calls this "broken world thinking": the world is always breaking, but such breaking can be regenerative and productive. This applies to technological cultures in the north as much as in the south and can be empirically substantiated by such varied studies as those focused on Swedish railway maintenance (Sanne 2014), ICT infrastructures in Namibia (Jackson, Pompe, and Krieshok 2012), or Californian repair cafés (Rosner 2014).

A focus on vulnerability thus opens new ways of thinking about the role of technoscience in development. First, it underlines that it is an illusion to think that a society without risk could be possible. Second, it shifts the agenda to pursuing resilience in societies while remaining attentive to the production of risk and unequal risk burden. Third, it underscores the idea that strategies to stimulate societal development in the direction of greater resilience exist in a broad spectrum of possibilities—from social to technical, and from governance to innovation, with an attitude of humility vis-à-vis our technoscientific capabilities.

Development as Practice

Development is not only a question of governance, politics, and policy; it is not only a matter of rethinking societies by finding more just and humble ways of dealing with vulnerabilities. Development is also practice, as manifested, for example, in dealing with immediate and practical problems, negotiating seemingly incompatible positions, and pragmatically accepting "good" when "excellent" is out of reach. For an analysis of development practice, STS offers an understanding of how entire sociotechnical ensembles or networks cohere, rather than the typical focus on decontextualized technical machines, scientific findings, and social institutions (Bijker 1995; Law and Hassard 1999). Analyses of technological cultures that focus on the construction of sociotechnical ensembles or networks and how these in turn shape society, like studies of postcolonial technoscience, help unsettle polarized oppositions between tradition and modernity.

Take, for instance, the case of handloom weaving in southern India (Mamidipudi 2016). Typical development discourses depict traditional craft activities as premodern, unproductive, and unsustainable, even as they continue to be the second most important source of livelihood in rural India. Empirical scrutiny of handloom weaving practices, however, reveals that some weaving communities have had economically sustainable livelihoods (Mamidipudi, Syamasundari, and Bijker 2012) while also being extremely innovative both in technological terms (e.g., advancing new loom technologies) and with regard to other elements of the ensemble (Mamidipudi 2016). Village-level weaving communities, for instance, have selectively espoused digitization as part of their craft: not to automate or mechanize the loom but to help coordinate certain aspects of the larger ensemble, such as estimating labor requirements, pricing cloth, designing cloth patterns, and visualizing such designs prior to installing them on the loom. Modern technologies thus complement traditional ones seamlessly in what becomes a highly innovative ensemble. However, weavers themselves tend to foreground the traditional (rather than innovative) aspects of their practices, not merely as a good marketing strategy but also to preserve sociocultural cohesiveness by articulating stable identities among craft communities. This analysis of development practice not only helps explain the resilience of some handloom communities but can also point to where interventions in the handloom ensemble are most likely to be effective.

These observations hold equally true of highly sophisticated technological systems as well. Sekhsaria (2013, 2016), for example, traces the development of nanotechnology innovations in India, highlighting the persistence of so-called traditional skills, practices, and knowledges even in cutting-edge technoscientific research. In this instance of cutting-edge physics research, however, such informal—but necessary—improvisation is frequently disavowed, with scientists and engineers going to great lengths to underscore the scientifically pure and objective nature of their work. Understood thus, tradition and modernity appear as strategically performative outcomes that help stabilize certain sociotechnical ensembles, rather than as the ideological opposites that the narrative of development-as-modernization presupposes (note here resonances with postcolonial studies; cf. Anderson 2002).

A related observation, then, is that ostensibly traditional communities often exhibit great resourcefulness and a very sophisticated understanding of how their entire sociotechnical ensemble works and can be adapted to changing circumstances. Indeed, as Pfaffenberger (1992, 512) points out, various traditional communities have time and again built highly sophisticated sociotechnical systems that have been all but invisible through the eyes of Western theory or policy. Often these were disrupted through the violent imposition of colonial domination that displaced "indigenous political, legal, and ritual systems" by outstripping the ability of local populations to make sense of and keep up with the societal transformations being induced (Lansing 1991). This observation challenges both proponents and opponents of technology-intensive models of development. Instead, the goal becomes one of creating the conditions under which communities can reflexively formulate actions toward achieving desired ends.

Furthermore, using sociotechnical ensembles as units of analysis in studying technological cultures focuses attention on the effective coordination of heterogeneous social, technical, and natural entities for such ensembles to function. STS scholarship enhances our understanding of how such coordination is achieved in a few different ways. Here, we highlight two important themes: standardization and knowledge brokerage.

Standardization through devices such as forms, censuses, and maps makes society legible and is hence indispensable to the functioning of the modern state (Scott 1998) and the sociopolitical imagination of modern nationhood (Anderson 1991). STS scholars have focused on how systems of classification help order society by "sorting things out" into various categories and how, over time, modern classificatory systems come to stand in for the objective truth (Bowker and Star 2000; Porter 1996). Such rationalization necessarily involves simplifications that do not map onto the realities that they seek to represent in their entirety. They nonetheless serve to orient action and devise collective systems within which such action becomes meaningful. Thus, for example, Loconto (2014) documents ways in which sustainability standards, such as organic and fair trade certifications, serve to orient practices of growing tea in rural Tanzania even as they signify different realities to variously situated actors. To draw on Delgado and Rodríguez-Giralt's (2014, 343) work again, such standards can be conceived of as "transient standardization devices" that are attuned to alternative developmental goals (e.g., novel forms of common property and maintaining diversity) in ways that allow for "trac[ing] back how, where, and by whom [particular objects] were made while they also remain open for transformation."

Scholars further modify this understanding by insisting that technical standardization (as carried out by bodies such as the International Organization for Standardization) is as much social as bureaucratic standardization is technical. Indeed, as Lawrence Busch (2011) argues, standards are "recipes for reality." Focusing on the processes of standardization, we contend, can therefore also provide new insights into development practice.

For example, to return to the case of handloom weaving, producing for emerging green markets at the end of the twentieth century was characterized by a tension: while consumption standards for green markets crystallized through certifications such as organic and fair trade, weavers themselves, for ease of weaving, preferred working with production standards of evenness of yarn (Mamidipudi 2016). This tension was eventually resolved by development practitioners by reference to an entirely different standard: that of the 'count.' The count of the yarn here is a number: this same number communicates staple length (of the cotton fiber) to the farmer, weight to the cotton trader, and thickness of thread to the spinner; it is also used to calculate weight in units of lengths (called hanks). The count indicates labor required of the weaver, color absorption for the dyer, resolution of the motif for the designer, and weight and drape for the consumer. To effect a change to organic cotton, this number has to be negotiated back and forth at every stage, through interactions between cotton trader and dyer, dyer and weaver, weaver and customer, and so on. As the new number travels across each stage of the work and part of the ensemble, the count emerges as a compatibility standard that allows for interoperability. It also allows for entities at different scales, such as individual weavers and large spinning mills, to interact with each other. Stated otherwise, this (new) standard coordinates action across the ensemble thereafter.

It follows then, also, that dissenting entities would find it difficult to function within the production and consumption cycle described by this particular ensemble. In other words, while the count brings together and standardizes innovations across the ensemble, it also serves to marginalize that which cannot be readily articulated within the given assemblage. Power is thus integral to any discussions of standardization: standards render certain qualities more visible and legitimate than others and thus codify certain values and visions for the future. Standards thus enact certain interests while rejecting others. They are, hence, always political and often fiercely contested. In its attention to standards in the making, STS scholarship stands to draw out which interests and futures are taken into account and which discounted; it stands to broaden the array of possibilities that exist within any given sociotechnical ensemble (Lampland and Star 2009; Timmermans and Berg 2003).

Making standards work requires skilled mediation that is capable of translating abstract demands, such as those of sustainability, into locally meaningful idioms. The central role of mediators in enabling development work is already richly documented in the development literature (Lewis and Mosse 2006). Examples we have used thus far also highlight the idea that standards that orient development practice, such as those of fair trade or organic production, often originate in cultural contexts different from where such production actually happens and are hence not self-evident to producers who are supposed to uphold them. The importance of skilled interpreters who can adapt such demands to the specificities of particular contexts is hence readily evident (Kilelu et al. 2011; Klerkx et al. 2012).

But mediation in developmental contexts can take other forms as well. A task of increasing importance, which STS scholars are particularly well-equipped to undertake, is that of "knowledge brokerage": facilitating the movement of knowledge and technology from one place to another, with the aim of strengthening individual and organizational capacity to learn and innovate (Bijker, Caiati, and d'Andrea 2012). Managing everyday lives in contemporary risk societies, as Giddens (1990) argues, necessarily involves creating and trusting abstract systems of expertise. However, such expertise is increasingly fragmented because of the complexity of modern life. For example, in his study of nanotechnology in India, Sekhsaria (2016) documents how even promising innovations such as point-of-use, low-cost water filters failed to take off in the marketplace due to apparent disconnects among technology researchers, developers, entrepreneurs, and end users. STS scholarship, with its rich understanding of how various forms of expertise come together and collaborate (Galison 1997; Traweek 2000), stands to cultivate cross-disciplinary "third spaces" where different forms of knowledge and expertise can interact toward creating more resilient technological cultures (Khandekar 2014).

Conclusion: Development for STS

In this chapter we have argued for the pertinence and timeliness of STS engaging with studies and practices of development. This plea builds on two arguments. The first is about extrapolating the political and intellectual agendas of each, and the second is about substantive conceptual and empirical gains for both. The first argument built on the increasing engagement of STS outside its traditional Euro-American contexts and on the recognition in development circles that science and technology play crucial roles in what we have called technological cultures. The second argument was made by first broadly reviewing literature in postcolonial studies of technoscience and by then elaborating three specific themes of engagement between STS and development.

Specific insights that STS has to offer to development, we suggested, are a new perspective on governance in technological cultures with associated new possibilities for political action, a broadening of the development agenda from risk to vulnerability with associated new entry points for promoting sustainable livelihoods, and finally the foregrounding of sociotechnical ensembles in development practice with associated possibilities for intervention and innovation. Underlying these three contributions are the epistemological and methodological underpinnings of contemporary scholarship shared by both constructivist and postcolonial traditions of STS: to move beyond the standard dichotomies of developed/developing, modern/traditional, global north/ global south, and West/East. This sensibility, in turn, opens up a broad and novel range of strategies for intervention toward effecting social change.

To conclude, we want to reverse the question and ask what such engagements may yield for STS. What lessons does development offer for STS scholarship? Sometimes constructivist approaches in STS have been read as downplaying "real" problems, such as the suffering of vulnerable populations (Winner 1993). In its effort to be symmetrical (also in analyzing the oppressor and the oppressed), constructivist STS is sometimes accused of not sufficiently questioning the power hierarchies that produce inequalities (Radder 1992, 1996). This limited reading has been refuted in scholarly terms by several authors (cf. special issue by Richards and Ashmore 1996), but what about the political and practical implications of such critiques?

Part of the strength of constructivist STS is to resist reifying categories that are unproductive and to use a symmetrical and impartial analysis that opens up new views and ways of intervention. However, sometimes STS researchers may experience a political urgency or ethical responsibility to take a stance, resulting in prioritizing ontology over expertise, justice over democracy, equality over symmetry, power over normativity, and politics over governance. For researchers, such explicitly political positioning and intervention is most effective when drawing on the aforementioned symmetrical analysis. It thus calls for a subtle balance between understanding and intervening, between symmetry and taking sides. In its symmetrical approach to understanding interaction of unequal entities (as is always the case in the contexts of development), STS may use its explanatory power to make explicit narratives that are otherwise not heard. The next step then could be to intervene by taking a political stance by amplifying otherwise marginalized narratives and dissenting with the usually dominant ones. Engagement with development theory and practice provokes STS researchers to move beyond mere observation to actively participate in the construction of new realities. This is certainly not an entirely new argument within STS (Cozzens and Woodhouse 1995; Jasanoff 2003; Martin and Richards 1995; Sismondo 2008), but it gains specificity and acumen in the context of development.

Let us try to explore this specificity. Development, as we have articulated in this chapter, has three strands: the evolving academic history of development studies, development policy and governance, and development practice that involves both powerful elites and vulnerable populations. Engagement with development in these three contexts can push STS to confront certain choices about its research agenda, its theories and methods, and its interventions and politics.

In its research agenda, STS is stimulated to choose sites and case studies that are not located in the geographical or scientific centers of the contemporary world system. Examples from development studies demonstrate that it is intellectually rewarding for scholarship to privilege relatively low-tech innovations in contexts that are not high on political agendas, such as those focused on tuberculosis, nonpesticidal agriculture, small hydroelectric plants, and handloom weaving (Engel 2015; Höffken 2012; Mamidipudi 2016; Quartz 2011). Furthermore, for drawing lessons about heuristics, theories, and methods, STS researchers could turn their analyses on themselves and reflect on how their engagements and value systems might fruitfully be adopted to contribute toward more robust developmental outcomes. Thus, STS researchers might locate themselves as activists in the sociotechnical ensembles that they study, as was the case for example, in the development of the Indian and African manifestos for science and technology.

Furthermore, engaging with development policy and governance invites STS to expand its conceptual bases. It pushes STS, for instance, to integrate more fully with contemporary social theories. Here, STS has as yet underinvested in locating itself within political-economic analyses. The relationships between development, the rise of consumerist new middle classes, and models of political organization such as democracy, for example, have hardly received attention within STS scholarship (Khandekar and Reddy 2015; Witsoe 2013; see Brown 2015 for emerging literature). Moreover, as Law and Lin (2015) suggest, in order to overcome their parochialism, STS scholars will have to recognize the privileging of English language and Euro-American formulations in their theoretical models (cf. Fu 2007). Engaging in an exercise of "postcolonial symmetry," Law and Lin (2015) hence draw attention to the extensive STS scholarship being produced in various academic ecologies across the globe (including those of East, South East, and South Asia, Africa, and Latin America) and advocate for cultivating a conceptually and linguistically plural intellectual space: "To think about STS in ways that are indeed Chinese- or Spanish- or Hindi-inflected" (ibid., 12).

Engaging with development practice pushes STS toward extending its mission from understanding to intervening. This implies the construction of new narratives that can help enroll not just academics but also stakeholders across cultures and domains, cutting through hierarchies, and contributing to a redistribution of power. Such new narratives require the development of new vocabularies, as well as an expansion of the conceptual bases of the field. In doing so, STS becomes more relevant to interventionists, both in policy and in practice, and is itself positioned to sometimes accept non-scholarly roles.

References

Abraham, Itty. 1998. *The Making of the Indian Atomic Bomb: Science, Secrecy and the Postcolonial State*. New York: Zed Books.

___. 2006. "The Contradictory Spaces of Postcolonial Techno-Science." *Economic and Political Weekly* 41 (3): 210–17.

Adams, Vincanne, Suellen Miller, Sienna Craig, Sonam, Nyima, Droyoung, Phuoc V. Le, and Micheal Varner. 2007. "Informed Consent in Cross-Cultural Perspective: Clinical Research in the Tibetan Autonomous Region, PRC." *Culture, Medicine and Psychiatry* 31 (4): 445–72.

Adas, Michael. 1989. *Machines as the Measure of Men: Science, Technology, and Ideologies of Western Dominance. Cornell Studies in Comparative History*. Ithaca, NY: Cornell University Press.

Akrich, Madeleine. 1992. "The De-scription of Technical Objects." In *Shaping Technology/Building Society: Studies in Sociotechnical Change*, edited by Wiebe E. Bijker and John Law, 205–24. Cambridge, MA: MIT Press.

Amir, Sulfikar. 2012. *The Technological State in Indonesia: The Co-constitution of High Technology and Authoritarian Politics*. New York: Routledge.

Anderson, Benedict. 1991. *Imagined Communities: Reflections on the Origin and Spread of Nationalism*. London: Verso.

Anderson, Warwick. 2002. "Introduction: Postcolonial Technoscience." *Social Studies of Science* 32 (6): 643–58.

———. 2006. *Colonial Pathologies: American Tropical Medicine, Race, and Hygiene in the Philippines*. Durham, NC: Duke University Press.

Arnold, David. 1993. *Colonizing the Body: State Medicine and Epidemic Disease in Nineteenth-Century India*. Berkeley: University of California Press.

Basch, Linda G, Nina Glick Schiller, and Cristina Szanton Blanc. 1994. *Nations Unbound: Transnational Projects, Postcolonial Predicaments, and Deterritorialized Nation-States*. Langhorne, PA: Gordon and Breach.

Bassett, Ross. 2009. "Aligning India in the Cold War Era: Indian Technical Elites, the Indian Institute of Technology at Kanpur, and Computing in India and the United States." *Technology and Culture* 50 (4): 783–810.

Bauchspies, Wenda K. 2012. "The Community Water Jar: Gender and Technology in Guinea." *Journal of Asian and African Studies* 47 (4): 392–403.

———. 2014. "Presence from Absence: Looking within the Triad of Science, Technology and Development." *Social Epistemology* 28 (1): 56–69.

Beumer, Koen. 2016. *Nanotechnology and Development: Styles of Governance in India, South Africa, and Kenya*. Ph.D. dissertation, Maastricht University, Maastricht, Netherlands.

Bhabha, Homi K. 1993. *The Location of Culture*. New York: Routledge.

Bijker, Wiebe E. 1995. *Of Bicycles, Bakelites, and Bulbs: Toward a Theory of Sociotechnical Change*. Cambridge, MA: MIT Press.

———. 2003. "The Need for Public Intellectuals: A Space for STS: Pre-presidential Address, Annual Meeting 2001, Cambridge, MA." *Science, Technology, & Human Values* 28 (4): 443–50.

Bijker, Wiebe E., Roland Bal, and Ruud Hendriks. 2009. *The Paradox of Scientific Authority: The Role of Scientific Advice in Democracies*. Cambridge, MA: MIT Press.

Bijker, Wiebe E., Giovanni Caiati, and Luciano d'Andrea. 2012. "Knowledge Brokerage for Environmentally Sustainable Sanitation: Position Paper and Guidelines from the EU-FP7 BESSE Project." Maastricht University, Maastricht, Netherlands.

Bijker, Wiebe E., Thomas Parke Hughes, and Trevor J. Pinch, eds. 2012. *The Social Construction of Technological Systems: New Directions in the Sociology and History of Technology*. Cambridge, MA: MIT Press.

Bowker, Geoffrey C., and Susan Leigh Star. 2000. *Sorting Things Out: Classification and Its Consequences*. Cambridge, MA: MIT Press.

Brown, Mark B. 2015. "Politicizing Science: Conceptions of Politics in Science and Technology Studies." *Social Studies of Science* 45 (1): 3–30.

Bucchi, Massimiano, and Federico Neresini. 2008. "Science and Public Participation." In *The Handbook of Science and Technology Studies*. 3rd ed., edited by Edward J. Hackett, Olga Amsterdamska, Michael Lynch, and Judy Wajcman, 449–72. Cambridge, MA: MIT Press.

Busch, Lawrence. 2011. *Standards: Recipes for Reality*. Cambridge, MA: MIT Press.

Chakrabarty, Dipesh. 2000. *Provincializing Europe: Postcolonial Thought and Historical Difference*. Princeton, NJ: Princeton University Press.

Chatterjee, Partha. 1993. *The Nation and Its Fragments: Colonial and Postcolonial Histories*. Princeton, NJ: Princeton University Press.

Cherlet, Jan. 2014. "Epistemic and Technological Determinism in Development Aid." *Science, Technology, & Human Values* 39 (6): 773–94.

Collins, Harry, and Robert Evans. 2002. "The Third Wave of Science Studies: Studies of Expertise and Experience." *Social Studies of Science* 32 (2): 235–96.

Conklin, Alice L. 1997. *A Mission to Civilize: The Republican Idea of Empire in France and West Africa, 1895–1930*. Palo Alto, CA: Stanford University Press.

Cozzens, Susan E. 2012. "Editor's Introduction: Distributional Consequences of Emerging Technologies." *Technological Forecasting and Social Change* 79 (2): 199–203.

Cozzens, Susan E., Sonia Gatchir, Kyung-Sup Kim, Gonzalo Ordóñez, and Anupit Supnithadnaporn. 2008. "Knowledge and Development." In *The Handbook of Science and Technology Studies*. 3rd ed., edited by Edward J. Hackett, Olga Amsterdamska, Michael Lynch, and Judy Wajcman, 787–811. Cambridge, MA: MIT Press.

Cozzens, Susan E., and Edward J Woodhouse. 1995. "Science, Government and the Politics of Knowledge." In *The Handbook of Science and Technology Studies*. 2nd ed., edited by Sheila Jasanoff, Gerald E. Markle, James C. Peterson, and Trevor J. Pinch, 533–53. Thousand Oaks, CA: Sage.

de Laet, Marianne, and Annemarie Mol. 2000. "The Zimbabwe Bush Pump: Mechanics of a Fluid Technology." *Social Studies of Science* 30 (2): 225–63.

Delgado, Ana, and Israel Rodríguez-Giralt. 2014. "Creole Interferences: A Conflict over Biodiversity and Ownership in the South of Brasil." In *Beyond Imported Magic: Essays on Science, Technology, and Society in Latin America*, edited by Eden Medina, Ivan da Costa Marques, and Christina Holmes, 331–48. Cambridge, MA: MIT Press.

Edwards, Paul N., and Gabrielle Hecht. 2010. "History and the Technopolitics of Identity: The Case of Apartheid South Africa." *Journal of Southern African Studies* 36 (3): 619–39.

Eglash, Ron, Jennifer L. Croissant, Giovanna di Chiro, and Rayvon Fouché, eds. 2004. *Appropriating Technology: Vernacular Science and Social Power*. Minneapolis: University of Minnesota Press.

Ehrlich, Paul R. 1971. *The Population Bomb*. New York: Ballantine Books.

Engel, Nora. 2015. *Tuberculosis in India: A Case of Innovation and Control*. Hyderabad: Orient BlackSwan.

Escobar, Árturo. 1995. *Encountering Development: The Making and Unmaking of the Third World*. Princeton, NJ: Princeton University Press.

Fanon, Frantz. 1963. *The Wretched of the Earth*. New York: Grove Press.

___. 1967. *Black Skin, White Masks*. New York: Grove Press.

Ferguson, James. 1990. *The Anti-Politics Machine: "Development," Depoliticization, and Bureaucratic Power in Lesotho*. Minneapolis: University of Minnesota Press.

Fortun, Kim. 2012. "Ethnography in Late Industrialism." *Cultural Anthropology* 27 (3): 446–64.

Fu, Daiwie. 2007. "How Far Can East Asian STS Go?" *East Asian Science, Technology and Society: An International Journal* 1 (1): 1–14.

Galison, Peter. 1997. *Image and Logic: A Material Culture of Microphysics*. Chicago: University of Chicago Press.

Giddens, Anthony. 1990. *The Consequences of Modernity*. Stanford, CA: Stanford University Press.

Grove, Richard. 1995. *Green Imperialism: Colonial Expansion, Tropical Island Edens, and the Origins of Environmentalism, 1600–1860*. Cambridge: Cambridge University Press.

Gupta, Akhil. 1992. "The Song of the Nonaligned World: Transnational Identities and the Reinscription of Space in Late Capitalism." *Cultural Anthropology* 7 (1): 63–79.

___. 1998. *Postcolonial Developments: Agriculture in the Making of Modern India*. Durham, NC: Duke University Press.

Hackett, Edward J. 2008. "Politics and Publics." In *The Handbook of Science and Technology Studies*. 3rd ed., edited by Edward J. Hackett, Olga Amsterdamska, Michael Lynch, and Judy Wajcman, 429–32. Cambridge, MA: MIT Press.

Halfon, Saul E. 2007. *The Cairo Consensus: Demographic Surveys, Women's Empowerment, and Regime Change in Population Policy*. Lanham, MD: Lexington Books.

Haraway, Donna J. 1988. "Situated Knowledges: The Science Question in Feminism and the Privilege of Partial Perspective." *Feminist Studies* 14 (3): 575–99.

Harding, Sandra G. 1998. *Is Science Multicultural? Postcolonialisms, Feminisms, and Epistemologies*. Bloomington: Indiana University Press.

Hayden, Cori. 2003. *When Nature Goes Public: The Making and Unmaking of Bioprospecting in Mexico*. Princeton, NJ: Princeton University Press.

___. 2005. "Bioprospecting's Representational Dilemma." *Science as Culture* 14 (2): 185–200.

Headrick, Daniel R. 1981. *The Tools of Empire: Technology and European Imperialism in the Nineteenth Century*. Oxford: Oxford University Press.

Hecht, Gabrielle. 2006. "Negotiating Global Nuclearities: Apartheid, Decolonization, and the Cold War in the Making of the IAEA." *Osiris* 21: 25–48.

Hess, David, Steve Breyman, Nancy Campbell, and Brian Martin. 2008. "Science, Technology, and Social Movements." In *The Handbook of Science and Technology Studies*. 3rd ed., edited by Edward J. Hackett, Olga Amsterdamska, Michael Lynch, and Judy Wajcman, 473–98. Cambridge, MA: MIT Press.

Höffken, Johanna I. 2012. *Power to the People? Civic Engagement with Small-Scale Hydroelectric Plants in India*. Den Bosch, Netherlands: Uitgeverij BOXpress.

Höhler, Sabine, and Rafael Ziegler. 2010. "Nature's Accountability: Stocks and Stories." *Science as Culture* 19 (4): 417–30.

Hommels, Anique, Jessica Mesman, and Wiebe E. Bijker. 2014. *Vulnerability in Technological Cultures: New Directions in Research and Governance*. Cambridge, MA: MIT Press.

Jackson, Steven J. 2014. "Rethinking Repair." In *Media Technologies: Essays on Communication, Materiality, and Society*, edited by Tarleton Gillespie, Pablo Boczkowski, and Kirsten Foot, 221–39. Cambridge, MA: MIT Press.

Jackson, Steven, Alex Pompe, and Gabriel Krieshok. 2012. "Repair Worlds: Maintenance, Repair, and ICT for Development in Rural Namibia." In *Proceedings of the 2012 Computer-Supported Cooperative Work Conference*, 11–15. Seattle, WA.

Jasanoff, Sheila. 2003. "Breaking the Waves in Science Studies: Comment on Harry Collins and Robert Evans, 'The Third Wave of Science Studies'." *Social Studies of Science* 33 (3): 389–400.

___. 2005. *Designs on Nature: Science and Democracy in Europe and the United States*. Princeton, NJ: Princeton University Press.

___. 2007. "Technologies of Humility." *Nature* 450 (7166): 33.

Khandekar, Aalok. 2013. "Education Abroad: Engineering, Privatization, and the New Middle Class in Neoliberalizing India." *Engineering Studies* 5 (3): 179–98.

___. 2014. "Third Spaces of Collaboration." Paper presented at Technologies for Development, Closing Conference for the NWO-Sponsored Project, "Nanotechnologies for Development in India, Kenya, and the Netherlands." Brussels.

Khandekar, Aalok, and Deepa S. Reddy. 2015. "An Indian Summer: Corruption, Class, and the Lokpal Protests." *Journal of Consumer Culture* 15 (2): 221–47.

Kilelu, Catherine W., Laurens Klerkx, Cees Leeuwis, and Andy Hall. 2011. "Beyond Knowledge Brokering: An Exploratory Study on Innovation Intermediaries in an Evolving Smallholder Agricultural System in Kenya." *Knowledge Management for Development Journal* 7 (1): 84–108.

Kim, Jongyoung. 2007. "Alternative Medicine's Encounter with Laboratory Science: The Scientific Construction of Korean Medicine in a Global Age." *Social Studies of Science* 37 (6): 855–80.

Klerkx, Laurens, Marc Schut, Cees Leeuwis, and Catherine Kilelu. 2012. "Advances in Knowledge Brokering in the Agricultural Sector: Towards Innovation System Facilitation." *IDS Bulletin* 43 (5): 53–60.

Lampland, Martha, and Susan Leigh Star. 2009. *Standards and Their Stories: How Quantifying, Classifying, and Formalizing Practices Shape Everyday Life*. Ithaca, NY: Cornell University Press.

Lansing, Stephen J. 1991. *Priests and Programmers: Technologies of Power in the Engineered Landscape of Bali*. Princeton, NJ: Princeton University Press.

Law, John, and John Hassard. 1999. *Actor Network Theory and After*. Malden, MA: Blackwell.

Law, John, and Wen-yuan Lin. 2015. "Provincialising STS: Postcoloniality, Symmetry and Method." Accessed at http://www.heterogeneities.net/publications/LawLinProvincialising STS20151223.pdf.

Leslie, Stuart, and Robert Kargon. 2006. "Exporting MIT: Science, Technology, and Nation-Building in India and Iran." *Osiris* 21: 110–30.

Lewis, David, and David Mosse. 2006. *Development Brokers and Translators: The Ethnography of Aid and Agencies*. Bloomfield, CT: Kumarian Press.

Li, Tania Murray. 2007. *The Will to Improve: Governmentality, Development, and the Practice of Politics*. Durham, NC: Duke University Press.

Loconto, Allison. 2014. "Sustaining an Enterprise, Enacting SustainabiliTea." *Science, Technology, & Human Values* 39 (6): 819–43.

Mamidipudi, Annapurna. 2016. *Reframing Sustainability: The Case of Handloom Weaving in India*. Ph.D. dissertation, Maastricht University, Maastricht, Netherlands.

Mamidipudi, Annapurna, B. Syamasundari, and Wiebe E. Bijker. 2012. "Mobilising Discourses: Handloom as Sustainable Socio-Technology." *Economic and Political Weekly* 47 (25): 41–51.

Marnie, Sheila, Millennium Development Goals Task Force, and United Nations Development Group. 2013. *A Million Voices: The World We Want. A Sustainable Future with Dignity for All*. New York: United Nations Development Group. Accessed at http://www.undp.org/content/undp/en/home/librarypage/mdg/a-million-voices--the-world-we-want/.

Marques, Ivan da Costa. 2014. "Ontological Politics and Latin American Local Knowledges." In *Beyond Imported Magic: Essays on Science, Technology, and Society in Latin America*, edited by Eden Medina, Ivan da Costa Marques, and Christina Holmes, 85–109. Cambridge, MA: MIT Press.

Marres, Noortje. 2007. "The Issues Deserve More Credit." *Social Studies of Science* 37 (5): 759–80.

Martin, Brian, and Evelleen Richards. 1995. "Scientific Knowledge, Controversy, and Public Decision Making." In *Handbook of Science and Technology Studies*. 2nd ed., edited by Sheila Jasanoff, Gerald E. Markle, James C. Peterson, and Trevor J. Pinch, 506–31. Thousand Oaks, CA: Sage.

McNeil, Maureen. 2005. "Introduction: Postcolonial Technoscience." *Science as Culture* 14 (2): 105–12.

Medina, Eden, Ivan da Costa Marques, and Christina Holmes, eds. 2014. *Beyond Imported Magic: Essays on Science, Technology, and Society in Latin America*. Cambridge, MA: MIT Press.

Moon, Suzanne. 2007. *Technology and Ethical Idealism: A History of Development in the Netherlands East Indies*. Leiden: CNWS Publications.

___. 2009. "Justice, Geography, and Steel: Technology and National Identity in Indonesian Industrialization." *Osiris* 24 (1): 253–77.

Mosse, David. 2013. "The Anthropology of International Development." *Annual Review of Anthropology* 42 (1): 227–46.

Nandy, Ashis, ed. 1990. *Science, Hegemony and Violence: A Requiem for Modernity*. New Delhi: Oxford University Press.

Pfaffenberger, Bryan. 1992. "Social Anthropology of Technology." *Annual Review of Anthropology* 21 (1): 491–516.

Phalkey, Jahnavi. 2013. *Atomic State: Big Science in Twentieth-Century India*. Ranikhet: Permanent Black.

Pigg, Stacy Leigh. 1992. "Inventing Social Categories through Place: Social Representations and Development in Nepal." *Comparative Studies in Society and History* 34 (3): 491–513.

Pigg, Stacy Leigh, and Linnet Pike. 2001. "Knowledge, Attitudes, Beliefs, and Practices: The Social Shadow of AIDS and STD Prevention in Nepal." *South Asia: Journal of South Asian Studies* 24 (S1): 177–95.

Pollock, Anne. 2014. "Places of Pharmaceutical Knowledge-Making: Global Health, Postcolonial Science, and Hope in South African Drug Discovery." *Social Studies of Science* 44 (6): 848–73.

Porter, Theodore M. 1996. *Trust in Numbers: The Pursuit of Objectivity in Science and Public Life*. Princeton, NJ: Princeton University Press.

Prahalad, Coimbatore K. 2005. *The Fortune at the Bottom of the Pyramid*. Upper Saddle River, NJ: Wharton School Publishers.

Prakash, Gyan. 1999. *Another Reason: Science and the Imagination of Modern India*. Princeton, NJ: Princeton University Press.

Quartz, Julia. 2011. *Constructing Agrarian Alternatives: How a Creative Dissent Project Engages with the Vulnerable Livelihood Conditions of Marginal Farmers in South India*. Maastricht, Netherlands: Maastricht University Press.

Radder, Hans. 1992. "Normative Reflexions on Constructivist Approaches to Science and Technology." *Social Studies of Science* 22 (1): 141–73.

___ 1996. *In and about the World: Philosophical Studies of Science and Technology*. Albany: State University of New York Press.

Raj, Kapil. 2007. *Relocating Modern Science: Circulation and the Construction of Knowledge in South Asia and Europe, 1650–1900*. New York: Palgrave Macmillan.

Richards, Evelleen, and Malcolm Ashmore. 1996. "More Sauce Please! The Politics of SSK: Neutrality, Commitment and Beyond." *Social Studies of Science* 26 (2): 219–28.

Rogaski, Ruth. 2004. *Hygienic Modernity: Meanings of Health and Disease in Treaty-Port China*. Berkeley: University of California Press.

Rosner, Daniela. 2014. "Making Citizens, Reassembling Devices: On Gender and the Development of Contemporary Public Sites of Repair in Northern California." *Public Culture* 26 (1): 51–77.

Rostow, Walt Whitman. 1960. *The Stages of Economic Growth: A Non-Communist Manifesto*. Cambridge: Cambridge University Press.

Rottenburg, Richard. 2009. *Far-Fetched Facts*. Cambridge, MA: MIT Press.

Roy, Arundhati. 1999. *The Cost of Living*. New York: Modern Library.

Said, Edward W. 1978. *Orientalism*. New York: Vintage Books.

Saidi, Trust. 2016. *Traveling Nanotechnologies*. Maastricht, Netherlands: Maastricht University Press.

Sanne, Johan M. 2014. "Vulnerable Practices: Organizing through Bricolage in Railroad Maintenance." In *Vulnerability in Technological Cultures: New Directions in Research and Governance*, edited by Anique Hommels, Jessica Mesman, and Wiebe E. Bijker, 199–216. Cambridge, MA: MIT Press.

Schiebinger, Londa. 2005. "Forum Introduction: The European Colonial Science Complex." *Isis* 96 (1): 52–55.

Schumacher, Ernst Friedrich. 1973. *Small Is Beautiful: Economics as If People Mattered*. New York: Harper and Row.

Scott, James C. 1998. *Seeing Like a State: How Certain Schemes to Improve the Human Condition Have Failed*. New Haven, CT: Yale University Press.

Sekhsaria, Pankaj. 2013. "The Making of an Indigenous Scanning Tunneling Microscope." *Current Science* 104 (9): 1152–58.

___. 2016. *Enculturing Innovation: Indian Engagements with Nanotechnology*. Maastricht, Netherlands: Maastricht University Press.

SET-DEV and ATPS. 2011. *The African Manifesto for Science, Technology and Innovation*. Brussels: ATPS.

SET-DEV and KICS. 2011. *Knowledge Swaraj: An Indian Manifesto on Science and Technology*. Hyderabad: University of Hyderabad.

Seth, Suman. 2009. "Putting Knowledge in Its Place: Science, Colonialism, and the Postcolonial." *Postcolonial Studies* 12 (4): 373–88.

Shah, Esha. 2009. "Manifesting Utopia: History and Philosophy of UN Debates on Science and Technology for Sustainable Development." 25. STEPS Working paper. Brighton: STEPS Centre.

Sharma, Aradhana. 2008. *Logics of Empowerment: Development, Gender, and Governance in Neoliberal India*. Minneapolis: University of Minnesota Press.

Shrum, Wesley, and Yehouda Shenhav. 1995. "Science and Technology in Less Developed Countries." In *The Handbook of Science and Technology Studies*. 2nd ed., edited by Sheila Jasanoff, Gerald E. Markle, James C. Peterson, and Trevor J. Pinch, 627–51. Thousand Oaks, CA: Sage.

Sismondo, Sergio. 2008. "Science and Technology Studies and an Engaged Program." In *The Handbook of Science and Technology Studies*. 3rd ed., edited by Edward J. Hackett, Olga Amsterdamska, Michael Lynch, and Judy Wajcman, 13–31. Cambridge, MA: MIT Press.

Storey, William Kelleher. 1997. *Science and Power in Colonial Mauritius*. Rochester, NY: University of Rochester Press.

___. 2008. *Guns, Race, and Power in Colonial South Africa*. New York: Cambridge University Press.

Tilley, Helen. 2010. "Global Histories, Vernacular Science, and African Genealogies; Or, Is the History of Science Ready for the World?" *Isis* 101 (1): 110–19.

___. 2011. *Africa as a Living Laboratory*. Chicago: University of Chicago Press.

Timmermans, Stefan, and Marc Berg. 2003. *The Gold Standard: The Challenge of Evidence-Based Medicine and Standardization in Health Care*. Philadelphia: Temple University Press.

Tironi, Manuel, and Javiera Barandiarán. 2014. "Neoliberalism as Political Technology: Expertise, Energy, and Democracy in Chile." In *Beyond Imported Magic: Essays on Science, Technology, and Society in Latin America*, edited by Eden Medina, Ivan da Costa Marques, and Christina Holmes, 305–29. Cambridge, MA: MIT Press.

Traweek, Sharon. 2000. "Faultlines." In *Doing Science + Culture: How Cultural and Interdisciplinary Studies Are Changing the Way We Look at Science and Medicine*, edited by Roddey Reid and Sharon Traweek, 21–48. New York: Routledge.

Turnbull, David. 2000. *Masons, Tricksters and Cartographers: Comparative Studies in the Sociology of Scientific and Indigenous Knowledge*. New York: Harwood Academic Publishers.

United Nations, and Department of Economic Affairs. 1951. *Measures for the Economic Development of Underdeveloped Countries*. New York: Department of Economic Affairs.

United Nations Development Group. 2014. "Delivering the Post-2015 Development Agenda: Opportunities at the National and Local Levels." Accessed at http://www.undp.org/content/dam/undp/library/MDG/Post2015-SDG/UNDP-MDG-Delivering-Post-2015-Report-2014.pdf.

United Nations Development Programme. 2003. *Human Development Report*. Oxford: Oxford University Press.

___. 2015. "Sustainable Development Goals." United Nations Development Programme. Accessed at http://www.undp.org/content/dam/undp/library/corporate/brochure/SDGs_Booklet_Web_En.pdf.

Vaughan, Scott. 2015. "Inclusivity and Integration: The New Sustainable Development Goals and a Second Chance for Bretton Woods: Commentary." International Institute for Sustainable Development. Accessed at https://www.iisd.org/sites/default/files/publications/inclusivity -integration-sustainable-development-goals-commentary_0.pdf.

Visvanathan, Shiv. 1997. *A Carnival for Science: Essays on Science, Technology and Development*. New Delhi: Oxford University Press.

___. 2009. "The Search for Cognitive Justice." *Seminar* 597: 40–50.

Watson-Verran, Helen, and David Turnbull. 1995. "Science and Other Indigenous Knowledge Systems." In *The Handbook of Science and Technology Studies*. 2nd ed., edited by Sheila Jasanoff, Gerald E. Markle, James C. Peterson, and Trevor J. Pinch, 115–39. Thousand Oaks, CA: Sage.

Winner, Langdon. 1993. "Upon Opening the Black Box and Finding It Empty: Social Constructivism and the Philosophy of Technology." *Science, Technology, & Human Values* 18 (3): 362–78.

Witsoe, Jeffrey. 2013. *Democracy against Development: Lower-Caste Politics and Political Modernity in Postcolonial India*. Chicago: University of Chicago Press.

Wynne, Brian. 2013. *Rationality and Ritual: Participation and Exclusion in Nuclear Decision-Making*. New York: Routledge.

IV Organizing and Governing Science

Laurel Smith-Doerr

The intersections of science and its social contexts have become more complex, leading to new kinds of work and institutional arrangements. Chapters in this section of the handbook shed light on work and institutional arrangements of science, which have undergone some major shifts. The amounts and sources of funding for science have shifted and diversified as higher education has reached an increasing (but still limited) number of people across the globe. The institutionalization of scientific training, and its internationalization, creates and reinforces class inequalities as well as other disparities, for example, across race and gender, among disciplines, between elite scientists and other research workers, across different research topics, and between industrial science firms and systems of higher education. The contingent and precarious nature of work for many scientists (in higher education and elsewhere) has called into question the legitimacy and prestige of science as a profession. Value systems also present complex contexts for science that are laden into knowledge production through a wide range of pathways, including the organization of research priorities, the legal and ethical regulation of scientific work, the integration of scientific research in technological innovation, and the development of new practices and expectations in the communication of science with broader public and policy audiences. This complex intersection between science and society is highlighted across the chapters in this section of the handbook with attention to both the social contexts and the work practices of science and technology.

Science and technology are conducted in social contexts, including a changing laboratory context. Early STS research on academic laboratories focused on the work of knowledge production and credibility within these spaces. In more recent research, including retrospective interviews, narratives now tell a story of the faculty principal investigators who were generously funded by government agencies in the industrialized world during the Cold War period. At the time, however, leading STS scholars like Bruno Latour and Karin Knorr Cetina conducted their now classic lab studies in the 1970s and 1980s and found a more complex story of credit pressures and rivalries. Still, narratives told now by scientists and engineers about changing funding contexts have a ring of truth. These changes (perceived and material) demand new research of laboratory life to understand the shifting landscape of science careers, practices, and processes. In the wake of these changes, STS has contributed research on university-industry partnerships and the threats and promises of patenting and commercial outcomes of academic knowledge production. Commercialization themes were perhaps more strongly present in the third edition (2008) of this handbook but nonetheless shape the concerns of the chapters in this volume's section, which develop themes around organizing and governing science and technology. Multiple chapters in this section discuss the growing emphasis on interdisciplinarity in the organizing of science and on the governance of ethics compliance, for example, which are related in complicated ways to the growing commercial landscape of science.

Science and technology are work, and the labor practices have changed in some ways but in other ways durable power inequalities do not seem to budge. Chapters in this section of the handbook provide ways to think about where there has been change in scientific work and where there is institutional inertia. The work activities of science and technology occur within organizations and under institutional constraints. The professionalization processes that individual scientists experience are shaped by the reward systems of universities and departments, rules and policies of funding agencies, and of wider social pressures like increasing hopes and expectations for science to solve economic and social problems. In current retrospective interviews with scientists, engineers, and science policy makers, narratives about increasing pressures on scientists to produce "broader impacts" make a contrast with earlier periods in the post–World War II era where fewer strings were attached to funding. At that time in the 1960s and 1970s, sociologists of science Merton and Zuckerman famously charted the norms of science, which indeed appeared to focus on more insular concerns within the scientific community and self-regulation. For example, the norm of universalism was/is the widespread narrative that scientists as a community subject truth claims to impersonal criteria and, therefore, that no external regulation of knowledge practices is needed. More recent research on particularistic biases, including gendered evaluation of scientific work, has called the universalistic self-regulation of science into question. STS has reflected on this kind of increasing concern of science and technology communities with social embeddedness, for example, analyzing struggles with how to institutionalize and make feasible the practices of ethical behavior for individual scientists. In developing discussion of the contexts and work of science, the chapters in this section begin broadly with an attention to organizing—analyzing structures and inequalities in science and engineering—and later chapters move toward attention to governance—analyzing the value-laden discussions in science policy making and in science communication.

Organizing

Chapter 24, by Mary Frank Fox, Kjersten Bunker Whittington, and Marcela Linková, describes recent research and theory on persistent gender inequalities in the workforce of academic scientists in the United States and Europe. Gender inequality in science is an old problem and one that has been studied since the earliest days of the field, and that was included in previous editions of the handbook. This chapter provides a fresh look at the problem by including a more reflexive perspective and discussing how research on gender inequity in science, including STS scholarship, has been engaged (or not) in U.S. and European policy interventions aimed at increasing the participation and leadership of women in science. Chapter 25, by Edward J. Hackett, John N. Parker, Niki Vermeulen, and Bart Penders, examines change and variation in the organization of science from discipline-based lone investigators to large interdisciplinary collaborations. The authors posit that the trends toward bureaucratization and commercialization, which go hand in hand with these changes in knowledge-production systems around the globe, create new kinds of work contexts and new forms of work for scientists that STS should study; however, we may be limited in effecting or resisting these changes. Chapter 26, by Harry Collins, Robert Evans, and Martin Weinel, is a deep dive into the concept of interactional expertise—or fluency in the language and practice of an area of science. The authors argue for further STS research using this approach, in addition to more constructivist approaches to scientific expertise.

Governing

Chapter 27, by Ruthanne Huising and Susan S. Silbey, charts how laboratory practices are increasingly regulated by law and legal notions of compliance, and discusses the implications of this increasing surveillance of the lab space. Notably, the shift to surveillance and audit by legal authorities outside of science is a shift away from trust and a relational approach to governing everyday lab practices. Chapter 28, by Stephen Hilgartner, Barbara Prainsack, and J. Benjamin Hurlbut, describes the ethics policies and funding for scholarship on the social implications of "unlocking the secrets" of DNA, which were attached to the massive amounts of human genomics science funding in the United States and Europe beginning in the 1990s. The chapter reflects on the ways that STS researchers have critically engaged with these policy programs in which ethics are posed as governance tools. Chapter 29, by Jack Stilgoe and David H. Guston, raises questions about the implications of living in an innovation society where higher stakes and the rapid pace of change weigh heavily in considerations of technological developments. The authors contend that responsibility in science and technology is being posed in new ways, with STS scholarship as part of that conversation on responsibility. Chapter 30, by Maja Horst, Sarah R. Davies, and Alan Irwin, analyzes the institutional practices for science communication and its related tensions and complexities. As science communicates, scientists begin to realize that creating space for more voices can diversify pathways of learning about science but also can open up science to contestation in media forums; in discussing the complexities, the authors outline a variety of themes in science communication requiring analysis from STS perspectives.

Occupying

This section of the handbook thus concerns science as an occupation and how science occupies a societal location. We might also reflect on the fact that these chapters point to how STS seeks to occupy science. There are normative claims—implicit and explicit—about how STS ought to intervene in discussions of gender inequality (Fox et al.), changes in science funding (Hackett et al.), expertise recognition (Collins et al.), laboratory regulation (Huising and Silbey), innovation policy (Stilgoe and Guston), ethics of genomics research (Hilgartner et al.), and science communication (Horst et al.). These discussions of how STS should occupy/occupies science and technology enrich each of the chapters in this section. The field has a complicated relationship with normative claims—often more comfortable with analyzing and critiquing those kinds of claims than making them—but still we would identify and claim common values toward democratizing science in contrast to elitism and toward reflecting on our practices.

Reflection on practices can reveal the implications for both science and society of involving new actors like publics in the discussion. Publics have a voice and can refuse or reject scientific frameworks in ways that deeply challenge traditional systems of science. Even when traditional systems of science seemingly remain unmoved, greater attention to publics still changes the work of scientists. As Hilgartner et al. note with the case of genomics, scientists have tried to grapple with societal implications; they may lack the language to speak about control of the products of their research but may also experience a new sense of responsibility that may or may not be related to new policies that require them to be certified in ethics training. One question is, how does

STS engage with the institutionalization of ethics? These kinds of questions about STS's relationship to institutionalization are similar to those raised in the chapter by Fox et al. about gender policies in different industrialized countries. STS researchers have always sought to go beyond policies to "count the women" and ask questions about gendered organizations and technology; yet despite years of research, we are still far from closing the gender gap. Taking a hard look at our field's practices (and cooptation?) in the institutions of science is important.

Another theme that crosses several of the chapters in this section is the call for communication between science and society, both as an object of STS study and as a practice in which STS scholars engage. Science communication is seen as a policy solution to dealing with ethical dilemmas, a way to create transparency, a responsibility of publicly funded scientists. STS criticizes the "communication as the answer" approach as being aligned with a more linear model of science (in which science creates and hands down knowledge in a straight line to society), which Horst et al. carefully outline in their chapter. Yet engaging with science communication is important for STS if we want to promote a democratic approach to science. Thinking about how to do science communication in new and different ways—to occupy science with democratic values and to do so in a reflexive way—is an important avenue by which the field can engage with publics and scientists. These themes cross into other sections of the *Handbook* as well, but a particular strength of this section is that the chapters raise questions about science as a workplace, governed by institutions. With this lens, viewing communication and ethics and responsibility as part of the work that scientists do, and as integrated with their expertise, allows us to reflect on engagement in a different way. Taking the idea of interactional expertise from the Collins et al. chapter, for example, we might conclude that STS needs to understand epistemic communities in a deep way in order to converse with them and for us to be recognized as experts. We invite readers to peruse the chapters of this section together in order to think through key dimensions of the questions of how, where, when, and by whom science is done, and who speaks for science.

24 Gender, (In)equity, and the Scientific Workforce

Mary Frank Fox, Kjersten Bunker Whittington, and Marcela Linková

Introduction

This chapter's focus on gender and scientific careers embodies a core tenet of science and technology studies (STS): science is deeply intertwined with processes of status building, power, and organizational change. Scientific fields are marked by immense inequalities in status, authority, and material and symbolic rewards (Harding 2006; Zuckerman 1988). This widespread inequality is evidenced in disparities between scientists in their rank and decision-making capacities, levels of research funding, equipment and materials available, discoveries made, and recognition received (Stephan 2012). Science also has powerful influences on, and connections with, central social institutions including higher education and the state (Jasanoff 2004).

Scientific activity is co-produced with relations of gender, as well as race and ethnicity, geopolitical positions, and other inequalities. The consequences are substantial. Scientific fields often define what is taken for granted by literally billions of people (Cozzens and Woodhouse 1995) and shape understandings of the world (Hackett 2008). Thus, to be in control of scientific research is to be involved with directing the future, and this is highly valued (Wajcman 1991).

As a relational category, gender operates for institutions and individuals (Harding 1986). Categories of women and men form systems of stratification, reflected in economies and markets, states and politics, religions, and social structures built on women's and men's unequal statuses (Lorber 1994). These hierarchies are constituted in social interactions and in material and discursive practices that signal and reinforce gender relations (Radtke and Stam 1994). Hierarchical gender relations prevail in scientific careers. These appear in women's, compared to men's, lower participation in scientific fields, ranks/positions, publication productivity and commercial activity, and recognition and rewards. Because science is influential and powerful and because gender stratification persists across fields, gender divisions within scientific careers exemplify, and also support and legitimate, the hierarchical relations of women and men in society (Fox 2001). For these reasons, science and technology are critical sites for the study of gender and other inequalities, such as race and ethnicity, that intersect with gender.[1]

The power that underlies science and gender relations is key to our analysis. We attend to ways that gender operates as a basis of subordination or exclusion within scientific institutions, and present the mechanisms and debates invoked as explanations of such inequality. This involves attention to reflexivity of science and technology studies and policy making and the ways STS understandings can help transform gender hierarchies (Morley 2015; Teelken and Deem 2013). Relations of power can also shape scientific knowledge through research questions asked, methods applied, and inferences and explanations formed (Keller 1995; Schiebinger 2008). While we acknowledge this, a more detailed treatment of the topic is beyond this chapter's scope and appears elsewhere in the volume.

We pay specific attention to doctoral-level scientists because they are central in the production of research, the training of scientists, and the leading of science policy. We also take a cross-national approach, concentrating on the United States and European Union.[2] Our approach is not global, yet reflects methodological and substantive considerations. Methodologically, challenges exist in securing and reconciling data and information across countries and thus meaningfully accounting for international contexts. At the same time, a focus on the United States and Europe stems from these countries' position as scientific centers with significant human and material resources that attract scientists worldwide and that affect the global pursuit of science. Scientific activity and standards, practices, and procedures in the United States and Europe thus influence the extent of equity[3] in the international scientific workforce.

Consequently, our aims are to

1. identify the nature and extent of gender disparities in the participation, performance, and rewards of scientists, pointing to key issues that merit explanation, while taking a critical perspective on metrics of productivity and performance;
2. characterize and appraise prevailing debates about explanations of gender disparities that contribute to understandings of gender and scientific careers; and
3. assess existing policy interventions and promising models within U.S. and European governments and organizations.

In doing this, we see that the issues and problems of gender and participation, performance, and rewards in scientific careers are complex, multifaceted matters. So too are policy prospects for advancing gender equity and futures of scientific careers.

Key Dimensions of Participation, Performance, and Rewards

Participation

Gender and Doctoral Degrees Awarded in Science/Engineering over Time   Although data on workforce participation over time are unavailable or unreliable, information on doctoral degrees awarded is more readily accessible. Documenting patterns of doctoral degrees awarded by gender over time is important because these patterns point to the potential for a trained pool of researchers, including women. Commonly assumed is that an increase in the number of women with doctoral degrees will translate into greater workforce participation; later we discuss the shortcomings of this assumption.

The total numbers of doctoral degrees have grown steadily in the United States and Europe, with substantial gains especially since the 1980s.[4] The distribution of women is uneven across fields. In the United States between 2000 and 2013, women received approximately 20 percent to 30 percent of doctoral degrees in engineering (20.3%), mathematics/statistics/computer science (24.9%), and physical/environmental/atmospheric sciences (30.5%), compared to 49.7 percent in life sciences and 58.7 percent in social sciences (National Science Foundation 2016, table A2-30). In Europe, a more or less balanced gender composition is found in the social sciences, business, and law (49%) and in agricultural and veterinary sciences (52%), compared to 40 percent in science, mathematics and computing, and 26 percent in engineering, manufacturing, and construction (European Commission 2013a, 54). However, differences exist across countries. For example, in 2001, Portugal, Lithuania, and Latvia had the highest proportions of women Ph.D. graduates in engineering (50%, 38%, and 36%, respectively) while Luxemburg (17%), Germany and Slovenia (15%) had the lowest.

Gender and Sectors of Employment   The proportion of women in the workforce is contingent upon historical, political, and sociocultural contexts, including the evolution of each country's gender regime (Walby 2004) and welfare policy. Cross-country comparisons can be challenging, as data collection and reporting standards vary. Yet trends exist across countries. Horizontal and vertical segregation continues, with the proportion of women the smallest in technical and natural sciences, decreasing at increasing ranks of the research career. The proportion of women among employed scientists and engineers in 2013 was 29 percent in the United States (National Science Foundation 2016, figure 3-27), and 32 percent in the EU 27 in 2010 (European Commission 2013a, 18) with significant variation by country (53% women in Poland and 18% in Switzerland) (ibid., 20).

In the 27 member states of the European Union in 2009, women represent 40 percent of researchers in higher education, 40 percent in the government sector, and 19 percent in the business enterprise sector, again with variation by country (ibid.).[5] In the United States in 2013, the academic sector had the highest proportion of women doctorate holders (35%), with 21 percent in the industrial sector, and 13 percent in government (National Science Foundation, Survey of Doctoral Recipients [SDR]). Differences across U.S. sectors are partially related to the fact that women are relatively absent from fields with the strongest ties to industry (computer science, engineering) (National Science Foundation 2016, table A3-13).

The percentage of scientists in industry has been changing as this sector gains in importance, particularly in countries with higher research and development (R&D) (Auriol, Misu, and Freeman 2013, 24). In the United States, the proportion of men in industry rose by one-third (25% to 34%) between 1973 and 2013, with an even greater proportional increase (7% to 21%) for women (Long 2001; National Science Foundation [SDR]). Across countries, the proportion of women in industrial science relates positively to the strength of a given country's industrial sector; the proportion of women in industry is lower in countries with stronger industrial sectors and higher R&D expenditures per researcher. This relationship is weakened among those countries with proactive policies of gender equity, however (European Commission 2008, 2009; Lipinski 2014).[6]

In academia, the percentage of women in tenured or tenure-track positions and as full/grade A[7] professors has increased in most years and countries, across fields. Variations also persist: women are more heavily concentrated in psychology, social sciences, and health disciplines, yet remain less than a third in tenured or tenure-track positions in computer science, mathematics, engineering, and physical sciences (European Commission 2013a; National Science Foundation 2015, table 9-24). At senior levels, women make up 20 percent of grade A positions in the European Union, with new member states having higher proportions than older ones (European Commission 2013a, table 3-1). In the United States, 24 percent of full professors are women, and men are more likely than women to attain the rank of full professor ten or more years after receipt of Ph.D. (National Science Foundation 2015, table 9-24). In most fields, women are overrepresented as instructors/lecturers. A concern is that the increasing percentages of women in the academic sector may be undercut by rising rates of temporary contracts (Auriol, Misu, and Freeman 2013).

Overall, these data reflect aggregate totals and are due only in part to fewer numbers of women in the doctorate pool at older cohorts. The fraction of women who have advanced to full-time, full professor positions remains smaller than the fraction of science/engineering doctorates earned in prior years.

Career Performance, Recognition, and Rewards

Understanding publication rates, and other dimensions of performance, is important in order to assess—and help redress—gender differences in rewards because these are partially explained by productivity (Fox 1999; Long 2001; Sonnert and Holton 1995a). "Partially" is a key term. The extent to which productivity governs rewards is a recurring debate (e.g., Ceci et al. 2014), and metrics of performance are infused with considerations of gender and dimensions of inequality, as we shall see. Further, women's records of performance do not automatically translate into rewards and recognition equivalent to men's (Wennerås and Wold 1997).

Publications and Commercial Activity   We focus on both publications and commercial activity in the academic sciences. While the former has long been regarded as the "coin of the realm" (Storer 1973), academic science has transformed in recent decades with an increase in patenting, licensing, start-up incubation, and founding of companies (Mowery et al. 2001). Commercial activity is not a requirement for tenure and promotion. However, the increased importance of commercial activity may heighten gender gaps in status because those who are already successful in the academy appear better able to capitalize on commercial success (Colyvas et al. 2012).

A range of studies point to the lower publication productivity of women compared to men in the United States (see Fox 2005; Leahey 2006; Long 1992) and in Europe (see Abramo, D'Angelo, and Caprasecca 2009; Barjak 2006; Prpić 2002), with cross-country variations among European countries (Naldi et al. 2005). Women scientists publish on average 40 percent to 70 percent the number of articles men do. The gender gap in publication productivity has been declining, especially within the life and social sciences (van Arensbergen, van der Weijden, and van den Besselaar 2012) and among the youngest cohorts of scientists (Mauleón, Bordons, and Oppenheim 2008). Women faculty members also receive fewer patents than men, and patent at lower rates, with some decrease in the gender disparity over time (Kugele 2010; Sugimoto et al. 2015). Compared to men, women faculty are also less likely to consult (Corley and Gaughan 2005), join private-sector scientific advisory boards (Ding, Murray, and Stuart 2013), receive U.S. National Institutes of Health Small Business Innovation Research awards (Tool and Czarnitzki 2005), and become company founders (Lowe and Gonzalez Brambilia 2007).

Extensive research documents that gender differences in productivity relate to women's (1) lower rank (Caprile et al. 2012; Long 2001; Rørstad and Aksnes 2015); (2)

location in less prestigious and resourced institutions and lower integration into scientific networks (Fox and Mohapatra 2007; Long 2001; Xie and Shauman 2003); and (3) family circumstances (Cole and Zuckerman 1987; Fox 2005; Hunter and Leahey 2010; Kyvik and Teigen 1996). We discuss these factors in subsequent sections.

Citations, Prizes, and Awards   Citations to research are a feature of formal scientific communication and a means of influence because they enhance circulation of knowledge, embed findings in a wider research context, and are a form of social capital. Citations are also the basis for calculating journals' impact factors and the Hirsh index, which increasingly figure in assessments of individuals, departments, and institutions. In addition, scientific prizes define and reward excellence, signaling valued performance and bestowing honor (Stephan 2012).

Women scientists have fewer citations than men (Larivière et al. 2013). In part, women publish more often with local coauthors, and men with international authors, and this garners more citations among men (Aaltojarvi et al. 2008; Aksnes et al. 2011). Women are also underrepresented in receipt of prizes and awards (Hedin 2014; Husu and Koskinen 2010; Lincoln et al. 2012), and this has not changed significantly with the increases in numbers of women scientists. Apart from the relationship to scientific networks, citations and awards are also subject to gender bias, as we address later.

Research Funding and Salary   Research funding and material resources are long-standing concerns for women's equity (e.g., Massachusetts Institute of Technology 1999). After receipt of start-up funds for newly hired scientists, U.S. universities expect faculty to fund their laboratories with competitive external funding, largely from federal agencies. European scientists compete for external research funding through the EC's Framework Programmes, including the European Research Council[8] (ERC), and through country-level support. Block grants to institutions are more common in Europe than in the United States, but competitive funding to individual scientists and institutions is on the rise in Europe (Felt 2009).

A study of early career life scientists in the United States reports that doctoral-level men receive over twice as much start-up funding than do women scientists, with gender gaps highest at the largest hospitals and universities (Sege, Nykiel-Bub, and Selk 2015). Data from the ERC indicate that women apply for grants much less than men, and often women's success rates are lower, especially in fields where their proportions are higher, such as life sciences (European Commission 2012a, 2013b, 2014; Vernos 2013).[9] Across career stages, men are more likely to apply and receive subsequent awards from the National Institutes of Health (NIH), the leading funder of scientific research in the United States (Pohlhaus et al. 2011); patterns vary for other fields and agencies (Hosek et al. 2005).

Studies of scientists' salaries often focus on academia because academic positions are more comparable across institutions than positions in other sectors, and the data more readily available. Among U.S. full-time employed scientists, women's <u>salaries</u> are 21 percent lower than men's (National Science Foundation 2016, figure 3-31), although the difference depends on average age, work experience, academic training, sector of employment, occupation, productivity, and demographic characteristics. Notably, taking publication productivity into account does not reduce appreciably the unexplained gender gap for full professors, the rank at which the gender difference in salary is greatest (Ginther 2004). In Europe, the gender pay gap in academia has been investigated unevenly across countries; however, persistent gender gaps are documented (Caprile et al. 2012; European Commission 2007, 2013c). Furthermore, means of payment beyond base salary, such as <u>bonuses and honoraria</u>, obscure transparency in salary and contribute to gender gaps (European Commission 2001). These gaps occur despite the fact that in some European countries, academic salaries are determined at national, rather than institutional, levels (Estermann, Nokkala, and Steinel 2011).

Limits of Available Data in Study of Gender and Scientific Careers

Despite many years of national and international data collection, a complete picture of women's and men's careers in science is restricted by limitations of data available. We highlight three areas here.

First, few research studies in the United States and Europe extend analysis of gender disparities in science to include the range of ways in which women and men of color and ethnic backgrounds are underrepresented or disadvantaged. Furthermore, very little is known about gay, lesbian, bisexual, and transgender (LGBT) populations in these fields. Studies show that career and productivity outcomes for these groups are especially challenged (see, for example, Farrell et al. 2015; Gutiérrez y Muhs et al. 2012; Williams, Phillips, and Hall 2015).

Second, more research is needed on scientific careers outside academia. Women and underrepresented groups are less likely than white men to indicate strong interest in careers at research-intensive institutions (Gibbs et al. 2014). Furthermore, despite the notable rise in industrial science employment in the United States and Europe, we know little about the nature of gender disparities in this sector. For example, some research highlights the unique challenges of women in technology fields (Simard et al. 2008), but needed are systematic and longitudinal career studies on industrial scientists. In addition, few assessments are available on gender disparities in government science or in science policy at national levels (Cozzens, Fealing, and Smith-Doerr 2012).

Third, comparisons across European countries are hampered by language barriers and the lack of coordinated data collection. Further challenges exist in drawing country-level comparisons amid differences in R&D expenditure, sociopolitical and economic development, gender and welfare regimes, research governance, and other factors.

Debates and Explanations

Pipeline Model

In 1983 the Rockefeller Foundation issued a report (Berryman 1983) that described trends in the representation of women and racial minorities among U.S. doctorate holders in science and mathematics with a metaphor of the "educational pipeline." Conjured were unidimensional links between educational stages and occupational outcomes. In the decades to follow, the pipeline metaphor came to constitute a common depiction of women's underrepresentation in science, characterized as "leaky" with fewer women than men arriving at the end (Blickenstaff 2005). Family formation is a frequently cited factor (Goulden, Mason, and Frasch 2011), together with the choices and interests of high school students, choices of college majors, and decisions to move on to graduate school and beyond.

Women are more likely than men to enter science tracks for the first time in college (Xie and Shauman 2003), and fewer women than men indicate an intention to major in scientific fields (Ceci et al. 2014). An argument of "demographic inertia" is that proportions of women at given academic ranks are subject to the existing gender and age distributions that affect the proportional representation of new Ph.D.s (see Shaw and Stanton 2012). By implication, potential lags exist between the timings of (1) the hiring of women and (2) their proportions among academic ranks within institutions. Indeed, computing and mathematics are two fields in which gender differences in transition from graduate school to an academic career are most pronounced. However, these differences persist even after taking demographic inertia into account (Shaw and Stanton 2012).

The pathways (or life course) model represents a contrasting view to that of the pipeline, emphasizing dynamic processes that include potential exits and reentry to science and technology over time. This perspective, represented by Xie and Shauman (2003) and others (Fealing, Lai, and Myers 2015), gives attention to the sequencing of educational and occupational events but emphasizes that transitions are "age-dependent, interrelated, and contingent on (but not determined by) earlier experiences and societal forces" (Xie and Shauman 2003, 12). While academic performance is influential for scientific careers, the character of the influence can change over time. Likewise, changes occur in the nature of family influences, and the anticipated conflict between childbearing and a demanding career may be salient long before women and men experience either children or careers. Furthermore, women constitute larger proportions of doctorate degree holders than proportions at advanced ranks, even after accounting for time since Ph.D. (National Science Foundation 2015, table 9-24).

In contrast to the pipeline model, a pathways perspective suggests that education and professional experience fail to uniformly translate into expected career outcomes for women compared to men. This brings forth alternative explanations on the influence of marriage and family, scientific culture and practices, and organizational-level arrangements, addressed below.

Marriage, Family, and Household

Some explanations of gender disparity in scientific careers focus on the influence of primary social institutions, and particularly, marriage and the family. The argument is that the demands of science are exceptionally high, and so are those of family, and that these two "greedy institutions" (Coser 1974) conflict. This is especially the case for women who often have greater responsibility for household demands and care work (Jacobs and Gerson 2004; Suitor, Mecom, and Feld 2001). A survey of academic scientists in nine U.S. research universities finds that while both women and men report that work interferes with family more than family interferes with work, women report a higher level of conflict in each direction (Fox, Fonseca, and Bao 2011).

Scientific work is revealing for work-family interactions because the expectations are that the "ideal typical" scientist's foremost commitment is to work—with few outside interests or responsibilities (Bailyn 2003). Scientists display strong personal association with work (Sonnert and Holton 1995a, 1995b), and their professional achievements, in turn, shape their perceptions of self and what they expect of others (Sharone 2004). Further, academic institutions tend to foster an increasingly competitive "culture of excellence" that intensifies work and goals for achievement (Hermanowicz 2009). Compounding this, faculty members frequently characterize their departments' and institutions' evaluative criteria as subjective or vague (Fox 2015). This raises work anxiety and contributes to a culture in which long hours at the lab are expected and desirable.

Furthermore, reductions in U.S. federal funding—most notably in the biomedical sciences—have lowered grant acceptance rates (especially for early-career investigators) and heightened the precariousness of university lab work (Daniels 2015). Europe has seen a similar trend: increases in competitive funding, falling acceptance rates, and increases in temporary contracts, especially in the postdoctoral phase (Felt 2009; Sigl 2015). These changes raise questions about the sustainability of research careers and bring specific types of risk for researchers taking time away for parenting (European Commission 2009).

The influence of marriage, households, and family has been the focus of extensive research. Studies have pointed to neutral or positive relationships between marriage and women's publication productivity, and to both positive and negative effects of the presence of children (Cole and Zuckerman 1987; Fox 2005; Kyvik and Teigen 1996; Whittington 2011). More nuanced and time-sensitive measures of parenthood and productivity appear in recent studies that include the ages of children in the home, the span of scientists' careers, and the timing of parenthood (Fox 2005; Hunter and Leahey 2010).

Further, in the United States, married women and women with children are less likely than the childless to transition from Ph.D. to a tenure-track job (Ginther and Kahn 2009; Wolfinger, Mason, and Goulden 2009); and compared to men, women scientists are more likely to stay single, have fewer children (or fewer children than desired), have children after tenure, and report missing family events (Drago et al. 2006; Ecklund and Lincoln 2011). In Europe, data are inconclusive on this as researchers—both women and men—are reported to have more children than the working population (European Commission 2013a, figure 3.8), while particular studies also report that men researchers have the same number of children as the working population but women have fewer (see Caprile et al. 2012). Furthermore, more women scientists are married to male scientists than the reverse with challenges in locating two jobs and managing national and international mobility (European Commission 2013d; Vohlídalová 2014). The magnitude of work-family conflict would likely be greater if studies could capture those who have left the career at earlier stages in scientific careers (Fox, Fonseca, and Bao 2011).

Culture and Practices of Science and Scientific Organizations

Science is an institution built on the ideal of merit as a basis for recognition and reward, and universalism[10] in who can participate. The traditional scientific belief system is that quality matters, not social characteristics such as gender, race and ethnicity, socioeconomic status and class, or geopolitical location. Despite these ideals and the increasing proportions of women in scientific fields, both women and men regard science as a masculine domain. Out of 5,000 school-aged children across three countries who participated in an open-ended "Draw a Scientist" test (Chambers 1983), only 28 girls (and no boys) drew female scientists. Since then, hundreds of studies have verified this gender divide, to greater or lesser degree (Steinke et al. 2007). Relatedly, half a million Implicit Association tests, completed by people from 34 countries, point to biases that associate science with men (Nosek et al. 2009). A recent survey of 5,000 people in several Western European countries revealed that 67 percent think that women lack the skills to reach top academic positions; and when asked which field is suitable for women, 89 percent selected the option "anything but science" (L'Oreal 2015). Such belief systems are especially salient for fields in which raw, innate talent is thought to be required, such as physics, math, or philosophy (Leslie et al. 2015).[11]

Feminist studies of science and technology and social psychologists provide insights into the ways that the culture, institutions, and practices of science are co-constituted by social, economic, and geopolitical forces. These studies unmask ways that gender and other types of bias operate in science, and ways that scientific participation is shaped by, and contributes to, broader societal inequities. They argue that power relations frame research questions and that exclusion of women from the making of knowledge contributes to unequal gender relations in science, academia, and society (Anderson 2015; Schiebinger 1989). This argument points to links among those who produce knowledge, the knowledge that is produced, and who and what get recognized (Haraway 1997). Early ethnographic work by Traweek (1988), for example, examined scientific work in a community of physicists, describing heroic, competitive individualism as a gendered model for recognition and success in the field. An implication is that men and women may be building their careers in discriminatory institutions; and this issue emerges with renewed strength and resonance in examinations of the gendered impacts of the neoliberal university in Europe (Morley 2003, 2015; Teelken and Deem 2013).

From this perspective, what constitutes excellence can be elusive and intersects with gender and age, race, notions of leadership (van den Brink and Benschop 2011), and other social characteristics. Men tend to be regarded as more competent than women in tasks with high social value, especially if these tasks involve assertiveness, agency, and instrumentality (Ridgeway 2011). These beliefs then create a higher standard for competence for women scientists because in male-typed tasks, a man's success reinforces the belief that men are competent, while a woman's success gets attributed to luck rather than ability (Foschi 2000). In addition, when women adopt behaviors typically associated with men and success, they are often penalized, and perceived as bossy, overly assertive, and not likeable (Williams 2005).

In keeping with this, the work of men—as students, colleagues, authors, and experts—is consistently judged as superior, even when the only thing that differs in the work evaluated is the name associated with it (Moss-Racusin et al. 2012; Reuben, Sapienza, and Zingales. 2014; Steinpreis, Anders, and Ritzke 1999); and thus women and men are held to different standards of performance (Wennerås and Wold 1997). Tenure assessments and letters of recommendation use dissimilar expressions to describe and stress the qualities of women and men (Marchant, Bhattacharya, and Carnes 2007; Trix and Psenka 2003), and award committees chaired by men (compared to women) are significantly less likely to give prizes to women (Hedin 2014; Lincoln et al. 2012). Men are also less likely to cite women, compared to men, authors (Hutson 2006; Maliniak, Powers, and Walter 2013) and are more likely than women to cite themselves (King et al. 2015).

Women and underrepresented groups also judge their own abilities and sense of belonging in science in accord with these cultural beliefs (Warmuth and Hanappi-Egger 2014). Women display less confidence, a lower sense of entitlement for success, and altered aspirations for scientific careers (Cech et al. 2011; Cheryan et al. 2009; Correll 2001; Faulkner 2006). These belief systems then contribute to attrition of women and marginalized groups from science (Beoku-Betts 2004; Gutiérrez y Muhs et al. 2012).

Organizational Perspectives and Forms of Organizations

Organizational perspectives on men's and women's participation in science emphasize the relationship between the organization and form of workplaces and gender inequities. The perspective is critical because scientific work is conducted through organizational practices and policies and relies on collaboration of students and colleagues (see Blau 1974; Fox 2001; Hargens 1975; Pelz and Andrews 1976). While the production of science in any setting involves the mobilization of many individuals to complete a project, organizational settings differ in the arrangements of power, methods of evaluation, and circumstances of rewards. Addressing organizational context shifts the emphasis beyond the relationship of individual characteristics and experience to the structural and environmental antecedents (such as institutional climate, formal and informal organization) of work (see Fox, Sonnert, and Nikiforova 2011).

Work environments have been found to be more supportive of gender equity when bureaucratic forms of governance (including written rules and guidelines) are in place (Baron et al. 2007; McGuire 2002; Roth and Sonnert 2011). The issue is that loosely defined criteria and processes of evaluation decrease the use of objective criteria for performance and allow for bias in assessment (Long and Fox 1995; McIlwee and Robinson 1992; Reskin 2003). Others, however, challenge the notion of bureaucracy as being able to overcome features of gendered organizations (Acker 1990; Morley 2003). They emphasize that bureaucratic organizational processes are reflections of masculine practices that derive from gendered structures and status differences.

Recently, scholars of work and organizations have become interested in the emergence of a new form of corporate organizing popular among science-based organizations (and other knowledge-based industries), one that is based on a network rather than hierarchical form of governance (Powell 1990). Hierarchical firms are more highly centralized, with defined roles, clear lines of authority, and formal decision-making procedures. In contrast, firms with a network form tend to maintain a more fluid, flat organization, have knowledge and direction more equally distributed across individuals, and establish more collective reward structures. Scientists in network firms report that they work toward a common goal, are less competitive internally, and have an increased emphasis on teamwork.

In a study of life science Ph.D.s in the United States, Smith-Doerr (2004) finds that women Ph.D.s were more likely to hold leadership positions in biotechnology firms than in the more hierarchical settings of academia and large drug companies. Women in biotech firms also have more equitable outcomes in patenting and publishing productivity than in other settings (Whittington 2007, 2009; Whittington and Smith-Doerr 2008). These findings suggest that the incentive structure and logic of organizing in these firms may support women's involvement by allowing for more inclusive collaboration across (and beyond) the organization. Formally specified rules and procedures support gender equity when transparency is in place and where women remain token minorities (see Ridgeway 2011); while the organization of work in networked, horizontally organized settings may serve to enhance women's (and men's) integration and involvement in science-based work.

Policy

Existing Interventions and Their Potential

Policy interventions for equity are important because they embody conceptions of what is wrong for women (as well as men) in science and technology and because they present solutions for what can or should be done toward improvement (Fox, Sonnert, and Nikiforova 2011). The arguments underlying policies are diverse, embodying a range of interactions between science and societies. First is an argument of justice, related to representative democracy and equitable distribution of resources, as well as belief systems based upon universalism (Long and Fox 1995). Second is a business case centered on efficiency where women scientists are a resource that cannot be wasted (see Pearson and Fechter 1994). Here, equity is seen as producing gains for economic advancement in global competition. Third, greater integration of gender in research and innovation and diversity of the scientific workforce are seen as fostering quality and robustness in science (making research more accountable to public interests and funding) (European Commission 2012b, 13).

Policy interventions are complex and not necessarily linear, involving multiple actors with differing levels of power in networks and roles that change and evolve. The case of Europe is especially intricate. The European Commission operates as a supranational actor along with national and regional policies and governments, research institutions and universities, funders, gender and feminist associations, scholars, experts, and activists. These actors strategically use one another to advance (as well as stall) their agendas on both the European Union and national levels (Linková and Červinková 2011). Thus, policies adopted and actions taken (or not) are a result of geopolitical histories of European countries, the specific nature of national gender and welfare regimes (including provisions for parental leave), women's and men's participation in the labor force, a country's R&D expenditures, the size of its business enterprise sector, and other factors (European Commission 2008; Tenglerová 2014). Despite EU efforts, the differences between countries with proactive gender equality policies and those that are comparatively inactive have widened since the new millennium (Lipinsky 2014).

Policies for equity tend to take individual or structural approaches. The former emphasize women's attitudes, values, aptitudes, and behaviors and their effect on participation and performance. Structural approaches emphasize potential characteristics of the settings that disadvantage women such as patterns of exclusion or inclusion in research groups, access to human and material resources, and practices of evaluation as part of gendered organization (Cronin and Roger 1999; Fox 2001, 2015; Sonnert and Holton 1995a, 1995b; Zellers, Howard, and Barcic 2008). In both the United States and Europe, programs and policies have shifted since the new millennium toward structural interventions on the basis of insights from policy (Rossiter 2012).[12]

A turning point toward structural approaches in the United States occurred in 2001 when the National Science Foundation launched a new program, called ADVANCE Institutional Transformation Awards. The ADVANCE awards constitute multiyear funding packages to institutions acknowledging that it would be difficult to advance women in science without changing the settings in which they work. Awardees' initiatives have ranged from those focusing on work-family arrangements, tenure and promotion practices, recruitment and retention of faculty, and internal networks of education, communication, and resources (Bilimoria and Liang 2012; Fox 2008). Similarly, since 2009, and building on the earlier structural approaches, the European Commission has funded structural projects emphasizing changes in institutions, as well as in the production of knowledge (European Commission 2012b, 2013e), and focusing on three priorities: research careers, leadership, and the dimension of gender in research.

Programs that take a structural approach have been shown to be more successful in advancing gender-sensitive workplace conditions. However, common (and available) actions actually implemented by program officials often target individual change because these are organizationally more feasible, less likely to interfere with existing arrangements, and less likely to face barriers from administrators and others invested in the status quo. These patterns reveal why programmatic efforts are challenging, and why numbers of organizational efforts can fail (Fox, Sonnert, and Nikiforova 2011).

Promising Interventions: Further Implications for Policy

Assessment of Merit, Transparency, and Clarity of Criteria Because evaluative practices are the means of assessment for advancement, a potential area for policy involves clarity in processes of evaluation (Bunton and Corrice 2011; European Commission 2012b). Decisions on research grants are often made through information solicited from (standardized) grant applications and peer-review meetings to evaluate them. Promising interventions suggest that attention be paid to equitable processes of solicitation and allocation. The ERC has recently eliminated self-evaluations and reports of applicants' greatest achievement. This change reduces the impact of women's tendency to undervalue their achievements. Others have recommended processes of awarding promotions, honorary positions, and prizes (Carnes et al. 2005) that attend to gender disparities in letters of nomination for women compared to men, as well as the composition of evaluation groups.

While clear criteria and reporting standards support gender equity through increased transparency, also needed are policies designed to address practices of evaluation that occur through informal practices. For example, the practice in the United States of raising senior faculty members' salaries based on outside offers may disadvantage women because they are more likely to have spouses with careers that affect whether they are moveable (Roos and Gatta 2009). Furthermore, when women do exercise outside offers, some evidence exists that the boosts of salary received are lower than those for men (Blackaby, Booth, and Frank 2005). These practices, and other taken-for-granted arrangements, are ripe for interventions.

Evaluation also involves individuals' perceptions of standards for success. Men scientists' clarity about criteria for tenure and promotion is linked to seniority as well as informal factors (frequency of speaking with colleagues and departmental climate); women's clarity is predicted only by the latter (Fox 2015). A policy implication is that women's clarity about criteria for tenure and promotion is predicted by social integration and cannot be expected to occur simply as a matter of time (seniority) in their institutions.

Finally, direct institutional action through targeted programs or incentives may also help override bias in science. For example, one study finds that life science women report less exposure to the commercial process, fewer invitations to collaborate, and different sources of support than men to initiative academic activities (close colleagues as opposed to senior advisers, and institutional assistance) (Murray and Graham 2007). The presence of targeted institutional support (for example, technology transfer office outreach or facilitated networking) could reduce gender gaps by providing women improved awareness of and access to these opportunities. Recent initiatives by a host of European countries have increased dramatically women's memberships on corporate boards through use of quotas (including those in Norway, France, Iceland, Italy, the Netherlands, Spain, Germany, and on their heels, possibly Sweden and the European Union).

Reevaluating Performance Metrics   Counts of publications and citations and other metrics of productivity, such as journal impact factors, are drivers in hiring and promotion decisions and in honorific recognitions in science. A growing consideration is that metrics-based assessments may not capture the quality of research and result in unintended conduct among researchers. Concerns include those that citation distributions are skewed; impact factors are field specific; potential exists for journals inappropriately boosting their impact factors (see Martin 2016); and data used for calculating impact factors are not always transparent and available to the public. In addition, scholars have questioned the extent to which counts and citations capture scientific impact. Various groups have called for rejection of impact factors as measures of quality (see, for example, the 2012 San Francisco Declaration on Research Assessment[13]). More recently, the Leiden Manifesto (Hicks et al. 2015) presents ten principles for accountable and meaningful use of metrics-based assessment.

Although these responses do not focus on gender-related concerns, implications exist for gender equity. Some evidence points to a pattern of a more cautious and comprehensive approach among women, compared to men, scientists (Fox and Mohapatra 2007; Sonnert and Holton 1995b). This caution may result in women having as high or higher citations per paper as do men, but lower numbers of articles and total citations. In addition, Leahey (2006) finds that women specialize less than men—that is, they are more diverse in their areas of research and less likely to publish repeatedly in one special area. This, she finds, relates to women's lower overall publication productivity; in parallel, it may also relate to lower overall citations.

In keeping with these issues, the NSF and NIH in the United States and the ERC in Europe present promising models for the use of performance metrics. These agencies now ask that applicants include on grant applications a limited number of publications (the most relevant five or ten papers, depending on career stage), rather than a full list. Other steps are inclusion of broader scientific influences, such as public and professional engagements, or other forms of scientific output, in assessing the influence of research. Requirements to discuss the broader impacts of proposed research now appear in grant applications for European projects, the NSF, and many national systems (such as the UK's Research Excellence Framework), although debate continues about the ability to assess impact on society.

Institutional Areas: Social and Family Policies   Policies vary widely by country with respect to parental leave, salaries obtained (or not) during leaves, and provisions for child care. In Europe, job protection in parenthood and parental leaves are provided by the law. The U.S. policy on these leaves lags considerably behind other developed countries, depends on the place of employment, and often operates as an employment benefit for certain groups of employees. Predominant ideologies of motherhood also vary widely across countries, shaping policy and the choices of women and men. For example, the socially enforced and nearly universal norm of a three-year parental leave for mothers, which has taken root in the Czech Republic since the 1990s, would be unimaginable in most countries of Europe, including countries with highly family-friendly social policies (Hašková and Saxonberg 2015).

Policies that anticipate parental leaves support advancement by keeping scientists engaged during leave periods. In addition, career reentry programs can provide funding for scientists to return to independent investigator status after a period away (see, for example, the NIH Research Supplement to Promote Reentry into Biomedical Careers funding initiative [NIH PA-15–321 2015]). In addition, funding programs that use academic or career age, rather than biological age, may better support equity by reducing the influence of parental leave when evaluating productivity. Similarly, the NIH, the ERC in Europe, and some European countries now invite explanations about periods of leave in the biographical section of principal investigators' grant applications in an effort to make life-course transitions more transparent.

Other promising policies address expectations that researchers be geographically (and in Europe, internationally) mobile, especially in their early-career stages. Two examples are the Academy of Finland (the Finnish National Research Council), which provides up to 20 percent extra funding if a mobile researcher is traveling with under-age dependents, and the ECs Marie Skodowska-Curie Fellowships allowance for mobility. These policies are useful only if scientists are comfortable taking advantage of them (Drago et al. 2006). Some studies suggest a disproportionate use of these policies by women inequities (Lundquist et al. 2012; Rhoades and Rhoades 2012) and seemingly gender-neutral policies may work to the advantage of men, rather than women, who use them (given gender differences in household divisions of labor). Thus, these patterns could perpetuate existing gender inequities. These findings emphasize the need for continued attention to the complex and nuanced nature of gendered organization of scientific work and careers.

Conclusions

Gender hierarchies in scientific careers reflect and reinforce relations of status and power in societies. Understanding these hierarchies is fundamental to science and technology studies and to policies for equity.

Despite increasing numbers of women in science in the past three decades, we find deep and persistent gender disparity in participation by field, and in performance, recognition, and rewards in scientific careers. Explanations for these disparities lie not simply in the individual interests, intentions, and choices of women and men to pursue scientific education and careers and thus remain in the pipeline. Rather, explanations lie within complex institutional influences, including marriage, family, and households, and in the culture and practices that connect science and masculinity and construct power relations that shape competence and success in ways that disadvantage women and other groups. Intersecting with these influences are organizational forms and arrangements that create disadvantageous and/or discriminatory systems of interactions, evaluations, and rewards.

Just as scientific careers are socially and organizationally shaped, so they can be reshaped through policies that support gender, as well as racial and ethnic, equities. Toward this end, promising policies are taking structural approaches that address features of the cultures, institutions, and organizations in which scientists live and work. Key are policies that (1) govern assessment of merit, transparency, and clarity of evaluation; (2) reevaluate metrics of performance; and (3) attend to parental leaves, expectations for geographic (and international) mobility, and research funding that takes into account the timing of family responsibilities. These policies reflect STS understandings of science as deeply intertwined with power and status building and with the ways and means that social and organizational change can empower social groups positively, proactively, and broadly across countries and societies.

Acknowledgments

Marcela Linková acknowledges support from the Ministry of Education, Youth and Sports project LE12003 and the support for long-term conceptual development of a research organization RVO:68378025.

Notes

1. In some European countries, intersectional research is neglected (Caprile et al. 2012, 191), while it is better developed in the United Kingdom, the Nordic countries, and German-speaking countries (Kerner 2012).

2. In the United States and the United Kingdom, *science* refers primarily to the natural, biological, physical, mathematical, and engineering sciences; in continental Europe, it often includes the social sciences and humanities.

3. *Inequality* refers to inequivalence; *inequity*, to unfairness, injustice, and/or bias.

4. In many countries, data on gender and field are available only since the 1990s or 2000s.

5. In some European countries—especially the former socialist ones—representation in the governmental sector is historically strong because of Academies of Sciences, government entities that act(ed) as primary sites for research.

6. European countries that have highly developed innovation systems and strong gender equity policy (Norway, Denmark, Austria, Switzerland, Netherlands, Germany) have a higher annual growth rate for women in recent years. Central and Eastern Europe have less developed systems of both innovation and gender policies.

7. In European data reporting, grades A, B, and C typically correspond to the positions of full, associate, and assistant professor, respectively.

8. ERC grants are the most prestigious funding instruments in Europe.

9. Further data are available at http://erc.europa.eu/sites/default/files/document/file/Gender _statistics_April_2014.pdf.

10. Universalism entails the requirement that contributions be assessed without considerations of the social attributes of those making scientific claims (Merton [1942] 1973, 270).

11. In this study, the same finding applies to African Americans but not Asian Americans.

12. In some European countries (especially in the Nordic region), such approaches were in evidence since the late 1970s (Bergman and Rustad 2013).

13. See http://www.ascb.org/dora/.

References

Aaltojarvi, Inari, Ilkka Arminen, Otto Auranen, and Hanni-Mari Pasanen. 2008. "Scientific Productivity, Web Visibility and Citation Patterns in Sixteen Nordic Sociology Departments." *Acta Sociologica* 51 (1): 5–22.

Abramo, Giovanni, Ciriaco A. D'Angelo, and Alessandro Caprasecca. 2009. "Gender Differences in Research Productivity: A Bibliometric Analysis of the Italian Academic System." *Scientometrics* 79 (3): 517–39.

Acker, Joan. 1990. "Hierarchies, Jobs, Bodies: A Theory of Gendered Organizations." *Gender & Society* 4 (2): 139–58.

Aksnes, Dag W., Kristoffer Rorstad, Fredrik Piro, and Gunnar Sivertsen. 2011. "Are Female Researchers Less Cited? A Large-Scale Study of Norwegian Scientists." *Journal of American Society for Information Science and Technology* 62 (4): 628–36.

Anderson, Elizabeth. 2015. "Feminist Epistemology and Philosophy of Science." In *The Stanford Encyclopedia of Philosophy*, edited by Edward N. Zalta. Accessed at http://plato.stanford.edu/archives/fall2015/entries/feminism-epistemology/.

Auriol, Laudeline, Max Misu, and Rebecca A. Freeman. 2013. "Careers of Doctorate Holders: Analysis of Labour Market and Mobility Indicators." *OECD Science, Technology, and Industry* Working Papers 2013/4. Paris: OECD Publishing.

Bailyn, Lotte. 2003. "Academic Careers and Gender Equity: Lessons Learned from MIT." *Gender, Work, and Occupations* 10 (2): 137–53.

Barjak, Franz. 2006. "Research Productivity in the Internet Era." *Scientometrics* 68 (3): 343–60.

Baron, James, Michael T. Hannan, Greta Hsu, and Ozgecan Kocak. 2007. "In the Company of Women: Gender Inequality and the Logic of Bureaucracy in Start-up Firms." *Work and Occupations* 34 (1): 35–66.

Beoku-Betts, Josephine A. 2004. "African Women Pursuing Graduate Studies in the Sciences: Racism, Gender Bias, and Third World Marginality." *NWSA Journal* 16 (1): 116–35.

Bergman, Solveig, and Linda Rustad. 2013. *The Nordic Region—A Step Closer to Gender Balance in Research? Joint Nordic Strategies and Measures to Promote Gender Balance among Researchers in Academia.* Copenhagen: Nordic Council of Ministers.

Berryman, Sue E. 1983. *Who Will Do Science?* New York: Rockefeller Foundation.

Bilimoria, Diana, and Xiangfen Liang. 2012. *Gender Equity in Science and Engineering: Advancing Change in Higher Education.* New York: Routledge.

Blackaby, David, Alison L. Booth, and Jeff Frank. 2005. "Outside Offers and the Gender Pay Gap: Empirical Evidence from the UK Academic Labour Market." *Economic Journal* 115 (501): F81–F107.

Blau, Peter. 1974. *The Organization of Academic Work*. 2nd ed. New Brunswick, NJ: Transaction Publishers.

Blickenstaff, Jacob C. 2005. "Women and Science Careers: Leaky Pipeline or Gender Filter?" *Gender and Education* 17 (4): 369–86.

Bunton, Sarah A., and April M. Corrice. 2011. "Perceptions of the Promotion Process: An Analysis of U.S. Medical School Faculty." *Analysis in Brief (American Association of Medical Colleges)* 11 (5): 4–5.

Caprile, Maria, Elisabetta Addis, Cecilia Castaño, Ineke Klinge, Marina Larios, Danièle Meulders, and Susana Vázquez-Cupeiro. 2012. *Meta-analysis of Gender and Science Research: Synthesis Report*. Luxembourg: Publications Office of the European Union.

Carnes, Molly, Stacie Geller, Evelyn Fine, Jennifer Sheridan, and Jo Handelsman. 2005. "NIH Director's Pioneer Awards: Could the Selection Process Be Biased against Women?" *Journal of Women's Health* 14 (8): 684–91.

Cech, Erin, Brian Rubineau, Susan Silbey, and Caroll Seron. 2011. "Professional Role Confidence and Gendered Persistence in Engineering." *American Sociological Review* 76 (5): 641–66.

Ceci, Stephen J., Donna K. Ginther, Shulamit Kahn, and Wendy M. Williams. 2014. "Women in Academic Science: A Changing Landscape." *Psychological Science in the Public Interest* 15 (3): 75–141.

Chambers, David W. 1983. "Stereotypic Images of the Scientist: The Draw-a-Scientist Test." *Science Education* 67 (2): 255–65.

Cheryan, Sapna, Victoria C. Plaut, Paul G. Davies, and Claude M. Steele. 2009. "Ambient Belonging: How Stereotypical Cues Impact Gender Participation in Computer Science." *Journal of Personality and Social Psychology* 97 (6): 1045–60.

Cole, Jonathan, and Harriet Zuckerman. 1987. "Marriage, Motherhood, and Research Performance in Science." *Scientific American* 256: 119–25.

Colyvas, Jeannette A., Kaisa Snellman, Janet Bercovitz, and Maryann Feldman. 2012. "Disentangling Effort and Performance: A Renewed Look at Gender Differences in Commercializing Medical School Research." *Journal of Technology Transfer* 37 (4): 478–89.

Corley, Elizabeth, and Monica Gaughan. 2005. "Scientists' Participation in University Research Centers: What Are the Gender Differences?" *Journal of Technology Transfer* 30 (4): 371–81.

Correll, Shelley J. 2001. "Gender and the Career Choice Process: The Role of Biased Assessments." *American Journal of Sociology* 106 (6): 1691–730.

Coser, Lewis A. 1974. *Greedy Institutions: Patterns of Undivided Commitment*. New York: Free Press.

Cozzens, Susan, and Edward Woodhouse. 1995. "Science, Government, and the Politics of Knowledge." In *The Handbook of Science and Technology Studies*. 2nd ed., edited by Sheila Jasanoff, Gerald E. Markle, James C. Petersen, and Trevor Pinch, 533–53. Thousand Oaks, CA: Sage.

○ Cozzens, Susan, Kaye Husbands Fealing, and Laurel Smith-Doerr. 2012. *Women in Science and Technology Policy*. Grant Award: National Science Foundation.

Cronin, Catherine, and Angela Roger. 1999. "Theorizing Progress: Women in Science, Engineering, and Technology in Higher Education." *Journal of Research in Science Teaching* 36 (6): 637–61.

Daniels, Ronald. 2015. "A Generation at Risk: Young Investigators and the Future of the Biomedical Workforce." *Proceedings of the National Academy of Sciences of the United States* 112 (2): 313–18.

Ding, Waverly, Fiona Murray, and Toby Stuart. 2013. "From Bench to Board: Gender Differences in University Scientists' Participation in Corporate Scientific Advisory Boards." *Academy of Management Journal* 56 (5): 1443–64.

Drago, Robert, Carol L. Colbeck, Kai D. Stauffer, Amy Pirretti, Kurt Burkum, Jennifer Fazioli, and Tara Habasevich. 2006. "The Avoidance of Bias against Caregiving: The Case of Academic Faculty." *American Behavioral Scientist* 49 (9): 1222–47.

Ecklund, Elaine, and Anne E. Lincoln. 2011. "Scientists Want More Children." *PLoS ONE* 6 (8): e22590.

Estermann, Thomas, Terhi Nokkala, and Monika Steinel. 2011. *University Autonomy in Europe II: The Scorecard*. Brussels: European University Association.

European Commission. 2001. *Science Policies in the European Union: Promoting Excellence through Mainstreaming Gender Equality*. Luxembourg: Publications Office of the European Union.

___. 2007. *The Study on the Remuneration of Researchers in the Public and Private Sectors*. Luxembourg: Publications Office of the European Union.

___. 2008. *Benchmarking Policy Measures for Gender Equality in Science*. Luxembourg: Office for Official Publications of the European Communities.

___. 2009. *The Gender Challenge in Research Funding: Assessing the European National Scenes*. Luxembourg: Office for Official Publications of the European Communities.

___. 2012a. *Annual Report on the ERC Activities and Achievements in 2011*. Luxembourg: Publications Office of the European Union.

___. 2012b. *Structural Change in Research Institutions: Enhancing Excellence, Gender Equality and Efficiency in Research and Innovation*. Luxembourg: Office for Official Publications of the European Communities.

___. 2013a. *She Figures 2012*. Luxembourg: Publications Office of the European Union.

___. 2013b. *Annual Report on the ERC Activities and Achievements in 2012*. Luxembourg: Publications Office of the European Union.

___. 2013c. *More 2. Remuneration—Cross Country Report.* Brussels: European Commission.

___. 2013d. *More 2. Final Report. Study on Mobility Patterns and Career Paths of Researchers.* Brussels: European Commission.

___. 2013e. *Gendered Innovations: How Gender Analysis Contributes to Research.* Luxembourg: Office for Official Publications of the European Communities.

___. 2014. *Annual Report on the ERC Activities and Achievements in 2013.* Luxembourg: Publications Office of the European Union.

Farrell, Stephanie, Rocio Chavela Guerra, Adrienne Minerick, Tom Waidzunas, and Erin Cech. 2015. *EAGER: Promoting LGBTQ Equality in Engineering through Virtual Communities of Practice.* Grant Award: National Science Foundation.

Faulkner, Wendy. 2006. *Genders in/of Engineering: A Research Report.* Edinburgh: Economic and Social Research Council.

Fealing, Kaye Husbands, Yufeng Lai, and Samuel L. Myers Jr. 2015. "Pathways vs. Pipelines to Broadening Participation in the STEM Workforce." *Journal of Women and Minorities in Science and Engineering* 21 (4): 271–93.

Felt, Ulrike, ed. 2009. *Knowing and Living in Academic Research: Convergence and Heterogeneity in Research Cultures in the European Context.* Prague: Institute of Sociology of the Academy of Sciences of the Czech Republic.

Foschi, Martha. 2000. "Double Standards for Competence: Theory and Research." *Annual Review of Sociology* 26: 21–42.

Fox, Mary Frank. 1999. "Gender, Hierarchy, and Science." In *Handbook of the Sociology of Gender,* edited by Janet S. Chafetz, 441–57. New York: Kluwer Academic/Plenum Publishers.

___. 2001. "Women, Science, and Academia: Graduate Education and Careers." *Gender & Society* 15 (5): 654–66.

___. 2005. "Gender, Family Characteristics, and Publication Productivity among Scientists." *Social Studies of Science* 35 (1): 131–50.

___. 2008. "Institutional Transformation and the Advancement of Women Faculty: The Case of Academic Science and Engineering." In *Higher Education: Handbook of Theory and Research,* vol. 23, edited by John C. Smart, 73–103. New York: Springer.

___. 2015. "Gender and Clarity of Evaluation among Academic Scientists in Research Universities." *Science, Technology, & Human Values* 40: 487–515.

Fox, Mary Frank, Carolyn Fonseca, and Jinghui Bao. 2011. "Work and Family Conflict in Academic Science: Patterns and Predictors among Women and Men in Research Universities." *Social Studies of Science* 41 (5): 715–35.

Fox, Mary Frank, and Sushanta Mohapatra. 2007. "Social-Organizational Characteristics of Work and Publication Productivity among Academic Scientists in Doctoral-Granting Departments." *Journal of Higher Education* 78 (5): 542–71.

Fox, Mary Frank, Gerhard Sonnert, and Irina Nikiforova. 2011. "Programs for Undergraduate Women in Science and Engineering: Issues, Problems, and Solutions." *Gender & Society* 25 (5): 589–615.

Gibbs, Kenneth, John McGready, Jessica Bennett, and Kimberly Griffin. 2014. "Biomedical Science PhD Career Interest Patterns by Race/Ethnicity and Gender." *PLOS One* 9 (12): e114736.

Ginther, Donna. 2004. "Women in Economics: Moving Up or Falling Off the Academic Career Ladder?" *Journal of Economic Perspectives* 18 (3): 193–214.

Ginther, Donna K., and Shulamit Kahn. 2009. "Does Science Promote Women? Evidence from Academia 1973–2001." In *Science and Engineering Careers in the United States* (National Bureau of Economic Research conference report), edited by Richard B. Freeman and Daniel F. Goroff, 163–94. Chicago: University of Chicago Press.

Goulden, Marc, Mary A. Mason, and Kari Frasch. 2011. "Keeping Women in the Science Pipeline." *Annals of the American Academy* 638: 141–62.

Gutiérrez y Muhs, Gabriella, Yolanda F. Niemann, Carmen G. González, and Angela P. Harris, eds. 2012. *Presumed Incompetent: The Intersections of Race and Class for Women in Academia.* Boulder: University Press of Colorado.

Hackett, Edward. 2008. "Politics and Publics." In *The Handbook of Science and Technology Studies.* 3rd ed., edited by Edward J. Hackett, Olga Amsterdamska, Michael Lynch, and Judy Wajcman, 429–32. Cambridge, MA: MIT Press.

Haraway, Donna J. 1997. *Modest_Witness@Second_Millennium.FemaleMan©_Meets_OncoMouse™: Feminism and Technoscience.* New York: Routledge.

Harding, Sandra. 1986. *The Science Question in Feminism.* Ithaca, NY: Cornell University Press.

____. 2006. *Science and Social Inequality: Feminist and Postcolonial Issues.* Champaign-Urbana: University of Illinois Press.

Hargens, Lowell. 1975. *Patterns of Scientific Research: A Comparative Analysis of Research in Three Scientific Fields.* Washington, DC: American Sociological Association.

Hašková, Hana, and Steven Saxonberg. 2015. "The Revenge of History—The Institutional Roots of Post-Communist Family Policy in the Czech Republic, Hungary and Poland." *Social Policy and Administration.* doi: 10.1111/spol.12129.

Hedin, Marika. 2014. "A Prize for Grumpy Old Men? Reflections on the Lack of Female Nobel Laureates." *Gender & History* 26 (1): 52–63.

Hermanowicz, Joseph C. 2009. *Lives in Science: How Institutions Affect Academic Careers.* Chicago: University of Chicago Press.

Hicks, Diana, Paul Wouters, Ludo Waltman, Sarah de Rijcke, and Ismael Rafols. 2015. "Bibliometrics: The Leiden Manifesto for Research Metrics." *Nature* 520 (7548): 429–31.

Hosek, Susan, Amy Cox, Bonnie Ghosh-Dastidar, Aaron Kofner, Nishal Ramphal, Jon S. Scott, and Sandra Berry. 2005. *Gender Differences in Major Federal External Grant Programs*. Santa Monica, CA: Rand Corporation.

Hunter, Laura A., and Erin Leahey. 2010. "Parenting and Research Productivity: New Evidence and Methods." *Social Studies of Science* 40 (3): 433–51.

Husu, Liisa, and Paula Koskinen. 2010. "Gendering Excellence in Technological Research: A Comparative European Perspective." *Journal of Technology Management and Innovation* 5 (1): 127–39.

Hutson, Scott R. 2006. "Self-Citation in Archaeology: Age, Gender, Prestige, and the Self." *Journal of Archaelogical Method and Theory* 13 (1): 431–35.

Jacobs, Jerry, and Kathleen Gerson. 2004. *The Time Divide: Work, Family, and Gender Inequality*. Cambridge, MA: Harvard University Press.

Jasanoff, Sheila. 2004. *States of Knowledge: The Co-production of Science and the Social Order*. New York: Routledge.

Keller, Evelyn F. 1995. "Gender and Science: Origin, History, and Politics." *Osiris* 10: 27–38.

Kerner, Ina. 2012. "Questions of Intersectionality: Reflections on the Current Debate in German Gender Studies." *European Journal of Women's Studies* 19 (2): 203–18.

King, Molly M., Shelley J. Correll, Jennifer Jacquet, Carl T. Bergstrom, and Jevin D. West. 2015. "Men Set Their Own Cites High: Gender and Self-Citation across Fields and over Time." Working paper. Palo Alto, CA: Stanford University

Kugele, Kordula. 2010. "Patents Invented by Women and Their Participation in Research and Development: A European Comparative Approach." In *Women in Engineering and Technology Research*, edited by A. Godfroy-Genin, 373–92. The PROMETEA Conference Proceedings No. 1. Zurich: Lit Verlag.

Kyvik, Svein, and Mari Teigen. 1996. "Child Care, Research Collaboration, and Gender Differences in Scientific Productivity." *Science, Technology, & Human Values* 21 (1): 54–71.

Larivière, Vincent, Chaoqun Ni, Yves Gingras, Blaise Cronin, and Cassidy R. Sugimoto. 2013. "Bibliometrics: Global Gender Disparities in Science." *Nature* 504 (7479): 211–13.

Leahey, Erin. 2006. "Gender Differences in Productivity: Research Specialization as a Missing Link." *Gender & Society* 20 (6): 754–80.

Leslie, Sarah-Jane, Andrei Cimpian, Meredith Meyer, and Edward Freeland. 2015. "Expectations of Brilliance Underlie Gender Distributions across Academic Disciplines." *Science* 347 (6219): 262–65.

Lincoln, Anne, Stephanie Pincus, Janet Koster, and Phoebe Leboy. 2012. "The Matilda Effect in Science: Awards and Prizes in the United States, 1990s and 2000s." *Social Studies of Science* 42 (2): 307–20.

Linková, Marcela, and Alice Červinková. 2011. "What Matters to Women in Science? Gender, Power and Bureaucracy." *European Journal of Women's Studies* 18 (3): 215–30.

Lipinsky, Anke. 2014. *Gender Equality Policies in Public Research*. Luxembourg: Office for Official Publications of the European Communities.

Long, J. Scott. 1992. "Measures of Sex Differences in Scientific Productivity." *Social Forces* 71: 159–78.

___. 2001. *From Scarcity to Visibility: Gender Differences in the Careers of Doctoral Scientists and Engineers*. Washington, DC: National Research Council.

Long, J. Scott, and Mary Frank Fox. 1995. "Scientific Careers: Universalism and Particularism." *Annual Review of Sociology* 21: 45–71.

L'Oreal. 2015. "The L'Oréal Foundation Unveils the Results of Its Exclusive International Study." Accessed at: http://www.loreal.com/media/press-releases/2015/sep/the-loreal-foundation-unveils -the-results-of-its-exclusive-international-study.

Lorber, Judith. 1994. *Paradoxes of Gender*. New Haven, CT: Yale University Press.

Lowe, Robert, and Claudia Gonzalez Brambilia. 2007. "Faculty Entrepreneurs and Research Productivity." *Journal of Technology Transfer* 32 (3): 173–94.

Lundquist, Jennifer H., Joya Misra, and KerryAnn O'Meara. 2012. "Parental Leave Usage by Fathers and Mothers at an American University." *Fathering: A Journal of Theory, Research, and Practice about Men as Fathers* 10 (3): 337–63.

Maliniak, Daniel, Ryan Powers, and Barbara F. Walter. 2013. "The Gender Citation Gap in International Relations." *International Organization* 67 (4): 889–922.

Marchant, Angela, Abhik Bhattacharya, and Molly Carnes. 2007. "Can the Language of Tenure Criteria Influence Women's Academic Advancement?" *Journal of Women's Health* 16 (7): 998–1003.

Martin, Ben. 2016. "Editors JIF-Boosting Strategem: What Are Appropriate and Which Not." *Research Policy* 45 (February): 1–7

Mauleón, Elba, María Bordons, and Charles Oppenheim. 2008. "The Effect of Gender on Research Staff Success in Life Sciences in the Spanish National Research Council." *Research Evaluation* 17 (3): 213–25.

McGuire, Gail. 2002. "Gender, Race, and the Shadow Structure: A Study of Informal Networks and Inequality in a Work Organization." *Gender & Society* 16 (3): 303–22.

McIlwee, Judith, and J. Gregg Robinson. 1992. *Women in Engineering: Gender, Power, and Workplace Culture*. Albany: State University of New York Press.

Merton, Robert K. [1942] 1973. "The Normative Structure of Science." In *The Sociology of Science*, edited by Robert K. Merton, 267–78. Chicago: University of Chicago Press.

Massachusetts Institute of Technology (MIT). 1999. "A Study on the Status of Women Faculty in Science at MIT." Accessed at: http://web.mit.edu/fnl/women/women.pdf.

Morley, Louise. 2003. *Quality and Power in Higher Education*. Philadelphia: Open University Press.

___. 2015. "Troubling Intra-actions: Gender, Neo-liberalism and Research in the Global Academy." *Journal of Education Policy* 31 (1): 28–45.

Moss-Racusin, Corinne A., John F. Dovidio, Victoria L. Brescoll, Mark J. Graham, and Jo Handelsman. 2012. "Science Faculty's Subtle Gender Biases Favor Male Students." *Proceedings of the National Academy of Sciences of the United States of America* 109 (41): 16474–79.

Mowery, David C., Richard R. Nelson, Bhaven N. Sampat, and Arvids A. Ziedonis. 2001. "The Growth of Patenting and Licensing by US Universities: An Assessment of the Effects of the Bayh-Dole Act of 1980." *Research Policy* 30 (1): 99–119.

Murray, Fiona, and Leigh Graham. 2007. "Buying Science and Selling Science: Gender Differences in the Market for Commercial Science." *Industrial and Corporate Change* 16 (4): 657–89.

Naldi, Fulvio, Daniela Luzi, Adriana Valente, and Ilaria V. Parenti. 2005. "Scientific and Technological Performance by Gender." In *Handbook of Quantitative Science: The Use of Publication and Patent Statistics in Studies of S & T Systems*, edited by Henk F. Moed, Wolfgang Glanzel, and Ulrich Schmoch, 299–314. New York: Kluwer Academic Publishers.

National Institutes of Health PA-15-321. *NIH Research Supplements to Promote Re-entry into Biomedical and Behavioral Research Careers*. Accessed at http://grants.nih.gov/grants/guide/pa-files/PA-15-321.html.

National Science Foundation. *Survey of Doctoral Recipients. 1973–2013*. Arlington, VA: National Center for Science and Engineering Statistics.

National Science Foundation. 2015. "Women, Minorities, and Persons with Disabilities in Science and Engineering: Special Report NSF 15-311." Arlington, VA: National Center for Science and Engineering Statistics. Available at http://www.nsf.gov/statistics/wmpd/.

___. 2016. *National Science Foundation Indicators*. Arlington, VA: National Center for Science and Engineering Statistics. Available at: http://www.nsf.gov/statistics/2016/nsb20161/#/.

Nosek, Brian A., Frederick L. Smyth, N. Sriram, Nicole M. Lindner, Thierry Devos, Alfonso Ayala, and Anthony G. Greenwald. 2009. "National Differences in Gender-Science Stereotypes Predict National Sex Differences in Science and Math Achievement." *Proceedings of the National Academy of Sciences of the United States of America* 106 (26): 10593–97.

Pearson, Willie, and Alan Fechter, eds. 1994. *Who Will Do Science? Educating the Next Generation*. Baltimore: Johns Hopkins University Press.

Pelz, Donald, and Frank Andrews. 1976. *Scientists in Organizations: Productive Climates for Research and Development*. Ann Arbor, MI: The Institute for Social Research.

Pohlhaus, Jennifer R., Hong Jiang, Robin M. Wagner, Walter T. Schaffer, and Vivian W. Pinn. 2011. "Sex Differences in Application, Success, and Funding Rates for NIH Extramural Programs." *Academic Medicine: Journal of the Association of American Medical Colleges* 86 (6): 759–67.

Powell, Walter W. 1990. "Neither Market Nor Hierarchy: Network Forms of Organization." *Research in Organizational Behavior* 12: 295–336.

Prpić, Katarina. 2002. "Gender and Productivity Differentials in Science." *Scientometrics* 55 (1): 27–58.

Radtke, H. Lorraine, and Henderikus J. Stam, eds. 1994. *Power/Gender: Social Relations in Theory and Practice*. London: Sage.

Reskin, Barbara. 2003. "Including Mechanisms in Our Models of Ascriptive Inequality." *American Sociological Review* 68 (1): 1–21.

Reuben, Ernesto, Paola Sapienza, and Luigi Zingales. 2014. "How Stereotypes Impair Women's Careers in Science." *Proceedings of the National Academy of Sciences of the United States of America* 111 (12): 4403–8.

Rhoades, Steven E., and Christopher H. Rhoades. 2012. "Gender Roles and Infant/Toddler Care: Male and Female Professors on the Tenure Track." *Journal of Social, Evolutionary, and Cultural Psychology* 6 (1): 13–31.

Ridgeway, Cecilia L. 2011. *Framed by Gender: How Gender Inequality Persists in the Modern World*. New York: Oxford University Press.

Roos, Patricia A., and Mary L. Gatta. 2009. "Gender (In)equity in the Academy: Subtle Mechanisms and the Production of Inequality." *Research in Social Stratification and Mobility* 27 (3): 177–200.

Rørstad, Kristoffer, and Dag W. Aksnes. 2015. "Publication Rate Expressed by Age, Gender and Academic Position—A Large-scale Analysis of Norwegian Academic Staff." *Journal of Infometrics* 9 (2): 317–33.

Rossiter, Margaret. 2012. *Women Scientists in America: Forging a New World since 1972*. Baltimore: Johns Hopkins University Press.

Roth, Wendy D., and Gerhard Sonnert. 2011. "The Costs and Benefits of 'Red Tape': Anti-bureaucratic Structure and Gender Inequity in a Science Research Organization." *Social Studies of Science* 41 (3): 385–409.

Schiebinger, Londa. 1989. *The Mind Has No Sex? Women in the Origins of Modern Science*. Cambridge, MA: Harvard University Press.

___, ed. 2008. *Gendered Innovations in Science and Engineering*. Stanford, CA: Stanford University Press.

Sege, Robert, Linley Nykiel-Bub, and Sabrina Selk. 2015. "Sex Differences in Institutional Support for Junior Biomedical Researchers." *JAMA* 314 (11): 1175–77.

Sharone, Ofer. 2004. "Engineering Overwork: Bell-Curve Management in a High Tech Firm." In *Fighting for Time: Shifting Boundaries of Work and Social Life*, edited by Cynthia F. Epstein and Arne Kalleberg, 191–218. New York: Russell Sage.

Shaw, Allison K., and Daniel Stanton. 2012. "Leaks in the Pipeline: Separating Demographic Inertia from Ongoing Gender Differences in Academia." *Proceedings of the Royal Society of London B: Biological Sciences* 279: 3736–41.

Sigl, Lisa. 2015. "On the Tacit Governance of Research by Uncertainty: How Early Stage Researchers Contribute to the Governance of Life Science Research." *Science, Technology, & Human Values*, doi: 0162243915599069.

Simard, Caroline, Andrea D. Henderson, Shannon K. Gilmartin, Londa Schiebinger, and Telle Whitney. 2008. *Climbing the Technical Ladder: Obstacles and Solutions for Mid-level Women in Technology*. Stanford, CA: Michelle R. Clayman Institute for Gender Research, Stanford University, and Anita Borg Institute for Women and Technology.

Smith-Doerr, Laurel. 2004. *Women's Work: Gender Equality vs. Hierarchy in the Life Sciences*. Boulder, CO: Lynne Rienner Publishers.

Sonnert, Gerhard, and Gerald Holton. 1995a. *Gender Differences in Science Careers*. New Brunswick, NJ: Rutgers University Press.

___. 1995b. *Who Succeeds in Science? The Gender Dimension*. New Brunswick, NJ: Rutgers University Press.

Steinke, Jocelyn, Maria Knight Lapinski, Nikki Crocker, Aletta Zietsman-Thomas, Yaschica Williams, Stephanie Higdon Evergreen, and Sarvani Kuchibhotla. 2007. "Assessing Media Influences on Middle School Aged Children's Perceptions of Women in Science Using the Draw-a-Scientist Test (DAST)." *Science Communication* 29 (1): 35–64.

Steinpreis, Rhea E., Katie A. Anders, and Dawn Ritzke. 1999. "The Impact of Gender on the Review of the Curricula Vitae of Job Applicants and Tenure Candidates: A National Empirical Study." *Sex Roles* 41 (7–8): 509–28.

Stephan, Paula. 2012. *How Economics Shapes Science*. Cambridge, MA: Harvard University Press.

Storer, Norman, ed. 1973. *The Sociology of Science: Theoretical and Empirical Investigations*. Chicago: University of Chicago Press.

Sugimoto, Cassidy R., Chaoqun Ni, Jevin D. West, and Vincent Larivière. 2015. "The Academic Advantage: Gender Disparities in Patenting." *PLoS One* 10 (5): e0128000. doi:10.1371/journal.pone.0128000.

Suitor, Jill, Dorothy Mecom, and Ilana Feld. 2001. "Gender, Household Labor and Scholarly Productivity among University Professors." *Gender Issues* 19 (4): 50–67.

Traweek, Sharon. 1988. *Beamtimes and Lifetimes: The World of High Energy Physicists*. Cambridge, MA: Harvard University Press.

Teelken, Christine, and Rosemary Deem. 2013. "All Are Equal, But Some Are More Equal than Others: Managerialism and Gender Equality in Higher Education in Comparative Perspective." *Comparative Education* 49 (4): 520–35.

Tenglerová, Hana. 2014. "The Policy of Inactivity: Doing Gender-Blind Science Policy in the Czech Republic 2005–2010." *Central European Journal of Public Policy* 8 (1): 78–106.

Tool, Andrew, and Dirk Czarnitzki. 2005. "Biomedical Academic Entrepreneurship through the SBIR Program." Working Paper 11450, National Bureau of Economic Research.

Trix, Frances, and Carolyn Psenka. 2003. "Exploring the Color of Glass: Letters of Recommendation for Female and Male Medical Faculty." *Discourse & Society* 14 (2): 191–220.

van Arensbergen, Pleun, Inge van der Weijden, and Peter van den Besselaar. 2012. "Gender Differences in Scientific Productivity: A Persisting Phenomenon?" *Scientometrics* 93 (3): 857–68.

van den Brink, Marieke, and Yvonne Benschop. 2011. "Gender Practices in the Construction of Academic Excellence: Sheep with Five Legs." *Organization* 19 (4): 507–24.

Vernos, Isabelle. 2013. "Research Management: Quotas Are Questionable." *Nature* 495 (7439): 39.

Vohlídalová, Marta. 2014. "Academic Mobility in the Context of Linked Lives." *Human Affairs* 24 (1): 89–102.

Wajcman, Judy. 1991. *Feminism Confronts Technology*. University Park: Pennsylvania State University Press.

Walby, Sylvia. 2004. "The European Union and Gender Equality: Emergent Variations of Gender Regime." *Social Politics* 11 (1): 4–29.

Warmuth, Gloria-Sophia, and Edeltraud Hanappi-Egger. 2014. "Professional Socialization in STEM Academia and Its Gendered Impact on Creativity and Innovation." In *Gender Considerations and Influence in the Digital Media and Gaming Industry*, edited by Julie Prescott and Julie E. McGurren, 156–74. Hershey, PA: IGI Global.

Wennerås, Christine, and Agnes Wold. 1997. "Nepotism and Sexism in Peer-review." *Nature* 387 (6631): 341–43.

Williams, Joan C. 2005. "The Glass Ceiling and the Maternal Wall in Academia." *New Directions for Higher Education* 130: 91–105.

Williams, Joan, Katherine Phillips, and Erika Hall. 2015. "Double Jeopardy: Gender Bias against Women of Color in Science." Center for WorkLife, UC Hastings College of Law.

Whittington, Kjersten Bunker. 2007. *Employment Sectors as Opportunity Structures: The Effects of Location on Men's and Women's Scientific Dissemination*. Dissertation, Stanford University.

___. 2009. "Patterns of Male and Female Dissemination in Public and Private Science." In *The New Market for Scientists and Engineers: The Science and Engineering Workforce in the Era of Globalization,* edited by Richard B. Freeman and Daniel F. Goroff, 195–228. Chicago: University of Chicago Press.

___. 2011. "Mothers of Invention? Gender, Motherhood, and New Dimensions of Productivity in the Science Profession." *Work and Occupations* 38 (3): 417–56.

Whittington, Kjersten Bunker, and Laurel Smith-Doerr. 2008. "Women Inventors in Context: Disparities in Patenting across Academia and Industry." *Gender & Society* 22 (2): 194–218.

Wolfinger, Nicholas, Mary A. Mason, and Marc Goulden. 2009. "Stay in the Game: Gender, Family Formation, and Alternative Trajectories in the Academic Life Course." *Social Forces* 87 (3): 1591–621.

Xie, Yu, and Kimberlee A. Shauman. 2003. *Women in Science: Career Processes and Outcomes.* Cambridge, MA: Harvard University Press.

Zellers, Darlene, Valerie Howard, and Maureen Barcic. 2008. "Faculty Mentoring Programs: Re-envisioning Rather than Reinventing the Wheel." *Review of Educational Research* 78 (3): 552–88.

Zuckerman, Harriet. 1988. "The Sociology of Science." In *Handbook of Sociology,* edited by Neil J. Smelser, 511–74. Newbury Park, CA: Sage.

25 The Social and Epistemic Organization of Scientific Work

Edward J. Hackett, John N. Parker, Niki Vermeulen, and Bart Penders

Science is work, and viewing science as a form of work demystifies it, eclipsing the quest for timeless truth with more mundane efforts to secure resources, conduct research, construct arguments, open (and protect) spheres of inquiry, and produce evidence convincing enough to pass through peer review into print. How scientific work is organized matters for science as knowledge and practice, shaping what is learned and how work is done. The work of scientists is organized into an overlapping and intersecting array of social and institutional arrangements. Scientists collaborate with one another and with citizens, students, technicians, practitioners, and other professionals in organizational settings that include disciplines, specialties, and research areas; for-profit, government, and nonprofit sectors; universities and departments; institutes and centers; invisible colleges and thought collectives; paradigms and epistemic cultures; research schools, groups, and teams; collaborations, laboratories, and collaboratories; social networks and social movements; boundary organizations, synthesis centers, and countless hybrids and variants of these. Demystifying scientific work and inquiring into its organizational form bring into focus three broad themes that have both analytic and normative aspects: First, how is scientific work organized at various scales, ranging from the institutional to the microsocial? Such questions have driven scholarship in science studies and cognate fields, and have practical and normative implications for science policy and management. Second, how do epistemic and social processes interact to form (and re-form) disciplines and specialties? The creation, diffusion, and application of scientific knowledge are causes and consequences of the organization of scientific work, and the alignment of science to social purposes that is built into scientific organizations shapes what research is done and what remains undone. Finally, the evolving epistemic and social patterns of scientific work—in particular, the essential ambiguity in the ever-tightening coupling of scientific inquiry to societal purposes—entail principles of scientific governance that bring power in all its forms to bear on the institution of science. Asking how science and its organization are co-produced has yielded sharp and enduring insights into the organization and dynamics of science.

The social organization of science has been a core concern of science studies *avant la lettre* for nearly a century (e.g., Fleck [1935] 1981; Merton [1938] 1970; Weber [1918] 1948) and continues to attract intense scholarly activity. This line of inquiry has generated various ways of understanding the interaction between science and its organization, such as epistemic cultures (Knorr Cetina 1999) and co-production (Jasanoff 2004), and extends beyond STS into such fields as the history of science, organizational behavior, network science, economics, sociology, informatics, and the nascent Science of Team Science ("the other STS": Hall et al. 2012; Paletz, Smith-Doerr, and Vardi 2011). This chapter engages with this vibrant and varied literature, describing current ideas about the social organization of science, how they have evolved, and what they mean for science, for society, and for social studies of science.

New ideas for organizing science are as old as organized science itself, and three venerable but still generative ideas orient our discussion. First, seventeenth-century utopias such as the New Atlantis, the City of the Sun, and Christianopolis prominently positioned imaginative new scientific organizations in their imagined landscapes, sometimes at the center and open to societal engagement, other times near the periphery and separate from quotidian affairs. The Accademia dei Lincei, an organization founded by Federico Cesi in 1603 and still active today, offered a grand humanistic vision of scientific collaboration guided by principles of justice and righteousness that would cross national and disciplinary lines to work for the greater good of humanity. The Future Earth initiative (launched in 2012), taking form under the aegis of the International Council for Science (ICSU), is a successor to this grand vision of science organized on a global scale for the benefit of all. The organization of scientific work and its place in the social order have been under negotiation and construction for centuries, and there is every reason to expect this to continue.

Weber's lecture on "Science as a Vocation," delivered at the University of Munich in 1918, offers a second point of orientation. At that watershed moment science existed simultaneously as a vocation or calling and in an emergent form that was more bureaucratic and less enchanted, conducted in alienating "state capitalist institutes ... of medicine or natural science" whose members endured "quasi-proletarian existences" (Weber 1948, 131). Looking backward, Weber saw the scientist of seventeenth-century London pursuing a gentleman's avocation alone or alongside his nearly invisible technicians (Shapin 1994). Looking forward, he saw the outlines of academic capitalism and the economic dependence of science that have created a marketplace where scientists sell their ideas and services at some cost to their calling and commitment (Hackett 1990; Mirowski 2011; Slaughter and Rhoads 2004; Stephan 2012).

The third reference point is the exponential growth and concomitant differentiation of science, first noticed by Price, which caused a dramatic rise in scientific collaboration that "has been increasing steadily and ever more rapidly since the beginning of the [twentieth] century," causing "one of the most violent transitions that can be measured in recent trends of scientific manpower and literature" (Price 1963, 77–79). This powerful transformation of scientific work has been driven by a sharp rise in public investment in research and a concomitant rising expectation that the investment will yield greater health, prosperity, and security.

Underlying these landmark reflections on the organization of science are three intertwined dynamics, and we will discuss each in turn. The first is *aggregation*, which ranges from the lone investigator to large, bureaucratic research enterprises. The second is *specialization*, the social and material processes that initiate and shape new areas of research. The third is interdisciplinary integration, or *synthesis*, which counterbalances specialization through the creative recombination of data, expertise, and ideas, often with practical applications in view. We close by discussing how aspects of the organization of science influence its purposes, politics, and place in society, and by offering ideas about the social organization of science in the future and the research challenges these will offer STS scholars.

Aggregation

The aggregation of science has been studied at scales ranging from the community or institution of science through disciplines and specialties to small groups, teams, laboratories, and individuals. Fleck and Merton first showed how groups of various sizes shape the organization of knowledge production. Fleck (1981, 1986) argues that science advances through the activities of "thought collectives": small, intensely interacting research groups that through persistent intellectual and emotional exchange develop a distinctive cognitive framework (or "thought style") that guides problem choice, evaluative standards, and literary styles (1981, 39, 99, 106). Shifting social relationships within the collective correlate with changes in research perspective and working style (1986, 74–75). In *Science, Technology and Society in Seventeenth Century England*, Merton maintains that the development of a sufficiently sizeable network of talented researchers, combined with cultural and material factors, provided the social matrix necessary for the development of the Royal Society of London and the rise of science as an independent social institution (1938, 78).

About twenty-five years later, Kuhn built on these ideas in *The Structure of Scientific Revolutions* (1962), proposing that the incremental growth of scientific knowledge is complemented by occasional episodes of revolutionary change brought on by the accumulation of anomalies that defy accepted explanations. Recognition of the profoundly social and emotional process of revolutionary change in science sparked a revolution in science studies, unleashing a torrent of research. Kuhn's initial focus on "scientific communities" (1962) was replaced in the second edition (1970) by "scientific specialties" of up to several hundred people competing with one another for intellectual hegemony over a substantive area. Scholars now understood that the all-too-human competition for recognition and reward drove epistemic change in science. But only with the work of Fleck and Kuhn, and their concepts of "thought collectives" and "paradigms," did interaction between the organization of science and scientific knowledge become visible, revealing how organizational transformations affect knowledge production, and how cognitive innovations have social implications.

During the same decade, Hagstrom's *The Scientific Community* (1965) demonstrated the importance of small scientific groups by arguing that solidarity and conformity within the scientific community are principally grounded in social control produced through informal interpersonal exchanges that take place in local contexts, such as the corridors and offices of academic departments and laboratories. Allocation of scholarly recognition and moral reproach by peer groups motivate creative scientific work and suppress scientific misconduct. Such interactions make the institutional purposes of science salient for everyday scientific work; complementarily, the social control of science as an institution emerges from its exercise within smaller scientific groups and networks.

"Big science"—the science of big instruments, big groups, and big money—was first investigated in the 1960s (Price 1963; Weinberg 1961). The concept was initially applied to large-scale physics research (e.g., the Manhattan Project), and later to astronomy, space research, ecology, and molecular biology. The term has taken on a variety of meanings (Capshew and Rader 1992), but common features include centralization around large and expensive instruments, industrialization, multidisciplinary collaboration, institutionalization, science-government relations, and internationalization. Such changes reflected the scaling-up of various processes during the modern era, and wonder at the scope of these transformations was tempered by concern about the diseases of "adminstratitis, moneyitis, and journalitis" (Weinberg 1961) and the decline of the intellectual mavericks (Price 1963). Debates about the appropriate scale for science continue (Vermeulen, Parker, and Penders 2010; Westfall 2003). The study of exponential growth in science stimulated other quantitative metrics, turning Price (a physicist and historian of science) into an acclaimed information scientist and a founder of bibliometrics (Elkana et al. 1978; Garfield 1984).

Price's quantitative approach, blended with a sociological reading of Kuhn's work and network analytic methods, sparked the first attempts to link scientific groups with changes in the content of scientific ideas. Ben-David and Collins (1966) showed how social relations among small groups of collaborators aided in the development of psychology, while Crane (1969, 1972) contended that science is mainly performed and evaluated within "invisible colleges"—small, dense, informal research networks. Building upon these foundations, Mullins (1972, 1973) and Griffith and Mullins (1972) proposed that "coherent groups"—small, highly emotive, intensely interacting research groups—are the primary drivers of scientific change, and seed the larger invisible colleges that develop around them. Together, these studies formed a specialty within science studies that focused on the development of scientific specialties (e.g., Chubin 1976; Cole and Zuckerman 1975; Edge and Mulkay 1976; Law 1976). Studies of scientific/intellectual movements (Frickel and Gross 2005) continue this tradition by joining the small-group dynamics that form the energetic kernel of scientific social movements to the large-scale dynamics of change.

Studies of laboratories first received systematic ethnographic study in Bruno Latour and Steve Woolgar's *Laboratory Life* (1979).[1] Conceptualizing lab scientists as members of tribes, they explored how researchers transform experiments into publications, arguing that technical facts are constructed rather than merely communicated. Focused on "science in the making" (Latour 1987), lab studies became a central concern of STS: scholars entered labs to explore their knowledge-producing practices. Traweek (1988), for instance, compared high-energy physics communities, while Knorr Cetina (1999) explored the epistemic cultures of molecular biology and particle physics, detailing scientists' collaborative patterns and ways of justifying knowledge claims. Lab studies nourished development of STS (Doing 2008), but the community's immersion in labs also made it difficult to see how science and technology interact with the world outside laboratory walls. Exceptions include studies of how institutional isomorphism brings business and professional standards and practices into the university (Hackett 1990), how articulation work aligns laboratory strategies and practices with the demands and interests of the wider world (Fujimura 1996), how asymmetrical convergence causes academic labs to resemble their for-profit counterparts (Vallas and Kleinman 2008), and how knowledge and social order are co-produced in practice (Jasanoff 2004).

Building on "big science" studies, the turn of the millennium witnessed increasing interest in scientific collaboration at scales ranging from groups through international research networks. Such collaborations may involve citizens, practitioners, and other professionals and may reach beyond scientific aims to societal purposes. They may be informally organized or firmly institutionalized, and their membership and aspirations may extend across institutional or national boundaries. In such arrangements collaborators share expertise, credibility, and resources (Hackett 2005a; Maienschein 1993) and are drawn together by funding programs, political motivations, pressure for societal relevance or simply because collaboration is viewed as good in and of itself. Overall, collaboration is driven by a variety of purposes, and while collaborations in different specialties have different characteristics (Vermeulen, Parker, and Penders 2013), they are seldom truly global (Wagner 2008).

Collaboratories are arrangements of information and communication technologies (ICTs) that support collective data analysis, remote operation of instruments, and collaboration at a physical distance (Glasner 1996; Olson, Zimmerman, and Bos 2008; Wulf 1989). Collaboratories widen access to research resources, promote interdisciplinarity, and may bridge the world's crippling knowledge divides by building research capacity in the global south (ICSU/ISSC 2010).

Alongside aggregation, scholarship also demonstrates individualizing aspects of research. For instance, before the era of large-scale molecular biology Knorr Cetina (1999) argued that knowledge production in that field depends on individual exchange, in contrast to the integrated collaboration required in particle physics. Additionally, Shapin (1994, 2008) maintains that the personal virtues of individual researchers, such as trustworthiness and honor, were critical for the development of science as a social institution. Further, he contends that the radical uncertainties of contemporary science make this ever more the case today. Investigation of the interaction between the individual and the collective is encapsulated in the concept of "epistemic living spaces" (Felt 2009), which shows how the personal and the institutional are intertwined and shaped by broader epistemic, symbolic, and political forces.

STS is currently moving simultaneously toward analyses of the very small and the very large. At one extreme are studies of small group collaborations that ask how the social and physical environments of groups interact with their composition to shape scientific knowledge. This includes investigations of research groups (Hackett 2005b; Hampton and Parker 2011) and "collaborative circles" (Farrell 2001; McLaughlin 2008), coherent groups (Parker and Hackett 2012, 2014) and the emerging "Science of Team Science" (Hall et al. 2012; Stokols et al. 2008). At the other extreme, bibliometricians have mapped the entire scientific enterprise to uncover relations between disciplines and track the broader influence of articles, research centers, and policy decisions on science writ large. Such efforts originated in the 1970s (Small and Griffith 1974), underwent an unsettled period in the 1980s (Leydesdorff 1987), and have emerged as one of the most technically sophisticated communities studying science today (Boerner 2010; Leydesdorff and Rafols 2009).[2]

Studies of the growth and aggregation of scientific organizations provide knowledge useful for improving science policy and management. The Science of Team Science, for example, enthusiastically offers itself as an instrument for improving the efficacy of collaboration, while maps of science reveal hot and cold regions and survey territory fertile for investment and investigation.

Specialization

STS's analysis of aggregations small and large is set against the backdrop of the history of specialization of scientific knowledge, punctuated by episodes of recombination or integration. The scientific community is a patchwork of overlapping groups, networks, and communities of practice drawn together by shared identities, problems, and methods, and separated by cultural, historical, and epistemic fissures and fault lines. The emergence of disciplines contributed to processes of specialization—beginning with natural philosophy and ramifying into the research areas we know today—and discipline formation, recombination, and branching became focal concerns of the social history of science in the late 1970s (Clarke 1998; Sturdy 2011; Suárez-Díaz 2009).[3] However, studies of disciplines and specialties are written in a highly variable vocabulary that ranges across the map of science: paradigms, social worlds, epistemic cultures, thought styles and cultures, ways of knowing, styles of scientific reasoning, and many more.[4]

Disciplines, specialties, and research areas arise for several reasons: new research apparatus or techniques illuminate the previously unknowable (Bechtel 1986; Clarke 1998; Mulkay, Gilbert, and Woolgar 1975); scientific roles and occupations form at the intersection of established ones (Ben-David and Collins 1966; Frickel 2004); coherent networks arise around potentially generative questions or phenomena (Griffith and Mullins 1972; Powell et al. 2005); research captures the attention of influential interest groups (Clarke 1998; Frickel 2004; Lenoir 1997); new uses are found for new scientific knowledge (Schweber 2006; Shostak 2005). The cumulative effects of several factors often enable the institutionalization of a new field.

The development of fields and specialties typically follows a pattern (Chubin 1976; Collins 1998; De May 1992; Parker and Hackett 2012). Initially, researchers at several locations begin exploring similar problems without knowledge of each other's efforts, and publication is widely dispersed across different disciplinary journals. Through such publications researchers gradually become aware of their common interests. Dense channels of formal and informal communication arise among researchers, and these networks thicken to become the coherent groups and invisible colleges of the nascent research area (Crane 1969; Farrell 2001; Mullins 1972). Improved communication occasions scientific debate and consensus gradually emerges about problems, definitions, techniques, and findings. As it stabilizes the community develops a characteristic *thought style*: a shared cognitive framework characterized by common perspectives, evaluative standards, methods, techniques, and literary styles (Fleck 1981, 99; Hacking 2002; Rose-Greenland 2013). Thought styles are emotive and cognitive, activating "a certain mood" that facilitates "directed perception, with corresponding mental and objective assimilation of what has been so perceived" (Fleck 1981, 99). The thought style gradually enforces a social and cognitive way of doing and seeing, stabilizes meanings, and reinforces the thought structure. The community now perceives the world differently, resulting in potentially incommensurable understanding between those within and outside the new, semi-autonomous scientific domain (cf. Knorr Cetina 1999; Kuhn 1962).

Communication and intellectual growth are inseparable from material aspects of the scientific process. Disciplines are also constellations of practices, instruments, and materials that structure work and shape social networks and thought styles (Collins 1994; Latour 1987; Law and Hassard 1999). Machines, tools, technologies, protocols, and institutes, as well as buildings and journals, are fundamental components of disciplines and epistemic cultures (Fujimura and Chou 1994; Knorr Cetina 1999; Schoenberger 2001). Experimental systems (Rheinberger 1997), platforms (Keating and Cambrosio 2003), and ensembles of research technologies (Hackett et al. 2004) position epistemic things near the center of the research process. The social networks that form new research areas coexist and interact with genealogies of research systems that are manipulated to produce new scientific phenomena and enable the high-consensus, rapid discovery science (Collins 1998; Hackett 2011; Schroeder 2007). Importantly, all these require resources to construct, coupling science more tightly to the interests of the state and capital, and each is itself a resource that imposes social control, promotes competition, and produces stratification.

Once established, research communities are socially stratified and characterized by internal solidarity and external conflict and competition among groups within the field (Becher 1989; Bourdieu 1988). They become arenas of competing emotional and intellectual alliances (Gieryn 1999; Soreanu and Hudson 2008) or "strategic action fields" in which multiple groups compete for status and attention (Bourdieu 1988; Collins 1998; Fligstein and McAdam 2012). But conflict and competition are not the only motivations: the demands of the local social milieu, attempts to realize one's intellectual potential, and the biographical and existential meaning provided by scientific work also propel disciplines (Camic and Gross 2008; Parker and Hackett 2012, 2014; Swedberg 2011).

Disciplinary growth and development resemble social movements (Frickel and Gross 2005). Subgroups within the scientific community attempt to develop and win acceptance for research programs that challenge the current state of scientific knowledge or supplant established scientific techniques, often in the face of tremendous resistance from other scientists. Such movements can succeed when they mobilize key material and emotional resources and high-status intellectuals and recruitment centers, and when they frame their research in ways that resonate with others in the field. After the intellectual beachhead has formed, the movement's ideas become accepted and gradually institutionalized into professional associations, conferences, journals, and (rarely) a new discipline (De May 1992).

Once established, disciplines become structured in ways that influence knowledge work. These include the types of resources needed to conduct research (Frickel and Gross 2005), the power of peer review (Crane 1972; Csíkszentmihályi 1999), and each discipline's relative degree of "attention space"—the number of creative contributions that can be accommodated in journals, conferences, and elsewhere (Collins 1998). Each factor shapes the social organization of scientific work. Consider, for instance, field-level consensus, or the level of agreement among researchers in a field about research questions, methods, and meaningful scientific contributions. Members of high-consensus fields enjoy greater funding, greater autonomy, and more collaboration (Beyer and Lodahl 1976; Fox 2008; Pfeffer and Langton 1993). In low-consensus fields systems for communicating scientific methods and results are more idiosyncratic, research coordination and control units are smaller and less powerful, and considerable theoretical pluralism exists (Fuchs 1992; Whitley 2001). The organization of research in scientific fields, structured by the investment and distribution of resources, shapes scientific knowledge and practice.

Integration

Disciplines and specialties accumulate knowledge by focusing research on specific topics addressed in characteristic ways that meet shared standards of evidence and closure (Jacobs 2014; Jacobs and Frickel 2009). Simultaneously, however, the narrowness of their subjects and methods, and the fact that many intellectual problems fall between or beyond research areas, mean that disciplinary horizons limit scientific advancement and the use of science to solve applied problems (Kostoff 2002). The increased burden of specialization is offset by further narrowing expertise, which limits one's ability to innovate through recombination (Jones 2009). Uneasiness about differentiation and the loss of unity in science can be traced back decades (Weingart 2010). As "researchers tend to work *on* problems, not *in* disciplines" (Klein 2000, 13), they mobilize elements derived from multiple styles, disciplines, or cultures of research (Radick 2000), and collect novel ideas that shape inquiry, collaboration, and the organization of communities (Barry and Born 2013; Strathern 2006).

Intellectual fragmentation stimulates efforts to re-integrate knowledge to produce holistic explanations, resulting in varieties of collaboration that extend beyond disciplinary borders (e.g., multidisciplinarity, interdisciplinarity, transdisciplinarity). These different modes of integration organize inquiry across the fault lines of scientific work and communities through methodological borrowing, theoretical enrichment, and convergent problem solving (Klein 1996). At the low end of integration, multidisciplinary collaboration draws methods, ideas, and theory from disparate academic disciplines and fits them together in much the same way that tiles form a mosaic: the tiles and their individual meanings remain identifiable parts of a new composite (Huutoniemi et al. 2010). Multidisciplinarity is thus additive rather than integrative: the disciplines and disciplinary frameworks remain unaltered and the relationships among disciplines are not well defined (Klein 2010). Interdisciplinary research, in contrast, achieves a qualitatively higher level of integration that dissolves the coherence of disciplines—the tiles are disintegrated and their constituent elements reconstituted into a coherent new whole. Integrative in process and outcome, interdisciplinarity is an epistemic accomplishment that reaches across the hierarchic division of disciplines to meet the challenges posed by complex questions and problems (Klein 2010). Transdisciplinarity, the most unusual and ambitious of the three collaborative varieties, transcends boundaries separating disciplines and professions (Felt et al. 2013; Nowotny, Scott, and Gibbons 2001), reaching into other sectors and communities to fashion coherent solutions and explanations from a diversity of expertise, evidence, and epistemic practices. Critical, transgressive, and synthetic, transdisciplinarity is "the contemporary version of the historical quest for systematic integration of knowledge" (Klein 2010, 24).

Beginning in the mid-1990s, multi-, inter- and transdisciplinary research has been accompanied by scientific synthesis, a new form of scientific collaboration that integrates disparate theories, methods, and data across disciplines, specialties, professions, and scales to produce explanations of greater generality or completeness (Carpenter et al. 2009; Rodrigo et al. 2013). Synthesis happens when theories, concepts, methods, and data are imported from quite different sources—within science and outside—in collaborations often catalyzed by an urgent problem or compelling intellectual challenge.

Synthesis produces emergent knowledge that extends beyond any one discipline, data-set, or method. It occurs both within and across disciplines, specialties, and sectors, and so differs in substance and scope from interdisciplinary and transdisciplinary research.

Specifically constructed for the purpose, synthesis centers (http://synthesis-consortium.org) convene small, intensely interacting working groups of about five to fifteen scientists, policy makers, and practitioners with complementary expertise and data to focus for several days on research problems contributing to fundamental understanding and practical problem solving (Hackett et al. 2008; Hampton and Parker 2011). Synthesis groups gather experts in different disciplines and professions and isolate them for a strictly delimited period of time. The emotional energy and social solidarity of the working group allow collaborators to overcome initial resistance to intersectoral or interdisciplinary collaboration, which, in turn, facilitates rising levels of trust and instrumental intimacy, productive alternation between creative and critical modes of scientific practice, and group flow (Hackett and Parker 2016; forthcoming). Consequently, synthesis groups engage in a highly creative and productive intellectual process, accomplishing in several days what takes other groups months or even years to achieve (Hackett et al. 2008). Synthesis offsets the negative effects of hyperspecial-ization, leverages massive and diverse data, and increases chances of serendipitous dis-covery and transformative science. As such, it is vital for a future in which increasingly specialized sciences face intellectual questions and real-world problems that demand rapid application of ideas and evidence drawn from a wide range of sciences and other bodies of expertise.

The Purposes and Politics of Scientific Organization

Characterizing science as work and examining it carefully exposes its political implica-tions and social inequalities. Just as politics arises in the social and epistemic tensions between specialization and integration, the politics of scientific organization become apparent in discussions about the purpose of science and its connection to other sec-tors in society, especially government and industry. Tensions also arise between the scientific collective and the individual scientist in such matters as the management of research, access to instruments and materials, accountability, job satisfaction, and stratification.

Fifty years ago the distinction between basic and applied research mattered a lot, with nonacademic scientists working on different problems and in different ways than their counterparts in academe (Mulkay 1977; Pelz and Andrews 1966). But in recent years new conceptualizations have arisen, grounded in new understandings of the social purposes of science. Underlying this changed relationship is a transformation from an *industrial* to a *postindustrial society*, as already suggested by Bell (1973) and Drucker (1969), which ushered in a knowledge society founded upon science and technology rather than industry. While Bell and Drucker offer similar analyses of this transformation, they envision quite different implications for the position of science (de Wilde 2001). Bell imagines a privileged place for theoretical knowledge and knowledge institutions, while Drucker predicts the industrialization and commoditization of knowledge. Seen from the vantage point of 2015, Bell's vision seems as idealistic as the utopians described in the chapter introduction, while Drucker's *realpolitik* echoes Weber's prescient warning.

More recently, those who maintain that there has been a transformation from *mode 1* to *mode 2 science* sense the emergence of a new form of knowledge production that displaces disciplinary and fundamental knowledge practices with a more reflexive, transdisciplinary, and heterogeneous "knowledge production" situated in the context of application (Gibbons et al. 1994). Accordingly, the authors identify and perhaps advocate the reform of established institutions, disciplines, practices, and policies. A later book (Nowotny, Scott, and Gibbons 2001) argues that increasing societal complexity, uncertainty, and reflexivity require that science become more thoroughly embedded in society to produce "socially robust knowledge." Similarly, Funtowicz and Ravetz (1993) identify a transformation from *normal* to *post-normal science* that arises in the context of growing uncertainty—when uncertainty becomes fundamental and risks are high, knowledge practices deviate from Kuhn's "normal science." Post-normal science blends scientific methods and principles of conduct with values and practices drawn from outside the scientific community, creating a more pluralistic form of inquiry.[5]

The rise of post-normal science has stimulated collaboration between academic science and private industry, which, in turn, has elicited a vibrant body of theorizing and empirical research. Growing emphasis on the commercialization of public-sector research and development has encouraged universities to enter into cooperative agreements with industry to transfer and develop technologies with commercial potential (Owen-Smith and Powell 2003). As such, the *triple-helix theory* signals an organizational change from separated domains in society to the entanglement of the domains of science, government, and industry (Leydesdorff and Etzkowitz 2001) and can be read as a recipe for innovation in the organization and outcomes of science. These new views on the relation between science and society are also accompanied by discussions about for-profit science and public-private partnerships, and new reflections, theories, and policies on science and its role in society (which at the time of this writing is called "responsible research and innovation"; see Stilgoe and Guston, chapter 29 this volume).

Boundary organizations—formal institutions that mediate interactions between the science and policy communities, bridging their diverse purposes, incongruent values, and mutual incomprehension—are an increasingly common means of linking scientific work to societal purposes (Guston 1999; Parker and Crona 2012). In a boundary organization, researchers, practitioners, and policy makers, often enabled by professional facilitators, use boundary objects to motivate, coordinate, guide, and reward collaboration while discouraging partisanship and imbalanced influence.

Theories of boundary organizations emerged from pathbreaking work on boundary objects and boundary work (Gieryn 1999; Star and Griesemer 1989), and were originally developed in a limited array of distinctive settings where it was generally assumed that science and policy communities were clearly delineated and had equivalent ability to exert power over the organization. Boundary organization theory also assumes that the organization can reconcile conflicting demands and achieve lasting stability between science and policy (Cash 2001; Guston 1999). Boundary organization theory was later modified to account for boundary management in more complex institutional environments wherein the organization serves multiple stakeholders with different levels of power whose competing demands incorporate scientific, political, and industrial agendas. This perspective allows for more realistic analyses of boundary management in complex social environments (e.g., Crona and Parker 2011; Parker and Crona 2012) and has been particularly influential for understanding the use of boundary organizations to promote environmental sustainability (e.g., Boezeman, Vink, and Leroy 2013; von Heland, Crona, and Fidelman 2014).

The role of government in directing scientific work is seen clearly in the rise of strategic research programs and priorities, as has recently been occurring in the form of "grand challenge" campaigns that direct scientific inquiry and hold it accountable to societal purposes (Calvert 2013; Rip 1998).[6] In the 2000s, grand challenges became "a tool for mobilizing an international community of scientists towards predefined global goals with socio-political as well as technical dimensions" (Brooks et al. 2009, 9). For example, following the Gates Foundation (2003) initiative Grand Challenges in Global Health, 400 prominent researchers and politicians stated in the Lund Declaration (2009) that "European research must focus on the Grand Challenges of our time moving beyond current rigid thematic approaches" (1). Today there are grand challenges in many sectors, created either top-down or in public consultations, directing us toward certain futures and away from others (Calvert 2013).

Grand challenges and similar large-scale, targeted research initiatives often require scientists to organize in complementary ways. For instance, the international ATLAS detector at CERN involves about 3,000 researchers (http://www.atlas.cern), while the Laser Interferometer Gravitational Wave Observatory (LIGO) and the U.S. National Ecological Observatory Network (NEON) each cost nearly $500 million to construct. Such massive instrument-centered projects structure the research agenda and commitments of scientists and funding agencies for decades. They also entail demanding reviews before construction and for the many years of active research. The recent downsizing of the $433 million U.S. National Ecological Observatory Network is an example of what happens when large-scale science outgrows decision makers' willingness to pay (Mervis 2015).

Most prominently, these big science initiatives have contributed to the integration of management practices in research: science has become project work. Rooted in large-scale, government-driven scientific efforts, such as the Manhattan Project and the Apollo space program, project design and management developed in fields of construction and engineering during the 1960s (Cicmil and Hodgson 2006; Hodgson 2004; Lock 2003). As part of the New Public Management (Boston et al. 1996; Ferlie et al. 1996), the 1990s saw the project mode expanding across industries and other sectors in a process aptly described as the "projectification of society" (Midler 1995).

Nowadays collaborations and individual research are also predominantly project work, from national and European research programs to the work of Ph.D. students (Torka 2009; Vermeulen 2009). The project format determines the structure of the research process, for example, through clear timeframes and preset deliverables that often go against the grain of the uncertainty and openness of knowledge creation processes. Project funding requires a predefined proposal for research that outlines the goals and outcomes, the research process and its schedule—including a clear beginning and end, as well as participants and their responsibilities.

While this pattern of organization makes research more legible to outsiders and more accountable through audits and evaluations, it also contributes to the bureaucratization of research and its associated red tape (Power 1997). For instance, the U.S. National Science Foundation supports the construction of large-scale research instrumentation through the Major Research Equipment and Facilities Construction account that imposes strong reporting requirements and places enduring demands on budgets to support research that uses the newly developed instrument. First applied to such major research projects, the culture of evaluation has spread to performance-based, research-funding systems that establish resource levels for large-scale institutions (Butler 2010; Hicks 2012; Martin and Whitley 2010). Such evaluation systems employ peer review and/or academic output indicators, including some that measure societal impacts beyond the academic. These measures have become highly performative or reactive, yet at the same time influence resource distribution and other dimensions of stratification in science (Good et al. 2015; Rushforth and de Rijcke 2015).

Access to research resources—instruments and facilities, students and collaborators, and that most precious resource, time—is both cause and consequence of stratification in science. Science is stratified along many dimensions, including publication and citation rates, resources, credibility, and participation by women and members of certain ethnic groups (particularly at higher academic ranks or in more prestigious institutions; Cole and Cole 1973). For example, a small proportion of researchers are responsible for a disproportionate share of scientific publications, and a relatively small pool of articles receives a disproportionate share of citations (Garfield 2006; Lotka 1926; Price 1986). A small number of influential researchers thus wield vastly disproportionate influences on their fields. And though their representation has improved in recent decades, women remain underrepresented in science (European Commission 2006; National Science Board 2008; see Fox, Whittington, and Linková, chapter 24 this volume). The same pattern holds true for non-Western scientists, particularly among scientific elites. Across disciplines between 40 percent and 90 percent of the world's most highly cited scientists live in the United States and Western Europe (Basu 2006; Parker, Vermeulen, and Penders 2010). STS scholarship is similarly skewed, and only in the past ten years has a truly global STS community begun to emerge as conferences are held in non-Western countries and new regional STS journals are established.

The professional expectations and organizational environments of university faculty influence their job satisfaction and dissatisfaction (self-doubt and anxiety) (Hermanowicz 1998). Different tiers within the university system (high, middle, and low) constitute different academic "social worlds" with characteristic norms regarding performance, and which provide differential access to the resources needed to meet them. Professional expectations interact with these local resource availabilities to shape scientists' job satisfaction. Middle-tier researchers experience the greatest levels of job satisfaction because their performative standards are flexible and they have ample resources to meet them. Lower-tier researchers lack the resources to conduct meaningful scientific work and so experience less job satisfaction. Top-tier researchers have ample resources, but unrelenting pressure leads to dissatisfaction and perpetual self-doubt. The personal and emotive aspects of scientific work are shaped by the stratification system of science (Hermanowicz 2003).

Looking Forward

"... the future is not what it used to be ..."
(Laura Riding and Robert Graves, [1937] 2001, 170)

In the first edition of the STS handbook (Spiegel-Rösing and Price 1978, 93–148), Michael Mulkay arrived at Riding and Graves's view of the future of science by first reviewing the "sociology of the scientific research community" in a long chapter that began with the distinction between pure (basic) and applied research, outlined the normative structure of science and the social processes (rewards, exchange) that supported it, and sketched the dynamics of discipline and specialty formation. His view of the future is a valuable counterpoint to our own. Mulkay continued by observing that growth brings differentiation (specialization) and interdisciplinary collaboration, and the resulting networks (invisible colleges) offer communication channels, recognition, and coordination (once competition-driven secrecy eased). Differentiation and growth demand resources, creating dependence on government funding, accompanied by increasing expectations for rapid and certain societal benefits (tighter coupling of science to social purposes). Through this process pure research would be supplanted by applied research, Mulkay thought, leading him to close with some observations about its social characteristics.

Mulkay's chapter identifies the seeds of many transformative forces that are restructuring science today, most notably the diminishing desire of governments to fund research for its own sake (Schuster and Finkelstein 2006), accompanied by rising concern for its measurable societal benefits (construed narrowly in the United States as national health, wealth, and security). Today, however, the dichotomy between pure and applied research would be accompanied by "scare quotes" to signal distancing from such an unqualified distinction. To the extent one would today distinguish between investigator-initiated or curiosity-driven research and research that is specified and delimited by a patron, that distinction would be discussed as an essential ambiguity or tension that reflects a shifting compromise in the societal contract that organizes science.

Collaboration in varied forms is rampant (Parker, Vermeulen, and Penders 2010; Penders, Vermeulen, and Parker 2015; Wuchty, Jones, and Uzzi 2007), and specialization has sparked counterbalancing efforts at creative recombination. The capitalist spirit in science, first identified by Weber, is currently folded into a regime of academic capitalism (Hackett 1990), economization (Berman 2014), or neoliberalism (Mirowski 2011) that has all but obliterated the distinction between pure and applied research (and some would say that science was never pure anyway; [Shapin 2010]). Concern for the normative structure of science has long passed: counternorms (Mitroff 1974) and sociological ambivalence (Merton 1973) have given way to high-resolution studies of strategies for managing the essential tensions of science to productive and creative effect (Hackett 2005b; Hackett and Parker 2016; Lee, Walsh, and Wang 2015; Uzzi et al. 2013). Growth, differentiation, and the intellectual structure of science are today analyzed on a massive scale (Uzzi et al. 2013) and depicted in exquisite detail (Boerner 2010; Wyatt et al., chapter 3 this volume). Limited resources and scientists' dependence on them remain as challenging as ever, exacerbated by macroeconomic volatility that wracks whole economies and governments.

Several emerging trends will shape the social organization of science in the years ahead, and these changes pose challenges for science studies scholars. Technologies transform sciences, and ICTs are the most powerful and transformative technologies of our day and will be the foundation and accelerant for all of the others. ICT-mediated collaborative research, or "e-science," uses "the Internet as an underlying research technology or infrastructure" to alter research practices across various disciplines and knowledge domains (Meyer and Schroeder 2015, 4). Internet-mediated research is practiced at unprecedented scale, scope, and speed, and the practice is spreading rapidly. This explicit and formalized network-building process requires public rules of membership and thus represents a transition from the notion of the invisible college to that of a quite visible college or network of collaborators (Crane 1972; Price 1963). But distance matters (Olson and Olson 2000; Olson, Zimmerman, and Bos 2008), as do other differences: scientists within a building whose paths overlap are more likely to collaborate and secure funding (Kabo et al. 2014), whereas collaborations that span institutions incur transaction costs that impair performance (Cummings and Kiesler 2005). And intense, isolated, and enduring interaction fuels sociality, trust, and the escalating intimacy that promote scientific integration or synthesis (Hackett and Parker 2016, forthcoming; Parker and Hackett 2012). The paradox of ICT-mediated science and the challenges for STS scholars is to understand how to achieve the velocity and intensity that promotes excellent science while working at a distance with diverse collaborators. Perhaps the promise of ICTs is not their ability to do old things in new ways—that is, to host interpersonal collaborations and distal analysis and operation of instruments—but to do something quite new: to create "knowledge machines" and an emerging form of e-science that is qualitatively different from traditional modes of inquiry (Meyer and Schroeder 2015).

While transforming the microsocial processes of collaboration, ICT also enables global scientific collaboration at unprecedented reach and scale. The promise of such collaboration is reflected in the Future Earth initiative of the International Council for Science, for example, which aspires to construct a "global research platform" to enable collaboration among scientists and diverse societal partners to develop the knowledge necessary to initiate transformations toward sustainability and sustainable development (future-earth_10-year-vision_web.pdf). The global organization of research for humanitarian purposes is a noble pursuit with some 400 years of history (remember the Lincei), but the social organization of global research will be shaped by global concerns and considerations (competitiveness, sustainability, capacity building—or exploitation—of talent in developing countries). In 2011 the International Council for Science undertook a foresight exercise to develop scenarios for the future of science (ICSU 2011). The report recognized the power of ICT, the intellectual opportunities of global interdisciplinary collaboration, and the desperate urgency of transdisciplinary collaboration to bring science and engineering to bear on wicked societal problems of every imaginable sort (health, resource use, poverty, urbanization, water, climate change, and such), but in the end the analysis distilled to two dimensions: would science be organized in ways that *engaged* societal problems or would it be *detached* from them? And would states' interests be parochially *national* or embracingly *global* (ICSU 2011, 18)? Aggressive nationalism, science for sale in the marketplace, or the prevalence of national interests over global interests would produce a pattern of collaboration unlikely to meet global needs, or even oppose them. Only a science engaged with social purposes and supported by national governments committed to the common good would form the collaborations necessary to produce knowledge equal to the challenges ahead. ICT may be necessary for us to realize the dreams of Federico Cesi and the Lincei, but they are not sufficient.

STS scholarship will have much to study but little to do that will likely influence the macroscopic dynamics that are shaping science. No amount of scholarship will shift states' interests from the national to the global, or from detached to engaged. But science engaged with global challenges, as it must be if it is to inform the pressing issues of our day, will offer much of interest to STS scholars and will demand much in return.

For example, ICTs and other new forms of research technology make possible closely monitored, carefully studied, and adaptively managed collaborations that will be shaped by social researchers who study them in real time and provide continuous feedback about their progress. The nascent Science of Team Science initiative, currently more aspiration than accomplishment, will advance in partnership with studies of individual and group creativity (e.g., Amabile 1996; Farrell 2001) and with the increasingly sophisticated use of sociometric sensors and other devices that capture rich and detailed (and massive) data about the process and outcomes of interpersonal collaboration, coupled with the emerging tools to analyze extremely large data sets.

Diverse publics are also increasingly engaged in research. The era of epistolary science has passed, and for the last century science has been conducted mainly in universities and industry, increasingly under a regime of capitalism, economization, and neoliberalism. But in recent years a countervailing practice has emerged—"citizen science"—"scientific work undertaken by members of the general public, often in collaboration with or under the direction of professional scientists and scientific institutions" (Conz 2006; ODE 2014). Amateur scientists have discovered new planets, assessed regional biodiversity, and uncovered intricate patterns of protein folding. There are also new trends toward crowdfunding of scientific research by the general public, particularly as governmental funding for science has receded in recent years (Meyer and Schroeder 2015). The rise of amateur science, citizen science, crowdfunded science, and other alternative patterns of organization call for us to reconsider the social contract of science and to reopen debates about who counts as a legitimate scientific collaborator, how scientific knowledge acquires practical relevance, and how public awareness and trust are achieved and sustained.

Finally, "open science" is on the rise, making the inner workings of scientific research accessible to all (Fecher and Friesike 2014; Meyer and Schroeder 2015, 175–186; The Royal Society 2012).[7] While open access is removing the cost and copyright barriers to published research, making it available to everyone with an Internet connection, open data also implies sharing data either before or soon after publication, which would transform the research process by opening it up. Such openness may redefine accountability in science by bringing unprecedented scrutiny to scientific claims and to the entire course of data collection, cleaning, management, analysis, and interpretation (Hartter et al. 2013). Still, these processes involve substantial costs and complications that may create new inequalities that derive from the need to pay for open access publication and/or data curation.[8] The Matthew effect is alive and well (Merton 1973).

In closing, we return to our points of reference: science is an increasingly organized, institutionalized, and managed form of professional work tightly woven into the fabric of society and tightly coupled to social purposes. Whose purposes? is the unavoidable but difficult question. Science organized to serve capital and the interests of the richest billion or so people in the world will look very different from science that is globally engaged and working in collaboration with the global south. Science of the first sort will become increasingly tightly coupled to the economic, defense, and well-being needs of a fraction of the world, whereas science of the second sort will engage diverse publics in varied places to collaborate on wicked problems that contribute to sustainability and social equity. Both forms will be imbued with values, but the values will differ sharply.

Whatever the driving purposes and guiding values, science will be an increasingly collaborative activity, with work organized in a wide spectrum of virtual and face-to-face formats. ICTs, collaboratories, synthesis centers, and forms of virtual collaboration that we can dimly imagine will likely shape the scientific future of the years ahead. Scientists may still feel a vocation for science, but that calling will probably lead to a highly structured and deeply capitalized and regulated workplace. The apparent freedom of ICT-mediated scientific work will be offset by the increased ease of surveillance.

Growth and specialization depend upon resources and purposes controlled by scientific decision makers and those in government who oversee their work. At this writing (January 2016) U.S. science budgets have increased modestly, while the cost of research—including the cost of oversight and accountability—continues to rise more rapidly, and research opportunities increase geometrically. With increasing specialization comes geometrically increasing opportunities for inter- and transdisciplinary integration or synthesis, adding an additional burden of organization and expense to budgets already strained. Priorities are inevitable, accompanied by increased accountability and demands for metrics of science that will evaluate progress and inform decisions. The structure of scientific knowledge and perhaps of scientific revolutions will be shaped by conscious decisions made outside the research context. And, of course, all this will offer myriad opportunities for studies of the social and epistemic organization of science.

Notes

1. Though historians had done so previously using documentary approaches (see Pickstone 2000).

2. While often working with different tools toward different ends, STS and bibliometrics have a long history of informing one another and are generally complementary enterprises (see Wyatt et al., chapter 3 this volume). The overlap and mutual relevance of these research areas is apparent.

3. For the study of disciplines, see Kohler (1982) and Lenoir (1997).

4. For paradigms, see Kuhn [1962] 1970); for social words, see e.g., Gerson (1983) and Clarke (1991); for epistemic cultures, see Knorr Cetina (1999); for thought styles and cultures, see Fleck ([1935] 1981); for ways of knowing, see Pickstone (2000); for styles of scientific reasoning, see Hacking (2002).

5. But these claims have also received important criticism (see e.g., Hessels and van Lente 2008; Tuunainen 2002).

6. See also the Fred Jevons lecture on "Fashions in Science Policy" given by Arie Rip on March 2, 2014, at the University of Manchester (https://www.youtube.com/watch?v=kKqX-5VOqLc).

7. Peer review itself is becoming more open through such sites as https://pubpeer.com.

8. For details, please see http://blogs.lse.ac.uk/impactofsocialsciences/2015/04/21/to-what-are-we-opening-science/.

References

Amabile, Teresa. 1996. *Creativity in Context: Update to the Social Psychology of Creativity.* Boulder, CO: Westview Press.

Barry, Andrew, and Georgina Born. 2013. *Interdisciplinarity: Reconfigurations of the Social and Natural Sciences.* London: Routledge.

Basu, Aparna. 2006. "Using ISI's 'Highly Cited Researchers' to Obtain a Country Level Indicator of Citation Excellence." *Scientometrics* 68 (3): 361–75.

Becher, Tony. 1989. *Academic Tribes and Territories: Intellectual Inquiry and the Cultures of Disciplines.* Milton Keynes: The Society for Research into Higher Education and Open University Press.

Bechtel, William, ed. 1986. *Integrating Scientific Disciplines.* Dordrecht, Netherlands: Martinus Nijhoff Publishers.

Bell, Daniel. 1973. *The Coming of Post-industrial Society: A Venture in Social Forecasting.* New York: Basic Books.

Ben-David, Joseph, and Randall Collins. 1966. "Social Factors in the Origins of a New Science: The Case of Psychology." *American Sociological Review* 31 (4): 451–65.

Berman, Elizabeth P. 2014. "Not Just Neoliberalism: Economization in US Science and Technology Policy." *Science, Technology, & Human Values* 39 (3): 397–431.

Beyer, Janice M., and Thomas M. Lodahl. 1976. "A Comparative Study of Patterns of Influence in United States and English Universities." *Administrative Science Quarterly* 21 (1): 104–29.

Boerner, Katy. 2010. *Atlas of Science.* Cambridge, MA: MIT Press.

Boezeman, Daan, Martinus Vink, and Pieter Leroy. 2013. "The Dutch Delta Committee as a Boundary Organization." *Environmental Science and Policy* 27: 162–71.

Boston, Jonathan, John Martin, June Pallot, and Pat Walsh. 1996. *Public Management: The New Zealand Model.* Oxford: Oxford University Press.

Bourdieu, Pierre. 1988. *Homo Academicus.* Stanford, CA: Stanford University Press.

Brooks, Sally, Melissa Leach, Henry Lucas, and Erik Millstone. 2009. *Silver Bullets, Grand Challenges and the New Philanthropy.* STEPS Working Paper 24. Brighton: STEPS Centre.

Butler, Linda. 2010. *Impacts of Performance-Based Research Funding Systems: A Review of the Concerns and Evidence.* Paris: OECD.

Calvert, Jane. 2013. "Systems Biology, Big Science and Grand Challenges." *BioSocieties* 8 (4): 466–79.

Camic, Charles, and Neil Gross. 2008. "The New Sociology of Ideas." In *The Blackwell Companion to Sociology*, edited by Judith R. Blau, 236–39. Malden, MA: Blackwell.

Capshew, James H., and Karen A. Rader. 1992. "Big Science: Price to Present." *Osiris* 7 (1): 2–25.

Carpenter, Stephen R., Virginia Armbrust, Peter W. Arzberger, F. Stuart Chapin, James J. Elser, Edward J. Hackett, Anthony R. Ives, et al. 2009. "Accelerate Synthesis in Ecology and Environmental Sciences." *BioScience* 59 (8): 699–701.

Cash, David W. 2001. "In Order to Aid in Diffusing Useful and Practical Information: Agricultural Extension and Boundary Organizations." *Science, Technology, & Human Values* 26 (4): 431–53.

Chubin, Daryl E. 1976. "The Conceptualization of Scientific Specialties." *Sociological Quarterly* 17 (4): 448–76.

Cicmil, Svetlana, and Damian Hodgson. 2006. *Making Projects Critical.* Basingstoke: Palgrave Macmillan.

Clarke, Adele. 1991. "Social Worlds/Arenas Theory as Organizational Theory." In *Social Organization and Social Process: Essays in Honor of Anselm Strauss*, edited by David R. Maines, 119–58. New York: Aldine de Gruyter.

___. 1998. *Disciplining Reproduction: Modernity, American Life Sciences, and the Problems of Sex.* Berkeley: University of California Press.

• Cole, Jonathan R., and Stephen Cole. 1973. *Social Stratification in Science.* Chicago: University of Chicago Press.

• Cole, Jonathan R., and Harriet Zuckerman. 1975. "The Emergence of a Scientific Specialty: The Self-Exemplifying Case of the Sociology of Science." In *The Idea of Social Structure: Papers in Honor of Robert K. Merton*, edited by Lewis A. Coser, 139–74. New York: Harcourt Brace Jovanovich.

Collins, Randall. 1994. "Why the Social Sciences Won't Become High-Consensus, Rapid-Discovery Science." *Sociological Forum* 9 (2): 155–77.

___. 1998. *The Sociology of Philosophies: Towards a Global Theory of Intellectual Change.* Cambridge, MA: Harvard University Press.

Conz, David B. 2006. "Citizen Technoscience: Amateur Networks in the International Biodiesel Movement." Ph.D. dissertation. Arizona State University, Tempe, AZ.

• Crane, Diana. 1969. "Social Structure in a Group of Scientists." *American Sociological Review* 36 (3): 335–52.

• ___. 1972. *Invisible Colleges.* Chicago: University of Chicago Press.

Crona, Beatrice I., and John N. Parker. 2011. "Network Determinants of Knowledge Utilization: Preliminary Lessons from a Boundary Organization." *Science Communication* 33 (4): 448–71.

• Cummings, Jonathan N., and Sara Kiesler. 2005. "Collaborative Research across Disciplinary and Organizational Boundaries." *Social Studies of Science* 35 (5): 703–22.

Csikszentmihályi, Mihalyi. 1999. "Implications of a Systems Perspective for the Study of Creativity." In *Handbook of Creativity*, edited by R. J. Sternberg, 313–38. Cambridge: Cambridge University Press.

De May, Marc. 1992. *The Cognitive Paradigm: An Integrated Understanding of Scientific Development.* Chicago: University of Chicago Press.

Doing, Park. 2008. "Give Me a Laboratory and I Will Raise a Discipline: The Past, Present, and Future Politics of Laboratory Studies in STS." In *The Handbook of Science and Technology Studies.* 3rd ed., edited by Edward J. Hackett, Olga Amsterdamska, Michael Lynch, and Judy Wajcman, 279–96. Cambridge, MA: MIT Press.

Drucker, Peter F. 1969. *The Age of Discontinuity: Guidelines to Our Changing Society.* London: Heinemann.

Edge, David O., and Michael J. Mulkay. 1976. *Astronomy Transformed: The Emergence of Radio Astronomy in Britain.* New York: Wiley.

Elkana, Yehuda, Joshua Lederberg, Robert K. Merton, Arnold Thackray, and Harriet Zuckerman, eds. 1978. *Toward a Metric of Science: The Advent of Science Indicators.* New York: Wiley.

European Commission. 2006. *She Figures 2006: Women in Science, Statistics and Indicators.* Directorate-General for Research, Science and Society. Brussels: European Commission.

Farrell, Michael P. 2001. *Collaborative Circles: Friendship Dynamics and Creative Work.* Chicago: University of Chicago Press.

Fecher, Benedikt, and Sascha Friesike. 2014. "Open Science: One Term, Five Schools of Thought." In *Opening Science: The Evolving Guide on How the Internet Is Changing Research, Collaboration and Scholarly Publishing*, edited by Soenke Bartling and Sascha Friesike, 17–47. Heidelberg: Springer Open.

• Felt, Ulrike, ed. 2009. *Knowing and Living in Academic Research: Convergence and Heterogeneity in Research Cultures in the European Context.* Prague: Academy of Sciences of the Czech Republic.

• Felt, Ulrike, Judith Igelsboeck, Andrea Schikowitz, and Thomas Voelker. 2013. "Growing into What? The (Un)disciplined Socialisation of Early Stage Researchers in Transdisciplinary Research." *Higher Education* 65 (4): 511–24.

Ferlie, Ewan, Lynn Ashburner, Louise Fitzgerald, and Andrew Pettigrew. 1996. *New Public Management in Action.* Oxford: Oxford University Press.

• Fleck, Ludwik. [1935] 1981. *Genesis and Development of a Scientific Fact.* Chicago: University of Chicago Press.

___. 1986. "Scientific Observation and Perception in General." In *Cognition and Fact: Materials on Ludwik Fleck*, edited by Robert S. Cohen and Thomas Schnelle, 59–78. Dordrecht: D. Reidel.

Fligstein, Neil, and Doug McAdam. 2012. *A Theory of Fields*. New York: Oxford University Press.

Fox, Mary F. 2008. "Collaboration between Science and Social Science: Issues, Challenges and Opportunities." In *Integrating the Sciences and Society: Challenges, Practices and Potentials*, edited by Harriet Hartman, 17–30. Bingley: Elsevier.

Frickel, Scott. 2004. *Chemical Consequences: Environmental Mutagens, Scientist Activism, and the Rise of Genetic Toxicology*. New Brunswick, NJ: Rutgers University Press.

Frickel, Scott, and Neil Gross. 2005. "A General Theory of Scientific/Intellectual Movements." *American Sociological Review* 70 (2): 204–32.

Fuchs, Stephan. 1992. *The Professional Quest for Truth: A Social Theory of Science and Knowledge*. Albany: State University of New York Press.

Fujimura, Joan H. 1996. *Crafting Science: A Sociohistory of the Quest for the Genetics of Cancer*. Cambridge, MA: Harvard University Press.

Fujimura, Joan H., and Danny Y. Chou. 1994. "Dissent in Science: Styles of Scientific Practice and Controversy over the Cause of AIDS." *Social Science in Medicine* 38 (8): 1017–36.

Funtowicz, Silvio, and Jerome Ravetz, 1993. "Science for the Post-Normal Age." *Futures* 25 (7): 735–55.

Garfield, Eugene. 1984. "The 100 Most-Cited Papers Ever and How We Select Citation Classics." *Current Contents* 23: 3–9.

___. 2006. "The History and Meaning of the Journal Impact Factor." *Journal of the American Medical Association* 295: 90–93.

Gates Foundation. 2003. *Grand Challenges in Global Health*. Accessed February 11, 2016, at http://grandchallenges.org/.

Gerson, Elihu M. 1983. "Scientific Work, Social Worlds." *Knowledge: Creation, Diffusion, Utilization* 4 (3): 357–77.

Gibbons, Michael, Camille Limoges, Helga Nowotny, Simon Schwartzman, Peter Scott, and Martin Trow. 1994. *The New Production of Knowledge: The Dynamics of Science and Research in Contemporary Societies*. London: Sage.

Gieryn, Thomas F. 1999. *Cultural Boundaries of Science: Credibility on the Line*. Chicago: University of Chicago Press.

Glasner, Peter. 1996. "From Community to 'Collaboratory'? The Human Genome Mapping Project and the Changing Culture of Science." *Science and Public Policy* 23 (2): 109–16.

Good, Barbara, Niki Vermeulen, Brigitte Tiefenthaler, and Erik Arnold. 2015. "Counting Quality? The Case of the Czech Evaluation Methodology." *Research Evaluation* 24 (2): 91–105.

Griffith, Belver C., and Nicholas C. Mullins. 1972. "Coherent Groups in Scientific Change: 'Invisible Colleges' May Be Consistent throughout Science." *Science* 177: 959–64.

Guston, David H. 1999. "Stabilizing the Boundary between US Politics and Science: The Role of the Office of Technology Transfer as a Boundary Organization." *Social Studies of Science* 29 (1): 87–111.

• Hackett, Edward J. 1990. "Science as a Vocation in the 1990s: The Changing Organizational Culture of Academic Science." *Journal of Higher Education* 61 (3): 241–79.

Hackett, Edward J. 2005a. "Introduction to the Special Guest-Edited Issue on Scientific Collaboration." *Social Studies of Science* 35 (5): 667–72.

___. 2005b. "Essential Tensions: Identity, Control, and Risk in Research." *Social Studies of Science* 35 (5): 787–826.

___. 2011. "Possible Dreams: Research Technologies and the Transformation of the Human Sciences." In *The Handbook of Emergent Technologies in Social Research*, edited by Sharlene Nagy Hesse-Biber, 25–46. New York: Oxford University Press.

Hackett, Edward J., David Conz, John N. Parker, Jonathan Bashford, and Susan DeLay. 2004. "Tokamaks and Turbulence: Research Ensembles, Policy and Technoscientific Work." *Research Policy* 33 (5): 747–67.

• Hackett, Edward J., and John N. Parker. 2016. "Ecology Reconfigured: Organizational Innovation, Group Dynamics and Scientific Change." In *The Local Configuration of New Research Fields: On Regional and National Diversity*, edited by Martina Merz and Philippe Sormani, 153-72. Switzerland: Springer.

Hackett, Edward J., and John Parker. Forthcoming. "From Salomon's House to Synthesis Centers." In *Intellectual and Institutional Innovation in Science: Historical and Sociological Perspectives*, edited by Thomas Heinze and Richard Muensch. London: Palgrave Macmillan.

Hackett, Edward J., John N. Parker, David Conz, Diana Rhoten, and Andrew Parker. 2008. "Ecology Transformed: The National Center for Ecological Analysis and Synthesis and the Changing Patterns of Ecological Research." In *Scientific Collaboration on the Internet*, edited by Gary M. Olson, Ann Zimmerman, and Nathan Bos, 277–96. Cambridge, MA: MIT Press.

Hacking, Ian. 2002. "'Style' for Historians and Philosophers." In *Historical Ontology*, edited by Ian Hacking, 178–99. Cambridge, MA: Harvard University Press.

Hagstrom, Warren. 1965. *The Scientific Community*. New York: Basic Books.

Hall, Kara, Daniel Stokols, Brooke A. Stipelman, Amanda L. Vogel, Annie Feng, Beth Masimore, Glen Morgan, Richard P. Moser, Stephen E. Marcus, and David Berrigan. 2012. "Assessing the Value of Team Science: A Study Comparing Center- and Investigator-Initiated Grants." *American Journal of Preventive Medicine* 42 (2): 157–63.

Hampton, Stephanie E., and John N. Parker. 2011. "Collaboration and Productivity in Scientific Synthesis." *BioScience* 61 (11): 900–910.

Hartter, Joel, Sadie J. Ryan, Catrina A. MacKenzie, John N. Parker, and Carly A. Strasser. 2013. "Spatially Explicit Data: Stewardship and Ethical Challenges in Science." *PLoS Biology* 11 (9): e1001634.

Hermanowicz, Joseph C. 1998. *The Stars Are Not Enough: Scientists—Their Passions and Professions.* Chicago: University of Chicago Press.

___. 2003. "Scientists and Satisfaction." *Social Studies of Science* 33 (1): 45–73.

Hessels, Laurens, and Harro van Lente. 2008. "Re-thinking New Knowledge Production: A Literature Review and Research Agenda." *Research Policy* 37 (4): 740–60.

Hicks, Diana. 2012. "Performance-Based University Research Funding Systems." *Research Policy* 41: 251–61.

Hodgson, Damian. 2004. "Project Work: The Legacy of Bureaucratic Control in the Post-Bureaucratic Organization." *Organization* 11 (1): 81–100.

Huutoniemi, Katri, Julie Thompson Klein, Henrik Bruun, and Janne Hukkinen. 2010. "Analyzing Interdisciplinarity: Typology and Indicators." *Research Policy* 39 (1): 79–88.

ICSU. 2011. *Foresight Analysis Report 1: International Science in 2031—Exploratory Scenarios.* Paris: ICSU.

ICSU/ISSC. 2010. *Earth System Science for Global Sustainability: The Grand Challenges.* International Council for Science, Paris.

• Jacobs, Jerry A. 2014. *In Defense of Disciplines: Interdisciplinarity and Specialization in the Research University.* Chicago: University of Chicago Press.

• Jacobs, Jerry A., and Scott Frickel. 2009. "Interdisciplinarity: A Critical Assessment." *American Review of Sociology* 35: 43–65.

Jasanoff, Sheila. 2004. *States of Knowledge: The Co-production of Science and the Social Order.* London: Routledge.

Jones, Benjamin F. 2009. "The Burden of Knowledge and the 'Death of the Renaissance Man': Is Innovation Getting Harder?" *Review of Economic Studies* 76 (1): 283–317.

Kabo, Felichism W., Natalie Cotton-Nessler, Yongha Hwang, Margaret C. Levenstein, and Jason Owen-Smith. 2014. "Proximity Effects on the Dynamics and Outcomes of Scientific Collaborations." *Research Policy* 43 (9): 1469–85.

Keating, Peter, and Alberto Cambrosio. 2003. *Biomedical Platforms, Realigning the Normal and the Pathological in Late-Twentieth-Century Medicine.* Cambridge, MA: MIT Press.

Klein, Julie Thompson. 1996. *Crossing Boundaries: Knowledge, Disciplinarities, and Interdisciplinarities.* Charlottesville: University Press of Virginia.

___. 2000. "A Conceptual Vocabulary of Interdisciplinary Science." In *Practising Interdisciplinarity*, edited by Peter Weingart and Nico Stehr, 3–24. Toronto: University of Toronto Press.

Klein, Julie Thompson. 2010. "A Taxonomy of Interdisciplinarity" In *The Oxford Handbook of Interdisciplinarity*, edited by Robert Frodeman, Julie Thompson Klein, and Carl Mitcham, 15–30. Oxford: Oxford University Press.

Knorr Cetina, Karin. 1999. *Epistemic Cultures: How the Sciences Make Knowledge*. Cambridge, MA: Harvard University Press.

Kohler, Robert E. 1982. *From Medical Chemistry to Biochemistry: The Making of a Biomedical Discipline*. Cambridge: Cambridge University Press.

Kostoff, Ronald N. 2002. "Overcoming Specialization." *Bioscience* 52 (10): 937–41.

Kuhn, Thomas S. [1962] 1970. *The Structure of Scientific Revolutions*. Chicago: University of Chicago Press.

Latour, Bruno. 1987. *Science in Action: How to Follow Scientists and Engineers through Society*. Cambridge, MA: Harvard University Press.

Latour, Bruno, and Steve Woolgar. 1979. *Laboratory Life: The Construction of Scientific Facts*. Princeton, NJ: Princeton University Press.

Law, John. 1976. "The Development of Specialties in Science: The Case of X-Ray Protein Crystallography." In *Perspectives on the Emergence of Scientific Disciplines*, edited by Gerard Lemaine, Roy MacLeod, Michael Mulkay, and Peter Weingart, 123–52. Chicago: Aldine.

Law, John, and John Hassard, eds. 1999. *Actor Network Theory and After*. Oxford: Blackwell and the Sociological Review.

Lee, You-Na., John P. Walsh, and Jian Wang. 2015. "Creativity in Scientific Teams: Unpacking Novelty and Impact." *Research Policy* 44 (3): 684–97.

Lenoir, Timothy. 1997. "The Discipline of Nature and the Nature of Disciplines." In *Instituting Science: The Cultural Production of Disciplines*, edited by Timothy Lenoir, 45–74. Stanford, CA: Stanford University Press.

Leydesdorff, Loet. 1987. "Various Methods for the Mapping of Science." *Scientometrics* 11 (5–6): 291–320.

Leydesdorff, Loet, and Henry Etzkowitz. 2001. "A Triple Helix of University-Industry-Government Relations: Mode 2 and the Globalization of National Systems of Innovation." In *Science under Pressure*, 7–33. Aarhus: The Danish Institute for Studies in Research and Research Policy.

Leydesdorff, Loet, and Ismael Rafols. 2009. "A Global Map of Science Based on the ISI Subject Categories." *Journal of the American Society for Information Science & Technology* 60 (2): 348–62.

Lock, Dennis. 2003. *Project Management*. 8th ed. Aldershot: Gower Publishing.

Lotka, Alfred. 1926. "The Frequency Distribution of Scientific Productivity." *Journal of the Washington Academy of Sciences* 16 (12): 317–24.

Lund Declaration. 2009. *Europe Must Focus on the Grand Challenges of Our Time*. Accessed at http://www.vinnova.se/upload/dokument/Verksamhet/UDI/Lund_Declaration.pdf.

Maienschein, Jane. 1993. "Why Collaborate?" *Journal of the History of Biology* 26 (2): 167–83.

Martin, Ben, and Richard Whitley. 2010. "The UK Research Assessment Exercise: A Case of Regulatory Capture?" In *Reconfiguring Knowledge Production: Changing Authority Relationships in the Sciences and Their Consequences for Intellectual Innovation*, edited by Richard Whitley, Jochen Glaeser, and Lars Engwall, 51–80. Oxford: Oxford University Press.

McLaughlin, Neil. 2008. "Collaborative Circles and Their Discontents: Revisiting Conflict and Creativity in Frankfurt School Critical Theory" *Sociologica* 2 (2): 1–35.

• Merton, Robert K. 1973. *The Sociology of Science*. Chicago: University of Chicago Press.

• ___. [1938] 1970. *Science, Technology and Society in Seventeenth Century England*. New York: Howard Fertig.

Mervis, Jeffrey. 2015. "The Dimming of NEON." *Science* 349: 574.

Meyer, Eric T., and Ralph Schroeder. 2015. *Knowledge Machines: Digital Transformations of the Sciences and Humanities*. Cambridge, MA: MIT Press.

Midler, Christophe. 1995. "Projectification of the Firm: The Renault Case." *Scandinavian Journal of Management* 11 (4): 363–75.

Mirowski, Philip. 2011. *Science-Mart: Privatizing American Science*. Cambridge, MA: Harvard University Press.

Mitroff, Ian I. 1974. "Norms and Counter-Norms in a Select Group of the Apollo Moon Scientists: A Case Study of the Ambivalence of Scientists." *American Sociological Review* 39 (4): 579–95.

Mulkay, Michael. 1977. "Sociology of the Scientific Research Community." In *Science, Technology and Society: A Cross-Disciplinary Perspective*, edited by Ina Spiegel-Rösing and Derek J. de Solla Price, 93–148. Beverly Hills, CA: Sage.

Mulkay, Michael, G. Nigel Gilbert, and Steve Woolgar. 1975. "Problem Areas and Research Networks in Science." *Sociology* 9 (2): 187–203.

Mullins, Nicholas C. 1972. "The Development of a Scientific Specialty: The Phage Group and the Origins of Molecular Biology." *Minerva* 10 (1): 51–82.

___. 1973. *Theories and Theory Groups in Contemporary American Sociology*. New York: Harper & Row.

National Science Board. 2008. *Science and Engineering Indicators 2008*. Arlington, VA: National Science Foundation (vol. 1, NSB 08–01; vol. 2, NSB 08–01A).

Nowotny, Helga, Peter Scott, and Michael Gibbons. 2001. *Rethinking Science: Knowledge in an Age of Uncertainty*. Cambridge: Polity.

Olson, Gary M., and Judith S. Olson. 2000. "Distance Matters." *Human-Computer Interaction* 15: 139–78.

Olson, Gary M., Ann Zimmerman, and Nathan Bos, eds. 2008. *Scientific Collaboration on the Internet*. Cambridge, MA: MIT Press.

Owen-Smith, Jason, and Walter W. Powell. 2003 "The Expanding Role of University Patenting in the Life Sciences: Assessing the Importance of Experience and Connectivity." *Research Policy* 32 (9): 1695–711.

Oxford Dictionary of English. 2010. Oxford: Oxford University Press

Paletz, Susannah, Laurel Smith-Doerr, and Itai Vardi. 2011. *Interdisciplinary Collaboration in Innovative Science and Engineering Fields*. National Science Foundation Workshop Report. Accessed at https://sites.google.com/site/interdisciplinary2010/.

Parker, John N., and Beatrice I. Crona. 2012. "On Being All Things to All People: Boundary Organizations and the Contemporary Research University." *Social Studies of Science* 42 (2): 262–89.

Parker, John N., and Edward J. Hackett. 2012. "Hot Spots and Hot Moments in Scientific Collaborations and Social Movements." *American Sociological Review* 77 (1): 21–44.

___. 2014. "The Sociology of Science and Emotions." In *The Handbook of the Sociology of Emotions*, vol. 2, chapter 25, edited by Jan E. Stets and Jonathan H. Turner. New York: Springer.

Parker, John N., Niki Vermeulen, and Bart Penders. 2010. *Collaboration in the New Life Sciences*. Farnham: Ashgate.

Pelz, Donald C., and Frank M. Andrews. 1966. *Scientists in Organizations: Productive Climates for Research and Development*. New York: Wiley.

Penders, Bart, Niki Vermeulen, and John N. Parker. 2015. *Collaboration across Health Research and Medical Care: Healthy Collaboration*. Farnham: Ashgate.

Pfeffer, Jeffrey, and Nancy Langton. 1993. "The Effect of Wage Dispersion on Satisfaction, Productivity, and Working Collaboratively: Evidence from College and University Faculty." *Administrative Science Quarterly* 38 (3): 382–407.

Pickstone, John V. 2000. *Ways of Knowing: A New History of Science, Technology and Medicine*. Manchester: Manchester University Press.

Powell, Walter W., Douglas R. White, Kenneth W. Koput, and Jason Owen-Smith. 2005. "Network Dynamics and Field Evolution: The Growth of Interorganizational Collaboration in the Life Sciences." *American Journal of Sociology* 110 (4): 1132–205.

Power, Michael. 1997. *The Audit Society: Rituals of Verification*. Oxford: Oxford University Press.

Price, Derek J. de Solla. 1963. *Little Science, Big Science*. New York: Columbia University Press.

___. 1986. *Little Science, Big Science and Beyond*. New York: Columbia University Press.

Radick, Gregory. 2000. "Two Explanations of Evolutionary Progress." *Biology and Philosophy* 15 (4): 475–91.

Rheinberger, Hans-Jorg. 1997. *Toward a History of Epistemic Things: Synthesizing Proteins in the Test Tube*. Stanford, CA: Stanford University Press.

Riding, Laura, and Robert Graves. [1937] 2001. "From a Private Correspondence on Reality." In *Essays from 'Epilogue' 1935–1937*, edited by Mark Jacobs, 163–73. Manchester: Carcanet Press.

Rip, Arie. 1998. "Fashions in Science Policy." In *Getting New Technologies Together: Studies in Making Sociotechnical Order,* edited by Cornelius Disco and Barend van den Meulen. Berlin: Walter de Gruyter.

Rodrigo, Allen, Susan Alberts, Karen Cranston, Joel Kingsolver, Hilmar Lapp, Craig McClain, Robin Smith, Todd Vision, Jory Weintraub, and Brian Wiegmann. 2013. "Science Incubators: Synthesis Centers and their Role in the Research Ecosystem." *PLoS Biology* 11: e1001468.

Rose-Greenland, Fiona. 2013. "Seeing the Unseen: Prospective Loading and Knowledge Forms in Archaeological Discovery." *Qualitative Sociology* 36 (3): 251–27.

Rushforth, Alexander D., and Sarah de Rijcke. 2015. "Accounting for Impact? The Journal Impact Factor and the Making of Biomedical Research in the Netherlands." *Minerva* 53 (2): 117–39.

The Royal Society. 2012. *Science as an Open Enterprise*. The Royal Society Science Policy Centre report 02/12, London.

Schoenberger, Erica. 2001. "Interdisciplinarity and Social Power." *Progress in Human Geography* 25 (3): 365–82.

Schroeder, Ralph. 2007. *Rethinking Science, Technology, and Social Change*. Stanford, CA: Stanford University Press.

• Schuster, Jack H., and Martin J. Finkelstein. 2006. *The American Faculty: The Restructuring of Academic Work and Careers*. Baltimore: Johns Hopkins University Press.

Schweber, Libby. 2006. *Disciplining Statistics: Demography and Vital Statistics in France and England, 1830–1835*. Durham, NC: Duke University Press.

Shapin, Steven. 1994. *A Social History of Truth*. Chicago: University of Chicago Press.

• ___. 2008. *The Scientific Life: A Moral History of a Late Modern Vocation*. Chicago: University of Chicago Press.

___. 2010. *Never Pure: Historical Studies of Science as If It Was Produced by People with Bodies, Situated in Time, Space, Culture, and Society, and Struggling for Credibility and Authority*. Baltimore: Johns Hopkins University Press.

Shostak, Sara. 2005. "The Emergence of Toxicogenomics: A Case Study of Molecularization." *Social Studies of Science* 35 (3): 367–403.

Slaughter, Sheila, and Gary Rhoades. 2004. *Academic Capitalism and the New Economy*. Baltimore: Johns Hopkins University Press.

Small, Henry, and Belver C. Griffith. 1974. "The Structure of Scientific Literature." *Science Studies* 4 (1): 17–40.

Soreanu, Raluca, and David Hudson. 2008. "Feminist Scholarship in International Relations and the Politics of Disciplinary Emotion." *Millennium—Journal of International Studies* 37 (1): 123–51.

Spiegel-Rösing, Ina, and Derek J. de Solla Price, eds. 1978. *Science, Technology and Society: A Cross-Disciplinary Perspective*. Beverly Hills, CA: Sage.

Star, Susan Leigh, and James R. Griesemer. 1989. "Institutional Ecology, 'Translations,' and Boundary Objects: Amateurs and Professionals in Berkeley's Museum of Vertebrate Zoology, 1907–1939." *Social Studies of Science* 19 (3): 387–420.

Stephan, Paula. 2012. *How Economics Shapes Science*. Cambridge, MA: Harvard University Press.

Stokols, Daniel, Kara L. Hall, Brandie K. Taylor, and Richard P. Moser. 2008. "The Science of Team Science: Overview of the Field and Introduction to the Supplement." *American Journal of Preventive Medicine* 35 (2): 77–89.

Strathern, Marilyn. 2006. "Knowledge on Its Travels: Dispersal and Divergence in the Makeup of Communities." *Interdisciplinary Science Reviews* 31 (2): 149–62.

Sturdy, Steve. 2011. "Looking for Trouble: Medical Science and Clinical Practice in the Historiography of Modern Medicine." *Social History of Medicine* 24 (3): 739–57.

Suárez-Díaz, Edna. 2009. "Molecular Evolution: Concepts and the Origin of Disciplines." *Studies in History and Philosophy of Biological and Biomedical Sciences* 40 (1): 43–53.

Swedberg, Richard. 2011. "Thinking and Sociology." *Journal of Classical Sociology* (11) 1: 31–49.

Torka, Marc. 2009. *Die Projektförmigkeit der Forschung*. Baden-Baden: Nomos.

Traweek, Sharon. 1988. *Beamtimes and Lifetimes: The World of High Energy Physicists*. Cambridge, MA: Harvard University Press.

Tuunainen, Juha. 2002. "Reconsidering the Mode 2 and Triple Helix: A Critical Comment Based on Case Study." *Science Studies* 15 (2): 36–58.

Uzzi, Brian, Satyam Mukherjee, Michael Stringer, and Ben Jones. 2013. "Atypical Combinations and Scientific Impact." *Science* 342: 468–72.

Vallas, Steven Peter, and Daniel Lee Kleinman. 2008. "Contradiction, Convergence and the Knowledge Economy: The Confluence of Academic and Commercial Biotechnology." *Socio-Economic Review* 6 (2): 283–311.

Vermeulen, Niki. 2009. *Supersizing Science: On the Building of Large-Scale Research Projects in Biology*. Maastricht: Maastricht University Press.

Vermeulen, Niki, John N. Parker, and Bart Penders. 2010. "Big, Small or Mezzo? Lessons from Science Studies for the Ongoing Debate about 'Big' versus 'Little' Science." *EMBO Reports* 11 (6): 420–23.

___. 2013. "Understanding Life Together: A Brief History of Collaboration in Biology." *Endeavour* 37 (3): 162–71.

von Heland, Franciska, Beatrice I. Crona, and Pedro Fidelman. 2014. "Mediating Science and Action across Multiple Boundaries in the Coral Reef Triangle." *Global Environmental Change* 29: 53–64.

Wagner, Caroline S. 2008. *The New Invisible College: Science for Development.* Washington, DC: Brookings Institution Press.

• Weber, Max. [1918] 1948. "Science as a Vocation." In *From Max Weber: Essays in Sociology*, edited by C. Wright Mills and Hans H. Gerth, 77–128. New York: Oxford University Press.

Weinberg, Alvin M. 1961. "Impact of Large-Scale Science on the United States: Big Science Is Here to Stay, But We Have Yet to Make the Hard Financial and Educational Choices It Imposes." *Science* 134 (3473): 61–164.

Weingart, Peter. 2010. "A Short History of Knowledge Formations." In *The Oxford Handbook of Interdisciplinarity*, edited by Robert Frodemann, Julie Thompson Klein, and Carl Mitcham, 3–14. Oxford: Oxford University Press.

Westfall, Catherine. 2003. "Rethinking Big Science: Modest, Mezzo, Grand Science and the Development of the Bevalac, 1971–1993." *Isis* 94 (1): 30–56.

Wilde, Rein de. 2001. *De kenniscultus: Over nieuwe vormen van vooruitgangsgeloof.* Maastricht: Maastricht University Press.

Whitley, Richard. 2001. *The Intellectual and Social Organization of the Sciences.* New York: Clarendon.

Wuchty, Stefan, Benjamin F. Jones, and Brian Uzzi. 2007. "The Increasing Dominance of Teams in Production of Knowledge." *Science* 316 (5827): 1036–39.

Wulf, William. 1989. "The National Collaboratory." In *Towards a National Collaboratory.* Unpublished report of a National Science Foundation invitational workshop. New York: Rockefeller University.

26 Interactional Expertise

Harry Collins, Robert Evans, and Martin Weinel

Introduction

Interactional expertise is a concept rooted in science and technology studies (STS). The term was first used in 2002 as part of a normative theory of expertise developed by Collins and Evans (2002) to address what they saw as a potential problem for STS, namely, how a constructivist sociology of knowledge should contribute to technological decision making in the public domain. Launched as "studies of expertise and experience" (SEE), their argument was that the relational model of expertise, which dominated STS and had been so successful in showing how the expertise attributed to actors depends on their social position, was creating a new "problem of extension" in which more inclusive forms of decision making would become increasingly unworkable unless a new way of distinguishing experts and nonexperts was developed.

So what is interactional expertise? Interactional expertise is fluency in the spoken language associated with a practice. It is more than mere mimicry as it requires the speaker to be able to initiate and hold new conversations. For similar reasons, namely, that it requires a deep understanding of substantive content, interactional expertise cannot be reduced to interactional competence either.[1] Instead, what distinguishes interactional expertise is the claim that, under the right social circumstances, fluency in a spoken language and a conceptual understanding of the domain to which it refers, can be acquired without experiencing the practice. To put the same point in more vernacular terms: if practitioners know how to walk the walk, interactional experts know how to walk the talk.

In this chapter we outline the ideas and debates that informed the development of interactional expertise, pulling together the developing body of literature on expertise in knowledge production and describing how the idea of interactional expertise has led to theoretical, pedagogic, and methodological innovations. In so doing we discuss the distinction between "realist" and "attributional" or "network" models of expertise;

the developing understanding of interactional expertise; and the relationship between language, practice, and culture. We conclude by examining how the concept of interactional expertise is applied in current research and point to future opportunities, including the use of a new social research method—the Imitation Game—that has been developed as a direct result of this STS work on expertise.

Intellectual History

Although the term *interactional expertise* first appeared in 2002 (Collins and Evans 2002), the debates that inform the idea precede its naming and an adequate discussion of its properties does not appear until 2004 (Collins 2004). In what follows, we set out the "hidden" intellectual history of interactional expertise and its relationship to these cognate literatures.

The Idea of Interactional Expertise

The ideas that underpin interactional expertise can be traced back to the Strong Program (Bloor 1991) in STS and its philosophical foundations in Wittgenstein's analysis of a "form of life" (Winch 1958; Wittgenstein 1953).[2] Central to these is the conception of language and practice as two sides of the same coin, with fluency in both language and practice being acquired by, and given meaning through, participation in a collective form of life.

The jump to interactional expertise involves two elements. The first is to recognize that the language component of a form of life and the physical activities that language describes are *both* social practices. The second is that fluency in the language that describes a practice—hereafter, a "practice language"—can be acquired independently of the corresponding physical practice. Although the first point is implicit in commonly used STS concepts like "paradigm" (Kuhn 1996), the second point—that a practice language can be acquired in the absence of physical practice—emerges from the debates between phenomenologists and AI researchers that occurred during the 1990s about whether a machine can reproduce the language of a human expert.

Insofar as the prehistory of interactional expertise is concerned, the crucial moment in these debates occurs with the publication of Hubert Dreyfus's (1992) *What Computers Still Can't Do* and concerns the ability of Madeleine, a severely disabled young woman who is nevertheless able to hold sophisticated conversations on all manner of topics, most of which she could never have experienced directly. Madeleine was initially invoked by AI researcher Douglas Lenat to refute the claims of phenomenologists like Dreyfus by showing that language can be acquired in the absence of embodied practice.

In a review of Dreyfus's book, Collins (1996) argued that Lenat's conclusion is only half right: it may be that embodied practice is not necessary to learn a language but this does not mean a machine will succeed. The reason is that, as noted above, language is a social practice and so, unless and until computers, humanoid or otherwise, are able to become social entities, they will be unable to use language in a humanlike way.

Although the term *interactional expertise* is being applied to these debates retrospectively, it is clear that its defining feature—that it is possible to acquire a language independently of the practice it describes—matches the way in which Madeleine's undisputed abilities were understood and explained by Collins. Of course, phenomenologists can—and do—argue that Madeleine's abilities can be explained without reference to interactional expertise. For example, Selinger, Dreyfus, and Collins (2007) contains a four-part exchange among Selinger, Dreyfus, and Collins which explores the importance of embodiment and socialization in understanding expertise. More recently, Ribeiro and Lima (2015) have offered a critique of interactional expertise that restates many of the core phenomenological concerns but which, in stressing that Madeleine is a social being, lends unintended support to the position it claims to reject.

Tacit and Explicit Knowledge

The second intellectual tradition that informs interactional expertise is work addressing the nature of tacit knowledge. First introduced by Polanyi, the term *tacit knowledge* derives from the insight that "we know more than we can tell" (Polanyi 1966) and is often contrasted with explicit knowledge. Thus, explicit knowledge is knowledge that can be codified and written down as facts and rules, whereas tacit knowledge refers to the unarticulated skills and knowledge that underpin all social activities but which are used without conscious thought.

Interactional expertise complicates this distinction between tacit and explicit knowledge because its key terms—language and practice—do not map onto the dichotomy of explicit and tacit knowledge. This is because, although language is used to make knowledge explicit, the act of speaking or writing is a social practice. Language speaking, like other practices, is developed through socialization into a collectivity of expert practitioners, with the performance judged by, and held accountable to, the standards of the relevant peer community. This suggests that language speaking has several important properties (see, e.g., Collins and Evans 2007), including the following.

• The same principle of socialization applies to the learning of all human languages, from the most general natural languages to the most specialized, niche or regional language.

• Successful language speaking requires the acquisition of tacit knowledge in order for the right words to be chosen and the infelicities of others to be accommodated and repaired.

• Fluency in a language, like any other practical skill, requires regular use if it is not to atrophy or become outdated as social norms change over time.

Where does this leave the explicit knowledge that language makes visible (or audible) for others to keep and inspect at some later point? Clearly, the emergence of a written language through which knowledge can be stored and transported has a transformative effect on those societies that develop the capability and its value should not be underestimated. Nevertheless, literate cultures retain a strong oral tradition in which face-to-face interactions provide the primary means of socialization (Goody and Watt 1968) and, as STS has shown, this is true for even that most formal of cultures, the scientific community (e.g., Merz 1998; Rowe 1986; Shapin 2007).

The Reality of Interactional Expertise

Turning to the practical application of interactional expertise, it is worth reemphasizing that the term was coined as part of normative theory of expertise (Collins and Evans 2002). The problem this theory attempted to solve can be seen by considering statements such as this:

[T]he collective societal definition of what the issues and concerns are which should enjoy priority public attention and attempted resolution [are] not unconnected with specialist technical expertises, and where appropriate it should be informed by these, but it does not at all reduce to this. (Wynne 2007, 108)

Despite the fact that this statement was written as part of a paper that criticizes the SEE approach, it is hard to disagree with its approach to technological decision making in the public domain. The difficult questions, and the arguments about realist-vs.-relational approaches, are only revealed when we examine the details. What do terms like "specialist technical expertises" and "where appropriate" actually mean in practice and does STS, as a field whose object of study is knowledge and expertise, have anything to say about how these judgments are made?

In order to make the argument that it is possible to distinguish levels of expertise, Collins and Evans (2002) introduced a simple tripartite model of specialist expertise illustrated by reference to sociological fieldwork:

1. *No expertise*: That is the degree of expertise with which the fieldworker sets out; it is insufficient to conduct a sociological analysis or do quasi participatory fieldwork.

2. *Interactional expertise*: This means enough expertise to interact interestingly with participants and carry out a sociological analysis.

3. *Contributory expertise*: This means enough expertise to contribute to the science of the field being analyzed. (Collins and Evans 2002, 254)

The idea was that readers of the article would recognize these different levels of expertise in their own fieldwork experiences and see that "having an expertise" could be distinguished from "not having an expertise." Perhaps more important, given what was then the relatively recent experience of the "science wars," in which the argument was often made that STS scholars could not understand the science involved because they were not practitioners (e.g., Koertge 2000), recognizing the existence of interactional expertise shows that the crucial difference in understanding occurs between levels 1 and 2 and not between levels 2 and 3. STS scholars do not have to be practitioners of, or contributors to, the domains of practice they study in order to justify their conclusions. Instead, the validity of STS comes from its immersion in the oral culture of those communities and the acquisition of the tacit knowledge needed to master the practice languages used within the relevant social groups.

Where STS researchers have developed interactional expertise, they have every right to apply their own (contributory) disciplinary expertise to that domain of science because their analysis builds on an understanding of the field that matches that of the practitioners. In some cases, these STS-inspired accounts will be critical, in which case interactional expertise can help to defend and legitimate their views. In other cases, the relationship will be less antagonistic and the acquisition of interactional expertise by STS researchers may lead to new collaborations in which the disciplinary expertise of STS plays a significant role (Whyte and Crease 2010). Of course, this does not mean that social science approaches that draw on other perspectives—for example the anthropological strangeness of the outsider—do not have value. Rather, the point is that these accounts will require a different kind of justification in the face of a complaint that the author "does not know what they are talking about."

Researching Expertise and Experience

In this section, we examine research that draws upon the studies of expertise and experience approach, explaining how it differs from other approaches and providing cross-references to other chapters.

Expertise and Expert Status

One aim of SEE is to open up expertise as an analysts' category. The approach is intended to complement those more descriptive studies that foreground the attribution or accomplishment of expert status (see Ottinger, Barandiarán, and Kimura, chapter 35 this volume; see also Jasanoff 2002, 2003) by providing an independent framework that can be used to describe the expertise held by social actors, regardless of how it is valued by others. In effect, there is an analytic choice to be made: expertise can be "something people do [or] something people have or hold" (Carr 2010, 18) and interactional expertise is something people have rather than something they do.

The ideal of pure description to which the SEE approach can be most clearly contrasted is found in ethnomethodological studies that examine how expertise is displayed and articulated in social contexts (e.g., Button and Sharrock 1998; Hoeppe 2014; Livingston 2006; Lynch 1997; Sormani et al., chapter 4 this volume).[3] In these studies, the idea of using analytic categories like "tacit knowledge" or "interactional expertise" to explain how social groups are (un)able to do things is rejected as a matter of principle in favor of an approach that focuses solely on explicating:

The *occasions* in which actual people (when they do) articulate and display tacit knowledge, explicit knowledge, and expertise as relevant for and in their undertakings. (Coopmans and Button 2014, 2)

One example of this approach is provided in the detailed analysis of courtroom interactions that occur when witnesses, whether designated experts or not, have their credibility called into question and so have to perform their expertise in order to have it recognized by the court (Matoesian 1999, 2008). While these studies reveal much about how expert status is practically accomplished though discursive strategies (e.g., upgrading, one-upping) and appropriate body language (e.g., hand gestures, facial expressions such as a thinking face), they cannot tell us if the outcome is justified against any other criteria than those used by the participants. For example, ethnomethodologists are not interested in showing whether a jury, or anyone else, has made a "mistake" as that would require some external criteria against which participants' actions could be judged. Instead, their analysis is concerned with describing the methods used by participants to assign or deny expert status and to render these judgments accountable in their own terms.

The dilemma this emphasis on expertise as a performance creates is nicely illustrated in the courtroom experiences of Simon Cole, an STS scholar who has studied fingerprint identification practices and who has been called as an expert witness in order to discredit fingerprint evidence. In a paper that reflexively examines his attempts to present his expertise as legitimate, Cole reviews the approach he used in *People v. Hyatt*, where a pretrial hearing ruled that he could not testify and his work was dismissed as "junk science," and says that:

So actually, I've changed my testimony in the most recent cases and tried to avoid using Daubert as the framework for it, and so now I just talk about reliability. (Lynch and Cole 2005, 278–9)[4]

The point to note is that while these different forms of presentation may be more or less effective in establishing his credibility in the courtroom and may tell us something about how expert status is constructed in U.S. courtrooms, without an independent metric for gauging Cole's expertise, they have no bearing on what he actually knows or whether the court was right to dismiss his work.

In the vast majority of STS work, however, the approach is rather different. Analytic categories such as "civic epistemology" (Jasanoff 2007), "obligatory passage point" (Callon 1986), "epistemic culture" (Knorr Cetina 1999), "boundary work" (Gieryn 1999), and "boundary objects" (Star and Griesemer 1989) are routinely used to theorize descriptions in ways that participants might neither understand nor accept. In addition, these accounts are often linked to an implicit normative agenda. As Sheila Jasanoff has written,

Many STS scholars think that the institutions, practices, and products of science and technology should be characterized in new ways not only for the sake of descriptive adequacy and analytic clarity, but also in order to reorder power relationships: for example, to make the exercise of power more reflexive, more responsible, more inclusive, and more equal. (Jasanoff 2013, 101)

For many in STS, the most challenging of these studies are the ones that focus on the interactions between lay citizens, special interest groups, and the scientific or policy elite and in which issues of expertise, power, and social justice are intertwined. They include, among others, topics such as environmental issues (Delborne 2008; Suryana-rayanan and Kleinman 2013), health and medicine (Epstein 1996; Popay and Williams 1996), and public engagement with science and technology (Kerr, Cunningham-Burley, and Tutton 2007). Many of these issues are discussed in other chapters of this handbook (see for example Breyman et al., chapter 10; Hess et al., chapter 11; Jasanoff, chapter 9 this volume), so here we focus on how the distinction between relational and realist approaches to expertise illuminates what is at stake in this work.

For those working with a relational model of expertise, the primary focus is on how the credibility of knowledge claims is negotiated and the status of expert awarded or withheld. Methodologically, the problem is how to identify the relevant social groups in what may be a diverse and porous social setting in which epistemic claims are being made alongside, and as part of, arguments about social justice, moral responsibility, and political rights. Thus, for example, the group struggling to have its experiences recognized as credible by elite institutions might be a low-income community that lives near a major industrial plant (Ottinger 2013), women who believe their health has been damaged by breast implants (Jasanoff 2002), social movements mobilizing public opposition to nuclear power (Welsh 2000), or an STS scholar attempting to give evidence as an expert witness (Lynch and Cole 2005).

Within this co-productionist approach, concerns with social justice and expertise come together because of the ways in which the unequal distribution of social and cultural resources determine which group is able to successfully claim the status of expert. In many cases this can be a potent critique of institutionalized privilege, but its power depends on the ability of the researcher to show that the marginalized group "really are" experts. In other words, the credibility of the argument depends on demonstrating that an epistemic injustice (Fricker 2007) has occurred. If this cannot be done, then the resulting critique is not based on expert status being wrongly denied but on a violation of democratic or other procedural principles that would apply regardless of expertise.

Classifying Expertise: The Periodic Table of Expertises

Showing that a particular group really does have expertise requires a change in perspective: expertise as an actors' category is no longer sufficient and expertise as an analysts' category is needed. The basic classification of specialist expertise (no expertise, interactional expertise, and contributory expertise) is one example of this approach.[5] It has since been developed into a more extensive "periodic table of expertises" in which different kinds of socialization give rise to different kinds of expertise (see table 26.1). The main features of this classification can be summarized as follows.

Working from the top down, the first two rows identify what is necessary for an individual to acquire any of the other expertises listed in the rest of the table. These are the *ubiquitous expertises*, needed to take part social life and two general *dispositions*—the ability to get on with others and the ability to reflect on one's own experience—that can assist socialization.

The next two rows (specialist expertises and meta-expertises) provide the core of the classification. The *specialist expertises* mark the progress made as an individual engages more intensively with a particular domain and have a transitive relationship to each other. The first three categories denote levels of understanding that can be achieved by the autodidact; the last two are the interactional and contributory expertises that can be developed only through immersion in the relevant domain of practice. In contrast, the *meta-expertises* row recognizes that no one can be an expert about everything and so lists the ways in which judgments about expert claims can be made in the absence of high-level specialist expertise. These include social judgments based on more or less ubiquitous knowledge that make no reference to the content of what is being judged and more informed judgments based on some limited familiarity with the expertise in question. Unlike the specialist expertises, there is no logical relationship between the different kinds of meta-expertise. The final row of the table (meta-creiteria) identifies some criteria that might be used for identifying or choosing among experts.

Table 26.1

UBIQUITOUS EXPERTISES					
DISPOSITIONS				Interactive Ability	Reflective Ability
SPECIALIST EXPERTISES	**UBIQUITOUS TACIT KNOWLEDGE**			**SPECIALIST TACIT KNOWLEDGE**	
	Beer-mat Knowledge	Popular Understanding	Primary Source Knowledge	Interactional Expertise	Contributory Expertise
				Polimorphic	Mimeomorphic
META-EXPERTISES	**EXTERNAL**		**INTERNAL**		
	Ubiquitous Discrimination	Local Discrimination	Technical Connoisseurship	Downward Discrimination	Referred Expertise
META-CRITERIA	Credentials	Experience			Track-Record

Source: Collins and Evans (2007).

Researching the different categories of expertise listed in the table can take many forms. Ubiquitous expertises are perhaps best explored via skilled ethnographic work that reveals how social orders are sustained through carefully coordinated, skillful practice. Given the emphasis on socialization, specialist expertises are distinguished from ubiquitous ones by their relative scarcity rather than by their content or "difficulty" (Collins 2013). This means that both the esoteric skills of the laboratory (e.g., Doing 2004) and the everyday skills of driving a car (e.g., Dreyfus and Dreyfus 1986) could be classed as specialist expertises. Such specialist expertises are a staple topic of STS and are, again, best revealed through careful interpretivist studies that document the shared assumptions and cooperative work needed for high-level skills to become established as a shared form of expertise (MacKenzie and Spears 2014; Wylie 2015).

What SEE and the theory of expertise represented by the periodic table of expertises add to these accounts is an explicit emphasis on tacit knowledge and the socialization needed to acquire it. Using this approach, it is possible to distinguish between the kinds of information that anyone might acquire by virtue of reading the academic or quasi-academic literatures and the skilled judgment that can be exercised only by domain experts (Collins 2014b; Reyes-Galindo 2014). The categories also make it possible to explore the ways in which scientists and others are able to contribute to technological decision making in the public domain by providing a language through which the expertise claimed by establishment figures can be compared and contrasted with that claimed by social movements and others (Evans and Plows 2007). This work is entirely compatible with constructivist accounts focusing on the discourse and identity work of participants but by treating expertise as the property of an individual or group, it is also possible to identify when legitimate claims to expertise—be it that of the scientist, social movement, or lay person—have been denied (e.g., Durant 2015).

The meta expertises and lower-level specialist expertises are more relevant to research on science communication and public participation where citizens participate *as* citizens, not experts, and so draw on ubiquitous rather than specialist expertises (Horst and Michael 2011). The challenge in such circumstances is twofold. On the one hand, science communicators have to convey a meaningful account of the science in question such that it can be grasped by those who have neither the time nor the opportunity to be fully immersed in the relevant social networks (Miller 2001). On the other hand, citizens themselves have to make trust decisions in order to choose between competing claims and must do so without having full access to the necessary tacit knowledge (Evans 2011). As with the study of specialist expertises, the use of the periodic table can provide a conceptual framework within which such work can take place, providing consistent descriptions between cases and enabling comparisons over time.

Expertise and Sociotechnical Integration

Recognizing expertise as the property of social collectivities has implications for the understanding of interdisciplinary work: interdisciplinary collaboration becomes an instance of multiculturalism, with the challenge being to find ways of communicating across the different expert cultures. Gorman (2002) has linked the tripartite division of specialist expertises to Peter Galison's metaphor of a trading zone (Galison 1997), in which distinct cultural groups coordinate their actions by developing a rudimentary shared language or pidgin before, in some cases, going on to develop richer forms of interaction based on creoles or even a fully fledged "interlanguage." According to Gorman, the type of trading zone that emerges depends on the kinds of expertise actors have about each other, with elite, top-down trading zones emerging where expertise is held by only one party and more collaborative versions developing as the mutual understanding of participants increases. More recent work (Collins, Evans, and Gorman 2007) has explored the relationship between expertise and type of trading zone in more detail, leading to a more general model in which there are four major types of trading zones defined by the extent to which the mixing of different groups is collaborative or coerced and the outcome is a heterogeneous mixture in which the original groups are still identifiable or a homogeneous group in which all participants share the same culture and language.

The sociological model of knowledge that informs interactional expertise can also shed light on, and support, other forms of research about interdisciplinary working, particularly where this aims to enhance research practice. In these contexts, the idea of interactional expertise can be used to explain how talk about methods, techniques, and values can generate sophisticated levels of understanding and enable more effective collaborative work. For example, the Toolbox Project (see, e.g., O'Rourke and Crowley 2013) has developed a modular set of surveys that are designed to elicit attitudes about different aspects of research practice. This data is then used to facilitate discussion groups in which the different assumptions of team members are explored in more detail. The outcome of this critical epistemic reflection is the development of some interactional expertise in the disciplines represented by others and, over time, an enhanced level of mutual understanding among the members of multidisciplinary research teams. Likewise, the Socio-Technical Integration Research (STIR; see Fisher 2007) program uses an "embedded humanist" based within the scientific team to probe and disrupt their taken-for-granted assumptions and so prompt reflection about their routine practices. Here the idea is that the embedded researcher, who typically comes from a humanities or social science discipline, will develop interactional expertise in the work of the host scientific team and then use this understanding to frame interventions and proposals that draw on concepts and ideas from their home discipline. In the Toolbox Project, STIR, and other, similar approaches, the underpinning idea is that the members of different scientific disciplines have been socialized into different contributory expertises and that learning to talk the language of these different practices will improve the quality of collaboration.

Expanding Interactional Expertise

Although the basic idea of interactional expertise has not changed since it was first proposed in 2002, the scope and scale of its application has expanded. When first introduced via the example of the social science fieldworker, the implication was that interactional experts were relatively rare. Seen in the context of the periodic table of expertises, however, in which all contributory experts are also interactional experts, it is clear that there are many more interactional experts than the initial example implied. More recently still, the research inspired by SEE has taken on ever-broader domains of contributory expertise, such as religion, gender, and sexuality, with the implication that there are interactional experts in these domains as well (Collins and Evans 2014). As a result, interactional expertise is now seen as more ubiquitous than previously thought.

In addition to these developments, which can be seen as following the logic of the original argument through to its conclusion, there have also been two attempts to expand the concept of interactional expertise in other ways. For example, Goddiksen (2014) suggests expanding the definition such that the kind of science learned in the classroom might also count as interactional expertise. Unfortunately, this would effectively merge interactional expertise with primary source knowledge and/or popular understanding and, in so doing, erase the distinction between specialist and ubiquitous tacit knowledge that is the key feature of interactional expertise, and indeed the whole typology of which it is a part (Reyes-Galindo and Duarte 2015). It also has the consequence of concealing the distinction between knowledge acquired through reading or listening to secondhand accounts and expertise developed through immersion in the community of practitioners, something that all social scientists know is important.

In contrast, Plaisance and Kennedy (2014) suggest expanding the notion of interactional expertise by putting more emphasis on what interactional expertise allows you to do rather than on how it is acquired. Their idea is based on the definition of interactional expertise as "enough expertise to interact interestingly with participants and carry out a sociological analysis" that was originally introduced in 2002 (Collins and Evans 2002, 254). Their rationale is that this more inclusive definition will make it easier to reap the benefits of interactional expertise, which, they say, include

identifying and legitimizing non-traditional experts; improving the uptake and application of scientific research by stakeholder communities; and improving the quality of scientific work by including new perspectives and social locations in knowledge production. (Plaisance and Kennedy 2014, 61)

Here the reasons for rejecting the proposal and insisting on defining interactional expertise as fluency in the practice language are more subtle than those used in the critique of Goddiksen and relate to two separate issues. First, there is a desire to avoid the ambiguity that the idea of "interesting interactions" has created. In what we might call the "fieldwork definition," the fact that a deep understanding of the subject is needed in order to have the kinds of interesting conversations that enable sociological analysis is not as clear as it should be. To see why this matters, one need only note that the threshold for being "interesting" is a lot lower than the threshold for "fully informed." Second, and perhaps more important, it is not clear that the goals Plaisance and Kennedy want to reach actually need interactional expertise in the first place. For example, the rights of stakeholders to be heard should not depend on their ability to engage with esoteric expertises but on the fact that they are stakeholders and, as such, do not need additional qualification or skill in order for their views to matter (for a more detailed response, see Collins and Evans 2015a; Collins, Evans, and Weinel 2015).

The Fractal Model of Society

In making sense of the expanded notion of interactional expertise described in the previous sections, Collins (2011) developed the "fractal model" of society, which starts from the sociological axiom that an individual is made up of the various social groups he or she participates in. Some of these groups will be large and heterogeneous, some will be small and homogeneous, and most will be somewhere in between. All will overlap to some extent and, most important of all, each group has nested within it smaller, more specialist subgroups that have the same basic form and properties as the larger group. As with many other sociological approaches, such as "taken-for-granted reality" (Schutz 1964), "culture" (Geertz 1973), "subculture" (Yinger 1982), and "microculture" (Fine 2010), this notion of social group assumes that the experiences of any given individual are defined by the combination of groups in which he or she actively participates.

The link between group membership and expertise is made via contributory expertise. Each social group, assuming it is distinctive in some way, has a practice language and a set of practices associated with it. Fully socialized members of the group learn both and so have both interactional expertise (the practice language) and contributory expertise (the practice). There are practice languages for esoteric expertises like theoretical physics and for more everyday activities like driving a car. In addition, because social groups are nested in a fractal-like way, members of lower-level groups will share the practice language of the higher-level groups from which they are derived. Thus, for example, both theoretical physicists and gravitational wave physicists will be part of the higher level group of physicists and, along with chemists and bioscientists, make up part of the still larger group of research scientists.[6]

The fractal model has two important consequences for our understanding of interactional expertise. The first is that any individual is fluent in the practice languages associated with all the social groups in which he or she actively participates. That is to say, all fully socialized group members have both contributory and interactional expertise, which means that the interactional expertise of the social scientist is remarkable only because it is acquired without the corresponding contributory expertise. For this reason, those who acquire their interactional expertise in this way—that is, without the obligation to also acquire any contributory expertise—are referred to as "special interactional experts" (Collins 2011).

The other consequence of the fractal model follows from this recognition that interactional experts are much more common than the early publications of Collins and Evans (e.g., 2002) implied and links the idea of interactional expertise to the division of labor. In interdisciplinary science, but also many other settings, members of different social groups must understand each other well enough to collaborate without each individual being able to do the practical work of every specialty and subspecialty represented in the team as a whole. The question is, how can this be done? The answer, using the fractal model, is that there is an interdisciplinary practice language, which exists at the level of the team, that "contains" enough interactional expertise in the relevant specialist practices for the team as a whole to coordinate their actions.

This relationship between interactional expertise and the fractal model is of particular importance for STS research as it means that qualitative sociological fieldwork can be much more secure in its understanding of a community than its critics imply. The reason is that because most practitioners will only have contributory expertise in their own specialist area, their understanding of what their colleagues do comes via the shared interactional expertise of the group's practice language. As this expertise is also what the social scientist is able to access through their social interactions, there is, as Collins puts it,

almost no difference between interactional experts and contributory experts as far as their relationship to the practice language is concerned. (Collins 2011, 278)

Of course, none of this should be taken to mean that the acquisition of interactional expertise is a free or easy pass to high-level technical understanding. Acquiring interactional expertise is hard and the idea in no way obviates the need for deep immersion in a specialist community as a precursor to technical debate.

Interactional Expertise in Action

By combining the transitive ladder of specialist expertise set out in the periodic table of expertises and the fractal model, it is possible to explore how expertise is developed, shared, and used among different social groups. Although there are many papers that cite the idea of interactional expertise, here we focus on work that either engages more deeply with the idea or draws on and develops it in some way.

Collaborations across Disciplines and Specialisms

Tiago Ribeiro Duarte (2013) has applied the fractal model to the work of paleo-oceanographers, exploring how information is exchanged within this multidisciplinary research area and also documenting the role played by technicians in supporting laboratory work. He finds that collaboration at different fractal levels requires different kinds of interaction. Interactional expertise being particularly salient with, for example, paleo-modelers attending more general paleo-oceanography conferences to pick up the tacit knowledge needed to understand the provenance of the data they use. In a similar way, Reyes-Galindo (2011, 2014) explores the relationships between theoretical and experimental physicists, showing how high trust relations between the two typically depend on close interpersonal contact that enables the sharing of tacit knowledge and the development of interactional expertise.

In contrast, much of the public concern expressed about the Climategate controversy, in which leaked private emails between climate scientists were interpreted as proving that the consensus position was the result of a conspiracy among scientists, can be explained by the absence of interactional expertise among the commentators; those more familiar with the nature of science in general, and climate science in particular, saw nothing unusual or unprofessional in what was said (Collins 2014a). Priaulx and Weinel (2014) examine a similar case where legal scholars attempt to use insights from behavioral psychology for regulatory purposes. They suggest that a lack of interactional expertise can explain why lawyers—in this particular instance—base their regulatory proposals on the highly controversial and widely rejected notion of "hedonic adaptation."

Health Care Professionals and Their Patients

Much of the STS research on the interactions between medical science and patients groups has emphasized the sophisticated understandings of biomedical science held by patients (Epstein 1996; Popay and Williams 1996). Using the idea of interactional expertise, it is also possible to reverse the question and examine the extent to which medical practitioners understand the lived experience of illness. Typically an entry point for critique (Irwin 1995; Layton et al. 1990), the calls for a "new medical conversation" (Mazur 2003) imply that the medical professional must learn to understand the patient perspective. Given that most practitioners will not suffer from the conditions they treat, this means that doctor-patient communication should also be seen as an opportunity for the therapist to go beyond taking a history of symptoms and improve their understanding of—that is to say, develop their interactional expertise in—the patient's lived experience. Milton (2014) uses the idea of interactional expertise to interrogate the relationship between autistic people and autism researchers, and argues that researchers need to develop greater interactional expertise in autistic cultures if they are to retain the trust and cooperation of participants. In other research, Evans and Crocker (2013) and Wehrens (2014) both use the Imitation Game method described below in order to explore the extent to which medical professionals whose work depends on providing advice to clients are able to demonstrate interactional expertise in the experience of living with the conditions they treat. This kind of understanding, which can be distinguished from emotional empathy, is particularly important in the treatment of chronic illnesses where research (Donaldson 2003; Greenhalgh 2009; Tyreman 2005) has shown that recovery is slower, treatment trajectories longer, and health costs higher when patients find the advice they are given difficult to implement.

Education and Training in Interdisciplinary Sciences

The emerging field of "service science" is becoming increasingly important in management and engineering. Drawing on the metaphor of "T-shaped" expertise (Glushko 2008), practitioners of service science describe their work as combining in-depth technical knowledge in a specific area (the vertical bar) with the ability to interact with, and contextualize that knowledge in, a wide range of other specialisms (the horizontal bar). But how can we understand, and if necessary problematize, the different kinds of expertise involved in the horizontal and vertical bars?

Using SEE, service science seems analogous to the clinician's need to understand the patient's experience: the vertical element is the contributory expertise of the technical skill, while the horizontal bar represents, and depends upon, the development of interactional—not contributory—expertise in the domains where it is being applied

(Gorman 2010; Gorman and Spohrer 2010). This explicit focus on developing interactional expertise has intriguing implications for education and training: whereas traditional degree courses focus on developing contributory expertise in one or more specialist domains, a degree in service science requires contributory expertise in service science to be complemented by interactional expertise in the fields where the core skills will be deployed. This, in turn, opens up the possibility of designing courses specifically to promote special interactional expertise in the language of the relevant specialism rather than contributory expertise in its practice (Berardy, Seager, and Selinger 2011).

The Public Communication of Science and Technology

Given that interactional expertise can only be acquired through sustained social interaction with the relevant specialist communities, it is most unlikely that laypeople are afforded the opportunity to develop this level of understanding for any esoteric practice. In such circumstances, laypeople must rely on others to sort the wheat from the chaff, with journalists often fulfilling this role via the interactional expertise developed through long-term interactions with their sources and contacts (Boyce 2006; Reich 2012). In practice, of course, journalistic accounts are shaped by news values, editorial priorities, and the availability of experts, so their interpretation of an expert debate may itself be challenged (Hinnant and Len-Rios 2009; Kruvand 2012). Similar problems also arise, for example, in political settings where policy makers may ignore consensual expert opinion when formulating policies, as happened in South Africa when Thabo Mbeki, after reading about potential dangers of the drug on the Internet, refused to approve the use of AZT to reduce the mother to child transmission of HIV (Collins, Weinel, and Evans 2010).[7] Nevertheless, to the extent that science communicators and others need to accurately represent the state of research in some scientific domain to their audience, having some level of interactional expertise in the relevant science would seem to be a prerequisite.

New and Future Directions

Looking to the future, interactional expertise has the potential to inform a wide range of research that draws on many of the different approaches within STS.

Expertise Research in STS

Within mainstream STS, interactional expertise and the wider typology of which it is the best known element, can inform a number of research approaches, some of which are more radical than others.

At the most conservative end of the spectrum, interactional expertise provides a useful explanatory resource that can be used in traditional STS case studies involving different expert communities. For example, nothing is lost if the AIDS treatments activists described by Epstein (1996), who successfully challenged the regulation and design of clinical trials, are said to have acquired interactional expertise in the science of HIV. On the other hand, much may be gained. For those concerned with STS as an academic discipline, there is value in using a consistent framework to describe the expertise of different groups and different case studies as this can only aid comparative research. In addition, for those who want to influence the ways in which technological decision making takes place, using the typology of expertise and, in particular, the idea of interactional expertise means that it is possible to argue on both political and epistemic grounds that a particular group has a right to contribute.

In a slightly more normative vein, interactional expertise, and the emphasis on tacit knowledge and socialization that go with it, can be used to inform work relating to interdisciplinarity and teamwork. One example of this is the emerging network of scholars interested in the nature of sociotechnical collaborations. A recent paper (Fisher et al. 2015) provides a synthesis of several approaches, with interactional expertise identified as a key theoretical resource for these more applied programs of work. Other related work, particularly that around service science and sustainability, has also started to integrate the idea of interactional expertise into its pedagogy and practice (Berardy, Seager, and Selinger 2011; Gorman and Spohrer 2010).

Finally, there is the debate about the relationship between expertise and democracy. The initial reactions to SEE typically highlighted its allegedly technocratic recommendations and contrasted them with a more participatory ethos that was said to characterize STS (see Wynne 2003; Jasanoff 2003 in particular; see Collins and Evans 2003; Collins, Weinel, and Evans 2010 for replies).[8] More recently, however, there are signs that this assumption in favor of democratization and/or increased participation is beginning to wane, with several recent articles now setting out a range of political positions and views, all of which are democratic in some sense, that can be used to enrich the conception of politics used within STS (Brown 2015; Durant 2011). In this new, more heterogeneous, context there will be less need for authors to distance themselves from a realist understanding of expertise even when it would suit their aims better not to (as Suryanarayanan and Kleinman 2013 do, for example). Thus, the future of legal and policy-related STS could see a more proactive and engaged stance by at least some members of the STS community in which it becomes increasingly legitimate to talk openly and explicitly about the expertise possessed by, and not merely attributed to, different social actors.

SEE and the Imitation Game

The Imitation Game has been developed to investigate the nature of interactional expertise. In essence, the Imitation Game is a more rigorous version of the parlor game that inspired Alan Turing's famous Turing Test for the intelligence of computers (Turing 1950). Each Imitation Game consists of three players who communicate using specialist software. One player, drawn from the target group, acts as the judge/interrogator and creates questions that are sent to the other two players. One of these, known as the nonpretender, is also drawn from the target group and answers naturally. The other, the pretender, is recruited from a different group and asked to answer as if he or she were a member of the target group. The judge then compares the answers and tries to work out which comes from the pretender and which from the nonpretender. The hypothesis is that, where the pretender has interactional expertise, the judge will be unable to distinguish between the two sets of answers (for more details of Imitation Game research, see Collins et al. 2006; Collins and Evans 2014).[9]

There are now several variations of the method and it has been used over a wide range of topics. The principal highlights of this research, and the questions still left open, include the following.

• **Scaling up for quantitative research:** Imitation Games initially took place with three players playing a single game and each game being played sequentially (e.g., Collins et al. 2006). It is now possible to play many three-player games simultaneously and to separate out different activities (e.g., generating questions, collecting answers, making judgments) in order to make the most efficient use of what might be limited populations. As a result, Imitation Games can now be used for both detailed qualitative research and large-scale quantitative research (Collins et al. 2015).

• **Expanding the theory of interactional expertise:** Initial Imitation Game research focused on "proof of concept" topics, such as color blindness and perfect pitch (Collins et al. 2006) and the success of long-term fieldwork (Giles 2006). More recent Imitation Game research has tackled topics that extend both interactional and contributory expertises to more widespread social groups and practices, including gender, religion, and sexuality (Collins and Evans 2014). This has led to a more fundamental analysis of the nature of representative samples in qualitative and quantitative research (Collins and Evans 2015b).

• **Applications within STS:** Imitation Game research has been used to explore the extent to which medical professionals are able to discursively take the perspective of patients with chronic conditions. Evans and Crocker (2013) have shown that dietitians are much more likely to "pass" as people living with Celiac disease than a control group of lay citizens, suggesting that the source of the expertise was interactions with patients rather than more general experiences of dietary choices. More recently, Wehrens (2014) has begun to use the Imitation Game to explore the relationship between therapists at an eating disorder clinic and their patients; here the Imitation Game is being used both as a research tool and as an intervention to facilitate post-game discussions among participants.

Looking to the future, we see a wide range of possibilities for Imitation Game research. Large-scale studies can deploy a range of quantitative measures based on the content of questions and answers, the reasons given by judges, and demographic data collected from participants to map the content, concerns, and boundaries of particular cultures. In-depth analysis of transcripts can explore how expertise and credibility are constructed and evaluated in the absence of visual clues. Cross-cultural research can explore how the "same" topic is experienced differently in different settings, while longitudinal studies can map how the practice language associated with a particular group or expertise changes over time (e.g., Kubiak and Weinel 2016).

Looking more specifically at how the Imitation Game might be used in STS, we believe there is scope for more research on the nature of medical communication and interdisciplinary working. In the case of medical contexts, possible questions include, how does the quality and quantity of interactional expertise needed vary between acute and chronic treatments? Are different kinds of medical practitioners more able than others to transcend the medical model? Can the Imitation Game be used to promote the development of interactional expertise needed to bridge the gap between biomedicine and the lived experience of patients? Similarly for interdisciplinarity, Imitation Games—either formal or informal—could be used by members of research teams to explore how far they understand each other's conceptual worlds and how such understanding might be promoted. More radically, we could even imagine the acquisition of interactional rather than contributory expertise being the aim of at least some modules in an interdisciplinary degree with, for example, thermodynamics being taught as a "second language" with competence assessed via Imitation Game–type tasks (as first mooted by Berardy, Seager, and Selinger 2011).

Conclusions

The typology of expertises of which interactional expertise is a central element provides a clear and consistent conceptual scheme that can be used to describe both elite and lay actors. For those more concerned with interventions, in the sense of either facilitating interdisciplinarity or improving technological decision making in the public domain, these categories provide a way of identifying relevant experts and arguing for their relevance independently of the way in which others choose to assign expert status. If, as seems likely, we are entering an age in which the appearance of controversy can be sustained in the public realm long after debate has ended within the expert community, then the realist approach that interactional expertise embodies will become even more important.

Notes

1. The literature cited in this chapter is drawn almost exclusively from the English language literature. This reflects the fact that work on interactional expertise is predominantly Anglophone. There is a more international literature on interactional competence ("compétence interactionnelle" in French and "Interaktionskompetenz" in German), but this does not address the questions that animate interactional expertise.

2. For an introduction to the Strong Program, see Barnes, Bloor, and Henry (1996) and Bloor (1991).

3. See also the ad hoc special section on ethnomethodological studies in *Social Studies of Science*, vol. 41 (December 2011).

4. For more details on the case, see *People v. James Hyatt*, Supreme Court of the State of New York, King County. Indictment no. 8852/2000, 10 October 2001.

5. Both interactional and contributory expertise are discussed in some detail in a pair of papers published in *Studies in History and Philosophy of Science* (Collins and Evans 2015a; Collins, Evans, and Weinel 2015). One outcome of this reappraisal is that contributions to the technical phase are seen to come from a much wider range of sources, with some implications for the ways in which the categories in the periodic table of expertises are understood and used.

6. There is also a time dimension in the sense that being socialized into a community is itself a social process. See, for example, Traweek (1992).

7. The point is that Mbeki intervened in what Wynne (2007, 108) calls "expert technical debate" about propositional questions on the basis of having read about the safety of AZT and other drugs on the Internet in isolation. Mbeki's knowledge of the safety of AZT in the context of the prevention of mother-to-child transmission of HIV can, at best, be described as primary source knowledge. The normative SEE framework argues, however, that actors need at least interactional expertise to contribute to "expert technical debate" (Weinel 2007).

8. For a similar exchange, see Durant (2008) and Wynne (2008).

9. Development of the Imitation Game method has benefitted from an Advanced Research Grant awarded to Harry Collins by the European Research Council (269463 IMGAME, 2011–2016). More information about the project, including links to download iOS and Android apps that allow informal Imitation Games to be played on mobile devices, can be found at http://blogs .cardiff.ac.uk/imgame.

References

Barnes, Barry, David Bloor, and John Henry. 1996. *Scientific Knowledge: A Sociological Analysis*. Chicago: University of Chicago Press.

Berardy, Andrew, Thomas P. Seager, and Evan Selinger. 2011. "Developing a Pedagogy of Interactional Expertise for Sustainability Education." In *IEEE International Symposium on Sustainable Systems and Technology (ISSST)*, 1–4. Chicago: IEEE.

Bloor, David. 1991. *Knowledge and Social Imagery*. 2nd ed. Chicago: University of Chicago Press.

Boyce, Tammy. 2006. "Journalism and Expertise." *Journalism Studies* 7 (6): 889–906.

Brown, Mark B. 2015. "Politicizing Science: Conceptions of Politics in Science and Technology Studies." *Social Studies of Science* 45 (1): 3–30.

Button, Graham, and Wes Sharrock. 1998. "The Organizational Accountability of Technological Work." *Social Studies of Science* 28 (1): 73–102.

Callon, Michel. 1986. "Some Elements of a Sociology of Translation: Domestication of the Scallops and the Fishermen of St. Brieuc Bay." In *Power, Action and Belief: A New Sociology of Knowledge?*, edited by John Law, 196–223. London: Routledge.

Carr, E. Summerson. 2010. "Enactments of Expertise." *Annual Review of Anthropology* 39 (1): 17–32.

Collins, Harry. 1996. "Embedded or Embodied? A Review of Hubert Dreyfus' What Computers Still Can't Do." *Artificial Intelligence* 80 (1): 99–117.

___. 2004. "Interactional Expertise as a Third Kind of Knowledge." *Phenomenology and the Cognitive Sciences* 3 (2): 125–43.

___. 2011. "Language and Practice." *Social Studies of Science* 41 (2): 271–300.

___. 2013. "Three Dimensions of Expertise." *Phenomenology and the Cognitive Sciences* 12 (2): 253–73.

___. 2014a. *Are We All Scientific Experts Now?* New Human Frontiers Series. Cambridge: Polity.

___. 2014b. "Rejecting Knowledge Claims Inside and Outside Science." *Social Studies of Science* 44 (5): 722–35.

Collins, Harry, and Robert Evans. 2002. "The Third Wave of Science Studies: Studies of Expertise and Experience." *Social Studies of Science* 32 (2): 235–96.

___. 2003. "King Canute Meets the Beach Boys: Responses to the Third Wave." *Social Studies of Science* 33 (3): 435–52.

___. 2007. *Rethinking Expertise*. Chicago: University of Chicago Press.

___. 2014. "Quantifying the Tacit: The Imitation Game and Social Fluency." *Sociology* 48 (1): 3–19.

___. 2015a. "Expertise Revisited, Part I—Interactional Expertise." *Studies in History and Philosophy of Science Part A* 54: 113–23.

___. 2015b. "Probes, Surveys and the Ontology of the Social." *Journal of Mixed Methods Research* Online First; December 15, 2015. doi:10.1177/1558689815619825.

Collins, Harry, Robert Evans, and Mike Gorman. 2007. "Trading Zones and Interactional Expertise." *Studies in History and Philosophy of Science Part A* 38 (4): 657–66.

Collins, Harry, Robert Evans, Rodrigo Ribeiro, and Martin Hall. 2006. "Experiments with Interactional Expertise." *Studies in History and Philosophy of Science Part A* 37 (4): 656–74.

Collins, Harry, Robert Evans, and Martin Weinel. 2015. "Expertise Revisited, Part 2: Contributory Expertise." *Studies in History and Philosophy of Science Part A*. doi:10.1016/j.shpsa.2015.07.003.

Collins, Harry, Robert Evans, Martin Weinel, Jennifer Lyttleton-Smith, Andrew Bartlett, and Martin Hall. 2015. "The Imitation Game and the Nature of Mixed Methods." *Journal of Mixed Methods Research*. doi:10.1177/1558689815619824.

Collins, Harry, Martin Weinel, and Robert Evans. 2010. "The Politics and Policy of the Third Wave: New Technologies and Society." *Critical Policy Studies* 4 (2): 185–201.

Coopmans, Catelijne, and Graham Button. 2014. "Eyeballing Expertise." *Social Studies of Science* 44 (5): 758–85.

Delborne, Jason A. 2008. "Transgenes and Transgressions: Scientific Dissent as Heterogeneous Practice." *Social Studies of Science* 38 (4): 509–41.

Doing, Park. 2004. "'Lab Hands' and the 'Scarlet O': Epistemic Politics and (Scientific) Labor." *Social Studies of Science* 34 (3): 299–323.

Donaldson, Liam. 2003. "Expert Patients Usher in a New Era of Opportunity for the NHS." *BMJ* 326 (7402): 1279–80.

Dreyfus, Hubert L. 1992. *What Computers Still Can't Do: A Critique of Artificial Reason*. Cambridge, MA: MIT Press.

Dreyfus, Hubert L., and Stuart E. Dreyfus. 1986. *Mind over Machine: The Power of Human Intuition and Expertise in the Era of the Computer*. New York: Free Press.

Duarte, Tiago Ribeiro. 2013. "Expertise and the Fractal Model: Communication and Collaboration between Climate-Change Scientists." Ph.D., Cardiff University, Cardiff, UK. Accessed at http://orca.cf.ac.uk/49632/.

Durant, Darrin. 2008. "Accounting for Expertise: Wynne and the Autonomy of the Lay Public Actor." *Public Understanding of Science* 17 (1): 5–20.

___. 2011. "Models of Democracy in Social Studies of Science." *Social Studies of Science* 41 (5): 691–714.

___. 2015. "The Undead Linear Model of Expertise." In *Policy Legitimacy, Science and Political Authority: Knowledge and Action in Liberal Democracies*, edited by Michael Heazle, John Kane, and Haig Patapan 17–38. Earthscan Science in Society. New York: Routledge.

Epstein, Steven. 1996. *Impure Science: AIDS, Activism, and the Politics of Knowledge*. Berkeley: University of California Press.

Evans, Robert. 2011. "Collective Epistemology: The Intersection of Group Membership and Expertise." In *Collective Epistemology*, edited by Hans Bernhard Schmid, Daniel Sirtes, and Marcel Weber, 177–202. Frankfurt: Ontos Verlag.

Evans, Robert, and Helen Crocker. 2013. "The Imitation Game as a Method for Exploring Knowledge(s) of Chronic Illness." *Methodological Innovations Online* 8 (1): 34–52.

Evans, Robert, and Alexandra Plows. 2007. "Listening without Prejudice? Re-discovering the Value of the Disinterested Citizen." *Social Studies of Science* 37 (6): 827–53.

Fine, Gary Alan. 2010. *Authors of the Storm: Meteorologists and the Culture of Prediction*. Chicago: University of Chicago Press.

Fisher, Erik. 2007. "Ethnographic Invention: Probing the Capacity of Laboratory Decisions." *NanoEthics* 1 (2): 155–65.

Fisher, Erik, Michael O'Rourke, Robert Evans, Eric B. Kennedy, Michael E. Gorman, and Thomas P. Seager. 2015. "Mapping the Integrative Field: Taking Stock of Socio-technical Collaborations." *Journal of Responsible Innovation* 2 (1): 1–23.

Fricker, Miranda. 2007. *Epistemic Injustice: Power and the Ethics of Knowing*. Oxford: Oxford University Press.

Galison, Peter Louis. 1997. *Image and Logic: A Material Culture of Microphysics*. Chicago: University of Chicago Press.

Geertz, Clifford. 1973. *The Interpretation of Cultures*. New York: Basic Books.

Gieryn, Thomas F. 1999. *Cultural Boundaries of Science: Credibility on the Line*. Chicago: University of Chicago Press.

Giles, Jim. 2006. "Sociologist Fools Physics Judges." *Nature* 442 (7098): 8.

Glushko, Robert J. 2008. "Designing a Service Science Discipline with Discipline." *IBM Systems Journal* 47 (1): 15–27.

Goddiksen, Mads. 2014. "Clarifying Interactional and Contributory Expertise." *Studies in History and Philosophy of Science Part A* 47 (September): 111–17.

Goody, Jack, and Ian Watt. 1968. "The Consequences of Literacy." In *Literacy in Traditional Societies*, edited by Jack Goody, 27–68. Cambridge: Cambridge University Press.

Gorman, Michael E. 2002. "Levels of Expertise and Trading Zones: A Framework for Multidisciplinary Collaboration." *Social Studies of Science* 32 (5–6): 933–38.

___. 2010. "Trading Zones, Normative Scenarios, and Service Science." In *Handbook of Service Science*, edited by Paul P. Maglio, Cheryl A. Kieliszewski, and James C. Spohrer, 665–75. Boston: Springer US.

Gorman, Michael E., and James C. Spohrer. 2010. "Service Science: A New Expertise for Managing Sociotechnical Systems." In *Trading Zones and Interactional Expertise: Creating New Kinds of Collaboration*, edited by Michael E. Gorman, 75–106. Cambridge, MA: MIT Press.

Greenhalgh, Trisha. 2009. "Patient and Public Involvement in Chronic Illness: Beyond the Expert Patient." *BMJ* 338: b49.

Hinnant, Amanda, and Maria E. Len-Rios. 2009. "Tacit Understandings of Health Literacy: Interview and Survey Research with Health Journalists." *Science Communication* 31 (1): 84–115.

Höppe, Gotz. 2014. "Working Data Together: The Accountability and Reflexivity of Digital Astronomical Practice." *Social Studies of Science* 44 (2): 243–70.

Horst, Maja, and Mike Michael. 2011. "On the Shoulders of Idiots: Re-thinking Science Communication as 'Event.'" *Science as Culture* 20 (3): 283–306.

Irwin, Alan. 1995. *Citizen Science: A Study of People, Expertise, and Sustainable Development*. London: Routledge.

Jasanoff, Sheila. 2002. "Science and the Statistical Victim: Modernizing Knowledge in Breast Implant Litigation." *Social Studies of Science* 32 (1): 37–69.

___. 2003. "Breaking the Waves in Science Studies: Comment on H. M. Collins and Robert Evans, 'The Third Wave of Science Studies'." *Social Studies of Science* 33 (3): 389–400.

___. 2007. *Designs on Nature Science and Democracy in Europe and the United States*. Princeton, NJ: Princeton University Press.

___. 2013. "Fields and Fallows: A Political History of STS." In *Interdisciplinarity: Reconfigurations of the Natural and Social Sciences*, edited by Andrew Barry and Georgina Born, 99–118. London: Routledge.

Kerr, Anne, Sarah Cunningham-Burley, and Richard Tutton. 2007. "Shifting Subject Positions: Experts and Lay People in Public Dialogue." *Social Studies of Science* 37 (3): 385–411.

Knorr Cetina, Karin. 1999. *Epistemic Cultures: How the Sciences Make Knowledge*. Cambridge, MA: Harvard University Press.

Koertge, Noretta, ed. 2000. *A House Built on Sand: Exposing Postmodernist Myths about Science*. New York: Oxford University Press.

Kruvand, Marjorie. 2012. "'Dr. Soundbite': The Making of an Expert Source in Science and Medical Stories." *Science Communication* 34 (5): 566–91.

Kubiak, Daniel, and Martin Weinel. 2016. "DDR-Generationen Revisited—Gibt es einen Generationszusammenhang der Wendekinder?" In *Die Generation der Wendekinder*, edited by Adriana

Lettrari, Christian Nestler, and Nadja Troi-Boeck, 107–29. Wiesbaden: Springer Fachmedien Wiesbaden.

Kuhn, Thomas S. 1996. *The Structure of Scientific Revolutions*. 2nd ed., vol. 14. Chicago: University of Chicago Press.

Layton, David, Edgar Jenkins, Sally Macgill, and Angela Davey. 1990. *Inarticulate Science? Perspectives in the Public Understanding of Science and Some Implications for Science Education*. Driffield, UK: Studies in Education.

Livingston, Eric. 2006. "The Context of Proving." *Social Studies of Science* 36 (1): 39–68.

Lynch, Michael. 1997. *Scientific Practice and Ordinary Action: Ethnomethodology and Social Studies of Science*. Cambridge: Cambridge University Press.

Lynch, Michael, and Simon Cole. 2005. "Science and Technology Studies on Trial: Dilemmas of Expertise." *Social Studies of Science* 35 (2): 269–311.

MacKenzie, Donald, and Taylor Spears. 2014. "'The Formula That Killed Wall Street': The Gaussian Copula and Modelling Practices in Investment Banking." *Social Studies of Science* 44 (3): 393–417.

Matoesian, Gregory M. 1999. "The Grammaticalization of Participant Roles in the Constitution of Expert Identity." *Language in Society* 28 (4): 491–521.

___. 2008. "Role Conflict as an Interactional Resource in the Multimodal Emergence of Expert Identity." *Semiotica 2008* (171): 15–49.

Mazur, Dennis John. 2003. *The New Medical Conversation: Media, Patients, Doctors, and the Ethics of Scientific Communication*. Lanham, MD: Rowman & Littlefield.

Merz, Martina. 1998. "'Nobody Can Force You When You Are across the Ocean'—Face to Face and E-Mail Exchanges between Theoretical Physicists." In *Making Space for Science: Territorial Themes in the History of Science*, edited by Crosbie Smith and Jon Agar, 313–29. London: Palgrave Macmillan.

Miller, Steve. 2001. "Public Understanding of Science at the Crossroads." *Public Understanding of Science* 10 (1): 115–20.

Milton, Damian. E. 2014. "Autistic Expertise: A Critical Reflection on the Production of Knowledge in Autism Studies." *Autism* 18 (7): 794–802.

O'Rourke, Michael, and Stephen J. Crowley. 2013. "Philosophical Intervention and Cross-Disciplinary Science: The Story of the Toolbox Project." *Synthese* 190 (11): 1937–54.

Ottinger, Gwen. 2013. *Refining Expertise: How Responsible Engineers Subvert Environmental Justice Challenges*. New York: New York University Press.

Plaisance, Kathryn S., and Eric B. Kennedy. 2014. "A Pluralistic Approach to Interactional Expertise." *Studies in History and Philosophy of Science Part A* 47 (September): 60–68.

Polanyi, Michael. 1966. *The Tacit Dimension*. Chicago: University of Chicago Press.

Popay, Jennie, and Gareth Williams. 1996. "Public Health Research and Lay Knowledge." *Social Science & Medicine* 42 (5): 759–68.

Priaulx, Nicky, and Martin Weinel. 2014. "Behaviour on a Beer Mat: Law, Interdisciplinarity and Expertise." *Journal of Law, Technology and Policy* (2): 361–91.

Reich, Zvi. 2012. "Journalism as Bipolar Interactional Expertise." *Communication Theory* 22 (4): 339–58.

Reyes-Galindo, Luis I. 2011. "The Sociology of Theoretical Physics." Ph.D. thesis, Cardiff, UK: Cardiff University. Accessed at http://orca.cf.ac.uk/15106/.

___. 2014. "Linking the Subcultures of Physics: Virtual Empiricism and the Bonding Role of Trust." *Social Studies of Science* 44 (5): 736–57.

Reyes-Galindo, Luis I., and Tiago Ribeiro Duarte. 2015. "Bringing Tacit Knowledge Back to Contributory and Interactional Expertise: A Reply to Goddiksen." *Studies in History and Philosophy of Science Part A* 49: 99–102.

Ribeiro, Rodrigo, and Francisco P. A. Lima. 2015. "The Value of Practice: A Critique of Interactional Expertise." *Social Studies of Science*. Online before print. doi:10.1177/0306312715615970.

Rowe, David E. 1986. "'Jewish Mathematics' at Gottingen in the Era of Felix Klein." *Isis* 77 (3): 422–49.

Schutz, Alfred. 1964. *Collected Papers. 2: Studies in Social Theory*, edited by Arvid Brodersen. The Hague: Martinus Nijhoff.

Selinger, Evan, Hubert L. Dreyfus, and Harry Collins. 2007. "Embodiment and Interactional Expertise." *Studies in History and Philosophy of Science* 38 (4): 722–40.

Shapin, Steven. 2007. *A Social History of Truth: Civility and Science in Seventeenth-Century England*. Chicago: University of Chicago Press.

Star, Susan Leigh, and James R. Griesemer. 1989. "Institutional Ecology, 'Translations' and Boundary Objects: Amateurs and Professionals in Berkeley's Museum of Vertebrate Zoology, 1907–39." *Social Studies of Science* 19 (3): 387–420.

Suryanarayanan, Sainath, and Daniel Lee Kleinman. 2013. "Be(e)coming Experts: The Controversy over Insecticides in the Honey Bee Colony Collapse Disorder." *Social Studies of Science* 43 (2): 215–40.

Traweek, Sharon. 1992. *Beamtimes and Lifetimes: The World of High Energy Physicists*. Cambridge, MA: Harvard University Press.

Turing, Alan. 1950. "Computing Machinery and Intelligence." *Mind* 59 (236): 433–60.

Tyreman, Stephen. 2005. "An Expert in What? The Need to Clarify Meaning and Expectations in 'The Expert Patient.'" *Medicine, Health Care and Philosophy* 8 (2): 153–57.

Wehrens, Rik. 2014. "The Potential of the Imitation Game Method in Exploring Healthcare Professionals' Understanding of the Lived Experiences and Practical Challenges of Chronically Ill Patients." *Health Care Analysis* 23 (3): 253–71.

Weinel, Martin. 2007. "Primary Source Knowledge and Technical Decision-Making: Mbeki and the AZT Debate." *Studies in History and Philosophy of Science Part A* 38 (4): 748–60.

Welsh, Ian. 2000. *Mobilizing Modernity: The Nuclear Moment*. International Library of Sociology. New York: Routledge.

Whyte, Kyle Powys, and Robert P. Crease. 2010. "Trust, Expertise, and the Philosophy of Science." *Synthèse* 177 (3): 411–25.

Winch, Peter. 1958. *The Idea of a Social Science and Its Relation to Philosophy*. Studies in Philosophical Psychology. London: Routledge & Kegan Paul.

Wittgenstein, Ludwig. 1953. *Philosophical Investigations*, translated by G. E. M Anscombe. Oxford: Blackwell.

Wylie, Caitlin Donahue. 2015. "'The Artist's Piece Is Already in the Stone': Constructing Creativity in Paleontology Laboratories." *Social Studies of Science* 45 (1): 31–55.

Wynne, Brian. 2003. "Seasick on the Third Wave? Subverting the Hegemony of Propositionalism: Response to Collins & Evans (2002)." *Social Studies of Science* 33 (3): 401–17.

___. 2007. "Public Participation in Science and Technology: Performing and Obscuring a Political–Conceptual Category Mistake." *East Asian Science, Technology and Society: An International Journal* 1 (1): 99–110.

___. 2008. "Elephants in the Rooms Where Publics Encounter 'Science'? A Response to Darrin Durant, 'Accounting for Expertise: Wynne and the Autonomy of the Lay Public.'" *Public Understanding of Science* 17 (1): 21–33.

Yinger, J. Milton. 1982. *Countercultures: The Promise and the Peril of a World Turned Upside Down*. New York: Free Press.

27 Surveillance and Regulation of Laboratory Practices

Ruthanne Huising and Susan S. Silbey

This chapter explores the confrontation between the authority of law and the authority of science by looking directly in the heart of science: the laboratory. We review the literature on how scientists respond to regulation of their laboratory practices, suggesting ways in which contemporary laboratory practices are shaped, or not, by law.

Law does not supply merely an environment surrounding the boundary of scientific spaces as happens when scientific creations become intellectual property or when scientific knowledge is introduced in courts as evidence. Law also inhabits science, materially and behaviorally. However, the role of legal regulation within the procedures as well as the spaces of science is perhaps one of the least explored or understood aspects of the relationship between law and science. This chapter will identify the range of scholarship exploring the intersections of law and science by focusing specifically on studies of the design and implementation of environmental, health, and safety regulations for academic research laboratories.

We examine academic research laboratories as examples of particularly difficult governance sites. These spaces often elude regulatory warnings and rules because of the professional status of faculty members, the opacity of scientific work to outsiders, and the generally loose coupling of policy and practice in most organizations, and especially so in academic organizations. We describe efforts to challenge faculty privilege, showing how academic scientists are buffered from the consequences of their activities, thus impeding the legal goals of responsibility and accountability. If scientists are difficult to manage, can we identify practices that nonetheless herd these privileged actors?

This chapter makes three key contributions. First, rather than focus on the transformation of science into legal work and discourse as most studies do—as evidence in court, as recommendations for policy, as proposed standards for best practices, or as intellectual property—we look at how and what happens when legal mandates are interpreted by scientists and translated into scientific practices. Thus, this chapter does not address how science purposely works to shape law but may nonetheless do so inadvertently by its responses to legal regulations. Second, by directing attention to the law where it is not obvious, nor closely observed, we contribute to studies of the everyday life of law that have been so productive in recent years. This also allows us to see what is often absent in the extensive library of science and technology studies (STS) laboratory studies (Doing 2008). Most scientists, despite some recalcitrance, end up complying with most regulations, although all do not, and some more often than others. When scientists comply, they often do so by delegating routine requirements to subordinates or relying on support staff who buffer them from the most intrusive regulations (Gray and Silbey 2014; Huising and Silbey 2011).

Third, this is also an inquiry into the practices of risk management and surveillance. Although contemporary surveillance technologies create their own distinctive implementation patterns (e.g., Marx 2016), they also reproduce the traditional gap between what law requires and what actually happens, a variably sized chasm between regulation and its implementation. Empirical studies show that successful regulators manage this inevitable gap with locally relevant, pragmatic adaptations. We call this relational regulation, an approach used by some front-line managers to keep laboratory practices within an acceptable range of variation close to regulatory specifications. Regulatory agents govern the gap between law and its implementation by acknowledging the impossibility of perfect compliance between abstract rules or norms and situated action. Thus, to successfully govern laboratory practices, regulators mimic some recognizable—if contested—norms of science, entertaining skepticism (Owen-Smith 2001) and governing through collegiality (Evans 2010), direct observation (Galison 1997; Knorr Cetina 1981; Lynch 1985), and local invention and tinkering within each laboratory (Clarke and Fujimura 1992; Gooding 1992; Knorr Cetina 1992; Pickering 1995).

This chapter addresses the implementation of regulations in American academic laboratories. Although there are historical and international cases that parallel the American example, there are substantial national variations, which cannot be responsibly covered in this chapter. While we concentrate on environmental, health, and safety regulations, other regulatory topics include human subjects (Stark 2012), animals (Birke, Arluke, and Michael 2007), and drugs (Abraham and Davis 2009; Carpenter 2010). How much compliance practices vary across these fields is worthy of investigation. Within these parameters, we provide students of science and technology studies with a window into often less well noticed phenomena to illustrate with one example long-standing competitions and collaborations between scientific and legal expertise. By substituting systems of audit and surveillance for relations of trust and collegiality, legal regulations participate in the reconstruction of everyday rituals of scientific practice. Examples of successful laboratory regulation are important not only because of the

increasing significance of scientific and educational institutions in contemporary societies but also because these institutions serve as models for emerging organizational
forms that depend on innovation, flexibility, and large knowledge bases.[1]

We begin with a quick review of the ways science enters legal processes and then
characterize science, not in terms of its methods or epistemology specifically but in
terms of trust among scientists. We suggest that the intrusion of legal regulation into
scientific spaces signals a subtle shift from a world of trust among peers to one of
surveillance by outside authorities. We then describe contemporary techniques of surveillance through information technologies working to standardize laboratory safety
practices and, before concluding, offer the notion of relational regulation, mimicking
conventional scientific norms, as a pragmatic approach to regulatory compliance.

Science in Law

In the contemporary literature describing the ways in which science intersects with
law, the largest portion of the research adopts what scholars sometimes call a "law first"
perspective (Sarat and Kearns 1993), examining the role of science when it enters formal legal arenas. This law first approach often assumes the institutional autonomy of
law, treating scientific questions and processes as problems the law needs to manage as
threats to the integrity of legal processes. This law first perspective regularly conceals,
because it rarely questions, the ways in which legality itself is formed by these engagements, constituted within and outside of formal legal institutions. Although a mature
conceptual framework is not yet in place for addressing the constitutional implications
of epochal changes in science and technology (Jasanoff 2011), evidence is accumulating about the ways in which "modes of authorization in science and the law build
upon, mimic or incorporate one another" (Jasanoff 2003, 164; Smith and Wynne 1989;
Wynne 1988). Although much modern culture attempts to keep separate the worlds
of nature and culture (e.g., science, politics, law), they are so intimately married in
our language and ordinary reasoning that we rarely take account of the implications
(Delaney 2001; Latour 1993). Despite the hypersensitivity of contemporary scholars to
this pervasive hybridity, we have not yet accounted "satisfactorily for the [distribution
of similarities and] divergences among our constitutional understandings [of science
and law] in different times and cultures" (Jasanoff 1999).

Science and technology studies have adopted the term "boundary work" to
describe the social transactions that mediate between science and other institutions
(Gieryn 1995, 1999). Scholars vary in the degree to which they see these boundaries as more or less given, or in the process of being built, maintained, defended, and

broken down (Miller 2000). Nonetheless, "recognizing that there is no unbridgeable chasm between science and non-science and that the flexibility of boundary work may threaten some important values and interests" (Guston 2001, 399), scholars have suggested that some objects and standardized packages stabilize boundary work by creating shared practices (Fujimura 1991; Miller 2001; Moore 1996). Boundary-spanning organizations may provide just those needed opportunities "for the creation and use of boundary objects and standardized packages; second they involve the participation of actors from both sides of the … frontier of the two relatively different social worlds of politics and science" (Guston 2001, 400f). Thus, we may conceive the introduction of environmental health and safety regulators into laboratories, as we do below, in terms of the boundary work these agents do translating legal mandates, expectations of hierarchical control, and aspirations for consistent conformity into the culture of frontier science. Because a boundary organization, such as an EHS (environment, health, and safety) office, serves two masters, Jasanoff (1996, 397) has described its work as simultaneously "co-producing" knowledge and social order.

For the moment, however, let us quickly look, as legal scholars often do, at science as an independent variable, perturbing legal processes, while offering assistance for establishing social and physical facts (e.g., Edmond and Mercer 2000; Faigman 1999; Jasanoff 1995). In this endeavor, science is distinctly secondary to legal norms and methods, a recalcitrant, hard-to-manage tool mobilized in service of legal decision making. When researchers look at how science enters legal disputes and the evidentiary quality of scientific data, they often focus specifically on the role of scientists as expert witnesses—emissaries from the scientific community to the halls of justice. In general, however, the legal process is challenged by scientific methods, as well as the epistemology and dialogic nature of scientific inquiry (e.g., for histories of expert testimony, see Golan 1999; Jones 1994). Issues concerning scientific proof have been central in legal debates concerning, for example, the reliability and interpretation of evidence regarding DNA "fingerprinting" (Lynch et al. 2008), battered women's syndrome (Faigman 1986), insanity pleas, rape trauma (Frazier and Borgida 1992), silicone implants, and the death penalty (Sarat 2001). Some research shows how evidentiary conundrums are fed, in part, by claims of irrefutable truth based on processes of visualization that nonetheless involve subtle, uncertain methods and interpretations, displaying visual evidence's status as human artifice rather than "natural" expression (Cole 1999; Dumit 1999; Moonkin 1998; Rafter 2001).

Although much of this research depicts science as an alien imported into the courtroom, subjected to foreign linguistic and interpretive conventions, burdens of proof, and authority structures, there are alternative voices. Jasanoff (2004), for example, argues instead of science disrupting law, the law constructs and reconstructs science for its own purposes and in its own image. In the courtroom, science's distinctive claims to knowledge and impartiality are challenged, perhaps even eroded. The probabilistic and statistical reasoning at the heart of much modern science generates frequent misunderstandings (Faigman and Baglioni 1998; Rubinfeld 1985). While some observers question the ability of legal actors, in particular juries, to understand and interpret expert testimony or its reliability (e.g., Cutler, Dexter, and Penrod 1989; Diamond and Casper 1992), others claim that the increased reliance on scientific testimony that is difficult to interpret and understand creates openings for "junk science," or a pseudoscience, that has been undertaken specifically to produce results desired by and supporting particular litigants (Faigman 1991; Foster, Bernstein ,and Huber 1993). Lynch et al. (2008), for example, provide an analysis of the kinds of incommensurable data presented in trial courts, and of the efforts in British trials to create commensurability between common sense and scientific evidence through Bayesian statistics.

If the focus on court trials exemplifies a law first perspective that ignores accounts of the mutual constitution and co-production of law and science (Jasanoff 1996), it also exaggerates the dominance of litigation and trials as a site of intersection of law and science processes (cf. Ewick and Silbey 1998). Law and science intersect in diverse ways, very much independent of courtroom drama. It bears remembering that modern science has been institutionalized with organizational and financial support from government as well as business, private philanthropy, and public-interest organizations (Guston 1999; Moore 1996; Owens 1990), in each instance secured through legal instruments. The institutionalization of modern science has been achieved not only through government and philanthropic largesse but also through legal techniques that transform the search for knowledge into a profit-making activity through the creation of intellectual property that makes markets of scientific knowledge (Silbey 2014). As with expert testimony, however, we also encounter heated controversies, for example, concerning title and property in scientific work because the right to benefit economically from scientific invention has become entangled with state funding and related concerns about public access to the fruits of public investment. The legal controversies persist because the right to turn knowledge into capital depends, doctrinally, on a demonstration of independent, discontinuous invention of new knowledge, which is never entirely independent or discontinuous (Landecker 1999; Voelkel 1999). Over the course of the twentieth century, hard-line opposition to this appropriation of a public good has weakened; patenting and technology licensing officers now exist as important and profitable administrative units within universities as well as firms (Owen-Smith 2005).

If the rituals of the trial exacerbate the distinctions between making facts in science and building cases in law (Fuchs and Ward 1994), and patenting and technology licensing obscure the boundaries between public and private benefits from science, public policy deliberations provide neither a more certain nor a problem-free arena for scientific discourse. Public policy debates often focus on the uncertain human, material, and environmental consequences of scientific innovations. During policy deliberations, members of the public, policy makers, and scientists compete to define both problems and relevant knowledge (Ewald and Utz 2002). Each group mobilizes disqualifying and legitimating discourses to "further their own interests" (ibid., 447) (Collins and Evans 2002; Gusterson 2000; Jasanoff 1990; Levidow 2001). Because "[k]nowledge claims are deconstructed during the rule-making process, exposing areas of weakness or uncertainty and threatening the cognitive authority of science ... the legitimacy of the final regulatory decision depends upon the regulator's ability to reconstruct a plausible scientific rationale for the proposed action" (Jasanoff 1987, 447).[2]

From Trust to Surveillance

At its core, because scientific authority is based upon a claim that seeing is believing, scientists are engaged in projects of self-presentation (Hilgartner 2000). This requires that methods and results be accessible to would-be observers and skeptics, distributed through public presentation, peer-review, and replication. Up until the nineteenth and twentieth centuries, that space of experimentation was a "truth spot" (Gieryn 1999), the place where empirical truths were revealed to and certified by a select audience of gentlemen (Shapin 1988). The particular location, configuration, ownership, and design of what served as a space of science were constructed by and available for privileged observers. The legitimacy of scientific claims depended on the idiosyncrasies of the place and the particular relationship that existed between scientist and audience, on the trust inscribed within the social ties that connected scientists, their laboratory spaces, and their publics.

In the contemporary world, this private, place-based method of observing and certifying scientific knowledge production has been replaced by claims of universalism and processes designed to promote disinterestedness. First, laboratory architecture (Galison and Thompson 1999) has changed dramatically over the last two centuries, no longer garden conservatories or kitchen sculleries. Laboratories have developed into vast, prototypical, universal products with interchangeable parts and equipment, unremarkable in the ease with which they are reproduced and installed in very different physical conditions, geographical climates, and cultural locations. What was private and personal, identified with and occupied by a specific, socially located individual became, with the collaboration of public authorities, more accessible, standardized, and relatively indistinguishable from other similar spaces. Their contents have been so architecturally standardized that contemporary laboratories are designed and built LEGO style: a pattern module is composed of stock materials, and then arranged in various configurations, most often in rows and bays, to fit a building's dimensions and each research group's desired social organization (Gieryn 1998).

Second, the laboratory is no longer the place where truth is lodged; it has been demoted to a backstage. The publication or text has become the new public space, open and accessible to all, in and through which the provenance of science is established. Rather than seeing the experiment, we see the report of it. We defer to the report because of the literary technology of the scientific journal including peer review and critique (Shapin 1988). Truth transcends its place of discovery.

In short, because the laboratory has become relatively standardized in its construction and composition, it can disappear as an epistemological marker; we can take it for granted because it is constant and universal. However, efforts to interrogate, test, and vet the quality and veracity of scientific claims published in peer-reviewed journals are nonetheless limited by the relative independence of laboratories and the global nature of science that despite aspirations to consistency and universality renders laboratory practices idiosyncratic and local (Knorr Cetina 1999; Traweek 1988). The extreme specialization and the tacit knowledge infused at all layers of the work make the essential criteria of replication tricky (Collins 1974, 1975; Latour and Woolgar 1979; Lynch 1982). Thus, in their daily activities and collaborations, scientists and scientific work still depend on trust.

Trust is "a willingness to accept vulnerability" based on expectations of "technically competent role performances from those involved with us in social relations and systems" (McEvily, Perrone, and Zaheer 2003, 92–93). Scientists routinely make decisions about whom, within their community, they will trust to perform as a credible scientist based on personal relations (Collins 2001), familiarity (Shapin 1994), reputation (Krige 2001), as well as formal markers of status such as institutional affiliations and past productivity. In other words, science proceeds in large part through relations and reputation rather than surveillance. Collins (2001), for example, describes how replication is not at all simple or straightforward, instead mediated by social knowledge and interactions among research groups. When a laboratory is known and familiar, even in the face of replication difficulties, other labs will work until replication is achieved rather than reject findings without testing the results. The assumption of reliability, grounding the investment in replication, is generated by intimate, repeated transactions among laboratories. Global dispersion of scientific inquiry across political boundaries sometimes impedes this kind of exchange and trust. While providing access to often unspoken tacit knowledge that is part of all science (Collins 2001; Polanyi 1958), regular transactions and exchange (Evans 2010) simultaneously demonstrate and create mutual trust in scientists' honesty and commitment to the norms of disinterested inquiry.

The fundamental trust relations are engendered by long years of training, habituating scientists to both the formal and informal norms and producing a form of governance as each member self-regulates in line with historic, professional, and local expectations (Van Maanen and Barley 1982). Of course, norms are not static (Evans 2010; Hackett 1990; Owen-Smith and Powell 2001); they change over time and vary by discipline (Sauermann and Stephan 2013). Moreover, not all norms are equally approved by all group members; contestation and conflict persists in all communities and professions. Nonetheless, norms structure interactions through widely circulating claims about how competent members of the profession should, and do, conduct themselves (Zucker 1986). Although Merton's ([1942] 1957) model of scientific norms may be more aspirational than realized, those aspirations circulate widely, providing a basis of trust that insulates scientists from some forms of regulation.

Even though individual scientists may have personal interests, they are expected to do their science in a disinterested manner. Disinterestedness names, in effect, the consequence of the internal policing of scientific claims to assure that base motives of profit, priority, and award are constrained by the epistemological insistence on replication. This is never perfect and thus controversies about replication and disinterest periodically flood public media. Trust, as the basis of much scientific exchange, persists to the degree that scientific communities practice skepticism, demonstrate and respond to critical inquiry, and balance competing drives for innovation and replication, or tradition versus subversion, relevance versus originality, conformity versus dissent (Bourdieu 1975; Foster, Rzhetsky, and Evans 2015; Lynch 1982; Merton [1942] 1957; Owen-Smith 2001; Whitley 2000).

As knowledge production becomes socially distributed, some argue that trust based upon reputation and credibility is not likely to be a useful marker of reliability and validity. New modes of knowledge production involving complex transdisciplinary collaborations across government, academia, and industry, such as mode 2 (Gibbons et al. 1994), triple-helix (Etzkowitz and Leydesdorff 2000), or post-academic (Ziman 2000) science, entail new modes of quality control. First, collaborators may not be academic scientists from one's discipline, sector, or geographic locale, and thus the collaborator's knowledge and status cannot be easily assessed, especially when teams contain members with vastly different expertise and epistemologies. Second, the collaborator may not be socialized to observe and operate in accordance with community and professional norms (which while generally shared, do vary cross-culturally and across fields of knowledge production). Finally, "scientific 'peers' can no longer be [easily] identified because there is no longer a stable taxonomy of codified disciplines from which 'peers' can be drawn" (Nowotny, Scott, and Gibbons 2003, 187). Lacking repeated, interpersonal transactions within familiar boundaries, it may be difficult to generate the trust conventionally used to assess who does good science.

Despite these predictions, recent analyses of scientific collaborations find that unstated, tacit arrangements and handshakes have not been entirely replaced by formal contracts or new forms of quality control (Hackett 2005). Trust and familiarity remain central in scientific collaboration. Collaborators continue to engage with each other on the basis of relations of respect, trust, and affinity generated through interpersonal transactions or familiarity (Jha and Welch 2010).

While trust relations and norms remain important factors in scientific collaboration, they no longer insulate science from external regulation. Law has entered the house of science by asserting jurisdiction over the spaces where experimentation is done. Scientific advances and technological innovations have created increased probabilities of material hazard that have raised concerns about the exposure of laboratory personnel or research subjects to diverse risks. In the age of risk management this creates openings for various techniques of surveillance—through bureaucratic accounting, record-keeping, and physical protections—to become habitual parts of the laboratory without appearing to dislodge the fundamental trust relations (O'Malley 1996; Parthasarathy 2004). As legal authorities and universities create demands for safer and more ethical science, new regimes of surveillance, inspection, and audit challenge the conventional self-policing cultures of laboratory science. The legal regulation of science becomes a task of controlling the spaces and materials in and with which scientists work; these regulations do not, however, challenge the scientists' control of truth, what constitutes a fact or knowledge. It remains, however, an open question to what degrees and in what ways the practices for establishing facts will be transformed incrementally by the regulation of materials, space, and human subjects.

Law in Science

What the law regulates, constrains, and enables may be influenced, if not determined, by science's methods, products, and conclusions identifying hazards and their consequences. *How* regulation takes place, through what sorts of procedures and sanctions, may be the law's specific prerogative (but here too law may collaborate with science). Thus we manage the dangers of radiation through an elaborate system of continuous surveillance that can lead to mandatory cessation of operation, personal exclusion from work, or heavy monetary fines. However, we respond to the dangers of smoking by requiring notices on cigarette packages, prohibiting advertising and sale to minors, but tax rather than prohibit consumption for adults. And, in most American states, we respond to the dangers of some contagious diseases not only by notices but also by mandated inoculation. In each of these instances, the dangers have been identified through scientific research (e.g., Abraham and Davis 2007), but the modes and forms of regulation are legal inventions.

By the late nineteenth century, chemistry laboratories, for example, began to install ventilation hoods to exhaust the chemical fumes. They were not required by law, at first, but were merely examples of good research practice; there was neither regular surveillance nor active enforcement. By the end of World War II, radioisotopes were familiar research substances, the control of which was delegated to the U.S. Atomic Energy Commission, which later became the Nuclear Regulatory Commission. Handling radioisotopes was not merely a matter of good research practice but subject to surveillance and enforcement of the licenses federal agencies issue to work with radioactive materials. The licenses specify safe conditions of operation through architecture, training, self-monitoring, and periodic inspections. Universities, national laboratories, and firms set up in-house offices to provide service to those using radioactive materials. These offices usually assume the role of consulting advisers supporting researchers' self-policing.

In the 1970s this unspoken bargain between regulators and scientists began to wither in the face of a series of scientific inventions as well as embarrassments that awakened public notice and political institutions to what seemed like unnecessarily risky, if not evil, scientific enterprises. As a consequence, demands for more direct regulation of the processes of doing science escalated, moving from concerns about humans as subjects of research and experimentation to more widespread public concerns about hazardous workplaces (which would eventually include laboratories), environmental damage, and sustainability.

Beginning with attention to human subjects of research, the Nuremberg Code following World War II established voluntary consent by human subjects as the fundamental principle of ethical scientific or medical research (Halperin 1950; Stark 2011). Nonetheless, for forty years (1932–1972) the U.S. Public Health Service had been conducting an experiment on four hundred African American men in the late stages of syphilis without the men's consent (Freimuth et al. 2001). The public health service did not tell the men from what disease they were suffering, nor were the subjects provided treatment. They were merely observed as their disease proceeded along its course. The men were told only that they were being treated for "bad blood." With the ambition to provide a complete model of the disease's ravages, doctors monitored the patients' progress until death. A newspaper story in the *Washington Star* on July 25, 1972, broke the story. Although the doctors and state officials are reported to have been unrepentant, claiming that someone was making a mountain out of a molehill, "under the glare of publicity, the government ended their experiment, and for the first time provided the men with effective medical treatment for syphilis."[3]

Other examples of irresponsible science became public. During the 1940s, 800 pregnant women, patrons of a prenatal clinic at Vanderbilt University, were given a cocktail including a tracer dose of radioactive iron as a means of establishing the iron requirements of pregnant women. Several universities participated in another experiment in which terminally ill patients were injected with plutonium to determine how the radioactive chemical would spread through the body. During the 1950s, mentally limited boys at the Fernald School in Waltham, Massachusetts, were fed radioactive calcium and iron with their breakfast cereal in exchange for extra milk and trips to baseball games. In the 1960s patients with incurable cancers were exposed to heavy doses of full-body radiation, a procedure that was largely abandoned by that time.

The 1974 National Research Act, creating the National Commission for the Protection of Human Subjects of Biomedical and Behavioral Research, signals the entry of law into science's unique domain: research design and methods. Social science as well as medical research was to be included under new rules for conducting scientific research.

Also in the early 1970s, Congress and the public responded to accounts of biologists unpacking and reconfiguring DNA. When rDNA (recombinant DNA) was first announced, the public mobilized to oppose the research and succeeded in securing a research moratorium in Cambridge, Massachusetts, while negotiations ensued between local authorities and researchers. Biologists organized themselves, following a call in *Science* magazine for a voluntary worldwide moratorium, to respond to the growing concerns as well as reservations by scientists themselves. One hundred and forty scientists, lawyers, and members of the press met at the Asilomar Conference center to develop voluntary guidelines to ensure the safety of DNA technology and research (Krimsky 2005; Weiner 1979). The scientists, who certainly had an interest in seeing that research resumed, were also concerned about the health of lab workers and the public's perception of their work. In February 1975 the biologists meeting at Asilomar recommended protocols that were subsequently established nationally,[4] with more stringent regulations imposed in Cambridge, Massachusetts, where citizens had first mobilized.

The recommendations sought to protect laboratory personnel, the general public, and the environment from any hazards that might be generated through biological experiments (Krimsky 1982). The protocols focused on laboratory architecture, relying on the laboratory workers' skill in handling biological material. The first principle for dealing with possible risks addressed containment, while the second principle asserted that containment should match the estimated risk as closely as possible. The two principles led to a categorization of hazards with commensurate physical containment through the use of hoods, limited access, or negative pressure laboratories. The ubiquitous signs labeling spaces and specifying procedures are now familiar sights along the corridors of scientific research buildings. The conference also recommended, but did not require, biological and physical barriers to limit spread of recombinant DNA, for example, using hosts unable to survive in natural environments and nontransmissible and fastidious vectors (plasmids, bacteriophages, and other viruses) able to survive only in specified hosts. The contemporary biology laboratory is now a space governed by a network of laws, regulations, and rules. Rather than addressing the behavior of scientists, through the spatial regulation of science, processes of social control are materialized in architecture, signs, autoclave machines, laboratory coats, and goggles, thus sustaining the scientists' authority and capacity for autonomy and self-governance.

For most of the twentieth century, universities acquiesced to national priorities for research and development, in exchange for which the government funded the research and deferred to the traditions of academic self-governance. By the end of the twentieth century, however, science's collaboration with the regulatory state led back, in an almost perfect feedback loop, into the sanctified spaces beyond biology through regulations concerning health, safety, environmental protection, ethical practice, and conditions of employment. Although such regulations were in principal universal, with very few exempt categories, scientists and academic institutions nonetheless retained a good measure of the organizational autonomy for which they contracted (Guston and Keniston 1994). Thus, in 1991, OSHA (Occupational Health and Safety Administration) enacted the lab standard exempting research facilities from regulations that had been designed for high-production enterprises. Relying on what was considered limited volumes and exposure, the rules of the lab standard delegated to the regulated organizations themselves authority to determine how these more generally applicable federal and state regulations would be adapted within universities and research laboratories. A compromise was struck between the regulators' interests in safety and the universities' interest in self-governance.

"The law is now an inescapable feature of the conditioning environment" that produces socially embedded science. However, law is not solely the context of science but inhabits the very procedures and routines of doing science. "Accounts of the development of science are incomplete without taking on board the shaping influence of legal imperatives and imaginations, and of necessity the work of legal practitioners and institutions" (Jasanoff 2008, 762). This influence is felt within the labs primarily by the universities taking on the responsibility as conduits, translators, and mediators of the legal mandate (Latour 2005).

Relational Regulation

Despite regulatory and managerial efforts to integrate concerns for the natural environment, scientists' safety, and public health into laboratory practices, these efforts, like the norms of science itself, are thus far largely aspirational. Although most scientists comply most of the time, not all comply all of the time and the rate of compliance varies by disciplines and departments. While most scientists complete mandatory safety training—often after multiply repeated requests—and respond to inspectors' requests to relabel waste or remove equipment blocking entrances, they often perceive these requests as impediments to their work (Gray and Silbey 2014) and frequently do not sustain the requested changes to behavior (Bruns 2009). One lab may have a perfect inspection at time 1 and noncompliant conditions at time 2. In general across organizations, decoupling of aspirations from practical on-the-ground achievements is not unusual and, in this regard, science is no different from most human endeavors. Eventually, ambitions for safer, environmentally sound laboratories *may* become like other legally sanctioned norms (such as nondiscriminatory employment and human subjects protections)—thoroughly institutionalized, part and parcel of the habituated and taken-for-granted conditions of everyday life (exemplified in safety standards for building materials, the organization of city streets and traffic lanes, or the security of drinking water). In 2016, however, this is a work in progress.

Regular disregard of external demands for environmentally safer laboratories is possible because universities are loosely coupled professional bureaucracies (Mintzberg 1979) and because of scientists' privileged position in the organization. Professional bureaucracies integrate two organizing logics. One side of the organization is made up of professionals who collegially and collectively govern. The other side of the organization is made up of supporting administrators who are organized as a Weberian, hierarchical, top-down bureaucracy with descending lines of authority and increasing specialization. This organizational structure facilitates differential interpretations of and responses to legal mandates and differential experiences of regulation and self-governance. What is regarded as academic freedom by scientists looks like mismanagement, if not anarchy, to regulators. What the regulators (in the government and the university staff functioning and interpreted by scientists as internal regulators) require—consistent conformity—faculty members abhor. While scientists welcome help from a staff for financial and material support, as well as response to accidents, emergencies, and breakdowns, they may resent rules and requirements about lab coats and safety glasses (Bruns 2009; Huising and Silbey 2011), material handling and storage, security measures, the disposal of contaminated and hazardous materials, and required documentation (Stephens, Atkinson, and Glasner 2011). They interpret such prescriptive rules as more than intrusions. They describe them as obstacles to scientific progress and the smooth operation of their laboratory (Evans 2014).

To the scientists, regulators (whether from the government or the university) are, like most "civilians," unaware of the real processes of science and the organization of laboratories. First, regulators and their rules cannot anticipate the emerging, novel materials that are created in laboratories. "We are inventing the processes that others' use as packaged tools." Second, regulators lack the deep expertise of the materials and processes used in laboratories. Oversight of laboratory materials should be left to scientists as "[we] understand chemistry, and also the health effects ... We understand why something's carcinogenic, we understand why something is corrosive" (Gray and Silbey 2014, 129). When external or internal overseers make shaky claims about the safety of materials or justify their requirements with thin scientific arguments, scientists are quick to point out that regulations are without scientific basis; they demand alterations (Huising 2014). Third, regulators and the rules are not relevant because of the low volumes of material and the low frequency with which they are used. If regulators were familiar with laboratory protocols and practices, scientists reason, they would understand there are minimal risks and dangers (Silbey and Ewick 2003). Given this, scientists often do not divert time from scientific work to attend to these issues (Gray and Silbey 2014).

Governing the gap between regulatory requirements and the daily operations of the laboratories is a complex process, made even more demanding by the vast array of materials, processes, and experimental apparatus that fill scientific laboratories as well as the status differences among the various actors. Perfect compliance between regulatory requirements and laboratory performance is a chimera, and always has been because the gap between regulatory expectations and performance is less a function of the distance between the source of prescriptions and the site of action than it is a product of the insufficiency of formalized, prescribed processes to handle the complex, situated demands faced daily in laboratories, or at most work sites for that matter. Training programs, manuals, inspection checklists, standard operating procedures, and databases cannot by themselves create compliance with risk management systems or laboratory prescriptions. Instead, "it is the uniquely human ability to vary and adapt actions to suit local conditions that preserves [production and creativity] in a dynamic and uncertain world" (Reason 2000, 4).

Although gaps between design and enactment of laboratory safety regulations may seem irreducible, they can be governed through relational regulation (Huising and Silbey 2011). Relational regulation can be practiced by workers who see themselves and their work as links in a complex web of interactions rather than as offices of delimited responsibilities and interests (Silbey, Huising, and Coslovsky 2009). Instead of focusing closely and only sporadically taking account of the larger connections and reverberations of their actions, these actors (sometimes called sociological citizens, ibid.) view their work, in this case supporting laboratory safety, as the outcome of human decisions, indecisions, trial and error, rather than formally rational action. In a dynamic organization, they reconceive their own role as insignificant by itself yet essential to the whole. Relational regulation governs the gap with a pragmatic, sociological appreciation of the ongoing production of social and material life, that is, by acknowledging the impossibility of perfect conformity between abstract rules and situated action but nonetheless managing to keep practices within a band of variation surrounding—but not perfectly coincident with—regulatory specifications.

Instead of integrating regulatory requirements into their daily routines, scientists and the universities that employ them rely on administrative offices and staff to monitor and police the boundaries between science and the world beyond the university (Gieryn 1995, 1998). Mediating between science and social institutions such as law, these boundary-spanning offices allow outsiders to cross into the sacred spaces of science while channeling those interventions in ways less likely to be resisted by independent-minded, resistant faculty (Guston 1999). These administrative offices facilitate compliance by insulating scientists and scientific practices from unwanted government intrusion (Huising and Silbey 2013; Smith-Doerr and Vardi 2015), becoming an essential resource, a condition of possibility for the continuing production of science. They support faculty members with resources that enable compliance with regulations concerning human subjects of research (institutional review board, or IRB), occupational health and safety, and animal care and handling, all conditions for external funding and internal licenses to practice with hazardous materials (e.g., Benner and Sandström 2000).

These boundary-spanning staff ensure that the organization has some degree of coupling between policies and practices (Weick 1976) so that most of the time, laboratories function safely, without incident, and more often than not in compliance with most regulations.[5] This compliance is achieved by the responsible staff in two ways: by buffering scientists (Huising and Silbey 2011, 2013) and by demonstrating to scientists that they—the hazard specialists and regulators—work according to the same logics (of skepticism, disinterest, collegiality, direct observation, and pragmatic adaptation), and developing trust through familiarity and personal relations just as scientists do among themselves (Huising 2015).

Although faculty members, along with university leadership, are ultimately responsible for compliance, along with university leadership, knowing faculty members are resistant, administrative staff may take on this responsibility. For example, some hazard specialists and regulators may put aside their techno-legal expertise and instead of instructing lab members or waiting for compliance, will help laboratories perform the day-to-day work of removing waste, checking equipment, cleaning up chemical or radioactive spills (Huising 2015). If they continue to help in laboratories, the regulatory staff demonstrate, through both their interactions and their deeds, that they understand where the legal regulatory requirements fit (or do not fit) into lab work and will alter them to accommodate the specific variations, lab by lab, that feeds scientific progress generally. Further, these in-house regulators may make explicit the trust they have in the scientists to follow through when staff are not present.

When regulators (whether government or in-house) both are familiar to scientists and demonstrate their knowledge of the laboratory, scientists are more likely to comply. By creating workarounds that buffer faculty members from the details of regulatory compliance, support staff defer to the authority of scientists, sustaining rather than eroding their privileged status (Huising and Silbey 2013). Thus, internal regulators can coax compliance from scientists on the basis of frequent interaction in laboratories (familiarity) and by demonstrating, in their prescriptions and enforcement, that they understand the priorities, routines, challenges, and stress of laboratory life (local knowledge) (Huising 2014). These findings run parallel with Sims's (2005) observation within a pulsed power laboratory, where scientists willingly complied when the safety regime reinforced fundamental logics of scientific work. Scientists incorporated safety features into their protocols when those regulations mimicked the physicists' efforts to create transparency and control of physical matter.

Those regulators who manage to seduce compliance from otherwise recalcitrant laboratory scientists deploy simple communicative, relational practices: they identify actors in a diverse network, search for information, synthesize and collate what they learn, and then craft pragmatic accommodations that recognize the interdependence of general policies and local variations, legal prescriptions and laboratory procedures. Such regulatory success is more likely to emerge where regulatory staff engage in face-to-face conversations, where they enjoy unscheduled work time, and finally where there is an organizational culture of macro, rather than micro, management. Those who are able to routinely and successfully govern the gap between regulation and implementation meet this challenge through their understanding of the organization's relational interdependencies (Huising and Silbey 2011).

Beyond external demands to modify practices, scientists themselves identify safety and security concerns that are direct or indirect consequences of their work. The "responsibility" movements in stem cell research (Evans 2012), nanotechnology[6] and synthetic biology,[7] and the green chemistry movement are examples of scientist-led efforts to integrate concerns for safety and security into the fabric of the science itself. Nanotechnologists acknowledge their obligation to protect their science and the wider society by routinely evaluating the safety of nanotech projects (McCarthy and Kelty 2010). This requires that scientists divert resources from producing new knowledge to producing knowledge of safety of new knowledge and products (Kelty 2009). A similar trajectory has been observed in green chemistry (Lynch 2015) where scientists take into account the environmental consequences of their traditional materials while designing new ways of doing chemistry (Woodhouse and Breyman 2005, 205). Thus, as scientists begin to reflect on and question the consequences of their work, they create new professions, forums, and knowledge related to these consequences.

These emergent efforts to challenge assumptions about the responsibility of bench scientists for the safety and security consequences of their work are examples of what is sometimes called "safety culture," the contemporary watchword of laboratory hazard experts and government regulators (National Research Council 2014). After more than a half century of concerted regulation, with diverse incentives and enforcement methods, safety culture has become a common explanation for accidents and a recipe for voluntary self-improvement in complex sociotechnical systems, and especially laboratories (Silbey 2005).

There are signs in current procedures that these locally emergent self-regulatory projects may become articulated into standardized best practices and codes of conduct adopted by organizations and regulators not as cultural norms but as official requirements. Like the biologists' self-regulation project that began at Asilomar and was subsequently required by the CDC and then the NIH, the responsibility movements in chemistry, nanotechnology, and synthetic biology may eventually become institutionalized. Future scientists may experience the required practices as externally imposed and may circle back to the need for relational approaches to implementing green chemistry and responsible nanotechnology as legally mandated laboratory regulations.

Digitizing Surveillance

Regulators (e.g., the U.S. Environmental Protection Agency) and funders (e.g., National Institutes of Health, National Science Foundation, and Department of Energy) have continued to shift responsibility to assess, monitor, and mitigate risk to the regulated parties themselves. This compromise has become materialized in the widespread use of management systems designed to coordinate the diverse informational and procedural transactions of large, complex organizations. What had been dispersed and often idiosyncratic forms of overseeing and reporting about organizational behavior have been brought together under the umbrella of what are called enterprise resource or management systems. These digital technologies bundle various software packages to centralize collection of information in common archives that, ostensibly, can be easily searched and analyzed for continuous observation and improvement toward organizational goals. These systems locate the design, standard setting, and implementation of legal regulation within the regulated organization itself (Coglianese and Nash 2001; Howard-Grenville, Nash, and Coglianese 2008). If used to their technological capacity, management systems move from regulating spaces to surveilling organizational members, groups, and practices, seeking to limit forms and aspects of risk, including the behavior of scientists themselves. Ultimately, however, digital technologies have not eliminated the human actors who provide training for working with hazardous substances, inspect laboratories, and respond to accidents.

The adoption of management systems as a coordinating and centralizing information device to implement legal regulations signals a transformation in the relative autonomy and status of scientists, no longer operating to such a great extent by disciplinary and professional norms or trust alone. Inside these governance arrangements, scientists become like workers in any other large organization, subject to a standardized model of responsibility and accountability (Silbey 2009). When implemented in laboratories, management systems attempt to reassign compliance responsibility from a staff of techno-legal experts to researchers working at the laboratory bench (including students, technicians, support staff of various ranks as well as postdocs and faculty). Formerly, laboratory safety was largely the task of specialists: for example, industrial engineers responsible for air quality; hazardous waste experts responsible for containing and removing chemical waste from labs and periodically from the university grounds; health physicists responsible for securing licenses for radioactive materials, training lab workers in handling them, and removing and storing the materials until they decay; biosafety officers responsible for approving experimental designs as well as overseeing the condition of autoclaves used to destroy biological matter. Researchers are, by the fact of the management system's existence, given responsibility for ensuring that their daily research practices comply with city, state, and federal regulations that had previously been the job of professional staff. This shift to expanded responsibility across the levels of personnel (university administrators, staff, lab scientists) is supposedly enabled by the creation of digitally available standard operating procedures, inspection checklists, and enhanced training. The expert staff now serve as coordinators, trainers, and *internal regulators* of the system with lab personnel sharing responsibility for achieving compliance (Huising 2014, 2015). Researchers are expected to inspect their own labs on a regular basis, even as internal regulators (staff) inspect on a less frequent schedule. Researchers are responsible for ensuring chemical waste is labeled following detailed prescriptions and collected within hoods for limited durations. Researchers are responsible for ensuring that the laboratory has signs specifying and providing the appropriate apparel for the lab's specific experimental procedures. Standard operating procedures (SOPs) are posted on bulletin boards and accessed through online repositories, for example, explaining the uses of double gloving, radiation dosage tags, waste water practices, open and closed doors, sharps collection, and spill response procedures, to name but six of what is often a repertoire of at least sixty SOPs.

The adoption of a management system is, we suggest, in the first instance, a tool for surveillance, routinely observing and recording performances and outcomes[8] because scientists' new regulatory responsibilities include continuous collection and submission of performance data to the central information systems. These data, their collection, storage, and analyses are intended to reinforce a local sense of accountability for continuous improvement in laboratory compliance. However, system efficacy—effectiveness and efficiency—lies in the *quality* of the information that circulates through the system and the capacity of the system to respond through feedback loops to affect negative, undesired, or noncompliant action. In essence, management systems improve safety to the degree that the regulations are well written and consistent with what are identified by system managers as best practices, with lab personnel adjusting their behavior in response to information about noncompliant practices. Management systems are, in this second sense, control systems that enable the use of observations to change behavior.

Management systems, adopted by organizations to implement practices compliant with legal regulations, work by simultaneously distributing the responsibility to the ground-level actors—to know how they are doing and change their action—while mechanizing these functions through centralized surveillance and storage of data. At the same time, management systems still rely on human action to animate transactions with the labs. Ironically, management systems turn out to be paradoxical and potentially disruptive tools for scientific laboratories conventionally governed through trust and peer relationships. Even where universities and other scientific organizations do not install digitized management systems, the functions encoded in the systems—training, observation, record keeping—must be performed by organizational personnel if scientific research using chemicals, biomatter, radiation, animals, or humans is to be conducted consistently with U.S. law.

Conclusion

Some of the most trenchant contemporary research conceptualizes law as a form of authority *in relation to* other forms of authority (Espeland 2003), never operating independently of the social fields to which it is applied (Ewick and Silbey 1998). Seeking to understand "how the setting for those competitions—very often an organization—shapes the outcome" (Edelman 2004a, 2004b; Heimer 1999), we have explored the tension, perhaps competition, between law and science through government efforts to regulate laboratory practices. Looking for law in the laboratory tells us much about law but it also offers a particularly sharp focus on routine practices of science, for example, on the ways in which scientific authority and status creates obstacles to accountability for environmental and safety hazards (Huising and Silbey 2013). Although legal rules and organizational procedures create obligations for nondiscretionary conformity, scientists often struggle to avoid, in the name of scientific progress, the bureaucratically imposed responsibilities. Nonetheless, most scientists—although not all—do comply with most if not all requirements, often by delegating compliance responsibilities to subordinates or relying on support staff who buffer scientists from these obligations. The chapter highlighted a particular practice of relational regulation, where administrators recognize their work as links in a complex web of interactions rather than as offices of delimited responsibilities. Taking account of the larger connections and reverberations of their actions, they also acknowledge the impossibility of perfect conformity and instead practice a form of pragmatic adaption to keep laboratory practices within a range of acceptable variation close to the desired norm. They do so by working with rather than for or against the laboratory scientists, often mimicking some of the same practices of careful observation, trial and error. The introduction of management systems—as a digitized technology used to respond to government regulations—may centralize information gathering and decentralize responsibility but does not eliminate the need for the human actors negotiating these boundaries between law and science. Responsibility is not taken away from top leadership but is dispersed as well to the ground-level actors.

These observations at the ground-level engagement of law and science suggest avenues for additional research. For example, are the characteristic privileges of contemporary lab scientists a generational phenomenon? Will future scientists enjoy the status and relative autonomy of their predecessors? How might the relations described here be amended by looking comparatively around the globe? How is responsibility allocated in other engagements between law and science?

Notes

1. We also do not discuss industrial laboratories, which function quite differently. We are grateful to a reviewer for emphasizing the limited attention to comparative and industrial examples and for identifying places where scientific autonomy and collegiality was differently experienced than in Britain and the United States, from whose histories we draw our analysis. See for example, Alder (2010), Josephson (1991), Kaiser (2009), and Olesko (1991). Of course, in social relations autonomy is never total and collegiality certainly varies. Our characterization of the scientific life is relative to empirical accounts of other contemporary professions (Abbott 1988).

2. Given the saturation of contemporary life with legal regulations, there are innumerable substantive topics for which scientific knowledge plays a role in shaping public policies. Because our focus is on the law regulating laboratory practices, in this chapter we cannot devote attention to the wide range of public policies influenced by science, a topic of extensive literature that should be explored independently. Law and science are intimately entwined in processes and subjects we do not discuss, for example, intellectual property law (Silbey 2014), technology transfer offices (Owen-Smith and Powell 2001), restrictions on experimental materials (e.g., animals, cells, viruses) (Evans 2014), standard setting (Yates and Murphy 2014), and metrology (O'Connell 1993), to name but a few. We cannot discuss them all.

3. The Tuskegee Syphilis Experiment, see http://www.infoplease.com/ipa/A0762136.html.

4. In the United States, the Centers for Disease Control and Prevention publishes criteria for laboratory architecture for various levels of biohazard. Federal agencies demand compliance with these recommendations when proposed experiments use biological matter.

6. http://crnano.org/.

7. http://www.synberc.org/safety-and-security-resources.

8. Surveillance refers not merely to direct visual observation but names the collection of various processes of capturing accounts—visual, documentary, and physical—of action in designated spaces (Marx 2005, 2016).

References

Abbott, Andrew. 1988. *The System of Professions: An Essay on the Division of Expert Labor.* Chicago: University of Chicago Press.

Abraham, John, and Courtney Davis. 2007. "Deficits, Expectations and Paradigms in British and American Drug Safety Assessments: Prising Open the Black Box of Regulatory Science." *Science, Technology, & Human Values* 32 (4): 399–431.

___. 2009. "Drug Evaluation and the Permissive Principle: Continuities and Contradictions between Standards and Practices in Antidepressant Regulation." *Social Studies of Science* 39 (4): 569–98.

Alder, Ken. 2010. *Engineering the Revolution: Arms and Enlightenment in France, 1763–1815.* Chicago: University of Chicago Press.

Benner, Mats, and Ulf Sandström. 2000. "Institutionalizing the Triple Helix: Research Funding and Norms in the Academic System." *Research Policy* 29 (2): 291–301.

Birke, Lynda, Arnold Arluke, and Mike Michael. 2007. *The Sacrifice: How Scientific Experiments Transform Animals and People.* West Lafayette, IN: Purdue University Press.

Bourdieu, Pierre. 1975. "The Specificity of the Scientific Field and the Social Conditions for the Progress of Reason." *Social Science Information* 14 (6): 19–47.

Bruns, Hille C. 2009. "Leveraging Functionality in Safety Routines: Examining the Divergence of Rules and Performance." *Human Relations* 62 (9): 1399–426.

Carpenter, Daniel. 2010. *Reputation and Power: Organizational Image and Pharmaceutical Regulation at the FDA.* Princeton, NJ: Princeton University Press.

Clarke, Adele E., and Joan H. Fujimura 1992. "What Tools? Which Jobs? Why Right?" In *The Right Tools for the Job: At Work in Twentieth-Century Life Sciences,* 3–46. Princeton, NJ: Princeton University Press.

Coglianese, Cary, and Jennifer Nash. 2001. *Regulating from the Inside: Can Environmental Management Systems Achieve Policy Goals?* Washington, DC: Resources for the Future Press.

Cole, Simon. 1999. "What Counts for Identity? The Historical Origins of the Methodology of Latent Fingerprint Identification." *Science in Context* 12 (1): 139–72.

Collins, Harry. 1974. "The TEA Set: Tacit Knowledge and Scientific Networks." *Science Studies* 4 (2): 165–86.

___. 1975. "The Seven Sexes: A Study in the Sociology of a Phenomenon, or the Replication of Experiments in Physics." *Sociology* 9 (2): 205–24.

___. 2001. "Tacit Knowledge, Trust and the Q of Sapphire." *Social Studies of Science* 31 (1): 71–85.

Collins, Harry M., and Robert Evans. 2002. "The Third Wave of Science Studies: Studies of Expertise and Experience." *Social Studies of Science* 32 (2): 235–96.

Cutler, Brian L., Hedy R. Dexter, and Steven D. Penrod. 1989. "Expert Testimony and Jury Decision Making: An Empirical Analysis." *Behavioral Sciences & the Law* 7 (2): 215–25.

Delaney, David. 2001. "Making Nature/Marking Humans: Law as a Site of (Cultura) Production." *Annals of the Association of American Geographers* 91 (3): 487–503.

Diamond, Shari Seidman, and Jonathan D. Casper. 1992. "Blindfolding the Jury to Verdict Consequences: Damages, Experts, and Civil Jury." *Law & Society Review* 26: 513–63.

Doing, Park. 2008. "Give Me a Laboratory and I Will Raise a Discipline: The Past, Present, and Future Politics of Laboratory Studies in STS." In *The Handbook of Science and Technology Studies*, edited by Edward J. Hackett, Olga Amsterdamska, Michael E. Lynch, and Judy Wajcman. 3rd ed., 279–98. Cambridge, MA: MIT Press.

Dumit, Joseph. 1999. "Objective Brains, Prejudicial Images." *Science in Context* 12 (1): 173–201.

Edelman, Lauren B. 2004a. "Rivers of Law and Contested Terrain: A Law and Society Approach to Economic Rationality." Presidential Address, *Law & Society Review* 38 (2): 181–98.

___. 2004b. "The Legal Lives of Private Organizations." In *The Blackwell Companion to Law and Society*, edited by Austin Sarat, 231–52. Malden, MA: Blackwell.

Edmond, Gary, and David Mercer. 2000. "Litigation Life: Law-Science Knowledge Construction in (Bendectin) Mass Toxic Tort Litigation." *Social Studies of Science* 30 (2): 265–316.

Espeland, Wendy Nelson. 2003. "Understanding Law in Relation to Other Forms of Authority." *Amici* 11 (1): 2.

Etzkowitz, Henry, and Loet Leydesdorff. 2000. "The Dynamics of Innovation: From National Systems and 'Mode 2' to a Triple Helix of University-Industry-Government Relations." *Research Policy* 29 (2): 109–23.

Evans, James A. 2010. "Industry Collaboration, Scientific Sharing and the Dissemination of Knowledge." *Social Studies of Science* 40 (5): 757–91.

Evans, Joelle. 2012. "Moral Frictions: Ethics, Creativity and Social Responsibility in Stem Cell Science." Ph.D. dissertation, Massachusetts Institute of Technology, Cambridge, MA.

___. 2014. "Resisting or Governing Risk? Professional Struggles and the Regulation of Safe Science." *Academy of Management Proceedings* 2014 (1): 16124.

Ewald, François, and Stephen Utz. 2002. "The Return of Descartes's Malicious Demon: An Outline of a Philosophy of Precaution." In *Embracing Risk: The Changing Culture of Insurance and Responsibility*, edited by Tom Baker and Jonathan Simon, 273–301. Chicago: University of Chicago Press.

Ewick, Patricia, and Susan S. Silbey. 1998. *The Common Place of Law: Stories from Everyday Life.* Chicago: University of Chicago Press.

___. 2003. "Narrating Social Structure: Stories of Resistance to Law." *American Journal of Sociology* 108 (6): 1328–72.

Faigman, David. 1986. "Note, the Battered Woman Syndrome and Self-Defense: A Legal and Empirical Dissent." *Virginia Law Review* 72 (3): 619–47.

___. 1999. *Legal Alchemy: The Use and Misuse of Science in the Law.* Houndmills: W. H. Freeman.

Faigman, D. L. 1991. "Struggling to Stop the Flood of Unreliable Expert Testimony." *Minnesota Law Review* 76: 877.

Faigman, David, and A. J. Baglioni Jr. 1998. "Bayes' Theorem in the Trial Process." *Law and Human Behavior* 12 (1): 1–17.

Foster, Jacob G., Andrey Rzhetsky, and James A. Evans. 2015. "Tradition and Innovation in Scientists' Research Strategies." *American Sociological Review* 80 (5): 875–908.

Foster, Kenneth R., David E. Bernstein, and Peter W. Huber, eds. 1993. *Phantom Risk: Scientific Inference and the Law.* Cambridge, MA: MIT Press.

Frazier, Patricia A., and Eugene Borgida. 1992. "Rape Trauma Syndrome: A Review of Case Law and Psychological Research." *Law and Human Behavior* 16 (3): 293–311.

Freimuth, Vicki S., Sandra Crouse Quinn, Stephen B. Thomas, Galen Cole, Eric Zook, and Ted Duncan. 2001. "African Americans' Views on Research and the Tuskegee Syphilis Study." *Social Science & Medicine* 52 (5): 797–808.

Fuchs, Stephan, and Steven Ward. 1994. "What Is Deconstruction, and Where and When Does It Take Place? Making Facts in Science, Building Cases in Law." *American Sociology Review* 59 (4): 481–500.

Fujimura, Joan. 1991. "Social Worlds/Arenas Theory as Organizational Theory." In *Social Organization and Social Processes: Essays in Honor of Anselm L. Strauss,* edited by David Maines, 119–58. Hawthorne, NY: Aldine de Gruyter.

Galison, Peter. 1997. *Image and Logic: A Material Culture of Microphysics.* Chicago: University of Chicago Press.

Galison, Peter, and Emily Thompson. 1999. *The Architecture of Science.* Cambridge, MA: MIT Press.

Gibbons, Michael, Camille Limoges, Helga Nowotny, Simon Schwartzman, Peter Scott, and Martin Trow. 1994. *The New Production of Knowledge: The Dynamics of Science and Research in Contemporary Societies.* London: Sage.

Gieryn, Thomas F. 1995. "Boundaries of Science." In *The Handbook of Science and Technology Studies,* edited by Sheila Jasanoff, Gerald E. Markle, James C. Petersen, and Trevor Pinch, 393–443. Thousand Oaks, CA: Sage.

___. 1998. "Biotechnology's Private Parts (and Some Public Ones)." In *Making Space for Science: Territorial Themes in Shaping of Knowledge,* edited by Crosbie Smith and Jon Agar, 281–312. New York: St. Martin's Press.

___. 1999. *Cultural Boundaries of Science: Credibility on the Line.* Chicago: University of Chicago Press.

Golan, Tal. 1999. "The History of Scientific Expert Testimony in the English Courtroom." *Science in Context* 12 (1): 7–32.

Gooding, David. 1992. "The Procedural Turn; Or, Why Do Thought Experiments Work?" In *Cognitive Models of Science*, edited by Ronald Giere and Herbert Feigl, 45–76. Minneapolis: University of Minnesota Press.

Gray, Garry, and Susan S. Silbey. 2011. "The Other Side of the Compliance Relationship." In *Explaining Compliance: Business Responses to Regulation*, edited by Christine Parker and Vibeke Lehmann Nielsen, 123–38. Northampton, MA: Elgar.

___. "Governing Inside the Organization: Interpreting Regulation and Compliance." *American Journal of Sociology* 120 (1): 96–145.

Gusterson, Hugh. 2000. "How Not to Construct a Radioactive Waste Incinerator." *Science, Technology, and Human Values* 25 (3): 332–51.

Guston, David H. 1999. "Stabilizing the Boundary between US Politics and Science: The Role of the Office of Technology Transfer as a Boundary Organization." *Social Studies of Science* 29 (1): 87–111.

___. 2001. "Boundary Organizations in Environmental Policy and Science: An Introduction." *Science, Technology, & Human Values* 26 (4): 399–409.

Guston, David H., and Kenneth Keniston. 1994. *The Fragile Contract: University Science and the Federal Government*. Cambridge, MA: MIT Press.

Hackett, Edward J. 1990. "Science as a Vocation in the 1990s: The Changing Organizational Culture of Academic Science." *Journal of Higher Education* 61 (3): 241–79.

___. 2005. "Essential Tensions: Identity, Control, and Risk in Research." *Social Studies of Science* 35 (5): 787–826.

Halperin, Sidney L. 1950. "A Study of the Personality Structure of the Prisoner in Hawaii." *Journal of Clinical and Experimental Psychopathology* 12 (3): 213–21.

Heimer, Carol A. 1999. "Competing Institutions: Law, Medicine, and Family in Neo-natal Intensive Care." *Law and Society Review* 33 (1): 17–66.

Hilgartner, Stephen. 2000. *Science on Stage*. Palo Alto, CA: Stanford University Press.

Howard-Grenville, Jennifer, Jennifer Nash, and Cary Coglianese. 2008. "Constructing the License to Operate: Internal Factors and Their Influence on Corporate Environmental Decisions." *Law & Policy* 30 (1): 73–107.

Huising, Ruthanne. 2014. "The Erosion of Expert Control through Censure Episodes." *Organization Science* 25 (6): 1633–61.

___. 2015. "To Hive or to Hold? Producing Professional Authority through Scut Work." *Administrative Science Quarterly* 60: 263–99.

Huising, Ruthanne, and Susan S. Silbey. 2011. "Governing the Gap: Forging Safe Science through Relational Regulation." *Regulation and Governance* 5 (1): 14–42.

___. 2013. "Constructing Consequences for Noncompliance: The Case of Academic Laboratories." *Annals of the American Academy of Political and Social Science* 649 (1): 157–77.

Jasanoff, Sheila. 1987. "Contested Boundaries in Policy-Relevant Science." *Social Studies of Science* 17 (2): 195–230.

___. 1990. *The Fifth Branch: Science Advisors as Policymakers*. Cambridge, MA: Harvard University Press.

___. 1995. *Science at the Bar: Law, Science, and Technology in America*. Cambridge, MA: Harvard University Press.

___. 1996. "Beyond Epistemology: Relativism and Engagement in the Politics of Science." *Social Studies of Science* 26: 393–418.

___. 1999. "Reframing Rights: Constitutional Implications of Technological Change." Proposal to the National Science Foundation. Accessed at http://grantome.com/grant/NSF/SES-9906834.

___. 2003. "In a Constitutional Moment: Science and Social Order at the Millennium," In *Social Studies of Science and Technology: Looking Back Ahead*, edited by Bernward Joerges and Helga Nowotny, 155–80. Netherlands: Kluwer Academic Publishers.

___, ed. 2004. *States of Knowledge: The Co-production of Science and Social Order*. London: Routledge.

___. 2008. "Representation and Re-presentation in Litigation Science," *Environmental Health Perspectives* 116 (1): 123–29.

___, ed. 2011. *Reframing Rights: Bioconstitutionalism in the Genetic Age*. Cambridge, MA: MIT Press.

Jha, Yamini, and Eric W. Welch. 2010. "Relational Mechanisms Governing Multifaceted Collaborative Behavior of Academic Scientists in Six Fields of Science and Engineering." *Research Policy* 39 (9): 1174–84.

Jones, Carol A. G. 1994. *Expert Witnesses*. Oxford: Clarendon Press.

Josephson, Paul R. 1991. *Physics and Politics in Revolutionary Russia*. Vol 7. Berkeley: University of California Press.

Kaiser, David. 2009. *Drawing Theories Apart: The Dispersion of Feynman Diagrams in Postwar Physics*. Chicago: University of Chicago Press.

Kelty, Christopher M. 2009. "Beyond Implications and Applications: The Story of 'Safety by Design.'" *Nanoethics* 3 (2): 79–96.

Knorr Cetina, Karin D. 1981. *The Manufacture of Knowledge: An Essay on the Constructivist and Contextual Nature of Science*. Exeter: A. Wheaton.

___. 1992. "The Couch, the Cathedral, and the Laboratory: On the Relationship between Experiment and Laboratory in Science." In *Science as Practice and Culture,* edited by Andrew Pickering, 113–38. Chicago: University of Chicago Press.

___. 1999. *Epistemic Cultures: How the Sciences Make Knowledge.* Cambridge, MA: Harvard College.

Krige, John. 2001. "Distrust and Discovery: The Case of the Heavy Bosons at CERN." *Isis* 93 (3): 517–40.

Krimsky, Sheldon. 1982. *Genetic Alchemy.* Cambridge, MA: MIT Press.

___. 2005. "From Asilomar to Industrial Biotechnology: Risks, Reductionism and Regulation." *Science as Culture* 14 (4): 309–23.

Landecker, Hannah. 1999. "Between Beneficence and Chattel: The Human Biological in Law and Science." *Science in Context* 12 (1): 203–25.

Latour, Bruno. 1993. *We Have Never Been Modern,* translated by Catherine Porter. Cambridge, MA: Harvard University Press.

___. 2005. *Reassembling the Social: An Introduction to Actor-Network Theory.* Oxford: Oxford University Press.

Latour, Bruno, and Steve Woolgar. 1979. *Laboratory Life: The Construction of Scientific Facts.* Princeton, NJ: Princeton University Press.

Levidow, Les. 2001. "Precautionary Uncertainty: Regulating GM Crops in Europe." *Social Studies of Science* 31 (6): 842–74.

Lynch, Michael E. 1982. "Technical Work and Critical Inquiry: Investigations in a Scientific Laboratory." *Social Studies of Science* 12 (4): 499–533.

___. 1985. *Art and Artifact in Laboratory Science: A Study of Shop Work and Shop Talk in a Research Laboratory.* London: Routledge & Kegan Paul.

Lynch Michael, Simon A. Cole, Ruth McNally, and Kathleen Jordan. 2008. *The Truth Machine: The Contentious History of DNA Fingerprinting.* Chicago: University of Chicago Press.

Lynch, William T. 2015. "Second-Guessing Scientists and Engineers: Post Hoc Criticism and the Reform of Practice in Green Chemistry and Engineering." *Science and Engineering Ethics* 21 (5): 1217–40.

Marx, Gary T. 2016. *Windows into the Soul: Surveillance and Society in an Age of High Technology.* Chicago: University of Chicago Press.

___. 2005 "Seeing Hazily (but Not Darkly) through the Lens: Some Recent Empirical Studies of Surveillance Technologies." *Law & Social Inquiry* 30 (2): 339–99.

McCarthy, Elise, and Christopher Kelty. 2010. "Responsibility and Nanotechnology." *Social Studies of Science* 40 (3): 405–32.

McEvily, Bill, Vincenzo Perrone, and Akbar Zaheer. 2003. "Trust as an Organizing Principle." *Organization Science* 14 (1): 91–103.

Merton, Robert K. [1942] 1957. "Science and Democratic Social Structure." In *Social Theory and Social Structure*, 550–61. Glencoe, IL: Free Press.

Miller, Clark. A. 2000. "The Dynamics of Framing Environmental Values and Policy: Four Models of Societal Processes." *Environmental Values* 9: 211–33.

___. 2001. "Hybrid Management: Boundary Organizations, Science Policy, and Environmental Governance in the Climate Region." *Science, Technology, & Human Values* 26 (4): 478–500.

Mintzberg, Henry. 1979. *The Structuring of Organizations*. Upper Saddle River, NJ: Prentice Hall.

Moonkin, Jennifer L. 1998. "The Image of Truth: Photographic Evidence and the Power of Analogy." *Yale Journal of Law and the Humanities* 10 (1): 1–74.

Moore, Kelly. 1996. "Organizing Integrity: American Science and the Creation of Public Interest Organizations, 1955–1975." *American Journal of Sociology* 101 (6): 1592–627.

National Research Council. 2014. *Safe Science: Promoting a Culture of Safety in Academic Chemical Research*. Washington, DC: National Academies Press.

Nowotny, Helga, Peter Scott, and Michael Gibbons. 2003. "Introduction: 'Mode 2' Revisited: The New Production of Knowledge." *Minerva* 41 (3): 179–94.

O'Connell, Joseph. 1993. "Metrology: The Creation of Universality by the Circulation of Particulars." *Social Studies of Science* 23 (1): 129–73.

Olesko, Kathryn Mary. 1991. *Physics as a Calling: Discipline and Practice in the Königsberg Seminar for Physics*. Ithaca, NY: Cornell University Press.

O'Malley, Pat. 1996. "Risk and Responsibility." In *Foucault and Political Rationality*, edited by Andrew Barry, Thomas Osborne, and Nikolas Rose, 189–208. London: UCL Press.

Owen-Smith, Jason. 2001. "Managing Laboratory Work through Skepticism: Processes of Evaluation and Control." *American Sociological Review* 66 (3): 427–52.

___. 2005. "Dockets, Deals, and Sagas: Commensuration and the Rationalization of Experience in University Licensing." *Social Studies of Science* 35 (1): 69–97.

Owen-Smith, Jason, and Walter W. Powell. 2001. "Careers and Contradictions: Faculty Responses to the Transformation of Knowledge and Its Uses in the Life Sciences." *Research in the Sociology of Work* 10: 109–40.

Owens, Larry. 1990. "MIT and the Federal 'Angel': Academic R & D and the Federal-Private Corporation before World War II." *Isis* 81 (2): 188–213.

Parthasarathy, Shobita. 2004. "Regulating Risk: Defining Genetic Privacy in the United States and Britain." *Science, Technology, & Human Values* 29 (3): 332–52.

Pickering, Andrew. 1995. *The Mangle of Practice: Time, Agency, and Science*. Chicago: University of Chicago Press.

Polanyi, Michael. 1958. *Personal Knowledge: Towards a Post-critical Philosophy*. Chicago: University of Chicago Press.

Rafter, Nicole. 2001. "Seeing Is Believing: Images of Heredity in Biological Theories of Crime." *Brooklyn Law Review* 67: 71–99.

Reason, J., 2000. "Safety Paradoxes and Safety Culture." *Injury Control and Safety Promotion* 7 (1): 3–14.

Rubinfeld, Daniel L. 1985. "Econometrics in the Courtroom." *Columbia Law Review* 85: 1048–97.

Sarat, Austin. 2001. *When the State Kills*. Princeton, NJ: Princeton University Press.

Sarat, Austin, and Thomas Kearns, eds. 1993. *Law in Everyday Life*. Ann Arbor: University of Michigan Press.

Sauermann, Henry, and Paula Stephan. 2013. "Conflicting Logics? A Multidimensional View of Industrial and Academic Science." *Organization Science* 24 (3): 889–909.

Shapin, Steven. 1988. "The House of Experiment in Seventeenth-Century England." *Isis* 79 (3): 373–404.

___. 1994. *A Social History of Truth: Civility and Science in Seventeenth-Century England*. Chicago: University of Chicago Press.

Silbey, Jessica. 2014. *The Eureka Myth: Creators, Innovators, and Everyday Intellectual Property*. Stanford, CA: Stanford University Press.

Silbey, Susan S. 2005. "After Legal Consciousness." *Annual Review Law & Social Science* 1: 323–68.

___. 2009. "Taming Prometheus: Talk of Safety and Culture." *Annual Review of Sociology* 35: 341–69.

Silbey, Susan S., and Patricia Ewick. 2003. "The Architecture of Authority: The Place of Law in the Space of Science." In *The Place of Law*, edited by Austin Sarat, Lawrence Douglas, and Martha Umphrey, 77–108. Ann Arbor: University of Michigan Press.

Silbey, Susan, Ruthanne Huising, and Salo Vinocur Coslovsky. 2009. "The 'Sociological Citizen': Relational Interdependence in Law and Organizations." *L'Année sociologique* 59 (1): 201–29.

Sims, Benjamin. 2005. "Safe Science: Material and Social Order in Laboratory Work." *Social Studies of Science* 35 (3): 333–66.

Smith, Roger, and Brian Wynne. 1989. *Expert Evidence: Interpreting Science in the Law*. London: Routledge.

Smith-Doerr, Laurel, and Itai Vardi. 2015 "Mind the Gap: Formal Ethics Policies and Chemical Scientists' Everyday Practices in Academia and Industry." *Science, Technology, & Human Values* 40 (2): 176–98.

Stark, David. 2011. *The Sense of Dissonance: Accounts of Worth in Economic Life*. Princeton, NJ: Princeton University Press.

Stark, Laura. 2012. *Behind Closed Doors: IRBs and the Making of Ethical Research*. Chicago: University of Chicago Press.

Stephens, Neil, Paul Atkinson, and Peter Glasner. 2011. "Documenting the Doable and Doing the Documented: Bridging Strategies at the UK Stem Cell Bank." *Social Studies of Science* 41 (6): 791–813.

Traweek, Sharon. 1988. *Beamtimes and Lifetimes: The World of High Energy Physicists*. Cambridge, MA: First Harvard University Press.

Van Maanen, John, and Stephen R. Barley. 1982. "Occupational Communities: Culture and Control in Organization." TR-10: Technical report. Cambridge, MA: Sloan School of Management.

Voelkel, James R. 1999. "Publish or Perish: Legal Contingencies and the Publication of Kepler's Astronomia Nova." *Science in Context* 12 (1): 33–59.

Weick, Karl E. 1976. "Educational Organizations as Loosely Coupled Systems." *Administrative Science Quarterly* 21 (1): 1–19.

Weiner, Charles. 1979. "The Recombinant DNA Controversy: Archival and Oral History Resources." *Science, Technology, & Human Values* 4 (1): 17–19.

Whitley, Richard. 2000. *The Intellectual and Social Organization of the Sciences*. Oxford: Oxford University Press.

Woodhouse, Edward J., and Steve Breyman. 2005. "Green Chemistry as Social Movement?" *Science, Technology, & Human Values* 30 (2): 199–222.

Wynne, Brian. 1988. "The Toxic Waste Trade: International Regulatory Issues and Options." *Third World Quarterly* 11 (3): 120–46.

Yates, JoAnne, and Craig Murphy. 2014. "The Role of Firms in Industrial Standards Setting: Participation, Process, and Balance." Unpublished working paper.

Ziman, John. 2000. *Real Science: What It Is, and What It Means*. Cambridge: Cambridge University Press.

Zucker, Lynne G. 1986. "Production of Trust: Institutional Sources of Economic Structure, 1840–1920." In *Research in Organizational Behavior*, edited by Barry M. Staw and Larry L. Cummings, 53–111. Greenwich, CT: JAI Press.

28 Ethics as Governance in Genomics and Beyond

Stephen Hilgartner, Barbara Prainsack, and J. Benjamin Hurlbut

Introduction

The Human Genome Project (HGP) is often described as the first large-scale scientific project to incorporate investigation of its own ethical and social dimensions into the "core" research itself. As the U.S. strand of the HGP took shape at the close of the 1980s, it instituted a program on "Ethical, Legal, and Social Implications," or ELSI. This "ambitious experiment" (Juengst 1994, 121), sometimes said to be more novel than the genome mapping and sequencing project itself, advanced what became known as the "'ELSI hypothesis': that combining scientific research funding with adequate support for complementary research and public deliberation on the uses of new knowledge will help our social policies about science evolve in a well-informed way" (National Institutes of Health 1993, 48). As the words *implications* and *uses* imply, the U.S. ELSI program tended to focus on downstream "impacts" rather than on the shaping of genome knowledge and technology itself. The European counterpart to ELSI used the acronym ELSA, with the "A" denoting *aspects*, to avoid the deterministic and narrow connotations of the term *implications*. (Below, we will use the term "ELS programs" when referring to these programs as a category; specific programs will retain their original designation.)

The ELS "experiment" continues today; by 2013, the U.S. ELSI program had dispersed $317 million of research support (McEwen et al. 2014). In many other countries, ELS programs were established (more or less temporarily, as in the Austrian case) at national levels, including in Austria, Canada, Finland, Germany, the Netherlands, Norway, the United Kingdom, South Korea, and Switzerland, and also at the EU level (Chadwick and Zwart 2013; Zwart, Landeweerd, and van Rooij 2014).[1] The influence of the ELS model extends far beyond genomics. It has served as a model in other fields of emerging science and technology, including nano and synthetic biology, where actors have sought to anticipate and manage societal issues, reassure publics, and maintain political legitimacy.[2] It has also been incorporated into methodologies of technology assessment and societal impact assessment (Kreissl, Fritz, and Ostermeier 2015). Advocates of these latter approaches often hasten to distinguish these programs from U.S. ELSI. But even if we treat these newer formations of engagement with societal, regulatory, and ethical aspects of science and technology as genuinely new, the emergence of ELS programs in genomics nevertheless was a touchstone in their development. Moreover, ELS programs figure importantly in the governance of emerging science and technology, shaping scholarly fields on the one hand, and governance agendas on the other. As McEwen et al. (2014, 491) argue, "the most consequential impact of ELSI research" may consist of subtle yet significant changes in the cultural milieu in which genomics research is conducted and implemented and in the use of scientific evidence in decision making.

The varied incarnations of the ELS model, including in such areas as nano and synthetic biology, represent a notable development in science-society relations: deploying "ethics" programs as subsidiary components of larger technoscientific projects. This form of ethics corresponds neither to an academic field nor to rational reflection on the foundations of moral life but is a new governance tool built on the prior institutionalization of "bioethics" as a way to manage problems of moral ambiguity and disagreement in biomedicine.

The ELS model, emphatically an interdisciplinary one, was also implicated in changes in the social sciences, for example, by drawing scholars to the study of normative questions associated with major technoscientific initiatives. Over the years, many types of scholars have hitched their wagons to these programs and/or tried to push them in new directions. As a result, ELS programs have typically included a heterogeneous collection of disciplines and diverse research agendas. Science and technology studies (STS) scholars have often been part of the mix. These programs have presented opportunities, not least by providing funding to STS researchers (ourselves included) and by making participation in these enterprises possible. STS scholars have engaged with these programs in a variety of contributory and critical modes, which we will review below. Others again have worked to reshape these programs by bringing STS sensibilities under their remit. They have also participated in ELS work with the goal of inventing new roles for the social sciences in the governance of emerging technology. At other times, they have treated ELS activities as a site not only for investigating traditional topics of STS research, such as the construction of scientific facts and technology, but also for symmetrical analysis of the interplay of knowledge and conceptions of public benefit, right reason, and democratic order.

Today, a quarter century after the founding of the first ELS programs, it seems appropriate to both take them as objects of STS analysis and reflect on the relationship of STS to them. This chapter reviews the literature relevant to several key questions: How did the people who launched ELS programs envision and present them, and how have critics described their limitations? What imaginaries of science and society underlie and are reinforced by these programs? How do ELS programs serve as modes of governance in different polities? How has STS research engaged with these programs, whether by contributing to them or critiquing them, and what questions remain salient? And how may theoretical approaches need to evolve as STS increasingly attends to the interplay of knowledge, technology, and democratic governance?

The Rise of ELS Programs

In 1988, at the press conference to announce his appointment as director of the fledgling genome program at the U.S. National Institutes of Health (NIH), James D. Watson (1990) made a public commitment to spend 3 percent (later 5 percent) of the HGP budget on ethical and social implications. The Department of Energy (DOE), NIH's partner in supporting the HGP, also decided to fund ELSI work. With the cost of the genome project estimated at $3 billion dispersed over fifteen years, the 3 percent budget translated into an unprecedented level of funding for a U.S. program on bioethics or science and society.

Over the next few years, the U.S. ELSI program took shape. According to its founders, three features of the program made it distinctive. First, it represented a case in which a research community made an unprecedented commitment to tackling the societal effects of its work. The second feature—unprecedented funding—underlined the strength of this commitment and also reflected the expectation, strongly promoted by advocates of the HGP, that genome science would revolutionize human genetics and biomedicine. Third, the ELSI program was a novel "science policy experiment" (Juengst 1994, 121): a coordinated attempt to "anticipate and address the implications for individuals and society," to "stimulate public discussion," and to "develop policy options that would assure that the information is used for the benefit of individuals and society" (National Institutes of Health and Department of Energy 1990, 66). NIH and DOE sought to achieve these goals primarily by providing grants to support investigator-initiated research (ibid., 65–73). The agencies began by constituting an interdisciplinary working group of external advisers to guide the program, sponsoring agenda-setting conferences, and issuing a call for proposals.

The keywords of the ELSI vision—*anticipate, address,* and *guide*—expressed the hope of at least partially escaping what in other contexts had become known as the "Collingridge dilemma," namely, the idea that an intractable double bind stymies efforts to exert societal control over technology (Liebert and Schmidt 2010). Early in the development of an emerging technology, Collingridge (1980) argued, too little is known about its eventual shape to predict its social effects; later, once the technology is in widespread use, its societal impact may be clear but the commitment to the technology has become irreversible. Perhaps ELSI—with its ample funding, supportive scientific community, and coordinated research program—could widen the window of choice.

U.S. ELSI during the Human Genome Project

The predictable challenges of moving from mission statement to operational program to policy outcome were especially complex in the case of ELSI. The NIH and DOE lacked experience organizing programs involving a mix of natural scientists, clinicians, social scientists, and humanities scholars, and implementation proceeded in what the first ELSI director, Eric T. Juengst (1994, 122), described as a "decidedly unusual manner." The deepest problems, however, pertained to the goals of ELSI and the prospects for achieving them. People projected a variety of visions into the space opened by the ambiguous promise to anticipate and address issues and guide policy. Various parties imagined everything from a source of research funds, to an ethical watchdog, to an advisory body that would "resolve" issues, to a politically motivated public relations machine.[3]

Some critics questioned the very idea of ELSI (Juengst 1996, 64). Like the HGP itself, ELSI was initially controversial. When James D. Watson, first director of the HGP, announced plans for the ELSI program, many biologists worried that the project would waste scarce funds on "routine" science (Hilgartner 2004). The U.S. Congress was not fully on board (Cook-Deegan 1994), and the fear that what then was called "predictive genetic testing" would cause psychosocial harm or lead to genetic discrimination had become a prominent issue in U.S. bioethical discourse (Holtzman 1989; Nelkin and Tancredi 1989). In this context, some observers perceived ELSI as "alarmist hype" about research that raised no truly novel issues. Many scientists, including some high-ranking NIH officials, questioned the wisdom of spending large sums to subsidize "the vacuous *pronunciamentos* of self-styled 'ethicists'" (Juengst 1996, 64, 66).[4]

Other observers saw ELSI as a political strategy to mitigate opposition to genome research. George Annas (1989, 21), a bioethicist and legal scholar, predicted that ELSI research calling for a slower pace or more deliberate consideration of the dangers of the HGP would be dismissed as "uninformed" or "anti-intellectual." Biologist Ruth Hubbard argued that ELSI funding was co-opting a growing number of potential critics of genome research (Hubbard and Wald 1993, 159). Even some people closely involved with ELSI saw its creation as an "adroit political maneuver." In a retrospective commentary, the first coordinator of the DOE's ELSI program described ELSI as a "low-risk mechanism to deflect political challenges" and secure public funding in the face of Congressional concern (Yesley 2008, 3). Some critics saw ELSI as severely limited by the premise that the HGP was a prima facie good and that ELS work should be limited to preventing misuses of knowledge, not challenging or redirecting genome research itself (Juengst 1996, 68–69). Historian of science Charles Weiner (1994) contended that ELSI was deficient because it failed to examine the HGP itself, arguing that a truly anticipatory program would study societal issues *before* beginning the scientific research, so the results could inform HGP policies and priorities.

Critics also argued that the U.S. ELSI program lacked the political authority to serve as an effective agent of oversight and reform (Juengst 1996, 67). They maintained that the proper mechanism for addressing ELSI issues was not an academic research program but a national commission with the authority to make influential recommendations (Hanna 1995; U.S. Congress, House 1992; Yesley 2008). To avoid conflict of interest, this commission should be fully independent of the HGP, with its own dedicated funding. Critics also argued a program focused on investigator-initiated grants could not ensure that its research portfolio covered the most important topics, lacked the means to transmit research results to the policy arena, and was failing to produce policy outcomes. They further charged that the existence of ELSI had the perverse effect of misleading the public into thinking that emerging problems were being addressed (McCain 2002). Conflict over the autonomy and purpose of the program came to a head in an internal dispute between HGP director Francis Collins and ELSI's external advisers. In response, Collins consolidated control over the ELSI agenda within the NIH staff (Beckwith 2009; ELSI Evaluation Committee 1996; Jasanoff 2005).[5]

Defenders of ELSI charged that the critics had misguided expectations. Juengst (1996, 86–87) argued that "no single program, commission, or initiative" could hope to resolve the issues raised by ongoing advances in genome research "during a fifteen-year span, any more than the larger purpose of the HGP—understanding all the genes— could be so achieved." ELSI, Juengst maintained, was successfully accomplishing what should be regarded as its main mission: creating the body of knowledge needed to anticipate issues and weigh policy options, and building a cadre of people committed to generating this body of knowledge and participating in a sustained policy-development process (see also ELSI Research Planning and Evaluation Group 2000). ELSI defenders also argued that the accusation that the program was uncritical of genomics did not hold up to scrutiny (Juengst 1996; see also Zwart, Landeweerd, and van Rooij 2014, 8).[6] They pointed out that ELSI research had already had practical results, including influencing professional practice in genetic testing, proposing model legislation, and shaping research ethics in gene-hunting studies (Burke et al. 2014; Juengst 1996).

As the HGP drew to a close in the early 2000s, the NIH and DOE reaffirmed their commitment to an ELSI program focused on research and education, although with an intensified stress on providing "reliable data and rigorous approaches" on which to base policy decisions (Collins et al. 2003, 843). Like the rest of the HGP, ELSI was officially celebrated as a success. However, a number of observers concluded that ELSI had failed, not least because it was predicated on an imaginary of rational policy making that framed the challenges of governance as resolvable through "reliable data and rigorous approaches" (ibid.). Philosopher Philip Kitcher (2001, 189) concluded that the program had been "doomed to fail from the beginning." ELSI, he wrote, falsely assumed that the problems raised by genomics could be resolved in "a politically neutral way. Instead, principled solutions to the problems of genomic research simply became part of broader political debate, coming out on the losing side."

ELSA in European Countries

Variations among countries make generalizing about ELSA in Europe difficult. One prominent difference between ELSA at the (supranational) European level and its U.S. counterpart is that the former was never limited to human genome research. The EU anchored its ELSA program under the funding umbrella of its Fourth Framework Programme (1994–1998).[7] While EU funding represents only a small proportion of research funding within EU member countries, the EU Framework Programmes play an important agenda-setting role, helping to establish informal understandings of what counts as valuable research. Funding for ELSA research and "related areas" amounted to roughly 2 percent of the funding allocated to Life Sciences and Technologies research (CORDIS 2001). In the Fifth EU Framework Programme (1998–2002), ELSA research was included in all research and technology development programs (European Commission 2001; Zwart, Landeweerd, and van Rooij 2014). EU-level research programs in turn also catalyzed the adoption of the "ELSA approach" at national and regional levels. In the late 1990s, national funding bodies in several European countries, such as Austria, the Netherlands, and the United Kingdom, set up ELSA programs (Nelis, Reddy, and Mulder 2008), some of which focused solely on life sciences and existed for a limited time. Since then, the ELSA model has influenced the way the relationship between social, ethical, and legal research and natural science research is conceived in many countries. Although not all major life science initiatives have established dedicated

ELSA funding, they tend at a minimum to treat social, ethical, and legal dimensions as worthy of consideration (Zwart and Nelis 2009).

ELSA research in some countries, most importantly the Netherlands, has been influenced by the concepts and methodologies of technology assessment, especially by its less technocratic variant, Constructive Technology Assessment (Kreissl, Fritz, and Ostermeier 2015; Schot and Rip 1997). In other countries, understandings of good research on societal and ethical aspects of science have drawn on different local and national strands of scholarship (Houdy, Lahmani, and Marano 2011). Throughout Europe (including in countries with no officially labeled ELSA programs) the idea that science is embedded in and influenced by society, that the public accountability of science should be increased, and that ethical and societal questions are part of any scientific endeavor, continues to gain acceptance (see also Felt et al. 2013).

In Europe, ELS programs sometimes faced allegations that they viewed the social sciences and humanities through a reductionist lens, treating societal research as an instrumental tool for reducing friction in the uptake of new genomic (and other) technologies (Nelis, Reddy, and Mulder 2008; Radstake and Penders 2008; Williams 2006). In the United Kingdom, for example, in the early days of ELS it was assumed "that the specifically *social* research agenda of genomics (and other related sciences) could only begin *after* the scientific research itself had been conceived and conducted" (Ommer et al. 2011, original emphasis). Critical scholarship thus saw ELS programs as the "'handmaiden' of genomics" (Zwart and Nelis 2009) and challenged their deference to science. Indeed to some analysts, ELS programs appeared to focus mainly on anticipating and managing public concern, an orientation that may reflect a "fear of publics" more than a response to public fears (Marris 2015).

Expansion and Change

The visibility of ELSI and the various ELSA programs created a recognizable policy template that sociotechnical vanguards have brought to such areas as nanotechnology (Roco and Bainbridge 2007), neuroscience (Greely 2005), global health (Singer et al. 2007), environmental genomics (Sharp and Barrett 2000), and synthetic biology (Calvert and Martin 2009), as well as a new, globally connected "ELSI 2.0" (Kaye et al. 2012). At times, debate has focused on the desirability of the "ELSIfication" (Stein 2010) of reasoning about science in society, and ELS programs have been criticized for becoming a forum to manage public acceptance (Rip 2009).[8] Regardless, the imprint of the original ELS programs remains powerful. Most often, discussions have treated them as touchstones to emulate or evolve beyond.

During development of the U.S. National Nanotechnology Initiative (NNI), to take one example, advocates of integrating societal concerns directly into early-stage research used ELSI as both a positive and negative model. Some witnesses at a U.S. congressional hearing testified that the substantial debate sparked by ELSI research explained why public opposition to genome science had been minimal (Colvin 2003, 82) and recommended a financial commitment of 3 to 5 percent to study societal issues in nano (Kurzweil 2003, 44). In contrast, Fisher (2005, 323) echoed the claim that ELSI failed to contribute to policy and warned that "an effective program for influencing nanotechnology policy could easily be sacrificed in favor of a sham program that merely gives the impression of doing so."[9] Criticism of ELSI from STS scholars played an important role in arguments for addressing the societal aspects of nano via "real-time technology assessment" (Guston and Sarewitz 2002), "anticipatory governance" (Guston 2014), and "upstream engagement" (Wilsdon and Willis 2004). These authors called for a combination of early-stage anticipatory work and an intensification of democratic engagement by prominently involving publics in early-stage discussions, while rejecting the focus on "applications" and "impacts" typical of the U.S. ELSI program.

While there appear to be no detailed historical accounts of the development of ELS programs as a mode of governance, the literature conveys the sense that they have moved through several stages. In the early 1990s, the first ELS programs tended to frame their mission as providing substantive ethical and social scientific expertise, whereas a decade later, a new emphasis on stakeholder participation and democratization took hold. The participation paradigm grew especially important in Europe, where policy makers looked for new modes of governance in the aftermath of the "mad cow" controversy and the unexpectedly strong rejection of genetically modified (GM) crops. This paradigm figures social scientists as authorized experts on public interaction, skilled at eliciting the perspectives of "citizens" and "stakeholders" and creating forums for dialogue.

New buzzwords accompanied these changes, as labels shifted from ELS predecessors, such as "bioethics" and "technology assessment," to "ELSI," "ELSA," "real-time technology assessment," "anticipatory governance," and most recently, "responsible research and innovation" (RRI) (see Stilgoe and Guston, chapter 29 this volume; see also Lindner et al. 2016). The participatory paradigm also generated controversy. As "upstream engagement" gained traction in the United Kingdom and on the European continent, it encountered critics: some found participatory exercises to be insufficiently democratic, arguing that more genuine and deeper engagement was needed (Wynne 2006). Others rejected the premises of the participatory turn, contending that upstream engagement has the potential to produce misleading representations of public attitudes and preferences (Tait 2009). The question of who should speak for "the public" on a particular topic, or who counts as a stakeholder, remained contentious (e.g., Collins and Evans 2007, 10) and epistemically problematic (Lezaun and Soneryd 2007).[10] Felt et al. (2009) contributed a sobering perspective on what a "bottom-up approach to ethics" can actually do in context where the linear model of innovation and its concomitant views of responsibility are basic cultural assumptions.

In the United States, participatory approaches were initially most strongly associated not with genomics but with nano, where innovative programs devoted to real-time technology assessment and anticipatory governance took shape (Barben et al. 2008). Meanwhile, the U.S. ELSI program intensified its earlier paradigm of providing expert knowledge for decision making.[11] The NIH continued to define ELSI's central mission as addressing the "consequences" of the "applications" of genomics, framing ELSI as "Genomics to Society"—despite new developments, especially involving human subjects (Henderson et al. 2012), that render problematic extant research practices and corollary constructions of the "right" relationships between science and society (Saha and Hurlbut 2011). Eschewing the linearity of ELSI's "implications" framework, European ELS programs tended to be more critical of the assumptions and agendas of genome research than was typical of U.S. ELSI (Zwart, Landeweerd, and van Rooij 2014, 8).

In the early 2010s, the concept of RRI started to grow prominent in EU science policy discourse (Chadwick and Zwart 2013; Owen et al. 2013; von Schomberg 2013; see Stilgoe and Guston, chapter 29 this volume). ELS has partly been assimilated into this notion (Forsberg 2015). Like earlier ELS programs, RRI enjoys no consensus about its raison d'être, but its proponents clearly intend it to encompass the "entire innovation process," including scrutiny of early-stage research (van Schomberg 2013). Zwart, Landeweerd, and van Rooij (2014, 3–4) argue that RRI also places much more emphasis on collaborations with industry, which "not only pushes researchers into close proximity to their private-public 'objects' of research, but may also infect them with the aims and ideologies involved, such as innovation, creating jobs and similar tangible socio-economic impacts." Another difference between ELS and RRI is that the latter has further broadened the variety of disciplines involved by adding fields such as management and innovation studies (ibid., 16).[12] Some commentators worry that a shift toward RRI may portend a "merger of the ELSA-agenda (bent on promoting morally justified research) with innovation and industrial agendas"—a development that could contribute to silencing critical voices (ibid., 17). The debate thus expresses not only widening hopes for these programs but also recapitulates (in a somewhat new form) aspects of the original criticisms of ELS programs. This similarity reflects the fact that ongoing controversy about the ELS approach is rooted in different notions of the nature of science-society relations and of the norms that should govern them.

Problematic Positions: STS as, in, or about ELS Research?

For a tangle of institutional, epistemic, and political reasons, the relationship between STS and various ELS programs is complex and has at times been rocky. Much ELS work remains innocent of the perspectives that inform STS. While variation within and among ELS programs makes it impossible to generalize, it is worth recognizing that to a significant degree, ELS programs reflect and reinscribe traditional imaginaries of orderly science-society relations. These imaginaries often rest on views of the nature of science, technology, and society that STS problematizes, such as the fact/value distinction, the self-evidence of power relations, and asymmetrical explanation of the social causes of truth and error. ELS programs generally do not analyze their own position as loci of political power, nor theorize their own political role vis-à-vis the publics that they construct. And as noted above, these programs have tended to focus more on defining norms of ethical conduct than on examining the social effects of those norms once they are put into practice. For instance, most ELS work treats the human subject as a normatively stable entity rather than analyzing how scientific and "ethical" practices configure patients, experts, and publics (Sunder Rajan 2010).

Another key area where STS and predominant ELS approaches differ is in views about the "neutrality" of science and technology. Whereas STS has long treated normative and political choices as embedded in the very process of knowledge making and innovation, a discourse that frames technological change as necessarily beneficial (at least in the long run and in the aggregate) remains well institutionalized in the legal and organizational apparatuses governing innovation (Hilgartner 2009), including ELS programs. STS insights about the ways normative features are built into knowledge and technology before they leave the laboratory—in designs calibrated to imagined uses and users, in judgments about safety and efficacy, and in underlying ideological commitments to certain conceptions of progress—have received relatively little attention in ELS research (Hurlbut 2015a).

Nevertheless, some STS scholars see the novel institutional form of ELS programs as offering an opening to contribute to ELS work while critiquing this dominant discourse, especially via "anticipatory governance" and "upstream engagement," which reject the notion of neutral innovation. Others remain wary that that these approaches may nevertheless slide back into the constraining pattern of grounding upstream ethical reflection in speculation about downstream consequences rather than in a sense of precaution in the face of the unknown (Nordmann 2014).

Other STS scholars have attempted work that is at once critical and contributory by "embedding" social scientists within centers of life science or nano research with the goal of creating new forms of engagement. Reflecting on their roles as social scientists embedded in a synthetic biology community, Rabinow and Bennett (2012) draw on Nowotny, Scott, and Gibbons's (2001) distinctions between types of knowledge production to describe three "modes" of engagement between social and natural scientists and to diagnose the "critical limitations" of each approach. Rabinow and Bennett themselves endorse a mode that they call "human practices," which they envision as a truly collaborative, rather than merely cooperative, interaction between social and natural scientists. Indeed, a number of STS scholars have recently envisioned and undertaken new collaborative relationships with their natural and life science counterparts, seeking to enhance reflexivity and mutual learning (Doubleday 2007; Fisher 2007; Stegmaier 2009; Wickson, Strand, and Kjølberg 2014). Some of these approaches are explicitly described as "post-ELSI" (Balmer and Bulpin 2013).

Edmond and Mercer (2009), however, are concerned that institutional and organizational barriers will discourage scientists from engaging in these sorts of truly reflexive practices. Moreover, they worry about the "possible dangers of diluting external regulation and existing forms of accountability for scientists and engineers" (ibid. 2009, 445). Joly (2015) also argues that upstream approaches have a tendency to underestimate the strength of institutionalized constraints and power relations. This point is consistent with what "embedded" social scientists have written about the demands that ELS sponsors have placed on them (Balmer et al. 2015; Rabinow and Bennett 2012; Viseu 2015). As their experiences show, by enforcing definitions of "social and ethical dimensions" as subsidiary to instrumental and technoscientific goals, ELS programs may silence critical questions, not least by pigeonholing ELS researchers into prevailing conceptions of their proper role. Additionally, some STS scholars question the tacit presumption that collaborative, interdisciplinary research will transcend power asymmetries among disciplines and thus produce "better" knowledge. Indeed, in some instances, collaborations between life science and ELS researchers seem to have strengthened disciplinary hierarchies (Albert et al. 2008; Fitzgerald and Callard 2014; Prainsack et al. 2010).

The challenges of achieving truly critical engagement are intensified by competing conceptions of the proper role of the social sciences in ELS programs. Many ELS participants (e.g., Caulfield et al. 2013) assume that social scientists should contribute research-based facts to help formulate "evidence-based policy." In this view, ELS programs should produce expert knowledge that stands outside of politics, providing a neutral capacity to offer rational social scientific and moral arguments. Still others see social sciences as supplying techniques for engaging publics and thereby reaching democratic solutions. Both of these contributory approaches are in some respects at odds with critical STS stances that question the fact/value distinction, approach power from a Foucauldian perspective, or problematize the construction of subjects and their interests. The tension between traditional approaches that emphasize relevance to previously structured policy agendas and STS approaches that emphasize reflexivity and open-ended deliberation is particularly strong.

This tension is captured in Jasanoff's (2003) distinction between "technologies of hubris" and "technologies of humility." The former—a wide range of methods of measurement, prediction, and control—are the central instruments of the policy sciences, and they are well represented in many ELS programs. Jasanoff argues that these methods suffer from important limitations: they "downplay what falls outside their field of vision, and ... overstate whatever falls within"; they tend to preempt political discussion; and they have a limited capacity "to internalize challenges that arise outside their framing assumptions" (ibid., 239). Restoring balance requires technologies of humility: "institutionalized habits of thought" that "confront 'head-on' the normative implications of our lack of perfect foresight" (ibid., 227) by inviting deliberation over ambiguities in collective experience that the policy sciences often occlude or ignore.[13]

ELS as a Mode of Governance

As part of a broader move toward employing "ethics" as a tool of governance, ELS programs constitute important forms of authority in democratic governance of science and technology, providing an alternative or a supplement to black-letter law and often exercising power in subtle and indirect ways. ELS programs may reframe and thereby help "close" controversies, for instance, by recasting uncertainties as matters for personal or organizational reflection rather than formal regulation. Bioethics expertise often plays a powerful role in classifying issues as warranting or not warranting deliberation, defining the terms of acceptable debate, and establishing frameworks for evaluation and oversight. ELS programs have played a central role in producing, deploying, institutionalizing, and naturalizing such frameworks, not least by cultivating particular forms of "ethical, legal, and social" expertise. This section reviews and suggests future directions for STS research on this topic.

Ethics Programs and Social Order

ELS programs represent an important site where the co-production of knowledge and social order take place (Jasanoff 2004). In the main, the critical literatures on bioethics and on science-society relations have developed separately (Braun et al. 2010), with the study of bioethics and society (e.g., Bosk 2008; Evans 2011; Kleinman, Fox, and Brandt, 1999; Stark 2011) growing more out of social studies of medicine than out of STS. One consequence is that even critical studies of bioethics have failed to interrogate the ways epistemic authority configures practices and parameters of ethics, for instance, by producing ontological accounts to which ethics then defers (Jasanoff 2011a). However, scholars in STS and related fields have increasingly begun to set bioethical expertise in social and political context (e.g., Felt et al. 2008; Hilgartner, Miller, and Hagendijk 2015; Jasanoff 2005; Prainsack 2006; Salter and Jones 2005; Sperling 2013).

Jasanoff's perspective on the co-production of epistemic and normative arrangements, with its focus on the role of institutions, discourses, identities, constitutions, and imaginaries, offers one idiom for examining the role of ELS programs in simultaneously making knowledge and making order (Hagendijk 2015). From this perspective, as well as from Foucauldian perspectives on governmentality (Rose 2007), the rise of ELS programs is a development of constitutional significance—as suggested by novel terms such as "biosocieties" (the name of a journal launched in 2006) and "bioconstitutionalism" (Jasanoff 2011b). "Constitutional" here refers not primarily to the anchoring of ELS concerns in constitutional law but to shared imaginaries that underwrite notions of the modes of reasoning and ruling appropriate to a well-ordered state. The biological plays an increasingly fundamental role as an arena for negotiation of relationships between the state and its citizens, informing conceptions of human life, identity, and flourishing, and shaping imaginations of the means through which the state can deliver on its responsibility to secure the well-being of the lives in its care. As a domain in which accounts of the right relationship between science and society take shape, ELS activities occupy constitutional territory.

Questions about the politics of knowledge making apply to ethics no less than they do to science. ELS programs play an important role in "boundary work" (Gieryn 1999, 5) that separates facts and values. In contrast to bioethics, which tends to take the boundary between fact and value as given in advance, STS scholars have asked how this boundary is constructed and consider how specific delineations of the boundary shape what questions are (un)asked and how authority and responsibility are allocated. Arguably, the structure of most ELS programs replicates the fact/value distinction. By constructing a body of scholarship to serve as a custodian of values, the programs simultaneously establish a sharp epistemological, disciplinary, and institutional separation between facts and values and, by extension, between scientific authority and the politics of public moral sense making.

ELS programs have helped produce a cadre of ethical experts who position themselves as standing outside of politics, allowing them to claim the capacity to offer rational moral arguments (Jasanoff 2011c) or identify moral views that appropriately comport with norms of secular reason (Bennett 2015; Evans 2011). Notions of scientific neutrality and legitimate ethical judgment may be co-produced as ethical experts draw on the putative neutrality of scientific authority to underwrite their own authority to delimit the forms of reasoning that can (and cannot) legitimately inform democratic governance (Hurlbut 2015b). ELS programs thus mediate between scientific autonomy and democratic governance, providing a source of authority for defining—and, at times, dismissing—public concerns. Conversely, these programs may insulate scientists from responsibility, for example, by constructing oversight procedures that render research "ethical" or by casting moral uncertainty as a politically irresolvable feature of pluralistic societies, and thus too multivocal to warrant constraining innovation. By interrogating the commitments—epistemic, political, and moral—that underlie ethics programs, STS scholarship is beginning to bring their constitutional dimensions to the surface.

Comparative research offers one method for studying these constitutional issues. Jasanoff's (2005) study of biotechnology regulation in Germany, the United Kingdom, and the United States examines how each country approached risks and ethical questions, showing how regulatory agendas, institutional structures, and modes of reasoning developed in nationally distinctive ways. In her account, each of these nations draws on a relatively stable "civic epistemology" that patterns the modes of "public reason" (Jasanoff 2012) it tends to employ. These patterns express themselves in the ways in which governance via ethics is imagined and enacted, thus shaping the framing of bioethical problems and the programs intended to address them. In the United States, for example, the prevailing pattern is to ground authority in the ideal of impersonal objectivity, as opposed to in specific trusted persons in the United Kingdom, and in a proper balancing of relevant constituencies in Germany (Jasanoff 2005).

Where Jasanoff notes profound differences among the countries she studied, and also other scholars discern variation—for example, in the rhetorical structures of regulatory discourses around genetic technologies (Gottweis and Prainsack 2006; Hauskeller 2010)—Braun et al. (2010) emphasize similarities in the ethics regimes instituted in the French, German, and U.K. contexts. In seeking to identify what sets the new "governmental ethics regime" apart from earlier visions of "speaking truth to power," Braun et al. (2010, 855) suggest that ethics regimes provide controlled forms of talk through which "struggles over the proper relationship between politics, science, and society are organized." They also argue that ethics commissions increasingly take on a "pastoral" role, acting almost like a therapist who helps the public better understand itself.

Imagining and Making Futures

Anticipation has always been one of the keywords of ELS discourse. STS scholars have written extensively about the performative aspects of anticipatory knowledge in the making of futures (e.g., Brown, Rappert, and Webster 2000; Fortun 2008; Gusterson 2008; Sunder Rajan 2006; see also Konrad et al., chapter 16 this volume). Contemporary bioethics—the starting point for ELSI and the wider ELS movement—almost invariably ties into a broader "sociotechnical imaginary" (Jasanoff and Kim 2015) that envisions science as the engine of change, casting the challenge of prediction as one that takes technoscientific novelties as its starting point and asks about the normative questions that those novelties raise. Imagining technology as the cause of sociotechnical change, Hurlbut (2015c, 128–29) argues, makes governance into "a matter of ontological discernment: of asking 'what is new here?'" Normative questions are thus "rendered subsidiary to expert assessments of novelty" (ibid.). This epistemic division of labor casts bioethics as necessarily reactive, with "its focus on downstream consequences and its preoccupation with the question of whether a given technological domain is sufficiently novel to engender new normative problems" (ibid.).

Hedgecoe (2010) perceives a similar allocation of epistemic authority, arguing that bioethics not only accepts but also reinforces the predictions and expectations of scientists and industry. Even when bioethicists react negatively to scientific developments, they may reinforce the expectations underlying them (Tutton 2011). To the extent that societal issues are treated as always lagging behind the science, ELS programs instantiate a "cultural cartography" (Gieryn 1999, 4) that parcels out authority in a particular way, positioning the creation of new knowledge (the domain of science) as institutionally separate from making societal choices about how to use knowledge (the domain of markets, law, policy, and ethics).

Recent European approaches to ethics (e.g., Grunwald 2012), and also anticipatory governance in the United States, seek to move beyond these institutional divisions by building new modes of preparedness for sociotechnical change that invite earlier and more democratic engagement. Critics of these "upstream" approaches question the ability of foresight techniques to provide sufficiently accurate predictions to underwrite revising the institutional order (e.g., Tait 2009). Tait also worries about upstream engagement preemptively restricting the choices available to future consumers in markets. This criticism is more than an epistemic point about the limits of foresight; it is also a constitutional one about the proper allocation of control over the development of science and technology. In claims about the limits of foresight, the temporality and institutional site of decision making are both at stake.[14] But for the advocates of anticipatory governance, the claim that foresight cannot make accurate predictions misses the point. The aim of anticipation is not to predict precisely how the future will unfold but to build the capacity to respond to the unpredicted and unpredictable, in part by seeking to broaden participation in shaping technological and societal agendas (Barben et al. 2008).

The promoters of anticipatory governance see the capacity to collectively explore alternative futures as providing the basis for integrating social concerns into scientific practice. They also argue that when social scientists engage with scientists in the laboratory, the interaction may change scientists' perceptions of their own role. Guston (2014) and his colleagues have found that scientists and engineers who do not initially conceive of themselves as engaged in "governance" come through repeated interactions to "capably reflect" on their roles in shaping the world beyond the laboratory. Such changes may be reconfiguring the concept of "responsibility" in science (McCarthy and Kelty 2010).[15] Joly (2015, 144–45), however, argues that despite its sophistication, anticipatory governance and related approaches adopt a problematic discourse of distributed governance that "tends to make invisible the asymmetries of power and of resource distribution potentially affecting technology development." These approaches also run the risk of capture by the technological projects with which they engage (ibid.). In light of this debate, studying anticipation-in-action is an important direction for future STS research (Rabinow and Stavrianakis 2013).

Soft Regulation and the Formation of Political Collectives

A number of commentators suggest that governance via "ethics" exerts relatively little regulatory force. For example, Huijer (2006) describes how a Dutch program intended to support symmetrical interaction between bioscientists and social scientists was instead made to smooth the integration of genomics into society. Similar critiques have been levied at participatory efforts to include publics "upstream." These exercises rarely engender significant change, critics contend, not least because of entrenched asymmetries of power. Writing from a Foucauldian perspective, López and Robertson (2007) provide an account of the Canadian ELS program that focuses on the constitution of citizens, arguing that the program produced demobilized Canadians. For some authors, systematic and explicit enrollment of ethics as a governance tool has both advantages and limitations. Braun et al. (2010, 858) argue that while governmental ethics programs may increase inclusiveness, allow for consideration of values beside efficiency, and eschew the assumption that knowledge lacks normative content, these advantages come at a price: they allow "no space for antagonistic political positions, long-term limits to certain technologies, or the question of whether certain technologies should be pursued and made available at all."

Governance via "ethics" also raises questions at the intersection of STS and political theory. Irwin (2006, 299) argues that the use of new governance machinery of the participatory paradigm by state actors should not only be criticized for its inadequacies but "should also be viewed as symptomatic of the state of science-society relations." In her work on institutionalized ethics programs in the EU since the 1990s, Tallacchini (2015) argues that EU "ethics" allows regulatory frameworks already in place to be analogically extended to new technologies more quickly than would be possible through a legislative process. She makes the point that ethics operates as "soft law"—a form of guidance that has normative content and practical effects but is not backed (at least directly) by the state power that underwrites black-letter law. This soft law approach may not control emerging technologies, but it facilitates their acceptance by conveying comforting rhetoric: "a message of flexibility and individual and corporate accountability to the market, a flavor of democracy to citizens, and a sense of (self-reassuring) political control" (ibid., 169–70) to institutions and member states. In Tallacchini's account, ethics has played a crucial role in EU integration, and studying it offers a way not only to investigate science-society relations but, more broadly, to understand the EU as a political project. Laurent (2016, 233) too observes that efforts to engage European publics in ethical evaluation of technological change function as performative sites for enacting a "politically perfected European collective."

Studying the role of these modes of regulation in forming and reconfiguring political collectives represents a promising line of STS inquiry which invites deeper engagement with such fields as critical legal studies and political theory that have tended to overlook the role of science and technology as a locus of political authority (Jasanoff, chapter 9 this volume; Kennedy 2016).

Conclusions

The rise of ELS programs represents an important development in contemporary politics of science and technology. The ambiguities of promises to "anticipate and address" issues, "engage" publics, and "guide policy" created a space into which diverse constituencies have projected a variety of hopes and fears. Accordingly, various parties have envisioned the ELS enterprise as everything from policy instrument, to research program, to regulatory watchdog, to public relations exercise, to experiment in novel modes of democratization. Observers have perceived it in conflicting ways: as an opportunity for funding research and a political "tax" on science, as an early warning system and an alarmist overreaction, and as a means to educate the public and a way to create the false impression that problems are being addressed. Underlying these varied perceptions are disparate visions of society, of the nature of science and technology, and of the appropriate way to govern rapid sociotechnical change.

STS scholars have explored a variety of stances toward ELS programs. They have sought to use their expertise in service of predefined ELS agendas, worked to transform programs by incorporating STS sensibilities, tried to invent new forms of mutual learning, and analyzed ELS programs as an emergent form of authority. While STS interventions have at times been welcomed, they have also encountered friction, not least in response to methodologically foundational STS sensibilities that treat science and governance as objects of critical scrutiny.

The studies of ethics and governance reviewed above notwithstanding, STS has undertaken relatively little systematic research on ELS programs. Much of the extant literature proceeds in a mode of celebration and/or criticism aimed at specific scientific and ELS agendas. There is not yet a large, theoretically informed literature on ELS as a form of governance and what its emergence means. Not only do shorter commentaries dominate STS literature on ELS, but most of these commentaries treat ELS programs as isolated, *de novo* phenomena that have evolved over the past quarter century without considering them as part of an array of governance techniques at play in specific historical and social contexts. Most studies do not develop *longue durée* histories, and most do not take up the leads provided by the existing sociological analyses of bioethics, historical accounts, or works in political theory on science and governance (e.g., Ezrahi 1990, 2012). Comparative research is also unusual, despite the fact it shows the importance of national differences (Felt et al. 2010; Jasanoff 2005; Macnaghten and Guivant 2011; Prainsack 2006). Clearly, there is a need for additional critical investigation of ELS and "ethics" as modes of governance, with attention to variation among cultural and political contexts.

ELS programs offer STS scholars a site not only for investigating traditional STS topics like the construction of facts or the shaping of technologies but also for symmetrical analysis of scientific knowledge and imaginaries of public benefit, right reason, and democratic order. A co-productionist approach that analyzes ELS programs in relation to institutions, discourses, identities, and extant dispositions of power offers promising directions for future work. Such research may help envision new roles for STS in the next generation of ELS undertakings, and illuminate the place of ELS programs in the evolving repertoires of governance of contemporary societies.

Notes

1. The U.K. Economic and Social Research Council (ESRC) allocated £41 million to five centers (from 2002 to 2012); the Netherlands provided €32 million to a center on genetics and society;

and Genome Canada dedicated 2.4 percent of its C\$537 million budget (2000 to 2008) to "ethical, environmental, economic, legal, and social implications" (Stegmaier 2009).

2. Programs with a specific technoscientific focus (e.g., on genomics or nano) differ greatly from broader programs, such as the U.S. National Science Foundation's (NSF) program on Science, Technology, and Society, which were launched well before ELSI (Hollander and Steneck 1990) and continue today.

3. The U.S. ELSI staff argue that the program began with "a broad and somewhat diffuse focus, which sometimes erroneously led to its being understood as having substantial responsibility for the development of policy solutions to the full range of complex ethical and societal issues raised by genomics" (McEwen et al. 2014, 482).

4. Beyond ensuring that "society learns to use the information only in beneficial ways," Watson (1990, 46) warned that a failure to act could lead to "abuses" and provoke "a strong popular backlash against the human genetics community."

5. Hilgartner was a member of the ELSI Evaluation Committee.

6. Juengst (1996, 75–76) notes that ELSI initially funded some research that critically examined the HGP itself, including studies by Dorothy Nelkin and Susan Lindee; Lily Kay; Sahotra Sarkar; and one of us (Hilgartner). Before long, support for interrogating the underpinnings of the HGP largely dried up.

7. "Framework Programmes" are the main device for allocating EU research funding. On the first ELSA program, see CORDIS 2001.

8. See Giordano and Olds (2010), Klein (2010), Singh (2010), and Stein (2010) for a slice of the debate over neuro-ELSI.

9. According to Fisher and Mahajan (2006, 9), Langdon Winner (2003) "had ELSI in mind" when he told the Congress to avoid setting up a nanoethics program that would "institutionalize a disconnection among societal concerns, research, and science and technology policy."

10. Discursive shifts among such terms as *publics*, *citizens*, and *stakeholders* are important here because they define the scope of democratic deliberation and allocate roles in it (Felt et al. 2007, 57–59).

11. Outreach programs were also important in ELSI, especially at the DOE. For an assessment, see Lewenstein and Brossard (2006).

12. The 2014 round of the British research excellence evaluation assessed the societal "impact" of research for the first time (see also Bauer 2012).

13. See also Stirling (2012), who calls for approaches to the governance of S&T aimed at "opening up" to inquiry rather than "closing down" a healthy, mature, accountable, democratic politics of technology choice.

14. Michael Polanyi's (1962) arguments about the unpredictability of science—with which he justified its autonomy—are a famous example, which Guston (2012) critiques.

15. Tallacchini (2015, 169) points out that "responsibility" in EU ethics does not mean liability or accountability but care and responsiveness, terms lacking any operational legal meaning.

References

Albert, Mathieu, Suzanne Laberge, Brian D. Hodges, Glenn Regehr, and Lorelei Lingard. 2008. "Biomedical Scientists' Perception of the Social Sciences in Health Research." *Social Science & Medicine* 66 (12): 2520–31.

Annas, George J. 1989. "Who's Afraid of the Human Genome?" *Hastings Center Report* 19 (4): 19–21.

Balmer, Andrew S., and Kate J. Bulpin. 2013. "Left to Their Own Devices: Post-ELSI, Ethical Equipment and the International Genetically Engineered Machine (iGEM) Competition." *BioSocieties* 8 (3): 311–35.

Balmer, Andrew S., Jane Calvert, Claire Marris, Susan Molyneux-Hodgson, Emma Frow, Matthew Kearnes, Kate Bulpin, Pablo Schyfter, Adrian Mackenzie, and Paul Martin. 2015. "Taking Roles in Interdisciplinary Collaborations: Reflections on Working in Post-ELSI Spaces in the UK Synthetic Biology Community." *Science & Technology Studies* 28 (3). Accessed January 15, 2016, at http://www.stis.ed.ac.uk/el/publications/project_publications/Balmer_et_al_2015_STS_Author _copy.pdf.

Barben, Daniel, Erik Fisher, Cynthia Selin, and David H. Guston. 2008. "Anticipatory Governance of Nanotechnology: Foresight, Engagement, and Integration." In *The Handbook of Science and Technology Studies*. 3rd ed., edited by Edward J. Hackett, Olga Amsterdamska, Michael E. Lynch, and Judy Wajcman, 979–1000. Cambridge, MA: MIT Press.

Bauer, Martin W. 2012. "Science Culture and Its Indicators." In *Communication in the World: Practices, Theories and Trends,* edited by Bernard Schiele, Michel Claessens, and Shunke Shi, 295–312. New York: Springer.

Beckwith, Jonathan R. 2009. *Making Genes, Making Waves: A Social Activist in Science*. Cambridge, MA: Harvard University Press.

Brown, Nik, Brian Rappert, and Andrew Webster, eds. 2000. *Contested Futures: A Sociology of Prospective Techno-science*. Aldershot: Ashgate.

Bennett, Gaymon. 2015. *Technicians of Human Dignity: Bodies, Souls, and the Making of Intrinsic Worth*. New York: Fordham University Press.

Bosk, Charles L. 2008. *What Would You Do? Juggling Bioethics and Ethnography*. Chicago: University of Chicago Press.

Braun, Kathrin, Svea Luise Herrmann, Sabine Konninger, and Alfred Moore. 2010. "Ethical Reflection Must Always Be Measured." *Science, Technology, & Human Values* 35 (6): 839–64.

Burke, Wylie, Paul Appelbaum, Lauren Dame, Patricia Marshall, Nancy Press, Reed Pyeritz, Richard Sharp, and Eric Juengst. 2014. "The Translational Potential of Research on the Ethical, Legal, and Social Implications of Genomics." *Genetics in Medicine* 17 (1): 1–9.

Calvert, Jane, and Paul Martin. 2009. "The Role of Social Scientists in Synthetic Biology." *EMBO Reports* 10 (3): 201–4.

Caulfield, Timothy, Subhashini Chandrasekharan, Yann Joly, and Robert Cook-Deegan. 2013. "Harm, Hype and Evidence: ELSI Research and Policy Guidance." *Genome Medicine* 5 (3): 21.

Chadwick, Ruth, and Hub A. E. Zwart. 2013. "From ELSA to Responsible Research and Promisomics." *Life Sciences, Society and Policy* 9 (1): 3.

Collingridge, David. 1980. *The Social Control of Technology*. London: Printer.

Collins, Francis S., Eric D. Green, Alan E. Guttmacher, and Mark S. Guyer. 2003. "A Vision for the Future of Genomics Research: A Blueprint for the Genomic Era." *Nature* 422 (6934): 835–47.

Collins, Harry, and Robert Evans. 2007. *Rethinking Expertise*. Chicago: University of Chicago Press.

Colvin, Vicki L. 2003. "Statement," *The Societal Implications of Nanotechnology*. Hearing before the U.S. House Committee on Science, 108th Congress. First session, April 9, 2003. Serial no. 108–13.

Cook-Deegan, Robert. 1994. *The Gene Wars: Science, Politics, and the Human Genome*. New York: W. W. Norton.

CORDIS. 2001. "ELSA—Ethical, Legal and Social Aspects" (29 January 2001), accessed January 25, 2016, http://cordis.europa.eu/elsa-fp4/.

Doubleday, Robert. 2007. "The Laboratory Revisited: Academic Science and the Responsible Development of Nanotechnology." *NanoEthics* 1 (2): 167–76.

Edmond, Gary, and David Mercer. 2009. "Norms and Irony in the Biosciences: Ameliorating Critique in Synthetic Biology." *Law & Literature* 21 (3): 445–70.

ELSI Evaluation Committee. 1996. "Report on the Joint NIH/DOE Committee to Evaluate the Ethical, Legal, and Social Implications Program of the Human Genome Project." Final report. U.S. National Center for Human Genome Research and the Department of Energy, accessed at http://www.genome.gov/10001745.

ELSI Research Planning and Evaluation Group. 2000. "A Review and Analysis of the Ethical, Legal, and Social Implications (ELSI) Research Programs at the National Institutes of Health and the Department of Energy." Final report. U.S. National Human Genome Research Institute and U.S. Department of Energy, accessed at http://www.genome.gov/10001727.

European Commission. 2001. ELSA, accessed June 14, 2015, http://ec.europa.eu/research/life/elsa/.

Evans, John H. 2011. *The History and Future of Bioethics: A Sociological View*. Oxford: Oxford University Press.

Ezrahi, Yaron. 1990. *The Descent of Icarus*. Cambridge, MA: Harvard University Press.

___. 2012. *Imagined Democracies: Necessary Political Fictions*. Cambridge: Cambridge University Press.

Felt, Ulrike, Daniel Barben, Alan Irwin, Pierre-Benoît Joly, Arie Rip, Andy Stirling, Tereza Stöckelová, and D. De La Hoz Del Hoyo. 2013. *Science in Society: Caring for Our Futures in Turbulent Times*. Strasbourg: European Science Foundation.

Felt, Ulrike, Maximilian Fochler, Astrid Mager, and Peter Winkler. 2008. "Visions and Versions of Governing Biomedicine: Narratives on Power Structures, Decision-Making and Public Participation in the Field of Biomedical Technology in the Austrian Context." *Social Studies of Science* 38 (2): 233–57.

Felt, Ulrike, Maximillian Fochler, Annina Müller, and Michael Strassnig. 2009. "Unruly Ethics: On the Difficulties of a Bottom-up Approach to Ethics in the Field of Genomics." *Public Understanding of Science* 18 (3): 354–71.

Felt, Ulrike, Maximilian Fochler, and Peter Winkler. 2010. "Coming to Terms with Biomedical Technologies in Different Technopolitical Cultures: A Comparative Analysis of Focus Groups on Organ Transplantation and Genetic Testing in Austria, France, and the Netherlands." *Science, Technology, & Human Values* 35 (4): 525–53.

Felt, Ulrike, Brian Wynne, Michel Callon, Maria Eduarda Gonçalves, Sheila Jasanoff, Maria Jepsen, Pierre-Benoît Joly, Zdenek Konopasek, Stefan May, Claudia Neubauer, Arie Rip, Karen Siune, Andy Stirling, and Mariachiara Tallacchini. 2007. *Taking European Knowledge Society Seriously*. Report of the Expert Group on Science and Governance to the Science, Economy and Society Directorate, Directorate-General for Research, European Commission. Brussels: European Commission.

Fisher, Erik. 2005. "Lessons Learned from the Ethical, Legal and Social Implications Program (ELSI): Planning Societal Implications Research for the National Nanotechnology Program." *Technology in Society* 27 (3): 321–28.

___. 2007. "Ethnographic Invention: Probing the Capacity of Laboratory Decisions." *NanoEthics* 1 (2): 155–65.

Fisher, Erik, and Roop L. Mahajan. 2006. "Contradictory Intent? US Federal Legislation on Integrating Societal Concerns into Nanotechnology Research and Development." *Science and Public Policy* 33 (1): 5–16.

Fitzgerald, Des, and Felicity Callard. 2014. "Social Science and Neuroscience beyond Interdisciplinarity: Experimental Entanglements." *Theory, Culture & Society* 32 (1): 3–32.

Forsberg, Ellen-Marie. 2015. "ELSA and RRI." *Life Sciences, Society and Policy* 11 (1): 1–3.

Fortun, Michael A. 2008. *Promising Genomics: Iceland and deCODE Genetics in a World of Speculation*. Berkeley: University of California Press.

Gieryn, Thomas F. 1999. *Cultural Boundaries of Science: Credibility on the Line*. Chicago: University of Chicago Press.

Giordano, James, and James Olds. 2010. "The Interfluence of Neuroscience, Neuroethics, and Legal and Social Issues: The Need for (N)ELSI." *AJOB Neuroscience* 1 (4): 12–14.

Gottweis, Herbert, and Barbara Prainsack. 2006. "Emotion in Political Discourse: Contrasting Approaches to Stem Cell Governance: The US, UK, Israel, and Germany." *Regenerative Medicine* 1 (6): 823–29.

Greely, Henry T. 2005. "Neuroethics and ELSI: Similarities and Differences." *Minnesota Journal of Law, Science & Technology* 7 (2): 599–637.

Grunwald, Armin. 2012. *Responsible Nanobiotechnology: Philosophy and Ethics*. Singapore: Pan Stanford Publications.

Gusterson, Hugh. 2008. "Nuclear Futures: Anticipatory Knowledge, Expert Judgment, and the Lack That Cannot Be Filled." *Science and Public Policy* 35 (8): 551–60.

Guston, David H. 2012. "The Pumpkin or the Tiger? Michael Polanyi, Frederick Soddy, and Anticipating Emerging Technologies." *Minerva* 50 (3): 363–79.

___. 2014. "Understanding 'Anticipatory Governance.'" *Social Studies of Science* 44 (2): 218–42.

Guston, David H., and Daniel Sarewitz. 2002. "Real-Time Technology Assessment." *Technology in Society* 24 (1): 93–109.

Hagendijk, Rob. 2015. "Sense and Sensibility: Science, Society and Politics as Coproduction." In *Science and Democracy: Making Knowledge and Making Power in the Biosciences and Beyond*, edited by Stephen Hilgartner, Clark A. Miller, and Rob Hagendijk, 220–38. London: Routledge.

Hanna, Kathi E. 1995. "The Ethical, Legal and Social Implications Program of the National Center for Human Genome Research: A Missed Opportunity?" In *Society's Choices: Social and Ethical Decision Making in Biomedicine*, 432–57. Washington, DC: National Academies Press.

Hauskeller, Christine. 2010. "How Traditions of Ethical Reasoning and Institutional Processes Shape Stem Cell Research in Britain." *Journal of Medicine and Philosophy* 29 (5): 509–32.

Hedgecoe, Adam. 2010. "Bioethics and the Reinforcement of Socio-technical Expectations." *Social Studies of Science* 40 (2): 163–86.

Henderson, Gail E., Eric T. Juengst, Nancy M. P. King, Kristine Kuczynski, and Marsha Michie. 2012. "What Research Ethics Should Learn from Genomics and Society Research: Lessons from the ELSI Congress of 2011." *Journal of Law, Medicine & Ethics* 40 (4): 1008–24.

Hilgartner, Stephen. 2004. "Making Maps and Making Social Order: Governing American Genome Centers, 1988–93." In *From Molecular Genetics to Genomics*, edited by Jean-Paul Gaudillière and Hans-Jörg Rheinberger, 113–28. London: Routledge.

___. 2009. "Intellectual Property and the Control of Technology: Inventors, Citizens, and Powers to Shape the Future." *Chicago-Kent Law Review* 84 (1): 197–224.

Hilgartner, Stephen, Clark A. Miller, and Rob Hagendijk, eds. 2015. *Science and Democracy: Making Knowledge and Making Power in the Biosciences and Beyond*. New York: Routledge.

Hollander, Rachelle D., and Nicholas H. Steneck. 1990. "Science- and Engineering-Related Ethics and Values Studies: Characteristics of an Emerging Field of Research." *Science, Technology, & Human Values* 15 (1): 84–104.

Holtzman, Neil A. 1989. *Proceed with Caution: Predicting Genetic Risks in the Recombinant DNA Era*. Baltimore: Johns Hopkins University Press.

Houdy, Philippe, Marcel Lahmani, and Francelyne Marano, eds. 2011. *Nanoethics and Nanotoxicology*. Heidelberg: Springer.

Hubbard, Ruth, and Elijah Wald. 1993. *Exploding the Gene Myth*. Boston: Beacon Press.

Huijer, Marli. 2006. "Between Dreams and Reality: The Dutch Approach to Genomics and Society." *BioSocieties* 1 (1): 91–95.

Hurlbut, J. Benjamin. 2015a. "Reimagining Responsibility in Synthetic Biology." *Journal of Responsible Innovation* 2 (1): 113–16.

___. 2015b. "Religion and Public Reason in the Politics of Biotechnology." *Notre Dame Journal of Law, Ethics and Public Policy* 29: 423–52.

___. 2015c. "Remembering the Future: Science, Law, and the Legacy of Asilomar." In *Dreamscapes of Modernity: Sociotechnical Imaginaries and the Fabrication of Power*, edited by Sheila Jasanoff and Sang-Hyun Kim, 126–51. Chicago: University of Chicago Press.

Irwin, Alan. 2006. "The Politics of Talk: Coming to Terms with the 'New' Scientific Governance." *Social Studies of Science* 36 (2): 299–320.

Jasanoff, Sheila. 2003. "Technologies of Humility: Citizen Participation in Governing Science." *Minerva* 41 (3): 223–44.

___. 2004. *States of Knowledge: The Co-production of Science and Social Order*. London: Routledge.

___. 2005. *Designs on Nature: Science and Democracy in Europe and the United States*. Princeton, NJ: Princeton University Press.

___. 2011a. "Making the Facts of Life." In *Reframing Rights: Bioconstitutionalism in the Genetic Age*, edited by Sheila Jasanoff, 59–83. Cambridge, MA: MIT Press.

___. 2011b. *Reframing Rights: Bioconstitutionalism in the Genetic Age*. Cambridge, MA: MIT Press.

___. 2011c. "Constitutional Moments in Governing Science and Technology." *Science and Engineering Ethics* 17 (4): 621–38.

___. 2012. *Science and Public Reason*. London: Routledge.

Jasanoff, Sheila, and Sang-Hyun Kim, ed. 2015. *Dreamscapes of Modernity: Sociotechnical Imaginaries and the Fabrication of Power*. Chicago: University of Chicago Press.

Joly, Pierre-Benoît. 2015. "Governing Emerging Technologies? The Need to Think Out of the (Black) Box." In *Science and Democracy: Knowledge as Wealth and Power in the Biosciences and Beyond*, edited by Stephen Hilgartner, Clark A. Miller, and Rob Hagendijk, 133–55. New York: Routledge.

Juengst, Eric T. 1994. "Human Genome Research and the Public Interest: Progress Notes from an American Science Policy Experiment." *American Journal of Human Genetics* 54 (1): 121.

___. 1996. "Self-Critical Federal Science? The Ethics Experiment within the Human Genome Project." *Social Philosophy & Policy* 13: 63–95.

Kaye, Jane, Eric M. Meslin, Bartha M. Knoppers, Eric T. Juengst, Mylène Deschênes, Anne Cambon-Thomsen, Donald Chalmers, Jantina De Vries, Kelly Edwards, Nils Hoppe, Alastair Kent, Clement Adebamowo, Patricia Marshall, and Kazuto Kato. 2012. "ELSI 2.0 for Genomics and Society." *Science* 336 (6082): 673–74.

Kennedy, David. 2016. *A World of Struggle: How Power, Law, and Expertise Shape Global Political Economy*. Princeton, NJ: Princeton University Press.

Kitcher, Philip. 2001. *Science, Truth, and Democracy*. Oxford: Oxford University Press.

Klein, Eran. 2010. "To ELSI or Not to ELSI Neuroscience: Lessons for Neuroethics from the Human Genome Project." *AJOB Neuroscience* 1 (4): 3–8.

Kleinman, Arthur, Renee C. Fox, and Allan M. Brandt. 1999. "Bioethics and Beyond." Special issue. *Daedalus: Journal of the American Academy of Arts and Sciences* 128 (4).

Kreissl, Reinhard, Florian Fritz, and Lars Ostermeier. 2015. "Societal Impact Assessment." In *International Encyclopedia of the Social and Behavioral Sciences*. 2nd ed., edited by James D. Wright, 873–77. Oxford: Elsevier.

Kurzweil, Raymond. 2003. "Statement," *The Societal Implications of Nanotechnology*. Hearing before the U.S. House Committee on Science, 108th Congress. First session, April 9, 2003. Serial no. 108–13.

Laurent, Brice. 2016. "Perfecting European Democracy: Science as a Problem of Technological and Political Progress." In *Perfecting Human Futures: Transhuman Futures and Technological Imaginations*, edited by J. Benjamin Hurlbut and Hava Tirosh-Samuelson, 218–37. Dordrecht: Springer.

Lezaun, Javier, and Linda Soneryd. 2007. "Consulting Citizens: Technologies of Elicitation and the Mobility of Publics." *Public Understanding of Science* 16 (3): 279–97.

Lewenstein, Bruce V., and Dominique Brossard. 2006. *Assessing Models of Public Understanding in ELSI Outreach Materials*. Ithaca, NY: Cornell University, accessed at http://www.osti.gov/scitech/biblio/876753.

Liebert, Wolfgang, and Jan C. Schmidt. 2010. "Collingridge's Dilemma and Technoscience: An Attempt to Provide a Clarification from the Perspective of the Philosophy of Science." *Poiesis & Praxis* 7 (1–2): 55–71.

Lindner, Ralf, Stefan Kuhlmann, Sally Randless, Bjorn Bedsted, Guido Gorgoni, Erich Griessler, Allison Loconto, and Niels Mejlgaard. 2016. "Navigating Towards Shared Responsibility." Karlsruhe, Germany: ResAGorA project report, accessed at https://indd.adobe.com/view/eaeb695e -a212-4a34-aeba-b3d8a7a58acc.

López, José, and Ann Robertson. 2007. "Ethics or Politics? The Emergence of ELSI Discourse in Canada." *Canadian Review of Sociology and Anthropology* 44 (2): 201–18.

Macnaghten, Phil, and Julia S. Guivant. 2011. "Converging Citizens? Nanotechnology and the Political Imaginary of Public Engagement in Brazil and the United Kingdom." *Public Understanding of Science* 20 (2): 207–20.

Marris, Claire. 2015. "The Construction of Imaginaries of the Public as a Threat to Synthetic Biology." *Science as Culture* 24 (1): 83–98.

McCain, Lauren. 2002. "Informing Technology Policy Decisions: The US Human Genome Project's Ethical, Legal, and Social Implications Programs as a Critical Case." *Technology in Society* 24 (1): 111–32.

McCarthy, Elise, and Christopher Kelty. 2010. "Responsibility and Nanotechnology." *Social Studies of Science* 40 (3): 405–32.

McEwen, Jean E., Joy T. Boyer, Kathie Y. Sun, Karen H. Rothenberg, Nicole C. Lockhart, and Mark S. Guyer. 2014. "The Ethical, Legal, and Social Implications Program of the National Human Genome Research Institute: Reflections on an Ongoing Experiment." *Annual Review of Genomics and Human Genetics* 15 (1): 481–505.

National Institutes of Health, National Center for Human Genome Research. 1993. *Progress Report: Fiscal Years 1991 and 1992*. NIH Publication No. 93–3550. Bethesda, MD: National Institutes of Health.

National Institutes of Health and Department of Energy. 1990. *Understanding Our Genetic Inheritance: The U.S. Human Genome Project: The First Five Years, FY 1991–1995*. Springfield, VA: National Technical Information Service.

Nelis, Annemiek, Collin Reddy, and Elles Mulder. 2008. *ELSA Genomics in the ERASAGE Consortium. Similarities and Differences in Approaches and Themes in the Study of the Relationship between Genomics and Society*. Nijmegen, The Netherlands: Centre for Society and Genomics.

Nelkin, Dorothy, and Laurence Tancredi. 1989. *Dangerous Diagnostics: The Social Power of Biological Information*. Chicago: University of Chicago Press.

Nordmann, Alfred. 2014. "Responsible Innovation, the Art and Craft of Anticipation." *Journal of Responsible Innovation* 1: 87–98.

Nowotny, Helga, Peter Scott, and Michael T. Gibbons. 2001. *Re-thinking Science: Knowledge and the Public in an Age of Uncertainty.* Cambridge: Polity.

Ommer, Rosemary, Brian Wynne, Robin Downey, Erik Fisher, Emily Marden, and Patricia Kosseim. 2011. *Pathways to Integration. Genome British Columbia GSEAC Subcommittee on Pathways to Integration.* Vancouver: Genome BC.

Owen, Richard, John Bessant, and Maggy Heintz, eds. 2013. *Responsible Innovation: Managing the Responsible Emergence of Science and Innovation in Society.* Chichester: John Wiley.

Polanyi, Michael. 1962. "The Republic of Science: Its Political and Economic Theory." *Minerva* 1: 54–74.

Prainsack, Barbara. 2006. "'Negotiating Life': The Regulation of Human Cloning and Embryonic Stem Cell Research in Israel." *Social Studies of Science* 36 (2): 173–205.

Prainsack, Barbara, Mette N. Svendsen, Lene Koch, and Kathryn Ehrich. 2010. "How Do We Collaborate? Social Science Researchers' Experience of Multidisciplinarity in Biomedical Settings." *BioSocieties* 5 (2): 278–86.

Rabinow, Paul, and Gaymon Bennett. 2012. *Designing Human Practices: An Experiment with Synthetic Biology.* Chicago: University of Chicago Press.

Rabinow, Paul, and Anthony Stavrianakis. 2013. *Demands of the Day: On the Logic of Anthropological Inquiry.* Chicago: University of Chicago Press.

Radstake, Maud, and Bart Penders 2008. "Inside Genomics: The Interdisciplinary Faces of ELSA." *Graduate Journal of Social Science* 4: 4–10.

Rip, Arie. 2009. "Futures of ELSA." *EMBO Reports* 10 (7): 666–70.

Roco, Mihail, and William Sims Bainbridge, eds. 2007. *Nanotechnology: Societal Implications I: Maximizing Benefits for Humanity.* Dordrecht: Springer.

Rose, Nikolas. 2007. *The Politics of Life Itself: Biomedicine, Power, and Subjectivity in the Twenty-First Century.* Princeton, NJ: Princeton University Press.

Saha, Krishanu, and J. Benjamin Hurlbut. 2011. "Research Ethics: Treat Donors as Partners in Biobank Research." *Nature* 478 (7369): 312–13.

Salter, Brian, and Mavis Jones. 2005. "Biobanks and Bioethics: The Politics of Legitimation." *Journal of European Public Policy* 12 (4): 710–32.

Schot, Johan, and Arie Rip. 1997. "The Past and Future of Constructive Technology Assessment." *Technological Forecasting and Social Change* 54 (2): 251–68.

Sharp, Richard R., and J. Carl Barrett. 2000. "The Environmental Genome Project: Ethical, Legal, and Social Implications." *Environmental Health Perspectives* 108 (4): 279–81.

Singer, Peter A., Andrew D. Taylor, Abdallah S. Daar, Ross E. G. Upshur, Jerome A. Singh, and James V. Lavery. 2007. "Grand Challenges in Global Health: The Ethical, Social and Cultural Program." *PLoS Medicine* 4 (9): e265.

Singh, Ilina. 2010. "ELSI Neuroscience Should Have a Broad Scope." *AJOB Neuroscience* 1 (4): 11–12.

Sperling, Stefan. 2013. *Reasons of Conscience: The Bioethics Debate in Germany*. Chicago: University of Chicago Press.

Stark, Laura. 2011. *Behind Closed Doors: IRBs and the Making of Ethical Research*. Chicago: University of Chicago Press.

Stegmaier, Peter. 2009. "The Rock 'n' Roll of Knowledge Co-production." *EMBO Reports* 10 (2): 114–19.

Stein, Donald G. 2010. "ELSI: Creating Bureaucracy for Fun and Profit." *AJOB Neuroscience* 1 (4): 21–22.

Stirling, Andy. 2012. "Opening Up the Politics of Knowledge and Power in Bioscience." *PLoS Biology* 10 (1): e1001233.

Sunder Rajan, Kaushik. 2006. *Biocapital: The Constitution of Postgenomic Life*. Durham, NC: Duke University Press.

___. 2010. "The Experimental Machinery of Global Clinical Trials: Case Studies from India." In *Asian Biotech: Ethics and Communities of Fate*, edited by Aihwa Ong and Nancy N. Chen, 55–80. Durham, NC: Duke University Press.

Tait, Joyce. 2009. "Upstream Engagement and the Governance of Science." *EMBO Reports* 10 (1S): S18–22.

Tallacchini, Mariachiara. 2015. "To Bind or Not to Bind? European Ethics as Soft Law." In *Science and Democracy: Making Knowledge and Making Power in the Biosciences and Beyond*, edited by Stephen Hilgartner, Clark A. Miller, and Rob P. Hagendijk, 156–75. New York: Routledge.

Tutton, Richard. 2011. "Promising Pessimism: Reading the Futures to Be Avoided in Biotech." *Social Studies of Science* 41 (3): 411–29.

U.S. Congress, House of Representatives, Committee on Government Operations. 1992. *Designing Genetic Information Policy: The Need for an Independent Policy Review of the Ethical, Legal, and Social Implications of the Human Genome Project*. Washington, DC: U.S. Government Printing Office.

Viseu, Ana. 2015. "Caring for Nanotechnology? Being an Integrated Social Scientist." *Social Studies of Science* 45 (5): 642–64.

von Schomberg, René. 2013. "A Vision of Responsible Research and Innovation." In *Responsible Innovation*, edited by Richard Owen, John Bessant, and Maggy Heintz, 51–74. London: John Wiley.

Watson, James D. 1990. "The Human Genome Project: Past, Present, and Future." *Science* 248: 44–49.

Weiner, Charles. 1994 "Anticipating the Consequences of Genetic Engineering: Past, Present, and Future." In *Are Genes Us? The Social Consequences of the New Genetics*, edited by Carl F. Cranor, 31–55. New Brunswick, NJ: Rutgers University Press.

Wickson, Fern, Roger Strand, and Kamilla Lein Kjølberg. 2014. "The Walkshop Approach to Science and Technology Ethics." *Science and Engineering Ethics* 21 (1): 241–64.

Williams, Robin. 2006. "Compressed Foresight and Narrative Bias: Pitfalls in Assessing High Technology Futures." *Science as Culture* 15 (4): 327–48.

Wilsdon, James, and Rebecca Willis. 2004. *See-through Science: Why Public Engagement Needs to Move Upstream*. London: Demos.

Winner, Langdon. 2003. "Statement." In *The Societal Implications of Nanotechnology*, Hearing before the Committee on Science, U.S. House of Representatives, 108th Congress, 1st Session, April 9. Serial No. 108–13, 57 Washington DC: U.S. Government Printing Office.

Wynne, Brian. 2006. "Public Engagement as Means of Restoring Trust in Science? Hitting the Notes, but Missing the Music." *Community Genetics* 3 (9): 211–20.

Yesley, Michael. 2008. "What's ELSI Got to Do with It? Bioethics and the Human Genome Project." *New Genetics and Society* 21 (1): 1–6.

Zwart, Hub A. E., Laurens Landeweerd, and Arjan van Rooij. 2014. "Adapt or Perish? Assessing the Recent Shift in the European Research Funding Arena from 'ELSA' to 'RRI.'" *Life Sciences, Society and Policy* 10 (1): 1–19.

Zwart, Hub A. E., and Annemiek Nelis. 2009. "What Is ELSA Genomics?" *EMBO Reports* 10 (6): 540–44.

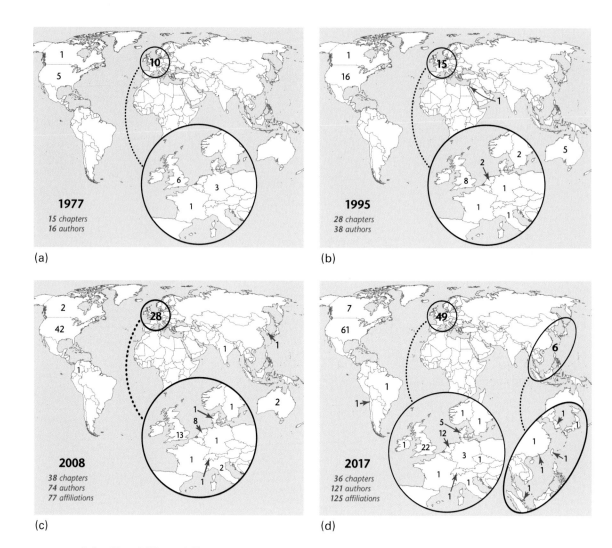

Color Plate 1 (Figure 0.1)

Geography of institutional affiliations of authors in the fourth (2017), third (2008), second (1995), and first (1977) editions of the *Handbook of Science and Technology Studies*. Map of World with Countries (http://freevectormaps.com/world-maps/WRLD-EPS-01-0013)—Outline by FreeVector-Maps.com.

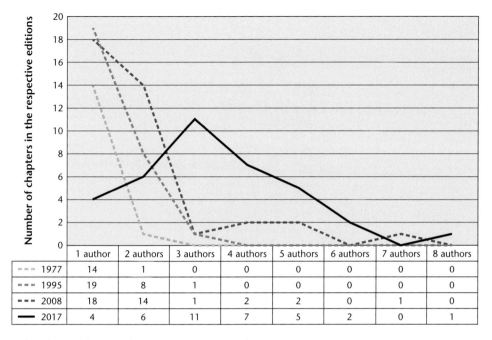

	1 author	2 authors	3 authors	4 authors	5 authors	6 authors	7 authors	8 authors
1977	14	1	0	0	0	0	0	0
1995	19	8	1	0	0	0	0	0
2008	18	14	1	2	2	0	1	0
2017	4	6	11	7	5	2	0	1

Color Plate 2 (Figure 0.2)

Changing coauthorship patterns in the editions of *The Handbook of Science and Technology Studies*.

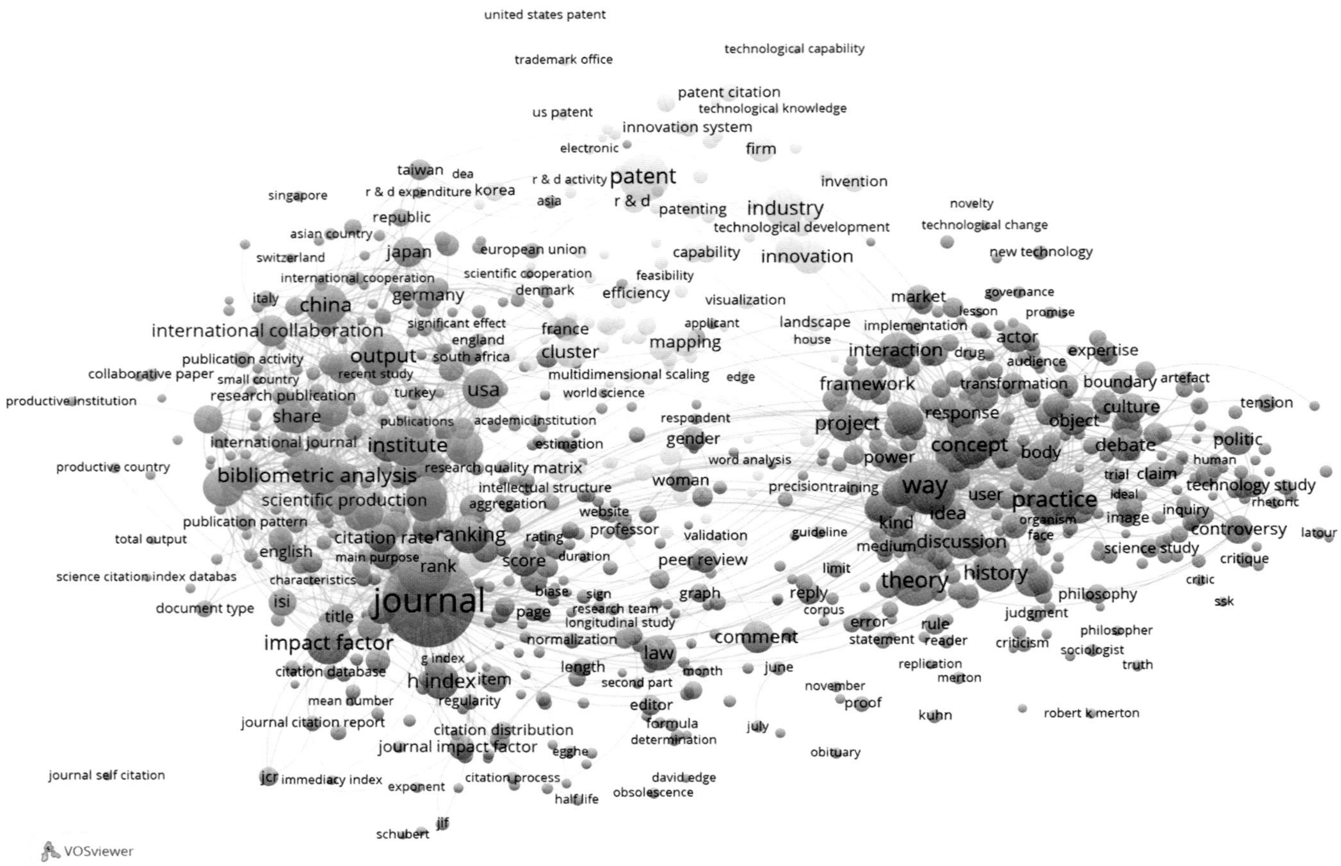

Color Plate 3 (Figure 3.1)

Semantic map of 896 most frequently used words in titles and abstracts of 5,677 publications in *Scientometrics* and *Social Studies of Science*. The map is available for web-starting at http://www.vosviewer.com/vosviewer.php?map=http://www.leydesdorff.net/sts_hbk/fig4map .txt&network=http://www.leydesdorff.net/sts_hbk/fig4net.txt&label_size_variation=0.5&n_lines=1000&zoom_level=1 .txt&network=http://www.leydesdorff.net/sts_hbk/fig4net.txt&label_size_variation=0.5&n_lines=1000&zoom_level=1

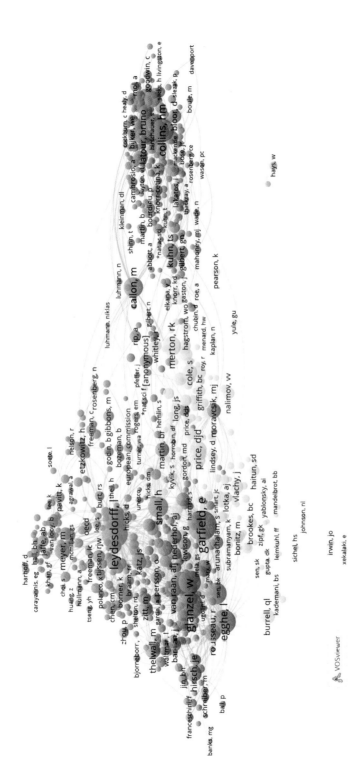

Color Plate 4 (Figure 3.2)

Co-citation analysis of 879 authors in *Scientometrics* and *Social Studies of Science* who are cited more than 20 times (6 October 2014). The map is available for web-starting at http://www.vosviewer.com/vosviewer.php?map=http://www.leydesdorff.net/sts_hbk/fig6map.txt&network=http://www.leydesdorff.net/sts_hbk/fig6net.txt&label_size_variation=0.4&n_lines=1000&visualization=1

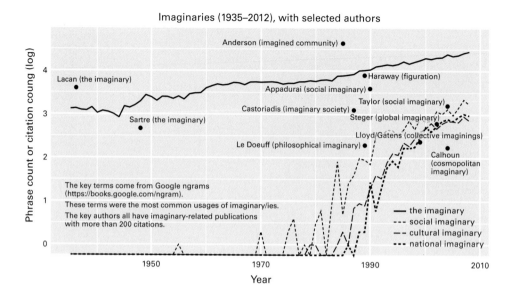

Color Plate 5 (Figure 15.1)

Chart of use of the concept of imaginaries (1935–2012) by selected authors.

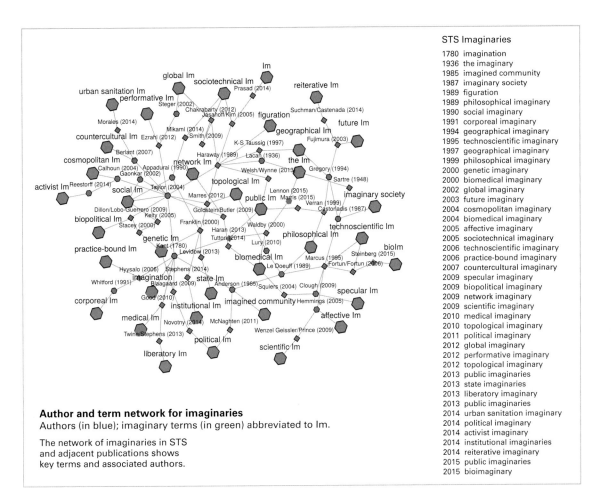

Author and term network for imaginaries
Authors (in blue); imaginary terms (in green) abbreviated to Im.

The network of imaginaries in STS
and adjacent publications shows
key terms and associated authors.

STS Imaginaries

1780	imagination
1936	the imaginary
1985	imagined community
1987	imaginary society
1989	figuration
1989	philosophical imaginary
1990	social imaginary
1991	corporeal imaginary
1994	geographical imaginary
1995	technoscientific imaginary
1997	geographical imaginary
1999	philosophical imaginary
2000	genetic imaginary
2000	biomedical imaginary
2002	global imaginary
2003	future imaginary
2004	cosmopolitan imaginary
2004	biomedical imaginary
2005	affective imaginary
2005	sociotechnical imaginary
2006	technoscientific imaginary
2006	practice-bound imaginary
2007	countercultural imaginary
2009	specular imaginary
2009	biopolitical imaginary
2009	network imaginary
2009	scientific imaginary
2010	medical imaginary
2010	topological imaginary
2011	political imaginary
2012	global imaginary
2012	performative imaginary
2012	topological imaginary
2013	public imaginaries
2013	state imaginaries
2013	liberatory imaginary
2013	public imaginaries
2014	urban sanitation imaginary
2014	political imaginary
2014	activist imaginary
2014	institutional imaginaries
2014	reiterative imaginary
2015	public imaginaries
2015	bioimaginary

Color Plate 6 (Figure 15.2)
Author and term network for imaginaries.

29 Responsible Research and Innovation

Jack Stilgoe and David H. Guston

Introduction: Caring for the Future

"Responsible innovation" and "responsible research and innovation" have become important themes for the governance of science and technology in the twenty-first century. But the ideas that sit behind these terms are perennial ones for the internal workings of the scientific community, its relationships with the outside world, and for science and technology studies (STS). Among the reasons for the current rearticulation of these terms, which may at first sound trite, is that discussions of social responsibility in science have fallen out of fashion, and innovation—unbound and unmodified—is ascendant. At least since World War II, recognition of the power of science and technology has forced reconsideration of the responsibilities that should follow such power. STS has helped build an understanding that the processes and consequences of science and technology should be governed responsibly. In addition, STS scholarship has pointed to the need to responsibly imagine the purposes of science and innovation and the framing of problems, although this broader responsibility has been less readily accepted.

In this chapter our aim is to explore and draw connections between current discussions of responsible innovation and STS. We begin by observing the (re-)emergence of the term *responsibility* as a response to perceived governance crises. We then retrace the STS concern with the politics of technology that provides a starting point for so much productive research. In the sections that follow, we chart the relationship between responsible innovation and public engagement with science, a debate in which STS had taken an interest and become similarly implicated. We look at the laboratory as a site for the conventional discussion of scientific controversy and at attempts to broaden the notions of responsibility being discussed there. We then look beyond, to the sociotechnical complexity of innovation in the world, to see new challenges to our understanding of responsible innovation and new challenges to STS. Finally, we observe some of the ways in which the language of responsible innovation has enabled STS researchers to play new roles in real-time discussions of emerging science and technology. Work taking place around responsible innovation is not just interested in asking what counts as responsible or irresponsible; it is also pointing to how things might be improved, building on a normative interest in democratizing and pluralizing debates and decisions about science, technology, innovation, and sociotechnical futures. Our conclusion is that these new, co-constructive relationships are important even if they are often neither straightforward nor comfortable.

The (Re-)emergence of Responsible Innovation

During the 1970s debates about social responsibility grew among U.S. scientists. The infamous Tuskegee syphilis study, which had over decades examined the progression of untreated syphilis in rural African American men, was shut down and lessons from this scandal were inscribed into the Belmont report and the Common Rule for the ethical treatment of human research subjects (Jones 1981). The U.S. Technology Assessment Act, drawing attention to the need to consider probable impacts of technological application, was finalized in 1972, with the Office of Technology Assessment (OTA) of the U.S. Congress emerging a couple of years afterwards. In 1975 molecular biologists descended on the Asilomar conference center in California to discuss the risks of recombinant DNA research, in a major attempt at self-described "self-regulation." Their hope was that, as they sought to control the potential of genetic modification, they could improve upon physicists' response to the creation of nuclear weapons (Kaiser and Moreno 2012).

These precedents, along with research on social and ethical aspects of genomics described below, were in mind during early efforts to institutionalize new forms of innovation governance. In this vein one of us proposed in 2002 that universities, as hubs of knowledge-based innovation, should establish "centers for responsible innovation." These would serve as counterweights to growing commercial norms in the university context and extend the concept of responsibility beyond contributing to the economy and not abusing research subjects (Guston 2004). While the term *responsible innovation* had been in modest circulation, it had not yet been core to STS theorizing and practice. The phrase and the ideas behind it, however, have well-established ancestry. In the first edition of the STS handbook, edited by Ina Spiegel-Rösing and Derek de Solla Price (1977), social responsibility was recognized as a central theme of what was then the emerging field of science, technology, and society. Toward the end of the twentieth century, ethics scholarship had begun an empirical turn, in which the realities and constraints of scientific practice had provided new research questions (Borry, Schotsmans, and Dierickx 2005; Kroes and Meijers 2000). Technology studies had explored questions such as how artifacts can produce "a specific geography of responsibilities" in allowing or constraining particular actions (Akrich 1992, 207). But links with governance and institutions had not yet been clearly drawn.

Since the mid-2000s, the terms *responsible innovation* (RI) and *responsible research and innovation* (RRI)—which, for the purposes of this chapter, we take to be synonymous—have entered intellectual and institutional discourse rapidly, if sometimes surreptitiously. STS researchers have variously been involved in encouraging the term's uptake, studying its usage, making sense of it in policy and pedagogy, critiquing its myriad simplistic or instrumental invocations, and catalyzing the creation of communities and projects. The *Journal of Responsible Innovation* was launched in 2014 as a way to map and advance these activities.

This process of scholarly and rhetorical blossoming has meant, inevitably, that actors have attached multiple meanings to *responsible innovation*. Before offering our own definition, it is important to note some of the motivations behind its recent usage in policy. In Europe, RRI has become a theme that cuts across Horizon 2020, the European Commission's (EC) latest and largest funding program. De Saille (2015) catalogs the spread of RRI at the European level, explaining it in terms of European policy makers seeking to get what they want from investments in science and innovation.[1]

The concern de Saille describes comes in part from European governments' desire to avoid such surprising and costly crises of public confidence as occurred with Europeans' rejection of genetically modified crops (see Wynne 2001 for an STS analysis of this controversy). More recently, the Dutch experience with the aborted rollout of smart energy meters, undermined by concerns about privacy that were realized too late, has given policy makers further cause for concern (European Commission 2013). RRI is also connected with the growing prominence of "grand challenges," which can be seen as policy recognition of three key pressures: first, the need to demonstrate and increase the external value of science; second, anxiety among scientists and engineers that the utility of their work is perceived in purely economic terms; and third, recognition that past efforts to increase impact through policies that emphasize the supply side of innovation have been a failure (ERA Expert Group 2008; Lund Declaration, 2009).[2]

René von Schomberg's (2011) characterization of RRI in Europe ties it explicitly to the values that should drive EU policy: scientific and technological advancement; the promotion of social justice, equality, solidarity, and fundamental rights; a competitive social market economy; sustainable development and quality of life. Von Schomberg (2013, 39) defines RRI as

[a] transparent, interactive process by which societal actors and innovators become mutually responsive to each other with a view to the (ethical) acceptability, sustainability and societal desirability of the innovation process and its marketable products in order to allow a proper embedding of scientific and technological advances in our society.

In the United States, responsible innovation received an early but indirect push with the inclusion of "responsible development" as one of the four strategic goals of the U.S. National Nanotechnology Initiative (NNI), which was launched in 2001. The NNI rhetoric on responsible development was informed by the successes and failures of the Human Genome Initiative's ethical, legal, and social implications (ELSI) program (Cook-Deegan 1994; Hilgartner, Prainsack, and Hurlbut, chapter 28 this volume), much of which maintained a studied distance from the research front. The NNI's post-ELSI orientation brought STS closer to science and engineering. This proximity focused responsible development in two ways: first, on the environmental implications and applications of nanotechnology and, second, on integrating social science, as a way of divining and making sense of societal concerns, into the nano-scale science and engineering research itself (Fisher, Mahajan, and Mitcham 2006).

Ideas about responsible innovation flourished in this domain, in part from a desire to broaden the deterministic language of responsible "development." A vision for the "anticipatory governance" (Barben et al. 2008) of nanotechnology and other emerging technologies was articulated, defined as the "capacity … to manage emerging knowledge-based technologies while such management is still possible" (Guston 2014b, 218). This work built upon a normative framework derived from Winner (1977, 323), who reasoned that because "technology in a true sense *is* legislation," it required elements of public participation akin to legislation to make it legitimate. Anticipatory governance proposes the development of *foresight*, *engagement*, and *integration* as a response to various pathologies of innovation that are conventionally realized only with hindsight.

Consistent with the vision of extending these capacities more broadly through society, Stilgoe, Owen, and Macnaghten (2013) describe a framework for responsible innovation aiming to improve the *anticipation*, *inclusion*, *reflexivity*, and *responsiveness* (AIRR) of research and innovation systems. As with the demand for integration, the highlighting of responsiveness follows a recognized need to better connect processes of "opening up" (Stirling 2008) with the practice, institutions, and cultures of the broader innovation system. One specific manifestation of this effort is with the United Kingdom's Engineering and Physical Sciences Research Council, which adopted a version of this framework, translating AIRR into AREA: anticipation, reflection, engagement, and action (Owen 2014). National research funders in Norway and the Netherlands have also created programs of research and engagement under the heading of responsible innovation.

The idea of responsibility that is enacted through such efforts is not legalistic and retrospective but prospective, recognizing profound uncertainties and encouraging and supporting researchers to join intellectual forces to explore them. In terms broad enough to capture the range of possible issues and targets of responsible innovation, a more concise definition of responsible innovation is "care for the future through responsive stewardship of science and innovation in the present" (Owen et al. 2013, 36). This definition not only captures the idea of anticipation that flows through most of these efforts, but it also implies the scrutiny of the societal goals to which emerging innovations are commonly offered as a response. The point is not to ask what emerging technologies can do to help us, but rather to ask the prior framing question of what we can do to help, or not, emerging technologies.

One important research agenda for STS is in continuing the study of the emergence of responsible innovation as a hybrid construction at the interface of science and policy. But this kind of approach has not been and, we would argue, should not be the only contribution made by STS researchers. As seen in the U.S. experience related above and as we describe later in this chapter, STS people are more than just researchers in these circumstances. In the STS tradition of what might be called "constructive constructivism" (Rip 1994), they are often aiming to ameliorate as well as analyze, to help reconstruct as well as deconstruct. The roots of this engagement go back to the STS diagnosis of the pathologies to which responsible innovation might offer some remediation. The depth of the STS agenda motivates much of the remainder of our discussion.

Technology as Means and Ends

The conventional approach to questions of technology governance, as developed over the second half of the twentieth century, has been one of technology assessment, risk assessment, and risk management, coupled with ethical attention to questions of implications and research practices (Macnaghten and Chilvers 2013; Wynne 2001). The growth of institutions of risk and ethics governance follows a narrative of responsibility with which the scientific community has grown comfortable. There have been controversies and crises in the governance of science, including prominent examples such as the Manhattan Project, the Nuremberg trials, the Tuskegee study, and accidents involving nuclear power, pharmaceuticals and other technologies, many of which have produced regulatory responses. Scientific cultures can justifiably claim some ability to self-correct in matters of ethics, even if lessons are often learned slowly and travel across cultures and scientific domains glacially (Briggle and Mitcham 2012).

The 1975 Asilomar conference, at which early recombinant DNA researchers met to draw up guidelines for an attempt at self-governance, is more than a case in point. Asilomar continues to be held by some scientists, such as those researching mutant influenza viruses, as a paradigmatic demonstration of the capacity of the scientific community to take care of an emerging issue (Kaiser and Moreno 2012), and those involved in emerging technologies from synthetic biology to geoengineering are constituting new, self-consciously Asilomar-style meetings (Hurlbut 2015). Yet STS scholars have explained how Asilomar failed to capture the full range of public concerns as they were expressed at the time or would emerge in the future (Nelkin 2001; Wright 2001). Jasanoff (2013) discusses the Asilomar meeting in terms of the "containment" not only of genetically engineered organisms but also of the controversies that they might trigger and thus the scope of public debate. Participants cast the debate as one of health and safety, with questions of the political economy of recombinant organisms—questions that became central as controversy emerged later—initially overlooked. For Krimsky (2005), the reductionism of the Asilomar debate failed to prepare either scientists or policy makers for impending controversies over industrial biotechnology. And Guston (2006) has shown that the very idea of "self"-governance is irreparably problematic, as scientists with diverse opinions invariably need something other than science to aid in decision making.

STS has thus helped to articulate the need to escape simple distinctions between self-governance and top-down regulation and between "good" and "bad" uses of a technology, even if it has not fully penetrated the veil around the scientific community. The field has built up historically and sociologically rich accounts of the plurality of the motivations for research and the sources of innovation. Just as the linear model of innovation is flawed for its presumption that science is the wellspring of technology, so we know also that necessity is not necessarily the mother of invention (see Williams and Edge 1996 for a survey). STS instead reads such stories of innovation as performances emanating from traditional assumptions about scientific authority, autonomy, and responsibility, and STS scholars have developed new frameworks of governance that get past a fixation on risk (Wynne 2002) to concentrate instead on "the governance of innovation itself" (Felt et al. 2007, 11).

Responsible innovation directs attention not only to the well-rehearsed risks, uncertainties, and unintended consequences of technology, but more importantly to an innovation system and its role in shaping futures. Furthering Winner's technology-as-legislation, STS has discussed the "social constitution" of emerging technologies, the social and political arrangements that particular technologies demand (Grove-White, Macnaghten, and Wynne 2000; Kearnes et al. 2006; Szerszynski et al. 2013). We can see, without resorting to technological determinism (i.e., the idea that social change is driven by technical change), how technologies variously open up or narrow choices. As Latour (2008, 5) describes when discussing tools of genetic modification,

[s]cience, technology, markets, etc. have amplified, for at least the last two centuries, not only the scale at which humans and nonhumans are connecting with one another in larger and larger assemblies, but also the intimacy with which such connections are made. Whereas at the time of ploughs we could only scratch the surface of the soil, we can now begin to fold ourselves into the molecular machinery of soil bacteria.

The growing disruptive power of technology to intervene not just in intimate ways with living systems but also at global scales (Beck 1992) magnifies the importance of scrutinizing emerging technologies. Once we get past technological determinism, we can see that there are choices to be made within innovation that do not just relate to its acceptance or rejection but rather to the multiplicity of mechanisms and arenas through which and within which innovation is constituted and governed. It is this recognition of multiplicity that keeps the emphasis on governance from being just another manifestation of neoliberal framing of research policy (Guston 2014b).

Some in neighboring disciplines have developed these insights in what could be seen as updates of Winner for the information age—focusing on black-boxed, inscrutable algorithms (Pasquale 2015) with a maxim that "code is law" to capture the growing power of software to inscribe new social rules (Lessig 1999). This descriptive approach may be engaged in something of a dialectic with the more normative question of how to democratize the politics of technology: In order to expand (democratize) and diversify (pluralize) participation in the aspects of innovation that legislate for the future, we may need to redescribe what we think the challenges may be. How we frame those challenges of course helps determine the kinds of normative and social responses elicited. The political philosophy behind many such approaches is derived from the realist democratic theorist E. E. Schattschneider (1960), who argued in the context of the American civil rights movement that the first step toward democratization was the creation of a conflict that would then attract more attention and offer the opportunity to reframe the issue at the heart of the conflict in light of the composition of the new set of participants.

The framing of responsible innovation includes in this dialectic how initial conditions are framed as problems such that knowledge-based innovation is presumed to be a solution. At first blush, the twenty-first century seems to have lost enthusiasm for this presumption, historically captured by the concept of the "technological fix."

Alvin Weinberg (1966) originally discussed (and endorsed) the term during the Cold War, reflecting an American reaction to what was perceived as Soviet "social engineering." Weinberg identified social problems as intrinsically complex (we would perhaps now describe them as "wicked" [Rayner 2012]) and saw technology as a way to cut through such Gordian knots rather than having to understand and disentangle their social threads. With the end of the Cold War, technological enthusiasm has, according to Morozov (2013), morphed rather than dimmed. "Solutionism," the idea that technology, especially information and communication technology, can cure our ills, has relocated to Silicon Valley and has begun to reframe a huge range of problems such that they become targets for ICT "solutions."

STS has not limited itself to revealing alternatives to and contingencies in predetermined technological trajectories. STS researchers have also sought to constructively engage with governance practices. Sarewitz (2011, 95) argues that, when it comes to technology assessment (TA), "current approaches are almost entirely reactive, ponderous and bureaucratic." STS has contributed to the development of new approaches to TA, particularly in the (well-funded) area of nanotechnology. Constructive Technology Assessment (Rip, Schot, and Misa 1995), Real Time Technology Assessment (Guston and Sarewitz 2002), Upstream Engagement (Wilsdon and Willis 2004; Wynne 2002), and Midstream Modulation (Fisher, Mahajan, and Mitcham 2006) all seek to open up consideration of alternative technological possibilities while they are still being imagined (Stirling 2008).

Among the ways such work contributes to the agenda of RI is through reconnecting STS to a perennial debate about the direction of science and innovation. The unevenness of technological progress was perhaps most succinctly described by Richard Nelson (1977) in his "Moon and the Ghetto" lectures in the 1970s when he asked why innovators and policy makers seemed willing and able to solve some problems—such as getting man on the moon and eradicating communicable disease—and not others—such as child illiteracy and drug addiction (for a more recent commentary, see Nelson 2011). Nelson's message spoke to a social movement, and associated literature, on appropriate technology and development (Breyman et al., chapter 10 this volume; Kaplinsky 2011; Leach and Scoones 2006; Schumacher 1973), as well as to a newer literature that has followed up on the promise of emerging technologies and found it wanting (Woodson 2012). A crucial contribution of STS is to delineate instances when problems might or might not be amenable to technological treatment. For example, Sarewitz and Nelson (2008) offer three criteria to help set priorities among problems to ascertain which might be more susceptible to technological problem solving. Comparing the relative failure of technologies to create positive outcomes in education to the relative success of technologies to do so for infectious disease, Sarewitz and Nelson (2008, 871) argue that for investments in R&D to lead to rapid social progress, (1) "the technology must largely embody the cause-and-effect relationship connecting problem and solution," (2) "the effects of the technological fix must be assessable using relatively unambiguous or uncontroversial criteria," and (3) R&D should be focused on a preexisting, "standardized technical core."

To say that technological fixes are good only under these conditions, however, may be another way of saying that they are not very good in general. We can add such an analysis to the typologies offered by Winner (1977) and Sclove (1995) for desirable technologies. Winner sees flexibility, intelligibility by nonexperts, and the avoidance of dependency as qualities that we should look for and nurture in "good" technology. Sclove's list of criteria for "democratic technologies" is longer, including, for example, ecologically sustainable technologies and those that promote global pluralism in technology choice while excluding in particular those that create transboundary ecological impacts. Similarly, de Laet and Mol (2000) describe the attractiveness of the Zimbabwe Bush Pump, an archetypical appropriate technology, in terms of its "fluidity," its adaptability across a variety of contexts. While we might be hard-pressed to disagree with these criteria pertaining to technologies themselves, STS research on responsible innovation also needs to concentrate on questions about what strategies and processes might encode a reliable path toward such technologies (e.g., Ganzevles, van Est, and Nentwich 2014; Kiran, Oudshoorn, and Verbeek 2015; Mampuys and Brom 2015) and the extent to which the articulation of principles is sufficient or necessary to establish an appropriate approach (e.g., Holbrook and Briggle 2014; Schroeder and Ladikas 2015; Ziegler 2015).

Responsible Innovation and Public Engagement

The imperative to engage in technology assessment at that moment when the technological and the social seem more formative and flexible has led to a growing STS interest in emerging technologies. Worldwide, significant funding for R&D on emerging technologies, as well as some mandates for research integrated with social sciences (Rodríguez, Fisher, and Schuurbiers 2013), has provided the opportunity. As Joly (2015) points out, the previous edition of this handbook contained an entire section on "emerging technosciences," and STS scholars—ourselves included—have made use of many of the types of technology assessment mentioned above in the study of nanotechnology, synthetic biology, geoengineering, robotics and artificial intelligence, personalized medicine, and other areas characterized by profound uncertainty, high stakes, and a robust politics of novelty (Guston 2014b).

The STS concern with emerging technologies as a site for "society in the making" (Callon 1987) is elucidated by Jasanoff (2004, 278–79) in terms of the co-production of natural and social orders:

Important normative choices get made during the phase of emergence: in the resolution of conflicts; the classification of scientific and social objects; the standardization of technological practices; and the uptake of knowledge in different cultural contexts. Once the resulting settlements are normalized (social order) or naturalized (natural order), it becomes difficult to rediscover the contested assumptions that were freely in play before stability was effected.

The recognition of the lack of determinacy and thus the presence of politics—as well as the structuring effects of innovations such that people live their lives in, with, and through new technologies—bring normative consequences to Winner's diagnosis. If Winner's cause is motivated by the identification of technology as legislation, its placards might perhaps read "No innovation without representation." If science is to be better aligned with public values, how might those values be articulated? This raises a familiar question for STS: How should publics be engaged and represented in science and innovation (Brown 2006; Chilvers and Kearnes 2016)?

Moves toward RRI can be understood as a further development of two-way public engagement with science. The last two decades have seen a blossoming of dialogic activities on issues involving science, at least in the United States and Northern Europe, but the motivations for this remain confused and contested (Stilgoe, Lock, and Wilsdon 2014; Lezaun, Marres, and Tironi, chapter 7 this volume). While one can divide rationales for public participation into three categories: *normative*, that participation is a good thing in itself; *instrumental*, that participation can build trust and smooth the implementation of decisions; and *substantive*, that participation produces better decisions (Fiorino 1990), many recent STS perspectives judge the institutional motives behind moves toward greater participation as primarily instrumental. Rayner (2004) and Wynne (2006) both make a compelling case that efforts at public engagement are stubbornly motivated by a deficit model that has mutated as rhetoric has shifted from public *understanding* to *engagement* to *upstream engagement*. Where the aim was once to remedy public deficits in scientific knowledge, engagement now often seems directed at a perceived deficit of public trust (see the chapters in Chilvers and Kearnes 2016 for further discussion).

The public is still routinely seen as a problem in science governance issues (e.g., Rip 2006b), and the practice of public engagement can exacerbate this imagined pathology, distracting from pathologies of innovation itself. Inasmuch as STS researchers have been involved in calling for, consulting on, and conducting deliberative processes, we are also implicated. As Latour (2007) and De Vries (2007) have discussed, social scientists, by taking their own calls to "open up" too literally, have been too quick to advocate public dialogue rather than focusing on faulty governance processes that could otherwise have been their target.

Some social scientists and practitioners of public engagement have prioritized analysis of the processes of public engagement rather than question its purposes (Marris and Rose 2010). A focus on the public and the means of their engagement may at times force consideration of new perspectives and questions, but it may equally impede the necessary *institutional reflexivity* (Wynne 1993) required for good governance. If the deficit model critique applies equally to the instrumental and substantive rationales for public engagement, then engagement can only come to be seen as self-evidently worthwhile, rather than as a means to an end. Thus, rather than opening up the possibility of revealing new conflicts, engagement practice is often directed at closure and consensus (Horst and Irwin 2010; Stirling 2008). This endpoint has led some STS researchers to describe participatory methods as themselves "technologies" (Lezaun and Soneryd 2007) or "experiments" (Laurent 2011; Stilgoe 2012), thereby facilitating STS deconstructions and reconstructions.

In a positive development for the co-constructive nature of responsible innovation, public engagement is diversifying beyond conventional deliberative practice (Davies et al. 2012; Guston 2014a; Selin 2014b). Rather than focusing on interlocution conditioned by reading and writing, these new engagements involve thinking and conversing around activities including taking photographs, crafting objects, and other forms of making. Not only do such forms of material deliberation provide broader opportunities for people to learn through different modalities, but they also provide concrete opportunities for people to engage in explicit co-creation, hypothetically preparing them for co-creative roles in the course of their everyday governance of innovation as well.

The Politics of Anticipation

Like the phrase "public engagement with science," "responsible innovation" implies a problem. Where previously the problem was seen as either unengaged science or unengaged publics, depending on one's perspective, tackling irresponsible innovation or what Beck (2000; also Adam and Groves, 2011) calls "organized irresponsibility" might be seen as a bigger issue. But while moves toward public engagement were actively challenged from some quarters (see, for example, Durodié 2003; Taverne 2005), the term *responsible innovation* is unlikely to incite direct opposition. Introducing the term to new audiences has prompted more than a few people to ask us, "Responsible innovation, who could be against *that*?" (Guston 2015). There is evidence that one can "steer with big words" such as responsible innovation (Bos et al. 2014), but there is a risk that responsible innovation could become a new label for business as usual, instrumentally deployed to smooth the path of innovation.

The pathologies of innovation to which STS researchers might easily point would include the unpredictability and uncontrollability of large sociotechnical systems (Krohn and Weingart 1987; Perrow 1984, 2011), institutionalized ignorance of early warnings (Harremoës et al. 2001), the altered nature of human action (Jonas 1985), the tendency toward hype (Borup et al. 2006; Simakova and Coenen 2013), and the various forms of lock-in, described below, that make sociotechnical changes of direction difficult. Some of the most important STS insights into public engagement come not from studies of explicit engagement but from studies of how "the public" is constructed in the practice of science and technology (Hill and Michael 1998; Maranta et al. 2003; Woolgar 1990; Wynne 1993). In a similar way, in engaging with notions of responsible innovation, we should look not just to the novel activities taking place under this name but also to the ways in which scientists and innovators imagine their own changing responsibilities.

The emergence of science as an organized activity from the seventeenth century onwards has been accompanied by "metascientific" (Ziman 2001) questions of social responsibility, mostly being posed within the scientific community itself (Glerup and Horst 2014). STS, including its prehistory from Merton onwards, has always maintained an interest in questions of what counts as "good" science and technology. The links with ethics, in its various forms (Hilgartner, Prainsack, and Hurlbut, chapter 28 this volume) are now well established.

Over the last few decades, a view in which good science can be cleanly separated from deviant, bad science, often labeled "fabrication, falsification and plagiarism" (FFP), has been questioned by those who observe scientific practice (Gieryn and Figert 1986; Steneck 2006). The more general term *questionable research practices* has provided a more realistic category of behaviors that ought to be discouraged and the more positive label *responsible conduct of research* has come to be adopted, including not just research *integrity*, evaluated according to professional standards, but also research *ethics*, in which society more broadly might have a say (Briggle and Mitcham 2012; Steneck 2006).

This discussion, however, still tends to maintain a rather tight definition of the responsibilities of scientists and engineers, leading to the distinction between their responsibilities to their professional colleagues, rendered as microethics, and their macroethical responsibilities to the broader society (Herkert 2005). Philosopher Heather Douglas (2003) has described how the responsibilities of scientists toward their professional colleagues (or, more grandly, toward the pursuit of truth) are often seen as trumping their general responsibilities toward society. This hierarchy is a dominant "division of moral labor" (Rip and Shelley-Egan 2010), in which scientific cultures have come to see social, ethical, and political issues as someone else's business, despite a long history of discussions about social responsibility led by scientists themselves. During the Cold War, for example, growing concerns about the use of science for military ends led to the creation of initiatives such as the Pugwash conferences on Science and World Affairs (which would go on to win the 1995 Nobel Peace Prize) (Sismondo 2011; Rip 2014). In the 1970s, in addition to the Asilomar meeting, others began debating the more general possibility and desirability of setting "limits of scientific inquiry" (Holton and Morison 1979).

Since Cold War discussions of scientific responsibility, however, the political economy of science has changed markedly. The emergence of what some have called "neoliberal science" (Lave, Mirowski, and Randalls 2010), with closer interweaving of public and private agendas, has made it even harder for scientists to defend (if it was ever defensible) an independent "republic of science" (Polanyi 1962) with an unfettered "right to research" (Brown and Guston 2009). A continued increase in the scale and scope of technoscience, coupled with the potential, captured by Latour in the quote above, to intervene in increasingly profound ways, changes the stakes of the debate on responsibility. In this regard, Latour's (1999) own focus on laboratories as sites of negotiation for such matters (even if we understand laboratories as extending into the outside world) starts to feel limiting. Indeed, if responsible innovation is to have purchase and STS is to contribute to challenging the dominant demarcation of responsibilities, we must remember the insight from literature in the social shaping of technology (Williams and Edge 1996), that innovation happens in use as well as in research and development.

For its defenders, the bulwarks of the autonomy of science protect against overreaching questions of scientific responsibility. But they also protect against scrutiny of and interference in the direction of scientific and technological development, as in Polanyi's (1962) argument that because the progress of science is unpredictable (and its societal consequences even more so), then it is ungovernable. The concept of anticipation as used in anticipatory governance and adopted by responsible innovation is meant to alter this logic by articulating a future-oriented disposition that can provide appropriate guidance for action in the present. Such anticipations would not be needed if scientific and technological development were truly predictable, as governance would then be transparent and certain, but studies in the social shaping of technology have systematically undermined the idea that innovation is inevitable. This process is tractable but challenging because even if technology is not autonomous it can, especially at scale, offer a convincing impression of autonomy (Hughes 1993; Winner 1977). Sociotechnical systems can build up "momentum" (Hughes 1993) as their trajectories are constrained by what has been variously described as "path dependency" (David 2001), "escalating commitment" (Staw 1976), "entrenchment" (Collingridge 1980), "entrapment" (Walker 2000), "lock-in" (Arthur 1989), and "obduracy" (Selin and Sadowski 2016). For Collingridge (1980, 19), the emergent intransigence of technological systems that these terms all in their way describe poses a dilemma of control:

[A]ttempting to control a technology is difficult, and not rarely impossible, because during its early stages, when it can be controlled, not enough can be known about its harmful social consequences to warrant controlling its development; but by the time these consequences are apparent, control has become costly and slow.

As described above, emerging technologies present insurmountable challenges for risk-based governance models. Some have argued, however, that focusing on the uncertain futures of these emerging technologies and their emergent properties is both impossible and undesirable. Nordmann (2014), for example, argues, first, that the world of the future is not the world of the present and the latter cannot see into the former; second, that we overlook history and the imagination of alternative worlds through paternalistic future projections of the present; and, third, that there are trivial and nontrivial versions of the future (e.g., easy extrapolations and discontinuities, respectively), and the anticipatory element of responsible innovation seems less able to deal with the latter.

Some, however, counter that anticipations provide appropriate interpretive orientation for future-oriented decision making (van der Burg 2014; Wilsdon 2014). More pointedly, one might recast the problem that Nordmann categorizes as "paternalism" rather as "care," as many environmental ethicists and some versions of RI do (Groves 2015). In a rhetorical anticipation, Selin (2014a, 106) argues, "The court of the future seems more likely to condemn us for negligence than for paternalism." Competing with the hubris of technoscientific visionaries, who claim resources and allegiance based on precisely the kinds of problematic claims that Nordmann attacks, would seem to demand the democratic articulation of alternative futures in ways that have been developed by STS over the last few decades.

Responsible Innovation, Emergence, and Technological Systems

Technologies that come to pose profound questions of governance may not initially seem problematic. In some cases, as with genetically modified crops or personal genetic tests, one can easily imagine how aspects of technical or contextual novelty might create new ethical dilemmas and political challenges. However, some problematic ramifications may only emerge at larger, systemic scales (Hellström 2003). Conventional biofuel crops, which may seem initially mundane or even environmentally benign, may be grown at a scale at which they put land-use pressure on food crops. Similarly, privacy concerns with particular social media may only emerge once a particular platform has reached a saturation point. As science and policy excitement grows around "big data," the rhetoric of which has been radically depoliticized so that value questions are hidden beneath those concerning the practical uses of data (Crawford, Gray, and Miltner, 2014), those interested in responsible innovation will need to pay particular attention to these dynamics of emergence.

In recognition of the impossibility of prediction and control and the limits of authoritative decision making, some have argued for "tentative governance."[3] Others have described the need for "meta-regulation" (Dorbeck-Jung and Shelley-Egan 2013). These authors argue that the task is one of "responsibilisation," constructing the conditions in which responsible actors are able to respond to surprises in the light of uncertain information. Before we can talk clearly about the allocation of responsibilities, we need to outline "second-order" (Illies and Meijers 2009) or "meta" responsibilities (Stahl 2013). These are the responsibilities upon actors to enable the possibility of making responsible choices in the future. So Collingridge's (1980) proposal for "corrigibility" in technology systems, or Winner's (1977, 326) principle that "technologies be built with a high degree of flexibility and mutability" could be taken as metagovernance recommendations.

In order for such approaches to have significant effect, however, there needs to be close engagement with the institutional, as well as the cultural and individual, practices of science and innovation. Rip (2006a) has discussed the importance of understanding the "de facto governance" of science and innovation, including what Pellizzoni (2004) calls the "logic of unresponsiveness"—the often-hidden interests, assumptions, and dynamics that steer innovation toward particular ends. A growing body of scholarship is concentrating on the political economy of universities and science (e.g., Berman 2011; Lave, Mirowski, and Randalls 2010; Tyfield 2012). Doing responsible innovation therefore demands a degree of engagement with dynamics beyond the practices of scientists and innovators.

Responsible Innovation in Action

So what does responsible innovation look like? How do we know it when we see it? And how might STS play a role in nurturing and evaluating its development? Recent years have seen the active, if still modest, involvement of STS researchers in the development of emerging technologies. As described above, nanotechnology has been a focus for the development of multiple models of engagement and technology assessment, all of which can be seen as advancing a broad agenda of responsible innovation, although their specific tools and aims would vary.

After nanotechnology, the emerging practice of synthetic biology has become the second major site for responsible innovation as a form of social innovation (Rip 2014). STS researchers have been involved in laboratories as well as policy rooms, seeking to shape trajectories in more responsible ways, and their perspectives have been sought by funding agencies to help construct sponsored research agendas (e.g., Brian 2015). At the same time, they have taken the STS principle of reflexivity to heart, putting their own roles under investigation (Calvert 2010; Calvert and Martin 2009; Rabinow and Bennett 2012; Sismondo 2007; Stemerding 2015). As synthetic biologists have adopted the language of responsible innovation, STS has been able to trace and critique the uptake of terms and ideas (Marris 2015). And scientometrics research has begun to map the integration of social sciences into various emerging technologies (Shapira, Youtie, and Li 2015). The emerging science of geoengineering has also seen STS researchers enter discussions with scientists and funding institutions that point to a more literal and more constructive sense of the word *collaboration* (Stilgoe 2015).

STS contributions have been, first, to argue that concerns with microethics apply also to macroethical questions around the power of innovation in shaping the future. Second, they have recognized a broader area of application, the innovation system, rather than science or research more narrowly conceived. The challenge is not just a conceptual one; it is also a methodological one. If innovation is a form of collective experimentation (Latour 2011), then responsible innovation has become a form of collective experimentation in which STS researchers have inveigled themselves.

Some of those most intimately involved have also criticized the ways in which STS has become embedded in emerging science and technology practices and policies. Wynne's (2007) argument is that the involvement of social scientists in emerging innovation risks occluding the politics that their role should be helping to reveal. However, Wynne has suggested some important roles that STS can and does play, including revealing the normative models of publics that are being enacted in science and policy, understanding and challenging "expert" presumptions about relevant public issues, describing the diversity of other cultures in which understandings might differ and be able to contribute, and, finally, putting this in the context of historical and philosophical perspectives on science as public knowledge (Wynne 2016). While these roles work to undermine the unitary framing of a relatively unreflexive, expert-driven, and even technocratic pursuit, they do not necessarily serve the function of diversifying the kinds of people who get to participate in future-making.

One way in which STS has sought to engage with institutionalized conversations about responsibility has been through bioethics. As discussed above, the current machinery of bioethics is in part a response to past failures of responsibility. As it has come to be institutionalized, however, it has attracted critique from STS because of its concentration on particular ethical concerns, such as the protection of human subjects' autonomy through informed consent, to the neglect of what has been called "public ethics" (Nuffield Council on Bioethics 2012). Macroethical concerns about the direction of innovation and the distribution of its benefits and risks have been largely overlooked in ethical governance. And the relationship of ethics to science tends to be oppositional. Rabinow and Bennett (2012, 35), in their reflections on engaging as collaborators in a large synthetic biology project, express their concerns that

[bioethics, as currently practiced in official settings, tends to undervalue the extent to which ethics and science can play a mutually formative role. More significant, it undervalues the extent to which science and ethics can collaboratively contribute to and constitute a good life in a democratic society.

Rabinow and Bennett here include the important role of democratic society, but their emphasis on the mutual constitution of the good life between scientists and even new bioethicists can, without grounding with public engagement, remain in the rarefied space of technocracy (Thompson 2013).

The experiences of some STS researchers who have been involved in experimental collaborations with scientists (some under the RI banner but many not) has led them to call for a "post-ELSI" social science (Balmer et al. 2012; Rabinow and Bennett 2012). The aim is to expand upon critique, in the familiar mode of articulating alternatives and revealing complexities, to develop new styles of engagement. Clearly, this move raises questions about the methodological and analytical challenges of being entangled in emerging science. In this sense, we are seeing a reworking of an older debate about the politics of STS (Radder 1998). Such a debate becomes unavoidable if we are to take seriously responsible innovation as a policy agenda and a set of ideas still in formation. There needs to be an urgent discussion of the opportunities and uncertainties of an approach that is explicitly ameliorative, one in which, rather than just studying co-production in action (Jasanoff 2004), STS researchers are themselves involved in co-producing knowledge and social order. An STS or a responsible innovation that serves only to expose technocratic framings but not to construct new, more expansive, diverse, and participatory ones is, in our view, not going far enough.

For STS, then, responsible innovation is a challenge and an opportunity, something old and something new, to be critically examined and to be reflexively practiced, to be concerned about and to be hopeful about. As a policy construction, it is, as we have described, still in the making, which means that it is too early to adjudicate on the value of the term or the activities that are carried out in its name. In this chapter, we have described the conceptual foundations and emergent structures that justify our own interest and involvement. Our conclusion is that it will remain an important agenda for STS research and engagement for years to come.

Acknowledgment

David H. Guston's contribution to this chapter is in part based upon work supported by the U.S. National Science Foundation cooperative agreement #0937591 and grant #1257246. Any findings, conclusions, or opinions are those of the author and do not necessarily represent NSF.

Notes

1. The European Commission has supported numerous RRI research and coordination projects. See, for example, Governance for Responsible Innovation (GREAT), the Global Model and Observatory for International Responsible Research and Innovation Coordination (Project Responsibility), Governance Framework for Responsible Research and Innovation (RES-AGORA), PROmoting Global REsponsible research and Social and Scientific innovation (ProGReSS), and the RRI-TOOLS initiative to develop a "tool kit" for responsible research and innovation (for which Stilgoe is a project partner).

2. In Europe the 2009 Lund Declaration stated that "Europe must focus on the grand challenges of our time" (see https://www.vr.se/download/18.249c421a1504ad6d28144942/1444391884365/Lund_Declaration_2009.pdf). In the United States in 2008, the National Academy of Engineering also articulated a set of fourteen "grand challenges for engineering in the twenty-first century."

3. See, for example, the 2010 conference on Tentative Governance in Emerging Science and Technology at the University of Twente, http://www.utwente.nl/igs/conferences/2010-tentative-governance/.

References

Adam, Barbara, and Chris Groves. 2011. "Futures Tended: Care and Future-Oriented Responsibility." *Bulletin of Science, Technology & Society* 31 (1): 17–27.

Akrich, Madeleine. 1992. "The De-scription of Technical Objects." In *Shaping Technology/Building Society: Studies in Sociotechnical Change*, edited by Wiebe E. Bijker and John Law, 205–24. Cambridge, MA: MIT Press.

Arthur, W. Brian. 1989. "Competing Technologies, Increasing Returns, and Lock-in by Historical Events." *The Economic Journal* 99 (394): 116–31.

Balmer, Andy, Kate Bulpin, Jane Calvert, Matthew Kearnes, Adrian Mackenzie, Claire Marris, Paul Martin, Susan Molyneux-Hodgson, and Pablo Schyfter. 2012. "Towards a Manifesto for Experimental Collaborations between Social and Natural Scientists," accessed at https://experimentalcollaborations.wordpress.com/2012/07/03/towards-a-manifesto-for-experimental-collaborations-between-social-and-natural-scientists/.

Barben, Daniel, Erik Fisher, Cynthia Selin, and David H. Guston. 2008. "Anticipatory Governance of Nanotechnology: Foresight, Engagement, and Integration." In *The Handbook of Science and Technology Studies*. 3rd ed., edited by Edward J. Hackett, Olga Amsterdamska, Michael Lynch, and Judy Wajcman, 979–1000. Cambridge, MA: MIT Press.

Beck, Ulrich. 1992. *Risk Society: Towards a New Modernity*, vol. 17. New York: Sage.

___. 2000. "Risk Society Revisited: Theory, Politics and Research Programmes." In *The Risk Society and Beyond: Critical Issues for Social Theory*, edited by Barbara Adam, Ulrich Beck, and Joost Van Loon, 211–29. Thousand Oaks, CA: Sage.

Berman, Elizabeth Popp. 2011. *Creating the Market University: How Academic Science Became an Economic Engine*. Princeton, NJ: Princeton University Press.

Borry, Pascal, Paul Schotsmans, and Kris Dierickx. 2005. "The Birth of the Empirical Turn in Bioethics." *Bioethics* 19 (1): 49–71.

Borup, Mads, Nik Brown, Kornelia Konrad, and Harro van Lente. 2006. "The Sociology of Expectations in Science and Technology." *Technology Analysis & Strategic Management* 18 (3–4): 285–98.

Bos, Colette, Bart Walhout, Alexander Peine, and Harro van Lente. 2014. "Steering with Big Words: Articulating Ideographs in Research Programs." *Journal of Responsible Innovation* 1 (2): 151–70.

Brian, Jenny Dyck. 2015 "Special Perspectives Section: Responsible Research and Innovation for Synthetic Biology." *Journal of Responsible Innovation* 2 (1): 78–80.

Briggle, Adam, and Carl Mitcham. 2012. *Ethics and Science: An Introduction*. Cambridge: Cambridge University Press.

Brown, Mark B. 2006. "Survey Article: Citizen Panels and the Concept of Representation." *Journal of Political Philosophy* 14 (2): 203–25.

Brown, Mark B., and David H. Guston. 2009. "Science, Democracy, and the Right to Research." *Science and Engineering Ethics* 15 (3): 351–66.

Callon, Michel. 1987. "Society in the Making: The Study of Technology as a Tool for Sociological Analysis." In *The Social Construction of Technological Systems: New Directions in the Sociology and History of Technology*, edited by Wiebe E. Bijker, Thomas P. Hughes, and J. Trevor Pinch, 83–103. Cambridge, MA: MIT Press

Calvert, Jane. 2010. "Synthetic Biology: Constructing Nature?" *The Sociological Review* 58 (S1): 95–112.

Calvert, Jane, and Paul Martin. 2009. "The Role of Social Scientists in Synthetic Biology." *EMBO Reports* 10 (3): 201–4.

Chilvers, Jason, and Matthew Kearnes, eds. 2016. *Remaking Participation: Science, Environment and Emergent Publics*. Abingdon: Routledge.

Collingridge, David. 1980. *The Social Control of Technology*. London: Open University Press.

Cook-Deegan. Robert M. 1994. *The Gene Wars: Science, Politics, and the Human Genome*. New York: W. W. Norton.

Crawford, Kate, Mary L. Gray, and Kate Miltner. 2014. "Critiquing Big Data: Politics, Ethics, Epistemology." Introduction to special section: Big Data, *International Journal of Communication* 8: 10.

David, Paul A. 2001. "Path Dependence, Its Critics and the Quest for 'Historical Economics.'" In *Evolution and Path Dependence in Economic Ideas: Past and Present*, edited by Pierre Garrouste and Stavros Ioannidis, 15–40. Cheltenham: Edward Elgar.

Davies, Sarah R., Cynthia Selin, Gretchen Gano, and Ângela Guimarães Pereira. 2012. "Citizen Engagement and Urban Change: Three Case Studies of Material Deliberation." *Cities* 29 (6): 351–57.

de Laet, Marianne, and Annemarie Mol. 2000. "The Zimbabwe Bush Pump: Mechanics of a Fluid Technology." *Social Studies of Science* 30 (2): 225–63.

de Saille, Stevienna. 2015. "Innovating Innovation Policy: The Emergence of 'Responsible Research and Innovation.'" *Journal of Responsible Innovation* 2 (2): 1–33.

De Vries, Gerard. 2007. "What Is Political in Sub-politics? How Aristotle Might Help STS." *Social Studies of Science* 37 (5): 781–809.

Dorbeck-Jung, Bärbel, and Clare Shelley-Egan. 2013. "Meta-regulation and Nanotechnologies: The Challenge of Responsibilisation within the European Commission's Code of Conduct for Responsible Nanosciences and Nanotechnologies Research." *NanoEthics* 7 (1): 55–68.

Douglas, Heather E. 2003. "The Moral Responsibilities of Scientists (Tensions between Autonomy and Responsibility)." *American Philosophical Quarterly* 40 (1): 59–68.

Durodié, Bill. 2003. "Limitations of Public Dialogue in Science and the Rise of New 'Experts.'" *Critical Review of International Social and Political Philosophy* 6 (4): 82–92.

ERA Expert Group. 2008. *Challenging Europe's Research: Rationales for the European Research Area (ERA)*. Brussels: European Commission.

European Commission. 2013. *Options for Strengthening Responsible Research and Innovation*. Luxembourg: European Union DG Research and Innovation.

Felt, Ulrike, Brian Wynne, Michel Callon, Maria Eduarda Gonçalves, Sheila Jasanoff, Maria Jepsen, Pierre-Benoît Joly, Zdenek Konopasek, Stefan May, Claudia Neubauer, Arie Rip, Karen Siune, Andy Stirling, and Mariachiara Tallacchini 2007. *Taking European Knowledge Society Seriously*. Luxembourg: European Union DG for Research.

Fiorino, Daniel J. 1990. "Citizen Participation and Environmental Risk: A Survey of Institutional Mechanisms." *Science, Technology, & Human Values* 15 (2): 226–43.

Fisher, Erik, Roop L. Mahajan, and Carl Mitcham. 2006. "Midstream Modulation of Technology: Governance from Within." *Bulletin of Science, Technology & Society* 26 (6): 485–96.

Ganzevles, Jurgen, Rinie van Est, and Michael Nentwich. 2014. "Embracing Variety: Introducing the Inclusive Modelling of (Parliamentary) Technology Assessment." *Journal of Responsible Innovation* 1 (3): 292–313.

Gieryn, Thomas F., and Anne E. Figert. 1986. "Scientists Protect Their Cognitive Authority: The Status Degradation Ceremony of Sir Cyril Burt." In *The Knowledge Society: The Growing Impact of Scientific Knowledge on Social Relations*, vol. 10, edited by Gernot Böhme and Nico Stehr, 67–86. Dordrecht: Springer Netherlands.

Glerup, Cecilie, and Maja Horst. 2014. "Mapping 'Social Responsibility' in Science." *Journal of Responsible Innovation* 1 (1): 31–50.

Groves, Christopher. 2015. "Logic of Choice or Logic of Care? Uncertainty, Technological Mediation and Responsible Innovation." *NanoEthics* 9 (3): 321–33.

Grove-White, Robin, Phil Macnaghten, and Brian Wynne. 2000. "Wising Up: The Public and New Technologies." Research report by the Centre for the Study of Environmental Change, Lancaster: IEPPP, Lancaster University.

Guston, David H. 2004. "Responsible Innovation in the Commercialized University." In *Buying In or Selling Out? The Commercialization of the American Research University*, edited by Donald G. Stein, 161–74. New Brunswick, NJ: Rutgers University Press.

———. 2006. "On Consensus and Voting in Science: From Asilomar to the National Toxicology Program." In *The New Political Sociology of Science*, edited by Scott Frickel and Kelly Moore, 378–404. Madison: University of Wisconsin Press.

———. 2014a. "Building the Capacity for Public Engagement with Science in the United States." *Public Understanding of Science* 23 (1): 53–59.

———. 2014b. "Understanding 'Anticipatory Governance.'" *Social Studies of Science* 44 (2): 218–42.

———. 2015. "Responsible Innovation: Who Could Be against That?" *Journal of Responsible Innovation* 2 (1): 1–4.

Guston, David H., and Daniel Sarewitz. 2002. "Real-Time Technology Assessment." *Technology in Society* 24 (1): 93–109.

Harremoës, Poul, David Gee, Malcolm MacGarvin, Andy Stirling, Jane Keys, Brian Wynne, and Sofia Guedes Vaz. 2001. *Late Lessons from Early Warnings: The Precautionary Principle 1896–2000. Environmental Issue Report 22*. Copenhagen: European Environment Agency.

Hellström, Tomas. 2003. "Systemic Innovation and Risk: Technology Assessment and the Challenge of Responsible Innovation." *Technology in Society* 25 (3): 369–84.

Herkert, Joseph R. 2005. "Ways of Thinking about and Teaching Ethical Problem Solving: Microethics and Macroethics in Engineering." *Science and Engineering Ethics* 11 (3): 373–85.

Hill, Alison, and Mike Michael. 1998. "Engineering Acceptance: Representations of the Public in Debates on Biotechnology." *The Social Management of Genetic Engineering*, edited by Peter Wheale, René Von Schomberg, and Peter Glasner, 201–17. Aldershot: Ashgate.

Holbrook, J. Britt, and Adam Briggle. 2014. "Knowledge Kills Action—Why Principles Should Play a Limited Role in Policy-Making." *Journal of Responsible Innovation* 1 (1): 51–66.

Holton, Gerald James, and Robert S. Morison, eds. 1979. *Limits of Scientific Inquiry*. New York: W. W. Norton.

Horst, Maja, and Alan Irwin. 2010. "Nations at Ease with Radical Knowledge on Consensus, Consensusing and False Consensusness." *Social Studies of Science* 40 (1): 105–26.

Hughes, Thomas P. 1993. *Networks of Power: Electrification in Western Society, 1880–1930*. Baltimore: Johns Hopkins University Press.

Hurlbut, J. Benjamin. 2015. "Remembering the Future: Science, Law, and the Legacy of Asilomar." In *Dreamscapes of Modernity: Sociotechnical Imaginaries and the Fabrication of Power*, edited by Sheila Jasanoff and Sang-Hyun Kim, 126–51. Chicago: University of Chicago Press.

Illies, Christian, and Anthonie Meijers. 2009. "Artefacts without Agency." *The Monist* 92 (3): 420–40.

Jasanoff, Sheila, 2004. "Afterword." In *States of Knowledge: The Co-production of Science and the Social Order*, edited by Sheila Jasanoff, 274–82. London: Routledge.

___ . 2013. "Epistemic Subsidiarity-Coexistence, Cosmopolitanism, Constitutionalism." *European Journal of Risk Regulation* 4 (2): 133–41.

Joly, Pierre-Benoît. 2015. "Governing Technologies—The Need to Think Outside the (Black) Box." In *Science and Democracy: Making Knowledge and Making Power in the Biosciences and Beyond*, edited by Stephen Hilgartner, Clark Miller, and Rob Hagendijk, 133–55. London: Routledge.

Jonas, Hans. 1985. *The Imperative of Responsibility: In Search of an Ethics for the Technological Age*. Chicago: University of Chicago Press.

Jones, James H. 1981. *Bad Blood: The Tuskegee Syphilis Experiment: A Tragedy of Race and Medicine*. New York: Free Press.

Kaiser, David, and Jonathan Moreno. 2012. "Dual-Use Research: Self-Censorship Is Not Enough." *Nature* 492 (7429): 345–47.

Kaplinsky, Raphael. 2011. "Schumacher Meets Schumpeter: Appropriate Technology below the Radar." *Research Policy* 40 (2): 193–203.

Kearnes, Matthew, Robin Grove-White, Phil Macnaghten, James Wilsdon, and Brian Wynne. 2006. "From Bio to Nano: Learning Lessons from the UK Agricultural Biotechnology Controversy." *Science as Culture* 15 (4): 291–307.

Kiran, Asle H., Nelly Oudshoorn, and Peter-Paul Verbeek. 2015. "Beyond Checklists: Toward an Ethical-Constructive Technology Assessment." *Journal of Responsible Innovation* 2 (1): 5–19.

Krimsky, Sheldon. 2005. "From Asilomar to Industrial Biotechnology: Risks, Reductionism and Regulation." *Science as Culture* 14 (4): 309–23.

Kroes, Peter, and Anthonie Meijers, eds. 2001. *The Empirical Turn in the Philosophy of Technology*. Oxford: Elsevier Science.

Krohn, Wolfgang, and Peter Weingart. 1987. "Commentary: Nuclear Power as a Social Experiment-European Political "Fall Out" from the Chernobyl Meltdown." *Science, Technology, & Human Values* 12 (2): 52–58.

Latour, Bruno. 1999. "Give Me a Laboratory and I Will Raise the World." In *The Science Studies Reader*, edited by Mario Biagioli, 258–75. New York: Routledge.

___. 2007. "Turning around Politics: A Note on Gerard De Vries' Paper." *Social Studies of Science* 37 (5): 811–20.

___. 2008. "It's Development, Stupid!" Or: How to Modernize Modernization." In *Postenvironmentalism*, edited by Jim Proctor, 1–13. Cambridge, MA: MIT Press.

___. 2011. "From Multiculturalism to Multinaturalism: What Rules of Method for the New Socio-Scientific Experiments?" *Nature and Culture* 6 (1): 1–17.

Laurent, Brice. 2011. "Technologies of Democracy: Experiments and Demonstrations." *Science and Engineering Ethics* 17 (4): 649–66.

Lave, Rebecca, Philip Mirowski, and Samuel Randalls. 2010. "Introduction: STS and Neoliberal Science." *Social Studies of Science* 40 (5): 659–75.

Leach, Melissa, and Ian Scoones. 2006. *The Slow Race: Making Technology Work for the Poor*. London: Demos.

Lessig, Lawrence. 1999. *Code and Other Laws of Cyberspace*. New York: Basic Books.

Lezaun, Javier, and Linda Soneryd. 2007. "Consulting Citizens: Technologies of Elicitation and the Mobility of Publics." *Public Understanding of Science* 16 (3): 279–97.

Macnaghten, Phil, and Jason Chilvers. 2013. "The Future of Science Governance: Publics, Policies, Practices." *Environment and Planning C: Government and Policy* 32 (3): 530–48.

Mampuys, Ruth, and Frans W. A. Brom. 2015. "Governance Strategies for Responding to Alarming Studies on the Safety of GM Crops." *Journal of Responsible Innovation* 2 (2): 201–19.

Maranta, Alessandro, Michael Guggenheim, Priska Gisler, and Christian Pohl. 2003. "The Reality of Experts and the Imagined Lay Person." *Acta Sociologica* 46 (2): 150–65.

Marris, Claire. 2015. "The Construction of Imaginaries of the Public as a Threat to Synthetic Biology." *Science as Culture* 24 (1): 83–98.

Marris, Claire, and Nikolas Rose. 2010. "Open Engagement: Exploring Public Participation in the Biosciences." *PLoS Biology* 8 (11): e1000549.

Morozov, Evgeny. 2013. *To Save Everything, Click Here: Technology, Solutionism, and the Urge to Fix Problems That Don't Exist.* London: Penguin Random House UK.

Nelkin, Dorothy. 2001. "Beyond Risk: Reporting about Genetics in the Post-Asilomar Press." *Perspectives in Biology and Medicine* 44 (2): 199–207.

Nelson, Richard R. 1977. *The Moon and the Ghetto.* New York: W. W. Norton.

___. 2011. "The Moon and the Ghetto Revisited." *Science and Public Policy* 38 (9): 681–90.

Nordmann, Alfred. 2014. "Responsible Innovation, the Art and Craft of Anticipation." *Journal of Responsible Innovation* 1 (1): 87–98.

Nuffield Council on Bioethics. 2012. *Emerging Biotechnologies: Technology, Choice and the Public Good.* London: Nuffield Council on Biotechnologies, accessed at http://nuffieldbioethics.org /sites/default/files/Emerging_biotechnologies_full_report_web_0.pdf.

Owen, Richard. 2014. "The UK Engineering and Physical Sciences Research Council's Commitment to a Framework for Responsible Innovation." *Journal of Responsible Innovation* 1 (1): 113–17.

Owen, Richard, Jack Stilgoe, Phil Macnaghten, Mike Gorman, Erik Fisher, and Dave Guston. 2013. "A Framework for Responsible Innovation." In *Responsible Innovation: Managing the Responsible Emergence of Science and Innovation in Society*, edited by Richard Owen, John Bessant, and Maggy Heintz, 27–50. Hoboken, NJ: Wiley.

Pasquale, Frank. 2015. *The Black Box Society.* Cambridge, MA: Harvard University Press.

Pellizzoni, Luigi. 2004. "Responsibility and Environmental Governance." *Environmental Politics* 13 (3): 541–65.

Perrow, Charles. 1984. *Normal Accidents: Living with High Risk Systems.* Princeton, NJ: Princeton University Press

___. 2011. "Fukushima and the Inevitability of Accidents." *Bulletin of the Atomic Scientists* 67 (6): 44–52.

Polanyi, Michael. 1962. "The Republic of Science: Its Political and Economic Theory." *Minerva* 1 (1): 54–73.

Rabinow, Paul, and Gaymon Bennett. 2012. *Designing Human Practices: An Experiment with Synthetic Biology*. Chicago: University of Chicago Press.

Radder, Hans. 1998. "The Politics of STS." *Social Studies of Science* 28: 325–31.

Rayner, Steve. 2004. "The Novelty Trap: Why Does Institutional Learning about New Technologies Seem So Difficult?" *Industry and Higher Education* 18 (6): 349–55.

___. 2012. "Uncomfortable Knowledge: The Social Construction of Ignorance in Science and Environmental Policy Discourses." *Economy and Society* 41 (1): 107–25.

Rip, Arie. 1994. "Science & Technology Studies and Constructive Technology Assessment." *EASST Newsletter* 13 (3): 11–16.

___. 2006a. "A Co-evolutionary Approach to Reflexive Governance—and Its Ironies." In *Reflexive Governance for Sustainable Development*, edited by Jan-Peter Voss and Dierk Bauknecht, 82–100. Cheltenham: Edward Elgar

___. 2006b. "Folk Theories of Nanotechnologists." *Science as Culture* 15 (4): 349–65.

___. 2014. "The Past and Future of RRI." *Life Sciences, Society and Policy* 10: 17.

Rip, Arie, Johan W. Schot, and Thomas J. Misa. 1995. "Constructive Technology Assessment: A New Paradigm for Managing Technology in Society." In *Managing Technology in Society: The Approach of Constructive Technology Assessment*, edited by Arie Rip, Thomas Misa, and Johan Schot, 1–12. London: Pinter.

Rip, Arie, and Clare Shelley-Egan. 2010. "Positions and Responsibilities in the 'Real' World of Nanotechnology." In *Understanding Public Debate on Nanotechnologies: Options for Framing Public Policies: A Working Document by the Services of the European Commission*, 31–38. Brussels: European Commission.

Rodríguez, Hannot, Erik Fisher, and Daan Schuurbiers. 2013. "Integrating Science and Society in European Framework Programmes: Trends in Project-Level Solicitations." *Research Policy* 42 (5): 1126–37.

Sarewitz, Daniel. 2011. "Anticipatory Governance of Emerging Technologies." In *The Growing Gap between Emerging Technologies and Legal-Ethical Oversight: The Pacing Problem*, vol. 7, edited by Gary E. Marchant, Braden R. Allenby, and Joseph R. Herkert, 95–105. Dordrecht: Springer Netherlands.

Sarewitz, Daniel, and Richard Nelson. 2008. "Three Rules for Technological Fixes." *Nature* 456 (7224): 871–72.

Schattschneider, Elmer E. 1960. *The Semisovereign People: A Realist's View of Democracy in America*. New York: Holt, Rinehart, & Winston.

Schroeder, Doris, and Miltos Ladikas. 2015. "Towards Principled Responsible Research and Innovation: Employing the Difference Principle in Funding Decisions." *Journal of Responsible Innovation* 2 (2): 169–83.

Schumacher, Ernst F. 1973. *Small Is Beautiful: Economics as If People Mattered*. London: Blond & Briggs.

Sclove, Richard. 1995. *Democracy and Technology*. New York: Guilford Press.

Selin, Cynthia. 2014a. "On Not Forgetting Futures." *Journal of Responsible Innovation* 1 (1): 103–8.

___. 2014b. "Merging Art and Design in Foresight: Making Sense of Emerge." *Futures* 70: 1–86.

Selin, Cynthia, and Jathan Sadowski. 2016. "Against Blank Slate Futuring: Noticing Obduracy in the City through Experiential Methods of Public Engagement." In *Remaking Participation: Science, Environment and Emergent Publics*, edited by Jason Chilvers and Matthew Kearnes, 218–37. Abingdon: Routledge.

Shapira, Philip, Jan Youtie, and Yin Li. 2015. "Social Science Contributions Compared in Synthetic Biology and Nanotechnology." *Journal of Responsible Innovation* 2 (1): 143–48.

Simakova, Elena, and Christopher Coenen. 2013. "Visions, Hype, and Expectations: A Place for Responsibility." In *Responsible Innovation: Managing the Responsible Emergence of Science and Innovation in Society*, edited by Richard Owen, John Bessant, and Maggy Heintz, 241–67. Hoboken, NJ: Wiley.

Sismondo, Sergio. 2007. "Science and Technology Studies and an Engaged Program." In *The Handbook of Science and Technology Studies*. 3rd ed., edited by Edward J. Hackett, Olga Amsterdamska, Michael Lynch, and Judy Wajcman, 13–32. Cambridge, MA: MIT Press.

___. 2011. *An Introduction to Science and Technology Studies*. Hoboken, NJ: Wiley.

Spiegel-Rösing, Ina, and Derek de Solla Price, eds. 1977. *Science, Technology and Society: A Cross-Disciplinary Perspective*. Beverly Hills, CA: Sage.

Stahl, Bernd Carsten. 2013. "Responsible Research and Innovation: The Role of Privacy in an Emerging Framework." *Science and Public Policy* 42 (6): 1–9.

Staw, Barry M. 1976. "Knee-Deep in the Big Muddy: A Study of Escalating Commitment to a Chosen Course of Action." *Organizational Behavior and Human Performance* 16 (1): 27–44.

Stemerding, Dirk. 2015. "iGEM as Laboratory in Responsible Research and Innovation." *Journal of Responsible Innovation* 2 (1): 140–42.

Steneck, Nicholas H. 2006. "Fostering Integrity in Research: Definitions, Current Knowledge, and Future Directions." *Science and Engineering Ethics* 12 (1): 53–74.

Stilgoe, Jack. 2012. "Experiments in Science Policy: An Autobiographical Note." *Minerva* 50 (2): 197–204.

___. 2015. *Experiment Earth: Responsible Innovation in Geoengineering*. Abingdon: Routledge.

Stilgoe, Jack, Simon J. Lock, and James Wilsdon. 2014. "Why Should We Promote Public Engagement with Science?" *Public Understanding of Science* 23 (1): 4–15.

Stilgoe, Jack, Richard Owen, and Phil Macnaghten. 2013. "Developing a Framework for Responsible Innovation." *Research Policy* 42 (9): 1568–80.

Stirling, Andy. 2008. "'Opening Up' and 'Closing Down' Power, Participation, and Pluralism in the Social Appraisal of Technology." *Science, Technology, & Human Values* 33 (2): 262–94.

Szerszynski, Bronislaw, Matthew Kearnes, Phil Macnaghten, Richard Owen, and Jack Stilgoe. 2013. "Why Solar Radiation Management Geoengineering and Democracy Won't Mix." *Environment and Planning A* 45 (12): 2809–16.

Taverne, Dick. 2005. *The March of Unreason: Science, Democracy, and the New Fundamentalism.* Oxford: Oxford University Press.

Thompson, Charis. 2013. *Good Science: The Ethical Choreography of Stem Cell Research.* Cambridge, MA: MIT Press.

Tyfield, David. 2012. "A Cultural Political Economy of Research and Innovation in an Age of Crisis." *Minerva* 50 (2): 149–67.

van der Burg, Simone. 2014. "On the Hermeneutic Need for Future Anticipation." *Journal of Responsible Innovation* 1 (1): 99–102.

von Schomberg, René. 2011. "The Quest for the 'Right' Impacts of Science and Technology: An Outlook towards a Framework for Responsible Research and Innovation." In *Les Nanotechnologies: Vers un Changement d'Échelle Éthique?*, edited by Céline Kermisch and Marie-Geneviève Pinsart, 269–88. Brussels & Paris: E.M.E. & InterCommunications.

———. 2013. "A Vision of Responsible Innovation" In *Responsible Innovation: Managing the Responsible Emergence of Science and Innovation in Society,* edited by Richard Owen, John Bessant, and Maggy Heintz, 27–50. Hoboken, NJ: Wiley.

Walker, William. 2000. "Entrapment in Large Technology Systems: Institutional Commitment and Power Relations." *Research Policy* 29 (7): 833–46.

Weinberg, Alvin M. 1966. "Can Technology Replace Social Engineering?" *Bulletin of the Atomic Scientists* 22 (10): 4–8.

Williams, Robin, and David Edge. 1996. "The Social Shaping of Technology." *Research Policy* 25 (6): 865–99.

Wilsdon, James. 2014. "From Foresight to Hindsight: The Promise of History in Responsible Innovation." *Journal of Responsible Innovation* 1 (1): 109–12.

Wilsdon, James, and Rebecca Willis. 2004. *See-Through Science: Why Public Engagement Needs to Move Upstream.* London: Demos.

Winner, Langdon. 1977. *Autonomous Technology: Technics-Out-of-Control as a Theme in Political Thought.* Cambridge, MA: MIT Press.

Woodson, Thomas. 2012. "Research Inequality in Nanomedicine." *Journal of Business Chemistry* 9 (3): 133–46.

Woolgar, Steve. 1990. "Configuring the User: The Case of Usability Trials." *The Sociological Review* 38 (S1): 58–99.

Wright, Susan. 2001. "Legitimating Genetic Engineering." *Perspectives in Biology and Medicine* 44 (2): 235–47.

Wynne, Brian. 1993. "Public Uptake of Science: A Case for Institutional Reflexivity." *Public Understanding of Science* 2 (4): 321–37.

___. 2001. "Creating Public Alienation: Expert Cultures of Risk and Ethics on GMOs." *Science as Culture* 10 (4): 445–81.

___. 2002. "Risk and Environment as Legitimatory Discourses of Technology: Reflexivity Inside Out?" *Current Sociology* 50 (3): 459–77.

___. 2006. "Public Engagement as a Means of Restoring Public Trust in Science—Hitting the Notes, but Missing the Music?" *Public Health Genomics* 9 (3): 211–20.

___. 2016. "Ghosts of the Machine: Publics Meanings and Social Science in a Time of Expert Dogma and Denial" *Remaking Participation: Science, Environment and Emergent Publics*, edited by Jason Chilvers and Matthew Kearnes, 218–37. Abingdon: Routledge.

Ziegler, Rafael. 2015. "Justice and Innovation—Towards Principles for Creating a Fair Space for Innovation." *Journal of Responsible Innovation* 2 (2): 184–200.

Ziman, John. 2001. "Getting Scientists to Think about What They Are Doing." *Science and Engineering Ethics* 7 (2): 165–76.

30 Reframing Science Communication

Maja Horst, Sarah R. Davies, and Alan Irwin

In current knowledge societies, where technoscience is seen as a key driver for economic growth and national prosperity, science communication represents a crucial activity. Certainly, public communication of science and technology—whether in the form of public engagement or less overtly deliberative formats, such as science media, museums, festivals, and websites—continues to be supported by national and international policies and funding programs. In Europe, for instance, the Horizon 2020 program's emphasis on Responsible Research and Innovation incorporates the aim of building "a more scientifically literate society able to actively participate in and support democratic processes, and science and technology developments" (European Commission 2014). In the United States, the National Science Foundation funds activities that "advance informal learning in science" (National Science Foundation 2015). Furthermore, science communication practice is undergoing a period of experimentation and change, with new formats such as science comedy (Riesch 2015), citizen science (Haklay 2013; Silvertown 2009), and Web 2.0 platforms (Ranger and Bultitude 2014; You 2014) creating a heterogeneous landscape for the public consumption of scientific knowledge (Bucchi and Trench 2014; Holliman et al. 2009a, 2009b; Yeo et al. 2015).

Given this level of interest and activity concerning science communication, but also the theoretical and policy challenges of communicating science, it is hardly surprising that such matters have been central to the development of science and technology studies (STS)—and that STS scholarship has contributed to the development of science communication as a field. Over the last two decades, STS work around what has become known as "public engagement with science" (PES) has been especially prominent, having an impact both upon scholarship in this field but also engagement practice.

However, and as we will discuss, science communication covers a broader area than is implied by PES alone. In that sense, PES is highly significant for science communication but is by no means coterminous with it. Furthermore, science communication brings together a number of classic themes within STS in sometimes novel and provocative ways, including the status of knowledge claims in contemporary society, the relationship between science and democracy, and the role of organizational cultures and institutional arrangements in sociotechnical innovation. Equally, research in science communication has been conducted within STS but also in other fields, for example, in connection with the international Public Communication of Science and Technology (PCST) network. While STS scholarship may have contributed substantially to the field of science communication, not all work within science communication would consider itself to be STS in conceptual and methodological orientation.

This chapter highlights the ways in which the study of science communication can benefit from, and add to, the methods and concerns of STS. Accordingly, it covers both work that has been conducted by STS scholars and research within neighboring fields. It is structured into three sections. First, a scene-setting introduction to science communication and some of the conceptual models that have been developed to describe it; second, a discussion of the key themes that have emerged within recent research and analysis; and third, a presentation of current issues and challenges, including some suggestions for how these might be addressed by future research.

Setting the Scene: Science Communication in Context

Science communication has been a part of science since its earliest days. As Steven Shapin and others have shown, some notion of the public has always been constitutive of the scientific process (Shapin 1990). Overt efforts at improving public awareness and understanding of scientific knowledge similarly have a long history: from Victorian Mechanics' Institutes (Cooter and Pumfrey 1994) and educational pamphlets (Keene 2014) to the activities of "visible scientists" (Goodell 1975) such as Joaquín Gallo, Jane Goodall, Neil deGrasse Tyson, Niels Bohr, and Brian Cox.

There is also a more contemporary history of science communication. As Brian Trench (2008, 119) put it, "Science communication has been telling a story of its own development, repeatedly and almost uniformly, for almost a decade." This dominant story tells of a transition from communication formats marked by "deficit model" approaches to those that emphasize dialogue, debate, and participation (Gregory and Lock 2008). In the United Kingdom, deficit model approaches to communication (understood as assuming a singular, cognitive deficit public whose trust in science will be ensured through the provision of scientific facts; Irwin and Wynne 1996) were often associated with the rise of the "public understanding of science" (or PUS) movement in the 1980s and 1990s (Bodmer 2010; Durant, Evans, and Thomas 1989). In the United States, discussion has rather tended to emphasize "scientific literacy" (Logan

2001; Miller 2004). In both cases, science communication practitioners were expected to respond to the problem of an uninformed—and distrustful—public by efficiently transferring scientific knowledge to them.

The fault line through which contemporary science communication has tended to define itself, then, is a contrast between *deficit* (one-way, elitist, fact-oriented) and *dialogue* (two-way or interactive, participatory, reflective upon technoscience's broader implications). The various conceptual models that have been developed for science communication tend to confirm this. Even where they include three or even four types of communication, the one-way/dialogue contrast is generally pivotal. One such model is provided by Maja Horst (2008), who distinguishes *diffusion* (where the emphasis is on the public listening to science), *deliberation* (with an emphasis on science listening to the public), and *negotiation* (focused on communication as constitutive of social relations) as ways of framing science communication. Other models have been proposed—by Trench (2008) and Palmer and Schibeci (2014), for example—based on a similar diffusion/deliberation (or deficit/dialogue) formulation.

These accounts (and other, similar models of the communication process; see Brossard and Lewenstein 2009; Rowe and Frewer 2005) thus classify science communication according to how knowledge, whether that is scientific or lay, is understood as traveling or being constructed. One noteworthy exception to this is Burns, O'Connor, and Stocklmayer (2003), who focus on the purpose(s) of a science communication activity, and its impacts at the level of individuals, rather than on information flow. Their definition of science communication includes issues of personal awareness with science, enjoyment, interest, and the forming of opinions about science-related matters as well as broader understanding of science's content, processes, and social characteristics.

For the purposes of this chapter, and at least as a starting point for discussion, we define science communication as *organized, explicit, and intended actions that aim to communicate scientific knowledge, methodology, processes, or practices in settings where nonscientists are a recognized part of the audiences*. There are four significant features of this definition.

First, this definition does not specify the sender or producer of the communication—in other words, who is involved aside from "nonscientist audiences." Though much of the academic discussion of, variously, PUS, engagement, and dialogue has tended to focus on those working in science as the key producers of science communication, it is important to note that, today, science communication involves many other kinds of professionals, from university outreach officers to museum "explainers." Science communication is no longer about (only) particular visible scientists but a whole raft of different professionals, organizations, and institutions—including activist groups, campaigners, citizen scientists, and NGOs.

Second, we do not wish to use the directionality or nature of knowledge exchange and production as the key means of defining or classifying science communication. In our view, communication, whether deficit model or dialogic in form, always contains an interactive element. Any kind of science communication, even the most static text or public lecture, will be actively received, reconstructed, and interpreted. In that sense, it is always co-produced. Some communication formats emphasize and make overt this co-production more than others. But the point is that the deficit model of a blank slate public has never held up: publics are never entirely passive, even when communicators assume that they are so.

Third, the definition is deliberately open with regard to the forms and formats it covers. As we will discuss in the next section, all sorts of activities (as defined above) *aim to communicate scientific knowledge, methodology, processes, or practices*: science journalism, blogging, museums, festivals, cafés, and events, certainly, but also deliberative exercises, public consultations, medical appointments, and, perhaps more incidentally, science fiction and advertising.

Fourth, and despite this relative openness, our approach here distinguishes science communication from more general communication. This is an important, but also contestable, point when so much communication in contemporary societies draws upon science-related material (a popular television series such as *Breaking Bad*, advice to households on waste recycling, many pro- or anti-abortion campaigns) but does not fit our definition in terms of the explicit and intended focus on science. One could extend the definition of science communication much more broadly to include all forms of communication which touch (or draw) upon science in some way—or else argue that science communication is simply communication with no special characteristics of its own since science is bound up in so many ways with the practices and experiences of everyday life. Our suggestion is that such moves would miss some important and distinctive features of science communication as an activity (as we will demonstrate in the following sections). However, it is important to stress that in practice the definition of science communication is always open to challenge and reconstruction, and that this definitional issue is an important aspect of the relationship between STS and science communication.

Important Strands of Science Communication

In this section, we introduce three key areas of science communication scholarship, focusing on how STS has contributed to each. We include academic accounts that imply a more or less linear understanding of communication as information transfer, as well as literature which understands communication as a constitutive activity for making sense of the world. For presentational purposes, this section distinguishes between science communication forms and formats (through which scientists and publics come into contact with each other); scientists as communicators; and publics as audiences. Of course, this analytical distinction breaks down as soon as we delve deeper into the research—an issue to which we return in the following section.

Forms and Formats of Science Communication

It has to be acknowledged that academic attention has not been evenly distributed among science communication's different formats. Science PR, for instance, has received rather little attention (but see Bauer and Bucchi 2008). Similarly, science festivals (Bultitude 2014; Jensen and Buckley 2012), science blogging (Riesch and Mendel 2014), and social media and online campaigning (Brossard and Scheufele 2013) are just starting to appear in the literature. However, there have been two particular foci for scholarly attention over the last decade: science in the mass media and activities that fall under the rubric of public engagement.

Some work on science in the mass media pertains to straightforward content analysis: how much, and what kind of, science is presented in science news, TV, or books. For instance, one extensive analysis of the British Broadcasting Corporation's 2009–2010 output found that science was mentioned in half of its TV news programming (Mellor, Webster, and Bell 2011). In another study, Boykoff (2010) reports that the coverage of climate change in fifty leading newspapers around the world increased five times in the period 2004–2009, pointing particularly to the important role of media coverage in the global south. There has been substantial interest in the ways the media frame science stories and particularly how such frames vary among national contexts (Massarani et al. 2013; Priest 2006; Reis 2008). Other analyses have focused specifically on the role of media coverage in public controversies, for instance, describing how such coverage supports the creation of scientific facts (Neresini 2000), influences public trust and risk perception (Durfee 2006; Varughese 2011), and supports certain national narratives (Kruvand and Hwang 2007). Drawing more generally on media studies, there has been an interest in theorizing the likely effects of media coverage of science on the publics who encounter it. Some of this work has focused on agenda setting, framing, priming, and issue arenas (Bauer 2002; Nisbet et al. 2002), while other studies have focused on the notion of medialization to describe the influence of the media logic on the system of science (Rödder, Franzen, and Weingart 2012).

An important and continuing focus within science communication has been public engagement with science, or PES. Developing out of the tradition of researching science-public interactions (for example, Irwin and Wynne 1996) but also from the body of academic work studying the relationship between technical expertise and political decision making (Brown 2015; Nelkin 1975), STS scholars and other social scientists have paid considerable attention to processes of public involvement with (especially) new areas of science and technology (Callon, Lascoumes, and Barthe 2011; Horlick-Jones et al. 2007). This academic field has been stimulated by (and indeed contributed to) a series of policy statements and institutional reports advocating greater societal scrutiny of emerging areas of science and technology (Felt et al. 2013; House of Lords 2000).

While some PES-related activities have been explicitly framed in the context of formal, scientific, decision-making processes (Hagendijk and Irwin 2006; Horlick-Jones et al. 2007), others have developed in connection with science communication institutions such as science museums, universities, and learned societies (Davies et al. 2009; Turney 2006). Again, the range of included activities is large just as there is considerable variation among national and regional contexts: science cafés, public events (such as consensus conferences run by academics), museum events, and university open days have all been discussed under the label of "public engagement" (Bell 2008; Grand 2009; Nishizawa 2005). Typically, such events involve live, face-to-face encounters between scientists and laypeople. Discussion may take place in small groups or in a public lecture or meeting, but the aim is to provide (carefully mediated) opportunities for laypeople to question or interrogate the views and work of scientists (Kerr, Cunningham-Burley, and Tutton 2007). As such, public engagement events often incorporate reflections on the wider implications of contemporary technoscience.

Understanding what happens in, and what results from, public engagement has been a key focus of STS over the last two decades, with research exploring questions of how the public is being defined for these purposes, the frameworks of meaning being imposed (not least in terms of the questions set for discussion; Felt and Fochler 2010), the institutional purposes of engagement (Hagendijk and Irwin 2006), the relationship between specific engagement initiatives, and questions of national identity and political culture (Horst and Irwin 2010; Wu and Qiu 2013; Zhang 2007), the status given to scientific evidence in relation to other forms of knowledge and interpretation (Leach, Scoones, and Wynne 2005; Tironi 2014), and the long-term effects on participants (Davies 2011; Kerr, Cunningham-Burley, and Tutton 2007). In more specific terms, STS treatments of PES have focused on (among other matters) questions of access to information and expertise (Chilvers 2008; Irwin 2006), the handling of doubt, uncertainty, and indeterminacy (Clark 2013; Wynne 1992), the specific form of interactions between scientists and citizens (Irwin 1995; Suryanarayanan and Kleinman 2012), and the design and conduct of deliberative processes (Burgess, Harrison, and Filius 1998; Pellizzoni 2001). Much of this analysis has been critical of institutional activities, presenting specific initiatives as centralized and relatively limited attempts to "capture" public opinions without any underlying commitment to deeper dialogue or principles of democratic governance (Marris 2015; Stilgoe, Lock, and Wilsdon 2014; Welsh and Wynne 2013).

The increasingly important phenomenon of citizen science must also be noted in the context of public engagement with science (Bonney et al. 2016; Ellis and Waterton 2004; Freitag and Pfeffer 2013). As Haklay (2013) has discussed, "citizen science" covers a range of participatory levels from basic crowdsourcing (or the use of citizens as sensors) through forms of "distributed intelligence" to "participatory science" and what he presents as "extreme" citizen science—where there is close collaboration between scientists and nonscientists with regard to such matters as problem definition, data collection, and analysis. As originally developed within STS scholarship, citizen science drew attention both to the manner in which groups of citizens could serve as knowledge generators (and not simply passive recipients of expert knowledge) and to the contextuality of all knowledge claims—whether based on close observation of everyday circumstances or formalized scientific methods (Irwin 1995). Beyond STS, citizen science has become a popular term to describe the engagement of members of the public in scientific projects, for example, surveys of bird populations, astronomical observations, and whale watching (Socientize Project 2014). In terms of our discussion of science communication, citizen science emphasizes both the diversity and distributed expertise of nonscientific audiences and the importance of two-way exchange between citizen and scientific groups. Even in its most "non-extreme" form, citizen science draws attention to the dynamic relationship, and increasingly open borders, between organized science and wider society—with important consequences for the form but also for the content of science communication.

Finally, it will be clear from the work cited above that the *type* of science that is communicated has also been a cross-cutting theme. Media content analysis, for instance, has explored long-term trends in the kinds of science covered in newspapers, finding a growth in medical and health-related news (Bauer et al. 2006). Other work has discussed the role of news values—such as novelty or geographical immediacy—in story selection (Henderson and Kitzinger 2007; Priest 2001). In the context of public engagement, the possibility of allowing public understandings, arguments, and values to shape research priorities has meant that the focus has very often been on new and emerging technologies. Nanotechnology, geoengineering, and synthetic biology have in that way provided important test cases for STS research (see, for example, Bowman and Hodge 2007).

Scientists as Communicators

Traditionally, science communication has been understood as an activity conducted by individual scientists ("visible scientists," in Goodell's terms; 1975). However, the last fifty years have also seen a growing interest in the role of other actors, for example, journalists and the media (Friedman, Dunwoody, and Rogers 1999; Lewenstein 1995), but also professional science communicators employed in consultancies, in research organizations, and working as freelancers (Bennett and Jennings 2011; Neresini and Bucchi 2011). It is nevertheless still the case that scientists are central to science communication and an understanding of their values, attitudes, and motivation structures is crucial (Bauer and Jensen, 2011). Thus, in an early analysis, Hilgartner (1990) pointed to the way in which the particular view of science communication as popularization serves as a political resource for scientists, since it lets them retain authority over what counts as knowledge. Hilgartner's conceptual analysis has been followed by more empirical analyses of how scientists think of their role in science communication (Casini and Neresini 2013; Jia and Liu 2014). These studies generally identify the dominance of a deficit model among scientists.

STS studies have also pinpointed the ways in which scientists employ various "discursive repertoires" (Gilbert and Mulkay 1984) to draw boundaries between science and nonscience (Burchell 2007). In particular, scientists have been described as dissatisfied with media coverage of science, which they tend to find too sensational and dramatic as well as overspeculative and with too much focus on risk (Gunter, Kinderlerer, and Beyleveld 1999). Petersen et al. (2009) point to the fact that this criticism is often coupled with a very poor understanding of the way the media function and scientists' own role in science coverage. However, recent studies have demonstrated that scientists' media contacts are more frequent and generally less problematic than previously thought (Dunwoody, Brossard, and Dudo 2009; Peters et al. 2008).

While the literature described above has focused on scientists' perceptions of the public, the media, and the purpose of science communication, a different set of studies (often drawing upon quantitative communication studies) has focused more directly on scientists' motivations for participating in public communication. In a secondary analysis of two large-scale surveys, Besley, Oh, and Nisbet (2012) found that scientists' attitudes toward public engagement were more significant than demographic factors such as gender and age. Other studies have demonstrated that scientists are motivated by a wish to improve public interest, awareness, understanding, and enthusiasm for science (see, for example, Royal Society 2006). Poliakoff and Webb (2007) found that scientists' previous behavior is the most important predictor for whether they were likely to engage in public communication. However, this study also concluded that scientists' perception of their colleagues' level of engagement influences their own participation in public communication. In a 2014 study, Johnson, Ecklund, and Lincoln investigated disciplinary differences and found that while physicists see public communication as a threat to their scientific reputation and credibility, biologists assign much greater value to a public communication profile.

This overview illustrates that despite various efforts, there is still no coherent understanding of the influence of collective values or disciplinary differences on scientists' decision to engage in science communication. Meanwhile, science communication is primarily presented as an activity carried out by individual scientists. This however might be changing. More recent studies have examined the role of PR departments (Lynch et al. 2014), professional science communicators (Borchelt and Nielsen 2014; Chilvers 2008), and the training of scientists for public communication (Besley, Dudo, and Storksdieck 2015; Miller and Fahy 2009).

Audiences of Science Communication

Research into the audiences of science communication basically falls into two main categories. There is, on the one hand, work exploring how various publics report on their knowledge and attitudes to science. On the other hand, research in STS traditions has investigated how science communication produces its audiences—focusing on the performative character of, for example, engagement processes. The first body of work has relied heavily on surveys of publics, whether general or specific, while the latter has emphasized qualitative analysis of the design and framing of particular engagement formats.

Studies of public understanding of science have a long history (Logan 2001), and general surveys attempting to identify public knowledge of and attitudes toward science remain central (see, for instance, Besley 2013; Castell et al. 2014). Importantly, these studies have demonstrated that there is no simple causal relation between knowing a lot about science and being positive toward it (Allum, Boy, and Bauer 2002; Castelfranchi et al. 2013). General surveys have also sought to uncover what "the public" think of certain technologies, and why (e.g., Pidgeon et al. 2012 for a study of public responses to geoengineering). In a similar vein, just over half of the respondents in a large UK study "think that the benefits of science outweigh any harmful effects"

(Castell et al. 2014, 26). Surveys often report that publics are interested in information about science and technology, and that they choose to consume forms of science communication such as science news, magazines, and museum exhibitions (ibid.; National Science Board 2014). More specifically, evaluative research on specific types of science communication has identified particular audiences for science communication and the different motivations they may express (Jensen and Buckley 2012). Thus, research on public lectures in science centers and museums has identified a number of types of visitor, including students working on the topic under discussion, individuals personally affected by that topic, and friends and family of the lecturer (Rennie and Williams 2006).

In her study of a gallery at the London Science Museum, Sharon Macdonald (1995) found that visitors' reasons for attending were often not much to do with science at all. Rather, they saw their visit as part of a "cultural itinerary" whereby they could, for instance, visit a key attraction in the UK capital or nostalgically re-create childhood visits with their own family. Similarly, she found that visitors' interpretations of the content of the gallery were shaped by their orientations and interests. They were, she writes, "reconfiguring the exhibition, sometimes in ways unanticipated by, or even explicitly contradicting, the makers" (Macdonald 1995, 20). This research suggests that science communication's audiences consume and use it in ways that go beyond simply satisfying an interest in science. Instead, audiences use science communication in the context of, and as part of, their personal, professional, and civic experiences.

The second major theme of research into audiences has emphasized the ways in which they are constituted by the structure of the engagement or communication format. Much of this work has focused on public engagement, exploring the ways in which publics are understood and thereby performed within ostensibly open and accessible formats (Irwin 2001). Braun and Schultz (2010), in reviewing this work, suggest four types of public that are commonly performed. They distinguish the *general* public, the *pure* public, the *affected* public, and the *partisan* public, with the pure and affected publics seen as most valuable within engagement exercises, since they either bring a neutral position or authentic expertise. Similarly, Felt and Fochler (2010) have discussed how the "machinery" of public engagement functions to frame participants, and their role in deliberation, in particular ways. Their discussion also emphasizes that participants do have scope to resist and reimagine those roles and that such imposed versions of the public, whether performed in public engagement or in scientific governance more generally, need not be final.

Current Issues in Science Communication

Taken overall, STS research has, then, served to trouble some of the assumptions which typically frame science communication—for instance, that its formats are innocent (and therefore do not act to constitute or constrain their participants), that it is primarily carried out by scientists, and that its users are motivated by a simple desire to know more science. Analysis of PES and other science communication has also destabilized the idea that science communication is an activity that can be grasped by one or another model of communication. Rather, any kind of science communication is a complex phenomenon and efforts to understand it through one model or analytical lens invariably produce a number of overflows. Even the most dialogically organized public engagement event might disappoint ideals of openness and egalitarianism, just as the most well-planned form of science information can be transformed through irony, appropriated by local social contexts, or simply ignored by audiences (Horst and Michael 2011).

Building on these insights, this section outlines some key issues for current and future research. It is structured around three major ways in which the study of science communication is important for understanding the role of science and technology in society: science communication's significance for the ability to make science, its connection to issues of organizational communication and identity formation, and finally, its importance for issues of democratic governance and citizenship.

Science Communication as Making Science

In Dorothy Nelkin's (1995) classic book about science communication, the fundamental point was that press coverage of science and technology was "selling science" rather than providing critical journalism. In particular, this was the case with new and emerging scientific fields. Building on a long tradition of studying controversies, she pointed to the fact that public communication about science is neither neutral nor objective—rather it is argumentation with the goal of convincing or persuading an audience of a particular state of affairs (Horst 2010).

In a related vein, the theme of futures and future making through communication has been central within STS and the area known as the sociology of expectations (Brown, Rappert, and Webster 2000, van Lente 1993). If actors are to secure a specific kind of future, then they must engage in a range of rhetorical, organizational, and material work in order to bring this future about (Brown, Rappert, and Webster 2000, 3). According to this literature, we should not focus on whether these expectations are justified or hyped in an effort to sell science but rather consider their performativity in the present (Michael 2000). Although visions and expectations are also formed through material and organizational efforts, communication and "storytelling" (Deuten and Rip 2000) represent a key medium through which these expectations are shaped. On this basis, Plesner and Horst (2013) have suggested that we speak of "innovation communication" as an integral part of the innovation process. Such communication aims to generate support for the sociotechnical development in question.

Especially when it comes to the communication of emerging science and technology, it is evident that companies and scientific organizations, as well as involved scientists, are invoking particular futures. For example, Hellsten (2002) demonstrates how life science companies have tried to turn images of fear into those of hope by focusing on health care and suggesting that humankind should take control of nature to make a better future. However, Guice (1999) has pointed to the fact that whereas all new technological trends have to be promoted, promotional work is often most effective when it is invisible. One of the most extreme examples of such unnoticed promotional work is found in the medical publication planning profession, whose job it is to orchestrate publications in order to support marketization of pharmaceutical drugs (Sismondo 2009).

Innovation communication highlights that science communication is not something separate from science but part of the process of generating resources and legitimacy in order to make and maintain it. Inspired by Deuten and Rip (2000), who discuss "product creation processes" as a genre, we suggest that it might be useful to think of innovation communication as something that takes place in many genres. Besides news and media coverage, we might also look at strategic reports (van Lente and Rip 1998), funding applications (Myers 1985), and patents (Packer and Webster 1996). In all these strategic forms of communication, actors aim to influence particular imaginations of technological possibilities and scientific futures in order to mobilize expectations. Pinch (2003) hinted at this when he argued that salespeople are "the missing masses" of technological development and that we should pay more attention to selling as an activity.

Work on the sociology of expectations has also pointed to the fact that expectations—stories about the future—do not just have one author. Rather, scientists and managers of research organizations are always coauthors of stories, which are simultaneously shaped by many other agents: "biotechnology firms (the small as well as the large variety), venture capitalists, retailers, consumer and environmental groups, all collude in creating a multi-actor—and multi-authored—story" (Deuten and Rip 2000, 84). Van Lente and Rip suggest that we think of these authors of visions as "promise champions" and remind us that they "need not be individuals. In fact, in the age of strategic science and technology, a plethora of collective actors has emerged to carry and protect new developments" (1998, 232). Science communication as the selling of science and innovation might therefore not be the result of individuals acting from specific interests but rather of a strategic authorial voice that emerges from an assemblage of relations between individual and collective actors and interests.

Science Communication as Organizational Communication

Simultaneously, organizations are becoming increasingly important to the conduct and understanding of science communication (Casini and Neresini 2013). Certainly, science communication is rapidly being professionalized, in part because (thanks to policy enthusiasm for public engagement) it is now a growth industry and in part because universities and other knowledge-based organizations have become much more focused on their public image and reputation (Bauer and Bucchi 2008). At the same time, science itself is no longer primarily the work of single individuals but a highly organized and collective phenomenon (Shapin 2009). Science communication therefore can no longer be interpreted only as the activity of individual scientists. It is also institutional and professional, and can be understood as a form of organizational communication in universities and other research organizations.

Brass and Rowe (2009) point out that universities' concern with their external reputation is focused around two perceived risks: damage to organizational reputation and damage to the legitimacy of knowledge and expertise. To avoid such damage, universities increasingly enforce prescriptive guidelines for academics' public statements and commentary. While these tendencies suggest that universities are moving toward stricter central control of science communication, it should also be recognized that traditionally there has been a large degree of freedom for individual scientists to speak publicly compared, for example, with employees of most commercial companies (Callagher, Horst, and Husted 2015). It is also important to recognize that scientists often have a dual allegiance (Anderson et al. 1980): to their organization and to their discipline. They might therefore consider disciplinary colleagues in competing research organizations as more a part of their internal organization than professional communication staff in their own organization.

Competition between universities and other research organizations has the potential to drive a celebrity culture (see Fahy 2015), in which high-profile scientists become important parts of the organizational brand. More generally, leading scientists can be seen as foci for the public image of, for instance, a university, acting as internal role models and interpreters of what science is and should be. As an example of the latter, Brosnan and Michael (2014) have shown how, for a research group, the promise of neuroscience being one day translated into clinical practice was embodied and vivi-fied—rendered alive—in the figure of a group leader, whose expertise spanned clinical and basic neurosciences.

Outside of STS, there has been substantial attention to the development of entre-preneurial universities and the motivations for scientists to take up commercial activi-ties. It is a common conclusion that socialization, group norms, and organizational support impact the degree to which individual scientists engage in commercial activi-ties and associated forms of communication (Bercovitz and Feldman 2008; Häussler and Colyvas 2011). STS-inspired analyses have demonstrated that academic and entre-preneurial activities are taking place alongside each other (Lam 2010). This can lead to complicated boundary maintenance arising "from the conflicting procedures and requirements of the two activities as well as from the double roles assumed by the actors involved" (Tuunainen and Knuuttila 2009, 684).

The importance of these perspectives and analyses stems from the fact that they point to the crucial connection between organizational communication and organiza-tional/individual identity formation. Within the symbolic perspective of organizational communication, a crucial point concerns the close links between internal perceptions of identity and external images of the organization (Hatch and Schultz 1997). Thus, the ways in which scientific organizations describe themselves—both externally and internally—are closely related to the ways in which scientists make sense of their own jobs. The branding of the organization, for instance, influences how scientists imagine their own careers (Duberley, Cohen, and Mallon 2006). Felt and Fochler (2012) have pointed to this as an increasingly important issue, since science communication regu-larly portrays science as a glamorous, business-like and result-focused venture, which junior scientists can struggle to align with their everyday experience of scientific work.

The professionalization of science communication therefore suggests a range of issues connected to changes in the wider academic landscape, the nature of scientific identities, and the way knowledge is branded and represented. Importantly, public communication associated with organizational and sectoral changes toward commer-cialization and marketization should not just be seen as new ways of packaging knowl-edge. Rather these organizational changes potentially influence the way in which scientists perceive their own role and that of the organization, thereby also shaping what science is and should be.

Science Communication for Scientific Citizenship

At least two foundational themes have run through STS analyses of public engage-ment. The first is *power*. STS has been interested in who—and what—is able to speak authoritatively and in which contexts. Research on public engagement activities has often indicated that—regardless of the language of dialogue and participation—such events all too often constrain, rather than enable, lay public involvement in active discussion and questioning of scientific knowledge (Kerr, Cunningham-Burley, and Tutton 2007). The second theme is, relatedly, that of *democracy*. Accounts of public engagement have often, implicitly or explicitly, linked such activities (generally in critical terms) to a larger democratic drive to open up science to public debate and, at least potentially, to empower citizens with regard to scientific decision making (Powell and Colin 2008). That science communication is politically as well as culturally important runs as an undercurrent through almost all discussion of it (Durant, Evans, and Thomas 1989). Yet this raises a series of broader questions. What is the role of science communication in a democratic society? Should it be understood as capacity building and citizen empowerment, or rather as the imposition of reductionist and scientistic forms of understanding (Wynne 2006)?

Stirling (2008) has emphasized the importance of opening up the governance of science and technology to plural and conditional policy advice, revealing alternative courses of action and different framing conditions. In a similar vein, Guston (2014, 234) has presented "anticipatory governance" as a means of "improving the societal capacity to articulate and apply public values in the context of emerging technologies." One key question for science communication then is whether it can adopt this "opening up" role rather than falling back into an older (and deficit-oriented) tradition of seeking to disseminate authoritative knowledge about science and society. In line with Welsh and Wynne (2013), part of this role would be the articulation of different "collective public meanings" and the questioning of normative social commitments to dominant models of science and progress. Seen in this light also, science communication is not simply a matter of dealing with what is conventionally defined as knowledge but is also bound up with questions of governance cultures, institutional learning, and relations of power and ownership—including the manner in which these affect communication structures and processes.

Developing further this citizenship role for science communication, a focus on justice in this context raises questions of democratic rights, of accountability, and of the responsibility to do what is best for the collectivity (including collectivities as yet unborn). Rather than separating questions of power, democracy, and knowledge into separate domains, ideas of cognitive/epistemic (in)justice precisely interconnect and intermingle these two, suggesting also that modes of citizenship and forms of knowledge flow together through many areas of social life (Fricker 2007; Visvanathan 2005). For science communication, this justice dimension suggests many challenging

questions of an ethical, political, and sociological kind: What are the societal responsibilities of the science communicator? How does science communication connect with
political decision making around science and innovation? How can science communication address vulnerable groups within society?

While questions of citizenship, democracy, and justice might seem to imply a rather
traditional model of both scientific governance and science-public relations, STS scholars have also explored alternative ways in which members of society engage with science and society—including the relationship between consumption and citizenship
and the manner in which public participation can be "materialized" through objects
and devices (Marres 2012; Michael 1998). Taken together with the growing body of
scholarship which identifies new governance forms (for example, mixtures of the public and private), fresh perspectives on agency, and the particular challenge of responding to global concerns such as climate change (Sundström, Soneryd, and Furusten
2010), one can identify new questions about the very nature of scientific citizenship
but also about the forms of science communication which are required in such settings.
Certainly, the theme of "science communication for scientific citizenship" raises profound issues concerning the changing nature of communication in this setting but also
its relationship to the problematic character of scientific citizenship.

Conclusion

One immediate lesson from the discussion in this chapter is that science communication can in no way be seen as a narrow topic or a neatly bounded activity. Instead, it
sits at a crucial point of intersection and influence between science, technology, and
society. As we have suggested, STS is an important part of this broader picture without
by any means covering it all. In this concluding section, we first address the implications of STS for science communication before turning the question around in order
to consider the linkage between research around science communication and larger
themes and issues within STS.

In terms of the contribution of STS to science communication, it seems that STS has
been especially valuable in challenging underlying assumptions concerning the nature
of science communication, drawing attention to some of the emerging forms and formats of science communication, and developing new themes and concerns (e.g., the
linkage between science communication and scientific citizenship). However, we have
also noted several areas where a stronger STS contribution could be made to science
communication, not least concerning important topics such as the influence of new

social media, organizational studies of science communication, and the relationship between science communication activities and the structure of scientific careers.

More generally, there is considerable further scope for STS scholars to bring tools, critical insights, and theoretical perspectives to the study and practice of science communication. Certainly, we are aware that we have only painted with the broadest of brushstrokes in this chapter. In particular, science communication could benefit from a more nuanced understanding of the nature of scientific knowledge and the ways in which it is produced. From our perspective, this might develop a more sophisticated treatment of what is being communicated (i.e., of the science itself) but also encourage greater reflection on the very idea of science communication and how it is to be defined, bounded, and enacted. Crucially, this should include the question of how and when institutions employ the language of science to deal with what might otherwise be presented as political, ethical, and cultural views and arguments concerning the direction of innovation or the desirability of new goods and processes.

Turning to the linkage between research in this particular field and larger themes within STS as a whole (i.e., of what science communication can do for STS), there are many possibilities. Certainly, the notions that science communication should be seen not as a side activity but as a crucial form of co-production, that the presentation of science can also be a matter of organizational identity building, and that questions of epistemic justice deserve larger scholarly and political attention have significant implications for STS as a whole.

Going further, research in science communication draws attention to the role that informal engagement with science can play in scientific citizenship. In that way it enables STS scholars to observe how lay citizens use museums, popular science, or the Internet as parts of their civic lives. Equally, the foregrounding by science communication research of emotional and aesthetic responses to science—such as pleasure, excitement, entertainment, wonder, and fear—brings a new dimension to the still largely epistemic orientation of STS. More broadly, science communication encourages STS to look beyond scientific organizations, industry, and the state as sites in which science is created and negotiated, and calls attention to science-related activism, citizen science, and "garage science" as general phenomena which may disturb conventional ideas of scientific institutions and scientific practice. As we have suggested above, science communication research regularly demonstrates not simply the public absorption of scientific messages but also the co-construction of new forms of knowledge, often in unconventional locations.

Science communication is both an important topic and also a complex, cross-cutting, and problematic activity. As we hope to have suggested in this chapter, science communication in many ways takes us to the heart of STS as a body of research and practice. As this chapter closes, we are very sure that the potential for further scholarship and reflective practice extends considerably beyond the crowded overview presented here. Already, we look forward to the treatment of science communication in the next edition of the STS handbook.

References

Allum, Nick, Daniel Boy, and Martin Bauer. 2002. "European Regions and the Knowledge Deficit Model." In *Biotechnology: The Making of a Global Controversy*, edited by Martin Bauer and George Gaskell, 224–43. Cambridge: Cambridge University Press.

Anderson, Robert Morris, Robert Perucci, Dan E. Schendel, and Leon E. Trachtman. 1980. *Divided Loyalties: Whistle-Blowing at BART*. West Lafayette, IN: Purdue University Press.

Bauer, Martin W. 2002. "Arenas, Platforms, and the Biotechnology Movement." *Science Communication* 24 (2): 144–61.

Bauer, Martin W., and Massimiano Bucchi 2008. *Journalism, Science and Society: Science Communication between News and Public Relations*. London: Routledge.

Bauer, Martin W., and Pablo Jensen. 2011. "The Mobilization of Scientists for Public Engagement." *Public Understanding of Science* 20 (1): 3–11.

Bauer, Martin W., Kristina Petkova, Pepka Boyadjieva, and Galin Gornev. 2006. "Long-Term Trends in the Public Representation of Science across the 'Iron Curtain': 1946–1995." *Social Studies of Science* 36 (1): 99–131.

Bell, Larry. 2008. "Engaging the Public in Technology Policy: A New Role for Science Museums." *Science Communication* 29 (3): 386–98.

Bennett, David J., and Richard C. Jennings, eds. 2011. *Successful Science Communication: Telling It Like It Is*. Cambridge: Cambridge University Press.

Bercovitz, Janet, and Maryann Feldman. 2008. "Academic Entrepreneurs: Organizational Change at the Individual Level." *Organization Science* 19 (1): 69–89.

Besley, John C. 2013. "The State of Public Opinion Research on Attitudes and Understanding of Science and Technology." *Bulletin of Science, Technology & Society* 33 (1–2): 12–20.

Besley, John C., Anthony Dudo, and Martin Storksdieck. 2015. "Scientists' Views about Communication Training." *Journal of Research in Science Teaching* 52 (2): 199–220.

Besley, John C., Sang Hwa Oh, and Matthew Nisbet. 2012. "Predicting Scientists' Participation in Public Life." *Public Understanding of Science* 22 (8): 971–87.

Bodmer, Walter. 2010. *Public Understanding of Science: The BA, the Royal Society and COPUS*. Notes and Records of the Royal Society. Accessed at http://rsnr.royalsocietypublishing.org/content /early/2010/06/16/rsnr.2010.0035.

Bonney, Rick, Tina B. Phillips, Heidi L. Ballard, and Jody W. Enck. 2016. "Can Citizen Science Enhance Public Understanding of Science?" *Public Understanding of Science* 25 (1): 2–16.

Borchelt, Rick, and Kristian Hvidtfelt Nielsen. 2014. "Public Relations in Science." In *Handbook of Public Communication of Science and Technology*. 2nd ed., edited by Massimiano Bucchi and Brian Trench, 58–69. London: Routledge.

Bowman, Diana M., and Graeme A. Hodge. 2007. "Nanotechnology and Public Interest Dialogue: Some International Observations." *Bulletin of Science, Technology and Society* 27 (2): 118–32.

Boykoff, Max. 2010. "Indian Media Representations of Climate Change in a Threatened Journalistic Ecosystem." *Climatic Change* (99): 17–25.

Brass, Kylie, and David Rowe. 2009. "Knowledge Limited: Public Communication, Risk and University Media Policy." *Continuum: Journal of Media & Cultural Studies* 23 (1): 53–76.

Braun, Kathrin, and Susanne Schultz. 2010. "'... A Certain Amount of Engineering Involved': Constructing the Public in Participatory Governance Arrangements." *Public Understanding of Science* 19 (4): 403–19.

Brosnan, Caragh, and Mike Michael. 2014. "Enacting the 'Neuro' in Practice: Translational Research, Adhesion and the Promise of Porosity." *Social Studies of Science* 44 (5): 680–700.

Brossard, Dominique, and Bruce V. Lewenstein. 2009. "A Critical Appraisal of Models of Public Understanding of Science." In *Understanding and Communicating Science: New Agendas in Communication*, edited by LeeAnn Kahlor and Patricia Stout, 11–39. New York: Taylor and Francis.

Brossard, Dominique, and Dietram A. Scheufele. 2013. "Science, New Media, and the Public." *Science* 339 (6115): 40–41.

Brown, Mark B. 2015. "Politicizing Science: Conceptions of Politics in Science and Technology Studies." *Social Studies of Science* 45: 3–30.

Brown, Nik, Brian Rappert, and Andrew Webster, eds. 2000. *Contested Futures: A Sociology of Prospective Techno-Science*. Aldershot: Ashgate.

Bucchi, Massimiano, and Brian Trench. 2014. *Handbook of Public Communication of Science and Technology*. 2nd ed. New York: Routledge.

Bultitude, Karen. 2014. "Science Festivals: Do They Succeed in Reaching beyond the 'Already Engaged'?" *JCOM: Journal of Science Communication* 13 (4): C01.

Burchell, Kevin. 2007. "Empiricist Selves and Contingent 'Others': The Performative Function of the Discourse of Scientists Working in Conditions of Controversy." *Public Understanding of Science* 16 (2): 145–62.

Burgess, Jacquie, Carolyn M. Harrison, and Petra Filius. 1998. "Environmental Communication and the Cultural Politics of Environmental Citizenship." *Environment and Planning A* 30 (8): 1445–60.

Burns, Timothy W., David J. O'Connor, and Susan M. Stocklmayer. 2003. "Science Communication: A Contemporary Definition." *Public Understanding of Science* 12 (2): 183–202.

Callagher, Lisa, Maja Horst, and Kenneth Husted. 2015. "Exploring Societal Responses towards Managerial Prerogative in Entrepreneurial Universities." *International Journal of Learning and Change* 8 (1): 64–82.

Callon, Michel, Pierre Lascoumes, and Yannick Barthe. 2011. *Acting in an Uncertain World: An Essay on Technical Democracy.* Cambridge, MA: MIT Press.

Casini, Silvia, and Federico Neresini. 2013. "Behind Closed Doors: Scientists' and Science Communicators' Discourses on Science in Society." *TECNOSCIENZA: Italian Journal of Science and Technology Studies* 3 (2): 37–62.

Castelfranchi, Yurij, Elaine Meire Vilela, Luciana Barreto de Lima, Ildeu de Castro Moreira, and Luisa Massarani. 2013. "Brazilian Opinions about Science and Technology: The Paradox of the Relation between Information and Attitudes." *História, Ciências, Saúde-Manguinhos* 20: 1163–83.

Castell, Sarah, Anne Charlton, Michael Clemence, Nick Pettigrew, Sarah Pope, Anna Quigley, Jayesh Navin Shah, and Tim Silman. 2014. *Public Attitudes to Science 2014.* London: Department for Business, Innovation and Skills. Accessed at https://www.gov.uk/government/uploads/system/uploads/attachment_data/file/348830/bis-14-p111-public-attitudes-to-science-2014-main.pdf.

Chilvers, Jason. 2008. "Deliberating Competence: Theoretical and Practitioner Perspectives on Effective Participatory Appraisal Practice." *Science, Technology, & Human Values* 33 (2): 155–85.

Clark, Lisa. 2013. "Framing the Uncertainty of Risk: Models of Governance for Genetically Modified Foods." *Science and Public Policy* 40: 479–91.

Cooter, Roger, and Stephen Pumfrey. 1994. "Separate Spheres and Public Places: Reflections on the History of Science Popularization and Science in Popular Culture." *History of Science* 32 (3): 237–67.

Davies, Sarah R. 2011. "How We Talk When We Talk about Nano: The Future in Laypeople's Talk." *Futures* 43 (3): 317–26.

Davies, Sarah, Ellen McCallie, Elin Simonsson, Jane L. Lehr, and Sally Duensing. 2009. "Discussing Dialogue: Perspectives on the Value of Science Dialogue Events That Do Not Inform Policy." *Public Understanding of Science* 18 (3): 338–53.

Deuten, J. Jasper, and Arie Rip. 2000. "Narrative Infrastructure in Product Creation Processes." *Organization* 7 (1): 69–93.

Duberley, Joanne, Laurie Cohen, and Mary Mallon. 2006. "Constructing Scientific Careers: Change, Continuity and Context." *Organization Studies* 27 (8): 1131–51.

Dunwoody, Sharon, Dominique Brossard, and Anthony Dudo. 2009. "Socialization or Rewards? Predicting US Scientist–Media Interactions." *Journalism & Mass Communication Quarterly* 86 (2): 299–314.

Durant, John, Geoffrey Evans, and Geoffrey P. Thomas. 1989. "The Public Understanding of Science." *Nature* 340: 11–14.

Durfee, Jessica L. 2006. "'Social Change' and 'Status Quo' Framing Effects on Risk Perception: An Exploratory Experiment." *Science Communication* 27 (4): 459–95.

Ellis, Rebecca, and Claire Waterton. 2004. "Environmental Citizenship in the Making: The Participation of Volunteer Naturalists in UK Biological Recording and Biodiversity Policy." *Science and Public Policy* 31 (2): 95–105.

European Commission, 2014. *Public Engagement in Responsible Research and Innovation.* Accessed at http://ec.europa.eu/programmes/horizon2020/en/h2020-section/public-engagement-responsible -research-and-innovation.

Fahy, Declan. 2015. *The New Celebrity Scientists.* London: Rowman & Littlefield.

Felt, Ulrike, Daniel Barben, Alan Irwin, Pierre-Benoît Joly, Arie Rip, Andy Stirling, and Tereza Stöckelová. 2013. *Science in Society: Caring for Our Futures in Turbulent Times.* Strasbourg: European Science Foundation.

Felt, Ulrike, and Maximilian Fochler. 2010. "Machineries for Making Publics: Inscribing and Describing Publics in Public Engagement." *Minerva* 48 (3): 219–38.

___. 2012. "Re-ordering Epistemic Living Spaces: On the Tacit Governance Effects of the Public Communication of Science." In *The Sciences' Media Connection: Public Communication and Its Repercussions,* edited by Simone Roedder, Martina Franzen, and Peter Weingart, vol. 28, 133–54. Sociology of the Sciences Yearbook. Dordrecht: Springer.

Freitag, Amy, and Max J. Pfeffer. 2013. "Process, Not Product: Investigating Recommendations for Improving Citizen Science 'Success.'" *PLoS ONE* 8 (5): e64079.

Fricker, Miranda. 2007. *Epistemic Injustice: Power and the Ethics of Knowing.* Oxford: Oxford University Press.

Friedman, Sharon M., Sharon Dunwoody, and Carol S. Rogers. 1999. *Communicating Uncertainty: Media Coverage of New and Controversial Science.* Mahwah, NJ: Lawrence Erlbaum.

Gilbert, Nigel, and Michael Mulkay. 1984. *Opening Pandora's Box: A Sociological Analysis of Scientists' Discourse.* Cambridge: Cambridge University Press.

Goodell, Rae. 1975. *The Visible Scientists.* Boston: Little, Brown.

Grand, Ann. 2009. "Cafe Scientifique." In *Practising Science Communication in the Information Age: Theorising Professional Practices,* edited by Richard Holliman, Jeff Thomas, Sam Smidt, Eileen Scanlon, and Elizabeth Whitelegg. Oxford: Oxford University Press.

Gregory, Jane, and Simon J. Lock. 2008. "The Evolution of 'Public Understanding of Science': Public Engagement as a Tool of Science Policy in the UK." *Sociology Compass* 2 (4): 1252–65.

Guice, Jon. 1999. "Designing the Future: The Culture of New Trends in Science and Technology." *Research Policy* 28: 81–98.

Gunter, Barrie, Julian Kinderlerer, and Deryck Beyleveld. 1999. "The Media and Public Under-standing of Biotechnology: A Survey of Scientists and Journalists." *Science Communication* 20 (4): 373–94.

Guston, David H. 2014. "Understanding 'Anticipatory Governance.'" *Social Studies of Science* 44 (2): 218–42.

Hagendijk, Rob, and Alan Irwin. 2006. "Public Deliberation and Governance: Engaging with Sci-ence and Technology in Contemporary Europe." *Minerva* 44 (2): 167–84.

Haklay, Muki. 2013. "Citizen Science and Volunteered Geographic Information: Overview and Typology of Participation." In *Crowdsourcing Geographic Knowledge*, edited by Daniel Sui, Sarah Elwood, and Michael Goodchild, 105–22. Berlin: Springer Verlag.

Hatch, Mary Jo, and Majken Schultz. 1997. "Relations between Organizational Culture, Identity and Image." *European Journal of Marketing* 31 (5/6): 356–66.

Häussler, Carolin, and Jeannette A. Colyvas. 2011. "Breaking the Ivory Tower: Academic Entre-preneurship in the Life Sciences in UK and Germany." *Research Policy* 40 (1): 41–54.

Hellsten, Iina. 2002. "Selling the Life Sciences: Promises of a Better Future in Biotechnology Advertisements." *Science as Culture* 11 (4): 459–79.

Henderson, Lesley, and Jenny Kitzinger. 2007. "Orchestrating a Science 'Event': The Case of the Human Genome Project." *New Genetics and Society* 26 (1): 65–83.

Hilgartner, Stephen. 1990. "The Dominant View of Popularization: Conceptual Problems, Politi-cal Uses." *Social Studies of Science* 20 (3): 519–39.

Holliman, Richard, Jeff Thomas, Sam Smidt, Eileen Scanlon, and Elizabeth Whitelegg. 2009a. *Practising Science Communication in the Information Age: Theorising Professional Practices.* Oxford: Oxford University Press.

Holliman, Richard, Elizabeth Whitelegg, Eileen Scanlon, Sam Smidt, and Jeff Thomas. 2009b. *Investigating Science Communication in the Information Age: Implications for Public Engagement and Popular Media.* Oxford: Oxford University Press.

Horlick-Jones, Tom, John Walls, Gene Rowe, Nick Pidgeon, Wouter Poortinga, Graham Murdock, and Tim O'Riordian. 2007. *The GM Debate: Risk, Politics and Public Engagement.* New York: Routledge.

Horst, Maja. 2008. "In Search of Dialogue: Staging Science Communication in Consensus Confer-ences." In *Communicating Science in Social Contexts*, edited by Donghong Cheng, Michel Claes-sens, Toss Gascoigne, Jenni Metcalfe, Bernard Schiele, and Shunke Shi, 259–74. Dordrecht: Springer.

———. 2010. "Collective Closure? Public Debate as the Solution to Controversies about Science and Technology." *Acta Sociologica* 53 (3): 195–211.

Horst, Maja, and Alan Irwin. 2010. "Nations at Ease with Radical Knowledge: On Consensus, Consensusing and False Consensusness." *Social Studies of Science* 40 (1): 105–26.

Horst, Maja, and Mike Michael. 2011. "On the Shoulders of Idiots: Re-thinking Science Communication as 'Event.'" *Science as Culture* 20 (3): 283–306.

House of Lords. 2000. *Third Report: Science and Society.* London: The Stationery Office, Parliament. Accessed at http://www.publications.parliament.uk/pa/ld199900/ldselect/ldsctech/38/3801.htm.

Irwin, Alan. 1995. *Citizen Science: A Study of People, Expertise and Sustainable Development.* London: Routledge.

___. 2001. "Constructing the Scientific Citizen: Science and Democracy in the Biosciences." *Public Understanding of Science* 10 (1): 1–18.

___. 2006. "The Politics of Talk: Coming to Terms with the 'New' Scientific Governance." *Social Studies of Science* 36 (2): 299–320.

Irwin, Alan, and Brian Wynne. 1996. *Misunderstanding Science? The Public Reconstruction of Science and Technology.* Cambridge: Cambridge University Press.

Jensen, Eric, and Nicola Buckley. 2012. "Why People Attend Science Festivals: Interests, Motivations and Self-Reported Benefits of Public Engagement with Research." *Public Understanding of Science* 23 (5): 557–73.

Jia, Hepeng, and Li Liu. 2014. "Unbalanced Progress: The Hard Road from Science Popularisation to Public Engagement with Science in China." *Public Understanding of Science* 23 (1): 32–37.

Johnson, David R., Elaine Howard Ecklund, and Anne E. Lincoln. 2014. "Narratives of Science Outreach in Elite Contexts of Academic Science." *Science Communication* 36 (1): 81–105.

Keene, Melanie. 2014. "Familiar Science in Nineteenth-Century Britain." *History of Science* 52 (1): 53–71.

Kerr, Anne, Sarah Cunningham-Burley, and Richard Tutton. 2007. "Shifting Subject Positions: Experts and Lay People in Public Dialogue." *Social Studies of Science* 37: 385–411.

Kruvand, Marjorie, and Sungwook Hwang. 2007. "From Revered to Reviled: A Cross-Cultural Narrative Analysis of the South Korean Cloning Scandal." *Science Communication* 29 (2): 177–97.

Lam, Alice. 2010. "From 'Ivory Tower Traditionalists' to 'Entrepreneurial Scientists'? Academic Scientists in Fuzzy University–Industry Boundaries." *Social Studies of Science* 40 (2): 307–40.

Leach, Melissa, Ian Scoones, and Brian Wynne, eds. 2005. *Science and Citizens: Globalization and the Challenge of Engagement.* London: Zed Press.

Lewenstein, Bruce V. 1995. "From Fax to Facts: Communication in the Cold Fusion Saga." *Social Studies of Science* 25 (3): 403–36.

Logan, Robert A. 2001. "Science Mass Communication: Its Conceptual History." *Science Communication* 23 (2): 135–63.

Lynch, John, Desire Bennett, Alison Luntz, Court Toy, and Eva VanBenschoten. 2014. "Bridging Science and Journalism: Identifying the Role of Public Relations in the Construction and Circulation of Stem Cell Research among Laypeople." *Science Communication* 36 (4): 479–501.

Macdonald, Sharon. 1995. "Consuming Science: Public Knowledge and the Dispersed Politics of Reception among Museum Visitors." *Media, Culture and Society* 17 (1): 13–29.

Marres, Noortje. 2012. *Material Participation: Technology, the Environment and Everyday Publics*. Basingstoke: Palgrave Macmillan.

Marris, Claire. 2015. "The Construction of Imaginaries of the Public as a Threat to Synthetic Biology." *Science as Culture* 24 (1): 83–98.

Massarani, Luisa, Dominique Brossard, Carla Almeida, Bruno Buys, and Emily Acosta Lewis. 2013. "Media Frame Building and Culture: Transgenic Crops in Two Brazilian Newspapers during the 'Year of Controversy.'" *E-Compós* 16 (1).

Mellor, Felicity, Stephen Webster, and Alice Bell. 2011. *Content Analysis of the BBC's Science Coverage: Review of the Impartiality and Accuracy of the BBC's Coverage of Science*. London: BBC.

Michael, Mike. 1998. "Between Citizen and Consumer: Multiplying the Meanings of 'Public Understanding of Science.'" *Public Understanding of Science* 7: 313–27.

———. 2000. "Futures of the Present: From Performativity to Prehension." In *Contested Futures: A Sociology of Prospective Techno-Science*, edited by Nik Brown, Brian Rappert, and Andrew Webster, 21–39. Aldershot: Ashgate.

Miller, Jon D. 2004. "Public Understanding of, and Attitudes toward, Scientific Research: What We Know and What We Need to Know." *Public Understanding of Science* 13 (3): 273–94.

Miller, Steve, and Declan Fahy. 2009. "Can Science Communication Workshops Train Scientists for Reflexive Public Engagement? The ESConet Experience." *Science Communication* 31 (1): 116–26.

Myers, Greg. 1985. "The Social Construction of Two Biologists' Proposals." *Written Communication* 2 (3): 219–45.

National Science Board. 2014. *Science and Engineering Indicators 2014*. Arlington, VA: National Science Foundation.

National Science Foundation. 2015. *Advancing Informal STEM Learning Program*. Accessed at http://informalscience.org/nsf-aisl.

Nelkin, Dorothy. 1975. "The Political Impact of Technical Expertise." *Social Studies of Science* 5 (1): 35–54.

———. 1995. *Selling Science: How the Press Covers Science and Technology*. New York: W. H. Freeman.

Neresini, Federico. 2000. "And Man Descended from the Sheep: The Public Debate on Cloning in the Italian Press." *Public Understanding of Science* 9: 359–82.

Neresini, Federico, and Massimiano Bucchi. 2011. "Which Indicators for the New Public Engagement Activities? An Exploratory Study of European Research Institutions." *Public Understanding of Science* 20 (1): 64–79.

Nisbet, Matthew C., Dietram A. Scheufele, James Shanahan, Patricia Moy, Dominique Brossard, and Bruce V. Lewenstein. 2002. "Knowledge, Reservations, or Promise? A Media Effects Model for Public Perceptions of Science and Technology." *Communication Research* 29 (5): 584–608.

Nishizawa, Mariko. 2005. "Citizen Deliberations on Science and Technology and Their Social Environments: Case Study on the Japanese Consensus Conference on GM Crops." *Science and Public Policy* 32 (6): 479–89.

Packer, Kathryn, and Andrew Webster. 1996. "Patenting Culture in Science: Reinventing the Scientific Wheel of Credibility." *Science, Technology, & Human Values* 21 (4): 427–53.

Palmer, Sarah E., and Renato A. Schibeci. 2014. "What Conceptions of Science Communication Are Espoused by Science Research Funding Bodies?" *Public Understanding of Science* 23 (5): 511–27.

Pellizzoni, Luigi. 2001. "The Myth of the Best Argument: Power, Deliberation and Reason." *British Journal of Sociology* 52: 59–86.

Peters, Hans Peter, Dominique Brossard, Suzanne de Cheveigne, Sharon Dunwoody, Monika Kallfass, Steve Miller, and Shoji Tsuchida. 2008. "Interactions with the Mass Media." *Science* 321 (5886): 204.

Petersen, Alan, Alison Anderson, Stuart Allan, and Clare Wilkinson. 2009. "Opening the Black Box: Scientists' Views on the Role of the News Media in the Nanotechnology Debate." *Public Understanding of Science* 18 (5): 512–30.

Pidgeon, Nick, Adam Corner, Karen Parkhill, Alexa Spence, Catherine Butler, and Wouter Poortinga. 2012. "Exploring Early Public Responses to Geoengineering." *Philosophical Transactions of the Royal Society A: Mathematical Physical and Engineering Sciences* 370 (1974): 4176–96.

Pinch, Trevor. 2003. "Giving Birth to New Users: How the Minimoog Was Sold to Rock and Roll." In *How Users Matter: The Co-construction of Users and Technology*, edited by Nelly Oudshoorn and Trevor Pinch, 247–70. Cambridge, MA: MIT Press.

Plesner, Ursula, and Maja Horst. 2013. "Before Stabilization." *Information, Communication & Society* 16 (7): 1115–38.

Poliakoff, Ellen, and Thomas L. Webb. 2007. "What Factors Predict Scientists' Intentions to Participate in Public Engagement of Science Activities?" *Science Communication* 29 (2): 242–63.

Powell, Maria C., and Mathilde Colin. 2008. "Meaningful Citizen Engagement in Science and Technology: What Would It Really Take?" *Science Communication* 30 (1): 126–36.

Priest, Susanna Hornig. 2001. "Cloning: A Study in News Production." *Public Understanding of Science* 10: 59–69.

___. 2006. "The Public Opinion Climate for Gene Technologies in Canada and the United States: Competing Voices, Contrasting Frames." *Public Understanding of Science* 15 (1): 55–71.

Ranger, Mathieu, and Karen Bultitude. 2016. "'The Kind of Mildly Curious Sort of Science Interested Person Like Me': Science Bloggers' Practices Relating to Audience Recruitment." *Public Understanding of Science* 25 (3): 361–78.

Rennie, Léonie J., and Gina F. Williams. 2006. "Adults' Learning about Science in Free-Choice Settings." *International Journal of Science Education* 28 (8): 871–93.

Reis, Raul. 2008. "How Brazilian and North American Newspapers Frame the Stem Cell Research Debate." *Science Communication* 29 (3): 316–34.

Riesch, Hauke. 2015. "Why Did the Proton Cross the Road? Humour and Science Communication." *Public Understanding of Science* 24 (7): 768–75.

Riesch, Hauke, and Jonathan Mendel. 2014. "Science Blogging: Networks, Boundaries and Limitations." *Science as Culture* 23 (1): 51–72.

Rödder, Simone, Martina Franzen, and Peter Weingart, eds. 2012. *The Sciences' Media Connection: Public Communication and Its Repercussions*, vol. 28. Sociology of the Sciences Yearbook. Dordrecht: Springer.

Rowe, Gene, and Lynn J. Frewer. 2005. "A Typology of Public Engagement Mechanisms." *Science, Technology, & Human Values* 30 (2): 251–90.

Royal Society. 2006. *Science and the Public Interest: Communicating the Results of New Scientific Research to the Public*. London: The Royal Society.

Shapin, Steven. 1990. "Science and the Public." In *Companion to the History of Modern Science*, edited by R. C. Olby, Geoffrey N. Cantor, John R. R Christie, and M. J. S. Hodge, 990–1007. London: Routledge.

___. 2009. *The Scientific Life: A Moral History of a Late Modern Vocation*. Chicago: University of Chicago Press.

Silvertown, Jonathan. 2009. "A New Dawn for Citizen Science." *Trends in Ecology & Evolution* 24 (9): 467–71.

Sismondo, Sergio. 2009. "Ghosts in the Machine: Publication Planning in the Medical Sciences." *Social Studies of Science* 39 (2): 171–98.

Socientize Project. 2014. *Green Paper on Citizen Science for Europe: Towards a Society of Empowered Citizens and Enhanced Research*. Socientize/European Commission.

Stilgoe, Jack, Simon J. Lock, and James Wilsdon. 2014. "Why Should We Promote Public Engagement with Science?" *Public Understanding of Science* 23 (1): 4–15.

Stirling, Andrew. 2008. "'Opening Up' and 'Closing Down': Power, Participation, and Pluralism in the Social Appraisal of Technology." *Science, Technology, & Human Values* 33 (2): 262–94.

Sundström, Göran, Linda Soneryd, and Staffan Furusten. 2010. *Organizing Democracy: The Construction of Agency in Practice*. Northampton, MA: Edward Elgar.

Suryanarayanan, Sainath, and Daniel L. Kleinman. 2012. "Be(e)coming Experts: The Controversy over Insecticides in the Honey Bee Colony Collapse Disorder." *Social Studies of Science* 43 (2): 215–40.

Tironi, Manuel. 2014. "Disastrous Publics: Counter-enactments in Participatory Experiments." *Science, Technology, & Human Values* 40 (4): 564–87.

Trench, Brian. 2008. "Towards an Analytical Framework of Science Communication Models." In *Communicating Science in Social Contexts*, edited by Donghong Cheng, Michel Claessens, Toss Gascoigne, Jenni Metcalfe, Bernard Schiele, and Shunke Shi, 119–35. Springer.

Turney, Jon. 2006. *Engaging Science: Thoughts, Deeds, Analysis and Action*. London: Wellcome Trust.

Tuunainen, Juha Kalevi, and Tarja Tellervo Knuuttila. 2009. "Intermingling Academic and Business Activities: A New Direction for Science and Universities?" *Science, Technology, & Human Values* 34 (6): 684–704.

van Lente, Harro. 1993. *Promising Technology: The Dynamics of Expectations in Technological Developments*. Ph.D. thesis. University of Twente, Enschede, Netherlands.

van Lente, Harro, and Arie Rip. 1998. "The Rise of Membrane Technology: From Rhetorics to Social Reality." *Social Studies of Science* 28 (2): 221–54.

Varughese, Shiju Sam. 2011. "Media and Science in Disaster Contexts: Deliberations on Earthquakes in the Regional Press in Kerala, India." *Spontaneous Generations: A Journal for the History and Philosophy of Science* 5 (1): 36–43.

Visvanathan, Shiv. 2005. "Knowledge, Justice and Democracy." In *Science and Citizens: Globalization and the Challenge of Engagement*, edited by Melissa Leach, Ian Scoones, and Brian Wynne, 83–100. London: Zed Books.

Welsh, Ian, and Brian Wynne. 2013. "Science, Scientism and Imaginaries of Publics in the UK: Passive Objects, Incipient Threats." *Science as Culture* 22 (4): 540–66.

Wu, Guosheng, and Hui Qiu. 2013. "Popular Science Publishing in Contemporary China." *Public Understanding of Science* 22 (5): 521–29.

Wynne, Brian. 1992. "Uncertainty and Environmental Learning: Reconceiving Science and Policy in the Preventive Paradigm." *Global Environmental Change* 2 (2): 111–27.

___. 2006. "Public Engagement as a Means of Restoring Trust: Hitting the Notes but Missing the Music?" *Community Genetics* 9: 211–20.

Yeo, Sara K., Michael A. Xenos, Dominique Brossard, and Dietram A. Scheufele. 2015. "Selecting Our Own Science How Communication Contexts and Individual Traits Shape Information Seeking." *The Annals of the American Academy of Political and Social Science* 658 (1): 172–91.

You, Jia. 2014. "Who Are the Science Stars of Twitter?" *Science* 345 (6203): 1440–41.

Zhang, Xiaoling. 2007. "Breaking News, Media Coverage and 'Citizen's Right to Know' in China." *Journal of Contemporary China* 16 (53): 535–45.

V   Engaging with Societal Challenges

Clark A. Miller

Science and technology are powerful human institutions that permeate contemporary societies. They influence how people are born, grow, live, work, and die. Even something as basic as what we eat is profoundly shaped by science via new breeds of plants and animals, new ways to process food and regulate its safety, new ideas about nutrition and health. Modern ideologies hold scientific reasoning, technological innovation, and efficient organization as essential to societal progress. Yet, historically, the pervasiveness of science and technology in modern societies has created as many challenges as it has solved. This trend continues. Many of the most significant and controversial insights to come from research in STS ultimately derived from revealing the not-so-happy truth that science and engineering are not merely benign forces for social good. They are also powerful sources and instruments for ordering social worlds, and with them, patterns of social inclusion and exclusion, fairness and inequality, risk and destruction. As societies face and seek solutions to the grand challenges of the twenty-first century, these insights are crucial. STS researchers are thus being drawn into and increasingly asked to play leadership roles in projects and debates with real impact. Their goal: to help communities and societies reconfigure the relationships among science, technology, and other important facets of social, economic, and political life, today and tomorrow, in the service of improving human lives and livelihoods.

The chapters in this section synthesize STS scholarship around a selection of significant issues currently confronting societies around the globe. The power of STS in these arenas lies in opening up the black boxes of science and technology; breaking down artificial boundaries that separate science, technology, and society; and opening up new possibilities for imagining and implementing alternative sociotechnical arrangements. As the chapters in earlier sections of the *Handbook* have illuminated in great depth, people inhabit both epistemic and material systems that entangle social identities, values, relationships, and institutions. These systems—dynamically structured ways of conceptualizing and organizing natural, human, and technological environments—are products of human cognitive and physical labor and organization. In the lexicon of STS, society and technoscience are *co-produced*: through their work, people simultaneously arrange both the world and what is known about it. As workers, citizens, scientists, consumers, managers, engineers, and leaders, people imagine, make, and govern worlds inhabited by both themselves and a great diversity of machineries. They imagine, make, and inhabit ways of knowing that make sense of those worlds. In the process, they imagine and reimagine, make and remake themselves and those around them.

The fact that people create the epistemic and material systems they inhabit emphasizes the significance of three key STS ideas—design, innovation, and expertise—that orient the chapters in this section. In STS, *design* is both feature and practice. It encompasses both the aesthetic and organizational patterning of an object or an entity—the sleek cool visual lines of an Apple iPhone and the ordering of its internal components—as well as the means of doing design work. The chapters in this section highlight the centrality of design—and redesign—to today's great challenges. The design of cities, Fortun and her colleagues illustrate in chapter 34, for example, shapes the patterns of destruction that flow from purportedly "natural" disasters. Likewise, as Vogel and her colleagues highlight in chapter 33, the design of intelligence-gathering systems determines what is known—and not known—about the security threats that nations face. For STS, design is never just technical but always *socio*technical, raising important questions about who participates, in what ways, and to what effect. Equally important are questions of how design choices ultimately intersect with the arrangements and dynamics of social networks and relationships, distributions of health, wealth, and well-being, and exercises of power and responsibility. How engineers practice design and organize design processes—and especially how they collect and incorporate insights from prospective users of technologies—impacts whether those technologies enable new modes of thriving among diverse communities, for example, the aging or the disabled, as Joyce and her colleagues explore in chapter 31. Design practices implicate who technologies empower, as well as whose lives and livelihoods are put at risk, subjected to new constraints, or otherwise excluded from benefiting from new technologies. In the same way, Beck and her colleagues remind us in chapter 36, the design (i.e., framing) of scientific and policy problems and expert institutions shapes who participates and therefore who has voice and power in processes of global governance.

STS is not merely concerned with unpacking existing designs and design processes. One of the most important insights in the chapters in this section and from STS research more broadly is that science and technology are remarkably flexible in design and meaning. The organization of science and technology is neither given nor fixed by nature but rather the result of tacit and explicit choices that could have been made otherwise. The decision to design to one set of standards, as opposed to another, reflects commitments to particular values and meanings that could change. This implies at least the potential for redesign, even if that potential may not be visible to those living inside particular technoscientific arrangements, who may experience the implications for design choices over long periods of time.

The prospect of redesign highlights the second key STS idea at work in this section: *innovation*. Societal discourses often define innovation in terms of scientific and engineering creativity, especially for discrete technological devices. For STS, however, this definition has proven too narrow. Innovation inevitably entails not just technical work but also a diverse array of people taking up and putting to use new ideas and technologies in service of many goals and objectives. This process readjusts not only the evolving design of devices but also the social identities, values, behaviors, relationships, networks, institutions, and infrastructures interwoven with them. Innovation is thus a process of transforming—via co-production—sociotechnological systems and arrangements, including both technologies and their accompanying forms of life and work throughout society.

The complexities of innovation in sociotechnological systems lie at the heart of many of the grand challenges confronting contemporary societies. Perhaps most obviously, as Iles and Beck and their colleagues show in chapters 32 and 36, respectively, efforts to prevent climate change and create more sustainable forms of agriculture require innovation that extends well beyond technologies for producing energy and food to the social, economic, and political institutions within which existing technologies are embedded. What is ultimately required in such efforts are global transformations of sociotechnological systems, including not only technologies but also social, economic, and political practices, identities, values, ways of knowing, relationships, and modes of organization. As an alternative yet equally powerful example, the dynamics of aging also illustrate the limitations of existing models of technological innovation that restrict its scope to the laboratory or the start-up company. As Joyce and her colleagues show in chapter 31, current innovation strategies often constrain the ability of scientists and engineers to design and create technologies that facilitate rather than complicate the ability of individuals to reconfigure their own sociotechnological arrangements in ways that allow them to continue to live as they'd like as their bodies and minds age.

Just as important, as Joyce and her colleagues remind us, as do Ottinger and her colleagues in chapter 35, is that when we understand and model innovation more broadly—as processes of change in the relationships among science and technology, their users, and larger societies and systems—the drivers of innovation are certainly not always scientists and engineers. In environmental justice movements, mobilized citizens can be powerful sources of innovation in how societies know and understand the problems they confront and how they develop workable solutions.

The relative significance of scientists, engineers, users, and citizens in fostering innovation in different contexts highlights the significance of *expertise* as the third focal idea that frames the chapters in this section. For STS, expertise is not just about who has knowledge. Rather, it is also about how the making of knowledge is organized; who participates, in what ways, and at what points in the process; who has rights and responsibilities to speak authoritatively about knowledge; and the norms and rules for both making and applying knowledge to important societal decisions. Crucially, therefore, as each of the following chapters explores, a central question for STS is about the relative power and authority of diverse knowers and forms of knowledge in governance processes. Thus, the construction of expertise and experts, as well as their differentiation from publics, is a political exercise entailing the allocation and restraint of power. As societies confront contemporary social and political challenges, the framing and analysis of problems and solutions is profoundly influenced by how societies define what counts as expertise, who counts as an expert, how expert advisory institutions are organized, and what authority experts are granted in relation to other participants in decision processes.

The chapters in this section highlight both the centrality of expertise to processes of decision making across a wide range of domains, from engineering to public policy, as well as the importance of understanding the design of expertise—how particular definitions of expertise are constructed and adopted and acquire stability or are challenged in particular decision processes. How do experts set criteria for the design of new technologies and the distributions of benefit, cost, and risk they create (e.g., in the discussion of technologies of aging in chapter 31 and agricultural technologies in chapter 32)? Whose knowledge and values should guide policy decisions (e.g., in discussions of agricultural policy in chapter 32 and environmental policy in chapter 35)? How are diverse forms of expertise organized, classified, put in relation to one another, or even excluded, with what implications for the construction and flow of knowledge within a policy domain (e.g., for security expertise in chapter 33 and climate change expertise in chapter 36)? Who is an expert within a particular domain and who is not, who decides, using what standards, and how might STS contribute to rethinking these definitions (e.g., in disaster research in chapter 34 and in environmental justice advocacy in chapter 35)?

Just as significant, innovation often requires reconfiguring prevailing notions of expertise in society and the institutional arrangements built on top of them. This is rarely easy. As Joyce and her colleagues argue in chapter 31, even opening up engineering design processes to the users of prospective technologies can be difficult to do effectively. Paradigms held in place by powerful communities of experts can be remarkably durable, as Iles and his colleagues demonstrate in chapter 32. Yet, as Ottinger and her colleagues emphasize in chapter 35—and Fortun and her colleagues anticipate in chapter 34—long-term engagement and collaboration has the potential, if done well, to create opportunities to reframe problems, respecify desired outcomes, and redesign solutions in ways that bridge diverse communities of experts and publics.

Contemporary societies confront significant social, policy, and economic challenges: rising inequality, threats to security, unsustainable markets, shifting population demographics, and many more. More often than not, science and technology lie in complex ways at the heart of both these challenges and proposed solutions. As the chapters in this section illuminate, STS offers a wealth of critical insights, methods, models, conceptual frameworks, and modes of analysis that can help disentangle societal problems and inform the design of solutions. The growing vibrancy and impact of the field's participation in these arenas is promising, even as it also challenges STS researchers to continue to reflect on how their scholarship and pedagogy may evolve as they become further entangled in the worries of the world. Opening up ideas and practices of design, innovation, and expertise to critical reflection and reconfiguration offers an important start on remaking future societies for the better.

31 Aging: The Sociomaterial Constitution of Later Life

Kelly Joyce, Alexander Peine, Louis Neven, and Florian Kohlbacher

Introduction

Populations across the globe are aging. The percentage of people older than 65 years is growing in many nations. As part of this trend, the number of the oldest old (people over 80) is also expected to increase (Kinsella and He 2009; United Nations 2013). Although the "graying of society" is happening at different rates and to differing degrees in various countries, most countries are expected to see increases in their elder populations.[1] Japan, because it has a large percentage of old people and has experience managing this trend, is considered a super aging society leading the way into the future (Kohlbacher and Rabe 2015; Muramatsu and Akiyama 2011).

While one could argue that lengthening the lifespan is a success—after all, in countries with aging populations it is increasingly less common to die from illness, wars, or accidents—aging populations are instead often interpreted or framed as a serious problem or even a threat (Johnson 2005; Kohlbacher 2011). Paradoxically, even as extending life is considered a highly desirable goal of emerging technologies and science, aging populations are considered a problem for social policy and social stability. Policy makers, companies, researchers, and lobby groups around the world commonly suggest that aging will lead to a global crisis for health-care systems, for pension schemes, for the innovative capacity of economies, and for the social relations between age groups (Neven 2011, 2015; Peine et al. 2015; Roberts and Mort 2009). This negative rhetoric about aging populations is frequently accompanied by a narrative that positions investment in technoscientific innovation as a "triple win" (Neven 2011, 2015): alleviating the societal consequences of demographic aging, providing a better life for old people (often by promising that elders will be able to live longer at home), and generating business activity and economic growth.

This grand narrative about science, technology, and aging, however, is also worrisome. It positions aging and old people as inherently problematic and proceeds on a naïve understanding of the relations between technological design, ideas about later life, and the lived realities of old people (Joyce and Loe 2010a; Neven 2011, 2015; Peine et al. 2015; Roberts and Mort 2009). So far, the triple-win narrative has largely perpetuated the idea that science and technology are separate from the meaning and experiences of aging. In such scenarios, aging is understood as a separate, distinct process, for which science and technology can provide improvements or solutions (Peine et al. 2015). As a consequence, ongoing debates have not elaborated an understanding of how we already age with technology, how new generations of elders are already users and consumers of technology, and how, therefore, science and technology have contributed to the current definitions and practices of later life. What falls out from ongoing debates is the more fundamental, and arguably increasing, intertwinement of later life with scientific practices and technology design and use. That is, the very definition of what counts as a normal later life, we argue, is bound up in technoscientific innovation and use and vice versa.

Against this background, science and technology studies (STS) scholars have identified the intersection of technology, science, and aging as a key issue for contemporary societies and are working to significantly improve understandings of how the lives and livelihoods of elders are bound up with technologies, science, and technological systems. Such interactions both shape and are shaped by a wide range of social, institutional, economic, policy, and material settings, relationships and practices that surround aging, constituting what we call the sociomaterial constitution of later life (Peine et al. 2015). This chapter synthesizes insights from STS scholarship on aging, science, and technology and argues for the importance of aging as a topic for future research in the field. In doing so, we highlight how STS can make a much needed contribution to the study of aging, science, and technology and enrich ongoing practices and discourses in aging research, social and technological innovation, and public policy.

We elaborate our synthesis along three lines. First, we focus on what STS research terms sociotechnical imaginaries, that is, core ideas about the roles of science and technology in society that permeate the social imagination of particular cultural contexts. In our case, these imaginaries include ageism, biomedicalization, successful aging, anti-aging medicine, and aging-in-place. Second, we address how these kinds of explicit and implicit ideas about later life are represented and subsequently built into technological design. We pay particular attention to how elders and sociotechnical imaginaries about elders are included (or not) in design practices. The third section moves from design to use where we emphasize the diversity and agency of old people. This section calls attention to elders as active agents involved in technoscientific innovation. Finally, the conclusion highlights directions for future research, emphasizing the creation of sociotechnical imaginaries that see later life as a time of creative, collective activities as well as suggestions for how STS and designers can collaborate. In all, we argue that STS insights into the nexus of aging, science, and technology can have broad impact, with the potential to contribute to sociotechnological innovation and design practices and priorities across a range of domains, local and national policies related to aging populations, and the quality of elders' daily lives.

Sociotechnical Imaginaries of Aging

Gerontologists have long demonstrated that aging is socially constituted (e.g., Dannefer and Daub 2009; Riley, Kahn, and Foner 1994). That is, the meaning and experiences of aging vary across time and contexts—aging is not an ahistorical, biological process. Drawing on Jasanoff and Kim's (2009) notion of sociotechnical imaginaries, this section examines the central ideas that socially constitute aging in contemporary life. Such ideas help us envision how we would like to age and what a "good" old age looks like. As defined by Jasanoff (2015, 7), sociotechnical imaginaries are "collectively held, institutionally stabilized, and publicly performed visions of desirable futures, animated by shared understandings of forms of social life and social order, attainable through, and supportive of, advances in technology and science." To understand the context in which scientific and technological innovation occurs in relation to aging, we need to understand the ideas and values that often drive this work (Joyce and Loe 2010a; Joyce et al. 2015). These visions for the future shape what is considered desirable for old age.

Ageism

Ageism is one of the key facets in the social constitution of aging. As a sociotechnical imaginary that pervades contemporary life, ageism imbues research and design efforts as well as how people imagine what good aging and a good later life could and should be. Although institutions and individuals challenge ageism, it still pervades visions for the future.

The term *ageism* was first coined by Robert Neil Butler (1969) to call attention to discrimination and bias against old people and aging. Ageism pervades society at macro (e.g., social structures, institutions, policies) and micro (e.g., interpersonal interactions, internalization) scales. It both shapes and is shaped by science and technology research, development, marketing, and use. The way ageism takes place varies across local and national contexts: there is not one form of ageism; its prevalence, forms, and intensity can differ by class, gender, sexuality, and ethnicity and by local and national contexts (Calasanti, Slevin, and King 2006; Cruikshank 2003; Gullette 2004).

For instance, an example from Japan highlights how context and culture shape ageism. There has been an influential body of literature claiming that Japanese people highly respect older people (e.g., Palmore and Maeda 1985) based on the traditional dominance of Confucianism, which demands respect toward old people (O'Leary 1993). The use of honorific language in Japan and priority seating for old people support this observation (Koyano 1989). However, there is substantial empirical evidence to the contrary that challenges the narrative of the "honorable elder" in Japan (Koyano 1989, 1997; Levy 1999; Tsuji 1997). Overall, it seems that negative images of old age prevail in Japan (Formanek 2008) or at least that both positive and negative views exist at the same time (Prieler et al. 2009, 2015).

In a study of robot development in the Netherlands, negative images of old age also emerged. Although robot designers intended their robot to provide diverse, healthy old people with practical assistance and emotional companionship, their final design was still seen as stigmatizing. The test users, who were in their sixties and seventies, interpreted the robot as a signifier of old age, which for them equaled frailty, isolation, and loneliness. The test users promptly stated that this was a robot for "old people" and actively disassociated themselves from both this category and the robot; they had internalized negative notions of "being old" and did not want to use the technology as a result (Neven 2010).

Ageism can take many forms, ranging from lack of representation and invisibility to negative stereotypes and slurs. Researchers around the globe who study the portrayal of old people in advertising, for example, find elders to be underrepresented (Prieler et al. 2009; Simcock and Sudbury 2006; Swayne and Greco 1987) and at times negatively or stereotypically represented (Hiemstra et al. 1983; Prieler et al. 2011, 2015; Zhou and Chen 1992). When old people are represented, ageist stereotypes highlight how old people are frail, sickly, forgetful, conservative, needy, technologically illiterate, and nearing death (Featherstone and Hepworth 1993, 2005; Johnson 2005; Neven 2011). As these ageist ideas are internalized, ageism becomes "a potent factor affecting the adoption of new technologies by older people" (Cutler 2005, 67). This is particularly significant since, unlike sexism and racism, which are widely recognized as discriminatory, ageism is a much more accepted form of inequality (Prieler et al. 2015; Wilkinson and Ferraro 2002, 341).

Ageism not only impacts how we understand old people and aging, it also shapes who counts as the imagined user of technologies and who is made invisible or marginalized. Although there are technologies and medicine aimed directly at elders, across cultural contexts, there is a tendency to imagine youthful bodies as the normal or ideal user for new technologies, which may make it harder for elders to adopt and use technoscientific innovations and interventions that are not directly aimed at them.

From Medicalization to Biomedicalization

The second sociotechnical imaginary that constructs aging is medicalization and the closely associated concept of biomedicalization. Medicalization, initially described by sociologists Irving Zola (1972), Eliot Friedson (1970), and Jesse Pitts (1968), calls attention to the increasing tendency to define mental, physical, and emotional processes as diseases. Medicalization pathologizes more and more processes previously considered normal, increasing the scope for medical knowledge and practice to define and intervene in everyday life. This trend has positive aspects, such as potentially destigmatizing behaviors previously thought to be under one's personal control and choices (e.g., addiction). Yet, this move is also of concern because it transforms issues from being perceived as social problems in need of societal responses to individual problems in need of medical solutions. For example, medicalization encourages the management of hyperactive children through pharmaceuticals and medical interventions instead of reimagining classroom enrollment sizes and educational practices and increasing the degree of physical activity and play in education and daily life (Conrad and Schneider 1992).

Given the changes in the organization and practice of medicine since the medicalization thesis was originally formulated in the 1970s, Clarke et al. (2003) suggest the term *biomedicalization* be used to demarcate contemporary medicalization processes. They outline five processes in particular that constitute biomedicalization: (1) the increasing privatization of biomedicine, with for-profit companies and hospitals gaining in prominence; (2) the extension of risk and surveillance categories resulting in more and more healthy conditions being labeled as predisease or at the very least risky; (3) the escalating use of technology and science in clinical practice and home care; (4) the availability of new computer technologies, such as the Internet, that change knowledge production and distribution; and (5) the production of new individual and collective health-related identities. In combination, these processes help produce the conditions needed to fuel the continued expansion of biomedical categories and treatments into additional aspects of life. This expansion enables new definitions of what counts as "normal aging" and the production of technoscientific interventions to help people achieve the new aging norms.

There is a rich body of anthropological and sociological research that documents the transformation of the emotional, mental, and physical changes associated with aging into "illnesses" (Cruikshank 2003; Estes and Binney 1989; Gubrium 1986; Lock 1993).

Memory loss, for example, was redefined into the medical category Alzheimer's disease in the 1960s and 1970s (Ballenger 2006; Gubrium 1986), and normal bodily processes such as menopause were transformed into pathology (Bell 1987; Utz 2011). Adding a productivity twist, individuals are now expected to work at maintaining cognitive health through exercises and nutrition (Williams, Higgs, and Katz 2012).

The rise of large-scale pharmaceutical industries has sped up and expanded the reach of biomedicalization into even more contours of daily life (Bell and Figert 2012; Conrad 2007; Dumit 2012). Pharmaceutical companies, working with physicians, patient foundations, government policy makers, and media, have led, for example, to the creation of new illnesses such as erectile dysfunction that have broad definitions that increase the number of people who might have the "disease" (Conrad 2007; Fishman 2007; Loe 2004) and the lowering of diagnostic thresholds for conditions such as high blood pressure and osteoporosis so that more people are now at risk (Greene 2008; Welch, Schwartz, and Woloshin 2011). In almost all cases, pharma is presented as the treatment for these new diseases. To be human in the twenty-first century is to be medicated.

The biomedicalization of aging is generative, creating new identities, norms, relationships, and work processes. Importantly, it may help bring attention to conditions previously neglected, provide relief for pain, and help people recover from illness in some situations (Lupton 1997, 98). Identities organized around biomedical categories may also provide the grounds for new group identities and social movements (Klawiter 2008). There is concern, though, that the widespread expansion of illness categories into everyday life means that elders will be increasingly managed and subjected to surveillance by medical professionals, tests, and clinical processes in ways that do not yield meaningful health or quality-of-life benefits at the individual level (Greene 2008; Loe 2004). Biomedicalization also leads to increased costs, both in terms of money and time, as well as feelings of unease about not being normal or healthy. The biomedicalization of life and aging, in other words, may have gone too far, producing negative financial and emotional effects.

Successful Aging and Anti-Aging Medicine

Even as societies reconstruct aging processes in terms of disease or pathology, two sociotechnical imaginaries—successful aging and anti-aging medicine—aim to stave off aging altogether. Successful aging challenges the idea that aging is about decline and focuses instead on strategies individuals can use to bolster physical, emotional, and mental health (see, e.g., Martin et al. 2015; Rowe and Kahn 1997).[2] This ideology has penetrated different countries and groups within countries to various degrees. In North

America critical scholarship on successful aging shows how it dovetails with the rise of sexual disease categories and drugs to change sexuality norms. Now, men and women are supposed to desire and perform penetrative sex until death (Calasanti and King 2005; Fishman 2007; Loe 2004; Mamo and Fishman 2001; Marshall 2006; Marshall and Katz 2002). Men, in particular, are expected to be virile and maintain erections to successfully embody and perform old manhood (Calasanti and King 2005). Rejecting notions of elders as passive consumers, elders also resist changing notions of sexuality and the pharmaceuticalization of sexuality (Hurd Clarke 2006; Loe 2004, 2015).

Anti-aging surgery is one strategy used to pursue youth and the status accorded to it (Bayer 2005; Brooks 2010; Kinnunen 2010). The decision to undergo anti-aging surgery is widely shaped by social and cultural contexts, varying by gender (Higgs and Jones 2009; Hurd Clarke and Korotchenko 2011), as well as across race, ethnicity, age, and national setting. Kinnunen (2010), for example, demonstrates how arguments for cosmetic surgery and anxieties about one's appearance are shaped by ageist discrimination, national contexts, and processes of globalization. Using in-depth interviews, Kinnunen shows how ageism combined with pressure to be more like Americans (who are constructed as happy and outgoing) creates a context in which Finnish people use surgical interventions to alter both their appearance and emotional selves. By transforming physical characteristics understood as Finnish (e.g., heavy foreheads, sagging eyelids, and bulbous noses), Kinnunen's respondents aim to make themselves look younger *and* more "white" and thus higher up in a perceived racial hierarchy within Europe.

The rise of anti-aging medicine, also called longevity medicine and regenerative medicine, offers another way to manage changes associated with aging. Anti-aging medicine gained momentum as a medical subfield in the 1990s in the United Kingdom, the United States, and Australia (Binstock 2004; Cardona 2009; Faulkner 2015; Fishman, Binstock, and Lambrix 2008; Moreira and Bond 2008; Mykytyn 2008, 2010). During this decade, physicians and other anti-aging proponents tried to gain scientific legitimacy by creating professional organizations such as the American Academy of Anti-Aging Medicine (A4M), the British Longevity Society and the World Anti-Aging Academy of Medicine and journals such as the *Journal of Anti-Aging Medicine*. By the 2000s, anti-aging societies and conferences were created nationally and internationally. The Japanese Society of Anti-Aging Medical Association and the European Society for Preventative, Regenerative and Anti-Aging Medicine, for example, were formed at this time. The A4M and other anti-aging organizations promote a variety of mechanisms to help people reach the triple-digit lifespan. Within this paradigm, stem cell research, nanotechnology research, supplements, caloric restriction, and off-label prescription drug use offer potential anti-aging interventions.

Interestingly, delaying or averting aging is only one facet of what draws physicians to anti-aging medicine. The motive for practicing anti-aging medicine comes in part from a desire to pursue personalized medicine, one that individualizes treatment for each patient and rejects the assembly-line organization of contemporary medicine (Fishman, Settersten, and Flatt 2010). Even as they focus on "the art of medicine," anti-aging practitioners aim to help clients achieve optimal fitness in mental, physical, and spiritual realms, accomplishing what Fishman, Settersten, and Flatt (ibid., 203) call "the full instantiation of a biomedicalization of lifestyle."

While the imaginaries of successful aging and anti-aging medicine resist many of the negative stereotypes associated with aging by recasting aging in a positive light, they also help produce ageism by underscoring both the undesirability of old age and the desirability of looking and acting like a youthful person even when people are in their 70s and 80s. In these scenarios, old people's abilities and ways of being in the world are rendered undesirable, transforming diverse ways of being into a narrower range of what counts as normal. Emphasizing the value of slower movements and changes in balance often associated with aging bodies, Joyce and Mamo (2006, 111) write:

The positioning of slowness and rest as undesirable has implications for us all at any age. In our caffeinated, sped-up world, aging people and bodies provide a potential reminder of the value and threat of slowness. This potential site of resistance (and perhaps respite) is lost when anti-aging ideals, put forth in advertisements for technologies, identify aging with the agility and energy associated with able-bodied youth.

Furthermore, these paradigms emphasize that it is *an individual's responsibility* to actively resist growing old and construct old age as a time of busyness and productivity (Katz 2000). Positioned as a technoscientifically mediated choice, people who do not pursue anti-aging interventions or practices may experience moral sanctions or stigma at work and at home (Brooks 2010). And, of course, such interventions often have limited effect—despite individuals' best efforts, bodies and minds do change. Attempts to maintain and achieve youth for one's lifespan set up expectations that are unlikely to be met.

Aging-in-Place

Finally, aging-in-place is a sociotechnical imaginary that drives current technoscience and policy interventions in Canada, the United States, and Europe (Johansson, Josephsson, and Lilja 2009; Sixsmith et al. 2014). Aging-in-place prioritizes the goal of keeping elders in their homes for as long as possible. The Centers for Disease Control and Prevention (CDC) (2015) defines aging-in-place as "the ability to live in one's own home and community safely, independently, and comfortably, regardless of age, income, or ability level." In contrast, keeping elders in their homes for as long as possible is not a primary objective in contexts where children are expected to care for their parents as they age.

The aging-in-place narrative fuels extensive research and innovation. Companion robots are being developed to provide company and support for elders so that they will have companionship and medical assistance (Joyce, Loe, and Diamond-Brown 2015; Kohlbacher and Rabe 2015; Kohlbacher, Herstatt, and Levsen 2015; Neven 2010; Peine and Moors 2015). Housing modifications, sensors, and communication technologies are also innovations aimed to support aging-in-place (Loe 2015). Since the focus on aging-in-place highlights individuals in their homes as imagined users, it may make it harder to imagine designing for groups of people or collective spaces or activities.

The emphasis on aging-in-place dovetails with a shift of care from hospitals and clinics to homes and communities (Dow and McDonald 2007). This shift, and the technological innovations that accompany it, have reconfigured care identities and care work, creating new identities and responsibilities for those who care for elders. Recent work on the ethical implications of telecare technologies, such as remote alarm or monitoring systems, demonstrates how domesticating these technologies reconfigures responsibilities, power, space, and sense of self in the homes of older persons (Callen et al. 2009; Mort, Finch, and May 2009; Mort et al. 2015; Mort, Roberts, and Milligan 2009; Mort, Roberts, and Callen 2012; Oudshoorn 2011; Pols and Willems 2011; Roberts, Mort, and Milligan 2012; Schillmeier and Domènech 2010).

In the United Kingdom, telecare includes systems in which an alarm is installed on one's body (via an alarm necklace or bracelet) and sensors in one's home; these monitoring systems send information back to a central hub where action can be taken as needed (Mort et al. 2015; Pritchard and Brittain 2015). Such systems divide care into three realms: monitoring, physical care, and social-emotional care (Roberts and Mort 2009). Although the rhetoric used to promote telecare systems promises to deliver interventions in all three realms, telecare primarily provides monitoring. Furthermore, telecare systems simplify the complexity of providing care and ignore the interconnection of monitoring, physical care, and social-emotional care. Roberts and Mort (2009, 147) provide an example of this when they write:

Automated dispensers, unlike face-to-face carers, cannot ensure that clients actually take their medication: they only "know" that the container has been turned over at a relevant time (a gesture that would tip the correct pills out). Telecare in this case, then, relies on the client taking the medication once reminded by the box, so is only suitable for some users. As one provider explained to us, however, even in these cases, telecare does not necessarily remove medication-related care work. Although daily visits may no longer be necessary, someone must regularly fill the pill dispenser: a task that can be done by a pharmacist or family member (who must sign a document accepting responsibility for this potentially risky action).

The importance of face-to-face interactions and the physical and emotional labor done by family, friends, and professional caregivers are often downplayed in imagined technological solutions. Despite the rhetoric of technological solutions, it is clear that technologies cannot manage all aspects of aging. And even if they could, a life managed primarily by robots and sensors may not create the type of life that is desired or enjoyed by elders.

Investigating the Design Process: How Ageism Becomes Embedded in Design

The previous section discussed the sociotechnical imaginaries that animate scientific and technological innovation aimed at aging populations. In this section, we move to a discussion of the design process itself. Using insights from material-semiotics—that is, understanding how innovators' ideas about users become embedded in technological design (Akrich 1995; Oudshoorn 2011)—STS research has begun to grasp how designers struggle to imagine the identities and practices of elder technology users and how such imagination is itself influenced by stereotypes of what it means to be old. This emerging, yet vibrant body of STS research highlights the co-production of ageism and technoscientific objects, critically investigates user involvement as a scripted process, and exposes the design paternalism that permeates many projects aimed at elder users.

Double Naturalization: Reciprocal Shaping of Ageist Imagery and Age Scripts

Using concepts such as user representations (Akrich 1995), scripts (Akrich 1992), and double naturalization (Neven 2011), we can see how stereotypical representations of aging inform technological design. User representations refers to the way designers imagine and represent possible users. As Akrich (1995, 168) argues, "[I]nnovators are from the very start constantly interested in their future users. They construct many different representations of these users, and objectify these representations in technical choices" (see also van den Scott, Sanders, and Puddephatt, chapter 17 this volume). Scripts are the end project of user representations in design; it is when innovators take their ideas about users and embed them into the actual design of a technology where the scripts then suggest how the technology should be used (Akrich 1992).

Although designers are also bound by technical limitations and the corporate or institutional context in which they work (Jelsma 2003; Mackay et al. 2000), how the prospective user is represented has a pivotal influence on design decisions. If the prospective elder user is, for instance, seen as frail, needy, and conservative or as a person with a low technological literacy and limited capacity to learn (Compagna and Kohl-bacher 2015; Neven 2011; Peine and Neven 2011), then the resulting design will likely vary when compared to a technology that was designed for a person seen as young, fit, and technologically savvy. Such "age scripts" consequently enable and constrain the users' scope for action (Neven 2010; Peine and Neven 2011).

We should, of course, not assume that users are powerless against these scripts (Oud-shoorn and Pinch 2008). Nevertheless, we should also not be blind to the possibility of forced use, the surveillance of old people to manage risks, and reduced agency for old people, especially when options to interact with technologies are taken away or when monitoring of compliance is a feature of the technology, as is, for instance, the case in many telecare technologies (Aceros, Pols, and Domènech 2015; Mort, Roberts, and Milligan 2009; Neven 2015). The resulting redistribution of power relationships can easily go unnoticed as design decisions are black-boxed. Technologies become a given, or naturalized, seemingly a part of "the way the world simply is." This move leads to what Neven (2011) calls double naturalization, that is, a process in which the user representation and then in turn the technological design are taken for granted and accepted as natural or normal.

In the case of elders, double naturalization refers to a reciprocal, mutually rein-forcing structure whereby designers' age scripts translate widely held ageist views into technologies. These technologies in turn become naturalized themselves as part of the normal and accepted life world of elders. As Neven (2011, 182) notes, this double move calls attention to how "stereotypical or ageist ways of talking about and treating older people may underlie designs of technologies for older people and, in turn, these technologies engrain these views into different (obdurate) media and thus subtly and obscurely, but potentially very forcefully, reinforce these views of older people."

How Elders Are Included and Imagined as Users

Against this background, design studies in STS have tried to understand how certain ideas about old people and later life come to prevail in design processes, while others are downplayed. It is not as if designers have not tried to involve and embed elders in design practices. In fact, in the past ten to twenty years user involvement has become a central feature in design practices, where the needs and practices of old people, their caregivers, and their families are positioned as an indispensable design input (Malanowski 2009; Östlund 2005; Peace and Hughes 2010). As a consequence, geron-technological design processes typically proceed on a collaborative basis (Botero and Hyysalo 2013), involving social scientists, designers, engineers, end users, and so forth. Moreover, they typically employ a wide range of social science and design methods to understand and engage with users. These methods range from surveys, laboratory tests, and field trials to meticulous, qualitative and interactive approaches like ethnographies, cultural probes, theater play, and co-creation (Peine and Neven 2011).

It is surprising, therefore, that the imagery of elders and later life that has emerged from such attempts has continued to naturalize a rather imbalanced imaginary of aging, focusing on frailty rather than competence, on passivity rather than activity, on overburdening rather than learning, and on coping with diseases rather than facilitating health (Joyce and Loe 2010b; Östlund 2004; Peine, Rollwagen, and Neven 2014). In other words, a double naturalization of ageism has remained influential, often despite initial intentions to avoid simplified understandings of old people and attempts to involve and understand elders in the design process (Peine and Moors 2015).

For STS scholars this is puzzling but not surprising. The material-semiotic tradition in STS user research has long demonstrated that user involvement does not provide an unfettered view on *the user* but is based on (explicit or implicit) premises, imaginations, choices, methods, and traditions that feed into representations and constructions of users, their needs, and their practices (Akrich 1995; Hyysalo 2006; Oudshoorn and Pinch 2008; Peine and Herrmann 2012; Woolgar 1990). From this perspective, the question is not so much *whether* users should be involved in design—a question an average gerontechnologist would answer with a loud and clear yes—but *how* they should be involved, under which premises, and who should control the outcome. There are four related key reasons why designers often end up constructing elders as passive, incompetent, and dependent.

First, collaborative projects need to define *doable problems* that can be addressed within the organization and timeline of the project (Lassen, Bønnelycke, and Otto 2015). User involvement thus involves a translational process by which often complex knowledge about the situated nature and diversity of lived realities needs to be conveyed into well-specified problems and user needs as a feasible target for design (Östlund 2005; Peine and Neven 2011). The reduction of complexity and diversity into specific problems is vulnerable to imbalanced power relations between users and designers (Compagna and Kohlbacher 2015; Neven 2010). Disabilities or illnesses seem to be the low-hanging fruit in this regard, which can be translated most conveniently into doable cues for design (Lassen, Bønnelycke, and Otto 2015; Malanowski 2009; Peine and Moors 2015). Having knowledge about the lived realities of old people is thus not a guarantee that this knowledge will drive the goal or outcome of a design project. Rather, the organization and timing of such projects, as well as the position of those engaged with knowledge about users, is important as well.

This leads to a related but more general second point: recent STS studies illuminate the closure mechanisms by which competing types of knowledge about old people are foregrounded or sidelined in different stages of user-centered design (Compagna and Kohlbacher 2015; Hyysalo 2006; Lassen, Bønnelycke, and Otto 2015; Lehoux 2006; Neven 2010; Oudshoorn, Neven, and Stienstra 2016; Peine and Moors 2015; Peine and Neven 2011). These studies thus question the naïve assumption that user involvement per se enriches the imagery of later life that comes to prevail in design. In STS, a key component of the closure process is called "generification work" (Pollock, Williams and D'Adderio 2007). Generification work highlights how designers need to build generic designs that address a wide diversity of user needs and practices and are robust across a variety of local circumstances. Even when knowledge about elders is included in design processes, the competing aim of generification can undermine this inclusion (Peine and Moors 2015).

Third, these elements of the design process matter because a secondary set of mental images of what old people are like as collaborators in innovation shapes how their input is understood. In this view, old people are, for instance, seen as people who have a short attention span, are quickly overburdened, cannot think conceptually, will try to please the designers with their answers, and so on (Neven 2010). Such representations provide ways in which designers can negate old people's input (Neven 2010). If elders are not included in later stages of design, this view can "speak" for them, impacting resulting technologies.

Finally, Peine, Rollwagen, and Neven (2014) argue that science and technologies that specifically target elders are designed from an implicitly paternalistic stance. They show how reifying elder technology users as a separate, static entity that preexists technology is itself an imaginary act with consequences. It positions old people as objects whose needs can be mapped and understood rather than as agents who learn about and develop new needs. Designers, then, are put in charge of mapping elders' needs and of defining their roles and practices. In a paternalistic stance, designers have to obtain an extensive understanding of old people's needs and then determine how to meet these needs. Good technology for old people is not seen as a potential source of playful encounters that might give rise to new practices and "needs." Rather, it is in elders' best interests, in terms of health, quality of life, and well-being, to follow what designers offer to them. Such design paternalism is intertwined with the double naturalization of ageist stereotypes: widespread, paternalistic framings of old people (Moody 1988) as dependent, decrepit, conservative, and technologically illiterate contribute to an environment in which designers speak *about* the needs of old people instead of providing old people the means to learn about and articulate their needs. In the triple-win narrative describe above, the designer or innovator is positioned as a type of hero who both helps needy people and solves societal problems, and who thus seemingly has a rightful place to do this.

To sum up, STS design studies highlight design paternalism as a key feature of how knowledge about elders is included in design and generification work as a key process to be understood better in order to fight ageism. Design paternalism is problematic not only because of the imagery of later life that it enacts but also in the specific way it takes agency away from old people—both in the design process itself (Compagna and Kohlbacher 2015; Neven 2010) and in the form of passive age scripts (Neven 2010, 2015; Peine, Rollwagen, and Neven 2014). Thus, design paternalism, with its basis in stereotypical views of old people, leads to technological designs that are suitable for only a small group of elders. Moreover, even elders who "belong" to this group may reject these technologies as they still see themselves as young and active and do not want technologies designed for "old people" (Barak 2009; Neven 2010; Sudbury-Riley, Kohlbacher, and Hofmeister 2015).

Theorizing Elders as Agents

To counter the practice of speaking for old people, viewing old people as passive users of technology, and the limits of the design process, we turn to gerontologists, feminists, and anthropologists and sociologists of aging who have a long history of studying old people as agents and of placing elders' embodied experiences in the center of analyses (Calasanti and Slevin 2001; Cruikshank 2003; Gullette 2004; Hurd Clarke 2010; King and Calasanti 2009; Loe 2004; Twigg 2000; Twigg and Buse 2013). Rejecting ideas that position old people as passive victims of social norms and practices—as well as of technological design—such work highlights elders' ability to meaningfully act upon and shape sociotechnical worlds. Activism by elders in the form of organizations such as the Gray Panthers, founded in 1970 by Maggie Kuhn, the Older Women's League (OWL), founded in 1980 by Tish Sommers and Laurie Shields, and AARP (formerly the American Association of Retired Persons), founded in 1958 by Dr. Ethel Percy Andrus, also stresses elders' agency and lived experiences with the aim to put such perspectives front and center in policy making and sociotechnical innovation.

Building on the focus of aging literatures on elders' agency and the STS emphasis on users as meaningful actors, recent STS work grays the cyborg (Joyce and Mamo 2006), taking note of how hybrids of machine and body are not only young people (Haraway 1991). Elders, too, are cyborgs in daily life. Moreover, old people are also active participants in technoscientific processes. Developing the concept of technogenarians to call attention to elders as agents, Joyce and Loe (2010a, 3) write, "Old people are not passive consumers of technologies such as walkers and drugs. Elders creatively utilize technological artifacts to make them more suitable for their needs even in the face of technological design and availability constraints. In this way they are technogenarians: individuals who create, use, and adapt technologies to negotiate health and illness in daily life." Understanding elders as technogenarians foregrounds elders' agency in deciding how and which technologies will or will not be used (Brittain et al. 2010; Joyce and Loe 2010b; Joyce and Mamo 2006; Loe 2015; Long 2011; Matsumoto 2011; Neven 2010; Wigg 2010). Such a perspective starts with elders' diverse, lived experiences and priorities, building on these to identify opportunities for research, development, and policy interventions; it also follows elders as they use (or don't use) technologies in daily life. This perspective argues for a richer understanding of elders that highlights elders' ability to act meaningfully upon the world: it emphasizes the plasticity of old age, and the importance of agency in adequately grasping lived realities of different stages during the life course (Dannefer 2008; Peine and Neven 2011).

A three-year longitudinal study of individuals aged 85 to 102 who are aging-in-place in upstate New York found that these elders had three relationships to technology and science: savvy tech operators, ambivalent users, and nonusers (Loe 2015). These nonagenarians relied on everyday technologies (e.g., phones, tea kettles, crockpots, mobility technologies) to negotiate daily life. When ambivalence was articulated, it was primarily in relation to pharmaceuticals and to recommendations for surgical interventions that would have little impact. As one respondent (Loe 2015, 144) noted, "I talk with [a friend] about old age and what they do to old people—medically I mean—these newfangled medical things. She told me she saw someone in the hospital, eighty-three, and she was in pain, and they were giving her something so she could live five months. We both agree, we don't want any part of that. We'd rather be comfortable. To me, the eye [injections] are worth doing. I can try that again, and see if it works. So far not. But that other stuff—I won't take it. It is for the doctors, not for us. Their pride." Starting with elders' embodied lives thus demonstrates how elders' agency contributes to decisions about which technologies matter and why.

Researching old people as technogenarians challenges design paternalism and its inherent assumption that old people are technologically illiterate along a number of dimensions. A recent set of European studies about the domestication of telecare technologies, for example, shows how such technologies obtain their identity only through the practices of elders and their caregivers (López Gómez 2015; Mort, Roberts, and Callen 2012; Oudshoorn 2011; Pritchard and Brittain 2015; Sánchez Criado et al. 2014). Numerous frictions emerge between telecare scripts and existing practices of care. Rather than identify them as threats for the implementation of telecare, these studies elaborate upon such frictions as moments of creative encounters in which a wide range of actors contribute to the renegotiation of care practices. Furthermore, STS-inspired research outlines how old people often are early adopters or innovators whose creative practices of tinkering with technoscientific objects can be an important design stimulus (Östlund 2011; Östlund and Linden 2011; Peine et al. 2015).

When theorizing elders as technogenarians, it is crucial to use an intersectional lens to produce a deep understanding of the diversity found in the category "old people." Intersectional approaches examine how dimensions such as race, class, gender, sexuality, and age intersect in and help produce people's lived experiences, aspirations, and life chances and inequalities (for an overview see Grzanka 2014). The intersection of age with difference produces unique local embodiments and meanings (Calasanti and Slevin 2001; Joyce and Mamo 2006; Oudshoorn, Neven, and Stienstra 2016). Examining inequality, intersectionality, and technology is important to understand which groups are privileged and which are made vulnerable in the process of the creation of sociotechnological infrastructures, their use, and meaning making (Oudshoorn, Neven, and Stienstra 2016).

Theorizing elders as technogenarians has yet to make it into the design literature. Bridging the two literatures opens up the space to talk about the design process pitfalls discussed earlier (where designers define doable problems, elders' views may be undermined by generification work and the prevalence of ageist stereotypes in later stages of design, and the type of knowledge about elders tends to be produced through design paternalism). Putting the two literatures in dialogue asks us to examine the innovation process at every step of the way. When and why are technological solutions privileged over other types of innovation (e.g., policy, social arrangements, economic arrangements)? Would understanding elders as agents change the types of innovation explored? What would happen if we expanded the innovation process to follow elders as they use and make sense of technologies in daily life? Would this help challenge ideas that construct old people as technologically challenged?

Future Directions

This chapter demonstrates how sociotechnical imaginaries such as ageism, biomedicalization, anti-aging medicine, and aging-in-place contribute to technoscientific interventions aimed at old people and to changing ideas of what counts as normal aging. As such, this chapter extends a long-standing insight from gerontological research—that aging is a social and cultural process—into the realm of science and technology, theorizing how science, technology, and aging are inextricably linked. The practices of designing and using technoscientific objects need to be studied as important arenas in which ideas and practices of later life take shape, even as they also shape technologies for elders. In this regard, STS scholarship shows how age scripts that emphasize elders as having a low technological literacy and limited capacity to learn have reinforced ageist stereotypes and culminated in design paternalism. This is at cross-purposes with empirically grounded theorizing of old people as active agents involved in technoscientific innovation. Starting with the idea of old people as technogenarians challenges assumptions about elders' technological literacy, puts elders' views and practices at the center of critical analysis, and sees creative interactions as crucial to the design process.

Critical STS reflections on the intersection of aging, science, and technology not only provide opportunities to prevent the incorporation of stereotypical representations of later life in the design of new technologies and scientific research. They also question the very notion of *technological solutions* itself, thus highlighting how we should reflect upon the sociotechnical networks that new technologies and scientific knowledge engender. Right now, visions for the future of aging highlight the isolated individual in his or her home and aim to stave off aging as much as possible. Instead of these sociotechnical imaginaries, what if we imagined old age as a place of exploration, one where people can explore creative and communal activities in collective or public spaces, and elders are imagined as subjects instead of objects of study? Such visions might lead to very different types of technological and scientific innovations and interventions: ones that put elders' views, social ties, and creative acts in the foreground instead of being bracketed or devalued.

Above all, this chapter highlights the need for STS scholars to engage with two groups. The first is social gerontologists and aging studies scholars (e.g., psychologists of aging, sociologists of aging, anthropologists of aging, physicians of gerontology). STS has the theoretical and methodological tools to critically study the sociomaterial practices that constitute the daily life for old people today and tomorrow, the technologies and material setups they rely on, the collectives including health-care services they are implicated in, and how new technologies enter the established sociomaterial assemblages that make up later life. Gerontologists bring decades of scholarship on the social construction and experiences of aging. Combined, STS and aging approaches can interrogate the social dimensions of technoscientific innovations, including the rhetoric that positions them as "solutions" or "interventions." That way, STS can also help overcome the idea that the experiences of and norms for aging are somehow separate from technology and science—an assumption that can often be found in the aging disciplines as well as policy and innovation discourses in the field.

Second, STS can play a role in foregrounding social science findings about old people, health, and well-being and can create or foster dialogues with social scientists, scientists, engineers, and designers so that a richer repertoire of imaginaries about later life drive sociotechnical innovation. Opening the black box of user involvement seems crucial in this regard. STS should critically engage with the conviction that more knowledge about elders and later life will automatically lead to interventions that help people live well in their later years. We need to ask, Which knowledge? For and by whom? To what end? What is the "good" life that is imagined? Who will have access to this "good" life? Transforming STS research findings into reflexive interventions that matter implies a more careful reflection about when and where new science or technology might not be the best "solution" (Peine and Moors 2015; Schot and Rip 1997).

STS should also critically engage with the notion of *gerontechnology* itself that defines a distinct domain of science and technology in relation to aging. This positions old people as peculiar and downplays that technologies aimed at old people often serve multiple markets (e.g., people living with disabilities) and may be beneficial to people of various ages and abilities. We believe that focusing on specific practices of later life, such as phasing out of work or managing extended, often global family networks, can be an important inspiration for design. But harnessing this inspiration, again, is a matter of framing: Do we frame old people as inherently problematic users of technologies to whom reduced and downsized versions of "normal" technologies need to be provided? Or, do we perceive the emerging silver market as a playing field for new technoscientific practices that can ultimately serve people of various generations and abilities (Joyce and Loe 2010b; Kohlbacher 2011; Peine, Rollwagen, and Neven 2014)?

For design and innovation processes, this chapter suggests the following: going forward, STS and designers can collaborate to co-create rich imagery of creative, meaningful later lives with technology that guides designing for the future. Such collaborations can identify the sociotechnical imaginaries that drive innovation, thereby avoiding simplistic notions of agency and old people, as well as questioning whether the imagined user is the isolated individual in a home or one who is actively involved in collective spaces and practices. Adding post-implementation studies that look at what actually happens to people and relationships once a product is used in everyday life to the innovation process will show what elders do as agents. Such a move will also illuminate how networks and relationships are changed (or not). Detailed and STS-inspired case studies of aging with technology and science will deliver not requirements, but inspirations.

Notes

1. For reports and data related to aging populations, see United Nations World Population Prospects, the 2015 Revision: http://esa.un.org/wpp/.

2. For an in-depth discussion of successful aging, see Martinson and Berridge (2015).

References

Aceros, Juan C., Jeannette Pols, and Miquel Domènech. 2015. "Where Is Grandma? Home Telecare, Good Aging and the Domestication of Later Life." *Technological Forecasting and Social Change* 93: 102–11.

Akrich, Madeleine. 1992. "The De-scription of Technical Objects." In *Shaping Technology Building Society: Studies in Sociotechnical* Change, edited by Wiebe E. Bijker and John Law, 205–24. Cambridge, MA: MIT Press.

___. 1995. "User Representations: Practices, Methods and Sociology." In *Managing Technology in Society: The Approach of Constructive Technology Assessment*, edited by Arie Rip, Thomas J. Misa, and Johan Schot, 167–84. London: Pinter Publishers.

Ballenger, Jesse. 2006. *Self, Senility and Alzheimer's Disease in Modern America: A History*. Baltimore: Johns Hopkins University Press.

Barak, Benny. 2009. "Age Identity: A Cross-Cultural Global Approach." *International Journal of Behavioral Development* 33 (1): 2–11.

Bayer, Kathryn. 2005. "Cosmetic Surgery and Cosmetics: Redefining the Appearance of Age." *Generations* 29 (3): 13–18.

Bell, Susan. 1987. "Changing Ideas: The Medicalization of Menopause." *Social Science and Medicine* 24 (6): 535–42.

Bell, Susan, and Anne Figert. 2012. "Medicalization and Pharmaceuticalization at the Intersections: Looking Backward, Sideways and Forward." *Social Science and Medicine* 75 (5): 775–83.

Binstock, Robert. 2004. "Anti-aging Medicine and Research: A Realm of Conflict and Profound Societal Implications." *Journal of Gerontology* 59A (6): 523–33.

Botero, Andrea, and Sampsa Hyysalo. 2013. "Ageing Together: Steps towards Evolutionary Co-design in Everyday Practices." *CoDesign* 9 (1): 37–54.

Brittain, Katie, Lynne Corner, Louise Robinson, and John Bond. 2010. "Aging in Place and Technologies of Place: The Lived Experience of People with Dementia in Changing Social, Physical and Technological Environments." *Sociology of Health & Illness* 32 (2): 272–87.

Brooks, Abigail. 2010. "Aesthetic Anti-aging Surgery and Technology: Women's Friend or Foe?" *Sociology of Health & Illness* 32 (2): 238–57.

Butler, Robert N. 1969. "Ageism: Another Form of Bigotry." *The Gerontologist* 9: 243–46.

Calasanti, Toni, and Neal King. 2005. "Firming the Floppy Penis: Age, Class, and Gender Relations in the Lives of Old Men." *Men and Masculinities* 8 (1): 3–23.

Calasanti, Toni, and Kathleen Slevin. 2001. *Gender, Social Inequalities, and Aging.* New York: AltaMira Press.

Calasanti, Toni, Kathleen Slevin, and Neil King. 2006. "Ageism and Feminism: From 'Et Cetera' to Center." *NWSA Journal* 18 (1): 13–30.

Callen, Blanca, Miquel Domènech, Daniel López, and Francisco Tirado. 2009. "Telecare Research: (Cosmo)politicizing Methodology." *ALTER—European Journal of Disability Research* 3: 110–22.

Cardona, Beatriz. 2009. "'Anti-aging Medicine' in Australia: Global Trends and Local Practices to Redefine Aging." *Health Sociology Review* 18 (4): 446–60.

Centers for Disease Control and Prevention (CDC). 2015. "Healthy Places Terminology." Accessed March 27, 2015, at http://www.cdc.gov/healthyplaces/terminology.htm.

Clarke, Adele, Laura Mamo, Janet Shim, Jennifer Fosket, and Jennifer Fishman. 2003. "Biomedicalization: Theorizing Technoscientific Transformations of Health, Illness, and U.S. Biomedicine." *American Sociological Review* 68 (2): 161–94.

Compagna, Diego, and Florian Kohlbacher. 2015. "The Limits of Participatory Technology Development: The Case of Service Robots in Care Facilities for Older People." *Technological Forecasting and Social Change* 93: 19–31.

Conrad, Peter. 2007. *The Medicalization of Society.* Baltimore: Johns Hopkins University Press.

Conrad, Peter, and Joseph Schneider. 1992. *Deviance and Medicalization: From Badness to Sickness.* Philadelphia: Temple University Press.

Cruikshank, Margaret. 2003. *Learning to Be Old: Gender, Culture, and Aging.* New York: Rowman and Littlefield.

Cutler, Steven. 2005. "Ageism and Technology." *Generations* 29 (3): 67–72.

Dannefer, Dale. 2008. "The Waters We Swim: Everyday Social Processes, Macrostructural Realities, and Human Aging." In *Social Structures and Aging Individuals—Continuing Challenges*, edited by K. Warner Schaie and Ronald P. Abeles, 3–22. New York: Springer.

Dannefer, Dale, and Antje Daub. 2009. "Extending the Interrogation: Life Span, Life Course, and the Constitution of Human Aging." *Advances in Life Course Research* 14 (1–2): 15–27.

Dow, Briony, and John McDonald. 2007. "The Invisible Contract: Shifting Care from the Hospital to the Home." *Australian Health Review* 31 (2): 193–202.

Dumit, Joseph. 2012. *Drugs for Life: How Pharmaceutical Companies Define Our Health.* Durham, NC: Duke University Press.

Estes, Caroll L., and Elizabeth A. Binney. 1989. "The Biomedicalization of Aging." *Gerontologist* 29 (5): 587–96.

Faulkner, Alex. 2015. "Usership of Regenerative Therapies: Age, Aging and Anti-aging in the Global Science and Technology of Knee Cartilage Repair." *Technological Forecasting and Social Change* 93: 44–53.

Featherstone, Mike, and Mike Hepworth. 1993. "Images in Ageing." In *Ageing in Society: An Introduction to Social Gerontology*, edited by John Bond, Peter Coleman, and Sheila M. Peace, 250–75. Thousand Oaks, CA: Sage.

___. 2005. "Images of Ageing: Cultural Representations of Later Life." In *The Cambridge Handbook of Age and Ageing*, edited by Malcom L. Johnson, Vern L. Bengtson, Peter G. Coleman, and Thomas B. L. Kirkwood, 354–62. Cambridge: Cambridge University Press.

Fishman, Jennifer. 2007. "Making Viagra: From Impotence to Erectile Dysfunction." In *Medicating Modern America: Prescription Drugs in History*, edited by Andrea Tone and E. Sigal Watkins, 292–352. New York: New York University Press.

Fishman, Jennifer, Robert Binstock, and Marcie A. Lambrix. 2008. "Anti-aging Science: The Emergence, Maintenance, and Enhancement of a Discipline." *Journal of Aging Studies* 22 (4): 295–303.

Fishman, Jennifer, Richard Settersten, and Michael Flatt. 2010. "In the Vanguard of Biomedicine? The 'Return' to the Art of Medicine in the Anti-aging Movement." *Sociology of Health & Illness* 32 (2): 197–210.

Formanek, Susan. 2008. "Traditional Concepts and Images of Old Age in Japan." In *The Demographic Challenge: A Handbook about Japan*, edited by Florian Coulmas, Harald Conrad, Annette Schad-Seifert, and Gabriele Vogt, 323–43. Leiden: Brill.

Friedson, Elliot. 1970. *The Profession of Medicine*. New York: Dodd, Mead.

Greene, Jeremy. 2008. *Prescribing by Numbers: Drugs and the Definition of Disease*. Baltimore: Johns Hopkins University Press.

Grzanka, Patrick R., ed. 2014. *Intersectionality: A Foundations and Frontiers Reader*. Boulder, CO: Westview Press.

Gubrium, Jaber. 1986. *Old Timers and Alzheimer's: The Descriptive Organization of Senility*. Greenwich, CT: Jai Press.

Gullette, Margaret. 2004. *Aged by Culture*. Chicago: University of Chicago Press.

Haraway, Donna J. 1991. *Simians, Cyborgs and Women: The Reinvention of Nature*. New York: Routledge.

Hiemstra, Roger, Maureen Goodman, Mary Ann Middlemiss, Richard Vosko, and Nancy Ziegler. 1983. "How Older People Are Portrayed in Television Advertising: Implications for Educators." *Educational Gerontology* 9 (2–3): 111–22.

Higgs, Paul, and Ian Rees Jones. 2009. *Medical Sociology and Old Age: Towards a Sociology of Health in Later Life*. London: Routledge.

Hurd Clarke, Laura. 2006. "Older Women and Sexuality: Experiences in Marital Relationships across the Life Course." *Canadian Journal on Aging* 25 (2): 129–40.

___. 2010. *Facing Age: Women Growing Older in Anti-aging Culture*. Lanham, MD: Rowman and Littlefield.

Hurd Clarke, Laura, and Alexandra Korotchenko. 2011. "Aging and the Body: A Review." *Canadian Journal of Aging* 30 (3): 495–510.

Hyysalo, Sampsa. 2006. "Representations of Use and Practice-Bound Imaginaries in Automating the Safety of the Elderly." *Social Studies of Science* 36 (4): 599–626.

Jasanoff, Sheila. 2015. "Future Imperfect: Science, Technology and the Imaginations of Modernity." In *Dreamscapes of Modernity: Sociotechnical Imaginaries and the Fabrication of Power*, edited by Sheila Jasanoff and Sang-Hyun Kim, 1–33. Chicago: University of Chicago Press.

Jasanoff, Sheila, and Sang-Hyun Kim. 2009. "Containing the Atom: Sociotechnical Imaginaries and Nuclear Regulations in the U.S. and South Korea." *Minerva* 47 (2): 119–46.

Jelsma, Jaap. 2003. "Innovating for Sustainability: Involving Users, Politics and Technology." *Innovation: The European Journal of Social Sciences* 16 (2): 103–16.

Johannson, Karin, Staffan Josephsson, and Margareta Lilja. 2009. "Creating Possibilities for Action in the Presence of Environmental Barriers in the Process of 'Ageing in Place.'" *Ageing and Society* 29 (1): 49–70.

Johnson, Malcolm L. 2005. "The Social Construction of Old Age as a Problem." In *The Cambridge Handbook of Age and Ageing*, edited by Malcolm L. Johnson, Vern L. Bengtson, Peter G. Coleman, and Thomas B. L. Kirkwood, 563–71. Cambridge: Cambridge University Press.

Joyce, Kelly, and Meika Loe. 2010a. "A Sociological Approach to Aging, Technology and Health." *Sociology of Health & Illness* 32 (2): 171–80.

___, eds. 2010b. *Technogenarians: Studying Health and Illness through an Aging, Science, and Technology Lens*. Chichester: Wiley-Blackwell.

Joyce, Kelly, Meika Loe, and Lauren Diamond-Brown. 2015. "Science, Technology and Aging." In *The Handbook of Cultural Gerontology*, edited by Julia Twigg and Wendy Martin, 157–64. New York: Routledge.

Joyce, Kelly, and Laura Mamo. 2006. "Graying the Cyborg: New Directions in Feminist Analyses of Aging, Science, and Technology." In *Age Matters: Realigning Feminist Thinking*, edited by Toni Calasanti and Kathleen Slevin, 99–122. New York: Taylor and Francis.

Katz, Stephen. 2000. "Busy Bodies: Activity, Aging and the Management of Everyday Life." *Journal of Aging Studies* 14 (2): 135–52.

King, Neal, and Toni Calasanti. 2009. "Aging Agents: Scholarly Imputations of Empowerment." *International Journal of Aging and Social Policy* 29 (1/2): 39–48.

Kinnunen, Taina. 2010. "A Second Youth: Pursuing Happiness and Respectability through Cosmetic Surgery in Finland." *Sociology of Health & Illness* 32 (2): 258–71.

Kinsella, Kevin, and Wan He. 2009. *U.S. Census Bureau, International Population Reports, P95/09–1, An Aging World: 2008*. Washington, DC: U.S. Government Printing Office.

Klawiter, Maren. 2008. *The Biopolitics of Breast Cancer: Changing Cultures of Disease and Activism.* Minneapolis: University of Minnesota Press.

Kohlbacher, Florian. 2011. "Business Implications of Demographic Change in Japan: Chances and Challenges for Human Resource and Marketing Management." In *Imploding Populations in Japan and Germany: A Comparison*, edited by Florian Coulmas and Ralph Luetzeler, 269–94. Leiden: Brill.

Kohlbacher, Florian, Cornelius Herstatt, and Nils Levsen. 2015. "Golden Opportunities for Silver Innovation: How Demographic Changes Give Rise to Entrepreneurial Opportunities to Meet the Needs of Older People." *Technovation* 39–40 (1): 73–82.

Kohlbacher, Florian, and Benjamin Rabe. 2015. "Leading the Way into the Future: The Development of a (Lead) Market for Care Robotics in Japan." *International Journal of Technology, Policy and Management* 15 (1): 21–44.

Koyano, Wataru. 1989. "Japanese Attitudes toward the Elderly: A Review of Research Findings." *Journal of Cross-cultural Gerontology* 4 (4): 335–45.

___. 1997. "Myths and Facts of Aging in Japan." In *Aging: Asian Concepts and Experiences—Past and Present*, edited by Susanne Formanek and Sepp Linhart, 213–27. Vienna: Verlag der Österreichischen Akademie der Wissenschaften.

Lassen, Aske Juul, Julie Bønnelycke, and Lene Otto. 2015. "Innovating for 'Aging' in a Public–Private Innovation Partnership: Creating Doable Problems and Alignment." *Technological Forecasting and Social Change* 93: 10–18.

Lehoux, Pascale. 2006. *The Problem of Health Technology: Policy Implications for Modern Health Care Systems*. New York: Routledge.

Levy, Becca R. 1999. "The Inner Self of the Japanese Elderly: A Defense against Negative Stereotypes of Aging." *International Journal of Aging and Human Development* 48 (2): 131–44.

Lock, Margaret. 1993. *Encounters with Aging: Mythologies of Menopause in Japan and North America.* Berkeley: University of California Press.

Loe, Meika. 2004. *The Rise of Viagra: How the Little Blue Pill Changed Sex in America*. New York: New York University Press.

___. 2015. "Comfort and Medical Ambivalence in Old Age." *Technological Forecasting and Social Change* 93: 141–46.

Long, Susan. 2011. "Tension, Dependency, and Sacrifice in the Relationship of an Elderly Couple." In *Faces of Aging: The Lived Experiences of the Elderly in Japan*, edited by Yoshiko Matsumoto, 60–86. Stanford, CA: Stanford University Press.

López Gómez, Daniel. 2015. "Little Arrangements That Matter: Rethinking Autonomy-Enabling Innovations for Later Life." *Technological Forecasting and Social Change* 93: 91–101.

Lupton, Deborah. 1997. "Foucault and the Medicalisation Critique." In *Foucault, Health and Medicine*, edited by Alan Petersen and Robin Bunton, 94–112. New York: Routledge.

Mackay, Hugh, Chris Carne, Paul Beynon-Davies, and Doug Tudhope. 2000. "Reconfiguring the User: Using Rapid Application Development." *Social Studies of Science* 30 (5): 737–57.

Malanowski, Norbert. 2009. "ICT-Based Applications for Active Ageing: Challenges and Opportunities." In *Information and Communication Technologies for Active Aging: Opportunities and Challenges for the European Union*, edited by Marcelino Cabrera and Norbert Malanowski, 107–27. Amsterdam: IOS Press.

Mamo, Laura, and Jennifer Fishman. 2001. "Potency in All the Right Places: Viagra as a Gendered Technology of the Body." *Body & Society* 7 (4): 13–35.

Marshall, Barbara. 2006. "The New Virility: Viagra, Male Aging and Sexual Function." *Sexualities* 9 (3): 345–62.

Marshall, Barbara, and Stephen Katz. 2002. "Sexual Fitness and the Aging Male Body." *Body & Society* 8 (4): 43–70.

Martin, Peter, Norene Kelly, Boaz Kahana, Eva Kahana, Bradley J. Willcox, D. C. Willcox, and Leonard W. Poon. 2015. "Defining Successful Aging: A Tangible or Elusive Concept?" *Gerontologist* 55 (1): 14–25.

Martinson, Marty, and Clara Berridge. 2015. "Successful Aging and Its Discontents: A Systematic Review of the Social Gerontology Literature." *Gerontologist* 55 (1): 58–69.

Matsumoto, Yoshiko, ed. 2011. *Faces of Aging: The Lived Experiences of the Elderly in Japan*. Stanford, CA: Stanford University Press.

Moody, Harry R. 1988. *Abundance of Life: Human Development Policies for an Aging Society*. New York: Columbia University Press.

Moreira, Tiago, and John Bond. 2008. "Does the Prevention of Brain Ageing Constitute Anti-ageing Medicine? Outline of a New Space of Representation for Alzheimer's Disease." *Journal of Aging Studies* 22 (4): 356–65.

Mort, Maggie, Tracy Finch, and Carl May. 2009. "Making and Unmaking Telepatients." *Science, Technology, & Human Values* 34 (1): 9–33.

Mort, Maggie, Celia Roberts, and Blanca Callen. 2012. "Ageing with Telecare: Care or Coercion in Austerity?" *Sociology of Health & Illness* 35 (6): 799–812.

Mort, Maggie, Celia Roberts, and Christine Milligan. 2009. "Ageing, Technology and the Home: A Critical Project." *ALTER—European Journal of Disability Research* 3 (2): 85–89.

Mort, Maggie, Celia Roberts, Jeannette Pols, Miquel Domènech, and Ingun Moser. 2015. "Ethical Implications of Home Telecare for Older People: A Framework Derived from a Multisited Participative Study." *Health Expectations* 18 (3): 438–49.

Muramatsu, Naoko, and Hiroko Akiyama. 2011. "Japan: Super-aging Society Preparing for the Future." *Gerontologist* 51 (4): 425–32.

Mykytyn, Courtney. 2008. "Medicalizing the Optimal: Anti-aging Medicine and the Quandary of Intervention." *Journal of Aging Studies* 22 (4): 313–21.

___. 2010. "A History of the Future: The Emergence of Contemporary Anti-aging Medicine." *Sociology of Health & Illness* 32 (2): 181–96.

Neven, Louis. 2010. "'But Obviously Not for Me': Robots, Laboratories and the Defiant Identity of Elder Test User." *Sociology of Health & Illness* 32 (2): 335–47.

___. 2011. *Representations of the Old and Aging in the Design of the New and Emerging: Assessing the Design of Ambient Intelligence Technologies for Older People.* Enschede, Netherlands: University of Twente.

___. 2015. "By Any Means? Questioning the Link between Gerontechnological Innovation and Older People's Wish to Live at Home." *Technological Forecasting and Social Change* 93: 32–43.

O'Leary, James S. 1993. "A New Look at Japan's Honorable Elders." *Journal of Aging Studies* 7 (1): 1–24.

Östlund, Britt. 2004. "Social Science Research on Technology and the Elderly: Does It Exist?" *Science Studies* 17 (2): 44–62.

___. 2005. "Design Paradigms and Misunderstood Technology: The Case of Older Users." In *Young Technologies in Old Hands: An International View on Senior Citizen's Utilization of ICT*, edited by Birgit Jaeger, 25–39. Copenhagen: DJOF Publishing.

___. 2011. "Silver Age Innovators: A New Approach to Old Users." In *The Silver Market Phenomenon: Marketing and Innovation in the Aging Society*, edited by Florian Kohlbacher and Cornelius Herstatt, 15–26. Berlin: Springer.

Östlund, Britt, and Kjell Linden. 2011. "Turning Older People's Experiences into Innovations: Ippi as the Convergence of Mobile Services and TV Viewing." *Gerontechnology* 10 (2): 103–9.

Oudshoorn, Nelly. 2011. *Telecare Technologies and the Transformation of Healthcare.* Basingstoke: Palgrave Macmillan.

Oudshoorn, Nelly, Louis Neven, and Marcelle Stienstra. 2016. "How Diversity Gets Lost: Age and Gender in Design Practices of Information and Communication Technologies." *Journal of Women and Aging* 28 (2): 170–85.

Oudshoorn, Nelly, and Trevor Pinch. 2008. "User-Technology Relationships: Some Recent Developments." In *The Handbook of Science and Technology Studies*. 3rd ed., edited by Edward Hackett, Olga Amsterdamska, Michael Lynch, and Judy Wajcman, 541–65. Cambridge, MA: MIT Press.

Palmore, Erdman, and Daisaku Maeda. 1985. *The Honorable Elders Revisited: A Revised Cross-Cultural Analysis of Aging in Japan.* Durham, NC: Duke University Press.

Peace, Sheila, and Jonathan Hughes, eds. 2010. *Reflecting on User-Involvement and Participatory Research.* London: Centre for Policy on Ageing.

Peine, Alexander, Alex Faulkner, Birgit Jaeger, and Ellen Moors. 2015. "Science, Technology and the 'Grand Challenge' of Aging: Understanding the Socio-material Constitution of Later Life." *Technological Forecasting and Social Change* 93: 1–9.

Peine, Alexander, and Andrea M. Herrmann. 2012. "The Sources of Use Knowledge: Towards Integrating the Dynamics of Technology Use and Design in the Articulation of Societal Challenges." *Technological Forecasting and Social Change* 79 (8): 1495–512.

Peine, Alexander, and Ellen H. M. Moors. 2015. "Valuing Health Technology: Habilitating and Prosthetic Strategies in Personal Health Systems." *Technological Forecasting and Social Change* 93: 68–81.

Peine, Alexander, and Louis Neven. 2011. "Social-Structural Lag Revisited." *Gerontechnology* 10 (3): 129–39.

Peine, Alexander, Ingo Rollwagen, and Louis Neven. 2014. "The Rise of the "Innosumer": Rethinking Older Technology Users." *Technological Forecasting and Social Change* 82: 199–214.

Pitts, Jesse. 1968. "Social Control: The Concept." In *International Encyclopedia of the Social Sciences*, edited by David L. Sills, 381–95. New York: Macmillan.

Pollock, Neil, Robin Williams, and Luciana D'Adderio. 2007. "Global Software and Its Provenance: Generification Work in the Production of Organizational Software Packages." *Social Studies of Science* 37 (2): 254–80.

Pols, Jeannette, and Dick Willems. 2011. "Innovation and Evaluation: Taming and Unleashing Telecare Technology." *Sociology of Health & Illness* 33 (3): 484–98.

Prieler, Michael, Florian Kohlbacher, Shigeru Hagiwara, and Akie Arima. 2009. "How Older People Are Represented in Japanese TV Commercials: A Content Analysis." *Keio Communication Review* 31: 5–21.

___. 2011. "Gender Representation of Older People in Japanese Television Advertisements." *Sex Roles: A Journal of Research* 64 (5/6): 405–15.

___. 2015. "The Representation of Older People in Television Advertisements and Social Change: The Case of Japan." *Aging and Society* 35: 865–87.

Pritchard, Gary W., and Katie Brittain. 2015. "Alarm Pendants and the Technological Shaping of Older People's Care." *Technological Forecasting and Social Change* 93: 124–32.

Riley, Matilda White, Robert L. Kahn, and Anne Foner, eds. 1994. *Age and Structural Lag.* New York: Wiley-Interscience.

Roberts, Celia, and Maggie Mort. 2009. "Reshaping What Counts as Care: Older People, Work and New Technologies." *ALTER—European Journal of Disability Research* 3 (2): 138–58.

Roberts, Celia, Maggie Mort, and Christine Milligan. 2012. "Calling for Care: 'Disembodied' Work, Teleoperators and Older People Living at Home." *Sociology* 46 (3): 490–506.

Rowe, John W., and Robert L. Kahn. 1997. "Successful Aging." *Gerontologist* 37 (4): 433–40.

Sánchez Criado, Tomás, Daniel López, Celine Roberts, and Miquel Domènech. 2014. "Installing Telecare, Installing Users: Felicity Conditions for the Instauration of Usership." *Science, Technology, & Human Values* 39 (5): 694–719.

Schillmeier, Michael, and Miquel Domènech. 2010. *New Technologies and Emerging Spaces of Care.* Farnham: Ashgate Publishing.

Schot, Johan, and Arie Rip. 1997. "The Past and Future of Constructive Technology Assessment." *Technological Forecasting and Social Change* 54: 251–68.

Simcock, Peter, and Lynn Sudbury. 2006. "The Invisible Majority? Older Models in UK Television Advertising." *International Journal of Advertising* 25 (1): 87–106.

Sixsmith, Judith, Andrew Sixsmith, Agneta Malmgren Faenge, Dorte Naumann, Csaba Kucsera, Signe Tomsone, Maria Haak, Sylvia Dahlin-Ivanoff, and Ryan Woolrych. 2014. "Healthy Aging and Home: The Perspectives of Very Old People in Five European Countries." *Social Science & Medicine* 106: 1–9.

Sudbury-Riley, Lynn, Florian Kohlbacher, and Agnes Hofmeister. 2015. "Baby Boomers of Different Nations: Identifying Horizontal International Segments Based on Self-Perceived Age." *International Marketing Review* 32 (3/4): 245–78.

Swayne, Linda E., and Alan J. Greco. 1987. "The Portrayal of Older Americans in Television Commercials." *Journal of Advertising* 16 (1): 47–54.

Tsuji, Yoko. 1997. "Continuities and Changes in the Conceptions of Old Age in Japan." In *Aging: Asian Concepts and Experiences: Past and Present*, edited by Susanne Formanek and Sepp Linhart, 197–210. Vienna: Verlag der Österreichischen Akademie der Wissenschaften.

Twigg, Julia. 2000. *Bathing: The Body and Community Care.* London: Taylor and Francis.

Twigg, Julia, and Christina E. Buse. 2013. "Dress, Dementia and the Embodiment of Identity." *Dementia* 12 (3): 326–36

United Nations, Department of Economic and Social Affairs, Population Division. 2013. *World Population Aging 2013.* ST/ESA/SER.A/348.

Utz, Rebecca. 2011. "Like Mother, (Not) Like Daughter: The Social Construction of Menopause and Aging." *Journal of Aging Studies* 25 (2): 143–54.

van den Scott, Lisa-Jo, Carrie Sanders, and Antony Puddephatt. 2016. "Reconceptualizing Users with Enriching Ethnography." In *The Handbook of Science and Technology Studies*. 4th ed., edited by Ulrike Felt, Rayvon Fouché, Clark Miller, and Laurel Smith-Doerr. Cambridge, MA: MIT Press.

Welch, H. Gilbert, Lisa Schwartz, and Steve Woloshin. 2011. *Overdiagnosed: Making People Sick in the Pursuit of Health*. Boston: Beacon Press.

Wigg, Johanna. 2010. "Liberating the Wanderers: Using Technology to Unlock Doors for Those Living with Dementia." *Sociology of Health & Illness* 32 (2): 288–303.

Wilkinson, Jody, and Kenneth Ferraro. 2002. "Thirty Years of Ageism Research." In *Ageism: Stereotyping and Prejudice against Older Persons*, edited by Todd Nelson, 339–58. Cambridge, MA: MIT Press.

Williams, Simon, Paul Higgs, and Stephen Katz. 2012. "Neuroculture, Active Aging and the 'Older Brain': Problems, Promises and Prospects." *Sociology of Health & Illness* 34: 64–78.

Woolgar, Steve. 1990. "Configuring the User: The Case of Usability Trials." *The Sociological Review* 38 (S1): 58–99.

Zhou, Nan, and Mervin Y. T. Chen. 1992. "Marginal Life after 49: A Preliminary Study of the Portrayal of Older People in Canadian Consumer Magazine Advertising." *International Journal of Advertising* 11: 343–54.

Zola, Irving. 1972. "Medicine as an Institution of Social Control." *American Sociological Review* 20 (4): 487–504.

32 Agricultural Systems: Co-producing Knowledge and Food

Alastair Iles, Garrett Graddy-Lovelace, Maywa Montenegro, and Ryan Galt

Introduction

Food is both a material out of which human cultures get made and a product of them. From ancient Egyptian grains to the bread riots of eighteenth century England to the fast-food worker strikes of 2015, food has long been the stuff of subsistence and social transformation. More recently, it has also been a central focus of science: making chemicals to kill crop pests and fertilize soil, creating plants through breeding and biotechnology, pasteurizing and testing samples of raw milk. The fields of agronomy, animal science, and horticulture on which early agricultural science was based are now witnessing the rise of gene chipping, climate modeling, and databases with terabytes of soil, temperature, and agrobiodiversity data. Yet food is supremely democratic and knowable. Particularly when it comes to food, humans from many cultures are opinionated and willingly proffer advice on eating and nutrition. Growing and eating food is also culturally symbolic, reflecting larger imaginaries and patterns of social meaning formation. The sorts of food that people consume speak not only to their preferences but also to the agricultures their cultures have developed or adopted and to their traditions and changing identities in a world of highly mobile capital, science, and trade. Food is thus a window into the many incredibly complex sociotechnological systems that span the globe, and on which humans depend.

For over 12,000 years, many generations of farmers have experimented with crops, animals, and processing methods, learning how to manage complex ecological and social conditions. Their early scientific trials gave birth to a remarkable variety of indigenous foods, exemplified in the hundreds of potato types found in the Peruvian and Bolivian Andes. Starting in Western Europe from the 1700s on, and accelerating across the twentieth century in particular, agricultural systems have undergone dramatic changes (Mazoyer and Roudart 2006). Farms have increased in their sizes, yields, and degrees of ecological simplification; they have become integrated into industrial production networks and markets that turn mass-produced crops and animals (often designed for higher yields) into processed foods. We argue that such productivist forms of agriculture need to be understood as the co-production of forms of agricultural knowledge, technology, organization, landscapes, politics, markets, consumers, eaters, and species. Understanding how all of these facets have been co-produced together—interweaving changes in agricultural science and technology with broader social, economic, and political changes—and how these transformations have made productivist agriculture so powerful vis-à-vis other forms of agriculture has been a particularly important site for research in the field of science and technology studies (STS).

Science and technology studies provides a suite of concepts, methods, and cases that reveal the ways in which productivist systems have arisen, acquired power and authority, and ultimately transformed agriculture (and its knowledges, systems, organization, and people) through encounters with science and technology. STS scholars have built a deep foundation of research that critically unpacks science, expertise, and other ways of knowing, such as local knowledge. They have investigated how scientists and industry have engaged in boundary work to make productivist agricultures more legitimate and powerful vis-à-vis "traditional" agricultures. These scholars have also learned about how the vast infrastructures of growing, processing, and distributing food have taken material shape in the everyday work needed to supply food to billions of people. STS has revealed the commonly shared epistemologies and ontologies hidden in seemingly disparate arenas of farming methods, food processing, and eating practices. STS has made visible, in other words, the politics and processes by which agriculture and food have been co-produced over the past century with material infrastructures, ecological landscapes, and social imaginaries, values, and institutions.[1]

At the same time, STS research on agriculture also looks toward the future of agriculture. Over the next few decades, ongoing processes of agricultural transformation will configure our collective human futures. Productivist trajectories in agricultural systems continue to develop new kinds of science and technology—such as flex crops for fuels, chemicals, and food, or artificial meat—that will shape the types of foods that humans eat and the kinds of bodies and diseases they develop. This knowledge may further lock human societies into highly intensive and environmentally damaging agricultural forms, despite pretensions to being sustainable. Yet, even as productivist forms of agriculture evolve, new modalities of knowledge and practice are emerging, such as agroecology and organic farming, to challenge them and offer alternatives to the world's farmers and eaters. In exploring these alternatives, the powerful critical methods of STS expose not just the limitations of productivist agriculture but how political struggles over epistemologies in growing vegetables open up new spaces for experimentation in other parts of the food system. Nonetheless, the work of advocates of change will be long and hard because they too must co-produce all the different facets of alternative agricultures.

We first discuss how taking a co-production approach can help make visible the mutual constitution of agricultural systems and technoscientific development. We then review STS research on the rise of productivist systems, followed by the challenges and transformations to such systems—underlining how this work has become more diverse in its geography, topics, methods, and participants. We conclude by arguing for the reconceptualization of food systems around diversity, not homogeneity. Many more insights are ready to be revealed through the work of new generations of researchers excited by the possibilities of doing STS work on agriculture.

Navigating Agricultural Systems through a Co-production Idiom

Agricultural systems are complicated forms of life. Unsurprisingly the "global food system" is a catchall for countless food systems that exist at and across local, subnational, national, and global levels. It consists of the metabolic processes and organisms that turn sunlight into plant sugars, the co-evolution of farmer knowledge and physical landscapes, and the millennia of seed selection that turned wild plants into edible food.[2] It comprises farm systems, processing systems, transportation systems, and marketing systems; it is trade and investment, public policy, research and development, and finance. These enfold knowledge and labor in each process and relationship. Industrialized agriculture dominates in some regions, while traditional and indigenous farming thrives in others. Much more commonly, agri-food systems feature farms and landscapes employing mixed conventional and traditional practices.

Nonetheless, industrialized variants of agriculture are imbued with particular political, economic—and *epistemic*—power in contemporary societies; over the course of the twentieth century, they have encroached on, destabilized, and displaced preexisting systems with less power, and they continue to do so in many parts of the world. The roots of productivism lie in Malthusian theory and the fear that population would outstrip world food supply. Contemporary productivist systems embed this logic, emphasizing the production of cheap, abundant food to meet projected population growth and caloric demand. This goal is achieved by continually boosting crop and animal yields, increasing scales of economy, and reducing production costs (Busch 2005). Starting in the Industrial Revolution in Europe and the United States, agricultural scientists and engineers, wealthy farmers, companies, and government agencies developed numerous technologies to allow farmers to intensify their production, reduce labor, and control nature (Mazoyer and Roudart 2006). Over the twentieth century, successive generations of new technologies, from barbed wire and tractors, pesticides and fertilizers, to hybrid seeds and genetically modified crops, served the goal of expanding food surpluses. Particularly in the United States and Europe, government policies, agricultural research systems, farming landscapes, and industrial food chains co-evolved around providing these surpluses. Supermarkets, food-processing firms, and distribution networks displaced farmer markets, local mills and abattoirs, and village stores. Such "productivist" forms of agriculture spread globally, to the point that China, Brazil, and other developing countries are now copying the meat and grain production practices of the United States (e.g., Schneider and Sharma 2014).

The dominance of productivist agricultural systems raises fundamental questions. Why does industrial farming hold greater legitimacy than traditional systems or alternatives such as organic agriculture? Why do companies and governments turn to technological solutions to repair agriculture's sustainability problems? How do scientists come to wield greater power than peasant farmers?

Since the late 1970s, agri-food studies and the social studies of agriculture have matured as an interdisciplinary field that connects agriculture and food to social structures and dynamics. These studies have arisen as a response to growing interest in the problems of industrial agriculture and international trade in food. Scholars have written on topics spanning food histories, producer-consumer relations, commodity chains research, and land politics. They have dealt with agrarian identities, farmer knowledge and practices, and (re)production of food as livelihood, political economy, worldview, and memory.[3] Many scholars have used STS concepts and methods—from boundary work and epistemic politics to actor-network theory and controversy studies—to enrich their agri-food analyses. As a consequence, scholars have gained new facility in unpacking how seemingly natural agricultural landscapes and seemingly inevitable technological innovations reflect human agency and politics. They are better able to trace how changes in knowledge underlie changes in industrial practices and supply chains. Reciprocally, studies of genetically modified (GM) crops and food-related regulatory science have proven especially important as core domains of STS research, generating significant insights into the politics of expertise, public understanding of science, and cross-comparative differences in regulation and politics of risk that are now widely applied across the field (see e.g., Jasanoff 2005; Kinchy 2012; Levidow and Carr 2009; Wynne 2001). Table 32.1 summarizes the major themes, ideas, and sites in STS scholarship on agriculture.

Table 32.1
Overview of Agricultural STS Themes, Sites, and Ideas*

	Established	Emerging
Major STS themes	Green revolution technologies Governance of food safety risks (e.g., BSE; more recently, leafy greens) Pesticide regulation The changing nature of public science in the biotechnology era Public understanding of GMOs, especially on a cross-national basis Standards for use in making foods and setting eco-labels Development of industrial agriculture technologies and organisms	Agrobiodiversity Seeds and IP Farmer knowledge politics Science in policy making aimed at markets and eaters
Key sites of STS work	Regulatory agencies Governments Agricultural departments Universities (especially land-grant universities in the United States) International S&T networks and institutions (e.g., CGIAR)	Food supply chains Food companies Farms Extension field days Social movements (e.g., the MST in Brazil)

Key ideas in STS	
Trust and deference to expertise Constructions of risk and risk society Lay/expert knowledge Boundary work around S&T Credibility of knowledge used in regulatory processes	Technological determinism Standardization and classification Local/indigenous knowledge Disciplining of actors through surveillance Embodied knowledge Community-based agrobiodiversity activities

*We composed the table based on our collective reviews and knowledge of the STS and agri-food literatures. By "emerging," we mean that a small but substantial amount of STS work has already been done.

Nonetheless, STS scholarship has not necessarily developed a synthesis of the human, cognitive, material, and structural patterns that agricultural systems contain. In this light, we suggest that using "co-production" as a storytelling idiom can help clarify how contemporary societies build agricultural systems through processes of knowledge-making and material work (Jasanoff and Wynne 1998). Many strands of co-production exist in the STS literature on agriculture, reflecting the field's diverse constructivisms. For example, "envirotech" scholars bridging environmental history and history of technology reject traditional views of technology driving deterministic impacts on plants and animals (e.g., Gorman and Mendelsohn 2011; Russell 2001; Vileisis 2011). Instead, these scholars trace (1) how agricultural technologies co-evolve with spatial landscapes and industrial systems and (2) how numerous organisms form interdependent relations with productive systems to the extent they can be seen as technologies in their own right. Russell et al. (2011) contend that the Industrial Revolution in Britain depended on workers fed by imported U.S. wheat, thereby co-producing agrarian transformation via more intensive farming and new shipping infrastructure.

We are particularly interested in how and why knowledge-making is incorporated into *institutions and practices of political economy making and governance*, and in reverse the influence of these practices on the making and use of knowledge (Jasanoff 2004). Scientific knowledge and political/legal/economic order are co-produced at multiple stages in their joint evolution, from legitimating findings in agricultural stations, to developing technologies based on this science, to creating legal and bureaucratic regimes to manage the technological applications. Ideas and practices, then, can reshape physical landscapes and living forms, as much as they can remold social norms and legal and political institutions. Simultaneously, these changes can loop back to change the terms in which people think about themselves and their place in the world (Jasanoff 2004).

The modern chicken exemplifies how agricultural organisms have been remodeled to suit the demands of mass production and consumption (Boyd 2001; Galusky 2010; Squier 2010). Chicken meat was once marginal in the U.S. diet. In 1948, a supermarket pioneer, A&P, and the USDA began running a recurrent competition to create the "Chicken of Tomorrow." They enlisted thousands of farmers to compete to breed "superior meat-type chickens," with the goals of achieving high feed-to-weight conversion and attracting consumers. Over the next fifty years, breeding programs supported by land-grant institutions across the United States created larger birds with more white, breast meat. The discovery of vitamin D enabled confinement rearing, while breakthroughs in antibiotics and hormones brought animal nutrition science and the pharmaceutical industry into a network of public and private broiler production interests. Increasingly consolidated "integrators," whose contract farming systems took hold around confinement chicken rearing in the rural South, formed a lynchpin of this network. Reengineering the modern bird, in sum, transformed labor organization, research and development, government policies, and consumer markets, not only for chicken but also for livestock agriculture more generally. In turn, the reciprocal pressures of consumer demand, favorable laws, and further science produced the modern bird: in 1957 broiler chickens weighed 900 grams when they were 56 days old; by 2005 they had swelled to 4.2 kilograms.[4]

Unpacking Industrial Agriculture

As the story of the chicken transformed suggests, STS scholars are particularly interested in exploring the relationships between agriculture and technoscience. How, for example, does scientific and technological knowledge influence the emergence of industrial agriculture? Why and how, in turn, does the legitimacy of industrial agriculture depend on science? One way to begin answering these questions is to examine how parts of industrial food systems developed. Much of this history pertains to the United States: it was in the U.S. South and Midwest, as well as in California, that many innovations first appeared and matured. These innovations were progressively exported to other regions in the United States, to other industrial countries, and across the world via the green revolution. Historians such as Mintz (1985), Cronon (1991), Stoll (1998), and Fitzgerald (2003) have sketched the broad historical patterns of industrializing agriculture. STS scholars have built on such histories to do their own tracing of technologies and sciences. These collective works suggest several key co-production pathways (Jasanoff 2004). For example, many farmers gained new social status as technologically advanced growers (identity making). Government officials subsidized industrial farming or passed laws to regulate food safety (institution making). Agricultural managers and extension scientists used scientific efficiency and economic cheapness to justify growth (discourse making). These pathways joined people, organisms, technologies, and supply chains into stable forms that now appear indispensable. Yet many people, ecosystems, and organisms lost as a result: workers and animals were exploited; consumers began eating unhealthy diets with higher levels of carbohydrates, sugars, and fats; and small farmers and wild bees faced extinction pressures.

In this section, we trace the emergence of productivism in three key arenas: the transformation of the farm into a factory, the rise of agricultural research institutions as core players in making industrial farms and foods (from an era in which agricultural research was not particularly significant), and the wider industrialization of agricultural food systems through the growth of manufacturing chains, processed foods, and supermarkets.

Conceiving the Factory Farm

Colonial sugarcane plantations in the U.S. South, the Caribbean, and Latin America pioneered many industrial techniques, such as monoculture crops and large-scale processing (Mintz 1985). Banana, cotton, rubber, coffee, and cacao plantations spawned some early multinational enterprises. Yet it was in the Midwest and California that industrialization penetrated more deeply into biological cycles. Cronon (1991) narrated the genesis of many industrial food practices in Chicago's metropolis from the 1850s onward. Initially, settlers turned the Midwest prairie into grain and livestock farms but remained small-scale producers for local towns. As the meat and grain industries used the railway to source and distribute foods across larger territories and adopted mass production techniques, growers were integrated into new supply chains. With the advent of more powerful grain traders, standards for measuring foods in storage and processing systems were needed to enable exchange. Physical wheat in sacks metamorphosed into abstract commodities in grain elevators. New discourses of efficiency and scales of economy were co-produced with technological innovations such as machines for dissecting cattle (the "disassembly line") and refrigerating meat in railway carriages.

It took decades before many farmers came to reimagine their farms as industrial factories instead of ecologically diverse landscapes. Fitzgerald (2003) explores how the industrial farm became more material during the 1920s, when machinery and electrical technologies were increasingly reliable and affordable. Many farmers converted to industrial methods because of cultural, economic, and sociotechnical pressures. While some small farmers preserved their traditional practices, others expanded their holdings to repay debt and survive in the capitalist political economy. Larger farms could only be managed more readily (and economically) by using tractors, harvesters, and other technologies that reduced labor. These technologies encouraged monoculture cropping to facilitate planting, weeding, and harvesting. Fitzgerald describes the efforts of some farmers to escape their "ignorant" social status by acquiring the identity of profitable business managers. As new agricultural sciences entered land-grant universities, the U.S. Department of Agriculture, and agribusiness firms, farmers often deferred to extension scientists and marketing agents, whose "superior" knowledge was articulated in scientific language. Discourses of "factory farms," in turn, were reflective of larger political economies: They began circulating in the hinterland just at a time when Fordism flourished. While mass production and assembly lines transformed the culture and organization of manufacturing, rural systems followed suit. Farm managers were coaxed into cultures of bookkeeping and quantifying yield/revenue. Equally important, early agribusiness companies nurtured cultures of using industrial inputs and specializing in high-yield crops and livestock, sending out armies of marketing agents to do demonstrations.

Nonetheless, as Fitzgerald (2003), Henke (2008), Busch (2005), and others show, productivism encountered many resistances. Many American farmers defied new-fangled machines for decades, stirred by their conservative philosophies and agrarian identities. Kline (2000) discusses how farmers selectively used technologies such as electricity, automobiles, and telephones—and developed their own conventions of use, reflecting their traditions of local experimentation. They used telephones communally, not individually. Car engines were used to drive farm equipment. Henke (2008) highlights ongoing debates in land-grant universities and cooperative extension systems about their role in advancing agriculture. Who were the farmers they were serving? What outcomes should they pursue? Even now, productivist regimes must be "constantly remade in innumerable, localized engagements" (Jasanoff 2004, 43) as disparate participants perform their social roles.

Creating the Agricultural Knowledge State

In the early twentieth century, land-grant universities established cooperative extension programs as a particularly important site in which agriculture knowledge and material systems were co-produced (Busch and Lacy 1983; Henke 2008). Progressive Era progenitors of extension drew on a study of rural America commissioned by President Theodore Roosevelt, which contended that small farmers were obsolete. To improve rural lives, they suggested, technical advisers should aid growers in making their farms larger and more technologically sophisticated. Following the Lever-Smith Act in 1914, the University of California built a large network of extension agents to translate agricultural science knowledge into practical forms that farmers could quickly adopt. Extension agents worked assiduously to spread productivist practices and assumptions. Henke (2008) traces how advisers helped develop discourses of efficiency and expertise that not only reinforced their identity as experts but remolded farmer identities. One crucial mode of knowledge-making was the "field trial." Farmers rejected new techniques as unsuitable for their local soil, water, and climate conditions. Accordingly, advisers developed a procedure of conducting field trials that simultaneously served as scientific experiments and as processes for enlisting farmers into new practices they were wary of. Such trials raised familiar STS questions of how to verify and represent the resulting experimental knowledge (Henke 2008).

Government agencies, such as departments of agriculture, regulatory bodies, and legislatures, also helped expand productivism through their discourses and institution building. As repositories of knowledge and power, they recognized knowledge as authoritative, conferred identities on people, and perpetuated tacit models of human agency through their bureaucratic procedures (Jasanoff 2004). Following—and reinforcing—discourses of rural improvement and poverty, Congress was instrumental in creating land-grant universities, cooperative extension, and agricultural research programs. In response to Dust Bowl and Depression era crises, the Roosevelt administration invented policies ranging from food stamps and food reserves to soil conservation programs and short-lived support for farmer-led learning networks. It also invested sizably in public agricultural science at land-grant universities and the U.S. Department of Agriculture (Kloppenburg 1988). Research within these facilities had far-reaching consequences for remaking the structure and function of agri-food systems. Seed hybridization, introduced in the 1930s, separated farmers from reproduction of their own planting material, permitting private corporations to benefit from marketing seed. The advent of hybrid corn also induced a pivotal agrarian identity shift. The process of creating hybrid corn was simple, but the number of crosses and the record keeping required to keep track of them excluded most farmers from the process (Fitzgerald 1993). Seeds were only the proverbial tip of the agri-food iceberg, as they fostered changes to machinery, chemical inputs, and feed-meat complexes that transformed agriculture.

Following World War II, the federal government endorsed large-scale industrial production through its research directions and policies (especially through the "farm bill"). Congress enacted laws that (1) created a federal research funding mechanism, (2) developed an international food aid program, and (3) established the basis of price floors and countercyclical payments to support commodity crops like corn and wheat. Farmers were encouraged to grow more food, reducing its cost and creating a surplus that could be exported overseas as well as used in animal feed and processed foods. Commodity subsidies were made contingent on farmers expressing specific identities—large-scale, monoculture, technologically advanced. USDA evolved into a particularly powerful government institution, whose leaders influenced imaginaries of American agriculture for decades (Busch and Lacy 1983; Busch et al. 1991). As Hamilton (2014) shows, USDA officials conflated technological progress with the growth of "agribusiness," a term first coined in 1955. Ezra Benson, Secretary of Agriculture, avowed: "Inefficiency should not be subsidized in agriculture or any other segment of our economy" (cited in Hamilton 2014, 565). Carbohydrates and sugars therefore became available in unprecedented quantities for companies to use in then-novel processed products.

The productivist worldview spread widely around the planet when U.S. institutions and their international counterparts—including the U.S. government, the Rockefeller Foundation, the World Bank, and many developing country governments—forged a network of research institutes that later became the Consultative Group for International Agricultural Research (CGIAR). As the first generation of agriculture-focused STS scholars noted, central to these alliances was a process of circulating scientific discourses through institutions and actor networks, particularly in developing countries, although this circulation was largely hidden behind a more public discourse of fostering a "green revolution" around the world. Educational institutions trained bureaucrats and technical extension experts in productivist tenets; government departments then designed favorable policies, which foundations reinforced through their project funding (Fitzgerald 1986; Jennings 1988). Visible evidence of the green revolution included the growth and diffusion of high-yielding varieties, harvesting machines, pesticides, chemical fertilizers, and irrigation technologies. Yet, it was the transmission of knowledge, from the 1940s through the 1980s, that helped change the character of agri-food systems throughout the world, such as Mexico, Brazil, India, the Philippines, and the Peruvian Andes (e.g., Shepherd 2005). Within programs built around the needs of larger-scale, wealthier farmers, cadres of technical experts extended the theory, practice, and input commodities of industrial agriculture. These experts, in turn, persuaded farmers to adopt intensification techniques. Smith (2009) notes that the process was hardly linear; rice research varied across time according to how the Rockefeller Foundation, CGIAR's International Rice Research Institute, and other actors framed agricultural and development problems.

Building Food Infrastructures

Downstream from the farm, an array of supply chains, processing plants, food science laboratories, and supermarkets have co-evolved as a vast system that now reliably delivers food to many millions of people. Largely invisible to consumers, this infrastructure transforms crops and animals, mass-grown in industrial farms, into standardized packaged products designed to travel worldwide and to appeal to human palates. "Foods produced on larger scales must be predictable in quality, quantity, content, safety, cost, flavor, texture and return on investment" (Jauho et al. 2014). To achieve this stability, food infrastructures produce and use S&T knowledge in the many steps between farm and dinner table. They are beneficiaries and advocates of the industrial farming model; they also influence the preferences of consumers toward favoring high-calorie, high-fat and high-carbohydrate diets (Otter 2015).

Over the decades, supply chain actors have changed markedly in their structural power and ability to mobilize S&T to produce foods. From the 1920s to 1960s, U.S. and European companies like Bayer and Dow gained ascendancy by manufacturing essential inputs into farming: Using a postwar surplus of chemicals, they remade norms of input use, entrenching fossil fuels at the heart of modern agriculture. United Fruit Company and Dole—the first multinational food firms—helped popularize the concept of monoculture cropping in the 1920s in the form of banana and other tropical fruits. Despite growing public and scientific awareness of pesticide problems, new pesticides entered the market and spraying rates escalated (e.g., Russell 2001). Agrarian sociologists noted that farmers became captive to the technological treadmill, forced to invest in costly technologies such as machinery, hybrid seeds, and fertilizer to keep up with the ever-growing demands for larger scales of economy (Cochrane 1979).

By the 1960s and 1970s, food processors such as Unilever, Nestle, Heinz, and General Foods were more commanding. A plethora of processed foods appeared, creating new demand for cheap, abundant crops and livestock. Food science was widely used to manufacture these foods: ingredients and process conditions were quantified precisely to enable more uniform outputs (Otter 2015). Ready-made or frozen meals gained popularity during the 1950s as eating identities and discourses began to favor "time-poor" convenience and "modern" diets featuring foods from packages, not in fresh form. In the United States, Cold War mentalities and atomic-age science were instrumental in these developments, engendering Jell-O molds and foods that were "bound" in mayonnaise or "imprisoned" in pepper rings (Adler 2015).

Simultaneously, branded product differentiation intensified, as companies pursued more specific, lucrative markets (Hamilton 2003). This explosion of processed products called for delicate manufacturer tinkering with package design, ingredients, preparation techniques, and compositions to work reliably and provide consistent, palatable tastes (Moss 2013; Otter 2015). Processing changes the nutrients, textures, and flavors of vegetables. Consequently, sugars, salts, fats, and additives such as colors, preservatives, and stabilizers were added to foods to manipulate consumer desires and increase durability. Fractionation technologies borrowed from the chemical industry brought about another sea change, enabling whole foods to be broken down into interchangeable food parts (Goodman, Sorj, and Wilkinson 1987). Agrarian products such as milk transformed into casein, lecithin, whey, and milk protein isolate, while corn and soy brought high-fructose corn syrup and partly hydrogenated oils into the constitution of "fabricated foods." As a result, anonymous food science laboratories quietly obtained more authority within agribusiness, since their specifications could profoundly reshape farming or processing practices. Mass media and advertising using consumer research enabled the remaking of eater identities around particular food discourses (e.g., Mudry 2009). Such branding has since undergone a pendulum swing, away from Jell-O molds and toward all things "natural." Nonetheless, convenience foods have continued to multiply in their seeming diversity and attractions.

In the 1980s and 1990s, food companies began restructuring supply chains to meet low-cost business models (Belasco and Horowitz 2011). Two supply-chain types grew dominant: supermarket chains and fast-food chains, which both benefited from scientific and technological innovations to grease flows of food from farm to dinner table (or car seat). Supermarkets first appeared in Britain around 1912, and fast-food chains such as McDonald's proliferated from the 1950s onward. STS scholars are fruitfully examining uses of S&T knowledge in aiding companies to design outlets, regulate supply chains, and sell food. One example is Hamilton's (2008) history of how trucking facilitated the growth of regional and nationwide retail distribution networks. Fast-food chains pioneered many distribution techniques and means of controlling agricultural production practices (Hockenberry 2014). Early consumer psychology enabled supermarkets to plan their stores to steer consumers toward buying processed foods through layout concepts such as the "central store" or the shelves that showcase such foods (Powell 2014). The advent of universal product codes (UPC), barcodes, and electronic data interchange protocols between 1973 and 1980 eventually enabled retailers like Walmart and Safeway to "industrialize" their logistics and intensify "lowest-cost" discourses aimed at consumers (Hockenberry 2014). Advanced electronics and logistics have since empowered retailers to take much greater control over supply chains.

Underlying these developments is the making and use of standards in food systems. Historically, standardization has enabled industrialization: it permits things to be interchanged and reduces variable human judgment. Standard cropping practices enable mechanized harvesting, standardized factories churn out identically packaged foods, and standard nutrition labels enable consumers to see if milk is low-fat. But standards also entail setting expectations and norms for actors along supply chains. What should a food look and taste like? What foods can be sold on the market? What counts as better-quality food? How should food be produced? STS scholars have pondered these questions at length (e.g., Busch 2011; Busch and Bingen 2006; Callon, Méadel, and Rabeharisoa 2002).

Standards only appeared in food systems during the 1930s. Now, every part of food production and consumption is subject to numerous standards from private, non-governmental, and state authorities. Busch (2011) explores how different types of standards can "design" food. For example, filters are used to decide whether food has passed safety requirements; they serve as rejections of the unacceptable. Multitiered filters can decide whether a grain is wheat and further sort it into soft, hard, or durum, determining its use for processing into specific products. Standards have the power to reconstruct nature to make it conform to human expectations—for example, by pressuring wheat breeders to change crop biology according to grain classifications. Standards are commonly based on scientific knowledge and coupled with technologies to measure compliance. Standards are not immutable or static: they are continually renegotiated as politics, knowledge, behavior, and ecological conditions change.

Busch et al. (2006) further suggest that standards convey shared knowledge and assumptions among disparate actor groups who may be widely distributed worldwide in modern supply chains: "Standards permit persons with little knowledge of each other's practices, and even less of each other's thoughts, to coordinate their action" (ibid., 138). Crucially, as Busch and Tanaka (1996) found for canola oil in Canada, actors in each part of the supply chain have diverging opinions about ideal oil characteristics. Other STS studies have looked at the actor-network underlying the development of standards for soybeans in Brazil (De Sousa and Busch 1998). Standards, then, are a means to discipline these disparate, competing interpretations into agreements that ensure product consistency. Even so, actors can still construe standards differently, thus calling for external monitoring and verification through audits. Shrimp farmers in Indonesia understood "sustainable" practices very differently than a certifying body in Europe (Konefal and Hatanaka 2011). Credibility in these negotiations is key. Like other facets of industrial agriculture, agri-food standards gain authority through appearing technically objective and neutral; they gain legitimacy through being represented as scientific. Nonetheless, as Busch (2011) demonstrates, standards embody corporate, technical expert, or government choices, thus reinforcing pervasive power asymmetries across modern food systems. Who makes standards, and with what procedures? These remain important questions.

Evolving Challenges and Transformations to Productivist Systems

Productivist agricultural systems remain dominant in many parts of the world. Yet, new forms of agricultural knowledge and production are also increasingly prevalent. Over the past decade, new generations of STS researchers studying agriculture have diversified their topics, sites, geography, and methods considerably, bringing these new countercurrents to productivism into focus. This development reflects the broader attention of the field not only to how dominant sociotechnical systems come to be but also to how they are challenged. In particular, scholars are investigating (1) the (re)emergence of alternative systems such as organic agriculture, agroecology, and agrobiodiversity, and (2) how agricultural industries are responding to sustainability challenges by modifying productivist systems. Co-production helps scholars and practitioners understand the ways in which productivist agriculture systems transform sustainability into technological solutions. Co-production also suggests that developing alternative forms of science, policy, and practice could eventually change the epistemic underpinnings of agriculture with the outcome of marginalizing productivism.

Participatory Expertise and Legitimating Sustainable Alternatives
Exciting STS research shows that not only are alternative systems emerging, agriculture across the planet has an historical foundation of diverse agrarian practices and

foodway traditions on which to build. The meanings of technology and science are not predetermined by industrial agricultural norms (Kloppenburg 1991). Anthropologists and agricultural historians have investigated knowledge production through the use of indigenous (or local scale) technologies including fiber baskets, living fences, livestock breeding, fishing traps, stone mills, storage cisterns, and pasture formation practices such as burning (e.g., Shah 2008). Such technologies predated industrialization and have continued to evolve with experimentation and learning. These technologies are co-produced with social institutions such as village irrigation cooperatives and with agricultural identities like *campesinos* or yeomen.

The fundamental place of farmer-made knowledge in agriculture is also garnering more recognition. Researchers have consistently shown that farmers, gardeners, and artisan food processors throughout the planet hold extensive practical skill and tacit knowledge as part of their cultural heritages. Indeed, ancient farmers were the first scientists as they built techniques based on empirical observations of weather and pest incidence and identified particular plants they wanted to retain (Busch et al. 1991). Such knowledge underlies the mutually constitutive relationships between ecologies and agriculture. Graddy-Lovelace (2013, 2014), Montenegro (2015), Carney (2009), and other political ecology/STS scholars are extending the work of earlier scholars in geography (e.g., Carl Sauer) in showing that farmers co-produce natural and cultural landscapes through their agroecological/traditional practices. In systems from vertical archipelagos in the Peruvian Andes to the rice/fish agriculture of China, farmers continue to devise clever technologies and management institutions that amount to biocultural knowledge systems. They have bred plants, livestock, and fish to live in agrobiodiverse complexes that provide life-sustaining ecological resources from soil fertility to pollination. Seeds are a particularly significant component; farming communities have developed many in situ seed conservation, saving, and sharing practices. In this, farmer knowledge and science exceeds modernist agricultural expertise in allowing this diversity to unfold and continuously adapt.

Within institutional contexts for agriculture research and education, however, farmer and local knowledges remain largely black-boxed. Warner (2007, 2008) critiques much conventional agricultural extension for imposing a divide between technical experts and laypeople and for relying on a transfer model in which farmers are passive, willing recipients of information rather than knowledge generators. This model has structured agricultural science institutions and defined the social roles that farmers play. By contrast, sustainable food movements worldwide increasingly emphasize participatory knowledge-making in a new politics of agricultural expertise. Instead of ceding scientific and technical capacity to those who are professionally trained and certified, organic food activists, biodynamic vintners, worm-composting gardeners, lentil growers, and artisanal fermented-food makers are claiming authority and legitimacy to make knowledge on an equal epistemic footing with agricultural scientists. Their activism and communities of practice pose new (yet familiar) STS questions: Who are experts in sustainable food? What criteria and processes are used to recognize and verify expertise? How do experts relate to those who are considered less expert?

Organic agriculture has blossomed since the 1980s. STS scholars have helped us understand why organic food has gradually grown more authoritative vis-à-vis industrial food. Gieryn (1999) explores the early struggles of organic farming. Much composting theory came from a developing world region: Indore, India, where soil scientist Albert Howard ran a field station for twenty years. There he blended indigenous knowledge and scientific tests. British agricultural scientists and government officials, by contrast, rejected such fluidity; they drew cognitive boundaries between "scientific" industrial and "unscientific" alternative practices. Accordingly, they disdained organic methods as primitive farming. Through early international networks of scientific expertise, such attitudes percolated throughout the world, affecting colonial and then postcolonial authorities' views of indigenous practices. They also entered U.S. agricultural culture. Carolan (2006b) notes that in 1951 politicians and scientists characterized J. I. Rodale—an American organic pioneer—as irrational during a congressional hearing on agriculture's future. In 1971 the U.S. Secretary of Agriculture Earl Butz said, "We can go back to organic farming if we must—we know how to do it. However, before we move in that direction, someone must decide which 50 million of our people will starve" (in Obach 2015, 35). Such boundary work had the effect of stifling research and government support in organic farming for decades.

Since the 1980s, however, organic agriculture has become more broadly accepted in industrial and developing countries. Much of its legitimacy stems from communities of practice coopting the prestige and power of science. As Obach (2015) notes, in the U.S. context, scientists and universities devoted more attention to organic agriculture in the mid-1980s as USDA and Congress inserted research-funding clauses into the farm bill in response to lobbying pressures from organic farmers. The Rodale Institute commissioned the first scientific appraisals and built its own long-term research station. This is an example of how co-production processes can "ratchet up" the legitimacy of alternative agricultures. Warner (2007, 2008) argues that organic farming has gained greater credibility because many proponents are embracing the language, experimental practices, and standards of proof prevalent in conventional agricultural science. Adopting quantification and scientific imagery confers authority on alternatives to conventional practice. Studying agroecological partnerships in California, Warner discusses how growers and scientists can be convinced that "organic farming works" through quantifying the amount of pesticides reduced. Similarly, Carolan (2006a, 2008) argues that organic agriculture benefits from protagonists aligning with like-minded university-based institutes and researchers and thereby acquiring scientific standing. Ingram (2007) notes that U.S. alternative agricultural movements rely on "immutable mobiles" such as reports, newsletters, and research findings, as well as on expert communities, to circulate knowledge in what Latour calls a capillary societal network (Latour 1999). The organic movement in particular has successfully created a discourse of "healthy soils leads to healthy bodies" over seventy years. With growing scientific, policy, and market support, Chinese, Indian, and Latin American scientists and governments are treating organic food with greater respect.

One of the core elements of many alternative agricultural movements is the argument that communities of living organisms—and the communities of agricultural knowledge and practice that work with them—lie at the heart of all agriculture. Brice (2014a) examines the practices of winemakers in South Australia in deciding when to harvest grapes. Using ethnographic analysis, he demonstrates that vintners build and use a sophisticated practical sense that drives the coordination of labor, trucks, and processing plants around the unpredictable ripening of grapes. A critical period of harvesting grapes exists: it must be done at the right time and must be finished quickly. Technical biochemical tests of grape sugar content helps vintners determine when to harvest, but they also rely on their field observations of weather and season conditions and of grape state (taste, color, sound). Brice underscores the agency of plants in shaping the behavior of farmers, who learn to monitor phenomena like soil texture, rainfall, and pest activity. Similarly, Grasseni (2005) studies the "skilled vision" of dairy cow farmers in northern Italy when they learn to identify their own cattle and choose promising individuals to breed. Expert breeder-farmers integrate sight, touch, and smell in appraising cattle, while referring to an ensemble of materials such as blueprints of the ideal cow and cattle registers. The ability to envisage human-plant/animal cohabitation is often stigmatized by modern technical experts as "preindustrial awareness of time," properly replaced by technological and calendar time. Yet this ability reveals the persistence of other ways of time reckoning that rely on farmer knowledge and observations that no mechanization can provide.

Particularly intriguing in this regard is scholarship exploring the persistence of life within apparently industrialized processes—and pointing to the unsuspected (by scholars and policy makers) importance of interspecies relations in growing food. Far from human agents controlling the process, farmers must work with diverse species (Carlisle 2015), from cows to fungi and from lentil plants to insects. Brice (2014b) ponders an example of wine that vintners reluctantly pasteurize to remove an enzyme remaining inside grapes from fungal infection but that destroys ongoing microbial community interactions that give wine its aroma and taste, which diminishes the wine's economic value. Similarly, as Paxson (2008) notes, many cheese makers and drinkers of raw milk complain that pasteurization results in "dead" milk where "healthy" microbes no longer can join the human microbiome. Innovative work on vermicomposting also suggests that urban gardeners are generating their own knowledge and expertise around soil amendments (Abrahamsson and Bertoni 2014). Many environmental conditions affect worm colonies in compost bins, so the success of vermicomposting depends on developing attunement to worm needs, namely, "learning to speak worm" (ibid., 134). Gardeners work with worms and microbes to create metabolic communities that transform food waste into fertilizer. Compost politics exist as well: "The worlds of earthworms, their external and internal digestive processes, the mites, the nematodes, the decomposing kitchen waste, [and] the vermicomposter" (ibid., 143) must be brought together. The idea of metabolic communities inside foods, soil-replenishing methods, and processing technologies shows different co-production processes at play that offer ways to think about agriculture differently.

STS researchers studying agroecology in Latin America take this scholarship a crucial step further, highlighting the political, institutional, and civic dimensions of agricultural communities of practice by showing how social movements can be co-produced with new forms of science and technology. Agroecology is "the application of ecological concepts and principles to the design and management of sustainable food systems" (Gliessman 2015, 18). For example, farmers can benefit from cultivating interactions between plants and insects so that pests are dispelled; they can intercrop plants to provide pollinator habitat, nitrogen fixation, and overyielding effects. Unique among alternative movements that challenge productivism, agroecology proponents assert that science, movement, and practice are already entwined in a more advanced form of agriculture (Méndez, Bacon, and Cohen 2013). Although organic agriculture has begun to gain credibility in scientific institutions, it remains the case that organic has succeeded more in marketing/retail and certification/labeling than in achieving a coherent set of scientific principles and practices (Obach 2015). Agroecology, by contrast, is an advanced science in its own right, based heavily on complex systems science and ecology. These forms of agricultural science pivot sharply away from technoscientific, reductionist approaches, characterizing agricultural systems in terms of nonlinear environmental change, complex biotic and abiotic interactions, and diversity across ecological, spatial, and temporal scales (e.g., Gliessman 2015). Importantly for STS, agroecologists approach this research with a step change in scientific praxis: for many proponents, agroecology is transdisciplinary work that challenges the epistemic boundaries of what science is and who participates. Vandermeer and Perfecto (2013) have described the synergies of Western science and indigenous knowledge in agroecology. The former is broad but shallow knowledge, the latter is deep, but narrow; both are complex traditions of equal, though differentiated, legitimacy.

In studying the *Campesino a Campesino* and Landless Workers movements in countries such as Cuba, Brazil, and Mexico, agroecology has much to offer STS in the way of emerging trends in the remaking of scientific communities and practices. These movements not only create space in which farmers can question industrial methods but make science explicitly political and civic. They are new sites for co-producing agrarian citizenship (Wittman 2009), educational processes, and agricultural science. Holt-Giménez (2006) details agroecological pedagogy in depth: he shows that farmer schools constitute sites of science-in-the-field. In these contexts, farmers are more likely to trust in, and learn from, their cultural peers than from technical advisers. Farmer schools devise their own experiments, make tools for visualizing problems, and test conventions. Participating in their own experiments can demonstrate to farmers that agroecology works. In effect, they are turned into witnesses who can attest to the merits of sustainable farming. Knowledge is treated as always politically and socially embedded—intuiting the constructivist moorings of STS.

Studying agroecology, however, underscores the significant tensions that exist regarding "experts" in social movements, particularly in developing countries. Delgado (2008, 2010) investigates knowledge production inside social movements, using the case of the Movimento Sem Terra (MST) in Brazil. She criticizes the Western-centric biases built into STS notions of expertise, such as assuming a sharp demarcation between laypeople and technical experts. Movimento dos Trabalhadores Sem Terra began in the 1970s as a landless laborer struggle for land access and grew into a large, dispersed movement with many communities across the country. Years of internal debate led to the movement endorsing agroecology. Delgado shows that agroecology has differing meanings and politics inside the Movimento dos Trabalhadores Sem Terra. The movement has built its own network of agricultural technicians and farming schools in rivalry with a state-sponsored system. Thus, it has gained the power to co-produce knowledge and institutions. For older technicians, agroecology is revolutionary politics to liberate farmers from Cartesian thought. For younger technicians, agroecology rethinks human-nature relations, through valuing indigenous wisdom, and creating a *dialogo de saberes* (dialogue of knowledges). Many MST farmers use conventional farming methods including pesticides and monocultures. Technicians must wrestle with how to persuade these laypeople to change their ways. Some technicians retain an information-deficit model and assume that farmers are ignorant and must be dissuaded from productivism. In many cases, technicians reproduce expert-lay distinctions, categorizing farmers without their say. Nonetheless, many farmers reject their categorization and insist on having their indigenous knowledge recognized.

Making Productivism More Sustainable?

Starting in the 1970s civil society actors, scientists, farmers, and governments increasingly questioned industrial agriculture's ecological and social costs, ranging from feedlot pollution and fertilizer runoff to animal suffering and worker abuses (e.g., Thompson 1995). In Europe, more stringent regulations and agri-conservation policies materialized, while in the United States, market pressures prevailed due to weak government intervention. Civil society opposition to industrial food, however, is appearing worldwide. Concerns about exposure to pesticide residues and nutrient deficiencies motivated growing consumer demand for organic food in industrialized countries from Italy to Japan. In China, many consumers now fear contaminated domestic milk and meat and seek safer imports from Australia, New Zealand, and Europe (Tracy 2010). In many countries of Africa, farmers and eaters are reluctant to embrace GM crops, which they see as being foisted on them by powerful companies and their own governments (e.g., Mwale 2006). In response, the agricultural industry has experimented with different ways to capture value from "sustainability."

For decades, STS scholars have helped to analyze disputes, at once ontological and epistemic, over what is and is not sustainable agriculture. Recently, for example, some scholars have taken interest in the processes and politics of measuring sustainability in supply chains. Wal-Mart, Unilever, and most multinational firms now buttress their sustainability claims by establishing new institutions such as sustainability consortia and standards and metrics to push their suppliers into adopting greener practices. For example, Freidberg (2014, 2015) considers how life cycle assessments are now acquiring power to govern globally sourced foods. Life cycle assessment quantifies the environmental impacts of producing food, often sidelining salient political and social debates. Freidberg finds that corporations continue to be powerful producers of (selective) knowledge along supply chains. In their embryonic development, agri-food life cycle assessments are gaining greater authority through being objective and comprehensive. Yet they also embody negotiations within companies and across industry communities over what will count as sustainable food. Agri-food evaluations often exclude worker health and other labor impacts, thus discounting their importance. Inside large multinational corporations, workforces are typically distributed across many countries and different national cultures; they may vary in their views of what sustainability means and whether it matters. Much political and cognitive work, therefore, must be done to make life cycle audits credible to these diverse audiences.

Other scholars are tracking new international calls for a "doubly green revolution"—a sustainable version of agricultural transformation in poor parts of the world that will better aid farmers in regions that ostensibly were previously untouched, such as Africa (Conway 1997). In 2009 the Royal Society in Britain published a prominent report that defined sustainable intensification as "agriculture in which yields are increased without adverse environmental impact and without the cultivation of more land." Sustainable intensification has quickly acquired cachet in guiding scientific research, philanthropy, and policy programs at national and global levels. Sustainable intensification is pointedly catholic in scope: genetically modified crops will be considered alongside agroecology. STS-inflected scholars (e.g., Kerr 2012) and environmental NGOs, however, criticize sustainable intensification as another salvo in industry's subtle efforts to disguise further intensification as sustainable. Sustainable intensification is discursively imbued with authority by being linked to what its proponents argue is the imperative of increasing global food production 70 percent to 100 percent by 2050 to feed a population of 9 billion people. However, Tomlinson (2013) has critically deconstructed this "feeding the world" claim by demonstrating its improvised origins in Food and Agricultural Organization policy statements and showing how numerous government and scientific actors have repeated it as though it were objective truth. Critics have also shown that investments in sustainable intensification appear to be geared toward biotechnology and industrial technologies. Sustainable intensification is being associated with climate-smart, precision, and eco-efficient agriculture practices. In precision agriculture, for example, industry is beginning to use drones, sensors, and GPS to economize irrigation, fertilizer, and planting strategies. Reflecting this new trend, in 2014 Monsanto bought a little-known company, Climate Corporation, to take control of its extensive database of U.S. field locations and climate conditions. Data-intensive forms of knowledge-making, in other words, are entering agriculture in gene banks and farm fields alike.

Food companies and researchers are also engaging in promissory, even evangelical, discourses surrounding envisioned technoscientific food futures: artificial meat, superfoods like açaí berry and einkorn grain, soylent meal replacements, probiotics, and other so-called functional foods. STS asks, Why is food such a key site for (especially American) experimentations with fantasies of the future? In large part, this is because industrial crops and livestock have become highly fungible—seen as capable of being manipulated into endless forms. Artificial meat is viewed as a solution to the many ethical, social, and environmental damages of meat production (Galusky 2014; Marcu et al. 2014). Researchers in Maastricht have worked for ten years to culture muscle cells in a scaffold and direct their growth into tissue through electrical stimulation and physical vibrations (Specter 2011). In October 2013 the first synthetic hamburger was eaten in London. Growing meat in vitro promises to alleviate the suffering of livestock as they are raced through feedlots to the slaughterhouse. Growing only muscle tissue obviates the energy squandered on growing other tissues such as bones and avoids the complications of tending livestock (Galusky 2014). Yet synthetic meat faces many challenges from high price to philosophical and aesthetic acceptance. Galusky argues that if humans start growing meat in vitro, they must provide biological inputs through technological interventions, thus becoming more dependent on machines. Artificial meat would uproot food production from its geographical, ecological, and social contexts. Producing meat in vitro, for example, ignores the role of animals in fertilizing diversified farms. In addition, artificial meat must succeed in winning consent from many reluctant participants across food systems. Such experiments are still nascent but indicate an important juncture. Will future food be sourced technologically or organically?

Still another industrial agriculture development is the beginning of a biobased economy. Goodman, Sorj, and Wilkinson (1987) forecast the "bioindustrialization" of agriculture: agribusiness would join chemical companies, biotechnology businesses, oil firms, and financial institutions to create new joint ventures crossing previously clearly bounded industry sectors. This prophecy has come true. Crops are increasingly being used not just for food and animal feed but also to make energy, fibers, and chemicals. As Borras et al. (2015) note, certain crops are now treated as fungible "flex crops." Since the early 2000s, biofuel production has grown in the United States and in developing countries from Brazil to Indonesia, partly because of government incentives to mitigate climate change (Levidow and Paul 2010). Similarly, biobased chemicals are materializing as a new agricultural output. Numerous companies—both giants and start-ups—are using corn, sugarcane, and cellulosic matter to produce chemicals. They are experimenting with fermentation and genetically modified organisms that can process raw materials or make the desired chemical. The biorefinery is developing as a factory that can transform crops into whatever outputs are desired or most profitable. Yet the multiple meanings of flexible crops are hardly settled (Borras et al. 2015). Farmed plants and trees often have multiple, though discrete, uses in traditional and indigenous agricultures. These specific uses are being erased as biotechnology models promise new fungible ones. Moreover, the future directions of flex crops are still malleable, as seen in switchgrass in the United States. In principle, switchgrass can be grown as part of diversified farms. In practice, researchers and companies are already trying to industrialize switchgrass by designing plantations, developing machines to plant and harvest the grass, devising agri-chemical protocols, and genetically modifying plants. In sum, the agricultural industry continues to devise technological adaptations but these are appearing at a time when societal scrutiny is more intense.

Toward Agriculture STS

We show that co-production idioms can help elucidate the often-obscure processes and pathways that have collectively built agricultural and food systems. The making of productivist regimes is founded on 150 years of the co-production of knowledges, technologies, institutions, cultures, organisms, humans, and markets. Paradoxically, the strength of the industrial system stems, in part, from its own heterogeneity. Industrial food is not one single beast but develops in multidirectional, locally and spatially variable ways while following an underlying, recurrent logic of productivism. Co-production processes have imbued industrial food with a strong legitimacy grounded in science and technology. Industrial food is credible because it appears efficient and inevitable. Yet the system depends on the massive day-to-day labor of workers across different sectors of the agri-food system. Much of the power of industrial food is due to its technological momentum and deeply entrenched structures and processes that reinforce one another, thereby locking human societies into productivist methods. We see this tendency in the attempts of productivist agricultures to become more sustainable, using technological fixes such as artificial meat, flex crops, and sustainable intensification. STS scholars can continue critically unpacking how and why productivist systems have been assembled, what they mean for human health and human rights, and the politics of how people authorize these systems with their beliefs and behavior.

We also want to highlight agri-food systems not merely as productive but as reproductive. Regenerating and renewing diversity can be helpfully illuminated in terms of co-productive ecological and social processes. As STS scholars are helping to reveal, agricultural systems worldwide still harbor substantial diversity—often in the forms of agrobiodiversity, heterogeneous landscapes, farmer and food-maker knowledge, customary laws, and inherited foodways. Alternative agricultures that promote diversity are appearing, or making a comeback, often with the aid of scientific authority as in organic food, and often through social movements as with agroecology. STS scholars can do much to interrogate the epistemic and political conditions for reconceptualizing food as diverse—not homogenous—systems. They can foster a dialogue of knowledges among different disciplines, ecologies, cultures, and sociotechnical worlds. We have provided some glimpses of the diverse texts and voices that can be brought to bear on this important work. Indigenous farmer-activists like to say, *"Se hace el camino al andar"* (We make our path as we walk it). STS scholars, then, have a responsibility to enable peoples worldwide to walk the paths they want.

Acknowledgments

We wish to thank Patrick Baur (Iles) and Kate Munden-Dixon and Jennifer Sedell (Galt) for very valuable research assistance and suggestions. We also wish to thank three searching reviewers and Clark Miller for their helpful feedback.

Notes

1. In this chapter we focus in depth more on the STS literature that emphasizes agricultural systems (and in particular industrial agricultural systems) than on the literature on food, nutrition, and health. Obesity, nutrition, culinary methods, and foodways are all deeply intertwined with productivist systems. But for reasons of coherence and brevity, we cannot adequately synthesize the entire food system.

2. By farmers, we also mean ranchers, pastoralists, livestock operators, fish farmers, and farmworkers.

3. Useful overviews of the complicated agri-food area can be found in Buttel (2001) and Goodman and Watts (1997). STS studies of food and agriculture offer important bridges to many other scholarly communities: business history, geography, economic sociology, history of medicine, medical anthropology, environmental studies, food studies, culinary histories, and so forth.

4. See https://www.washingtonpost.com/news/wonk/wp/2015/03/12/our-insatiable-appetite-for -cheap-white-meat-is-making-chickens-unrecognizable/.

References

Abrahamsson, Sebastian, and Filippo Bertoni. 2014. "Compost Politics: Experimenting with Togetherness in Vermicomposting." *Environmental Humanities* 4: 125–48.

Adler, Tamar. 2015. "Betty Crocker's Absurd, Gorgeous Atomic-Age Creations." *New York Times Magazine*, October 27, 2015.

Belasco, Warren, and Roger Horowitz. 2011. *Food Chains: From Farmyard to Shopping Cart*. Philadelphia: University of Pennsylvania Press.

Borras, Saturnino, John Franco, Ryan Isakson, Les Levidow, and Paul Vervest. 2015. "The Rise of Flex Crops and Commodities: Implications for Research." *Journal of Peasant Studies* 43 (1): 93–115.

Boyd, William. 2001. "Making Meat: Science, Technology, and American Poultry Production." *Technology and Culture* 42 (4): 631–64.

Brice, Jeremy. 2014a. "Attending to Grape Vines: Perceptual Practices, Planty Agencies and Multiple Temporalities in Australian Viticulture." *Social & Cultural Geography* 15(8): 942–65.

___. 2014b. "Killing in More-Than-Human Spaces: Pasteurisation, Fungi, and the Metabolic Lives of Wine." *Environmental Humanities* 4: 171–94.

Busch, Lawrence. 2005. "Commentary on 'Ever since Hightower: The Politics of Agricultural Research Activism in the Molecular Age.'" *Agriculture and Human Values* 22 (3): 285–88.

___. 2011. *Standards: Recipes for Reality.* Cambridge, MA: MIT Press.

Busch, Lawrence, and Jim Bingen. 2006. "Introduction: A New World of Standards." In *Agricultural Standards*, edited by Lawrence Busch and Jim Bingen, 3–28. Amsterdam: Springer.

Busch, Lawrence, and William Lacy. 1983. *Science, Agriculture, and the Politics of Research.* Boulder, CO: Westview Press.

Busch, Lawrence, William Lacy, John Burkhardt, and Laura Lacy. 1991. *Plants, Power and Profit: Social, Economic and Ethical Consequences of the New Biotechnologies.* Oxford: Basil Blackwell.

Busch, Lawrence, Elizabeth Ransom, Tonya Mckee, Gerald Middendorf, and John Chesebro. 2006. "Paradoxes of Innovation: Standards and Technical Change in the Transformation of the US Soybean Industry." In *Agricultural Standards*, edited by Lawrence Busch and Jim Bingen, 137–55. Amsterdam: Springer.

Busch, Lawrence, and Keiko Tanaka. 1996. "Rites of Passage: Constructing Quality in a Commodity Subsector." *Science, Technology, & Human Values* 21 (1): 3–27.

Buttel, Fred. 2001. "Some Reflections on Late Twentieth Century Agrarian Political Economy." *Sociologia Ruralis* 41 (2): 165–81.

Callon, Michel, Cécile Méadel, and Vololona Rabeharisoa. 2002. "The Economy of Qualities." *Economy and Society* 31 (2): 194–217.

Carlisle, Liz. 2015. "Making Heritage: The Case of Black Beluga Agriculture on the Northern Great Plains." *Annals of the Association of American Geographers* 106 (1): 1–15.

Carney, Judith. 2009. *Black Rice: The African Origins of Rice Cultivation in the Americas.* Cambridge, MA: Harvard University Press.

Carolan, Michael. 2006a. "Social Change and the Adoption and Adaptation of Knowledge Claims: Whose Truth Do You Trust in Regard to Sustainable Agriculture?" *Agriculture and Human Values* 23 (3): 325–39.

___. 2006b. "Do You See What I See? Examining the Epistemic Barriers to Sustainable Agriculture." *Rural Sociology* 71 (2): 232–60.

___. 2008. "Democratizing Knowledge: Sustainable and Conventional Agricultural Field Days as Divergent Democratic Forms." *Science, Technology, & Human Values* 33 (4): 508–28.

Cochrane, Willard. 1979. *The Development of Industrial Agriculture: A Historical Analysis*. Minneapolis: University of Minnesota Press.

Conway, Gordon. 1997. *The Doubly Green Revolution: Food for All in the Twenty-First Century*. New York: Penguin Books.

Cronon, William. 1991. *Nature's Metropolis: Chicago and the Great West*. New York: W. W. Norton.

Delgado, Ana. 2008. "Opening Up for Participation in Agro-Biodiversity Conservation: The Expert–Lay Interplay in a Brazilian Social Movement." *Journal of Agricultural and Environmental Ethics* 21 (6): 559–77.

___. 2010. "Activist Trust: The Diffusion of Green Expertise in a Brazilian Landscape." *Public Understanding of Science* 19 (5): 562–77.

De Sousa, Ivan, and Lawrence Busch. 1998. "Networks and Agricultural Development: The Case of Soybean Production and Consumption in Brazil." *Rural Sociology* 63 (3): 349.

Fitzgerald, Deborah. 1986. "Exporting American Agriculture: The Rockefeller Foundation in Mexico, 1943–53." *Social Studies of Science* 16 (3): 457–83.

___. 1993. "Farmers Deskilled: Hybrid Corn and Farmers' Work." *Technology and Culture* 34 (2): 324–43.

___. 2003. *Every Farm a Factory: The Industrial Ideal in American Agriculture*. New Haven, CT: Yale University Press.

Freidberg, Susanne. 2014. "Footprint Technopolitics." *Geoforum* 55: 178–89.

___. 2015. "It's Complicated: Corporate Sustainability and the Uneasiness of Life Cycle Assessment." *Science as Culture* 24 (2): 157–82.

Galusky, Wyatt. 2010. "Playing Chicken: Technologies of Domestication, Food, and Self." *Science as Culture* 19 (1): 15–35.

___. 2014. "Technology as Responsibility: Failure, Food Animals, and Lab-grown Meat." *Journal of Agricultural and Environmental Ethics* 27 (6): 931–48.

Gieryn, Thomas. 1999. *Cultural Boundaries of Science: Credibility on the Line*. Chicago: University of Chicago Press.

Gliessman, Stephen. 2015. *Agroecology: The Ecology of Sustainable Food Systems*. Baton Rouge, LA: CRC Press.

Goodman, David, Bernardo Sorj, and John Wilkinson. 1987. *From Farming to Biotechnology*. Oxford: Basil Blackwell.

Goodman, David, and Michael Watts. 1997. *Globalising Food: Agrarian Questions and Global Restructuring*. New York: Routledge.

Gorman, Hugh, and Betsy Mendelsohn. 2011. "Where Does Nature End and Culture Begin? Converging Themes in the History of Technology and Environmental History." In *The Illusory*

Boundary: Environment and Technology in History, edited by Martin Reuss and Stephen Cutcliffe, 265–90. Charlottesville: University of Virginia Press.

Graddy-Lovelace, Garrett. 2013. "Regarding Biocultural Heritage: In Situ Political Ecology of Agricultural Biodiversity in the Peruvian Andes." *Agriculture and Human Values* 30 (4): 587–604.

___. 2014. "Situating In Situ: A Critical Geography of Agricultural Biodiversity Conservation in the Peruvian Andes and Beyond." *Antipode* 46 (2): 426–54.

Grasseni, Cristina. 2005. "Designer Cows: The Practice of Cattle Breeding between Skill and Standardization." *Society & Animals* 13 (1): 33–50.

Hamilton, Shane. 2003. "The Economies and Conveniences of Modern-Day Living: Frozen Foods and Mass Marketing, 1945–1965." *Business History Review* 77 (1): 33–60.

___. 2008. *Trucking Country: The Road to America's Wal-Mart Economy*. Princeton, NJ: Princeton University Press.

___. 2014. "Agribusiness, the Family Farm, and the Politics of Technological Determinism in the Post–World War II United States." *Technology and Culture* 55 (3): 560–90.

Henke, Christopher. 2008. *Cultivating Science, Harvesting Power*. Cambridge, MA: MIT Press.

Hockenberry, Mark. 2014. Elements of Food Infrastructure. *Limn* 4, accessed at http://limn.it /elements-of-food-infrastructure/.

Holt-Giménez, Eric. 2006. *Campesino a Campesino: Voices from Latin America's Farmer to Farmer Movement for Sustainable Agriculture*. Oakland, CA: Food First Books.

Ingram, Mrill. 2007. "Biology and Beyond: The Science of Back to Nature Farming in the United States." *Annals of the Association of American Geographers* 97 (2): 298–312.

Jasanoff, Sheila. 2004. *States of Knowledge: The Co-production of Science and the Social Order*. New York: Routledge.

___. 2005. *Designs on Nature: Science and Democracy in Europe and the United States*. Princeton, NJ: Princeton University Press.

Jasanoff, Sheila, and Brian Wynne. 1998. "Science and Decision-Making." In *Human Choices and Climate Change*, edited by Steve Rayner and Elizabeth Malone, 1–77. Columbus, OH: Battelle Press.

Jauho, Mikko, David Schleifer, Bart Penders, and Xaq Frohlich. 2014. "Preface: Food Infrastructures." *Limn* 4, accessed at http://limn.it/preface-food-infrastructures/.

Jennings, Bruce. 1988. *Foundations of International Agricultural Research. Science and Politics in Mexican Agriculture*. Boulder, CO: Westview Press.

Kerr, Rachel Bezner. 2012. "Lessons from the Old Green Revolution for the New: Social, Environmental and Nutritional Issues for Agricultural Change in Africa." *Progress in Development Studies* 129 (2–3): 213–29.

Kinchy, Abby. 2012. *Seeds, Science, and Struggle: The Global Politics of Transgenic Crops*. Cambridge, MA: MIT Press.

Kline, Ronald. 2000. *Consumers in the Country: Technology and Social Change in Rural America*. Baltimore: Johns Hopkins University Press.

Kloppenburg, Jack. 1988. *First the Seed: Political Economy of Agricultural Biotechnology*. Madison: University of Wisconsin Press.

___. 1991. "Social Theory and the De/reconstruction of Agricultural Science: Local Knowledge for an Alternative Agriculture." *Rural Sociology* 56 (4): 519–48.

Konefal, Jason, and Mari Hatanaka. 2011. "Enacting Third-Party Certification: A Case Study of Science and Politics in Organic Shrimp Certification." *Journal of Rural Studies* 27 (2): 125–33.

Latour, Bruno. 1999. *Pandora's Hope: Essays on the Reality of Science Studies*. Cambridge, MA: Harvard University Press.

Levidow, Les, and Susan Carr. 2009. *GM Food on Trial: Testing European Democracy*. New York: Routledge.

Levidow, Les, and Helena Paul. 2010. "Global Agrofuel Crops as Contested Sustainability, Part I: Sustaining What Development?" *Capitalism Nature Socialism* 21 (2): 64–86.

Marcu, Afrodita, Rui Gaspar, Pieter Rutsaert, Beate Seibt, David Fletcher, Wim Verbeke, and Julie Barnett. 2014. "Analogies, Metaphors, and Wondering about the Future: Lay Sense-Making around Synthetic Meat." *Public Understanding of Science* 24 (5): 547–62.

Mazoyer, Marcel, and Laurence Roudart. 2006. *A History of World Agriculture: From the Neolithic Age to the Current Crisis*. New York: New York University Press.

Méndez, Ernesto, Christopher Bacon, and Roseann Cohen. 2013. "Agroecology as a Transdisciplinary, Participatory, and Action-oriented Approach." *Agroecology and Sustainable Food Systems* 37 (1): 3–18.

Mintz, Sidney. 1985. *Sweetness and Power*. New York: Viking.

Montenegro, Maywa. 2015. "Are We Losing Diversity? Navigating Ecological, Political, and Epistemic Dimensions of Agrobiodiversity Conservation." *Agriculture and Human Values* 32 (2): 1–16.

Moss, Michael. 2013. *Salt, Sugar, Fat: How the Food Giants Hooked Us*. New York: Random House.

Mudry, Jessica. 2009. *Measured Meals: Nutrition in America*. Stony Brook: State University of New York Press.

Mwale, Peter. 2006. "Societal Deliberation on Genetically Modified Maize in Southern Africa: The Debateness and Publicness of the Zambian National Consultation on Genetically Modified Maize Food Aid in 2002." *Public Understanding of Science* 15 (1): 89–102.

Obach, Brian. 2015. *Organic Struggle: The Movement for Sustainable Agriculture in the United States*. Cambridge, MA: MIT Press.

Otter, Christopher. 2015. "Industrializing Diet, Industrializing Ourselves: Technology, Energy, and Food, 1750–2000." In *The Routledge History of Food*, edited by Carol Helstosky, 220–49. New York: Routledge.

Paxson, Heather. 2008. "Post-Pasteurian Cultures: The Microbiopolitics of Raw-Milk Cheese in the United States." *Cultural Anthropology* 23 (1): 15–47.

Powell, Michael. 2014. "All Lost in the Supermarket." *Limn* 4, accessed at http://limn.it/all-lost-in-the-supermarket/.

Russell, Edmund. 2001. *War and Nature: Fighting Humans and Insects with Chemicals from World War I to Silent Spring.* Cambridge: Cambridge University Press.

___. 2010. "Can Organisms Be Technology?" In *The Illusory Boundary: Environment and Technology in History*, edited by Martin Reuss and Stephen Cutcliffe, 249–64. Charlottesville: University of Virginia Press.

Russell, Edmund, James Allison, Thomas Finger, John Brown, Brian Balogh, and Bernand Carlson. 2011. "The Nature of Power: Synthesizing the History of Technology and Environmental History." *Technology and Culture* 52 (2): 246–59.

Schneider, Mindi, and Shefali Sharma. 2014. "China's Pork Miracle? Agribusiness and Development in China's Pork Industry." *Global Meat Complex: The China Series.* Minneapolis, MN: Institute of Agricultural and Trade Policy, accessed at http://www.iatp.org/documents/china%E2%80%99s-pork-miracle-agribusiness-and-development-in-china%E2%80%99s-pork-industry.

Shah, Esha. 2008 "Telling Otherwise: A Historical Anthropology of Tank Irrigation Technology in South India." *Technology and Culture* 49 (3): 652–74.

Shepherd, Christopher. 2005. "Imperial Science: The Rockefeller Foundation and Agricultural Science in Peru, 1940–1960." *Science as Culture* 14 (2): 113–37.

Smith, Elta. 2009. "Imaginaries of Development: The Rockefeller Foundation and Rice Research." *Science as Culture* 18 (4): 461–82.

Specter, Michael. 2011. "Test Tube Burgers: How Long Will It Be before You Can Eat Meat That Was Made in a Lab?" *The New Yorker*, May 23, 2011, 32–38.

Squier, Susan. 2010. *Poultry Science, Chicken Culture: A Partial Alphabet.* New Brunswick, NJ: Rutgers University Press.

Stoll, Steven. 1998. *The Fruits of Natural Advantage: Making the Industrial Countryside in California.* Berkeley: University of California Press.

Thompson, Paul. 1995. *The Spirit of the Soil: Agriculture and Environmental Ethics.* New York: Routledge.

Tomlinson, Isobel. 2013. "Doubling Food Production to Feed the 9 Billion: A Critical Perspective on a Key Discourse of Food Security in the UK." *Journal of Rural Studies* 29: 81–90.

Tracy, Mehan. 2010. "The Mutability of Melamine: A Transductive Account of a Scandal." *Anthropology Today* 26 (6): 4–8

Vandermeer, John, and Ivette Perfecto. 2013. "Complex Traditions: Intersecting Theoretical Frameworks in Agroecological Research." *Agroecology and Sustainable Food Systems* 37 (1): 76–89.

Vileisis, Ann. 2011. "Are Tomatoes Natural?" In *The Illusory Boundary: Environment and Technology in History*, edited by Martin Reuss and Stephen Cutcliffe, 211–32. Charlottesville: University of Virginia Press.

Warner, Keith. 2007. *Agroecology in Action: Extending Alternative Agriculture through Social Networks.* Cambridge, MA: MIT Press.

___. 2008. "Agroecology as Participatory Science: Emerging Alternatives to Technology Transfer Extension Practice." *Science, Technology, & Human Values* 33 (6): 754–77.

Wittman, Hannah. 2009. "Reworking the Metabolic Rift: La Vía Campesina, Agrarian Citizenship, and Food Sovereignty." *Journal of Peasant Studies* 36 (4): 805–26.

Wynne, Brian. 2001. "Creating Public Alienation: Expert Cultures of Risk and Ethics on GMOs." *Science as Culture* 10 (4): 445–81.

33   Knowledge and Security

Kathleen M. Vogel, Brian Balmer, Sam Weiss Evans, Inga Kroener,
Miwao Matsumoto, and Brian Rappert

For the past hundred years, the pursuit of national and international security has been largely synonymous with the use of science and technology to design and deploy powerful weapons. From the battleships and chemical weapons of World War I to the radar systems and atomic bombs of World War II to the intercontinental ballistic missiles and nuclear arsenals of the Cold War, scientists and engineers transformed the nature of warfare during the twentieth century. Not surprisingly, given their focus on the social and political contexts and dynamics of science and technology, researchers in science and technology studies (STS) have made extensive contributions to analyzing this transformation and understanding its complex consequences for the organization and funding of science, technology, and the military, as well as the relationships between national security and the social, economic, and political dynamics of modern societies. Today, science and technology remain central to contemporary national and international security. STS scholars, working within and outside of security establishments, are making important contributions regarding how to think about and respond to current global security challenges.

Security landscapes have fundamentally changed in the twenty-first century and with them STS contributions to security studies. Science and technology are no less integral to national and international security today, but their influence is changing. New weapons, such as drones and computer viruses, remain important, but increasingly the focus of security is shifting into the realms of knowledge and information. Just as the national security states of the Soviet Union and Eastern Europe built massive information architectures to serve their needs, so, too, the rise of information societies is reconfiguring the power of states across the world to gather information as they pursue expanded security.

The processes of knowledge making have been identified as crucial to both the making of security and the broader implications of security mechanisms (such as regimes, frameworks, technologies, practices, and materialities). Although the topics studied in relation to security have altered significantly in past decades (Rappert, Balmer, and Stone 2008; Sapolsky 1977; Smit 1995), the common strengths of those working in the STS-security interface have been (1) to provide counter-narratives on security compared to traditional understandings and approaches found in other disciplines; (2) to innovate and adapt methods, tools, frameworks, and ideas to emerging security concerns; (3) to contextualize how these concerns can be understood in terms of a broader historical and discursive context; and (4) to inform and critique policy responses.

The STS-security literature that comprises this chapter is grouped around four main themes and questions:

1. *Imagining security: the scope, boundaries, and discourse of security*: How do societies define, imagine, frame, scope, and otherwise know "security" as a field of human activity and how do these knowledges of security shape and get shaped by the security enterprise?

2. *Knowledge, non-knowledge, secrecy, and ignorance*: What are the practices for managing knowledge (and access to knowledge) within security enterprises, how do they work, and how do they impact the relation between security enterprises and larger societies?

3. *Knowing citizens: surveillance, whistleblowing, and big data*: What do security enterprises know about citizens (especially in democratic societies), how is that knowledge produced and consumed, and with what consequences?

4. *Reflexive knowing: tensions for scholars investigating security matters*: How are STS's own knowledge-making practices shaped by its engagements?

In the remainder of the chapter, we examine how research since the time of the previous edition of this handbook has addressed these themes and questions.

Imagining Security: The Scope, Boundaries, and Discourse of Security

Security is frequently invoked to justify state action and is often taken as a clear-cut "public good" that must be defended by the state (Loader and Walker 2007). It is an umbrella term, however, that "enables and conceals a diverse array of governing practices, budgetary practices, political and legal practices, and social and cultural values and habits" (Valverde 2001, 90). How security gets framed, therefore, and the kinds of knowledges that are brought to bear on it, matter enormously for what security means and the power and influence of security enterprises.

In the past decade, for example, public health and security agendas have converged, particularly in relation to the bioterrorist threat. This has led to the overflow of "biosecurity" into numerous areas of social life, as health and other institutions become increasingly "securitized" (Hinchliffe et al. 2013; Kelle 2007; Lakoff 2008). These issues connect to far wider interests in STS that focus on biopower in diverse settings such as biodiversity, genomics, and synthetic biology (e.g., Lemke 2011). At the same time, a critique of the new bioeconomy has emerged within STS that is focused on its more complex political-economic character and how this is bound up with social expectations (Birch and Tyfield 2013). Additional work in STS has conceptualized the convergence of security and public health agendas with respect to disease (e.g., Cooper 2008). Several alternative framings of security issues have been explored, including "biosocial" over biosecurity (Vogel 2013a), defining cyber threats in economic or market terms rather than technical (Cavelty 2007), and framing concerns as ones of vulnerability rather than security (Hommels, Mesman, and Bijker 2014).

Security issues have also been explored in the context of building broader societies. Kosek (2010) draws out how the honeybee's existence is tightly tied to the empires of the twentieth century and has in the process become militarized and a source of military metaphor through interspecies relationships. Jasanoff and Kim (2009) show that wrapping a technology such as nuclear energy in the mantle of security seems logical only when seen within the context of a particular sociotechnical imaginary (e.g., the United States during the Cold War; for a more extensive treatment of imaginaries in STS, see McNeil et al., chapter 15 this volume), while in others (such as South Korea) it can play a strong role in helping the state and its citizens build and achieve a vision around development.

Security scopes, defined as the analytic categories that encompass security, have multiplied in recent years, with their displacement, crossover, and rearrangement suggesting new frameworks for investigating hitherto dismissed problem areas. For example, human security may be the broadest security concept because its universal concern with the security of people subsumes multiple scopes, including economic, food, health, environmental, personal, community, and political dimensions of security as evidenced in policy reports (United Nations Development Programme 1994). Scopes therefore maintain a distance from "the traditional prioritization of the state" created by identifying "people as the primary referent for understanding security" (Hobson, Bacon, and Cameron 2014, 2). Such an extended concept, however, could decrease its analytical power because of definitional expansion and ambiguity (Paris 2001), and lead a problematic politicization of securitization (O'Manique 2006).

Security scopes and boundaries do not necessarily overlap; they may deviate from each other or a even be mutually conflicting, thereby leading to tensions in how to address security and policy concerns. For example, enriching plutonium is a considerable technological hurdle, which if further developed as a nuclear weapon can cause a significant shift in national security. Yet, environmental security can be destroyed by the same radioactive material, the disposal of which could entail lasting effects on humans (Huang, Gray, and Bell 2013; Macfarlane 2012). Thus, the same object can enact multiple possibilities. This situation reflects the post–Cold War complexities of the science-technology-security interface, for example, such as the intricate relationships between climate change and global politics (Edwards 2013). Although one of the best options for maximizing national security, the nuclear energy option could also generate human and global environmental insecurity (Hecht 2012; Jobin 2013). Japan's experience with its Sunshine Project reveals that even renewable energy options, such as ocean thermal energy conversion technology, can generate insecurity through the unexpected creation of stratospheric ozone depletion (Matsumoto 2005). Slayton's work on electrical grids has illustrated how "smart grids" have conflicting scopes involving efficiency, reliability, and security (Slayton 2013a). All of these examples show the trade-offs between the different scopes of security and the characteristics common to them.

Another area of recent attention is how infrastructures underpin and shape security, particularly in the cyber domain. Ribes and Lee (2010) discuss the need to consider the integrated social (users, institutions, agencies) and technical systems (computers, databases, computer networks, fiber optic cables, grids, automated sensors, and imaging technologies) that comprise cyber-infrastructures and how these heterogeneous entities are made to relate to one another and work together. Comfort, Boin, and Demchak (2010) expand on this framework to discuss the importance of designing new sociotechnical mechanisms (including considering the tacit knowledge resident within organizations) for creating more resilient "socio-cyber" infrastructure systems that can survive against global cyber threats.

In the post-9/11 security environment, resilience and preparedness have emerged as central motifs in security discourse (e.g., Collier and Lakoff 2008; Lakoff 2008; Walker and Cooper 2011). Sims (2011) discusses how the concept of "resilience" has emerged and become dominant in homeland security rhetoric due to its ability to serve as a "boundary object" (Star and Griesemer 1989), providing a common rubric for the national security communities to communicate with each other and the public, avoiding potential controversies while responding to collective security anxieties. The framing of the problem as resilience means that only certain types of technological problems and solutions become visible for policy interventions, whereas others (e.g., public health or social problems and solutions) are ignored.

The relation between scientific discourse and security has also received continuing attention. This work has led to a focus for STS on how security matters are intimately bound up with how we might "know" about social and technological complexities and risks inherent in large-scale technical systems. In her study of the Chernobyl nuclear accident, Schmid's (2015) research reveals how a variety of narratives exist to explain the disaster; these reflect distinct co-construction of ideas about high-risk technologies, human-machine interactions, and organizational methods for ensuring safety and productivity of nuclear energy. Slayton's (2013b) work on missile defense has compared how physicists and computer scientists developed distinctive "disciplinary repertoires" about technology to construct different authoritative arguments about the risks associated with weapons systems (for complementary work that examines expertise debates and missile defense, see Spinardi 2014). Sims and Henke's (2012) analysis of U.S. nuclear weapons scientists' discourse in the post–Cold War period reveals that they espouse a discourse of "sociotechnical repair" with respect to computer simulations, experimental science, and new nuclear weapons designs that has maintained their credibility in the absence of testing. Spinardi (2008) has also examined the broader epistemic politics of testing surrounding the U.S. missile defense program even among supporters of the technology, who dispute not only whether the tests are representative but also what type of knowledge is considered most credible, what real-world use would involve, and what type of system design is best.

Discourses on security issues in many scientific and technological areas have recently focused on the "dual-use" aspects of innovation and knowledge production (Marris, Jefferson, and Lentzos 2014; Rabinow and Bennett 2012; Spinardi 2012). While initially characterized as turning military technology into civilian, or "spin-off," technology, dual-use also refers to "spin-on," turning civilian/benevolent research into military/hostile uses to create a complex "combination of spin-on and de facto spin-off" (Matsumoto 2006, 111). In addition, the usefulness of common conceptualizations of dual use as an analytic category has been questioned for the manner in which it is based on simplistic notions of the link between scientific knowledge and technological innovation (e.g., McLeish 2007).

Perhaps more important, the character of the dual-use dilemma has begun to shift. Where once the primary concern was for the dual utility of research in military and civilian settings, today dual-use conversations focus instead on how security enterprises should know when (and when not) to classify research, objects, or even people as security threats. The most prominent of these in recent policy discussions has been on the "dual-use dilemma" of the life sciences (National Research Council 2004; Tucker 2012), where benign research might be misused for malevolent purposes. Concerns about the creation of new genetically engineered viruses and other scientific and engineering developments have ushered in calls for more government oversight and regulation over science. In this context, scientific advising, scientific autonomy, public engagement, and the state are in continual tension (Brown 2007; Evans and Valdivia 2012).

One possibility for addressing the dual-use dilemma is through upstream and participatory governance of security concerns (Caduff 2010; Jefferson, Lentzos, and Marris 2014). Many scholars have argued that broader public input involving a diverse array of expertise is needed on these contentious issues in order to have a more holistic understanding of the issues, problems, stakeholders, values, and agendas at play (in general security matters, see Berling and Bueger 2015; in synthetic biology, see Evans 2014; in American military software development, see Lindsay 2010; in terrorism, see Stampnitsky 2013; in bioterrorism, see Vogel 2013b; in nuclear waste storage, see Macfarlane and Ewing 2006). Despite the need for technical input, scientists' engagement with the dual-use dilemma is variable and sometimes poor (Rappert and Selgelid 2013). Moreover, assumptions underpinning the drive to educate scientists on the ethics of dual use are not necessarily shared in developing-country environments where more pressing problems exist, ranging from local research priorities to expectations about laboratory safety and maintaining equipment (Bezuidenhout 2014).

Beyond these dual-use discussions, STS-inspired work in the so-called turn to ontology contributes to an expanded capacity to understand how objects become dangerous. McLeish and Balmer (2012) chart the history of nerve gas, showing that converting a discovery of a pesticide in the civilian sector into a nerve gas for military use was not a simple handing over of technology. Instead, a network of different secrecy regimes (commercial and military), institutional goals, and research practices had to be actively connected and reconfigured to enact the transfer. Furthermore, interviews with scientists involved in weapons programs and potential dual-use research have demonstrated that, while dual uses are frequently represented as a simple act of replication, tacit knowledge and other important micro- and macro-level sociotechnical factors are important in conducting science and turning the resulting knowledge into dangerous weapons (see Ben Ouagrham-Gormley 2014; Dennis 2013; Hymans 2012; Lentzos 2014; Revill and Jefferson 2014; Vogel 2008). Moving away from obvious weapons, Neyland (2008) used the idea of "mundane objects of terror" to explore the ontological reworking of everyday objects (bin bags, letters, water bottles) into objects of concern for security personnel and publics. Again, this is not just a matter of relabeling something as inherently harmful but rather an active reconfiguration of networks of ordinary, everyday things and people to reestablish them as a dangerous letter, a suspicious bin-bag, or a threatening bottle of water. We can also think of these moves as "security theater" (Schneier 2003), superficial measures introduced in the name of security to allay public fears or to convince them that something is being done.

Overall, the dual-use discussion rests on central questions in STS about the politics of knowledge—in this instance, both how we know something is benign or malign as well as what it is in the first place (Hecht 2010). This problem is significant not only for weapons but also for people. In the context of recent debates about strikes by unmanned aerial vehicles (UAVs, or drones) against civilians in countries such as Afghanistan, for example, Suchman (2015, 4) asks how bodies are made "subject to categorization as friend or enemy, and the latter are rendered a military target?" The nature of the difficulty is neatly illustrated by the Pentagon Defense Science Board: "Enemy leaders look like everyone else; enemy combatants look like everyone else; enemy vehicles look like civilian vehicles; enemy installations look like civilian installations; enemy equipment and materials look like civilian equipment and materials" (Defense Science Board Summer Study 2004 cited in Suchman 2015, 5).

Not surprisingly, these and other aspects of the development and use of drones and other kinds of military robots raise profound ethical questions such as, is it permissible to even build lethal robots or should we only build "pacifist" robots? Does possessing the ability to kill via robots lead to ethical slippage? Who should be held responsible for the deaths caused by autonomous robots? (Altmann et al. 2013; Asaro 2013; Gusterson 2016; Noorman and Johnson 2014; Sparrow 2009). Drones enact a present and future vision of warfare in which automated war asymmetrically reduces human casualties (Packer and Reeves 2013). This ability to enhance human capacity and create the "soldier of the future" has only been lightly touched on by STS (Viseu and Suchman 2010), despite the rich potential for analysis of this topic stemming from Haraway's (1991) seminal work on cyborgs. Some recent studies reflect on the proliferation of militarized cyborg bodies and how related imaginaries, such as sleep reduction, redefine what it is to be human in military settings (e.g., C. H. Gray 2003; Wolf-Meyer 2009). Enhancement, however, is not only a question for humans as Johnson (2015) observes how a U.S. ethology laboratory received Department of Defense (DoD) funding to create robotic lobsters to disarm underwater explosives. Having flagged these various studies of futuristic military innovation, it is a useful rejoinder to note some recently produced studies of low-tech, low-profile military innovation and of why apparently outmoded military technologies, such as the bayonet and gun trucks, have persisted (Kollars 2014; Stone 2012).

Knowledge, Non-knowledge, Secrecy, and Ignorance

For STS, a central question about any arena of human activity is how the production of knowledge is organized and regulated—and how this in turn creates areas of non-knowledge and ignorance (for discussions of this phenomenon, see Hess et al., chapter 11 this volume; and Ottinger, Barandiarán, and Kimura chapter 35 this volume). These questions have come to the fore in recent STS research on security knowledges, especially as they concern problems of secrecy and disclosure. Although secrecy is not unique to security studies, the practices, mechanisms, and institutions of secrecy in security matters are different from how secrecy works in the commercial sector and therefore worth special attention. While some level of intentional concealment about activities undertaken in the name of national and international security is often granted, what is appropriate in practice can be contested. In military conflicts in the twenty-first century, the openness (or not) of those planning for, undertaking, or counting the costs of military actions has been central to disputes about the merits of those actions. Because access to information is unidirectional and individuals are compartmentalized in classified spaces, secrecy and disclosure are central to the contests over who is able to speak with authority (Barak 2011) and how democratic control can be exercised over military and intelligence entities.

Recent studies have highlighted the extensive efforts undertaken in the past to hide the secrets of war and statecraft, for example, maps of the "New World" (Portuondo 2009), as well as place attention on debates about where the proper balance lies between security and openness, for example, for export control regimes (Reppy and Felbinger 2011) and patents and classification of nuclear weapons during the Manhattan Project, Cold War, and post–Cold War periods (Wellerstein 2008). With this work has come reflection on how actors and analysts struggle to make sense of the capabilities of technology in conditions of partial information (e.g., Boudeau 2007). Emphasis has also shifted from military secrecy and security to other ways secrecy is implicated in securing a population. For example, the experience of extreme events such as the Fukushima nuclear accident and its aftermath have motivated a new conceptual framework to reveal structurally deep-rooted secrecy created by participatory communication practices, such as science cafés, which did nothing to communicate the negative aspects of nuclear power plants to the people concerned (Matsumoto 2014).

Extending previous approaches that have focused on geographies of knowledge (Dennis 1999), secrecy has been understood not simply as a negative act of blocking information but rather as a generative force that constructs particular power relationships and identities (for a nuclear weapons example, see Masco 2010). Another focus has been how attempts to conceal can result in disclosure as well as how actions notionally undertaken for transparency can conceal (for a comparison of the tensions in secrecy versus transparency regimes in the Obama and Bush administrations, see Birchall 2011). Likewise, Paglen (2009) has sought out the traces left by secrecy, such as redacted "blank spots" on aerial maps. Such efforts to spatially and temporally hide objects illustrate the tensions and contradictions of secret keeping, as blanking out simultaneously reveals the act of and boundaries of concealment, as did redaction of documents prior to public release during the Cold War (Rindzeviciute 2015). The implications for scholarly writing regarding the interplay of secret keeping and secret telling have also received attention. With particular reference to the requirements for the deliberate withholding of information imposed on those who study national security, Rappert (2009) has used writing with concealment and absence as a means of examining the conditions of knowledge production in security diplomacy, as well as scholars' cultivating sensitivities for investigating secrecy.

Secrecy generally also entails ignorance. In line with a wider turn to ignorance in the social sciences (Gross and McGoey 2014; Proctor and Schiebinger 2008), scholars working on security topics have developed this theme in several directions. Those seeking to guard information often seek to maintain a state of ignorance, as well as induce one. Intelligence, counterintelligence, international diplomacy, covert operations, and other activities associated with the maintenance of secrets are characterized by moves and countermoves intended to deceive, disguise, and beguile. Agencies of the state can go to extraordinary lengths to hide their capabilities (or the lack of them) from others. For example, in Cold War British chemical weapons programs, authorities were particularly concerned that the Soviets would discover not what they possessed but that Britain had no stockpile of chemical weapons (Balmer 2010). And a 1968 brochure for "open day" at Britain's main biological defense establishment "hid" its secrets in plain sight by including sections entitled "what you will see" but also "what you will not see" (Balmer 2015).

In closed security communities, classification and compartmentalization of information can create ignorance that can create problems for how these communities (and those on the outside) assess their work (for an example related to the U.S. national lab complex, see Bussolini 2011). Vogel (2013a) has built on Eden's (2004) approach to examine the practices whereby U.S. government and nongovernment agencies produce narrow bioweapon threat assessments, as in the case of the intelligence justifying the 2003 Iraq War. In a study of a Swedish military intelligence unit, Minna Räsänen and James Nyce (2013) find that raw intelligence is never simply raw because of the multiple practices by which it is produced. Macrakis (2008) has argued the belief in technology by the East German Stasi ultimately led it to overestimate what intelligence could establish. Conditions of ignorance can create a "surplus of ambiguity" in which performativity and attempts to construct sources of authority figure highly, as seen in debates about nuclear weapons development (Gusterson 2008). Building in institutional forgetting has forced certain communities, like the victims of the Gujarat riots in India, to live in states of perpetual vulnerability (Visvanathan and Setelvad 2014). Along these lines, it is noteworthy that actors can embrace their own ignorance. For instance, when counting the deaths from conflict, belligerents often claim the facts of the matter cannot be known (Andreas and Greenhill 2010; Rappert 2012; Stone 2007). Some are even able to enhance their credibility and authority through acknowledging ignorance: scientific advisers to the military frequently draw attention to knowledge gaps and draw on their epistemic authority to claim "we don't know" without the risk of being called ignorant in order to secure funding and other resources (Balmer 2012).

Furthermore, it has become difficult to distinguish uncertainty from the production of "non-knowledge" as higher degrees of uncertainty increasingly become involved. Extreme events such as the 2011 Fukushima nuclear accident epitomized this dilemma, as the politics of secrecy crept into the uncertainty of allocating responsibility and achieving human security in an emergency (Downer 2014). This coupling of non-knowledge production with uncertainty provides one of the promising problem areas for studies on various scopes of security such as for the homeland (Comfort 2005), food (Mooney and Hunt 2009), and genetic security (Bonneuil, Foyer, and Wynne 2014).

Researching non-knowledge raises particular problems for scholars. A tempting asymmetrical move is for STS analysts to assume the role of unmasking non-knowledge through their claims to knowledge, without realizing the full extent of their ignorance (Proctor and Schiebinger 2008). Against this temptation, Rappert and Balmer (2015) brought together a broad set of scholars to debate how, when, and by whom matters become or are absented, and to debate the role of scholars' own research in making concerns absent and present. Another possibility for analysts is to seek ways of intervening into discussions infused with secrecy and ignorance. Through doing so it has been possible to make prevalent presumptions delimiting possibilities into active topics for questioning (in health security, see Heimer 2012; on the international ban on cluster munitions, see Rappert 2012).

Knowing Citizens: Surveillance, Whistleblowing, and Big Data

The structures and practices of state-sponsored surveillance have long been a topic of interest in STS (Foucault 1975). The rise of interconnected, large-scale information technology systems and accompanying big data analytics as an integral element of contemporary societies has given new rise to new concerns over the scope, scale, and intrusiveness that now characterize and create new opportunities for modern state surveillance. Arguably one of the most significant events to raise attention about these issues involved former defense contractor Edward Snowden's 2013 release of thousands of classified documents containing details of numerous global surveillance programs, run by the U.S. National Security Agency (NSA) in conjunction with various telecommunications companies and other governments. What became clear through the documents was that the NSA had been making use of big data analytics in order to mine metadata and analyze online behavior. *Big data* is a complex term that covers a whole host of data analytics activities. Often in security discussions, it refers to processing personal information and activities in order to extract information that might lead to detecting terrorist or criminal activities and networks. In the case of the NSA, these practices amounted to the ability to gather large quantities of data and the ability to repurpose that data for different uses at a later date (Andrejevic and Gates 2014).

Elements of the U.S. security community have tended to emphasize, in a Janus-face manner, the benefits of big data collection and analysis to improve security while at the same time raising new security risks related to the use of big data and its vulnerabilities to cyber intrusion (e.g., Berger and Roderick 2014). What this focus tends to obscure is how big data practices and tools are inherently social and, therefore, how big data analytics designed and developed by particular users ultimately influence the kinds of data collected and used (boyd and Crawford 2012; Gillespie 2014). Further, it also fails to understand how a narrow technical and material focus on big data (and its associated data-mining practices) can exclude considerations of how social factors, such as tacit knowledge, remain within emerging security threats (Vogel 2014) and how individuals, groups, and institutions are integral to the design, development, and use of big data.

The Snowden revelations triggered a flurry of stories about how intelligence agencies were putting virtually everyone under surveillance and intruding upon privacy in many different ways. They also sparked debates in policy circles across many countries about the need for a reform of surveillance laws, rights to privacy, and data protection.[1] For some, this practice of large-scale surveillance by the NSA and its counterparts must be understood as an "indicator of a much larger transformation affecting the way the boundaries of national security function" (Bauman et al. 2014, 126). Others have begun to examine the larger impact and consequences of big data practices: (1) to intensify surveillance by expanding interconnected data sets and analytical tools and (2) to create a specific form of abstracted, fragmented elements of data that focuses attention on a future orientation to justify more intrusive data-gathering practices while severing ties of surveillance practices from historical critique (Amoore 2011; Lyon 2014). Even before the Snowden leaks, Clarke (2009) observed the growing interest of policy makers in privacy issues for information technologies, with the construction of novel privacy assessments and privacy regimes for computer usage for business processes in the public and private sectors to mitigate privacy concerns and business risks.

These have not been the only examples of prominent national security big data disclosures. In 2010 hundreds of thousands of military reports and diplomatic communications were released through WikiLeaks that revealed secret information about state activities, illustrating how the public could also surveil the state. In purportedly "lifting the lid" on backstage dealings, the released material was often positioned as core bedrock facts that could resolve disputes about what took place, about how many people actually died, what officials really thought, and so on, rather than as knowledge claims by specific individuals and institutions that also needed to be scrutinized (Andreas and Greenhill 2010; Rappert 2012). Beyond WikiLeaks and the Snowden revelations, the idea that in order to obtain security one must give up privacy to counter terrorist attacks began to appear regularly as a notion in political and media discourse (for further discussion, see Solove 2011). This has resulted in a variety of policy measures by the George W. Bush and Obama administrations to justify more intrusive surveillance efforts on foreigners and citizens that are approved in secret court hearings outside of public awareness or purview.

Surveillance and security technologies are meant to reassure people and give them confidence, yet people's perceptions of how these technologies work and their impacts on their own lives can lead them to conclude that their security may not be better and that their freedoms and basic rights as citizens are threatened. For example, to deal with security concerns, an array of security and surveillance technologies are ushered in, such as fences/borders, gated communities, the widespread use of remote video surveillance cameras, and the privatization of public space (e.g., Bennett et al. 2014; Doyle, Lippert, and Lyon 2011). Research suggests that rather than managing peoples' fear of crime and enhancing feelings of safety, this range of measures may have the opposite effect (for an example of security fences, see Leuenberger 2014; in airport security, see Woolgar and Neyland 2013; for facial recognition technologies, see M. Gray 2003). Some argue, for instance, that the presence of surveillance cameras does not affect an individual's perception of public space in a positive way; rather these technologies work to emphasize existing negative feelings. For others, the lack of control over our own security and our reliance on surveillance technologies rather than other people to keep us safe has meant that feelings of insecurity have risen over recent years (Bauman and Lyon 2012).

In conversation with Ulrich Beck's work (1992), recent scholarship has illustrated how technological developments and geopolitics are bound up with the understandings and classification of risks (e.g., in relation to pandemic outbreaks, see Abeysinghe 2013; in predictive data mining in the global war on terrorism, see Guzik 2009), and surveillance and security technologies are coopted into risk management strategies. In addition, the varying sociotechnical practices of surveillance can have a variety of wide-ranging impacts on domestic and international populations to include concerns surrounding racial profiling and policing, human rights abuses, population control, consumer monitoring and manipulation, and the creation of new surveillance-industrial complexes (Ball, Haggerty, and Lyon 2012).

Given such concerns, questions need be asked about how and by whom surveillance and security systems should be regulated and how these systems shift the dynamics and exercise of power. Tensions also arise over regulation and the needs of intelligence-led operations across many domains, such as the security of urban spaces (Pieri 2014) and the sharing of viruses and public health information systems (Stephenson 2011). Echoing past themes, recent analyses have underscored the need to approach surveillance systems as sociotechnical networks requiring considerable maintenance (de Vasconcelos Cardoso 2012). The bigger the data demands, the more reliance on the infrastructure for collecting and storing these massive quantities of data, for the maintenance and security of this infrastructure, as well as for understanding the techniques and technologies that can make sense of this data (Andrejevic and Gates 2014).

Reflexive Knowing: Tensions for Scholars Investigating Security Matters

Building on important past work of those who have studied diverse security communities (Dennis 1999; Gusterson 1996; MacKenzie and Spinardi 1995; McNamara 2001), STS scholars have increasingly studied security through close engagement with members of security establishments. Epistemological issues regarding secrecy and related practices in security settings, and how these relate to the production of knowledge on contemporary and historical security threats, have been of particular interest (for diverse range of perspectives, see Masco 2010 for a discussion in the increasing scope of classification from the Cold War and post 9/11; Maret and Goldman 2008, 2011 who examine various facets of U.S. government secrecy regimes).

The calls made after 9/11 for greater political engagement from the STS community in security topics (Bijker 2003) continue to be salient (Vogel 2006). This has resulted in research with varying types of engagements. These include the direct involvement of scholars from different academic disciplines positioned within security institutions (in the U.K. intelligence community, see Aldrich, Cormac, and Goodman 2014; in the U.S. Counterterrorism Center, see Nolan 2013; in the U.S. military, see Kusiak 2008). STS researchers have also served as observers and commentators on the outside (for observations of security institutions, see Berling and Bueger 2015; for observations and analysis of a Swedish military unit, see Räsänen and Nyce 2013; for analysis of a U.S. biosecurity intelligence unit, see Vogel 2013/2014; for analysis of historical U.K. chemical and biological weapons cases, see Balmer 2012; for observations of the government spin involved in the Iraq, Afghanistan, and Libyan wars, see Rappert 2012). In addition, a collection of current and former intelligence practitioners have published articles discussing facets of knowledge production in intelligence and how such scholarly inquiry can suggest improvements for intelligence collection, analysis, and policy making, and they have brought these perspectives to STS audiences (for a range of perspectives on this, see Fingar 2011; Johnston 2005; Jones 2011; Kerr et al. 2005; Miller 2008; Nolan 2013).

Practitioner engagement can help scholars to appreciate secret communities and how they produce security knowledge, to learn about the hidden history of science and technology in these domains, to observe and document the effects of security on public spheres, and to propose novel interventions. Yet, such relations require scholars to be reflexive about how and to what extent they choose to engage and to take into account the normative, methodological, ethical, and epistemological dimensions of their choices as researchers. The breadth of such reflexive consideration has arguably been a defining hallmark of recent STS studies (e.g., more extended discussions of reflexive method in STS, see Law, chapter 1 this volume; Lezaun, Marres, and Tironi, chapter 7 this volume) compared to traditional security studies.

Debate continues regarding whether, how, and to what extent scholars can/should/will maintain critical distance in their engagement, scholarly inquiry, and interventions.[2] These are questions that have been long discussed within the related discipline of psychology, in which tensions have existed between those who have sought to bring psychology to help military activities and policy making and those who have sought to create an alternative psychology that would derail military aggression and be oriented to creating peaceful economic development and political stability (Herman 1995). In anthropology, there has long been a critique within the discipline against engaging in any form of collaboration with the military or intelligence agencies, given a variety of abuses and deaths of local populations that occurred during the Cold War; yet the post-9/11 environment has ushered in a new generation of anthropologists who seek to engage with the defense establishment (Price 2011). There are emerging discussions in the constructivist domains of international relations and security studies for need to have ideas about the securitization of politics infiltrate into a variety of national and international policy organizations (e.g., Berling and Bueger 2015). Within STS, these discussions have identified the need to be reflexive on how the closed social worlds of national security shape what those on the outside know about security threats and how knowledge is produced in these communities.

Beyond the issue of whether one should engage with security practitioners, and whether to do so as an insider or outsider, there remain additional challenging methodological issues once a researcher has decided to pursue their subject of inquiry. This would involve being self-aware of a researcher's position vis-à-vis his or her research subjects in security, as well as investigating the pros and cons of obtaining a security clearance to do this kind of work. While ethical concerns—such as procedural ethics, situational ethics, and relational ethics (Ellis 2007)—are apparent in many research settings, studying security issues involves its own flavor of these concerns, which overlaps with other issues that involve secrecy (such as dealing with intellectual property or repressed minorities). For each of these concerns, future work could usefully consist of a series of studies.

Procedural ethics deals with matters of confidentiality and informed consent, for example, how to gather informed consent in interviews with security practitioners. If informed consent is not possible (or desirable) by practitioners, alternatives or workarounds to the formal informed consent process need to be considered. These arrangements, however, can pose risks for the informants given that some security communities are quite small and knowledgeable insiders can determine identities.

Situational ethics deals with dilemmas and discomforts that can arise during fieldwork. Sociologists have documented the unique moral relationships that researchers have with their subjects given that what they reveal can be off-limits to others (Gunn 2005). New STS studies could examine these relationships for whether they raise new and different concerns from other kinds of social science research. Moreover, for those who obtain security clearances to do their research, how do they deal with their co-participation in security events and institutional structures? Rather than serving as mere observers, these researchers must deal with issues of closeness and critical distance and the roles of participant and spectator. How do these individuals navigate the complexity of their situated position in serving roles as institutional ethnographers and official historians (e.g., see Berling and Bueger 2013)? Within the field of anthropology, several studies have outlined these concerns for the academic community, including whether social science researchers should be engaged at all in research that condones violence (for a range of anthropological discussions on this issue, see Albro et al. 2012; American Anthropological Association 2007; Gusterson 2015; McNamara and Houtman 2007; McNamara and Rubenstein 2011; Price 2008, 2012). Within the umbrella of STS, besides anthropologists who self-identify with STS, there has been little discussion published on how STS scholars should approach their study of security communities the same, or differently, from anthropologists. Those working in these areas need to be aware of the varied ways of acknowledging and working with an uncertain and negotiated status and position in such situations, and more work needs to be published that addresses these issues in security.[3]

Relational ethics deals with assuming responsibility for our actions as researchers and the consequences of our stories on those being researched. This ethical concern deals with issues such as what is, and ought to be, left missing from narratives and accounts from research about security matters. How does secrecy affect access to and interpretation of primary sources? How does it influence historical analysis and writing, for example by "constructing transparency" through the impression that access to secret sources in and of itself means a complete and "real" account can be told (Balmer 2012)? Also, given security practitioners' concerns about revelation of identity and information disclosures, the role of the researcher in selectively deleting particular details from the public account becomes an important issue for reflection (Rappert 2009; Vogel 2015). Confidential spaces can facilitate open discussion on security issues but raise new ethical issues about how concealment, absences, and revelation shape what is known.

Other problems arise with attempts to triangulate findings from different methods of data collection due to limited source information. In this context of limited information, problems arise in terms of validation of findings given small sample sizes, limited access to informants or materials, or other documents that might offer counter claims. Here, as stated above, researchers must be aware of how certain information may be concealed by the actors studied (even information that is not classified).[4] Scholars also need to attend to how their methods themselves selectively disclose only some matters, often in a manner that parallels those being studied (Perkins and Dodge 2009; Rappert and Balmer 2015).

Finally, direct engagement with security organizations raises epistemological challenges. In order to capture the rich, social dimensions of technical work, STS research typically involves studies that consist of detailed historical or ethnographic data. These qualitative analytic methods focus on obtaining rich, in-depth understandings of the why and how of a particular case. These approaches can elucidate important contextual factors and understandings that other methods cannot capture. At least some security communities, however, have been described by practitioners as being composed of multiple, largely autonomous, often idiosyncratic units (Johnston 2005). Therefore, using a case study approach, what is to be learned about security work? What is fore-grounded, backgrounded, or rendered invisible by this approach? What types of generalizations and group identities are possible regarding the production of knowledge in security—and how does compartmentalization of security practices shape the organization of knowledge communities in security regimes? Alternatively, what do the case studies highlight or exemplify about the social processes, structures, values, and features that we might recognize as present, absent, or interconnected, in analogous cases (Galison 2008)? These issues highlight the distinctive challenges posed by secrecy for case study methodology that need concerted attention and explication.

Beyond engagement with security practitioners, there are also opportunities for greater engagement between STS and the field of security studies. In recent years, some scholarship in security studies has explored the breakdown of traditional boundaries around technology and capability, construction of the enemy and the state, conceptions of security and threats, and distinctions between theory and praxis (e.g., Cavelty 2007; Smith III 2014). STS-inflected analysis could be advanced to engage with these works by explicating how boundaries break down and how new boundaries are defined and defended. Bourne (2012) has recently called for engagement between critical security studies and STS. He argues that actor-network theory's emphasis on materiality and the emergence of identity and agency from assemblages of humans and nonhumans offers a means to move beyond debates on arms control within which "weapons are often encountered as mere artefacts of deeper and prior autonomous logics of anarchy and technological development and diffusion" (Bourne 2012, 163).

Conclusion

Two trends in the history of STS-security literature emerge from our review. First, the thinking on security issues in STS has remained closely linked to the changing analytic techniques of STS as a whole. Second, security-related events and developments, such as the end of the Cold War, September 11, and the progression of the war on terror, have also shifted the focus of questioning in the field toward what we have broadly labeled as the reimagination of security, non-knowledges, ways of knowing citizens, and reflexive knowing about studying security. Given these considerations, and instead of trying to predict future security challenges, we argue that what is needed now is to continue to build an STS-security community and literatures that can provide critical perspectives, as well as be responsive to new developments so as to remain relevant to changes in the field and the world. At the same time, this scholarship needs to develop longer-term research agendas to challenge commonplace presumptions about science and technology, and it needs to cultivate self-reflexive attention to the commitments entailed by our own research collaborations and interventions.

One area, however, that needs critical attention are non-Western studies. In surveying historical and recent STS literature, we noticed a persistent lack of non-Western studies of security. Although there has been important past scholarship on the Indian and Asian nuclear programs (Abraham 1998; DiMoia 2010; Flank 1993) and some limited work on the nuclear programs of other autocratic regimes (Hymans 2012), there is little recent work about non-Western security issues that go beyond the Fukushima nuclear disaster or the Soviet weapons complex. Glaring deficiencies on important contemporary security issues remain in the geographical areas of China, South Asia, the Koreas, and the Middle East. We need more STS-security scholars trained in Chinese, Korean, Arabic, and South Asian languages who could conduct detailed historical and contemporary case studies of security discourse and issues in these domains. This would allow a better understanding of how security and policy practitioners, as well as nonstate actors, in these countries think about and shape security threats, and how these perspectives might challenge and inform Western notions on security.

Notes

1. This initial surge in media coverage, however, quickly abated and research suggests that the debates have not translated into tangible policy reforms (Davies 2014).

2. Starting in 2004 there have been periodic panels at the annual Society for Social Studies of Science conference that raise issues of scholarly engagement with security. See 2004 4S Panel, "War on Terror," in Paris, France. Also, see the 2009 meeting in Washington, D.C., "Rethinking Expertise in Defense and Intelligence," available at http://www.4sonline.org/files/4S _Program2009_lg.pdf; see the 2013 meeting in San Diego, CA, "Ethnography of Engagement between S&TS and U.S. Intelligence: Lessons to Date," and "S&TS and Security: Training and Engaging," available at http://convention2.allacademic.com/one/ssss/4s13/index.php?click_key =1&cmd=Multi+Search+Load+Person&people_id=3679670&PHPSESSID=t21lmuequf51lh2udp4u d68oc6.

3. STS scholars working in other research domains (e.g., synthetic biology) have also raised methodological questions about close collaboration with their research subjects, see Calvert (2013).

4. The issue of actors selectively revealing and hiding information, and focusing on their style of presentation, is a topic that has received much attention in the wider social science literature, especially building on the work of Erving Goffman (1959).

References

Abeysinghe, Sudeepa. 2013. "When the Spread of Disease Becomes a Global Event: The Classification of Pandemics." *Social Studies of Science* 43 (6): 905–26.

Abraham, Itty. 1998. *The Making of the Indian Atomic Bomb: Science, Secrecy and the Postcolonial State*. London: Zed Books.

Albro, Robert, George Marcus, Laura A. McNamara, and Monica Spoch-Spana, eds. 2012. *Anthropologists in the SecurityScape: Ethics, Practice, and Professional Identity*. Walnut Creek, CA: Left Coast Press.

Aldrich, Richard J., Rory Cormac, and Michael S. Goodman, eds. 2014. *Spying on the World: The Declassified Documents of the Joint Intelligence Committee, 1936–2013*. Edinburgh: Edinburgh University Press.

Altmann, Jürgen, Peter Asaro, Noel Sharkey, and Robert Sparrow. 2013. "Armed Military Robots: Editorial." *Ethics and Information Technology* 15 (2): 73–76.

American Anthropological Association. 2007. "Commission on the Engagement of Anthropology with the US Security and Intelligence Communities." Final report, 4 November, accessed at http://www.policyarchive.org/handle/10207/20640.

Amoore, Louise. 2011. "Data Derivatives on the Emergence of a Security Risk Calculus for Our Times." *Theory, Culture & Society* 28 (6): 24–43.

Andreas, Peter, and Kelly M. Greenhill. 2010. Introduction. In *Sex, Drugs, and Body Counts*, edited by Peter Andreas and Kelly M. Greenhill, 1–22. Ithaca, NY: Cornell University Press.

Andrejevic, Mark, and Kelly Gates. 2014. "Big Data Surveillance: Introduction." *Surveillance & Society* 12 (2): 185–96.

Asaro, Peter M. 2013. "The Labor of Surveillance and Bureaucratized Killing: New Subjectivities of Military Drone Operators." *Social Semiotics* 23 (2): 196–224.

Ball, Kirstie, Kevin Haggerty, and David Lyon, eds. 2012. *Routledge Handbook of Surveillance Studies*. London: Routledge.

Balmer, Brian. 2010. "Keeping Nothing Secret: United Kingdom Chemical Weapons Policy in the 1960s." *Journal of Strategic Studies* 33 (6): 871–93.

___. 2012. *Secrecy and Science: A Historical Sociology of Biological and Chemical Warfare*. Farnham: Ashgate.

___. 2015. "An Open Day for Secrets: Biological Warfare, Steganography and Hiding Things in Plain Sight." In *Absence in Science, Security and Policy: From Research Agendas to Global Strategies*, edited by Brian Rappert and Brian Balmer, 34–54. Basingstoke: Palgrave.

Barak, Eitan. 2011. *Deadly Metal Rain: The Legality of Flechette Weapons in International Law*. Leiden: Brill.

Bauman, Zygmunt, and David Lyon. 2012. *Liquid Surveillance: A Conversation*. Cambridge: Polity.

Bauman, Zygmunt, Didier Bigo, Paulo Esteves, Elspeth Guild, Vivienne Jabri, David Lyon, and R. B. J. Walker. 2014. "After Snowden: Rethinking the Impact of Surveillance." *International Political Sociology* 8 (2): 121–44.

Beck, Ulrich. 1992. *Risk Society: Towards a New Modernity*. London: Sage.

Bennett, Kevin Haggerty, David Lyon, and Val Steeves, eds. 2014. *Transparent Lives: Surveillance in Canada*. Edmonton: Athabasca University Press.

Ben Ouagrham-Gormley, Sonia. 2014. *Barriers to Bioweapons: The Challenges of Expertise and Organization for Weapons Development*. Ithaca, NY: Cornell University Press.

Berger, Kavita M., and Jennifer Roderick. 2014. "National and Transnational Security Implications of Big Data in the Life Sciences." *Joint AAAS-FBI-UNICRI Project*, Washington, DC: American Association for the Advancement of Science. Accessed at http://www.aaas.org/sites/default/files /AAAS-FBI-UNICRI_Big_Data_Report_111014.pdf.

Berling, Trine Villumsen, and Christian Bueger. 2013. "Practical Reflexivity and Political Science: Strategies for Relating Scholarship and Political Practice." *PS: Political Science & Politics* 46 (1): 115–19.

___, eds. 2015. *Security Expertise. Practices, Power, Responsibility*. London: Routledge.

Bezuidenhout, Louise. 2014. "Moving Life Science Ethics Debates beyond National Borders: Some Empirical Observations." *Science and Engineering Ethics* 20 (2): 445–67.

Bijker, Wiebe E. 2003. "The Need for Public Intellectuals: A Space for STS." *Science, Technology, & Human Values* 28 (4): 443–50.

Birch, Kean, and David Tyfield. 2013. "Theorizing the Bioeconomy: Biovalue, Biocapital, Bioeconomics or ... What?" *Science, Technology, & Human Values* 38 (3): 299–327.

Birchall, Clare. 2011. "There's Been Too Much Secrecy in This City: The False Choice between Secrecy and Transparency in US Politics." *Cultural Politics: An International Journal* 7 (1): 133–56.

Bonneuil, Christophe, Jean Foyer, and Brian Wynne. 2014. "Genetic Fallout in Bio Cultural Landscapes: Molecular Imperialism and the Cultural Politics of (Not) Seeing Transgenes in Mexico." *Social Studies of Science* 44 (6): 901–29.

Boudeau, Carole. 2007. "Producing Threat Assessments: An Ethnomethodological Perspective on Intelligence on Iraq's Aluminium Tubes." In *Technology and Security: Governing Threats in the New Millennium*, edited by Brian Rappert, 66–88. Basingstoke: Palgrave.

Bourne, Mike. 2012. "Guns Don't Kill People, Cyborgs Do: A Latourian Provocation for Transformatory Arms Control and Disarmament." *Global Change, Peace and Security* 24 (1): 141–63.

boyd, danah, and Kate Crawford. 2012. "Critical Questions for Big Data: Provocations for a Cultural, Technological, and Scholarly Phenomenon." *Information, Communication & Society* 15 (5): 662–79.

Brown, Phil. 2007. *Toxic Exposures: Contested Illnesses and the Environmental Health Movement*. New York: Columbia University Press.

Bussolini, Jeffrey. 2011. "Los Alamos as Laboratory for Domestic Security Measures: Nuclear Age Battlefield Transformations and the Ongoing Permutations of Security." *Geopolitics* 16 (2): 329–58.

Caduff, Carlo. 2010. "Public Prophylaxis: Pandemic Influenza, Pharmaceutical Prevention, and Participatory Governance." *BioSocieties* 5 (2): 199–218.

Calvert, Jane. 2013. "Collaboration as a Research Method? Navigating Social Scientific Involvement in Synthetic Biology." In *Early Engagement and New Technologies: Opening Up the Laboratory*, edited by Neelke Doorn, Daan Schuurbiers, Ibo van De Poel, and Michael E. Gorman, 175–94. Dordrecht: Springer.

Cavelty, Myriam Dunn. 2007. "Cyber-Terrorism-Looming Threat or Phantom Menace? The Framing of the US Cyber-Threat Debate." *Journal of Information Technology & Politics* 4 (1): 19–36.

Clarke, Roger. 2009. "Privacy Impact Assessment: Its Origins and Development." *Computer Law & Security Review* 25 (2): 123–35.

Collier, Stephen J., and Andrew Lakoff. 2008. "Distributed Preparedness: The Spatial Logic of Domestic Security in the United States." *Environment and Planning D: Society and Space* 26 (1): 7–28.

Comfort, Louise K. 2005. "Risk, Security, and Disaster Management." *Annual Review of Political Science* 8: 335–56.

Comfort, Louise K., Arjen Boin, and Chris Demchak, eds. 2010. *Designing Resilience: Preparing for Extreme Events*. Pittsburgh: University of Pittsburgh Press.

Cooper, Melinda E. 2008. *Life as Surplus: Biotechnology and Capitalism in the Neoliberal Era*. Seattle: University of Washington Press.

Davies, Simon, ed. 2014. "A Crisis of Accountability: A Global Analysis of the Impact of the Snowden Revelations." Accessed at https://citizenlab.org/wp-content/uploads/2014/06/Snowden -final-report-for-publication.pdf.

Dennis, Michael Aaron. 1999. "Secrecy and Science Revisited: From Politics to Historical Practice and Back." In *Secrecy and Knowledge Production. Occasional Paper No. 23*, edited by Judith Reppy, 1–16. Ithaca, NY: Cornell University Peace Studies Program.

___. 2013. "Tacit Knowledge as a Factor in the Proliferation of WMD: The Example of Nuclear Weapons." *Studies in Intelligence* 57 (3): 1–9.

de Vasconcelos Cardoso, Bruno. 2012. 'The Paradox of Caught-in-the-Act Surveillance Scenes: Dilemmas of Police Video Surveillance in Rio de Janeiro." *Surveillance & Society* 10 (1): 51–64.

DiMoia, John. 2010. "Atoms for Sale? Cold War Institution-Building and the South Korean Atomic Energy Project, 1945–1965." *Technology and Culture* 51 (3): 589–618.

Downer, John. 2014. "Disowning Fukushima: Managing the Credibility of Nuclear Reliability Assessment in the Wake of Disaster." *Regulation and Governance* 8 (3): 287–309.

Doyle, Aaron, Randy Lippert, and David Lyon., eds. 2011. *Eyes Everywhere: The Global Growth of Camera Surveillance*. London: Routledge.

Eden, Lynn. 2004. *Whole World on Fire: Organizations, Knowledge and Nuclear Weapons Devastation*. Ithaca, NY: Cornell University Press.

Edwards, Paul N. 2013. *A Vast Machine: Computer Models, Climate Data, and the Politics of Global Warming*. Cambridge, MA: MIT Press.

Ellis, Carolyn. 2007. "Telling Secrets, Revealing Lives: Relational Ethics in Research with Intimate Others." *Qualitative Inquiry* 13 (1): 3–29.

Evans, Sam Weiss. 2014. "Synthetic Biology: Missing the Point." *Nature* 510 (7504): 218.

Evans, Samuel A. W., and Walter D. Valdiva. 2012. "Export Controls and the Tensions between Academic Freedom and National Security." *Minerva* 50 (2): 169–90.

Fingar, Thomas. 2011. *Reducing Uncertainty: Intelligence Analysis and National Security*. Stanford, CA: Stanford University Press.

Flank, Steven M. 1993. "Exploding the Black Box: The Historical Sociology of Nuclear Proliferation." *Security Studies* 3 (2): 259–94.

Foucault, Michel. 1975. *Discipline & Punish: The Birth of the Prison*. New York: Vintage Books.

Galison, Peter. 2008. "Ten Problems in History and Philosophy of Science." *Isis* 99 (1): 111–24.

Gillespie, Tarleton. 2014. "The Relevance of Algorithms." In *Media Technologies: Essays on Communication, Materiality, and Society*, edited by Tarleton Gillespie, Pablo J. Boczkowski, and Kirsten A. Foot. Cambridge, MA: MIT Press.

Goffman, Erving. 1959. *The Presentation of Self in Everyday Life*. Garden City, NY: Doubleday.

Gray, Chris Hables. 2003. "Posthuman Soldiers in Postmodern War." *Body and Society* 9 (4): 215–26.

Gray, Mitchell. 2003. "Urban Surveillance and Panopticism: Will We Recognise the Facial Recognition Society?" *Surveillance & Society* 1 (3): 314–30.

Gross, Matthias, and Linsey McGoey, eds. 2014. *Routledge International Handbook of Ignorance*. London: Routledge.

Gunn, Joshua. 2005. *Modern Occult Rhetoric: Mass Media and the Drama of Secrecy in the Twentieth Century*. Tuscaloosa: University of Alabama Press.

Gusterson, Hugh. 1996. *Nuclear Rites: A Weapons Laboratory at the End of the Cold War*. Berkeley: University of California Press.

___. 2008. "Nuclear Futures: Anticipatory Knowledge, Expert Judgment, and the Lack That Cannot Be Filled." *Science and Public Policy* 35 (8): 551–60.

___. 2015. "Ethics, Expertise and Human Terrain." In *Security Expertise: Practice, Power, Responsibility*, edited by Trine Villumsen Berling and Christian Bueger, 204–27. London: Routledge.

___. 2016. *Drone: Remote Control Warfare*. Cambridge, MA: MIT Press.

Guzik, Keith. 2009. "Discrimination by Design: Data Mining in the United States' 'War on Terrorism.'" *Surveillance and Society* 7 (1): 3–20.

Haraway, Donna J. 1991. *Simians, Cyborgs, and Women: The Reinvention of Nature*. New York: Routledge.

Hecht, Gabrielle. 2010. "The Power of Nuclear Things." *Technology and Culture* 51 (1): 1–30.

___. 2012. *Being Nuclear: Africans and the Global Uranium Trade*. Cambridge, MA: MIT Press.

Heimer, Carol A. 2012. "Inert Facts and the Illusion of Knowledge: Strategic Uses of Ignorance in HIV Clinics." *Economy and Society* 41 (1): 17–41.

Herman, Ellen. 1995. *The Romance of American Psychology*. Berkeley: University of California Press.

Hinchliffe, Steve, John Allen, Stephanie Lavau, Nick Bingham, and Simon Carter. 2013. "Biosecurity and the Topologies of Infected Life: From Borderlines to Borderlands." *Transactions of the Institute of British Geographers* 38 (4): 531–43.

Hobson, Christopher, Paul Bacon, and Robin Cameron, eds. 2014. *Human Security and Natural Disasters*. London: Routledge.

Hommels, Anique, Jessica Mesman, and Wiebe E. Bijker, eds. 2014. *Vulnerability in Technological Cultures: New Directions in Research and Governance*. Cambridge, MA: MIT Press.

Huang, Gillan Chi-Lun, Tim Gray, and Derek Bell. 2013. "Environmental Justice of Nuclear Waste Policy in Taiwan: Taipower, Government, and Local Community." *Environment, Development and Sustainability* 15 (6): 1555–71.

Hymans, Jacques E. C. 2012. *Achieving Nuclear Ambitions: Scientists, Politicians, and Proliferation*. New York: Cambridge University Press.

Jasanoff, Sheila, and Sang-Hyun Kim. 2009. "Containing the Atom: Sociotechnical Imaginaries and Nuclear Power in the United States and South Korea." *Minerva* 47 (2): 119–46.

Jefferson, Catherine, Filippa Lentzos, and Claire Marris. 2014. "Synthetic Biology and Biosecurity: How Scared Should We Be?" Workshop report. London: King's College. Accessed at http://www.kcl.ac.uk/sspp/departments/sshm/research/Research-Labs/CSynBI@KCL-PDFs/Jefferson-et-al-%282014%29-Synthetic-Biology-and-Biosecurity.pdf.

Jobin, Paul. 2013. "Radiation Protection after 3.11: Conflicts of Interpretation and Challenges to Current Standards Based on the Experience of Nuclear Plant Workers." Paper presented at Forum on the 2011 Fukushima/East Japan Disaster, 13 May. Berkeley. Accessed at https://fukushimaforum.wordpress.com/workshops/sts-forum-on-the-2011-fukushima-east-japan-disaster/manuscripts/session-3-radiation-information-and-control/radiation-protection-after-3-11-conflicts-of-interpretation-and-challenges-to-current-standards-based-on-the-experience-of-nuclear-plant-workers/.

Johnson, Elizabeth. 2015. "Of Lobsters, Laboratories, and War: Animal Studies and the Temporality of More-Than-Human Encounters." *Environment and Planning D* 33 (2): 296–313.

Johnston, Rob. 2005. "Analytic Culture in the U.S. Intelligence Community: An Ethnographic Study." Washington, DC: Center for the Study of Intelligence, U.S. Central Intelligence Agency.

Jones, Brian. 2011. *Failing Intelligence: The True Story of How We Were Fooled into Going to War in Iraq*. London: Biteback.

Kelle, Alexander. 2007. "The Securitization of International Public Health: Implications for Global Health Governance and the Biological Weapons Prohibition Regime." *Global Governance* 13 (2): 217–35.

Kerr, Richard, Thomas Wolfe, Rebecca Donegan, and Aris Pappas. 2005. "Collection and Analysis on Iraq: Issues for the U.S. Intelligence Community." *Studies in Intelligence* 49 (3): 47–54.

Kollars, Nina. 2014. "Military Innovation's Dialectic: Gun Trucks and Rapid Acquisition." *Security Studies* 23 (4): 787–813.

Kosek, Jake. 2010. "Ecologies of Empire: On the New Uses of the Honeybee." *Cultural Anthropology* 25 (4): 650–78.

Kusiak, Pauline. 2008. "Sociocultural Expertise and the Military: Beyond the Controversy." *Military Review* 88 (6): 65–76.

Lakoff, Andrew. 2008. "The Generic Biothreat, or, How We Became Unprepared." *Cultural Anthropology* 23 (3): 399–428.

Lemke, Thomas. 2011. *Biopolitics: An Advanced Introduction*. New York: New York University Press.

Lentzos, Filippa. 2014. "The Risk of Bioweapons Use: Considering the Evidence Base." *BioSocieties* 9 (1): 84–93.

Leuenberger, Christine. 2014. "Technologies, Practices and the Reproduction of Conflict: The Impact of the West Bank Barrier on Peace Building." In *Borders, Fences and Walls: State of Insecurity?*, edited by Elisabeth Vallet, 211–30. Burlington, VT: Ashgate.

Lindsay, Jon R. 2010. "'War upon the Map': User Innovation in American Military Software." *Technology and Culture* 51 (3): 619–51.

Loader, Ian, and Neil Walker. 2007. *Civilizing Security*. Cambridge: Cambridge University Press.

Lyon, David. 2014. "Surveillance, Snowden and Big Data: Capacities, Consequences, Critique." *Big Data & Society* 1 (1): 1–13.

Macfarlane, Allison. 2012. "The Nuclear Fuel Cycle and the Problem of Prediction." *Japanese Journal for Science, Technology & Society* 21: 69–85.

Macfarlane, Allison, and Rodney C. Ewing. 2006. *Uncertainty Underground: Yucca Mountain and the Nation's High-Level Nuclear Waste*. Cambridge, MA: MIT Press.

MacKenzie, Donald, and Graham Spinardi. 1995. "Tacit Knowledge and the Uninvention of Nuclear Weapons." *American Journal of Sociology* 101 (1): 44–99.

Macrakis, Kristie. 2008. *Seduced by Secrets: Inside the Stasi's Spy-Tech World*. Cambridge: Cambridge University Press.

Maret, Susan L., and Jan Goldman, eds. 2008. *Government Secrecy: Classic and Contemporary Readings*. Westport, CT: Libraries Unlimited.

___, ed. 2011. *Government Secrecy*. Vol. 19. Bingley, UK: Emerald Group Publishing.

Marris, Claire, Catherine Jefferson, and Filippa Lentzos. 2014. "Negotiating the Dynamics of Uncomfortable Knowledge: The Case of Dual Use and Synthetic Biology." *Biosocieties* 9 (4): 393–420.

Masco, Joseph. 2010. "Sensitive but Unclassified: Secrecy and the Counterterrorist State." *Public Culture* 22 (3): 433–63.

Matsumoto, Miwao. 2005. "The Uncertain but Crucial Relationship between a 'New Energy' Technology and Global Environmental Problems." *Social Studies of Science* 35 (4): 623–51.

___. 2006. *Technology Gatekeepers for War and Peace*. Basingstoke: Palgrave Macmillan.

___. 2014. "The 'Structural Disaster' of the Science-Technology-Society Interface: From a Comparative Perspective with a Prewar Accident." In *Reflections on the Fukushima Daiichi Nuclear Accident*, edited by Joonhong Ahn, Cathryn Carson, Mikael Jensen, Kohta Juraku, Shinya Nagasaki, and Satoru Tanaka, 189–214. Heidelberg: Springer.

McLeish, Caitriona. 2007. "Reflecting on the Problem of Dual Use." In *A Web of Prevention: Biological Weapons, Life Sciences, and the Governance of Research*, edited by Brian Rappert and Caitríona McLeish, 189–208. Sterling, VA: Earthscan.

McLeish, Caitriona, and Brian Balmer. 2012. "Discovery of the V-Series Nerve Agents." In *Innovation, Dual-Use, and Security: Managing the Risks of Emerging Biological and Chemical Technologies*, edited by Jonathan Tucker, 273–87. Cambridge MA: MIT Press.

McNamara, Laura A. 2001. *Ways of Knowing about Weapons: The Cold War's End at the Los Alamos National Laboratory*. Ph.D. dissertation, University of New Mexico, accessed at http://fas.org/programs/ssp/nukes/new_nuclear_weapons/cwendlosalamos.pdf.

McNamara, Laura A., and Gustaaf Houtman. 2007. "Culture, Critique, and Credibility: Speaking Truth to Power during the Long War." *Anthropology Today* 23 (2): 20–21.

McNamara, Laura A., and Robert A. Rubinstein, eds. 2011. *Dangerous Liaisons: Anthropologists and the National Security State*. Santa Fe, NM: School for Advanced Research Press.

Miller, Bowman H. 2008. "Improving All-Source Intelligence Analysis: Elevate Knowledge in the Equation." *The International Journal of Intelligence and Counterintelligence* 21 (2): 337–54.

Mooney, Patrick H., and Scott A. Hunt. 2009. "Food Security: The Elaboration of Contested Claims to a Consensus Frame." *Rural Sociology* 74 (4): 469–97.

National Research Council, Committee on Research Standards and Practices to Prevent the Destructive Application of Biotechnology. 2004. *Biotechnology Research in an Age of Terrorism*. Washington, DC: National Academies Press.

Neyland, Daniel. 2008. "Mundane Terror and the Threat of Everyday Objects." In *Technologies of (In)security*, edited by Katja Franko Aas, Helene Oppen Gundhus, and Heidi Mork Lomell, 21–40. London: Routledge.

Nolan, Bridget, 2013. *Information Sharing and Collaboration in the United States Intelligence Community: An Ethnographic Study of the National Counterterrorism Center*. Ph.D. dissertation, University of Pennsylvania.

Noorman, Merel, and Deborah G. Johnson. 2014. "Negotiating Autonomy and Responsibility in Military Robots." *Ethics and Information Technology* 16 (1): 51–62.

O'Manique, Coleen. 2006. "The Securitization of HIV AIDS in Sub-Saharan Africa: A Critical Feminist Lens." In *A Decade of Human Security*, edited by David R. Black, Timothy M. Shaw, and Sandra J. MacLean, 161–78. Aldershot: Ashgate.

Packer, Jeremy, and Joshua Reeves. 2013. "Romancing the Drone: Military Desire and Anthropophobia from SAGE to Swarm." *Canadian Journal of Communication* 38 (3): 309–31.

Paglen, Trevor. 2009. *Blank Spots on the Map: The Dark Geography of the Pentagon's Secret World*. New York: Dutton.

Paris, Roland. 2001. "Human Security: Paradigm Shift or Hot Air?" *International Security* 26 (2): 87–102.

Perkins, Chris, and Martin Dodge. 2009. "Satellite Imagery and the Spectacle of Secret Spaces." *Geoform* 40 (4): 546–60.

Pieri, Elisa. 2014. "Emergent Policing Practices: Urban Space Securitisation in the Aftermath of the Manchester 2011 Riots." *Surveillance and Society* 12 (1): 38–54.

Portuondo, Maria M. 2009. *Secret Science: Spanish Cosmography and the New World*. Chicago: University of Chicago Press.

Price, David H. 2008. *Anthropological Intelligence: The Deployment and Neglect of American Anthropology in the Second World War*. Durham, NC: Duke University Press.

___. 2011. *Weaponizing Anthropology: Social Science in Service of the Militarized State*. Oakland, CA: AK Press.

___. 2012. "Counterinsurgency and the M-VICO System: Human Relations Area Files and Anthropology's Dual-Use Legacy." *Anthropology Today* 28 (1): 16–20.

Proctor, Robert N., and Londa L. Schiebinger, eds. 2008. *Agnotology: The Making and Unmaking of Ignorance*. Stanford, CA: Stanford University Press.

Rabinow, Paul, and G. Gaymon Bennett. 2012. *Designing Human Practices: An Experiment with Synthetic Biology*. Chicago: University of Chicago Press.

Rappert, Brian. 2009. *Experimental Secrets: International Security, Codes, and the Future of Research.* New York: University Press of America.

___. 2012. *How to Look Good in a War: Justifying and Challenging State Violence.* London: Pluto.

Rappert, Brian, and Brian Balmer, eds. 2015. *Absence in Science, Security and Policy: From Research Agendas to Global Strategy.* London: Palgrave.

Rappert, Brian, Brian Balmer, and John Stone. 2008. "Science, Technology and the Military: Priorities, Preoccupations and Possibilities." In *The Handbook of Science and Technology Studies.* 3rd ed., edited by Edward J. Hackett, Olga Amsterdamska, Michael Lynch, and Judy Wajcman, 719–40. Cambridge, MA: MIT Press.

Rappert, Brian, and Michael J. Selgelid, eds. 2013. *On the Dual Uses of Science and Ethics: Principles, Practices, and Prospects.* Canberra: The Australian National University.

Räsänen, Minna, and James M. Nyce. 2013. "The Raw Is Cooked: Data in Intelligence Practice." *Science, Technology, & Human Values* 38 (5): 655–77.

Reppy, Judith, and Jonathan Felbinger. 2011. "Classifying Knowledge, Creating Secrets: Government Policy for Dual-Use Technology." In *Government Secrecy: Special Issue of Research in Social and Public Policy Perspectives*, edited by Susan Maret, 277–99. Bingley, UK: Emerald Group Publishing.

Revill, James, and Catherine Jefferson. 2014. "Tacit Knowledge and the Biological Weapons Regime." *Science and Public Policy* 41 (5): 597–610.

Ribes, David, and Charlotte P. Lee. 2010. "Sociotechnical Studies of Cyberinfrastructure and e-Research: Current Themes and Future Trajectories." *Computer Supported Cooperative Work* 19: 231–44.

Rindzeviciute, Egle. 2015. The Overflow of Secrets: A Study of the Disclosure of Soviet Repression in the Museum." *Current Anthropology* 56 (S12): S276–S285.

Sapolsky, Harvey. 1977. "Science, Technology and Military Policy." In *Science, Technology, and Society: A Cross-Disciplinary Perspective*, edited by Ina Spiegel-Rösing and Derek de Solla Price, 443–71. London: Sage.

Schmid, Sonja D. 2015. *Producing Power: The Pre-Chernobyl History of the Soviet Nuclear Industry.* Cambridge, MA: MIT Press.

Schneier, Bruce. 2003. *Beyond Fear: Thinking Sensibly about Security in an Uncertain World.* New York: Copernicus Books.

Sims, Benjamin. 2011. "Resilience and Homeland Security: Patriotism, Anxiety, and Complex System Dynamics." *Limn* 1 (1): 6–8

Sims, Benjamin, and Christopher R. Henke. 2012. "Repairing Credibility: Repositioning Nuclear Weapons Knowledge after the Cold War." *Social Studies of Science* 42 (3): 324–47.

Slayton, Rebecca. 2013a. "Efficient, Secure Green: Digital Utopianism and the Challenge of Making the Electrical Grid 'Smart.'" *Information and Culture* 48 (4): 448–78.

___. 2013b. *Arguments That Count: Physics, Computing, and Missile Defense, 1949–2012*. Cambridge, MA: MIT Press.

Smit, Wim. 1995. "Science, Technology and the Military: Relations in Transition." In *Handbook of Science and Technology Studies*, ed. S. Jasanoff, G. E. Markle, J. C. Petersen, and T. Pinch, 598–626. London: Sage.

Smith, Frank L, III. 2014. *American Biodefense: How Dangerous Ideas about Biological Weapons Shape National Security*. Ithaca, NY: Cornell University Press.

Solove, Daniel J. 2011. *Nothing to Hide: The False Tradeoff between Privacy and Security*. New Haven, CT: Yale University Press.

Sparrow, Robert. 2009. "Building a Better WarBot: Ethical Issues in the Design of Unmanned Systems for Military Applications." *Science and Engineering Ethics* 15 (2): 169–87.

Spinardi, Graham. 2008. "Ballistic Missile Defence and the Politics of Testing: The Case of the US Ground-Based Midcourse Defence." *Science and Public Policy* 35 (10): 703–15.

___. 2012. "The Limits to 'Spin-off': UK Defence R&D and the Development of Gallium Arsenide Technology." *British Journal for the History of Science* 45 (1): 97–121.

___. 2014. "Technical Controversy and Ballistic Missile Defence: Disputing Epistemic Authority in the Development of Hit-to-Kill Technology." *Science as Culture* 23 (1): 1–26.

Stampnitsky, Lisa. 2013. *Disciplining Terror: How Experts and Others Invented "Terrorism."* Cambridge: Cambridge University Press.

Star, Susan Leigh, and James Griesemer. 1989. "Institutional Ecology, 'Translations' and Boundary Objects: Amateurs and Professionals in Berkeley's Museum of Vertebrate Zoology, 1907–1939." *Social Studies of Science* 19 (3): 387–420.

Stephenson, Niamh. 2011. "Emerging Infectious Disease/Emerging Forms of Biological Sovereignty." *Science, Technology, & Human Values* 36 (5): 616–37.

Stone, John. 2007. "Technology and the Problem of Civilian Casualties in War." In *Technology and Security: Governing Threats in the New Millennium*, edited by Brian Rappert, 133–51. Basingstoke: Palgrave.

___. 2012. "The Point of the Bayonet." *Technology and Culture* 53 (4): 885–908.

Suchman, Lucy. 2015. "Situational Awareness: Deadly Bioconvergence at the Boundaries of Bodies and Machines." *MediaTropes* 5 (1): 1–24.

Tucker, Jonathan B., ed. 2012. *Innovation, Dual Use, and Security: Managing the Risk of Emerging Biological and Chemical Technologies*. Cambridge, MA: MIT Press.

United Nations Development Programme (UNDP). 1994. *Human Development Report 1994*. New York: Oxford University Press.

Valverde, Mariana. 2001. "Governing Security, Governing through Security." In *The Security of Freedom: Essays on Canada's Anti-terrorism Bill*, edited by Patrick Macklem, Ronald Joel Daniels, and Kent Roach, 83–92. Toronto: University of Toronto Press.

Viseu, Ana, and Lucy Suchman. 2010. "Wearable Augmentations." In *Technologized Images, Technologized Bodies*, edited by Jeanette Edwards, Penelope Harvey, and Peter Wade, 161–84. Oxford, NY: Berghahn Books.

Visvanathan, Shiv, and Teesta Setelvad. 2014. "Narratives of Vulnerability and Violence: Retelling the Gujarat Riots." In *Vulnerability in Technological Cultures: New Directions in Research and Governance*, edited by Anique Hommels, Jessica Mesman, and Wiebe E. Bijker, 109–30. Cambridge, MA: MIT Press.

Vogel, Kathleen M. 2006. "Bioweapons Proliferation: Where Science Studies and Public Policy Collide." *Social Studies of Science* 36 (5): 659–90.

___. 2008. "Framing Biosecurity: An Alternative to the Biotech Revolution Model?" *Science and Public Policy* 35 (1): 45–54.

___. 2013a. *Phantom Menace or Looming Danger? A New Framework for Assessing Bioweapons Threats.* Baltimore: Johns Hopkins University Press.

___. 2013b. "Necessary Interventions: Expertise and Experiments in Bioweapons Intelligence Assessments." *Science, Technology & Innovation Studies* 9 (2): 61–88.

___. 2013/2014. "Expert Knowledge in Intelligence Assessments: Bird Flu and Bioterrorism." *International Security* 38 (3): 39–71.

___. 2014. "Big Data and the Invisible, Social Dimensions of Science." *HCBDR '14 Proceedings of the 2014 Workshop on Human Centered Big Data Research.* New York: ACM. Accessed at http://dl.acm.org/citation.cfm?id=2609877.

___. 2015. Project Jefferson: Technological Surprises and Critical Omissions. In *Absence in Science, Security and Policy: From Research Agendas to Global Strategies*, edited by Brian Rappert and Brian Balmer, 114–31. Basingstoke: Palgrave.

Walker, Jeremy, and Melinda Cooper. 2011. "Genealogies of Resilience: From Systems Ecology to the Political Economy of Crisis and Adaptation." *Security Dialogue* 42 (2): 143–60.

Wellerstein, Alex. 2008. "From Classified to Commonplace: The Trajectory of the Hydrogen Bomb Secret." *Endeavour* 32 (2): 47–52.

Wolf-Meyer, Matthew. 2009. "Fantasies of Extremes: Sports, War and the Science of Sleep." *BioSocieties* 4 (2–3): 257–71.

Woolgar, Steve, and Daniel Neyland. 2013. *Mundane Governance.* Oxford: Oxford University Press.

34 Researching Disaster from an STS Perspective

Kim Fortun, Scott Gabriel Knowles, Vivian Choi, Paul Jobin, Miwao Matsumoto, Pedro de la Torre III, Max Liboiron, and Luis Felipe R. Murillo

Disasters are not new and have long garnered the attention of researchers, especially those able to help direct practical disaster response. Policy and operations-oriented disaster research has developed in many national contexts, usually organized around hazards with special local significance—earthquakes, hurricanes, or landslides, for example. In most contexts, natural and industrial hazards have been dealt with separately, further complicating the disaster research field. Disaster research has also developed at the international level—with increasing scope as climate change adaptation has come to be seen as disaster mitigation (Hannigan 2013; Lavell et al. 2012). Though diverse in focus and organization, policy and operations-oriented research has shared a general goal: practical effectiveness in reducing death and injury in disaster contexts.

Science and technology studies (STS) disaster research has taken shape largely in parallel, usually focusing on expansive historical and structural conditions productive of disaster vulnerability, dynamics, and response. Informed by theory from many disciplines and traditions of thought, STS disaster research has been less likely to theorize (and help respond to) disaster itself than to work through disaster to extend conceptualizations of power, legitimacy, inequality, change, and other phenomena at the center of debates in social theory. STS disaster researchers have also—as might be expected—focused especially on ways science and technology simultaneously produce risk in modern industrial societies and provide tools to assess and manage it.

Increasingly, different communities of disaster researchers are coming together, all looking for refreshed perspective and relevance.[1] The time for such collaboration is ripe. Disasters have made headlines in recent years due to staggering losses and controversies surrounding both causes and failures to recover in ways observers and impacted communities expect—drawing in the press, governments, insurance agencies, and other stakeholders. Climate change, the transition to unconventional oil and gas energy resources, neoliberalism, and other structural shifts portend still more disaster in the future. Further, critical observers are pointing to important shifts in the way disasters are conceptualized and managed. John Hannigan, for example, describes how "a humanitarian aid model for dealing with disasters became widely accepted in international affairs during the 1970s and 1980s; faltered in the 1990s; and is currently being challenged by a new approach to disaster management wherein risk management and insurance logic replace humanitarian concern as guiding principles" (2013, 1). The emergence of the Anthropocene as an organizing concept for environmental analysis—highlighting the enduring earth systems impact of human activities (associated with industrialization in particular)—has also implicated disaster research (Clark 2014).

STS can play an important role in emerging, collaborative disaster research, helping to draw out the many, cumulative failures productive of disaster, fostering expansive historical sensibilities. Operational definitions of disaster (reproduced in many media and policy accounts of disaster) emphasize the suddenness of the event and measurable harms, often assuming that disaster begins and ends with relief operations. STS researchers have challenged both this temporal framing and the measurability of disaster, emphasizing arrays of causes and effects (extending both backward and forward in time), and how disaster research itself (and associated ways of identifying and categorizing injuries as disaster-related, for example) is a sociopolitical process. A useful STS definition of disaster thus foregrounds *failures of diverse, nested systems, producing injurious outcomes that cannot be straightforwardly confined in time or space, nor adequately addressed with standard operating procedures and established modes of thought.* Such a definition implicates STS itself as well as what STS studies, calling for critical reflection on established methods, theoretical frameworks, and ways of representing and sharing research results.

Within STS, disaster research calls for reconsideration of ways theory is related to practice, in turn calling for new habits and modes of collaboration across disciplines. Disaster research also calls for reconsideration of well-established theoretical frameworks in STS and throughout the social sciences. Disaster involves system *dysfunction,* for example, upsetting long-standing social theoretical assumptions that systems tend to be functional, with components working interactively to produce order and stability. Functionalism in social theory emphasizes the positive operations of systems, and the ways these operations produce broadly shared perspectives across social groups. Disaster is more complicated, drawing into high relief how systems fail, through complex interactions that almost inevitably exacerbate the marginal positioning and inordinate vulnerability of particular social groups. STS disaster research thus has potential to reorient the field as a whole, with grounding in critical perspectives on structural functionalism and related ways of conceiving social order, with significant political implications (Bhabha 2015; Williams 2006).[2] STS disaster research thus contributes to the extended project of unsettling and decolonizing STS (Anderson 2009; Harding 2011; Law and Lin 2015; Seth 2009).

This chapter thus has double purpose, arguing that STS disaster research can make a significant contribution to emerging, collaborative disaster research, bringing together diverse research communities and modes of analysis, and that STS disaster research poses potentially transformative challenges to STS itself.

Following an introductory section on genealogies of disaster research, the chapter highlights three themes in STS disaster research: (1) expansive time frames for understanding disaster, (2) the role of design and expertise in disaster, and (3) denaturalizing disaster disparities. In different ways, research addressing these themes draws out how modern sociotechnological systems produce risk and harm, often without accountability—suggesting how STS disaster research can realize a core aim of critical theory, accounting for what dominant modes of expertise, representation, and political economy discount and reproduce. Concluding sections of the essay describe the methodological challenges of STS disaster research and its contemporary significance, theoretical and political.

Genealogies of Disaster Research

The United Nations General Assembly designated the 1990s as the International Decade for Natural Disaster Reduction, focusing primarily on earthquakes, volcanoes, landslides, and droughts. But the 1990s were also shaped by efforts to respond to a series of high-profile technological disasters—in Bhopal, Chernobyl, and Three Mile Island, among other sites. In the 2000s, disaster became associated with terrorism and securitization. Both Hurricane Katrina in 2005 and Japan's triple disaster (earthquake, tsunami, and nuclear) in 2011 provoked awareness of compound disaster involving the failure of multiple, interconnected systems. As could be expected, these developments brought disaster research to new prominence in many fields. The roots of disaster research are, however, much deeper. Indeed, disaster could be said to be foundational to the emergence of the social sciences in the nineteenth century.

Industrial accidents as well as other routine outcomes of industrialism like unemployment, suicide, and poverty were at the center of early social science research. As researchers have documented in different ways, the social sciences became formalized in the context of crying needs for social reform. Social science methods and disciplines grew up alongside growing government and business needs for social information, and new private and social insurance programs built around the identification and mitigation of risk (Backhouse and Fontaine 2014; Lipset 1969; Skocpol and Ruesche-meyer 1996). Scientific and theoretical interests also drove early developments in the social sciences, in ways clearly tethered to concrete albeit ever-changing, real-world conditions, complicating what "theory" in the social sciences would become. Disaster research continues to bear the imprint of these foundations, powerfully shaped—in productive but challenging ways—by simultaneous, cross-cutting demands for sharp social criticism and theory, as well as research-enhanced operational capacity.

Reviews of social science and humanities literatures on disaster give some sense of the scope of work already available, including in anthropology (Oliver-Smith 1986, 1996), sociology (Calhoun 2004; Tierney 2007, 2012), political science (Comfort 2005), public health (Glik 2007), human ecology (Susman et al. 1983), and at the interface of the social sciences and engineering (Kendra and Nigg 2014).[3] Further, disaster research goes by many names and has developed from an array of intellectual traditions. There are strong, coherent threads of research on humanitarianism (de Waal 2010; Redfield 2011) that are clearly about disaster although sometimes without explicit reference to the term. There is also a large literature about the structural violence that produces gross disparities in health around the world—at a scale most would concede to be disastrous (e.g., Farmer 1996, 2003); this literature, too, usually runs parallel to self-defined disaster research. Additionally, as further elaborated below, the meaning of the term *disaster* has been extensively debated, with acute recognition of implications for politics and practice (Erikson 1994; Law and Singleton 2009; Oliver-Smith 1986; Quar-antelli 1998). Work on disaster has developed in different ways in different contexts, with challenging implications for collaboration across both disciplinary and natural boundaries during disasters. A key reflexivity challenge for disaster research going forward is to recognize and build on this rich and complex history.

From Classic to Disaster-Focused STS

Foundational STS scholarship has informed the field's research on disaster in important ways. For example, studies of the ways knowledge is produced and legitimated in different contexts (Shapin and Schaffer 1985), of changing ways of representing scientific findings (Daston and Galison 2007), and of different scientific cultures and forms of collectivity (Knorr Cetina 1999; Traweek 1988) laid the groundwork for studies of the production of disaster knowledge and expertise. Long-running STS concerns with risk (Jasanoff 1986, 1994) also provide an important backdrop.

Extending Charles Perrow's (1984) well-known characterization of "normal acci-dents" in complex industrial systems, for example, STS disaster researchers have ana-lyzed how risk and injury are produced not only by technical systems themselves but also by the cumulative interaction of different kinds of systems—of knowledge, representation, politics, economics, and so on—showing how these systems function *and dysfunction*. In this emphasis on normal dysfunctionality, STS disaster researchers extend an important thread of theory in STS and beyond—in postcolonial, feminist, and race studies, in particular—that attend to what dominant systems marginalize and discount (Franklin 1995; de Lauretis 1987). They also extend important threads of work within STS on "agnotology" and the production of ignorance (Centemeri 2014; Frickel 2008; Frickel and Vincent 2011; Kuchinskaya 2014; Proctor and Schiebinger 2009), which interlace with threads of work on slow rather than acute disasters, resulting from toxic contamination and other routine, often discredited sources of risk and injury.[4]

In the last decade or so—moved by disaster at sites around the world—STS research on disaster has accelerated and deepened. Tironi, Rodriguez, and Guggenheim's (2014) edited volume *Disasters and Politics: Materials, Experiments, Preparedness* is exemplary, foregrounding how disaster configures politics through an array of processes, at different scales. The collection is impressively international in scope, including comparative case studies covering Europe, the United States, Latin America, and Australia. Important for our argument here is Mike Michael's perspective in the afterword of the volume. Michael describes how disaster narratives have long shared linear temporality—dating at least to the 1930s in sociology, when L. J. Carr[5] "advance[d] a typology which classified disaster according to the combination of the disaster's temporal form (whether it is instantaneous or progressive) and its aftermath (whether it results in social disarray, or leaves social organization by and large intact)" (ibid., 237). Michael lays out how the essays in the edited collection can be arranged in this way, with each essay positioned on a timeline in accord with the time of disaster on which it focuses. Michael also describes alternative ways to think about what disaster is and how it can be analyzed and represented—inspired, for example by STS theorist Donna Haraway's (1994) notion of knots in the children's game cat's cradle—with futures, pasts, and presents tied intricately together in dense "involution" (Michael 2014, 238).

Work responsive to Japan's triple disaster in 2011 also illustrates the momentum and shape of contemporary STS research on disaster. Working to deal with the intersecting impacts of the earthquake, tsunami, and nuclear disaster, STS researchers came together across areas of specialization in new ways. STS researchers have also led development of related teaching resources and of scholarly presence in both the mainstream media and policy arenas.[6] Importantly, Japan-based STS researchers have used Japan's triple disaster (often referred to simply as "3-11") to build and sustain a dialogue with STS researchers elsewhere, making an important contribution to internationalizing the STS community beyond the United States and Europe.

Miwao Matsumoto's (2013, 2014) work on "structural disaster" is exemplary. Focusing first on a hidden but critical accident caused by failure of a standard military technology in Japan just before World War II, Matsumoto argues that the resulting military-industrial-university complex in Japan has a telling structural similarity to the behavior patterns of heterogeneous agents involved in the Fukushima nuclear disaster—and that this complex helps explain why crucial information on radiation dispersal from the Fukushima Daiichi plant (indispensable for proper evacuation) was restricted to government insiders. Building from the Fukushima case, Matsumoto highlights the limits inherent in lines of explanation for disaster that privilege either conflicts of interest or cultural essentialism. Instead he argues for an understanding of how these factors interconnect, coupling structural integration with functional disintegration in both prewar military and postwar nuclear regimes in Japan. Matsumoto's methodology and argument have provided important guidance in efforts to understand the multiple factors, histories, and long-term outcomes of Japan's 2011 disaster—a disaster Sara Pritchard (2012) has termed an "envirotechnical" disaster because of interlinkages between its environmental and technological dimensions.

The work of David Slater at Sophia University in Tokyo also illustrates some of the best characteristics of contemporary STS disaster research. Slater developed an online oral history archive of the 3-11 disaster and has published extensively, in both Japanese and English, describing both the sudden brutality of 3-11 and the protracted, cumulative anxiety about radiation, displacement, compensation, and discrimination that followed (illustrating how fast disaster often turns into slow disaster). Slater also writes about the methodological challenges of "urgent ethnography," laying out the unique demands of research during times of emergency, and the operational and ethical challenges of disaster relief and representation (Slater 2013; Slater, Keiko, and Kindstrand 2012; Slater, Morioka, and Danzuka 2014).

Contemporary STS disaster research is diversely rooted and oriented but stands out both for its capacity to deal with the interlocking complexes of factors contributing to disaster and for recognition of the ways that thinking about disaster itself has profound operational and political consequences. As we describe in more detail in the next section of the chapter, this positions STS researchers to collaborate with people and organizations involved in operational response to disaster. It also challenges STS researchers to think in new ways about their own methods, theory, modes of representation, and politics—and the interrelationship of these.

Themes in STS Disaster Research

Drawing on diverse threads of STS research, interlaced with theory from many traditions, STS researchers are poised to make an array of contributions to disaster research, in ways that feedback into development of STS writ large. Here, we describe three themes that have developed so far, each illustrative of different scales and types of analysis.

Expanding Time Frames of Disaster

At the height of the Cold War, U.S. sociologist Charles Fritz wrote what became a foundational definition of disaster for the social sciences. Fritz (1961, 655) found disaster to be "an event, concentrated in time and space, in which a society or a relatively self-sufficient subdivision of a society undergoes severe danger and incurs such losses to its members and physical appurtenances that the social structure is disrupted and the fulfillment of all or some of essential functions of the society is prevented." Over the following decades an enormous volume of work across the social sciences has challenged and refined this definition, an analytical challenge stoked by Fritz himself through his role in founding the long-standing Disaster Roundtable of the U.S. National Academies. At issue was when disasters were conceived to begin and end, powerfully implicating what types of phenomena would count under the definition.

Temporally delimited constructs of disaster have also exerted powerful effects on disaster operations. In the standard practice of emergency management, disasters are generally broken into four phases: mitigation, preparedness, response, and recovery. In this way, practitioners acknowledge that a disaster is more than a single event happening in one time and place. Nevertheless, the communities of disaster practice (and most disaster research until recently) have focused overwhelmingly on the "preparedness" and "response" phases of disaster, with mitigation and recovery left as indeterminate poles receding into the past and stretching into the future. Expanding analyses out to these time scales reveals structural factors that foster risk and point to recovery pathways. The bounding of "disaster time" is thus a subject of great interest for STS disaster research, with entwined analytic and political implications.

The 1984 Union Carbide chemical plant disaster in Bhopal, India, provides a telling example, as Kim Fortun and her collaborator T. R. Chouhan (who worked in the Carbide plant) report. Union Carbide located the cause of the disaster in the actions of one supposedly disgruntled worker who attached a water hose directly to Storage Tank 610, starting a reaction process that resulted in the total release of the contents of the tank (40,000 tons of toxic chemicals) into the atmosphere. Thousands were killed, hundreds of thousands more exposed. Workers locate the cause of the disaster further back in the piping configuration—in a routine maintenance exercise involving water-washing of clogged pipes, from which water was introduced into Tank 610—because of a missing slip bind that should have isolated the water used in this routine exercise. The implications of these different accounts are significant. In locating the direct cause of the Bhopal disaster just a little further out in the piping configuration, the workers' account points to a chain of failures leading to disaster, reaching back over years and to the headquarters and engineering offices of Union Carbide in the United States. In the workers' account, a missing slip bind was the proximate cause of the Bhopal disaster, but the real story was the inadequate design of the plant, poor maintenance, and declining investment in worker training as plans were made to decommission the plant (because its market promise failed) (Chouhan 1994; Fortun 2001). Here, as in many other disasters, an expanded time frame provides important critical purchase.

Bhopal also illustrates how acute disasters often become drawn-out, slow disasters. Pollution from the abandoned Bhopal plant has now seeped into local water supplies, for example, creating routine (rather than exceptional or explosive) exposures to toxic chemicals. Such slow exposures are often even harder to deal with—scientifically and politically—than fast exposures. "Slow disaster" has thus become a way to conceptualize the injury produced by everyday operations of the many entangled systems people live within and depend on today (Bleicher and Gross 2011; Liboiron 2012), injuries that are often difficult to account for because cultural, legal, and technical "regimes of imperceptibility" work against such accounting (Murphy 2006).

Renowned disaster sociologist Kai Erikson (1976), best known for his seminal account of the horrifically fast flood disaster in Buffalo Creek, West Virginia, in 1972 (caused by the bursting of a coal slurry impoundment dam), called for recognition of slow disaster in the mid-1990s. Erikson (1994, 20) emphasized that "chronic conditions as well as acute events can induce trauma, and this, too, belongs in our calculations" of disaster. Theorizing a "new species of trouble," Erikson described a "chronic dread" and sense of helplessness among people subject to daily exposure—or potential exposure—to radiation and toxic chemicals. Erikson argues that these exposures create a new, insidious type of trauma, one deserving scholarly and political attention.

Historian Kate Brown extends this approach in arguing that plutonium production at both Hanford in the United States and Mayak in the former Soviet Union have created "slow-motion disasters" because, over time, each has released more curies of ionizing radiation into the environment than the Chernobyl disaster (Brown 2013). At both sites, the releases were not primarily the result of meltdowns or catastrophic events (although there have been major accidents at Mayak) but rather of normal operating procedures, aging infrastructure, and, in some cases, experiments. Unlike the Chernobyl disaster, releases took place over decades instead of days.

Literature scholar Rob Nixon's (2011) conceptualization of "slow violence" has also proven influential. Slow violence in Nixon's terms is "a violence that is neither spectacular nor instantaneous, but rather incremental and accretive, its calamitous repercussions playing out across a range of temporal scales" (2011, 2). Nixon emphasizes the "need to engage the representational, narrative, and strategic challenges posed by the relative invisibility of slow violence" (ibid.). STS researchers have taken up this challenge, with many insistently expansive in what they deem disaster (Jobin 2013a; 2013b, forthcoming; Liboiron 2012), recognizing that crises and disasters are not first-order observations but distinctions made with analytic and political purpose (Bond 2013; Roitman 2013).

STS researchers have also shown how imagining and planning for disaster powerfully shapes the future, in ways often outside political accountability (Choi 2015; Collier and Lakoff 2014; Lakoff 2010; Masco 2014). Basic political matters are at stake, such as divisions of responsibility between private and public sectors, what counts as a security issue, and who should be involved in and lead humanitarian response in disaster contexts. Decisions about such matters are built into disaster response plans but rarely deliberated publicly. Sometimes these plans are never implemented but still reshape priorities, resource flows, and what is considered legitimate. When implemented, the urgency of disaster response often leaves no space for questions. The stakes are thus high; the time frames of disaster research need to be expanded forward into the future as well as back in history to capture them.

Disaster through Design and Expertise

In 1984, the year of the Union Carbide chemical plant disaster in Bhopal (and just two years before the nuclear disaster in Chernobyl), organizational sociologist Charles Perrow (1984) published what would become a critical text for disaster studies. *Normal Accidents* describes industrial facilities—nuclear power plants, chemical processing plants, air transport networks—as tangles of sociotechnical systems so tightly coupled that run-away incidents exceeding what experts can understand, much less control, must be considered routine. Perrow's emphasis is on the way industrial systems almost inevitably harbor the potential for disaster, even when operating under optimal conditions and designed for safety. From this perspective, accidents are normal rather than exceptional; the industrial systems of modern societies are inherently dangerous and should be governed as such.

Normal Accidents was also methodologically significant, directing scholarly attention to the role of technical and organizational design in disaster. Diane Vaughan's examination of the U.S. *Challenger* disaster (1996) followed this path. Her analysis of risk modeling and decision making at the National Aeronautics and Space Administration (NASA) both drew upon and advanced a stream of sociological work focused on "the dark side of organizations." Vaughan defined organizational deviance as "an event, activity, or circumstance, occurring in and/or produced by a formal organization, that deviates from both formal design goals and normative standards or expectations, either in the fact of its occurrence or in its consequences, and produces a suboptimal outcome" (1999, 273).

The methodological move to focus on the dark side of organizations (and modernity writ large) has been foundational for disaster research, providing grounds for many important studies. Sonja Schmid's (2015) *Producing Power: The Pre-Chernobyl History of the Soviet Nuclear Industry*, for example, describes how the design of nuclear reactors in the Soviet Union was shaped by strong personalities, desire for international recognition, bureaucratic jockeying, and political culture, all in turn shaping what would happen in Chernobyl. Schmid's core argument is that structural conditions and design decisions led to catastrophic technological failure—complicating claims that human error by plant operators was *the* cause of the Chernobyl disaster. *Producing Power* is a deeply Soviet story while also pointing to dynamics of disaster that recur across cases.

Schmid's account of nuclear power technology draws out the historical conditions that shape hazardous technology design and the ways such design is beyond the full control of system designers and practitioners. Scott Snook also extends these insights in his book *Friendly Fire*, about the 1994 accidental shooting down of two U.S. army helicopters by U.S. F-15 fighter aircraft in the no-fly zone over Iraq, with 26 United Nations personnel on board. His description of "practical drift" in technology development complicates risk management even further, describing "the slow, steady uncoupling of practice from written procedure" (Snook 2001, 194) in supposedly high-reliability organizations. Drift happens over time, in different ways in different subunits of an organization, often with increasing detachment of subunits and thus decreasing understanding of what is happening outside any particular subunit. Further, when disaster occurs, the response is often overcorrection rather than improved understanding of where risk comes from and how it plays out. This, in turn, largely results from the forms of expertise cultivated and respected within an organization. Experts both build and maintain technical and organizational systems that harbor risk and are called on to make sense of their failures—with inevitably limited perspective (given that expertise, by definition, focuses attention in particular aspects of a problem, through particular frames of reference).

Constance Perin's (2005) work on the nuclear industry elaborates on this, drawing out how organizational context shapes and sometimes undercuts different forms of expertise (see also chapters 35 and 36 this volume). In *Shouldering Risks: The Culture of Control in the Nuclear Power Industry*, Perin argues that reducing and managing risk in nuclear power stations—and, by extension, other complex and high-risk industries—is hampered by the hierarchical ordering of command-and-control approaches over real-time logics. The former emphasizes adherence to procedure, compliance-oriented regulatory and management frameworks, measurable outcomes and probability estimates, design, and compartmentalization. The latter stresses processes of doubt and discovery, judgment, experience and professionalization, contextual and qualitative factors, as well as maintenance and reconfiguration. This conceptual (and social) hierarchy tends to limit what can be seen, since the procedures produced by calculated logics deal only with the expected and quantifiable and often fail to see nuclear power plants as "peopled" systems.

The inevitable limits of all forms of professional expertise—as well as their inevitable political positioning—has many consequences, not least the ways it often provokes the emergence of lay, often oppositional forms of expertise. As long recognized in STS, "public understanding of science" is often deep, locally grounded, and enhanced with other forms of knowledge (Wynne 1992, 1996). Sometimes, "lay" experts emerge from the grassroots, as in the case of mothers concerned about food safety in the wake of the Fukushima disaster (Sternsdorff-Cisterna 2015). Other lay (e.g., unofficial) experts also bring unexpected forms of expertise into disaster response. Murillo (2015), for example, describes the mobilization of a wide network of computer experts to address the lack of public information on radiation levels in Japan after the Fukushima disaster. This network included volunteers from hackerspaces, Free and Open Source software and hardware projects, companies, nonprofit organizations, and educational institutions in Japan and the United States—self-organized into a highly distributed and effective collective that came to be relied on by local governments as well as citizens. Volunteers also played an enormous role in the wake of Hurricane Sandy, which became the largest relief effort in New York City's history. Sixty thousand people volunteered, from all walks of life, more than four times the number deployed by the Red Cross (FEMA 2013). These volunteers brought many kinds of expertise into their work and acquired new forms of expertise in the process. At street level, it was all too obvious to most of these people how years of infrastructure neglect in New York City—particularly in poor neighborhoods—set the stage for disaster (Liboiron and Wachsmuth 2013).

Disaster researchers themselves are yet another kind of expert that shapes disaster planning and response, helping delineate what counts as hazards, as disaster-related, and as possible pathways to recovery. Sometimes "disaster researchers" are researchers never before involved in disaster response who are brought in because of relevant specialized expertise. Hurricane Katrina and the Deepwater Horizon oil spill drew in both regulatory and academic scientists, for example, many with little previous experience with disaster, reconfiguring careers and entire research fields in the process (Bond 2013; Frickel and Vincent 2011). The ways these researchers focused their work had clear social impacts, delineating, for example, what areas of New Orleans would be considered contaminated and what would be classified as ready or not for rehabitation (Frickel 2008; Frickel and Vincent 2011). Experts of many kinds are also called into rebuilding programs, simultaneously negotiating technical issues (*where* homes should be rebuilt, for example), financial and commercial pressures, and the many challenges of participatory planning and action (Felt and Fochler 2010; Vaughn forthcoming).

Disaster researchers also come from research fields and schools of thought long codified to work specifically on disaster. As noted previously, these have developed in many different national contexts and internationally, including many fields of engineering and science as well as more operational fields such as emergency management. Scott (2001) describes how a cadre of disaster experts was built in the United States, turning a diverse group of professionals—insurance inspectors, engineers, scientists, journalists, public officials, civil defense planners, and emergency managers—into authorities with both expansive and limited power. Knowles (2011) describes how these experts helped create elaborate fire and building codes—and about how these codes didn't block construction of New York's Twin Towers, despite their clearly risky, experimental design. He also describes increasing investment in preparation for terrorist attacks, even though floods, fires, and earthquakes pose far more consistent threats. The story is about becoming prepared for some things while remaining profoundly unprepared for others. It is also a story about learning from and managing disasters once they occur, often with crossed purposes. Building on Steve Hilgartner's (2007) analysis of official investigations of Hurricane Katrina, Knowles (2014a) describes how disaster investigations are an expected, necessary stage in the life cycle of a technological disaster—a normal outgrowth of the very technoscientific mode of thinking that brings high-risk technological systems into existence in the first place. Disaster investigations work to soothe public fears and restore faith in experts; yet, investigations may reveal negligence that opens the door to sustained critiques of corporate, regulatory, and/or governmental leadership (Hilgartner 2007; Knowles 2013).

Investigation and reflection in the wake of disaster is complex, culturally coded, and often weighted by vested interests. There is much to learn through comparison. In the late 1990s, for example, the European Environmental Agency undertook a study of "late lessons from early warnings of disaster" to advance understanding of how the precautionary principle could have been used to anticipate and mitigate an array of hazards, including asbestos, PCBs, benzene, and sulfur dioxide (all highly hazardous pollutants), DES (a pharmaceutical that produced birth defects), and practices resulting in calamitous crash of fish stocks (Harremoës et al. 2001). Similar studies have not been undertaken in many other contexts. Indeed, a key purpose of the European study and report was to contribute to debates within and between Europe and the United States about the values of (and problems with) precautionary approaches. The results of the European effort are well worth attention and further development, emphasizing the need for regulatory independence while retaining an inclusive approach, to "take full account of the assumptions and values of different social groups," to "acknowledge and respond to ignorance as well as uncertainty and risk," and to ensure that real-world conditions are adequately accounted for in regulatory appraisals (ibid., 169). All these recommendations have the potential to bring STS perspectives directly into policy domains. In this report itself, for example, leading STS researcher Brian Wynne was an editor. The report thus demonstrates the potential role of STS *as* disaster expertise, while foregrounding deep problems with conventional practices and expectations of expertise.

Denaturalizing Disaster Disparities

Thirty years ago German sociologist Ulrich Beck (1987) argued that industrialized nations have come to spend as much time and effort managing risks and disasters as they used to spend managing the creation and operation of cities and factories. For Beck, the relationship was causal and reflexive: risks today "are risks *of modernization*. They are a *wholesale product* of industrialization, and are systematically intensified as it becomes global" (1992, 21; emphasis in original). Beck's central observation was that "risk society" allows no escape—for anyone—from global calamities like nuclear radiation and climate. This was a critical insight, highlighting how most if not all people today have what historian Brett Walker (2010) has called "industrialized bodies."

Beck's thesis has nonetheless been the subject of extended debate, with some arguing that it undercuts recognition of the way environmental problems and disaster disproportionately impact poor and minority communities. To address this, researchers have drawn on critical race, feminist, and postcolonial theory in different ways, describing "intersectionalities" that compound bias and disadvantage (Crenshaw 1991,

2010), sometimes in ironic ways.[7] Marjaana Jauhola (2010), for example, shows how radio programs that were part of recovery initiatives in Indonesia following the 2004 Indian Ocean tsunami idealized and reinforced conventional gender figurations in the very process of promoting gender equity. Jauhola also shows how neoliberal ideals of "working hard" and "making progress" are reinforced, silencing other ways of making sense of everyday life and the future.

Another stream of work concerned with disaster disparities focuses on vulnerability—structural positioning that makes particular people, social groups, and regions (often inordinately) subject to harm. Poverty creates vulnerability, as does gender, ethnicity, and oppositional political standing. Like the concept of risk society, the concept of vulnerability has also been subject to debate, with some arguing that it naturalizes susceptibility to harm, undercutting recognition of the way such susceptibility is produced by historical, cultural, and political economic forces. Extending critiques in human ecology and complexity theory (Susman et al. 1983), for example, Greg Bankoff (2003), has long called for deeper historical perspective in thinking about vulnerability, arguing for historical, natural, and social generation of hazards themselves (floods, for example) and for recognition of ways vulnerability (of women, for example) is socially produced rather than inherent. Bankoff's (2010) key argument is that there is no such thing as natural disaster.

Debates about disaster "resilience" also implicate social disparities. Disaster resilience has become a "national imperative" in many contexts, with governments investing (sometimes extensively) "to prepare and plan for, absorb, recover from, and more successfully adapt to adverse events" (U.S. National Academies 2012, 16). In many contexts, the Hyogo Framework for Action (adopted by the United Nations World Conference for Disaster Reduction in 2005) provides orientation, defining disaster resilience as "the ability of individuals, communities, organizations, and states to adapt to and recover from hazards, shocks, or stresses without compromising long-term prospects for development" (Combaz 2014, 2). Importantly, the Hyogo Framework emphasizes how resilience involves peoples' capacity to organize themselves (at local, national, regional, and international levels) to learn from past disasters so as to reduce their risks for future ones. The Hyogo Framework acknowledges both increasing disaster risks and the insufficiency of existing response capacity. It recognizes structural problems—related to development, poverty, gender relations, governance, and climate change—as key risk factors, emphasizing the need for special commitment to both gender equality and vulnerable social groups and countries (Manyena 2006).[8]

Debates about resilience have nonetheless continued, with many lines of argument. Some researchers have argued that focusing on resilience can distract both from work to *prevent* disaster and from recognition of how resilience ideals map onto the ideals and disavowals of neoliberalism (Alexander 2013; Dowty and Allen 2011; Weichselgartner and Kelman 2014). Researchers have also argued that thinking about resilience tends to neglect symbolic dimensions of social standing and capacity. Especially important for our argument here is the way resilience, and the Hyogo Framework in particular, works toward recovery and stability in ways that can reproduce previously entrenched disadvantage. In the terms of the Hyogo Framework, the goal is to create capacity to adapt to stress "by resisting or changing in order to reach and maintain an acceptable level of functioning and structure" (UNISDR 2005, 4). With such an orientation, the ways functional systems inevitably have and produce margins (and thus disadvantage) is easily obscured. There are also lines of argument focused on the insufficiencies of the frameworks used for research on resilience. Weichselgartner and Kelman, for example, argue against technical reductionism and "the quantitative streamlining of resilience into one index," emphasizing the need for "integrating different knowledge types and experiences to generate scientifically reliable, context-appropriate and socially robust resilience-building activities" (2014, 1). STS researchers have many ways to contribute to such an effort.

Conclusion: Challenges and Significance of STS Disaster Research

In this chapter we have argued that STS can make important contributions to efforts to expand the interdisciplinarity and sophistication of disaster research in different national contexts and internationally, attentive to political-economic and ecological shifts that could dramatically increase disaster vulnerability and injury. We have also argued that STS disaster research challenges STS itself in fundamental ways, unsettling functionalist assumptions through focus on the injurious dysfunction of diverse, nested systems (ecological, technological, discursive, and so on). Such dysfunction produces outcomes that can't be explained or addressed with standard operating procedures and established modes of thought. In the terms of postcolonial theory, disasters produce "enunciatory disorder" and nonsense (Bhabha 1994, 123), demanding new forms of knowledge and practice, recognizing interlacing power dynamics and systemic tendencies to marginalize critical voices. Disaster thus calls for especially persistent effort to multiply perspectives, recognizing the situatedness and partiality of knowledge (Haraway 1988; Harding 1986) and the value of explanatory pluralism (Keller 2002; Turkle and Papert 1990). This, however, is a special challenge in disaster contexts given the often urgent need for operational efficacy. Disasters are thus difficult, conceptually and practically, in complex and interlinked ways. In closing, we want to reiterate and elaborate on this.

Going forward, disaster research will need to be methodologically and theoretically inventive, empirically rich, and expressed in ways attentive to the acute problems of representations that beset disaster (and disaster studies) (Jhala 2004; Kuchinskaya 2013, 2014; Sontag 1965). STS disaster research should also be ethically and politically engaged in new ways, acknowledging a special need for collaboration among researchers, and between researchers, survivors, first responders, disaster managers, journalists, and other stakeholders, on an international scale.

Often, for example, it is important to have a positive societal impact *during* data collection and fieldwork, not just with finished results. Following Hurricane Sandy's landfall in New York City, for example, STS researchers helped establish the Superstorm Research Lab, in what came to be seen as a mutual aid research model. Members worked in solidarity with research subjects, doing pro bono work in exchange for interviews, compensating collaborators and contributors, and sharing economic, food, and labor resources whenever possible. Based on the experience of working in a disaster zone as a researcher, one member of Superstorm Research Lab, STS researcher Max Liboiron, created a Memorandum of Understanding (MOU) for researcher-community partnerships after watching other research groups create "second disasters" by parachuting into devastated communities, mining knowledge, claiming survivors' analytical insights as their own, and then walking out again. The MOU is based heavily on ideas of research ethics developed by tribal communities who have been managing the slow disaster of colonial research for generations (Liboiron 2013, 2015).

Sara Wylie and colleagues' work to develop do-it-yourself environmental monitoring technology also extends research ethics in STS. They use and sometimes themselves build inexpensive, open source technology that communities in slow disaster zones, such as fracking fields, can use to monitor their environments to keep themselves safer and to bring forward proof of environmental injustice (see also Ottinger, Barandiarán, and Kimura, chapter 35 this volume). They aim to build research capacity in the regions where they work so that local communities can continue research long after professional researchers have left the scene (Wylie et al. 2014).

Dealing with compound disasters also poses special challenges, ethically and methodologically. Compound disasters (a term that came into wide usage after the 3-11 disasters in Japan, as noted earlier) involve chains of destructive events with cumulative consequences—as when an earthquake causes a gas leak and fire, contaminating water supplies, thus undercutting hygiene, and then contributing to the outbreak of waterborne diseases. Extreme weather also leads to compound disasters, as with the storm surge flooding, levee breaches, and evacuation failures of Hurricane Katrina. Armed conflicts have also been productively described as compound disasters (Choi 2015; Wachira 1997). Dense urbanization and settlement in high-risk zones (coastal areas, for example) harbor significant potential for compound disaster, creating serious planning and governance challenges in geographies with closely interconnected technological and natural systems. In all contexts of compound disaster, people are in way over their head—figuratively, and sometimes literally. Compound disaster drives people to take on responsibilities and make decisions outside previously prescribed roles, usually without adequate preparation. STS researchers can be in such conundrums themselves, or can be called on to help other experts make sense of their shifting and expanding responsibilities in disaster.

Another direction for collaborative work could expand comparative analyses of disaster, learning to think together across disaster cases. Across fields, researchers can map and question what is generalizable and what is not, generating questions with both practical and theoretical relevance. Many cross-cutting questions have already been considered (Button 2010; Frickel and Fortun 2013; Knowles 2014a, 2014b), pointing to ways that a comparative body of STS disaster research could come together. Such research would depend on inventive study designs, building comparative perspectives and potential in from the outset, redirecting tendencies to study one disaster at a time. Such research could enable productive tracking between thick empirical descriptions of particular disasters and identification of patterns recurring across disasters—advancing conceptualizations of inequality, expertise, legitimacy, and other phenomena of enduring social theoretical concern, keeping STS research in scholarly conversations that extend far beyond both STS and disaster studies.

The challenges of STS disaster research are considerable but have engaged a diverse and dynamic group of researchers committed both to their own, personal contributions to disaster research and to development of a strong community of researchers able to work collaboratively and comparatively, responsive to both long-running and emergent problems. The Disaster-STS Research Network, which came together in the wake of the Fukushima disaster, is illustrative, aiming to bring together STS researchers from around the world, creating forums for theory development as well as for interface with policy and operations-oriented practitioners. The digital platform supporting the Disaster-STS Network also provides a place to archive, access, and collaboratively analyze research materials, building a strong base for future disaster research.[9] STS disaster research thus advances an historically and politically attuned thread of STS scholarship, with potential to help remake both the world and STS itself.

Notes

1. A well-articulated call for more intensive collaboration between more or less "applied" researchers was made by prominent disaster sociologist (and Director of the Natural Hazards Center at the University of Colorado, Boulder) Katherine Tierney (2007) in the review article "From the Margins to the Mainstream: Disaster Research at the Crossroads."

2. Michael Guggenheim makes a related argument in explaining a remarkable surge in social science research on disasters in recent years—that can't be explained either by naturalization (pointing to an actual increase in disasters) or by focus on increasing *perception* of disaster (what he calls a Zeitgeist argument). Instead, Guggenheim offers a third explanation, which he calls "politicization," associated with the emergence of theoretical frames that make us more sensitive to particular phenomena. Disasters, Guggenheim argues, have come to the center of social scientific attention because of an increasing interest in breaks and ruptures rather than in continuity and structure, in turn linked to an "idea of politics as problematization of the composition of the world." Disasters are thus seen as politics, and politics are disaster (Guggenheim 2014).

3. It is also useful to recognize diverse theoretical traditions that undergird disaster research. In the French context, for example, "pragmatic sociology" (building off Dewey and James), developed in seminal work by Luc Boltanski and Laurent Thévenot ([1991] 2006), is an important genealogy. Laura Centemeri (2014, 2015) extends this genealogy in her studies of the Sevesco disaster and of environmental valuation more generally. In the United States an important thread of disaster research stems from feminist and postcolonial concerns with the many ways social hierarchy and marginality are continually reproduced (Bhabha 1994; Chatterjee 1993; Cornell 1992; de Lauretis 1987; Spivak 1987). Notably, studies in this vein focus on both fast and slow disaster—in Bhopal and Fukushima, in the wake of the 2004 Indian Ocean tsunami, and in the study of ocean plastics, the shale gas boom and endocrine disruption, for example (Choi 2015; Fortun 2001; Liboiron 2012; Murillo 2015).

4. See Ottinger, Barandiarán, and Kimura, chapter 35 this volume, on "Environmental Justice: Knowledge, Technology, and Expertise."

5. For original publication, see Carr (1932).

6. See http://teach311.org/.

7. One important stream of work focuses on the ways high-risk technologies often produce labor disasters long before they produce more visible disasters "beyond the fence line." Gabrielle Hecht (2012), Paul Jobin (2013a, 2013b), and Markowitz and Rosner (2002), for example, have drawn this out in sobering detail.

8. Manyena, among others, has usefully described the history and cross-disciplinary travels of the concept of resilience, originating in nineteenth-century study of materials, coming into psychology in the 1940s, ecology in the 1970s, the social sciences and development aid in the 1990s, and economics and the study of organizations most recently (Manyena 2006; McAslan 2010).

9. See http://disaster-sts-network.org/.

References

Alexander, David E. 2013. "Resilience and Disaster Risk Reduction: An Etymological Journey." *Natural Hazards and Earth System Science* 13: 2707–16.

Anderson, Warwick. 2009. "From Subjugated Knowledge to Conjugated Subjects: Science and Globalization, or Postcolonial Studies of Science?" *Postcolonial Studies* 12 (4): 389–400.

Backhouse, Roger E., and Philippe Fontaine, eds. 2014. *A Historiography of the Modern Social Sciences*. Cambridge: Cambridge University Press.

Bankoff, Greg. 2003. "Constructing Vulnerability: The Historical, Natural and Social Generation of Flooding in Metro Manila." *Disasters* 27 (3): 224–38.

___. 2010. "No Such Thing as Natural Disasters." *Harvard International Review*, August 23, 2010, accessed on June 15, 2015, http://hir.harvard.edu/no-such-thing-as-natural-disasters.

Beck, Ulrich. 1987. "The Anthropological Shock: Chernobyl and the Contours of the Risk Society." *Berkeley Journal of Social Sociology* 32: 153–65.

___. 1992. *Risk Society: Towards a New Modernity*. London: Sage.

Bhabha, Homi K. 1994. *The Location of Culture*. New York: Routledge.

___. 2015. "The Beginning of Their Real Enunciation: Stuart Hall and the Work of Culture." *Critical Inquiry* 42 (1): 1–30.

Bleicher, Alena, and Matthias Gross. 2011. "Response and Recovery in the Remediation of Contaminated Land in Eastern Germany." In *Dynamics of Disaster: Lessons in Risk, Response, and Recovery*, edited by Rachel A. Dowty and Barbara L. Allen, 187–202. London: Earthscan.

Boltanski, Luc, and Laurent Thévenot. [1991] 2006. *On Justification: Economies of Worth*. Princeton, NJ: Princeton University Press.

Bond, David. 2013. "Governing Disaster: The Political Life of the Environment during the BP Oil Spill." *Cultural Anthropology* 28 (4): 694–715.

Brown, Kate. 2013. *Plutopia: Nuclear Families, Atomic Cities, and the Great Soviet and American Plutonium Disasters*. Oxford: Oxford University Press.

Button, Gregory. 2010. *Disaster Culture: Knowledge and Uncertainty in the Wake of Human and Environmental Catastrophe*. Walnut Creek, CA: Left Coast Press.

Calhoun, Craig. 2004. "A World of Emergencies: Fear, Intervention, and the Limits of Cosmopolitan Order." *Canadian Review of Sociology and Anthropology* 41 (4): 373–95.

Carr, Lowell Juilliard. 1932. "Disaster and the Sequence Pattern Concept of Social Change." *American Journal of Sociology* 38 (2): 207–18.

Centemeri, Laura. 2014. "What Kind of Knowledge Is Needed about Toxicant-Related Health Issues? Some Lessons Drawn from the Seveso Dioxin Case." In *Powerless Science? Science and*

Politics in a Toxic World, edited by Soraya Boudia and Nathalie Jas, 134–51. Oxford, NY: Berghahn Books.

___. 2015. "Reframing Problems of Incommensurability in Environmental Conflicts through Pragmatic Sociology: From Value Pluralism to the Plurality of Modes of Engagement with the Environment." *Environmental Values* 24 (3): 299–320.

Chatterjee, Partha. 1993. *Nationalist Thought and the Colonial World: A Derivative Discourse.* Minneapolis: University of Minnesota Press.

Choi, Vivian Y. 2015. "Anticipatory States: Tsunami, War and Insecurity in Sri Lanka." *Cultural Anthropology* 30 (2): 286–389.

Chouhan, T. R. 1994. *Bhopal: The Inside Story.* New York: Apex Press

Clark, Nigel. 2014. "Geo-politics and the Disaster of the Anthropocene." *The Sociological Review* 62 (S1): 19–37.

Collier, Stephen, and Andrew Lakoff. 2014. "Vital Systems Security: Reflexive Biopolitics and the Government of Emergency. *Theory, Culture, and Society* 32 (2): 19–51.

Combaz, Emilie. 2014. *Disaster Resilience: Topic Guide.* Birmingham, UK: GSDRC, University of Birmingham.

Comfort, Louise K. 2005. "Risk, Security, and Disaster Management." *Annual Review of Political Science* 8: 335–56.

Cornell, Drucilla. 1992. *The Philosophy of the Limit.* London: Routledge.

Crenshaw, Kimberle. 1991. "Mapping the Margins: Intersectionality, Identity Politics, and Violence against Women of Color." *Stanford Law Review* 43 (6): 1241–99.

___. 2010. "Close Encounters of Three Kinds: On Teaching Dominance, Feminism, and Intersectionality." *Tulsa Law Review* 46 (1): 151–89.

Daston, Lorraine, and Peter Galison. 2007. *Objectivity.* New York: Zone Books.

de Lauretis, Teresa. 1987. *Technologies of Gender: Essays on Theory, Film, and Fiction.* Bloomington: Indiana University Press

de Waal, Alex. 2010. "An Emancipatory Imperium? Power and Principle in the Humanitarian International." In *Contemporary States of Emergency: The Politics of Military and Humanitarian Interventions*, edited by Didier Fassin and Mariella Pandolfi, 295–316. Cambridge, MA: MIT Press.

Dowty, Rachel, and Barbara Allen, eds. 2011. *Dynamics of Disaster: Lessons in Risk, Response, and Recovery.* London: Earthscan.

Erikson, Kai. 1976. *Everything in Its Path: Destruction of Community in the Buffalo Creek Flood.* New York: Simon and Schuster.

___. 1994. *A New Species of Trouble: Explorations in Disaster, Trauma, and Community.* New York: W. W. Norton & Company.

Farmer, Paul. 1996. "On Suffering and Structural Violence: A View from Below." *Daedalus* 125 (1): 261–83.

___. 2003. Pathologies of Power: Health, Human Rights, and the New War on the Poor. Berkeley: University of California Press.

Federal Emergency Management Agency (FEMA). 2013. "Hurricane Sandy: Youthful Energy and Idealism Tackles Real World Disaster Response, Lessons Learned." *Information Sharing Report.* Washington, DC: FEMA.

Felt, Ulrike, and Maximilian Fochler. 2010. "Machineries for Making Publics: Inscribing and Describing Publics in Public Engagement." *Minerva* 48 (3): 219–38.

Fortun, Kim. 2001. *Advocacy after Bhopal: Environmentalism, Disaster, New Global Orders.* Chicago: University of Chicago Press.

Franklin, Sarah. 1995. "Science as Culture, Cultures of Science." *Annual Review of Anthropology* 24: 163–84.

Frickel, Scott. 2008. "On Missing New Orleans: Lost Knowledge and Knowledge Gaps in an Urban Hazardscape." *Environmental History* 13 (4): 634–50.

Frickel, Scott, and Kim Fortun. 2013. "Making a Case for Disaster-STS." *An STS Forum on the East Japan Disaster.* Accessed at https://fukushimaforum.wordpress.com/online-forum-2/online -forum/making-a-case-for-disaster-science-and-technology-studies/.

Frickel, Scott, and M. Bess Vincent. 2011. "Katrina's Contamination: Regulatory Knowledge Gaps in the Making and Unmaking of Environmental Contention." In *Dynamics of Disaster: Lessons in Risk, Response, and Recovery,* edited by Rachel A. Dowty and Barbara L. Allen, 11–28. London: Earthscan.

Fritz, Charles E. 1961. "Disasters." In *Contemporary Social Problems,* edited by Robert K. Merton and Robert A. Nisbet, 651–94. New York: Harcourt.

Glik, Deborah C. 2007. "Risk Communication for Public Health Emergencies." *Annual Review of Public Health* 28: 33–54.

Guggenheim, Michael. 2014. "Introduction: Disasters as Politics—Politics as Disasters." *The Socio-logical Review* 62 (S1): 1–16.

Hannigan, John. 2013. "Coping with Disasters without Borders." *European Financial Review.* December 29, 2013. Accessed March 2, 2016, at http://www.europeanfinancialreview. com/?p=616.

Haraway, Donna. 1988. "Situated Knowledges: The Science Question in Feminism and the Privilege of Partial Perspective." *Feminist Studies* 14 (3): 575–99.

___. 1994. "A Game of Cat's Cradle: Science Studies, Feminist Theory, Cultural Studies." *Configurations* 2 (1): 59–71.

Harding, Sandra. 1986. *The Science Question in Feminism*. Ithaca, NY: Cornell University Press.

___. 2011. *The Postcolonial Science and Technology Studies Reader*. Durham, NC: Duke University Press.

Harremoës, Poul, David Gee, Malcolm MacGarvin, Andy Stirling, Jane Keys, Brian Wynne, and Sofia Guedes Vaz. 2001. *Late Lessons from Early Warnings: The Precautionary Principle 1896–2000*. Brussels: European Environment Agency.

Hecht, Gabrielle. 2012. *Being Nuclear: Africans and the Global Uranium Trade*. Cambridge, MA: MIT Press.

Hilgartner, Stephen. 2007. "Overflow and Containment in the Aftermath of Disaster." *Social Studies of Science* 37 (1): 153–58.

Jasanoff, Sheila. 1986. *Risk Management and Political Culture*. New York: Sage.

___, ed. 1994. *Learning from Disaster: Risk Management after Bhopal*. Philadelphia: University of Pennsylvania Press.

Jauhola, Marjaana. 2010. "When House Becomes Home"—Reading Normativity in Gender Equality Advocacy in Post-Tsunami Aceh, Indonesia." *Gender, Technology and Development* 14 (2): 173–95.

Jhala, Jayasinhji. 2004. "In a Time of Fear and Terror: Seeing, Assessing, Assisting, Understanding and Living the Reality of the Consequences of Disaster." *Visual Anthropology Review* 20 (1): 59–69.

Jobin, Paul. 2013a "Beyond Uncertainty: Industrial Hazards and Class Actions in Taiwan and Japan." In *Environmental History in East Asia: Interdisciplinary Perspectives*, edited by Ts'ui-jung Liu, 341–85. New York: Routledge.

___. 2013b. "The Criminalization of Industrial Disease: Epidemiology in a Japanese Law Suit." In *Disease and Crime: A History of Social Pathologies and the New Politics of Health*, edited by Robert Peckham, 129–50. New York: Routledge.

___. Forthcoming. "Nuclear Gypsies in Fukushima before and after 3/11." In *Nuclear Portraits: People, Communities and the Environment*, edited by Laurel MacDowell. Toronto: University of Toronto Press.

Keller, Evelyn Fox. 2002. *Making Sense of Life: Explaining Biological Development with Models, Metaphors, and Machines*. Cambridge, MA: Harvard University Press.

Kendra, James, and Joanne Nigg. 2014. "Engineering and the Social Sciences: Historical Evolution of Interdisciplinary Approaches to Hazard and Disaster." *Engineering Studies* 6 (3): 134–58.

Knorr Cetina, Karin. 1999. *Epistemic Cultures: How the Sciences Make Knowledge*. Cambridge, MA: Harvard University Press.

Knowles, Scott Gabriel. 2011. *The Disaster Experts: Mastering Risk in Modern America*. Philadelphia: University of Pennsylvania Press.

___. 2013. "Investigating 3.11: Disaster and the Politics of Expert Inquiry." *An STS Forum on the East Japan Disaster.* Accessed at https://fukushimaforum.wordpress.com/online-forum-2/online-forum/making-a-case-for-disaster-science-and-technology-studies/.

___. 2014a. "Engineering Risk and Disaster: Disaster-STS and the American History of Technology." *Engineering Studies* 6 (3): 227–48.

___. 2014b. "Learning from Disaster? The History of Technology and the Future of Disaster Research." *Technology and Culture* 55 (4): 773–84.

Kuchinskaya, Olga. 2013. "Twice Invisible: Formal Representations of Radiation Danger." *Social Studies of Science* 43 (78): 78–96.

___. 2014. *The Politics of Invisibility: Public Knowledge about Radiation Health Effects after Chernobyl.* Cambridge, MA: MIT Press.

Lakoff, Andrew, ed. 2010. *Disaster and the Politics of Intervention.* New York: Columbia University Press.

Lavell, Allan, Michael Oppenheimer, Cherif Diop, Jeremy Hess Robert Lempert, Jianping Li, Robert Muir-Wood, and Soojeong Myeong. 2012. "Climate Change: New Dimensions in Disaster Risk, Exposure, Vulnerability, and Resilience." In *Managing the Risks of Extreme Events and Disasters to Advance Climate Change Adaptation: A Special Report of Working Groups I and II of the Intergovernmental Panel on Climate Change (IPCC),* 25–64. Cambridge: Cambridge University Press. Accessed at http://www.ipcc.ch/pdf/special-reports/srex/SREX-Chap1_FINAL.pdf.

Law, John, and Vicky Singleton. 2009. "A Further Species of Trouble?" In *The Cultural Meaning of the 2001 Outbreak of Foot and Mouth Disease in the UK,* edited by Martin Doering and Brigitte Nerlich, 229–42. Manchester: Manchester University Press.

Law, John, and Wen-yuan Lin. 2015. "Provincialising STS: Postcoloniality, Symmetry and Method." Accessed at http://www.heterogeneities.net/publications/LawLinProvincialising STS20151223.pdf.

Liboiron, Max. 2012. "Redefining Pollution: Plastics in the Wild." Ph.D. dissertation, New York University.

___. 2013. "Plasticizers: A Twenty-First Century Miasma." In *Accumulation: The Material Politics of Plastics,* edited by Jennifer Gabrys, Gay Hawkins, and Mike Michael, 22–44. New York: Routledge.

___. 2015. "Disaster Data, Data Activism: Grassroots Responses to Representations of Superstorm Sandy." In *Extreme Weather and Global Media,* edited by Diane Negra and Julia Leyda, 144–62. New York: Routledge.

Liboiron, Max, and David Wachsmuth. 2013. "The Fantasy of Disaster Response: Governance and Social Action during Hurricane Sandy." *Social Text Periscope* (October 29).

Lipset, Seymour M., ed. 1969. *Politics and the Social Sciences.* Cambridge: Oxford University Press.

Manyena, Siambabala Bernard. 2006. "The Concept of Resilience Revisited." *Disasters* 30 (4): 434–50.

Markowitz, Gerald, and David Rosner. 2002. *Deceit and Denial: The Deadly Politics of Industrial Pollution.* Berkeley: University of California Press.

Matsumoto, Miwao. 2013. "'Structural Disaster' Long before Fukushima: A Hidden Accident." *Development & Society* 42 (2): 165–90.

___. 2014. "The "Structural Disaster" of the Science-Technology-Society Interface from a Comparative Perspective with a Prewar Accident." In *Reflections on the Fukushima Daiichi Nuclear Accident,* edited by Joonhong Ahn, Cathryn Carson, Mikael Jensen, Kohta Juraku, Shinya Nagasaki, and Satoru Tanaka, 189–214. Heidelberg: Springer.

Masco, Joseph. 2014. *The Theater of Operations: National Security Affect from the Cold War to the War on Terror.* Durham, NC: Duke University Press.

McAslan, Alastair. 2010. *The Concept of Resilience: Understanding Its Origins, Meaning and Utility.* Adelaide: Torrens Resilience Institute.

Michael, Mike. 2014. "Afterward: On the Topologies and Temporalities of Disaster." *The Sociological Review* 62 (S1): 236–45.

Murillo, Luis Felipe Rosado. 2015. "Transnationality, Morality and Politics of Computing Expertise." Ph.D. dissertation, University of California, Los Angeles.

Murphy, Michelle. 2006. *Sick Building Syndrome and the Problem of Uncertainty: Environmental Politics, Technoscience, and Women Workers.* Durham, NC: Duke University Press.

Nixon, Rob. 2011. *Slow Violence and the Environmentalism of the Poor.* Cambridge, MA: Harvard University Press.

Oliver-Smith, Anthony. 1986. "Disaster Context and Causation: An Overview of Changing Perspectives in Disaster Research." In *Natural Disasters and Cultural Responses*, 1–38. Williamsburg, VA: College of William and Mary.

___. 1996. "Anthropological Research on Hazards and Disasters." *Annual Review of Anthropology* 25: 303–28.

Perin, Constance. 2005. *Shouldering Risks: The Culture of Control in the Nuclear Power Industry.* Princeton, NJ: Princeton University Press.

Perrow, Charles. 1984. Normal Accidents: Living with High-Risk Technologies. New York: Basic Books.

Pritchard, Sara B. 2012. "An Envirotechnical Disaster: Nature, Technology, and Politics at Fukushima." *Environmental History* 17 (2): 219–43.

Proctor, Robert N., and Londa Schiebinger, eds. 2009. *Agnotology: The Making and Unmaking of Ignorance.* Stanford, CA: Stanford University Press.

Quarantelli, Enrico Louis, ed. 1998. *What Is a Disaster? A Dozen Perspectives on the Question.* London: Routledge.

Redfield, Peter. 2011. *Life in Crisis: The Ethical Journey of Doctors without Borders.* Berkeley: University of California Press.

Roitman, Janet. 2013. *Anti-Crisis.* Durham, NC: Duke University Press.

Schmid, Sonja. 2015. *Producing Power: The Pre-Chernobyl History of the Soviet Nuclear Industry.* Cambridge, MA: MIT Press.

Seth, Suman. 2009. "Putting Knowledge in Its Place: Science, Colonialism, and the Postcolonial." *Postcolonial Studies* 12 (4): 373–88.

Shapin, Steven, and Simon Schaffer. 1985. *Leviathan and the Air-Pump: Hobbes, Boyle and the Experimental Life.* Princeton, NJ: Princeton University Press.

Skocpol, Theda, and Dietrich Rueschemeyer. 1996. *States, Social Knowledge, and the Origins of Modern Social Policies.* Princeton, NJ: Princeton University Press.

Slater, David H. 2013. "Urgent Ethnography." In *Japan Copes with Calamity: Ethnographies of the Earthquake, Tsunami and Nuclear Disasters of March 2011*, edited by Tom Gill, Brigitte Steger, and David Slater, 25–50. Bern: Peter Lang AG, International Academic Publishers.

Slater, David H., Nishimura Keiko, and Love Kindstrand. 2012. "Social Media, Information, and Political Activism in Japan's 3.11 Crisis." *The Asia-Pacific Journal* 10 (24): 1.

Slater, David H., Rika Morioka, and Haruka Danzuka. 2014. "Micro-Politics of Radiation: Young Mothers Looking for a Voice in Post-3.11 Fukushima." *Critical Asian Studies* 46 (3): 485–508.

Snook, Scott. 2001. *Friendly Fire.* Princeton, NJ: Princeton University Press.

Sontag, Susan. 1965. "Imagination of Disaster." *Commentary* 65: 42–48.

Spivak, Gayatri Chakravorty. 1987. *In Other Worlds: Essays in Cultural Politics.* London: Methuen.

Sternsdorff-Cisterna, Nicolas. 2015. "Food after Fukushima: Risk and Scientific Citizenship in Japan." *American Anthropologist* 117 (3): 455–67.

Susman, Paul, Phil O'Keefe, and Ben Wisner. 1983. "Global Disasters: A Radical Interpretation." In *Interpretations of Calamity from the Viewpoint of Human Ecology*, edited by K. Hewitt, 264–83. Boston: Allen and Unwin.

Tierney, Kathleen J. 2007. "From the Margins to the Mainstream? Disaster Research at the Crossroads." *Annual Review of Sociology* 33: 503–25.

___. 2012. "Disaster Governance: Social, Political and Economic Dimensions." *Annual Review of Environment and Resources* 37: 341–63.

Tironi, Manuel, Israel Rodríguez-Giralt, and Michael Guggenheim, eds. 2014. *Disasters and Politics: Materials, Experiments, Preparedness.* Oxford: Wiley-Blackwell.

Traweek, Sharon. 1988. *Beamtimes and Lifetimes: The World of High Energy Physicists*. Cambridge, MA: Harvard University Press.

Turkle, Sherry, and Seymour Papert. 1990. "Epistemological Pluralism and the Revaluation of the Concrete." *Signs: Journal of Women in Culture and Society* 16 (1): 128–57.

United Nations International Strategy for Disaster Risk Reduction (UNISDR). 2005. *Hyogo Framework for Action 2005–2015: Building the Resilience of Nations and Communities to Disasters*. World Conference on Disaster Reduction. Kobe, Hyogo, Japan. January 18–22.

U.S. National Academies, Committee on Increasing National Resilience to Hazards and Disasters. 2012. *Disaster Resilience: A National Imperative*. Washington DC: National Academies Press. Available at http://www.nap.edu/read/13457.

Vaughan, Diane. 1996. *The* Challenger *Launch Decision: Risky Technology, Culture, and Deviance at NASA*. Chicago: University of Chicago Press.

___. 1999. "The Dark Side of Organizations: Mistakes, Misconduct and Disasters." *Annual Review of Sociology* 25: 271–305.

Vaughn, Tyson Earl. Forthcoming. *Reconstructing Communities: Participatory Recovery Planning in Post Disaster Japan*. Cambridge, MA: MIT Press.

Wachira, George. 1997. "Conflicts in Africa as Compound Disasters: Complex Crises Requiring Comprehensive Responses." *Journal of Contingencies and Crisis Management* 5 (2): 109–17.

Walker, Brett L. 2010. *Toxic Archipelago: A History of Industrial Disease in Japan*. Seattle: University of Washington Press.

Weichselgartner, Juergen, and Ilan Kelman. 2014. "Geographies of Resilience Challenges and Opportunities of a Descriptive Concept." *Progress in Human Geography* 1–19. doi:10.1177/0309132513518834.

Williams, Christine. 2006. "Still Missing? Comments on the Twentieth Anniversary of 'The Missing Feminist Revolution in Sociology.'" *Social Problems* 53 (4): 454–58.

Wylie, Sara Ann, Kirk Jalbert, Shannon Dosemagen, and Matt Ratto. 2014. "Institutions for Civic Technoscience: How Critical Making Is Transforming Environmental Research." *The Information Society* 30 (2): 116–26.

Wynne, Brian. 1992. "Misunderstood Misunderstanding: Social Identities and the Public Uptake of Science." *Public Understanding of Science* 1 (3): 281–304.

___. 1996. "May the Sheep Safely Graze? A Reflexive View of the Expert–Lay Knowledge Divide." In *Risk, Environment, and Modernity: Towards a New Ecology*, edited by Scott Lash, Bronislaw Szerszynski, and Brian Wynne, 27–83. London: Sage Publications.

35   Environmental Justice: Knowledge, Technology, and Expertise

Gwen Ottinger, Javiera Barandiarán, and Aya H. Kimura

Introduction

On the Gulf Coast of the United States, a historic African American community lives within 100 feet of a Shell Chemical plant. Blaming its emissions for illness, residents campaign for relocation to someplace safer. When regulators and industry scientists deny that Shell's emissions are to blame, residents measure air quality with inexpensive, homemade samplers—and ultimately win relocation (Lerner 2005).

In Fukushima, Japan, radiation from the 2011 nuclear disaster threatens the livelihoods of rural organic farmers. In deciding whether to remain on their farms or move elsewhere, tensions arise among the farmers, their customers, and their families over whether radiation risks can be controlled and normalized—a view espoused by political elites—or whether they ought to be avoided—a view branded as irrational, irresponsible, and feminine (Kimura and Katano 2014).

In Huasco, Chile, a gold mine threatens to destroy glaciers that maintain a downstream agricultural valley. Tired of broken promises for jobs and rising ecological distress, rural workers successfully challenge the company and state's insistence that the glaciers can be safely "relocated." However, the state's approach to decision making marginalizes the community's and scientists' concerns about what mining and climate change mean for Chile's glaciers and farms (Barandiarán 2015a).

These are just a few examples of the environmental justice (EJ) cases analyzed by science and technology studies (STS) scholars. They come from all over the world but share several characteristics: they all involve situations where environmental burdens fall heavily on a group already structurally marginalized by virtue of race, gender, indigeneity, class, and/or rurality. They also feature contests over knowledge: what are the effects of pollution or industrial operations on the environment? On health? How should they be determined? Whose knowledge is credible? Depending on larger social and political contexts, the role, status, and even meaning of "science" or "expert" can vary greatly.

STS's contribution to social science research on environmental justice issues has been a better understanding of their epistemic dimensions—not only how knowledge claims are advanced and negotiated in particular cases but also how the conditions of knowledge production structure the information that is or is not available to environmentally burdened communities. Research has historically focused on North American cases; however, there is more to be learned from comparative work across national contexts. In addition, other core concepts from STS suggest directions for further research and advocacy: theories of sociotechnical systems suggest how the possibilities for environmental justice are structured and constrained not just in specific locales but globally; and research into the roles of scientists and engineers in sociotechnical controversies offer a better understanding of how technical practitioners do, and should, participate in fostering environmental justice.

We begin by clarifying the meanings of environmental justice, arguing that it is a frame for understanding inequity with respect to the environment that includes the distribution of hazards, participation in decision making, and recognition of both cultural identity and the unique knowledge of marginalized groups. The second section then describes STS's contributions to understanding the epistemic dimensions of EJ issues, especially the unevenness in knowledge production around environmental health and quality issues, and the political consequences of differences in lay and expert ways of knowing, and we argue for the need to expand those contributions through more comparative research. The third section turns to the global dimensions of environmental injustice, suggesting that an STS approach could usefully reveal how worldwide flows of knowledge, expertise, and technology structure environmental injustices. Finally, we highlight two areas of STS insight that, if mobilized more extensively, could significantly further our understanding of environmental justice: the role of sociotechnical systems in creating, reinforcing, and potentially undermining environmental inequities; and the roles and responsibilities of scientists and engineers in the struggle for environmental justice.

What Is Environmental Justice?

Environmental justice is a particular way of understanding race- and class-based inequities in pollution, facility siting, and enforcement of environmental laws. Although these inequities have a long history, the environmental justice "frame"—a way of looking at a problem that diagnoses it and points to solutions—first emerged in the United

States in the 1980s. In 1982 an African American community attempting to stop the construction of a hazardous waste dump in Warren County, North Carolina, dubbed the practice of targeting minority communities for waste and other hazardous sites "environmental racism" (Bullard 1994). Around the same time, largely white activists started calling attention to poor and working-class communities disproportionately exposed to toxic chemicals, such as at Love Canal, New York (Brulle and Pellow 2006). By the 1990s, the "environmental racism" and "antitoxics" framings of environmental inequities had largely been supplanted by an "environmental justice" frame to talk about the environmental burdens suffered by communities of color and poor white communities alike (Capek 1993; McGurty 2000).

The environmental justice frame was taken up by activists in other countries later, with the term becoming widespread only after 2000 (Walker 2012). On the world stage, the discourse of environmental justice refers to injustices at two levels. The unequal treatment of communities marginalized due to their social positionality along the lines of indigeneity, race, class, and gender *within* a given country is a central concern for the EJ movement internationally; cases of EJ activism include, for example, indigenous and peasant communities in Ecuador fighting state and multinational petroleum companies (Sawyer 2004) and the struggles of First Nations peoples in Canada to participate in environmental governance (Mascarenhas 2014; Tsosie 2012). *Global* inequities are an equally important category of environmental injustice. Global capital is implicated in the economic marginalization and environmental despoliation of developing countries (e.g., Fortun 2001; O'Rourke and Connolly 2003; Raman 2013; Sawyer 2004), with inequities especially visible on issues of hazardous waste disposal and the effects of global climate change (Iles 2004; Okerke 2008; Schlosberg and Collins 2014). Intergenerational equity, both within and across nations, is another justice issue brought to the fore by climate change. This tends to be framed by scholars and activists as "climate justice" rather than EJ; although, as we suggest below, STS analysis offers some tools for bridging the two framings.

As used by actors involved in the EJ movement, the concept of environmental justice is, on the one hand, expansive. The National People of Color Environmental Leadership Summit's seventeen Principles of Environmental Justice include considerations of equity, self-determination, and freedom from particular environmental harms (e.g., nuclear testing) alongside concern for worker safety, education, health care, and ecological systems themselves. On the other hand, contemporary uses of the term tend to be used more narrowly to refer to specific, place-based communities grappling with environmental problems—including food deserts and unfair housing practices but most often revolving around corporate polluters. Moreover, it has become common for not only marginalized communities but also communities who do not appear to be structurally disadvantaged to invoke the ideal of environmental justice in their campaigns, which leads geographer Gordon Walker (2009) to wonder whether all unevenness in environmental distributions should really be considered injustice.

Like most researchers working at the intersection of EJ and STS, we draw heavily, though not exclusively, on the imaginary of a disadvantaged community up against a powerful polluter in identifying relevant cases, and we use the term "EJ" in a way that mirrors actors' usage. However, our analysis is informed by work done by scholars to unpack the moral basis of calls for environmental justice. In particular, Schlosberg (2007) suggests that EJ has four interconnected components: distributive justice, the idea that environmental burdens and benefits will be shared equitably; procedural justice, or the rights of all people to participate meaningfully in decisions that affect their environment; recognition, the acknowledgment of the cultural identities of marginalized groups by dominant institutions; and capabilities, or the rights of every group to the conditions necessary to live a good life. We argue that epistemic justice—people's right to be respected in their capacities as knowers (Fricker 2007)—is also a central component of EJ, intertwined with but not reducible to procedural justice and recognition: in demanding that their testimony about local environmental conditions carry the same weight in decision making as that of credentialed experts, EJ activists are demanding recognition not just for their cultural identities but for their identities as knowers as well.

Comparing Epistemic Aspects of EJ Struggles across Nations

The most important contributions of STS to the study of environmental justice highlight the epistemic aspects of environmental injustices. STS sees science as a contestable, contingent set of socially produced knowledge claims that are intertwined with relations of power rather than as a body of fact that exists separate from the political sphere. The science that is produced reflects the interests of the political and economic elite and the worldviews of Western scientists in two ways that are particularly consequential for environmental justice. Money and status flow to researchers investigating topics of concern to military and industrial interests at the expense of topics important to less powerful groups (Hess 2007). In addition, the procedural and interpretive choices involved in scientific research itself necessarily entail value judgments that scientists too often make without an understanding of how their values might differ from those of marginalized communities (Douglas 2009; Elliot 2011). Communities living with environmental hazards thus face large areas of "undone science" and the persistent marginalization of their ways of knowing, as we describe below, while noting that the consequences of these dynamics for environmental justice vary cross-nationally according to societies' different political cultures (Jasanoff 2005).

Undone Science

Because resources for scientific research flow overwhelmingly to areas of interest to political and economic elites (Hess 2007; Mitman 2007; Murphy 2006), communities working for environmental justice confront "the systematic non-production of knowledge" (Frickel et al. 2010, 446), and even the active cultivation of ignorance (Proctor 2008) related to the environmental and health effects of industrial pollution. Indeed, *not* investigating these issues may be a strategic move on the part of elites, for fear that findings could jeopardize lucrative businesses (Proctor 2008, 2012).

Several areas of undone science are consequential for environmental justice, including the lack of research into synergistic effects of chemical exposures (Allen 2003; Tesh 2000) and the paucity of air quality data in neighborhoods nearest to industrial polluters (Ottinger 2010). In other parts of the world, the absences appear to be even greater. Recent surveys of environmental health research in China, for example, suggest that despite growing interest in pollution-related illness, little data about exposures or disease is available, and the attendant uncertainties shape how exposed communities experience pollution (Holdaway 2013; Lora-Wainwright 2013; Tilt 2013).

Knowledge gaps arising from undone science exist also in countries of the global south for systemic, political-economic reasons. In Latin America, dismantled state bureaucracies employ few if any scientists (e.g., Carey 2010; Mathews 2011), and funding for science is low and erratic (Carruthers 2008b). In the United States, well-funded environmental NGOs may help to fill in gaps, undertaking scientific research in areas neglected by mainstream academic scientists (Hess 2007). However, in less wealthy nations, there is also less funding available for activists (Carruthers 2008b). The scarcity of local funding opportunities leads many scientists in Latin America—and likely other less-developed regions, as well—to collaborate with foreign organizations or participate in research projects defined abroad. In those collaborations, they may work on problems of interest to foreigners rather than on questions that address the needs of the region, accentuating the region's lack of data (Kreimer 2006; Vessuri 2007).

The potential for undone science to aggravate environmental injustices has also an increasingly global dimension. In some places, multinational corporations and their spin-off foundations (e.g., the Gates Foundation) may play a central role in determining what science does and does not get done. Other times, actors whose interests seem marginalized in the United States, like communities concerned about environmental causes of cancer, may function as elites on the global stage, directing resources away from problems that afflict communities in poorer nations (e.g., Halfon 2007). Alternatively, elites in the global south often have interests that run counter to those of elites in the north; this sometimes spurs investment in science that fills local needs or fosters south-south science-based cooperation (e.g., Centellas 2014). While more science does not necessarily produce more just outcomes, to the extent public issues are increasingly subject to science-inflected political processes, research should examine worldwide patterns in what questions are being studied and what knowledge gaps are allowed to persist.

Local Knowledge and Lay Ways of Knowing

A second issue confronted by EJ activists around the world is the persistent marginalization of local knowledge and lay ways of knowing. STS shows that people without scientific credentials incorporate into their understandings of the environment their cultural identities, their assessments of the political and social contexts of scientific claims, and their own local, vocational, and experiential knowledge (Brown 1992; Irwin 1995; Irwin, Dale, and Smith 1996; Wynne 1996), coming to ways of making sense of the world that might be quite different from that of scientists.

In the environmental health realm, local knowledge has been pivotal in taking policy action on hazards that threaten marginalized communities, including groundwater contamination and subsistence anglers' exposure to high levels of toxins in fish (Corburn 2005; Kuehn 1996; Tesh 2000). Local knowledge and experiential knowledge often serve as the basis for community-initiated investigations of pollution and health issues. Communities often enlist credentialed scientists, engineers, and health experts in their knowledge production activities, which, depending on their specifics, have been labeled "popular epidemiology," "citizen science," "participatory monitoring," and "community-based participatory research," among others. In these activities, communities' approaches to understanding environmental issues, or lay ways of knowing, usually depart from those of the scientific mainstream: community groups use different standards of proof than professional epidemiologists; they prefer false positives to false negatives; and they refuse to divorce scientific study from the need for action (Allen 2003; Brown 1992; Corburn 2005). They frequently ask different questions, looking for ways to represent ongoing, systemic hazards rather than assessing regulatory compliance or identifying the root cause of a particular chemical release (Ottinger 2009) and insisting on the importance of cumulative impacts and synergistic effects (e.g., Bryant 2011). And they have in some cases developed scientific instruments, including air monitors and kite-based cameras, that speak directly to communities' questions and let communities' need for accuracy and precision drive the cost and complexity of the instruments used (Hemmi and Graham 2014; Ottinger 2010; Wylie et al. 2014).

The majority of STS research on local ways of knowing and environmental justice has focused on communities in the United States and United Kingdom. However, the dynamics described above are likely to be widespread: studies of indigenous communities also show, for example, how the local knowledge and ways of knowing of marginalized groups in the global south pose substantial challenges to Western science but are nonetheless largely dismissed as potential contributors to scientific understandings (de la Cadena 2005; Lave 2012; Philip 2004).

Nonetheless, we know relatively little about how differences across countries or regions shape how lay ways of knowing function in EJ contestation. Nations differ in their "civic epistemologies," which define how evidence is received and incorporated into regulatory processes. These are sure to condition how lay epistemologies are advanced and received in different regulatory contexts (Barandiarán 2015a; Jasanoff 2005). They may also condition the ways that evidence-gathering tools like popular epidemiology, community monitoring, and participatory mapping come to have meaning and influence in political processes.

The effectiveness of citizen science interventions also calls for further research. EJ activists' knowledge production efforts are typically meant to bring about concrete policy actions or change the terms of environmental policy debates, but there is a great deal of variability in the success of these efforts in U.S. communities (Brown 1992; Jalbert and Kinchy 2015; O'Rourke and Macey 2003; Ottinger 2010). In other nations, citizen science activities do not necessarily translate into political mobilization by marginalized communities—and thus do not become a tool for structural change (Auyero and Swistun 2009; Kimura 2016; Lora-Wainwright 2013).

The efficacy of citizen scientists' knowledge claims is also likely to be affected by who the citizen scientists are, in regards to gender, ethnicity, class, and nationality. The power and status of individuals within communities varies, with implications for how local knowledge is produced and taken up (Epstein 1996). For example, women are more likely to take leadership roles in fights against environmental injustices (e.g., Rocheleau et al. 1996), yet access to scientific and public spheres is often restricted more for women than men even within the same community. It would thus be instructive to know where and why low-cost, community-based scientific instruments are being adopted most frequently, by whom within the community, and with what impacts.

Finally, scholars have shown that calling attention to pollution in one's neighborhood may be dangerous for communities invested in preserving their sense of their homes as good places to live (Ottinger 2013a; Phillimore and Moffatt 2004). Even where citizen science is tightly linked to political mobilization, studies have shown that its effects are double-edged: while it may amplify EJ activists' grievances and result in increased regulatory action (O'Rourke and Macey 2003), direct engagement with scientific arguments may push EJ activists to reframe their issues in ways more aligned with mainstream science (Kinchy 2012; Ottinger 2010). This may limit their ability to push for systemic change, including change in how data are collected and interpreted. If STS research is to contribute to the EJ movement's development of effective strategies, then we must pay attention not only to the possibilities and obstacles to citizen science in different cultural contexts but also to when and where lay engagements with science make sense at all.

Epistemic Barriers to Procedural Justice

STS's focus on the epistemic aspects of environmental injustice adds an important element to social scientists' understanding of procedural injustices; however, there appears to be wide variation across nations in how scientific authority structures individuals' ability to participate meaningfully in political processes. In general, social science research on environmental justice tends to black-box science when thinking about procedural justice, assuming that there is authoritative information to be brought to bear, and that the challenges of structuring fair processes lie elsewhere. The contribution of STS research has been to point out how procedural justice is hampered by the growing trend to seek scientific solutions to contentious political questions, thereby taking them out of the realm of public debate. For example, in the developed world, experts' voices are privileged in public hearings, to the particular disadvantage of women and members of racial or ethnic minorities (Cole and Foster 2001; Gregory 1998; Kimura 2016). Comments from the public are also routinely reinterpreted using scientific frameworks, and community concerns that challenge those frameworks are marginalized, misunderstood, or overlooked as a result (Aitken 2009; Alatout 2008; Ottinger 2013b).

Most of the literature on the epistemic aspects of procedural justice, especially that in developed countries, thus shows science to be a privileged discourse in public processes. As a result, scholars have tended to frame community mobilization as, in part, a struggle against the power afforded to technical experts in political decision making (e.g., Fischer 2000). But research from the developing world suggests that the privileged position of science cannot be assumed everywhere. In Latin America scientists are not typically members of the governing elite and have limited access to regulators (Tironi and Barandiarán 2014; Vessuri 2007). Scientists' collaborations with foreign organizations, a result of resource constraints discussed above, undermines their credibility and may further lower their social status. State officials, further, may strategically produce confusion and disagreements about scientific information to limit their responsibility for community and the environmental well-being. For example, in Mexico, state officials regarded information about forest cover as dangerous because it could be used to hold them accountable for decisions that they could not control, made in far-off Mexico City and at the World Bank (Mathews 2011). And in Chile, where communities, industry, and the government hold very different ideas of what counts as credible science, the authority of science to speak at all on certain environmental conflicts is contested (Barandiarán 2015a).

While scientific discourses can play an important role in shaping the possibilities for procedural justice, they do not necessarily. More work remains to be done to understand why science is authoritative in some cultural and political contexts and not in others, extending research on civic epistemologies to a broader range of cultures than those of advanced, liberal democracies (Jasanoff 2005; e.g., Macnaghten and Guivant 2011; Zhang 2015). Comparative international work could help us understand the circumstances under which science is not an instrument of state power, and it would attune us to the possibility that there may be circumstances under which the state's embrace of science could be to the advantage of marginalized communities. We also face the challenge of developing a vision for procedural justice that would take seriously the problem of missing knowledge and the need to incorporate experiential knowledge into the decision-making process; the little existing work in this area (e.g., Ottinger 2013a) needs to be extended to account for variation across national contexts.

Global Aspects of Environmental Injustice

While our understanding of environmental justice would benefit from comparative research, investigations of global flows of science, technology, and expertise are also sorely needed. Research on the global dimensions of environmental justice have tended to focus on distributive injustices, focusing especially on the extraction of raw materials from developing nations for the benefit of developed ones (Carruthers 2008b; O'Rourke and Connolly 2003; Raman 2013), the movement of waste from wealthy nations to poor ones (Iles 2004), and the disproportionate impacts of climate change beginning to be felt in the global south despite those nations' relatively minor contributions to atmospheric CO_2 concentrations (Carey 2010; Joshi 2013; Miller 2006; Norgaard 2008). But global flows of raw materials, pollution, and waste are intertwined with the uneven circulation of knowledge and expertise, which combine with material flows to structure environmental injustices and environmental justice contestation even at the local level.

International Flows of Knowledge and Expertise

Most case studies of lay ways of knowing and EJ-oriented citizen science portray non-scientists' knowledge-production efforts as highly local; however, lay engagements with science, especially in the global south, are shaped by international forces as well. Indigenous knowledge in particular has been a target of international appropriation. Corporations and scientists in industrialized nations reap commercial benefits from plants harvested by communities in the global south by arguing that, although indigenous communities may have used a plant for its pharmaceutical properties, it is scientists who have "really" discovered the biochemically active ingredients (Hayden 2003; Shiva 2007)—a narrative that dismisses the possibility that indigenous people could be knowers in their own right. Indigenous peoples have tried to counter exploitation by appropriating the concept of indigenous knowledge, gaining recognition in some international treaties and reaching benefit-sharing agreements with multinational corporations (Martin and Vermeylen 2005).

A countervailing trend in international flows of knowledge and expertise is the tendency of lay knowers to seek the support of scientists in other countries. Creating international alliances has long been a strategy of the EJ movement: activists have forged ties across peer communities fighting the same multinational company (Hecht 2012; Khagram 2004) and linked their local campaigns to global environmentalist discourses (Carruthers 2008a; Rothman and Oliver 1999; Urkidi and Walter 2011) to increase their political power. Experts may play a particularly important role in international alliances, serving as advocates for marginalized communities and translating their grievances into scientific language (Fortun 2001). Even where they are conducting their own environmental monitoring or health studies, NGOs and community groups often bring in sympathetic scientists from other countries to make their case more authoritatively to decision makers (Kimura 2016; Kinchy 2012).

International alliances for citizen science raise a number of questions amenable to further STS investigation. As part of mapping out the possibilities for epistemic justice, especially increasing the resources available for lay knowers to have their epistemologies recognized by mainstream science, it would be worthwhile to ask whether these alliances support local epistemologies or merely help communities translate their issues into the terms of Western science—and with what consequences? Further study of participating experts is also warranted. Frickel (2010, 2011) finds that scientists' activism on behalf of environmental groups is enabled by under-the-radar professional networks, or "shadow mobilizations," based on research on U.S.-based expert-activists. Does that same model apply internationally? What additional work is required to create and maintain such networks across borders? Empirical approaches to these kinds of questions could aid scientists and grassroots groups in creating more effective collaborations.

Just Participation in Nonstate Decision Making

In STS and the social sciences more generally, research on community participation in environmental decision making has tended to focus on governmental processes. However, nation-states can no longer be considered sole—or, arguably, even primary—decision makers on environmental issues: corporations, treaty organizations, and international nonprofit organizations all play significant roles in defining and acting on environmental problems. Their activities structure the possibilities for marginalized communities to participate in decision making and, in general, limit their ability to advance alternatives to scientific ways of knowing or to object on nontechnical grounds such as equity or justice.

At the company level, benchmarks for environmental performance are increasingly being determined by corporations themselves, as nations move to incentive-based programs and voluntary initiatives as means of environmental governance. Regardless of the effectiveness of such regulatory approaches (see Liverman and Vilas 2006), they shift the mechanisms for participation by affected communities away from state-sponsored hearings and deliberations to industry-sponsored consultations and consumer choice. The chemical industry's Community Advisory Panels (CAPs), one example of industry-sponsored consultation, are understood by corporations as helping to maintain their "social license to operate." As vehicles for meaningful participation in industry's environmental decisions, however, CAPs have been shown to be inadequate (Allen 2003; Fortun 2001; Ottinger 2013b), as has the broader idea of social license (Miller 2014) because both enable corporations to structure their interactions with the public in a way that limits community influence.

Corporations often use voluntary and "green" initiatives to distinguish themselves from competitors, creating the possibility for public input through consumer choices. As a model for public participation, consumer choice shifts attention from the collective to the individual and the locus of decision making from civic institutions to the market. Defined as consumers rather than members of a political community, individuals have been impelled to treat air pollution–related illnesses by buying pharmaceuticals (Mitman 2007; Murphy 2006). Similarly, biomonitoring technologies have shifted risk regulation from industrial facilities to exposed bodies while turning our attention to the chemical hazards associated with consumer products.

International organizations also play an important role in defining the extent of corporations' and countries' environmental responsibilities. The International Standards Organization (ISO), for example, sets voluntary standards for environmental management and energy efficiency, among others; similarly, the World Health Organization (WHO) publishes guidelines for safe levels of air toxins. International standards can serve to legitimate both the performance of companies and decisions made by governments. However, by deleting any contestation or uncertainty that may have surrounded expert judgments made in the standard-setting process, international standards help insulate decisions about acceptable levels of environmental protection from public scrutiny and debate.

International organizations have also set standards for environmental decision making that impose norms from the developed world on the developing world. In particular, the World Bank and other global financiers now require environmental impact assessments (EIAs) for projects that they fund, making them effectively a global procedural standard (Pope et al. 2010). The EIA regime is justified by the authority of science and not only represents a reductionist approach to scientific knowledge; it assumes that the necessary knowledge exists in every country, is credible on the same terms, and can simply be plugged into standardized institutional relationships (Barandiarán 2015a; Li 2015). In practice, EIAs are implemented differently in different places, reflecting varying local regulatory requirements, community engagements, and political legacies (Kolhoff, Driessen, and Runhaar 2013; Richardson and Cashmore 2011). In all cases, though, the process shapes democratic practices in ways that distract from or marginalize political concerns and local values that do not fit within the EIA paradigm[1] and, by privileging reductionist science over alternative ways of knowing, perpetuate epistemic injustices (Goldman 2005).

The standards and practices of international organizations have not gone unopposed. Community groups frequently call into question the adequacy of standards for environmental quality, and EIA processes have met with significant social movement opposition both at policy meetings and at project sites. But international norms often push opponents to choose between translating their demands into dominant reductionist, scientific terms and trying to influence policy from their position on the margins—making it difficult for them to have a meaningful say in environmental decisions, as procedural justice would require.

Supranational institutions and social movement organizations have also participated in elevating "sustainability," "conservation," and the like to moral virtues that can be invoked to legitimate a wide variety of projects. However, projects justified in the name of sustainability may ignore or marginalize indigenous and local knowledge and as a result may ultimately have deleterious effects on vulnerable communities. For instance, conservation projects in the developing world have often taken a "fence and fine" approach that precluded local people from accessing forest resources (Anderson and Grove 1989); aquaculture, similarly justified in the name of conserving world fisheries, in some places is practiced at such an intensive scale that it has disrupted artisanal fisheries and degraded the marine environment (Barton 1997). And global markets for "sustainable" agricultural products, in some cases explicitly created to favor the products of farmers in the global south, may in fact be difficult for them to survive in (Loconto 2014).

The same patterns of injustice can be seen in efforts to promote so-called green energy as a strategy for mitigating climate change: the costs of hydropower and nuclear energy are borne disproportionately by local communities, as in the case of the Mekong dam (Yong and Grundy-Warr 2012). The production of rare earth metals critical for renewable energy technologies, including wind turbines, is also notorious for environmental and social damages such as radioactive byproducts, water pollution, and human resettlement for mine development, especially in China, the source of 85 percent of the global supply (Raman 2013). The specific decisions made in the name of conservation, sustainability, or climate-change mitigation occur in a variety of contexts: some are made in corporations or supranational organizations with few opportunities for public participation or the assertion of lay ways of knowing, as described above; others are made in public processes run by democratic states, where citizens presumably have the potential to influence the outcome. Even in the latter case, though, research suggests that discourses of sustainability act as a trump card, and the state's interest in promoting, for example, green energy can squeeze out meaningful public participation (Davies and Selin 2012; Ottinger, Hargrave, and Hopson 2014; Phadke 2013).

Environmental injustices are thus shaped by a set of global processes and international institutions that also perpetuate epistemic injustices by promulgating narrow versions of science and insulating them from critique based in alternative epistemologies. These institutions are not subject to democratic accountability, and while they don't go unopposed, the agonistic processes through which they are confronted favor those who have or can muster political power, for example, through social movement organizing. For STS scholars, the challenge is to theorize new modes of participation, deliberation, and regulation that can hold nonstate actors accountable and give voice to marginalized groups. The field's rich body of work on consensus conferences, technology assessment, and anticipatory governance (e.g., Davies and Selin 2012; Irwin

2001; Phadke 2013; Sclove 2010) provide a starting point, in that these STS approaches to public participation refuse to black-box science and technology and assert the legitimacy of lay ways of knowing—seeking, in effect, epistemic and procedural justice simultaneously. But more remains to be done to explore how these models could translate to corporate-community interactions, to supranational policy making, and to developing countries.

Beyond the Epistemic: Technology and Experts in EJ

While the primary focus of STS scholars interested in EJ issues has been knowledge-making practices of various sorts and their intersections with environmental decision making, two additional domains of STS have important implications for environmental justice studies: the technological infrastructures that create, monitor, and remediate environmental hazards, and the scientists and engineers whose professional practices, identities, and ethical codes shape the resources available to marginalized communities in any given EJ struggle. More attention to each of these areas would further not only social studies of technology and expertise but also social scientific understandings of environmental injustices.

Technological Politics

Among the core premises of STS are the ideas that technological choices can have unequal effects, benefiting some people to the exclusion of others and promoting some values or knowledge systems at the expense of others, and that the consequences of technology can only be understood by examining how technologies are inextricably intertwined with social and political institutions in what have been called "sociotechnical systems" (Akrich 1992; Hughes 1983; Sclove 1995; Winner 1980). These ideas contribute to a robust understanding of how structures of environmental inequality are produced and sustained. For example, decisions to dam, divert, and extract water from rivers have historically favored intensive and industrial economic activities over rural and low-impact livelihoods. Large dam projects also motivated setting up new state agencies that promoted and then perpetuated such inequalities (Alatout 2008; Carse 2012; White 1996; Worster 1992). Likewise, in the nineteenth century, the construction of railroads and then pipelines to carry energy to large cities created inequities between energy-producing regions and energy-consuming regions (Jones 2014).

In contemporary cases, we see similar patterns. Raman (2013) suggests that rare earth metals crucial to renewable energy technology are being extracted and governed in ways that mirror the sociotechnical system surrounding fossil fuels, highlighting the need to scrutinize supposedly green technologies and the sociotechnical systems in which they are embedded with an eye to their implications for environmental and social justice. Similarly, the electric car, often seen as an important element of a green lifestyle may or may not be compatible with visions of sustainability that take environmental justice seriously, depending on where the electricity comes from, among other factors (Iles 2013; see also Mulvaney 2013). This research has led to the identification of specific characteristics of technology that may stand in the way of one or more principles of EJ, such as the scale of large wind farms, which limit affected individuals' ability to benefit from and participate in decisions about them (Ottinger 2013c), and the one-way nature of most transportation systems for fuel, which concentrate harms at the point of extraction and benefits at the point of use (Jones 2013, 2014).

While STS offers insights into how technology helps create and sustain environmental inequities, scholars are also using STS theory and analysis to identify positive examples of technological innovation fostering environmental justice. In "technology- and product-oriented" social movements, for example, innovators deliberately organize to create technologies with less environmental impact—and potentially fewer environmental inequalities (Hess 2005, 2007). Similarly, in the citizen science realm, innovations in monitoring technology and the application of web-based crowdsourcing to local environmental problems have been shown to amplify community claims about pollution and influence regulatory responses (McCormick 2012; O'Rourke and Macey 2003).

While these studies offer a starting point, much more remains to be done to understand how technology and technological infrastructures may be cementing environmental inequalities and making progress toward environmental justice very difficult. Analyses of sociotechnical systems make clear that EJ struggles, although typically thought of in terms of specific locales, are not isolated but connected across nations and continents by technological infrastructures and political institutions. The development of nuclear weapons and energy technology, for example, resulted in some states and communities "becoming nuclear," and thus able to wield economic and political power, while others were denied that status and as a result marginalized from nuclear geopolitics and markets (Hecht 2012). Further, the denial and obfuscation of the ways that local conditions are shaped by global systems, and/or the subversion of activists' attempts to make connections across local struggles, may be deliberate strategies by powerful groups interested in maintaining environmental injustices (Fortun 2001; Ottinger 2013b). Examples that show the interconnectedness of local struggles underscore the need for a whole-system approach with a global purview to understand patterns of environmental inequity.

Focusing on sociotechnical systems could also bridge the gap between environmental justice and climate justice, to the extent that technological systems have intergenerational consequences. Transportation infrastructures for fossil fuels built in the nineteenth century not only created new inequities at the time; they demanded the ongoing expansion of fossil fuel consumption and continue to shape the landscape of benefits and harms (Jones 2014). Under the threat of climate change, discussions of both intergenerational equity and responsible innovation would be enriched by keeping in view the long-term impacts of design choices made in transforming sociotechnical systems.

The sociotechnical systems being built to understand and mitigate climate change are also amenable to STS analyses (e.g., Edwards 2010) that show the consequences of particular technological and bureaucratic regimes for inter- and intra-generational equity. The making of carbon and "ecosystem services" into commodities that can be traded on the free market each depend on technical and economic assumptions that privilege particular ethical positions (Roberts and Parks 2007; Srang-Iam 2012). Privileging "scientific" approaches may also be to the disadvantage of future generations, as well as today's marginalized populations: drawing on particular assemblages of data sets and simulations, climate models have been criticized for obfuscating complex intersectional impacts of climate change (Tuana 2013), and "objective" calculations of the costs of climate change versus those of remedial action draw on a discounting technique that gives little weight to the well-being of future generations (Nelson 2008).

Experts' Participation

STS research has shown that scientists can be an important resource to the EJ movement, and a growing body of literature examines the structural conditions under which it is possible for them to act as allies, as well as the effects of their participation in the movement. More work is needed, however, to understand international variations in scientists' activism and how the changing economics of scientific work alter the possibilities for scientists to serve as advocates. Because scientists are encouraged to seek private-sector funding and collaboration in lieu of government support, they are often discouraged from challenging corporate environmental practices. Research on the participation of experts in the EJ movement deserves to be extended to include engineers and to consider the ethical obligations of technical experts to environmental justice.

Case studies of EJ contestation show that scientists often play a decisive role, substantiating or amplifying grassroots groups' claims about the effects of local hazards (Allen 2000; Brown 1992). Some scholars advocate for new forms of inquiry on EJ issues that allow scientists and community groups to be equal contributors (e.g., Corburn

2005). There is some evidence that scientists' involvement with environmental justice campaigns has in fact changed scientific practice, as scientists try to incorporate communities' social demands and political goals into their research questions and methods (Cohen and Ottinger 2011; Kinchy 2012; Morello-Frosch et al. 2011; Ottinger and Cohen 2012). At the same time, the extent of these transformations appears to be limited by the ways that traditional models of science and scientific authority have been internalized by scientists and institutionalized by their organizations (Hoffman 2011; Johnson and Ranco 2011; Shilling, London, and Lievanos 2009).

The conditions under which scientists might become advocates, as well as the consequences of doing so, appear to vary from place to place. Research based in the United States shows that some scientists manage to maintain their professional reputations while supporting environmental movements by developing networks of like-minded scientists and keeping their advocacy under the radar (Frickel 2010). However, maintaining professional credibility while participating in environmental conflicts appears to be a more difficult task for scientists outside the United States, where creating such stable, professional networks may be an even greater challenge (Barandiarán 2015a). International collaboration among scientists can also be crippled by distrust among scientists from countries in the global north and south that is rooted in legacies of colonialism and power asymmetries (Hecht 2012; Lahsen 2004), although these hierarchies can be challenged by organizing science in specific ways (Barandiarán 2015b). Additional comparative research could help establish whether and to what extent activist scientists in the global south face graver consequences for acting politically than do scientists in the United States and could tease out the underlying causes of the variation, including the greater diversity of employment opportunities for scientists in the developed world and the size of the professional networks available to them, among others.

Worldwide, scientists' participation in EJ issues is taking place in the context of neoliberal reforms that privilege market-based mechanisms for funding and organizing science (Carey 2010; Mirowski 2011; Popp-Berman 2012; Shapin 2008). These shifts have reshaped the methods, organization, and content of science, even in the United States, where the state is still a major funder of research (Lave, Doyle, and Robertson 2010; Lave, Mirowski, and Randalls 2010; Moore et al. 2011)—and potentially altered the possibilities for scientists to gear their work to environmental justice goals. For example, in Peru, Chile, and the United States, environmental scientists often work as consultants in commercial organizations, which tends to result in regulatory standards and practices that promote market ideals and are used by environmental regulators to justify controversial government decisions (Barandiarán 2015a; Carey 2010; Lave, Doyle, and Robertson 2010; Li 2015; Tironi and Barandiarán 2014). Although in general neoliberalism appears to work against environmental justice (Lievanos, London, and Sze 2011; Ottinger 2013b; Sze 2007), it may also offer new opportunities (Moore et al. 2011), for example, for decentralized or nonprofit-led scientific research that could lend itself to greater community participation. More work is needed to understand the concrete consequences of neoliberal configurations of knowledge production for communities.

Most of the research on the role of technical professionals in EJ issues has focused on scientists. But engineers also participate, including by making industrial and development projects more socially and environmentally just (Barrington et al. 2012; Nieusma 2011; Nieusma and Riley 2001). At the same time, the persistent distinction made in engineering cultures between social and technical aspects of environmental issues tend to hamper progress toward environmental justice, especially when engineers draw boundaries that reflect values and unspoken assumptions at odds with community values and ways of looking at the world (Karwat, Eagle, and Wooldridge 2014; Ottinger 2013b). EJ research and advocacy would be well served by understanding better the factors that enable or limit engineers' participation in EJ issues, including how those factors may differ internationally and from the factors shaping scientists' participation.

Discussions of engineering and environmental or social justice often shade into discussions of ethics (e.g., Vallero and Vesilind 2006): how ought engineers be incorporating justice concerns into their practice? What does it mean to be a socially responsible engineer? We suggest that more work is needed to formulate an expanded science *and* engineering ethics that incorporates technical practitioners' responsibilities to environmental justice. In particular, STS research suggests that technical experts' responsibilities to EJ include promoting epistemic justice by taking seriously lay knowers' contributions and working with them to develop new tools for making sense of environmental inequities.

Conclusion

STS research on EJ has shed light on the multiple strategies that communities and the scientists who have worked with them have used to contest environmental injustice. STS scholars have analyzed the myriad ways in which knowledge claims gain authority and are used to exclude some and privilege others. They have identified that knowledge gaps are often strategic, not just an oversight. And they have identified ways in which communities have mobilized scientific knowledge to protect their health, challenge corporate and political elites, and redefine notions such as public health risks, toxicity, and pollution. STS scholarship on EJ issues has put beyond dispute the value and importance of local knowledges that challenge mainstream science, including citizen science, lay knowledges, and indigenous knowledges. As noted above, much work remains to be done, particularly with regard to the technologies used in the production of local knowledges, the international flows of information and knowledge, and in thinking about EJ contests comparatively but attuned to possible differences in the status of science across places.

But STS scholarship on EJ issues also raises questions about the limits of science as a positive political strategy: the impact of citizen science is unclear; the national and international norms that structure scientific analysis of EJ issues are weighty, opaque, and difficult to contest; and scientists working in increasingly privatized environments are unreliable allies. In a political context where contentious issues of equity and justice are frequently removed from public debate by transforming them into narrower scientific questions, EJ activists' efforts to mobilize science to contest environmental injustices may simply reinforce larger patterns of scientization without giving them any strategic advantage (Kinchy 2012).

For EJ advocates, STS analyses of how science, technology, and technical expertise can contribute to, reinforce, or help dismantle environmental injustices potentially offers strategic insight: when is confronting scientific authorities on their own terms likely to pay off? How can data be mobilized in ways that advance community perspectives and resist reinterpretation in expert frames? What changes in decision-making procedures or science-based standards could open up the most space for lay ways of knowing? Our answers to these and other questions will be improved by developing cross-cultural comparisons, bringing global flows into our analytical frames, and cultivating attention to technology and engineering as important shapers of the technoscientific terrain.

Note

1. There is an emerging field of "critical EIA studies" with scholarship examining EIA conflicts from an STS perspective. This includes work by Mark Gardiner on Namibia, by Nicolas Baya-Laffite on Argentina and Uruguay, by Amelia Fiske on Ecuador, and by Katherine Fultz on Guatemala.

References

Aitken, Mhairi. 2009. "Wind Power Planning Controversies and the Construction of 'Expert' and 'Lay' Knowledges." *Science as Culture* 18 (1): 47–64.

Akrich, Madeline. 1992. "The De-scription of Technical Objects." In *Shaping Technology/Building Society: Studies in Sociotechnical Change*, edited by Wiebe E. Bijker and John Law, 205–24. Cambridge, MA: MIT Press.

Alatout, Samer. 2008. "'States' of Scarcity: Water, Space, and Identity Politics in Israel, 1948–59." *Environment and Planning D: Society and Space* 26 (6): 959–82.

Allen, Barbara L. 2000. "The Popular Geography of Illness in the Industrial Corridor." In *Transforming New Orleans and Its Environs: Centuries of Change*, edited by Craig E. Colten, 178–201. Pittsburgh, PA: University of Pittsburgh Press.

___. 2003. *Uneasy Alchemy: Citizens and Experts in Louisiana's Chemical Corridor Disputes*. Cambridge, MA: MIT Press.

Anderson, David, and Richard H. Grove. 1989. *Conservation in Africa: Peoples, Policies and Practice*. Cambridge: Cambridge University Press.

Auyero, Javier, and Debora A. Swistun. 2009. *Flammable: Environmental Suffering in an Argentina Shantytown*. Oxford: Oxford University Press.

Barandiarán, Javiera. 2015a. "Chile's Environmental Assessments: Contested Knowledge in an Emerging Democracy." *Science as Culture* 24 (3): 251–75.

___. 2015b. "Reaching for the Stars? Astronomy and Growth in Chile." *Minerva* 53 (2): 141–64.

Barrington, Dani J., Stephen Dobbs, and Daniel I. Loden. 2012. "Social and Environmental Justice for Communities of the Mekong River." *International Journal of Engineering, Social Justice, and Peace* 1 (1): 31–49.

Barton, Jonathan. 1997. "Environment, Sustainability and Regulation in Commercial Aquaculture: The Case of Chilean Salmonid Production." *Geoforum* 18: 313–28.

Brown, Phil. 1992. "Popular Epidemiology and Toxic Waste Contamination: Lay and Professional Ways of Knowing." *Journal of Health and Social Behavior* 33 (3): 267–81.

Brulle, Robert J., and David N. Pellow. 2006. "Environmental Justice: Human Health and Environmental Inequalities." *Annual Review of Public Health* 27: 103–24.

Bryant, Bunyan. 2011. "Environmental Crisis or Crisis Epistemology? Working for Sustainable Knowledge." In *Environmental Crisis or Crisis of Epistemology? Working for Sustainable Knowledge*, edited by Bunyan Bryant, 1–34. New York: Morgan James Publishing.

Bullard, Robert D. 1994. "Environmental Justice for All." In *Unequal Protection: Environmental Justice and Communities of Color*, edited by Robert D. Bullard, 3–22. San Francisco: Sierra Club Books.

Capek, Stella M. 1993. "The 'Environmental Justice' Frame: A Conceptual Discussion and an Application." *Social Problems* 40 (1): 5–24.

Carey, Mark. 2010. *In the Shadow of Melting Glaciers: Climate Change and Andean Society*. Oxford: Oxford University Press.

Carruthers, David. 2008a. "Where Local Meets Global: Environmental Justice on the U.S.-Mexico Border." In *Environmental Justice in Latin America: Problems, Promise, and Practice*, edited by David Carruthers, 137–60. Cambridge, MA: MIT Press.

___, ed. 2008b. *Environmental Justice in Latin America: Problems, Promise, and Practice*. Cambridge, MA: MIT Press.

Carse, Ashley. 2012. "Nature as Infrastructure: Making and Managing the Panama Canal Watershed." *Social Studies of Science* 42 (4): 539–63.

Centellas, Kate. 2014. "'Cameroon Is Just Like Bolivia!' Southern Expertise and the Construction of Equivalency in South-South Scientific Collaborations." *Information & Culture: A Journal of History* 49 (2): 177–203.

Cohen, Benjamin R., and Gwen Ottinger. 2011. "Introduction: Environmental Justice and the Transformation of Science and Engineering." In *Engineers, Scientists, and Environmental Justice: Transforming Expert Cultures through Grassroots Engagement*, edited by Gwen Ottinger and Benjamin R. Cohen, 1–18. Cambridge, MA: MIT Press.

Cole, Luke W., and Sheila R. Foster. 2001. *From the Ground Up: Environmental Racism and the Rise of the Environmental Justice Movement*. New York: New York University Press.

Corburn, Jason. 2005. *Street Science: Community Knowledge and Environmental Health Justice*. Cambridge, MA: MIT Press.

Davies, Sarah R., and Cynthia Selin. 2012. "Five Dilemmas of the Practice of Anticipatory Governance." *Environmental Communication* 6 (1): 119–36.

de la Cadena, Marisol. 2005. "The Production of Other Knowledges and Its Tensions: From Andeanist Anthropology to *Interculturalidad*." In *World Anthropologies: Disciplinary Transformations within Systems of Power*, edited by Lins Ribeiro and Arturo Escobar, 201–24. Oxford: Berg Publishers.

Douglas, Heather E. 2009. *Science, Policy, and the Value-Free Ideal*. Pittsburgh, PA: University of Pittsburgh Press.

Edwards, Paul N. 2010. *A Vast Machine: Computer Models, Climate Data, and the Politics of Global Warming*. Cambridge, MA: MIT Press

Elliot, Kevin C. 2011. *Is a Little Pollution Good for You? Incorporating Societal Values in Environmental Research*. Oxford: Oxford University Press.

Epstein, Steven. 1996. *Impure Science: AIDS, Activism, and the Politics of Knowledge*. Berkeley: University of California Press.

Fischer, Frank. 2000. *Citizens, Experts, and the Environment: The Politics of Local Knowledge*. Durham, NC: Duke University Press.

Fortun, Kim. 2001. *Advocacy after Bhopal: Environmentalism, Disaster, New Global Orders*. Chicago: University of Chicago Press.

Frickel, Scott. 2010. "Shadow Mobilization in Environmental and Health Justice." In *Social Movements and the Transformation of U.S. Health Care*, edited by Jane Banaszak-Holl, Sandra R. Levitsky, and Mayer N. Zald, 171–87. New York: Oxford University Press.

___. 2011. "Who Are the Expert Activists of Environmental Health Justice?" In *Engineers, Scientists, and Environmental Justice: Transforming Expert Cultures through Grassroots Engagement*, edited by Gwen Ottinger and Benjamin Cohen, 21–39. Cambridge, MA: MIT Press.

Frickel, Scott, Sahra Gibbon, Jeff Howard, Joanna Kempner, Gwen Ottinger, and David Hess. 2010. "Undone Science: Charting Social Movement and Civil Society Challenges to Research Agenda Setting." *Science, Technology, & Human Values* 35 (4): 444–73.

Fricker, Miranda. 2007. *Epistemic Injustice: Power and the Ethics of Knowing*. Oxford: Oxford University Press.

Goldman, Michael. 2005. *Imperial Nature*. New Haven, CT: Yale University Press.

Gregory, Steven. 1998. *Black Corona: Race and the Politics of Place in an Urban Community*. Princeton, NJ: Princeton University Press.

Halfon, Saul E. 2007. *The Cairo Consensus: Demographic Surveys, Women's Empowerment, and Regime Change in Population Policy*. Lanham, MD: Lexington Books.

Hayden, Cori. 2003. *When Nature Goes Public: The Making and Unmaking of Bioprospecting in Mexico*. Princeton, NJ: Princeton University Press.

Hecht, Gabrielle. 2012. *Being Nuclear: Africans and the Global Uranium Trade*. Cambridge, MA: MIT Press.

Hemmi, Akiko, and Ian Graham. 2014. "Hacker Science versus Closed Science: Building Environmental Monitoring Infrastructure." *Information, Communication, and Society* 17 (7): 830–42.

Hess, David J. 2005. "Technology- and Product-Oriented Movements: Approximating Social Movement Studies and Science and Technology Studies." *Science, Technology, & Human Values* 30 (4): 515–35.

___. 2007. *Alternative Pathways in Science and Industry: Activism, Innovation, and the Environment in an Era of Globalization*. Cambridge, MA: MIT Press.

Hoffman, Karen. 2011. "From Science-Based Legal Advocacy to Community Organizing: Opportunities and Obstacles to Transforming Patterns of Expertise and Access." In *Technoscience, Environmental Justice, and the Spaces Between: Transforming Expert Cultures through Grassroots Engagement*, edited by Gwen Ottinger and Benjamin Cohen, 41–61. Cambridge, MA: MIT Press.

Holdaway, Jennifer. 2013. "Environment and Health Research in China: The State of the Field." *The China Quarterly* 214: 255–82.

Hughes, Thomas P. 1983. *Networks of Power: Electrification in Western Society, 1880–1930*. Baltimore: Johns Hopkins University Press.

Iles, Alastair. 2004. "Mapping Environmental Justice in Technology Flows: Computer Waste Impacts in Asia." *Global Environmental Politics* 4 (4): 76–107.

___. 2013. "Choosing Our Mobile Future: Degrees of Just Sustainability in Technological Alternatives" *Science as Culture* 22 (2): 164–71.

Irwin, Alan. 1995. *Citizen Science: A Study of People, Expertise, and Sustainable Development*. London: Routledge.

___. 2001. "Constructing the Scientific Citizen: Science and Democracy in the Biosciences." *Public Understanding of Science* 10 (1): 1–18.

Irwin, Alan, Alison Dale, and Denis Smith. 1996. "Science and Hell's Kitchen: The Local Understanding of Hazard Issues." In *Misunderstanding Science? The Public Reconstruction of Science and Technology*, edited by Alan Irwin and Brian Wynne, 47–64. Cambridge: Cambridge University Press.

Jalbert, Kirk, and Abby Kinchy. 2015. "Sense and Influence: Environmental Monitoring Tools and the Power of Citizen Science." *Journal of Environmental Policy & Planning*, doi:10.1080/15239 08X.2015.1100985.

Jasanoff, Sheila. 2005. *Designs on Nature: Science and Democracy in Europe and the United States*. Princeton, NJ: Princeton University Press.

Johnson, Jackie, and Darren Ranco. 2011. "Risk Assessment and Native Americans at the Cultural Crossroads: Making Better Science or Redefining Health?" In *Technoscience, Environmental Justice, and the Spaces Between: Transforming Expert Cultures through Grassroots Engagement*, edited by Gwen Ottinger and Benjamin Cohen, 179–99. Cambridge, MA: MIT Press.

Jones, Christopher F. 2013. "Building More Just Energy Infrastructure: Lessons from the Past." *Science as Culture* 22 (2): 157–63.

___. 2014. *Routes of Power: Energy and Modern America*. Cambridge, MA: Harvard University Press.

Joshi, Shangrila. 2013. "Understanding India's Representation of North–South Climate Politics." *Global Environmental Politics* 13 (2): 128–47.

Karwat, Darshan M. A., W. Ethan Eagle, and Margaret S. Wooldridge. 2014. "Are There Ecological Problems That Technology Cannot Solve? Water Scarcity and Dams, Climate Change and Biofuels." *International Journal of Engineering, Social Justice, and Peace* 3 (1–2): 7–25.

Khagram, Sanjeev. 2004. *Dams and Development: Transnational Struggles for Water and Power*. Ithaca, NY: Cornell University Press.

Kimura, Aya H. 2016. *Radiation Brain Moms and Citizen Scientists: The Gender Politics of Food Contamination after the Fukushima Nuclear Accident*. Durham, NC: Duke University Press.

Kimura, Aya H., and Yohei Katano. 2014. "Farming after the Fukushima Accident: A Feminist Political Ecology Analysis of Organic Agriculture." *Journal of Rural Studies* 34: 108–16.

Kinchy, Abby. 2012. *Seeds, Science, and Struggle: The Global Politics of Transgenic Crops*. Cambridge, MA: MIT Press.

Kolhoff, Arend J., Peter P. J. Driessen, and Hens A. C. Runhaar. 2013. "An Analysis Framework for Characterizing and Explaining Development of EIA Legislation in Developing Countries—Illustrated for Georgia, Ghana and Yemen." *Environmental Impact Assessment Review* 38(C): 1–15.

Kreimer, Pablo. 2006. "¿Dependientes o Integrados? La Ciencia Latinoaméricana y la Nueva División Internacional del Trabajo." *Nómadas* 24: 199–212.

Kuehn, Robert R. 1996. "The Environmental Justice Implications of Quantitative Risk Assessment." *University of Illinois Law Review* 1: 103–72.

Lahsen, Myanna. 2004. "Transnational Locals: Brazilian Experiences of the Climate Regime." In *Earthly Politics: Local and Global in Environmental Governance*, edited by Sheila Jasanoff and Marybeth L. Martello, 151–72. Cambridge, MA: MIT Press.

Lave, Rebecca. 2012. "Neoliberalism and the Production of Environmental Knowledge." *Environment and Society: Advances in Research* 3 (1): 19–38.

Lave, Rebecca, Martin Doyle, and Morgan Robertson. 2010. "Privatizing Stream Restoration in the US." *Social Studies of Science* 40 (5): 677–703.

Lave, Rebecca, Philip Mirowski, and Samuel Randalls. 2010. "Introduction: STS and Neoliberal Science." *Social Studies of Science* 40 (5): 659–75.

Lerner, Steve. 2005. *Diamond: A Struggle for Environmental Justice in Louisiana's Chemical Corridor*. Cambridge, MA: MIT Press.

Li, Fabiana. 2015. *Unearthing Conflict: Corporate Mining, Activism, and Expertise in Peru*. Durham, NC: Duke University Press.

Lievanos, Raoul S., Jonathon K. London, and Julie Sze. 2011. "Uneven Transformations and Environmental Justice: Regulatory Science, Street Science, and Pesticide Regulation in California." In *Engineers, Scientists, and Environmental Justice: Transforming Expert Cultures through Grassroots Engagement*, edited by Gwen Ottinger and Benjamin Cohen, 201–28. Cambridge, MA: MIT Press.

Liverman, Diana M., and Silvina Vilas. 2006. "Neoliberalism and the Environment in Latin America." *Annual Review of Environment and Resources* 31: 327–63.

Loconto, Allison. 2014. "Sustaining an Enterprise, Enacting SustainabiliTEA." *Science, Technology, & Human Values* 39 (6): 819–43.

Lora-Wainwright, Anna. 2013. "The Inadequate Life: Rural Industrial Pollution and Lay Epidemiology in China." *The China Quarterly* 214: 302–20.

Macnaghten, Phil, and Julia S. Guivant. 2011. "Converging Citizens? Nanotechnology and the Political Imaginary of Public Engagement in Brazil and the United Kingdom." *Public Understanding of Science* 20 (2): 207–20.

Martin, George, and Saskia Vermeylen. 2005. "Intellectual Property, Indigenous Knowledge, and Biodiversity." *Capitalism Nature Socialism* 16 (3): 27–48.

Mascarenhas, Michael. 2014. *Where the Waters Divide: Neoliberalism, White Privilege, and Environmental Racism in Canada*. Lanham, MD: Lexington Books.

Mathews, Andrew. 2011. *Instituting Nature: Authority, Expertise and Power in Mexican Forests*. Cambridge, MA: MIT Press.

McCormick, Sabrina. 2012. "After the Cap: Risk Assessment, Citizen Science, and Disaster Recovery." *Ecology and Society* 17 (4): 31–41.

McGurty, Eileen Maura. 2000. "Warren County, NC, and the Emergence of the Environmental Justice Movement: Unlikely Coalitions and Shared Meanings in Collective Action." *Society and Natural Resources* 13 (4): 373–87.

Miller, Clark A. 2006. "The Design and Management of International Scientific Assessments: Lessons from the Climate Regime." In *Assessments of Regional and Global Environmental Risks: Designing Processes for the Effective Use of Science in Decisionmaking*, edited by Alexander E. Farrell and Jill Jaeger, 187–205. Cambridge, MA: MIT Press.

___. 2014. "Globalization and Discontent." *Social Epistemology* 28 (3–4): 385–92.

Mirowski, Philip. 2011. *Science-Mart: Privatizing American Science*. Cambridge, MA: Harvard University Press.

Mitman, Gregg. 2007. *Breathing Space: How Allergies Shape Our Lives and Landscapes*. New Haven, CT: Yale University Press.

Moore, Kelly, Daniel Lee Kleinman, David J. Hess, and Scott Frickel. 2011. "Science and Neoliberal Globalization." *Theory and Society* 40 (5): 505–32.

Morello-Frosch, Rachel, Phil Brown, Julia Green Brody, Rebecca Gasior Altman, Ruthann A. Rudel, Ami Zota, and Carla Perez. 2011. "Experts, Ethics, and Environmental Justice: Communicating and Contesting Results from Personal Exposure Science." In *Engineers, Scientists, and Environmental Justice: Transforming Expert Cultures through Grassroots Engagement*, edited by Gwen Ottinger and Benjamin Cohen, 94–118. Cambridge, MA: MIT Press.

Mulvaney, Dustin. 2013. "Opening the Black Box of Solar Energy Technologies: Exploring Tensions between Innovation and Environmental Justice." *Science as Culture* 22 (2): 230–37.

Murphy, Michelle. 2006. *Sick Building Syndrome and the Problem of Uncertainty: Environmental Politics, Technoscience, and Women Workers*. Durham, NC: Duke University Press.

Nelson, Julie A. 2008. "Economists, Value Judgments, and Climate Change: A View from Feminist Economics." *Ecological Economics* 65 (3): 441–47.

Nieusma, Dean. 2011. "Middle-Out Social Change: Expert-Led Development Interventions in Sri Lanka's Energy Sector." In *Engineers, Scientists, and Environmental Justice: Transforming Expert*

Cultures through Grassroots Engagement, edited by Gwen Ottinger and Benjamin Cohen, 119–46. Cambridge, MA: MIT Press.

Nieusma, Dean, and Donna Riley. 2001. "Designs on Development: Engineering, Globalization, and Social Justice." *Engineering Studies* 2 (1): 29–59.

Norgaard, Richard. 2008. "Finding Hope in the Millennium Ecosystem Assessment." *Conservation Biology* 22 (4): 862–69.

Okerke, Chukwumerije. 2008. *Global Justice and Neoliberal Environmental Governance: Ethics, Sustainable Development, and International Co-operation.* London: Routledge.

O'Rourke, Dara, and Sarah Connolly. 2003. "Just Oil? The Distribution of Environmental and Social Impacts of Oil Production and Consumption." *Annual Review of Environment and Resources* 28: 587–617.

O'Rourke, Dara, and Gregg Macey. 2003. "Community Environmental Policing: Assessing New Strategies of Public Participation in Environmental Regulation." *Journal of Policy Analysis and Management* 22 (3): 383–414.

Ottinger, Gwen. 2009. "Epistemic Fencelines: Air Monitoring Instruments and Expert-Resident Boundaries." *Spontaneous Generations* 3 (1): 55–67.

___ . 2010. "Buckets of Resistance: Standards and the Effectiveness of Citizen Science." *Science, Technology, & Human Values* 35 (2): 244–70.

___. 2013a. "Changing Knowledge, Local Knowledge, and Knowledge Gaps: STS Insights into Procedural Justice." *Science, Technology, & Human Values* 38 (2): 250–70.

___. 2013b. *Refining Expertise: How Responsible Engineers Subvert Environmental Justice Challenges.* New York: New York University Press.

___. 2013c. "The Winds of Change: Environmental Justice in Energy Transitions." *Science as Culture* 22 (2): 222–29.

Ottinger, Gwen, and Benjamin R. Cohen. 2012. "Environmentally Just Transformations of Expert Cultures: Toward the Theory and Practice of a Renewed Science and Engineering." *Environmental Justice* 5 (3): 158–63.

Ottinger, Gwen, Timothy J. Hargrave, and Eric Hopson. 2014. "Procedural Justice in Wind Facility Siting: Recommendations for State-Led Siting Processes." *Energy Policy* 65: 662–69.

Phadke, Roopali. 2013. "Public Deliberation and Geographies of Wind Justice." *Science as Culture* 22 (2): 247–56.

Philip, Kavita. 2004. *Civilizing Natures: Race, Resources, and Modernity in Colonial South India.* New Brunswick, NJ: Rutgers University Press.

Phillimore, Peter, and Suzanne Moffatt. 2004. "'If We Have Wrong Perceptions of Our Area, We Can't Be Surprised If Others Do as Well': Representing Risk in Teeside's Environmental Politics." *Journal of Risk Research* 7 (2): 171–84.

Pope, Jenny, Alan Bond, Angus Morrison-Saunders, and Francois Retief. 2010. "Advancing the Theory and Practice of Impact Assessment: Setting the Research Agenda." *Environmental Impact Assessment Review* 41: 1–9.

Popp-Berman, Elizabeth. 2012. *Creating the Market University: How Academic Science Became an Economic Engine*. Princeton, NJ: Princeton University Press.

Proctor, Robert N. 2008. "Agnotology: A Missing Term to Describe the Cultural Production of Ignorance (and Its Study)." In *Agnotology: The Making and Unmaking of Ignorance*, edited by Londa Schiebinger and Robert N. Proctor, 1–34. Stanford, CA: Stanford University Press.

___. 2012. *Golden Holocaust: Origins of the Cigarette Catastrophe and the Case for Abolition*. Berkeley: University of California Press.

Raman, Sujatha. 2013. "Fossilizing Renewable Energies." *Science as Culture* 22 (2): 172–80.

Richardson, Tim, and Matthew Cashmore. 2011. "Power, Knowledge and Environmental Assessment: The World Bank's Pursuit of 'Good Governance.'" *Journal of Political Power* 4 (1): 105–25.

Roberts, J. Timmons, and Bradley C. Parks. 2007. *A Climate of Injustice: Global Inequality, North-South Politics, and Climate Policy*. Cambridge, MA: MIT Press.

Rocheleau, Dianne E., Barbara P. Thomas-Slayter, and Esther Wangari, eds. 1996. *Feminist Political Ecology: Global Issues and Local Experiences*. London: Routledge.

Rothman, Franklin Daniel, and Pamela E. Oliver. 1999. "From Local to Global: The Anti-Dam Movement in Southern Brazil, 1979–1992." *Mobilization: An International Quarterly* 4 (1): 41–57.

Sawyer, Suzana. 2004. *Crude Chronicles: Indigenous Politics, Multinational Oil, and Neoliberalism in Ecuador*. Durham, NC: Duke University Press.

Schlosberg, David. 2007. *Defining Environmental Justice: Theories, Movements, and Nature*. Oxford: Oxford University Press.

Schlosberg, David, and Lisette B. Collins. 2014. "From Environmental to Climate Justice: Climate Change and the Discourse of Environmental Justice." *Wires Climate Change* 5 (3): 359–74.

Sclove, Richard. 1995. *Democracy and Technology*. New York: Guilford Press.

___. 2010. *Reinventing Technology Assessment: A 21st Century Model*. Washington, DC: Woodrow Wilson International Center for Scholars.

Shapin, Steven. 2008. *The Scientific Life: A Moral History of a Late Modern Vocation*. Chicago: University of Chicago Press.

Shilling, Fraser M., Jonathon K. London, and Raoul S. Lievanos. 2009. "Marginalization by Collaboration: Environmental Justice as a Third Party in and beyond CALFED." *Environmental Science and Policy* 12 (6): 694–709.

Shiva, Vandana. 2007. "Bioprospecting as Sophisticated Biopiracy." *Signs* 32 (2): 307–13.

Srang-Iam, Witchuda. 2012. "Local Justice, Global Climate Injustice? Inequality and Tree Planting in Thailand." *Development* 55 (1): 112–18.

Sze, Julie. 2007. *Noxious New York: The Racial Politics of Urban Health and Environmental Justice.* Cambridge, MA: MIT Press.

Tesh, Sylvia Noble. 2000. *Uncertain Hazards: Environmental Activists and Scientific Proof.* Ithaca, NY: Cornell University Press.

Tilt, Bryan. 2013. "Industrial Pollution and Environmental Health in Rural China: Risk, Uncertainty and Individualization." *The China Quarterly* 214: 283–301.

Tironi, Manuel, and Javiera Barandiarán. 2014. "Neoliberalism as Political Technology: Expertise, Energy and Democracy in Chile." In *Beyond Imported Magic: Studying Science and Technology in Latin America*, edited by Eden Medina, Christina Holmes, and Ivan da Costa Marques, 305–30. Cambridge, MA: MIT Press.

Tsosie, Rebecca. 2012. "Indigenous Peoples and Epistemic Injustice: Science, Ethics, and Human Rights." *Washington Law Review* 87: 1133–202.

Tuana, Nancy. 2013. "Gender Climate Knowledge for Justice: Catalyzing a New Research Agenda." In *Research, Action and Policy: Addressing the Gendered Impacts of Climate Change*, edited by Margaret Alston and Kerri Whittenbury, 17–31. New York: Springer.

Urkidi, Leire, and Mariana Walter. 2011. "Dimensions of Environmental Justice in Anti-Gold Mining Movements in Latin America." *Geoforum* 42 (6): 683–95.

Vallero, Daniel A., and P. Aarne Vesilind. 2006. *Socially Responsible Engineering: Justice in Risk Management.* Hoboken, NJ: Wiley.

Vessuri, Hebe. 2007. *"O Inventamos o Erramos": La Ciencia como Idea-Fuerza en América Latina.* Bernal, Argentina: Universidad Nacional de Quilmes Editorial.

Walker, Gordon. 2009. "Environmental Justice and Normative Thinking." *Antipode* 41 (1): 203–5.

___. 2012. *Environmental Justice: Concepts, Evidence, and Politics.* New York: Routledge.

White, Richard. 1996. *The Organic Machine: The Remaking of the Columbia River.* New York: Hill and Wang.

Winner, Langdon. 1980. "Do Artifacts Have Politics?" *Daedalus* 109 (1): 121–36.

Worster, Donald. 1992. *Under Western Skies: Nature and History in the American West.* Oxford: Oxford University Press.

Wylie, Sara Ann, Kirk Jalbert, Shannon Dosemagen, and Matt Ratto. 2014. "Institutions for Civic Technoscience: How Critical Making Is Transforming Environmental Research." *The Information Society* 30 (2): 116–26.

Wynne, Brian. 1996. "Misunderstood Misunderstandings: Social Identities and Public Uptake of Science." In *Misunderstanding Science? The Public Reconstruction of Science and Technology*, edited by Alan Irwin and Brian Wynne, 19–46. Cambridge: Cambridge University Press.

Yong, Ming Li, and Carl Grundy-Warr. 2012. "Tangled Nets of Discourse and Turbines of Development: Lower Mekong Mainstream Dam Debates." *Third World Quarterly* 33 (6): 1037–58.

Zhang, Joy Yueyue. 2015. "The 'Credibility Paradox' in China's Science Communication: Views from Scientific Practitioners." *Public Understanding of Science* 24 (8): 913–27.

36 The Making of Global Environmental Science and Politics

Silke Beck, Tim Forsyth, Pia M. Kohler, Myanna Lahsen, and Martin Mahony

Introduction

Global environmental change is one of the most visible and worrying concerns of the twenty-first century. Scientific projections of catastrophic climate change—disintegrating ice sheets, violent storms, and declining food production—have combined with anxieties about other worldwide environmental problems such as biodiversity loss, land degradation, distortion of phosphorous cycles, persistent organic pollutants, and the deterioration of ecosystems to generate intense political outcries about the possible consequences of human-induced environmental change and demands for a rapid and comprehensive response.

This chapter reviews how science and technology studies (STS) have contributed to the increasingly vital debates over how humans should respond to global environmental change. In particular, STS investigates how knowledge about the global environment gets made and deployed—for example, by whom, using which frames, and based on what kinds of models or other knowledge infrastructures—because these factors give rise to additional uncertainties and carry significant implications for global environmental governance. STS researchers have examined the fundamental role of science and scientific institutions in identifying and monitoring global environmental change, assessing impacts, and shaping responses. Global environmental monitoring and modeling have grown rapidly since the 1980s, and expert organizations such as the Intergovernmental Panel on Climate Change (IPCC)—the world's leading scientific authority on climate change—have acquired new political authority. These activities and actors have immense influence on how policy is made, plus on how members of the public perceive environmental problems and put pressure on governments to act. Indeed, the deep significance of scientific knowledge to humanity's future was acknowledged in the early 1990s when Mostafa Tolba (1991), the head of the United Nations Environment Programme (UNEP) predicted that responses to climate change—a problem that is essentially undetectable absent scientific instruments—would influence the choices of "every individual on this planet."

The chapter begins by reviewing STS research on the making of the global environment, and of knowledges about it, with a particular focus on climate change and biodiversity loss. The social sciences in general have a crucial role in explaining the causes and consequences of global environmental change and informing more effective, equitable, and durable solutions to today's sustainability challenges (Hackmann, Moser, and Clair 2014). STS extends this work uniquely by focusing on how scientific knowledge reflects social influences and institutional choices, how the social shaping of knowledge in turn impacts how problems are known and which solutions are proposed and adopted, and how knowledge production and policy making can make these influences more transparent and governable. Understanding these influences on scientific knowledge—and science's corollary role in shaping social and policy outcomes—allows a better awareness of the nature of risks posed by global environmental change and encourages more flexible, more inclusive and, arguably, more effective approaches to policy making.

The Making of Global Environmental Science

In this first section we summarize the evolution of global environmental sciences and the social, political, and institutional contexts within which these sciences arose. These factors influence how we know about "global" things. The section has three stages: the making of the global environment as an ontological object, that is, an object about which humans think and reflect; the generation of scientific infrastructures for gathering and analyzing knowledge; and the making of environmental policy problems as frames for how knowledge is generated.

The Making of the "Global Environment"

Understanding how the global environment came to be an object of scientific and political concern can help us make sense of the social values and practices that influence explanations of global environmental change. These insights can then be used to indicate how alternative influences might lead to complementary insights about environmental change; or how labeling problems as "global" within some contexts might emphasize certain values and desired outcomes more than others.

Conceptions of the world as a globally connected system ordered by physical, chemical, and biological laws have a long history, animated not just by abstract theoretical advances but by processes of European expansion and the imperial thirst for both facts and resources. However, it was arguably during the Cold War when the imperialism and occasional internationalism of the earth sciences gave way to more concerted efforts to understand, predict, and perhaps control something called the "global environment" (Edwards 2006). Although the postwar rise of environmental science research is often attributed to the emergence of ecological concerns in the 1960s, Doel (2003) argues that military patronage was key to the institutionalization of the earth and environmental sciences in the United States. In 1961 the Pentagon declared: "[T] he environment in which the Army, Navy, Air Force, and Marine Corps will operate covers the entire globe and extends from the depths of the ocean to the far reaches of interplanetary space" (quoted in Doel 2003, 656–57). The military's desire to "have an understanding of, and an ability to predict and even control" this global environment (quoted in Doel 2003, 636) saw funds poured into the geophysical sciences, arguably at the expense of disciplines like ecology and biology, which lacked the promises of precise mathematical prediction and utilitarian control (Masco 2010).

Rich resources were devoted to projects like numerical weather prediction, with new, global computer models developed at places like Princeton University with strong state and military backing. Although military interests in *control* did not always mesh with their scientists' primary concern with *prediction* (Harper 2003), the earth sciences enjoyed a Cold War renaissance that recalled their blossoming in the context of U.S. and European expansionism in the late nineteenth century. This time though, the *global* was the geopolitical space of interest, and through emerging computational techniques, the earth sciences as a field of applied geophysics rapidly established the global environment as a new scientific and political object. The atmosphere, the climate system, the biosphere, plate tectonics—during the second half of the twentieth century all these systems came to be understood in newly global terms through novel infrastructures of environmental knowledge production.

Related to this global emphasis, risks portrayed by global assessments also tended to focus on the globe. During the 1980s and 1990s, most global environmental assessments had little systematic regional or subnational analysis, and accordingly tended to project risk in universal terms that appeared to be most appropriately conceptualized, analyzed, and managed on scales no smaller than the planet itself (Miller 2004; Miller and Edwards 2001; Takacs 1996). Takacs (1996), for example, also argues that this globalizing trend served a discursive political objective. He argues that by relabeling nature as "biodiversity," scientists hoped to build links between images of nature in any one location and nature in its global variety—and so to foster concern for environmental protection not just at home but everywhere around the world (Miller and Edwards 2001).

Knowledge Infrastructures for the Global Environment

The idea of knowledge infrastructures refers to the social systems and ways of organizing that developed in order to generate and order global knowledges. Paul Edwards in particular has summarized how large-scale sociotechnical systems have been constructed over long periods of time to make objects like the atmosphere knowable on a global scale. The "knowledge infrastructure" is made up of technologies of observation and remote sensing, data standards, disciplined observers, global institutions, data-processing algorithms, and computer models (Edwards 2006).

The military's patronage of the earth sciences notwithstanding, Edwards (2006) shows how a successful global knowledge infrastructure could only come to fruition through international cooperation. In meteorology, the voluntary internationalism of the nineteenth and early twentieth centuries proceeded in a patchwork fashion based on countries' shared interests. After World War II this shifted to a "quasi-obligatory globalism" whereby infrastructural collaboration and meteorological practice became subject to governmental control and international negotiation, facilitated by new global political organizations such as the United Nations (Miller 2001, 2004).

This growth of global knowledge infrastructures also occurred amid substantial social changes in societies where global environmentalism became a visible political force. Indeed, social scientists have applied preexisting work on democratic impulses against technocracy in Europe and the United States in the 1960s (involving civil rights, antiwar, and antinuclear movements) to global environmental politics. Ulrich Beck (1992), for example, wrote about the "democratic triumph" that can grow from public awareness of environmental risk as an unavoidable reality. Beck argued that this process both builds up and undermines scientific authority because society becomes used to a chronic condition of uncertainty about these risks. Yet, this new "risk society" also represents an opportunity for people's hopes, ideas, and interests to triumph over "the hard—the organizations, the established, the powerful, and the armed" (ibid., 117) because the uncertainties would demonstrate the limits of knowledge or indeed the inevitability of risk. He saw this as an opportunity to liberate politics, law, and the public sphere from "the patronization by technocracy" (ibid., 109).

While the U.S. military was populating outer space with earth-observing satellites and funding the development of global predictive models, different kinds of anti-technocratic environmental knowledge were emerging closer to the ground. Rachel Carson's *Silent Spring*, published in 1962, castigated government support for ecologically damaging pesticides by drawing on a very different knowledge infrastructure— social networks of amateur field observers, concerned scientists, classified government documents, and expert interviews—to present a compelling and influential picture of environmental decline. Carson's work inspired a new generation of activists who united concerns for local environmental conservation with new, countercultural forms of "planetary" consciousness (Russill 2016).

Although geophysics overshadowed biological and ecological concerns for local environments and population dynamics in the early Cold War, questions of population and "carrying capacity" soon found a technoscientific home. The infamous *Limits to Growth* report (Meadows, Randers, and Meadows 1972) emerged at a time when global predictive models were starting to establish themselves as scientific tools but had yet to achieve great traction as tools of political persuasion (Ashley 1983). While organizing early assessments of the science of global environmental change, Carroll Wilson of the Massachusetts Institute of Technology (MIT) raised funds to conduct new work into global, systemic problems—the "world *problématique*"—using emerging computer simulation techniques. Jay Forrester, a computing pioneer interested in the modeling of urban and industrial dynamics, joined Wilson along with his former student Dennis Meadows. Forrester and Meadows modeled nonlinear systems with multiple interacting feedback loops, arguing that conventional planning relied on misguided assumptions of linear causality and the application of simple fixes to symptoms rather than address more elusive causes. Relatively simple dynamic models of natural resources, population, pollution, capital, and agriculture were connected to build a global picture, and the resulting conclusion that exponential growth rates were unsustainable on a finite planet echoed Forrester's earlier simulations of "overshoot and collapse" in his work on urban and industrial systems (Edwards 2010).

The *Limits* authors faced criticism and even derision in the scientific community, where a lack of reliable supporting data and questionable underlying assumptions about how individuals respond in the face of scarcity (Taylor and Buttel 1992) hobbled their work's credibility. But the predictions of world collapse had an electric effect on international environmentalism at a time when computers were still new for most consumers of environmental science—and when concerns about pollution (such as at Love Canal, Lake Erie), the ongoing Vietnam War, and the first photographs of planet Earth from the moon had already made global environmentalism a common concern (Jasanoff 2004). Indeed, debate about the underlying veracity of the *Limits* report continues today, almost a half century later (Turner 2014). The impact of the *Limits to Growth* study was thus a product of a confluence of technological, scientific, environmental, and political factors—the very interactive factors that are the central focus of STS.

Framing Global Environmental Change

The above discussion indicates that different modes of environmental knowledge making have framed the issue of global change in different ways. It is therefore important to understand how knowledge infrastructures have influenced framings of global
change, and vice versa. Analyzing framings shows how different aspects of problems
are given priority, and which solutions are rendered thinkable.

Chris Russill (2016) argues that the dominance of geophysical approaches to environmental change in the United States shaped the way in which the climate change
debate was framed in the 1980s. Displacing earlier, ecological conceptions of climate-
society interactions, which emphasized local conditions and extreme events, global
"trend detection" techniques, supported by extensive observational and computational infrastructures, rendered climate change as something linear, gradual, and thus
manageable through conventional economic and political means. Russill links the new
dominance of trend-detection framings to early 1980s U.S energy security debates,
where the issue of human interference in the climate system was parceled off as a
question of carbon dioxide pollution and the analysis of incremental changes and subsequent "impacts." However, rather than emphasize global trends and impacts, a new
group of ecologists and complexity scientists, such as William C. Clark, lobbied for
framings that emphasized society-ecology-climate interactions at multiple scales. Climate change necessitated a "broader perspective that could help to locate the carbon
dioxide question within the context of related economic trends, political agendas, and
environmental problems" (Clark 1985, 3). Amid this complexity, "no-one's needs will
be served by single 'bottom line' assessments that purport to speak for all people and
all times" (ibid., 2). Clark and others advocated a risk assessment and management
approach that could account for uncertainties as well as the possibility of extreme,
nonlinear changes. Although overshadowed in the 1980s by a geophysical emphasis on
global, linear trends, this risk-based framing has recently come back into favor (Russill
2016).

In the early 1990s similar concerns about the framing of climate change through
global, universalizing metrics were expressed in a famous confrontation between the
World Resources Institute (WRI) in the United States and the Indian Centre for Science
and Environment (CSE). This controversy concerned the relationship between scientific representations and different visions of appropriate levels of social and economic
development (Agarwal and Narain 1991). In a report which drew on U.S. pollution policy framings and annual inventory techniques, the WRI listed Brazil, China, and India
as among the top six countries contributing to climate change based on current rates of
deforestation and fossil-fuel emissions. However, as CSE pointed out, WRI had ignored

the vast differences in per capita emissions between richer and poorer countries. It matters, they stressed, whether fuels are burnt for subsistence or luxury lifestyles, and whether rich countries had cut down forests or burnt fossil fuels in the past. For the CSE, the priority was not the annual calculation of global emissions but the (uneven) global distribution of responsibility for such emissions in the past, present, and future. CSE's criticisms did not deny environmental change or the need for ameliorative policies. Instead, their analysis of how WRI's index was constituted showed how scientific representations might exclude and even penalize different visions of appropriate levels of development, in addition to engendering geopolitically inflected distrust.

The implications of these episodes are that drawing on different representations of global environmental science aligns observers with different sets of values and concerns. Through close, often ethnographic study of the social life of scientific modeling, STS scholars have performed fine-grained analyses of scientific reports and assessments since the mid-1990s to interrogate the meaning associated with scientific projections as well as the values and objectives that give rise to them. Shackley and Wynne (1995, 1996), for example, studied communities of climate modelers and their interactions with policy makers in a crucial period of political agenda setting. They observed a "mutual construction" of scientific knowledge for policy, whereby "politics" directly influenced the practices of modeling (van der Sluijs et al. 1998). For example, policy makers might have a large say in defining the content and structure of modeling research programs or scientists might concentrate on particular facets of a problem they believed policy makers would or should be concerned about. Likewise, the policy world of climate change was heavily structured by scientific knowledge, focused on the technocratic management of a global system, and promising reduced scientific uncertainty as research advanced (Shackley et al. 1998).

STS research has highlighted how different forms of knowledge are fashioned, the strengths and limitations of those different ways of knowing, and the ways in which they arise in and embed assumptions from diverse social and cultural contexts. STS scholars have pointed out how geophysical framings of climate change, based on the physical properties of greenhouse gases (such as atmospheric residence time, radiative signature, and photochemical reactivity), give little insight into the social context, meaning, or ability to adapt to the risks posed by changes in these physical properties (Jasanoff and Wynne 1998; Lahsen 2010). These wider contexts and meanings, it is argued, must be identified through different analytical processes than the modeling of climatic change alone (Demeritt 2001; Jasanoff 2010). Against this, however, some historians of global science have argued that climate models should not be accused of "reductionism" because they have, in fact, adopted increasing levels of holism and complexity (Dahan-Dalmedico 2008, 2010). Nonetheless, the norms by which models are "verified," rendered useful and authoritative can also vary among national contexts and over time (Guillemot 2010; Lahsen 2013).

A wider implication of this work is the observation that science and politics are not separate fields. Rather, examining exactly *how* science and politics are intertwined opens up new spaces for thinking about how knowledge-making and decision making might be organized in effective, legitimate, and democratic ways. The language of "co-production" has been used to make sense of the mutual construction of science and politics. Jasanoff (2004) positions co-production as a metaconcept that has, in various guises, guided how STS scholars have thought about the co-evolution of nature and society, science, and politics. For scholars like Miller (2004), the lens of co-production encourages analysis both of how particular scientific framings of problems call forth certain kinds of response and of how such framings themselves do not emerge in a social vacuum but through social processes, which shape the content and direction of scientific work. Scientific knowledge and social order are co-produced.

The seminal work of Bruno Latour (1987, 1993) has likewise urged new thinking about *whose* values are included or excluded in framings and translations of science and policy rather than just pointing to the presence of values as such. Actor-network theory (ANT) is another way in which the connection between knowledge and social structure has been newly theorized. ANT studies objects such as the ozone hole or additional greenhouse gas concentrations as functions of how human and nonhuman actors build material-semiotic networks through which visions of the material and social world emerge and stabilize (Latour 2005, 2015). ANT can help identify how the shared objectives and worldviews of policy actors contribute toward seeing nonhuman artifacts such as climate or biodiversity in fixed and unquestioned terms. ANT theorists have argued, for example, that climate change has forced political debate to be reconfigured around a new, and apparently external concern about carbon, but where the representation of the alleged "natural" aspects of carbon are the result of calculative apparatuses through which physical facts are made visible or useful to different political interests (Blok 2010; Callon 2009). The value of STS in this respect is in showing the constitution of the physical parameters of risk that frame social responses.

Further studies have employed Foucault's concept of governmentality to make similar arguments about how the representation of apparently apolitical, technical things can serve political purposes by legitimizing policy outcomes such as carbon offsetting or emissions trading (Lövbrand and Stripple 2011; Lövbrand, Stripple, and Wiman 2009).

STS concepts have also been applied in relation to framings of global biodiversity. Historical research in India, for example, has linked the emergence of ideas of equilibrium and fragility largely through the activism and influence of scholars from North America in the 1960s, resulting in conservation policy that emphasized large parks (often excluding people) rather than diverse and smaller reserves (Lewis 2004, 140). More recently, scholars have shown how counting or mapping concepts such as ecosystem services and natural capital are increasingly imbued with an economic logic that translate the diversity of nature into a single financial measure (Turnhout, Neves, and de Lijster 2014; Waterton, Ellis, and Wynne 2013). This STS research has sought to demonstrate how accounting systems exercise disciplinary power by adopting practices of standardization and codification that erase local differences and alternative framings. But research also shows how public resentment of standardizations can mobilize resistance and counterexpertise (Gupta, Lövbrand, et al. 2012).

The Making and Organization of Expertise in Global Affairs

The former section demonstrates that global environmental "science" is not just information that has emerged to describe and explain the world but that it has particular history and context. This new section now extends these discussions about how knowledge gets made toward the making of scientific authority used by scientists to speak about global environmental knowledge. We argue that science is organized as a form of expertise in global affairs, but that the analysis of this scientific expertise is so far lacking in political studies outside of STS. In particular, it is important to understand how processes of producing expertise and policy-relevant knowledge are organized, designed, and work.

The first part of this section discusses how scientific authority is made through processes of contextual decision making. It focuses on design choices that define who gets to participate in defining which knowledge matters, through what processes, and at what times. These choices shape how knowledge is generated and subsequently adopted into policy. The second part then focuses on two particularly important examples of expert organizations, comprising the IPCC and the Intergovernmental Platform on Biodiversity and Ecosystem Services (IPBES). Again, the focus is on the design choices that determine the governance of knowledge-making and communication and that, in turn, have implications for making expertise authoritative by defining whose voices or framings are heard and, consequently, which knowledge is incorporated into decision making. This analysis has special significance for building global consensus across different contexts and spatial scales, a topic we pick up in the section "Entanglement between the Global and the Local."

Epistemic Authority in the Making

How do claims of epistemic authority come into being? To answer this question, there is a need to consider how different societies or political communities organize scientific expertise in order to gain legitimacy and trust in different societies, organizations, or jurisdictions.

A number of important analytical concepts have been used to capture these politics of expertise. "Boundary work" has demonstrated how new, and frequently international, scientific and advocacy networks establish epistemic authority and gain influence within policy arenas (Miller 2009). Gieryn (1983, 782) defined boundary work as the "ideological efforts by scientists to distinguish their work and its products from nonscientific intellectual activities." According to Jasanoff (2015), boundary work is the determination of who belongs within the relevant expert collective and, hence, is entitled to speak for it, as well as who does not belong and, hence, lacks such authority. In turn, boundary *organizations* seek to manage the flow of information between science and politics (Guston 2001). Providing policy-relevant knowledge and advice is deeply implicated; experts combine scientific reasoning with social and political judgments—even when their formal role is to assess science (Jasanoff 1990; Miller 2001).

Analyzing boundary work, therefore, provides fundamental insights into the dynamic processes by which expertise becomes authoritative. "Authoritative" here can mean unchallenged, accepted unquestionably as true, or as of more value than expertise or knowledge from other sources. Some typical practices of boundary work include rules of membership for an expert organization, its lines of accountability, the standards by which it defines evidence, and its procedures for review and certification (Chilvers and Kearnes 2016; Miller 2009). Boundary work can also include patterns of behavior—such as gender bias, as claimed by feminist analysts concerning expertise about breast cancer (Batt 1994). Whether consciously or not, design choices such as limiting participation to selected invited scientists, making strict distinctions between "scientific" and "traditional" knowledge (Martello 2004), and regulating the discussion of scientific uncertainty (Shackley and Wynne 1996) can serve to give the appearance of a stable boundary and a clear-cut line of demarcation between actors with different levels of authority and between their respective realms of responsibility. Research on boundary work shows how expert organizations establish design rules that create different responsibilities and procedures for producing and legitimizing knowledge, which in turn influence how (and by whom) problems are framed. Making this boundary work transparent allows stakeholders or other observers to challenge these framings and seek alternatives that involve or define stakeholders in different ways—a process sometimes called "reflexive politicization" (Hajer 2009; Stirling 2008).

Work in STS has shown that the persuasiveness of scientific knowledge depends on both the institutional dynamics of its production and the context of its uptake and spread in society (Shapin 1995). Not only do processes and design matter, but also cultural circumstances aid or impede public acceptance of its claims. These conditions include diverse "civic epistemologies"—or public ways of knowing (Jasanoff 2005). The concept of civic epistemology indicates the "institutionalized practices" and "collective knowledge-ways" through which "members of a given society test and deploy knowledge claims used as a basis for making collective choices" (Jasanoff 2005, 255). This concept offers the ability to understand the contexts in which expert claims are made and considered legitimate and the different ways in which the boundary between science and politics is drawn in different places. Accordingly, this concept does not just refer to the agency of "experts" or scientific organizations but also to the contexts and cultures in which they are considered legitimate. For example, information issued by one expert organization is unlikely to be legitimate simply because of the method used to generate it. Instead, different policy makers and their constituencies have different priorities in terms of framing and seeking information (an example is the disagreement between WRI and CSE above). Locally prominent discourses and social concerns can act as filters for identifying which "facts" should be collected and investigated further—a process named "problem closure" (Hajer 1995, 62). Hence, expertise does not simply lie in specific exercises of knowledge-making (such as models or laboratory experiments), but also the social conditions and institutional dynamics in which knowledge is rendered authoritative and unchallengeable. This realization allows analysts to ask why, for example, different models are considered appropriate for specific policy concerns (Guillemot 2010) or how experts and expert organizations are considered authoritative and trustworthy in different places (Beck 2012a; Mahony 2014). These approaches also help us to understand how and why particular knowledges are publicly challenged as legitimate claims to truth about the global environment. The credibility of knowledge claims and trust in experts relates to long-established, culturally situated practices of interpretation and reasoning that function differently in different national and transnational contexts.

Climate Change: The IPCC as Boundary Organization

The Intergovernmental Panel on Climate Change is widely considered the most significant environmental boundary organization. It was created in 1988 to inform political negotiations toward the United Nations Framework Convention on Climate Change (UNFCCC). Much of the boundary drawing between legitimate and illegitimate climate science has occurred through the very processes of producing integrated climate assessments, above all within the IPCC, which functions as a global "thought collective" for climate science (Jasanoff 2015). It has pioneered the development of procedures for producing and evaluating knowledge, but also illustrates many of the dilemmas faced by expert organizations (Beck 2012b; Hulme 2008; Miller 2004). It has faced the challenge to maintain political relevance and scientific integrity despite intense political pressures, tight deadlines, and a continually evolving, multidisciplinary scientific field. The design choices that the IPCC made mirror these inevitable tensions. For example, the IPCC deliberately includes both government representatives and scientists in order to satisfy concerns of member states. Government representatives participate and vote in plenary sessions that approve the IPCC work program and governance structure, as well as elect the IPCC chair and bureau members. The IPCC's first chairman, Bert Bolin, defended this "global representation" design involving countries' national decision makers and scientists as necessary to consolidate trust in the assessments (*cf.* Bert Bolin, cited in Schneider 1991). Unsurprisingly, this structure was one reason that U.S. Senator Chuck Hagel claimed that "summaries of IPCC reports are written not by the scientists, but by U.N. environmental activists. ... The IPCC summaries aren't science, they're U.N. politics" (Hagel 2001, 4). Research has also suggested that the governments of Brazil and India have likewise questioned the IPCC's trustworthiness (Lahsen 2009; Mahony 2014).

As a response to these criticisms, the IPCC decided to adopt a style of building authority by claiming neutrality and balance in scientific and geopolitical representation. The IPCC's role as a key boundary organization is shown partly by its attempt to make a clear separation between science and politics. In 1990 the IPCC leadership decided to no longer put forward policy recommendations and to base its authority primarily on scientific rigor, claiming that "[t]he work of the organization is [...] policy-relevant and yet policy-neutral, never policy-prescriptive."[1] This decision has subsequently been reflected in design choices that have major implications for the design of the organization—for example, the IPCC frequently asserts that it does not conduct research but rather produces regular assessment reports that synthesize research conducted and published by scientists. It is also mirrored by the way assessments are conducted by three working groups: Working Group I, the chief focus of attention to date, assesses the science of climate change; Working Groups II and III follow with their impact studies and mitigation options. This division of tasks among working groups serves to separate fundamental science (WG I) from risk assessment (WG II) and risk management (WG III) and thus to protect the "science" of climate change from political choices and normative considerations.

This choice is also reflected in the problem framing chosen for IPCC reports (and especially Working Group I), which defines the risks posed by climate change in terms of additional atmospheric greenhouse gas concentrations rather than in terms of what those concentrations might mean in different contexts worldwide (see discussion in the section "Entanglement between the Global and the Local" and also the discussion of the WRI and CSE dispute above). To date, knowledge-making for climate change has been framed and organized primarily around facets of the natural environment put at risk by human affairs. It has been difficult, if not impossible, for the IPCC to break away from the early framing of climate change around global average temperature as the preeminent indicator of risk.

In turn, these choices also influence the organization of IPCC reports and membership. For example, the IPCC's emphasis on peer-reviewed publications and traditional modes of scientific authorization result in distinctive disciplinary and geographical patterns, primarily involving a predominance of geophysical scientists from more developed countries (Edwards 2010). Important alternative sources of knowledge, such as legal and indigenous or other "local" knowledge, are excluded. Climate scientists rather than energy engineers have played the principle roles in organizing and prioritizing knowledge to address climate change, even as energy-sector emissions of carbon dioxide dominate the driving processes that contribute to the problem (Bjurström and Polk 2011).

Moreover, the IPCC has adopted a series of design choices concerning the communication of decisions and information from its different chapters and working groups. For example, much emphasis has been given to global temperature and to the policy instruments to secure the target of limiting the growth in temperature to 2°C (~ 3.6°F). This framing has been widely justified in order to provide a simple and transparent objective for global climate change policy. The design choice to emphasize consensus procedures, however, has been called selective and unnecessarily exclusive by some analysts because the resulting analyses reflect only the "lowest common denominator" areas of agreement rather than covering what they consider the most pressing questions of climate science and policy (Hansen 2007; Oppenheimer et al. 2007). Other scholars have argued that consensus procedures significantly simplify the epistemological, political, and ethical complexities of climate change (Hulme 2011), which can lead to unresolved political differences when trying to implement this outcome (O'Reilly, Oreskes, and Oppenheimer 2012). There is, indeed, much debate about the range of policy options that could be hidden when focusing on the outcome alone. Moreover,

STS scholars have argued that seeking to increase public trust by strengthening claims about global temperature and its associated risks can also be problematic because it exposes debates about climate change to the analysis of new statistics about warming and reduces the possibility of attributing responsibility to agents at lesser scales, such as specific nations or forms of consumption, rather than embracing more normative concerns about what to do under conditions of increasing climatic complexity (Beck 2012b; Guillemot 2010; Jasanoff 2010).

Even if the IPCC's claim is to be neutral, it acts as a politically powerful agent in politics. More than anything else, however, the Nobel Peace Prize awarded jointly to the IPCC and Al Gore in 2007 is also seen as an acknowledgment of the political impact of the IPCC: by providing "sound" scientific evidence and a science-based narrative of human-induced climate change, the panel has succeeded simultaneously in creating an awareness of the risks of climate change among both politicians and the wider public and in instigating the political activities required to address it. At the same time, the Nobel Prize added to the IPCC's legitimacy and further enhanced its public visibility.

One consequence of the IPCC's growing political influence and visibility has been greater attention to its activities by a wide variety of political actors. As we discuss below, national responses to public controversies over IPCC science show that the organization enjoys different levels of credibility in different nation-states and communities (Beck 2012a; Jasanoff 2011). Analyses of the production and circulation of climate change knowledges suggest that "climate science" itself should not be considered paradigm-driven or "normal" in the sense of being grounded in central theories and techniques, or the product of a single community of scientists, but instead as a composite of many different forms of observing and recording, each with its distinctive criteria of what constitutes good or bad work (Jasanoff 2015). Hence, international organizations such as the IPCC face significant challenges to develop scientific objects and rules of procedures that bridge different styles of knowledge-making and cultures.

Biodiversity: The MA and IPBES

The challenge of biodiversity loss has also seen different styles of establishing epistemic authority. Key expert organizations include the Millennium Ecosystem Assessment (MA) and Intergovernmental Platform on Biodiversity and Ecosystem Services (IPBES).

The MA was based on the idea that a purely global assessment is not sufficient to address problems related to biodiversity but they require a multiscale assessment. Whereas the IPCC primarily provides input into international climate policy, the MA and IPBES seek to address decision making at national and subnational levels, which are of great importance for biodiversity conservation and ecosystem management. The

MA decided to explicitly address the issue of scale and to conduct a multiscale assessment; as a result of this design choice, it triggered debates about how to bridge both different scales and different epistemologies (Miller and Erickson 2006).

STS scholars have taken a particular interest in this tension between epistemic pluralism and the push for unity and how it is reflected by particular design choices. For example, the MA's conceptual framework, while recognizing the importance of the local, was guided by "the need to establish a global form of expert consensus" (Filer 2009, 88). As the above discussion of the IPCC demonstrates, consensus can mean exclusions, of either unauthorized voices or uncertain knowledge. In the case of the MA, the push for a unitary scientific voice produced a situation in which local knowledge had to be translated into "scientific language" in order to be mediated through global, unitary categories (Brosius 2006, 133). The organization itself emphasized that "[t]he choice of scale is not politically neutral, because the selection may intentionally or unintentionally privilege some groups" (Millennium Ecosystem Assessment 2003, 122). Many contributors too came to the conclusion that the political dimension is an inherent feature of assessment design that deserves to be recognized more explicitly in future assessments (Carpenter et al. 2009).

Established in 2012 under the auspices of UNEP, the IPBES builds upon the conceptual and practical achievements of the MA and IPCC. In the early negotiation phase, prominent scientists argued in favor of a globally orchestrated, centralized organization akin to the IPCC (Loreau et al. 2006). But in contrast to the IPCC's globalism, and informed by the experiences of the MA and also by the input of STS scholars (Larigauderie 2015), IPBES decided to pursue a multiscalar assessment, with representatives of "local" and "indigenous" knowledge invited to the process (Beck, Esguerra, and Görg 2014; Turnhout et al. 2012). Indeed, IPBES's design choices seek, notionally at least, to take seriously the claim that global environmental knowledge-making demands diverse voices and openness to diverse forms of knowledge.

STS scholars might interpret IPBES as a participatory experiment at the global scale (Chilvers and Kearnes 2016), and as an attempt to reconcile "local" voices and knowledges with the global organization of scientific advice. By taking stakeholder engagement seriously, IPBES has faced the challenge of navigating difficult issues concerning how to account for expertise while also allowing for political representation. This experimentation with a new kind of epistemic pluralism resulted in the agreement of a guiding conceptual framework that encompasses both "ecosystem services" and a normative attachment to "Mother Earth" (Borie and Hulme 2015). As a boundary organization, IPBES faces inevitable challenges such as deciding who is included or excluded, with distinct patterns of boundary work observable as actors struggle to define legitimate participation. Categories such as "stakeholder" (and the design of classifications of subcategories) themselves become sites of political conflict, as they crucially influence the basis of representation inside the organization and bring forth numerous political and normative disagreements (for example "indigenous" groups may be included, while other agricultural smallholders might not be).

Despite these actions, however, the IPBES process is still progressing, and it remains unclear whether and how the competing demands of scientists, politicians, and diverse stakeholders will be incorporated into the rules and procedures of IPBES. The case of IPBES presents at least one example of a more inclusive approach to the institutional design and dissemination of expertise, where new forms of knowledge are co-produced with diverse social norms concerning ways of approaching global, yet multiscalar, environmental problems.

Entanglement between the Global and the Local

Co-producing Scale and Epistemic Authority

As the discussion above demonstrates, authoritative knowledge is made authoritative not just in global expert organizations but also in the national and local contexts in which such knowledge often originates and, importantly, in which it is put to use. The STS interest in science and culture *in the making* has therefore developed new perspectives on the *politics of scale* and the ways in which interactions between global and local forms of knowledge and political action are conducted and contested (Beck, Esguerra, and Görg 2014; Jasanoff and Martello 2004).

"Politics of scale" can refer to both the sociopolitical impacts of spatial structures (such as the distribution of legislative power between the European Union and one member state) and the politics of how specific scales are portrayed as legitimate sites for sociopolitical action (Bulkeley 2005). The latter resonates more directly with STS because, for STS scholars, categories like "global" and "local" should not to be taken for granted but rather considered as *outcomes* of particular forms of knowledge-making.

Indeed, the way that environmental problems are framed can have scalar consequences. Some analysts, for example, have invoked the concept of governmentality to indicate how representing anthropogenic climate change as essentially a systemic, atmospheric change makes some climate mitigation strategies such as global carbon trading seem rational or even natural, while marginalizing other policy options such as building local adaptive capacity (Gupta, Andresen, et al. 2012; Lövbrand and Stripple 2011; Oels 2005). ANT analysts have added to these arguments by highlighting new spatialities (or topologies) of political organization, such as transnational carbon markets, which are additionally strengthened by forms of accounting that conform to scientific standards but which overflow with contestable normative assumptions, giving rise to uneven geographies of political resistance at multiple scales (Blok 2010; Fogel 2004; Lahsen 2009).

Similarly, standardized templates of risk derived from singular measures such as additional greenhouse gas concentrations reduce attention to the local contexts in which additional gases are considered damaging, including the historical, socio-economic factors underlying social vulnerability to global environmental change (Hulme 2011; Lemos and Boyd 2010). As an alternative, more localized and deliberative forms of environmental governance, such as community-based adaptation to climate change, can offer means of adding local experiences and norms to global assessments of risk (Forsyth 2014). This more normative form of multilevel governance also implies that the spatial scales of environmental policies can be driven by how, and where, problems are defined in terms of meaning rather than in asocial terms such as physical scale alone.

Applying Global Environmental Knowledge between Contexts

Understanding the scalar and "ontological politics" (Carolan 2004) of global objects like carbon dioxide molecules or the "green economy" (Forsyth and Levidow 2015) means understanding how different civic epistemologies or local contexts can reframe and infuse a supposedly global concern with preexisting local interests and, consequently, change the matter under concern. The challenges of applying global environmental science and politics in different sociopolitical contexts require analysis of the conditions under which different articulations of knowledge and policy are considered legitimate and authoritative.

Public scientific controversies offer analysts a way into thinking about these themes. Around the time of the 15th Conference of the Parties to the UN Framework Convention on Climate Change (COP 15), held in Copenhagen in December 2009, a media storm ("Climategate") arose over the illegal publication of e-mails written by leading climate researchers and over the discovery of errors in the 2007 IPCC Working Group II report. Comparative studies have shown that disagreement expressed in public disputes about IPCC science is rooted in fundamental differences over values or about the role of science in policy making. In India and the United Kingdom, responses to Climategate saw discussion focused on the trustworthiness of scientists and the appropriate national response (Jasanoff 2011; Mahony 2014). In Germany Climategate did little to challenge widespread worry about climate change (Grundmann and Scott 2012) while in the United States, where public challenges to scientific authority are more common

(Jasanoff 2005), Climategate resonated with those who have long attacked climate science and who see climate policy as an anti-individualistic form of politics (Latour 2015; Wynne 2010). These divergent responses show that disputes about climate change are not solely functions of the quality of the science, as it is the same science—or the same controversies—producing very different politics in different places.

The case of India further demonstrates that the comparative politics of global environmental change cannot be understood in isolation from the ways in which political collectives produce and contest authoritative knowledge. The aforementioned study by Agarwal and Narain (1991) is often taken as a starting point for perceptions of climate knowledge politics in India (Dubash 2013). Simultaneously, India has also been criticized for showing low levels of participation and a lack of trust in the IPCC (Biermann 2001; Kandlikar and Sagar 1999).

More recently, the so-called Glaciergate scandal ensued when it emerged that the IPCC's Fourth Assessment Report (2007) contained the widely criticized claim that Himalayan glaciers could melt by 2035. Mahony (2014) argued the incident revealed a distinctive civic epistemology in India that identifies scientific knowledge as one among many sources of political authority within contentious politics (Lele 2011). Simultaneously, a new Indian Network for Climate Change Assessment (INCCA) was established and was seen by some as an "Indian IPCC" with a tacit remit to reclaim epistemic sovereignty. This all occurred amid the efforts of Jairam Ramesh, the Environment Minister, to reorder Indian climate politics away from a rejection of binding emissions targets and toward the notion of voluntary emissions pledges, to be monitored by outside actors (Stevenson 2011).

This rescaling of political authority and visibility over India's material flows of greenhouse gases led to new forms of territorial knowledge and assertions of epistemic sovereignty in public debates about climate policy. This was a case not of science linearly informing policy, or of politics manipulating science, but of the intended co-production of new forms of knowledge and social order through an enactment of the national as a jointly epistemic and political object (Mahony 2014). By studying the scalar politics of epistemic and political authority, STS scholars can contribute not only to making sense of public controversies but also to bridging "the gap between co-production as an analytic approach and co-production as a strategic instrument in the hands of knowledgeable social actors" (Jasanoff 2004, 281). This is an acknowledgment that appreciation of the complexities of science-politics relationships does not reside exclusively within STS, and it is a call for STS scholars to build connections with new, like-minded actors in new places in order to diversify the field's empirical reach and to enhance its "real-world" impact.

Analyzing civic epistemologies indicates that trust in science is related to the persuasive power of the people and institutions who speak for science (Hajer 2012) and that these forms of persuasion and authority vary among different national and transnational contexts. The public controversies surrounding Climategate show that convergence on a shared set of values or universally valid norms, as suggested by the epistemic communities model of Haas (1992), cannot be taken for granted. From an STS perspective, the controversies surrounding global environmental change can be seen as evidence of both existing civic epistemologies and of *civic epistemologies in the making*, as national and international contexts of knowledge-making interact in novel ways. These factors reveal why the IPCC faces more resistance in the United States and India, for example, than in Germany. A new question for STS scholars and science-policy practitioners is not about how to implement standardized rules about political relevance, transparency, and accountability but about what these broad concepts actually mean in different political cultures (Gupta, Andresen, et al. 2012; Hulme 2010; Mahony 2015). This new objective requires STS to broaden its own geographical scope and reach out to new empirical sites, building on existing work in places like Brazil (Cerezo and Verdadero 2003; Lahsen and Nobre 2007), Mexico (Mathews 2011), and Thailand (Forsyth and Walker 2008). STS ideas are also influencing work in sociology (Hajer 2009; Lidskog and Sundquist 2015), interpretative political science (Lövbrand 2014; Stevenson 2011), international development (Leach, Scoones, and Stirling 2010), and international relations (Siebenhüner 2008), where scholars are taking a renewed interest in the politics of knowledge production and in the inevitable tensions between "global" and "local" understandings of environmental change.

Conclusion: STS Engagement and Global Environmental Research

If there is one takeaway message about STS research on global environmental problems, it is that knowledge about these problems is inescapably political. Contestations over the categories used to frame climate change or biodiversity demonstrate that our knowledge of these problems is made by particular representations of change, devices, infrastructures, institutional dynamics, and cultures. Being aware of these factors helps improve scientific understanding of these complex problems, as well as cope with challenges such as uncertainties and ignorance. Looking critically at how knowledge is made—and who and what has been included and excluded—also improves the likelihood of successful responses among diverse stakeholders (Turnhout et al. 2012). Knowledge about global environmental change is not only partial and growing; it is *performative* by co-creating causes, effects, potential solutions, and affected constituencies—for example, by defining simultaneously the mechanisms by which risk occurs and who is vulnerable (Jasanoff 2010). STS research adds to other social sciences by demonstrating the social influences within, and of, environmental knowledge and the possibilities thus created of remaking knowledge and potential social outcomes.

Connected to this observation is the realization that scientific expertise about global environmental problems does not exist outside of the places and histories that made it—or without social and political meaning to the locations where it is applied. Accordingly, STS research has demonstrated that—contrary to traditional ideas of epistemic communities—scientific knowledge alone is rarely effective in compelling public policy (Lidskog and Sundquist 2015). This suggests a need to shift focus from the production of expert knowledge and truth claims to the ways knowledge resonates with and is reframed in politics. STS scholars have also argued that scientific knowledge needs to be analyzed for its political implications: if scientific evidence is the only basis on which global policy action is justified, then policy commitment and public consent are reduced merely to a question of whether the science is right or wrong (Hulme 2009). This idealized notion of neutral expertise and evidence-based policy makes experts vulnerable to backlash campaigns based on simplified approaches to knowledge (Wynne 2010).

Instead, there is a need to understand the social and political settings in which science and expertise are embedded, used, and presented as authoritative. First, in general terms, there is a need to acknowledge that knowledge-making about the global environment cannot be interpreted only as a neutral input into policy; rather it needs to be understood as partial, but improvable, insights into environmental problems. Second, concerning the generation of expertise, there is a need to interrogate the accountability and representativeness of organizations such as the IPCC and IPBES, and to appreciate that these questions increase their legitimacy rather than detract from their purpose to produce neutral expertise.

Demonstrating that these steps will lead to a more complex understanding of environmental problems and potential solutions is already a challenge for contemporary policy making. These steps also pose questions for STS. Rather than cling to the optimistic idea that more engagement and diversity will automatically achieve reflexivity, trust, and better outcomes, there is a need to acknowledge that diverse ideas of inclusivity and appropriate representation will always exist (Chilvers and Kearnes 2016). This implies there will be no simple transition away from frameworks that establish global models based on single, unitary, and consensus-driven processes of defining and communicating expertise. Rather, the means of achieving a more deliberative and questioning manner of understanding global change in more contextual and self-aware ways remains an important topic for exploration—but one where there will also be no simple or unitary model.

STS research makes us aware of how using knowledge in fixed or unquestioned ways closes down political options, can frequently be socially exclusive, and lead to public mistrust in knowledge and expert organizations (Beck et al. 2014). STS research contributes to social and physical sciences by offering a critical analysis of how knowledge about environmental change engages with social influences, as well as opening up space for competing ways of seeing and knowing nature and society (Lövbrand et al. 2015). These objectives contribute not only toward understanding more meaningful trajectories of environmental risks but also toward investigating a wider (and more inclusive) set of policy choices and visions for society in the future.

Note

1. See http://ipcc.ch/organization/organization.shtml.

References

Agarwal, Anil, and Sunita Narain. 1991. *Global Warming in an Unequal World: A Case of Environmental Colonialism.* New Delhi: Centre for Science and Environment.

Ashley, Richard. 1983. "The Eye of Power: The Politics of World Modeling." *International Organization* 37 (3): 495–535.

Batt, Sharon. 1994. *Patient No More: The Politics of Breast Cancer.* Edinburgh: Scarlet.

Beck, Silke. 2012a. "The Challenges of Building Cosmopolitan Climate Expertise—with Reference to Germany." *Wiley Interdisciplinary Reviews: Climate Change* 3 (1): 1–17.

___. 2012b. "Between Tribalism and Trust: The IPCC under the 'Public Microscope.'" *Nature and Culture* 7 (2): 151–73.

Beck, Silke, Maud Borie, Alejandro Esguerra, Jason Chilvers, Katja Heubach, Mike Hulme, Rolf Lidskog, Eva Lövbrand, Elisabeth Marquard, Clark Miller, Tahani Nadim, Carsten Nesshoever, Josef Settele, Esther Turnhout, Eleftheria Vasileiadou, and Christoph Goerg. 2014. "Towards a Reflexive Turn in Governing Global Environmental Expertise—The Cases of the IPCC and the IPBES" *GAIA* 23 (2): 80–87.

Beck, Silke, Alejandro Esguerra, and Christoph Görg. 2014. "The Co-production of Scale and Power: The Case of the Millennium Ecosystem Assessment and the Intergovernmental Platform on Biodiversity and Ecosystem Services." *Journal of Environmental Policy and Planning,* 1–16. doi:10.1080/1523908X.2014.984668.

Beck, Ulrich. 1992. *Risk Society: Towards a New Modernity*. London: Sage.

Biermann, Frank. 2001. "Big Science, Small Impacts—in the South? The Influence of Global Environmental Assessments on Expert Communities in India." *Global Environmental Change* 11 (4): 297–309.

Bjurström, Andreas, and Merritt Polk. 2011. "Physical and Economic Bias in Climate Change Research: A Scientometric Study of IPCC Third Assessment Report." *Climatic Change* 108: 1–22.

Blok, Anders. 2010. "Topologies of Climate Change: Actor-Network Theory, Relational-Scalar Analytics, and Carbon-Market Overflows." *Environment and Planning D: Society and Space* 28 (5): 896–912.

Borie, Maud, and Mike Hulme. 2015. "Framing Global Biodiversity: IPBES between Mother Earth and Ecosystem Services." *Environmental Science & Policy* 54: 487–96.

Brosius, J. Peter. 2006. "What Counts as Local Knowledge in Global Environmental Assessments and Conventions?" In *Bridging Scales and Knowledge Systems: Concepts and Applications in Ecosystem Assessment*, edited by Walter V. Berkes, Fikret Wilbanks, and Thomas J. Reid, 129–44. Washington, DC: Island Press.

Bulkeley, Harriet. 2005. "Reconfiguring Environmental Governance: Towards a Politics of Scales and Networks." *Political Geography* 24 (8): 875–902.

Callon, Michel. 2009. "Civilising Markets: Carbon Trading between in Vitro and in Vivo Experiments." *Accounting, Organizations and Society* 34 (3–4): 535–48.

Carolan, Michael. 2004. "Ontological Politics: Mapping a Complex Environmental Problem." *Environmental Values* 13 (13): 497–522.

Carpenter, Stephen, Harold Mooney, John Agard, Doris Capistrano, Ruth DeFries, Sandra Diaz, Thomas Dietz, Aanatha Duraiappah, Alfred Oteng-Yeboah, Henrique Pereira, Charles Perrings, Walter Reid, José Sarukhan, Robert Scholes, and Anne Whyte. 2009. "Science for Managing Ecosystem Services: Beyond the Millennium Ecosystem Assessment." *PNAS* 106 (5): 1305–12.

Carson, Rachel. 1962. *Silent Spring*. New York: Fawcett Crest.

Cerezo, José López, and Carlos Verdadero. 2003. "Introduction: Science, Technology and Society Studies—From the European and American North to the Latin American South." *Technology in Society* 25 (2): 153–70.

Chilvers, Jason, and Matthew Kearnes, eds. 2016. *Remaking Participation: Science, Environment and Emergent Publics*. London: Routledge.

Clark, William C. 1985. "On the Practical Implications of the Carbon Dioxide Question." *IIASA working papers*. Laxenburg: International Institute for Applied Systems Analysis.

Dahan-Dalmedico, Amy. 2008. "Climate Expertise: Between Scientific Credibility and Geopolitical Imperatives." *Interdisciplinary Science Reviews* 33 (1): 71–81.

___. 2010. "Putting the Earth System in a Numerical Box? The Evolution from Climate Modeling toward Global Change." *Studies in History and Philosophy of Science Part B: Studies in History and Philosophy of Modern Physics* 41 (3): 282–92.

Demeritt, David. 2001. "The Construction of Global Warming and the Politics of Science." *Annals of the Association of American Geographers* 91 (2): 307–37.

Doel, Ronald. 2003. "Constituting the Postwar Earth Sciences: The Military's Influence on the Environmental Sciences in the USA after 1945." *Social Studies of Science* 33 (5): 635–66.

Dubash, Navroz. 2013. "The Politics of Climate Change in India: Narratives of Equity and Cobenefits." *Wiley Interdisciplinary Reviews: Climate Change* 4 (3): 191–201.

Edwards, Paul. 2006. "Meteorology as Infrastructural Globalism." *Osiris* 21 (1): 229–50.

___. 2010. *A Vast Machine: Computer Models, Climate Data, and the Politics of Global Warming.* Cambridge, MA: MIT Press.

Filer, Colin. 2009. "A Bridge Too Far: The Knowledge Problem in the Millennium Ecosystem Assessment." In *Virtualism, Governance and Practice: Vision and Execution in Environmental Conservation*, edited by James G. Carrier and Paige West, 84–111. New York: Berghahn.

Fogel, Cathleen. 2004. "The Local, the Global, and the Kyoto Protocol." In *Earthly Politics: Local and Global in Environmental Governance*, edited by Sheila Jasanoff and Marybeth L. Martello, 103–25. Cambridge, MA: MIT Press.

Forsyth, Tim. 2014. "Climate Justice Is Not Just Ice." *Geoforum* 54: 230–32.

Forsyth, Tim, and Les Levidow. 2015. "An Ontological Politics of Comparative Environmental Analysis: The Green Economy and Local Diversity." *Global Environmental Politics* 15 (3): 140–51.

Forsyth, Tim, and Andrew Walker. 2008. *Forest Guardians, Forest Destroyers: The Politics of Environmental Knowledge in Northern Thailand.* Seattle: University of Washington Press.

Gieryn, Thomas F. 1983. "Boundary-Work and the Demarcation of Science from Non-Science: Strains and Interests in Professional Ideologies of Scientists." *American Sociological Review* 48 (6): 781–95.

Grundmann, Reiner, and Mike Scott. 2012. "Disputed Climate Science in the Media: Do Countries Matter?" *Public Understanding of Science* 23 (2): 220–35.

Guillemot, Hélène. 2010. "Connections between Simulations and Observation in Climate Computer Modeling: Scientists' Practices and 'Bottom-Up Epistemology' Lessons." *Studies in History and Philosophy of Science Part B* 41 (3): 242–52.

Gupta, Aarti, Steinar Andresen, Bernd Siebenhüner, and Frank Biermann. 2012. "Science Networks." In *Global Environmental Governance Reconsidered*, edited by Frank Biermann and Philipp Pattberg, 69–93. Cambridge MA: MIT Press.

Gupta, Aarti, Eva Lövbrand, Esther Turnhout, and Marjanneke J. Vijge. 2012. "In the Pursuit of Carbon Accountability: The Politics of REDD+ Monitoring, Reporting and Verification." *Current Opinion in Environmental Sustainability* 4 (6): 726–31.

Guston, David. 2001. "Boundary Organizations in Environmental Policy and Science: An Introduction." *Science, Technology, & Human Values* 26 (4): 399–408.

Haas, Peter. 1992. "Introduction: Epistemic Communities and International Policy Coordination." *International Organization* 46 (1): 1–35.

Hackmann, Heide, Susanne C. Moser, and Asunción Lera St. Clair. 2014. "The Social Heart of Global Environmental Change." *Nature Climate Change* 4 (8): 653–55.

Hagel, Chuck. 2001. "Intergovernmental Panel on Climate Change (IPCC) Third Assessment Report." In *Hearing before the Committee on Commerce, Science, and Transportation, United States Senate One Hundred Seventh Congress, first session, May 1 2001*. Washington DC: U.S. Government Printing Office.

Hajer, Maarten. 1995. *The Politics of Environmental Discourse: Ecological Modernization and the Policy Process*. Oxford: Oxford University Press.

___. 2009. *Authoritative Governance: Policy Making in the Age of Mediatization*. Oxford: Oxford University Press.

___. 2012. "A Media Storm in the World Risk Society: Enacting Scientific Authority in the IPCC Controversy (2009–10)." *Critical Policy Studies* 6 (4): 452–64.

Hansen, James E. 2007. "Scientific Reticence and Sea Level Rise." *Environmental Research Letters* 2: 024002.

Harper, Kristine. 2003. "Research from the Boundary Layer: Civilian Leadership, Military Funding and the Development of Numerical Weather Prediction (1946–55)." *Social Studies of Science* 33 (5): 667–96.

Hulme, Mike. 2008. "Geographical Work at the Boundaries of Climate Change." *Transactions of the Institute of British Geographers* 33 (1): 5–11.

___. 2009. *Why We Disagree about Climate Change*. Cambridge: Cambridge University Press.

___. 2010. "Problems with Making and Governing Global Kinds of Knowledge." *Global Environmental Change* 20 (4): 558–64.

___. 2011. "Reducing the Future to Climate: A Story of Climate Determinism and Reductionism." *Osiris* 26 (1): 245–66.

Intergovernmental Panel on Climate Change. 2007. *Climate Change 2007: Impacts, Adaptation and Vulnerability. Contribution of Working Group II to the Fourth Assessment Report of the Intergovernmental Panel on Climate Change*. Cambridge and New York: Cambridge University Press.

Jasanoff, Sheila. 1990. *The Fifth Branch: Science Advisers as Policymakers*. Cambridge, MA: Harvard University Press.

___. 2004. Afterword. In *States of Knowledge: The Co-production of Science and Social Order*, edited by Sheila Jasanoff, 274–82. London: Routledge.

___. 2005. *Designs on Nature: Science and Democracy in Europe and the United States*. Princeton, NJ: Princeton University Press.

___. 2010. "Testing Time for Climate Science." *Science* 328 (5979): 695–96.

___. 2011. "Cosmopolitan Knowledge: Climate Science and Global Civic Epistemology." In *Oxford Handbook of Climate Change and Society*, edited by John Dryzek, Richard B. Norgaard, and David Schlosberg, 129–43. Oxford: Oxford University Press.

___. 2015. "Science and Technology Studies." *Research Handbook on Climate Governance*, edited by Karin Baeckstrand and Eva Lövbrand, 36–48. Cheltenham: Edward Elgar.

Jasanoff, Sheila, and Marybeth L. Martello, eds. 2004 *Earthly Politics: Local and Global in Environmental Governance*. Cambridge, MA: MIT Press.

Jasanoff, Sheila, and Brian Wynne. 1998. "Science and Decision Making." *Human Choice and Climate Change*, vol. 1, edited by Steve Rayner and Elizabeth L. Malone, 1–87. Columbus, OH: Battelle Press.

Kandlikar, Milind, and Ambuj Sagar. 1999. "Climate Change Research and Analysis in India: An Integrated Assessment of a South-North Divide." *Global Environmental Change* 9 (2): 119–38.

Lahsen, Myanna. 2009. "A Science–Policy Interface in the Global South: The Politics of Carbon Sinks and Science in Brazil." *Climatic Change* 97 (3–4): 339–72.

___. 2010. "The Social Status of Climate Change Knowledge: An Editorial Essay." *Wiley Interdisciplinary Reviews: Climate Change* 1 (2): 162–71.

___. 2013. "Anatomy of Dissent: A Cultural Analysis of Climate Skepticism." *American Behavioral Scientist* 57 (6): 732–53.

Lahsen, Myanna, and Carlos A. Nobre. 2007. "Challenges of Connecting International Science and Local Level Sustainability Efforts: The Case of the Large-Scale Biosphere–Atmosphere Experiment in Amazonia." *Environmental Science & Policy* 10 (1): 62–74.

Larigauderie, Anne. 2015. "The Intergovernmental Platform on Biodiversity and Ecosystem Services (IPBES): A Call to Action." *GAIA* 24 (2): 73.

Latour, Bruno. 1987. *Science in Action: How to Follow Scientists and Engineers through Society*. Milton Keynes: Open University Press.

___. 1993. *We Have Never Been Modern*. New York: Harvester Wheatsheaf.

___. 2005. *Reassembling the Social: An Introduction to Actor-Network Theory*. Oxford: Oxford University Press.

___. 2015. "Telling Friends from Foes in the Time of the Anthropocene." In *The Anthropocene and the Global Environment Crisis: Rethinking Modernity in a New Epoch*, edited by Clive Hamilton, Christophe Bonneuil, and François Gemenne, 145–55. London: Routledge.

Leach, Melissa, Ian Scoones, and Andrew Stirling. 2010. *Dynamic Sustainabilities: Technology, Environment, Social Justice.* London: Earthscan.

Lele, Shamchchandm. 2011. "Climate Change and the Indian Environmental Movement." In *Handbook of Climate Change and India,* edited by Navroz K. Dubash, 208–17. London: Earthscan.

Lemos, Maria, and Emily Boyd. 2010. "The Politics of Adaptation across Scales: The Implications of Additionality to Policy Choice and Development." In *The Politics of Climate Change: A Survey*, edited by Max Boykoff, 96–110. London: Routledge.

Lewis, Michael. 2004. *Inventing Global Ecology: Tracking the Biodiversity Ideal in India, 1947–1997.* Athens: Ohio University Press

Lidskog, Rolf, and Göran Sundqvist. 2015. "When Does Science Matter: International Relations Meets Science and Technology Studies." *Global Environmental Politics* 15 (1): 1–20.

Loreau, Michel, Alfred Oteng-Yeboah, Mary T. K. Arroyo, Didier Babin, Robert Barbault, Michael Donoghue, Madhav Gadgil, Christoph Häuser, Carlo Heip, Anne Larigauderie, Keping Ma, Georgina Mace, Harold Mooney, Charles Perrings, Peter Raven, Jon Sarukhan, Peter Schei, Robert J. Scholes, and Robert T. Watson. 2006. "Diversity without Representation." *Nature* 442 (7100): 245–46.

Lövbrand, Eva. 2014. "Knowledge and the Environment." In *Advances in International Environmental Politics.* 2nd ed., edited by Michele Betsill, Kathryn Hochstetler, and Dimitris Stevis, 161–84. Basingstoke: Palgrave Macmillan.

Lövbrand, Eva, and Johannes Stripple. 2011. "Making Climate Change Governable: Accounting for Carbon as Sinks, Credits and Personal Budgets." *Critical Policy Studies* 5 (2): 187–200.

Lövbrand, Eva, Johannes Stripple, and Bo Wiman. 2009. "Earth System Governmentality." *Global Environmental Change* 19 (1): 7–13.

Lövbrand, Eva, Silke Beck, Jason Chilvers, Tim Forsyth, Johan Hedrén, Mike Hulme, Rolf Lidskog, and Eleftheria Vasileiadou. 2015. "Who Speaks for the Future of Earth? How Critical Social Science Can Extend the Conversation on the Anthropocene." *Global Environmental Change* 32: 211–18.

Mahony, Martin. 2014. "The Predictive State: Science, Territory and the Future of the Indian Climate." *Social Studies of Science* 44 (1): 109–33.

___. 2015. "Climate Change and the Geographies of Objectivity: The Case of the IPCC's 'Burning Embers' Diagram." *Transactions of the Institute of British Geographers* 40 (2): 153–67.

Martello, Marybeth L. 2004. "Negotiating Global Nature and Local Culture: The Case of Makah Whaling." In *Earthly Politics: Local and Global in Environmental Governance*, edited by Sheila Jasanoff and Marybeth L. Martello, 263–84. Cambridge, MA: MIT Press

Masco, Joseph. 2010. "Bad Weather: On Planetary Crisis." *Social Studies of Science* 40 (1): 7–40.

Mathews, Andrew S. 2011. *Instituting Nature: Authority, Expertise, and Power in Mexican Forests.* Cambridge, MA: MIT Press.

Meadows, Donella H., Jorgen Randers, and Dennis Meadows. 1972. *The Limits to Growth.* New York: Universe Books.

Millennium Ecosystem Assessment (MA). 2003. *Ecosystems and Human Well-Being: A Framework for Assessment.* Washington, DC: Island Press.

Miller, Clark. 2001. "Hybrid Management: Boundary Organizations, Science Policy, and Environmental Governance in the Climate Regime." *Science, Technology, & Human Values* 26 (4): 478–500.

___. 2004. "Climate Science and the Making of a Global Political Order." In *States of Knowledge: The Co-production of Science and Social Order,* edited by Sheila Jasanoff, 46–66. London: Routledge.

___. 2009. "Epistemic Constitutionalism in International Governance: The Case of Climate Change." In *Foreign Policy Challenges in the 21st Century*, edited by Michael Heazle, Martin Griffiths, and Tom Conley, 141–63. Cheltenham: Edward Elgar.

Miller, Clark, and Paul Edwards. 2001. *Changing the Atmosphere: Expert Knowledge and Environmental Governance.* Cambridge, MA: MIT Press.

Miller, Clark, and Paul Erickson. 2006. "The Politics of Bridging Scales and Epistemologies: Science and Democracy in Global Environmental Governance." In *Bridging Scales and Knowledge Systems: Concepts and Applications in Ecosystem Assessment*, edited by W. V. Berkes, Fikret Wilbanks, and Thomas Reid, 297–314. Washington, DC: World Resources Institute.

Oels, Angela. 2005. "Rendering Climate Change Governable: From Biopower to Advanced Liberal Government?" *Journal of Environmental Policy and Planning* 7 (3): 185–207.

Oppenheimer, Michael, Brian C. O'Neill, Mort Webster, and Shardul Agrawala. 2007. "Climate Change: The Limits of Consensus." *Science* 317 (5844): 1505–6.

O'Reilly, Jessica, Naomi Oreskes, and Michael Oppenheimer. 2012. "The Rapid Disintegration of Projections: The West Antarctic Ice Sheet and the Intergovernmental Panel on Climate Change." *Social Studies of Science* 42 (5): 709–31.

Russill, Chris. 2016. "The Climate of Communication: From Detection to Danger." In *Reframing Climate Change: Constructing Ecological Geopolitics*, edited by Shannon O'Lear and Simon Dalby, 31–51. London: Routledge.

Schneider, Stephen. 1991. "Three Reports of the Intergovernmental Panel on Climate Change." *Environment* 33: 25–30

Shackley, Simon, and Brian Wynne. 1995. "Global Climate Change: The Mutual Construction of an Emergent Science-Policy Domain." *Science and Public Policy* 22 (4): 218–30.

___. 1996. "Representing Uncertainty in Global Climate Change Science and Policy: Boundary-Ordering Devices and Authority." *Science, Technology, & Human Values* 21 (3): 275–302.

Shackley, Simon, Peter Young, Stuart Parkinson, and Brian Wynne. 1998. "Uncertainty, Complexity and Concepts of Good Science in Climate Change Modelling: Are GCMs the Best Tools?" *Climatic Change* 38 (2): 159–205.

Shapin, Steven. 1995. "Cordelia's Love: Credibility and the Social Studies of Science." *Perspectives on Science* 3 (3): 255–75.

Siebenhüner, Bernd. 2008. "Learning in International Organizations in Global Environmental Governance." *Global Environmental Politics* 8 (4): 92–116.

Stevenson, Hayley. 2011. "India and International Norms of Climate Governance: A Constructivist Analysis of Normative Congruence Building." *Review of International Studies* 37 (3): 997–1019.

Stirling, Andrew. 2008. "'Opening Up' and 'Closing Down' Power, Participation, and Pluralism in the Social Appraisal of Technology." *Science, Technology, & Human Values* 33 (2): 262–94.

Takacs, David. 1996. *The Idea of Biodiversity: Philosophies of Paradise*. Baltimore: Johns Hopkins University Press.

Taylor, Peter J., and Frederick H. Buttel. 1992. "How Do We Know We Have Global Environmental Problems? Science and the Globalization of Environmental Discourse." *Geoforum* 23 (3): 405–16.

Tolba, Mostafa. 1991. "Address by Dr. Mostafa K. Tolba." In *Climate Change: Science, Impacts and Policy. Proceedings of the Second World Climate Conference*, edited by J. Jaeger and H. L. Ferguson. Cambridge: Cambridge University Press.

Turner, Graham. 2014. "Is Global Collapse Imminent?" In *MSSI Research Paper No. 4*. Melbourne Sustainable Society Institute: University of Melbourne.

Turnhout, Esther, Bob Bloomfield, Mike Hulme, Johannes Vogel, and Brian Wynne. 2012. "Listen to the Voices of Experience." *Nature* 488: 454–55.

Turnhout, Esther, Katja Neves, and Elisa de Lijster. 2014. "'Measurementality' in Biodiversity Governance: Knowledge, Transparency, and the Intergovernmental Science–Policy Platform on Biodiversity and Ecosystem Services (IPBES)." *Environment and Planning A* 46 (3): 581–97.

van der Sluijs, Jeroen P., Josée van Eijndhoven, Simon Shackley, and Brian Wynne. 1998. "Anchoring Devices in Science for Policy: The Case of Consensus around Climate Sensitivity." *Social Studies of Science* 28 (2): 291–323.

Waterton, Claire, Rebecca Ellis, and Brian Wynne. 2013. *Barcoding Nature: Shifting Cultures of Taxonomy in an Age of Biodiversity Loss*. New York: Routledge.

Wynne, Brian. 2010. "When Doubt Becomes a Weapon." *Nature* 466: 441–42.

Contributors

Morana Alač is associate professor in communication and the Science Studies program at the University of California, San Diego. She conducts ethnographic research of scientific laboratories and other settings of technology production and use, and she works with video to focus on the dynamics of embodied social interaction.

Sulfikar Amir is associate professor of sociology at the School of Social Sciences, Nanyang Technological University. He is the author of *The Technological State in Indonesia: The Co-constitution of High Technology and Authoritarian Politics* and has produced documentary films. In addition to development, technological nationalism, and nuclear politics, his recent works cover risk, disaster, and resilience studies.

Michael Arribas-Ayllon is senior lecturer at the School of Social Science, Cardiff University, United Kingdom. His research interests include medical sociology, science and technology studies, and critical psychology. He is the lead author of *Genetic Testing: Autonomy, Responsibility and Blame* (Routledge).

Brian Balmer is professor of science policy studies in the Department of Science and Technology Studies, University College London. His research combines historical and sociological approaches to understanding the nature of expertise, particularly in the life sciences. He has published extensively on the nature of secrecy and on the history and sociology of biological and chemical warfare.

Javiera Barandiarán is assistant professor in the Global Studies program at the University of California, Santa Barbara. She is working on a book on the politics of scientific credibility in Chilean environmental conflicts since the transition to democracy. Her research has been published in *Science as Culture, Minerva*, and the edited volume *Beyond Imported Magic: Essays in Science, Technology and Society in Latin America* (MIT Press, 2014), among others.

Silke Beck is senior researcher in the Department of Environmental Politics, Helmholtz Centre for Environmental Research, Leipzig, Germany, and the chair of the interdisciplinary working group "governance and institutions." Her research focuses on the relationship between science and governance in global environmental politics.

Koen Beumer works on emerging science and technology in the global south and is currently a postdoctoral researcher at the University of Groningen, the Netherlands. He received his Ph.D. from Maastricht University and has previously worked at the University of Amsterdam and the Dutch National Advisory Council for Science and Technology Policy.

Wiebe E. Bijker is professor in the Department of Technology and Society Studies at Maastricht University, the Netherlands, and the Norwegian University of Science and Technology. More details on his research and publications can be found at www .maastrichtuniversity.nl/web/Profile/w.bijker.htm.

Anders Blok is associate professor in the Department of Sociology, University of Copenhagen, Denmark. His current research focuses on the intersections of science, technology, and environmental politics in cities across Europe and East Asia. He is the coauthor (with Torben E. Jensen) of *Bruno Latour: Hybrid Thoughts in a Hybrid World* (2011) and the coeditor (with Ignacio Farías) of *Urban Cosmopolitics: Agencements, Assemblies, Atmospheres* (2016).

Alain Bovet is professor of communication at the HEG-Arc, University of Applied Sciences and Arts of Western Switzerland. He is currently working on a video analysis of repair and maintenance practices in urban housing, conducted at ETH Wohnforum, ETH Zurich.

Geoffrey C. Bowker is professor of informatics at the University of California, Irvine. He is the coauthor with Susan Leigh Star of *Sorting Things Out: Classification and Its Consequences* and author of *Memory Practices in the Sciences* (both from MIT Press). He directs the Infrastructures Series at MIT Press with Paul Edwards.

Steve Breyman is associate professor in the Science and Technology Studies Department at Rensselaer Polytechnic Institute. He writes about social movements, foreign policy, environmental politics, and political economy. His books include *Why Movements Matter* (SUNY Press, 2001) and *Movement Genesis* (Westview, 1997).

Regula Valérie Burri is professor of science and technology studies at HafenCity University in Hamburg, Germany. Her research centers on topics such as visual knowledge, science and art, and the governance of science and technology. She is the author of *Doing Images: Zur Praxis medizinischer Bilder* (2008), the founder of artLAB, an experimental research and teaching format involving art practice, and has been a codirector of a postgraduate program on artistic research in Hamburg.

Nancy D. Campbell is professor in the Science and Technology Studies Department at Rensselaer Polytechnic Institute. She writes about feminist health social movements and the history of drug and alcohol treatment. Her books include *Gendering Addiction: The Politics of Drug Treatment in a Neurochemical World* (Palgrave, 2011), *Using Women: Gender, Drug Policy, and Social Justice* (Routledge, 2000), and *Discovering Addiction: The Science and Politics of Substance Abuse Research* (University of Michigan Press, 2007).

Vivian Choi is assistant professor of anthropology at St. Olaf College. Her research and teaching focus on technologies of disaster and risk management, insecurity, disaster nationalism, and climate change in Sri Lanka and the Indian Ocean basin. Her forthcoming book is entitled *Disaster Nationalism: Tsunami and Civil War in Sri Lanka*. She is also coeditor of the Online Photo Essay Initiative for the journal *Cultural Anthropology*.

Marisa Leavitt Cohn is assistant professor at the IT University of Copenhagen, where she is a member of the Technologies in Practice and Interaction Design research groups and co-head of the ETHOS lab. She is also a research fellow in critical design and engineering at the Madeira Interactive Technologies Institute. With a Ph.D. in information and computer science from the University of California, Irvine, she combines approaches from HCI, anthropology, and STS to examine the politics of computational work in the design, maintenance, and repair of sociotechnical systems.

Harry Collins is Distinguished Research Professor and director of the Centre for the Study of Knowledge, Expertise and Science (KES). He is fellow of the British Academy and winner of the Bernal prize for social studies of science. His eighteen books cover sociology of scientific knowledge, artificial intelligence, the nature of expertise, and tacit knowledge. He is continuing his research on the sociology of gravitational wave detection, expertise, fringe science, science and democracy, technology in sport, and a new technique—the "Imitation Game"—for exploring expertise and comparing the extent to which minority groups are integrated into societies.

Sarah R. Davies is based in the Department of Media, Cognition and Communication at the University of Copenhagen, where her research and teaching focus on science communication and public engagement with science. Her publications include the edited volumes *Science and Its Publics* (2008) and *Understanding Nanoscience and Emerging Technologies* (2010) and articles in journals such as *Social Studies of Science, Science Communication,* and *Science as Culture.*

Pedro de la Torre III is a Ph.D. candidate in the Department of Science and Technology Studies at Rensselaer Polytechnic Institute and a graduate of the New School for Social Research's master's program in anthropology. He is currently working on a dissertation about the chronopolitics of legacy waste and nuclear remediation at the Hanford Site in Eastern Washington (USA).

Gary Lee Downey is Alumni Distinguished Professor of Science and Technology Studies and affiliated professor in the Women's and Gender Studies program at Virginia Tech. Contributions to STS making and doing include critical participation in engineering formation, infrastructural scholarship in engineering studies, and serving as 4S president (2013–2015). He is author of *The Machine in Me,* coauthor of *Engineers for Korea,* and coeditor of *Cyborgs and Citadels* and *What Is Global Engineering Education For?* (www.downey.sts.vt.edu).

Joseph Dumit is the chair of Performance Studies and professor of science and technology studies and of anthropology at the University of California, Davis. He is the author of *Drugs for Life: How Pharmaceutical Companies Define Our Health* (2012) and *Picturing Personhood: Brain Scans in Biomedical America* (2004). He is a cofounder of ModLab (modlab.ucdavis.edu), a co–principal investigator on a grant funding the KeckCAVES, working on improvisation in art and anthropology, and currently designing a game about fracking.

Virginia Eubanks is associate professor in the Department of Women's Studies at the State University of New York, Albany. She writes about digital technology and social justice. Her books include *Ain't Going to Let Nobody Turn Me Around* (SUNY Press, 2014) and *Digital Dead End: Fighting for Social Justice in the Information Age* (MIT Press, 2011).

Robert Evans is professor of sociology at the Cardiff School of Social Sciences. With Harry Collins he is the coauthor of "The Third Wave of Science Studies" (*Social Studies of Science,* 2002) and *Rethinking Expertise* (University of Chicago Press, 2007), as well as a number of papers on interactional expertise and the Imitation Game method. His current work continues this emphasis on the nature of expertise and, in particular, the relationship between expertise and democracy.

Sam Weiss Evans focuses on the construction and governance of security concerns in emerging technology, particularly in relation to the social construction of ignorance and the organizational dynamics around who gets to decide when we need to be worried about what. He teaches in the STS programs at Harvard University and Tufts University and is a research associate at MIT's program on emerging technology.

Ignacio Farías is assistant professor in the Munich Center for Technology in Society and the Department of Architecture at the Technical University of Munich, Germany. His research focuses on political and democratic challenges resulting from current transformations of urban infrastructures in Europe and Latin America. He is the coeditor (with Thomas Bender) of *Urban Assemblages: How Actor-Network Changes Urban Studies* (2009) and (with Anders Blok) of *Urban Cosmopolitics: Agencements, Assemblies, Atmospheres* (2016).

Andrew Feenberg is Canada Research Chair in Philosophy of Technology in the School of Communication, Simon Fraser University, where he directs the Applied Communication and Technology Lab. He also serves as program director at the Collège International de Philosophie. His recent books include *Between Reason and Experience: Essays in Technology and Modernity* (MIT Press) and *The Philosophy of Praxis: Marx, Lukács and the Frankfurt School* (Verso).

Ulrike Felt is professor in the Department of Science and Technology Studies and currently dean of the Faculty of Social Sciences at the University of Vienna, Austria. Her research focuses on issues of governance, technoscience and democracy, public participation, and changing research cultures, as well as the role of time/future in science and society issues. Her work has covered the life sciences, biomedicine, nanotechnologies, nuclear power, and sustainability research. She has been involved in policy advice to the European Commission, to the European Science Foundation, as well as to national policy bodies. She has previously served as a 4S and EASST council member. From 2002 to 2007 she was editor in chief of *Science, Technology, & Human Values*. In 2014 she was, together with a team of STS researchers, winner of the John Ziman award of EASST for the European Science Foundation policy brief *Science in Society: Caring for Our Futures in Turbulent Times*.

Jennifer R. Fishman (Ph.D., sociology, University of California, San Francisco) is associate professor in the Social Studies of Medicine Department and Biomedical Ethics Unit at McGill University. Her research analyzes the often unexamined and presumptive values and ethical frameworks within new biomedical and scientific enterprises and how these impact research trajectories and clinical medicine. Her recent work has been published in venues including *PLoS ONE*; *CMAJ*; *CMAJ Open*; *Trends in Genetics*; *Social Science and Medicine*; *Sociology of Health and Illness*; *Science, Technology and Human*

Values; *Personalized Medicine*; and *New Genetics and Society*. She is the coeditor of *Biomedicalization: Technoscience, Health, and Illness in the U.S.* (Duke University Press, 2010).

Laura Forlano is assistant professor of design at the Institute of Design and affiliated faculty in the College of Architecture at Illinois Institute of Technology where she is director of the Critical Futures Lab. Her research is focused on the intersection of emerging technologies, material practices, and the future of cities. She received her Ph.D. in communications from Columbia University.

Tim Forsyth is professor of environment and development at the London School of Economics and Political Science. He has conducted research on local experiences and responses to environmental change and on technology transfer within climate change policy, with special reference to Southeast Asia.

Kim Fortun is a cultural anthropologist and professor in the Department of Science and Technology Studies at Rensselaer Polytechnic Institute. Her research and teaching focus on environmental risk and disaster and on experimental ethnographic methods and research design. Her research has examined how people in different geographic and organizational contexts understand environmental problems, uneven distributions of environmental health risks, developments in the environmental health sciences, and factors that contribute to disaster vulnerability.

Laura Foster is assistant professor in the Department of Gender Studies at Indiana University, where she is also affiliate faculty in the IU Maurer School of Law and African Studies Program. She is also senior research associate in the Intellectual Property Unit at University of Cape Town Faculty of Law. Her current book project examines how contestations over patent ownership rights, indigenous San knowledge, and Hoodia plants in South Africa present emerging sites of struggle over who does and does not belong.

Rayvon Fouché is director of the American Studies Program and associate professor in the School of Interdisciplinary Studies at Purdue University. His work explores the multiple intersections of cultural representation, racial identification, and technological design. He has published or edited three books including *Black Inventors in the Age of Segregation* (Johns Hopkins University Press), *Appropriating Technology: Vernacular Science and Social Power* (University of Minnesota Press), and *Technology Studies* (Sage Publications). Johns Hopkins University Press will publish his forthcoming book, *Game Changer: Technoscience and the Fate of Athletic Competition*.

Mary Frank Fox is an ADVANCE professor in the School of Public Policy and codirector of the Center for the Study of Women, Science, and Technology at Georgia Institute of Technology. Her research has introduced and established ways in which the participation and performance of women and men reflect and are affected by social and organizational features of science and academia; and her publications appear in more than fifty scholarly and scientific journals, books, and collections. She is a member of the Council of the Section on Science, Knowledge, and Technology of the American Sociological Association, prior cochair of the Social Science Advisory of the National Center for Women and Information Technology, and member of advisory boards for the U.S. National Science Foundation.

Scott Frickel is associate professor of sociology at the Institute at Brown (University) for Environment and Society, where his research investigates intersections of knowledge, environment, and politics. He is author of *Chemical Consequences: Environmental Mutagens and the Rise of Genetic Toxicology* and coeditor of *The New Political Sociology of Science* (with Kelly Moore), *Fields of Knowledge* (with David J. Hess), and *Investigating Interdisciplinary Collaboration* (with Mathieu Albert and Barbara Prainsack).

Joan H. Fujimura moved to the University of Wisconsin, Madison to teach sociology and to build the Science and Technology Studies Program and the Holtz Center for Research in STS, after working as the Henry R. Luce Professor of Biotechnology and Society at Stanford University and teaching sociology at Harvard University. Fujimura has studied developments in cancer research, molecular genetics, bioinformatics, genomics, and now new developments in epigenetics and systems biology in conjunction with developing theoretical concepts that include doability, standardization, bandwagons, boundary work, theory-methods packages, sociomaterial analysis, continuities, awkward surpluses, and genome geography. Her current research is on race and genomics, interdisciplinary and collaborative research, and notions of "mixed race" or "multiracialisms" in the sciences and social sciences.

Ryan Galt is associate professor and MacArthur Foundation Endowed Chair in Global Conservation and Sustainability in the Department of Human Ecology at the University of California, Davis. His work focuses on agrifood system governance and sustainability from the perspectives of political ecology, agrarian political economy, and critical realism.

Garrett Graddy-Lovelace is assistant professor, teaches, and researches agricultural policy and agrarian politics at American University's School of International Service in Washington, DC. A critical geographer, she draws upon political ecology and postcolonial studies in current research on agricultural biodiversity conservation and on international and domestic impacts of U.S. farm policies.

Christian Greiffenhagen is assistant professor in the Department of Sociology at the Chinese University of Hong Kong and senior lecturer in the Department of Social Science at Loughborough University. He is working in the areas of science and technology studies (STS) and human–computer interaction (HCI).

Christopher Groves's work focuses on how people and institutions negotiate and deal with an intrinsically uncertain future—one increasingly imagined against the backdrop of global environmental change and accelerating technological innovation. Along with the ethical and political implications of a range of future-oriented discourses and practices (e.g., risk management, precautionary regulation, land-use planning, building resilience), he examines how our ideas about what it means for individuals and whole societies to take responsibility for their futures are being changed by emerging technologies (such as bio- and nanotechnology and climate engineering).

Patrick R. Grzanka (Ph.D., American Studies, University of Maryland) is assistant professor of psychology and affiliate faculty in the Women's Studies and American Studies programs at the University of Tennessee, Knoxville. His interdisciplinary research, which has been supported by the National Science Foundation, explores the co-construction of psychological science and social inequalities at the intersections of race, gender, and sexuality. He is the editor of *Intersectionality: A Foundations and Frontiers Reader* (Westview, 2014).

David H. Guston is professor and founding director of the School for the Future of Innovation in Society at Arizona State University. From 2005 to 2016 he directed the Center for Nanotechnology in Society at ASU, funded by the U.S. National Science Foundation to pursue a vision of the anticipatory governance of nanotechnology and other emerging technologies. He has also been the director of the NSF-funded Virtual Institute for Responsible Innovation and was founding editor in chief of the *Journal of Responsible Innovation*.

Edward J. Hackett is vice provost for research at Brandeis University and a professor in the Heller School of Social Policy and Management and in the Department of Sociology. He has written about the social organization and dynamics of science, science policy, and environmental justice, and is editor of *Science, Technology, & Human Values*.

David Hakken was an information ethnographer who until recently directed the Social Informatics program in the School of Informatics and Computing at Indiana University, Bloomington. Trained in Americanist anthropology, his research focused on how cultures shape digital information and communications technologies and their uses, as well as, in turn, how social spaces are shaped by these technologies. The purpose of this work, as in his most recent book, *Beyond Capital: Values, Commons, Computing, and the Search for a Viable Future* (Routledge), is to promote technologies that expand rather than undermine human capabilities. David died in April 2016. His coauthors dedicate his chapter to his memory.

Joan Haran holds a Marie Skłodowska-Curie Global Fellowship at the University of Oregon and Cardiff University. Previously, she was a fellow at the Hanse-Wissenschaftskolleg (HWK) in Delmenhorst, Germany, affiliated with the *Fiction Meets Science* program, and working on her monograph, *Genomic Fictions: Genes, Gender and Genre*. Her research is conducted at the intersections of gender studies, cultural studies, and science and technology studies.

Sandra Harding is the author or editor of sixteen books and special journal issues on topics in feminist and postcolonial philosophy and social studies of science. She received the John Desmond Bernal Award from 4S in 2013, and coedited *Signs: Journal of Women in Culture and Society* from 2000 to 2005. She is a Distinguished Research Professor of Education and Gender Studies at the University of California, Los Angeles.

David J. Hess is professor of sociology at Vanderbilt University, where he is also the James Thornton Fant Chair in Sustainability Studies, the director of the Program in Environmental and Sustainability Studies, and the associate director of the Vanderbilt Institute for Energy and Environment. His most recent book is *Undone Science: Social Movements, Mobilized Publics, and Industrial Transitions* (MIT Press, 2015). He works on the politics of sustainability transition policies and on mobilized publics, science, and technology.

Stephen Hilgartner is professor of science and technology studies in the Department of Science and Technology Studies, Cornell University. He studies the social dimensions and politics of contemporary and emerging science and technology, especially in the life sciences. His research focuses on situations in which scientific knowledge is implicated in establishing, contesting, and maintaining social order.

Maja Horst is professor in the Department of Media, Cognition and Communication at the University of Copenhagen, where her research focuses on science communication, research management, and the social responsibility of science. She has published, among other places, in *Social Studies of Science Science, Communication,* and *Public Understanding of Science*. She has also experimented with interactive science communication installations, for which she received the Danish Science Minister's Communication Prize.

Ruthanne Huising is an ethnographer of work and organizations. She studies how organizations respond to external pressures to change and the implications of these changes for professional control and expertise. Across her various projects she has observed how organizations accommodate regulatory change (Human Pathogens and Toxins Act), auditing fads (environmental management systems), and efficiency efforts (Ontario perioperative coaching program), and the complex responses of scientists, biosafety officers, health physicists, surgeons, nurses, and administrators. Huising is an associate professor in the Desautels Faculty of Management at McGill University. She received her Ph.D. from the Sloan School of Management at MIT.

J. Benjamin Hurlbut is assistant professor in the School of Life Sciences at Arizona State University. He studies the governance of morally and technically complex domains in the biosciences, examining the interplay of developments in science and technology with shifting notions of democracy, of religious and moral pluralism, and of public reason. He holds a Ph.D. in the history of science from Harvard University.

Alastair Iles is associate professor in the Department of Environmental Science, Policy, and Management at the University of California. Trained in law and policy, he researches the knowledge politics and governance of sustainable agriculture systems. He is also co-faculty director of the Berkeley Food Institute.

Alan Irwin is professor in the Department of Organization at Copenhagen Business School. His previous books include *Citizen Science* (1995), *Sociology and the Environment* (2001), and (with Mike Michael) *Science, Social Theory and Public Knowledge* (2003). With Brian Wynne, he was the editor of *Misunderstanding Science?* (1996). His current research focuses on science–industry relations, the role of business schools in society, and scientific governance.

Sheila Jasanoff is Pforzheimer Professor of Science and Technology Studies at Harvard University's John F. Kennedy School of Government. Her research centers on the engagements of science and technology with law, politics, and policy in modern democratic societies, with a particular focus on the role of science in cultures of public reason. Her books include *The Fifth Branch* (1990), *Science at the Bar* (1995), *Designs on Nature* (2005), and *The Ethics of Invention* (2016).

Paul Jobin is associate researcher at the Institute of Sociology, Academia Sinica, Taiwan, and associate professor in the Department of East Asian Studies, Université Paris

Diderot. His research addresses industrial hazards in Japan and Taiwan, with a particular focus on the interaction between workers and residents, mobilization for the recognition of damages through litigation and compensation systems, and the role of medicine in these conflicts.

Kelly Joyce is professor in the Department of Sociology and director of the Center for Science, Technology and Society at Drexel University. Joyce is the author of the *Magnetic Appeal: MRI and the Myth of Transparency* (Cornell University Press, 2008) and is coeditor of *Technogenarians: Studying Health and Illness through an Aging, Science, and Technology Lens* (Wiley-Blackwell, 2010). Her research takes up the intersections of technology, science, and aging, emphasizing sociotechnical support for collective, creative later lives.

Aalok Khandekar received his Ph.D. in the Department of Science and Technology Studies at Rensselaer Polytechnic Institute and has subsequently held postdoctoral and lecturer positions at the Department of Technology and Society Studies, Maastricht University, the Netherlands. His current research interests are in cultivating capacity to address global challenges of development and sustainability, especially as they relate to science, technology, and engineering. Previously, he worked on understanding the formation of technoscientific subjectivities and the new middle class in India under contemporary conditions of globalization.

Aya H. Kimura is associate professor in the Department of Women's Studies at the University of Hawai'i, Mānoa. She is the author of an award-winning book, *Hidden Hunger: Gender and Politics of Smarter Food* (Cornell University Press, 2013), and *Radiation Brain Moms and Citizen Scientists: The Gender Politics of Food Contamination after Fukushima* (Duke University Press, forthcoming). She also coedited *Food and Power: Visioning Food Democracy in Hawai'i* (University of Hawai'i Press, forthcoming).

Abby Kinchy is associate professor in the Science and Technology Studies Department at Rensselaer Polytechnic Institute. She writes about political controversies over food production and energy. She is author of *Seeds, Science, and Struggle: The Global Politics of Transgenic Crops* (MIT Press, 2012).

Daniel Lee Kleinman is associate dean of the graduate school at the University of Wisconsin, Madison, where he is also a professor in the Department of Community and Environmental Sociology. He is the inaugural editor of the Society for Social Studies of Science's open access journal, *Engaging Science, Technology, and Society*. His most recent book, coauthored with Sainath Suryanarayanan, *Vanishing Bees: Science, Politics, and Honey Bee Health*, is forthcoming from Rutgers University Press.

Scott Gabriel Knowles is associate professor of history at Drexel University. He is also a research fellow of the Disaster Research Center, University of Delaware. His most recent book is *The Disaster Experts: Mastering Risk in Modern America*. Knowles is coeditor (with Kim Fortun) of *Critical Studies in Risk and Disaster*, a book series published by the University of Pennsylvania Press.

Florian Kohlbacher is associate professor of marketing and innovation in the International Business School Suzhou (IBSS) at Xi'an Jiaotong-Liverpool University (XJTLU), as well as the founding director of the XJTLU Research Institute on Aging and Society (RIAS). He is coeditor of *The Silver Market Phenomenon: Marketing and Innovation in the Aging Society*, 2nd edition (Springer, 2011), and coauthor of *Advertising in the Aging Society: Understanding Representations, Practitioners, and Consumers in Japan* (Palgrave Macmillan, 2016).

Pia M. Kohler is assistant professor in the Environmental Studies Program at Williams College. Her research examines expert institutions providing science advice to multilateral environmental agreements. She has also studied negotiations for the global governance of chemicals, including under the Stockholm Convention on Persistent Organic Pollutants and the Minamata Convention on Mercury.

Kornelia Konrad is assistant professor in the Department of Science, Technology and Policy Studies at the University of Twente. Her work focuses on the role of anticipation in science, technology, and innovation, in particular on the dynamics and implications of hype-disappointment cycles, different modes of future orientation, constructive technology assessment, and foresight. A further interest lies with sociotechnical change at the field and sector level.

Inga Kroener works as a senior research analyst at Trilateral Research Ltd. Her research interests lie in the area of governance, policy, and ethics in relation to surveillance and security technologies. Prior to joining Trilateral, she worked as a lecturer in science and technology studies at University College London. She has also worked as a senior research associate at Lancaster University, for the RSA, and for Defra. Kroener is the author of the book *CCTV: A Technology under the Radar* (Ashgate, 2014).

Myanna Lahsen is senior associate researcher in the Earth System Science Center of the Brazilian National Institute for Space Research (INPE) and is adviser to Nature Climate Change and Executive Editor of WIREs Climate Change's domain on the Social

Status of Climate Change Knowledge and of *Environment* magazine. A cultural anthropologist by Ph.D. (Rice University), she is a recipient of numerous awards and fellowships, including a U.S. EPA "STAR" fellowship and three postdoctoral fellowships, one at the U.S. National Center for Atmospheric Research and two at Harvard University, where she also was lecturer on environmental science and public policy.

John Law is emeritus professor in sociology at the Open University and honorary professor in the Centre for Science Studies, Lancaster University. Author of *After Method: Mess in Social Science Research*, he uses material semiotic tools to explore methods, environmental topics, human-animal interactions, postcolonial knowledge relations, and alternative postcolonial forms of STS. (See www.heterogeneities.net.)

Loet Leydesdorff (Ph.D., sociology; M.A., philosophy; and M.Sc., biochemistry) is professor emeritus at the Amsterdam School of Communications Research (ASCoR) of the University of Amsterdam. He has published extensively in scientometrics, social network analysis, systems theory, and the sociology of innovation (see www.leydesdorff .net/list.htm).

Javier Lezaun is James Martin Lecturer at the Institute for Science, Innovation and Society (InSIS), and associate professor in the School of Anthropology and Museum Ethnography, University of Oxford. He holds a Ph.D. in science and technology studies from Cornell University. His work focuses on the relationship between technoscientific and political change.

Max Liboiron is assistant professor of sociology and environmental sciences at Memorial University of Newfoundland. Her research focuses on how harmful yet invisible threats such as disasters, toxicants, and marine plastics become visible in science and activism, and how these methods of representation relate to action. She directs the Civic Laboratory for Environmental Action Research (CLEAR), which creates citizen science technologies for environmental monitoring of plastic pollution, and manages *Discard Studies*, an online hub for research on waste and pollution.

Marcela Linková is the head of the Centre for Gender and Science at the Institute of Sociology of the Czech Academy of Sciences. Her research focuses on sociology of gendered organizations, research careers, governance of research, and research assessment from a gender perspective. She also examines the material-discursive practices through which gender equality policies and initiatives are adopted and implemented at the European and Czech country level. She is a member of the European Commission's

Helsinki Group on Gender in Research and Innovation and has served as a member of expert and advisory bodies for the European Commission and Czech government.

Yanni Loukissas is assistant professor of digital media in the School of Literature, Media and Communication at Georgia Tech, where he directs the Local Data Design Lab. He is the author of *Co-Designers: Cultures of Computer Simulation in Architecture* (Routledge, 2012). He holds a bachelor's degree in architecture from Cornell and a Ph.D. in design and computation from MIT.

Adrian Mackenzie (professor in Technological Cultures, Department of Sociology, Lancaster University) has published work on technology: *Transductions: Bodies and Machines at Speed* (2002), *Cutting Code: Software and Sociality* (2006), *Wirelessness: Radical Empiricism in Network Cultures (*2010), and *Into the Data: An Archaeology of Machine Learning* (2016). He is currently working on the circulation of data-intensive methods across science, government, and business. He codirects the Centre for Science Studies, Lancaster University, UK.

Martin Mahony is research fellow in the School of Geography at the University of Nottingham, UK. He is interested in the political histories of the atmospheric sciences and has published widely on climate change and the science-policy interface. His current research concerns the historical relationships of meteorology, climate, and the state in Britain's colonial empire.

Annapurna Mamidipudi was trained as an engineer and has recently completed her doctoral thesis in STS from the Maastricht University Science, Technology and Society (MUSTS) program. Her research is focused on theorizing innovation in handloom weaving and is grounded in her fifteen-year-long experience working in an NGO that supports livelihoods of vulnerable handloom weavers in rural India.

Laura Mamo (Ph.D., sociology, University of California, San Francisco) is professor of health education and associate director of the Health Equity Institute at San Francisco State University. Her research explores the intersections of biomedical knowledges, gender and sexuality, and health inequalities. Her current work examines biomedical and social controversies surrounding the emergence of viral cancer knowledge and practice, the production and meanings of the concept of sexual health, and the ways LGBTQ sexuality circulates in U.S. schools. She is the author of *Queering Reproduction: Achieving Pregnancy in the Age of Technoscience* (Duke University Press, 2007) and coeditor of *Biomedicalization: Technoscience, Health, and Illness in the U.S.* (Duke University Press, 2010).

Noortje Marres is associate professor in the Centre for Interdisciplinary Methodologies (CIM) at the University of Warwick (UK). She studied sociology and philosophy of science and technology at the University of Amsterdam, and much of her research focuses on problems of participation in technological societies. Her first book, *Material Participation: Technology, the Environment and Everyday Publics* (Palgrave), came out in paperback in 2015, and she is currently completing her second, *Digital Sociology: The Reinvention of Social Research* (Polity, forthcoming).

Miwao Matsumoto is professor and chair of the Department of Sociology, the University of Tokyo. His research focuses on the "structural disaster" of the science-technology-society interface, the complex and emergent relationships between technological trajectory and global environmental problems, and the formation and transformation of the military-industrial-university complex. He is currently exploring theoretical frameworks to sociologically analyze the path-dependent dynamics of the social decision-making process with reference to high-level radioactive waste disposal issues.

Maureen McNeil is emeritus professor, affiliated with the Department of Sociology, Lancaster University, UK. Most of her research is at the intersection of gender studies, cultural studies, and science and technology studies, with a particular interest in reproductive narratives, theories, and practices. Her publications include *Feminist Cultural Studies of Science and Technology* and, with Joan Haran, Jenny Kitzinger, and Kate O'Riordan, *Human Cloning in the Media*.

Clark A. Miller is associate director of the School for the Future of Innovation in Society and the Consortium for Science, Policy and Outcomes at Arizona State University. He is an avid advocate for science and technology studies and chairs ASU's doctoral program in the human and social dimensions of science and technology. His publications include *Changing the Atmosphere: Expert Knowledge and Environmental Governance*; *Arizona's Energy Future*; *Nanotechnology, the Brain, and the Future*; *Science and Democracy*; and *The Practices of Global Ethics*.

Staša Milojević is associate professor in the School of Informatics and Computing at Indiana University, Bloomington. Her research focuses on applying computational and statistical methods for understanding the processes that underpin modern science as a social and intellectual activity, especially using artifacts of scholarly communication and large bibliographic data sets.

Maywa Montenegro is a Ph.D. candidate at the University of California, Berkeley. With a background in molecular biology and science journalism, she now combines political ecology, geography, and STS in studies of the knowledge politics underpinning access to crop genetic resources. This work informs her broader interests in agroecology and theories of transition to sustainable food systems.

Kelly Moore is associate professor of sociology and director of graduate studies at Loyola University Chicago. She is the author of *Disrupting Science: Scientists, Social Movements and the Politics of the Military, 1945–1969* (Princeton University Press, 2008), coeditor of *The Routledge Handbook of Science, Technology and Society* (2014), and coeditor of *The New Political Sociology of Science* (University of Wisconsin Press, 2006).

Luis Felipe R. Murillo is an anthropologist and affiliate researcher at the Berkman Center for Internet and Society, Harvard University. His research interests include computing, politics, and language studies. He is currently conducting a multisited project on computer hacking outside the Euro-American axis with a focus on Free Software, Open Hardware, and Open Data projects.

Alondra Nelson is Dean of Social Science and professor of sociology and gender studies at Columbia University. Her research and teaching explore the intersections of science, medicine, technology, and social inequality. Her books include *The Social Life of DNA: Race, Reparations, and Reconciliation after the Genome; Body and Soul: The Black Panther Party and the Fight against Medical Discrimination; Genetics and the Unsettled Past: The Collision of DNA, Race, and History* (edited with Keith Wailoo and Catherine Lee); and *Technicolor: Race, Technology, and Everyday Life* (edited with Thuy Linh N. Tu).

Louis Neven is broadly interested in the ways in which technology and aging are co-constructed and in how the match between (geron)technologies and the wants, needs, practices, and identities of older people can be improved. Louis is currently a lector (research professor) and leads the Active Aging research group at Avans University of Applied Sciences in Breda, the Netherlands.

Casey O'Donnell is associate professor in the Department of Media and Information at Michigan State University. His research examines the creative collaborative work of video game design and development. His first book, *Developer's Dilemma*, was published by MIT Press in 2014.

Carsten Østerlund is associate professor at the School of Information Studies, Syracuse University. He earned a Ph.D. in management from MIT and is a former student at the University of California, Berkeley, University of Aarhus, and University of Copenhagen, Denmark. He has been affiliated with the Work Practice and Technology Group at Xerox PARC.

Gwen Ottinger is associate professor in the Department of Politics and the Center for Science, Technology, and Society at Drexel University. She is coeditor of *Technoscience and Environmental Justice: Expert Cultures in a Grassroots Movement* and author of *Refining Expertise: How Responsible Engineers Subvert Environmental Justice Challenges*, which won the 2015 Rachel Carson Prize.

Han Woo Park is a full professor at Yeungnam University, South Korea. He was a pioneer in network science of open big data in the early 2000s (often called *Webometrics*). He is currently the president of the World Association for Triple Helix and Future Strategy Studies.

John N. Parker is an honors fellow at Barrett Honors College at Arizona State University and an associate research scientist at the University of California, Santa Barbara. He is an expert in the sociology of science, creativity, emotions, and organizations. His research examines social dimensions of scientific work, scientific social movements, scientific elites, and scientific creativity.

Alexander Peine holds a tenured position as assistant professor of science, technology and innovation studies at Utrecht University. Peine's research explores sociotechnical change at the intersection of ICT innovation, the built environment, and aging covering diverse themes such as smart homes, eBikes, robots, and nanotechnology. In terms of theory, he is particularly interested in bridging STS and social gerontology to explore the sociomaterial constitution of later life.

Bart Penders is assistant professor in biomedicine and society at Maastricht University, the Netherlands, where he studies how scientists collaborate to create knowledge, how they render such knowledge credible, and how nonscientists are involved in knowledge production and credibilization. He held visiting and honorary fellowships at Bielefeld University, the University of Canterbury, and Harvard University.

Hector Postigo is associate professor in the Department of Media Studies and Production at Temple University. He holds a Ph.D. in science and technology studies from Rensselaer Polytechnic Institute. He has published articles and books about AOL digital labor, hacking copyright in the digital age, and amateur video game–modification labor processes.

Barbara Prainsack is a political scientist working at the Department of Social Science, Health & Medicine at King's College London. Her work explores social, ethical, and regulatory aspects of biomedicine and bioscience. Her current projects explore participatory practices in biomedicine and bioscience and the role of solidarity in guiding practice and policy in medical research and practice.

Alex Preda is a professor at King's College London. His most recent book is *Noise: Living and Trading in Electronic Finance* (University of Chicago Press, 2016). Currently he conducts ethnographic research on competitions.

Antony J. Puddephatt is associate professor and chair of the Department of Sociology at Lakehead University. He has published on a variety of themes to do with the social theory of George Herbert Mead, as it relates to science and technology studies, environmental sociology, global social work practice, and modern theories about language and thought. He has also been interested in sociology in Canada, how to encourage better theorizing in ethnographic research, and the organizational culture of competitive chess. Most recently, he is principal investigator on an insight development grant to study open access scholarly publishing in Canada, funded by the Social Sciences and Humanities Research Council of Canada.

Ramya M. Rajagopalan is research associate of life sciences at CHF (Chemical Heritage Foundation) and visiting fellow at the University of Wisconsin, Madison. Her research examines epistemic practices and the dynamics of interdisciplinary collaboration in contemporary life sciences research and biomedicine, focusing on human genomics, personalized medicine, epigenetics, and systems biology. Her coauthored article with Joan Fujimura titled "Different Differences: The Use of 'Genetic Ancestry' versus Race in Biomedical Human Genetic Research" won the 2013 David Edge Best Article Prize from the 4S.

Brian Rappert is professor of science, technology, and public affairs at the University of Exeter (UK). His long-term interest has been the examination of the strategic management of information, particularly in relation to armed conflict. More recently he has been interested in the social, ethical, and political issues associated with researching and writing about secrets, as in his books *Experimental Secrets* and *How to Look Good in a War*.

David Ribes is associate professor in the Department of Human Centered Design and Engineering (HCDE) at the University of Washington. He is a sociologist of science and technology who focuses on the development and sustainability of research infrastructures (i.e., networked information technologies for the support of interdisciplinary science), their relation to long-term changes in the conduct of science, and epistemic transformations in objects of research. His methods are ethnographic, archival-historical, and comparative. See davidribes.com for more.

Deboleena Roy is associate professor of women's, gender, and sexuality studies and neuroscience and behavioral biology at Emory University. She received her Ph.D. in reproductive neuroendocrinology and molecular biology from the Institute of Medical Science at the University of Toronto. Her fields of interest include feminist theory, feminist science and technology studies, neuroscience, molecular biology, postcolonial theory, and reproductive justice movements.

Chris Salter is an artist, University Research Chair in New Media, Technology and the Senses at Concordia University in Montreal and codirector of the Hexagram Network for Research-Creation in Media Art, Design, Technology, and Digital Culture. His artistic work has been shown all over the world, and he is the author of *Entangled: Technology and the Transformation of Performance* (MIT Press, 2010) and *Alien Agency: Experimental Encounters with Art in the Making* (MIT Press, 2015).

Carrie B. Sanders, Ph.D., is associate professor of criminology at Wilfrid Laurier University in Canada. She specializes in social constructionism and qualitative research methodologies. Her primary research interests, which have received funding by the Social Sciences and Humanities Research Council of Canada, examine the integration and utilization of information technologies, analytics, and big data in police practices.

Pankaj Sekhsaria recently completed his doctorate in STS from Maastricht University, the Netherlands. Titled *Enculturing Innovation: Indian Engagements with Nanotechnology*, his dissertation investigates innovation and technoscientific practices in nanotechnology labs in India. He is a writer, photographer, and author/editor of four books on the environment in India. His debut novel based in the Andaman Islands, *The Last Wave*, was published in 2014 by HarperCollins Publishers India.

Cynthia Selin is a scholar of future-oriented deliberation working at the intersection of responsible innovation, future studies, and sustainability. With a joint appointment in Arizona State University's School for the Future of Innovation in Society and the School of Sustainability, Selin investigates and invents methodologies for clarifying uncertainty and explores more theoretical questions about anticipation in society. She is an associate fellow at the University of Oxford, where she co-teaches the Oxford Scenarios Programme.

Kalpana Shankar is lecturer at the School of Information and Communication Studies, University College Dublin, Ireland. She earned her Ph.D. in library and information studies from the University of California, Los Angeles. Her current research focuses on sustainability of data archives, open data in Ireland, and related data practices.

Susan S. Silbey is Leon and Anne Goldberg Professor of Humanities, professor of sociology and anthropology, and professor of behavioral and policy sciences, Sloan School of Management at MIT. Silbey is interested in the governance, regulatory, and audit processes in complex organizations, currently studying the introduction of environment, safety, and health regulations in research laboratories. She is the author of *The Common Place of Law: Stories from Everyday Life*; *In Litigation: Do the 'Haves' Still Come Out Ahead?*; *Law and Science: (I) Epistemological, Evidentiary, and Relational Engagements* and *(II) Regulation of Property, Practices, and Products*.

Stephen C. Slota is a doctoral candidate in the Department of Informatics at the University of California, Irvine, and has an MLIS from the University of Pittsburgh. He is currently studying knowledge infrastructures and science policy with a particular focus on knowledge production in new scientific cyberinfrastructure. He was a founding member of the EVOKE lab and is a fellow of the Newkirk Center for Science and Society.

Laurel Smith-Doerr is the director of the Institute for Social Science Research and professor of sociology at the University of Massachusetts Amherst. She is author of *Women's Work: Gender Equity v. Hierarchy in the Life Sciences*. Her work investigates how science is organized in contemporary knowledge-based communities in order to understand persistent inequalities. She conducts research on topics such as interorganizational collaboration among biotech firms and implications of different organizational contexts for gender equity in science, scientific collaboration, and tensions in the institutionalization of science policy related to ethics, and gender pay gaps at government science agencies. She has previously served as a 4S council member and as an NSF program officer and believes in advocating for STS and social science in organizations.

Philippe Sormani is head of the research program at the Swiss Institute in Rome and associate researcher of the Institut Marcel Mauss, EHESS, Paris. Ethnography, ethnomethodology, and science and technology studies are his main fields of research.

Jack Stilgoe is senior lecturer in the Department of Science and Technology Studies, University College London. His teaching and research interests are in science and innovation policy and the governance of emerging technologies. He is the author of *Experiment Earth: Responsible Innovation in Geoengineering* (Routledge, 2015)

Banu Subramaniam is a professor in the Department of Women, Gender, Sexuality Studies at the University of Massachusetts Amherst. Trained as a plant evolutionary biologist, she seeks to engage the feminist studies of science in the practices of experimental biology. She is author of *Ghost Stories for Darwin: The Science of Variation and the Politics of Diversity* (2014) and coeditor of *Feminist Science Studies: A New Generation* and *Making Threats: Biofears and Environmental Anxieties*. Spanning the humanities and the social and natural sciences, she works at the intersections of biology, women's studies, ethnic studies, and postcolonial studies.

Kim TallBear, author of *Native American DNA: Tribal Belonging and the False Promise of Genetic Science*, is associate professor, Faculty of Native Studies, University of Alberta. TallBear is a citizen of the Sisseton-Wahpeton Oyate in South Dakota. She blogs on indigeneity and technoscience at www.kimtallbear.com.

Manuel Tironi is associate professor in the Instituto de Sociología at Pontifica Universidad Católica de Chile, where he convenes the Controversies, Environment and Society Group. Tironi's research sits at the intersection of science and technology studies, environmental studies, and politics. His research on engagement in envirotech disasters has been published in *Science, Technology, and Human Values*, *The Sociological Review*, and *Geoforum*, among other journals, and his coedited volume *Disasters and Politics: Materials, Experiments, Preparedness* (Wiley-Blackwell) won the 2014 Amsterdamska Award by the European Association for the Study of Science and Technology.

Richard Tutton is senior lecturer in the Department of Sociology and Centre for Science Studies at Lancaster University, UK. His research interests are in the exploration and conceptualization of visions, imaginaries, and futures in contemporary technoscience and medicine. His publications include *Genomics and the Reimagining of Personalized Medicine* (Ashgate, 2014).

Lisa-Jo K. van den Scott, Ph.D., is assistant professor of sociology at Memorial University of Newfoundland in Canada. Her area of specialization is the sociology of walls. Her current work explores the introduction of permanent walls to the Inuit of Arviat, Nunavut, Canada, at the intersection of symbolic interactionism and the social construction of technology approach.

Harro van Lente is full professor in the Department of Science and Technology Studies at Maastricht University, the Netherlands. He was trained in physics and philosophy, and he studies the dynamics of emerging technologies. He has published widely on the sociology of expectations, technology assessment, foresight, sustainability, and the politics of knowledge production.

Niki Vermeulen is a Dutch STS scholar working in the Science, Technology and Innovation Studies Unit of the University of Edinburgh as a lecturer in history/sociology of science. She specializes in the organization and institutionalization of research, with a special focus on scientific collaboration across the life sciences. Next to her academic work, she has been active as a science policy adviser and consultant, trying to improve conditions for research.

Janet Vertesi is assistant professor in the Sociology Department at Princeton University. She is the author of *Seeing Like a Rover: How Robots, Teams and Images Craft Knowledge of Mars* (University of Chicago Press, 2015) and coeditor of *Representation in Scientific Practice Revisited* (MIT Press, 2014). In addition to her ethnographic studies of scientific collaboration and representation in scientific practice on NASA's robotic spacecraft teams, she has published about seventeenth-century astronomy and contributes to the human-computer interaction and computer-supported cooperative work communities.

Kathleen M. Vogel is associate professor at North Carolina State University in the Department of Political Science and serves as director of the Science, Technology, and Society Program. Vogel has also spent time as a faculty member in Cornell's STS Department, as a visiting scholar at Sandia National Laboratories, and as a William C. Foster Fellow in the U.S. Department of State. Her research focuses on studying the social and technical dimensions of bioweapons threats and the production of knowledge in intelligence assessments.

Martin Weinel is research associate at the Centre for the Study of Knowledge, Expertise and Science (KES) at Cardiff University. In collaboration with others, most notably Harry Collins, Rob Evans, and Nicky Priaulx, he has written a number of papers on aspects of expertise, science policy, interdisciplinarity, science communication, and the Imitation Game. He is currently working on the ERC-funded Imitation Game project.

Kjersten Bunker Whittington is associate professor of sociology at Reed College. Her research addresses how men's and women's career trajectories and innovative output are influenced by network dynamics and collaborative relationships, the organization of work in firms and universities, and gendered dimensions of science and technology. She has served on expert and advisory bodies for the National Science Foundation and the American Sociological Association, and most recently as special assistant to the NIH associate director for research on women's health through the AAAS Science Policy Fellowship Program.

Logan D. A. Williams is assistant professor of history, philosophy, and sociology of science in Lyman Briggs College and the Department of Sociology at Michigan State University. The journals *Minerva, Technology in Society*, and *Perspectives on Global Technology and Development* have published her research on innovation, appropriation, and technology choices. Her book manuscript about the global circulation of science, technology, and management practices in ophthalmology is under contract with the University of North Carolina Press.

Sally Wyatt is professor of digital cultures in development at Maastricht University and director of the Netherlands Graduate Research School for Science, Technology and Modern Culture (WTMC). Her research focuses on the use of digital technologies, both in the creation of knowledge in the humanities and the social sciences, and the ways in which people incorporate such technologies in their health information–seeking practices.

Teun Zuiderent-Jerak is LiU Research Fellow and Senior Lecturer in science and technology studies at the Department of Thematic Studies—Technology and Social Change of Linköping University, Sweden. His research focuses on standardization and quality improvement practices in health care, the construction of markets for public values, and sociological research that explicitly aims to intervene in the practices it studies. His book *Situated Intervention: Sociological Experiments in Health Care* was published in 2015 by MIT Press.

Name Index

Subject Index

aboriginal communities
 in Australia, 32, 47, 69–70, 436, 447, 448
 in Canada, 237
academic-military-industrial complex, 291,
 1008
accountability
 of artists' work in scientific contexts, 152
 boundary work on organizational lines of,
 1068
 citizenship role for science communication
 and, 895
 disaster research on, 1005, 1011
 documentation for, 69–70, 73, 77
 of ELSA research, 829, 833
 ethnomethodology and, 125, 770
 global environmental issues and, 1041
 governance via ethics (soft law) and, 839
 laboratory management systems for, 810,
 811
 meaning of, in different political cultures,
 1077, 1078
 objectivity of scientists and, 35
 open science and, 751
 of research programs, 76, 714, 743, 745, 746,
 752, 793
 science democracy relationship and, 259,
 267, 270
 scientific authority in environmental disputes and, 1037
 scientific control of environmental and
 safety hazards and, 812

activism. *See also* social movements
 antinuclear, 88, 264, 266, 291, 297, 299, 771,
 857, 1043, 1062
 origin of STS and, 87
actor networks, 65, 117, 180, 270, 328, 595,
 952
actor-network theory (ANT), 7
 analysis of methods of power using, 43
 architectural practice and, 148, 564, 565, 567
 climate change debate and, 1066
 concept of agency and, 659n9
 co-production in, 641, 643–44
 Critical Engineering Manifesto and, 178
 critical theory of technology and, 636–37,
 638, 640–41, 644–45, 652–53, 659n8
 documents in, 66–67
 ethnomethodological tradition and, 533
 governmentality and, 1074–75
 interchangeability of human and nonhuman
 actors in, 533, 640, 1066
 ordering methods and, 42
 parliaments of things and, 276
 politics of technology and, 636
 practices of scientists and, 268, 270
 situated constitution of, 118, 119
 STS use of, 41–42, 90, 635, 946
 success of heterogeneous networks and, 328
 urban politics and, 497, 573
 urban studies using, 557, 569, 570, 571–72,
 574
admixture mapping, 359–60